Shock Compression of Condensed Matter — 2003

Previous Proceedings in the Series of
Conferences of the American Physical Society Topical Group on
Shock Compression of Condensed Matter

Year	Held in	Publisher	ISBN
2001	Atlanta, Georgia, USA	AIP Conference Proceedings 620	0-7354-0068-7
1999	Snowbird, Utah, USA	AIP Conference Proceedings 505	1-56396-923-8
1997	Amherst, Massachusetts, USA	AIP Conference Proceedings 429	1-56396-738-3
1995	Seattle, Washington, USA	AIP Conference Proceedings 370	1-56396-566-6
1993	Colorado Springs, Colorado, USA	AIP Conference Proceedings 309	1-56396-219-5
1991	Williamsburg, Virginia, USA	North-Holland	0-444-89732-1
1989	Albuquerque, New Mexico, USA	North-Holland	0-444-88271-5
1987	Monterey, California, USA	North-Holland	0-444-87097-0
1985	Spokane, Washington, USA	Plenum Press	0-306-42276-X
1983	Santa Fe, New Mexico, USA	North-Holland	0-444-86904-2
1981	Menlo Park, California, USA	AIP Conference Proceedings 78	0-88318-177-0

Other Related Titles from AIP Conference Proceedings

712 Materials Processing and Design: Modeling, Simulation and Applications; NUMIFORM 2004; Proceedings of the 8th International Conference on Numerical Methods in Industrial Forming Processes
Edited by Somnath Ghosh, Jose C. Castro, and June K. Lee, June 2004, 2 vol. hard cover set, 0-7354-0188-8; CD-ROM: 0-7354-0189-6

700 Review of Progress in Quantitative Nondestructive Evaluation: Volume 23
Edited by Donald O. Thompson and Dale E. Chimenti, February 2004, 2 vol. hard cover set, CD-ROM included, 0-7354-0173-X

To learn more about these titles, or the AIP Conference Proceedings Series, please visit the webpage
http://proceedings.aip.org/proceedings

Shock Compression of Condensed Matter — 2003

Proceedings of the Conference of the American Physical Society
Topical Group on Shock Compression of Condensed Matter
held in Portland, Oregon, July 20–25, 2003

PART ONE

Edited by:

MICHAEL D. FURNISH
Sandia National Laboratories
Albuquerque, New Mexico, USA

YOGENDRA M. GUPTA
Washington State University
Pullman, Washington, USA

JERRY W. FORBES
University of Maryland
Port Tobacco, Maryland, USA

SPONSORING ORGANIZATION
American Physical Society

◎ **CD-ROM INCLUDED**

Melville, New York, 2004
AIP CONFERENCE PROCEEDINGS ■ 706

EDITORS

Michael D. Furnish
Sandia National Laboratories
MS 1168, P.O. Box 5800
Albuquerque, NM 87185-1168
USA
E-mail: mdfurni@sandia.gov

Yogendra M. Gupta
Washington State University
Dept. of Physics and Institute for Shock Physics
Pullman, WA 99164-2814
USA
E-mail: ymgupta@wsu.edu

Jerry W. Forbes
University of Maryland
6535 Chelsea Way
Port Tobacco, MD 20677
USA
E-mail: jfor@direcway.com

L.C. Catalog Card No. 2004107953
ISBN 0-7354-0181-0
ISSN 0094-243X
Printed in the United States of America

CONTENTS

PART ONE

CHAPTER I

PLENARY

CHAPTER II

EQUATION OF STATE: NONENERGETIC MATERIALS

CHAPTER III

EQUATION OF STATE: ENERGETIC MATERIALS

CHAPTER IV

PHASE TRANSITIONS

CHAPTER V

MODELING, SIMULATION, AND THEORY: NONREACTIVE MATERIALS

CHAPTER VI

MOLECULAR DYNAMICS MODELING: NONREACTIVE MATERIALS

CHAPTER VIII

SPALL, FRACTURE, AND FRAGMENTATION

CHAPTER IX

CONSTITUTIVE AND MICROSTRUCTURAL PROPERTIES OF METALS

CHAPTER X

MECHANICAL PROPERTIES OF POLYMERS

CHAPTER XI

MECHANICAL PROPERTIES OF COMPOSITES

CHAPTER XII

MECHANICAL PROPERTIES OF CERAMICS, GLASSES, IONIC SOLIDS, AND LIQUIDS

PART TWO

CHAPTER XIII

MECHANICAL PROPERTIES OF REACTIVE MATERIALS

CHAPTER XIV

DETONATION AND BURN PHENOMENA

CHAPTER XV

EXPLOSIVE AND INITIATION STUDIES

CHAPTER XVI

SHOCK-INDUCED MODIFICATIONS AND MATERIAL SYNTHESIS

CHAPTER XVII

INSTRUMENTATION

CHAPTER XVIII

EXPERIMENTAL TECHNIQUES

CHAPTER XIX

ISENTROPIC COMPRESSION

CHAPTER XX

OPTICAL AND ELECTRICAL MEASUREMENTS

CHAPTER XXI

IMPACT PHENOMENA, BALLISTICS, HYPERVELOCITY STUDIES, AND EXOTIC SHOCK CONFIGURATIONS

CHAPTER XXII

LASER-DRIVEN SHOCKS

CHAPTER XXIII

GEOPHYSICS, STRUCTURES, AND MEDICAL APPLICATION

PREFACE

The Thirteenth Biennial International Conference of the APS Topical Group on Shock Compression of Condensed Matter (SCCM) was held at the Doubletree Hotel (Lloyd Center) in Portland, Oregon, U.S.A. from July 20 through July 25, 2003. The purpose of this conference, as with earlier conferences, was to provide a forum for scientists and engineers studying the physics, chemistry, mechanics, and materials science of condensed matter at high dynamic pressures and compressions to exchange ideas and technical information. Presentations at the conference included contributed papers, in addition to invited and plenary lectures on selected topics with visionary themes. Theoretical, computational and experimental results describing the physical, chemical, structural, electro-mechanical and optical responses of condensed materials to shock wave loading were discussed.

Five hundred and nineteen scientists and engineers from 14 countries registered at the conference, including 69 students from 9 countries. The countries represented included: Canada (9), France (20), Germany (7), Israel (16), Italy (3), Japan (12), Netherlands (2), Norway (1), Portugal (2), People's Republic of China (1), Russia (18), South Korea (4), United Kingdom (43), and the United States (381). Twenty-three of the students were supported by the SCCM Topical Group: 16 from U.S. universities, 2 from U.S. National Laboratories, and 5 from other nations – UK, Israel and Japan. There were 510 scheduled technical presentations (390 oral and 120 poster), with 350 appearing as written contributions in this volume.

Jim Asay received the 2003 APS Shock Compression Science Award at the conference. He presented a plenary talk entitled, "Wave Structure Studies in Condensed Matter Physics - Single Crystals to Magnetic Effects."

A special session was organized in memory of George E. Duvall, a pioneer in the field and recipient of the second APS Shock Compression Science Award in 1989, who passed away in 2003. Brief comments were made by several prominent scientists who had interacted with George Duvall during four decades of his active involvement in the field.

Neil Holmes was the Master of Ceremonies at the conference banquet, which included a piano concert by Dr. Margaret Brink, a faculty member in the School of Music and Theater Arts at Washington State University. Alita Roach and William Deal of the Los Alamos National Laboratory were recognized at the conference banquet for their dedicated efforts and contributions, spanning twenty years, to the SCCM conferences.

The Organizing Committee for the conference was comprised of the following people:
Yogendra Gupta, Washington State University, Conference Chair
Jerry Forbes, Lawrence Livermore National Laboratory, Technical Program Coordinator
J. Michael Boteler, Naval Surface Warfare Center – Indian Head, Conference Treasurer
Ricky Chau, Lawrence Livermore National Laboratory, Poster Co-Chair
Clint Hall, Sandia National Laboratories, Poster Co-Chair
Paulo Rigg, Los Alamos National Laboratory, Audio-Visual Aids
Michael Furnish, Sandia National Laboratories, Publications Chair
Vitali Nesterenko, University of California at San Diego, Educational Outreach
Paul Urtiew, Lawrence Livermore National Laboratory, International Advisor
James Asay, Washington State University, Conference Advisor
Robert Hixson, Los Alamos National Laboratory, Conference Advisor
Neil Holmes, Lawrence Livermore National Laboratory, Conference Advisor
Alita Roach, Los Alamos National Laboratory, Conference Coordinator
William Deal, Los Alamos National Laboratory, Conference Coordinator

The Technical Program Committee was comprised of the following people:

Program Chair: Jerry Forbes, Lawrence Livermore National Laboratory
Rebecca Brannon (Sandia National Laboratories): Computational Mechanics of Materials
Robert Cauble (Lawrence Livermore National Laboratory): Warm Dense Matter, EOS
Ricky Chau (Lawrence Livermore National Laboratory): Planetary Science, Geophysics
Dana Dlott (University of Illinois): Optical Studies, Biological Systems
Marcus Knudson (Sandia National Laboratories): EOS, Mesoscale
Philip Miller (Naval Surface Warfare Center – Indian Head): Energetic Materials, Detonation
David Moore (Los Alamos National Laboratory): Chemical Reactions, Optical Studies
Vitali Nesterenko (UC/San Diego): Materials Related Phenomena (Mechanics), Nonlinear Waves
Paulo Rigg (Los Alamos National Laboratory): Phase Transitions
Carter White (Naval Research Laboratory): Condensed Matter Theory, M.D. Simulations
Choong-Shik Yoo (Lawrence Livermore National Laboratory): Static Pressure Studies
Anna Zurek (Los Alamos National Laboratory): Materials Related Phenomena, Fracture

During 2003, the American Physical Society Topical Group Officers were:

Stephan Bless, University of Texas, Austin, Chair
Brad L. Holian, Los Alamos National Laboratory, Chair-Elect
Craig Tarver, Lawrence Livermore National Laboratory, Vice Chair
Michael Furnish, Sandia National Laboratories, Past Chair
J. Michael Boteler, Naval Surface Warfare Center – Indian Head, Secretary/Treasurer
Ray Lemar, Naval Surface Warfare Center-Indian Head, Webmaster

American Physical Society Topical Group Executive Committee Members were: Michael P. Desjarlais, Sandia National Laboratories, Marcus D. Knudson, Sandia National Laboratories, William J. Nellis, Lawrence Livermore National Laboratory, and Naresh N. Thadhani, Georgia Institute of Technology.

The Organizing Committee would like to express its appreciation to the many people who contributed to the success of the Conference: the session chairs, the authors, the plenary and invited speakers, and the participants. The editors are especially grateful to the session chairs and other people who reviewed the manuscripts.

Special thanks are due to the many people who willingly gave their time, talent, support, and encouragement in the various aspects of the organization of this conference. Alita M. Roach and William E. Deal invested large amounts of time and effort before, during, and after the meeting. These two individuals deserve special thanks for their dedication toward the success of the conference. Others who made special efforts to make things work for the meeting include Rosella Atencio-Gerst, Justin Brown, Fran Chavez, Amy Lucero, Janet Neff-Shampine, and Mary Wong. The editors are particularly appreciative of the energy expended by Stacey Hanks in the months following the conference in the continued collection, editing and filing of papers.

MICHAEL FURNISH
YOGENDRA GUPTA
JERRY FORBES

FOREWORD

Shock wave compression represents a unique approach to understanding condensed matter response to extreme conditions and to provide insight into nonlinear wave propagation. During the last sixty years, shock wave compression of condensed matter has developed into an important research endeavor that cuts across many disciplines in science and engineering.

The 2003 conference emphasized shock wave phenomena, a broad range of applications, and fundamental understanding of condensed matter through innovations in theory/computations and experiments. The breadth of topics covered in the plenary lectures of Neil Ashcroft, James Asay, Donald Curran, and Russell Hemley demonstrated the scientific maturity and the broad applicability of this challenging, multidisciplinary field. James Asay's lecture associated with his APS Shock Compression Science Award for 2003 emphasized the importance of wave propagation methods to probe physical states of matter and mechanical response under dynamic loading.

The contributed, invited, and plenary papers presented in these proceedings represent the latest advances in shock wave research. Included are papers on equation of state developments, phase transitions, chemical reactions, warm dense matter, inelastic deformation and fracture, materials-related phenomena, geophysics and planetary science, energetic materials and detonation, *ab-initio* and molecular dynamics calculations, experimental developments, and nonlinear wave propagation. A special session on biological applications was also held at the conference. The scientific and technical advances reported in these proceedings make this two-volume set the latest authoritative reference for this field. As with the previous proceedings, the diversity of material presented here is difficult to organize into well-developed categories. However, an extensive index provides a convenient entry into various scientific and technical topics.

In 2003 and 2004, two of the most eminent individuals in the field, George E. Duvall and Lev V. Al'tschuler, passed away. They were the second and third recipients of the APS Shock Compression Science Award, respectively, and made pioneering contributions to shock wave research. These two remarkable physicists played a major role in the development of this field and mentored a large number of scientists for more than four decades. Their scientific contributions will always be remembered and acknowledged.

The APS Topical Conference on Shock Compression of Condensed Matter held in Portland, Oregon was indeed a team effort and I want to express my sincere appreciation to everyone who contributed to the success of this conference and to the publication of these proceedings. In particular, four individuals deserve special acknowledgement: Alita Roach, William Deal, Michael Furnish and Jerry Forbes. I am very grateful to these four colleagues for their hard work, dedication, and good cheer.

Yogendra M. Gupta

Topical Group on Shock Compression of Condensed Matter
2002 - 2003 APS Fellows

The APS Fellowship Program was created to recognize members who may have made advances in knowledge through original research and publication or made significant and innovative contributions in the application of physics to science and technology. They may also have made significant contributions to the teaching of physics or service and participation in the activities of the Society. Each year, no more than one-half of one percent of the then current membership of the Society are recognized by their peers for election to the status of Fellow in the American Physical Society.

Congratulations to the following members of the APS who have been elevated to fellowship through the Topical Group on Shock Compression of Condensed Matter since the 2001 biennial meeting.

2002:

Vladimir E. Fortov, Russian Academy of Sciences

For pioneering work on the physical properties of hot dense plasmas at multimegabar pressures and very high temperatures achieved with shock compression.

Brad Lee Holian, Los Alamos National Laboratory

For pioneering use of large-scale atomistic computer simulations (massively parallel nonequilibrium molecular dynamics) in studying shock waves in condensed matter..

2003:

Michael Cowperthwaite, SRI International (retired)

For seminal contributions to shock wave propagation in reactive materials, detonation science, analysis of unsteady waves, and thermochemical equilibrium calculations.

Vitali Federovich Nesterenko, University of California at San Diego

For pioneering contribution to strongly nonlinear wave propagation in granular materials, through the discovery of a new solitary wave, and to shock (localized shear) mesomechanics in porous and heterogeneous media.

Toshimori Sekine, National Institute for Materials Science, Japan

For his pioneering work in shock synthesis of cubic Si_3N_4 and spinel phases in the Si_3N_4AlN-Al_2O_3 system, and for experimental studies elucidating the shock metamorphism of minerals and meteorites.

James R. Asay

Recipient of the APS Shock Compression Science Award, 2003

"For pioneering personal research in shock compression science, for leadership in developing programs and tools that have strongly impacted the field, and for leadership in the technical community."

CONFERENCES OF THE AMERICAN PHYSICAL SOCIETY TOPICAL GROUP ON SHOCK COMPRESSION OF CONDENSED MATTER

1979	First Biennial APS SCCM Meeting	Pullman, Washington
1981	Second Biennial APS SCCM Meeting	Menlo Park, California
1983	Third Biennial APS SCCM Meeting	Santa Fe, New Mexico
1985	Fourth Biennial APS SCCM Meeting	Spokane, Washington
1987	Fifth Biennial APS SCCM Meeting	Monterey, California
1989	Sixth Biennial APS SCCM Meeting	Albuquerque New Mexico
1991	Seventh Biennial APS SCCM Meeting	Williamsburg, Virginia
1993	Eighth Biennial APS SCCM Meeting	Colorado Springs, Colorado
	Jointly with the International Association for Research and Advancement of High Pressure Science and Technology (AIRAPT)	
1995	Ninth Biennial APS SCCM Meeting	Seattle, Washington
1997	Tenth Biennial APS SCCM Meeting	Amherst, Massachusetts
1999	Eleventh Biennial APS SCCM Meeting	Snowbird, Utah
2001	Twelfth Biennial APS SCCM Meeting	Atlanta, Georgia
2003	Thirteenth Biennial APS SCCM Meeting	Portland, Oregon

The Fourteenth Biennial APS SCCM International Conference will be held from July 31 through Aug. 5, 2005, at the Hyatt Regency Inner Harbor, Baltimore MD, with Carter White (NRL), Mark Elert (USNA) and Tom Russell (NSWC/IH) serving as Co-Chairs.

CHAPTER I

PLENARY

CP706, *Shock Compression of Condensed Matter - 2003*
edited by M. D. Furnish, Y. M. Gupta, and J. W. Forbes
© 2004 American Institute of Physics 0-7354-0181-0/04/$22.00

WAVE STRUCTURE STUDIES IN CONDENSED MATTER PHYSICS - SINGLE CRYSTALS TO MAGNETIC EFFECTS

James R. Asay

Institute for Shock Physics, Washington State University, Pullman, WA, 99164

Abstract. Wave structure methods have played an important role in probing mechanical and physical states of matter under dynamic loading. Applications cover a broad spectrum of research, including dynamic yielding; shock-induced phase transformations; energetic reactions, tensile and compressive strength; and viscoplastic deformation. A large variety of experimental configurations have been developed to explore these phenomena using an extensive range of time-resolved diagnostics. These methods were developed on single-stage light gas guns for the most part, but extended to higher-pressure capabilities, including explosive loading, propellant guns and two-stage light gas guns. More recently, peak pressures accessible with these methods have been extended to even higher impact velocities and pressures through novel experimental platforms, including a modified two-stage light gas gun that increases impact velocities to about 15 km/s, magnetically driven flyer plates that extend the velocities above 20 km/s, and laser-induced shock loading which increases peak pressures even further. In addition to shock compression studies, magnetic loading enables a new application of wave structure studies using large amplitude ramp waves to probe shockless, or nearly isentropic compression, to pressures exceeding 3 Mbar. Furthermore, the use of time-resolved diagnostics to measure the structure of magnetically induced ramp waves provides off-Hugoniot data unachievable with other methods. In this presentation, I will give a brief summary of wave structure techniques for studying thermomechanical and physical properties and discuss several examples from the research that my colleagues and I have performed using these methods.

INTRODUCTION

My research has covered a range of technologies and experimental methods. However, one underlying theme involves use of shock structure methods to probe material response on sub-microsecond timescales. In this paper, I decided to focus on this topic, including research for a variety of studies ranging from the study of condensed matter physics to the development of new experimental methods.

A variety of loading techniques and diagnostics have been developed for studying the material dynamic response [1]. The techniques include gas guns and the newly developed magnetic loading technique, but there are others that I don't have time to discuss. A large variety of time-resolved diagnostics have also been developed [2]. Most of the applications I'll discuss involve using velocity interferometry [3]. These discussions will occur in chronological order, starting with my early career at Washington State University, followed by research projects initiated at Sandia National Laboratories and extending to the most recent developments using pulsed power systems at Sandia. These capabilities allow studies of material properties at significantly higher pressures than possibly with conventional techniques.

Typical shock wave structures contain information important to studying a variety of effects, including initial yielding, viscoplasticity in steady shock waves, and structural phase transitions. Unloading from the shock state reveals additional information about local sound speeds, dynamic material strength, tensile failure, and details of reverse phase transitions.

When I first began research on shock compression in the late sixties, impact velocities and the corresponding impact pressures were limited to pressures available with single stage light gas guns. Since then, impact velocities and the associated impact pressures have steadily increased. At Sandia, we modified a two-stage gas gun to a three-stage configuration, referred to as the Hypervelocity Launcher, which increased impact velocities to about 12 km/s [4] for EOS measurements. Starting in about 1996, I led a research group that developed a new launcher capability that used magnetic loading with the Z Accelerator to shocklessly compress planar samples to Mbar pressures and to also launch thin flyer plates to velocities exceeding 20 km/s [5, 6].

HIGH-PRESSURE MECHANICAL RESPONSE

The area of high-pressure mechanical properties has been one of my long-standing interests. For mechanical pressure measurements, there are basic features of shock and unloading wave structures that can be used to probe mechanical states. In an elastic-plastic material, initial uniaxial strain response is assumed to be elastic to the onset of yielding at the Hugoniot Elastic Limit, or HEL [7]. Further loading is plastic to a peak stress, which is offset from a mean stress or hydrostatic state of stress by two thirds the compressive yield stress. Upon unloading, initial response is again elastic, followed by plastic or hydrodynamic response when the shear stress achieves a value equal to the reverse critical shear stress. This mechanical response results in a specific wave structure consisting of an elastic wave, followed by a plastic wave during compression and elastic, followed by plastic waves during unloading. For reloading from the shocked state, the simple elastic-plastic theory predicts plastic response.

My early experience using wave structure techniques began at WSU in my thesis research under the direction of Professor George Duvall and involved dynamic yielding of single crystal LiF. Prof. Duvall initially asked me to examine steady wave propagation in LiF since these analyses had not been previously applied to single crystals. I performed shock wave experiments on various thicknesses of LiF crystals in the <100> direction and found that the elastic precursor amplitude (HEL) varied substantially from one crystal to another. I immediately concentrated on why the HEL varied so significantly, since this was a prerequisite to obtaining reproducible results. Resolving this difficult issue turned out to change the goal of the thesis work, which became identification of deformation mechanisms for dynamic yielding of LiF.

Initially, the controlling mechanism for this variation was a mystery. Early theoretical work [8-11] indicated that decay of the elastic precursor was related to the plastic strain rate which could be expressed in terms of the initial dislocation density, the Burgers vector, and the dislocation velocity. In this formulation, the dislocation velocity was controlled through a drag coefficient that could possibly be related to impurities or some other effect, but there was very little theoretical guidance to go by in identifying specific mechanisms responsible for the large variations in precursor amplitude. Also, there were a range of impurities present in the crystals, so the critical defects were not immediately apparent. After considerable investigation, I discovered that divalent impurities, specifically Mg impurities, were the main problem and that very small variations in the concentration of these defects resulted in extremely large variations in precursor amplitude. Reported Mg^{++} concentrations varied from about 50-150 ppm. I initially didn't expect any influence on mechanical properties for this range of variation, but this assumption turned out to be wrong, in that changes of a few ppm had an enormous effect on precursor amplitude. Once I realized this, it was possible to maintain constant impurity levels and to obtain reproducible results. We were also able to independently check this hypothesis by irradiating LiF crystals to introduce radiation-induced divalent defects that had similar effects on precursor decay.

Since this early work, there have been a large number of follow-on experiments under the direction of George Duvall and later Yogendra Gupta at WSU to clarify the micromechanical deformation mechanisms of LiF, including the impurity state, crystal orientation, atomic rearrangements occurring during stress relaxation, and other effects.

Upon joining Sandia, I continued an interested in the mechanical response of materials under shock loading but with an emphasis on the issue of high-pressure material strength. I was especially interested in developing experimental configurations that would critically evaluate the elastic-plastic model developed by Richard Fowles [7]. Strength effects under shock loading are often neglected in high-pressure EOS studies and had not received significant attention at that time. As Fowles had shown, compressive strength under shock loading could be estimated by comparing the Hugoniot state to a hydrostatic pressure at the corresponding volume. However, this approach does not give accurate results at high stresses because of the generally small difference between the longitudinal stress and the hydrostat and the fact that hydrostatic response is not accurately known at high pressure. Because of this, we explored other ways of estimating strength in the shocked state.

By making rather basic assumptions about the state of stress under shock compression and the high-pressure behavior of the yield surface, Joel Lipkin and I [12] and later Lalit Chhabildas and I [13] developed a method for estimating strength properties at high pressure. These measurements provided an estimate of the yield surface and therefore determination of both the shear stress and the shear strength in the shocked state. In most models of elastic-plastic response, these are assumed to be equal. A surprising result was that these values differed by up to a factor of three at the higher shock pressures. A related effect is that the Hugoniot state appeared to lie within the yield envelope, indicating shock-induced softening. In later experiments using lateral and longitudinal

stress gauges Lev Al'tshuler and colleagues [14] obtained data that agreed with our results. Other experiments conducted on a variety of materials, including Be, Ta, and W, showed a similar effect of shock softening and a elastic precursor from the shocked state.

Several hypotheses were forwarded for these observations [13, 15, 16]. After critically examining several possibilities and the microscopic-based dislocation models in particular, we explored a meso-scale based description that assumed a distribution of shear stresses in the shocked state. The basic idea is that multiple wave reverberations occurring between grains, or relaxation of shear stresses at local heterogeneities within grains, results in a distribution of shear stress states and a mean value differing from the critical shear strength. This is not a new idea and several investigators [17, 18], have discussed similar effects. However, I think we were the first to use this concept to interpret the recompression and unloading results. We assumed a Gaussian distribution function to describe these effects and incorporated the distribution into a standard elastic-plastic model, which was able to describe both the quasi-elastic response for recompression and unloading from the shocked state. Recent work by Tracy Vogler shows that a distribution model can be self-consistently incorporated into an elastic-plastic model and used to treat the full wave propagation problem [19]. As with any continuum model, this model cannot be shown to be unique and other data are needed to confirm it.

Similar effects of shock-induced softening, followed by strength recovery, have been observed in a variety of other materials, suggesting a general effect during shock compression. As mentioned, several models have been proposed to explain these effects. The explanation I favor includes the distribution effects discussed here, along with incorporation of microscopic effects of dislocation flow. I believe that our present models need to be carefully assessed in terms of the existing data base and possibly changed to accurately represent these effects. This is one of the current challenges for the community.

PHASE TRANSITIONS

I'll now turn to a study of phase transitions using wave profile techniques. My first job at Sandia was to develop an experimental research program to study shock-induced melting. My interest was to apply shock wave structure methods to this problem and to develop a sensitive method for detecting melting. At the time, existing Hugoniot studies had shown that traditional $U_s - u_p$ measurements were not a sensitive indicator of melting in shock wave experiments. Generally, only a subtle change is observed in the slope of the shock velocity-particle-velocity curve at the onset of melting.

Bismuth was chosen for this work because its low-pressure phase diagram was well known, and calculations showed that compression waves should undergo a discontinuity in structure when the Hugoniot intersects the melt boundary [20, 21]. For initial pressures of about 25 kbar, intersection of the Hugoniot from the Solid I phase with the melt boundary produces a three-wave structure if thermodynamic equilibrium prevails. A large number of experiments were performed to explore this possibility, but these conclusively showed that a three-wave structure did not occur. I later found that collapse of deviator stresses near the melt boundary produced a smearing of the expected wave structure. However, these experiments did give evidence for melting. This was achieved by adjusting the final pressure in each experiment to exceed the triple point pressure. If melting did not occur under shock compression, a two-wave structure would occur at different initial temperatures, corresponding to intersection of the Solid I – Solid II phase boundary at different pressure levels. However, it was found that this transition always occurred at the triple point, independent of starting temperature, indicating that melting did occur.

A drawback was that it was not possible to determine the actual shock pressure where melting occurred. I therefore extended the wave structure experiments to include both compression and unloading in the vicinity of the melt boundary. This was successful in detecting the onset of melting [22]. In these experiments, the initial temperature was held constant at 473 K, and the shock pressure systematically increased until the melt pressure was exceeded. The initial unloading velocity from the shocked state is sensitive to the onset of melting, since it corresponds to the elastic longitudinal velocity for unloading in the solid and the bulk velocity for unloading in the liquid phase. This behavior was clearly observed as the calculated liquid mass fraction increased to 15%.

Analysis of the full unloading wave profiles also provided preliminary evidence that refreezing occurred [22]. For the highest pressure studied, unloading to near zero stress produced a final density corresponding to the initial solid phase. In addition, the unloading wave structure for this experiment showed evidence of transitioning from a rarefaction fan to a rarefaction shock, as expected for reverse phase transitions. Dynamic freezing is a topic of current scientific interest as evidenced by the several presentations on this topic at this conference.

I intended to extend the shock melting technique to other materials. However, in the early seventies, high-pressure windows were not available for time-resolved interferometer studies at Mbar pressures. Furthermore, techniques for performing full unloading wave with two-stage gas guns had not yet been worked out. Because of this, I designed shock and release experiments on porous aluminum with a very fine pore structure to establish thermal equilibrium in a few tens of ns after shock loading. We prepared porous samples that were about 60% of solid density, with average pore sizes of 1.3 microns. This allowed study of shock-induced melting at about 70 kbar instead of the 1.2 Mbar required for solid Al [23].

As in the case of bismuth, a decrease in the release wave speed from elastic longitudinal to bulk velocities was observed at the onset of melting [23]. Since the phase boundary for Al was also well established from static values, we felt that the studies on both Bi and Al served to establish this method as a reliable technique for detecting melting. The work on shock-induced melting evolved into studies of shock-induced vaporization [24]. This topic cannot be covered in the present article because of space limitations.

MAGNETIC LOADING TECHNIQUES

My most recent work at Sandia involved use of the Z Accelerator to produce shockless loading of condensed materials to peak pressures of several Mbar. This is a revolutionary new capability that opens up a new area of research for wave structure techniques. The magnetic approach produces quasi-isentropic loading, which is useful for evaluating off-Hugoniot EOS data [5, 25]. A spin off is that it can also be used to launch very high-velocity flyer plates under nearly shockless loading conditions [6]. This maintains the impact region of the flyer plate at near ambient conditions useful for performing symmetric impact experiments and for obtaining near-absolute Hugoniot measurements. Both of these capabilities have been long-standing goals of the shock community.

This new technique has been referred to as the "Isentropic Compression Experiment" or ICE after terminology first used by Lynn Barker. In this method, the sample is located on a planar surface of an anode directly connected to a cathode. Flow of a current between the two planar conductors produces a nearly planar magnetic field that loads the sample over a time interval of about 200 ns. Samples are mounted behind the planar anode conductor and experience a shockless pressure through the increasing magnetic field at the surface of the conductor. The method can be used to study up to twelve samples simultaneously. A large number of studies have been conducted with this technique, several of which are discussed in this conference.

Analysis techniques have been developed for analyzing the ramp waves produced by this technique, including a Lagrangian analysis approach [26] and a finite-difference analysis approach [27]. The method has been used to study the quasi-isentropic response of several metals and ceramics to pressures exceeding 3 Mbar; phase transitions, including dynamic freezing; effects of aging, grain size changes in metal alloys; the optical properties of laser window materials; and the effects of radiation damage. An inherent feature is the ability to detect very small changes in material properties, since the loading path of the material is directly studied, versus determination of end states produced by shock loading. For example, David Reisman and colleagues were able to easily detect changes in dynamic mechanical properties over the stress range of a couple hundred kbar in a radiated steel sample with radiation-induced atomic porosity on the order of 1% [28]. This would be extremely difficult with standard shock wave techniques because many separate experiments would be required. The typical error bars obtained in shock wave experiments are on the order of differences observed so that it would be difficult to clearly identify mechanical differences due to radiation damage.

One of the major spin-offs of the ICE technique involves launching of flyer plates at near ambient density to velocities exceeding 20 km/s [6]. A laboratory capability for doing this has been a long-standing goal of the shock wave community. The experimental configuration is similar to that for shockless loading except the flyer plate is designed so that it remains uniaxial in its central region for sufficiently long times to achieve the desired impact velocity, typically about 300 ns. Furthermore, the magnetic field must be confined during this time to states far from the impact region so that the density at the impact surface remains near its ambient value. In the experiments conducted on Z, flyer plates nearly 1 mm thick and 10 mm or so in lateral dimension were launched with a magnetic field, accelerated across a gap of about 3 mm and impacted onto a sample of comparable dimensions. The velocity of the flyer was measured to 1% or less and the shock speed in the sample was measured with similar accuracies, resulting in near absolute Hugoniot data can for symmetric impact experiments, with a slight density correction.

One of the major issues in these experiments concerned diffusion of the magnetic field through the flyer plate before impact. This effect produces joule heating that lowers the flyer density and can cause vaporization of the flyer. Traditional wave structure techniques turned out to be useful in diagnosing this problem. A series of 3-D magneto-hydrodynamics (MHD) simulations of the experiment indicated that the high-pressure conductivity of aluminum is crucially important to the successful launching of flyer plates, since it controls the rate of magnetic diffusion. Effects of magnetic diffusion could be detected through time-resolved measurement of the flyer plate velocity history and used to evaluate MHD simulations of the experiment [29]. These experiments allowed validation of the conductivity model, which was then used to optimize experimental configurations. In addition, measurement of the velocity history at an aluminum sample-LiF window interface after impact of the flyer plate with an aluminum sample provided information about the density gradient in the flyer plate, which is needed to make density corrections for Hugoniot experiments.

The magnetic flyer technique was used to determine the Hugoniot of aluminum [30], which is well known in the pressure range of a few Mbar, in order to qualify the technique, and also to determine the Hugoniot of liquid deuterium [31] to about 1 Mbar. EOS data on liquid deuterium recently obtained with high intensity lasers had suggested a softer response than expected from existing models in the high-pressure regime [32].

For the aluminum experiments, symmetric impact was used to obtain near absolute Hugoniot data from impactor and shock velocity measurements. For liquid deuterium, an impedance matching technique was used to determine the Hugoniot from measurement of shock velocities and the measured Hugoniot and unloading response of aluminum.

One of our goals was to ensure that the flyer plate technique produced a constant pressure drive in order to ensure that the experimental data were of high quality. At the pressures produced in these experiments, typical particle velocity measurements are not possible because laser windows do not maintain transparency at these pressures. Based on work at Lawrence Livermore National Laboratory, we found that the shock velocity in deuterium could be directly measured, since D_2 becomes conductive for shock pressures exceeding about 400 kbar [32]. These measurements indicated that the impact pressure was constant to better than 2% for times on the order of at least 30ns after impact. These measurements thus ensured that high-quality drives could be obtained with the flyer plate method. These measurements also emphasize that for ultra-high-pressure EOS measurements, using

VISAR or other interferometric techniques, it will be necessary to rethink how to probe the shock state. Direct measurements of the shock velocity in conductive windows with a VISAR is one way to gather this information.

A recent result from the magnetic flyer experiments illustrates the advantage of the high velocities and relatively long loading times possible with magnetic flyer plates for studying expanded states, now being referred to as warm dense matter [33].

In this experiment, aluminum was first shocked to varying pressure states up to 5 Mbar and released into either a silica aerogel buffer plate or into a vacuum. Impedance matching experiments with aerogel allowed determination of release to states of aluminum to about 700 kbar. In addition, full unloading into vacuum allowed study of the vaporous products for states exceeding the critical point. In agreement with data obtained on other materials, such as lead and cadmium [24], these results suggested a two-phase expansion of the liquid-vapor products, illustrating that both two-phase flow and kinetic effects of vaporization can be addressed. However, considerable work remains to work out the details for separating kinematic effects and material property information.

A final point illustrated by the unloading experiments is that the magnetic drive technique provides an opportunity for studying a broad range of the EOS surface, including ramp loading which approximates isentropic response, shock Hugoniot data, and unloading experiments that allow study of expanded states. The combination of these is powerful for developing broad-based EOS theories.

FUTURE OPPORTUNITIES

I've shown a few examples from my personal research and the research performed by colleagues I've worked with, which illustrate the breadth of problems that wave structure measurements have been applied to. These are only but a few of the large body of research that exists, but I hope they convey the usefulness of these techniques in probing material response. I'll close by discussing a few research areas, which I think, hold considerable promise for wave structure measurements.

In prior studies, shock wave structure measurements have been used to probe continuum material response for the most part, but one of the major challenges is study of dynamic material response at the mesoscopic or grain scale [18]. Initial results by Yuri Mescheryakov and coworkers [34] have shown that time-resolved interferometer measurements can provide important information about deformation processes occurring at the mesoscale. They have also begun development of theoretical models that relate mesoscale response to continuum properties. This is another exciting area, although systematic studies are necessary to obtain useful data at the mesoscale that can be directly correlated with continuum properties.

Another area of growing importance for both practical and scientific applications concerns the dynamic response of vapor and dense plasma states. The initial results obtained by several investigators illustrate that wave structure measurements are useful for studying time-dependent effects of expansion products and for providing information on kinetics and mixed phase flow.

A final example concerns the use of wave structure techniques to investigate magnetohydrodynamic phenomena, which have significant scientific and programmatic applications, but which have not been a predominant research area in the shock physics community. Initial results obtained with the magnetic flyer technique that connect hydrodynamic response to magnetic effects is encouraging and suggests that this MHD phenomena can be studied with these techniques.

ACKNOWLEDGMENTS

It is a great honor and I am extremely pleased to receive the Topical Group Shock Compression Science Award. I would like to express my gratitude to the many colleagues that I've worked with throughout the years who have made it possible for me to receive the award. Any successes I've achieved are really due to the many people I've been honored to work with. Although I can't name them all here, I am especially indebted to Arthur Guenther and George Duvall for providing the initial interest and direction in shock wave physics. I am especially grateful to my family and to my wife Pat who has provided support throughout my career.

REFERENCES

1. Davison, L. and R.A. Graham, in *Physics Reports* **55**, 257, 1979.
2. Graham, R.A. and Asay, J.R. High-Temp.-High Press. 10, 355, 1978.
3. Barker, L.M., and R.E. Hollenbach, J. Appl. Phys. **45**, 4872, 1974.
4. Chhabildas, L.C., Barker, L.M., Asay, J.R., Trucano, T.G., Kerley, G.I., and Dunn, J.E., **Shock Compression of Condensed Matter 1991**, ed. By S.C. Schmidt, R.D. Dick, J.W. Forbes, and D.G. Tasker, Elsevier Science Publishers, 1992.
5. Hall, C.A., Asay, J.R., Knudson, M.D., Stygar, W.A., Spielman, and Pointon, T.D., Rev. Scien. Instru. **72**, 1, 2001.
6. Hall, C.A., Knudson, M.D., Asay, J.R., Lemke, R., and Oliver, B., Int'l J. Impact Engin'g **26**, 275, 2001.
7. Fowles, G.R., J. Appl. Phys. **32**, 1475, 1961.
8. Johnson, J.N., Jones, O.E., and Michaels, T.E., J. Appl. Phys. **41**, 2330, 1970.
9. Asay, J.R., Fowles, G.R., Duvall, G.E., Miles, M.H., and Tinder, R.F., J. Appl. Phys. **45**, 2132, 1975.
10. Gilman, J.J. **Micromechanics of Flow in Solids**, McGraw Hill Book Co., 1969.
11. Taylor, J.W., J. Appl. Phys. **36**, 3146, 1965.
12. Asay, J.R., and Lipkin, J., J. Appl. Phys. **49**, 4242, 1978.
13. Asay, J.R., and Chhabildas, L.C., in **Shock Waves and High-Strain-Rate Phenomena in Metals**, M.A. Meyers and L.E. Murr (eds.), Plenum Press, NY, 1981.
14. Al'tshuler, L.V., Pavlovskii, M.N., Komissarov, V.V., and P.V. Makarov, Combustion, Explosion and Shock Waves 35, 92, 1999.
15. Lipkin, J., and Asay, J.R., J. Appl. Phys. **48**, 182, 1977.
16. Johnson, J.N., "Micromechanical considerations in shock compression of solids", in **High-Pressure Shock Compression of Solids**, J.R. Asay and M. Shahinpoor (eds.), Springer-Verlag, NY, 1993.

17. Meyers, M.A.., and Carvalho, M.S., "Shock front irregularities in polycrystalline metals", Mat. Sci. and Eng. **24**, 131, 1976.

18. Asay, J.R., and Chhabildas, L.C., "Paradigms and Challenges in Shock Wave Research", in **High-Pressure Shock Compression of Solids VI**, Y. Horie, L. Davison, and N.N. Thadhani (eds.), Springer-Verlag, NY, 2003.

19. Vogler, T.J., and Asay, J.R., this conference.

20. Johnson; J.N., Hayes, D.B., Asay, J.R., J. Phys. and Chem. Solids **35**, 501, 1974.

21. Asay, J.R., J. Appl. Phys. **45**, 4441, 1974.

22. Asay, J.R., J. Appl. Phys. **48**, 2832, 1977.

23. Asay, J.R., and Hayes, D.B., J. Appl. Phys. **46**, 4789, 1975.

24. Trucano, T. G. and Asay, J. R., J. Impact Eng'g; **5**, 645, 1986.

25. Asay, J.R., in **Shock Compression of Condensed Matter-1999**, ed. by M.D. Furnish, L.C. Chhabildas, and R.S. Hixson, AIP Conf. Proc. 505, 261, 2000.

26. Reisman, D.B., Toor, A., Cauble, R.C., Hall, C.A., Asay, J.R., Knudson, M.D., and Furnish, M.D., J. Appl. Phys. 89, 1625, 2001.

27. Hayes, D.B. and Hall, C.A., **Shock Compression of Condensed Matter-2002**, ed. by M.D. Furnish, N. Thadhani, and Y. Horie, AIP Conf. Proc. 620, 1177, 2002.

28. Reisman, D.B., Wolfer, W.G., Elsholz, A., Furnish, M.D., J. Appl. Phys. **93**, 8952, 2003.

29. Lemke, R.W., Knudson, M.D., Hall, C.A., Haill, T.A., Desjarlais, M.P., Asay, J.R., and Mehlhorn, T.A., Phys. Plasmas, April, 2003.

30. Knudson, M.D., Lemke, R.W., Hayes, D.B., Hall, C.A., Deeney, C., Asay, J.R., to be published, Appl. Phys., 2003.

31. Knudson, M.D., Hanson, D.L., Bailey, J.E., Hall, C.A., and Asay, J.R., Phys. Rev. Lett. **87**, 225501, 2001.

32. Collins, G.W., DaSilva, L.B., Celliers, P., Gold, D.M., Foord, M.E., Wallace, R.J., Ng, A., Weber, S.V., Budil, K.S., Cauble, R., Science **281**, 1178, 1998.

33. Knudson, M.D., private communication, 2003.

34. Meshcheryakov, Yu. I., and Divakov, A.K., Sov. Phys. Tech. Phys. **30**, 348, 1985.

CP706, *Shock Compression of Condensed Matter - 2003*
edited by M. D. Furnish, Y. M. Gupta, and J. W. Forbes
© 2004 American Institute of Physics 0-7354-0181-0/04/$22.00

DYNAMIC FRACTURE AND FRAGMENTATION

D. R. Curran

SRI International, 333 Ravenswood Ave., Menlo Park, CA 94025

Abstract. Modeling dynamic fracture and fragmentation has many applications ranging from space debris impact to armor penetration. Modeling scales range from microscopic descriptions at the grain level, to mesomechanical models that average the microprocesses over many grains, to standard continuum models, and up to engineering models at the structural scale. Designers and interpreters of experiments prefer to work at the highest scale possible to minimize computational time, but the higher scale models often have weak links to the underlying physics. This paper discusses current activity to tie the various scale models together by computationally simulating selected experiments at various scales and comparing the "knob settings," thereby linking the higher scale models to the lower scale ones. Examples are given from the community working on shrapnel hazards in large laser and pulsed power facilities and from the community working on earthquake propagation in the earth's crust.

INTRODUCTION

Modeling dynamic fracture and fragmentation is of importance in many applications. One application of current interest is the evaluation of shrapnel hazards in large laser or pulsed power facilities caused by the disassembly of irradiated components in the target chamber.

The field has grown tremendously since the beginning of fracture mechanics in the 1950s. One can divide the field into four modeling size scales: (1) microscopic modeling of failure at the grain or atomic level, (2) mesomechanical modeling in a relevant volume element (RVE), (3) continuum modeling, and (4) "global" modeling at the scale of engineering structures.

A comprehensive review of the field would be beyond the scope of this paper. Instead, this paper focuses on a specific question: How do we approach relating the models that arise from research conducted at the above four scales?

The main advantage of working with models at the largest appropriate scale is the saving in computational time. The micromodels are detailed at the atomic or grain level and can require days to weeks to run in a large-scale application. The mesomodels can often run in hours per problem. They are themselves continuum models but have the constraint that they can only be used in cases where the macro stress and strain fields vary slowly over an RVE. Simpler continuum models can often run faster than the mesomodels because they depend on fewer variables and require less iterative subcycling. Finally, the "global" models are analytic in nature, do not necessarily require a hydrocode, and are truly interactive, running in seconds on a desktop (or laptop) computer.

To add confidence to the use of the higher scale models, we need to relate the model parameters, or "knobs," at each level to each other. We also need to assess the hazards inherent in using a higher scale model, i.e., how much physics have we lost?

Another side to the story involves the experimental data that are used to validate and calibrate the models. Often the experiments are confined to the same scale as the models themselves. Clearly, we also need to relate the various scale experiments to each other.

In the remainder of this paper we first propose a general approach to this problem and then show two examples: one in which "micro" refers to events at the micrometer scale and one in which "micro" refers to events that may occur at the meter scale!

APPROACH

Figure 1 outlines a general approach to relating the models and associated experiments.

The plan is to perform overlapping calculational simulations of selected experiments. Some of the experiments are labeled "virtual," which refers to thought experiments in which simple loading conditions and material properties can be exactly specified, whereas in real experiments and materials they may not be easily measured.

There are a number of activities under way in the community that more or less are following the above program. These include penetration of ceramic armor, rock, and concrete; penetration of fabrics; fragmentation of irradiated metals; and propagation of macroscopic cracks in the earth's crust (where "micro cracks" may be meters in size!).

As specific examples for discussion, we have chosen the latter two cases: fragmentation of irradiated metals and propagation of macroscopic cracks in the earth's crust. Furthermore, in the first case we restrict ourselves to fragmentation of plastically deforming (ductile) metals, although fragmentation of elastic solids and melted material is also of interest.

FRAGMENTATION OF DUCTILE METALS

The microscopic experimental picture often shows void nucleation at grain boundaries, impurities, or second phase particles (1). Void growth over a wide range of void sizes has been found in post-test analyses of impacted specimens to obey a viscous growth law (1, 2). The observations also show a wide variety of void coalescence mechanisms, including ligament stretching, shear localization between voids, and void clustering.

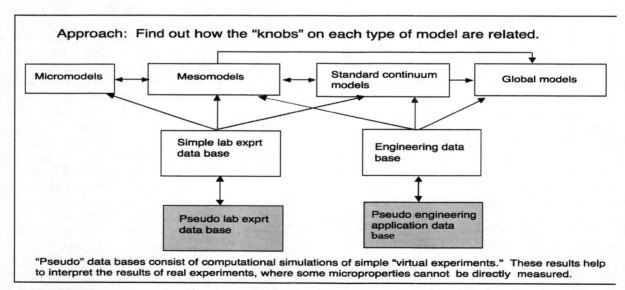

FIGURE 1. Proposed procedure for the transfer of micro- and mesomodel results to continuum and global models.

A number of papers in the present conference deal with such experiments and with molecular dynamics (MD) and other microscopic computations of such processes.

The corresponding meso parameters are found in the meso nucleation rate functions, void growth functions, and void coalescence algorithms. Overlapping calculations with micro- and mesomodels must be made to relate the micro parameters to the meso ones. In accordance with the above proposed procedure, simple lab experiments with well-characterized initial and boundary conditions are best suited for this task.

We have chosen, as an example, to begin with the simplest possible "virtual experiment," namely, imposed triaxial tensile strain histories.

In this paper we focus on the next step: linking mesomodel results to those of a continuum model. The mesomodel chosen is an updated version of the DFRACT model (1). The continuum/global model chosen is the energy balance (EB) model of Grady and colleagues (3-5).

The updated DFRACT model adds a void coalescence algorithm to the original version. The algorithm is based on three key assumptions: (a) coalescence is triggered at a critical relative void volume, ε_{FC} , (b) each void is surrounded by a fragment, and (c) the local porosity is equal to the overall porosity.

These assumptions lead to the result that the fragment size distribution is attainable directly from the void size distribution at the onset of coalescence:

$$N_g(R) = N_t(\text{coalesc.})\exp(-R/R_1) \qquad (1)$$

where $N_g(R)$ is the number of voids per unit volume with radii greater than R, $N_t(\text{coalesc.})$ is the void density at the onset of coalescence, and R_1 is the average void size in this assumed Poisson distribution. An equivalent spherical fragment with the same volume as the actual fragment has a radius R_F given by

$$R_F = R(\varepsilon_{FC}^{-1} - 1)^{1/3} \qquad (2)$$

The fragment size distribution corresponding to Eq.(1) is

$$N_g(R_F) = N_t(\text{coalesc.})\exp(-R_F/R_1) \qquad (3)$$

The "average fragment size" D is the average spacing between voids at coalescence:

$$D = N_t(\text{coalesc.})^{-1/3} \qquad (4)$$

Also,

$$\varepsilon_{FC} = 8\pi\, N_t(\text{coalesc.})R_1^3 = 8\pi\, (R_1/D)^3 \qquad (5)$$

Thus, the values of ε_{FC} and $N_t(\text{coalesc.})$ determine the fragment size distribution.

In DFRACT [1], the evolution of the void size distribution of Eq.(1) is given by nucleation and growth laws:

$$\partial N_t/\partial t = \omega(\exp[(\sigma_m - \sigma_{no})/\sigma_1] - 1), \;\; \sigma_m > \sigma_{no} \quad (6)$$

$$\partial R_1/\partial t = R_1(\sigma_m - \sigma_{go})/4\eta, \;\; \sigma_m > \sigma_{go} \qquad (7)$$

where ω is a nucleation rate coefficient, σ_m is the tensile mean stress, σ_{no} is the nucleation threshold stress, σ_1 is a stress sensitivity parameter, σ_{go} is the growth threshold stress (the growth rate is zero at lower stresses), and η is viscosity. Finally, the voids are nucleated with a size R_o , which in the calibrated DFRACT models was set at the easily observable value 1 μm, even though actual nucleation may occur at smaller sizes.

The DFRACT meso fragmentation model thus has seven input parameters (in addition to the density and modulus): ε_{FC} , R_o , ω. σ_{no} , σ_1 , σ_{go} , and $\tilde{\eta}$

We performed a number of calculations with the above mesomodel, using aluminum parameters from Ref. 1: Bulk modulus = 75 GPa, density = 2.87 g/cm³, R_o = 1 micron, ω = 3×10^9/cm³-s, σ_{no} = 300 MPa, σ_{go} = 150 MPa, σ_1 = 40 MPa, and η = 75 poise. In addition, we chose ε_{FC} = 0.05, although this value is arbitrary; an appropriate value must come from micro experiments and calculations.

We first chose an imposed strain history given by

$$\varepsilon = (t/\tau)^{1/4} \qquad (8)$$

where $\tau = 1$ s. This history expands the material to a strain of 0.05 in about 10 μs, and the strain rate ranges from about 10^6 s^{-1} at 10 ns to about 10^3 s^{-1} at 10 μs, i.e., over three orders of magnitude. The result is shown in Figure 2 below.

The calculated average fragment size D was 21 μm.

The calculation showed a burst of nucleation around the stress peak, during which the order of magnitude of D was established, followed by continuous nucleation of voids until coalescence occurred at a strain of 0.05. The later nucleation "fine-tuned" (reduced) the value of D by about 30%.

Similar calculations were done for constant imposed strain rates ranging from 10^4 s^{-1} to 10^8 s^{-1}.

We also performed a parameter sensitivity study in which the nucleation and growth parameters were varied. An interesting result was that D was only weakly dependent on the nucleation rate. Decreasing ω by seven orders of magnitude only increased the peak stress and D by about 30%. These depend mostly on the value of N_t and not on how fast it is attained.

This welcome result means that the number of necessary parameters in the EB model may be expected to be less than the seven of the mesomodel. Of course, more complicated load histories may yield a different conclusion. Nevertheless, we proceed to construct an EB model.

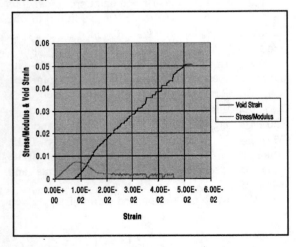

FIGURE 2. Calculated stress/modulus and relative void volume vs. total strain.

The original EB model proposed by Grady et al. (3-5) shows poor agreement with the DFRACT model, but this is to be expected because the two have different assumptions regarding the energy partition. For plastically deforming solids, Grady's original assumption was that the stored elastic energy is converted to plastic work $Y\varepsilon_c$, where Y is the yield strength and ε_c is a critical plastic strain.

To be consistent with the DFRACT model, we replace $Y\varepsilon_c$ with the viscoplastic work, which depends on the viscosity η. In addition, Grady's "horizon condition" is replaced by the assumption that the stress peak occurs at a strain equal to $8\pi N_t R_o^3 = \square\pi(R_o/D)^3$, where R_o is the void nucleation size (assumed to be 1 mm in DFRACT).

We thus replace $Y\varepsilon_{FC}$ with the integral of $(\sigma_m - \sigma_{go})\partial\varepsilon_F$ from ε_o to ε_{FC}, where ε_o is the strain at which the stress reaches its maximum. By assuming that the nucleation rate has become negligible at ε_o and that the strain rate is mostly due to void growth (assumptions only roughly correct), we can obtain

$$\sigma_m - \sigma_{go} = (4/3)\eta(\partial\varepsilon_F/\partial t)/\varepsilon_F \quad (9)$$

Then the viscoplastic work integral becomes $(4/3)\eta(\partial\varepsilon_F/\partial t)\ln(\varepsilon_{FC}/\varepsilon_o)$. The stored elastic energy is $(1/2)K\varepsilon_o^2$, where we assume that the voids have their initial nucleation size R_o, so

$$\varepsilon_o \approx 8\pi N_t R_o^3 = 8\pi(R_o/D)^3 \quad (10)$$

The energy balance then gives

$$(D/R_o)\{\ln[(\varepsilon_{FC}/8\pi)(D/R_o)^3]\}^{1/6}=[237K/\eta(\partial\varepsilon/\partial t)]^{1/6} \quad (11)$$

Because $\{\ln[(\varepsilon_{FC}/8\pi)(D/R_o)^3]\}^{1/6}$ is a very slowly varying function of (D/R_o) and is of the order of unity, the original energy balance result is changed to

$$D/R_o \approx [237K/\eta(\partial\varepsilon/\partial t)]^{1/6} \quad (12)$$

Comparison of the DFRACT results with (12) is shown in Figure 3.

It can be seen that the continuum model gives rough agreement with the meso model, at least for

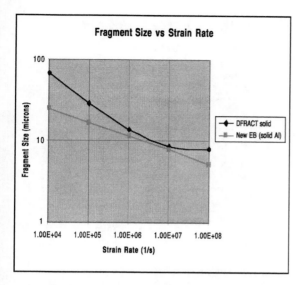

FIGURE 3. Comparison of DFRACT meso and EB continuum model results for fragment size vs. strain rate.

the higher strain rates. Furthermore, the number of parameters has been reduced from seven to two.

We conclude that linking mesomodels to continuum and global models for ductile fracture appears possible, although more work is needed to confirm this preliminary conclusion. In particular, it may well be that for more complex loading histories, the above EB simplifications will be inadequate.

PROPAGATION OF MACROCRACKS IN THE EARTH'S CRUST

The next example has been chosen at the other extreme of scales: the propagation of kilometer-long cracks in the earth's crust. Such cracks are of interest to the earthquake community. They advance by the nucleation, growth, and coalescence of microcracks in a "process zone" at the macrocrack tip. The "microcracks" range in size from centimeters to meters. The confining pressure ranges from zero at the earth's surface to about 0.3 GPa at 15 km depth.

The most common continuum model used in classical fracture mechanics is based, for fracture in tension (Mode I), on the initiation toughness K_{IC} and the crack propagation toughness K_{ID}, where

K_{ID} is a function of crack velocity. Under shear (Mode II), the corresponding parameters are K_{IIC} and K_{IID}.

These parameters are a measure of the energy absorbed in the process zone as the macrocrack advances by driving the microcracks to coalescence.

A key problem is that typical laboratory measurements of the above Ks are performed on centimeter-sized specimens, where the grains and microcracks are micrometers in size. The challenge is to extrapolate these results six orders of magnitude in scale!

A way to begin is to perform laboratory experiments at varying confining pressures and to carefully characterize the microcrack distributions in the process zone. It is hoped that if we can develop good micro- and mesomodels of the process zone at laboratory scale, we will get valuable clues as to how to extend the models to full scale.

A key laboratory experiment is the edge impact experiment performed by Kalthoff and colleagues (6). The main features of the experiment are shown in Figure 4.

Measurements with this and other techniques on Solnhofen limestone gave results that are unusual for brittle materials, namely, the energy absorbed in the process zone increased with macrocrack velocity, with the result that the macrocrack propagated for unusually great distances before reaching its limit velocity (governed by sound speed) (6).

The explanation must lie in microcrack activity in the process zone. To study this question further and to lay the groundwork for extrapolating the laboratory results to full scale, Kalthoff and colleagues are planning edge impact experiments at varying levels of confining pressure. Extensive post-test characterization of the microcrack orientations and sizes in the process zone will be carried out.

These results should form the basis for the development of micro- and mesomodels and for establishing the desired links to the continuum fracture toughness values.

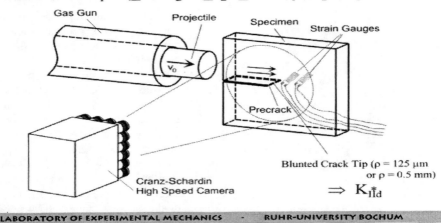

FIGURE 4. Edge impact experiment. Courtesy of Prof. Joerg Kalthoff, Ruhr-University, Bochum

CONCLUSIONS

The proposed program for relating models at different scales seems feasible. Recent advances in MD calculational capabilities appear capable of putting mesomodels on firmer ground, but the mesomodels may have to be made more complex to describe the micro void or crack coalescence behavior.

Improved mesomodels may be able to put continuum and global models on firmer ground, but the latter two will probably have to add knobs to handle complex loading conditions.

Many calculations with the models on overlapping problems will be necessary.

ACKNOWLEDGMENTS

Thanks are due to Professor Joerg Kalthoff and Mr. Andreas Bertram, Ruhr – University Bochum, for permission to discuss some of their preliminary experimental results and to Dr. James Belak, Lawrence Livermore National Laboratory, for providing information regarding recent advances in MD calculations.

REFERENCES

1. Curran, D. R., Seaman, L., and Shockey, D. A., *Physics Reports* **147**, 253-388 (1987).
2. Antoun, T., Seaman, L., Curran, D. R., Kanel, G. E., Razorenov S. V., and Utkin, A. V., *Spall Fracture*, Springer, New York, 2003, pp. 78-81, 225-236.
3. Grady, D. E., "The Spall Strength of Condensed Matter," *J. Mech. Phys. Solids* **3**, 322-384 (1988).
4. Grady, D. E., "Spall and Fragmentation in High-Temperature Metals," in *High-Pressure Shock Compression of Solids II – Dynamic Fracture and Fragmentation*, edited by L. Davison et al., Springer, New York, 1996, pp. 219-236.
5. Kipp, M. E., and Grady, D. E., "Experimental and Numerical Studies of High-Velocity Impact Fragmentation," in *High-Pressure Shock Compression of Solids II – Dynamic Fracture and Fragmentation*, edited by L. Davison et al., Springer, New York, 1996, pp. 282-339.
6. Bertram, A., and Kalthoff, J. F., "Bruchenergie laufender Risse in Gestein," 34. *Materialprüfung* **45**, No. 3, Carl Hanser Verlag, Munchen, Germany (2003).

CP706, *Shock Compression of Condensed Matter - 2003*
edited by M. D. Furnish, Y. M. Gupta, and J. W. Forbes
© 2004 American Institute of Physics 0-7354-0181-0/04/$22.00

NEW FINDINGS IN STATIC HIGH-PRESSURE SCIENCE

Russell J. Hemley and Ho-kwang Mao

*Geophysical Laboratory, Carnegie Institution of Washington,
5152 Broad Branch Road NW, Washington, DC 20015*

Abstract. Recent static high P-T experiments using diamond anvil cell techniques reveal an array of phenomena and provide new links to dynamic compression experiments. Selected recent developments are reviewed, including new findings in hot dense hydrogen, the creation of new metals and superconductors, new transitions in molecular and other low-Z systems, the behavior of iron and transition metals, chemical changes of importance in geoscience and planetary science, and the creation of new classes of high-pressure devices based on CVD diamond. These advances have set the stage for the next set of developments in this rapidly growing area.

INTRODUCTION

Accelerating developments in static high-pressure techniques over the last decade have created myriad new opportunities for research on materials under extreme conditions [1]. Specifically, the advances in dynamic ultrahigh pressure techniques such as x-ray driven shocks produced from intense laser sources, gas-guns, and magnetic compression methods [2] parallel numerous recent developments in static high-pressure physics, principally as a result of the continued refinements of diamond-anvil cell (DAC) techniques. Indeed, recent advances in static high P-T methods provide new links to dynamic compression experiments. These include major improvements in both laser heating and resistive heating experimentation coupled with a broad range of analytical probes such as x-ray, optical, and transport techniques. The results provide crucial tests of theory. Here we review selected recent findings and developments, focusing on areas of overlap between static and dynamic compression, both areas of agreement as well areas where there are differences and thus where the two approaches are leading to new questions.

HOT DENSE HYDROGEN

The high P-T behavior of the isotopes of hydrogen is central to our understanding of dense matter. This knowledge includes accurate determinations of equations of state (EOS), phase transitions, electronic transitions, and thermodynamic properties, in both the solid and fluid state to multimegabar pressures (>300 GPa) over a wide temperature range, and ultimately to the high-density plasma [3,4] and the volume collapse reported at <100 GPa but higher temperature [5,6] (see also Ref. [7,8]). Studies of solid hydrogen at very high densities during the past decade have revealed marked changes in optical response and a complex phase diagram from 100 to 300 GPa (corresponding to >7 to 13-fold compression of the solid [9,10]). A recent experiment has confirmed our earlier observations of visible absorption in hydrogen [11] at the highest pressures reached so far on the material under static compression (~300 GPa) [12]. It has also been proposed that the melting line drops at this pressure, giving possible reentrant melting on compression at low temperature (<300 K) [13-15]. Accurate high P-T studies are required to examine these issues.

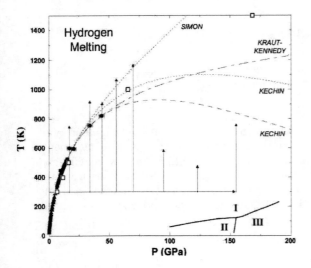

FIGURE 1. Hydrogen melting line. The solid symbols are static compression data and the arrows indicate representative P-T paths followed from Ref. [16]; the open squares are predictions from classical simulations [17]. Various extrapolated melting laws fit to experimental data are shown, including the two recent Kechin law analyses [14,16], along with the I-II-III solid-solid phase boundaries [10].

Recent measurement of the sound velocity as a function of pressure and temperature have provided improved determination of the EOS, as well as thermodynamic properties, of the fluid. Previously, fluid and crystalline hydrogen have been measured to above 20 GPa at high temperature using Brillouin scattering techniques [18]. The measurements have led to the construction of new effective intermolecular potentials that were consistent with all static and shock-wave data available at the time [19]. Fluid perturbation theory calculations performed with the potential were used to predict sound velocities and the EOS of fluid H_2 at higher P-T conditions, including those found within the molecular layer of Jovian planets. This work is particularly important for comparison with Z-accelerator and laser shock determinations.

By using different scattering geometries, the Brillouin technique can be used to constrain the refractive indices as well, which can be cross-checked, and for VISAR measurements in laser shock experiments.

Recent developments have made it possible to contain hydrogen in resistively heated DACs to still higher P-T for Raman measurements [16]. Specifically, measurements of the solid and fluid phases have been obtained to above 1100 K near 70 GPa and to above 650 K in 150-GPa range, conditions previously inaccessible by static compression experiments. The magnitude of the vibron frequency temperature derivative increases by a factor of ~30 over the measured pressure range, indicating an increase in intrinsic anharmonicity and weakening of the molecular bond. Moreover, the data give a direct measure of the melting curve that extends previous optical investigations by up to a factor of four in pressure. Use of the Kechin law gives an even lower melting maximum, however, other fits do not. Interestingly, the RRY model provides a reasonably reliable match to the melting line in the lower pressure range but does not give a melting maximum [17].

Although a conducting state has been observed in shocked high-temperature fluid hydrogen at 140 GPa and temperatures estimated to be 3000 K [3], at twice the pressure (285 GPa) and low temperatures (20-140 K), hydrogen remains an insulator, as shown by optical spectroscopy [20]. Direct measurements of electrical conductivity up to 210 GPa also confirm that low-temperature solid hydrogen is non-conducting (at and below 300 K) [21]. The pressure dependence of the electronic spectrum (*e.g.*, band gap and excitons), which are largely outside the optical observation window of the DAC, can be accessed using inelastic x-ray scattering (IXS) [22].

NOVEL METALS AND SUPERCONDUCTORS

One of the most important findings in studies of metals under pressure in recent years has been the surprising structural complexity that can obtain in ostensibly simple metals. The complexity of lithium has been reviewed, and is a case in point.

Direct electrical measurements to 30 GPa [23] and combined electrical and magnetic susceptibility studies to 80 GPa [24] revealed that the material superconducts under pressure, the latter study indicating a maximum T_c of 16 K. More recent measurements carried out in a helium medium are in accord although no T_c could be measured in the higher pressure range with that technique [25]. Raman measurements in ^{7}Li and ^{6}Li to 123 GPa at 180 K reveal high-frequency band above 60-70 GPa, which can be interpreted as a result of the phase transition to the "paired" state in fair agreement with theoretical predictions. The spectral changes coincide with the phase transition above 70 GPa in measurements of the pressure dependence of T_c. However, the observed frequency is too high compared to the calculations. The reported results should stimulate further theoretical and experimental work on the unusual "paired" state in Li at high densities.

Another novel low-Z metal and superconductor is oxygen [26,27]. Raman measurements in oxygen under high static pressures (up to 134 GPa) clearly demonstrate that the material remains molecular through the transition to metallic phase [28]. Contrary to previous studies, recent work reveals only a moderate decrease in intensity of the Raman vibron mode at the transition. Moreover, a rich low-frequency spectrum is observed. The transition may be accompanied by a small reorientation of the molecules accompanied by relatively weak changes in molecular bonding. A large pressure range of phase coexistence and pressure hysteresis in the transformation is observed.

TRANSITION METALS: EXAMPLE OF IRON

Common metals exhibit intriguing and sometimes unexpected phenomena over the range of P-T conditions now accessible experimentally. An excellent example is Fe, a material of obvious primary importance in geophysics. But it is also a metal that has been representative of the confluence of static and dynamic compression for the past 50 years. The α-ε transition in Fe was first clarified in early shock-wave experiments, but it continues to be of interest. In particular, the relationship between the mechanism of the transition and the texture development on static compression in ε-Fe has been a topic of current interest experimentally and theoretically. This is also important for understanding seismological evidence for texture in the Earth's inner core.

By the use of x-ray transparent gasketing techniques for radial x-ray diffraction, the ellipticity of the diffraction rings induced by non-hydrostatic strains can be measured. From the intensities, we obtain information on the development of texture, which can be recast in terms of inverse pole figures. The d-spacings as a function of ϕ measured by radial diffraction reveal the strength, and the intensities of each reflection *(hkl)* as a function of ϕ give the texture, which can be monitored as a function of time at a given pressure. Studies at high temporal, spatial, and wavelength resolution require a dedicated high-energy synchrotron facility. One method for approaching this problem derives moduli explicitly from the functional behavior of d-spacings in various directions [29]. A second method involves reproducing d-variations for different coefficients with FEM polycrystal plasticity models and finding an optimal fit [30], revealing the active slip systems. For example, we could determine that in ε-Fe basal slip is active based on energy dispersive diffraction data to 220 GPa [31]. More recently, we have obtained much higher resolution data based on angle-dispersive diffraction in combination with laser annealing. These results confirm the earlier results but provide more accurate determination of both the deformation mechanisms and single-crystal elastic moduli [32].

The higher P-T regime has also been an important meeting ground for static and dynamic compression experiments (Fig. 2). The presence of an additional high P-T phase was suggested in the Hugoniot study of Brown and McQueen [33]. Definitive resolution of this question should come from high P-T x-ay diffraction measurements. The melting line of Fe has been determined, with no evidence for additional phases. The dataset, plus its thermodynamically constrained extension, may be compared with shock data. Interestingly, these data and their analysis give a melting temperature of 5800 K at the pressure of the inner core boundary, which is a constraint on the temperature at depth within the planet (Fig. 2). Despite the wide stability field of ε-Fe and the absence of many proposed or inferred phases (see Ref. [34]), the possibility of additional high P-T phases remains.

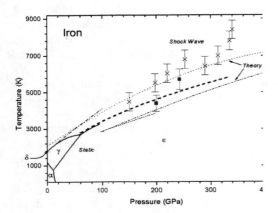

FIGURE 2. High P-T phase diagram of iron [34]. The solid lines were determined from static compression experiments; the thick lines is the thermodynamically consistent extrapolation of the ε-Fe melting line; the squares and crosses are Hugoniot data in the vicinity of melting from Refs. [33] and [35], respectively; the thin dashed lines are first-principles calculations of melting (see Ref. [34]).

This work has been complemented by spectroscopic studies of Fe. Nuclear resonant inelastic x-ray scattering (NRIXS) gives detailed information on thermodynamics and elasticity through the phonon density of states (DOS). The results may be compared with other studies of the phonons of Fe under pressure, including non-resonant IXS [36] and Raman spectroscopy [37]. Most recently, NRIXS has been extended to high temperature with laser heating techniques [38]. As discussed above, the structure of liquids can be directly examined with new techniques. High P-T diffraction studies of liquid Fe to 2500 K in large volume (Paris-Edinburgh, P-E) cells [39] reveal combined P-T induced structural changes in the liquid. The new generation of large volume DACs will allow the extension of these methods to beyond the maximum pressures of P-E cells (which so far are limited to below ~25 GPa).

The phonon DOS can be obtained by IXS measurement of polycrystalline samples integrated through the entire Brillouin zone. The NRIXS method developed for the Mössbauer isotope ^{57}Fe has produced DOS measurements of α-Fe at ambient conditions [40,41] identical to results obtained from inelastic neutron scattering (INS) [42]. The NRIXS technique carried out to 153 GPa [43] has been improved by a factor of ten in sensitivity, and a well-defined DOS of Fe in a helium medium has been obtained to 50 GPa. These properties have previously eluded direct experimental study and have been dependent solely upon theoretical calculations. Very recent developments have extended high-pressure NRIXS of Fe to 1500 K with resistive and laser heating. NRIXS spectroscopy has been successfully applied to other Mössbauer-active nuclei.

ALTERED CHEMISTRY AND CHEMICAL BONDING

We return to recent findings in molecular systems and examples of changes in chemistry and chemical bonding induced by high pressures and temperatures. Investigations of the melting of ice VII has been extended to higher pressure using x-ray diffraction techniques, extending considerably previous work. Notably, there is a large disagreement with the results of Ref. [14] though we obtained excellent agreement with their work in the region of overlap for H_2. The results provide a constraint on the EOS of the fluid at high pressure, extending the measurements of Ref. [44]. The P-V-T EOS of supercritical fluid H_2O was been extended near the ice VII melting curve using monochromatic x-ray tomography methods [45]. The method is currently applicable over the P-T range of resistively-heated DACs for determination of fluid densities with ±1% accuracy.

The chemistry of hydrocarbons under very high pressure is of fundamental interest, as well as being important for combustion, geoscience, and planetary science. Static compression experiments carried out nearly 40 years ago established that hydrocarbons subjected to high temperatures above 15 GPa break down to form diamond [46]. Subsequent shock-wave compression experiments showed a large increase in density along the Hugoniot at these pressures that were interpreted in terms of the breakdown transition [47]. These observations suggested that diamond and possibly related refractory high-density, low-Z materials may form the cores of large planets (*e.g.*, Ref. [48]). Later first-principles calculations predicted that methane condenses to form heavier hydrocarbons (*i.e.*, ethane), that is, prior to the presumed complete dissociation into the elements [49,50]. Infrared experiments establish that methane can persist to pressures well above a megabar on compression at 300 K [51] (Fig. 3a). In contrast, laser heating of methane at pressures at pressures below 15 GPa results in breakdown of the material [52] (see also Ref. [53]). Raman measurements reveal the presence of higher hydrocarbons and the formation of free hydrogen (Fig. 3b), in qualitative agreement with theoretical predictions (but occurring at much lower pressures).

A long-standing question is the formation and stability of methane in the Earth's mantle, and the extent to which the mantle could be a source of methane and even heavier hydrocarbons in reservoirs in the crust. In fact, whether or not natural gas and heavier hydrocarbons can be created on the Earth and high *P-T* condition has been the subject of debate for about fifty years. Surprisingly, very few experiments have been done to examine this. Recently, we have looked into this question using high *P-T* laser heating and Raman spectroscopy techniques. Methane readily forms on heating $CaCO_3$, FeO, and H_2O at 5 GPa. The results are consistent with thermodynamic calculations, which also predict the formation of higher hydrocarbons [54].

There is evidence for new kinds of covalent interactions that develop under high compression, including the formation of high energy density materials such as polymeric nitrogen and polynitrogen. Prompted by seminal theoretical work [55,56] and subsequent experimental studies [57], great progress was made during the past year with the synthesis of polymeric nitrogen above 150 GPa and its low-temperature recovery at ambient pressure [58,59]. Recent experiments have uncovered an array of examples of new high *P-T* chemistry in low-Z materials, including new bonding patterns in CO_2 [60-62], CO [63], and N_2O [64-66], and HMX reactions [67]. The above static high-pressure chemical studies provide an important benchmark for the interpretation of shock data, as demonstrated for the high *P-T*

breakdown of CO_2 [62,68] as well as CH_4 described above [47,52,53]. The mixing and diffusion of active species and the thermal conductivity of the mixtures remain major experimental and theoretical challenges.

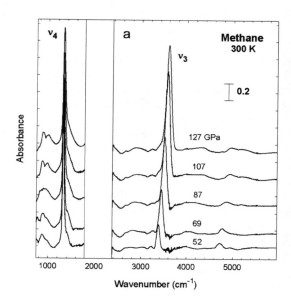

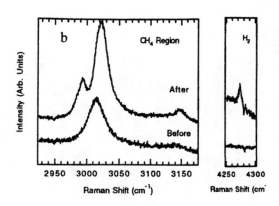

FIGURE 3. High-pressure behavior of methane. (a) Synchrotron infrared measurements at 300 K [51]. (b) Raman measurements at 15 GPa before and after heating with a CO_2 laser [52].

In principle, spectroscopic features near core electron absorption edges measured by x-ray absorption (XANES) reveal rich information about such changes in bonding character. These types of measurements for low-Z elements have not been possible high-pressure, as the pressure vessels completely block the soft x-ray (<4 keV). The near edge spectra can also be obtained by IXS but has been limited to ambient pressure studies. With recent advances in inelastic x-ray scattering spectroscopy, this limitation has been removed. The inelastic K-edge scattering of Be (111 eV), B (188 eV), C (284 eV), N (410 eV), and O (543 eV) under pressure has been observed.

The technique has been used to probe bonding changes in the pressure-induced transition in graphite near 15 GPa [69]. A transition in a graphitic h-BN has been the subject of a related study [70]. The experiment started with graphitic h-BN, the near edge peaks corresponding to the $\pi*$ and $\sigma*$ bonding of B and N [71,72] observed at low pressure. Above 12 GPa, the σ features grow at the expense of those associated with π, indicating that bonding changes are associated with the transition. The high-pressure phase is a framework superhard material. High-pressure near K-edge spectroscopy of the second row elements (Li to Ne) and near L-edge spectroscopy from Na to transition elements is now open to investigation.

FIGURE 4. Single crystal diamond produced by chemical vapor deposition [73]. The diamond is approximately 2.5 mm high and was grown in one day starting with a conventional type 1b as seed, which forms the table (base) of the anvil.

CVD DIAMOND AND NEW HIGH-PRESSURE DEVICES

Continued developments in high-pressure science depend critically upon advances in instrumentation. Among the most essential needs is the enlargement of sample volume without compromising P-T range and accessibility for *in situ* measurements. This means producing large gem anvils. One route is to employ other anvil materials that are more easily enlarged; this includes single-crystal SiC (moissanite), which has been used as an anvil material to reach pressures above 65 GPa [74,75]. Considerable progress has been made in scaling up this device with supported (belt-type) assemblies in neutron diffraction experiments[76].

Another approach is to stay with diamond. During the past two years we have had considerable success producing single crystal diamond by microwave plasma chemical vapor disposition (CVD; Fig. 4) [77]. Recent experiments show that these hybrid high P-T/CVD diamonds are at least as strong as the conventional diamonds we use, reaching at least 245 GPa based on x-ray diffraction of metal standards [78] (*e.g.*, Pt whose EOS determined by shock-wave techniques [79]). On high P-T annealing these yellow diamonds become transparent. Moreover, we have found all of these single crystals have very high fracture toughness relative to natural and synthetic diamond. But more remarkable is the very high hardness of these same diamonds that have been subjected to high P-T processing. These have Vickers hardnesses above 160 GPa [73]. In addition to its importance for static pressure experiments, this new single-crystal material should enable new classes of shock-compression studies.

DISCUSSION AND OUTLOOK

The central theme of this paper has been the variety of new findings in materials subjected to the combination of extreme P-T conditions. The P-V-T EOS is one of the most basic pieces of information needed for predicting material behavior, verifying theoretical calculations, and pressure calibration. A decade of co-development of synchrotron x-ray techniques for accurate EOS studies of metals to >300 GPa has set the stage for the next level of development. Determination of melting curves of metals has been controversial because of uncertainties in *P-T* measurements and melting criteria. With the improvements in high *P-T* methods described above, accurate determinations of melting curves are now possible.

Characterization by combined static and dynamic experiments provides the opportunity for major advances in our understanding of many kinds of high *P-T* material behavior. For example high *P-T* rheological studies have important implications for a wide range of static pressure work, as uncertainties in the strength of materials are one of the most important sources of systematic errors in static pressure calibrations. Understanding the behavior of materials at high plastic strain ($\varepsilon_P >$ 100%), and high strain rates (up to 10^5) is crucial for many problems. Constitutive properties such as plasticity, strength, failure, fracture, anisotropy, texture, and preferred orientation under pressure are intimately associated with sample strain. Importantly, the texture provides data for the interpretation of active deformation mechanisms, and phase transformations induced by pressure and temperature produce systematic changes in texture patterns.

The confluence of developments described above is also providing new synergy between experiment and first-principles theory. Just as in theoretical multiscale treatments [80], we have for the first time an opportunity to explore multiscale phenomena experimentally, starting with local structure (atomic level spectroscopic probes), multiple nanoscale effects (vibrational/magnetic excitations, diffraction), to grain aggregates and extended defects (imaged *in situ* by x-ray topography), to entire sample volumes (x-ray radiography, optical image analysis).

The future is particularly bright for new classes of high-pressure synchrotron and neutron scattering experiments. In addition to the non-resonant inelastic scattering studies discussed above, resonant inelastic x-ray scattering (RIXS) techniques been used to probe specific electronic excitations (*e.g.*, in NiO [81]). Detailed study of phonon dynamics in highly compressed materials has eluded study until recently. Emerging techniques have demonstrated the potential of

enabling full phonon dynamics studies at high *P-T*. The results can be used to bridge the gap between the shock-wave and DAC experiments and to validate theory. Fourth generation (coherent) synchrotron radiation sources, as well as the next generation neutron scattering facilities are coming on line.

Finally, the high *P-T* properties of the materials described above are closely interrelated. With the growing versatility of high-pressure devices, multiple measurements can be performed and multiscale properties determined on the same sample under well characterized conditions of *P, T*, stress, and sample history. New large volume megabar techniques should provide uniquely powerful experiment input to multiscale modeling, which requires examination of properties over the full spectrum of length scales from the atomic to the macroscopic. With new large and "smart" anvils, combined susceptibility, electrical, optical, x-ray, and neutron experiments measurements are now possible. Also exciting is the direct combination of static and dynamic compression techniques, for example, in shock studies on precompressed samples [82]. Moreover, detailed experimental and theoretical studies of the new anvils is setting the stage for reaching still higher static pressures.

ACKNOWLEDGEMENTS

We are grateful to numerous colleagues for their input into the work described above. We also thank S. Gramsch for comments on the manuscript. This was supported by DOE (CDAC), NSF, NASA, and the W. M. Keck Foundation.

REFERENCES

1. Hemley, R. J., Chiarotti, G., Bernasconi, M. & Ulivi, L. (eds.) *High-Pressure Phenomena, Proceedings of the International School of Physics "Enrico Fermi"* (North Holland, New York).
2. Asay, J. R. *et al.* Use of Z-pinch sources for high-pressure equation-of-state studies. *Int. J. Impact Engin.* **23**, 27-38 (1999).
3. Weir, S. T., Mitchell, A. C. & Nellis, W. J. Metallization of fluid molecular hydrogen at 140

GPa (1.4 Mbar). *Phys. Rev. Lett.* **76**, 1860-1863 (1996).

4. Nellis, W. J., Weir, S. T. & Mitchell, A. C. Minimum metallic conductivity of fluid hydrogen at 140 GPa (1.4 Mbar). *Phys. Rev. B* **59**, 3434-3449 (1999).

5. Collins, G. W. *et al.* Measurements of the equation of state of deuterium at the fluid insulator-metal transition. *Science* **281**, 1178-1181 (1998).

6. Mostovych, A. N., Chan, Y., Lehecha, T., Schmitt, A. & Sethian, J. D. Reflected shock experiments on the equation-of-state properties of liquid deuterium at 100-600 GPa (1-6 Mbar). *Phys. Rev. Lett.* **85**, 3870-3873 (2000).

7. Knudson, M. D., Hanson, D. L., Bailey, J. E., Hall, C. A. & Asay, J. R. Equation of state measurements in liquid deuterium to 70 GPa. *Phys. Rev. Lett.* **87**, 225501 (2001).

8. Militzer, B. *et al.* Calculation of a deuterium double shock Hugoniot from ab initio simulations. *Phys. Rev. Lett.* **87**, 275502 (2001).

9. Mao, H. K. & Hemley, R. J. Ultrahigh-pressure transitions in solid hydrogen. *Rev. Mod. Phys.* **66**, 671-692 (1994).

10. Hemley, R. J. & Mao, H. K. Progress in cryocrystals to megabar pressures. *J. Low Temp. Phys.* **122**, 331-344 (2001).

11. Mao, H. K. & Hemley, R. J. Optical observations of hydrogen above 200 gigapascals: evidence for metallization by band overlap. *Science* **244**, 1462-1465 (1989).

12. Loubeyre, P., Ocelli, F. & LeToullec, R. Optical studies of hydrogen to 320 GPa and evidence for black hydrogen. *Nature* **416**, 613-617 (2002).

13. Ashcroft, N. W. in *High-Pressure Phenomena, Proceedings of the International School of Physics "Enrico Fermi"* (eds. Hemley, R. J., Chiarotti, G., Bernasconi, M. & Ulivi, L.) in press (North Holland, New York).

14. Datchi, F., Loubeyre, P. & LeToullec, R. Extended and accurate determination of the melting curves of argon, helium, ice (H_2O), and hydrogen (H_2). *Phys Rev. B* **61**, 6535-6546 (2000).

15. Scandolo, S. Liquid-liquid phase transition in compressed hydrogen from first-principles simulations. *Proc. Nat. Acad. Sci.* **100**, 3051-3053 (2003).

16. Gregoryanz, E., Goncharov, A. F., Matsuishi, K., Mao, H.-K. & Hemley, R. J. Raman spectroscopy of hot dense hydrogen. *Phys. Rev. Lett.* **90**, 175701 (2003).

17. Ross, M., Ree, F. H. & Young, D. A. The equation of state of molecular hydrogen at very high density. *J. Chem. Phys.* **79**, 1487-1494 (1983).

18. Matsuishi, K., Gregoryanz, E., Mao, H.-K. & Hemley, R. J. Equation of state and intermolecular interactions in fluid hydrogen from Brillouin scattering at high pressures and temperatures. *J. Chem. Phys.* **118**, 10683-10695 (2003).

19. Duffy, T. S., Vos, W., Zha, C. S., Hemley, R. J. & Mao, H. K. Sound velocity in dense hydrogen and the interior of Jupiter. *Science* **263**, 1590-1593 (1994).

20. Goncharov, A. F., Gregoryanz, E., Hemley, R. J. & Mao, H. K. Spectroscopic studies of the vibrational and electronic properties of solid hydrogen to 285 GPa. *Proc. Nat. Acad. Sci.* **98**, 14234-14237 (2001).

21. Hemley, R. J., Eremets, M. I. & Mao, H. K. in *Frontiers of High Pressure Research II:* (eds. Hochheimer, H. D., Kuchta, B., Dorhout, P. K. & Yarger, J. L.) 201-216 (Kluwer, Amsterdam, 2002).

22. Mao, H. K., Kao, C. C. & Hemley, R. J. Inelastic scattering at ultrahigh pressures. *J. Phys. Condens. Matter* **13**, 7847-7858 (2001).

23. Shimizu, K., Ishikawa, H., Takao, D., Yagi, T. & Amaya, K. Superconductivity in compressed lithium at 20 K. *Nature* **419**, 597-599 (2002).

24. Struzhkin, V. V., Eremets, M. I., Gan, W., Mao, H. K. & Hemley, R. J. Superconductivity in dense lithium. *Science* **298**, 1213-1215 (2002).

25. Deemyad, S. & Schilling, J. S. Superconducting phase diagram of Li metal in nearly hydrostatic pressures up to 67 GPa. *Phys. Rev. Lett.* **91**, 167001 (2003).

26. Desgreniers, S., Vohra, Y. K. & Ruoff, A. L. Optical response of very high density oxygen to 132 GPa. *J. Phys. Chem.* **94**, 1117-1122 (1990).

27. Shimizu, K., Suhara, K., Ikumo, M., Eremets, M. I. & Amaya, K. Superconductivity in oxygen. *Nature* **393**, 767-769 (1998).

28. Goncharov, A. F., Gregoryanz, E., Hemley, R. J. & Mao, H. K. Molecular character of the metallic high-pressure phase of oxygen. *Phys. Rev. B* **68**, 100102 (2003).

29. Singh, A. K., Mao, H. K., Shu, J. & Hemley, R. J. Estimation of single-crystal elastic moduli from polycrystalline x-ray diffraction at high pressure: Applications to FeO and iron. *Phys. Rev. Lett.* **80**, 2157-2160 (1998).

30. Dawson, P., Boyce, D., MacEwen, S. & Rogge, R. Residual strains in HY100 polycrystals: Comparisons of experiments and simulations. *Metall. Mater. Trans.* **31A**, 1543-1555 (2000).

31. Wenk, H.-R., Matthies, S., Hemley, R. J., Mao, H. K. & Shu, J. The plastic deformation of iron at pressures of the Earth's inner core. *Nature* **405**, 1044-1047 (2000).

32. Merkel, S., Wenk, H. R., Gillet, P., Mao, H.-K. & Hemley, R. J. Deformation of polycrystalline iron

up to 30 GPa and 1000 K. *Phys. Earth and Planet. Inter.* **in press** (2003).

33. Brown, J. M. & McQueen, R. G. Phase-transitions, gruneisen-parameter, and elasticity for shocked iron between 77-GPA and 400 GPA. *Geophys. J. R. Astron. Soc.* **91**, 7485-7494 (1986).

34. Hemley, R. J. & Mao, H. K. In situ studies of iron under pressure: new windows on the Earth's core. *Internat. Geol. Rev.* **43**, 1-30 (2001).

35. Yoo, C. S., Holmes, N. C. & Ross, M. Shock temperatures, melting and phase diagram of iron at earth core conditions. *Phys. Rev. Lett.* **70**, 3931-3934 (1993).

36. Fiquet, G., Badro, J., Guyot, F., Requardt, H. & Krisch, M. Sound velocities in iron to megabar pressures. *Science* **291**, 468-471 (2001).

37. Merkel, S., Goncharov, A. F., Mao, H. K., Gillet, P. & Hemley, R. J. Raman spectroscopy of iron to 152 gigapascals: implications for Earth's inner core. *Science* **288**, 1626-1629 (2000).

38. Lin, J., Sturhahn, W., Mao, H. K. & Hemley, R. J. Sound velocities of iron at high pressures and temperatures.

39. Sanloup, C. *et al.* Structural changes in liquid Fe at high pressures and high temperatures from synchrotron x-ray diffraction. *Europhys. Lett.* **52**, 151-157 (2000).

40. Seto, M., Yoda, Y., Kikuta, S., Zhang, X. W. & Ando, M. Observation of nuclear resonant scattering accompanied by phonon excitation using synchrotron radiation. *Phys. Rev. Lett.* **74**, 3828-3831 (1995).

41. Sturhahn, W. *et al.* Phonon density of states measured by inelastic nuclear resonant scattering. *Phys. Rev. Lett.* **74**, 3832-3835 (1995).

42. Minkiewicz, V. J., Shirane, G. & Nathans, R. Phonon dispersion relation for iron. *Phys. Rev.* **162**, 528-531 (1967).

43. Mao, H. K. *et al.* Phonon density of states of iron up to 153 GPa. *Science* **292**, 914-916 (2001).

44. Eggert, J. H., to be published.

45. Guo, Q., Hu, J., Mao, H. K. & Hemley, R. J. *to be published.*

46. Wentorf Jr., R. H. The behavior of some carbonaceous materials at very high pressures and high temperatures. *J. Phys. Chem.* **69**, 3063-3069 (1965).

47. Nellis, W. J., Ree, F. H., van Thiel, M. & Mitchell, A. C. Shock compression of liquid carbon monoxide and methane to 90 GPa (900 kbar). *J. Chem. Phys.* **75**, 3055 (1981).

48. Ross, M. The ice layer in Uranus and Neptune -- diamonds in the sky. *Nature* **292**, 435-436 (1981).

49. Ancilotto, F., Chiarotti, G. L., Scandolo, S. & Tosatti, E. Dissociation of methane into

50. hydrocarbons at extreme (planetary) pressure and temperature. *Science* **275**, 1288-1290 (1997).

50. Kress, J. D., Bickham, S. R., Collins, L. A., Hoilan, B. L. & Goedecker, S. Tight-binding molecular dynamics of shock waves in methane. *Phys. Rev. Lett.* **83**, 3896-3899 (1999).

51. Goncharov, A. F., Badro, J., Liu, Z., Mao, H. K. & Hemley, R. J. *to be published.*

52. Schindelback, T., Somayazulu, M., Hemley, R. J. & Mao, H. K. Breakdown of methane at high pressures and temperatures. *MRS 1997 Fall Meet. Prog*, 583 (1997).

53. Benedetti, L. R. *et al.* Dissociation of CH_4 at high pressures and temperatures: diamond formation in giant planet interiors? *Science* **286**, 100-102 (1999).

54. Scott, H., Hemley, R. J., Mao, H. K., Herschbach, D. & Fried, L. to be published.

55. McMahan, A. K. & LeSar, R. Pressure dissociation of solid nitrogen under 1 Mbar. *Phys. Rev. Lett.* **54**, 1929-1932 (1985).

56. Mailhiot, C., Yang, L. H. & McMahan, A. K. Polymeric nitrogen. *Phys. Rev. B* **46**, 14419-14435 (1992).

57. Lorenzana, H. E., Yoo, C. S., Barbee III, T. W. & McMahan, A. K. Phase transition in high-pressure nitrogen. *Bull. Am. Phys. Soc.* **39**, 816 (1994).

58. Goncharov, A. F., Gregoryanz, E., Mao, H. K., Liu, Z. & Hemley, R. J. Optical evidence for a nonmolecular phase of nitrogen above 150 GPa. *Phys. Rev. Lett.* **85**, 1262-1265 (2000).

59. Eremets, M. I., Hemley, R. J., Mao, H. K. & Gregoryanz, E. Electrical conductivity of semiconducting nitrogen to 240 GPa and its low pressure stability. *Nature* **411**, 170-174 (2001).

60. Iota, V., Yoo, C. S. & Cynn, H. Quartzlike carbon dioxide:an optically nonlinear extended solid at high pressures and temperatures. *Science* **283**, 1510-1513 (1999).

61. Yoo, C. S. *et al.* Crystal structure of carbon dioxide at high pressure: "superhard" polymeric carbon dioxide. *Phys. Rev. Lett.* **83**, 5527-5530 (1999).

62. Tschauner, O., Mao, H. K. & Hemley, R. J. New transformations in CO_2 at high pressures and temperatures. *Phys. Rev. Lett.* **87**, 075701 (2001).

63. Lipp, M., Evans, W. J., Garcia-Baonza, V. & Lorenzana, H. E. Carbon monoxide: Spectroscopic characterization of the high-pressure polymerized phase. *J. Low Temp. Phys.* **111**, 247-256 (1998).

64. Somayazulu, M. *et al.* Novel broken symmetry phase from N_2O at high pressures and high temperatures. *Phys. Rev. Lett.* **87**, 135504 (2001).

65. Song, Y., Somayazulu, M., Mao, H. K., Hemley, R. J. & Herschbach, D. R. High pressure structure and equation of state of nitrosonium nitrate from

synchrotron x-ray diffraction. *J. Chem. Phys.* **118**, 8350-8356 (2003).

66. Song, Y. *et al.* High-pressure stability, transformations and vibrational dynamics of nitrosonium nitrate from synchrotron infrared and Raman spectroscopy. *J. Chem. Phys.* **119**, 2232-2240 (2003).

67. Yoo, C. S. & Cynn, H. Equation of state, phase transition, decomposition of beta-HMX (octahydro-1,3,5,7-tetranitro-1,3,5,7-tetrazocine) at high pressures. *J. Chem. Phys.* **111**, 10229-10235 (1999).

68. Nellis, W. J. *et al.* Equation of state of shock-compressed liquids: carbon dioxide and air. *J. Chem. Phys.* **95**, 5268-5272 (1991).

69. Mao, W. L. *et al.* Bonding changes in compressed superhard graphite. *Science* **302**, 425-427 (2003).

70. Meng, Y. *et al.* to be published.

71. Galambosi, S., Soininen, J. A., Hämäläinen, K., Shirley, E. L. & Kao, C.-C. Nonresonant inelastic x-ray scattering study of cubic boron nitride. *Phys. Rev. B* **64**, 024102-1 (2001).

72. Watanabe, N., Hayshi, H., Udagawa, Y., Takeshita, K. & Kawata, H. Anisotropy of hexagonal boron nitride core absorption spectra by x-ray spectroscopy. *Appl. Phys. Lett.* **69**, 1370-1372 (1996).

73. Yan, C. *et al.* Ultrahard single-crystal diamond from chemical vapor deposition. *to be published.*

74. Xu, J. & Mao, H. K. Moissanite: a window for high-pressure experiments. *Science* **290**, 783-785 (2000).

75. Xu, J., Mao, H. K., Hemley, R. J. & Hines, E. Moissanite anvil cell: a new tool for high-pressure research. *J. Phys.: Condensed Matter* **14**, 11543-11548 (2002).

76. Xu, J., Mao, H. K., Hemley, R. J. & Hines, E. Large volume high pressure cell with supported moissanite anvils. *Rev. Sci. Instrum.*, submitted.

77. Yan, C., Vohra, Y. K., Hemley, R. J. & Mao, H. K. Very high growth rate chemical vapor deposition of single-crystal diamond. *Proc. Nat. Acad. Sci.* **99**, 12523-12525 (2002).

78. Mao, W. L. *et al.* Generation of ultrahigh pressure using single-crystal chemical-vapor-deposition diamond anvils. *Appl. Phys. Lett.*, in press.

79. Holmes, N. C., Moriarty, J. A., Gathers, G. R. & Nellis, W. J. The equation of state of platinum to 660 GPa (6.6 Mbar). *J. Appl. Phys.* **66**, 2962 (1989).

80. Moriarty, J. A. *et al.* Quantum-based atomistic simulation of materials properties in transition metals. *J. Phys.: Condens. Matter* **14**, 2825 (2002).

81. Shukla, A. *et al.* Charge transfer at very high pressure in NiO. *Phys. Rev. B* **67**, 081101(R) (2003).

82. Loubeyre, P. *et al.*, to be published.

CHAPTER II

EQUATION OF STATE:
NONENERGETIC MATERIALS

CP706, *Shock Compression of Condensed Matter - 2003*
edited by M. D. Furnish, Y. M. Gupta, and J. W. Forbes
© 2004 American Institute of Physics 0-7354-0181-0/04/$22.00

MEASUREMENTS OF SOUND SPEED IN ZINC IN THE NEGATIVE PRESSURE REGION

G.S. Bezruchko[1], G.I. Kanel[2], and S.V. Razorenov[1]

[1]*Institute of Problems of Chemical Physics, Chernogolovka, Moscow region, 142432 Russia*
[2] *Institute for High Energy Densities, IVTAN, Izhorskaya 13/19, Moscow, 125412 Russia*

Abstract. In the paper we present the results of measurements of sound speed in zinc single crystals with orientation <0001> at initial temperatures 17°C and 322°C over the stress range from 13 GPa of shock compression down to −2 GPa of tension. Within this stress range zinc of given orientation is elastic. The method is based on the measurement and analysis of wave reverberation in a plate, one surface of which is free whereas other surface contacts with a high-impedance material (molybdenum). The results were confirmed by computer simulation. Nonlinearity of longitudinal compressibility of zinc exceeds that of the bulk compressibility.

INTRODUCTION

Tensile stresses up to 15–20 GPa are available now in spall experiments at nanosecond load durations.[1,2] This makes it actual to expand the equations of state (EOS) into the negative pressure region. Sin'ko and Smirnov[3] performed the first-principle calculations of EOS in the negative pressure region. In this paper we discuss one of the possible ways of experimental study of compressibility of a matter by means of measurements of sound speed both in shock compressed state and under tension.

METHOD

In order to explain the idea, let us consider the wave dynamics after collision of the thick flyer plate of high-impedance material with the thin target of softer material. Figures 1 and 2 present the time–distance and the particle velocity – pressure diagrams which explain reflections and interactions of the compression and rarefaction waves.

Reflection of the shock wave from the target free surface creates rarefaction wave presented in Fig. 1

by a fan of C_--characteristics R_h to R_t. In this wave the state of matter is changing along the Riemann's isentrope AB in Fig. 2. On the interface between the target and the impactor the reflection of the wave occurs again and new rarefaction wave C_+ appears. Unloading of the impactor occurs along the Riemann's isentrope An. In the region ABi (Fig. 1) of interaction of two opposite waves the pressures and particle velocities are changing along characteristics in accordance to Riemann's integrals. Since the state in the intersection point k should satisfy the conditions along both C_--characteristic $FBki$ and C_+-characteristic $nklE$, one can see in Fig. 2 that the negative pressures are generated near the impactor/target interface. This interface cannot support tensile stresses, therefore the influence of impactor on the wave process in target is ceased when the pressure at interface dropped to zero (point n in Figs. 1 and 2). Residual part of the C_- rarefaction wave reflects from appeared free surface and produces a compressive C_+ wave.

Thus, the reflection of a rarefaction wave from the interface with material of higher impedance produces new rarefaction wave. As a result, a short tensile pulse is generated in the unloaded target, which, in turn, should produces a pullback in the free surface velocity history $u_{fs}(t)$. If the $u_{fs}(t)$ profile has been measured, the time–distance diagram can be recovered and thus the sound speeds in shock-compressed and stretched target material can be determined.

The sequence of analysis may be the following. The triangle FBD in Fig. 1 is bounded by the tail C_- characteristic FB of the unloading wave and the head C_+ characteristic BD of the re-reflected tensile pulse, which slope corresponds to the sound speed at zero pressure c_0. Hence, the position of point B is determined through the time interval between the moments t_F and t_D of appearances of the shock front and the rarefaction front in the free surface velocity history. Since the characteristics C_+ and C_- are symmetrical in Lagrangian coordinates, the line AB obeys to the equation

$$\frac{dx}{dt} = \frac{h-x}{t-t_F} \qquad (1)$$

where h is the target thickness and x is substantional (Lagrangian) coordinate in the target. Integration of Eq. (1) gives the relationship:

$$t_A - t_F = (t_B - t_F)\frac{h - x_B}{h} \qquad (2)$$

Lagrangian sound speed a_A in shock-compressed matter is then determined as

$$a_A = c_A \rho/\rho_0 = h/(t_A - t_F) \qquad (3)$$

where $c_A = \sqrt{(\partial p/\partial \rho)_s}$ is thermodynamic (Eulerian) sound speed.

In order to recover C_+ characteristic $nklE$ and to determine the sound speed at ultimate tension we need first to determine the time moment t_n when the pressure dropped to zero at the interface. This means that we have to recover the C_- characteristic Fmn with states varying along Riemann's isentrope mn in Fig. 2. Between the points F and m on this characteristic sound speed is constant and particle velocity is $u_m = (u_B + u_n)/2$. The appropriate sound speed a_m can be found using linear relationship between the Lagrangian sound speed a and the particle velocity u_p

$$a = c_0 + 2bu_p \qquad (4)$$

This relationship ia based on the assumption that, in the pressure – particle velocity coordinates, Riemann's unloading isentropes are symmetrical to the Hugoniot. Thus, $a_m = c_0 + 2b(u_B - u_m)$. Small curvilinear part mn of the characteristic Fmn can be replaced by a straight line with a slope of average sound speed $-(a_m + c_0)/2$ without essential loss of the accuracy

The first estimation of sound speed at ultimate tension is: $a_k = h/(t_E - t_n)$, where t_E is the time moment of appearance of minimum in the free surface velocity. After that the curvilinear parts nk and lE of the characteristic $nklE$ are revised by further approximations. Finally, the sound speed at ultimate tension is determined as

$$a_k = (x_l - x_k)/(t_l - t_k). \qquad (5)$$

MATERIAL

Zinc has h.c.p. crystal structure with abnormally large axial c/a ratio of 1.856. As a result, its

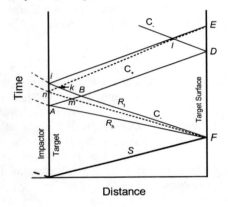

FIGURE 1. Time–distance diagram of wave interactions after collision of a high-impedance impactor plate with thin target.

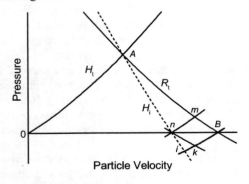

FIGURE 2. Variations of states in the target and the impactor plates along characteristics.

longitudinal compressibility along c-axis is very close to bulk compressibility.[4] Due to that the Hugoniot elastic limit at shock compression in the direction <001> is as high as 21 GPa.[5] The spall strength of zinc in this direction is 2 GPa and remains that at the temperature increasing at least up to 370°C (melting temperature is 419°C). Hugoniot data for zinc is well approximated by linear relationship $U_s = 3.0 + 1.57u_p$ ($\pm 1\%$).

Zinc samples were cut from single crystals of 99.99% purity, grown by a directed crystallization technique.[6] The samples were plates parallel to the (001) plane with dimensions in plane of 15×10 mm and thickness varied from 0.25 to 0.8 mm. The measured longitudinal sound speed in the <001> direction is $c_l = 2.98\pm0.03$ km/s.

EXPERIMENTAL PROCEDURE

Scheme of the experiments is shown in Fig. 3. Aluminum impactor plates of thickness 4 mm were launched by explosive facility up to velocity

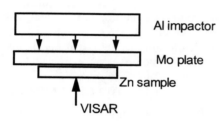

FIGURE 3. Scheme of the experiments.

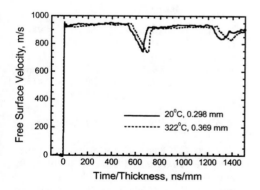

FIGURE 4. The free surface velocity histories of zinc single crystals at normal and elevated temperatures. The time is normalized by the sample thickness at 20°C.

1.45±0.05 km/s. Shock wave was introduced into sample via 2 mm molybdenum base plate, separated from sample by a 0.02 mm gap. The free-surface velocity profiles $u_{fs}(t)$ were recorded by VISAR. Shock compression occured within the elastic deformation region.

The measurements have been done at the room temperature and 320°C. The details of arrangement for the shock-wave measurements at elevated temperatures were discussed in [5,7].

RESULTS

Fig. 4 presents free surface velocity histories of zinc crystals measured at normal and elevated temperatures. For better comparison the time is normalized by the sample thickness. As expected, the wave profiles exhibit short velocity drops in vicinity of $6\cdot10^{-4}$ s/m as a result of reflection of unloading wave from the zinc/molybdenum interface.

At room temperature, the front of tensile pulse came to the surface at $t_D/h = 5.27\cdot10^{-4}$ s/m of normalized time after the shock front. For $c_0 = 2980$ m/s we obtains $t_B/h = t_D/2h = 2.635\cdot10^{-4}$ s/m and $x_B/h = 1 - c_0t_B/h = 0.2148$. Eq. (2) gives us $t_A/h = 2.069\cdot10^{-4}$ ns/mm, that results in the Lagrangian sound speed $a_A = h/t_A = 4833$ m/s at stress $\sigma = 13.2$ GPa. Assuming that the free surface velocity $u_{fs} = 945$ m/s equal to doubled particle velocity u_A, we obtains for the relationship (4) $b = (a_A - c_0)/2u_A = 1.96$. Thus, we determined the dependence of sound speed on particle velocity which can be transformed to the stress-particle velocity, $\sigma(u_p)$ - (σ) relationship using Riemann's

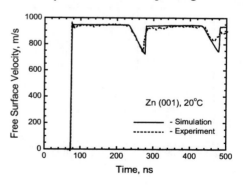

FIGURE 5. Comparison of measured free surface velocity history of zinc single crystal 0.298 mm in thickness with the result of computer simulation.

invariant

$$\sigma = \rho_0 (c_0 + bu)\,u\,, \quad a = \sqrt{c_0^2 + \frac{4b\sigma}{\rho_0}} \quad (6)$$

The expected free surface velocity in the minimum of tensile pulse $u_n = 710$ m/s is less than the measured value of 745 m/s that is obviously a result of stress relaxation at nucleation of spall fracture near the interface. For this reason, in order to recover C_+ characteristic nlE, we used not the measured time $t = 6.56\cdot10^{-4}$ ns/mm of the free surface velocity minimum, but the value $t_E = 6.80\cdot10^{-4}$ ns/mm, fond by extrapolation of $u_{fs}(t)$ to $u_n = 7.10\cdot10^{-4}$ m/s. The time moment t_n (Fig. 1) was determined using the value of $a_m = 3.44$ km/s, calculated from the relationship (4).

So the sound speed 2.53 km/s at ultimate tensile stress $\sigma_k = -2.3$ GPa was found, whereas its expected value is 2.52 km/s in accordance to Eqs. (4). The real tensile stress is limited by the spall strength of the Zn (-2.0 GPa in this shot).

The comparison of the measured free surface velocity history with the results of computer simulation is presented in Fig. 5. In the simulation the equation of state

$$\sigma = \sigma(V) + (\Gamma/V)[E - E(V)], \quad (7)$$

was taken, where $\sigma(V)$ is the reference isentrope

$$\sigma_c = \frac{\rho_0 c_0^2}{4b}\left[\exp\left(4b\frac{V_0 - V}{V_0}\right) - 1\right] \quad (8)$$

with $c_0 = 2.99$ km/s and $b = 1.93$. An ability of fracture was not taken into account. In the same

manner we have found $c_0 = 2.88$ km/s and $b = 1.73$ at the density $\rho_0 = 6.941$ g/cm^3 for the test at 322°C

DISCUSSION

The $a(\sigma)$ dependence (6) shows that the sound speed turns to zero at $\sigma \approx -8.3$ GPa both at room temperature and at 322°C. This value can be considered as an estimation of ideal strength of zinc of given orientation. Hence, about 25% of ultimately possible tensile stress has been realized in the experiments. In this range zinc has the same dependence of sound speed on stress as for compression.

Fig. 6 presents isentropes of longitudinal, $\sigma(V)$, and bulk, $p(V)$, compression/tension of a zinc crystal which were calculated with Eq. (8), using the results of present measurements. As it should be, under compression $\sigma(V) > p(V)$ and the discrepancy grows with compression. At 13 GPa of shock compression this difference reaches 1.2 GPa. However, relative position of these curves in the region of tension looks anomalous. Normally these curves should cross each other only in the initial point. Remaining of the equality of pressure and longitudinal stress is equivalent to shear modulus turning to zero, that means instability of the crystal structure.

ACKNOWLEDGEMENTS

The work was supported by Russian Foundation for Basic Research under grant number 03-02-16379.

REFERENCES

1. G.I. Kanel, S.V. Razorenov, et al. *J. Appl. Phys.*, **74**(12), 7162-7165 (1993).
2. G.I. Kanel, S.V. Razorenov, et al. In: *High-Pressure Science and Technology - 1993*. Ed.: S.C.Schmidt et al., AIP Conference Proceedings **309**, p.1043 (1994).
3. G. V. Sin'ko and N. A. Smirnov. *JETP Letters*, **75**(4), 184-186 (2002)
4. H.M. Ledbetter. *J. Phys. Chem. Ref. Data*, **6**(4), 1181-1203 (1977).
5. A.A. Bogach, G.I. Kanel, S.V. Razorenov, et al. *Physics of the Solid State*, **40**(10), 1676-1680 (1998).
6. A. Antonov, C.V. Kopetskii, L.S. Shvindlerman, V. Sursaeva *Sov. Phys. Doklady*, **18**, 736-738 (1974).
7. G.I. Kanel, S.V. Razorenov, A.A. Bogach, et al. *J.Appl.Phys.*, **79**(11), 8310-8317 (1996).

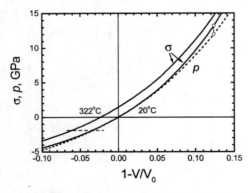

FIGURE 6. Isentropes (8) of longitudinal (along C-axis) and bulk compression of zinc. Horizontal dashed line shows achieved tension.

CP706, *Shock Compression of Condensed Matter - 2003*
edited by M. D. Furnish, Y. M. Gupta, and J. W. Forbes
© 2004 American Institute of Physics 0-7354-0181-0/04/$22.00

HUGONIOTS AND SHOCK TEMPERATURE OF DENSE HELIUM UNDER SHOCK COMPRESSION

Qifeng Chen[1,2] Lingcang Cai[2] Zizheng Gong[3] Fuqiang Jing[2]

[1]*Lab. of Computational Physics, Institute of Applied Physics and Computational Mathematics, P. O. Box 8009-26,Beijing 100088, P R China.*
[2] *Lab. for Shock Wave and Detonation Physics Research, Institute of Fluid Physics, P.O. Box 919-102, Mianyang, Sichuan, 621900 P R China*
[3]*Institute of Physics, Southwest Jiaotong University, Chengdu, Sichuan 610031, P R China.*

Abstract.   The Hugoniot data and shock temperatures of gaseous helium were measured in the pressure range 500-700MPa and temperature 6.7-10.8kK generated by flyer accelerated up to ~6km/s with a two-stage light-gas gun. Gaseous specimens were shocked from initial pressure 5MPa at 285K. Spectral radiance histories from the shocked helium sample were studied experimental by using pyrometer. Shock velocity was measured and particle velocity was determined by the shock impedance matching method. The equation of state and degree of ionization has been calculated by applying the Saha model with Debye-Hüchel correction. In comparison of the experiment with theoretical results, we found that shock pressure versus particle velocity is in agreement with each other, but for the shock temperatures. It implies that the theoretical degree of ionization should be lower than that of the experiments due to their lower values of shock temperature, and thus the ionization energies and level populations of the compressed atoms are also changed.

INTRODUCTION

Helium is the simplest and abundant element in the universe and also main constituents of some planets, such as Jupiter and Saturn. The high-pressure equation of state (EOS) of helium has been the subject of considerable interest, principally due to importance of the EOS to such areas as planetary astrophysics, and our fundamental understanding of warm dense matter. In past several years, some significant advances have been made in the experimental and theoretical Hugoniot equation of state of liquid helium [1-2]. However, until recently, there is dearth of measured the Hugoniots and shock temperatures of dense gaseous helium. In addition to completing the equation of state information, the shock temperatures are important in the study of phenomena such as the thermophysical properties

of dense plasmas, opacity, and ionizing. This type of thermodynamic information is often a much more sensitive test of theoretical EOS models than pressure/volume information alone.

The method for measuring shock temperature exceeding a few thousand degrees Kelvin is time resolved optical pyrometer. The spectral radiance emitted from the shocked dense gas is measured as a function of time at several wavelengths. This method has been used successfully to measure the temperature of transparent materials under shock compression[3]. In this work, we have applied this approach to the shocked dense helium.

EXPERIMENTAL METHOD

Strong shock was generated in dense gas samples by impact of projectiles driven by a two-stage light-gas gun. The helium gas samples were condensed in a cavity. The sample cavity was a cylinder, 35mm in diameter and 4.0mm thick. A cylinder chamber of aluminum encloses the gas having an initial pressure 5.0MPa. A two-stage gas gun accelerates a 30mm in diameter and 3mm thick 93W (4.2Ni2.45Fe0.35C_OW) disk projectile to speeds up to velocity of 4.2-5.2km/sec. The flyer velocity was measured by magneto-flier velocity system to an accuracy of 0.5%. Impact of the disk on a 4mm thick aluminum base-plates generates a planar shock wave which enters the gas chamber, heating the helium gas to an incandescent plasma, and the rear was formed by a synthetic sapphire (Al_2O_3) window 8mm thick. The window was used to allow optical access for the absolute spectral radiance measurements. Spectral radiance histories from the shocked dense helium were recorded using a seven-channel fiber-optic-coupled pyrometer as shown Fig.1.

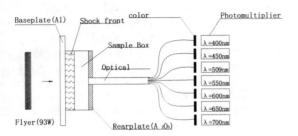

Figure.1 A layout of experimental arrangement

After the flyer impact onto a 4mm thick aluminum base-plate, a strong planar shock wave is generated and propagates in the base-plate and then the gas sample, and, consequently, heats the gas to be become an incandescent plasma. Fig.2 is a typical spectral radiance history recorded by pyrometer for shocked He (shot 1) at a central wavelength channel of λ =700nm, from which the shock transit time, Δt, across sample and the spectral radiance, I(ε,λ,T), emitted from the shock-induced equilibrium incandescent plasma can be separately obtained. Where T is the temperature, λ the wavelength, ε the effective emissivity, providing the grey-body radiation model holds for this thermal plasma.

Shock temperatures were fit to use a grey-body Plank spectrum model. The shock wave velocity was accurately determined by the thick of chamber and transit time of the single shock wave in samples. Fig.1 is an example of output of one of the photomultiplier at 509nm(shot No.1). Using the flyer velocity, initial densities of helium, flyer, and base-plate, the known equation of state for the 93W flyer and the Al base-plate, the pressure, density, internal energy can be determined by impedance matching and Hugoniot relations.[2]

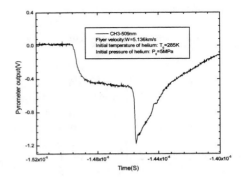

Figure.2 A measured spectral radiance history at λ=700nm for experiment He No.1.

From the measured Δt the shock wave velocity, u_s, the particle velocity, u_p, and the shock pressure, P_H, of the shocked gas can be successively deduced through following procedure. Since the thickness, d, of the gas sample is well known before experiment, so u_s could be immediately computed by

$$u_s = d / \Delta t \qquad (1)$$

Now let us begin to calculate the values of u_p and P_H in baseplate. The impact velocity of flyer, W, is measured by magneto-flier velocity method. Upon impact by the flyer, two shock waves are created, one travels backwards in the flyer with a particle velocity $u_{p,f}$, and shock pressure $P_{H,f}$, and the other travels forwards in the baseplate with particle velocity $u_{p,b}$ and shock pressure $P_{H,b}$. Therefore, P_H and u_p in both flyer and base-plate could be written as[4]

$$P_{H,f} = \rho_{0f}[-C_{0f} + \lambda_f(u_f - W)](u_f - W), \qquad (2)$$

$$P_{H,b} = \rho_{0b}(C_{0b} + \lambda_b \cdot u_b) \cdot u_b, \qquad (3)$$

with the help of pressure and particle velocity continuity condition across the flyer/baseplate interface, i.e.

$$P_H = P_{H,f} = P_{H,b}, \qquad (4)$$

$$u_p = u_{p,f} = u_{p,b}, \qquad (5)$$

combining Eqs.(2)-(5), then yields

$$u_p = \frac{-B + \sqrt{B^2 - 4AC}}{2A}, \qquad (6)$$

where $\quad A = \rho_{0f}\lambda_f - \rho_{0b}\lambda_b$

$B = -\rho_{0f}C_{0f} - 2W\rho_{0f}\lambda_f - \rho_{0b}C_{0b}$

$C = \rho_{0f}W(C_{0f} + \lambda_f W)$,

here λ and C_0 are the two material constants in the linear $u_s = C_0 + \lambda u_p$ relation which could be determined by separate experiments, ρ_0 is the initial density, and the subscripts f and b refer to the flyer and the baseplate, respectively. Once u_p is computed from Eq.(6), then the value of P_H can be immediately calculated from Eq.(3).

When the shock wave arrives at the baseplate/sample interface, a release wave reflected from the interface and a shock wave transmitting in the sample will be created. The release pressure, P_s, in baseplate can be computed by

$$P_S = \exp[\gamma_0(\eta - \eta_i)]\left\{P_i + \rho_0 C_0^2 \times \int_{\eta_i}^{\eta} \frac{1 + \lambda x - \gamma_0 x}{(1 - \lambda x)^3} \exp[\gamma_0(\eta_i - x)]dx\right\}$$

$$(7)$$

where γ_0 is the Grüneisen parameter under ambient condition (see Table I), P_i is the initial isentropic release pressure. $\eta = 1 - V_{S,b}/V_0$, $\eta_i = 1 - V_{i,b}/V_0$, where V_0, $V_{i,b}$ and $V_{S,b}$ are the initial sample, the initial release, and a certain specific volumes, respectively. The shock pressure in gas sample, $P_{H,g}$, can be written as

$$P_{H,g} = \rho_{0,g}u_{s,g}u_{p,g}, \qquad (8)$$

here the subscript g refers to the gas sample. Combining Eqs. (7)-(8) and with the pressure and particle velocity continuity condition across the baseplate/gas interface and the useful Hugoniot relation connecting specific volume with particle velocity, we may get the Hugoniot data of the shocked gas which are summarized in Table 1.

TABLE 1. Experimental Hugoniot data and shock temperature for dense helium

Shot No.	He#1	He#2	He#3
P_0, MPa	5.0	5.0	5.0
W, km/s	5.136	4.524	4.208
U_p, km/s	7.860	7.019	6.493
U_s ,km/s	10.28	9.08	8.80
P, MPa	700	546	497
T,kK	10.7	8.4	6.7
ΔT, K	300	300	300

As to the experimental shock-temperature T_H, it could be computed by fitting the measured spectral

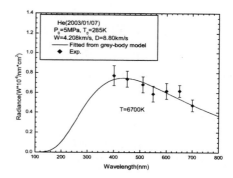

Figure 3 Results of least-squares fits to the radiance data for experiment He No.3.

radiance histories recorded by pyrometer for that shot, a typical one is shown in Fig 2, to a grey-body Plank radiance formula

$$I_{exp}(\varepsilon, \lambda, T) = \varepsilon I_{Pl}(\lambda, T) = \frac{\varepsilon C_1}{\lambda^5} \frac{1}{\exp(C_2/\lambda T) - 1},$$

$$(9)$$

by nonlinear least-square method. Fig.3 shows the result He gas sample of the shot no.1. All values of the experimental T_H are also listed in Table 1

THEORETICAL MODEL

A plateau-like trace appearing on the later half of the measured spectral radiance history, displaying in Fig.2 as an example, can be considered as a demonstration that the shock-induced incandescent plasmas is under thermal equilibrium in our experiments. Based upon the above-mentioned equilibrium manifestation and the accompanying lower density nature, we use Saha model to analyze the experimental results.

$$\frac{N_{i+1}N_e}{N_i} = \frac{\sum_j g_{(i+1)j}e^{-E_{(i+1)j}/kT}}{\sum_j g_{ij}e^{-E_{ij}/kT}} \frac{2V}{\lambda_e^3} \quad (i=0,1) \quad (10)$$

Pressure (P) and energy (E) of the dense gases can be computed by

$$P = \sum_i P(N_i), \quad (11)$$

$$E = \sum_i E_i(N_i), \quad (12)$$

where the sum runs over all values of atoms, ions and electrons. The final pressure, specific volume, and internal energy under shock compression may be calculated by combining following Hugoniot relations

$$E_H - E_0 = (P_H + P_0)(V_0 - V_H)/2, \quad (13)$$

$$u_p - u_0 = (V_0 - V_H)\sqrt{\frac{P_H - P_0}{V_0 - V_H}}, \quad (14)$$

$$u_s - u_0 = V_0\sqrt{\frac{P_H - P_0}{V_0 - V_H}}, \quad (15)$$

and with Eqs.(11)-(12) and $E=E_H$ condition.

RESULTS AND DISCUSSION

The experimental single-shock Hugoniot data and shock temperature for the gaseous helium at initial pressures 5MPa and room temperature condition are listed in Table 1.

In comparison of the experimental with theoretical results, we found that shock pressure (P_H) –particle velocity (u_p) are in agreement with

each other as shown Fig.4, but for the shock temperatures as shown Fig.5. It implies that the theoretical degree of ionization should be lower than that of the experiments due to their lower values of shock temperature (T_H), and thus the ionization energies and level populations of the compressed atoms are also changed.

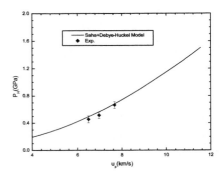

Figure 4 Shock pressure vs. particle velocity

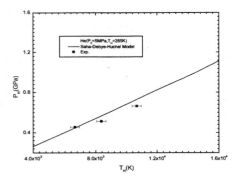

Figure 5 Shock pressure vs. shock temperature

REFERENCES

1. Nellis W J, Holmes N C, Mitchell A C et al., *Phys. Rev. Lett.* **53** 1248-1251 (1984)
2. Louebeyre P., Letoullec R., and Pinceaux J. P., *Phys. Rev. Lett.* **69** 1216-12191 (1992)
3. Holmes N C, Ross M and Nellis W J, *Phys. Rev. B* **52** 15835 (1995)

CP706, *Shock Compression of Condensed Matter - 2003*
edited by M. D. Furnish, Y. M. Gupta, and J. W. Forbes
© 2004 American Institute of Physics 0-7354-0181-0/04/$22.00

STATIC X-RAY DIFFRACTION STUDY OF CERIUM: THE STANDARD APPROACH & THE MAGIC-ANGLE APPROACH

Gary N. Chesnut[1], Becky D. Streetman[1], David Schiferl[1], William W. Anderson[1], Malcolm Nicol[2], and Yue Meng[3]

[1]*Los Alamos National Laboratory, Los Alamos, NM 87545*
[2]*University of Nevada, Las Vegas, NV 89154*
[3]*Argonne National Laboratory, HP-CAT, Argonne, Il 60439*

Abstract. Cerium, a member of the rare earth metals, has been studied up to 300 kilobars in a diamond-anvil cell using energy- and angular-dispersive x-ray diffraction with a synchrotron source. The purpose of this experiment was to examine the electronic and structural behavior of cerium and to examine the effects of deviatoric stress due to non-hydrostatic conditions within the sample environment. Using the standard sample orientation and data from various orientations, the effects of deviatoric stress are shown.

INTRODUCTION

Cerium exhibits several unique electronic and structural properties, which have prompted many experimental and theoretical studies with respect to pressure and temperature. Despite the abundance of literature available on cerium, many issues regarding its structural behavior are unresolved. For example, the phase diagram up to 140 kilobars and 1000 K has several unknown regions, and many of the reported phase transitions have poorly defined phase boundaries. One common problem reported in the literature is unreliable pressure sensors [1,2] which has lead to large errors in pressure-temperature-volume calculations and thus, equation-of-state values one hopes to extract. Another unresolved issue is the determination of the correct low-symmetry structure between 50 and 130 kilobar.

Under ambient pressure and temperature, cerium is stable in the γ-fcc phase [3]. At ambient temperature, this structure remains stable to approximately 7 kilobars. γ-fcc possesses an unusual property, such that the derivative of the bulk modulus has been shown to be negative [4]. In other words, cerium becomes more elastic as it is compressed. The first phase transformation, γ-fcc → α-fcc, occurs below 10 kilobars; this is the only iso-structural phase transformation observed in the lanthanide series. The most interesting aspect of this phase transition is that it occurs with a large volume collapse between 13% and 16%.

This is an indication that the 4f electrons are becoming delocalized with the application of pressure. The α-fcc phase has been shown to be stable to about 50 kilobars where a phase transformation occurs to a low-symmetry orthorhombic or monoclinic structure. There has been much debate over these two structures. Some have suggested that sample purity is an issue [4]. The predominant theory is that the preparation of the sample is actually what determines which structure exists [5]. Finally, the ultra-high pressure phase, observed between 0.13 Mbar and 2.08 Mbar [6], is reported to be body-centered tetragonal.

Cerium has many unresolved issues with respect to pressure and temperature, which require closer examination. Also, it is an excellent material to study the effects of deviatoric stresses due to non-hydrostatic conditions since it possesses high- and low-symmetry structures. In this discussion, the effects of quasi-hydrostatic pressure on cerium using the standard experimental geometry with a Merrill-Bassett diamond-anvil cell are compared to experiments using a newly designed diamond-anvil cell, capable of accessing the diffraction data with the standard geometry and all geometries covering the entire stress region of the sample including the magic angle geometry.

THE STANDARD APPROACH

The standard approach to this type of experiment can be performed with energy- or angle-dispersive x-ray diffraction. The geometry of the standard approach is shown in Fig. 1. The feature to notice is, Ψ, the angle between the compression axis of the diamonds and the reciprocal lattice vector, $g(hkl)$. In a standard experiment, Ψ is approximately 90°.

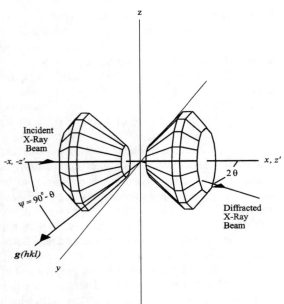

FIGURE 1. The geometry of the standard approach with Ψ approximately equal to 90°. The stress axis z' is parallel to the x-direction.

Base-line x-ray diffraction experiments were performed on cerium using the standard approach. Several Merrill-Bassett diamond-anvil cells (DAC) were loaded with polycrystalline cerium. The samples ranged from 50 to 100 microns in diameter. Ruby was used as a pressure calibrant, using the R_2 line [7], and liquid argon was used as a pressure medium with all samples except those at the lowest pressures. The samples were studied at beamline X17-C, the National Synchrotron Light Source (NSLS) at Brookhaven National Laboratory using energy-dispersive x-ray diffraction (EDXD). The size of the beam was 50 x 60 microns, the diffraction angle was 15°, and a germanium solid-state detector was used.

The γ-fcc phase was observed up to 7.3 kilobars with an initial volume (volume/atom) of 34.367 Å^3/atom. At 7.3 kilobars, a phase transition from γ-fcc to α-fcc was observed with a volume collapse of 15%. The α-fcc phase remained stable up to $\approx$50 kilobars. At 51 kilobars, a new phase, best indexed as the monoclinic C2/m structure (4 atoms/cell), was observed and remained stable to the highest pressure of 72 kilobars. All the phase transformations were reversible.

THE MAGIC-ANGLE APPROACH

We now consider non-hydrostatic stress effects, which are the cause of most of the large discrepancies between different sets of static data taken with the standard approach as well as of differences of such data from that of dynamic experiments.

Truly hydrostatic conditions are possible for only a small part of pressure range accessible with diamond-anvil cells. The best we can create is a quasi-hydrostatic environment that reduces the effects of deviatoric stresses and thus provides somewhat more accurate results. However, deviatoric stresses still exist, causing significant errors in pressure, volume, etc. In order to adequately examine the affects of the deviatoric stresses, radial x-ray diffraction must be done and the angle Ψ must be well defined.

The stress conditions in the diamond-anvil cell have been examined for several decades [8-15]. Singh et al [10] derived the equations for the lattice strains of each crystal system along with some supporting experimental data. Listed below is the equation for the measured d-spacing, $d_m(hkl)$:

$$d_m(hkl) = d_p(hkl)[1 + (1- 3\cos^2\Psi)Q(hkl)].$$

$d_p(hkl)$ is the hydrostatic d-spacing, and $Q(hkl)$ is a complicated function that depends on the material. This equation is rather complex and difficult to solve. However, a particularly simple solution occurs when $1-3\cos^2\Psi = 0$. For this to be true, Ψ = 54.74 degrees (the magic angle), and $d_m(hkl) = d_p(hkl)$. By choosing the magic angle, the effects of the deviatoric stresses have been removed from the data. This geometry is shown in Fig. 2. Preliminary experiments have demonstrated this to be true, indicating that sample orientation has significant affects on the data [16].

In this set of experiments, two newly designed DAC's were loaded with polycrystalline cerium. The samples were 110 microns in diameter. No pressure calibrant was used in order to generate cleaner spectra containing only cerium, and liquid argon was used as a pressure medium in one of the two cells. The experiments were performed at beamline X17-C at NSLS using EDXD; and at HPCAT located in the Advanced Photon Source (APS) at Argonne National Laboratory using angle-dispersive x-ray diffraction with λ = 0.37378 Å and a CCD detector. By using an area detector we were able to examine the entire stress region between Ψ = 0 and 360 degrees simultaneously (the standard experimental geometry is approximately Ψ = 90 degrees).

The purposes of this experiment were to illustrate the ease of magic angle x-ray diffraction experiments and to show the effects of deviatoric stress. A P-V curve was not generated for these experiments. However, data were obtained for each structure up to the bct structure at room temperature. Around 300 kilobars (the highest achieved pressure in this set of experiments), cerium showed systematic errors in volume of about 1.5 % between Ψ = 10 and 90 degrees. The major issue with these experiments is not systematic error, but the fact that d-spacings may cross since the value of each d_{hkl} is affected by different amounts for the same amount of stress, which is determined by the elastic constants, C_{ij}, of the material. This is most likely to occur in low-symmetry structures as is found in cerium. At approximately 60 kilobars, the low-symmetry

phase of cerium did appear to exhibit effects from deviatoric stress as shown in Fig. 3. The two EDXD spectra were collected at the same pressure

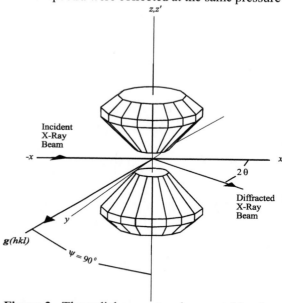

Figure 2. The radial geometry shown enables data collection at all values of Ψ from 0° to 360° either by rotating the stress axis z' through the y-z plane or by measuring different points around the powder diffraction ring, which is equivalent to changing the angle between the reciprocal lattice vector $g(hkl)$ and the stress axis z'.

using methodology stated previously with the only difference being the sample orientation. The lower spectrum is at 90 degrees and the upper spectrum is at the magic angle. The major differences in the spectra are indicated. It is also interesting that the spectrum at 90 degrees fits the same monoclinic structure found using the standard approach, and that the spectrum at the magic angle (essentially free of deviatoric stress effects) fits an orthorhombic structure. More x-ray diffraction data are needed to determine the structure unambiguously, but this result already shows the effects of deviatoric stress upon structural identification.

DISCUSSION

The current x-ray diffraction experiments on cerium demonstrate significant effects from deviatoric stress in non-hydrostatic environments. Systematic errors, as much as a few percent, have been observed in high- and low-symmetry

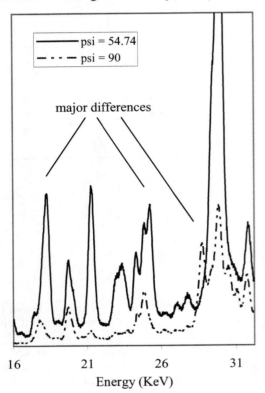

FIGURE 3. EDXD spectra of Ce at 60 kilobars at Ψ = 54.74 and 90 degrees with newly designed diamond-anvil cells. 2θ = 15 degrees.

structures. As shown in Fig. 3, low-symmetry structures can easily be solved incorrectly and generate significant errors in volume. Pressure media may alleviate some of these errors, but will not remove them, especially in the ultra high-pressure regime.

Considering the fact that the deviatoric stresses are inherently part of the data, they cannot be removed by using analytical techniques such as Rietveld Refinement. Nor can they be determined without great difficulty, even for the simplest of structures. The only way to solve high-symmetry structures like cubics and hexagonals with any precision, and to definitively solve low-symmetry structures is to use magic-angle x-ray diffraction. This will ultimately lead to better agreement between static and dynamic data, and lead to more accurate equations-of-state.

ACKNOWLEDGEMENTS

These experiments were done at X17-C (NSLS, Brookhaven National Lab) and HPCAT (APS, Argonne National Lab) with contributions from Rob Hixson, Jingzhu Hu, Bob Sander, Rachel Hixson, and Rush Davidson. DOE-BES, DOE-NNSA, NSF, DOD-TACOM, and the W.M. Keck Foundation support HPCAT. DOE-BES supports the NSLS. Use of APS is supported by the DOE-BES, Office of Science, under Contract No. W-31-109-Eng-38. LANL and UNLV are supported by DOE-NNSA. This work was, in part, supported by the US DOE under contract # W-7405-ENG-36.

REFERENCES

1. W.H. Zachariasen and F.H. Ellinger, Acta. Cryst. A 33, 155 (1977)
2. Y. Zhao and W.B. Holzapfel, J. Alloys Comp. 246, 216 (1997)
3. K.A. Gschneidner, Jr. and V. K. Pecharsky, J. Phase Equilibria 20, 612 (1999)
4. A.K. Singh, High Temp.-High Press. 12, 47 (1978)
5. M.I. McMahon and R.J. Nelmes, Phys. Rev. Lett. 78, 3884 (1997)
6. Y.K. Vohra, S.L. Beaver, J. Akella, C.A. Ruddle and S.T. Weir, J. Appl. Phys. 85, 2451 (1999)
7. Y. M. Gupta and X. A. Shen, Appl. Phys. Lett. 58 (6), 583 (1991)
8. G. L. Kinsland and W. A. Bassett, Rev. Sci. Instrum. 47 (1), 130 (1976)
9. A. K. Singh and C. Balasingh, J. Appl. Phys. 48, 5338 (1977)
10. A. K. Singh, C. Balasingh, H. K. Mao, R. J. Hemley, and J. Shu, J. Appl. Phys. 83, 7567 (1998)
11. H. K. Mao, J. Shu, G. Shen, R. J. Hemley, B. Li, and A. K. Singh, Nature 396, 741 (1998)
12. S. R. Shieh, T. S. Duffy, and B. Li, Phys. Rev. Lett. 89, 255507-1 (2002)
13. Y. Meng, D. J. Weidner, and Y. W. Fei, Geophys. Res. Lett. 20 (12), 1147 (1993)
14. T. Uchida, N. Funamori, and T. Yagi, J. Appl. Phys. 80 (2), 739 (1996)
15. T. S. Duffy, G. Shen, J. Shu, H. K. Mao, R. J. Hemley, and A. K. Singh, J. Appl. Phys. 86 (12), 6729 (1999)
16. Unpublished work by G. N. Chesnut, B. D. Streetman, D. Schiferl, W. W. Anderson, M. Nicol, and Y. Meng

CP706, *Shock Compression of Condensed Matter - 2003*
edited by M. D. Furnish, Y. M. Gupta, and J. W. Forbes
© 2004 American Institute of Physics 0-7354-0181-0/04/$22.00

VARIATION OF THERMAL AND COLD CURVE CONTRIBUTIONS TO THERMODYNAMIC FUNCTIONS ALONG THE HUGONIOT

Eric Chisolm, Scott Crockett, and Duane Wallace

Theoretical Division, Los Alamos National Laboratory, Los Alamos, New Mexico 87545

Abstract. We have developed a technique for constructing two-phase EOS for simple metals using lattice dynamics, liquid dynamics, and electronic structure theory, and we have tested this technique by constructing an EOS for Aluminum valid up to compressions over two and temperatures up to five times melting temperature [Chisolm, Crockett, and Wallace, to appear in Phys. Rev. B]. Here we investigate the predictions of this EOS for the pressure, energy, and entropy along the Hugoniot up to roughly 5 Mbar, showing the relative contributions of the cold curve ($T = 0$ isotherm) and thermal part of the EOS to each function. We also comment on the possibility of taking data from different regions of the Hugoniot as tests of different terms in the EOS.

INTRODUCTION

A material's principal Hugoniot, the set of all equilibrium states accessible by a single shock from room temperature and pressure, is determined by the Rankine-Hugoniot relations together with the material's equation of state (EOS), which in most studies is composed of three parts: the cold curve ($T = 0$ isotherm), the thermal contribution from the nuclei, and the thermal contribution from the electrons. The importance of each of these parts in determining the Hugoniot varies with pressure; standard shock physics lore makes the following (certainly correct) assertions:

1. The cold curve makes the dominant contribution to the Hugoniot at low pressures.
2. At high pressures, the thermal components dominate over the cold curve, and at very high P the electronic thermal component dominates over all the rest, as more and more electrons are ionized and the electronic degrees of freedom greatly outnumber the nuclear ones.

What is not so clear, however, is where the low pressure regime ends, where the high pressure regime begins, and exactly what happens in between. In particular, it is not obvious that the thermal nuclear contribution enters significantly at any point.

Recently, we have developed a technique for constructing two-phase EOS using results from lattice dynamics, liquid dynamics, and electronic structure theory, and we have illustrated the technique by constructing an EOS for Aluminum, valid to compressions just over two and temperatures up to five times melting [1]. Here, we will show the predictions for the energy, entropy, and pressure along the Hugoniot due to each part of the EOS, and we will comment qualitatively on how they vary relative to one another as pressure increases. We will emphasize qualitative trends because we expect all metals across the periodic table to follow similar trends as functions of compression, differing primarily by a scale factor of around two for the heavier metals. Equivalently, we will see how much each contribution to the pressure could be changed without ruining the agreement between the EOS and available $P - \rho$ Hugoniot data, thus giving us an idea of how accurately Hugoniot data in different regions test different parts of the EOS. Finally, we will discuss the relevance of these results for future theoretical work on EOS.

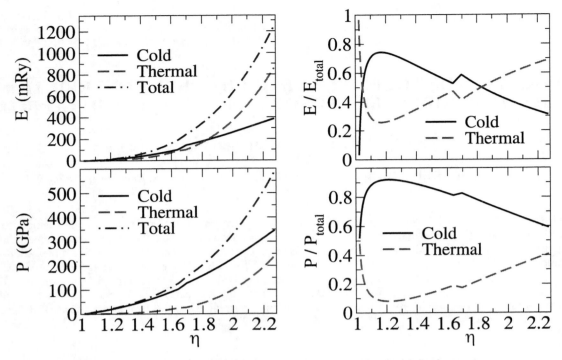

FIGURE 1. Left top: Total energy, cold curve energy, and thermal energy along the Hugoniot. Left bottom: Total, cold curve, and thermal pressure along the Hugoniot. Right top: Cold curve and thermal energy as fractions of total energy along the Hugoniot. Right bottom: Cold curve and thermal pressure as fractions of total pressure along the Hugoniot.

VARIATIONS ALONG THE HUGONIOT

The range of validity of our EOS allowed us to calculate the Hugoniot to a density of 6.15 g/cm^3, corresponding to a compression $\eta = \rho/\rho_{ref} = 2.28$. Over this range, Fig. 1 shows the total energy and pressure, respectively, along with the separate cold curve and total thermal (nuclear plus electronic) contributions. (The total energy, and each contribution to the energy, have been normalized to zero at room temperature and pressure.) Each of the left hand graphs shows all three parts (cold, thermal, total) separately, while the right hand graphs show the cold and thermal parts as fractions of the total. It's initially surprising that at very low pressures, the thermal energy and pressure dominate their cold curve counterparts, but one easily shows that for very small shocks the cold curve contributions to both quantities actually decrease, while the thermal parts always grow larger; then the cold curve parts turn around very rapidly and come to dominate at $\eta = 1.04$. Very soon after that, the thermal parts begin to catch up; the melting tran-

sition in the region from $\eta = 1.64$ to $\eta = 1.7$ reduces their contributions relative to the cold curve briefly, but by $\eta = 1.83$ ($P = 225$ GPa) the energy is dominated by the thermal contribution, and the pressure is on track to suffer a similar fate by $\eta = 2.5$ or so. Thus we see that by moderate compressions the thermal contribution to the EOS is non-negligible along the Hugoniot.

To see how the two thermal contributions compare, in Fig. 2, we show the thermal energy, entropy, and pressure, respectively, along with the separate nuclear and electronic contributions. (We did not discuss entropy in the previous paragraph because it is purely thermal.) Again, each of the left hand graphs shows all three parts (nuclear, electronic, thermal) while each right hand graph shows the nuclear and electronic parts as fractions of the thermal contribution. (The steep rise in the nuclear entropy at melt results because the configuration space available to the nuclei expands dramatically in the liquid phase.) Here the trends are obvious: The electronic part is almost entirely absent at low compressions, grow-

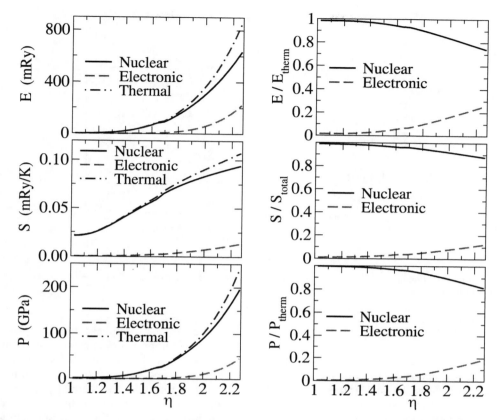

FIGURE 2. Left: Thermal, nuclear, and electronic energy, entropy, and pressure along the Hugoniot. Right: Nuclear and electronic energy, entropy, and pressure as fractions of thermal contribution along the Hugoniot.

ing very slowly as pressure increases, and the nuclear part clearly dominates through the entire range, dominating most in entropy and least in energy. Thus in this compression range, unsurprisingly, it is the nuclear thermal contribution, not the electronic contribution, which is most important.

Finally, in [1] we saw that the $P - \rho$ Hugoniot predicted by this EOS agrees with available data to an accuracy of 1% or so, which suggests the following question: How much could one change any one contribution to the pressure along the Hugoniot without spoiling the agreement with experiment? The answer is found in Fig. 3, which shows as a function of compression the fractional change in the cold curve, nuclear thermal, and electronic thermal pressure along the Hugoniot required to shift the total pressure by 1%. Clearly the Hugoniot is sensitive to very small changes in the cold curve pressure, on the order of

1 to 2% at all compressions, but it is sensitive also to the nuclear contribution at the 13% level throughout, and for $\eta \geq 1.8$ the sensitivity drops below 5%. Even the electronic contribution is significant at the higher pressures; at compressions above 2.1, a 20% change in the electronic part is sufficient to shift the total pressure out of agreement with experiment. Thus it is clear that while the cold curve contribution dominates, the nuclear thermal contribution still contributes significantly throughout the domain of validity of this EOS, and at the high end even the electronic part should be determined accurately to achieve acceptable agreement with available data.

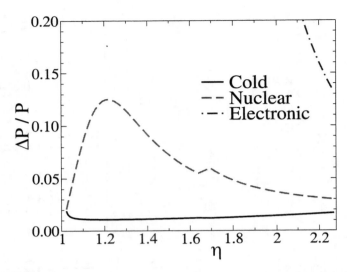

FIGURE 3. Fractional change in cold curve, nuclear thermal, or electronic thermal pressure required to shift total pressure along the Hugoniot by 1%.

CONCLUSIONS

While the cold curve clearly makes the largest contribution to energy and pressure along the Hugoniot at low compressions, the nuclear thermal part of the EOS becomes indispensable long before a compression of 2, and even the electronic thermal contribution must be treated properly in a regime that is (probably) well below the high-pressure region where it can be expected to approach the TFD form (see below). This makes it clear that an accurate account of the nuclear thermal EOS is required even at fairly low compressions, and that the intermediate pressure region of the Hugoniot is sensitive to its accuracy.

There is some ambiguity about where TFD becomes an accurate description of the thermal electronic behavior; traditions dating back to [2] suggest pressures around 10 Mbar, but other calculations [3] suggest a lower limit of 100 Mbar, at least for some materials. Further work to extend the region of applicability of this EOS and compare its predictions with TFD will help us understand more quantitatively where TFD can be expected to take over, and we may find a significant range of pressures where the thermal electronic contribution dominates the Hugoniot but TFD does not determine this contribution with sufficient accuracy.

Again, we expect these trends, expressed as functions of compression, to persist throughout the metals, with possible changes of scale needed for the heavier ones, but this should not affect our conclusions that both the thermal nuclear and thermal electronic contributions to the EOS need to be understood to some accuracy to properly account for Hugoniot data at compressions of 2 and higher.

ACKNOWLEDGMENTS

This work was supported by the U. S. Department of Energy through contract W-7405-ENG-36.

REFERENCES

1. Chisolm, E. D., Crockett, S. D., and Wallace, D. C., to appear in *Phys. Rev. B* (2003).
2. Feynman, R. P., Metropolis, N., and Teller, E., *Phys. Rev.* **75**, 1561 (1949).
3. Zittel, W. G., Meyer-ter-Vehn, J., Boettger, J. C., and Trickey, S. B., *J. Phys. F: Met. Phys.* **15**, L247 (1985).

CP706, *Shock Compression of Condensed Matter - 2003*
edited by M. D. Furnish, Y. M. Gupta, and J. W. Forbes
© 2004 American Institute of Physics 0-7354-0181-0/04/$22.00

A COMPARISON OF THEORY AND EXPERIMENT OF THE BULK SOUND VELOCITY IN ALUMINUM USING A TWO-PHASE EOS

Scott Crockett, Eric Chisolm, and Duane Wallace

Theoretical Division, Los Alamos National Laboratory, Los Alamos, New Mexico 87545

Abstract. We compute the bulk sound speed along the Hugoniot using a new solid-liquid two-phase equation of state (EOS) for aluminum [Chisolm, Crockett, and Wallace, to appear in Phys. Rev. B] and compare with experimental sound speeds from various sources. The experiment extends from the crystal through the entire solid-liquid two-phase region. The EOS and data closely agree on where the Hugoniot passes through the two-phase region, which corresponds to where aluminum melts. The bulk sound speed in the crystal region is consistent with the data, given the uncertainty in the experimental procedure. We also estimate shear moduli by using the experimental longitudinal sound speed data and the calculated bulk modulus. The shear modulus satisfies the approximation G_S/B_S=constant, within experimental error bars, throughout the crystal region on the Hugoniot.

INTRODUCTION

An aluminum equation of state was created recently using results from lattice dynamics, liquid dynamics and electronic structure theory [1]. With this EOS, one can calculate the adiabatic bulk modulus and bulk sound speed. In this work, we compare the results of these calculations to experimental bulk sound speed measurements along the Hugoniot. The comparison of theory and experiment is an additional guide to testing the validity of an EOS, given that the error bars and scatter in the experiment are reasonable. The data covers the part of the Hugoniot that crosses the solid-liquid coexistence region, allowing us to test the EOS in the two phase region. Using longitudinal sound speed data and the bulk modulus from the EOS, a reasonable estimate of the shear modulus can be computed, and this estimate can be used to test approximations for the shear modulus.

BULK AND LONGITUDINAL SOUND VELOCITIES

The longitudinal sound velocity c_l and the bulk sound velocity c_b are commonly measured in shock experiments. For the solid phase, the leading edge of the rarefaction wave is presumed to travel at the velocity c_l. This leading edge is observed as an initial drop in the wave profile. The velocity c_b is identified by a second drop in the wave profile where the material response changes from elastic to plastic. For the liquid phase, longitudinal waves do not propagate and the leading edge of the rarefaction wave travels at the velocity c_b. Experimental data for c_l and c_b are shown in Figure 1. These data are for 2024 aluminum [2, 3, 4, 5, 6, 7, 8, 9, 10, 11, 12, 13, 14, 15], except for one measurement on 6061 [16].

To calculate the bulk velocity c_b from our equation of state we first calculate the adiabatic bulk modulus

$$B_S = -V \left. \frac{\partial P}{\partial V} \right)_S, \qquad (1)$$

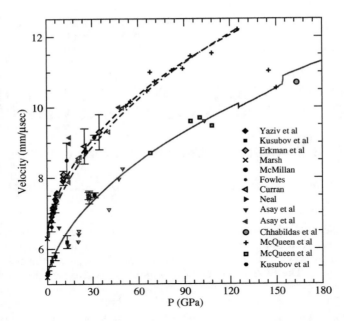

FIGURE 1. c_l and c_b vs P on the Hugoniot. The solid line is the calculated c_b from our EOS. The dashed line is a numerical fit to c_l, Eq.(1), and the dot-dashed line results from setting G_S/B_S equal to a constant. The data at $P = 0$ from Marsh are from ultrasonic measurements.

where the volume V is related to the density ρ by $V = 1/\rho$. Then c_b is obtained from

$$\rho c_b^2 = B_S. \qquad (2)$$

Our theoretical c_b for pure aluminum along the Hugoniot is compared with experiment for 2024 aluminum in Fig 1. At a given pressure and density, we expect the bulk sound velocity to be essentially the same for pure aluminum and 2024 aluminum. Notice that the theory shows a slight decrease in c_b through the solid-liquid two phase region. This reflects the fact that the $P - \rho$ adiabats cross the solid-liquid two-phase region at a slightly smaller slope than in the separate solid and liquid phases. The Figure illustrates that our theory is in excellent agreement with experiment for c_b, or equivalently for B_S.

Our equation of state does not provide a theoretical prediction for c_l, since the equation of state is in terms of isotropic pressure P. However, we can learn something about the shear modulus from the experimental data for c_l. For this purpose we show two curves representing c_l in Figure 1. The dashed line is simply the numerical fit

$$c_l = 6.36 + 0.3633P^{2/3} - 0.02631P. \qquad (3)$$

Note that the c_l fit goes through the ultrasonic data [3] at $P = 0$. The dot-dashed line is a calculation of c_l using an approximation for the adiabatic shear modulus G_S. We shall discuss this model for G_S in the next section.

SHEAR MODULUS

The shear modulus is an important material property and is needed to calculate plastic flow processes in hydrodynamic codes. In explosively driven processes, the shear modulus is needed over a wide range of the equation of state surface, and not just in the vicinity of the Hugoniot. An estimate of G_S for the solid phase at all temperatures up to melt, and to pressures of a megabar or so, reliable to an accuracy of 25%, would be most useful for practical hydrodynamical calculations.

We have previously developed an approximation for G_S based on setting $G_S/B_S = constant$, where the constant is to be determined separately for each metal [17]. This allows us to calculate G_S from an equation of state at any pressure and density, since B_S can be calculated from the equation of state. The

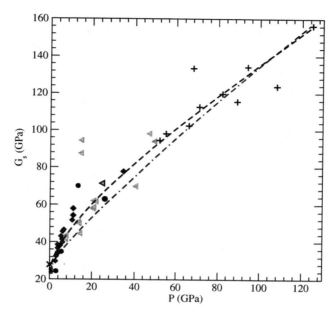

FIGURE 2. G_S vs P on the Hugoniot.

approximation was developed from a study of ultrasonic data, and it was used in making qualitative estimates of strength and dissipative effects in overdriven shocks in metals. While the ultrasonic data are highly accurate, they are limited to modest pressures, not more than 10 GPa. With the data we have collected for aluminum we can now test this approximation up to 125 GPa along the Hugoniot, the point where aluminum starts to melt.

From the theory for an isotropic material, we have

$$\rho c_l^2 = B_S + \frac{4}{3} G_S. \tag{4}$$

We used this equation, together with the experimental data for c_l from Figure 1 and with ρ and B_S determined from our EOS, to calculate G_S along the Hugoniot. The results are shown by the points in Figure 2. Notice these points do not precisely represent 2024 aluminum, since they incorporate some information for pure aluminum. The ratio G_S/B_S, from the G_S points of Figure 2, and again with B_S from our EOS, is shown by the points in Figure 3.

In Figure 3, the dashed line is a result of using the numerical fit for c_l, Eq.(3), and the equation of state for B_S, and Eq(4) to calculate G_S. To approximate G_S/B_S by a constant, we might choose any one of three values: (*a*) 0.37, the mean value of the points

in Figure 3, (*b*) 0.36, the ultrasonic value for 2024 aluminum at $P = 0$ and 295 K [3], or (*c*) 0.35, the ultrasonic value for pure aluminum at $P = 0$ and 295 K [18]. At the level of accuracy we are concerned with, all these constants are equally acceptable. Further, to an accuracy of 10% in G_S, any of these constants for G_S/B_S is as good as the fitted curve in Figure 3.

To test the approximation $G_S/B_S = constant$, let us take $G_S/B_S = 0.35$. The corresponding result in each figure is shown by a dot-dash line. The corresponding G_S and c_l curves agree with experimental data within error bars, as shown respectively in Figures 1 and 2.

CONCLUSION

Our equation of state formulation is based on well developed condensed matter theory, and is expected to be highly accurate for solid and liquid phases at all densities, and at temperatures to around five times the melting temperature[19]. This expectation was confirmed by comparing theory and experiment for the aluminum Hugoniot to 600 GPa and 35,000 K [1]. Here we compare theory and experiment for sound velocities in aluminum, and reach the following-

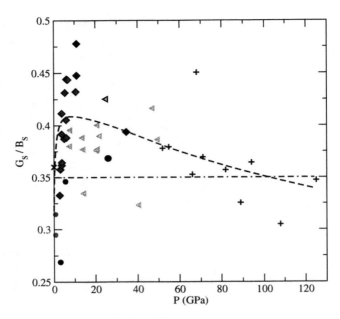

FIGURE 3. G_S/B_S vs P on the Hugoniot.

ing conclusions.

(a) The bulk sound velocity c_b, or equivalently the bulk modulus B_S, from our aluminum EOS is in excellent agreement with experiment on the Hugoniot, throughout the solid and two-phase regions.

(b) The approximation $G_S/B_S = constant$ agrees with experiment within error bars, on the Hugoniot through the solid phase. We expect this approximation to hold equally well away from the Hugoniot, throughout the entire solid phase region.

ACKNOWLEDGMENTS

This work was supported by the U. S. Department of Energy through contract W-7405-ENG-36.

REFERENCES

1. Chisolm, E. D., Crockett, S. D., and Wallace, D. C., to appear in *Phys. Rev. B* (2003).
2. Curran, D. R., *J. Appl. Phys.* **34**, 2677 (1963).
3. Marsh, S., *LASL Shock Hugoniot Data*, University of California Press, 1980.
4. Vorob'ev, A. A., Dremin, A., and Kanel, G., *Zh. Prikl. Mekh. Tekh. Fiz.* **5**, 94 (1974).
5. Yaziv, D., Rozenberg, Z., and Partom, Y., *J. Appl. Phys.* **53**, 353 (1982).
6. Fuller, P. J. A., and Price, J. H., *Brit. J. Appl. Phys. (J. Phys. D.)* **2, Sec. 2**, 275 (1969).
7. Neal, T., *J. Appl. Phys.* **46**, 2521 (1975).
8. Fowles, G. R., *J. Appl. Phys.* **32**, 1475 (1961).
9. Erkman, J. O., and Christensen, A. B., *J. Appl. Phys.* **38**, 5395 (1967).
10. Kusubov, A. S., and van Thiel, M., *J. Appl. Phys.* **40**, 3776 (1969).
11. Kusubov, A. S., and van Thiel, M., *J. Appl. Phys.* **40**, 893 (1969).
12. McQueen, R. G., Fritz, J. N., and Morris, C. E., *Shock Waves in Condensed Matter*, Elsevier, Science Publishers, 1984, pp. 95–98.
13. Chhabildas, L. C., Furnish, M. D., and Reinhart, W. D., *Shock Compression of Condensed Matter*, AIP, 2000, pp. 97–100.
14. Asay, J. R., and Chhabildas, L. C., *Shock waves in High-Strain-Rate Phenomena in Metals*, Plenum Press, 1981, p. 417.
15. Bushman, A. B., Kanel', G. I., Ni, A. L., and Fortov, V. E., *Intense Dynamic Loading of Condensed Matter*, Taylor & Francis, 1993.
16. McMillan, A. R., *Bull. Amer. Phys. Soc.* **13**, 1680 (1968).
17. Wallace, D. C., *Phys. Rev. B* **24**, 5607 (1981).
18. J. F. Thomas, J., *Phys. Rev.* **175**, 955 (1968).
19. Wallace, D., *Statistical Physics of Crystals and Liquids*, World Scientific, 2002.

CP706, *Shock Compression of Condensed Matter - 2003*
edited by M. D. Furnish, Y. M. Gupta, and J. W. Forbes
© 2004 American Institute of Physics 0-7354-0181-0/04/$22.00

WIDE RANGE EQUATION OF STATE OF WATER TAKING INTO ACCOUNT EVAPORATION, DISSOCIATION AND IONIZATION

V.V. Dremov, A.T. Sapozhnikov, M.A. Smirnova

Russian Federation Nuclear Centre-Institute of Technical Physics, P.O. Box 245, Snezhinsk, Chelyabinsk region 456770, Russia.

Here we present new wide range equation of state for water. It has been constructed by sewing together a number of local models describing the matter in different regions of the phase diagram. At the temperatures under dissociation and moderate densities semiempirical equations of state describing water with high accuracy and taking evaporation into account have been used. To construct thermodynamic model describing properties of water in the region covered by shock data Variational Perturbation Theory has been applied. Dissociation reactions have also been introduced in the model. In this region water is considered as a mixture of molecular fluids. Some peculiarities of intermolecular potential for water and their effect upon parameters of shock compression have been investigated. Results of calculation have been compared with experimental data on shock compression of porous ice and snow. At high densities the matter is considered as homogeneous mixture of atoms and Thomas-Fermi model with quantum corrections and nuclei treatment by Kopyshev is applied. At low densities and high temperatures the model of weakly non-perfect dissociating gas and Saha model of ionized gas have been used. The EOS has been converted into tabular form to make it efficient when using in hydrodynamic codes.

INTRODUCTION

Water from the one hand is one of the wide-spread nature substances and from the other hand is widely used as a working fluid in technology. That is why the experimental and theoretical investigations in the many fields of physics including the physics of shock waves require the equation of state of water capable of precise calculation of its thermodynamic properties in the wide range of temperatures and densities.

In the past twenty five years a number of wide range water EOSs have been constructed. Tabular EOS [1] describes water at densities $10^{-6} < \rho(g/cm^3) < 6$ and energies $0 < E(kJ/g) < 10^6$ The EOS [1] uses precise

approximations of the experimental data on water and vapour [2] and approximations [3] of calculation data [4] at high temperatures and densities obtained with Thomas-Fermi model. In the region covered by shock experiments EOS [1] uses modified EOS [5] and semiempirical approach from [3] to describe dissociation.

Wide range EOS [6] gives analytical forms for pressure and energy as the functions of density and temperature. It takes into account evaporation, oscillation excitations, dissociation (semiempirical approach) and hydrogen bonding. At high temperatures and pressures this EOS approximates calculation data [4]. EOS [6] provides for satisfactory description of experimental data [2] at $\rho > 0.01$ g/cm^3 and experimental data on shock compression of water but it has poor accuracy when describing experimental data on shock compression of porous ice and snow [7,8].

EOS [9] also has an analytical form and uses a specific volume and an entropy as the independent thermodynamic variables. This EOS is in satisfactory agreement with the experimental data on shock compression of water and porous ice but at pressures below 10 GPa and temperatures below 1000 K it fails when describing precise experimental data on static compression [2]. Again, an entropy as an independent variable hampers the use of the EOS in the hydrocodes.

The experimental data on shock compression of water into Megabar range (0.5–31.5 Mbar) [10-13] and new theoretical data on thermodynamic properties of water as a homogeneous mixture of oxygen and hydrogen obtained with Thomas-Fermi model with quantum corrections [9] led to construction of the tabular EOS GLOBUS [14]. The EOS works in the ranges $10^{-3} \leq \rho$ (g/см³)≤ 64, $300 \leq T(\text{K}) \leq 1.6 \cdot 10^8$ and has more sophisticated mathematical form comparably to EOS [1]. Potential pressure at densities lower and higher than the density at ambient conditions is presented on the uniform and logarithmic grids respectively. Potential energy is found from the equation of thermodynamic compatibility. The EOS domain of definition is subdivided into a number of rectangular subdomains corresponding to the models valid in the given ranges of density and temperature. Thermal pressure and energy are presented within the subdomains on the logarithmic grids. Between the grid nodes bicubic interpolation is used. Beyond the EOS's domain of definition a simple extrapolation is applied.

GLOBUS-3 EQUATION OF STATE

Here we present GLOBUS-3 EOS for water which is the modification of GLOBUS EOS [14]. The main difference of the new EOS is that the subdomains corresponding to different physical models are no more rectangular (See Fig.1).

When tabulating this EOS more comprehensive set of models and local EOSs comparably to [14] have been used:

1) the model of weakly non-perfect dissociating gas;

2) variational perturbation theory;

3) calculation data on ionized gas of low density – Saha model [15];

4) calculation data obtained with Thomas-Fermi model with quantum corrections [16] and nuclei treatment by Kopyshev [17] for homogeneous mixture of oxygen and hydrogen.

GLOBUS-3 EOS for water is superior to earlier EOSs when describing regions of dissociation and ionization, experimental data on shock compression of porous ice and snow [7, 8] as well as water properties at extremely high pressures on the Hugoniot [13]. The ranges of the applicability of the EOS are $10^{-4} \leq \rho$ (g/см³) ≤ 100; $273 \leq T$ (K) $\leq 6 \cdot 10^7$.

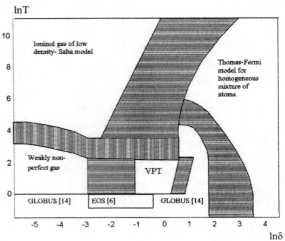

FIGURE 1. Layout of models sewed together in the frame of the wide range GLOBUS-3 EOS of water. Shading denotes interpolation regions between models.

APPLICATION OF VARIATIONAL PERTURBATION THEORY (VPT) TO WATER

To construct thermodynamic model describing properties of water in the region covered by shock data obtained in experiments with porous ice and snow [7-8] VPT [18] has been applied. Dissociation reactions have also been introduced in the model according to [19].

We found that at densities corresponding to condensed matter the major reaction affecting the parameters of water shock compression is

$$2H_2O \leftrightarrow H_3O^+ + OH^- \qquad (1)$$

This reaction is responsible for appearance of the conductivity of water when shock compression [21].

In the paper by Ree [20] the following form of the intermolecular potential for water was proposed (exp-6):

$$\varphi(r) = \varepsilon(T) \left\{ \left(\frac{6}{\alpha-6} \right) \exp[\alpha(1-r/r_0^*)] - \left(\frac{\alpha}{\alpha-6} \right) \left(\frac{r_0^*}{r} \right)^6 \right\}, \quad (2)$$

where parameter ε depends upon temperature (due to water molecule dipole moment). This formula being a simple approximation of the results of ab-initio calculations [21] provides for excellent agreement with the experimental data on the Hugoniot of water.

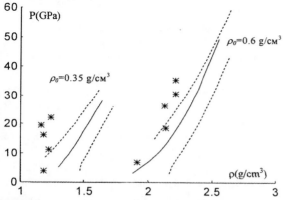

FIGURE 2. Hugoniots of porous ice. Stars – experimental data [7], solid lines – calculation with the potential [20], dashed lines bound the region where the Hugoniot lies when applying different possible approximations of the ab-initio data [21].

However calculations carried out for porous ice (see Fig.2) showed strong disagreement with the experimental data. Dissociation at the pressures P<20 GPa for these Hugoniots is <1% and hence does not affect their positions. Conclusion is that the approximation (2) is unsatisfactory. Indeed, not the minimum depth only but the minimum position is also temperature dependent - it shifts right as the temperature increases (see [21]). This leads to the compressibility decrease. So as at the given pressure on the Hugoniot the temperature increases with initial porosity the compressibility decrease reveals itself stronger as the initial porosity grows.

Calculations carried out with limiting approximations corresponding to the lower (1000 K) and the upper curve (10000 K) from [21] bound the region where the Hugoniot lies when

applying different possible approximations of the ab-initio data [21] (see fig.2). The experimental data go out of this region as pressure grows. So, the matter is not only in successful approximation. We suppose the following reasons for the disagreement between the calculation and the experiment. Ab initio calculations [21]

1) do not take into account multiparticle interactions;

2) the multipolar and other depending upon mutual orientations interactions have been taken into account indirectly, namely, via the procedure of averaging over the orientations at the given temperature.

To improve the model two simple steps have been done. As it has been already mentioned above more accurate approximation of the ab-initio data [21] requires temperature dependence of r_0' parameter (characteristic radius) which has been taken in the form:

$$r_0' = r_{00}'((1.0 + \zeta \exp(-T_0/T)) \quad (3)$$

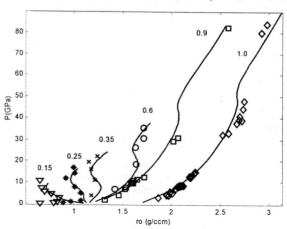

FIGURE 3. Hugoniots of water and porous ice and snow. Solid lines - calculation with modified potential, Markers — experimental data [7,8,10-13,23]. Digits near the curves are initial densities. Data for ρ_0=1.0 g/см^3 have been added +0.5 g/см^3.

Then we tried to take into account effectively the multiparticle effects. Remaining in the frame of pair potential we suppose that the characteristic radius growth with a rise of temperature (i.e. spherical symmetrization of the interaction) is suppressed as the density increases i.e. assume that T_0 parameter in (3) is a function of density

$$T_0 = T_{00}(1 + \nu\rho) \qquad (4)$$

The calculations with modified potential and experimental data on shock compression of water, porous ice and snow [7-8] are compared on the fig.3. One can see very good agreement.

A similar approach to water EOS has been applied in the paper [22] where intermolecular interaction is described by spherically symmetric exp-6 potential and an angular dependent multipolar contribution. Unfortunately in [22] there is no comparison with experimental data for porous ice and snow.

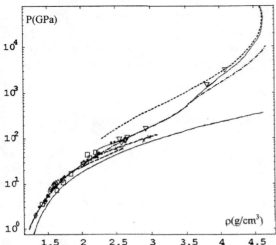

FIGURE 4. Hugoniot of water. Solid line –GLOBUS-3 EOS, dashed and dotted line – GLOBUS EOS [14], markers – exp. data [7-8, 10-13, 23], dashed line – Thomas-Fermi model with quantum corrections for homogeneous mixture of atoms, long dashed line – double compression calculated by GLOBUS-3, x – exp. data on double compression [7]. Lower solid line – potential pressure.

The principal Hugoniot of water calculated with GLOBUS-3 EOS in which different physical models have been sewed together is presented on the fig.4.

ACKNOWLEDGEMENTS

The authors thank E. Mironova for fruitful discussions and software support.

REFERENCES

1. Kovalenko, G.V., Sapozhnikov, A.T., *VANT, Ser.: Met. Prog. Chis. Resh. Zad. Mat. Fiz.*, 4(**6**), 40-46, (1979).

2. Vukalovich, M. P., Ryvkin, S.L., Alexandrov A.A., *Tables of thermodynamic properties of water and water vepour*, M.: Standart publishers, 1969.

3. Zamyshlyaev, B.V., Menzhulin, M.G., *Rus. Appl. Mech. and Tech. Phys.*, **3**, 113-118, (1971).

4. Lateer, R., *J.Phys.Rev.*, **99**, 1854–1859, (1955).

5. Sapozhnikov, A.T., Pershina, A.V., *VANT, Ser.: Met. Prog. Chis. Resh. Zad. Mat. Fiz.*, 4(**6**), 47-52, (1979).

6. Bobrovsky, S.V., Gogolev, V.M., Menzhulin, M.G., Shylova, R.V., *Rus. Appl. Mech. and Tech. Phys.*, **5**, 130-139, (1978).

7. Bakanova, A.A., Zubarev, V.N, Sutulov, Yu.N., Trunin, R.F., *ZHETF*, **68**, 1099–1107, (1975).

8. Trunin, R.F., Simakov, G.V., Zhernokletov, M.V., Dorokhin, V.V., *Teplofizika Vysokih Temperatur*, **37**, 732-737, (1999).

9. Kalitkin, N.N., Kuzmina, L.V., Sharipjanov, I.I., *Constraction of equations of state for chemical compounds*, Preprint #43. M.: IPM, USSR Acad. Sci., 1976.

10. Volkov, L.P., Voloshin, N.P., Mangasarov, P.A., et.al., *Pis'ma v ZHETF*, **31**, .564–548, (1980).

11. Mitchel, A.C., Nellis, W.J., *J.Chem.Phys.*, **76**, 6373–6281, (1982).

12. Poduretz, M.A., Simakov, G.V., Trunin, R.F., et.al., *ZHETF*, **62**, 710–712, (1972).

13. Avrorin, E.N., Vodolaga, B.K., Voloshin N.P., et.al., *ZHETF*, **93**, 613–626, (1987).

14. Sapozhnikov, A.T., Kovalenko, G.V., Gershchuk, Mironova, E.E., *VANT, Ser.: Mat. Mod. Fiz. Proc.*, **2**, 15–19, (1991).

15. Kalitkin, N.N., Ritus, I.V., Mironov, A.M., *Ionization equilibrium taking into account degeneration of electrons*, Preprint №46. M.: IPM, USSR Acad. Sci., 1983.

16. Kalitkin, N.N., Kyz'mina, L.V., *Tables of thermodynamic functions at high energy concentration*, Preprint №35, M.: IPM, USSR Acad. Sci., 1975.

17. Kopyshev, V.P., *Rus. Chis. Met. Mech. Splosh. Sred.*, **8**, 54–67, (1977).

18. Ross, M., *J.Chem.Phys.*, **71**, 1567, (1979).

19. Dremov, V.V., Modestov, D.G., *Chem. Phys. Reports*, **17**, 781-790, (1998).

20. Ree, F.H., *J.Chem.Phys.*, **81**, 1251-1262, (1984).

21. Ree, F.H., *J.Chem.Phys.*, **76**, 5287, (1982).

22. Jones, H.D., "Theoretical equation of state for water at high pressures", in *Shock Compression of Condensed Matter-2001*, edited by M.D. Furnish et. al., AIP Conference Proceedings 620, New York, 2002, pp.103-106.

23. Walsh, J.M., Rice, M.H., *J.Chem.Phys.*, **76**, 6273, (1982).

CP706, *Shock Compression of Condensed Matter - 2003*
edited by M. D. Furnish, Y. M. Gupta, and J. W. Forbes
© 2004 American Institute of Physics 0-7354-0181-0/04/$22.00

ON PHASE TRANSITION IN STRONGLY COUPLED HYDROGEN PLASMA

Vladimir S. Filinov*, Pavel R. Levashov*, Michael Bonitz[†], Vladimir E. Fortov* and Werner Ebeling**

Institute for High Energy Densities, RAS, Izhorskaya 13/19, Moscow 125412, Russia
[†]*Fachbereich Physik, Universität Rostock, D-18051 Rostock, Germany*
**Institute für Physik, Humboldt-Universität Berlin, Invalidenstrasse 110, D-10115 Berlin, Germany*

Abstract. Plasma phase transitions in dense hydrogen and electron-hole plasmas are investigated by direct path integral Monte Carlo (DPIMC) method. The results are compared with existing theoretical and experimental data. The phase boundary of the electron-hole liquid in germanium is calculated in agreement with the known experimental results. High-density hydrogen exhibits similar behavior. In both cases the phase transition is accompanied by the conductivity rise and the internal energy is lowered due to the formation of droplets. The high-density part of the deuterium shock Hugoniot is computed using the simulation results.

INTRODUCTION

The thermodynamics of strongly correlated Fermi systems at high pressures are of growing importance in many fields, including shock and laser plasmas, astrophysics, solids and nuclear matter [1–3]. Among the phenomena of current interest is also plasma phase transition (PPT), which occurs in situations where interaction and quantum effects are relevant. PPT has been predicted theoretically on the basis of the chemical picture of a partially ionized plasma [4–6]. However chemical models are unreliable for high-density quantum plasma close to pressure ionization and dissociation, therefore, there is a great interest in first-principle calculations of these systems which avoid such approximations.

DIRECT PATH INTEGRAL MONTE CARLO

The description and discussion of our computational scheme can be found elsewhere [7, 8]. The idea of the DPIMC is as follows: all thermodynamic properties of a two-component plasma with N_e electrons and N_p

protons at a temperature T and volume V are defined by the partition function $Z(N_e, N_p, V, T)$:

$$Z(N_e, N_p, V, T) = \frac{1}{N_e! N_p!} \sum_{\sigma} \int_V dq \, dr \, \rho(q, r, \sigma; T),$$

where q — coordinates of protons, r — coordinates of electrons, σ — spins of protons and electrons, ρ is the density matrix of the system. Taking into account the spin of the system and Fermi statistics the density matrix itself is then expressed by a path integral [7]. The physical meaning of such expression is that all the particles are represented by fermionic loops with n coordinates (beads). In our simulations we used the effective finite at zero distance quantum pair potential [8] which has been obtained by Kelbg as a result of a first-order perturbation theory solution of a Bloch equation.

HYDROGEN PLASMA SIMULATION

The simulation has been performed at temperatures from 10^4 K and higher in the wide range of particle densities. Under these conditions the exchange effects for protons are negligible. Earlier thermodynamic properties and pair distribution functions of degenerate strongly coupled hydrogen plasma (including the conditions of partial dissociation and ionization) have been calculated [9]. At high temperatures and small degeneracy parameters $\chi = n\lambda^3$ ($\lambda^2 = 2\pi\hbar^2\beta/m_e$ — thermal wavelenght of electron) we obtained a good agreement with independent restricted path integral Monte-Carlo calculations (RPIMC) [10]. But at lower temperatures and higher densities the agreement became worse. If the average interparticle distance is close to the size of hydrogen molecule we observed a bad convergence to the equilibrium state and the formation of many-particle clusters. As a result the pressure values can be negative under such conditions [11, 12]. Such effects are apparently excluded in the RPIMC calculations [10] by the additional assumptions (on the nodes of the density matrix) used to reduce the region of integration and the sum over permutations to even (positive) contributions only.

In Fig. 1 the dependency of full plasma energy vs. particle density on $T = 10^4$ K isotherm is shown. For ideal degenerate plasma (curve 1) the energy monotonically increases with particle density rise; at high densities the behavior of the energy of nonideal plasma is the same (curve 6) because of the degeneracy of electrons. But at small and medium densities the energy of nonideal plasma decreases and becomes negative. In the range of densities near the energy minimum we observed a bad convergency to the equilibrium state and energy values were calculated with low accuracy. Exactly in this region strong Coulomb correlations and quantum effects lead to the formation of bound states. The curve 3 in Fig. 1 corresponds to the model PACH (Padé approximations with chemical picture) [13], and the curve 4 shows the density functional theory (DFT) computations [14]. At high density limit for fully ionized plasma PACH and DFT results coincide. On the other hand, these data are systematically higher our DPIMC points for $n > 10^{24}$ cm^{-3} because of insufficient accuracy of calculations at high degeneracy parameter χ as well as small number of beads and particles.

The minimum on the dependency of energy vs. particle density is reproduced by all methods, but the minimum width and depth are different for every model. RPIMC method gives higher energy value in the minimum on the isotherm 10^4 K, whereas at

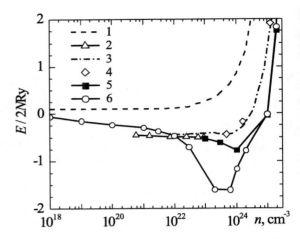

FIGURE 1. Internal energy of hydrogen for $T = 10^4$ K. 1 — ideal plasma, 2 — [10], 3 — PACH, 4 — [14], 5 — this work, clusters are prohibited, 6 — this work

$5 \cdot 10^4$ K the simulation results coincide within the accuracy of calculations. The analysis showed that in the region of energy minimum on the isotherm $T = 10^4$ K plasma uniform state is energetically unfavorable. This leads to the separation of the system and the formation of clusters (droplets) with high density on the background of low density plasma [12]. We observed such formations in the Monte-Carlo cell which is typical in simulations inside first-order phase transition regions. To prove that the uniform state is unstable in these conditions we reduced the region of integration in our simulations to exclude configurations with many-particle clusters, namely configurations, in which distances between three and more particles are smaller than a value d_{min} varying from 1.3 to $2a_B$, where a_B — Bohr radius. The energy of this artificially formed system rises significantly in a good agreement with RPIMC data [10] (in the density interval where RPIMC results are available).

The DFT computations [14] in a particle density range 10^{23}–10^{24} cm^{-3} revealed the existance of strong fluctuations of ionic density which was explained by the beginning of phase transition from the high-conductivity (quasi-metal) into low-conductivity state at the reduction of particle density. This assumption is confirmed by an abrupt conductivity rise (by 4-5 orders) in the density range from 0.3 to 0.5 g/cm^3 measured in experiments on shock and quasi-isentropic compression of hydrogen and deuterium [15].

Deuterium Shock Hugoniot

The calculated thermodynamic properties of hydrogen allowed us to compute the shock Hugoniot of deuterium. The Hugoniot equation determines the compressed state (P, V, E) of matter through its initial state (P_0, E_0, V_0):

$$H = E - E_0 + \frac{1}{2}(V - V_0)(p + p_0) = 0.$$

Following the work [16] we chose $p_0 = 0$, $\rho_0 = 0.171$ g/cm³, $E_0 = -15.886$ eV per atom. Then we computed the values of pressure P_i and energy E_i at a given constant temperature T (from 10^4 K to 10^6 K) and four values of volume V_i corresponding to $r_s = 1.7$, 1.86, 2 and 2.15, where $r_s = \bar{r}/a_B$, $\bar{r} = (3/4\pi n_p)^{1/3}$, n_p — particle density. Substituting the obtained values P_i, E_i and V_i into the Hugoniot equation we determine the volume range V_1, V_2 where the function $H(P,V,E)$ changes its sign. The value of density on the Hugoniot is calculated by linear interpolation of function H between V_1 and V_2.

The results of calculations together with selected theoretical and experimental data are shown in Fig. 2. At high pressures the RPIMC deuterium Hugoniot [16] and the DPIMC curve from the current work are very close. But at lower pressures the DPIMC Hugoniot deviates to higher densities and lies between Z-pinch experimental points [17] and laser shock wave data [18] near recently obtained registration by convergent geometry technique [19]. It is worth to mention that at temperatures below $2 \cdot 10^4$ K there is a big influence of phase separation to the thermodynamic properties. This obstacle leads to unreliable values on the shock Hugoniot at pressures below 2 Mbar.

ELECTRON-HOLE PLASMA SIMULATION

Since the PPT in dense hydrogen is still hypothetical and has not been observed experimentally, it is reasonable to look for other systems where similar conditions exist. A suitable example is electron-hole plasma in low-temperature semiconductors, for which droplets formation was well established and

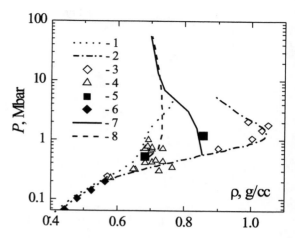

FIGURE 2. Shock Hugoiot of deuterium. Calculation: 1 — SESAME, 2 — [18], 7 — this work, 8 — [16]. Experiment: 3 — [18], 4 — [17], 5 — [19], 6 — [20]

observed experimentally three decades ago [21]. We, therefore, performed DPIMC simulations of isotherms of electron-hole plasma. At relatively low temperatures the properties of electrone-hole plasma can be described by the degenerate system of electrons and holes with constant effective masses. Below a critical temperature at low densities plasma is uniform and only electron-hole pairs (exitons) exist. At higher densities the simulations exhibit anomalously large fluctuations and an unstable behaviour of the pressure. The e-h plasma is found to separate and form large droplets and the photoconductivity rise is observed [21]. At very high densities plasma becomes uniform again.

The phase boundary of the electron-hole liquid in germanium obtained by our DPIMC method is presented in Fig. 3 in a reasonable agreement with the experimental data [22].

DISCUSSION AND CONCLUSION

Our interpretation of the facts described above is that they are a direct indication of a first-order phase transition [4–6]. There are also obvious similarities between hydrogen and electron-hole plasmas. The main point of debate of the present DPIMC results is the low value of the energy in the region of droplet formation, which at minimum is about three times

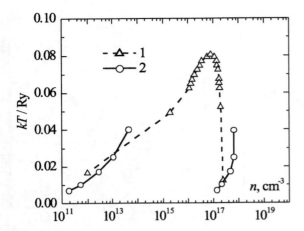

FIGURE 3. Phase boundary of electrone-hole plasma in bulk germanium. Experiment: 1 — [22]. Calculation: 2 — this work. Temperature is shown in units of the exiton binding energy

lower than that of molecular hydrogen. There are two possible factors which determine the simulation accuracy and which may be responsible: the limited number of particles ($N_e = N_i \lesssim 100$) and the finite number n of beads ($n = 20$ in the simulations). These factors may certainly affect the calculated values of the total energy at high densities, but has no influence on the general existence of the droplets. There is also a problem of extrapolation of the results to a macroscopic system because of finite-size effects. The surface effects of a small inhomogeneous system are significant; this, of course, leads to a lowering of the total energy. Therefore, in order to obtain more accurate data for the internal energy of a macroscopic two-component plasma at ultrahigh compression, an increase in the simulation size (CPU time) by at least a factor 10 is desirable which should become feasible in the near future.

REFERENCES

1. Kalman, B., editor, *Strongly Coupled Coulomb Systems*, Pergamon, Oxford, 1988.
2. Bonitz, M., editor, *Progress in Nonequilibrium Greens Functions*, World Scientific, Singapore, 2000.
3. Kraeft, W. D., and Schlanges, M., editors, *Proc. Int. Conf. on Strongly Coupled Plasmas*, World Scientific, Singapore, 1996.
4. Norman, G. E., and Starostin, A. N., *Sov. Phys. High Temp.*, **6**, 410 (1968).
5. Saumon, D., and Chabrier, G., *Phys. Rev. A*, **44**, 5122 (1991).
6. Schlanges, M., Bonitz, M., and Tschttschjan, A., *Contrib. Plasma Phys.*, **35**, 109 (1995).
7. Zamalin, V. M., Norman, G. E., and Filinov, V. S., *The Monte-Carlo Method in Statistical Thermodynamics*, Nauka, Moscow, 1977.
8. Filinov, V. S., Bonitz, M., Levashov, P. R., Fortov, V. E., Ebeling, W., Schlanges, M., and Koch, S. W., *J. Phys. A.: Math. Gen.*, **36**, 6069–6076 (2003).
9. Filinov, V. S., Bonitz, M., and Fortov, V. E., *JETP Letters*, **72**, 245–248 (2000).
10. Militzer, B., and Ceperley, D. M., *Phys. Rev. E*, **63**, 066404 (2001).
11. Filinov, V. S., Fortov, V. E., Bonitz, M., and Levashov, P. R., *JETP Letters*, **74**, 384 (2001).
12. Levashov, P. R., Filinov, V. S., Fortov, V. E., and Bonitz, M., "Thermodynamic Properties of Nonideal Strongly Degenerate Hydrogen Plasma," in *Shock Compression of Condensed Matter — 2001*, edited by M. D. Furnish, N. N. Thardhani, and Y. Horie, AIP, New York, 2002, pp. 119–126.
13. Kraeft, W. D., Kremp, D., Ebeling, W., and Röpke, G., *Quantum Statistics of Charged Particle Systems*, Akademie, Berlin, 1986.
14. Xu, H., and Hansen, J.-P., *Phys. Rev. E*, **57**, 211–223 (1998).
15. Weir, S. T., Mitchell, A. C., and Nellis, W. J., *Phys. Rev. Lett.*, **76**, 1860–1863 (1996).
16. Militzer, B., and Ceperley, D. M., *Phys. Rev. Lett.*, **85**, 1890–1893 (2000).
17. Knudson, M. D., Hanson, D. L., Bailey, J. E., Hall, C. A., and Assay, J. R., *Phys. Rev. Lett.*, **90**, 035505–1 (2003).
18. Da Silva, L. P., Celliers, P., Collins, G. W., Budil, K. S., Holmes, N. C., Barbee, T. W., Jr., Hammel, B. A., Kilkenny, J. D., Wallace, R. J., Ross, M., and Cauble, R., *Phys. Rev. Lett.*, **78**, 483–486 (1997).
19. Belov, S. I., Boriskov, G. V., et al., "Compression of solid deuterium by megabar pressures of shock waves," in *Substances, Materials and Constructions under Intense Dynamic Influences*, edited by A. L. Mikhailov, VNIIEF, Sarov, 2003, pp. 100–104.
20. Nellis, W. J., Mitchell, A. C., van Theil, M., Devine, G. J., Trainor, and R. J., N., Brown, *J. Chem. Phys.*, **79**, 1480 (1983).
21. Jeffries, C. D., and Keldysh, L. V., editors, *Electron-Hole Droplets in Semiconductors*, Nauka, Moscow, 1988.
22. Thomas, G. A., Rice, T.-M., and Hensel, J. C., *Phys. Rev. Lett.*, **33**, 219–222 (1974).

CP706, *Shock Compression of Condensed Matter - 2003*
edited by M. D. Furnish, Y. M. Gupta, and J. W. Forbes
© 2004 American Institute of Physics 0-7354-0181-0/04/$22.00

DFT CALCULATIONS OF STRUCTURAL AND THERMODYNAMIC PROPERTIES OF MOLTEN SN: ZERO-PRESSURE ISOBAR

S. M. Foiles

Sandia National Laboratories, Albuquerque NM 87145

Abstract. The dynamic compression of molten metals including Sn is of current interest. In particular, experiments on the compression of molten Sn by Davis and Hayes will be described at this conference. Supporting calculations of the equation of state and structure of molten Sn as a function of temperature and pressure are in progress. The calculations presented are ab initio molecular dynamics simulations based on electronic density functional theory within the local density approximation. The equation of state and liquid structure factors for zero pressure are compared with existing experimental results. The good agreement in this case provides validation of the calculations.

INTRODUCTION

This paper presents the results of an *ab initio* molecular dynamics study of the properties of molten Sn as a function of temperature. The work presented here represents the first phase of a study of the properties as a function of both temperature and pressure. The current work focuses on the results for the zero-pressure isobar where comparison with experiment is possible. Sn is an interesting material to study due to the potential for unusual behavior. There is an experimental indication of non-simple-liquid behavior in the form of the observed static structure factor, $S(q)$[1]. At low temperatures a small shoulder is observed on the high q side of the first peak in $S(q)$. This shoulder disappears at higher temperatures. It is of interest to know how this shoulder evolves as a function of pressure in addition to temperature. Sn is also an interesting material in terms of the nature of its bonding and structure. The bonding is intermediate between favoring close-packed and open structures. The elements above Sn in the periodic table, C, Si, and Ge, are all diamond structure semiconductor crystals at ambient conditions. The element below Sn, Pb, is a close-packed FCC metal

under ambient conditions. Sn itself is stable in the diamond structure at low temperature and pressure, but at higher temperatures and/or higher pressures, it is stable in a densely packed metallic structure. This structural polymorphism suggests that there may be interesting variations in the short-range order in the liquid state.

The structural polymorphism of Sn also suggests that it is important to perform *ab initio* based simulations. The effective interatomic interactions may well be substantially density and local environment dependent. The current simulations determine the energetics and forces from a density functional treatment of the quantum mechanics of the electrons while simulations based on interatomic potentials may not be reliable over wide range of densities and temperatures. An alternative approach to the one in the current work is to develop effective interatomic potentials for each density and temperature under consideration. Bernard and Maillet[2] have recently applied this approach to Fe as well as to Sn.

This paper presents two types of calculations. First, the zero-temperature energy-volume relations of various crystal structures of Sn will be presented. These results should correspond to the observed low-temperature, high-pressure phases. The second sets of results are simulations of the molten state at zero pressure. In this case, the simulation results can be directly compared to experimental information. In particular, the static structure factors and the densities as a function of temperature will be computed and compared to experiment. These results will then provide confidence in the future results for the properties at high pressure.

COMPUTATIONAL DETAILS

The energy calculations are based on density functional theory calculations of the electronic structure. The calculations are performed in the local density approximation (LDA) to the exchange and correlation energies. The electron-ion interaction is treated by a norm-conserving pseudopotential that was developed by Wright following the approach due to Hamann.[3] The pseudopotential treats the 5s and 5p electrons. The 4d electrons are treated as part of the ionic core. The calculations are performed using a plane-wave basis set and for the solid phase calculations, the k-space integrations were performed using the Monkhorst-Pack scheme. The results are convedrged with respect to grid size. For the liquid state calculations, the energy and forces are based on the gamma point. The calculations are performed using the SOCORRO code developed at Sandia National Laboratories.

The structure and thermodynamics of the liquid state are computed via completely *ab initio* molecular dynamics. The equations of motion of the atoms are computed using the Verlet algorithm with a thermostat. The simulations use a time step of about 5 fs and run for about 1000 time steps or 5 ps. The average internal potential energy, the pair correlation function and static structure factor are computed by a time average over the simulation. The static structure factor, $S(q)$, is computed for wave vectors that correspond to the reciprocal lattice of the simulation cell and then binned according to the magnitude of q to obtain the spherically symmetric result.

An important quantity of interest is the pressure. It is possible to compute the stress tensor, and so the pressure, directly for each configuration in the MD simulation. However, there are two numerical limitations to this approach. First, the calculation of stress in an electronic structure calculation converges slower in both k-point sampling and plane-wave cut-off than does the calculation of the energy. This dramatically increases thhe computational requirements. The second issue is that long time averages are required due to the instantaneous pressure fluctuations in an MD simulation.

In order to avoid these issues, an alternative approach was used here in order to determine the pressure. For a given temperature, simulations are performed at a variety of volumes. This is not extra work since the long-range goal is to study the properties as a function of pressure or alternatively density. From these simulations, the volume derivative of the internal potential energy can be obtained. The pressure, though, is related to the volume derivative of the free energy. Thus the entropy must be known. The entropy is estimated from an equivalent hard sphere liquid. In order to determine the packing fraction of the equivalent hard sphere liquid, the hard sphere diameter is adjusted to fit the computed static structure factor to the the Percus-Yevick solution[4]. This approach ignores the contribution of the variation of the electronic entropy to the pressure. Simple estimates show that this is small compared to the other terms at these temperatures.

RESULTS FOR SOLID Sn

The energy versus volume relationship was computed for various structures of Sn. The low energy structures should correspond to those observed at low temperatures as a function of pressure[5]. At low pressure, the equilibrium structure is diamond cubic (A4). This is the 'gray tin' phase. It has an open atomic structure with tetrahedral coordination. The next stable phase with increasing pressure is the metallic 'white tin' or A5 phase. This phase is formed from two interpenetrating body-centered tetragonal lattices. With increasing pressure, another metallic phase with a simple body centered tetragonal (BCT) structure becomes stable. Finally at very high pressures, the tetragonal distortion of the BCT phase is lifted and the BCC phase becomes stable.

In Figure 1, the computed energy versus volume is plotted for a variety of simple crystal structures. For the white Sn and BCT structures, the energy plotted is the minimum with respect to the c/a ratio for that fixed volume. In the case of white Sn, the c/a ratio was found to be 0.54 over the range of volumes considered here. This is in agreement with the experimentally observed ratio. For the BCT structure, the c/a ratio is a function of the volume with the c/a ratio increasing for decreasing volumes (increasing pressure).

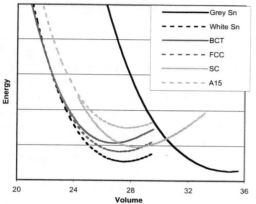

Figure 1: Computed energy with a grid spacing of 0.1 eV versus volume per atom in $\mathring{A}^3$ of Sn for selected crystal structures. Where necessary the energy has been minimized with respect to the c/a ratio.

The computed lowest energy structures correspond to the experimentally observed stable structures. The transition pressures at zero temperature are determined from the Maxwell common tangent construction. The computed transition from grey Sn to white Sn occurs at 10 kbars. The transition from white Sn to the BCT occurs at 140 kbars. These results are consistent with experiment[5].

The equilibrium lattice constants provide another test. For grey Sn, the experimental lattice constant corrected to zero temperature is 6.482Å. The calculations give 6.54Å. For the white Sn phase, the zero temperature experimental lattice constants are a = 5.812Å and c = 3.158Å. The calculated lattice constants are a = 5.89Å and c = 3.18Å. Note that the experimental and calculated values of (c/a) are both 0.54. Overall, the computed lattice constants are approximately 1% higher than the experimental values.

RESULTS FOR MOLTEN Sn

A simple, yet important, property of a molten metal is its density as a function of temperature.

The ability to predict this quantity provides a good test of the ability to predict the equation of state of the melt. The calculated densities are 6% lower than the experimental values[6] for the temperatures considered. This is consistent, though larger, to the larger lattice constants predicted from the density functional calculations compared with experiment. This reflects the approximations inherent in the pseudopotential and local density approximation treatment of the electronic structure and accounts for about half of the discrepancy in the predicted density. The other half presumably results from the approximate treatment of the liquid entropy. The thermal expansion of the liquid is predicted quite accurately. This indicates that relative densities predicted from the calculations should be quite accurate. The better agreement for the relative densities is not surprising since the contributions to the errors in the density would be expected to cancel when looking at relative densities.

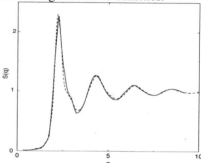

Figure 2: The static structure factor of molten Sn at 525 K and zero pressure. The solid line is the current result and the dashed line is the experimental data from reference 7.

The nature of the short-range order in the molten state as a function of temperature and pressure is of interest in general and for Sn in particular. The most common experimental measure of the local ordering in a liquid is the static structure factor, S(q), which can be measured by x-ray or neutron diffraction. There have been many experimental determinations of the structure factor for Sn near the melting point. In Figure 2, the computed structure factor is compared with that obtained by Waseda[7]. The data of Waseda was chosen since it is presented in a tabular fashion, which facilitates the comparison. The overall agreement between the calculations and the experiement is excellent.

In particular, the position and amplitudes of the maximums and minimums are reproduced. Note that there is a small shoulder on the high-q side of the first maximum in the S(q). This has been observed in all of the experimental observations performed near the melting point. This is of interest since it indicates a deviation of the local ordering from that obtained in simple hard sphere models. It is believed to be a reflection of tetrahedral ordering in the liquid. This shoulder is also present in the calculated structure factor.

The variation of the structure with temperature has also been observed experimentally and computed here. With increasing temperature, the amplitude of the oscillations in S(q) decrease as is expected. It is important to note that the shoulder on the first peak of S(q) that is observed near the melting point is not present in the higher temperature results in either the experimental or computational results.

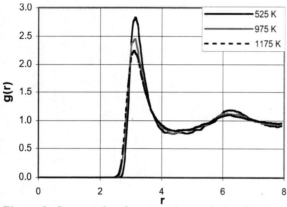

Figure 3: Computed real space pair correlation function for molten Sn at 525, 975 and 1175 K.

The real-space pair correlation function, g(r), can also be determined from the simulation. This is plotted for three temperatures in fig. 3. The computed variation of the correlation function with temperature is consistent with the variation of the pair correlation functions deduced from the inversion of the experimental structure factors. The short-range behavior of g(r) can also be determined experimentally using extended x-ray absorption fine structure (EXAFS)[8]. This has be done for Sn at a variety of temperatures. These experiments yield the initial rise and first peak of g(r). The trends in the calculations agree with the experi-

mental EXAFS data. Both show a shift of the initial rise of g(r) to smaller distances with increasing temperatures. The EXAFS results indicate a somewhat larger variation of the amplitude of the first peak of g(r) than is observed in the present calculations.[8]

CONCLUSIONS

The results presented here validate the approach to be used to study the equation of state and structure of molten Sn at elevated pressures. The electronic structure calculations produce the correct relative energies of solid phases. The thermal expansion and structure of the liquid along the zero-pressure isobar are in agreement with available experiments. Future work will extend the calculations for the molten state to higher pressures.

ACKNOWLEDGEMENTS

Sandia is a multiprogram laboratory operated by Sandia Corporation, a Lockheed Martin Company, for the United States Department of Energy's National Nuclear Security Administration under contract DE-AC04-94AL85000.

REFERENCES

1. Waseda, Y., "The Structure of Non-Crystalline Materials: Liquids and Amorphous Solids", McGraw-Hill, N.Y., (1980).
2. Bernard, S., Maillet, J.B., Phys. Rev. B66, 012103 (2002).
3. Wright, A. F., private communication; Hamann, D. R., Phys. Rev. B40, 2980 (1989).
4. Wertheim, M., Phys. Rev. Lett. 10, 321 (1963); Thiele, J. Chem. Phys. 39, 474 (1963).
5. Liu, L-G., Bassett, W. A., "Elements, Oxides, and Silicates: High Pressure Phases with Implications for the Earth's Interior", Oxford, N.Y., (1986).
6. Nasch, P.M., Steinemann, S. G., "Density and thermal expansion of molten manganese, iron, nickel, copper, aluminum and tin by means of the gamma-ray attenuation technique", Phys. Chem. Liq. 29, 43 (1995).
7. Y. Waseda, "The Structure of Non-Crystalline Materials: Liquids and Amorphous Solids", McGraw-Hill, New York (1980).
8. DiCicco, A., Phys. Rev. B53, 6174 (1996).

CP706, *Shock Compression of Condensed Matter - 2003*
edited by M. D. Furnish, Y. M. Gupta, and J. W. Forbes
© 2004 American Institute of Physics 0-7354-0181-0/04/$22.00

AN EMPIRICAL MATERIAL CONSTANT AND EQUATION OF STATE ON THE SOLIDS HUGONIOT

Zizheng Gong[1,2], Fu Dai [1], Li Zhang[1] and Fuqian Jing[1,3]

[1]*Institute of Physics, Southwest Jiaotong University, Chengdu 610031, P.R. China.*
[2]*Geophycical Laboratory, Carnegie Institution of Washington, Washington DC 20015, USA.*
[3]*Laboratory for Shock Wave and Detonation Physics Research, Institute of Fluid Physics, P.O.Box 919, Mianyang, Sichuan 621900, China..*

ABSTRACT: A new material parameter β: $\beta = (\rho_0 - \rho_{00}\, P_H / P_H') / \rho_0 \rho_{00} (1 - P_H / P_H')$, where ρ is density and subscript 0 and 00 represent different initial density, and P_H and P_H' represent Hugoniot pressure of ρ_0 and ρ_{00} which compressed to the same density ρ, was find out to keep in constant along Hugoniot. For metal $\beta_{metal} = 1.217\rho_0^{-0.884}$, where ρ_0 is the ideal crystal density. By using of β, Hugoniot data of different initial density samples can be simply converted by: $P_H = P_H' \rho_0 (\beta\rho_{00} - 1) / \rho_{00}(\beta\rho_0 - 1)$, and a new empirical Equation to express Hugoniot State was obtained: $p_H V^n = p_0 V_0^n$, where n is constant. The properties and limitation of this empirical material parameter to be constant with variations of pressure and porosity were discussed.

INTRODUCTION

Hugoniot data of porous samples are very important and interesting for establishing the Equation of State (EOS) and trying to understand the behavior of materials under high pressure and high temperature conditions. In the past over three decades, great amounts of experimental data have been measured for porous samples of all kinds of materials. In this article, from the systemically comparative study of previous experimental solid Hugoniot data of different porous samples, a new material parameters, which is independent of Hugoniot pressure for all kinds of solid materials was revealed, and the conveniences of introducing this constant were shown. The properties and the limitation of this empirical material parameter to be constant with variations of pressure and porosity were discussed.

A NEW MATERIAL CONSTANT ON SOLIDS HUGONIOT

For the two kinds of samples with different initial density ρ_0 and ρ_{00} of the same material, assumed the shock pressure at the same specific volume V ($V=1/\rho$) in the two samples are P_H and P_H', respectively. According to the Grüneisen EOS and Hugoniot EOS, a well-known formula to calculate Grüneisen parameter γ according to Hugoniot data of porous samples was obtained easily:

$$\gamma = \frac{2V(P_H' - P_H)}{P_H'(V_{00} - V) - P_H(V_0 - V)} \tag{1}$$

where subscript "*H*" represents Hugoniot state. Equation (1) can be rewritten as:

$$\gamma = \frac{2}{\rho\beta - 1} \tag{2}$$

where:

$$\beta = \frac{\rho_0 - \rho_{00} P_H / P_H'}{\rho_0 \rho_{00} (1 - P_H / P_H')} \quad (3)$$

The value of β at different shock pressure for several typical kinds of solid materials with different initial density were calculated and reviewed infra.

Because metals are very commonly used materials in research, and there are plenitudinous experimental Hugoniot data for different porosity metals, the value of β of three groups of metals were calculated. The results of aluminum (2024-Al), one of the typical low shock impedance metal; iron, cobalt and copper, the typical medium shock impedance metals; tantalum and tungsten, the typical high shock impedance metals, were listed in Table 1. Figure 1 shows β of iron, cobalt and copper at different pressure. Hugoniot data in calculation were cited from Reference [1] for aluminum (2024-Al), iron and copper. The others were cited from Reference [2]. There is little scattering on the calculated value of β at different pressure for the experimental error, β given in Table1 are the average value. In order to show the Detail of calculation, β of iron at different shock pressure for different initial density samples were given in Table 2. The calculated value of β for ionic crystal CsI and KCl is also given in Table 1; the experimental data used in the calculation were

TABLE 1. Average value of for some typical metals, ionic crystals, molecular crystal and minerals[*].

Material	$\overline{\beta}$ cm^3/g	Material	$\overline{\beta}$ cm^3/g
Al	0.39	Mg (OH)$_2$	0.367
Fe	0.23	CaSiO$_3$	0.354
Cu	0.19	MgO	0.254
Co	0.198	TiO$_2$	0.225
Ta	0.12	MgSiO$_3$	0.433
W	0.1	CO$_2$, ice	1.09
KCl	1.388	H$_2$O, ice	0.97
CsI	0.495	Polythene	1.19

[*]. The maximum porosity ($m = V_{00}/V_0$) of the samples is less then 1.8.

from Ref. [2]. For minerals, we choice the typical MgO (periclase), TiO$_2$ (rutile), MgSiO$_3$ (enstatite), CaSiO$_3$ and Mg(OH)$_2$ to calculate. The results were also given in Table 1. Hugoniot data of enstatitewere from Ref. [1], others from Ref. [2].

TABLE 2. The value of β of iron[*].

ρ_0 g/cm^3	ρ_{00} g/cm^3	P_H GPa	P_H' GPa	g/cm^3	β cm^3/g
7.864	6.988	47.43	59.36	9.83	0.21
7.856	6.959	118.69	148.01	11.02	0.21
7.861	6.933	100.72	117.35	10.7	0.24
7.861	6.055	100.20	153.06	10.8	0.23
7.850	6.050	29.43	51.05	9.20	0.22
7.850	5.919	42.51	79.69	9.62	0.22
7.840	4.770	34.99	120.48	9.40	0.24
6.960	4.770	35.51	120.48	9.40	0.24
6.050	4.880	51.05	84.56	9.20	0.26

[*]. Data from Ref. [1]。

The value of β for polythene was also calculated according to Ref. [1] and given in Table 1. Otherwise, The value of β for Bronzitite was calculated to be 0.321, but after phase transition it is 0.324, according to experimental data of Ref. [2].

Obviously, from the above calculation and review, such a conclusion can be obtained: (i) There exist a material parameter defined by equation (3), which is independence of shock pressure and shock temperature, whether for elementary substance, compound or minerals, whether for high density or low-density materials. (ii) For different materials, the value of is different. So we call material constant. (iii) The value of is different at different phases for one material.

It was very interesting that the value of for metal decrease monotonously with the increasing of metal initial density ρ_0 (crystal density), and there exist a universal relation between material constant and initial density ρ_0 for metal. Seeing Fig. 2, the relation can be described by:

$$\beta_{metal} = 1.217 \rho_0^{-0.884} \quad (4)$$

the unit of is cm^3/g, g/cm^3 of ρ_0. Equation (4) is very useful to estimate the Hugoniot both for por-

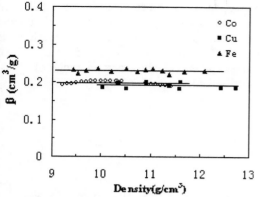

FIGURE 1. Relationship between   and   of aluminum iron, copper and cobalt.

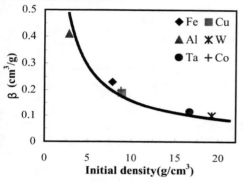

FIGURE 2. The universal relation between β and initial crystal density of metals (data are from Table 1).

ous and non-porous metal, which lack of enough experimental data.

SOME APPLICATIONS OF β

(1) A New Method To Convert Hugoniot Data Between Different Porous Samples

Equation (1) was often used to modify Hugoniot data for porosity, but there is some inconvenience for Gruneisen parameter is γ uncertain. However, by using of β, this problem becomes very simple. When compressed to the same density, shock pressure in the no-porous sample P_H and in the porous sample P'_H can be related by β according to equation (3):

$$P_H = P_H' \frac{\rho_0(\beta\rho_{00}-1)}{\rho_{00}(\beta\rho_0-1)} \qquad (5)$$

The experimental Hugoniot data of porous sample can be converted into no-porous data, or Hugoniot,

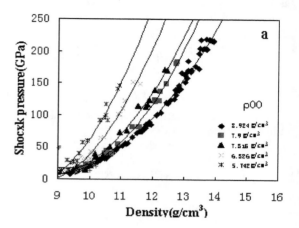

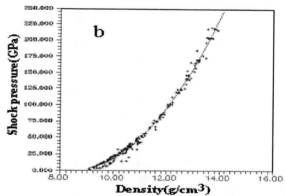

FIGURE 3. (a) Shock pressure vs. density of several porous Coppers. (b) Comparison of the porosity corrected P~ρ Hugoniot from Fig, 2(a) using eq. (4) to no-porous with experimental Hugoniot of no-porous Copper (solid line).

data of different initial density samples can be converted each other by using of equation (5) if only the material constant   is known. For example we calculated the Hugoniot of copper of non-porous sample (crystal density) based on the samples using equation (5), and were compared with the experimental data of non-porous sample in Fig.3. It can be seen from Fig.3 that the calculated Hugoniot data have a very good consistent with the experiments curve. This also verified the rationality of the constant .

(2) An Empirical Equation of Hugoniot State

From equation (3), because is constant, P_H/P'_H is constant likewise. So anyway, we can educe a constant n from:

$$\frac{p_H}{p'_H} = \frac{p_H V^n}{p'_H V^n} = \frac{p_0 V_0^n}{P_0 V_{00}^n} = \frac{V_0^n}{V_{00}^n} = \varpi \,,$$

so as to:

$$p_H V^n = p_0 V_0^n \qquad (6)$$

The form of Equation (6) is very similar to the adiabatic exponent EOS of gas and gives us a very simple way to describe shock wave EOS.

DISCUSSION

In which condition that is almost constant? Assumed porosity $m=V_{00}/V_0$, $\varpi=P_H/P'_H$, equation (3) can be rewritten as:

$$\beta=V_0(m-\varpi)/(1-\varpi) \qquad (7)$$

then the derivative differential of to m and ϖ are:

$$\partial\beta/\partial m=V_0/(1-\varpi) \qquad (8)$$

and

$$\partial\beta/\partial\varpi=(V_{00}-V_0)/(1-\varpi)^2 \qquad (9)$$

respectively. These two derivative differentials are all positive, for $(V_{00}-V_0) >0$ and $\varpi< 1$. The possible variety of with porosity and pressure were shown in Fig.4. It can been seen that the pressure region which ρ_0 keep constant (the left region to the dashed line in Fig. 4 will decrease with the increasing of porosity m. In other words, the pressure range which keep constant will narrow with the increasing of porosity. So we should point out that the material constant works merely in a certain range both of pressure and porosity.

It has been proved from thermodynamics and the experimental data analysis that [3,4], at certain pressure and temperature conditions, the total pressure in the equation of state of solids can be written as two independent functions:

$$P(V,T) = P(V,0) + P_{TH}(V_0,T)$$

where $P(V, 0)$ is the pressure on the zero temperature isotherm (cold pressure) and it is a function only of V, $P_{TH}(V, T)$ is the thermal pressure, which keep volume constant as temperature is increased, it is a function only of T. keep constant, i.e. $\varpi=P_H/P'_H$ is constant, this means that the relative contributions of cold pressure is dominative to the total pressure and the thermal pressure is negligible in a certain range of shock pressure and porosity.

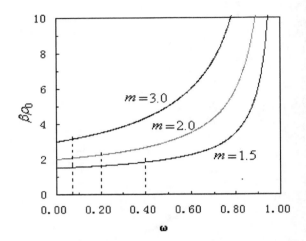

FIGURE 4. Relationship between $\beta\rho_0$ and P_H/P'_H (in this figure, $\varpi= P_H/P'_{H,}$ $m=V_{00}/V_0$).

ACKNOWLEDGEMENTS

This research is supported by the Natural Science Foundation of China under Grant No. 10299040.

REFERENCES

1. Stanley P. Marsh ed., *LASL Shock Hugoniot Data*, Berkeley, University of California Press, 1979.
2. Ahrens,T,J., edt, *Mineral Physics & Crystallography: a handbook of physical constants,* 2000 Florida Avenue, N. J., Washington, DC, Published by American Geophysical Union, 1995.
3. Anderson, O. L., *Phys. Erath Planet. Inter.,* **112**, 267-283 (1999).
4. Gong, Z., *et al.,* submitted and under published,(2003)

Correspondence author: gongzz@263.net
Or: z.gong@gl.ciw.edu

CP706, *Shock Compression of Condensed Matter - 2003*
edited by M. D. Furnish, Y. M. Gupta, and J. W. Forbes
© 2004 American Institute of Physics 0-7354-0181-0/04/$22.00

HIGH-PRESSURE DEBYE-WALLER AND GRÜNEISEN PARAMETERS OF GOLD AND COPPER

Matthias J. Graf, Carl W. Greeff and Jonathan C. Boettger

Los Alamos National Laboratory, Los Alamos, New Mexico 87545, USA

Abstract. The lattice vibrations are determined in the quasi-harmonic approximation for elemental Au and Cu to twice their normal density by first-principles electronic band-structure calculations. It is found for these materials that the important moments of the phonon density of states can be obtained to high accuracy from short-ranged force constant models. We discuss the implications for the Grüneisen parameters on the basis of calculated phonon moments and their approximations by using bulk moduli and Debye-Waller factors.

Accurate and reliable high-pressure standards are fundamental for the study of matter under extreme conditions and in the earth sciences. With the advent of neutron and x-ray diffraction measurements up to several hundred GPa in diamond anvil cells it becomes important to be able to determine precisely the pressure. At high pressures the equation of state (EOS) of elemental metals are commonly used as pressure scales. However, these scales are based on either extrapolation of low-pressure data or reduction from the Hugoniot to room temperature. The reduction of the Hugoniot curve onto $P - V$ isotherms requires model assumptions about the Grüneisen parameter, which are experimentally not well founded. Our ability to calculate accurate moments of the phonon distribution from first-principles electronic structures enables us to strongly constrain the Grüneisen parameter and to establish an accurate EOS and high-pressure standard for elemental gold and copper. This is an important step toward high-precision high-pressure experiments.

Since recent experiments raised concerns about the consistency of these standards [1], we revisited the problem of high-pressure standards using a semi-empirical approach combined with first-principles electronic structure calculations [2]. This method is different from the tight-binding approach that was earlier applied to solid copper [3]. We determined the lattice vibrational contribution to the EOS in the quasi-harmonic approximation of elemental *face centered cubic* Au and Cu by first-principles electronic structure calculations. We found that the important moments of the phonon density of states can be obtained to high accuracy from short-ranged force constant models.

In this paper, we present the implications for the Grüneisen parameter based on those calculations and their approximations by using bulk moduli and Debye-Waller factors which, in principle, can be obtained from ultrasound and neutron/x-ray diffraction measurements. The accuracy of our results is based on the accuracy of first-principles electronic structure calculations, which were performed in the local-density approximation (LDA) and generalized-gradient approximation (GGA) of density functional theory. From these results elastic moduli and zone boundary phonon frequencies were obtained. Fitting simultaneously a short-ranged, second nearest-neighbor Born-von Kármán (BvK) force matrix model to the zone boundary frequencies and the elastic moduli enabled us to generate phonon dispersion curves in the entire Brillouin zone and to compute any phonon moment needed, which is typically accurate within a few percent.

At high temperatures, i.e., in the classical limit, the temperature dependence of the pressure of a solid is dominated by the contribution from lattice vibrations, P^{vib}. Its temperature derivative at constant volume, $(\partial P^{\mathrm{vib}}/\partial T)_V$, is proportional to the Grüneisen parameter γ, which is defined by

$$\gamma = -\sum_{\mathbf{k}} \frac{d \ln \omega_{\mathbf{k}}}{d \ln V} = -\frac{d \ln \omega_0}{d \ln V} , \qquad (1)$$

where the phonon frequencies $\omega_{\mathbf{k}}$ are functions of

TABLE 1. Bulk modulus B, log-moment ω_0, thermal mean-square displacement $\langle u^2 \rangle$, and Grüneisen parameter γ for Au and Cu at ambient conditions were obtained from the theoretical data in Figs. 1(a) and 2(a) using Eq. (4). Experimental results are listed where available.

		B [GPa]	$\omega_0/2\pi$ [THz]	$\langle u^2 \rangle$ [pm^2]	γ	γ_B	γ_{DW}
Au	theo.	167.7	3.49	83.2	2.99	2.76	3.03
	expt.	167 Ref. [4]	3.65 Ref. [5]	63 Ref. [6]	2.95 Ref. [7]		
Cu	theo.	149.5	6.86	61.7	1.85	2.31	1.69
	expt.	137 Ref. [4]	6.43 Ref. [5]	76 Ref. [8]	2.02 Ref. [7]		

volume only and the summation is over all eigenmodes. Eq. (1) defines the logarithmic phonon moment ω_0.

For practical reasons one often uses an interpolating Debye phonon model for calculating the EOS or for analyzing diffraction data, instead of the more elaborate lattice dynamical models. Since in a Debye phonon model the Debye frequency ω_D is identical to all other phonon moments, the approximation $\omega_D \approx \omega_n$, with $n \geq -3$, is widely used for computing lattice vibrational properties and Grüneisen parameters. The moments ω_{-2} and ω_{-3} are related to the Debye-Waller factor at high temperatures and the sound velocity, respectively [7]. Using the theoretical relationships for the thermal mean-square displacement of an atom, $\langle u^2 \rangle \propto T/\omega_{-2}^2$, and the sound speed, $c \propto \omega_{-3} \propto \sqrt{BV^{1/3}}$, we find the following approximations to the Mie-Grüneisen theory for the high-temperature Grüneisen parameter,

$$\gamma \approx \gamma_B = -\frac{1}{6} - \frac{1}{2}\frac{d\ln B}{d\ln V}, \qquad (2)$$

$$\gamma \approx \gamma_{DW} = \frac{1}{2}\frac{d\ln\langle u^2\rangle}{d\ln V}. \qquad (3)$$

The derivation of Eq. (2) requires a constant Poisson ratio (Slater approximation). Moruzzi et al. [9] studied extensively its application to the 4d transition elements and found good agreement with experiment at ambient conditions. The advantage of using Eq. (2) for estimating the Grüneisen parameter is that it is readily accessible from ultrasound and diffraction measurements. On the other hand, the expression for the thermal parameter in Eq. (3) has not seen much application, because of the extreme difficulties of measuring accurate thermal mean-square displacements in high-pressure diffraction experiments [10]. Since these different approximations of the Grüneisen parameter emphasize different phonon

frequencies in the Brillouin zone compared to the log-moment ω_0, we do expect to find deviations from γ by at least several percent, when using these approximate formulas, reflecting the differences between different phonon moments.

The calculation of the phonon moments ω_n requires knowledge of the phonon frequencies for all **k** points in the Brillouin zone [7]. The direct first-principles calculation of frequencies on a dense **k** mesh is computationally intensive, while the phonon dispersions of most elements can be easily parameterized using lattice dynamical models like a generalized BvK force matrix model. Often it suffices to use a short-ranged force model to compute the low order moments accurately within a few percent [11]. In particular, for elemental Au and Cu the log-moment ω_0 is converged to less than 1% with a 2nd nearest-neighbor interatomic shell model at ambient conditions. Thus, we chose to calculate four zone boundary phonon frequencies corresponding to the transverse and longitudinal eigenmodes at the X and L points of the Brillouin zone. These are computed with standard frozen-phonon methods. Additionally, three elastic moduli are computed using the method by Söderlind et al. [12]. We fitted these results to a 2nd nearest-neighbor BvK force model, which then allows the evaluation of the frequencies $\omega_\mathbf{k}$ in the entire Brillouin zone. Details of this calculation will be published elsewhere [2].

We fitted the theoretical ω_0, B, and $\langle u^2 \rangle$ results to a functional form that gives a realistic V-dependence of the Grüneisen parameter, which has been used in many EOS calculations [13],

$$\gamma(V) = \gamma^\infty + A_1(V/V_0) + A_2(V/V_0)^2, \qquad (4)$$

where V_0 is the volume at ambient pressure, γ^∞ is the infinite density limit of γ, and A_1 and A_2 are fit parameters. Recently, it has been argued for the value $\gamma^\infty = 1/2$, instead of the commonly used $\gamma^\infty = 2/3$

TABLE 2. Fit parameters and ω_0, B, and $\langle u^2 \rangle$ for Au and Cu at ambient conditions were obtained from fitting the theoretical data in Figs. 1(a) and 2(a) using Eq. (5). The reference values γ_{ref} are from [7] and were derived using the thermodynamic definition of $\gamma = (V/C_V)(\partial P^{\text{vib}}/\partial T)_V$, with the specific heat C_V.

	γ_{ref}	$\gamma(V_0)$	$\gamma(V_0)_B$	$\gamma(V_0)_{DW}$	q	q_B	q_{DW}	$\omega_0/2\pi$ [THz]	B [GPa]	$\langle u^2 \rangle$ [pm^2]
Au	2.95	2.95	2.72	3.00	1.229	1.064	1.481	3.48	166.9	83.8
Cu	2.02	1.85	2.29	1.68	0.445	0.774	0.623	6.86	149.1	61.3

[14]. However, our results are insensitive to this difference and hence we chose $\gamma^\infty = 2/3$. The corresponding fitted values are listed in Table 1.

In order to test the robustness of the calculated γ values using Eq. (4), we also fitted ω_0, B, and $\langle u^2 \rangle$ to the widely used expression

$$\gamma(V) = \gamma(V_0)\,(V/V_0)^q \,, \qquad (5)$$

where the fit parameters $\gamma(V_0)$ and q are listed in Table 2; see also the inserts of Figs. 1(b) and 2(b).

In Figs. (1) and (2) we show the normalized log-moments, bulk moduli, and thermal mean-square displacement parameters of elemental Au and Cu based on our first-principles calculations. To emphasize the very similar scaling behavior of these lattice dynamical properties, we normalized their values by their corresponding ambient condition values.

We used the functional forms for the Grüneisen parameter given in Eqs. (4) and (5) to integrate Eqs. (1) through (3). This allowed us to fit the theoretical log-moments, bulk moduli, and thermal parameters and to extract the fitting parameters necessary for calculating the corresponding Grüneisen parameters shown in Figs. (1b) and (2b).

We found for the volume compressions considered here that γ of Au and Cu are quite well described by either expression (4) or (5). The differences in the calculated γ values are only a few percent. A detailed discussion of the accuracy of the different functional forms for $\gamma(V)$ and its consequences for the equation of state and the Hugoniot is given in Ref. [2]. In the case of Au, the approximate expressions for the Grüneisen parameter in Eqs. (2) and (3) give good agreement with the correct high-temperature γ obtained from Eq. (1). The deviations of γ_B and γ_{DW} from γ are mostly less than 8%. In the case of Cu, the deviations of γ_B and γ_{DW} from γ are generally bigger than for Au, but never more than 15% for compressions in the range $0.5 < V/V_0 < 1.0$. A possible explanation of the larger deviations for Cu may be that the Debye temperature and equivalently the

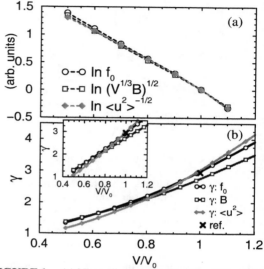

FIGURE 1. (a) Normalized (to ambient conditions) log-moment, bulk modulus, and thermal mean-square displacement of Au at $T = 296$ K from electronic structure calculations at $V/V_0 = 0.5, 0.6, 0.7, 0.8, 0.9, 1.0, 1.1$. (b) Corresponding Grüneisen parameters using Eq. (4). $\gamma(V_0) = 2.95$ [7] (cross) is shown for reference. Insert: Corresponding Grüneisen parameters using Eq. (5).

log-moment of Cu are almost twice as large as for Au. Since the typical temperature of lattice vibrations of Cu, $\hbar\omega_0/k_B \approx 330$ K, is slightly above room temperature (300 K), one may have to go beyond the classical approximation for accurately calculating the Grüneisen parameter at ambient conditions.

Comparing γ_{DW} and γ_B with γ, we find in the range of compressions $0.8 < V/V_0 < 1.0$ that the Debye-Waller approximation γ_{DW} results generally in better agreement with γ than the approximation using the bulk modulus γ_B. This situation is reversed below $V/V_0 \sim 0.8$. At such high compression γ_B is in very good agreement with γ of Au and in good agreement for Cu, while γ_{DW} deviates the most. A simple explanation of this very different behavior of γ_B and γ_{DW}

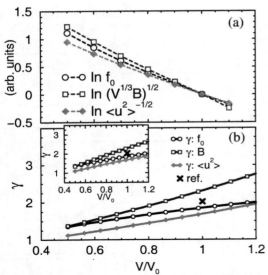

FIGURE 2. (a) Normalized (to ambient conditions) log-moment, bulk modulus, and thermal mean-square displacement of Cu at $T = 296$ K from electronic structure calculations at $V/V_0 = 0.5, 0.6, 0.7, 0.8, 0.9, 1.0, 1.1$. (b) Corresponding Grüneisen parameters using Eq. (4). $\gamma(V_0) = 2.02$ [7] (cross) is shown for reference. Insert: Corresponding Grüneisen parameters using Eq. (5).

the bulk modulus and the Debye-Waller factor, we found that for low compression, $0.8 < V/V_0 < 1.0$, the approximation using the thermal mean-square displacement is more accurate than the one using the bulk modulus, while at higher compression the bulk modulus gives generally better agreement with γ. Therefore, a combination of γ_B and γ_{DW}, which can be obtained from ultrasound and diffraction measurements, can give an estimate of the volume dependence of the Grüneisen parameter γ within approximately 10%. This provides a useful alternative for determining the Grüneisen parameter besides using the thermodynamic relation, which depends on the knowledge of the bulk modulus, thermal expansion, and specific heat.

This work was supported by the Los Alamos National Laboratory under the auspices of the U.S. Department of Energy.

at high compression, i.e., below $V/V_0 \sim 0.8$, is due to the drastic stiffening of the phonon frequencies and the simultaneous increase of the bulk modulus with decreasing V. At such high pressures and room temperature the excited phonons probe mostly the linear part of the phonon dispersion. The slope of the dispersion near the zone center is crudely proportional to the bulk modulus, while the Debye-Waller factor averages all frequencies weighted by the temperature dependent occupation factor of each mode. Therefore, we expect Eq. (3) to fail when the temperature becomes comparable to the high-temperature Debye-Waller phonon moment. Roughly speaking, for T of the order of the log-moment, $k_B T \sim \hbar \omega_0$.

In summary, we have successfully computed with high accuracy phonon moments of elemental gold and copper from first-principles electronic structure calculations combined with a Born-von Kármán force model. From the logarithmic phonon moments we calculated the volume dependence of the Grüneisen parameter up to twice the normal density of Au and Cu at ambient conditions. Comparing the Grüneisen parameter with approximations based on

REFERENCES

1. Akahama, Y., Kawamura, H., and Singh, A.K., J. Appl. Phys. **92**, 5892 (2002).
2. Greeff, C.W., and Graf, M.J., (arXiv.org/cond-mat/0311614), Phys. Rev. B in press.
3. Rudin, S., Jones, M.D., Greeff, C.W., and Albers, R.C., Phys. Rev. B **65**, 235114 (2002).
4. deLaunay, J., Solid State Phys. **2**, 220 (1956).
5. Dederichs, P.H., Schober, H., and Sellmyer, D.J., in Landolt-Börnstein, *Metalle: Phononzustände, Elektronenzustände und Fermiflächen*, eds. Hellwege, K.-H., and Olsen, J.L., Springer, Berlin, 1981, Vol. 13.
6. Killean, R.C.G, and Lisher, E.J., J. Phys. F: Metal Phys. **5**, 1107 (1975).
7. Wallace, D.C., *Statistical Physics of Crystals and Liquids*, World Scientific, New Jersey, 2002.
8. *International Tables for X-Ray Crystallography*, Kynoch, Birmingham, England, 1974, Vol. 3.
9. Moruzzi, V.L., Janak, J.F., and Schwarz, K., Phys. Rev. B **37**, 790 (1988).
10. Zhao, Y., Lawson, A.C., Zhang, J., Bennet, B.I., Von Dreele, R.B., Phys. Rev. B **62**, 8766 (2000).
11. Graf, M.J., Jeong, I.-K., Starr, D.L., and Heffner, R.H., Phys. Rev. B **68**, 064305 (2003).
12. Söderlind, P., Eriksson, O., Wills, J.M., and Boring, A.M., Phys. Rev. B **48**, 5844 (1993).
13. Abdallah, J., *User's manual for GRIZZLY*, Los Alamos National Laboratory report LA-10244-M.
14. Burakovsky, L., and Preston, D.L., (arXiv.org/cond-mat/0206160).

CP706, *Shock Compression of Condensed Matter - 2003*
edited by M. D. Furnish, Y. M. Gupta, and J. W. Forbes
© 2004 American Institute of Physics 0-7354-0181-0/04/$22.00

MEASUREMENTS AND SIMULATIONS OF WAVE PROPAGATION IN AGITATED GRANULAR BEDS

Stephen R. Hostler and Christopher E. Brennen

Department of Mechanical Engineering, California Institute of Technology, Pasadena, CA 91125

Abstract. Wave propagation in a granular bed is a complicated, highly nonlinear phenomenon. Yet studies of wave propagation provide important information on the characteristics of these materials. Fundamental nonlinearities of the bed include those in the particle contact model and the fact that there exists zero applied force when grains are out of contact. The experimental work of Liu and Nagel showed the strong dependence of wave propagation on the forming and breaking of particle chains. As a result of the nonlinearities, anomalous behavior such as solitary waves and sonic vacuum have been predicted by Nesterenko. In the present work we examine wave propagation in a granular bed subjected to vertical agitation. The agitation produces continual adjustment of force chains in the bed. Wave propagation speed and attenuation measurements were made for such a system for a range of frequencies considerably higher than that used for the agitation. Both laboratory experiments and simulations (using a two-dimensional, discrete soft-particle model) have been used. The present paper is progress report on the simulations.

INTRODUCTION

The complexities of wave propagation in a granular material are well documented. [1,2] Experiments have shown the sensitivity of wave propagation to the state of the material's microstructure which could even be changed by the wave itself [2]. Theoretical studies have predicted the existence of solitary waves and other interesting phenomena in these highly nonlinear materials [1]. Experiments have verified the existence of solitary waves in particle chains by comparing observed wave shapes to those predicted by the theory [3]. Despite the complexities, wave propagation in granular materials warrants further study as it is relevant in both industrial and natural settings and it also provides a promising, minimally invasive means of detecting buried objects [4] or investigating the state of the material itself.

Previous numerical studies have examined the properties of waves in systems with minimal disorder such as 1-D particle chains or 2-D regular packings. For a 2-D system, solitary waves have be

observed to have very little dispersion while propagating vertically through the bed [4]. Other work focused on the depth dependence of the wave speed in a 2-D, triangular lattice [5]. In this paper, we look at the effect of irregularity in particle packing on the characteristics of wave propagation. Our simulations include dissipation so the waves can only be considered semi-permanent.

This paper will exclusively focus on simulations with pulse inputs. Simulations have also been run with continuous (sinusoidal) excitation. Results from these simulations display many qualitative features seen in our experimental investigation of 3-D wave propagation. In particular, the 2-D simulations show the nonlinear beating and signal structure seen in the agitated bed experiments [6].

SIMULATIONS

The simulations are based on the discrete element method [7] in which each individual particle is tracked and its motion is determined by integrating the equations of motion. Interactions between particles are determined via a contact model that dictates normal and tangential forces due to contacts or collisions. We use a contact model that has a nonlinear, Hertzian spring in parallel with a viscous dashpot in the normal direction and a linear spring and frictional sliding element in the tangential direction [8]. As usual, the Hertzian spring yields the following non-linear relation between the normal force, F, and the compression, δ, of the particle of the form $F=K_2\delta^{3/2}$.($\delta>0$) and $F=0$ ($\delta<0$). The dashpot and frictional slider introduce a mechanism of energy dissipation in the system.

The computational domain consists of a rectangular, 2-D cell with fixed walls on the top and right side and movable walls at the bottom and left side. All simulations were performed in the presence of gravity. The box contains a few thousand particles that are initially randomly placed but settle into a packed bed during a preliminary phase of the calculation. A pulse is then initiated by movement of the left wall and, in some cases, movement of the floor is used to agitate the bed. The excitation introduced at the left wall is measured at the right wall by one or more detectors of specified size and depth in the bed. This procedure emphasizes the influence of disorder in packing and agitation on wave propagation by creating a unique microstructure for each test.

The pulses consist of a period of constant positive acceleration followed by an equal portion of the negative of this acceleration. This pulse shape has the advantage of not including any infinite acceleration (and corresponding infinite force) whereas a pulse in velocity or displacement would apply such a force. Some simulations were run with other pulse shapes. The propagating pulse shapes showed little dependence on the shape of the input pulse.

The simulation variables were non-dimensionalized using the mean particle diameter, the mean particle mass, and $(E/\rho)^{1/2}$ where E and ρ are the Young's modulus and density of the particle material. This scaling leads to a formulation in which the main parameters are a Froude number, the coefficient of restitution, Poisson's ratio, sliding friction coefficients, the number of particles, and geometric factors. In the contact model, this scaling reveals that the non-dimensional

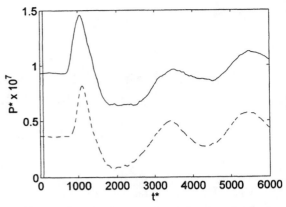

FIGURE 1. Waves shapes measured at two depths in a bed that is 50 particle diameters wide. The solid and dashed lines correspond to a sensor centered at 56 and 26 particle diameters below the free surface, respectively.

multiplying factor in the nonlinear normal spring relation is $K_2^* = 2/\pi(1-v^2)$ and is only a function of Poisson's ratio.

RESULTS

Static Bed

A typical received signal is shown in Fig. 1 for a case in which the ratio of the total pulse displacement to the static deformation of a particle is of the order of ten. The first peak is the semi-permanent wave with the subsequent peaks corresponding to its reflections off of the cell walls. The duration of the input pulse for this case is 80 time units as shown at the left side of figure. This comparatively short pulse leads to a wave with a width roughly ten times larger. Further simulations show that this wave length is quite insensitive to simulation parameters including the duration, amplitude, and shape of the input pulse.

A similar simulation with 10x the input pulse duration also produces a wave with a width of around 1000 time units. These periods correspond to a pulse width of 40 particles which is much larger than the ~5 particle diameters predicted by theory for the width of a solitary wave in a particle chain [1].

Despite the noticeable attenuation, the wave appears to maintain its shape through the reflections suggesting that the wave may be a solitary wave.

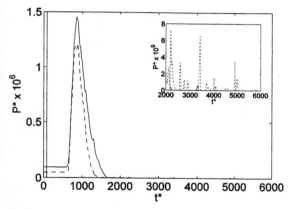

FIGURE 2. Wave shapes with 10x the acceleration input. The solid and dashed lines correspond to a sensor centered at 56 and 26 particle diameters below the free surface, respectively. The inset is a vertically magnified view of the collisional behavior for $t^* > 2000$.

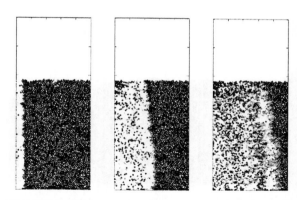

FIGURE 3. The wave front as it propagates across the bed. The particles are shaded according to the magnitude of the force on the particle in the x-direction. The time between frames is 300 time units. The cell is 50 particles wide and the free surface is at a height of 76 particles.

Early calculations tend to show the velocity of the wave increases slightly as it traverses the box. This may be explained in that the wave is still forming over the relatively small size of the box.

The received signal for a case with ten times the input pulse acceleration is shown in Fig. 2. The pulse duration is the same as before and again displayed at the left of the figure. The period of the waves is nearly identical to that in Fig. 1, but the leading edges here appear to be steeper. The steepening of the wave front is indicative of a shock-like structure such has been observed in simulations of an elastic wave propagating vertically in a granular bed [9]. Additionally, there are no reflections seen as the pressure bottoms-out behind the wave. The expansion wave that follows the compression wave appears to pull the particles out of contact with the right walls, disrupting contact with the detectors. This theory is reinforced by the insert plot in Fig. 2 that shows a close-up of the wake region. The multiple peaks in this plot have durations that correspond to the contact time of a collision of a single particle with the wall. This region behind the compression wave displays collisional behavior until lasting contact is restored.

Figure 3 shows the propagation of the wave in the bed in more detail. Each of the particles are plotted with the net force on the particle in the horizontal direction represented by the shading. Black corresponds to zero force, white corresponds

to maximum force. The wave front, seen as the white band in the first frame, is relatively flat just after the pulse has been applied. By the time the wave has traveled halfway across the box, there is already quite a bit of curvature to the wave front. The wave moves faster deeper in bed due to the dependence of the wave speed on pressure. Disorder in the packing of the bed leads to some corrugation of the front, but, in general, the trend is that of increasing velocity with depth. This disorder coupled with dissipation leads to a less sharply defined front by the third frame. Judging by Fig. 1, despite the apparent spreading in Fig. 3, the shape of the wave seems to remain coherent. This spreading of the wave front may actually be the process of formation of the semi-permanent wave. Even with the less clearly defined front, the curvature is still evident in the third frame.

Agitated Bed

Here we will show some typical results from the simulation with floor agitation. Since the width of the pulse is much shorter than the period of the vertical shaking, the pulse is able to probe the state of the bed at an instant in time.

For the wave depicted in Figs. 4 and 5, the base of the bed has just passed through its highest elevation and is on its way to its mean agitation position as shown in the insert of Fig. 5. At this point on its trajectory, one would expect the bed to

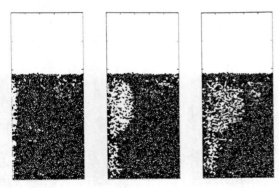

FIGURE 4. Pulse sent through an agitated bed in which the floor acceleration is ±1.5g. The particles are shaded by the magnitude of the force on the particle in the x-direction.

be expanding from the bottom. An expansion wave starts at the bottom and propagates to the top of the bed. Figure 4 shows the pulse passing through the top half of the cell where the particles are still in contact. In the lower half of the bed, the particles have come out of contact and the pulse cannot be transmitted. This description is corroborated by looking at the wave shapes in Fig. 5. The upper sensor detects a wave of similar shape to those seen in the static cases. A couple of waves that may have followed a slightly longer path obscure the back edge of the wave, but the width of the primary peak is consistent with the widths seen in Figs. 1 and 2. As for the sensor located 56 particle diameters from the free surface, no wave is measured at all.

The wave speed measured in the top section is 30% lower than that measured for an identical static case, suggesting that, though the particles are in contact, they are held by a much weaker confining force. This is supported by the much lower static pressure seen in Fig. 5 as compared to Fig. 1.

DISCUSSION AND CONCLUSIONS

We observe some form of semi-permanent wave that has a fixed width of roughly 40 particle diameters over large ranges of system parameters. Neglecting attenuation, the waves appear to maintain their shape, but show some variation in wave speed as they propagate.

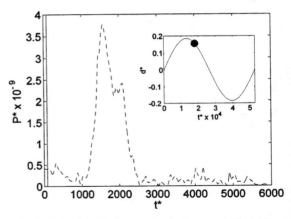

FIGURE 5. Wave shapes for pulses sent through an agitated bed. The dashed line corresponds to a sensor centered at 26 particle diameters below the free surface. The inset shows the displacement of the bed's floor when the pulse was sent.

Such pulses prove useful in diagnosing the state of an agitated bed due to their short duration relative to the period of shaking. The wave speed of a pulse can be determined at any instant of the shaking trajectory of the bed so long as particles remain in contact. This provides some insight into the current nature of the contacts and force chains in the bed as they undergo rearrangement due to agitation.

REFERENCES

1. Nesterenko, V.F., *Dynamics of Heterogeneous Materials*, Springer, New York, 2001.
2. Liu, C.H. and Nagel, S.R., *Phys. Rev. Lett.* **68**, 2301-2304 (1992).
3. Coste, C., Falcon, E., and Fauve, S., *Phys. Rev. E* **56**, 6104-6117 (1997).
4. Sen, S., Manciu, M., Sinkovits, R.S., Hurd, A.J., *Granular Matter* **3**, 33-39 (2001).
5. Melin, S., *Phys. Rev. E* **49**, 2353-2361 (1994).
6. Hostler, S.R. and Brennen, C.E., *Proc. Mat. Res. Soc. Symp.* **757**, 73-78 (2003).
7. Cundall, P.A. and Strack, O.D.L., *Geotechnique* **29**, pp. 47-65 (1979).
8. Wassgren, C.R., *Vibration of Granular Material*, Ph.D. Thesis, Calif. Inst. of Tech., 1997, pp. 16-37.
9. Potapov, A.V. and Campbell, C.S., *Phys. Rev. Lett.* **77**, 4760-4763 (1996).

CP706, *Shock Compression of Condensed Matter - 2003*
edited by M. D. Furnish, Y. M. Gupta, and J. W. Forbes
© 2004 American Institute of Physics 0-7354-0181-0/04/$22.00

QUASI-ISENTROPIC COMPRESSIBILITY OF GASEOUS DEUTERIUM IN PRESSURE RANGE UP TO 300 GPA

R.I.Il'kaev[1], V.E.Fortov[2], A.S.Bulannikov[1], V.V.Burtsev[1], V.A.Golubev[1], A.N.Golubkov[1], N.B.Davydov[1], M.V.Zhernokletov[1], S.I.Kirshanov[1], S.F.Manachkin[1], A.B.Medvedev[1], A.L.Mikhaylov[1], M.A.Mochalov[1], V.D.Orlov[1], V.V.Khrustalev[1], V.V.Yaroshenko[1].

[1] *RFNC – VNIIEF, Sarov*
[2] *IHED IVTAN RAS, Moscow*

Abstract. The authors discuss results of experiments with measurement of compression of gaseous deuterium at initial pressure of 25MPa in the pressure range up to 300GPa in cylindrical steel chambers capable to transform shock compression into quasi-isentropic compression. To record time trajectories of motion of steel shells compressing gas, two independently-operating sources of hard gamma-radiation with beams crossed at angle of 135° were used.

INTRODUCTION

Of special fundamental and pragmatic interests is presently the study of properties of hydrogen that is the simplest and most the abundant natural element. This interest is stimulated by increase of activities devoted to study of laser thermonuclear synthesis, study of structure and evolution of astrophysical objects. Experimental measurements of electric conductivity of liquid hydrogen [1] and gaseous hydrogen [2, 3] under quasi-isentropic compression revealed high level of specific electric conductivity close to that of metal at pressures of 50-1500 GPa. Experimental studies of deuterium compressibility at laser generation of shock waves revealed its abnormally large compression at pressures of ~100 GPa [4]. These data are not described by equation of state of hydrogen [5] based on data from [6-7]. Independent measurements of shock-wave compressibility of liquid deuterium performed in [8] at facility [9] in the pressure range up to 100 GPa did not confirmed results of the experiments from [4].

The studies performed by the most advanced devices and diagnostics cannot be, however, considered as completed studies. Interpretation of the data meets some principal difficulties. So, it needs additional experimental verification.

EXPERIMENTAL RESULTS

Contrary to [6-7], where the researchers studied compressibility of gaseous hydrogen in a steel spherical shell convergent to the center under effect of explosion products (EP), in this work, we measured compression of gaseous deuterium at initial pressure of ~25 MPa in the pressure range of ~ 300 GPa in cylindrical steel chamber allowing to transform shock compression to quasi-isentropic compression. For these studies, the design suggested in [10] was developed and manufactured. Scheme of the experimental device is presented in figure 1.

Compression of gaseous deuterium (1) in such device is performed by the system of direct and reflected cylindrical shock waves having small amplitudes and by steel cylindrical shells (2) and (3) convergent to the center under effect of explosion products (EP) of high explosive (HE) (4) surrounding the external shell (2). During multiple shock-wave compression in such device, it is possible to reduce thermal heating and to provide the regime of "cold" compression of gas. Initiation of external surface of HE by special focusing system of HE was simultaneously performed in 640 points. Compression of gas in the internal cavity was performed through a layer of gaseous deuterium. This cavity was protected against direct effect of HE on material of the shell (3) that excluded metal particles release into the gas cavity.

Thermal desorption sources of deuterium [11-12] were used in order to fill constructions having volume of ~3500 cm^3 with gas up to pressures of ~ 25 MPa. Powder of electrolytic vanadium is a deuterium carrier in gas source. This powder is used to prepare vanadium deuteride with atomic ratio D/V of ~1.9 by special technology. Extraction of deuterium and absorption of it by vanadium in the applied source have reversible character.

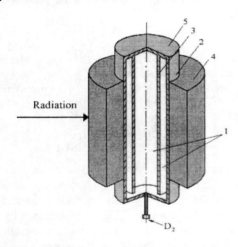

1 – gaseous deuterium; 2 – external shell, and 3 – internal shell; 4 – HE; 5 – flange

FIGURE 1. Experimental device for study of compressibility of gaseous deuterium.

To provide estimation of possible distortions of cylindrical shells of the chamber, gasdynamic code «EGAK» [13] was used for performing two-dimensional gasdynamic calculations of the compression device. Fragment of the chamber in the

initial state for one of the devices is depicted in figure 2.

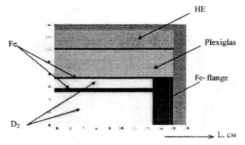

FIGURE 2. Fragment (1/4 part) of experimental device in initial state.

It follows from results of this calculation that high symmetry of internal cylindrical shell is kept in the central part of the chamber at length L = 2ΔL ≈ 160 mm. Distortions are observed only at the ends of the experimental design that is not an obstacle for X-ray recording of the central cavity of the shell. In the same cavity, it is also possible to place electrodes for measurement of electric conductivity.

Two sources of hard gamma-radiation [14] operating independently with beams crossed at angle of 135° were used to study time trajectory of motion of steel shell (fig. 3) compressing gas. The minimum radius of deuterium compression was recorded. Compression of gaseous hydrogen was earlier studied in similar experiments in [6-7].

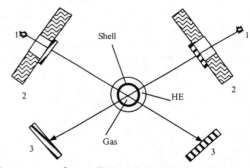

1 – sources of γ - radiation; 2 – protection; 3 – recorders.

FIGURE 3. Scheme of experiment for measurement of compressibility of gaseous deuterium.

R(t) diagram of shell motion (3) was determined in series of consequent experiments. The diagram was used to estimate sizes of the cavity at time of maximum compression of gas (at the moment of "stop"). With supposition of conservation of mass of compressed deuterium, its average density for cylindrical device was determined from the relation:

$$\rho = \rho_o \cdot (R_o/R_t)^2 \qquad (1)$$

where R_o – size of the shell in the initial state; R_t - size of the shell at time of X-ray recording; ρ_o - initial density of substance. The thermodynamic parameters, namely, pressure P and temperature T of compressed deuterium were determined from gasdynamic calculations, where equation of state of hydrogen from [5] was used; for the other materials – EOSs commonly applied in RFNC-VNIIEF were used. Deuterium pressures in the both cavities (fig. 1) were similar; the initial density was determined basing on the initial pressure of gas and temperature of medium during experiment.

To provide deuterium compression, three HE units having weights of 16, 24 and 32 kg were used in the experimental device. X-ray images of the shells for the experimental device with HE weight of 16 kg in the initial state and at the moment of maximum compression are presented in figure 4. The initial state of gas was the following: Po = 255 atm, To = 241K, ρ_0 = 0.0407 g/cm^3. As one can see in the obtained photos, the employed equipment allows to record satisfactorily the location of boundaries of the steel shells compressing gaseous deuterium.

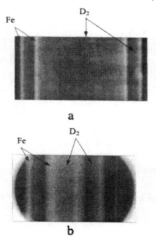

FIGURE 4. X-rayograms of shells in experimental device with HE weight of 16 kg: in initial state (a) and at time of maximum compression of deuterium (b).

DISCUSSION OF RESULTS

. Calculated distributions of density and pressure at the maximum compression of deuterium for this device are presented together with experimentally measured density in figure 5. Value of the calculated average density ρ_{calc} = 1.204 g/cm^3, which corresponds to pressure of compressed deuterium P_{calc} = 80 GPa, is in good agreement with the experimental value ρ_{exp} = (1.22 ± 0.05) g/cm^3, which corresponds to P = 82 GPa.

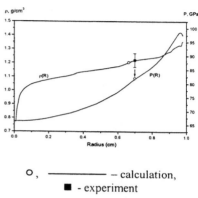

O, —————— – calculation,
■ - experiment

FIGURE 5. Distribution of density and pressure in compressed deuterium on cavity radius at stop of internal boundary of shell.

Fragments of X-rayograms obtained in the other experiment with charge having weight of 24 kg are presented in figure 6.

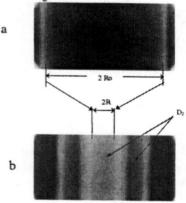

a – shadgraph of internal shell (3, see fig. 2) in initial state; b – shadgraphs of shells (2 and 3) at time of gammagraphy.
FIGURE 6. X-ray images of cylindrical shells in experiment with HE weight of 24 kg.

Results concerning compressibility of gaseous deuterium obtained in this work are given in figure 8. Also the figure shows the experimental and calculated data on quasi-isentropic compression of gaseous hydrogen in the pressure range up to 900 GPa basing on data from [6-7], and the theoretical dependence calculated in [15].

P, GPa

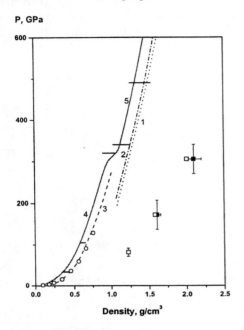

Deuterium (present work): ■ - experiment (deuterium), □ - calculation;

Hydrogen: ⌐ - experiment [6-7]; Calculation [15]: 1 – P_x atomic phase; isotherms T=0 K: 2 – atomic phase, 3 – molecular phase; isentropes: 4 – molecular phase, 5 – atomic phase

FIGURE 7. Quasi-isentropic compressibility of gaseous hydrogen and deuterium.

It follows from the experiments (figure 7) that compression of gaseous deuterium in the researched pressure range is similar to hydrogen compression from [6-7]; the ratio between densities of deuterium and hydrogen in compressed state is ≈ 2 with precision up to experimental error, i.e. it equals to the ratio of their densities in the initial state.

Results of the work testify that it is actually possible to measure compressibility of gaseous deuterium in the megabar range of pressures in the devices of the cylindrical geometry. The obtained data are used to verify physical models of equation of state of hydrogen and its isotopes.

REFERENCES

1. Nellis W.J., Weir S.T., Mitchell A.C., *Phys. Rev.* **B**. **59**, 5, 3434-3449 (1999). Weir S.T., Mitchell A.C., Nellis W.J., *Phys. Rev. Lett.* **76**, 11, 1860-1863 (1996). Nellis W.J., 48, 671-677 (2000).
2. Fortov V.E., Ternovoi V.Ya., Kvitov S.V. et al., *Letters to ZhETF* **69**, 12, 874-878 (1999).
3. Mochalov M.A., Kuznetsov O.N., *Bulletin of the American Physical Society.* *12TH APS Topical Group Meeting on Shock Compression of Condensed Matter*, edited by Donna M. Baudrau, CMP, Atlanta, Georgia, June 2001, **46**, 4, p.101.
4. Cauble R., Silva L., Perry T. et al., *Phys. Plasmas* **4**, 5, 1857-1861 (1977).
5. Kopyshev V.P., Khrustalev V.V., *PMTF* **1**, 122 – 128 (1980).
6. Grigoriev F.V., Kormer S.B., Mikhailova O.L. et al., *Letters to ZhETF* **16**, 286 (1972).
7. Grigoriev F.V., Kormer S.B., Mikhailova O.L. et al., *ZhETF* **75**, 1683 (1978).
8. Knudsen M.D., Hanson D.L., Bailey J.E. et al., *Phys. Rev. Lett.* **87**, 22, 225501-1 (2001).
9. Hanson D.L., Asay J.R., Hall C.A. et al., *Shock Compression of Condensed Matter-1999*, ebited by M.D.Furnish, L.C.Chhabildas, R.S.Hixson, 1175-1178.
10. Al'tshuler L.V., Dynin E.A., Svidinskii V.A., *Letters to ZhETF* **17**, 1, 20-22 (1973).
11. Golubev V.A., Zagrai V.D., Sotskov V.I., Khodalev V.F., Russian patent № 2155156, 7S 01 V 3/00 dated 18.01.99, Bul. № 24 dated 27.08.2000.
12. Golubkov A.N., Golubev V.A., Vedeneev A.I., Yaroshenko V.V., Russian patent № 2171784, 7C 01, G 31/00, S 01 V 6/02 dated 12.05.99, Bul. № 22 dated 10.08.2001.
13. Darova N.S., Dibirov O.A., Zharova G.V. et al., *VANT*, ser. MMFP **2**, 51-58 (1994).
14. Pavlovskii A.I., Kuleshov G.D., Sklizkov G.V. et al., *Papers of USSR AS* **160**, 68 (1965).
15. Kopyshev V.P., Urlin V.D., "Shock waves and experimental states of substances", edited by Academician V.E.Fortov, L.V.Al'tshuler, R.F.Trunin, A.I.Funtikov., M., "Nauka", 315-341 (2000).

CP706, *Shock Compression of Condensed Matter - 2003*
edited by M. D. Furnish, Y. M. Gupta, and J. W. Forbes
© 2004 American Institute of Physics 0-7354-0181-0/04/$22.00

HUGONIOT MEASUREMENTS OF HIGH PRESSURE PHASE STABILITY OF TITANIUM-SILICON CARBIDE (Ti_3SiC_2)

J. L. Jordan[1*], T. Sekine[2], T. Kobayashi[2], X. Li[2], N. N. Thadhani[1], T. El-Raghy[3], M.W. Barsoum[4]

[1] *School of Materials Science and Engineering, Georgia Institute of Technology, Atlanta, GA 30332-0245*
[2] *Advanced Materials Laboratory, National Institute for Materials Science, Tsukuba, Japan 305-0044*
[3] *3-ONE-2 LLC, 4 Covington Place, Voorhees, NJ 08043*
[4] *Department of Materials Engineering, Drexel University, Philadelphia, PA 19104*

Abstract. Hugoniot measurements of the high-pressure phase stability of titanium-silicon carbide (Ti_3SiC_2) were performed in this study. Ti_3SiC_2 is a unique ceramic having high stiffness, but low hardness. Time-resolved measurements employing plate-impact geometry were conducted on Ti_3SiC_2 samples in the pressure range of 50 to 120 GPa using the NIMS two-stage light-gas-gun. Experiments performed in the lower pressure range followed the continuous pressure-volume compressibility trend reported by Onodera, et al. [1] in static high-pressure experiments. At pressures around 80-120 GPa, deviation in pressure-volume compressibility to a more compressed state was observed indicating evidence of a possible phase change. Streak camera records of the free surface velocity measured using the inclined mirror method also showed discontinuous slope, indicating a possible pressure-induced phase transformation.

INTRODUCTION

Titanium-silicon ternary carbide (Ti_3SiC_2) is a unique material maintaining the high-temperature properties and wear resistance typical of ceramics while also demonstrating metal-like properties, including high electrical and thermal conductivity, and easy machinability [1]. It is an elastically stiff but soft material. While in most materials hardness typically scales with the elastic modulus, Ti_3SiC_2 shows an anomalous behavior. The unique properties of this ternary ceramic are attributed to its layered structure, which consists of TiC octahedra separated by layers of silicon atoms. The incompressibility and the metal-like deformation response of this ternary ceramic make it a potentially interesting material for studying its dynamic mechanical behavior and high pressure phase stability.

Although the structural characteristics and mechanical properties of this material have been extensively studied, its dynamic mechanical behavior and high pressure phase stability have not been characterized. The static high pressure compressibility properties were measured by Onodera, et al. [2]. They found a continuous pressure – volume compressibility behavior, indicating no phase transformation or decomposition during static application of pressure up to 60 GPa at room temperature. This paper will present results of dynamic high pressure compressibility and phase stability behavior determined from Hugoniot measurements performed at pressures up to 120 GPa.

*Present Address: Air Force Research Laboratory, AFRL/MNME, Eglin AFB, FL 32542

EXPERIMENTAL PROCEDURE

The Ti$_3$SiC$_2$ samples used in the present work were synthesized using the method described in Reference 1. They were obtained from 3-One-2, LLC, in the form of 62.5-mm diameter by 5-mm thick discs, from which 10-mm x 12-mm rectangles of 2.75-mm thickness were electo-discharged machined (EDM) for performing the Hugoniot measurements. Hugoniot experiments for equation of state measurements were conducted using a two-stage light gas gun at the National Institute for Materials Science (NIMS), in Tsukuba, Japan. The experimental set-up [3] shown in Figure 1 was used to measure the shock velocity based on time of travel through the sample thickness, given by the extinction times of mirrors M1/M4 and M2/M3. The particle velocity was determined from the free surface velocity, which was measured using the inclined mirror method. Signals from mirrors M1, M2, M3, M4, and the inclined mirror were recorded using a streak camera. The flyer and driver plates were stainless steel in all experiments, except T-177, where the flyer plate was aluminum. Values of the particle velocity were also calculated using the impedance matching method and the measured impact velocity.

RESULTS AND DISCUSSION

Hugoniot equation of state experiments involved measurements of shock and particle velocities recorded on a streak camera. The numerical values of the measured parameters (shock (U$_S$) and particle velocity (Up)) for experiments performed at different impact

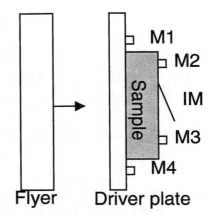

FIGURE 1. Schematic showing inclined mirror method of measuring shock and free surface velocity. The inclined mirror is indicated by IM, and the flat mirrors are indicated by M1 – M4. Extinction of reflectivity of mirrors M1 and M4 indicate arrival time of shock wave at front surface of sample. Extinction of reflectivity of M2 and M3 indicates arrival of shock wave at back surface of the sample. The inclined mirror gives a profile of the free surface velocity with time [3].

velocities are given in Table 1. The particle velocity calculated from the impedance matching method is also listed. It should be noted that the calculated particle velocities are lower than those obtained from free surface velocity measurements for experiments T-175, T-176, T-177, and T-190. The difference in the measured and calculated particle velocities is attributed to the presence of tilt in these experiments. While the streak camera records of free surface velocity measurements show interesting features (to be discussed later), the calculated particle velocity values are used for

Table 1. Summary of results from equation of state experiments.

Shot No.	Velocity (km/s)	Measured Us (km/s)[a]	Measured Up (km/s) [b]	Calc. Up (km/s) [c]	Calc. P (GPa)	Calc. V/V$_0$	Calc.T (K)
T-177	3.61	8.53 ± 0.1	1.88	1.27	49	0.85	610
T-190	3.24	8.78 ± 0.1	1.78	1.85	73	0.79	700
T-183	3.55	8.78 ± 0.1	2.09	2.05	81	0.77	800
T-175	4.02	9.31 ± 0.2	3.18	2.31	96	0.75	1470
T-176	4.87	9.67 ± 0.2	2.31	2.81	122	0.71	2170
2 slopes		9.21 ± 0.2	3.67	2.86	118	0.69	2050

[a] Shock velocity measured based on the extinction times of mirrors M1/M4 and M2/M3 in Figure 1 (a).

[b] Particle velocity determined from the free surface velocity, which was measured using the inclined mirror method

[c] Particle velocity calculated from the impedance matching method.

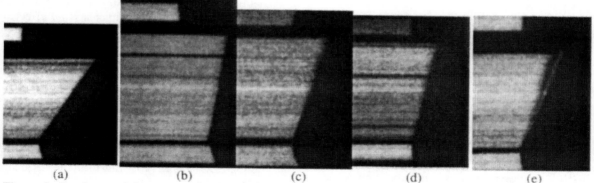

Figure 2. Streak record from experiment T-177; (b) streak record from experiment T-190; (c) streak record from experiment T-183; (d) streak record from experiment T-175; and (e) streak record from experiment T-176.

subsequent data analysis. However, there was good agreement between the measured and calculated values of the particle velocity in experiment T-183. Values of the shock pressure and relative volume (V/V_0) calculated using jump conditions and the measured U_S and calculated U_P values are also listed in Table 1. The shock temperature was also calculated from the Hugoniot experiments based on the method described in Reference 4.

The streak records of the three inclined mirror experiments are shown in Figure 2 (a-e). Figure 2 (a) shows a clear record of free surface velocity with a single slope. Figure 2 (b-e) shows blurred edges in the records of the free surface velocity. The blurred edges may be indicative of the presence of several phases having been formed due to shock-induced structural or chemical changes. The higher pressure record (Figure 2 (e)) shows a distinct change in slope illustrating an obvious change in the state of the material as a result of a shock-induced transformation.

Figure 3 shows the static pressure-volume compressibility data of Onodera, et al. [2] extended to higher pressures by curve fitting the data. The pressure-volume data obtained from the measured shock velocity and calculated particle velocity in the present experiments (listed in Table 1), are also shown in Figure 3. It can be seen that while the 49 GPa data point falls on the same curve as that of the low pressure (static) data of Onodera, et al. [2], data obtained from the present experiments at two higher pressures show deviation indicating possible phase change to an increased compressibility state. The static high pressure data from Onodera, et al. was also converted to shock and particle velocity data using jump conditions. The linear fit to the data yielded the shock velocity-particle velocity relationship given by the following:

$$U_S = 6.70 + 1.34 * U_P \qquad (1)$$

Figure 4 shows the $U_s - U_p$ data obtained from the current shock experiments at high pressures plotted along with the static high-pressure data of Onodera, et al. [2]. Values of longitudinal, shear, and bulk sound speed of the Ti_3SiC_2 samples measured using an ultrasound technique are also presented. It can be seen that the low pressure (49 GPa) data point from the present work falls on the same $Us - Up$ linear trend extrapolating on Onoderas' data, while the higher pressure data points show a deviation from linearity to an increased compressibility state. The deviation is again indicative of a possible phase change from a lower to higher density-state, occurring in the range of 50 – 75 GPa. Evidence of phase change is further confirmed by the blurred edge of the inclined mirror record in experiments T-190, T-183, and T-175 (at ~73 – 96 GPa), shown in Figure 2 (b-d), and the presence of two distinct slopes in the streak record of experiment 176 (at ~ 122 GPa), shown in Fig. 2 (e).

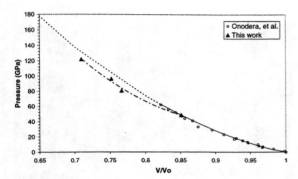

FIGURE 3. Pressure versus volume plot with Onodera, et al. [2] data and data from the three equation of state experiments.

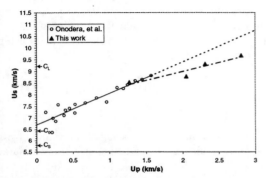

FIGURE 4. Particle velocity versus shock velocity with Onodera's data and data from the Tsukuba experiments. Onodera's data was converted from the reported pressure-volume data using $U_s = \dfrac{C_0 V_0}{V_0 - S(V_0 - V_1)}$ and

$U_p = \dfrac{C_0}{2S}\left(\sqrt{1 + \dfrac{4S}{\rho_0 C_0^2} P} - 1\right)$. The particle velocities in the Tsukuba experiments were calculated from the measured shock velocity using an impedance matching method.

It should be noted that the thermal decomposition of Ti_3SiC_2 at *ambient* pressure has been reported to be greater than 2300 °C, with the decomposition products believed to be Ti_3C_2 and silicon [5,6]. The calculated bulk temperature produced during shock compression in the experiments conducted in the present work is lower than the decomposition temperature and the time at temperature is not sufficient for diffusion-type

reactions to occur. Hence, the phase change observed in Hugoniot experiments is an effect of pressure-induced phase transformation.

CONCLUSIONS

Time-resolved measurements of pressure-volume compressibility and Hugoniot equation of state were conducted on titanium-silicon carbide (Ti_3SiC_2) samples in the pressure range of 50 to 120 GPa. At pressures around 80-120 GPa, Ti_3SiC_2 was found to transform to a more condensed state.

ACKNOWLEDGEMENTS

This work was accomplished while Dr. Jennifer Jordan was participating in the Summer Program for Graduate Students in Japan sponsored by the National Science Foundation. Funding for research was provided in part by the National Institute for Materials Science, Tsukuba, Japan, and ARO Grants No. DAAG55-98-1-0454 and DAAG55-98-1-0161.

REFERENCES

1. Barsoum, M. and T. El-Raghy, *Journal of the American Ceramic Society*, **79** [7], 1953-1956 (1996).
2. Onodera, A., H. Hirano, T. Yuasa, N.F. Guo, and Y. Miyamoto, "Equation of State of a Soft Ceramic: Ti_3SiC_2," in: *Science and Technology of High Pressure*, edited by M.H. Manghnani, W.J. Nellis, and M.F. Nicol, Universities Press, India 2000.
3. Sekine, T., S. Tashiro, T. Kobayashi, and T. Matsumura, *Shock Compression of Condensed Matter – 1995*, Eds. W.C. Tao, AIP, Woodbury, NY, 1996, 1201-1204.
4. Gupta, S.C., S.G. Love, T.J. Ahrens, *Earth and Planetary Science Letters*, **201**, 1-12 (2002).
5. M.W. Barsoum, "The $M_{N+1}AX_N$ Phases: A New Class of Solids; Thermodynamically Stable Nanolaminates," *Progress in Solid State Chemistry*, **28**, p. 201-281 (2000).
6. Y. Du, J.C. Schuster, H. Seifert, and F. Aldinger, "Experimental Investigation and Thermodynamic Calculation of the Ti-Si-C System," *J. of American Ceramic Society*, **83**, p. 197-203 (2000).

CP706, *Shock Compression of Condensed Matter - 2003*
edited by M. D. Furnish, Y. M. Gupta, and J. W. Forbes
© 2004 American Institute of Physics 0-7354-0181-0/04/$22.00

EQUATION OF STATE MEASUREMENTS IN LIQUID DEUTERIUM TO 100 GPA

M.D. Knudson, D.L. Hanson, J.E. Bailey, C.A. Hall, C. Deeney, and J.R. Asay

Sandia National Laboratories, Albuquerque, NW 87185-1181

Abstract. Using intense magnetic pressure, a method was developed to launch flyer plates to velocities in excess of 20 km/s. This technique was used to perform plate-impact, shock wave experiments on cryogenic liquid deuterium (LD_2) to examine its high-pressure equation of state (EOS). Using an impedance matching method, Hugoniot measurements were obtained in the pressure range of 22-100 GPa. The results of these experiments disagree with previously reported Hugoniot measurements of LD_2 in the pressure range above ~40 GPa, but are in good agreement with first principles, *ab-initio* models for hydrogen and its isotopes. Additionally, a novel approach was developed using a wave reverberation technique to probe density compression of LD_2 along the principal Hugoniot. Relative transit times of shock waves reverberating within the sample are shown to be sensitive to the compression due to the first shock. Results in the range of 22-75 GPa corroborate the ~4 fold density compression inferred from the impedance matching Hugoniot measurements, and provide data to differentiate between proposed theories for hydrogen and its isotopes.

INTRODUCTION

Prior to 1997, Hugoniot measurements of liquid hydrogen and deuterium had been limited to the pressure range below approximately 20 GPa [1], which is accessible by conventional gas-gun, plate-impact experiments. These measurements are in good agreement with widely used and accepted Sesame EOS model for liquid deuterium (LD_2) [2]. However, recent measurements from laser-driven experiments [3] at pressures of 20-300 GPa suggest that liquid deuterium is much more compressible than previously thought. The results from these laser-driven experiments suggest a maximum compression in excess of $\rho/\rho_0 = 6$, which deviates significantly from the Sesame EOS that predicts a maximum compression of approximately 4.2.

Despite efforts to model this apparent increase in compressibility, several theoretical models based on first principles, *ab-initio* methods [4-6] have

been unable to describe the experimental results above ~40 GPa. Rather, as the approximation schemes in these *ab-initio* methods have improved, these models [4-6] have converged, yielding results that are closer to the Sesame EOS at high shock pressures, and disagree with the phenomenological linear mixing model of Ross [7] that is in relatively good agreement with the laser-driven data.

Given the significant discrepancy between theory and experiment, it is desirable to obtain independent EOS measurements of LD_2. Recently, a new capability has been developed to isentropically compress materials to high pressures [8] using the intense magnetic pressure produced by the Sandia Z accelerator [9]. This new capability has been used to launch relatively large flyer plates to velocities about three times higher than that possible using conventional gas gun technology. The flyer plate technique for performing high-pressure shock wave experiments is particularly attractive for several reasons. First, the experiments are plate-impact experiments, and thus produce a well-defined shock loading of the sample, with a substantial duration of constant pressure (to 30 ns). Second, relatively large sample diameters and thicknesses are possible, thus increasing the accuracy of the EOS data. Finally, the large sample sizes allow for multiple and redundant diagnostics to be fielded, which further enhance the accuracy and confidence of the data.

In this paper we report the results of two independent sets of experiments performed on LD_2 using this new flyer plate technique. First, Hugoniot measurements using a traditional impedance matching technique are discussed. The result of these experiments disagree with the previously reported Hugoniot measurements using laser techniques for pressures above ~40 GPa. Second, we describe a novel experimental approach to infer density compression along the principal Hugoniot that is more discriminating than the inference from measurements of shock velocity, U_s, and mass or particle velocity, u_p. This technique monitors the relative arrival time of shock waves at the interface of the LD_2 sample and a sapphire window as the shock reverberates between the aluminum drive plate and the window. The relative timing in these experiments can be used to infer the compression of LD_2 in this higher-pressure regime with better precision than the more traditional density inference from U_s and u_p.

DEUTERIUM EXPERIMENTS

The experimental configuration used to obtain Hugoniot data with the magnetically driven impact technique can be seen in Ref [10]. The necessary cryogenics are provided by an expendable cryocell connected to a survivable cyrostat [11]. The cavity of the cryocell is defined by a stepped aluminum (6061-T6) pusher plate and a z-cut sapphire window, with cavity dimensions of approximately 5 mm in diameter and 300 and 600 μm in thickness. LD_2 samples are condensed in the cryocell by filling the cavity with high purity deuterium gas at 18 psi, cooling the cryocell to a temperature of 16-18 K, and then warming the cell to 22.0 ± 0.1 K [20]. This produces a quiescent LD_2

sample below the boiling point of about 25 K, with nominal initial density of 0.167 g/cm^3.

Shock waves are generated by planar impact of either an aluminum (6061-T6) or a titanium (Ti-Al6V4) flyer plate onto the aluminum pusher plate at the front of the cryocell. The rectangular flyer plate, ~12x25 mm in lateral dimension and ~200-300 μm in thickness [12], was accelerated across a nominal 3 mm vacuum gap by the magnetic field. Titanium flyer velocities as high as ~22 km/s have been achieved, which are capable of generating shock states to ~700 GPa in the aluminum drive plate and transmitting up to ~100 GPa shock waves into LD_2. Conventional velocity interferometry [13] (VISAR) is used to directly measure the velocity history of the flyer plate from launch to impact with an accuracy of ~0.5%.

The shock response of LD_2 is diagnosed with a number of fiber-optic coupled diagnostics. Typically, several optical fiber bundles of 100 and 200 μm diameter fibers are used, allowing multiple, redundant diagnostics, including (i) conventional VISAR, (ii) fiber-optic shock break out (FOSBO), and (iii) temporally and spectrally resolved spectroscopy. Fig. 1 shows sample data obtained from a typical LD_2 experiment. In all, up to 16 channels of data are obtained for each experiment, allowing up to 16 independent measurements of U_s in LD_2 and up to 4

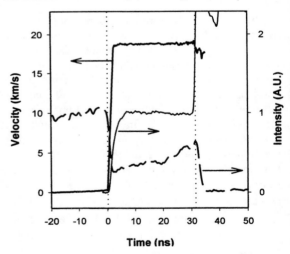

Figure 1. Typical data obtained in LD_2 experiments; (i) VISAR record of the shock front (solid black line), (ii) FOSBO record (dashed line), and (iii) self-emission record (solid gray line).

independent measurements of U_s in the aluminum drive plate. The uncertainty in U_s is ~2-3% from the measured transit time through the cell and the initial cell dimensions. Since the uncertainties were due to random errors, statistical techniques can be used to decrease the uncertainty in U_s to approximately 1% and 2% for the LD_2 sample and the aluminum drive plate, respectively [14].

The VISAR records for the higher-pressure experiments confirm the constancy of the pressure drive obtained from the flyer plate-impact, as shown in Fig. 1. In this case the VISAR velocity is indicative of U_s in the LD_2 because at shock pressures above ~30 GPa LD_2 becomes reflective [3]. From these records it is determined that the shock pressure is constant to better than 1% as the shock traverses the cryocell.

For the lower-pressure experiments, the shock front is not sufficiently reflective to obtain VISAR measurements. However, U_s is obtained for all experiments using the FOSBO and self-emission data. As seen in Fig. 1, both of these measurements provide a clear signature of shock arrival at the aluminum/LD_2 and the LD_2/sapphire interfaces. Also, in all experiments, high quality spectra are obtained over the continuous wavelength region between 250 and 700 nm. The detailed analysis of the spectral dependence of the self-emission, which provides a measure of the temperature of the shocked LD_2, will be discussed in a future publication. We emphasize that the constancy of the emission signal during the traversal of the shock through the cryocell further verifies the constancy of the pressure states achieved with the flyer plate-impact, as the intensity of emission is proportional to the pressure of the LD_2 to the ~1.75 power [15].

After the shock traverses the cell, wave interactions at the LD_2/sapphire interface result in a transmitted and a reflected shock. We exploit this by using the reflected shock to probe the location of the aluminum/LD_2 interface. The velocities of the shock (U_s) and the aluminum/LD_2 interface (u_p) determine the time that the reflected shock from the aluminum/LD_2 interface reaches the LD_2/sapphire interface. Thus, the ratio of the time between the first and second shock arrival at the LD_2/sapphire interface $(t_2 - t_1)$ to the original transit time across the cell $(t_1 - t_0)$, referred to as

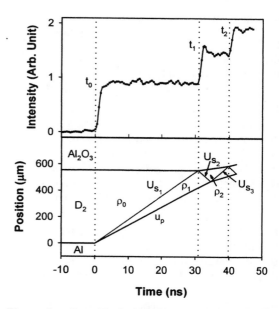

Figure 2. Top: Typical self-emission measurement indicating shock arrival at the aluminum/LD_2 interface (t_0), the first shock arrival at the LD_2/sapphire interface (t_1), and the second shock arrival at the LD_2/sapphire interface (t_2). Bottom: Position-time diagram indicating trajectories of the shock fronts and interfaces.

the reverberation ratio, is directly related to the density of the LD_2 in the Hugoniot state.

Qualitatively one can see from the position-time plot in Fig. 2 (drawn to scale for an initial shock of ~45 GPa) that due to the substantial compression of LD_2 upon first shock, the exact behavior of LD_2 upon re-shock has a relatively small influence on the reverberation ratio.

One can also show quantitatively the strong dependence of the reverberation ratio on ρ_1. Analysis of the position-time plot reveals that the ratio of the reverberation time, t_r, to the original transit time, t_i, is given by

$$\frac{t_r}{t_i} \equiv \frac{(t_2 - t_1)}{(t_1 - t_0)} = \rho_0 U_{s_1} \left(\frac{1}{\rho_1 U_{s_2}} + \frac{1}{\rho_2 U_{s_3}} \right) \quad (1)$$

where ρ_0 is the initial LD_2 density; ρ_1 and ρ_2 are the LD_2 densities due to the first and second shock, respectively; and U_{s_1}, U_{s_2}, and U_{s_3} are the velocities of the first, second, and third shock, respectively. Eq. 1 indicates that apart from the measured quantities ρ_0 and U_{s_1}, the ratio t_r/t_i depends on ρ_1, ρ_2, U_{s_2}, and U_{s_3}. However, model predictions over the pressure range examined in this study indicate that to a very good approximation $\rho_2 \approx 1.9\rho_1$ and $U_{s_3} \approx U_{s_2} \approx 1.1U_{s_1}$ for LD_2; Sesame [1], tight-binding (TB) [4], GGA-MD [5], and Young [16] predictions were compared, and the variations from these relations were found to be less than 10% for each of the models. Given this very similar behavior upon re-shock for the various models for LD_2, one can show that to a good approximation $t_i/t_r \approx \rho_1/1.39\rho_0 = 4.24\rho_1$, where ρ_1 is expressed in units of g/cc. Thus, the inverse of the reverberation ratio is roughly proportional to the density compression along the Hugoniot. It is to be emphasized, however, that when comparing experimental measurements with the various models for LD_2 the above approximation is not needed since the models will be used to determine ρ_1, ρ_2, U_{s_2}, and U_{s_3}.

The reverberation time is obtained from the time-resolved spectroscopy measurement; a typical spectroscopy measurement is shown in Fig. 2. The self-emission from the LD_2 sample provides a clear indication of shock arrival at the aluminum/LD_2 interface (at time t_0) and the first and second shock arrivals at the LD_2/sapphire interface (at times t_1 and t_2, respectively).

EXPERIMENTAL RESULTS

An impedance matching method, utilizing the Hugoniot jump conditions [17], was used to obtain Hugoniot points for the shocked LD_2. The initial shocked state of the aluminum drive plate is described in the pressure-particle velocity (P-u_p) plane by the point labeled A, and the shocked state of LD_2 is constrained to lie on a straight line, with the slope of the line given by $\rho_0 U_s$, where ρ_0 is the initial density of the LD_2 sample. An EOS model for aluminum [18] was used to calculate the release isentrope from state A in the aluminum drive plate. The intersection of the calculated release isentrope and the line defined by the LD_2 shock velocity determines u_p of the shocked

deuterium sample. The uncertainty in u_p for LD_2, typically 2-3%, was determined from the uncertainty in the shocked state of the aluminum drive plate, and thus from the uncertainty in U_s for the aluminum drive plate. The density compression was then determined from the jump conditions using the expression $\rho/\rho_0 = U_s/(U_s - u_p)$.

The accuracy in the impedance matching technique depends upon two factors: the accuracy in the measurement of U_s in the LD_2 and the accuracy of the calculated release isentrope for aluminum. The quality of the data shown in Fig. 1, and the multiplicity of the U_s measurement for each experiment indicates that U_s in the LD_2 is determined quite accurately. To determine the accuracy of the calculated release isentrope, release experiments in aluminum using a low density (200 mg/cm^3) silica aerogel were performed. This technique is similar to that used by Holmes, et al., to measure the aluminum release from ~80 GPa [19]. Direct impact experiments were performed to generate Hugoniot data for the aerogel between 30 and 75 GPa. Experiments were then performed in which a shock was transmitted from the aluminum drive plate into the silica aerogel, which simulates unloading to the LD_2 state. The measured U_s for the aerogel in the release experiment, along with the measured aerogel Hugoniot, determines a point in P-u_p space that the aluminum release isentrope must pass through. These measurements, which agreed with the model predictions to within experimental uncertainty, confirm the validity of the release calculations in aluminum over the pressure range of interest, and make a strong case for the procedure outlined above to obtain the LD_2 Hugoniot results reported in the present work.

The pressure-density compression states determined in this way for a total of nineteen experiments are displayed in Fig. 3. The lowest pressure experiment was found to be in good agreement with the results reported from the earlier gas gun experiments and the lower pressure laser experiments. However, at higher pressures, particularly the data centered around 70 GPa and the data point at ~100 GPa, there is a distinct deviation between the present results and those reported from the laser-driven experiments. Further, the data obtained from our study are in quite good agreement with the predictions from the

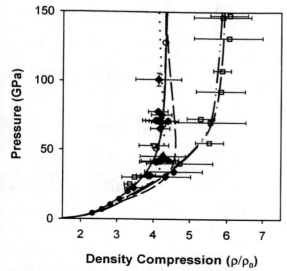

Figure 3. Deuterium Hugoniots. Theoretical models: Sesame (solid line [2]); Tight Binding (dotted line [4]); GGA-MD (dashed line [5]); PIMC (open circles [6]); Ross (dot-dot-dashed line [7]); Young (dot-dashed line [16]). Experiments: Gas gun (filled circles [1]); Laser-driven (open squares [3]); This work (gray diamonds); Convergent geometry (gray triangle [20]).

ab-initio models throughout the entire range of pressures investigated.

Also shown in Fig. 3 are the results of recent LD_2 Hugoniot measurements obtained using a convergent geometry technique [20]. The results, shown as a weighted average point in Fig. 3, appear to confirm the stiffer response observed in the present experiments. Given the fact that these experiments utilize completely independent experimental configurations, the agreement of the inferred density compression makes a strong case for a ~4-fold limiting compression for the equilibrium response of LD_2 along the principal Hugoniot.

Results of reverberation measurements at several initial pressure states are plotted in Fig. 4, along with predictions of various LD_2 EOS models. We chose to plot the initial shock speed in the LD_2, U_{s_1} (increase in U_{s_1} correlates to an increase in P), as a function of the inverse of the reverberation ratio, t_i/t_r (increase in t_i/t_r correlates to an increase in ρ_1), to allow for a clearer comparison with the P-ρ principal Hugoniot shown in Fig. 3.

As shown in Fig. 4, the measured reverberation ratios are in much better agreement with the various stiffer *ab-initio* and Sesame EOS models. In particular, the data at the highest shock velocities, corresponding to ~70 GPa, are in excellent agreement with the Sesame [2] and tight-binding (TB) [4] models. This agreement corroborates the principal Hugoniot results obtained through the impedance matching experiments. If the density compression was ~6-fold along the Hugoniot, as inferred from the laser driven experiments [3], t_i/t_r would continue to increase with increasing pressure. The fact that t_i/t_r is observed to decrease slightly from ~45 GPa to ~70 GPa is a strong indication that ~4-fold compression is not exceeded along the Hugoniot, and that the Hugoniot begins to stiffen at pressures above ~45 GPa. Furthermore, the maximum in t_i/t_r observed at ~45 GPa implies that a maximum in the density compression along the Hugoniot occurs at ~45 GPa, which is also consistent with the impedance matching results.

These reverberation measurements also discriminate between the stiffer *ab-initio* and Sesame EOS models in the pressure range of 25-50 GPa. The impedance matching Hugoniot measurements over this pressure range are unable to distinguish between the various stiffer EOS

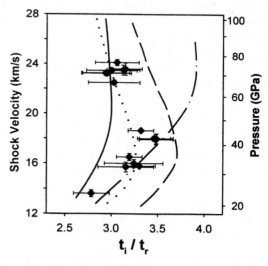

Figure 4. Measured reverberation timing compared with theoretical predictions. Lines and symbols as in Fig. 3.

models, all of which fall within the scatter and uncertainty of the measurements. However, as illustrated in Fig. 4, the differences in the model predictions for the reverberation ratio in this pressure range are significantly larger and exceed the measurement scatter and uncertainty.

CONCLUSIONS

In conclusion, we have performed high-velocity plate-impact, shock wave experiments to investigate the high-pressure EOS of LD_2. The results of impedance matching experiments are in agreement with theoretical models based upon first principles, *ab-initio* methods, and corroborate the stiff shock response at pressures up to ~100 GPa predicted by the Sesame EOS. Further, the present results disagree with earlier results reported from laser-driven experiments at pressures above ~40 GPa.

We have also exploited a novel reverberation technique to probe the density compression of LD_2 along the principal Hugoniot. The reverberation measurements confirm the stiffer Hugoniot response at high pressures inferred from impedance matching Hugoniot experiments and predicted by the various *ab-initio* and Sesame EOS models [4-6,2]. The results are consistent with a maximum density compression along the Hugoniot of ~4.2. Furthermore, the reverberation technique was found to be sufficiently sensitive to discriminate between the stiffer EOS models in the pressure range investigated. Best agreement with the measured reverberation ratios was found with the TB model [4].

While the present impedance matching Hugoniot and reverberation results may not completely resolve the discrepancy in the high-pressure response of LD_2, these results along with the recent Hugoniot measurements using convergent geometry techniques [20] make a strong case for a ~4-fold limited shock-compression. The consistency of these measurements lends confidence to each set of results, and suggests that systematic errors are likely not present. In particular, the agreement between results obtained using two completely independent loading techniques strengthens this argument.

ACKNOWLEDGEMENTS

Sandia is a multiprogram laboratory operated by Sandia Corporation a Lockheed Martin Company, for the U.S. DOE under contract DE-AC04-94AL8500.

REFERENCES

1. W.J. Nellis, *et al.*, *J. Chem. Phys.* **79**, 1480 (1983).
2. G.I. Kerley, *Molecular Based Study of Fluids* (ACS, Washington, D.C., 1983), p. 107; private communication (1998).
3. L.B. Da Silva, *et al.*, *Phys. Rev. Lett.* **78**, 483 (1997); G.W. Collins, *et al.*, *Science* **281**, 1178 (1998).
4. T.J. Lenosky, *et al.*, *Phys. Rev. B* **56**, 5164 (1997); L. Collins, *et al.*, *Phys. Rev. E* **52**, 6202 (1995); I. Kwon, *et al.*, *Phys. Rev. B* **50**, 9118 (1994).
5. T.J. Lenosky, *et al.*, *Phys. Rev. B* **61**, 1 (2000).
6. B. Militzer and D.M. Ceperley, *Phys. Rev. Lett.* **85**, 1890 (2000).
7. M. Ross, *Phys. Rev. B* **58**, 669 (1998); **54**, R9589 (1996).
8. C.A. Hall, *et al.*, *Rev. Sci. Instrum.* **72**, 1 (2001).
9. M.K. Matzen, *Phys. Plasma* **4**, 1519 (1996).
10. M.D. Knudson, *et al.*, *Phys. Rev. Lett.* **87**, 225501-1, (2001).
11. D.L. Hanson, *et al.*, in *Shock Compression of Condensed Matter-1999*, edited by M.D. Furnish, L.C. Chhabildas, and R.S. Hixson (AIP Press, New York, 2000), p. 1175.
12. The initial flyer thickness was nominally 800 μm. The ~200-300 μm thickness refers to portion of the flyer at impact that remained unaffected by magnetic diffusion.
13. L.M Barker and R.E. Hollenbach, *J. Appl. Phys.* **43**, 4669 (1972).
14. J.R. Taylor, *An Introduction to Error Analysis, 2nd edition* (University Science Books, California, 1982), p. 173.
15. The power varies as a function of wavelength; at 400 and 600 nm the power is approximately 1.9 and 1.5 respectively.
16. D. A. Young, private communication (2001).
17. G.E. Duvall and R.A. Graham, *Rev. Mod. Phys.* **49**, 523 (1977).
18. G.I. Kerley, Kerley Publishing Services Report No. KPS98-1, 1998 (unpublished).
19. N.C. Holmes, in *High-Pressure Science and Technology-1993*, edited by S.C. Schmidt, *et al.* (AIP Press, New York, 1994), p.153.
20. S.I. Belov, *et al.*, *JETP Lett.* **76**, 433 (2002).

CP706, *Shock Compression of Condensed Matter - 2003*
edited by M. D. Furnish, Y. M. Gupta, and J. W. Forbes
© 2004 American Institute of Physics 0-7354-0181-0/04/$22.00

DATABASE ON SHOCK-WAVE EXPERIMENTS AND EQUATIONS OF STATE AVAILABLE VIA INTERNET

Pavel R. Levashov*, Konstantin V. Khishchenko*, Igor V. Lomonosov† and Vladimir E. Fortov*

**Institute for High Energy Densities, RAS, Izhorskaya 13/19, Moscow 125412, Russia*
†*Institute of Problems of Chemical Physics, RAS, Chernogolovka, Moscow region 142432, Russia*

Abstract. The information on thermodynamic properties of matter at extremely high pressures and temperatures is very important both for fundamental researches and applications. We have collected about 14000 experimental points on shock compression, adiabatic and isobaric expansion and measurements of sound velocities behind the shock front for more than 400 substances. The database with graphical user interface containing experimental data, approximation modules, and caloric equations of state models has been worked out. One can search the information in the database in two different ways and obtain the experimental points in tabular or plain text formats directly via the Internet using common browsers. Registered users can remotely add new data into the database. It is also possible to draw the experimental points on graphs in comparison with different approximations and results of equation-of-state calculations. Recently we have added an ability to make calculations of shock Hugoniots, isentropes, isobars, and other curves using semiempirical equations of state for more than 100 substances via the Internet. One can present the results of calculations in text or graphical forms and compare them with any experimental data available in the database.

INTRODUCTION

Experimental data on investigations of thermodynamic properties in shock compression and release waves play an important role in modern physics of high pressures and temperatures. In particular, this information is commonly used for the creation of wide-range semiempirical equations of state (EOS). During last 50 years of shock-wave investigations about 15000 experimental registrations for more than 500 substances have been published.

Experimental data obtained at high energy densities are the result of numerous complex and expensive experiments. There are several compendiums of shock wave data available [1–4], but none of them is complete. For example, USA compendiums [1, 2] contain information obtained up to 1980, and don't include data on adiabatic expansion. In the relatively new handbooks [3, 4] there are only experimental registrations mainly obtained in Arzamas-16, Russia.

Therefore there is a requirement for a handbook including all available experimental data.

Rapid evolution of computer engineering and communications allows one to develop electronic handbooks with opportunities to search and represent information in any suitable format. In this work we present the public database which allows one to search necessary experimental data by different ways, represent the information in tabular and textual formats, and show data in graphs together with different approximations. Registered users can fulfil various editing operations in the database. Moreover there is an opportunity to calculate different thermodynamic functions for a big number of substances using several models of semiempirical equations of state. The database can be accessed via the Internet by address `http://teos.ficp.ac.ru/rusbank/`.

EXPERIMENTAL DATA AVAILABLE IN THE DATABASE

There are 4 types of experimental information in the database: on shock compression, adiabatic unloading, measurements of sound velocity behind the shock front and isobaric expansion. The database contains the complete compendium [1] and partialy the handbooks [2–4]. Experimental points are stored in the database as records. Each record includes measured parameters, reference and can have a comment.

The most numerous group of points is on shock compression of solid and porous samples. Every record in the database contains the initial density of substance (in g/cm^3), shock wave and particle velocities (in km/s). Today we have more than 13000 such experimental points.

Experiments on adiabatic expansion of substances allow one to study properties of materials in a wide range of densities and pressures from highly-compressed heated matter to expanded gas or plasma states. For all points on adiabatic expansion the database contains the initial density of substance, pressure (in GPa) and particle velocity on the isentrope (final state of substance after unloading). At present there are about 500 experimental points of this type in the database.

Measurements of sound velocity in shocked-compressed matter allow one to obtain the information about phase transitions in it. Every sound velocity record in the database contains the initial density of substance, particle and shock-wave velocities as well as the sound velocity value. As such experiments are very difficult there are only 139 registrations of this type now.

Experiments on isobaric expansion of thin wires by intense current pulse allow one to obtain a unique thermodynamically complete information about properties of substances near the liquid-gas phase boundary and the critical point at high temperatures. Each of 535 records of this type in the database can contain pressure, temperature (in kK), density, enthalpy (in kJ/g) and sound velocity depending on experimental conditions. Thus today there are more than 14000 records for about 450 substances combined into 11 groups in the database.

DATABASE DESCRIPTION AND REALIZATION

Despite experimental points the database also includes several data treatment modules: approximations of experimental dependencies and equations of state.

There are three different types of approximations of shock-wave data in the database [4–6]. In [6] experimental dependencies of shock velocity vs. particle velocity on shock Hugoniots of metals were approximated by piece-linear or piece-quadratic functions (up to 3 parts). The specific character of the dependence was determined by a statistical analysis. In the handbook [4] the approximations were constructed as piece-rational functions for a big number of substances; coefficients of dependencies are determined by the least-squares method. In [5] the calculated by quantum-statistical model of atom shock Hugoniots for elements were approximated by analitic quadratic functions. The coefficients of the functions depend on the porosity of substances. These approximations are applicable only at very high pressures accessible in nuclear underground explosions.

The database also contains semiempirical equations of state based on 2 caloric models [7, 8]. These equations of state are in a good agreement with experimental data on shock compression of solid and porous samples, double shock compression, adiabatic expansion and sound velocity measurements behind the shock front. The model [7] contains rather complex expressions for cold and thermal components of pressure and specific energy vs. density. This model was applied to the substances well-studied experimentally (metals, inorganic compounds). To create equations of state for poorly investigated substances (organic compounds) the simpler model was used [7]. Caloric equations of state doesn't take into account phase transitions and physics-chemical transformations of materials at high dynamic loading. In such cases equations of state for the original substance and products of transformation were constructed. In the database we have modules for shock Hugoniots, release isentropes, isobars, isokhors, isoenergy curves calculations to verify the agreement with experimental data. Now there are equations of state for 36 substances based on the model [7] and for 103 substances based on the model [8].

The database is functioning under Linux operating system. We use PostgresQL database server and Apache Web server. One can access the database via the Internet using any modern browser. PHP and Perl scripting languages have been used for textual data treatment and Web interface creation. Calculations using equations of state are accomplished by special FORTRAN modules.

The graphical user interface provides the search capabilities of necessary information by successive choices from the series of convergent subcategories or by the substance name. Data can be presented in textual or tabular formats. It is also possible to represent experimental dependencies and their approximations in graphical form. In Fig. 1 presented are shock Hugoniots of molybdenum samples with different initial densities in pressure-density plane along with quantum-statistical approximations [5] of these data and principal Hugoniot low-pressure part approximation [6]. One can change limits and scales on graph axes and put experimental points from specific references in the graph. Shock-wave data can be presented in different variables (the conversion between variables is fulfilled by means of conservation laws); there is also a possibility to represent only shock Hugoniots with given initial densities of samples and approximations chosen by user. As an example of graphical representaion of adiabatic expansion data in Fig. 2 shown are the release isentropes of magnesium in pressure-particle velocity plane.

In 2002 we have worked out the graphical user interface for the calculation of various thermodynamic curves using different equations of state models via the Internet. Now it is relatively easy to compute shock Hugoniots, release isentropes, isobars, isokhors and isoenergy curves for any suitable parameters and a wide circle of substances with the help of several equations of state models. The results can be shown in graph (in comparison with experimental points) or in a tabular form. The user interface provides for possibilities to correct graph parameters, enlarge or diminish its size, change limits and scales on axes, view the results in different variables, add new experimental or calculation data etc. One graph can contain curves calculated by means of several equations of state models for different substances together with any experimental data from the database. For example, in Fig. 3 shown are shock Hugoniots and release isentropes for polystyrene, computed us-

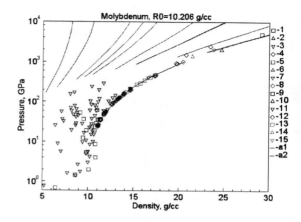

FIGURE 1. Shock Hugoniots of molybdenum samples with different initial densities: 1–15 — experimental data; a1 and a2 — approximations [6] and [5]

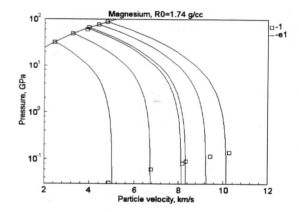

FIGURE 2. Release isentropes of magnesium: 1 — experiment; 2 — equation of state [8] calculation

ing the caloric equation of state [8] in comparison with experimental points.

CONCLUSION AND FUTURE WORK

The database presented in this work is a development of the database of shock-wave experiments available in the Internet since 1996 [9]. Today the number of experimental points is increased by a factor of 4, the approximations and graphical representation modules are added, the graphical user interface is elaborated for thermodynamic modeling with the help of

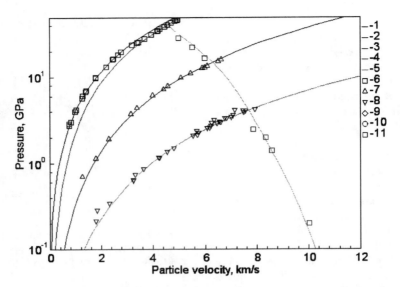

FIGURE 3. Shock Hugoniots of polystyrene samples with different initial densities and release isentrope of polystyrene in comparison with experimental data: 1 — principal Hugoniot calculated by the equation of state of original compound [8]; 2–4 — shock Hugoniots of samples with different initial densities, 5 — release isentrope, calculation using the equation of state of products of physical-chemical transformation [8]; 6–11 — experimental data

equations of state. Today the database is a flexible and modular system and can be adapted to the solution of different problems. In the nearest future we plan to add new types of experimental data and equations of state as well as to provide for the capability of modeling of typical shock-wave experiments via the Internet.

ACKNOWLEDGEMENTS

This work was supported by the Russian Foundation for Basic Research, Grant No. 01-07-90307.

REFERENCES

1. March, S. P., editor, *LASL Shock Hugoniot Data*, Univ. California Press, Berkeley, 1980.
2. Van Thiel, M., Shaner, J. W., and Salinas, E., *Compendium of Shock Wave Data. Livermore Lawrence Laboratory Report UCRL-50108*, Livermore, 1977.
3. Trunin, R. F., Gudarenko, L. F., Zhernokletov, M. V., and Simakov, G. V., *Experimental Data on Shock-Wave Compression and Adiabatic Expansion of Condensed Substances*, VNIIEF, Sarov, 2001.
4. Zhernokletov, M. V., Zubarev, V. N., Trunin, R. F., and Fortov, V. E., *Experimental Data on Shock Compression and Adiabatic Expansion of Condensed Substances at High Energy Densities*, Chernogolovka, 1996.
5. Kalitkin, N. N., and Kuz'mina, L. V., *Mat. Model.*, **10**, 111–123 (1998).
6. Al'tshuler, L. V., Bakanova, A. A., Dudoladov, I. P., Dynin, E. A., Trunin, R. F., and Chekin, B. S., *Zh. Prikl. Mekh. Tekhn. Fiz.*, **2**, 3–34 (1981).
7. Lomonosov, I. V., Bushman, A. V., Fortov, V. E., and Khishchenko, K. V., "Caloric equations of state of structural materials," in *High Pressure Science and Technology — 1993*, edited by S. C. Schmidt, J. W. Shaner, G. A. Samara, and M. Ross, AIP Press, New York, 1994, pp. 133–136.
8. Lomonosov, I. V., Fortov, V. E., and Khishchenko, K. V., *Chem. Phys. Rep.*, **14**, 51 (1995).
9. Levashov, P. R., Fortov, V. E., Khishchenko, K. V., Lomov, I. N., and Lomonosov, I. V., "Shock wave data base," in *Shock Compression of Condensed Matter — 1997*, edited by S. C. Schmidt, D. P. Dandekar, and J. W. Forbes, AIP Press, New York, 1998, pp. 47–50.

CP706, *Shock Compression of Condensed Matter - 2003*
edited by M. D. Furnish, Y. M. Gupta, and J. W. Forbes
© 2004 American Institute of Physics 0-7354-0181-0/04/$22.00

THEORETICAL INVESTIGATION
OF SHOCK WAVE STABILITY IN METALS

I. V. Lomonosov[1], V. E. Fortov[2], K. V. Khishchenko[2], and P. R. Levashov[2]

[1]*Institute of Problems of Chemical Physics, Chernogolovka, 142432 Moscow reg., RUSSIA*
[2]*Institute of High Energy Densities, Izhorskaya 13/19, 127412 Moscow, RUSSIA*

Abstract. Shock adiabats of metals with different initial volume and pressure have been calculated with the use of multi-phase equation of state. The stability of shock wave has been investigated with the use of known criteria. We found two types of instabilities occurring in the vicinity and inside two-phase liquid-gas region. They are the specific sound instability arising as a spontaneous sound emission from the shock discontinuity and two-wave configuration. The position of the instability region and its dependence on initial pressure and volume has been analyzed. Discussed are general regularities obtained on the base of the analysis for 30 metals.

INTRODUCTION

The shock wave is a unique tool in modern physics of high pressure. It produces the uniform distribution of pressure, density, energy and temperature behind the shock front for short period of time. The thermodynamic properties of shocked matter can be obtained for stationary 1D shock wave with the use of Hugoniot equations [1].

The gas dynamic analysis of the Riemann problem allows one to obtain criteria of shock wave stability. It can be done either with the usage of linearized gas dynamic equations [2-9] or on the basis of the general theory of decay and branching of arbitrary discontinuity [10]. The theory predicts two situations. In the first case the shock wave is unstable [2-4] and decays into other stable "elements" (stable shock waves, isentropic rarefaction or compression waves, tangential discontinuities, see for details Ref. [10] and references therein). The second case corresponds to a possibility of spontaneous emission of sound from the shock front [5-10].

Such instabilities can occur when the shock adiabat passes through specific regions of the phase diagram, characterized by thermodynamic anomalies. The two-phase liquid-gas region is the most attractive object for investigating the problem of shock wave stability. The basic reasons are that there are parts of the shock Hugoniot with $(\partial V / \partial p)_H > 0$ (index H means that the derivative is taken along the shock adiabat) inside the liquid–gas region, the isentropic sound velocity C_s has an anomalously low value near the boundary with the liquid phase and presence of other anomalies in thermodynamic functions.

The goal of the work is theoretical investigation of shock wave stability in metals.

CRITERIA OF SHOCK WAVE STABILITY

The stability of plane shock waves in media with arbitrary equation of state (EOS) has been studied by S. P. Dyakov [2] and then in subsequent works [3-8,10]. It is important to note that the recent analysis done on the basis of the general theory of decay and branching of an arbitrary discontinuity [10] leads to the same formulas obtained previously by use of linearized gas dynamic equations [2-9]. Omitting theoretical calculations, one can write in thermodynamic variables these criteria as

$$(\partial V / \partial p)_H < (V - V_0)/(p - p_0), \qquad (1)$$

$$(\partial V / \partial p)_H > [(V_0 - V)/(p - p_0)] \times$$
$$\times (1 + 2\beta^{-1}(D/c_S)), \qquad (2)$$

$$\frac{V_0 - V}{p - p_0} \times \frac{1 - \beta^{-2}(D/c_S)^2 - \beta^{-1}(D/c_S)^2}{1 - \beta^{-2}(D/c_S)^2 + \beta^{-1}(D/c_S)^2} <$$
$$< \left(\frac{\partial V}{\partial p}\right)_H < \frac{V_0 - V}{p - p_0}\left(1 + 2\beta^{-1}\frac{D}{c_S}\right). \qquad (3)$$

Here V_0, p_0 are referred to an initial state, $\beta = V_0 / V$, and D is shock velocity.

The case of absolute shock wave instability corresponds to Eqs. (1),(2). The condition (1) leads to gas dynamic flow with several shock discontinuities, which are separated by rarefaction waves. The condition (2) is much stronger then (1) and, according to [10], leads to decay of a shock discontinuity into several ones. A criterion of spontaneous sound emission (SSE) is defined by Eq. (3). Note that for fulfillment of this criterion the presence of region on shock adiabat with $(\partial V / \partial p)_H > 0$ is desired.

Equations (1)-(3) have been applied in theoretical investigations with the use of EOS for specific substances. Calculations done with the use of multi–phase EOS [11] for copper at p_0=1 bar (10^{-4} GPa) revealed SSE in two–phase liquid–gas region [12]. EOS for water–vapor system made it possible to examine regions of shock wave stability [13]. In accordance with [13], there are SSE regions and absence of absolute instability in water –vapor system. Analogous results have been obtained for shock adiabats of metals inside liquid–gas region [14]. SSE's region behind strong ionizing shock propagating in inert gases has been reported recently [15]. An application of quasi-classic Thomas – Fermi theory revealed SSE's region in tungsten at extreme high pressure [16].

CALCULATIONS AND RESULTS

Calculations of shock adiabats for ca. 30 metals have been done with the use of modified version [17] of multi–phase EOS [11]. The initial state has been changed from compressed solid and liquid metal to plasma and gas. These calculations correspond to matter that is in thermodynamic equilibrium.

The anomalous behavior has been found for shock adiabats starting inside two–phase liquid–gas region. Two different types of instabilities have been obtained.

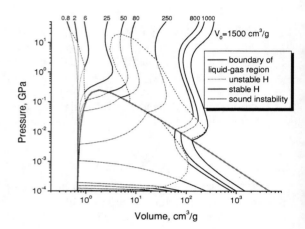

FIGURE 1. Shock adiabats of Mg at p_0=1 bar with different initial volumes V_0 in two–phase liquid–gas region.

Figure 1 illustrates the evolution of shock adiabats of magnesium started at initial pressure of 1 bar inside two-phase liquid–gas region. The check of criterion (3) gives parts on these adiabats having the sound instability both inside and outside the liquid–gas region. The corresponding dashed line ("instability" line) forms the boundary for the SSE–instability region so that each shock adiabat crossing this line has unstable parts with respect to SSE. Another situation takes place for shock adiabats with initial volumes V_0=0.8 – 250 cm³/g. The shock velocity D becomes less then isentropic sound velocity C_s when these shock adiabats cross the phase boundary of evaporation and come to liquid state. In accordance with general principles [1,18], such a shock transforms into an isentropic compression wave under this condition. The further increasing of pressure restores the condition of shock compression. The corresponding instability dashed line on Fig. 1 shows the position of this mechanic instability region. Note, that shock adiabats with $V_0=0.8 - 25$ cm³/g remain in liquid state while Hugoniots with $V_0=50 - 1000$ cm³/g cross again the evaporation region and exit to gas state. The subcritical adiabats have no anomalies in liquid, while supercritical Hugoniots have SSE parts in gas.

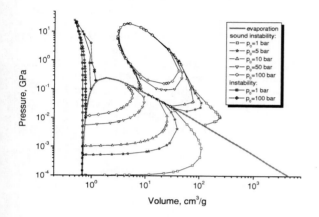

FIGURE 2. Evolution of instability lines in Mg depending on initial shock pressure p_0.

The variation of initial shock pressure p_0 changes the position of instability regions on the phase diagram as it is seen on Fig. 2. The mechanic instability region occurs closer to the critical point at higher initial pressures p_0. The SSE regions inside liquid–gas mixture remain similar geometrically. The geometry of SSE regions outside the liquid–gas mixture is a little bit different – these domains are separated from the evaporation boundary at pressures $p_0>10$ bar. It is interesting to note, that all these SSE regions are very close at high pressures, see Fig. 2.

As it is seen on Fig. 1, some shock adiabats (for example, with $V_0=1500$ cm³/g) have no anomalies. Only adiabats, which initial volume V_0 satisfies to an inequality $V_{min}(p_0) < V_0 < V_{max}(p_0)$ (the limiting volumes do not coincide with boundaries of liquid–gas region), cross corresponding instability line. These limiting volumes at different initial pressure p_0 are drawn on Fig. 3, so all

adiabats, which start under the line, have unstable parts. Let's consider the region under the line as an "instability" region.

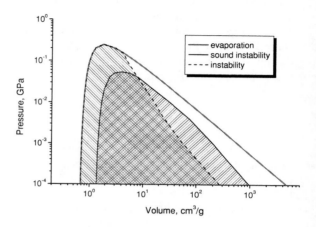

FIGURE 3. Instability regions in Mg.

Fig. 3 shows that shock adiabats with initial states inside liquid-gas region can have different types of instabilities. The first type is the transformation of shock wave to isentropic compression wave. In the second case this mechanic instability is supplemented by the sound instability inside and outside the liquid-gas mixture. The third situation corresponds to shock adiabats that cross the evaporating boundary only from the gas side and have only sound instability inside and outside the liquid-gas region.

DISCUSSION

Note that adiabats crossing the liquid–gas boundary from the liquid side lie below the wave ray (see Fig. 1). Along these parts $\left(\partial^2 p/\partial V^2\right)_H < 0$, while entropy is increasing. In this situation metal is compressed in the centered adiabatic wave $s=const$ which is separated from moving flow by surface of weak discontinuity [18].

Like previous analysis done only for shock adiabats inside liquid-gas region [14], the position of instability lines and instability region is individual for all metals with some similarity in groups of the Periodic Table. The influence of thermal electrons contribution in the EOS on the geometry of instability lines and instability region is strong, as it was found previously [14]. Nevertheless, the situation mentioned above for Mg with different type of instabilities of shock adiabats is analogous for all investigated metals. The physical explanation of anomalies of shock compression for the liquid-gas mixture is namely the low value of isentropic sound velocity C_s in this region of the phase diagram.

CONCLUSIONS

The stability of shock waves has been studied for different initial pressures and volumes with the use of advanced multi-phase EOS for 30 metals. Different types of instabilities have been found for shock adiabats that starts inside liquid-gas region. These mechanic and sound instabilities arise inside and outside liquid-gas region. The dependence of typical regions of these instabilities on initial pressure and volume has been investigated.

ACKNOWLEDGEMENTS

Funding was provided by the Programs of Presidium of Russian Academy of Sciences "Thermophysics and Mechanics of Intense Energy Loadings" and "Mathematical Modeling". Authors (IVL, KVK and PRL) also thank Russian Foundation for National Science Support.

REFERENCES

1. Zeldovich, Ya. B., and Raizer, Yu. P., *Physics of Shock Waves and High-Temperature Hydrodynamic Phenomena*, Academic Press, New York, 1966.
2. Dyakov, S. P. Zh. Eksp. Teor. Fiz. **27(9)**, 288, 1954.
3. Kontorovich, V. M. Sov. Phys. JETP **6**, 1179, 1957.
4. Iordanskii, S. V. Zh. Prikl. Matem. Mekh. **21(4)**, 465, 1957.
5. Kontorovich, V. M. Zh. Eksp. Teor. Fiz. **33(6)**, 1527, 1957.
6. Kontorovich, V. M. Akust. Zh. **5(3)**, 314, 1959.
7. D'yakov, S. P. Zh. Eksp. Teor.Fiz. **33(4)**, 948, 1957.
8. Fowles, G. R. Phys. Fluids **24**, 220, 1981.
9. Landau, L. D., and Lifshitz, E. M. *Fluid Mechanics,* Pergamon, Oxford, 1987.
10. Kuznetsov, N. M. Zh. Eksp. Teor. Fiz. **88(2)**, 470, 1985.
11. Bushman, A. V., Kanel, G. I., Ni, A. L., and Fortov, V. E., *Intense Dynamic Loading of Condensed Matter*, Taylor&Francis, London, 1993.
12. Bushman, A.V. In: *Dokl. II Vsesoyuz. Simp. Po Imp. Davleniyam,* Moskva, 39, 1976 [in Russian].
13. Kuznetsov, N.M., Davidova, O.N. Teplofiz .Visokh. Temp. **26(3)**, 567, 1988.
14. Lomonosov I.V., Fortov V. E., Khishchenko K. V., Levashov P. R. "Shock wave stability in metals", in Shock Compression of Condensed Matter, 1999 (M.D. Furnish, L.C. Chhabildas, R.S. Hixson, eds.), part I, pp.85-88.
15. Mond, M., Rutkevich, I. M. J. Fluid Mech. **275**, 121, 1994.
16. Fortova, T. N., Dremin, A. N., Fortov, V. E. Chisl. Met. Mekh. Splosh. Sredi **4**, 143, 1972.
17. Fortov V.E., Khishchenko K.V., Levashov P.R., Lomonosov I.V., Nucl. Instrum. Meth. in Phys. Res. A. vol. 415. p.604-608, 1998.
18. Fortov V. E. *Tepl. Visokh. Temper.* **19(1)**, 168, 1972.

CP706, *Shock Compression of Condensed Matter - 2003*
edited by M. D. Furnish, Y. M. Gupta, and J. W. Forbes
© 2004 American Institute of Physics 0-7354-0181-0/04/$22.00

TIME-RESOLVED X-RAY DIFFRACTION INVESTIGATION OF SUPERHEATING-MELTING OF CRYSTALS UNDER ULTRAFAST HEATING

Sheng-Nian Luo,[1] **Damian C. Swift,**[1] **Thomas Tierney,**[1]
Kaiwen Xia,[2] **Oliver Tschauner,**[3] **and Paul D. Asimow**[4]

[1] *P-24 Plasma Physics, Los Alamos National Laboratory, Los Alamos, NM 87545*
[2] *Graduate Aeronautical Laboratories, California Institute of Technology, Pasadena, CA 91125*
[3] *Department of Physics, University of Nevada, Las Vegas, NV 89154*
[4] *GPS Division, California Institute of Technology, Pasadena, CA 91125*

Abstract. The maximum superheating of a solid prior to melting depends on the effective dimensionless nucleation energy barrier, heterogeneities such as free surfaces and defects, and heating rates. Superheating is rarely achieved with conventional slow heating due to the dominant effect of heterogeneous nucleation. In present work, we investigate the superheating-melting behavior of crystals utilizing ultrafast heating techniques such as exploding wire and laser irradiation, and diagnostics such as time-resolved X-ray diffraction combined with simultaneous measurements on voltage and current (for exploding wire) and particle velocity (for laser irradiation). Experimental designs and preliminary results are presented.

INTRODUCTION

Superheating is the state where the long-range order of a solid persists above its equilibrium melting temperature (T_m). It is related to the kinetics of melting − to overcome the energy barrier for solid-liquid transitions. Thus superheating is supposed to be common in melting process where heating rate (Q) also plays an important role. Appreciable amount of superheating is seldom observed in conventional low heating rate experiments ($Q \sim 1$ K/s), due to the heterogeneous nucleation at free surfaces, defects and impurities. As heating rate and heterogeneous nucleation are competing factors, superheating could be marked for ultrafast heating as in light-gas gun loading and laser irradiation. To study the materials properties (in particular melting behavior) under extreme conditions, e.g., to interpret the results of shock melting experiments ($Q \sim 10^9 - 10^{12}$ K/s), understanding the

kinetics of melting (in time t, pressure P and temperature T) is of immediate importance.

If a solid with a dimensionless nucleation energy barrier (β) is superheated at heating rate Q to a maximum temperature T_c, the maximum superheating achieved is $\theta_c = T_c/T_m$.[1] β depends on solid-liquid interfacial energy γ_{sl}, heat of fusion ΔH_m and T_m: $\beta \equiv 16\pi\gamma_{sl}^3/3\Delta H_m^2 kT_m$. The systematics[1] of maximum superheating was established as

$$\beta = (A_0 - b \lg Q)\theta_c(\theta_c - 1)^2 \qquad (1)$$

where $A_0 = 59.4$, $b = 2.33$ and Q is normalized by 1 K/s. The values for parameter β are documented[1], e.g., β is about 1.5 and 8.2 for Cu and Ga, which correspond to θ_c of ~1.19 and 1.43 at $Q \sim 10^{12}$ K/s, respectively. Systematic molecular dynamics simulations[1] yielded results consistent with such systematics. Previous superheating experiments compare favorably to the systematics, but the amount of superheating data is very limited.

The above systematics simply supply an upper bound to superheating. To resolve the degree of superheating in real experiments, we need to know T and T_m, and phases of the heated sample. The persistence or breakdown of the long-range order of solid can best be resolved from time-resolved (or transient) X-ray diffraction (TXD). Accurate temperature measurement is critical but challenging. In this work, we utilize exploding wire[2] and laser irradiation techniques to investigate superheating-melting behavior under various heating rates with such diagnostics as TXD, line-imaging velocity interferometry (VISAR), and voltage and current measurements. Experimental designs and preliminary results are presented. The main purpose of this work is to point out the possible directions for experimental investigation of superheating-melting behavior. The preliminary results are mostly utilized to illustrate idea rather than draw any solid conclusions.

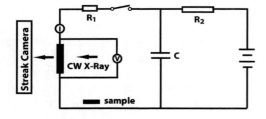

FIGURE 1. Schematic of exploding wire[2] circuit with TXD diagnostics. Metallic sample wire is subjected to ohmic heating with voltage (V) and current (I) evolutions recorded. *In-situ* TXD pattern of the sample is recorded continuously by the streak camera. X-ray source can be continuous wave (CW, e.g. from synchrotron) or pulsed.

EXPLODING WIRES

When a metallic wire of sample is subjected to discharging of a capacitor (Fig. 1), the temperature of the wire before it becomes melted, vaporized or plasma-ized, can rise to $\sim 10^3$ K in ~ 10-100 μs (average $Q \sim 10^7 - 10^8$ K/s) or less by ohmic heating. The high heating rate itself may induce significant superheating. A wide

range of heating rates can be obtained by varying capacity, applied voltage and resistance (Fig. 1), and the effect of heating rates can be investigated.

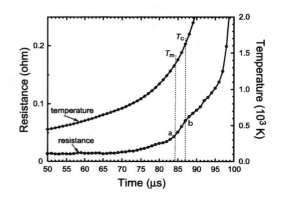

FIGURE 2. Time evolution of resistance (R, calculated from measured $V(t)$ and $I(t)$, not shown) and temperature (T) of the Cu wire during ohmic heating. This wire was 143 mm long and 0.3556 mm in radius.[3] Values of T assume solid state and Dulong-Petit limit for C_p. T_m (1356 K for Cu) is the equilibrium melting point, and T_c (~ 1600 K) the predicted maximum temperature at superheated state. Pressure is reasonably assumed to be 0 at solid states.

In the schematic of exploding wire, the current through (I) and voltage (V) across the exploding wire can be measured continuously. Thus the energy increase (ΔE) due to ohmic heating at any instant t_1 can be calculated as $\Delta E = \int_0^{t_1} V(t)I(t)dt$. Heat conduction and optical radiation to its environment, and strain energy changes can be neglected, so T_1 in the solid wire can be solved from $\int_0^{t_1} V(t)I(t)dt = \int_{T_0}^{T_1} C_p(T)dT$ where C_p is heat capacity. Temperature in the solid state (including superheated states if there are any) can be constrained (e.g. Fig. 2 for exploding Cu wire). To tell whether it is melted, one possible way is to examine the evolution of resistance $R(t) = V(t)/I(t)$ along with $T(t)$, as the electrical conductivity of liquid metal may differ from that of the solid. For example, T_c, the predicted maximum temperature of Cu (from Eq. 1 at $Q \sim 10^8$ K/s) at the superheated state, corresponds to R_b where the curvature of R starts to change, i.e. melting could occur at the superheated state b rather than a (corresponding to T_m). Upon catastrophic melting at b, the temperature excess cannot fully compensate the required latent heat to melt the whole sample, thus only partial melting occurs at b ($\sim 46\%$) and less pronounced superheating exists thereafter.

To infer the melting instant from $R(t)$, accurate temperature dependences of C_p and $R(t)$ for solid and liquid metals are needed. There still exists possibility that equilibrium melting occurs along ab — a more direct and definitive structure information can resolve such an uncertainty. Thus we propose to measure the X-ray diffraction patterns from continuous X-ray source simultaneously with $V(t)$ and $I(t)$ (Fig. 1). We expect that TXD, $R(t)$ and $T(t)$ would allow us to examine the details of melting process in a variety of metals.

We pointed out that catastrophic melting at the limit of superheating (if there is any) at b may be only partial. The possibility of continuous equilibrium melting (also partial melting) exists as well. Partial melting raises the issue — to what extend TXD can resolve melting — which remains to be investigated. Although only melting behavior near ambient pressure (the pressure from magnetic field is negligible) is investigated with exploding wire technique, it is directly relevant to high-pressure melting.

LASER IRRADIATION

Materials such as Si, Ge, Ga, Bi and water ice have negative Clausius-Clapeyron slopes at low pressures (e.g. Fig. 3). If we preheat (or precool) such solids to a certain T slightly below ambient T_m, and shock-load to a certain P in the negative dT_m/dP regime, and samples remain in solid states, then the samples are superheated. Fig. 4 is a schematic of laser irradiation (direct laser drive or laser-driven flyer) with TXD and line-imaging velocity interferometry (VISAR) diagnostics.[4] Phase change may be registered in the particle-velocity (u_p) history (Fig. 5) reduced from VISAR record. Two streak cameras (denoted as Bragg and Laue) measuring diffraction pattern in two orthogonal directions from shocked sample (Fig. 4).

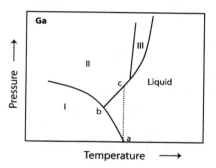

FIGURE 3. Schematic of Ga phase diagram. a denotes ambient condition ($T = 298$ K and $P = 0$. Note T_m is 303 K at zero pressure). b is the Ga I-II-liquid triple point at $T \sim 277$ K and $P \sim 1.2$ GPa. ac denotes an isotherm (298 K) to which the low-P end Hugoniot is close. If Ga sample remains crystalline at pressure below $P_c \sim 2.2$ GPa, it is certainly superheated for states on ac.

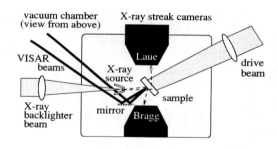

FIGURE 4. Schematic of laser irradiation with TXD and line-VISAR diagnostics. A laser beam shone on the sample surface (sample itself or coating) induces plasma, and expansion of the plasma cloud induces shocks propagating into the sample. TXD patterns are recorded in two directions by Bragg streak camera (parallel to shock wave propagation direction) and Laue camera (parallel to shock wave front).

We conducted VISAR and TXD measurements on Ga to illustrate the idea for similar experiments. Single-crystal Ga samples were grown and mounted for direct laser drive (initial temperature is ~ 298 K) . VISAR (Fig. 5) and TXD (Fig. 6) measurements were separate in the preliminary shots. An elastic precursor and indication of possible phase change (melting or solid-solid) are displayed on particle-velocity history (Fig. 5). Such a possible phase change is more

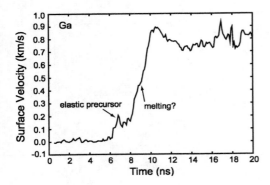

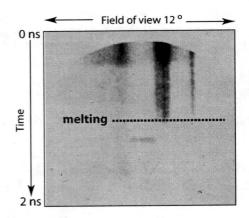

FIGURE 5. Free surface velocity history at the center of the sample (Ga) reduced from line-imaging VISAR record in a preliminary shot. No TXD was measured. The kink denoted as possible melting is more pronounced in original VISAR film record.

FIGURE 6. TXD pattern of laser-irradiated Ga from a separate preliminary shot. VISAR measurement was not conducted. The disappearance of diffraction pattern for the two lines to the right at ~1 ns (relative time) is interpreted as melting. The intensity for the diffraction line to the left is significantly dim from which no melting appears to be indicated. Further investigation is needed.

pronounced on un-reduced film record of VISAR. As only u_p is measured, P can only be obtained by assuming an equation of state. Laser-driven flyer experiment with measurements of flyer velocity (u_{fp}) and u_p may resolve P and shock velocity (U_s) from impedance match. In another shot, melting appears to occur on TXD at ~1 ns (relative time). Simultaneous measurement of u_p, diffraction pattern and u_{fp} or U_s (using stepped sample) could in principle resolve the details of melting and other phase changes.

As in the case of exploding wire, the temperature excess might not be enough to compensate the latent heat required to melt the bulk crystal at the limit of superheating (if there is any), thus only partial melting occurs. This concerns with the resolution of VISAR and TXD, i.e. the minimum portion of melting which can be resolved in TXD and VISAR recordings. In the case of Ga ($\Delta H_m = 5.6$ kJ/mol), bulk melting requires equivalent superheating (excess in T) of ~220 K which seems to be difficult to achieve for low P loading. But significant portion of melting could be achieved for materials with triple point at higher pressures, such as Si.

CONCLUSIONS

We present experimental designs of exploding wire and laser irradiation with time-resolved X-ray diffraction diagnostics to investigate superheating-melting behavior of crystals under ultrafast heating. Preliminary results suggest that such techniques would potentially allow us to examine systematically the detailed superheating-melting process.

ACKNOWLEDGMENTS

S.-N. Luo is sponsored by a Director's Post-doctoral Fellowship at Los Alamos National Laboratory (P-24 and EES-11). We appreciate the kind permission of QinetiQ for using their exploding wire data.

REFERENCES

1. Luo, S.-N., and Ahrens, T.J., Appl. Phys. Lett. **82**, 1836 (2003); Luo, S.-N. and Ahrens, T.J., Phys. Earth Planet. Int. (in press) (2003); Luo, S.-N., Ahrens, T.J., Çağın, T., Strachan, A., Goddard III, W.A., and Swift, D.C., Phys. Rev. B **68**, 134206 (2003).

2. Chace, W.G., *Exploding Wires*, Plenum Press, 1959.

3. Swift, D.C., private communication.

4. Swift, D.C., Paisley, D.L., Kyrala, G.A., and Hauer, A., in *Shock Compression of Condensed Matter-2001*, edited by M.D. Furnish et al., AIP Conference Proceedings 620, Melville, New York, 2002, pp 1192-5.

CP706, *Shock Compression of Condensed Matter - 2003*
edited by M. D. Furnish, Y. M. Gupta, and J. W. Forbes
© 2004 American Institute of Physics 0-7354-0181-0/04/$22.00

THE SHOCK HUGONIOT OF HYDROXY-TERMINATED POLYBUTADIENE

Y. Meziere, J. Akhavan, G.S. Stevens, J.C.F. Millett, N.K. Bourne

Royal Military College of Science, Cranfield University, Shrivenham, Swindon, SN6 8LA. United Kingdom.

Abstract. The response of polymers to shock loading is becoming of increasing importance, both as binder systems in plastic–bonded explosives (PBXs) and as structural materials in their own right. In this paper, we report on the shock Hugoniot of hydroxy-terminated polybutadiene (HTPB), which is commonly used as a binder system in PBXs, but whose shock response has yet to be presented in the open literature. Results indicate that the shock velocity – particle velocity relationship is linear, similar to some but not all polymer-based materials.

INTRODUCTION

The need to understand the shock response of polymeric materials to shock loading, both from a mechanical and microstructural standpoint is becoming of increasing interest. In particular, those polymers that find application as the binder phases in plastic-bonded explosives (PBXs) such as polychloro-trifluoroethylene (Kel-F) and poly-urethane based materials such as estane have received attention [1, 2]. However, the shock response of one polymer that has been little studied is hydroxy-terminated polybutadiene (HTPB). Whilst a number of studies have considered it as part of a composite system in PBXs and propellants [3], it's individual response to high-rate loading has not been considered. In contrast, its response to quasi-static loading rates has been examined. Wingborg [4] has shown that the mechanical properties of HTPB are influenced by the choice of hardener. For example, using the hardener dicyclohexylmethane 4,4'-diisocyanate ($H_{12}MDI$) resulted in a material with a tensile strength of *ca.* 9 MPa, compared to *ca.* 4 GPa when using the more usual isophorone diisocyanate (IPDI).

In addition to the IPDI hardener, most HTPB binders also contain additional chemicals such anti-oxidants and plasticizers that will also have an effect upon their mechanical response. Therefore, in this paper, we measure the Hugoniots of two HTPB compositions, one with a plasticizer (supplied by Royal Ordnance in the United Kingdom), and one without, manufactured by ourselves at the Royal Military College of Science. In a parallel paper, we have also recovered this second material for chemical and microstructural analysis [5].

EXPERIMENTAL

Both materials were cast as 10 mm plates onto 1 mm thick plates of either dural (aluminium alloy 6082-T6) or copper, to which manganin stress gauges (MicroMeasurements LM-SS-125CH-048) had previously been fixed. These were insulated from the metallic plates with 25 μm mylar with a slow setting epoxy adhesive. HTPB 1 was prepared by Royal Ordnance, Glascoed, to a proprietary composition. HTPB 2 was prepared in house to a similar composition but without the plasticizer. Once cured, an addition gauge was supported on the back of the target assemblies using a 12 mm block of polycarbonate. In this way, not only stress and particle velocity (through impedance matching), but timing information through known positions of both gauges would yield the shock velocity. Shock stresses were induced by the impact of 10 mm dural and copper flyer plates. Gauge records were converted to stress-time traces using the methods of Rosenberg *et al* [6]. Particle velocities (u_p) were determined using impedance matching. Specimen configurations and gauge placements are shown in figure 1.

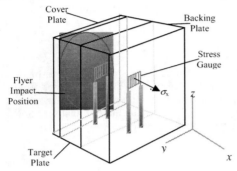

Figure 1. Specimen configuration and gauge placement.

HTPB1 had a density (ρ_0) of 0.85±0.01 g cm^{-3}, and a longitudinal sound speed (c_L) of 1.46 ±0.03 mm μs^{-1}, whilst HTPB2 has a density of 1.06±0.01 g cm^{-3}, and a sound speed of 1.43±0.03 mm μs^{-1}. The sound speeds were measured using quartz transducers operating at 5 MHz, using a Panametrics 500PR pulse receiver.

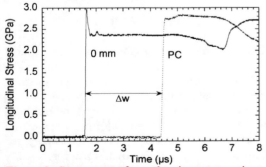

Figure 2. Gauge traces from plate impact experiment on HTPB1. Impact conditions are 10 mm copper flyer at 769 m s^{-1}. 0 mm trace from gauge at Cu/HTPB interface, PC – trace from HTPB/polycarbonate interface.

RESULTS AND DISCUSSION

In figure 2, we present typical gauge records from a plate impact experiment upon hydroxy-terminated polybutadiene, in this case, HTPB1.

A number of features in this figure are worthy of note. Firstly the trace labeled 0 mm records a sharp rise in signal, reaching a peak before settling to a steady value of *ca.* 2.4 GPa. This peak is due to the capacitive linking due to the fast-rising nature of the stress pulse and is discussed in more detail elsewhere [7]. The temporal spacing between the traces (Δw), in combination with the known separation of the gauges is used to determine the shock velocity, (U_s). Finally, the trace labeled PC, that is from the gauge supported on the back of the HTPB with a polycarbonate block, also shows a rapid rise in signal with a comparatively flat top. Due to the close impedance matching between polycarbonate, the epoxy adhesive and the gauge backing, the rise time is due to the thickness of the manganin gauge element, and thus the trace will be a good indication of the shape of the stress pulse as it travels through the sample. Even though the stress amplitude at 2.4 GPa is high, there is no evidence of a break in slope that would indicate the presence of an HEL. Whilst the HEL of this and similar materials is not known, it would be expected to be relatively low in comparison to other polymers (PMMA for example has an HEL quoted at 0.75 GPa [8]). Also note that the top of the pulse is relatively flat, suggesting a linear relationship between shock velocity and particle velocity. This is in contrast to PMMA [8], where a pronounced rounding of the pulse was observed above the HEL, where the U_s-u_p curve was also seen to be non-linear due to the high rate sensitivity of the material.

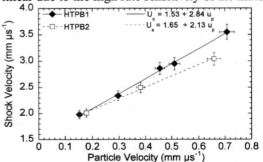

Figure 3. Shock Hugoniot of HTPB in shock velocity (U_s) – particle velocity (u_p) space.

This issue is explored further in figure 3. The plots are both materials are broadly similar, in that both show a linear response, as was suggested previously. The values of the shock parameters c_0 and S are different for each material. In particular, the value of c_0 in HTPB2 is higher than in HTPB1. This suggests that the compressibility of the latter is greater. Given that this material has had an addition of plasticizer, which will have the effect of increasing the mechanical compliance, this observation from the shock velocities makes sense.

One feature that is common to both materials concerns the relationship of the measured longitudinal sound speeds to the values of c_0 determined from measurements of the shock velocity. In most metallic systems, for example copper or tantalum, the value of c_0 corresponds to the ambient pressure bulk sound speed [9]. However, in both HTPBs investigated here, c_0 is greater than the measured ambient longitudinal sound speed. This is a feature that has been observed both by ourselves, for example in polyether ether ketone [10], an epoxy resin [11] and polychloroprene [12], and others, for example Carter and Marsh [13]. Indeed in that work, the authors investigated the shock response of 22 different polymers, and in all but four was it observed that c_0 was greater than c_L. Therefore, we can explain why no elastic precursor was observed the trace labeled PC in figure 2. The results from figure 3 show that the shock wave will always be faster than the elastic precursor, hence it will not be observed.

The relationship between shock velocity and particle velocity has been used to determine the shock stress (σ_x) through the relation,

$$\sigma_x = \rho_0 U_s u_p. \qquad 1.$$

Note that this does not take into account the strength of the material. However, comparison with the measured stresses from the gauges is a useful exercise, giving insights into the materials response to shock loading. The results are presented in figure 4.

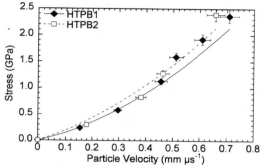

Figure 4. Shock Hugoniots of HTPB.

The measured values of stress from both materials are effectively identical to each other, in contrast to the shock velocity plots. The comparison of the stresses calculated from equation 1 (using the relevant values of density, c_0 and S) to the measured stresses in both materials is more revealing. In both materials, the agreement is good at lower stresses, but at higher levels, the curves diverge, with the measured stresses being the higher. However, as we have already stated, equation 1 does not allow for the materials strength and its variation with impact strength. The longitudinal stress generated during shock loading can be expressed as a function of the hydrostatic pressure, P and the materials shear strength, τ, thus,

$$\sigma_x = P + \frac{4}{3}\tau. \qquad 2.$$

Therefore, failure to take the effects of shear strength into account in this calculation could result in differences between calculated and measured stresses. We would point out that the 'stress' calculated from equation 1 is the hydrodynamic response of the material. whilst P is the hydrostatic pressure, and thus the two will be slightly different. Therefore, while we would not suggest that values of the shear strength could be determined from a combination of equations 1 and 2, differences between the measured Hugoniot stress and equation 1 will still reveal overall trends in materials response. Similar behaviour has been noted in other materials, including polyether ether ketone (PEEK) [10] and epoxy based resins [14], where these differences were correlated with an increasing shear strength with impact stress amplitude, determined from experimental measurements [15]. More interestingly, we have also noticed in materials where the measured and calculated stresses agree (for example polychloroprene [12]), the measured shear strength was observed to remain at a constant level with increasing impact stress [16], therefore supporting the hypothesis that the observed differences in figure 4 indicate that HTPB has a positive dependence of shear strength on the impact stress.

CONCLUSIONS

The shock response of two compositions of hydroxy-terminated polybutadiene has been determined in terms of the shock velocity, shock stress and particle velocity. Results in both materials show a linear relationship between U_s and u_p, with the pure material having a greater value of c_0. This would appear consistent with the fact that this material has no added plasticizer, which would have the effect of reducing the compressibility of the material. The Hugoniots in terms of stress and particle velocity in both materials are similar, but comparisons of measured and calculated stresses show differences at higher stresses, suggesting that these materials have an increasing shear strength with impact stress. Further work is in progress to determine if this is correct.

ACKNOWLEDGMENTS

We would like to thank Dr. Ian Murray and Dr. Ron Hollands of Royal Ordnance, Glascoed for provision of samples. We are also grateful to Gary Cooper, Matt Eatwell and Paul Dicker of RMCS for technical assistance.

REFERENCES

1. Anderson, M.U., in *Shock Compression of Condensed Matter - 1991*, S.C. Schmidt, et al., Editors. (1992), Elsevier: Amsterdam. p. 875.
2. Bourne, N.K. Gray III, G.T., in *Plasticity 99: Constitutive and Damage Modeling of Inelastic Deformation and Phase Transformation*, A.S. Khan, Editor. (1998), Neat Press: Fulton, Maryland. p. 619.
3. Sutherland, G.T., Forbes, J.W., Lemar, E.R., Ashwell, K.D., Barker, R.N., in *High Pressure Science and Technology 1993*, S.C. Schmidt, et al., Editors. (1994|), American Institute of Physics: New York. p 1381.
4. Wingborg, N., Polymer Testing, (2002), **21** 283.
5. Akavhan, J., Millett, J.C.F., Bourne, N.K. (2003) in *These Proceedings*
6. Rosenberg, Z., Yaziv, D., Partom, Y. J. Appl. Phys., (1980), **51** 3702.
7. Bourne, N.K., Rosenberg, Z. Meas. Sci. Technol., (1997), **8** 570.
8. Barker, L.M., Hollenbach, R.E. J. Appl. Phys., (1970), **41** 4208.
9. Marsh, S.P., *LASL Shock Hugoniot data*. (1980), Los Angeles: University of California Press. Los Angeles
10. Millett, J.C.F., Bourne, N.K., Gray III, G.T. J. Appl. Phys., (2003) Submitted.
11. Millett, J.C.F., Bourne, N.K., Barnes, N.R. J. Appl. Phys., (2002), **92** 6590.
12. Millett, J.C.F. Bourne, N.K., J. Appl. Phys., (2001), **89** 2576.
13. Carter, W.J., Marsh, S.P. *Hugoniot equation of state of polymers*. (1995), Los Alamos National Laboratory. LA-13006-MS
14. Barnes, N., Millett, J.C.F., Bourne, N.K., in *Shock Compression of Condensed Matter - 2001*, M.D. Furnish, N.N. Thadhani, Y. Horie Editors. (2002) American Institute of Physics: Melville, NY. p.135
15. Bourne, N.K., Millett, J.C.F., Barnes, N., Belcher I. in *Shock Compression of Condensed Matter - 2001*, M.D. Furnish, N. Thadhani, and Y. Horie, Editors. 2002, American Institute of Physics: Melville, NY. 649.
16. Bourne, N.K. Millett, J.C.F. Proc. R. Soc. Lond. A, (2002). **459** 567.

CP706, *Shock Compression of Condensed Matter - 2003*
edited by M. D. Furnish, Y. M. Gupta, and J. W. Forbes
© 2004 American Institute of Physics 0-7354-0181-0/04/$22.00

NON-LINEARITY OF POLYETHYLENE HUGONIOT UP TO 1 GPa AND ITS INTERPRETATION BY GRÜNEISEN PARAMETER ESTIMATION

Yasuhito Mori* and Kunihito Nagayama

*Department of Aeronautics and Astronautics, Faculty of Engineering, Kyushu University,
Hakozaki, Higashiku, Fukuoka 812-8581 Japan*
(** present address: Department of General Education, Sasebo National College of Technology,
Okishin, Sasebo, Nagasaki 857-1193 Japan*)

Abstract. Based on an optical or a gauge-and-optical method, shock Hugoniot curves for two kinds of polyethylene (PE) specimens, i.e., high-density polyethylene (HDPE) and linear low-density polyethylene (LLDPE), have been measured in the shock stress region up to 1 GPa. It is found that the shock-versus-particle velocity Hugoniot curves for these PE specimens are non-linear in this shock stress region. As one of the reasons for this non-linearity, it is plausible that the molecular chains of polymers have anharmonic potential in the direction of intermolecular motion and intramolecular motion. The Grüneisen parameter of representing the anharmonic properties of the potential has been calculated from the obtained Hugoniot curves for these PE specimens. Calculated values of the parameter change rapidly from large to small values with shock compression up to 1 GPa shock stress. Large values of the parameter correspond to the selective excitation of the intermolecular vibrational modes of with lower phonon frequency. It is revealed that the non-linearity of the Hugoniot curves for the PE specimens results from the drastic increase of the number of the excited vibrational modes of intramolecular motion in case the value of shock stress exceeds some level.

INTRODUCTION

Hugoniot curves for various polymers have been known in the relatively high shock stress region up to about 100 GPa. [1,2] In this region, shock velocity (u_s) versus particle velocity (u_p) Hugoniot curves for polymers tend to have almost linear relationship as is seen in other materials. In most cases, however, value of the extrapolated shock velocity for infinitesimally small particle velocity does not coincide with the measure ultrasonic longitudinal velocity of the specimen. This means that the $u_s - u_p$ Hugoniot curves for polymers may deviate from the high-pressure Hugoniot with some non-linearity in the lower shock stress region less than 1 GPa.

In order to discuss the functional form of the Hugoniot curves for polymers up to 1 GPa precisely, we have developed two experimental methods of using a single-stage compressed gas gun. One is an optical method of using a high-speed streak camera. [3,4,5,6] The other is a gauge-and-optical method of using a polyvinylidene-fluoride (PVDF) stress gauge and an optical prism pin. [6,7,8,9]

By these two methods, the Hugoniot data up to 1 GPa have been measured for two kinds of PE specimens. In this article, the effective thermodynamic Grüneisen parameters are calculated by the measured Hugoniot data to explain the measured Hugoniot function. It is found that Hugoniot curve shows non-linearity in this narrow stress range. It is also revealed that the non-linearity of the u_s-u_p Hugoniot curves for the PE specimens can be explained by the non-equilibrium excitation of intermolecular and intramolecular vibrational modes.

GRÜNEISEN PARAMETER FOR POLYMER

Polymers have molecular chain structures, which are bonded by strong covalent bonds in the intramolecular direction and by very weak van der Waals force in the intermolecular direction. Therefore, the molecular chains of polymers have anharmonic potential in the direction of intermolecular motion and intramolecular motion. These two quite different kinds of potential energy lead to the vibrational modes of relatively low and high eigen-frequencies, which are far apart. As explained later, this situation might be a main cause of the non-linearity of the Hugoniot curve for polymers especially in the low shock stress region. Behavior of these frequency spectra will be adequately described by the Grüneisen parameter, since it is the parameter representing the anharmonic properties of the potential. [10,11,12]

The frequency change of the lattice vibration with increasing volumetric strain can be defined as a microscopic Grüneisen parameter, γ_j,

$$\gamma_j = -\frac{V}{\omega_j}\left(\frac{\partial \omega_j}{\partial V}\right)_T = -\left(\frac{\partial \ln \omega_j}{\partial \ln V}\right)_T, \quad (1)$$

where ω_j, V and T are the frequency of the jth vibrational mode, specific volume, and temperature, respectively.[13] Values of γ_j for several molecular bonds such as CH_2 radical with high frequencies, e.g., C-C stretching mode, CH_2 rocking mode, etc. have been measured.[14] They are known to have very small values (0-0.09). It is known that the values of γ_j for van der Waals crystals becomes greater than that for ionic and covalent crystals.[15] Therefore, it is expected that the values of γ_j for the vibrational modes with very low frequencies for intermolecular motion become greater than that with very high frequency for intramolecular modes.

The thermodynamic Grüneisen parameter γ is defined as

$$\gamma = V\frac{\alpha K_T}{C_V} = V\left(\frac{\partial p}{\partial \varepsilon}\right)_V, \quad (2)$$

where C_V, α, K_T, p, and ε denote specific heat at constant volume, coefficient of thermal expansion,

isothermal bulk modulus, pressure, and specific internal energy, respectively.[10,11,12,13] It is found that γ gives the average value of all vibrational modes. Anharmonic properties in solids are generally evaluated by using the thermodynamic Grüneisen parameter γ defined by Eq. (2). In the case of metals, the value of γ equals to about 2.[16] On the other hand, in the case of polymers, the value of γ becomes the relatively small, e.g., about 0.7-1.1 for PE [17,18] and about 0.5 for PMMA. [19]

In the case of polymers under the hydrostatic pressure, Grüneisen parameters had been obtained from acoustic data of the polymers.[14] For example, This value is about 5 for PE.[14] This means that only the intermolecular modes of the chain potential is exited selectively when the acoustic wave propagates into PE specimen. In the excitation of vibrational modes, the propagation of weak shock wave is similar to that of the acoustic wave. Therefore, it is expected that only the intermolecular vibrational modes with very low frequencies are exited also in the weak shock propagation.

Effective thermodynamic Grüneisen parameters in the weak shocked polymers can be expressed by using a thermodynamic model and a Slater model. [10,11,12,20] The Grüneisen parameter γ derived by these models along the Hugoniot curves can be expressed by only the shock wave parameter except for the sound velocity at the shocked state a_H.

For simplicity, it is assumed that the Hugoniot curves for polymers can be expressed by the following linear relationship as

$$u_s = A + Bu_p, \quad (3)$$

where A- and B-coefficient are material constants. By using Eq. (3), the effective Grüneisen parameter in the weak shocked state for polymers can be derived as follows, [10]

$$\gamma = \frac{(u_s + Bu_p)(u_s - u_p)\left(\frac{1}{3} + \frac{B-1}{A}u_s\right) + \frac{B}{A}u_s(u_s - u_p)^2}{(u_s + Bu_p)(u_s - u_p) + \frac{B}{A}u_su_p(u_s - u_p)}. \quad (4)$$

From the measured Hugoniot data for HDPE and LLDPE up to 1 GPa, the Grüneisen parameters can be calculated by Eq. (4). Then the non-linearity of the u_s-u_p Hugoniot curves can be discussed.

SHOCK REGISTRATION SYSTEM

All of the shock experiments have made by a single-stage compressed gas gun, whose launch tube has 40 mm bore diameter and 2 m length. [21,22] A projectile of 20-250 g mass with a flyer plate is accelerated up to about 100-500 m/s by helium gas.

In the optical method, a very compact high-speed streak camera has been used, whose writing arm length is only about 150 mm. [3,4,5,6] As a light source, a homemade long-pulsed dye laser with 25-30 μs in pulse duration is used.

In the gauge-and-optical method, a PVDF stress gauge (Dynasen Inc., PVF$_2$-11-.125-EK) [23] and an optical prism pin are used.[6,7,8,9] A shock stress profile in the position embedded the gauge can be recorded. The optical prism pin is located on the free surface of the target specimen to detect the weak shock wave sensitively.

RESULTS AND DISCUSSION

Hugoniot curves for HDPE and LLDPE

Figure 1 shows the obtained $u_s - u_p$ Hugoniot curves for HDPE and LLDPE for the particle velocity region less than 0.5 km/s. Open symbols show the data by the optical method, and solid symbols show the data by a gauge-and-optical method. Two sets of data with different experimental methods are found to be in good harmony with each other as is seen in Fig. 1.

It is found that these Hugoniot curves are non-linear such that the gradient of the curves degreases gradually with compression. Only for the convenience of analysis, we assumed here that these Hugoniot curves have a kink in the particle velocity of about 0.17 km/s, and can be divided into two linear relationships as shown in Fig. 1. These lines can be expressed as

HDPE: $\quad u_s = 2.439 + 2.459 u_p \qquad (5)$
$\qquad\qquad u_s = 2.537 + 1.871 u_p \qquad (6)$
LLDPE: $\quad u_s = 1.992 + 2.990 u_p \qquad (7)$
$\qquad\qquad u_s = 2.220 + 1.787 u_p \qquad (8)$

By using these linear relationships, the values of γ in each part should be calculated by using Eq. (4).

Consideration of Non-linear Hugoniot Curves by using Grüneisen parameter

Figure 2 shows the relationship between the calculated Grüneisen parameter γ and the particle velocity u_p in the region of $0 < u_p < 0.5$ km/s. It is found that the value of γ becomes small suddenly in the particle velocity of about 0.17 km/s. The value of γ is almost determined by the gradient of the $u_s - u_p$ Hugoniot curve (B-coefficient in Eq. (3)). As shown in Fig. 1, the change of the gradient around $u_p = 0.17$ km/s for LLDPE is larger than that for HDPE. Therefore, as shown in Fig. 2, the change in value of γ for LLDPE is larger in comparison with that for HDPE.

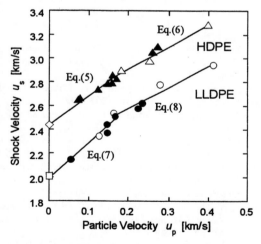

FIGURE 1. The obtained $u_s - u_p$ Hugoniot curves for HDPE and LLDPE. The data at $u_p = 0$ km/s are the measured longitudinal sound velocity for each specimen.

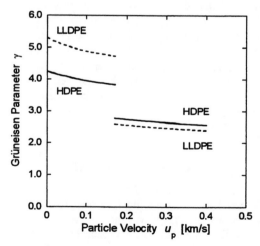

FIGURE 2. The calculated Grüneisen parameters for HDPE and LLDPE.

In the very weakly shocked region ($0 < u_p < 0.17$ km/s), it is found that the values of γ are relatively large, i.e., about 4-5. As described in the previous section, this means that the weakly shocked state is in thermodynamic non-equilibrium. These values of γ equal almost to the value evaluated by using the acoustic data of PE specimen.[15] In other words, present estimation of γ based on the Hugoniot data is consistent with those based on the pressure dependence of sound velocity. In this particle velocity region for HDPE and LLDPE, therefore, it is regarded that only the intermolecular vibrational modes with very low frequencies are selectively exited by the shock wave propagation.

In the particle velocity range of $0.17 < u_p < 0.4$ km/s, the value of γ is reduced to about 2.5-2.8. These values are still far from that in the thermal equilibrium state for PE specimen.[17,18] This result can then be explained as follows: with increasing pressure or shock temperature, anharmonicity of molecular vibration increases, which will gradually enhance mode coupling between intermolecular to intramolecular vibrational modes. After all, part of the vibrational energy is imparted to excite the intramolecular modes. Even so, the value of γ suggests that the shocked state in this stress region is still not in thermal equilibrium.

By using the previous Hugoniot data in higher pressure region for PE specimens [1], the Grüneisen parameter can be calculated. As a consequence, it is found that the value of γ for HDPE and LLDPE decreases monotonically from about 5 to about 2 in the region of $0 < u_p < 0.5$ km/s, and then approaches slowly to the value less than unity. Therefore, it is suggested that the non-linearity of the $u_s - u_p$ Hugoniot curves in the weakly shocked region up to 1 GPa arose by the increasing exitation of the intramolecular vibrational modes with higher frequencies. [24]

ACKNOWLEDGEMENT

The authors wish to thank Dr. Koichi Kitao of Materials and Processing Research Center of NKK Corporation for providing us with HDPE and LLDPE. We also wish to thank Mr. K. Hidaka for the measurement of the Hugoniot data for these PE specimens.

REFERENCES

1. *LASL Shock Hugoniot Data*, edited by S. P. Marsh (Berkeley, University of California Press, 1980).
2. W. J. Nellis, F. H. Ree, R. J. Trainor, A. C. Mitchell, and M. B. Boslough, *J. Chem. Phys.*, **80**, 2789 (1984).
3. Mori, Y, Tamura, T., and Nagayama K., *Rev. Sci. Instrum.*, **69**, 4, 1730-34 (1998).
4. Mori, Y., and Nagayama, K., *Shock Compression of Condensed Matter-1997*, 875-878 (1998).
5. Y. Mori and K. Nagayama, *Rev. High Pressure Sci. Technol.*, 7, 841-843 (1998).
6. Nagayama, K., Mori, Y., and Hidaka, K., *J. Materials Processing Technol.*, **85**, 20-24 (1999).
7. Mori, Y., Hidaka, K., and Nagayama, K., *Rev. Sci. Instrum.*, **71**, 6, 2492-96 (2000).
8. Y. Mori and K. Nagayama, *Shock Compression of Condensed Matter-2001*, 673-676 (2002).
9. Nagayama, K., Mori, Y., and Hidaka, K., *Rev. High Pressure Sci. Technol.*, 7, 858-860 (1998).
10. K. Nagayama and Y. Mori, *J. Phys. Soc. Jpn.*, **63**, 4070-77 (1994).
11. K. Nagayama, *J. Phys. Chem. Solids*, **58**, 271 (1997).
12. K. Nagayama and Y. Mori, *J. Appl. Phys.*, **84**, 6592-99 (1998).
13. E. Grüneisen, *Handbuch der Physik* ed. H. Greiger and K. Scheel (Springer, Berlin, 1926) **10**, pp. 1-59.
14. Y. Wada, A. Itani, T. Nishi, and S. Nagai, *J. Polym. Sci. A-2*, 7, 201 (1969).
15. Y. Wada, A. Itani, T. Nishi, and S. Nagai, *J. Polym. Sci. A-2*, 7, 201 (1969).
16. Ya. B. Zel'dovich and Yu. P. Raizer, *Physics of Shock Waves and High-Temperature Hydrodynamic Phenomena* (New York and London, Academic Press, 1966) Chap. XI.
17. C. K. Wu, G. Jura, and M. Shen, *J. Appl. Phys.*, **43**, 4348 (1972).
18. K. I. Tsirule and E. L. Tyunina, *High-Pressure Chemistry and Physics of Polymers* (Boca Raton, CRC Press, 1994), Chap.1.
19. I. Gilmour, A. Trainor, and R. N. Haward, *J. Polym. Sci.: Polym. Phys. Ed.*, **16**, 1291 (1978).
20. J. C. Slater, *Introduction to Chemical Physics* (McGraw-Hill, New York, 1939) Chap. XIII.
21. Nagayama, K., "*Shock Waves in Material Science*", Springer-Verlag, Tokyo, 1993, Chap. 9, pp.195-224.
22. Mori, Y., and Nagayama, K., *Proc. 2nd Symposium on High Speed Photography and Photonics*, Tohoku University, Sendai, Japan, 159-193 (1996).
23. J. A. Charest and C. S. Lynch, *Shock Compression of Condensed Matter-1989* (North-Holland, Amsterdam, 1990), pp.797-800.
24. Y. Mori, *Doctor's thesis* (in Japanese) (2002).

CP706, *Shock Compression of Condensed Matter - 2003*
edited by M. D. Furnish, Y. M. Gupta, and J. W. Forbes
2004 American Institute of Physics 0-7354-0181-0/04/$22.00

A MULTI-PHASE EQUATION OF STATE FOR SOLID AND LIQUID LEAD

C. M. Robinson

AWE, Aldermaston, Reading, Berks. RG7 4PR. U.K.

Abstract. This paper considers a multi-phase equation of state for solid and liquid lead. The thermodynamically consistent equation of state is constructed by calculating separate equations of state for the solid and liquid phases. The melt curve is the curve in the pressure, temperature plane where the Gibb's free energy of the solid and liquid phases are equal. In each phase a complete equation of state is obtained using the assumptions that the specific heat capacity is constant and that the Grüneisen parameter is proportional to the specific volume. The parameters for the equation of state are obtained from experimental data. In particular they are chosen to match melt curve and principal Hugoniot data. Predictions are made for the shock pressure required for melt to occur on shock and release.

INTRODUCTION

This paper considers an empirical thermodynamically consistent multi-phase equation of state (EoS) for solid and liquid lead.

The construction of the EoS follows the method used by Johnson et al (1) to determine a multi-phase EoS for bismuth. Different equations of state are used for the solid and liquid phases. The low pressure fcc to hcp solid-solid phase transition at ~15 GPa (2) is neglected, so it is assumed that there is only one solid phase. The phase that exists at any pressure and temperature is the phase that has the lowest Gibb's free energy, the melt curve is the locus of points with the same Gibb's free energy in the pressure, temperature plane. Along the melt curve the solid and liquid phases have different energies and densities, therefore there is a latent heat of melting and a density change when melt occurs. Between the solidus and liquidus the equation of state is obtained from the solid and liquid EoS's using the method of mixtures.

An empirical EoS is used for each phase, and experimental data is used to determine the parameters of the EoS model, particular attention is made to matching the available principal Hugoniot and melt curve data.

Predictions are made for the shock pressures required to melt lead on shock and on release from shock.

EQUATION OF STATE

A different EoS is used for the solid and liquid phases. In each phase it is assumed that the specific heat capacity at constant volume is constant. Then thermodynamic consistency requires that the Grüneisen parameter is a function of the specific volume only, here the Grüneisen parameter is assumed to be proportional to the specific volume. Then the Helmholtz free energy F is the following function of temperature T and specific volume v,

$$F(T,v) = F_0 + (c_v - S_0)(T - T_0) - c_v T \log\left(\frac{T}{T_0}\right) + \Gamma_0 c_v \varepsilon (T - T_0) + K_0 v_0 \phi(\varepsilon) \quad (1)$$

where F_0, S_0, c_v, Γ_0, K_0, v_0, and T_0 are constants and ϕ is a function of the strain $\varepsilon = 1 - v/v_0$ with $\phi(0) = \phi'(0) = 0$, $\phi''(0) = 1$. Then at zero pressure and temperature T_0 the Helmholtz free energy, entropy, Grüneisen parameter, isothermal bulk modulus and specific volume are given by F_0, S_0, Γ_0, K_0, and v_0 respectively. c_v is the specific heat capacity at constant volume. The pressure p, entropy S and Gibb's free energy G may be obtained from the Helmholtz free energy from the equations

$$p = -\frac{\partial F}{\partial v}\bigg|_T, \quad S = \frac{\partial F}{\partial T}\bigg|_v, \quad G = F + pv. \quad (2)$$

Note that $K_0\phi'(\varepsilon)$ is the pressure as a function of strain along the $T = T_0$ isotherm. The constants and the function ϕ are determined by fitting to experimental data.

The melt curve is determined by calculating the curve in the pressure temperature plane where the Gibb's free energies of the two phases are equal. In the mixed phase region between the solidus and liquidus the EoS is determined from the solid and liquid EoS using the method of mixtures, at a given pressure and temperature the specific volume, Helmholtz free energy and entropy are given by

$$\begin{aligned} v &= \alpha_s v_s + \alpha_l v_l \\ F &= \alpha_s F_s + \alpha_l F_l \\ S &= \alpha_s S_s + \alpha_l S_l \end{aligned} \quad (3)$$

where the subscripts s and l refer to the solid and liquid phases at the same pressure and temperature, and α_s and α_l are the solid and liquid mass fractions which must add up to one,

$$\alpha_s + \alpha_l = 1. \quad (4)$$

Hence the EoS in the mixed phase is completely determined from the EoS's of the solid and liquid phases.

PARAMETERS FOR THE EQUATION OF STATE

For the solid EoS zero pressure and temperature T_0 are taken to be room temperature and pressure (RTP). The constants F_0 and S_0 are chosen so that the Helmholtz free energy and entropy are zero at RTP, the remaining constants in the solid EoS may be obtained from RTP experimental data, e.g. from (3). The function ϕ is calculated so that the principal Hugoniot has a linear shock velocity u_s, particle velocity u_p curve,

$$u_s = c_0 + s u_p \quad (5)$$

where c_0 and s are constants. c_0 is the bulk sound speed at RTP and s is obtained from fitting to Hugoniot data (4) up to ~200 GPa. Substituting equation (5) and the EoS (from equations (1) and (2)) into the Rankine-Hugoniot equations results in an ordinary differential equation for ϕ, see (1), which is solved numerically.

For the liquid phase the constant T_0 is taken to be the room pressure (RP) melting temperature (5). The difference between the energies of the solid and liquid EoS's is expected to be small, therefore a small change in either EoS is likely have a significant effect on the position of the melt curve. Therefore the liquid EoS is chosen to match the experimental melt curve data. The melt temperature is assumed to be given by the equation

$$T = T_{m0} + \frac{T'_{m0}p + (\tfrac{1}{2}T''_{m0} + CT'_{m0})p^2}{1 + Cp} \quad (6)$$

where the constants T_{m0}, T'_{m0}, T''_{m0}, and C are obtained by fitting to experimental melt curve data (6,7,8) which is available to ~100 GPa. Note that T_{m0}, T'_{m0}, and T''_{m0} are respectively the temperature, gradient, and curvature of the melt curve at zero pressure. The function ϕ is chosen so that the Gibb's free energy of the liquid is equal to the Gibb's free energy of the solid on the melt curve, this results in an algebraic equation for ϕ.

The constants F_0 and S_0 may be determined by ensuring that T_0 is the melt temperature at RP (i.e. the Gibb's free energy of the solid and liquid are equal at temperature T_0, zero pressure) and that the latent heat of melting at RP matches the experimental value (5). The remaining constants in the EOS may be obtained from the available specific heat capacity data (5), density data (9,10) and sound speed data (11,12) at the RP melt temperature. However these constants must be chosen to be consistent with the chosen melt curve. In particular Clausius Clapeyron's equation must be satisfied at RP,

$$\frac{dT}{dp} = \frac{v_l - v_s}{S_l - S_s} = \frac{T}{L}(v_l - v_s), \qquad (7)$$

where the subscripts s and l refer to the solid and liquid phases respectively and L is the latent heat. By differentiating Clausius Clapeyron's equation with respect to pressure along the melt curve the following equation is obtained for the curvature of the melt curve,

$$\frac{d^2T}{dp^2} = \frac{T}{L}\left[\frac{v_l}{K_{Tl}}\left(1 - \frac{c_{vl}\Gamma_l}{v_l}\frac{dT}{dp}\right)^2 - \right.$$
$$\left. \frac{v_s}{K_{Ts}}\left(1 - \frac{c_{vs}\Gamma_s}{v_s}\frac{dT}{dp}\right)^2 + \frac{(c_{vl} - c_{vs})}{T}\left(\frac{dT}{dp}\right)^2\right] \qquad (8)$$

where K_T is the isothermal bulk modulus and the subscripts s and l refer to the solid and liquid phases respectively. The constants chosen for the liquid EoS and the equation for the melt curve are chosen so that equations (7) and (8) are satisfied at RP and the EoS agrees as closely as possible with experimental melt curve data (6,7,8) and experimental data at the RP melting point (5,9,10,11). The constants used are given in table 1.

RESULTS

The predicted principal Hugoniot, solidus and liquidus are plotted in the pressure, specific volume plane up to 200 GPa in Fig. 1 together with the

TABLE 1. Constants Used in the EoS Model.

	Solid	Liquid
T_0 (K)	288	600.6
F_0 (J kg^{-1})	0	-1.7050x10^4
S_0 (J kg^{-1} K^{-1})	0	138.17
v_0 (m^3 kg^{-1})	8.82x10^{-5}	9.395x10^{-5}
K_0 (Pa)	4.569x10^{10}	3.657x10^{10}
Γ_0	2.928	2.943
c_v (J kg^{-1} K^{-1})	120	113.9
s	1.47	-
T_{m0} (K)	-	600.6
T'_{m0} (K Pa^{-1})	-	7.885x10^{-8}
T''_{m0} (K Pa^{-2})	-	-4.349x10^{-18}
C (Pa^{-1})	-	3.441x10^{-11}

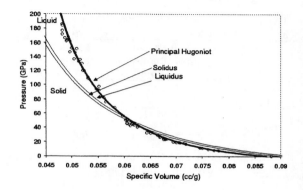

FIGURE 1. The Principal Hugoniot, Solidus and Liquidus in the Pressure, Specific Volume Plane. Circles Show Experimental Hugoniot Data (4).

experimental Hugoniot data (4). It can be seen that the Hugoniot matches the experimental data (4) in this pressure range, in fact the Hugoniot matches experimental data (13) up to 1 TPa. The sound speed on the Hugoniot is ~5% lower than the experimental data (14).

Note that there is a finite range of shock pressures that result in the lead being shocked to a partially molten state between the solidus and liquidus curves. There is a small discontinuity in the gradient of the Hugoniot as it crosses the solidus and liquidus curves, but this is not apparent in Fig. 1.

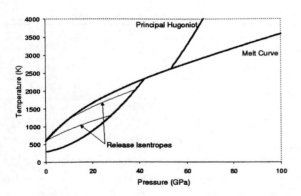

FIGURE 2. The Predicted Melt Curve and Principal Hugoniot in the Pressure, Temperature Plane. The Two Release Isentropes Shown Both Pass Through the Room Pressure Melting Point. At this Point one Isentrope is Solid, the other Liquid.

TABLE 2. Points Where the Principal Hugoniot Intercepts the Melt Curve.

	Solid/Mixed Phase Boundary (Solidus)	Liquid/Mixed Phase Boundary (Liquidus)
Shock Pressure	42.2 GPa	53.3 GPa
Shock Temperature	2349 K	2624 K

TABLE 3. Principal Hugoniot Pressures that Release to the Room Pressure Melting Temperature.

	Solid at Release	Liquid at Release
Shock Pressure	28.0 GPa	38.4 GPa

The melt curve and principal Hugoniot are shown in the pressure, temperature plane in Fig. 2. The Hugoniot is deflected along the melt curve for a short range of shock pressures, these states are the partially molten states between the solidus and liquidus. The shock pressures where the Hugoniot enters and leaves the melt curve are given in table 2. Also shown in Fig. 2 are two release isentropes. Both isentropes release to the melt temperature at RP, but the colder isentrope is solid at RP while the hotter isentrope releases to liquid. The shock pressures for the two release isentropes are given in table 3. Between these two shock pressures the lead releases to a partially molten state.

CONCLUSIONS

An EoS for solid and liquid lead has been constructed that matches the available experimental data for the melt curve (up to 100 GPa) and principal Hugoniot (up to 1 TPa). This allows the shock pressure required to melt lead on shock and release from shock to be determined.

REFERENCES

1.   Johnson, J.N., Hayes, D.B., and Asay, J.R. *J. Phys. Chem. Solids* **35**, 501-515 (1974).
2.   Young, D.A., Phase *Diagram of the Elements*, University of California Press, Berkley and Los Angeles, California, 1991, pp. 107-108.
3.   Swan, G.W., and Thornhill, C.K. *J. Mech. Phys. Solids* **22**, 349-370 (1974).
4.   Marsh, S.P., *LASL Shock Hugoniot Data*, University of California Press, Berkeley and Los Angeles, California, 1980, pp. 100-102.
5.   Hultgren, R., Orr, R.L., Anderson, P.D., and Kelley, K.K., *Selected Values of Thermodynamic properties of Metals and Alloys*, John Wiley & Sons, Inc., New York, 1963, pp. 206-210.
6.   Akella, J., Ganguly, J., Grover, R., and Kennedy, G., *J. Phys. Chem. Solids*, **34**, 631-636 (1973).
7.   Mirwald, P.W., and Kennedy, G.C., *J. Phys. Chem. Solids*, **37**, 795-797 (1976).
8.   Godwal, B.K., Meade, C., Jeanloz, R., Garcia, A., Liu, A.Y., and Cohen, M.L., *Science*, **248**, 462-464 (1990).
9.   Kirshenbaum, A.D., Cahill, J.A., and Grosse, A.V., *J. Inorg. Nucl. Chem.*, **22**, 33-38 (1961).
10. Lucas, L.D. *Mem. Sci. Rev. Met.*, **69**, 395-409 (1972).
11. Gitits, M.B., and Mikhailov, I.G., *Soviet Physics-Acoustics*, **11**, 372-375 (1966).
12. Hixson, R.S., Winkler, M.A., Shaner, J.W., *Int. J. Thermophysics*, **7**, 161-165 (1986).
13. Rothman, S.D., Evans, A.M., Horsfield, C.J., Graham, P., and Thomas, B.R., *Physics of Plasmas*, **9**, 1721-1733 (2002).
14. Boness, D.A., Brown, J.M., and Shaner, J.W., "Rarefaction Velocities in Shocked Lead," *in Shock Waves in Condensed Matter 1987*, edited by. Schmidt, S.C., and Holmes, N.C., Elsevier Science Publishers B.V., 1988, pp. 115-118.

CP706, *Shock Compression of Condensed Matter - 2003*
edited by M. D. Furnish, Y. M. Gupta, and J. W. Forbes
© 2004 American Institute of Physics 0-7354-0181-0/04/$22.00

SHOCK WAVES AND SOLITONS IN COMPLEX (DUSTY) PLASMAS

D.Samsonov, S.Zhdanov, G.Morfill

*Max-Planck-Institute for extraterrestrial Physics, Centre for Interdisciplinary Plasma Science, Garching
85741, GERMANY*

Abstract. Shock waves and solitons were obtained in 2D complex plasmas i.e. plasmas mixed with micron-sized particles. These particles acquire large $(1000 - 50000e)$ negative charges and strongly interact with each other. The particle cloud can form ordered structures and exist in a solid, liquid, or gaseous state. A monolayer hexagonal lattice was formed from monodisperse plastic micro-spheres in the sheath of an rf discharge and excited with an electrostatic pulse. It was found that weak pulses produced solitons, which did not change the phase state of the lattice. Stronger excitation created shock waves. The lattice melted behind the shock front and later recrystallized. Two shock regimes were distinguished. One had a stable, thin, well defined front, the other had an unstable front. The shocks were analyzed by tracking individual particles with a video camera and calculating their velocity, number density, kinetic temperature, and defect density. Molecular dynamics simulation reproduced the experimental results. Shock waves were also observed in a 3D complex plasma on board of the International Space Station. Those shocks were accompanied by a potential drop across the front. In this case not only the dust component but also ions, electrons and neutrals had to be considered.

INTRODUCTION

Complex (dusty) plasmas consist of submicron to sub-millimeter sized particles immersed into an ion-electron plasma. These particles collect ions and electrons and can emit secondary or photo-electrons and therefore charge up. Charging can be explained by the theory of electrostatic probes [1, 2]. In a static case a sum of all currents to a particle is equal to zero and it is equivalent to a floating probe. In the absence of emission, particles acquire a large negative charge, since the flux of light and fast electrons is much higher than that of heavy and slow ions.

Negatively charged particles can be confined by an external electric field. They are affected by the gravity and friction force due to collisions with neutrals. If there is an ion flow or a temperature gradient, the ion drag and thermophoretic forces should be taken into account.

Charged particles interact with each other electrostatically via a screened Coulomb (Yukawa) poten-

tial [3] $\phi = (Q/r) \exp(-r/\lambda_D)$, where Q is the particle charge, r is the interparticle distance and λ_D is the Debye length of the electron-ion plasma. Complex plasmas can be characterized by a coupling parameter Γ which is the ratio of the particle interaction energy to their kinetic energy. It was first predicted theoretically [4] and then observed experimentally [5, 6, 7] that complex plasmas with $\Gamma \gg 1$ can form ordered or crystalline structures. Complex plasmas can also be in gaseous ($\Gamma \ll 1$) or liquid ($\Gamma \simeq 1$) states [8, 9]. It was also suggested that complex plasmas can be used as macroscopic model systems to model phase transitions, waves, shocks, Mach cones, solitons, etc. The advantage is that they are easily observed with a video camera in real time at the kinetic level.

Shock waves were observed in complex plasmas before. Ion acoustic shocks were observed experimentally in Ref. [10]. A three-dimensional dust-acoustic shock was observed in a micro-gravity experiment [11]. Here we report an experimental obser-

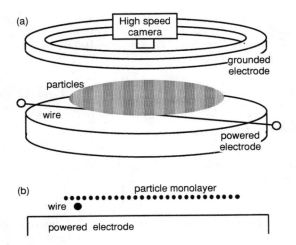

FIGURE 1. Sketch of apparatus. (a) Oblique view. Spherical monodisperse particles charged negatively and formed a monolayer levitating in the plasma sheath above the lower electrode. (b) Side view. The grounded wire was placed below the particles. Short negative pulses were applied to the wire to excite shocks. Reprinted from Ref. [12]

vation of shocks in a strongly coupled (crystalline) monolayer complex plasma.

EXPERIMENTAL PROCEDURE

We performed the experiments in a 13.56 MHz capacitively coupled rf discharge in a setup (Fig. 1) identical to that of Ref. [12]. A powered lower electrode and a ring upper electrode were placed in a vacuum chamber. The upper electrode and the chamber were grounded. A rf power of 10 W (measured as forward minus reverse) was applied to the lower electrode. The working pressure of 1.8 Pa was maintained by a flow of argon at a rate of 0.5 sccm. Monodisperse plastic micro-spheres 8.9±0.1 μm in diameter were levitated in the sheath $\simeq$ 9 mm above the lower electrode. They were confined radially in a bowl shaped potential formed by a rim on the outer edge of the electrode and formed a monolayer hexagonal lattice. The particle cloud was about 6 cm in diameter. The particle separation in the lattice was 650 μm at the excitation edge, 550 μm in the middle and 720 μm at the outer edge. A horizontal thin (0.2-0.3 mm) sheet of light from a doubled Nd:YAG diode pumped laser (532 nm) illuminated the parti-

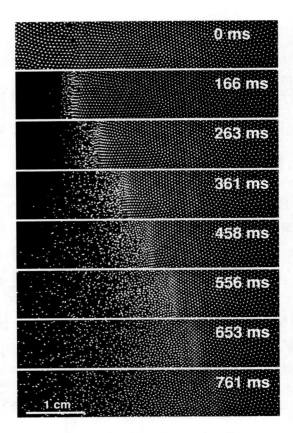

FIGURE 2. Shock wave propagating in a two-dimensional complex (dusty) plasma. Initially unperturbed lattice (at 0 ms) had hexagonal symmetry with some defects. The shock swept the particles (at 166 ms) and propagated into the crystal melting its structure. The amplitude of the shock reduced with time due to the neutral damping until it turned into a soliton (at 653 ms) and stopped melting the lattice. The particle cloud fully recrystallized after about 10 s.

cles, which were imaged by a top view digital video camera at 102.56 frames/s. The field of view was 1024 × 512 pixels or 4.42 × 2.21 cm and it contained about 3000 particles.

In order to excite shock waves, a horizontal tungsten wire 0.1 mm in diameter was stretched above the lower electrode, 4 mm below the particle layer and roughly half way between the center and the edge of the electrode (Fig. 1). The wire was normally grounded and it did not affect the particles. A short negative pulse applied to the wire repelled the negatively charged micro-spheres breaking the lattice above the wire and creating a pulsed one-dimensional compressional disturbance, propagating horizontally perpendicular to the wire. The particle cloud also oscillated in the vertical direction with a small amplitude which caused a periodic change of brightness of the particles. The parameters of the excitation pulse were selected stronger than in the experiments of Ref. [12]. We used long pulses (-100 V, 50 ms) to make stable shocks, and short pulses (-100 V, 5 ms) to make unstable shocks. We allowed about 1 min cooling time between the experimental runs.

The recorded video sequences were analyzed by a program which identified particle positions. Tracing the particles from one frame to the next yielded their velocity. It was then averaged in 50 narrow bins parallel to the wire. The kinetic temperature was calculated from the standard deviation of the particle velocity in the bins which depends only on the particle random motion. We used triangulation to determine the nearest neighbors for each particle. The local number density was determined as the inverse area of Voronoi cells. The compression factor was computed as a ratio of the number density to the unperturbed number density. All measured quantities were binned in order to reduce the influence of random fluctuations.

We performed a molecular dynamics simulation in order to support the experimental results. A monolayer lattice of 721 particles was formed in a three-dimensional confining potential. The lattice was strongly confined in the vertical direction and weakly in the horizontal. The particles interacted via a screened Coulomb (Yukawa) potential and their motion was damped by friction. Initially placed at random positions, the particles were equilibrated by running the code, until a round hexagonal lattice was formed and cooled down. Then a pulsed excitation force was applied to produce shocks.

EXPERIMENTAL RESULTS

The unperturbed lattice was in a crystalline state with hexagonal symmetry. It had 8% defects, mostly pairs of five- and seven-fold cells. The mean particle separation ($a = 570$ μm) was found from the first peak of the pair correlation function. The translational correlation length was $3a$ and the orientational correlation

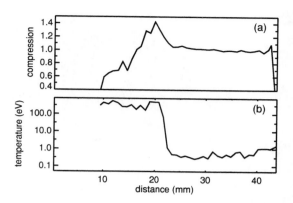

FIGURE 3. Compression factor (a) and kinetic temperature (b) in the shock wave versus distance to the excitation source at 390 ms. The shock front was manifested by a compression peak which coincided with a temperature jump.

length was $10a$. This indicated a relatively good crystalline structure.

After a short negative pulse (-100 V, 50 ms) was applied for the duration of about 5 frames, the particles were swept away from the wire (Fig. 2). A shock with a Mach number of 2.4-2.7 propagated into the lattice melting it. The amplitude of the shock reduced due to the neutral damping of the particle motion, until it turned into a supersonic compressional soliton [12], that did not melt the crystal. The cloud recrystallized after about 10 s.

The structure of the shock was revealed by plotting the compression factor and kinetic temperature (Fig. 3). A narrow peak of compression (Fig. 3a) coincided with a jump in kinetic temperature (Fig. 3b). The compression decreased after the shock had passed (left side of the plot) due to the cloud back-flow which re-expanded to its equilibrium position.

It was found that two shock regimes are possible. Long excitation pulses (50 ms) produced shocks with a stable linear front (Fig. 4). The roughness scale of the front was of the order of the distance between the particles. Short excitation pulses(5 ms) produced, on the other hand, shocks with an unstable front (Fig. 5) with a very rough structure. Its roughness scale was much larger than the interparticle distance.

Molecular dynamics simulation reproduced both stable and unstable shocks and helped explain how they are formed. The excitation force rapidly decreases with the distance to the wire. Particles do not

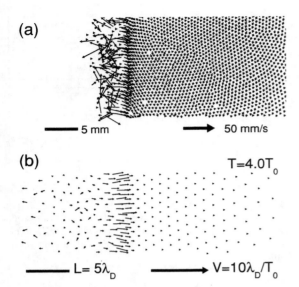

(a)

━━━ 5 mm   ➡ 50 mm/s

(b)   T=4.0T$_0$

━━━ L= 5λ_D ➡ V=10λ_D/T$_0$

FIGURE 4. Velocity vector map of a shock with a stable front, (a) experiment, (b) molecular dynamics simulation. The shock has a sharp linear front with a roughness scale smaller than the particle separation. All the particles in the front move at the same speed.

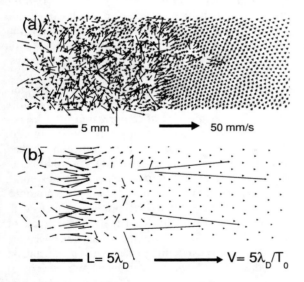

(a)

━━━ 5 mm   ➡ 50 mm/s

(b)

━━━ L= 5λ_D ➡ V= 5λ_D/T$_0$

FIGURE 5. Velocity vector map of a shock with an unstable front, (a) experiment, (b) molecular dynamics simulation. The shock has a rough front with a roughness scale much larger than the particle separation. The particles in the front move at different speeds, breaking the front.

move significantly during a short excitation pulse. Those far from the wire are weakly accelerated and only those few close to the excitation region get strongly accelerated and acquire large speeds. The fast particles penetrate the lattice, get scattered and break the lattice in an irregular pattern. Long excitation pulses allow the particles to move significantly during the excitation. They end up far from the wire where the force is weak. The overall effect is that more particles have nearly the same speed as they form a linear shock front.

SUMMARY

Shock waves with a Mach number of 2.4-2.7 were obtained in a two-dimensional complex plasma. They had a narrow compression peak at the front which coincided with a jump in kinetic temperature. The shocks decayed due to the neutral damping and formed supersonic compressional solitons. Shocks with a stable and an unstable fronts were observed.

REFERENCES

1. Whipple, E., *Rep. Prog. Phys.*, **44**, 1197–1250 (1981).
2. Bernstein, I., and Rabinowitz, I., *Phys. Fluids*, **2 (2)**, 112–121 (1959).
3. Konopka, U., Ratke, L., and Thomas, H. M., *Phys. Rev. Lett.*, **79**, 1269–1272 (1997).
4. Ikezi, H., *Phys. Fluids*, **29 (6)**, 1764–1766 (1986).
5. Thomas, H., Morfill, G., Demmel, V., Goree, J., Feuerbacher, B., and Möhlmann, D., *Phys. Rev. Lett.*, **73 (5)**, 652–655 (1994).
6. Hayashi, Y., and Tachibana, K., *Jpn. J. Appl. Phys.*, **33**, L804–L806 (1994).
7. Chu, J., and Lin, I., *Physica A*, **205 (1-3)**, 183–190 (1994).
8. Thomas, H., and Morfill, G., *Nature*, **379 (6568)**, 806–809 (1996).
9. Thomas, H., and Morfill, G., *J. Vac. Sci. Technol. A*, **14 (2)**, 501–505 (1996).
10. Nakamura, Y., Bailung, H., and Shukla, P. K., *Phys. Rev. Lett*, **83**, 1602–1605 (1999).
11. Samsonov, D., Morfill, G., Thomas, H., Hagl, T., Rothermel, H., Fortov, V., Lipaev, A., Molotkov, V., Nefedov, A., Petrov, O., Ivanov, A., and Krikalev, S., *Phys. Rev. E*, **67**, 036404 (2003).
12. Samsonov, D., Ivlev, A., Quinn, R., Morfill, G., and Zhdanov, S., *Phys. Rev. Lett.*, **88**, 095004 (2002).

CP706, *Shock Compression of Condensed Matter - 2003*
edited by M. D. Furnish, Y. M. Gupta, and J. W. Forbes
© 2004 American Institute of Physics 0-7354-0181-0/04/$22.00

SHOCK COMPRESSION OF DEUTERIUM AT MBAR PRESSURES AND THE INTERIOR OF JUPITER

D. Saumon[*] and T. Guillot[†]

[*]*Los Alamos National Laboratory, MS F699, Los Alamos, NM 87545*
[†]*Observatoire de la Côte d'Azur, BP 4229, 06304 Nice CEDEX 04, FRANCE*

Abstract. It is of great interest to planetary science to understand how the current experimental uncertainty on the hydrogen EOS affects the inferred structure of Jupiter. In particular, the mass of a core of heavy elements (other than H and He) and the total amount and distribution of heavy elements are very sensitive to the EOS of hydrogen and constitute important clues to its formation process. We present a study of the range of structures allowed by the current uncertainty in the hydrogen EOS. We show that an improved experimental understanding of hydrogen at Mbar pressures is necessary to put firm limits on the internal structure of Jupiter.

INTRODUCTION

Jupiter is a fluid planet in hydrostatic equilibrium that is composed of about 70% hydrogen by mass. At about 80% of the radius of the planet, corresponding to $P \sim 1 - 3\,\mathrm{Mbar}$, hydrogen goes from an insulating molecular fluid to an atomic metallic fluid. This transition has been the subject of much theoretical and experimental work but remains poorly understood. Recent shock compression experiments of deuterium in the Mbar range show a significant disagreement [1, 2, 3].

The internal structure of Jupiter is not directly accessible to observations and is inferred indirectly. Its rapid rotation results in a noticeable deformation and a non-spherical gravitational field that can be expressed as an expansion in spherical harmonics. The first three non-zero coefficients in this expansion (the "gravitational moments") have been measured during spacecraft flybys. Combined with the total mass, radius, and the rotation period of the planet, these provide integral constraints on the density profile of the planet. These constraints cannot be inverted to obtain the density profile, however. Instead, a simple model is assumed for the structure and composition of the rotating planet which is then subjected

to the rotational perturbation. Model parameters are adjusted to fit the observed constraints. The structure thus inferred is sensitive to the EOS adopted for hydrogen in the models.

Of great astrophysical interest are the mass of the core of heavy elements (all but H and He) in Jupiter and the amount of heavy elements distributed in the outer envelope. The total amount of heavy elements and their distribution inside Jupiter bears directly on its formation process by accretion of both gaseous (H and He) and solid material from the protoplanetary nebula.

For the first time, there are experimental data that allow us to constrain the EOS of hydrogen under the conditions found inside Jupiter. While the shock compression experiments do not agree, it is reasonable to assume that they bracket the actual Hugoniot of hydrogen in the Mbar range. Thus, we can determine the range of interior models of Jupiter allowed by the current uncertainties in the EOS of hydrogen.

EOS'S FOR HYDROGEN

Jupiter's interior follows an adiabat in the EOS for a H/He mixture with a specific entropy determined

TABLE 1. Equations of state compared to various data sets.

EOS	Sandia P–V [2, 3]	NOVA P–V [1, 4]	NOVA P–T [5]	Gas Gun 2nd Shock T [6]
LM (A)	no	yes	yes	yes
LM-SOCP	yes	no	$+1\sigma$	+30%
LM-H4	no	yes	yes	+20%
SESAME	yes	no	$+1\sigma$	+50%
SCVH-I	yes/no	yes/no	yes	+50%

from observations of the surface. On the other hand, shock-compression experiments follow Hugoniots that are typically much hotter than the Jupiter adiabat for the pressures of interest. For example, at 1 Mbar the temperature inside Jupiter is ~ 6000 K, while the principal Hugoniot reaches ~ 20000 K. The gas-gun reshock experiments overlap Jupiter's adiabat up to 0.8 Mbar, however (see Fig. 1).

Since the adiabat and the Hugoniot overlap minimally, two EOS's that predict nearly identical Hugoniots may produce different adiabats. For the purposes of this study, we computed interior models of Jupiter using 5 different hydrogen EOS's. These have been chosen as representative EOS's that reproduce selected subsets of data and realistically bracket the actual EOS of hydrogen. For the sake of simplicity, we consider only the following experiments:

- The (P, V) principal Hugoniot measured with the Z-machine at Sandia National Laboratory [2] and the Hugoniot point achieved by spherical shock wave compression [3]. These data indicate a maximum compression of ~ 4 for deuterium along the Hugoniot.
- The (P, V) principal Hugoniot measured at the NOVA facility at Lawrence Livermore National Laboratory [1, 4]. These data indicate a maximum compression of ~ 6 for deuterium along the Hugoniot.
- The (P, T) data along the principal Hugoniot measured at the NOVA facility [5].
- The single and double-shock gas-gun T measurements [6]. The double-shock temperatures may be systematically underestimated due to unquantified thermal conduction into the window upon shock reflection. We consider this data set as a lower limit on the reshock temperatures.

Description of the EOS's

Ross [7] developed a simple linear mixing (hereafter, LM) model to reproduce the unexpectedly low gas-gun reshock temperature measurements [6]. It was found that this model also agrees well with the subsequent NOVA Hugoniot [1, 4]. The linear mixing model is based on a linear interpolation in composition between a molecular fluid and a metallic fluid. An entropy term is introduced in the metallic EOS and adjusted to fit the data. The LM model is particularly useful for the present study as it can be easily modified to fit various data sets. We found that while it reproduces Hugoniot data well, the original LM model predicts an anomalous adiabat for Jupiter. We use 3 EOS's based on the LM model:

- **LM-A:** This EOS is identical to the original mixing model [7] except that the anomalous behavior of the adiabat has been corrected by replacing the entropy term (Eq. (7) of [7]: $\delta_e = -2.7$) by a temperature- and density-dependent term that is adjusted to fit the Hugoniot data.
- **LM-SOCP:** In this model, the metallic fluid EOS used by Ross (basically the OCP) is replaced by a screened OCP [8]. In addition, the entropy term is adjusted to reproduce the Sandia Hugoniot with $\delta_e = 0.5$. It is also a good representation of DFT-GGA simulations of deuterium [9] for $T \gtrsim 1$ eV and $\rho \gtrsim 0.6$ g/cm^3.
- **LM-H4:** In this EOS model[10], the need for a fitting parameter in the original LM model is removed by introducing D_4 chains as a new species.

The **SESAME** deuterium EOS [11] provides a fair representation of the Sandia Hugoniot and we adopt it here. For modeling Jupiter, we adopt the SESAME EOS 5251 for hydrogen, which is a deuterium EOS scaled in density.

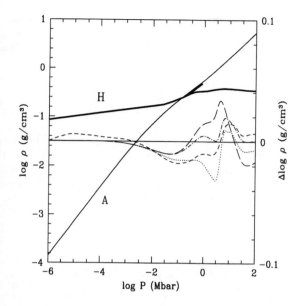

FIGURE 1. Jupiter adiabats for hydrogen. The curves labeled 'A' and 'H' show the SCVH-I adiabat and the first and second-shock Hugoniots (SESAME), respectively. The other curves and the scale on the right show *differences* in density between Jupiter adiabats computed with various EOS, relative to the SCVH-I: SESAME (short dashes), LM (A) (long dashes), LM-SOCP (dotted), and LM-H4 (dot-dash). The central pressure of Jupiter is about 70 Mbar.

Finally, we computed models with the **SCVH-I** EOS [12]. Like the SESAME EOS, it predates all 4 data sets. It agrees better with the Sandia (P, V) Hugoniot for $P \lesssim 0.7$ Mbar and then shifts toward the NOVA Hugoniot at higher pressures. This EOS has been used extensively in modeling the interiors of Jupiter [13, 14, 15, 16, 17] and serves as a basis for comparison.

A summary of how each EOS compares with the various experiments is given in Table 1. "Yes" indicates a good fit within the error bars of the measurements, "+1σ" in the fourth column indicates that the model Hugoniot overestimates the measurements by the reported experimental error bars. In the last column, the difference between the calculated temperatures and the data is given as a percentage (several times larger than the internal uncertainties). Jupiter adiabats for each of these EOS's are shown in Fig. 1.

The interior of Jupiter is represented by a 3-layer model with a core, an inner envelope and an outer envelope [14]. The core is assumed to be predominantly composed of heavy elements, while the envelope is dominated by hydrogen (about 70% by mass) and helium, with a small admixture of heavy elements. Because Jupiter formed in a region where heavy elements had condensed, the latter are divided in two groups according to their condensation temperatures. The most refractory elements form mostly silicates and solid iron ("rocks") at relatively high temperatures. The other condensates, the "ices," form at much lower temperatures and are mainly H_2O, CH_4 and NH_3. The relative ratio of rocks to ices in Jupiter is not known.

The free parameters of this model are the mass of the core and the mass of heavy elements homogeneously mixed in the H/He envelope. We also account for the following sources of uncertainty, by decreasing order of significance: (i) the unknown interior rotation (as a solid body or on cylinders tied to the surface rotation) (ii) the unknown rock/ice ratio; (iii) the unknown composition of the central core (from pure rocks to pure ices). We further include a $\pm 2\%$ uncertainty on the density profile along the adiabat for $P \geq 1$ Mbar, independently of the EOS chosen. This *ad hoc* variation is useful to estimate the effects of other uncertainties inherent to each EOS and from the additive volume rule used to obtain the EOS of H/He/rock/ice mixtures. The full parameter space is explored and models that do not fit the observed constraints within the 2σ error bars are rejected [16].

The results are summarized in Fig. 2. For each EOS considered, a range of core masses and masses of heavy elements in the envelope is obtained after varying the other parameters. The domain of acceptable models for a given hydrogen EOS is enclosed in a box.

All EOS's lead to models that are significantly enriched in heavy elements ($M_{\text{core}} + M_Z = 12$ to $36 M_\oplus$) compared to solar abundances (which would give $\sim 6 M_\oplus$). The total range of solutions for all EOS's considered is much larger than the allowed range for any given EOS. This means that the current uncertainties on the hydrogen EOS remains the largest source of uncertainty in interior models of Jupiter. The minimal overlap between the boxes for the EOS considered here implies that it is not yet possible

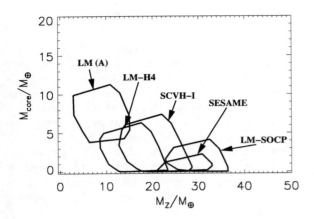

FIGURE 2. Jupiter's core mass M_{core} and the mass of heavy elements mixed in the H/He envelope M_Z (in Earth masses, $M_\oplus$). The total mass of Jupiter is $318 M_\oplus$. Each box represents the range of models that satisfy the observational constraints for a given choice of hydrogen EOS.

to determine with any confidence the core mass of Jupiter and its total content in heavy elements. All EOS's but LM (A), which is fitted to the NOVA Hugoniot and the gas-gun reshock temperatures, allow solutions with no core at all. The EOS's that reproduce the Sandia Hugoniot (SESAME, LM-SOCP, and to some extent, SCVH-I) predict small cores and a relatively large amount of heavy elements in the envelope. Most formation models of Jupiter require an initial core mass of 10 – 20 Earth masses [18]. The small core masses we find would imply significant mixing of the core and the envelope *after* the bulk of the planet had formed [19]. Alternatively, Jupiter may have formed without a core by a direct instability of the gas [20], but it is then difficult to explain its large enrichment in heavy elements.

Proposed space missions to Jupiter would lead to a tenfold reduction of the uncertainty on its gravitational moments, greatly reducing the range of acceptable models (*i.e.* smaller boxes in Fig. 2). Such improved measurements would most likely lead to rather distinct solutions for different choices of hydrogen EOS. Furthermore, the ability of a given EOS to give acceptable models depends somewhat on the assumptions for the model structure. In general, more complex models (with more parameters) can accommodate a wider range of EOS's. Our understanding

of the interior of Jupiter and of its formation process therefore relies rather critically on a better determination of the EOS of hydrogen in the 0.5 to 10 Mbar range at temperatures below 10^4 K.

ACKNOWLEDGMENTS

This work was supported in part by NASA Planetary Geology & Geophysics grant NAG5-8906, by the United States Department of Energy under contract W-7405-ENG-36, and by the *Programme National de Planétologie* (France).

REFERENCES

1. Collins, G. W. et al., *Science* **281**, 1178–1181 (1998).
2. Knudson, M. D., Hanson, D. L., Bailey, J. E., Hall, C. A., and Asay, J. R., *Phys. Rev. Lett.* **87**, 225501-1 – 225501-4 (2001); *ibid* **90**, 035505-1 – 035505-4 (2003).
3. Belov, S. I., et al., *JETP Lett.* **76**, 433-435 (2002).
4. Da Silva, L. B. et al., *Phys. Rev. Lett.* **78**, 483-486 (1997).
5. Collins, G. W. et al., *Phys. Rev. Lett.* **87**, 165504-1 – 165504-04 (2001).
6. Holmes, N. C., Ross, M. and Nellis, W. J., *Phys. Rev. B* **52**, 15835 –15845 (1995).
7. Ross, M. *Phys. Rev. B* **58**, 669 – 677 (1998); *ibid*, **60**, 6923 (1999).
8. Chabrier, G., *J. de Phys.*, **51**, 1607 – 1632 (1990).
9. Lenosky, T. J., Bickman, S. R., Kress, J. D., and Collins, L. A., *Phys. Rev B* **61**, 1 – 4 (2000)
10. Ross, M. and Yang, L. H., *Phys. Rev. B* **64**, 134210-1 – 134210-8 (2001)
11. Kerley, G. I., *Los Alamos Scientific Laboratory, Technical Report LA-4776*, 1 – 28 (1972).
12. Saumon, D. and Chabrier, G. and Van Horn, H. M., *Astrophys. J. Suppl.* **99**, 713 – 741 (1995).
13. Chabrier, G., Saumon, D., Hubbard, W. B. and Lunine, J. I., *Astrophys. J.* **391**, 817 – 826 (1992).
14. Guillot, T., Chabrier, G., Morel, P., and Gautier, D., *Icarus* **112**, 354 – 367 (1994).
15. Guillot, T., Gautier, D., and Hubbard, W. B., *Icarus* **130**, 534 – 539 (1997).
16. Guillot, T., *Plan. & Space Sci.* **47**, 1183-1200 (1999).
17. Gudkova, T. V. and Zharkov, V. N., *Plan. & Space Sci.* **47**, 1201 – 1210 (1999).
18. Pollack, J. B., Hubickyj, O., Bodenheimer, P., Lissauer, J. J., Podolak, M. and Greenzweig, Y., *Icarus* **124**, 62 – 85 (1996).
19. Guillot, T., Stevenson, D. J., Hubbard, W. B., and Saumon, D., in *Jupiter*, edited by F. Bagenal, U. of Arizona Press, Tucson, in press (2003).
20. Boss, A. P., *Astrophys. J. Lett.*, **536**, L101–104 (2000).

CP706, *Shock Compression of Condensed Matter - 2003*
edited by M. D. Furnish, Y. M. Gupta, and J. W. Forbes
© 2004 American Institute of Physics 0-7354-0181-0/04/$22.00

EQUATION OF STATE MEASUREMENTS FOR BERYLLIUM IN THE ICF CAPSULE REGIME

Damian Swift,[1] Dennis Paisley,[1] and Marcus Knudson[2]

[1]*P-24 Plasma Physics, Los Alamos National Laboratory, Los Alamos, NM 87545*
[2]*Sandia National Laboratories, Albuquerque, NM 87185*

Abstract. The dynamic response of beryllium on nanosecond time scales is important for controlling symmetry during the implosion of the fuel capsule in inertial confinement fusion. Particularly important is the behavior up to about 200 GPa, covering the foot of the implosion drive. We have performed experiments to measure the equation of state (EOS) and flow stress of beryllium, and to investigate solid-solid phase transitions and melting, using flyer impact and isentropic compression by pulsed electromagnetic fields at Z, and shocks induced by direct laser irradiation at TRIDENT. The principal diagnostic was VISAR velocimetry; transient x-ray diffraction was also used on some TRIDENT experiments. The Hugoniot and isentrope data were consistent with previously-reported EOS. The flow stress was inferred from elastic precursor waves to be about 6 GPa in the (0001) direction on these time scales, with significant sensitivity to orientation. Possible evidence was observed of the hex-bcc transition and of melting.

INTRODUCTION

Be-based capsules will be used to contain the fuel for studies of inertially-confined fusion (ICF). Capsule design requires accurate knowledge of the dynamic properties of Be, to ensure an adequate implosion symmetry. Relevant properties include the equation of state (EOS), orientation-dependent strength, melt, and solid-solid phase transformations, on nanosecond time scales. Laser-based drives can readily generate a diverse range of loading conditions, so it is important to determine the EOS over a conservatively wide range of states.

We have performed shock and isentropic loading experiments using the Z pulsed power facility at Sandia, and shock experiments at Los Alamos' TRIDENT laser.(1) Here we present EOS data on the principal Hugoniot and STP isentrope, and dynamic strength data obtained on nanosecond time scales.

FLYER IMPACT

On Z shot 825, Cu flyer plates were used to induce shock waves in Be and Al samples, using the discharge of the capacitor banks through coaxial conductors in the 'square short' configuration(3) to accelerate the flyers by magnetic pressure and ablation. Eight flyers were accelerated simultaneously on a single experiment, the thickness of each flyer being used to control its speed. Flyers were 9 mm in diameter and 200, 300, and 400 μm thick, giving impact speeds 8.4, 7.3, and 6.2±0.1 km/s respectively. The target assembly consisted of a Cu plate 225 μm thick (omitted in one experiment), the sample 525 μm thick, and a LiF window. The target assembly covered roughly half of the flyer. Laser Doppler velocimetry (point VISAR) was used to record the acceleration of the flyer and the velocity history at the interface between the sample and the LiF. An additional LiF window was mounted flush with the impact surface of the Cu plate as a measurement of flyer arrival time. (Fig. 1).

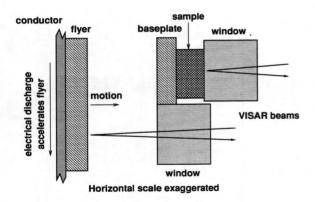

FIGURE 1. Schematic of flyer impact experiments.

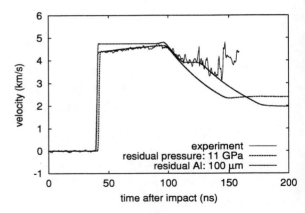

FIGURE 2. Comparison of experimental and simulated interface velocity history, including different phenomena. (Z shot 825, Be experiment 4: flyer at 7.2 ± 0.1 km/s.)

On this shot, four experiments were performed on Be samples (Alfa Aesar foil, 99.5% pure) and an additional four reference experiments on Al samples (Alfa Aesar foil, 99.998% pure). The thickest Cu foils apparently spalled and recollected during acceleration; this did not seem to affect the quality of the data. All flyers were still accelerating slightly on impact, at a rate equivalent to a residual pressure ~10 GPa. The effect of the residual pressure could be modeled using continuum mechanics simulations, but it contributed to uncertainty in Hugoniot states deduced from the data. The continuum mechanics simulations used published EOS for all components,(4,5) and strength was neglected.

Generally, quite good agreement was obtained between experimental and continuum mechanical interface velocity histories, using the observed flyer speed as input. The records diverged after the passage of the shock, presumably because of damage accumulating in the LiF. There was evidence of incremental release structures, consistent with ~100 μm of Al from the conductor remaining attached to the back of the flyer. Eliminating likely effects such as these, several features remained in the velocity histories which might indicate melt on shocking or on release; these will be discussed elsewhere. (Fig. 2.)

Given the flyer speed on impact, Hugoniot states were estimated from the shock speed in the sample (deduced from the transit time) or

from the peak interface velocity (deduced using pressure-particle speed analysis). There was generally greater uncertainty in the impact time, partly because of reflections between the flyer and the impact window. The Hugoniot states were consistent with previous data and models (Fig. 3).

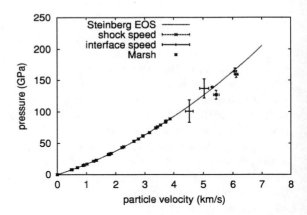

FIGURE 3. Principal shock Hugoniot for Be: experimental data and EOS models.

ISENTROPIC COMPRESSION

On Z shot 843, quasi-isentropic compression waves were induced in samples of Be of different thickness attached to a 'rectangular short' similar to the square short used to accelerate flyers and for previous isentropic compression studies,[3] but with a smaller cross-section hence inducing a higher magnetic field. LiF windows were mounted on each sample and on a representative region of the surface of the conductor; point VISAR records were obtained of the velocity history at each interface and hence of the evolution of the compression waves. Two records each were obtained of the conductor, 250 μm Be, and 500 μm Be. (Fig. 4.)

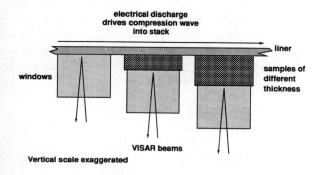

FIGURE 4. Schematic of magnetically-driven quasi-isentropic compression experiments.

If the EOS of the sample material is unknown, the evolution of the compression wave can be used to deduce the locus of the isentrope through the initial state,[6] which requires corrections to be made for the effect of the surface (window or free) in contact with the sample. As reasonable EOS are available for Be, we instead used a sequence of continuum mechanical simulations forward in time, adjusting the drive conditions (a pressure history applied to the inside of the Al conductor) until the velocity history at the surface of the conductor was reproduced, and then comparing the velocity history at the surface of each sample. The pressure in the sample was estimated to reach 195 GPa.

Once the drive conditions had been reproduced, the velocity history predicted at the surface of the 250 μm sample matched the experimental record to within its scatter. This indicates that the published EOS are accurate along the STP isentrope to ∼200 GPa, despite being fitted primarily to the empirical shock Hugoniot.

In the samples 500 μm thick, the velocity history indicated a region of shock rather than isentropic compression. This was not reproduced by the simulations. As the range of states explored in each sample was essentially the same, the main difference being difference of the order of 2 in strain rate, it seems likely the shock feature was caused by an uncontrolled experimental difference such as a gap in the assembly of the target stack. (Fig. 5.)

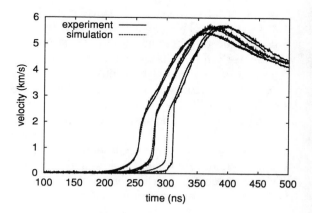

FIGURE 5. Comparison of experimental and simulated interface velocity history, for different thicknesses of Be (left to right: zero, 250, 500 μm) under quasi-isentropic loading.

LASER-INDUCED SHOCK

Further experiments were performed similar to those reported previously[1] in which the TRIDENT laser was used to induce shock waves in samples of Be between 12 and 125 μm thick by direct irradiation. The drive pulse was 1.0 to 2.5 ns long of 527 nm light, and spatial variations in the beam were smoothed using a Fresnel zone plate. The velocity history at the opposite side of the sample was recorded using a line-imaging VISAR, and transient x-ray diffraction records were obtained on some experiments which confirmed compressions inferred from the VISAR records.[2] The drive energy was ∼30 to 200 J, giving a shock pressure of up to 50 GPa. In addition to the four experiments on (0001) crystals, 26 experiments were performed on rolled foils.

A common feature of all these experiments

was an elastic precursor of amplitude 700 m/s for crystals and 400 m/s (with some sample to sample variation) for foils. The precursor for crystals exhibited structure: an initial peak, followed by a decrease to ~500 m/s. The precursor for foils was somewhat smeared out, partly because the reflectivity of the foils was far lower so the temporal resolution obtained from the VISAR analysis scheme was poorer. The precursor amplitudes are equivalent to a flow stress of 6.0±0.1 GPa (crystal) and 2.2±0.2 GPa (foil). Published flow stresses for polycrystalline Be range from 0.33 to 1.31 GPa (Y_0 to Y_{max}).(4) Given that the rolled foils had been subjected to a large amount of plastic work in manufacture and the texture was found to be dominated by (0001) planes within ~ 10° of the surface of the foil, 2.2 GPa is a reasonable value for flow stress: the difference between this and Y_{max} is likely to be a combination of texture and rate dependence. The precursor structure for the crystals suggests that an activation process occurs in plastic flow, such as generation of dislocations or the onset of a quantized flow process such as twinning. (Fig. 6.)

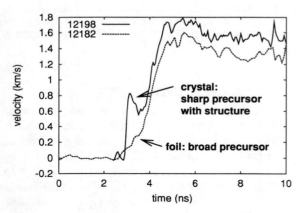

FIGURE 6. Comparison of free surface velocity histories for (0001) crystal (shot 12198) and rolled foil (shot 12182) of Be, subjected to shocks induced by laser irradiation. Shot 12198: ~3.3 PW/m² for 1.8 ns, giving ~17.5 GPa; shot 12182: ~2.8 PW/m² for 1.8 ns, giving ~13.1 GPa.

CONCLUSIONS

EOS data were obtained for Be up to about 200 GPa along the principal Hugoniot and the STP isentrope, from experiments at Z using samples 250 to 500 μm thick. Previously published EOS were found to be consistent with both types of measurement. Features were observed in velocity histories from flyer impact experiments which were suggestive of melting on shock or release.

Velocity histories were obtained from Be samples shocked to about 50 GPa by direct laser irradiation at TRIDENT using samples 12 to 125 μm thick. Hugoniot states were estimated from this data; these were also consistent with the published EOS. The velocity histories exhibited strong elastic precursors, equivalent to a flow stress of 2.2 GPa. In single crystals, the precursors had structure suggesting flow associated with a barrier of 6.0 GPa.

ACKNOWLEDGMENTS

We would like to thank the staff of Z and TRIDENT, the Sandia target fabrication team, Scott Evans for helping to prepare the Be samples, and John Bingert for making x-ray texture maps. This work was performed under the auspices of the US Department of Energy, contract W-7405-ENG-36.

REFERENCES

1. Swift, D.C., Paisley, D.L., Kyrala, G.A., and Hauer, A., "Simultaneous VISAR and TXD measurements on shocks in beryllium crystals," in *Shock Compression of Condensed Matter-2001*, edited by M.D. Furnish et al, AIP Conference Proceedings 620, Melville, New York, 2002, pp 1192-5.

2. D.C. Swift, *Analysis of TXD and VISAR experiments on shock waves in beryllium crystals*, Los Alamos National Laboratory report LA-UR-01-3410 (2001).

3. Reisman, D.B., Toor, A., Cauble, R.C., Hall, C.A., Asay, J.R., Knudson, M.D., and Furnish, M.D., J. Appl. Phys. **89**, 3, pp 1625-1633 (2001).

4. Steinberg, D.J., *Equation of state and strength properties of selected materials*, Lawrence Livermore National Laboratory report UCRL-MA-106439 change 1 (1996).

5. K S Holian (Ed.) "T-4 Handbook of Material Property Data Bases, Vol 1c: Equations of State," Los Alamos National Laboratory report LA-10160-MS (1984). SESAME EOS #2023 was used.

6. Aidun, J.B., and Gupta, Y.M., J. Appl. Phys. **69**, 6998 (1991).

CP706, *Shock Compression of Condensed Matter - 2003*
edited by M. D. Furnish, Y. M. Gupta, and J. W. Forbes
2004 American Institute of Physics 0-7354-0181-0/04/$22.00

AB-INITIO 0 K ISOTHERM FOR ORGANIC MOLECULAR CRYSTALS

Frank J. Zerilli[1] and Maija M. Kuklja[2]

[1]Research and Technology Department, Naval Surface Warfare Center, Indian Head, MD 20640
[2]Division of Materials Research, National Science Foundation, Arlington, VA 22230

Abstract. The 0 K isotherm for the organic molecular crystal 1,1-diamino-2,2-dinitroethylene is calculated reasonably accurately using the Hartree-Fock approximation to the solutions of the many-body Schrödinger equation for a periodic system as implemented in the computer program CRYSTAL. The equilibrium lattice parameters are predicted to within one percent of the values reported by Bemm and Östmark. Pressure values on the isotherm agree extremely well with the values measured by Peiris and co-workers. The key to obtaining such accuracy is the relaxation of all the molecular coordinates as well as the lattice parameters under a fixed volume constraint. It was found that Density Functional Theory (DFT) calculations gave much poorer results, but optimizations were not as extensively investigated in this case.

INTRODUCTION

In recent years there has been a large body of research devoted to the determination of optical, mechanical, and reactive properties of materials from numerical solutions of the many body Schrödinger equation. In particular, the equation of state of many solids has been determined in this manner: metals such as iron, copper, aluminum; elementary solids such as silicon; metal oxides; ionic solids such as NaCl. Not much work has been done on the equation of state of more complex materials such as organic molecular crystals. However, state of the art computer power has progressed to the point where it becomes practical to contemplate studying these more complex materials.

In this article, we describe a reasonably accurate calculation for the 0 K isotherm for 1,1-diamino-2,2-dinitroethylene ($C_2H_4N_4O_4$). This material is representative of a large class of organic molecular crystals referred to as energetic materials because of their propensity for rapid explosive decomposition under shock conditions. Bemm and Östmark[1]

determined the crystal structure of this material, which they named FOX-7, to be monoclinic with space group $P2_1/n$ and four molecules per unit cell. They reported the values of the lattice parameters a=6.9410 Å, b=6.5690 Å, c=11.315 Å, β=90.55°. Gilardi[2] has also studied the structure of FOX-7 and found the values a=6.9396 Å, b=6.6374 Å, c=11.3406 Å, and β=90.611°.

In previous work[3] we reported the results of calculations for a rigid molecule approximation as well as for several degrees of molecular structure optimization including optimization of the fractional atomic coordinates.

It turns out that the Hartree-Fock approximation gives excellent results for this type of calculation while density functional methods that have the advantage of including electron correlation energy were simply unable to come close, predicting unit cell volumes from 10 to 20 percent smaller than experimentally determined values. The Hartree-Fock calculations predict the equilibrium lattice parameters within about 1% and the corresponding unit cell volume within 3%.

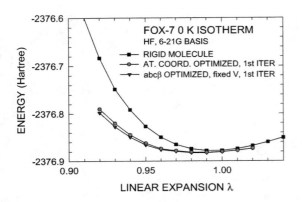

FIGURE 1. Energy along the 0 K isotherm for FOX-7.

In order to obtain this degree of accuracy, it is necessary not only to optimize the atomic coordinates of the molecules within the unit cell, but also to optimize the lattice parameters under a fixed volume constraint. The basic quantum calculations for a periodic structure were performed with the CRYSTAL98[4] computer program, developed by the Theoretical Chemistry Group at the University of Torino. Optimizations of the atomic coordinates and lattice parameters were done separately and iteratively. The atomic coordinates were first optimized using Zicovich-Wilson's LoptCG script[5], which calls CRYSTAL98 to calculate the energy for each configuration. The lattice parameters then were optimized under a fixed volume constraint using a locally written program based on the downhill simplex method of Nelder and Mead[6,7]. This was repeated until satisfactory convergence was achieved.

LoptCG is a flexible script, allowing the choice of the method of steepest descents or the Polak-Ribiere conjugate gradient method or some combination of these methods to find the minimum energy[6,8]. LoptCG calculates derivatives numerically and will automatically adjust the self-consistent field (scf) convergence criterion to the precision required to calculate the first derivative. It also takes advantage of the ability of CRYSTAL to start an scf calculation with the initial guess taken from density matrices calculated on a previous run. In the calculations reported here, the conjugate gradient method was used exclusively, with a convergence criterion of 10^{-3} in the gradient norm.

The simplex optimization routine, SOPT, which was used for the lattice parameter optimization under a fixed volume constraint, also calls CRYSTAL to determine the energy of a configuration and has the advantage of not requiring the calculation of derivatives. SOPT also starts successive energy computations from the density matrix of the previous computation, but does not alter CRYSTAL's scf convergence tolerances. The convergence criterion used for lattice parameter optimization is based on the square root of the sum of the squares of the lattice parameter change during one cycle. This was set to 1×10^{-5} Å.

CRYSTAL uses a set of basis functions composed of linear combinations of gaussian type atomic orbitals. Kunz[9] determined that a 6-21G split valence basis set affords a good compromise between accuracy of the computations and required computer resources in the case of the energetic solids RDX and TATB. Our own tests confirm this, and we have chosen to utilize this basis set in the computations reported here. As these basis functions are optimized for isolated, non-periodic systems, scaling factors are introduced which reduce the range of the outer orbitals to better adapt them for use in periodic systems. The scaling factors (1.05 for C, N, H; 1.00 for O) for Hartree-Fock computations were determined by minimizing the energy in test calculations and are similar to values obtained previously[9,10].

FOX-7 0 K ISOTHERM

The energy along the 0 K isotherm for FOX-7 is shown in Fig. 1 as a function of linear expansion. In the figure, a linear expansion λ of 1 corresponds to Gilardi's experimental values for the lattice parameters and an associated unit cell volume V_0 of 522 Å^3. For comparison, the energy calculated under uniform compression of the lattice with the assumption of rigid molecules* is shown. A considerable improvement in the energy occurs when the molecules are no longer treated as rigid and the fractional atomic coordinates are relaxed. In this relaxation, only the coordinates of the 14 irreducible atoms in the unit cell are considered,

---

*The rigid molecule structure is that for which the 13 bond lengths have been optimized for lattice parameters equal to 99% of Gilardi's parameters.

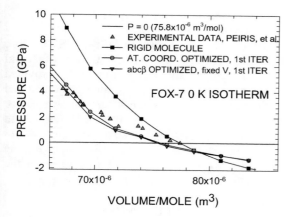

FIGURE 2. Pressure along the 0 K isotherm for FOX-7.

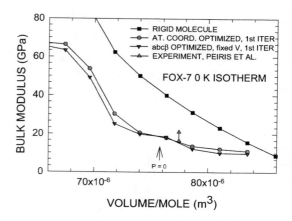

FIGURE 3. Bulk modulus along the 0 K isotherm for FOX-7.

preserving the symmetry of the space group. A further, small, but significant, improvement in the energy occurs when the lattice parameters, a, b, c, and β are relaxed under a fixed volume constraint, where the volume $V = \lambda^3 V_0$. Relaxation under the fixed volume constraint corresponds to true hydrostatic compression.

The experimental observations[1,2] are done at room temperature (300 K). The thermal expansion coefficient for FOX-7 has not yet been experimentally determined, but it was calculated by Sorescu, et al[11]. using molecular dynamics. They determined an average linear expansion coefficient of 6.8×10^{-5} K^{-1} at 273 K. A similar material, TATB ($C_6H_6N_6O_6$) has a measured linear thermal expansion coefficient[12] of the order of 5×10^{-5} K^{-1} at 300 K. The thermal expansion decreases to zero at 0 K. Thus, the correction for the temperature difference should amount to about 1%. This would shift the calculated equilibrium point up to 100.5% of Gilardi's value.

The pressure-volume dependence is calculated from the usual thermodynamic relation

$$P = -\left(\frac{\partial F(V,T)}{\partial V}\right)_T \qquad (1)$$

where F is the Helmholtz Free energy. At 0 K, this reduces to $P = -dE_0/dV$ where E_0 is the energy along the 0 K isotherm. The pressure, calculated from the energy curves in Fig. 1, is shown in Fig. 2, along with experimental data reported by Peiris, et al.[13] It can be seen that the rigid molecule, uniform

compression calculations predict much too high a pressure, while the calculations in which the atomic coordinates and lattice parameters have been relaxed are in excellent agreement.

In order to make the comparison, it is necessary to take into account the fact that the experimental data corresponds to room temperature, about 300 K. In the future we plan to extend the work to determine the equation of state for non-zero temperatures. At present there is limited theoretical information and no experimental data for thermal properties of FOX-7 but some estimates are possible. The pressure difference may be determined from the thermodynamic identity

$$\left(\frac{\partial P}{\partial T}\right)_V = \rho c_V \gamma \qquad (2)$$

where ρ is the density, c_V is the specific heat at constant volume, and γ is the Grüneisen parameter. The Grüneisen parameter may be determined from the relation

$$\gamma = \frac{\alpha \kappa_T}{\rho c_V} \qquad (3)$$

where α is the volume coefficient of thermal expansion and κ_T is the isothermal bulk modulus. With a density of 1900 kg/m^3, assuming a specific heat of 1000 J/kg-K and a Grüneisen parameter of 1.0, the pressure at 300 K would be about 0.6 GPa greater than the pressure at 0 K. This is very consistent with the results shown in Fig. 2.

The calculated bulk modulus is shown in Fig. 3, along with the zero pressure, room temperature range of values of the bulk modulus determined experimentally by Peiris, et al[13]. The calculated value is a little smaller than the measured value. Further, the measured value at room temperature is presumably smaller than the 0 K value, but we do not know by how much at this point. If we assume that the room temperature value of the bulk modulus is not much smaller than the 0 K value, then the calculated value agrees well with the experimental value. However, this may not be a good assumption. There is some evidence that the 0 K value of the bulk modulus could be as much as 50% larger than the room temperature (300 K) value according to Monte Carlo calculations for RDX ($C_3H_6N_6O_6$) reported by Sewell and Bennett[14]

CONCLUSIONS

Hartree-Fock calculations give good results for the 0 K isotherm for the organic molecular crystal for 1,1-diamino-2,2-dinitroethylene when the entire structure is relaxed under a fixed volume constraint. We believe that this conclusion will apply to organic molecular crystals generally. Calculations recently performed for solid nitromethane (orthorhombic, $P2_12_12_1$) lend support to this conclusion. They yield the equilibrium lattice parameters a=5.0877 Å, b=6.2385 Å, c=8.4270 Å, with a unit cell volume of 267.5 Å^3. This is to be compared to the experimental values at 4.2 K, a=5.1832 Å, b=6.2357 Å, c=8.5181 Å, with a unit cell volume of 275.3 Å^3, reported by Trevino, Prince, and Hubbard[15]. The calculated unit cell volume is 3% lower than the measured value.

Up to this point we have been unable to achieve similarly good results with density functional methods.

The comparison with experimental data in the case of FOX-7 depends on estimates of differences between the 0 K isotherm and the 300 K isotherm. Work is now underway to calculate the full T > 0 equation of state by constructing the free energy from the phonon frequency spectrum of the crystal.

ACKNOWLEDGMENTS

This work was supported by the NSWC Core Research Program and by the Office of Naval Research. Computational resources were provided by the Aeronautical Systems Center Major Shared Resource Center at Wright-Patterson Air Force Base, Ohio, under the Department of Defense High Performance Computing Initiative. M. K. is grateful to the Division of Materials Research of the National Science Foundation for support under the Independent Research and Development Program.

REFERENCES

1. U. Bemm and H. Östmark, *Acta Crystall.* **C54**, 1997-1999 (1998).
2. R. Gilardi, Naval Research Laboratory (private communications, 2001).
3. M. M. Kuklja, F. J. Zerilli, and S. M. Peiris, *J. Chem. Phys.* **118**, 11073 (2003).
4. R. Dovesi, V. R. Saunders, C. Roetti, M. Causà, N. M. Harrison, R. Orlando, and C. M. Zicovich-Wilson, CRYSTAL 98 User's Manual, University of Torino, Torino, 1998.
5. C. M. Zicovich-Wilson, LoptCG: http://www.-chimifm.unito.it/teorica/crystal/crystal.html
6. J. A. Nelder and R. Mead, *Computer J.* **7**, 308 (1965).
7. W. H. Press, B. P. Flannery, S. A. Teukolsky, and W. T. Vetterling, *Numerical Recipes* (Cambridge University Press: New York ,1989).
8. P. G. Mezey, in *Potential Energy Hypersurfaces* (Elsevier, New York: 1987).
9. A. B. Kunz, *Phys. Rev.* **B53**, 9733 (1996).
10. M. M. Kuklja and A. B. Kunz, *J. Appl. Phys.* **87**, 2215 (2000).
11. D. C. Sorescu, J. A. Boatz, and D. L. Thompson, *J. Phys. Chem.A* **105**, 5010 (2001).
12. J. R. Kolb and H. F. Rizzo, *Propellants and Explosives* **4**, 10-16 (1979).
13. S. M. Peiris, C. P. Wong, and F. J. Zerilli, (unpublished).
14. T. D. Sewell and C. M. Bennett, *J. Appl. Phys.* **88**, 88 (2000).
15. S. F. Trevino, E. Prince, and C. R. Hubbard, *J. Chem. Phys.* **73**, 2996 (1980).

CP706, *Shock Compression of Condensed Matter - 2003*
edited by M. D. Furnish, Y. M. Gupta, and J. W. Forbes
© 2004 American Institute of Physics 0-7354-0181-0/04/$22.00

HUGONIOT-MEASUREMENT OF GGG (Gd$_3$Ga$_5$O$_{12}$) IN THE PRESSURE RANGE UP TO OVER 100 GPA

Y. Zhang[1], T. Mashimo[1], K. Fukuoka[2], M. Kikuchi[2], T. Sekine[3], T. Kobayashi[3], R. Chau[4], W. J. Nellis[4]

[1] *Shockwave and Condensed Matter Research Center, Kumamoto University, Kumamoto 860, Japan*
[2] *Institute for Materials Research, Tohoku University, Sendai 980, Japan*
[3] *National Institute for Materials Science, Tsukuba, Ibaraki 305, Japan*
[4] *University of California Lawrence Livermore National Laboratory 7000 East Ave., Livermore, U.S.A.*

Abstract. Gadolinium gallium garnet (Gd$_3$Ga$_5$O$_{12}$: GGG) is expected to have very high shock-impedance compared with sapphire (Al$_2$O$_3$), etc., and thus to be used as a potential new anvil material in shock reverberation experiments on hydrogen and other low-Z materials. In this study, the Hugoniot-measurement experiments were performed using both a powder gun and two-stage gas guns in the pressure range to 100 GPa by means of the inclined-mirror method. The HEL stress was measured to be larger than 30 GPa. A kink was observed on the Hugoniot in the pressure range higher than 60 GPa, which might be caused by structural phase transition or decomposition.

INTRODUCTION

Gadolinium gallium garnet (Gd$_3$Ga$_5$O$_{12}$: GGG) with garnet structure has been used for the substitute of diamond, and the substrate material of magnetic garnet film. GGG is expected to have very high shock-impedance, and to be used as a potential new anvil material in shock reverberation experiments on hydrogen and other low-Z materials, instead of Al$_2$O$_3$. The high-pressure behavior of garnet structure materials is important in earth science, etc. It was reported that the garnet phase transformed to an amorphous phase under static compression around 84 GPa (1). In this study, Hugoniot-measurement experiments are performed on GGG single crystals by the inclined-mirror method in the pressure range up to over 100 GPa to study the Hugoniot-elastic limit (HEL), phase transition (PT) and equation of state (EOS).

EXPERIMENTAL PROCEDURE

In this study, the plate-shaped specimens parallel to the (111), with thickness of about 2.4 mm and diameter of about 13 or 18 mm were used. And the density of GGG specimens was measured to be 7.099 g/cm^3 by the Archimedean method, which was good agreement with the theoretical one, 7.10 g/cm^3. Shock-wave experiments were conducted using a keyed-powder gun (27 mm in basic bore diameter) (2) and two-stage gas guns (20 or 25 mm in bore diameter) (3). The impact velocities were in the range of 1.3-4.5 km/s. The Hugoniot parameters were measured by the inclined-mirror method using the rotating-mirror type streak cameras (4) (5), an image converter camera, xenon flash lamp and long-pulsed dye laser (6). The setting of angle of the inclined mirror was 4-5 degrees in this study.

RESULTS

Figure 1 shows streak photograph obtained by the inclined-mirror method at an impact velocity of 2.501 km/s (tungsten impact plate, tungsten driver plate). At points 1 and 2, the elastic shock wave arrived at the rear surfaces of the driver plate and the specimen, respectively. The kink due to the elastoplastic transition is observed at point 3 on the inclined image. The HEL stress of GGG was larger than 30GPa, which is much higher than those of Al_2O_3 (15-21 GPa) (7), Y_2O_3-ZrO_2 (25 GPa) (8), B_4C (17 GPa) (9), SiC (15 GPa) (10) etc.

A further kink was clearly observed at point 4 in addition to the kink of the elastoplastic transition. The pressure at this kink was higher than 60GPa. It may correspond to structural phase transition or decomposition reaction. This may be related to the amorphization under static compression reported by H. Hua et al. (1). The P (pressure) - U_p (particle velocity) Hugoiot relation showed that the shock-impedance of this material is greater than that of Al_2O_3 in the pressure range up to 200 GPa.

ACKNOWLEDGEMENTS

A part of this work was performed under the inter-university research program of IMR, Tohoku University.

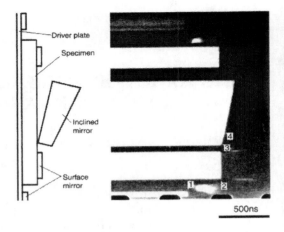

500ns

(time-expanded graph)

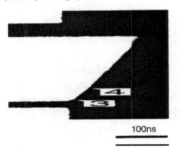

100ns

Figure 1. Streak photograph of GGG(111) by the inclined-mirror method (Impact velocity(W): 2.501 km/s).

REFERENCES

1. H. Hua, S. Mirov, and Y. K. Vohra, "Hihg-pressure and high-temperature studies on oxide garnets", Phys. Rev. B. 54, 6200, 1996.
2. T. Mashimo, S. Ozaki, and K. Nagayama, "Keyed-powder gun for the oblique-impact shock study of solids in several 10s of GPa region", Rev. Sci. Instr. 55, 226 , 1984.
3. Y. Syono, and T. Goto: Sci. Rep. Res. Inst. Tohoku Univ. "A Two-Stage Light Gas Gun for Shock Wave Research", A-29, pp.17-31, 1980.
4. M. Uchino, M. Kaetsu and T. Mashimo, "Shock-wave measurements of solids using the long-pulsed laser", SPIE-2869, 597 (1996).
5. T. Goto, Y. Syono, in: I. Sunagawa (ED.), "Technical Aspect of Shock Compression Experiments Using the Gun Method", Materials Science of the Earth's Interior, Terra, Tokyo, , 605 (1984).
6. T. Mashimo, A. Nakamura, Y. Hamada, "A compact high-speed streak camera for impact-shock study of solids", SPIE-1801, 170 (1992).
7. T. Mashimo, K. Tsumoto, K. Nakamura, "High-pressure phase transformation of corundum (α-Al_2O_3) observed under shock compression", Geophysical Res. Let. Vol. 27, No. 14, 2021 (2000).
8. T. Mashimo, A. Nakamura, M. Kodama, K. Kusaba, K. Fukuoka, Y. Syono, "Yielding and phase transition under shock compression of yttria-doped cubic zirconia single crystal and polycrystal", Advances In Ceramics, 24, 329 (1988).
9. T. Mashimo, M. Uchino, "Heterogeneous free-surface profile of B_4C polycrystal under shock compression", Journal of Applied Physics, 7064 (1997).
10. Unpublished data.

CP706, *Shock Compression of Condensed Matter - 2003*
edited by M. D. Furnish, Y. M. Gupta, and J. W. Forbes
© 2004 American Institute of Physics 0-7354-0181-0/04/$22.00

EXPERIMENTAL MEASUREMENT OF COMPRESSIBILITY AND TEMPERATURE IN SHOCK-COMPRESSED LIQUID XENON IN PRESSURE RANGE UP TO 350 GPa.

M.V.Zhernokletov[1], R.I.Il'kaev[1], S.I.Kirshanov[1], T.S.Lebedeva[1], A.L.Mikhaylov[1], M.A.Mochalov[1], A.N.Shuikin[1], V.E.Fortov[2]

[1] RFNC – VNIIEF, Sarov,
[2] IHED IVTAN RAS, Moscow

Abstract. Using generator of shock waves of the hemispherical geometry, densities of liquid xenon were measured at pressure of shock compression of 175 GPa in direct shock wave and at pressure of ~350 GPa in shock wave reflected from a sapphire window. High-velocity four-channel pyrometer was used to measure temperatures up to ~ 33000 K in pressure range up to 230 GPa at wavelengths of 406, 498, 550 and 600 nm. The photochronographic method was used to measure temperature of ~ 20000 K of shock-compressed liquid xenon in plane-wave devices in pressure range up to 95 GPa at wavelength of 430 nm. The obtained results were compared to available experimental and calculated data.

INTRODUCTION

In shock-wave experiments, xenon was studied in [1-4] up to pressures of 130 GPa, temperature was also measured in [3-4]; electric conductivity behind front of plane shock wave and quasi-isentropic compression up to density of 13 g/cm^3 were measured in [4]. In [5], the area of study of liquid xenon under quasi-isentropic compression was extended up to density of ~ 20 g/cm^3 at pressure of ~ 720 GPa. Experimental data from [1-5] and the structural transition revealed in [6-8] allowed to create a wide-range two-phase (Xe 1 and Xe 2) equation of state of xenon [4]. In this work, the area of study of properties of liquid xenon under shock compression is extended by use of a cryogenic construction manufactured basing on hemispherical device MZ-4 [9]. We measured density, temperature and pressure much higher than the parameters of shock-compressed liquid xenon in the earlier performed experiments.

EXPERIMENTAL DEVICE

Experiments with liquid xenon were performed in pressure range up to ~ 96 GPa with use of cryogenic construction of the plane type, which had been earlier applied for study of liquid argon in [10]. The obtained data on density, pressure, brightness temperature in the red area of spectrum (λ = 670 nm) and electric conductivity of shock-compressed liquid xenon are presented in [4-5].

For experiments at pressures higher than 100 GPa, a cryogenic device of hemispherical type with geometrical sizes similar to the MZ-4 design [9] was developed. The basic measuring unit of the experimental device is depicted in figure 1. A steel hemispherical impactor was accelerated by explosion products of powerful high explosive (HE). At impact of the impactor against a hemispherical baseplate (**1**), a shock wave was formed in it, and then the shock wave arrived to liquid xenon. The construction includes 6 couples of gauges (2-4) for measurement of average velocity of shock wave in liquid xenon (Xe1) and 6 couples of gauges (2-3) for measurement in baseplate (1).

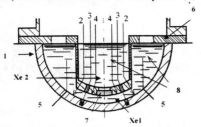

FIGURE 1. Hemispherical cryogenic device for study of properties of liquefied inert gases: 1 – baseplate; 2, 3, 4 – contacts; 5 – gauges for W measurement; 6 – insertion; 7 – sapphire window; 8 – xenon.

Along the normal to its surface, 5 gauges (5) for measurement of velocity of free surface W of steel impactor are fastened symmetrically on baseplate (1). Sapphire window (7) is glued into the central hole in insertion (6) in order to record velocity of shock wave and radiation of front in liquid xenon.

EXPERIMENTAL DATA

In this work, we performed an experiment at shock-wave loading, where we measured the kinematical parameters in shock-compressed liquid xenon. They are: D = 6.95 km/s, U = 4.65 km/s at pressure P = 95.7 GPa. Basing on data of this experiment and results from [2, 4, 5] in the area of particle velocities $1.5 \leq U \leq 5.5$ km/s, the generalized approximated dependence D(U) was obtained for liquid xenon:

$$D = 1.663 + 1.149 \, U \qquad (1)$$

Hereafter the measurement unit for D and U is km/s. With use of dependence (1), the equation of state of liquid xenon in the Mie-Gruneisen form was created. It was used in the gasdynamic calculation for modeling the shock-wave processes in the hemispherical experimental device. When choosing parameters of the equations of state of the other materials, the following D(U) dependencies were used: for aluminum

$$D = 5.378 + 1.348 \, U \qquad (2)$$

in the range $1 \leq U \leq 6.1$ km/s ($\rho_0 = 2.732$ g/cm^3 at To=165 K); for sapphire D(U) - the dependence from [11]:

$$D = 8.74 + 0.957 \, U \qquad (3)$$

Processing of experimental R(t) data obtained by measuring velocity of free surface of impactor at

impact against the baseplate gives the value W = 8.14 km/s. For this velocity, the following state at the external boundary of the baseplate (R_{ext} = 35mm) is determined: pressure P = 185 GPa and velocity of shock wave D = 12.66 km/s. For processing the data, the Hugoniot adiabat for iron [12] was used: $D_{Fe} = 3.664 + 1.790 \cdot U - 0.0342 \cdot U^2$ ($1.4 \leq U \leq 8$ km/s). The measured average velocity of shock wave in Al-baseplate (thickness of 3.98 mm) on the measurement basis of 3.5 mm was D_{Al} = 13.0 ($\pm$ 6%) km/s. The large error of measurement is mostly associated with quality of the contacts arrangement.

Velocity of shock wave in liquid xenon was measured by the electrocontact and optical methods. The optical method is based on recording of front luminescence at arrival of shock wave from the baseplate to liquid xenon and radiation «cutting off» at the interface with sapphire window. Luminescence was recorded by a four-channel pyrometer of visible range of spectrum [13] together with oscillographs Tektroniks TDS 3054. Three oscillographs with different sensitivities were simultaneously used in each spectral channel. During the experiment, 12 oscillograms were obtained. They allowed to measure times of shock wave motion in liquid xenon and sapphire window, as well as radiations amplitudes at wavelengths of 406, 498, 550, and 600 nm. The original oscillogram of shock wave front pulse radiation at wavelength of 406 nm is presented in fig. 2. In this figure, one can also see characteristic marks of time used for measurement of time of shock wave motion in liquid xenon (p.1–p.2) and in sapphire window (p.2-p.3). Processing of all data obtained in the experiment gives the following value for the average velocity of SW in liquid xenon: D_{Xe} = 9.3 km/s. The measurement accuracy is $\pm$ 1%. The average velocity of SW in sapphire measured by such a way is equal to D = 14.6 km/s ($\pm$ 1.2%).

Turning from the measured average values of shock wave velocity to instantaneous values at the borders of break decay was performed by the following method. Basing on results of gasdynamic calculation, for each of the areas, relative difference of velocities at the measurement radius R_{meas} and the borders was calculated: $\delta D = (R_{bord} - D_{meas})/D_{meas}$; and values of shock wave velocities measured in the experiment were changed for the obtained value.

Correction was ± 2.44% for the baseplate, ± 3% - for the area Xe1, ± 1.5% - for sapphire.

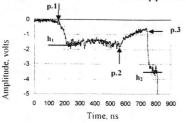

FIGURE 2. Oscillogram of shock wave front luminescence in liquid xenon: $\lambda = 406$ nm

Shock wave velocity changed with account for the correction at the baseplate-xenon interface was equal to $D_{bord} = 13.3$ km/s. With account for D_{bord} and shock adiabat of aluminum (2), the following parameters were obtained in shock-compressed baseplate at the border with xenon: $U_{Al} = 5.88$ km/s and $P_{Al} = 213.65$ GPa. Being brought to the boundaries with Al-screen and sapphire window with account for the corrections, shock wave velocities in liquid xenon have values of 9.0 km/s and 9.58 km/s, respectively. Basing on these values and solution of the problem on decay of arbitrary break, the following parameters in liquid xenon on direct shock wave were obtained: $P_1 = 175$ GPa, $D_1 = 9.0$ km/s, $U_1 = 6.58$ km/s, $\rho_1 = (11.0 \pm 0.5)$ g/cm^3, $\rho_o = 2.96$ g/cm^3. The corrected value of SW velocity in sapphire brought to the boundary with liquid xenon is equal to $D_{sapph-Xe1} = 14.4$ km/s. With account for (3), it determines state in sapphire: $P_2 = 343$ GPa and $U_2 = 5.91$ km/s. The corrected velocity in sapphire brought to the external boundary of the window was $D_{sapph-Xe2} = 14.8$ km/s at pressure P=377GPa and particle velocity U = 6.33 km/s.

Shock adiabat

Data from [1-4] together with the experimental values obtained in this work, as well as the new approximation dependence of all D(U) results:

$$D = 1.447 + 1.311\,U - 0.0248\,U^2, \quad (4)$$

are presented in figure 3. The lowest point at U = 1.05 km/s from [4] was not taken into account during approximation. C with account for (4), the following parameters are determined in liquid xenon (Xe1) at boundary with sapphire: D_1 = 9.58 km/s, $U_1 = 7.15$ km/s and $P_1 = 203$ GPa, and

state in liquid xenon in shock wave reflected from sapphire window: $P_2 = 343$ GPa, $U_2 = 5.91$ km/s, $\rho_2 = (13.5 \pm 1.3)$ g/cm^3.

Experimental results of this work together with dependencies calculated in [4] are presented in figure 4.

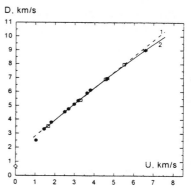

FIGURE 3. Hugoniot adiabat of liquid xenon: ● - [4], □ – [2], ◆ - present work, ◇ - sound velocity; 1 – approximation of (1), 2 – approximation of (4).

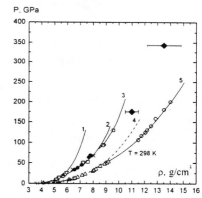

FIGURE 4. Phase diagram of liquid xenon. Experiment: ◆ - present work, ∇ - [1], □ – [2], × - [3], ● - [4], ○ – [7], + - [14], Δ - [15]; Calculation from [4]: 1- without account for excitation of electrons. 2 – with account for that, 3 – shock adiabat for the phase Xe 2.4 – isotherm T = 298 K for the phase Xe1. 5 – isotherm T = 298 K for the phase Xe 2.

Temperature of shock-compressed liquid xenon

The brightness temperature of shock-compressed liquid xenon in the red area of spectrum ($\lambda = 670$ nm) was measured in [4] by the photochronographic method in the pressure range up to 100 GPa. The same method was used in this work to measure temperatures in the blue area of the spectrum (λ = 430 nm). The temperature was evaluated using the value of experimental amplitude of radiation with account for preliminary calibration of measurement channel by the reference lamp. Results of these experiments are given in figure 5.

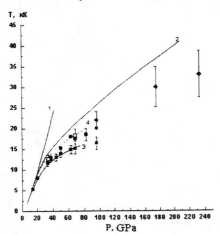

FIGURE 5. Temperature of shock-compressed liquid xenon versus pressure. Experiment: • - this work (λ = 430 nm), ◆ - this work (multichannel pyrometer); ■ - [4] (λ = 670 nm). □ –[3]. Calculation [4]: 1 – without account for excitation of electrons. 2 – with account for that, 3 – effective temperature in red area of spectrum.

At pressures of 95.6 GPa and higher, in this work, we used a four-channel pyrometer of visible range of spectrum [13] for temperature measurements. Radiation of shock wave front in liquid xenon was recorded at wavelengths λ = 406, 498, 550 and 600 nm separated by interferential light filters with passage amplitude of ~50% and the band $\Delta\lambda \approx$ 10 nm at the level of half of the maximum value. The problem of determination of temperature and radiation capability in four spectral flows measured in experiments was solved by the non-linear method of the smallest squares for the model with two parameters: T and Э. To calculate temperatures we used the model equation for the radiation capability in the following form:

$$Э = \exp{(A_o \cdot \lambda)}, \qquad (5)$$

and the flow of thermal radiation for measured brightness (spectral) temperatures was calculated by the Planck formula:

$$N(\lambda) = Э \cdot C_1 \lambda^{-5} [\exp{(C_2/\lambda T)} - 1]^{-1} =$$
$$= C_1 \lambda^{-5} [\exp{(C_2/\lambda T_{br})} - 1]^{-1} \qquad (6)$$

Here Э – radiation capability of body, λ - wavelength, T – actual temperature, T_{br} - brightness temperature; $C_1 = 1.19 \cdot 10^{-16}$ Wm^2/Sr and C_2 = 0.0144 mK are constants.

Then such temperature value T was chosen that the curve $N(\lambda)$ gave the best description of flows corresponding to the values of experimental temperatures. The obtained results are presented in figure 5

REFERENCES

1. Keeler R.N., M.van Thiel. Alder B.J.. *Physica* **31**. 1437 (1965).
2. Nellis W.J., M.van Thiel. Mitchel A.C.. *Phys. Rev. Lett.* **48**. 816 (1982).
3. Radousky H., Ross M., *Phys. Lett.* **A: 129**. 43 (1988).
4. Urlin V.D., Mochalov M.A., Mikhailova O.L., *Mathematical modeling* **3**. 42 (1991). Urlin V.D., Mochalov M.A., Mikhailova O.L.. *High Pressure research* **8**, 595 (1992).
5. Urlin V.D., Mochalov M.A.. Mikhailova O.L.. *TVT* **38**. 2. 227-231 (2000).
6. Jephcoat A.P.. Mao H-K. Finger L.W. et al.. *Phys.Rev.Lett.* **59**. 2670 (1987).
7. Goettel K.A.. Eggert J.H.. Silvera I.F., *Phys. Rev.Lett.* **62**. 665 (1989).
8. Reichlin R.. Brister K.E.. Mc.Mahan A.K.. et al., *Phys.Rev.Lett.* **62**. 6691 (1989).
9. Al'tshuler L.V.. Trunin R.F.. Krupnikov K.. Panov N.V.. *UFN* **166**. 5. 575-581 (1996).
10. F.V.Grigoriev. S.B.Kormer. O.L.Mikhailova et al.. *Sov. Phis. JETP.* **61**. 751 (1985).
11. Nellis W.J.. Radousky H.B.. Hamilton D.C. et al.. *J. Chem. Phys.* **93**. 3. 2244 (1991).
12. Experimental data on shock-wave compression and adiabatic expansion of condensed substances. Edited by Sc. Dr. R.F.Trunin. Sarov: RFNC-VNIIEF, 2001. 446 p.
13. Zhernokletov M.V.. Lebedeva T.S.. Medvedev A.B. et al., *In. Shock Compression of Condensed Matter– 2001*. edited by M.D.Furnish. N.N.Thadhani. and Y.Horie. AIP, 763-766.
14. Syassen K.. Holzapfer W.B.. *Phys.Rev.* **18**. 5826 (1978).
15. Jisman A.N.. Aleksandrov I.V.. Stishov S.M., *Phys. Rev.* **32**. 484 (1985).

CHAPTER III

EQUATION OF STATE:
ENERGETIC MATERIALS

CP706, *Shock Compression of Condensed Matter - 2003*
edited by M. D. Furnish, Y. M. Gupta, and J. W. Forbes
© 2004 American Institute of Physics 0-7354-0181-0/04/$22.00

PRESSURE IONIZATION OF CONDENSED MATTER UNDER INTENSE SHOCK WAVES AT MEGABARS

V. E. Fortov[1], V.K.Gryaznov[2], R.I.Il'kaev[3], A.L.Mikhaylov[3], V.B.Mintsev[2], M.A.Mochalov[3], A.A.Pyalling[2], V.Ya.Ternovoi[2], M.V.Zhernokletov[3]

[1]*Institute for High Energy Density IVTAN RAS, Moscow 127412 Russia*
[2] *Institute of Problems of Chemical Physics RAS, Chemogolovka, 142432 Russia*
[3] *Institute for Experimental Physics, VNIIEF, Sarov, 607190 Russia*

Abstract. Physical properties of hot dense matter at megabar pressures are considered. The new experimental results on pressure ionization of hot matter generated by multiple shock compression of hydrogen and noble gases are presented. The low-frequency electrical conductivity of shock compressed hydrogen, helium and xenon plasmas was measured in the megabar range of pressures. To reduce effects of irreversible heating and to implement a quasi-isentropic regime strongly compressed matter was generated by the method of multiple shock compression in planar and cylindrical geometries. As a result, plasma states at pressures of the megabar range were realized, where the electron concentration could be as high $n_e \sim 2 \times 10^{23}$ см$^{-3}$, which may correspond to either a degenerate or a Boltzmann plasma characterized by a strong Coulomb and a strong inter-atomic interaction. A sharp increase (by three to five orders of magnitude) in the electrical conductivity of a strongly nonideal plasma due to pressure ionization was recorded, and theoretical models were invoked to describe this increase. Opposite effect was observed for lithium compressed by multiple shock up to pressures ~ 200 GPa, where electrical conductivity was sharply decreased as pressure increased.

INTRODUCTION

The behavior of plasma under the conditions of a strong heating and compression is of considerable interest from the general physical point of view; it is also of practical interest for astrophysics, the physics of giant planets, and promising applications in power engineering [1]. Particular attention is being given to the ionization composition of plasma, since this provides a basis for calculating its thermodynamic, transport, and optical properties. It is well known that plasma can be generated not only via a strong heating up to temperatures comparable with the ionization potential, $k_B T \sim I$, but also via a strong compression to a state in which the inter-particle spacing becomes close to atomic sizes, $r_a \sim n_a^{-1/3}$; the second way is referred to as cold ionization or

pressure ionization. While thermal ionization processes have to date received quite an adequate study [1], investigation of pressure-produced ionization is much more complicated since one deals here with a cold ($k_B T < I$) compression of a plasma to pressures of a megabar range and densities that considerably exceed solid-state values. Under such conditions, the interaction between particles becomes strong, the electron shells of atoms and molecules overlap, and a typical level of electrical conductivity is comparable with that in metals. A considerable number of studies (see [2] and references therein) of this region motivated by searches for metallic hydrogen [2] in connection with its possible high temperature superconductivity in a metastable medium [3] have been devoted to the metallization ($T = 0$) of dielectrics at high pressures. By using the technique of strong shock waves to ensure a compression and an irreversible heating of matter, one can obtain very high pressures up to several Gbar, they are limited only by the intensity generation source. The viscous dissipation of the kinetic energy of the flux in the shock-wave front, along with compression, leads to a considerable heating of matter, and this stimulates the thermal $(T \sim I)$ ionization of a plasma, whose kinetics and thermodynamics have by now received a detailed study both for an ideal and for a strongly nonideal case (see [1, 4]). It should be noted that a number of theoretical models loose thermodynamic stability upon extrapolation to the region of strong nonideality and that this is attributed in [1, 4] to the occurrence of a first order plasma phase-transition. In order to separate density and thermal effects of ionization, one must naturally try to suppress the effects of irreversible heating $(k_БT << I)$ by implementing a quasi-isentropic compression. For this purpose, the compression of substance in this study was accomplished by means of a sequence of direct and reflected shock waves that emerge upon their reverberation in planar and cylindrical geometries. For the source of generation, we employed explosive devices of end-face and cylindrical throwing. By using processes of multiple shock compression, it proves to be possible to implement an order of magnitude reduced heating and an approximately tenfold increased compression of a plasma in relation to that in a direct shock.

EXPERIMENTAL PROCEDURE

A typical experimental assembly for multiple shock compression of condensed hydrogen and inert gases in planar geometry is shown in Fig. 1. Shock waves were generated by an impact of a steel impactor (2) 1–3 mm thick and 30–40 mm in diameter accelerated by detonation products of a condensed high explosive (1) to velocities of 3–8 km/s with the aid of the gradient-cumulation effect [1]. The absence of melting and evaporation of a shock-worker material, as well as the absence of mechanical fracture of the impactor during dynamic acceleration, was tested in methodological experiments. The transition of a shock wave from a metallic screen (3) of thickness 1–1.5 mm to the substance under study (4) having an initial

thickness of 1 to 5 mm generated, in it, the first shock wave of amplitude pressure $P_1 = 0.02-0.8$ Mbar; upon being reflected from a transparent sapphire window (5) 4–5 mm thick and 20 mm in diameter, this wave excited a repeated-compression shock wave. A further rereflection of shock waves between the screen 3 and the window 5 led to multiple shock compression of the sample to maximum pressures of $P \approx 1-2$ Mbar, whose level was determined by the velocity of the impinging impactor, its thickness, and the dimensions of the substance being studied.

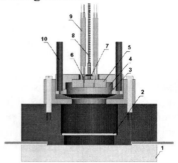

Figure 1. Experimental assembly for multiple shock compression of condensed hydrogen and inert gases in planar geometry: *1* - high-explosive, *2* - steel plate, *3*- bottom of the experimental assembly; *4* - substance under study, *5* - sapphire window, *6*- indium electrodes, *7* - shunting resistance, *8* - quartz– quartz light guide, *9* - coaxial electric cables, and *10* - gas-supplying pipes.

The initial states of the explored substances for a further multiple compression were either in the gas region of the phase diagram at pressure and temperature values of $P_0 = 5-35$ МПа, $T_0 = 77.4-300$ K, respectively, or in its liquid region at $P_0 \sim 0.1-1$ МПа, $T_0 \sim 20.4-160$ K. In the last case, liquefaction was performed from high-purity gases supplied to the assembly through pipes (*10*). Initial temperature was measured by thermocouples and platinum resistance thermometers. The process of multiple compression was observed by means of fast optic-electronic converters and five-channel fiber-optic-coupled pyrometer of time resolution 2–5 ns (*8*). Shock-compressed sapphire of the optic window 6 retained transparency up to $P \approx 20$ GPa and made it possible to record from five to six reverberations. In these experiments, the compression and irreversible heating of the substance under study were implemented by series of shock waves arising upon successive reflections from the sapphire window and the steel screen. A hydrodynamic analysis of the process revealed that, following the propagation of the first two waves through the compressed layer, a further compression proceeded in a quasi-isentropic way. This made it possible to advance to the region of higher densities ($\rho/\rho_0 \sim 10\text{-}100$) in comparison with single wave compression and to reduce the final temperature.

The electrical conductivity of a shock-compressed plasma was determined by a probe method. An electric current was supplied to the shock-compressed plasma under study by means of electrodes (6) that were arranged orthogonally to the plane of the shock-wave front. Further, the current propagated along the shock compressed sample and then arrived at the surface of the steel screen 3, whereupon it left the compressed region through a grounding electrode. The arising electric signals transferred by high-frequency coaxial cables (9) were recorded by multi-channel digital oscilloscopes whose transmission-band width was 500 MHz. Use was made of two- or three-electrode schemes for recording resistance.

The second series of measurements was performed by employing shock compression under the conditions of cylindrical geometry [5] (Fig. 2). A cylindrical charge of a high explosive (an alloy formed by trotyl and hexogen in the ratio 40 : 60), its 1 1 2 outer diameter being 30 cm, was initiated over the outer surface at 640 points that generated, at the inner surface of the charge, a highly symmetric detonation wave (the difference in time of arrival was not greater than 100 ns).

The arrival of this wave at the inner surface caused the centripetal motion of the steel impactor at an initial velocity of W ~ 5 km/s. The deceleration of this cylindrical impactor against the metallic surface of the chamber filled with the gas under study at an initial pressure of up to 70 MPa generated a converging shock wave. Successive reflections of the shock wave from the center of symmetry and from the inner surface of the chamber gave rise to multiple shock compression, which, as in the case of planar geometry, proved to be close to isentropic compression. At each instant of time, the profiles of thermodynamic parameters of multiple compression were determined on the

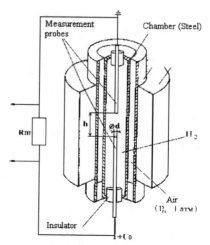

Figure 2. Scheme of cylindrical compression.

basis one- or two-dimensional gas-dynamic calculations.

Fig. 3 displays an oscillogram that was obtained in one of the experiments at an initial

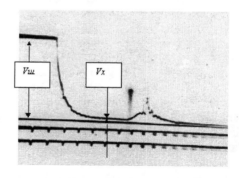

FIGURE 3. Oscillogram of the experiment where the initial hydrogen pressure was 70 MPa (the time-scale division here is 1 μs).

hydrogen pressure of $P_0 = 70$ GPa. The electrical conductivity was measured according to the classic two-point circuit diagram [53] involving a reference resistance connected in parallel with the resistance of the sample under study.

The resistance estimated by using this oscillogram is $R_x \approx 0.2\ \Omega$. The error in the electrical- conductivity values found in this way was estimated at 50%. The characteristic plasma parameters obtained in some experiments are quoted in Table 1.

EXPERIMENTAL RESULTS AND DISCUSSION

Experiments on multiple shock compression of hydrogen and inert gases make it possible to obtain physical information in a new region of phase diagram. For hydrogen pressures are up to 15 Mbar, temperatures of 3000 to 7000 K and densities one order of magnitude higher than those of solid state. This region is of interest since strong interaction both between atoms (molecules) ($\Gamma_a = r_a n_a^{-1/3} \sim 1$ - that is, molecular or the atomic size r_a is comparable with the interparticle spacing $n_a^{-1/3}$) and Coulomb particles (the mean interaction energy of charged particles E_C, is much greater than the mean kinetic energy of thermal motion, E_T ($\Gamma_D = E_C/E_T \sim 10$).). The situation is additionally complicated electrons become degenerated, ($n_e \lambda_e^3 \sim 200$, λ_e is the thermal de Broglie wavelength).

From Fig.4-5 one can see that the most prominent feature of the electrical conductivity of strongly nonideal plasma is sharp increase of the electrical conductivity at final stages of compression (by three to five orders of magnitude) in a narrow range of densities ($\rho \approx 0.3$–1 g/cm³ for hydrogen, and $\rho \approx$ 8–10 g/cm³ for xenon) at megabar pressures, reaching values of about 10^2–$10^3 \, \Omega^{-1} \, \text{cm}^{-1}$, which are peculiar to alkali metals.

Our measurements exhibit a pronounced threshold effect in density and are therefore in a qualitative contradiction with models of weakly nonideal plasma [1], which predict a monotonic decrease in the electrical conductivity in response to its isothermal compression [1]. Indeed, it is well known that, at low degrees of the ionization of plasma, its electrical conductivity is determined by the scattering of electrons on neutral particles and is qualitatively described by the Lorentz formula [4]

$$\sigma_{ea} = \frac{2\sqrt{2}}{3\sqrt{\pi}} \frac{e^2}{m_e^{1/2}(k_B T)^{1/2}} \frac{n_e}{n_a} \frac{1}{q_{ea}^*(T)}$$

where q_{ea} is the averaged cross section for electron scattering by atoms:

$$\frac{1}{q_{ea}^*(T)} = \frac{1}{(k_B T)^2} \int exp(-E/k_B T) \frac{dE}{q_{ea}(E)}$$

In turn, the composition of a plasma is described by Saha-like ionization-equilibrium equation [1]

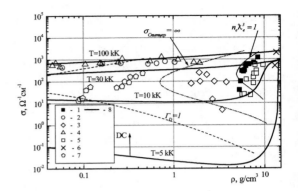

Figure 4. Electrical conductivity of xenon as a function of density. Experiment - *1-7* [6 and references herein]. ($n_e \lambda_e^3 = 1$) - the electron-degeneracy line, the line on which $\Gamma_D = 1$, and the line on which the electrical conductivity calculated by Spitzer's $\sigma_{\text{Spitzer}} = \infty$., 8 – calculation of present work.

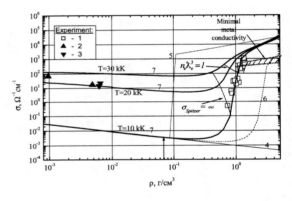

FIGURE 5. Electrical conductivity of helium as function of density: 1, 2, and 3 experimental [6 and references herein] calculations of electrical conductivity with composition (4) - ideal plasma; (5) –Debye–Hückel model [1]; (6) - the confined-atom model [7] with fixed radius of the helium atom ($r_a = 1.3a_0$); and (7) - the present work.

$$\frac{n_e n_i}{n_a} = \frac{2Q_i}{Q_a} \lambda_e^{-3} exp[-(I - \Delta I(n_e, n_o, T))/k_B T]$$

where Q_a and Q_i are the partition functions for atoms and ions, and ΔI is ionization potential lowering due to interparticle interaction. Thus, one can see that, at $\alpha_i \ll 1$, it follows from equations above that, in the absence of interaction ($\Delta I = 0$, and Q_i, Q_a = const), the electrical conductivity under isothermal compression corresponds to the curves for an ideal plasma in Fig 5. Nonideality, which must be included for determining the composition of plasma, is taken here into account by introducing a density- dependent quantity ΔI, Q_i and Q_a, that leads to nonthermal growth of ionization degree and the electrical conductivity upon isothermal compression. On the curve representing the electrical conductivity as a function of density at T = const, there appears a minimum, its depth being greater for lower temperatures. With increasing temperature, this minimum levels out as soon as thermal-ionization effects (at $k_B T \sim I$) become more pronounced than effects associated with pressure- ionization, which are significant at $k_B T << I$. As the density increases further at a given temperature, ionization processes are completed.

We can see that the exponential growth of the number of carriers due to the ionization potential lowering ΔI because of strong interparticle interaction in plasmas of condensed densities is the main reason behind a sharp increase in the measured electrical conductivity. Thus, the data obtained here for the electrical conductivity at $k_B T << I$ provide a unique possibility for adequately choosing thermodynamic models that would describe ΔI. For example, Figs. 4-5 shows that the standard Debye–Hückel model (curve 5 in Fig. 5) strongly overestimates effects of Coulomb interaction, leading to pressure ionization at densities that are two orders of magnitude lower than their experimental counterparts.

Quantitative calculations of physical parameters of dense plasma is based on quasichemical representation of nonideal plasma [1,2]. Following to this picture the free energy of a quasineutral mixture of electrons, ions, atoms and molecules can splitted into the contribution of the ideal-gas component and the term that takes into account interparticle interaction; that is,

$$F \equiv F_i^0 + F_e^0 + F_{ii,ie,ee,...}^{int}$$

It is assumed that heavy particles (atoms, ions, molecules) obey Boltzmann statistics, their contribution having the standard form

$$F_i^0 = \sum_j N_j k_B T (ln \frac{n_j \lambda_j^3}{Q_j} - 1)$$

where Q_j are partition functions of atoms, molecules and ions.

Contribution of free electrons F_e^0 corresponded to partially degenerated ideal Fermi gas. For a description of Coulomb effects we applied a version of the pseudopotential model for multiple ionization [8]. A key point of this model is that, formation of bound states leads to deviation of the interaction of free charges at short distances, from a Coulomb form, and this deviation results in noticeable positive shift as potential energy of free charges so their mean kinetic energy.

The contribution of the short-range repulsion of molecules, atoms, and ions is described within the soft-sphere approximation [9] generalized to the case of a multicomponent mixture; that is,

$$\frac{\Delta F_{SS}}{N k_B T} = C_s y^{s/3} (\varepsilon_{SS} / k_B T) + \frac{s+4}{6} Q y^{s/9} (\varepsilon_{SS} / k_Б T)^{1/3};$$

$$y = \frac{3Y\sqrt{2}}{\pi}; \quad Y = \frac{4\pi r_c^3}{3} = \frac{\pi \sigma_c^3}{6}; \quad r_c = \left[\sum n_j r_j^3 / \sum n_j\right]^{1/3}$$

Radii of each sort of particles were determined from Hartree-Fock calculations [7] of their electron structures. Note that corresponding corrections for short-range repulsion to the chemical potential are different for particles of different radii, and this determines the decrease in the ionization (dissociation) energy with increasing matter density. Results of calculations of principal Hugoniot for xenon are demonstrated in Fig.6.

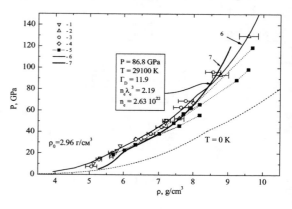

FIGURE 6. Shock compressed xenon. Experiment: *1 – 5* - [6] and references herein,. Theoretical results *6* - [5] and (*7*) - the present work. The dashed curve represents the "cold" curve from [5].

139

The electrical conductivity was described on the base of τ-approximation [10],

$$\sigma = \frac{4e^2 (kT)^{-3/2}}{3\sqrt{\pi}\, m_e} \frac{2}{\lambda_e^3} \int_0^\infty \varepsilon^{3/2} \tau(\varepsilon) \left(-\frac{\partial f_0}{\partial \varepsilon} \right) d\varepsilon$$

modified for the case of electron degeneracy. Here f_0 is the electron distribution function, τ – is the relaxation time,

$$\tau(\varepsilon)^{-1} = \sqrt{\frac{2\varepsilon}{m_e}} \left[\sum_j \gamma_j n_j Q_{ej}(\varepsilon) + n_a Q_{ea}(\varepsilon) \right]$$

Q_{ea} and Q_{ei} are the transport cross sections for, electron–atom and electron–ion scattering, γ_j is a correction for electron–electron scattering

$$\gamma_j = \gamma_j^0 + (1 - \gamma_j^0) \frac{T_F}{\sqrt{T_F^2 + T^2}}$$

T_F is the Fermi temperature.

Expression for the electrical conductivity takes into account the fact that, in the region of high compressions, free electrons are partly or fully degenerate. In this case, the Boltzmann distribution function of electrons is replaced by the Fermi–Dirac one. The effect of pressure ionization is pronounced in Fig. 3, 4. It is seen that an extrapolation of the simplest plasma models to the region of strong nonideality leads to the thermodynamic instability of Debye–Hückel models (Coulomb collapse)—arrow DC and to the divergence of Spitzer's formula.

densities yields satisfactory results in this region, which is not traditional for the model in question; nonetheless, experiments on direct measuring of free electrons density could help to understand in details process of pressure ionization. The first studies devoted to such measurements in plasmas have already been performed.

ACKNOWLEDGEMENTS

This work was supported by the Russian Foundation for Basic Research (project nos. 00-15-96738, 00-02-17550), by the Ministry for Industry and Science of the Russia (contract no. 40.009.1.11192, project no.1938.2003.02), and by the Presidium of RAS (Program for Scientific Research into Physics and Chemistry of Extreme States of Matter).

REFERENCES

1. *Encyclopedia of Low-Temperature Plasma*, Ed. by V. E. Fortov , Nauka, Moscow 2000.
2. E. G. Maksimov, Phys.–Usp. **42**, 1121 1999.
3. N. W. Ashkroft, Phys. Rev. Lett. **21**, 1748 1968.
4. V. E. Fortov and I. T. Yakubov, *Nonideal Plasma* Énergoatomizdat, Moscow 1994.
5. V. D. Urlin, M. A. Mochalov, and O. L. Mikhailova, High Press. Res. **8**, 595 1992.
6. Fortov V.E., Ternovoi V.Ya.., M.V.Zhernokletov et al., JETP **97**, 259 2003.

TABLE 1.

Substance	Initial state.	Final state.	P, GPa	ρ, g/cm^3	T,10^3 K	σ, Ω^{-1}cm^{-1}
		Planar compression				
H$_2$	P$_0$=25.6 MPa T$_0$=77.4 K	Max. compression	227	0.94	5.3	1600
He	P$_0$=28 MPa, T$_0$=77.4 K	Max. compression	126	1.37	15	1080
Xe	P$_0$=0. MPa, T$_0$=160 K	Max. compression	126	10	25	500
		Cylindrical compression				
H$_2$	P$_0$=50 MPa, T$_0$=293 K	Max. compression	1440	2.4	14	550
	P$_0$=70 MPa, T$_0$=293 K	Max. compression	1250	2	12.5	1100

CONCLUSIONS

Thus, it can be considered that the main reason of sharp increase in the measured electrical conductivity with compression .is connected with the exponential growth of the number of carriers. The chemical model used to describe the equation of state for plasmas in the region of ultrahigh

7. W. Ebeling, A. Förster, V. Fortov, V. Gryaznov, and A. Polishchuk, *Thermophysical Properties of Hot Dense Plasmas* (Teubner, Stuttgart, 1991).
8. I. L. Iosilevski, Teplofiz. Vys. Temp. **18**, 355 (1980).
9. D. A. Young, UCRL-52352, LLNL (1977).
10. V. K. Gryaznov, Yu.V.Ivanov, A.N.Starostin, V.E.Fortov., Teplofiz. Vys. Temp. **14**, 643 1976.

CP706, *Shock Compression of Condensed Matter - 2003*
edited by M. D. Furnish, Y. M. Gupta, and J. W. Forbes
© 2004 American Institute of Physics 0-7354-0181-0/04/$22.00

P,V,E,T EQUATION OF STATE FOR TATB-BASED EXPLOSIVES

K.F. Grebyonkin, A.L. Zherebtsov, V.V. Popova, M.V. Taranik

Russian Federation Nuclear Centre-Institute of Technical Physics, P.O. Box 245, Snezhinsk, Chelyabinsk region 456770, Russia.

Abstract. The equation of state (EOS) for TATB-based explosives has been built using complex information - the experimental data (Hugoniots of PBX9502 for different initial temperatures) as well as the results of *ab initio* and molecular dynamic calculations of the thermodynamic parameters (heat capacity at constant volume and isochoric pressure coefficient). The main features of these thermodynamic coefficients were identified, namely the strong temperature dependence of heat capacity and the strong volume dependence of isochoric pressure coefficient. Shock wave heating calculations for explosive with different initial temperatures and densities were performed. The analysis of calculated dependences T(P) was done.

INTRODUCTION

Recently, the new models for shock initiation and detonation of heterogenious insensitive high explosives (IHE's) with temperature as a main parameter have appeared [1,2]. The actual problem is to obtain a *P,V,E,T* EOS for the accurate calculation of temperature in a wide range of the dynamic loading and initial conditions.

This paper is considered as an attempt to derive a complete EOS for TATB and TATB-based explosives (with wt% of binder less than 10%). Theoretical description of thermodynamic properties of HE molecular crystals is a complicated task, therefore we used complex information, i.e. experimental and calculated data to build EOS suitable for hydrodynamic codes. PBX 9502 (95 wt%TATB, 5wt% Kel-F) was taken as a prototype of TATB-based explosives. This IHE contains low binder fraction, and experimental Hugoniots for different initial temperatures (ambient [3], -55°C, 75°C, 252°C [4]) are available. First of them was used while EOS building, others – for EOS verification.

EQUATION OF STATE

The form of EOS is as following :

$$p = p_C\left(V\right) + p_T\left(V,T\right) = p_C\left(V\right) + \left(\frac{dp}{dT}\right)_V \cdot T \; ;$$

$$\varepsilon(V,T) = \varepsilon_C(V) + \varepsilon_T(T) \; ;$$

$$\varepsilon_C(V) = -\int_{V_0}^{V} P_C\left(V^{'}\right)dV^{'} \; ; \; \varepsilon_T(T) = \int_{T_0}^{T} C_V\left(T^{'}\right)dT^{'} \; ,$$

where V - specific volume, T – temperature, p_C, ε_C - "cold" curve (zero Kelvin isotherm), p_T, ε_T - thermal pressure and internal energy ; C_V -heat capacity at constant volume, $(\partial P/\partial T)_V$ - isochoric pressure coefficient, V_0- specific volume of solid substance at T_0.

The approximation used in this paper is the following: C_V is a function of only of temperature, $(\partial P/\partial T)_V$ is a function of only of specific volume. Such behavior of C_V and $(\partial P/\partial T)_V$ corresponds to the thermodynamic consistency of these parameters $(\partial/\partial V\left[\left(C_V/T\right)_T\right] = \partial/\partial T\left[\left(\partial P/\partial T\right)_V\right])$ and is appropriate for high-molecular liquids and crystals. For example, temperature dependence of C_V of nitromethane [5] is stronger than its volume dependence and vice versa for $(\partial P/\partial T)_V$. It is possible to expect that for HE liquids and crystals (20-30 atoms in molecule) this feature will be more pronounced.

HEAT CAPACITY AT CONSTANT VOLUME

For high-molecular crystal, the contribution of intramolecular vibrations into heat capacity is dominant. The approximation used consists of the fact that the internal vibration modes of the TATB molecule in the solid slightly differ from those in the free molecule, and pressure of an order of $10\div20$ GPa influences insignificantly mode values. The vibration frequencies of isolated TATB molecule were obtained from *ab initio* calculation in Hartree-Fock approximation. Total heat capacity with regard for temperature dependent contribution of intramolecular vibrations was presented in the following form (T>200K):

$$C_V(T)= C_V^m \cdot \left[1 - \left(1 - \frac{C_V^0}{C_V^m} \right) \cdot \exp\left(\frac{T - T_0}{T_c} \right) \right],$$

where reference points (T_0=293K, C_V^0=1.0 J/g/K) and C_V^m=72R$\approx$2.32 J/g/K were used, parameter T_C was equal to 550K.

ISOCHORIC PREESSURE COEFFICIENT

Estimates of $(\partial P/\partial T)_V$ were obtained by classical molecular dynamic (MD) simulation. Molecular modeling of crystalline TATB was based on the atom–atom scheme, containing EXP6-potential and electrostatic interactions between atoms of various molecules. The parameters of EXP6-potential and intramolecular force constants were taken from [6] with slight corrections. Due to the results of *ab initio* calculations, the following values were taken in MD calculations: charges on atoms, which reproduced electrostatic potential of isolated molecule; bond strengths and bond lengths; barriers of rotation of NH_2 and NO_2 groups. *NVT*-ensemble and periodic super cell (3x3x3 unit cells, 1296 atoms) were used. The experimental values of volume and shape of unit cell [7] and their extrapolation into the range of slightly higher pressure were put in MD calculations. Temperature was varied within the range 300$\div$800K, time step was about 1 fs, time required to achieve the equilibrium values of pressure and internal energy at given V, T was 15$\div$25 ps. At first, we compared the calculated and experimental isothermal (293K) compression of TATB. Agreement was good enough to make estimates of $(\partial P/\partial T)_V$. Temperature dependence of $(\partial P/\partial T)_V$ was not determined using ensemble described above. Isochoric pressure coefficient was obtained as $\Delta P/\Delta T$ for each pair of (P,T) at constant V. Fig.1 shows $(\partial P/\partial T)_V$ calculated in such a way.

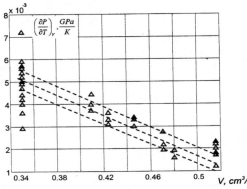

FIGURE 1 Isochoric pressure coefficient as a function of specific volume $((\partial P/\partial T)_V \rightarrow \Delta P/\Delta T$ for each pair of (P,T) at constant V, *NVT*-ensemble simulation).

Volume dependence is strong, $(\partial P/\partial T)_V$ increases several times in the specific volume range under study. To match the value of $(\partial P/\partial T)_V$ with its value at 1 atm isobar, $(\partial P/\partial T)_V$ was presented by linear function of compression $\delta = \rho/\rho_0(T_0)=V_0(T_0)/V$ ($\delta \geq 1$):

$$(\partial P/\partial T)_V = g_0 + g_1(\delta - 1) \text{ GPa/K},$$

where T_0 – initial temperature, g_0, ρ_0, V_0 - values of $(\partial P/\partial T)_V$, density (TMD) and specific volume, respectively at T_0, P_0=1atm. Parameter g_1 was equil to $1.2\cdot10^{-2}$ GPa/K .

It is necessary to notice that calculated pressure strongly depends on force constants of intramolecular degrees of freedom (torsion angle, barriers of rotation of functional groups), especially at elevated temperatures. Isochoric pressure coefficient $(\partial P/\partial T)_V$ strongly depends on volume only when these degrees of freedom are taking into account (for ensemble with rigid molecules, volume dependence of this coefficient is much weaker). Isochoric pressure coefficient of crystalline TATB is anisotropic. This coefficient in direction perpendicular to molecular layers is by an order higher than $(\partial P/\partial T)_V$ in the plane of aromatic rings.

THERMODYNAMIC PARAMETERS OF TATB AT ATMOSPHERIC PRESSURE

EOS uses thermodynamically consistent parameters of TATB at 1 atmosphere isobar. While obtaining these parameters for the temperature range 200÷600K, experimental data on temperature dependencies of heat capacity at costant pressure [8] and volumetric coefficient of thermal expansion (CTE) [9] were used as well as volumetric sound velocity and density at normal conditions [8]. Calculated $C_V(T)$ and estimates of $(\partial P/\partial T)_V$ at low pressure were also used. As the result CTE was given in this paper in the following form:
$\alpha(T)=7.0554 \cdot 10^{-7} \cdot T$ $(T \leq 160\ K)$;
$\alpha(T)=1.1333 \cdot 10^{-4}-5.7359 \cdot 10^{-7} \cdot T+3.5676 \cdot 10^{-9} \cdot T^2$,
$(T>160K)$, where α expressed in K^{-1}. Crystalline specific volume at temperature T^* may be easily obtained using

$$V\left(T^*\right)=V_0 \exp\left[\int_{T_0}^{T^*} \alpha\left(T\right)dT\right]$$

and reference point $\rho_0 =1.937$ g/cm^3 ,$T_0=293K$ [7].
C_P/C_V ratio was taken as following:
$C_P/C_V(T)=9.9978 \cdot 10^{-1}-5.3162 \cdot 10^{-6} \cdot T+4.6700 \cdot 10^{-7} \cdot T^2$,
$(T \geq 200K)$. Other thermodynamic coefficients can be determined from $C_V(T)$,$C_P/C_V=f(T)$, $\alpha(T)$, $V(T)$ using identities $c_0 = \sqrt{\left(C_P/\alpha^2 T\right) \cdot \left(C_P/C_V -1\right)}$,
$\Gamma = \alpha c_0^2 / C_P$, $(\partial P/\partial T)_V = \Gamma C_V /V$.
In the first approximation, there is no difference in $C_V(T)$,$C_P/C_V=f(T)$, $\alpha(T)$ dependencies between TATB-based explosives and pure TATB. The only thing that is to be changed is reference point (ρ_0,T_0) for a certain HE.

"COLD" CURVE

Analytical dependencies $C_V(T)$, $(\partial P/\partial T)_V=f(V)$, and PBX 9502 Hugoniot ($T_0=25°C$, $\rho_0=1.89$ g/cm^3) [3], allow to reproduce "cold" curve (zero Kelvin isotherm) for this HE for the pressure up to 25 GPa. "Cold" curve is a mirror image of $p_T(V)$ in the range of negative values, i.e. $p_C(V) \approx -p_T(V)$, where $P_T(V)$ is the volume dependence of thermal pressure at 1 atm isobar.

Estimate of "cold" curve for pure TATB was done using simplest mechanical equilibrium model, PBX 9502 and Kel-F Hugoniots. TATB "cold" curve is approximately 5÷10% steeper than that of PBX 9502. In the same manner, it is possible to correct "cold" curve for other TATB-based explosives.

RESULTS

EOS verification was done by comparison of the calculated and experimental Hugoniots of PBX 9502 for initial temperatures -55°C, 75°C, 252°C. Agreement with experimental data was good enough. The results for Hugoniot with initial temperature 252°C are presented in Fig.2.

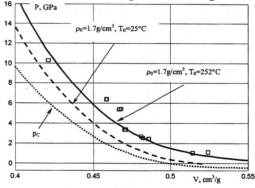

FIGURE 2 Calculated Hugoniots of PBX 9502 for initial density 1.7g/cm^3: dash line – ambient initial temperature (~15% porosity); solid line - $T_0=252°C$ (~2% porosity), squares-experimental data [4]; p_C – "cold" curve.

Thermal pressure reaches 40% of total pressure at the Hugoniot. For comparison, the Hugoniot of high-porous PBX 9502 with the same density, but for ambient initial temperature is shown the same figure. One can see essentially different shock compressibility of HE in these two cases.

EOS reproduces also experimental data on shock compression of TATB (ρ_0=1.806 g/cm^3) [8] using $p_c(V)$ for TATB.

Shock compression temperatures of PBX 9502 for different initial states are presented in Fig.3.

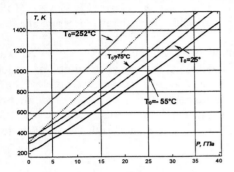

FIGURE 3 Calculated temperature along the Hugoniots of PBX 9502. Solid lines: T_0=-55°C (ρ_0=1.92g/cm^3); T_0=25°C (ρ_0=1.89g/cm^3); T_0=75°C (ρ_0=1.857g/cm^3); T_0=252°C (ρ_0=1.7g/cm^3). Dash line: T_0=25°C (ρ_0=1.7g/cm^3).

Fig.3 shows strong nonlinear effect: HE with higher initial temperature has stronger dependence T(P). The temperature about 800÷900K is achieved at different pressures depending on initial conditions: 17÷20 GPa for T_0=25°C, ρ_0=1.89g/cm^3; 7÷9 GPa for T_0=252°C, ρ_0=1.7g/cm^3; 9÷13 GPa for T_0=25°C, ρ_0=1.7g/cm^3. From this point of view, the significant decrease of initiation pressure in case of preheated HE or in case of high porosity samples can be explained.

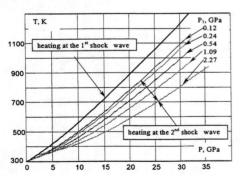

FIGURE 4 Calculated temperature at the 2nd shock wave of PBX 9502 (ρ_0=1.89g/cm^3, T_0=25°C) which was precompressed by the 1st weak shock wave. The amplitudes of the first shock wave are shown.

Decrease of shock-wave sensitivity of precompressed HE (for example, by a weak air shock wave or in case when the samples are heated in a metal confinement) is related to different factors including temperature (e.g. see Fig.4).

The analysis of calculated dependences T(P) shows that besides structure factors (porosity, grain size of HE), the temperature of shock heating to a high extent determines IHE decomposition rate.

Proposed EOS is valid at P>0.5 GPa. At lower pressure it is necessary to account for tensor and strength character of HE which are beyond the paper purposes.

REFERENCES

1. C.M.Tarver "Next generation experiments and models for shock initiation and detonation of solid explosives" *in Proc. of Int. Conf. Shock Compression of condensed matter*, p.873, 1999.
2. K. Grebyonkin "Physical model of shock wave detonation initiation in compressed fine-crystal high explosives", *Pis'ma v Jour. Tech. Phys.*, **20**, pp.1-2, 1998.
3. J.J. Dick, C.A. Forest, J.B. Ramsay, W.L. Seitz "The Hugoniot and shock sensitivity of a plastic-bonded TATB explosive PBX 9502", *J. Appl. Phys.*, **63**, pp.4881-4888, 1988.
4. J.C. Dallman, and J. Wackerle "Temperature-dependent shock initiation of TATB-based high explosives" *in Proc. of the 10th Symposium (Int.) on Detonation*, pp.130-137, 1993.
5. P.C.Lysne, D.R.Hardesty "Fundamental equation of state of liquid nitromethane to 100 kbar", *J. Chem. Phys.*, **59**, pp.6512-6523, 1973.
6. S.Mayo, B.Olafson, W.Goddard "DREIDING: A general force field for molecular simulation", *J. Phys. Chem.*, **94**, pp.8897-8909, 1990.
7. B.Olinger, H.Cady "The hydrostatic compression of PETN, TATB, CO_2 and H_2O to 10 GPa" *in Proc. of the 6th Symposium (Int.) on Detonation*, pp. 700-709, 1976.
8. B.Dobratz "The Insensitive High Explosive TATB : Development and Characterization – 1888 to 1994", Report LANL UC-741, Los Alamos, New Mexico,pp.53-85, 1995.
9. D.J.Pastine, R.R.Bernecker "*P,V,E,T*-equation of state for 1,3,5-triamino-2,4,6-trinitrobenzene", *J. Appl. Phys.*, **45**, pp.4458-4468, 1974.

CP706, *Shock Compression of Condensed Matter - 2003*
edited by M. D. Furnish, Y. M. Gupta, and J. W. Forbes
© 2004 American Institute of Physics 0-7354-0181-0/04/$22.00

THE ISENTROPE OF UNREACTED LX-04 TO 170 kbar

D.E. Hare[1], D.B. Reisman[1], F. Garcia[1], L.G. Green[1], J.W. Forbes[1], M.D. Furnish[2], Clint Hall[2], R.J. Hickman[2]

[1]Lawrence Livermore National Laboratory, Livermore, CA 94550
[2]Sandia National Laboratory, Albuquerque, NM 87185

Abstract. We present new data on the unreacted approximate isentrope of the HMX-based explosive LX-04, measured to 170 kbar, using newly developed long pulse isentropic compression techniques at the Sandia National Laboratories Z Machine facility. This study extends in pressure by 70% the previous state of the art on unreacted LX-04 using this technique. This isentrope will give the unreacted Hugoniot from thermodynamic relations using a Gruneisen equation of state model. The unreacted Hugoniot of LX-04 is important in understanding the structure of the reaction front in the detonating explosive. We find that a Hugoniot given by $U_S = 2.95$ km/s + 1.69 u_P yields for an isentrope a curve which fits our LX-04 ICE data well.

INTRODUCTION

LX-04 is a secondary high explosive (HE) formulation composed of 85% HMX (octahydro-1,3,5,7-tetranitro-1,3,5,7-tetrazocine) and 15% Viton-A (vinylidene fluoride/hexafluoropropylene copolymer) [1]. The unreacted Hugoniot of LX-04 is important for a detailed understanding of the impulse it will deliver during detonation. It is difficult to measure the properties of unreacted explosives by conventional shock compression techniques, which tend to induce reaction. On the other hand, ramp wave compression, such as is achieved in Isentropic Compression Experiments (ICE), is believed to be inherently a lower temperature dynamic loading process than shock compression. The ICE technique has already been demonstrated to be very effective at dynamically loading HEs without reaction, and experiments have already been performed on LX-04 as well as various other explosives [2,3]. The previous ICE work on LX-04 extended to 100 kbar [2], but recent improvements in waveform shaping motivated us to attempt to remeasure the LX-04 isentrope to higher pressures.

Our ICE experiments were conducted at Sandia National Laboratories Z Machine facility. The ICE technique [4,5] generates a large-amplitude (hundreds of kbar) compression wave of a few hundreds of nanosecond rise time (by comparison, shock wave rise times can be sub-nanosecond). This ramp wave is launched into the sample by magnetic pressure developed by an enormous current density in a conductive "floor" upon which the sample is mounted (For all our samples this floor was 800 μm of 6061- aluminum.). This current creates a time-dependent pressure boundary condition on the interior floor interface (the a-b interface of Fig. 2). References 4 and 5 contain excellent descriptions of this versatile technique.

PROCEDURE

Full density LX-04 samples of six different thicknesses and two NaCl (100) samples were mounted on four panels made from 6061 aluminum. Each panel is precision machined to 800 μm thick in three places. Thus each panel will accommodate three samples. All samples used a

NaCl (100) VISAR window. In addition, four LiF (100) windows were mounted, one per panel, directly on the Al floor to monitor the drive uniformity between panels. NaCl is a better acoustic impedance match to HE than the more generally used LiF. Commercially available VALYN VISAR equipment was used. The refractive index-density relations of Wackerle and Stacy [6] were used for the NaCl windows. Figure 1 shows the four panels with mounted samples. Current flow on each panel is along the same direction as the row of three windows. Figure 2 is a diagram illustrating an individual sample assembly.

wave of amplitude u at the VISAR interface of sample 1. ρ_0 and ρ are initial density and density (variable) respectively, and c_E is the Eulerian sound speed. Equation 1 is nothing

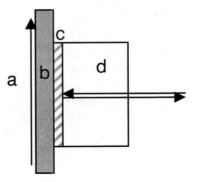

FIGURE 2. Side cut-away view of sample assembly. a) The assembly interior. It can be thought of as the interior of a short-circuited waveguide where a large magnetic field is developed by the current. b) the aluminum floor. The large surface current (vertical arrow) flows at the boundary between "a" and "b". The return conductor (known as the "stem") is to the left of "a" and is not shown. The floor serves both as a conductor for the current and to drive the ramp wave in the sample. c) the sample. d) the window. The interface between "c" and "d" is coated with 250 nm of silver to efficiently reflect the VISAR probe light, represented by the horizontal arrows.

FIGURE 1. The four panels with samples and windows mounted (twelve sample locations total). To give an idea of size, the windows are 6.00 mm diameter.

RESULTS/DISCUSSION

The ICE load curve in P-V space was computed based on the work of Fowles and Williams [7]:

$$c_L(u) = \frac{x_{01} - x_{02}}{t_1(u) - t_2(u)} \quad \text{(Eq. 1)}$$

$$dP = \rho_0 c_L(u)\, du \quad \text{(Eq. 2)}$$

$$\left(\frac{\partial P}{\partial \rho}\right)_S = c_E^2 = \frac{\rho_0^2}{\rho^2}(c_L(P))^2 \quad \text{(Eq. 3)}$$

In these equations: c_L is the Lagrange sound speed. u is the sample particle velocity. x_{01} is the inital thickness of sample 1, t_1 is the arrival time of the

more than a recipe for calculating c_L from VISAR data. Equation 2 allows us to eliminate the variable u in favor of the more useful P. Equation 3 is then integrated to get the isentrope. We do not consider material strength and do not distinguish between longitudinal wave stress and pressure P. A minimal data set for an analysis is two VISAR velocity histories from samples of the same material but of different initial thicknesses. Such an "EOS (equation of state) pair" of histories is represented by Fig. 3, for example. All LX-04 histories showed no clear evidence of reaction and looked quite similar to the histories of the (nonreactive) NaCl.

Equations 1-3 are appropriate to plane isentropic simple wave loading, and finite reflection from the sample-window interface spoils the simple wave assumption. Our window material NaCl is well (but not perfectly) matched to LX-04. We corrected u using NaCl and LX-04 Lagrange sound speeds measured in this experiment. Hayes

[8] elaborates on the zero-order nature of this type of correction, which according to his Fig. 1 leads to a 10% error in the worst case scenario (i.e. free surface). However, given that our zero-order correction increased u by less than 10% over its full range, we estimate 1-2% error in neglecting the higher order corrections for our specific case.

Unfortunately all our LX-04 waveforms shocked up slightly at the base. We treated the shock as setting up an initial state to a pressure given by the average strength shown in the two waveforms used to create a data set. This average initial shocked state is the starting point for the integration of Eqs. 2 and 3. The results of such a computation performed on our data are displayed in Fig. 4. Also shown in Fig. 4 is a linear (in U_S, u_P) Hugoniot (dashed line) and the associated isentrope (solid line) derived from this Hugoniot using a Gruneisen gamma model with the constant gamma/volume assumption. We used a Γ_0 of 1.25 based on thermodynamic data [1]. Based on the good agreement between the derived isentrope and our data we conclude that the Hugoniot given by $U_S = 2.95$ km/s + 1.69 u_P represents the unreacted Hugoniot data for full density LX-04 between 0 and 170 kbar. Our data are in good agreement with the previous ICE work on LX-04 [2] within their range of mutual overlap (up to 100 kbar). However our extended data range shows a less stiff response beyond 100 kbar than the extrapolated Hugoniot derived from the previous work would suggest. Hydrodynamic simulations using the Trac II code further support this conclusion.

Figure 5 shows a comparison of the room temperature isotherm of pure HMX [9] to our LX-04 results. The isotherm appears to be slightly stiffer than our Hugoniot. We claim that the difference is due to the 15 % binder composition of LX-04. We do not know of any EOS data for Viton-A in the literature but we did plot the Hugoniot for the somewhat similar Teflon [10]. Clearly the incorporation of 15 % of a soft Teflon-like material will soften the mixture's P-V curve relative to pure HMX.

There are numerous potential contiributions to the uncertainty in our results. There are those contributions which would exist even if Eqs 1-3 strictly and exactly applied to the data analysis: uncertainties in inital density, floor thickness,

sample thickness, particle velocity, and time measurement. There are also contributions related to the fact that the application of Eqs. 1-3 to the data is an

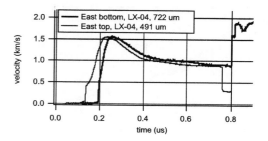

FIGURE 3. An EOS pair of LX-04 samples. The lack of amplitude growth at the peak and in the following rarefaction is evidence for insignificant reaction.

approximation. Partial shock-up, the approximation of isentropic loading by the ramp wave, finite chemical reaction, and the imperfect impedance match of the window to the sample will fall in this category. We have not yet conducted a detailed analysis of the uncertainty, but based on the following factors: 1) the spread between the data generated from three seperate LX-04 EOS pairs, 2) the relative difference between the computed isentrope and computed Hugoniot, 3) the good agreement with previous LX-04 data in the range of mutal overlap, and 4) the agreement between NaCl data taken with this same equipment in this same experiment and previously published NaCl Hugoniot data [12], we assess the (LX-04) uncertainty to be roughly +/- 5% in pressure at a given relative volume. We emphasize that the reader should consider this a rough guide only. A more accurate analysis will be presented in a future publication.

SUMMARY AND FUTURE WORK

We used the ICE technique and the improved pulse shaping capabilities of the Z machine at SNL to measure the unreacted approximate isentrope of full density LX-04 to 170 kbar, 70 % higher than previous work. From our data we compute an unreacted Hugoniot for LX-04 given by $U_S = 2.95$ km/s + (1.69) u_P. This data is reasonably consistent with existing HMX isotherm data.

We are hopeful that future advances will allow us to extend the pressure range on this and other HEs.

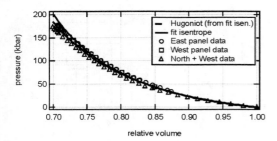

FIGURE 4. Results of this work: LX-04 isentrope data, fitted isentrope, and Hugoniot derived from this fitted isentrope. The derived Hugoniot parameters are A = 2.95 km/s, B = 1.69. Our estimate of uncertainty is +/- 5% in P at a given relative volume.

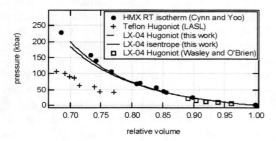

FIGURE 5. Comparison of HMX room temperature isotherm (Ref. 9), Teflon Hugoniot (Ref. 10), the fitted LX-04 isentrope and its derived Hugoniot (this work), and the shock compression data of Wasley and O'Brien (Ref. 11)

ACKNOWLEDGEMENTS

This work would not have happened without the assistance of Scott Humphery, Kevin Vandersall, and Allen Elsholz from LLNL; and from SNL: Jean-Paul Davis, Chuck Hardjes, Dave Bliss, Josh Mason and the outstanding technical staff of the Z Machine facility. DEH would like to acknowledge helpful discussions with Dennis Hayes of SNL, and Ed Lee, Craig Tarver, and Paul Urtiew of LLNL.

This work was performed under the auspices of the U. S. Department of Energy by the University of California, Lawrence Livermore National Laboratory under Contract No. W-7405-Eng-48.

REFERENCES

1. Owens, C., Nissen, A., and Souers, P.C., "LLNL Explosives Reference Guide" UCRL-WEB-145045 (2003).
2. Reisman, D.B., et.al., in "Shock compression of condensed matter-2001, edited by M.D. Furnish, N.N. Thadhani, and Y. Horie. (AIP, Melville NY, 2002) p.849.
3. Reisman, D.B., et. al., in "12th International Detonation Symposium" San Diego CA, 2002.
4. Hall, C.A., Phys. Plasmas **7**, 2069 (2000).
5. Reisman, D.B. et al, J. Appl. Phys. **89**, 1625 (2001).
6. Wackerle, J. and Stacy, H.L., "Shock waves in condensed matter, 1987" edited by Schmidt, S.C. and Holmes, N.C. (North Holland, New York, 1988).
7. Fowels, R. and Williams, R.F., J. Appl. Phys. **41**, 360 (1970).
8. Hayes, D.B., "Backwards Integration of the Equations of Motion to Correct for Free Surface Perturbations", Sandia National Laboratories Report, SAND2001-1440(2001).
9. Yoo, C.S. and Cynn, H., J. Chem Phys. **1ll**, 10229 (1999).
10. "LASL Shock Hugoniot Data", edited by S.P. Marsh (U. California Press, Berkeley, 1980). p.467
11. Wasley, R.J., and O'Brien, J.F., UCRL-12422-t (1965).
12. "LASL Shock Hugoniot Data", edited by S.P. Marsh (U. California Press, Berkeley, 1980). p.335

CP706, *Shock Compression of Condensed Matter - 2003*
edited by M. D. Furnish, Y. M. Gupta, and J. W. Forbes
2004 American Institute of Physics 0-7354-0181-0/04/$22.00

EQUATION OF STATE FOR LIQUID NITROMETHANE AT HIGH PRESSURES

Hermenzo D. Jones

Research & Technology Department
Naval Surface Warfare Center, Indian Head Division
Indian Head, MD 20640-5035

Abstract. A perturbation technique is used to calculate the thermodynamic properties of nitromethane for high pressures. The intermolecular interaction is described by a spherically symmetric, exponential-6 potential and an angular dependent, multipolar contribution. Isothermal compression and the shock Hugoniot for nitromethane are well characterized by the theory. Molecular dynamics calculations, with the intermolecular forces constructed from interatomic interactions, also describe the experimental observations under consideration. There is a considerable difference in the predicted shock temperatures from the two methods

INTRODUCTION

Liquid nitromethane, a simple homogeneous explosive, is a widely studied material. However, it has a large dipole moment, nearly twice that of water, which presents difficulty in modeling its thermodynamic behavior. Earlier theoretical efforts[1,2] made assumptions about particular thermodynamic properties, such as the specific heat, to formulate an equation of state (EOS). A few years ago, Winey, et al[3] successfully constructed an entire fundamental surface based on experimental data. Recently, there have been two significant molecular dynamic (MD) works[4,5] on liquid nitromethane, which provide a complete description in the low pressure regime.

In this work a complete EOS for liquid nitromethane based on intermolecular interactions is constructed from liquid-state perturbation theory and MD computations. For the perturbation approach, the intermolecular potential is written as the sum of a spherically symmetric, exponential-6 (exp-6) interaction and angular dependent, multipole corrections. Further details can be found in an earlier work.[6]

Molecular dynamics (MD) provides the solution of the classical equations of motion (Newton's equations) for a system of molecules, with the use of periodic boundary conditions. It offers a practical approach to complex systems, which cannot be described by analytical techniques. DL_POLY[7] is a computer program for performing molecular dynamics simulations of condensed matter. It can handle any assembly of flexible polyatomic molecules, atoms or ions and any mixture thereof. It uses as many as three different potential cutoffs to calculate short-range forces and the Ewald sum technique to handle long-range electrostatic forces. Simulations may be performed either in the usual NVE, NVT, NST or NPT ensembles. The detailed treatment of the nitromethane molecule given by Sorescu, et al[8] is utilized here. Both intra- and intermolecular forces are considered in the MD calculations, so the molecule is fully flexible. They found that their potential function accurately described the structural parameters for the crystal for the temperature and pressure range of interest.

As a simple test of the liquid-state perturbation theory, MD calculations for modified Stockmayer potential, which consists of an exp-6 potential and a dipole moment, are performed with MOLDY.[9] With liquid-state perturbation theory and MD calculations performed with DL_POLY[7], a room temperature isotherm for nitromethane and a shock Hugoniot are calculated and compared to observations.[10,11]

THEORY

The perturbation theory to be discussed here is based on the assumption that the repulsive intermolecular forces provide the dominant characteristics of the material. This technique, originated by Weeks, et al[12], should be quite applicable to the high pressure domain. The interaction between two molecules is taken as

$$\zeta(\vec{r}_1, \vec{r}_2) = \phi(r) + \delta(\vec{r}_1, \vec{r}_2) \qquad (1)$$

where the $\vec{r}_i$'s are the position vector of the particles. In the above, $\phi(r)$ contains the electronic repulsion and van der Waals attraction, while $\delta(\vec{r}_1, \vec{r}_2)$ consists of the multipolar interactions.

In this work the spherically, symmetric potential is given by

$$\phi(r) = \frac{\varepsilon\alpha}{\alpha-6}[\frac{6}{\alpha}\exp[\alpha(1-\frac{r}{r_m})]-(\frac{r_m}{r})^6] \ (2)$$

Formulation of the EOS proceeds with the construction of a reference system from the repulsive part of $\phi(r)$ and calculation of its free energy and radial distribution function. Free energies from the remaining molecular interactions are then computed via perturbation theory and detailed in Ref. 6.

For the MD calculations for nitromethane the potential is taken as

$$V^{total} = \sum_{i=1}^{N}(V_i^{intramolecular}$$
$$+ \frac{1}{2}\sum_{j}^{N}V_{ij}^{intermolecular}) \qquad (3)$$

The intermolecular potential consists coulombic interactions and exp-6 potentials. The intramolecular potential includes stretching, bending and torsional terms. All the specific parameters are given in Ref. 8.

RESULTS AND DISCUSSION

For the numerical computations of the modified Stockmayer potential system, the exp-6 potential parameters are taken as $\alpha=13.0$, $r_m'=3.15E-10$ m and $\varepsilon/k=78$. The dipole moment is chosen so that the reduced moment, $\mu/\sqrt{\varepsilon r_m^3} = 1$. The energy as a function of pressure for the reduced temperature, $kT/\varepsilon=1.15$, is shown in Fig.1. The perturbative values and the numerical results from MOLDY[9] compare favorably. In this pressure regime the dipolar energy is about 30 per cent of the total.

For the perturbative calculations for nitromethane the model exp-6 potential parameters are taken as $\alpha=16.0$, $r_m=5.10E-10$ m and $\varepsilon/k=225.0$. The pertinent multipole moments are

$$\mu_z = 3.317E-18(esu)$$
$$Q_{xx} = -3.0E-26(esu)$$
$$Q_{yy} = 7.0E.-26(esu)$$
$$Q_{zz} = -4.0E-26(esu)$$

With these parameters established, the thermodynamic behavior of nitromethane is now investigated.

For the room temperature isotherm, the pressures obtained from perturbative predictions and those from MD results are in close agreement with the diamond anvil cell data.[10] The energy associated with the multipolar contribution is comparable to that from the short range potential.

The initial conditions for the shock Hugoniot of liquid nitromethane are material are taken as $T_0 = 293.15$ K and $\rho_0 = 1129.$ kg/m^3. Inspection of the shock Hugoniot in the pressure-volume plane in Fig. 3 shows that there is good agreement between the theory and experimental data[11] from 0.1 Gpa to 9.0 GPa. For a pressure of 10.0 Gpa, the angular effects are still substantial. Inspection of Fig. 4, reveals that the predictions for the shock temperature from the two techniques differ by approximately twenty per cent at P=5.0 GPa. The disparity is readily understood by a comparison of the configuration energies along the T=500. K isotherm exhibited in Fig 5. The main

difficulty arises from the short range potential contribution. Evidently, the point model of a molecule with seven atoms is just inadequate.

SUMMARY

Thermodynamic properties for high pressures have been calculated with MD and liquid state perturbation theory for a fluid with angular dependent intermolecular interactions. Isothermal compression at room temperature is well described by both approaches. The calculated shock Hugoniot

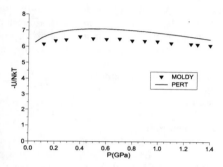

Figure 1. Configuration energy vs pressure for an modified Stockmayer potential for a reduced temperature of kT/ε=1.

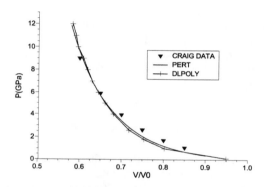

Figure 3. Pressure vs. compression for the shock Hugoniot of liquid nitromethane. The experimental points from Craig[11] are represented by the inverted triangles.

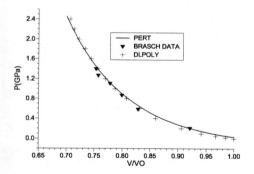

Figure 2. Pressure vs compression for liquid nitromethane for a temperature of 298.15 K. The experimental points from Brasch[10] are represented by the inverted triangles.

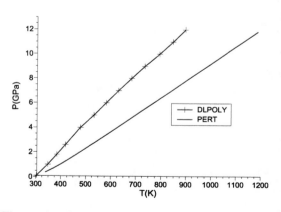

Figure 4. Pressure vs. temperature for the shock Hugoniot of liquid nitromethane.

151

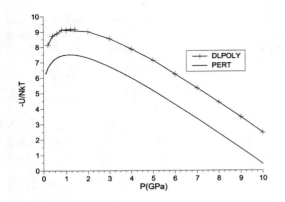

Figure 5. Configuration energy vs. pressure for liquid nitromethane for a temperature of 500. K

for nitromethane is in close agreement with the P vs volume data. However, the difference of the shock temperature indicates that the perturbation technique does not provide a sufficiently accurate description of a complex fluid, such as nitromethane.

ACKNOWLEDGMENTS

The Independent Research Program of the Naval Surface Weapons Center, Indian Head Division, supported this work. The author wishes to express sincere appreciation to Dr. Betsy Rice for many helpful discussions and supplying several files for DL_POLY.

REFERENCES

1. Cowperthwaite,M. and Shaw,R., J. Chem. Phys. 53, 555 (1970).
2. Enig, J.W and Petrone, F.J., Phys. Fluids, 9, 398 (1966).
3. Winey, J.M., Duvall, G.E., Knudson, M.D. and Gupta, Y.M., J. Chem. Phys. 113, 7492 (2000).
4. Sorescu, D.C., Rice, B.M. and Thompson, D.L., J. Phys. Chem. A 105, 9336 (2001).
5. Alper, H.E., Abu-Awwad, F. and Politzer, P., J. Phys. Chem. B 103, 9738 (1999).
6. Jones, H.D, "Theoretical Equation of State for Water at High Pressures", in Shock Compression of Condensed Matter-2001 edited by M.D. Furnish, N.N. Thadhani and Y. Horie 2002 American Institute of Physics 0-7354-0068-7
7. DL_POLY is a package of molecular simulation routines written by W. Smith and T.R. Forester, copyright The Council for the Central Laboratory of the Research Councils, Daresbury Laboratory at Daresbury, Nr. Warrington, 1996.
8. Sorescu, D.C., Rice, B.M. and Thompson, D.L., J. Phys. Chem. B, 104, 8406 (2000).
9. MOLDY is a molecular dynamics program written by Keith Refson, copyright 1989, 1993-1996 the Free Software Foundation, Inc.
10. Unpublished results from J.W. Brasch.
11. Craig, B.G., Los Alamos Scientific Laboratory Report Number GMX-8-MR-62-4, 1962.
12. Weeks, J.D., Chandler, D. and Anderson, H.C., J. Chem. Phys. 54, 5237 (1971).

CP706, *Shock Compression of Condensed Matter - 2003*
edited by M. D. Furnish, Y. M. Gupta, and J. W. Forbes
© 2004 American Institute of Physics 0-7354-0181-0/04/$22.00

CONSTANT VOLUME SPECIFIC HEAT CAPACITY OF THE CJ STATE OF NITROMETHANE

Julius Roth

Retired, Portola Valley, California

Abstract. The objective of the present study is the evaluation of c_v, the constant volume heat capacity along a locus of CJ states of nitromethane. Nitromethane (NM) was chosen because reliable measurements of its T_{CJ} are available, and there are requisite other data at varied initial densities ρ_0. Acquisition of the latter is accomplished by changing the initial temperature, but this also changes the initial internal energy E_0 and creates some problems. An equation for c_v was developed which contains the experimental T_{CJ}, assumed to be constant over the rather narrow range of ρ_0 of the other input data, and $\gamma = c_P/c_v$, which still needs to be evaluated. Thus only provisional values of c_v are presented.

INTRODUCTION

Reliable CJ variables are needed to validate EOS for CJ detonations. Presently only the detonation velocity D and its dependence on initial density ρ_0 have been measured precisely and reproducibly. Our previous studies (Ref 1, 2, 3) showed that the adiabatic exponent $\kappa = -(\partial \ln P / \partial \ln v)_S$ together with D can provide some of these needed values such as $\beta = (\partial \ln D / \partial \ln \rho_0)_{E_0}$, P, v, and $(\partial P / \partial v)_{E_0}$, $(\partial E / \partial v)_{E_0}$, $(\partial S / \partial v)_{E_0}$, $(\partial P / \partial v)_S$ the Jones $\alpha = P / (\partial E / \partial v)_P$, and the Grüneisen coefficient $\Gamma = v / (\partial E / \partial P)_v$. The present study attempts to extend the above methods to the use of κ with measured CJ temperatures T to estimate c_v, the CJ specific heat capacity at constant volume. The explosive examined is nitromethane (NM). It was chosen because there are measured T's for NM (Ref 4), and JCZ-3 calculations (Ref 5) indicate that T for NM is nearly constant over the narrow range of ρ_0 (and of v), for which experimental data are available. To avoid confusion, note that all symbols without subscripts are CJ state variables. All Greek letter symbols, except ρ, are dimensionless.

DEVELOPMENT OF EQUATIONS

The First Law provides a starting point to the numerical evaluation of c_v, namely: $(\partial S / \partial T)_v = T / c_v$. This used in the identity

$$(\partial T / \partial v)_{P_0} = (\partial T / \partial v)_S + (T / c_v)(\partial S / \partial v)_{P_0} \qquad (1)$$

advances the derivation (as shown in Appendix A) to give:

$$(\partial T / \partial v)_{P_0} = (\kappa / \Gamma)(P / c_v)(1 - \lambda) - \Gamma T / v \qquad (2)$$

To evaluate $(\partial T / \partial v)_{P_0}$ and c_v another equation is necessary, e.g. $\gamma = c_p / c_v = -(\kappa P / v)(\partial P / \partial v)_T^{-1}$, which, as shown in Appendix B results in

$$c_v = (1 - 1/\gamma)(\kappa / \Gamma)(Pv / \Gamma T) \qquad (3)$$

To get c_v in cal/(gK), P must be in cal/cm³. Inserting c_v into Eqn 1, as shown in Appendix B, gives $(\partial T/\partial v)_{P_0} = (\Gamma T/v)[(1-\lambda)/(1-1/\gamma)-1]$ (4)

Numerical values of κ, λ, etc. are shown below in "Input Data." The CJ temperature is assumed to be constant at 3430K (3). This leaves γ still undetermined. Thermochemical calculations (6) (7) provide γ. They require thermodynamic data for the expected dissociation of CJ products at high temperatures and pressures of several hundred bars, as well as computer codes which I do not have. Estimates of γ are shown subsequently.

INPUT DATA

The equations of the previous section have $v = \kappa/(\kappa+1)\rho_0$ as the independent variable, with ρ_0 and v varied by changing the initial temperature. This was done by Davis et al (8) and adapted to our purposes in (3). We used the Los Alamos data and some of their techniques in order to obtain $D_\infty = 6346 - 4.081T_0$ m/s where -4.081 m/s°C is $(\partial D/\partial T_0)_{P_0}$ m/s°C (3). Linear regression analysis gives $D_\infty = 2617.5 + 3207.3\rho_0$ m/s at a correlation of 0.99998. This then provides $\beta = d\ell nD / d\ell n\rho_0$, and subsequently λ (cf. Appendix A). Jones' $\alpha = (\kappa - 1 - 2\mu)/(1+\mu-\varepsilon)$ (3), is used to get $\Gamma = \kappa\alpha/(\alpha+1)$.

Numerical values of λ, κ, θ are shown as linear functions of v:

$\lambda = 0.865291 - 0.0813823v$, 0.99994
$\kappa = 3.130683 - 0.784344v$; 0.999998
$\theta = 0.024867 + 0.165231v$; 0.99998

The correlation coefficients are based on 21 individual entries for v, ranging from 0.60635 through 0.64537 cm³/g. This range of v corresponds to a ρ_0 range of 1.198 through 1.122 g/cm³, as a consequence of a T_0 range of -27.9 through 31.7°C. The range of T_0 is limited because NM freezes at -29°C, and because there is an abrupt change in the relation between ρ_0 and T_0 at about 30°C. P is best represented as follows:

P = 2259.647 $\rho_0^{2.038428}$; 0.999997.

The Grüneisen coefficient Γ decreases slowly from 0.47897 to 0.47789 over the above v range.

It can be approximated by an average $\overline{\Gamma} = 0.47845 \pm 0.00028$. However, over the range 0.60635 through 0.62399 cm³/g a fair fit is $\Gamma = 0.42935 - 0.022051v$, 09994. All of the above are empirical interpolation formulas and should not be used for extrapolations.

PROVISIONAL VALUES OF C_v

Additional confirmation of Eqn 3 is provided in Appendix C, where it is shown that, at large v, Eqn 3 becomes the expected $c_v = n R/(\gamma-1)$. With T ≡ 3430K, γ is the only undetermined element of the RHS of Eqn 3. JCZ-3 calculations (5) indicate that $dT/d\rho_0 \cong 0$ for NM at $\hat{\rho}_0 \cong 1.2$ g/cm³. Since

$d\rho_0/dv = -\rho_0^2 (\kappa+1)/(\kappa-\theta)$,

$dT/dv = -\rho_0^2 (\kappa+1)/(\kappa-\theta)(dT/d\rho_0)$, and then dT/dv is also zero. According to Eqn 4, $dT/dv = 0$ if $\hat{\lambda} = 1/\gamma$ at $\hat{\rho}_0 \cong 1.2$ g/cm³. We have chosen $\hat{\lambda} \cong 0.8519$ for $\hat{\rho}_0 = 1.195$ g/cm³ at $T_0 = -26°C$, because $\lambda = 0.8160$, at our largest ρ_0 of 1.198 g/cm³ and $T_0 = -27.9°C$, is too close to the freezing point of NM. With the above estimate of γ, we obtain the following provisional values of c_v. The following interpolation formula for c_v is not valid over the entire v-range. For v between 0.61656 and 0.63727 cm³/g, $c_v = 2.66481 - 2.36927v$ cal/gK (.99996). For $v > 0.63727$ cm³/g, $c_v = 1.1508$, 1.1461, 1.1375 cal/gK at $\rho_0 = 1.13320$, 1.12983, 1.122 g/cm³ respectively. For $v \le 0.61565$ g/cm³, $c_v = 1.2307$, 1.2276, 1.2071 cal/gK at $\rho_0 = 1.1980$, 1.19556, 1.17902 g/cm³ respectively.

All these c_v values are provisional pending appropriate calculations of γ.

THE DILEMMA OF $(\partial S/\partial v)_{P_0}$

We are concerned about the validity of Equations 2 and 4, because $(\partial S/\partial v)_{P_0}$ in Eqn 1 may have been improperly evaluated. In Appendix A, we used partial derivatives of S, which Davis et al

(8) derived for conditions where E_0 was constant. Thus we could have obtained $(\partial S/\partial v)_{E_0}$ and not $(\partial S/\partial v)_{P_0}$. It is readily shown that the identity $(\partial E/\partial v)_{E_0} = (\partial E/\partial v)_S + (\partial E/\partial S)_v (\partial S/\partial v)_{E_0}$ yields $(\partial S/\partial v)_{E_0} = (\kappa/\Gamma)(P/T)(1-\gamma)$ whose RHS is identical to the RHS of Eqn 1A of our Appendix A. Does this signify that $(\partial S/\partial v)_{P_0} = (\partial S/\partial v)_{E_0}$, or, as is more likely, that we employed the wrong partial derivative in Appendix A? In a later publication Davis (9) shows that $(\partial P/\partial S)_v = \Gamma T/v$, which we used in :

$(\partial P/\partial v)_{P_0} = (\partial P/\partial v)_S + (\partial P/\partial S)_v (\partial S/\partial v)_{P_0}$ to get the ubiquitous $(\partial S/\partial v)_{P_0} = (\kappa/\Gamma)(P/T)(1-\lambda)$. The above dilemma still holds. If $(\partial P/\partial S)_v = \Gamma T/v$ was obtained for conditions where E_0 is not constant, then $(\partial S/\partial v)_{P_0} = (\partial S/\partial v)_{E_0}$, but if $(\partial P/\partial S)_v = \Gamma T/v$ is a consequence of the derivations of (8), then our Equations 2 and 4 need correction.

The uncertainty in the form of Eqn 4 could affect the estimate of c_v shown in the preceding section. However, when γ is established by thermochemical calculations these estimates become superfluous and unequivocal values of c_v can be obtained.

CONCLUSION

This study proved to be more complicated than anticipated. To determine the locus of the required CJ parameters, one needs to vary the initial density of NM by changing its initial temperature. For most detonation studies T_0 is the ambient temperature and $dE_0/dv \cong 0$. In the present study the T_0 range is about 60°C and $dE_0/dv \neq 0$. How this affects dE_0/dv is still undetermined, as is its influence on $(\partial T/\partial v)_{P_0}$. As discussed in a preceding section, changes in $(\partial T/\partial v)_{P_0}$ will affect the current estimates of c_v, but when γ is properly evaluated, these estimates become unnecessary and c_v becomes well established.

APPENDIX A

Definitions

The following dimensionless terms have been evaluated previously: $\kappa = -(\partial \ell n P/\partial \ell n v)_S$ (1); $\alpha = P/(dE/dv)_P$ (3); $\theta = (\rho_0 d\kappa/d\rho_0)/(\kappa+1)$ (1). Equivalent forms of the Grüneisen coefficient: $\Gamma = -(\partial \ell n T/\partial \ell n v)_s = \kappa \alpha/(1+\alpha)$. A new term is $\lambda = (1+2\beta-\theta)/(\kappa-\theta)$ and $\beta = d\ell n D_\infty/d\ell n \rho_0$. β was called μ in (3), where α was shown to be $\alpha = (\kappa-1-2\mu)/1+\mu-\varepsilon$.

Derivations

Since T_0 was varied, use of partial derivatives at constant E_0 should be avoided. This was not always done in (3). In Eqn 1 the term $(\partial S/\partial v)_{P_0}$ can be evaluated as follows: $(\partial S/\partial v)_{P_0} = (\partial S/\partial v)_P + (\partial S/\partial P)_v (\partial P/\partial v)_{P_0}$. Davis (Ref 8, pp 2181-82) in a number of steps, derives $(\partial S/\partial v)_P = (P/T)(\kappa/\Gamma)$ and $(\partial S/\partial P)_v = (v/\kappa T)(\kappa/\Gamma)$, where we have replaced his $(1+\alpha)/\alpha$ by κ/Γ. $(\partial P/\partial v)_{P_0} = -(\kappa P \lambda)/v$, $[\lambda = (1+2\beta-\theta)/(\kappa-\theta)]$ was derived in (1) but appears there in an equivalent but different-looking form. $(\partial S/\partial v)_{P_0} = (P/T)(1-\lambda)(\kappa/\Gamma)$ (1A)

Putting Equation 1A into Equation 1 gives: $(\partial T/\partial v)_{P_0} = (\kappa/\Gamma)(P/c_v)(1-\lambda)$, which is Eqn 2.

APPENDIX B
Derivation of Equations 2 and 3

At large v, $\kappa = \gamma = c_P/c_v$ and according to (9) $\gamma = -(\kappa P/v)(\partial P/\partial v)_T^{-1}$ or $1/\gamma = -(v/\kappa P)(\partial P/\partial v)_T$ (1B)

Davis (10) in his Eqn 4.6 defines $g = (Pv)/(c_v T)$ and in Eqn 4.18 et seq. indicates that $\kappa - \Gamma^2/g = -(v/P)(\partial P/\partial v)_T$ or $1/g = [\kappa/\Gamma + (v/P\Gamma)(\partial P/\partial v)_T](1/\Gamma)$ and $(\partial P/\partial v)_T = [(c_v T\Gamma/Pv) - \kappa/\Gamma](P\Gamma)/v$ (2B)

155

and substituting (2B) into (1B) and rearranging we get $1/\gamma = -\Gamma/\kappa\left[-\kappa/\Gamma - (\Gamma T/v)(c_v/P)\right]$ and $c_v = (1 - 1/\gamma)(\kappa/\Gamma)(P_v/\Gamma T)$, which is Eqn 3. It is convenient to use Eqn 3 in this form since we had previously evaluated κ/Γ as well as $(Pv)/(\Gamma T)$. Now divide the first RHS term of Eqn 2 by c_v to get Eqn 4.

APPENDIX C

The equation for a perfect gas is $Pv = nRT$; $\kappa = \gamma = c_P/c_v$ and $\alpha = \Gamma = \gamma - 1$, where n is the number of moles per unit mass. The Jones equation gives $\beta = (\kappa + 1)/(2 + \alpha) - 1$, which becomes $\beta = (\gamma + 1)/(\gamma + 1) - 1 = 0$. Since both β and θ $= 0$, $\lambda = (1 + 2\beta - \theta)/(\kappa - \theta) - 1$ becomes $1/\gamma$. Equation 3 now reduces to: $c_v = nR/(\gamma - 1)$.

ACKNOWLEDGEMENTS

The author wishes to thank Mrs. Barbara Roth for her heroic efforts in putting the writer's messy manuscript into proper format. He also acknowledges helpful discussions with Dr. Michael Cowperthwaite concerning the "$(\partial S/\partial v)_{P_0}$ dilemma."

REFERENCES

1. Roth, J., "The Adiabatic Exponent in Steady Detonation," in *Proceedings of the Twentieth International Pyrotechnics Seminar,* Colorado Springs, CO, 1994, pp. 845-854.1

2. Roth, J., "Do JCZ and BKW EOS Need Modification?", in *Shock Compression of Condensed Matter, 1999,* edited by M.D. .Furnish *et al* , Snowbird, UT, 1999, pp. 219-222.

3. Roth, J. "CJ Parameters of Liquid Explosives," in *Proceedings of 27th International Pyrotechnics Seminar,* Grand Junction, CO, 2000.

4. Burton, J. T. A. *et al*, "Detonation Temperature of Some Liquid Explosives," *Seventh Symposium (International) on Detonation,* Annapolis, MD, 1981, pp. 759-767.

5. Hardesty, D. R. and Kennedy, J. E., "Thermochemical Estimation of Explosive Energy Output," *Combust. Flame,* 28:48 (1977), pp. 45-59.

6. Taylor, J., *"Detonation in Condensed Explosives,"* Oxford Press, London, 1952, pp. 82-83.

7. Stesik, L. N. and Shvedova, N. S., "Detonation of Condensed Explosives at Low Initial Densities," (in Russian), *PMTF* No. 4, 1964, pp. 124-126.

8. Davis, W. C. *et al*, Failure of the Chapman-Jouguet Theory for Liquid and Solid Explosives," *Physics of Fluids,* 8:12, Dec. 1965, pp. 2181-2182.

9. AMCP 706-180, "Principles of Explosive Behavior," U.S. Army Material Command, Washington, DC, 1972, p. 2-14, Eq. 2-73.

10. Davis, W. C., "Equation of State for Detonation Products," in *Eighth Symposium (International)on Detonation,* Albuquerque, NM, 1985, p. 789.

CP706, *Shock Compression of Condensed Matter - 2003*
edited by M. D. Furnish, Y. M. Gupta, and J. W. Forbes
© 2004 American Institute of Physics 0-7354-0181-0/04/$22.00

COMPLETE EQUATION OF STATE FOR β-HMX AND IMPLICATIONS FOR INITIATION

Thomas D. Sewell* and Ralph Menikoff*

Theoretical Division, MS-B214, Los Alamos National Laboratory, Los Alamos, NM 87545

Abstract. A thermodynamically consistent equation of state for β-HMX, the stable ambient polymorph of HMX, is developed that fits isothermal compression data and the temperature dependence of the specific heat computed from molecular dynamics. The equation of state is used to assess hot-spot conditions that would result from hydrodynamic pore collapse in a shock-to-detonation transition. The hot-spot temperature is determined as a function of shock strength by solving two Riemann problems in sequence: first for the velocity and density of the jet formed when the shock overtakes the pore, and second for the stagnation state when the jet impacts the far side of the pore. For a shock pressure below 5 GPa, the stagnation temperature from the jet is below the melt temperature at ambient pressure and hence insufficient for rapid reaction. Consequently, for weak shocks a dissipation mechanism in addition to shock heating is needed to generate hot spots. When the stagnation temperature is sufficiently high for rapid reaction, the shock emanating from the hot spot is computed, assuming a constant volume burn. For initial shocks below 20 GPa, the temperature behind the second shock is below 1000 K and would not propagate a detonation wave.

INTRODUCTION

Mesoscale simulations, which resolve the explosive grains and hot spots within a plastic-bonded explosive (PBX), are being used to better understand the initiation process. These simulations require as input the constitutive properties of the components of the PBX. A thermodynamically consistent equation of state (EOS) for the stable ambient polymorph β-HMX is developed below based on data from quasistatic compression experiments [1, 2] for the mechanical properties and molecular dynamics calculations of the crystal density of states for the thermal properties [3]. The specific heat is a function of temperature. This is particularly important for simulations that resolve hot spots since the chemical reaction rate is very sensitive to temperature.

The equation of state can be used to scope out hot-spot conditions in order to ascertain the dominant physical processes that a mesoscale simulation of initiation must incorporate. Assuming that hot spots are generated on a fast time scale behind a lead shock,

their rapid heating will result in reaction that can be approximated by a constant-volume burn. The burn results in a high pressure that generates a secondary shock wave emanating from the hot spot. The pressure of the constant-volume burn and temperature of the second shock depend on the strength of the lead shock wave. These quantities are determined by the equation of state.

FREE ENERGY

A thermodynamically consistent equation of state can be defined by the Helmholtz free energy. We assume that the free energy has the form of a generalized Hayes EOS [4]

$$F(V,T) = -\int_{V_0}^{V} dV\, P_0(V)$$
$$- \theta(V) \int_0^{T/\theta} \frac{d\widetilde{T}}{\widetilde{T}} \left(\frac{T}{\theta(V)} - \widetilde{T} \right) \widehat{C}_V(\widetilde{T}). \quad (1)$$

A Birch-Murnaghan form is used for the cold curve

$$P_0(V) = \tfrac{3}{2}K_0 \left[(V/V_0)^{-7/3} - (V/V_0)^{-5/3} \right]$$
$$\times \left(1 + \tfrac{3}{4}\left[K_0' - 4 \right]\left[(V/V_0)^{-2/3} - 1 \right] \right) . \quad (2)$$

The specific heat is assumed to have the form

$$\widehat{C}_V(\widetilde{T}) = \frac{\widetilde{T}^3}{c_0 + c_1\widetilde{T}^2 + c_2\widetilde{T}^2 + c_3\widetilde{T}^3} , \quad (3)$$

where $\widetilde{T} = T/\theta(V)$ is a scaled temperature. This form has the asymptotically correct limits: $\widehat{C}_V(T) \propto T^3$ for small T and $\widehat{C}_V(T) \to$ const for large T. The Debye temperature is assumed to have the form

$$\theta(V) = \theta_0 \left(\frac{V_0}{V} \right)^a \exp\left[b(V_0 - V)/V \right] . \quad (4)$$

Consequently, the Grüneisen coefficient is given by

$$\Gamma(V) = -\frac{V}{\theta}\frac{d\theta}{dV} = a + b\frac{V}{V_0} . \quad (5)$$

The pressure and internal energy are determined by the standard thermodynamic relations $P = -\partial_V F$ and $e = F - T\partial_T F$.

The EOS parameters for HMX are listed in table 1. These are fit to isothermal compression data [1, 5] and the temperature dependence of the specific heat from molecular dynamic calculations [3, p. 96]. Figures 1, 2 and 3 show a comparison between the data and the model EOS. The discrepancy of the thermal expansion coefficient is due to the neglect of the temperature dependence of either the bulk modulus or Grüneisen coefficient.

The sensitivity of the shock temperature to the specific heat is shown in figure 4. For dissipative mechanisms other than shock heating, a constant specific heat can have a large effect on hot spot temperature, and hence on chemical reaction rates.

ESTIMATES FOR INITIATION

One of the proposed mechanisms for generating hot spots is known as hydrodynamic pore collapse [9, §3.3]; the impact of a shock wave on a pore forms a micro-jet which subsequently stagnates on the far

TABLE 1. Model EOS parameters for β-HMX.

ρ_0	1.9	g/cm^3
K_0	16.5	GPa
K_0'	8.7	—
a	1.1	—
b	-0.2	—
c_0	5.265×10^{-1}	K·kg/MJ
c_1	3.073×10^2	K·kg/MJ
c_2	1.831×10^5	K·kg/MJ
c_3	4.194×10^2	K·kg/MJ

side of the pore, converting kinetic energy into thermal energy. This results in a localized region of high temperature, *i.e.*, a hot spot.

The peak hydrodynamic hot-spot temperature can be determined from the EOS as follows. The initial state (density, temperature and velocity) of the jet is determined by solving a Riemann problem between the shocked state of the explosive and the pore (gas or void). Due to convergence the tip of the jet will then accelerate. High resolution hydro simulations [10, these proceedings] show that the velocity of the jet tip can increase by a factor of 2. A second Riemann problem between the jet state and the ambient explosive then determines the hot-spot temperature.

For HMX based plastic-bonded explosives, temperatures over 1000 K are needed for a hot-spot induction time (tens of ns) short compared to the time-to-detonation (hundreds of ns) from Pop-plot data of shock initiation. When this occurs, the fast reaction in the hot spot can be approximated with a constant-volume burn. The pressure rise from the burn then gives rise to a secondary shock emanating from the hot spot. The temperature of the second shock can be calculated from the EOS. The burn states and shock states are shown in figures 5 and 6.

From figure 6 we see that a hot-spot temperature of at least 1000 K requires an incident shock pressure of about 10 GPa. Moreover, for shocks up to 20 GPa the second shock emanating from fast burning hot-spot has a temperature below 1000 K and is insufficient by itself to initiate a detonation wave. Hence, initiation of a detonation wave must be a collective phenomenon involving many hot spots.

Finally, we note that when the incident shock pressure is below 5 GPa, the hydrodynamic hot-spot temperature is below the melt temperature of HMX, ~ 550 K at ambient pressure. Hence, initiation from weak shocks requires an additional dissipative mechanism to increase the hot-spot temperature.

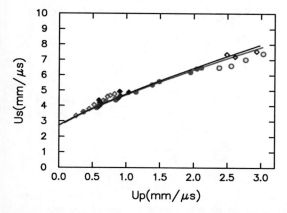 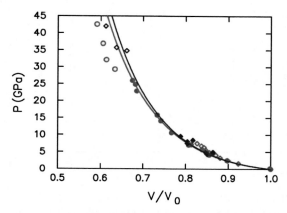

FIGURE 1. Mechanical response of HMX. Red and blue curves are isotherm and Hugoniot loci, respectively, from EOS model. Blue symbols are Hugoniot data [6, pp. 595–596] and red symbols are isothermal data [1, 2]. Open circles are above the pressure induced phase transition reported in [1] and are not used in the calibration of the EOS model.

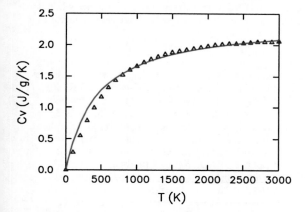 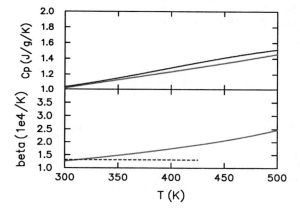

FIGURE 2. Specific heat. Symbols are from molecular dynamics calculations [3, p. 96] and red curve is fit, Eq. (3), used for the EOS model.

FIGURE 3. Specific heat [7] and the volumetric coefficient of thermal expansion [8] at atmospheric pressure. Blue curve is experimental data and red curve is model EOS.

CONCLUDING REMARKS

To gain an understanding of initiation from hot spots, mesoscale simulations should use chemical reaction rates. For weak shock (pressure below 5 GPa), a dissipative mechanism in addition to shock heating must be included. Proposed mechanisms include frictional heating across closed cracks or grain boundaries, and viscous heating which approximates rate-dependent plasticity. In addition, to evaluate reaction rates reliably, a simulation must accurately compute the temperature. This requires an EOS that accounts for the temperature dependence of the specific heat.

For HMX the reaction rate is only significant above melting. We note that the latent heat for melting is equivalent to $\Delta T \sim 200\,\text{K}$, and would have a large effect on the reaction rate. Furthermore, when HMX melts the volume change is substantial $\sim 8\%$ and confinement would affect pressure and hence the melt temperature. Constructing an EOS that accounts for melting requires data on the liquid state. Experimentally this would be difficult to obtain because the reaction rate becomes substantial above melting. The best chance for obtaining the needed liquid EOS data is from molecular dynamics simulations.

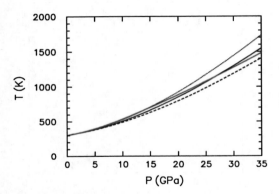

FIGURE 4. Temperature on Hugoniot locus. Red curve is model with temperature dependent specific heat. Blue curves have constant specific heat; $C_V = 1.2, 1.4, 1.6 \times 10^{-3}$ MJ/(kg·K) for dotted, solid and dashed lines, respectively. These values of C_V correspond to $T \simeq 360$K for dotted curve, temperature near β-δ transition ($T \simeq 440$K) for solid curve and near the melt temperature at ambient pressure ($550\,K$) for dashed curve.

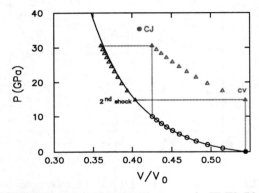

FIGURE 5. Loci relevant to hot spots in HMX. Black line is principal shock Hugoniot. Black circles indicate first shock. Red and blue triangles correspond to constant-volume burn from first shock and second shock to the burn pressure, respectively.

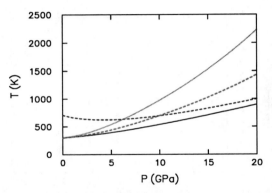

FIGURE 6. Temperature along loci in figure 5 parameterized by the pressure behind the first shock. Black line is temperature on principal shock Hugoniot. Dashed blue line is temperature of second shock at constant volume burn pressure from first shock. Red lines are estimates for hot-spot temperature based on a hydrostatic pore collapse; dashed line assumes no effect from convergence while dotted line assume velocity of jet tip increased by factor of 2.

ACKNOWLEDGMENTS

This work was carried out under the auspices of the U. S. Dept. of Energy at LANL under contract W-7405-ENG-36.

REFERENCES

1. Yoo, C., and Cynn, H., *J. Chem. Phys.*, **111**, 10229–10235 (1999).

2. Olinger, B., Roof, B., and Cady, H. H., "The linear and volume compression of β-HMX and RDX," in *Proc. Symposium (Intern.) on High Dynamic Pressures*, C.E.A., Paris, France, 1978, pp. 3–8.

3. Goddard, W. A., Meiron, D. I., Ortiz, M., Shepherd, J. E., and Pool, J., Annual technical report, Tech. Rep. 032, Center for Simulation of Dynamic Response in Materials, Calif. Inst. of Tech. (1998), http://www.cacr.caltech.edu/ASAP/onlineresources/publications/cit-asci-tr/cit-asci-tr032.pdf.

4. Sheffield, S. A., Mitchell, D. E., and Hayes, D. B., "The Equation of State and Chemical Kinetics for Hexanitrostilbene (HNS) Explosive," in *Sixth (International) Symposium on Detonation*, 1976, pp. 748–754.

5. Menikoff, R., and Sewell, T. D., *High Pressure Research*, **21**, 121–138 (2001).

6. Marsh, S., editor, *LASL Shock Hugoniot Data*, Univ. Calif. press, 1980, http://lib-www.lanl.gov/books/shd.pdf.

7. Koshigoe, L. G., Shoemaker, R. L., and Taylor, R. E., *AIAA Journal*, **22**, 1600–1601 (1984).

8. Herrmann, M., Engel, W., and Eisenreich, N., *Zeitschrift für Kristallographie*, **204**, 121–128 (1993).

9. Mader, C. L., *Numerical Modeling of Explosives and Propellants*, CRC Press, Baca Raton, FL, 1998, second edn.

10. Menikoff, R., "Porous Collapse and Hot Spots in HMX," in *Shock Compression of Condensed Matter*, 2003, these proceedings.

CHAPTER IV

PHASE TRANSITIONS

CP706, *Shock Compression of Condensed Matter - 2003*
edited by M. D. Furnish, Y. M. Gupta, and J. W. Forbes
© 2004 American Institute of Physics 0-7354-0181-0/04/$22.00

ISENTROPIC COMPRESSION EXPERIMENTS ON DYNAMIC SOLIDIFICATION IN TIN

Jean-Paul Davis, Dennis B. Hayes

Sandia National Laboratories, * *Albuquerque, NM 87185*

Abstract: Isentropic compression experiments were performed on molten tin (initial temperature 500-600 K), using the Sandia Z Accelerator to generate magnetically driven, planar ramp waves compressing the tin across the equilibrium liquid–solid phase boundary. Velocity interferometry measured time-resolved wave profiles at the tin/window interface. The experiments exhibit a departure from expected liquid response, time-dependent behavior above 8 GPa, and, at higher pressure, reduced wave speed relative to calculations using a nonequilibrium phase-mixture model. These phenomena may be due to a nonequilibrium solidification process, but verification of this conjecture will require further work.

INTRODUCTION

An isentropic compression loading technique (ICE) has been developed at Sandia Labs using fast pulsed power to provide ramped, planar, magnetic pressure loading of material samples over durations of several hundred nanoseconds.[1] The method is useful for the study of structural phase transitions under dynamic compression, particularly the solidi-fication transition not typically accessible by Hugoniot measurements due to the temperature rise under shock loading. Tin was chosen for the present study because it has a relatively simple phase diagram, a low melt temperature, and its high-pressure melting transition has previously been studied under dynamic release from Hugoniot states.[2]

Figures 1-2 show Hugoniot curves and various ICE loading paths in T-p and p-ρ space, computed using a three-phase model for tin described in the next section. The 600-K Hugoniot enters the mixed-phase region, but never completely transforms to the γ-solid phase, instead returning to the liquid at about 24 GPa. The metastable curves are

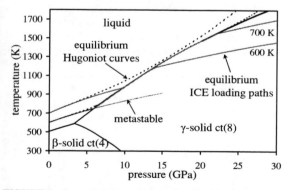

FIGURE 1. Computed phase diagram for tin with equilibrium Hugoniot curves (dashed) and both equilibrium (solid gray) and metastable (dotted gray, 600 K only) ICE loading paths for initial temperatures of 600 K and 700 K.

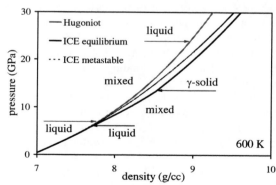

FIGURE 2. Computed compression curves for tin at an initial temperature of 600 K; equilibrium Hugoniot (gray), equilibrium (solid black) and metastable (dotted) ICE loading paths.

* Sandia is a multi-program laboratory operated by Sandia Corporation, a Lockheed Martin Company, for the United States Department of Energy under Contract DE-AC04-94AL85000.

given by extrapolation of the liquid equation-of-state (EOS) to over-compressed states. Unlike the Hugoniot curve, the equilibrium quasi-isentropic ICE loading path in Fig. 2 shows a small discontinuity in sound speed upon entering the mixed-phase region. Further computations (not shown) indicate that this will produce a signature in the stress-wave profile, but only for a short propagation distance due to rapid steepening of the wave above the transition. Thus, for dimensions typical of the present experiments, solidification can only be detected if the transformation rate is greater than about 10^7 s^{-1}.

THREE-PHASE MODEL FOR TIN

Calculations in the present work use a three-phase homogeneous mixture model for tin based on assumptions introduced by Horie and Duvall.[3] The method was developed by Andrews,[4] extended to N phases by Hayes,[5] and implemented in the WONDY hydrocode.[6] Each of the pure phases (α-solid, γ-solid, and liquid) has its own EOS, and extensive properties of the mixture are mass-averaged over all phases. For the EOS of each phase, a Birch isotherm is assumed with constant c_v and Γ/v. EOS parameters and reference-state values are based on those given by Mabire and Héreil.[2]

Transformation kinetics is introduced by a simple phenomenological model wherein the rate of change of mass fraction for a given phase is proportional to the weighted sum of the differences in Gibbs free energy from each other phase;[4]

$$\dot{x}_i = \sum_j \frac{G_j - G_i}{A_{ij}},$$

where A_{ij} is a constant of proportionality for transformation from phase i to phase j. For each transformation, a constant Gibbs energy offset ΔG_{ij} can be specified to represent metastable and hysteretic behaviors; for a given loading rate, this implicitly defines a characteristic time having the effect of a nucleation time. The goal of this macroscopic thermodynamic approach is not to illuminate underlying physics of dynamic freezing, but rather to connect experimental data to continuum modeling.

EXPERIMENTAL APPARATUS

A resistive band heater was used to heat a thin liquid cell of 6.8 mm diameter and Mo side walls clamped between a sapphire insulator and a top-hat

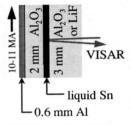

stepped sapphire or LiF interferometer window. Vapor-deposited Ni films promoted wetting of these substrates. Redundant fine-gauge thermocouples bonded into grooves machined on the insulator surface measured the temperature of the molten tin.

FIGURE 3. Sketch of the experimental configuration.

The insulator was bonded to the anode drive panel. The sample preparation process caused contamination of the Sn by up to 0.2% Cu and Fe-Cr alloy.

The experiment generated magnetic loading on the Al anode panel, launching a centered compression wave into the series of material layers shown in Fig. 3. Time-resolved particle velocity at the tin/window interface was measured using Valyn VISAR velocity interferometers with bare-fiber probes. At a separate location on the anode panel, in-situ particle velocity was measured at the interface between two sapphire discs in order to estimate the stress-wave profile in the sapphire insulator.

EXPERIMENTAL CONDITIONS

The experiments consisted of four anode panels arranged in a square configuration and subjected to a pulse of approximately 10-11 MA peak current and 400 ns rise-time, which, according to the present EOS for liquid tin and neglecting transformation to the solid state, produced approximately 11-13 GPa peak stress in the molten tin samples.

The difference in sample and window acoustic impedances and resulting ramp-wave interactions

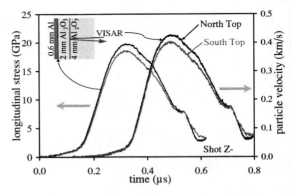

FIGURE 4. Particle velocity measurements in sapphire from two panels on shot Z-1008, and effective stress loading histories computed by backward integration through 2 mm of sapphire.

preclude direct Lagrangian analysis of the experimental data. If the isentropic compression response of sapphire is known, a time-resolved measurement of its in-situ particle velocity can be used to compute an effective loading profile by numerically integrating the equations of motion backward in Lagrangian coordinate from the measurement position,[7] as shown in Fig. 4. VISAR window corrections were taken the same as for shock loading.[8] Sapphire was modeled using the Mie-Grüneisen EOS (with Γ_0=1.5) referenced to the elastic Hugoniot.[9] It should be emphasized that backward integration is valid strictly for isentropic flow, i.e., for centered compression waves where sound speed is a unique function of particle velocity (or pressure).

If this backward-computed stress history lies outside the causal region of influence of the measured velocity history, then it can be used as a boundary condition in forward wave-propagation calculations, neglecting magneto-hydrodynamics, for a configuration where the in-situ measurement window is replaced by another material. This allows an indirect analysis of experimental data by comparison to computations using various models for tin.

PRELIMINARY RESULTS

Figures 5-6 show comparisons between experiment and computation using the above method. Thermal softening of the sapphire longitudinal modulus was negligible for the present temperature,[10] while that of the LiF bulk modulus was accounted for using the Los Alamos SESAME tabular EOS. Sample thickness was estimated from the room-temperature sample cell geometry and constant thermal expansion coefficients for the materials involved. In some cases (Fig. 6), calculations suggested the sample was thicker than estimated by up to 25 μm, possibly due to assembly issues.

Equilibrium calculations used $A_{\text{liq-}\gamma} = A_{\gamma\text{-liq}} = 0.01$ erg-s/g, resulting in only small departures from an equilibrium phase transition (note the rounded peak of the first wave). Frozen calculations used $A_{\text{liq-}\gamma} = A_{\gamma\text{-liq}} = 10^6$ erg-s/g, causing the tin to remain in the liquid state during the time of interest. The frozen case more accurately represents the experimental data, suggesting that the solidification process in tin is slow relative to the time scale of the experiment.

In an attempt to more closely match two of the experimental profiles, nonequilibrium calculations

were performed using intermediate $A_{\text{liq-}\gamma}$ to slow the transformation and non-zero ΔG_{ij} to delay onset of the transformation to higher pressure. By adjusting parameters, it was possible to reproduce the initial departure from the frozen velocity profile, but not the subsequent apparent decrease in wave speed.

The backward integration method, used above to compute forward-calculation boundary conditions, was also applied directly to the molten tin velocity data. Using an iterative technique,[7] the measurements were fit by an isentrope expressed in terms of Lagrangian sound speed c_L as a polynomial in pressure. The results at low pressure in Fig. 7 are consistent with propagation of a centered compression wave, within experimental uncertainty. At high pressure, however, the isentrope derived from the thicker sample deviates significantly from the thin-

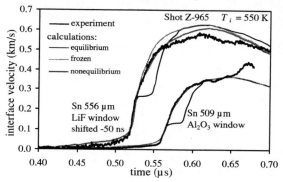

FIGURE 5. Experimental and calculated window-interface velocity profiles for two molten tin samples on shot Z-965. The terms equilibrium and frozen refer to the kinetics of solidification. Nonequilibrium calculation used $A_{\text{liq-}\gamma}$=8.0 erg-s/g and $\Delta G_{\text{liq-}\gamma}$=4×$10^7$ erg/g. Time origin is tied to a machine fiducial.

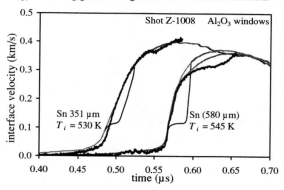

FIGURE 6. Window-interface velocity profiles for two samples on shot Z-1008; see Fig. 5 for legend. Nonequilibrium calculation used $A_{\text{liq-}\gamma}$=5.0 erg-s/g and $\Delta G_{\text{liq-}\gamma}$=1.2×$10^7$ erg/g. Thickness of second sample was adjusted to match wave arrival time.

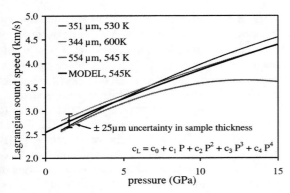

FIGURE 7. Results from backward-integration fit to quartic $c_L(p)$ isentrope for three different samples on shot Z-1008, compared to the present model for liquid tin. Deviation of the thicker sample suggests time-dependent behavior above 10 GPa.

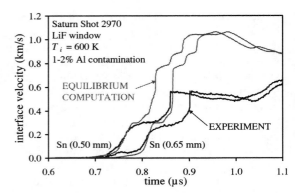

FIGURE 8. Experimental and calculated velocity profiles from an earlier experiment with contaminated molten tin, suggesting a strong initiation of freezing and transition to a metastable phase.

ner samples. In other words, the assumptions required for backward integration are violated, suggesting a time-dependent response above 8 GPa.

CONCLUSIONS

Preliminary results are inconsistent with extrapolation of the liquid tin EOS to higher pressure, and analysis of the measured velocity profiles shows time-dependent behavior in liquid tin above 8 GPa. A simple phenomenological model for transformation kinetics successfully reproduces initial departure from expected liquid tin response, but fails to capture subsequent behavior. Though not conclusive, this evidence suggests that the solidification transition in tin can initiate during ramped compressions of less than 100 ns duration, but does not follow an equilibrium path to a fully transformed state of γ-solid. The cause for reduced wave speed at higher pressure in the experiments relative to computations is unknown. Possible explanations include formation of an amorphous solid phase instead of the crystalline γ-solid phase, changes in phase velocity due to time-dependent behavior, or ramp-wave interactions between the window interface and a nonequilibirum transformation front. Further work is required to develop these ideas.

Interestingly, an earlier experiment with more strongly contaminated samples indicated rapid freezing, as shown in Fig. 8. Differences from the present experiments included a glass ceramic insulator, copper anode panels, an aluminum sample holder, and an uncoated split LiF window. The tin samples were contaminated with 1-2% aluminum,

some of which probably remained in solid form since the temperature during remelt for the experiment was lower than the temperature during sample preparation. Comparison of Fig. 8 with Figs. 5-6 suggests that the presence of solid crystallites could have a profound effect on the initiation kinetics of solidification. Figure 8 also shows a low wave speed above the transition (relative to computation) and a late-time velocity rise that appears to be due to response of the tin itself (edge effects and reverberation have been eliminated as possible causes).

REFERENCES

1. Hall, C. A. et al, *Rev. Sci. Instrum.* **72**, 3587–3595 (2001).
2. Mabire, C. and Héreil, P. L., *J. Physique IV* **10**, Pr9/749–754 (2000).
3. Horie, Y. and Duvall, G. E., *Behavior of Dense Media under High Dynamic Pressure*, IUTAM Symposium on High Dynamic Pressure, Gordon and Breach, New York, 1968, pp. 355–359.
4. Andrews, D. J., *J. Comp. Phys.* **7**, 310–326 (1971).
5. Hayes, D. B., *J Appl. Phys.* **46**, 3438–3443 (1975).
6. Kipp, M. E. and Lawrence, R. J., *WONDY V – A One-Dimensional Finite-Difference Wave Propagation Code*, Sandia National Laboratories Report SAND81-0930, 1982.
7. Hayes, D. B., *Backward Integration of the Equations of Motion to Correct for Free Surface Perturbations*, Sandia National Laboratories Report SAND2001-1440, 2001.
8. Hayes, D. B., *J Appl. Phys.* **89**, 6484–6486 (2001).
9. Barker, L. M. and Hollenbach, R. E., *J Appl. Phys.* **41**, 4208–4226 (1970).
10. Tefft, W. E., *J. Res. National Bureau of Standards* **70A**, 277–280 (1966).

CP706, *Shock Compression of Condensed Matter - 2003*
edited by M. D. Furnish, Y. M. Gupta, and J. W. Forbes
© 2004 American Institute of Physics 0-7354-0181-0/04/$22.00

TIME DEPENDENT FREEZING OF WATER UNDER MULTIPLE SHOCK WAVE COMPRESSION

D.H. Dolan and Y.M. Gupta

Institute for Shock Physics and Department of Physics, Washington State University, Pullman, WA 99164-2816

Abstract. Using shock wave reverberation experiments, water samples were quasi-isentropically compressed between silica and sapphire plates to peak pressures of 1–5 GPa on nanosecond time scales. Real time optical transmission measurements were used to examine changes in the compressed samples. Although the ice VII phase is thermodynamically favored above 2 GPa, the liquid state was initially preserved and subsequent freezing occurred over hundreds of nanoseconds only for the silica cells. Images detailing the formation and growth of the solid phase were obtained. These results provide unambiguous evidence of bulk water freezing on such short time scales.

INTRODUCTION

At thermodynamic equilibrium, a material assumes the configuration or phase that minimizes the Gibbs free energy at a particular pressure (P) and temperature (T) [1]. When P,T conditions are altered to favor a new phase, the system requires some time to reach its new configuration. Transformation times are often long enough (> 1 s) for the properties of the metastable state to be studied [2]. Phase transition studies require the proper P,T conditions and appropriate observational times.

Shock wave experiments produce rapid ($10^{-12} - 10^{-9}$ s) changes in the P,T state of a material, permitting fast phase change kinetics to be probed [3]. Several polymorphic and melting transitions have been studied previously under shock wave loading [4]. However, shock-induced freezing remains a challenge because it presents some fundamental difficulties. Shock wave compression is an irreversible, adiabatic process [4], and liquids undergo large volume changes ($>15\%$) at modest pressures (1-2 GPa). These factors result in large temperature increases that can prevent the P,T conditions required for freezing. Even if proper P,T conditions for freezing can be achieved, shock wave experiments are typically limited to 10^{-6} s durations. Given the long relaxation times observed in freezing [2], it is not clear that freezing can be observed in shock compression experiments.

Previous shock-induced freezing attempts have focused largely on ordinary liquid water [4–9]. Apart from the general importance of water [10], the large specific heat of the liquid phase moderates the temperature rise during adiabatic compression. Fig. 1 shows two calculated [11] limiting cases for adiabatic compression of liquid water initially at 1 bar and 25°C; the dark lines indicate experimentally determined phase boundaries [12]. Single shock compression, which represents the high temperature limit, approaches the ice VII phase boundary [12] near 4 GPa. Whether water actually freezes under single shock compression remains an open question [4]. Isentropic compression represents the low temperature limit of adiabatic loading; this curve passes deeply into the ice VII region for pressures above 2 GPa. If adiabatic freezing is possible, it will most likely occur along an isentrope.

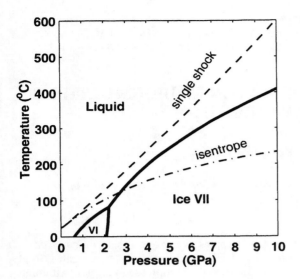

FIGURE 1. Calculated bounding P-T curves for adiabatic loading of liquid water.

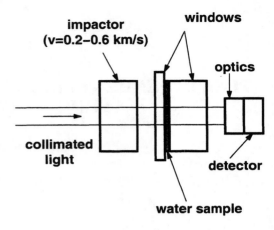

FIGURE 2. Schematic view of plate impact experiments on liquid water.

EXPERIMENTS

Quasi-isentropic compression was achieved through multiple shock wave loading [13] using the experimental configuration shown in Fig. 2. A thin (15-130 μm) water sample was confined between two optical windows (typically, quartz or sapphire) and struck with an impactor accelerated by a light gas gun [14]. The shock wave produced upon impact traversed the front window and reached the water sample at a time denoted as $t = 0$. The mechanical impedance mismatch between the liquid sample and the windows produced a series of reverberating shock waves that progressively compressed the liquid to the peak state. The pressure in the peak state was determined solely by the impact velocity and the choice of impactor/window materials [13]. In the limit of infinitesimal shock wave steps, the loading process becomes exactly isentropic. Numerical simulations [15] of the impact configuration in Fig. 2 indicated that the temperature rise during shock reverberation closely follows the isentropic path shown in Fig. 1.

The compressed water sample was optically transparent while it remained in a pure liquid state. Freezing was detected by a loss of transparency from optical scattering [16] due to the coexistence of liquid and solid phases; there is no optical absorption in water at the pressures of interest. These changes were probed in real time by passing collimated visible light from a pulsed xenon flash lamp through the water sample (Fig. 2). The illuminated region (1/4" diameter) was far from the sample edges ($\geq$ 1" diameter) to ensure a state of uniaxial strain for times of interest ($\sim$1 μs). Light passing through the sample was recorded with a fast photodiode detector (ns resolution) to measure an average transmission in the visible spectrum ($\approx$400-700 nm). Sample transmission was defined by the ratio of light intensity measured during the impact experiment and the intensity measured at ambient conditions. A pure liquid sample thus has a transmission of unity; any value less than unity indicates freezing. Images of the water samples were also acquired using a high speed framing camera. Snapshots (25 ns exposure time) were taken during the impact experiment and compared to images acquired at ambient conditions. Liquid regions were denoted by bright areas of the image; frozen regions appeared dark. Optical transmission and imaging data were obtained for water compressed to peak pressures ranging from 1.1 to 5.1 GPa.

RESULTS AND DISCUSSION

A typical optical transmission measurement is shown in Fig. 3. In this experiment, a 132 ± 5 μm thick

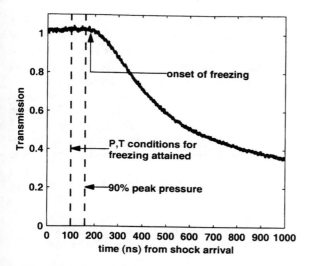

FIGURE 3. Typical optical transmission history for a water experiment with z-cut quartz windows.

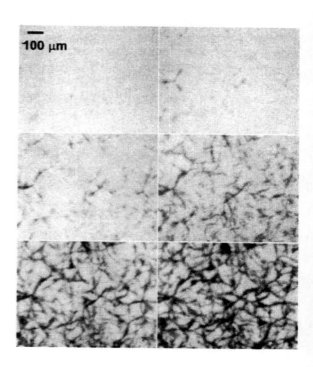

FIGURE 4. Images of water freezing under multiple shock compression. Beginning from the top left corner, these images were acquired at $t = 25, 175, 225, 275, 325,$ and 375 ns (moving left to right). Each image was exposed for 25 ns.

water sample confined within z-cut quartz windows was compressed to a peak pressure of 3.6 GPa. The sample reached 90% of the peak pressure in about 160 ns. The necessary P,T conditions for freezing were attained approximately 100 ns after shock arrival, but optical loss was not observed until about 80 ns later. After this incubation time, transmission dropped continuously and approached a steady state value at t>1000 ns. Similar transmission losses were observed in other water experiments when the peak pressure was greater than 2 GPa; below that pressure, the sample remained transparent. Transparency was expected at low pressures because the sample stays in the liquid phase below 2 GPa (Fig. 1). The rate of transmission loss was related to the pressure in the peak state. As this pressure increased, the incubation time became shorter; transmission loss after incubation also occurred more rapidly at high pressure. The initial water sample thickness also played a role in the transition. Thin samples lost transparency faster than thick samples because the reverberation process required less time. However, thick samples ultimately reached lower transmission values than thin samples. Lower transmission values in thick samples is an important result because it points out that the phase transformation spans the thickness of the compressed water sample.

The images shown in Fig. 4 were acquired during the same experiment, described in Fig. 3. The first frame was taken just after the shock entered the water. The onset of freezing was evident on the second frame, which was taken shortly after 90% peak pressure was reached. An intermediate frame (not shown) indicated that the sample was clear until at least $t = 150$ ns. The water sample did not lose transparency all at once, but instead transformed into a complex network of frozen regions. Freezing began from a number of independent sites distributed throughout the sample ($\sim 100~\mu m$ separation). Over time, the frozen regions became larger, covering more of the sample, although residual liquid was always present in the sample. As pressure was increased, more independent freezing sites were observed and less of the sample remained liquid.

The experiment shown in Fig. 3 and 4 was performed using distilled water in z-cut quartz windows. Similar results were obtained for water compressed in fused silica and soda lime glass windows. Freezing was also observed in water samples purified by different deionization, filtration, and distillation treatments, so the transformation did not depend on the details of sample preparation. It was necessary, however, for water to be in contact with at least one silica window for freezing to occur. Transmission and imaging experiments performed with a-cut sapphire windows did not show the optical changes observed using crystalline and amorphous silica windows. If one of the sapphire windows was replaced with z-cut quartz, freezing was observed, although the transmission loss was smaller than for experiments using two quartz windows.

These results can be understood qualitatively in terms of classical nucleation theory [2]. Freezing is preceded by a metastable period where stable ice nuclei are formed within the liquid phase. The nuclei form at a rate per unit volume , J, which depends upon the Gibb's free energy difference $\Delta g = g_l - g_s$ between the liquid and solid phases.

$$J \propto \exp\left(-\frac{B}{(\Delta g)^2}\right) \qquad (1)$$

B is determined by the geometry and surface energies of the solid nucleus. Since optical changes were observed only in the presence of silica windows, it can be concluded that B is much greater than $(\Delta g)^2$ in most of the water sample. Near a silica surface, however, B is reduced by interactions that favor the solid phase. Thus freezing is heterogeneously nucleated in these experiments. The fact that freezing does not begin without a silica window suggests that bulk impurities, such as dissolved gases, are of secondary importance on these time scales. A variety of surface effects have been reported at water-silica interfaces in other types of studies [17, 18], and the present results may be a manifestation of such phenomena. These interactions are presumably absent at sapphire interfaces, and freezing, dominated by homogeneous nucleation, may be too slow to be observed. The details of surface effects in these experiments are still under investigation at this time. When a silica surface is present, J is strongly tied to the value of Δg. As reverberation pressure increases, liquid water moves further into the solid domain (Fig. 1), so Δg becomes larger. Nucleation thus becomes more likely, shortening the metastable lifetime of the liquid phase and reducing the observed incubation time. Each of the independent features observed in Fig. 4 must have originated from a separate nucleation event, so it would be expected that the number of these features scales with the number of nucleation events. The average number of nucleation events in a fixed volume is proportional to J, so increases in peak pressure should lead to more independent features as observed.

Once ice nucleates at the water-silica interface, it grows both along and away from the windows. The former process can be seen directly in the imaging experiments (Fig. 4); the latter process can be inferred from the decrease in transmission for thicker water samples. The growth dynamics of freezing is strongly tied to the dissipation of latent heat, which leads to a complex pattern formation [19]. The sharp, irregular features shown in Fig. 4 are consistent with freezing in a metastable liquid, where growth instabilities form to maximize the heat dissipation.

SUMMARY

Under quasi-isentropic compression, time dependent freezing of water was observed. The transition occurred when the liquid phase was metastable with respect to ice VII. These measurements do not provide structural details about the particular solid phase formed, so it is unclear whether ice VII was formed. Wave profile measurements are being carried out to verify the details of this phase transition.

Freezing was initiated by heterogeneous nucleation sites located on amorphous and crystalline silica windows. Sapphire lacks these nucleation sites and did not induce freezing in compressed water.

ACKNOWLEDGMENTS

The authors wish to thank Kurt Zimmerman, Dave Savage, and Gary Chantler for their assistance in performing the experiments. Jim Johnson is thanked for discussions regarding the development of the equation of state. This work was supported sponsored by DOE Grant DE-FG03-97SF21388.

REFERENCES

1. H.B. Callen, *Thermodynamics*, Wiley, New York, 1985.
2. P.G. Debenedetti, *Metastable Liquids*, Princeton University Press, Princeton, 1996.

3. M.D. Knudson, Y.M. Gupta, *Phys. Rev. Lett.*, **81**, 2938 (1998).
4. G.E. Duvall, R.A. Graham, *Rev. Mod. Phys.*, **49**, 523 (1977).
5. J.M. Walsh, M.H. Rice, *J. Chem. Phys.*, **26**, 815 (1957).
6. L.V. Al'tshuler, A.A. Bakanova, R.F. Trunin, *Sov. Phys. Doklady*, **3**, 761 (1958).
7. Ya.B. Zel'dovich, S.B. Kormer, M.V. Sinitsyn, K.B. Yushko, *Sov. Phys. Doklady*, **6**, 494 (1961).
8. S.B. Kormer, K.B. Yushko, G.V. Krishkevich, *Sov. Phys. JETP*, **27**, 879 (1968).
9. A.P. Rybakov, *J. App. Mech. Tech. Phys.*, **37**, 629 (1996).
10. D. Eisenberg, W. Kauzmann, *The Structure and Properties of Water*, Oxford University Press, New York, 1969.
11. D.H. Dolan, Ph.D. thesis, Wasington State University, WA (2003).
12. C.W. Pistorius, E. Rapoport, J.B. Clark, *J. Chem. Phys.*, **48**, 5509 (1968).
13. K.M. Ogilvie, G.E. Duvall, *J. Chem. Phys.*, **78**, 1077 (1983).
14. G.R. Fowles, G.E. Duvall, J. Asay, P. Bellamy, F. Feistmann, D. Grady, T. Michaels, R. Mitchell, *Rev. Sci. Instrum.*, **41**, 984 (1970).
15. Y.M. Gupta, COPS wave propagation code (Stanford Research Institute, Menlo Park, CA, 1978), unpublished.
16. H.C. van de Hulst, *Light scattering by small particles*, Dover, New York, 1957.
17. K. Klier, A.C. Zettlmoyer, *J. Colloid Interface Sci.*, **58**, 216 (1977).
18. Q. Du, E. Freysz, Y.R. Shen, *Phys. Rev. Lett.*, **72**, 238 (1994).
19. J.S. Langer, *Rev. Mod. Phys.*, **52**, 1 (1980).

CP706, *Shock Compression of Condensed Matter - 2003*
edited by M. D. Furnish, Y. M. Gupta, and J. W. Forbes
© 2004 American Institute of Physics 0-7354-0181-0/04/$22.00

MELTING AT THE LIMIT OF SUPERHEATING

Sheng-Nian Luo,[1] Thomas J. Ahrens,[2] and Damian C. Swift[1]

[1]*P-24 Plasma Physics, Los Alamos National Laboratory, Los Alamos, NM 87545*
[2]*Seismological Laboratory, California Institute of Technology, Pasadena, CA 91125*

Abstract. Theories on superheating-melting mostly involve vibrational and mechanical instabilities, catastrophes of entropy, volume and rigidity, and nucleation-based kinetic models. The maximum achievable superheating is dictated by nucleation process of melt in crystals, which in turn depends on material properties and heating rates. We have established the systematics for maximum superheating by incorporating a dimensionless nucleation barrier parameter and heating rate, with which systematic molecular dynamics simulations and dynamic experiments are consistent. Detailed microscopic investigation with large-scale molecular dynamics simulations of the superheating-melting process, and structure-resolved ultrafast dynamic experiments are necessary to establish the connection between the kinetic limit of superheating and vibrational and mechanical instabilities, and catastrophe theories.

INTRODUCTION

Melting and freezing as first-order phase changes and their related kinetics, are of ubiquitous theoretical and experimental interest in condensed matter physics, materials science and engineering, geophysics and planetary sciences.[1] Metastable superheating and undercooling are inherent in melting and freezing processes. Determining the degree to which a solid can be superheated and a liquid undercooled, is a fundamental and challenging issue. Experimental investigation of the maximum superheating is particularly difficult due to the existence of heterogeneous nucleation sites (e.g. free surfaces and defects), and the difficulty in achieving high heating rates while making sensible measurements. Theoretical efforts in understanding superheating-melting have been seriously undermined by the paucity in superheating data. Molecular dynamics (MD) simulations have been utilized to probe melting and freezing processes at atomic level, and serve an important complementary approach to theoretical and experimental techniques.

Previous superheating-melting theories[1] including Lindemann and Born's criteria, order-disorder transition, catastrophes of entropy, volume and rigidity, and nucleation-based kinetic models are briefly reviewed. We present certain details on the recently developed systematics for the maximum superheating and undercooling.[1] Experimental, theoretical and simulation directions for future investigations of superheating-melting process are presented.

SUPERHEATING-MELTING THEORIES

Solids differ distinctly from liquids in both their long-range order and ability to resist shearing. The definitions and criteria for melting mostly involve vibrational and mechanical instabilities and order-disorder transitions.[1] Lindemann's vibrational criterion[1] states that melting occurs at the onset of an instability when the atomic displacements (e.g. the root-mean-squared displacements) during thermal vibrations exceed a certain threshold. Born's mechanical criterion[1] states that the stability against shearing stress vanishes (e.g. for cubic lattice $c_{44} = 0$ where c_{44} is the elastic constant in

Voigt's notation) and such shearing instability is essentially melting. Melting is also interpreted as structure transition from order to disorder as proposed by Lennard-Jones and Devonshire.[1] Such a order-disorder transition is arguably attributed by Cahn[1] to the spontaneous production of intrinsic lattice defects.

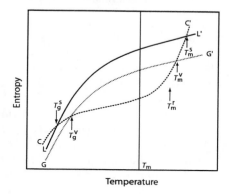

FIGURE 1. Schematic of entropy vs. temperature: A hierarchy of catastrophes as a succession of stability limits for the crystalline state. After Tallon.[1]

These instabilities in thermal vibration and resistance to shearing, and breakdown of long-range order are definitions of melting which inadequately describe the mechanism of melting, i.e. the kinetics of melting. Similarly, equilibrium thermodynamics simply states that liquid has lower Gibbs free energy than solid above melting temperature, T_m. Without considering nucleation process, catastrophes of certain physical quantities (e.g. molar entropy s, molar volume v and rigidity r) are employed by Tallon[1] to define a hierarchy of the limit of superheating (Fig. 1). The entropy of solid (along CC') intersects that of liquid (LL') at T_g^s and T_m^s ($T_g^s < T_m^s$). According to Kauzmann,[1] the glass transition preempts at the catastrophe point T_g^s, an undercooling state. Fecht and Johnson[1] thus proposed that melting preempts at T_m^s, the counterpart of T_g^s. The maximum superheating occurs at the T_m^s where the entropies of solid and liquid are equal. By correcting the liquid entropy for communal entropy of the liquid (e.g. $R\ln 2$ for certain monatomic systems where $R \equiv$ gas constant), the entropy line GG' intersects that of solid at T_g^v and T_m^v ($T_g^v < T_m^v$) where the

molar volumes of liquid (glass) and crystal are equal. Beyond T_m^v, liquid would become denser for normal materials, thus melt preempts at T_m^v. Isochoric melting defines a smaller superheating T_m^v at the volume catastrophe. Tallon further argued that rigidity instability occurs where the density of the superheated crystal equal to that of the liquid at the freezing point. This defines a superheating state T_m^r at the catastrophe of rigidity. The hierarchy of catastrophes of a succession of stability limits for crystalline state is elastic rigidity (T_m^r), volume (T_m^v) and entropy (T_m^s).[1]

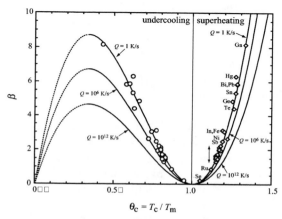

FIGURE 2. The systematics[1] of maximum superheating and undercooling for elements: $\beta = (A_0 - b \lg Q)\theta_c (1 - \theta_c)^2$. Circles are experimental value of undercooling at cooling rate $Q \sim 1$ K/s, and diamonds are calculated superheating at $Q \sim 1$ K/s. Solid and dotted curves are plots with $Q = 1$, 10^6 and 10^{12} K/s, respectively. Dotted curves denote the undercooling portions for $\theta_c = 0 - 1/3$. The maximum of β for undercooling occurs at $\theta_c = 1/3$ for each Q. The elements within the double-headed arrow are Ti, Al, Au, Cu, Hf, Cd, Pd, Ag, Co, Pt, Ta, Rh, Zr, Mn, Si, Sb, Ni, In and Fe in β-increasing order.

The concept of defining the limit of superheating by a hierarchy of catastrophes has clear physical implications but is oversimplified, and cannot be regarded as universal and is of little practical use. A large range of superheating is predicted by the catastrophes.[1] Furthermore, nucleation is inherent in melting and freezing process and depends on an integral of several physical parameters such as solid-liquid interfacial energy (γ_{sl}), heat of fusion per unit volume

173

(ΔH_m) and T_m, and heating rate Q. An appropriate estimation of superheating should be nucleation-based. Given γ_{sl} from undercooling experiments and assuming nucleation rate $I = 1$ s^{-1}cm^{-3}, Lu and Li[1] predicted less amount of superheating than Tallon's hierarchy.[1] Rethfelder et al.[1] estimated superheating assuming a critical volume of nucleation is formed during a given time scale. Although both studies are nucleation-based, their studies did not reveal the systematic nature of melting, and heating rate was not included.

We recently developed a framework[1] for the systematics of maximum superheating and undercooling which are based on undercooling experiments and classical nucleation theory, and incorporate heating rates. Systematic MD simulations and dynamic melting experiments demonstrate significant consistency with the systematics. Next we discuss the systematics for maximum superheating.

SYSTEMATICS FOR MAXIMUM SUPERHEATING

The technical challenges of achieving homogeneous nucleation in melting experiments limit the amount of data of superheating which in turn limits development of a practical superheating-melting theory. But a significant number of freezing experiments have been conducted where appreciable undercooling has been observed with homogeneous nucleation of crystals in liquids. As γ_{sl}, ΔH_m and T_m are common to both melting and freezing, undercooling experiments would allow us to make predictions on superheating.

Based on classical nucleation theories, the nucleation rate[1] for both melting and freezing can be expressed and approximated as

$$I = M(m, T) \exp\{-\frac{\Delta G_c}{kT} g(\phi)\} \approx I_0 f(\beta, \theta) \quad (1)$$

where M is a function of material properties (m) and temperature (T). ΔG_c is the critical Gibbs free energy for nucleation, k Boltzmann's constant, and $g(\phi)$ a geometrical factor depending on the wetting angle ϕ of a heterogeneous nucleant. For homogeneous nucleation, $g(\phi) = 1$, the case assumed in the following discussions. I_0 is a

constant prefactor.[1] We define the energy barrier for nucleation, β, as a dimensionless quantity,

$$\beta(\gamma_{sl}, \Delta H_m, T_m) = \frac{16\pi\gamma_{sl}^3}{3\Delta H_m^2 k T_m} \quad (2)$$

and the reduced temperature as $\theta = T/T_m$, and

$$f(\beta, \theta) = \exp\{-\frac{\beta}{\theta(\theta - 1)^2}\}. \quad (3)$$

Thus, the nucleation process is essentially dependent on β and θ. We denote the maximum superheating and undercooling as $\theta_c = T_c/T_m$. Previous undercooling experiments yielded values of θ_c^- (− denotes undercooling), γ_{sl}, ΔH_m and T_m (thus β) for elements and compounds.[1] θ_c obviously depends on heating (or cooling) rate (Q). Normally the reported experimental values of θ_c^- are for $Q_0 \sim 1$ K/s.

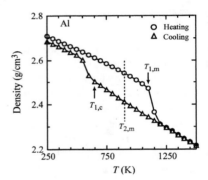

FIGURE 3. Typical single- and two-phase MD simulations of the melting and refreezing behavior: density vs. T. A complete hysteresis of density forms during stepped heating-cooling process for Al. $T_{1,m}$ and $T_{1,c}$ are the single-phase melting and freezing temperature at the superheated and undercooled states, respectively. $T_{2,m}$ is the equilibrium melting temperature from the two-phase simulations. Thus, $\theta_c^+ = T_{1,m}/T_{2,m}$ and $\theta_c^- = T_{1,c}/T_{2,m}$.

As β is common to both melting and freezing, values of β and $\theta_c^-(Q_0)$ allow us to predict the maximum undercooling $\theta_c^-(Q)$ and superheating $\theta_c^+(Q)$ at certain cooling and heating rate Q. For steady-state homogeneous nucleation of crystals from liquid (or melt in solid), Kelton[1] proposed that the probability x for a given amount of parent phase of volume v containing no new phase under certain cooling (or heating) rate Q is

$$x = \exp\{\pm \frac{vT_m I_0}{Q} \int_{\theta_c}^{1} f(\beta, \theta)d\theta\} \qquad (4)$$

where $+$ refers to superheating and $-$ to undercooling. The parameters for undercooling experiments at $Q \sim 1$ K/s, such as γ_{sl}, ΔH_m, T_m (thus β), and v can be regarded as equal to those for superheating and undercooling at different heating and cooling rates. By assuming x and I_0 is approximately equal for the undercooling and superheating cases, the maximum superheating and undercooling under any Q can be calculated from experimental value of $\theta_c^-(Q_0)$ (Fig. 2). The numerical relationship[1] between β, θ_c and Q is fitted as

$$\beta = (A_0 - b \lg Q)\theta_c(\theta_c - 1)^2 \qquad (5)$$

where $A_0 = 59.4$, $b = 2.33$, and Q is normalized by $Q_0 = 1$ K/s. Eq. (5) is referred to as the $\beta - \theta_c - Q$ systematics for the maximum superheating and undercooling.

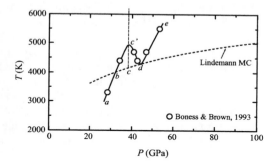

FIGURE 4. Shock-melting experiments on CsBr[1] demonstrate simultaneous drop in shock temperature and sound-speed (not shown), signaling melting of shocked crystal at higher shock pressures than P_c (the long dashed curve). Solid curves indicate the Hugoniot states. The dashed curve is the Lindemann melting curve (MC).[1] bc' segment denotes superheated states.

The $\beta - \theta_c - Q$ systematics for maximum superheating and undercooling are empirical in nature. An independent verification is MD simula-

tions of superheating and undercooling. Superheating was observed previously in a few studies.[1] We conducted systematic MD simulations with single- and two-phase techniques on fcc metals (Al, Ni, Cu, Rh, Pd, Ag, Ir, Pt, Au and Pb) and Be.[1] A typical example is shown in Fig. 3 for Al where superheating and undercooling can be determined. Current and previous simulations yielded values consistent with the $\beta - \theta_c - Q$ systematics at $Q \sim 10^{12}$ K/s. Thus the empirical systematics are validated at atomic level from MD simulations.[1]

Superheating has been observed in planar impact experiments with light-gas gun loading and intense laser irradiation on silicates, alkali halides and metals.[1] A representative example of shock-induced superheating ($Q \sim 10^{12}$ K/s) is shown in Fig. 4 for CsBr. Experimental superheating values compare favorably to the prediction of the systematics.[1]

DISCUSSION

We have established the $\beta - \theta_c - Q$ systematics for the maximum superheating and undercooling consistent with MD simulations and dynamic experiments. Future experimental efforts will employ in-$situ$ structure-resolved melting experiments with exploding wire and shockwave techniques.[2] MD simulations and theoretical efforts are needed to establish a universal relationship between kinetic limit of superheating and various definitions of melting, and catastrophe theories. The effects of heterogeneous nucleation sites at high heating rates, low dimensions and anisotropy, are also of interest.

ACKNOWLEDGMENTS

This work has been supported by U.S. NSF Grant EAR-0207934. S.-N. Luo is sponsored by a Director's Post-doctoral Fellowship at Los Alamos National Laboratory (P-24 and EES-11).

REFERENCES

1. Luo, S.-N., and Ahrens, T.J., Appl. Phys. Lett. **82**, 1836 (2003); Luo, S.-N. and Ahrens, T.J., Phys. Earth Planet. Int. (in press) (2003); Luo, S.-N., Ahrens, T.J., Çağın, T., Strachan, A., Goddard III, W.A., and Swift, D.C., Phys. Rev. B **68**, 134206 (2003), and references therein.

2. Luo, S.-N., Swift, D.C., Tierney, T., Xia, K., Tschauner, O., and Asimow, P.D., this conference.

CP706, *Shock Compression of Condensed Matter - 2003*
edited by M. D. Furnish, Y. M. Gupta, and J. W. Forbes
© 2004 American Institute of Physics 0-7354-0181-0/04/$22.00

FROM FERROELECTRIC TO QUANTUM PARAELECTRIC: $KTa_{1-x}Nb_xO_3$ (KTN), A MODEL SYSTEM

G. A. Samara

Sandia National Laboratories, Albuquerque, NM 87185

Abstract: The mixed perovskite oxides $KTa_{1-x}Nb_xO_3$, or KTN, are a model system for studying ferroelectric behavior and phase transitions under pressure. Crystals with $x > 0.1$ exhibit ferroelectric soft-mode behavior and a sequence of phase transitions, while for $x \leq 0.02$ a pressure-induced ferroelectric-to-relaxor crossover occurs. The system also exhibits a pressure-induced crossover from classical-to-quantum behavior ultimately leading to the complete suppression of the phase transition and the formation of a quantum paraelectric state.

INTRODUCTION

The ABO_3 perovskite oxides constitute an important family of ferroelectrics whose relatively simple chemical and crystallographic structures have contributed significantly to our understanding of ferroelectricity. They are among the most technologically important ferroelectrics finding a broad range of applications such as transducers, actuators and positioners at ambient pressure and as shock-actuated sensors and pulse power sources. They readily undergo structural phase transitions involving both polar and non-polar distortions from the high-temperature ideal cubic lattice.

In this paper we focus on the mixed perovskite system $KTa_{1-x}Nb_xO_3$, or KTN, which has turned out to be a model system for pressure studies. While the end members $KTaO_3$ and $KNbO_3$ might be expected to be similar, in reality they exhibit very different properties. Their mixed crystals, which can be grown over the whole composition range, exhibit a rich set of phenomena whose study has added greatly to our current understanding of the phase transitions and dielectric properties of these materials. Included among these phenomena are soft mode response, pressure-induced ferroelectric (FE)-to-relaxor (R) crossover, quantum mechanical suppression of the

transition and the appearance of a quantum paraelectric state. Each of these phenomena and other properties of KTN compositions were recently discussed elsewhere.[1] In this paper we restrict our presentation to some highlights.

CONTRAST BETWEEN $KTaO_3$ and $KNbO_3$ AND PHASE DIAGRAMS OF KTN

Despite the fact that the Ta and Nb ions have comparable sizes and the same valence, $KTaO_3$ and $KNbO_3$ are very different.[1] $KTaO_3$ remains cubic paraelectric down to the lowest temperatures, whereas the isomorphous $KNbO_3$ exhibits on cooling a sequence of three FE phase transitions, cubic to tetragonal to orthorhombic to rhombohedral (Fig. 1).

First-principles electronic structure and total energy calculations have shed light on this problem.[2,3] Specifically, it has been shown that strong hybridization of O $2p$ states and the d states of the transition metal ion reduces the short-range repulsion leading to off-center ionic displacements and the onset of the sequence of phase transitions in $KNbO_3$. In the case of $KTaO_3$, on the other hand, this hybridization is weak leading to suppression of ionic displacements and stabilization of the cubic phase.

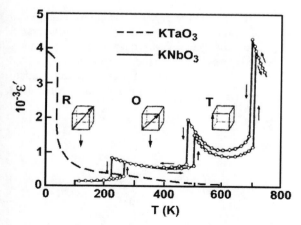

FIGURE 1. Contrast of the temperature dependences of the dielectric constants of $KTaO_3$ and $KNbO_3$.

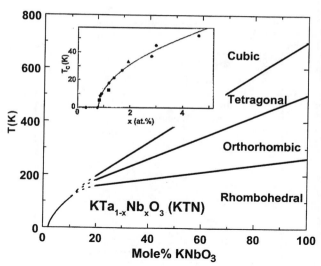

FIGURE 2. Temperature-composition phase diagram for $KTa_{1-x}Nb_xO_3$ at 1 bar. The inset shows details for $x \lesssim 0.04$ where the data come from different sources.

The broad features of the temperature-composition phase diagram of $KTa_{1-x}Nb_xO_3$ at 1 bar are shown in Fig. 2. Over most of the composition range ($\sim 0.1 \leq x \leq 1$) the system exhibits on cooling the indicated sequence of three structural phase transitions. These are equilibrium first-order phase transitions for which the dielectric properties and transition temperature are independent of measuring frequency below microwave frequencies. It is seen that the range of stability of the tetragonal and orthorhombic phases decreases, with increasing $KTaO_3$ content leading to a single phase transition from cubic to rhombohedral symmetry for $x \lesssim 0.1$. The inset in Fig. 2 shows the details of the phase diagram for $x \lesssim 0.04$. This region has attracted considerable attention in relation to a possible FE-to-R crossover and to the onset of quantum fluctuations.[4,5] There is considerable evidence for relaxor behavior for $x \lesssim 0.02$,[4,5] and the "transition" vanishes for $x < 0.008$.

We illustrate the influence of pressure on compositions with $x \geq 0.2$ by the results on a single crystal with $x = 0.36$. The cubic-tetragonal FE transition in this composition range is driven by the softening of a long wavelength TO phonon.[1] This soft mode consists of a vibration of the undistorted oxygen octahedron against the

remainder of the unit cell with the heavy Ta and Nb ions effectively fixed at the center of mass. Figure 3 shows the T dependence of the dielectric constant, ε', measured on heating at 0, 4 and 8 kbar. All three transitions exhibit large thermal hysteresis. The inset in Fig. 3 shows the pressure dependences of the transition temperatures. The lack of pressure dependence of the Ortho-Rh transition temperature reflects the fact that the volume strain, $\Delta V/V$, at this transition is very small,[6] since according to the Clasius-Clapeyron equation $dT/dP = \Delta V/\Delta S$, where ΔS is the change entropy at the transition temperature. The decrease of the Curie temperature with pressure is well understood in terms of soft mode theory.[1]

It is clear that the temperature ranges of stability of the tetragonal and orthorhombic phases decrease with increasing pressure. Extrapolation of the data to higher pressures (dashed lines) suggests that these two phases should vanish around ~ 25 kbar, and at higher pressures the crystal should transform directly from the cubic to the rhombohedral phases. Clearly, there is a strong analogy between increasing $KTaO_3$ content at 1 bar and increasing pressure for a fixed composition as can be seen on comparing Figs. 2 and 3.

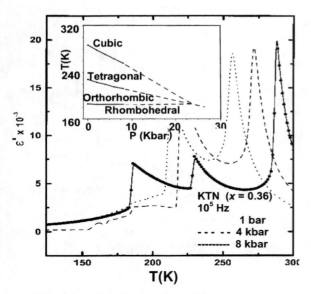

FIGURE 3. Temperature dependence of the dielectric constant of KTN ($x = 0.36$) at different pressures. The inset shows the pressure dependences of the various transition temperatures.

FERROELECTRIC-TO-RELAXOR CROSSOVER

One of the more interesting pressure effects observed in perovskites with random site disorder, as in KTN with small x, is the F-to-R crossover. This phenomenon has been observed in a variety of perovskites and appears to be a general feature of FE soft-mode materials.[5] Whereas ferroelectrics are characterized by macro-size polar domains, sharp equilibrium phase transitions at their Curie temperatures, T_c, and frequency independent dielectric properties (up to microwave frequencies), relaxors are characterized by nano-size polar domains associated with the disorder, the absence of a macroscopic phase transition and strong frequency dispersion in the dielectric properties. The frequency dependent peak (T_m) in $\varepsilon'(T)$ is simply a manifestation of the slowing downs of dipolar motion below T_m. The crossover is illustrated for KTN with $x = 0.02$ in Fig. 4. The 1-bar $\varepsilon'(T)$ response is independent of frequency and characteristic of FE behavior. At high

pressures the character of the response changes to that of a relaxor with strong frequency dispersion T_m and absence of a macroscopic symmetry change across T_m, i.e., the crystal remains cubic down to the lowest temperatures. The mechanism for the FE-to-R crossover has been discussed elsewhere.[1,5]

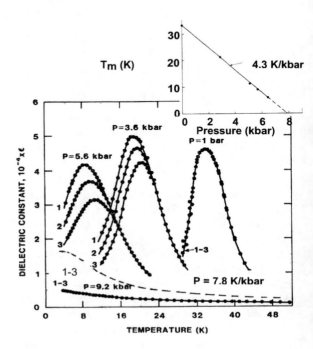

FIGURE 4. The dielectric response of KTN ($x = 0.02$) showing the pressure-induced FE-to-R crossover and the suppression of the relaxor state at $P \geq 7.8$ kbar. The labels 1, 2, 3 refer to data at 10^2, 10^4 and 10^6 Hz, respectively. There is no frequency dispersion at 1 bar, 7.8 kbar and 9.2 kbar. The inset shows that $T_m \rightarrow$ 0K with finite slope.

PRESSURE AS A PROBE OF PROPERTIES IN THE QUANTUM REGIME

The occurrence of transitions in solids is determined by a competition between cooperative, long-range forces, which try to order the system and fluctuations, which favor disruption of this order. When the transition occurs at high temperature (T), i.e., in the classical regime, thermal fluctuations are at work. These fluctuations dominate in the high T phase and there is no ordering; but, on lowering T the fluctuations decrease and eventually the ordering forces win out, and the system orders at a transition temperature, T_c. On the other hand, if the transition occurs at sufficiently low T, quantum fluctuations, or zero-point motions, come into play, and they can strongly influence the response of the system. Among these manifestations are new critical exponents, the suppression of T_c below its classical value, and ultimately the complete suppression of T_c (or T_m).[5]

One of the consequences of the suppression of the phase transition is the formation of a quantum paraelectric state. This state is characterized by a high static dielectric susceptibility (or low soft-mode frequency), which is constant at low T over an extended range of temperatures. $KTaO_3$ exhibits such a state at atmospheric pressure and other systems exhibit it at high pressure.[5,7]

As there are no known ABO_3 FE compounds with T_cs sufficiently low to make the full range of quantum effects discernible, experimental study of these effects necessitates that T_c be shifted down to the appropriate range by application of external fields. Ideally such fields should not change the symmetry of the high temperature phase. Hydrostatic pressure is an excellent such field, and we have used it successfully earlier to study these effects in a variety of crystals.[5,7] Another "external" field used by others is chemical substitution which yields mixed crystals.[5] However this substitution introduces randomness, compositional fluctuations, and clustering and breaks translational symmetry. These effects undoubtedly change some of the interaction parameters of the system, and it is not clear that they can be neglected.

The results in Fig. 4 show the complete suppression of the transition, or relaxor state, for P $\geq$ 7.8 kbar. The large values and temperature independence of ε' below 5K are the signatures of the quantum paraelectric state.

Finally, an interesting feature of the data in Fig. 4 is the fact that $T_m \rightarrow 0K$ with a finite slope as shown in the inset in Fig. 4. This result, which is qualitatively different from that observed for T_c (P) for equilibrium phase transitions, is a unique manifestation in pressure experiments of the non-equilibrium nature of the glass (relaxor) transition. It is attributed to the presence of residual configurational entropy in the relaxor phase near 0K.[5] By contrast, for equilibrium phase transition, the third law of thermodynamics requires that T_c vanishes with infinite slope, i.e., $dT_c / dP \rightarrow -\infty$ as $T_c \rightarrow 0K$.

ACKNOWLEDGEMENT

This work was supported in part by the Division of Materials Sciences, Office of Basic Energy Sciences, United States Department of Energy and by Sandia's Research Foundations under Contract No. DE-AC04-AL85000.

REFERENCES

1. Samara, G. A., *Mat. Res. Soc. Symp. Proc.*, Vol. **718**, 281 (2002) and references therein.
2. Cohen, R. E., *J. Phys. Chem. Solids* **61**, 139 (2000) and references therein.
3. Singh, D. J., *Phys. Rev.* B **52**, 12559 (1995); B **53** (1996).
4. Samara, G. A., *Ferroelectrics* **117**, 347 (1991).
5. Samara, G. A., in *Solid State Physics*, Vol. **56**, edited by Ehrenreich, H. and Spaepen, F., Academic Press (2001), Chapter 3, p. 239 and references therein.
6. Wang, M., Yang, Z. H., Wang, J. Y., Lin, Y. G., Guan, Q. C. and Wei, J. Q., *Ferroelectrics* **132**, 55 (1992).
7. Samara, G. A., *Physica B* **150**, 179 (1988).

CP706, *Shock Compression of Condensed Matter - 2003*
edited by M. D. Furnish, Y. M. Gupta, and J. W. Forbes
© 2004 American Institute of Physics 0-7354-0181-0/04/$22.00

MICROSTRUCTURAL EFFECTS ON THE SHOCK RESPONSE OF PZT 95/5

R. E. Setchell, B. A. Tuttle, J. A. Voigt

Sandia National Laboratories, Albuquerque, NM, 87185

Abstract. Shock-induced depoling of the ferroelectric ceramic PZT 95/5 is utilized in pulsed power devices. The bulk density and corresponding porous microstructure can be varied by adding different types and quantities of organic pore formers prior to bisque firing and sintering. In previous studies, a baseline material having a particular microstructure was examined in detail. Comparative experiments with a second material having a common density but a very different porous microstructure showed only subtle differences in mechanical and electrical shock properties. However, large differences in these properties were observed using materials prepared over a range of bulk densities. Recent studies have examined three new materials that were prepared at a common density matching that of the baseline material. Each was made using spherical pore formers having diameters within a narrow range, with nominal diameters varying from 15 μm to 140 μm. Normally poled samples of each material were used in identical planar impact experiments that produced peak stresses of 2.5 GPa under high-field conditions, or 4.5 GPa under short-circuit conditions. Unlike previous comparisons of common-density materials, consistent trends were evident in depoling currents, wave rise times and amplitudes, and yielding thresholds. Overall differences between two of these materials and the baseline material were relatively small, but the material made with the smallest pore former showed significant differences.

INTRODUCTION

A lead zirconate titanate ceramic having a Zr:Ti ratio of 95:5 and modified with 2% niobium, denoted as PZT 95/5, is used in shock-driven pulsed power devices. At ambient conditions this material is ferroelectric, and poling results in remanent polarization with associated bound charge. An antiferroelectric phase boundary exists at pressures of a few hundred MPa, and shock compression into this phase provides a fast mechanism for releasing the bound charge. Different processing methods have been used to produce PZT 95/5, with current materials made using oxide powders precipitated from chemical solutions. An important processing variable is the final material density and the associated porous microstructure. These are controlled through the

addition of organic pore formers prior to a bisque (lower temperature) firing and subsequent sintering, resulting in voids that reflect the morphology of the pore former. A material at a nominal bulk density of 7.30 g/cm^3 has about 9% total void volume fraction (porosity), composed of about 4% intergranular porosity and 5% porosity resulting from pore former addition.

The current study is part of a continuing program to determine an optimum porous microstructure for PZT 95/5, as well as to develop improved models for its mechanical and electrical behavior during shock compression. Previously, a nominal density baseline material was extensively characterized (1,2). The pore formers used in preparing this particular material were PMMA spheres having diameters from 50-100 μm. Shock Hugoniot states and transmitted wave profiles were obtained in unpoled and poled material for shock stresses varying from 0.9 to 4.6 GPa. Depoling currents from normally poled samples were measured under short-circuit and high-field conditions over the same range in stress. In a subsequent study, the baseline material was compared to new materials that were made with varying amounts of pore former (2,3). The pore former used in these materials was microcrystalline cellulose in the form of rods 5-15 μm in diameter and ≥ 20 μm long. A material having a density matching that of the baseline material showed only small differences in wave profiles and in depoling currents over a wide range of shock conditions. However, large differences were seen with materials having bulk densities that varied from 6.94 to 7.66 g/cm^3. In particular, the threshold stress for dynamic yielding varied by more than a factor of two over this density range.

In the current study, comparative shock experiments were performed using three new materials prepared at a common density matching that of the baseline material. Each of these materials was made using spherical pore formers having diameters within a tightly controlled range. The first material, identified by the Sandia designation HF963, used PMMA spheres that were 14-16 μm in diameter. This is comparable to the nominal grain size in sintered material (~10 μm). The second material, HF967, was made using polystyrene spheres 77-83 μm in diameter. The final material, HF969, was made using polystyrene spheres 135-145 μm in diameter. These materials had nominal densities of 7.29, 7.29, and 7.28 g/cm^3, respectively. Two particular experimental conditions were used. In the first, 2.5 GPa shocks were introduced into normally poled samples connected to a high external load so that a strong electrical field would develop during the shock motion. In the second condition, 4.5 GPa shocks were introduced into normally poled samples under short-circuit conditions. An unexpected result in these experiments prompted two additional high-stress experiments using unpoled samples.

EXPERIMENTAL CONFIGURATION

The configuration used for examining the shock response of normally poled samples in gas-gun experiments is shown in Fig. 1. The active PZT 95/5 sample is poled normal to the direction of shock motion, and is sandwiched between two similar but shorted samples that provide a shock-impedance match at their boundaries, as well as electrical connections to an external circuit. Depoling currents generated during shock transit pass through a load resistor chosen for either short-circuit or high-field conditions. Transmitted wave profiles at the sample/window interface are obtained using laser interferometry (VISAR). Additional details on this experimental configuration can be found elsewhere (2).

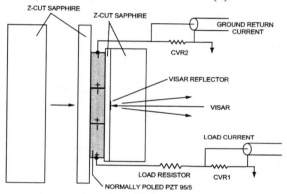

FIGURE 1. Configuration for gas-gun experiments on normally poled PZT 95/5 samples.

RESULTS

The first set of comparative experiments were conducted using 1150 Ω load resistors and an impact velocity of 0.154 km/s, which results in an impact stress of 2.50 GPa using the Hugoniot properties established for the baseline material (2). Figure 2 shows the transmitted wave profiles recorded in these experiments, together with a profile recorded earlier in an identical experiment with the baseline material. All samples were 4.0 mm thick. Two trends are apparent in these profiles. As the pore former size decreases, the rise time decreases and the final amplitude increases in the corresponding waves. The profile recorded with the baseline material has a slower rise time and ends below the level predicted by its Hugoniot. The observed level corresponds to the threshold stress for yielding in this material, and the available test time is insufficient to observe this profile slowly rise to the predicted final state (2). The profile for

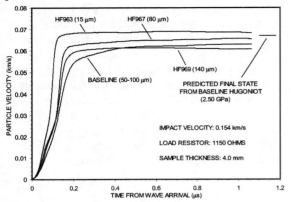

FIGURE 2. Transmitted wave profiles for 2.5 GPa shock inputs under high-field conditions.

the material made with the smallest pore formers ends above this state.

Figure 3 shows the load currents recorded on these experiments. The slow rise time results from the depoling current being divided between passing through the load resistor and being retained on the electrodes to account for the sample capacitance. A similar trend is seen in the load currents, with the highest current corresponding to the material made with the smallest pore former. No dielectric breakdowns occurred during these experiments, in which fields exceeded 36 kV/cm.

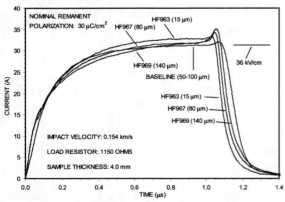

FIGURE 3. Load currents measured in 2.5 GPa shock experiments under high-field conditions.

The second set of comparative experiments were conducted using 10 Ω load resistors and an impact velocity of 0.330 km/s, which results in an impact

stress of 4.53 GPa using the baseline Hugoniot properties. The small load resistor results in

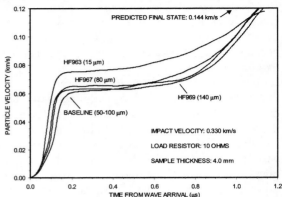

FIGURE 4. Transmitted wave profiles for 4.5 GPa shock inputs under short-circuit conditions.

essentially short-circuit conditions. Figure 4 shows the wave profiles recorded in these experiments, along with a profile from an identical baseline material experiment. In each profile, the plateau following the initial rise corresponds to the threshold for dynamic yielding. This level is followed by a slow rise towards a final Hugoniot state. The same trends are apparent, with the material made using the smallest pore former showing a surprisingly high yield threshold. Comparing these plateau values to the final profile values for the 2.5 GPa experiments (Fig. 2), it is apparent that the 2.5 GPa cases showed yield thresholds except for the smallest-pore case.

Figure 5 shows the load currents recorded in these experiments. An initial spike followed by a relaxation to a steady value is characteristic of a shock input well above the yield threshold (1). The same trend with pore former size is weakly apparent, with the smallest-pore case again showing the largest difference.

Because the threshold for yielding was so much higher for the smallest-pore case (Fig. 4), two additional experiments were conducted to confirm this result. Figure 6 shows transmitted wave profiles recorded in 4.6 GPa shock experiments on unpoled samples of the materials made with the smallest and largest pore formers. Very similar results for the yielding thresholds were found. The values shown for these thresholds were determined using the Hugoniot properties of sapphire and the baseline PZT 95/5 material in impedance-matching

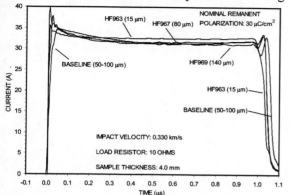

FIGURE 5. Load currents measured in 4.5 GPa shock experiments under short-circuit conditions.

calculations. These are also shown in Fig. 7, along with other values found for PZT 95/5 materials over a range of densities. The smallest-pore material shows a significant departure from what previously had been a smooth trend of yield stress increasing with initial density.

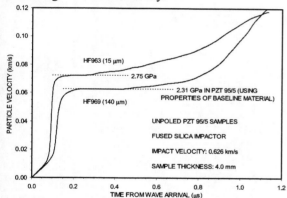

FIGURE 6. Transmitted wave profiles in unpoled materials for 4.6 GPa shock inputs.

SUMMARY

Results with these new common-density materials were in contrast to previous findings, with consistent trends evident in wave rise times and amplitudes, depoling currents, and yielding thresholds. Overall differences between two of these materials and the baseline material were rela-

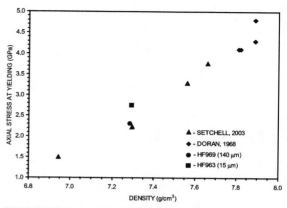

FIGURE 7. Dynamic yielding thresholds for PZT 95/5 at different initial densities.

tively small, but the material made with the smallest pore former showed significant differences. The microstructure of this material is unique in that the added porosity looks like missing single grains. Continuing studies will examine this material in more detail.

ACKNOWLEDGMENTS

The authors would like to thank David E. Cox for skillfully preparing and conducting the experiments. Sandia is a multiprogram laboratory operated by Sandia Corporation, a Lockheed Martin Company, for the United States Department of Energy's National Nuclear Security Administration under Contract DE-AC04-94AL85000.

REFERENCES

1. Setchell, R. E., "Recent Progress in Understanding the Shock Response of Ferroelectric Ceramics", in *Shock Compression of Condensed Matter – 2001*, edited by M. D. Furnish et al., AIP Conference Proceedings 620, New York, 2002, pp. 191-196.
2. Setchell, R. E., J. Appl. Phys. **94**, 573-588 (2003).
3. Setchell, R. E., Tuttle, B. A., Voigt, J. A., and Venturini, E. L., "Effects of Initial Porosity on the Shock Response of Normally Poled PZT 95/5," in *Shock Compression of Condensed Matter – 2001*, edited by M. D. Furnish et al., AIP Conference Proceedings 620, New York, 2002, pp. 209-212.

CHAPTER V

MODELING, SIMULATION, AND THEORY: NONREACTIVE MATERIALS

CP706, *Shock Compression of Condensed Matter - 2003*
edited by M. D. Furnish, Y. M. Gupta, and J. W. Forbes
© 2004 American Institute of Physics 0-7354-0181-0/04/$22.00

THE MATERIAL POINT METHOD AND SIMULATION OF WAVE PROPAGATION IN HETEROGENEOUS MEDIA

S. G. Bardenhagen[*], D. R. Greening[†] and K. M. Roessig[**]

[*]*Group T–14, MS B214, Los Alamos National Laboratory, Los Alamos, NM 87545*
[†]*Group X–7, MS F699, Los Alamos National Laboratory, Los Alamos, NM 87545*
[**]*Fraunhofer Institut für Kurzzeit Dynamik, Ernst–Mach–Institute, 4 Eckerstrasse, D–79104 Freiburg*

Abstract. The mechanical response of polycrystalline materials, particularly under shock loading, is of significant interest in a variety of munitions and industrial applications. Homogeneous continuum models have been developed to describe material response, including Equation of State, strength, and reactive burn models. These models provide good estimates of bulk material response. However, there is little connection to underlying physics and, consequently, they cannot be applied far from their calibrated regime with confidence. Both explosives and metals have important structure at the (energetic or single crystal) grain scale. The anisotropic properties of the individual grains and the presence of interfaces result in the localization of energy during deformation. In explosives energy localization can lead to initiation under weak shock loading, and in metals to material ejecta under strong shock loading. To develop accurate, quantitative and predictive models it is imperative to develop a sound physical understanding of the grain–scale material response.

Numerical simulations are performed to gain insight into grain–scale material response. The Generalized Interpolation Material Point Method family of numerical algorithms, selected for their robust treatment of large deformation problems and convenient framework for implementing material interface models, are reviewed. A three–dimensional simulation of wave propagation through a granular material indicates the scale and complexity of a representative grain–scale computation. Verification and validation calculations on model bi–material systems indicate the minimum numerical algorithm complexity required for accurate simulation of wave propagation across material interfaces and demonstrate the importance of interfacial decohesion. Preliminary results are presented which predict energy localization at the grain boundary in a metallic bicrystal.

INTRODUCTION

Many aspects of mechanical response of materials are fundamentally tied to localized events on scales determined by the material. This scale is often referred to as the mesoscale, and refers to structure much smaller than an engineering component. Examples include particle spacing in a particulate composite or average grain size in a polycrystalline metal, both of which are considered further here. While for some aspects of mechanical response mesostructural effects may be averaged out to give a representative bulk response, many aspects cannot. Examples include damage nucleation, shear banding and mechanical initiation of explosives, all of which are strongly dependent on the heterogeneity of material response at the mesoscale. While bulk engineering models can be developed which fit a particular data set, they cannot be extrapolated with confidence without an understanding of the relevant mesoscale physics.

This view is generally accepted in the materials community and drives both experimental efforts to characterize and resolve material response on shorter length scales, and theoretical efforts to develop mesoscale physics models and incorporate them in averaging schemes and numerical simulations. However, these remain very difficult tasks, largely on account of the myriad of physics possible at the mesoscale. Examples include anisotropy, spatially dependent material properties, defects, and interfacial response. If, in addition, material response is desired under shock loading, the short time scales and necessarily nonlinear material response further complicates matters.

Despite the extensive complications, progress has been made, particularly for model materials. However, even for model materials, experimental, theoretical, and computational efforts are generally unconnected. Two major stumbling blocks are: (1) using data which resolves mesoscale heterogeneity (experimental or computational) in engineering models, and (2) applying theory, computation and experiment to the same material, i.e. characterizing material mesostructure in three dimensions. Until approaches to these problems mature enough to consider "real" materials, it is unlikely that much confidence in extrapolating engineering models beyond the regimes in which they have been calibrated will develop.

Although some progress on one of the items enumerated above may be reported, in fact many smaller challenges need to be overcome in the field of numerical simulation of mesoscopic material response. Numerical simulation of solids with multiple phases and interfaces, particularly when undergoing large deformations, is an area of active research. The challenges divide into two areas, development of mesoscale physics models and numerical solution accuracy. Here the focus is primarily on the latter. Particle–In–Cell (PIC) methods, which have inherent advantages in tracking material response and interfaces, are briefly discussed. A more general setting for the Material Point Method (MPM), a particular PIC method, is reviewed and encouraging results for various algorithm variants are presented. Simulations of the dynamic response of several heterogeneous model materials are presented, focusing on specific physics and verification and validation calculations. Applications to plastic bonded explosives (PBXs) and polycrystalline metals are discussed.

THE GENERALIZED INTERPOLATION MATERIAL POINT METHOD

The MPM is one of the latest developments in PIC methods, originally used in computational fluid mechanics to model highly distorted fluid flow. Subsequent developments advanced the understanding of the algorithm and brought modifications to reduce numerical diffusion. A fundamental aspect of PIC methods is the interpolation of information between a grid and particles. How this is done and precisely which solution variables are ascribed to the grid, and which to the particles distinguishes various PIC algorithms. The general trend has been toward keeping more properties on the particles. This has been continued with the development of MPM, where the ability of the particles to advect constitutive model state variables has been exploited in the application to computational solid mechanics [1]. In MPM the grid may be viewed as a temporary computational scratch pad, used to advance material point states.

Particle methods seem well suited for computational solid mechanics, where it is natural to have material properties which are a function of location in a reference configuration. Solid mechanics constitutive models generate stress based on both the history and current mechanical state. These models are often complex and require the calculation of "internal variables" which represent history dependence. Lagrangian particles allow easy implementation of these constitutive models, and straight–forward advection of internal variables.

The derivation of the MPM algorithm has recently been cast in variational, or weak form [1], providing a standard setting for the discretization of the governing equations. In addition, however, this setting provides a venue for generalizing the MPM discretization technique. Here it is outlined how the variational form of the governing equations provides a consistent framework for generalizing MPM. The use of smoother representations of particle data allows an entire family of methods to be developed. These have been named the Generalized Interpolation Material Point (GIMP) methods [2]. An important result of the generalization is the availability of smoother representations of particle data on the computational grid, which can remove numerical artifact noise inherent in the original MPM formulation. Additionally, the relationship to other "meshless methods" can be identified.

The variational form for conservation of momentum may be written as in [2]. The essence of the discretization procedure is to represent a solid material continuum as a collection of body fixed (Lagrangian) particles, or "material points". Particles may be defined by "particle characteristic functions", $\chi_p(\mathbf{x})$, which specify the space occupied, perhaps only partially, by a given particle in the current configuration. Given material point data, f_p, a continuous representation may be constructed

$$f(\mathbf{x}) = \sum_p f_p \chi_p(\mathbf{x}).　\quad (1)$$

Eqn. 1 is used to substitute for the trial functions (namely density, stress, and rate of change of momentum density) in the variational form. To complete the discretization procedure, approximations to the test functions are introduced. The continuous representation, $g(\mathbf{x})$, of grid data, g_v, is taken to be

$$g(\mathbf{x}) = \sum_v g_v S_v(\mathbf{x}). \qquad (2)$$

Here $S_v(\mathbf{x})$ is a computational grid shape function, which takes unit value at node v and zero value at the other nodes. The shape functions are required to be a partition of unity. Substitution of the grid shape function representation for the admissible velocity fields, as in Eqn. 2, into the variational form, and use of the arbitrariness of the admissible velocity fields, yields the discrete governing equations. The fact that the approximations to trial and test functions use different basis functions identifies the discretization procedure as Petrov–Galerkin and highlights the similarity to other meshless–methods.

Inspection of the governing equation reveals they are identical in form to those presented for MPM [1]. The only difference is that the weighting functions, $\overline{S}_{vp}$, and gradient weighting functions, $\overline{\nabla S}_{vp}$, used to interpolate information between the particles and the grid, have been generalized

$$\overline{S}_{vp} = \frac{1}{V_p} \int_{\Omega_p \cap \Omega} \chi_p(\mathbf{x}) S_v(\mathbf{x})\, d\mathbf{x}, \qquad (3)$$

$$\overline{\nabla S}_{vp} = \frac{1}{V_p} \int_{\Omega_p \cap \Omega} \chi_p(\mathbf{x}) \nabla S_v(\mathbf{x})\, d\mathbf{x}. \qquad (4)$$

where V_p is the current volume of particle p, Ω_p is the support of χ_p, and Ω denotes the current volume of the material to be discretized. These generalized weighting and gradient weighting functions may be easily calculated for any combination of particle characteristic functions and grid shape functions. This superset of MPM has been named the Generalized Interpolation Material Point (GIMP) method [2].

The MPM may be recovered as a specific case of the GIMP methods by appropriate selection of the particle characteristic functions. Specifically, for

$$\chi_p(\mathbf{x}) = \delta(\mathbf{x} - \mathbf{x}_p) V_p, \qquad (5)$$

where $\delta(\mathbf{x} - \mathbf{x}_p)$ is the Dirac delta function, MPM [1] is recovered exactly.

The simplest extension to finite particles, i.e. particle characteristic functions with finite support, is to consider contiguous regions of non–overlapping support Ω_p ("contiguous particles"). The particle characteristic functions in this case may be written,

$$\chi_p(x) = 1 \quad if \quad x \in \Omega_p. \qquad (6)$$

This generalization replaces particle mass points with mass volumes. The result, for the bilinear grid shape functions typically employed, is weighting functions with larger, particle size dependent support. It has the advantage that it develops weighting functions in C^1 with a minimal amount of additional complexity, both theoretically and computationally, and is referred to as the "contiguous particle GIMP method" [2].

APPLICATIONS

Wave propagation through heterogeneous media is simulated in this section. The emphasis is on verification and validation calculations, i.e. demonstrating the ability to match analytical and experimental results for well characterized, generally idealized, materials. Stewardship of the nuclear weapons stockpile places stringent demands on the understanding of materials and their behavior under extreme conditions. Ultimately of interest is a sound physical understanding of the mesoscale material response in "real" materials, and a predictive capability.

Applications of interest include explosives and polycrystalline metals, both of which have grain scale structure. The properties of the individual grains and the presence of interfaces result in the localization of energy during deformation. In explosives energy localization can lead to initiation under weak shock loading, and in metals to material damage and ejecta under strong shock loading.

EXPLOSIVES

PBXs are composed of energetic grains embedded in a polymeric matrix, i.e. they are particulate composites. Accidental initiation of PBXs due to mild impact, or "weak-shocks" is not well understood. Low levels of average stress sometimes result in violent reactions. The inhomogeneity of the deformation at the scale of the grains clearly contributes to energy localization and the development of "hot spots", which may coalesce and lead to violent reaction. Candidate hot spot mechanisms include stress concentrations resulting in large material deformation and frictional sliding between grains.

In previous work[3] attention was focused on an idealized dry granular material. A 1mm x 1mm x 2mm box was packed with 1603 spheres, where the distribution of sphere diameters was chosen consistent with experimental measurements for a particular PBX. As the microstructure is inherently three-dimensional, resolving grain deformation and interactions in a repre-

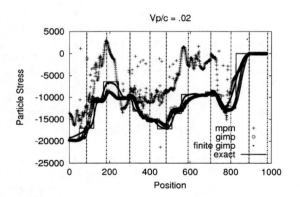

FIGURE 1. Verification of GIMP algorithm performance for a weak shock traveling through an elastic layered medium

sentative volume of the microstructure requires tremendous computational power. The calculations were performed using the parallel computing infrastructure developed at the University of Utah's Accelerated Strategic Computing Initiative (ASCI) center, the Center for the Simulation of Accidental Fires and Explosions (C-SAFE), and was run on 1000 processors. Stress propagation through the granular compact was simulated. The strongly heterogeneous structure of the stress wave is in qualitative agreement with work in 2D, both numerical and experimental, where the development of preferential load paths, or force chains, was examined.

These calculations raise interesting questions which are difficult to answer, in part due to the size of the computations (and data processing) needed. How many grains are required such that bulk response is accurately simulated, i.e. what is the minimum "representative volume element" size? Given that the bulk wave structure is determined by the collection of tortuous paths, grain to grain, through which the stress travels, will a steady wave develop or will it attenuate? How are these answers affected by details such as packing density, grain size distribution and shape? These results are reviewed here to suggest some of the remaining challenges as well as the scale and complexity of representative calculations. Other work in this area has recently begun to focus on data interrogation for computations of this scale, and incorporation of statistical information in engineering models [4].

Before pursuing simulations on more realistic representations of PBXs it is necessary to investigate algorithmic and physics model performance for material interfaces. Specifically of interest is wave transmission across an interface between different materials and the possibility of separation or debonding at an interface.

Confidence in modeling these phenomena is required in order to consider the plastic matrix in PBXs, as well as wave transmission between grains in a metallic polycrystal.

A convenient verification problem is the propagation of a stress wave through an elastic layered medium. If the wave propagates perpendicular to the material interfaces, and the layers are isotropic, then the problem is one–dimensional. Further, if the constitutive relationship for each layer is taken to be

$$\sigma = E(F - 1) \tag{7}$$

where σ is the Cauchy stress, E is the constant (within a layer) uniaxial strain modulus and F the deformation gradient, the problem can be easily solved exactly for finite deformations. The magnitude of transmitted and reflected waves at each interface may be calculated, the usual formulas for small deformations apply [5], and it is simply a bookkeeping exercise to track the current number and position of each of the stress waves.

To examine the performance of various GIMP algorithms, wave transmission through a periodically layered material was simulated. Ten alternating layers were considered, all with the same density. The stiffness of the "odd" layers (the layer in contact with the piston and then every other one beyond) was taken to be four times that of the "even" layers. The layers are of equal thickness and each is initially discretized by 100 grid cells with one particle per cell.

One end of the material is compressed by a rigid piston. The ratio of piston velocity, v_p, to the maximum material wavespeed in the problem, c, gives a measure of the shock strength. The simulation is run for one wave transit time. A Richtmyer–VonNeumann artificial viscosity term is added to smear the shock over several computational cells and a linear term is added to damp out ringing behind the shock. For the calculations presented here, a linear term coefficient of 0.4 and a Richtmyer–VonNeumann coefficient of 4.0 are used.

Figure 1 depicts particle stresses at the same time after impact as calculated using various GIMP algorithms for a weak shock case, $v_p/c = .02$. Vertical lines indicate current positions of the layer interfaces. The exact solution is depicted using a solid line which can be distinguished from the point data by its "blocky" character. The exact solution is simply a superposition of many left and right traveling square waves.

Three algorithm variants are depicted. Results for the MPM algorithm are depicted with crosses. Because material points are represented as discrete mass points, numerical noise can develop when particles fail to reg-

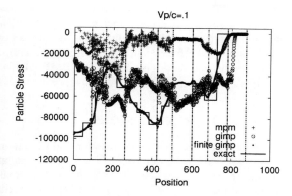

FIGURE 2. Verification of GIMP algorithm performance for a shock traveling through an elastic layered medium

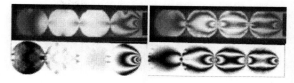

FIGURE 3. Comparison of experiment (top frames) and simulation (bottom) for wave propagation through photoelastics disks in binder.

ister uniformly on the computational grid. For finite deformations (in practice defined as deformations for which particles cross cell boundaries) this noise can overwhelm the solution, as is the case here. Results for the contiguous particle GIMP method are also depicted and labeled "finite gimp". Because it is a significant simplification in practice, results for the case where current particle volumes are not tracked (i.e. $V_p = const.$) are also presented, labeled "gimp". Both of these algorithms are substantially more accurate than the MPM and not much difference may be discerned between them in the weak shock case.

Figure 2 depicts particle stresses as before, but for a stronger shock, $v_p/c = 0.1$. For this case the only algorithm which successfully tracks the exact solution is the contiguous particle GIMP method in which current particle volumes are tracked, i.e. "finite gimp". For further discussion of these algorithms see [2].

This verification exercise should be interpreted as suggesting a minimal complexity required for successful simulation of shocks using GIMP methods. It remains to be determined how these simulations compare with other numerical techniques. In particular, there is substantial attenuation of the initial shock rise during the course of the calculations (from a few cells to tens of cells), possibly due to numerical diffusion, which bears further investigation.

More complex, although still idealized, material configurations have also been examined. Collaborators at Eglin AFB investigated stress waves in quasi two-dimensional, dry granular media. A Hopkinson bar was used to dynamically load collections of photoelastic disks. Using high-speed photography, the stress state was resolved as an impulse traveled through various assemblies. This data was used in successful frictional

contact algorithm validation efforts [6].

More recent efforts have examined the effect of an interstitial material or "binder" on wave propagation through simple configurations of Plexiglas disks. The binder is the same material, and has been cast around the disks everywhere except in the vicinity of the loading piston. Two cases were examined. In one case the binder was allowed to adhere to the disks and in the other the interface was greased to completely prevent adhesion. Results for these cases have been used to qualitatively evaluate the performance of an interfacial debonding algorithm.

Experimental and computational results for the two cases are depicted in Fig. 3. The stress wave propagates to the left. All frames correspond to a time shortly before the wave reaches the left boundary. The bonded case comparison is presented in the left frames, the greased case in the right frames. Experimental results are the top frames. For the bonded case simulations the interface is not allowed to separate (perfect bonding). For the greased case interpenetration is prevented but separation is allowed (no bond strength). The striking dependence on interfacial properties is captured in the simulations, and reasonable qualitative agreement with the experiments is found.

POLYCRYSTALLINE METALS

Engineering models provide good estimates of the bulk constitutive response of polycrystalline metals, as it is the collective response of many randomly oriented grains. However, the heterogeneity of the material properties on the mesoscale, due to the orientation of the individual, anisotropic grains, and the presence of grain boundaries, may result in the localization of shock wave energy. The occurrence of many important phenomena, such as void nucleation, fracture, shear banding and ejecta formation, are likely determined by the details of material heterogeneity.

A research program has been initiated at Los Alamos National Laboratory (LANL) to better understand grain scale mechanics of polycrystalline metals under shock

FIGURE 4. Reverse ballistics calculation on a bicrystal showing localization of velocity at the free surface.

loading via a combined program of experiment and simulation. NiAl was chosen as a model material because of its availability and degree of single grain anisotropy. Single grain and bicrystal samples have been grown and characterized. Plate impact experiments have been performed to drive a shock through the samples. Diagnostics have been developed to record the deformation of the free surface with sufficient spatial and temporal resolution to determine variations in the vicinity of the grain boundary. These topics are covered in more detail in other papers in this proceedings [7, 8, 9, 10, 11, 12].

The initial work on this project has been on bicrystals, i.e. two NiAl grains and their common grain boundary. It was predicted, using a Lagrangian hydrocode, that for certain orientations of the grain boundary relative to the impact surface, a localization of energy occurs at the intersection of the grain boundary and the free surface. This calculation was for perfectly bonded, elastic grains. Grain anisotropy was approximated by isotropic grains with different material properties. Here the focus is on verifying that a GIMP code obtains this result.

Figure 4 shows results from a contiguous particle GIMP calculation in which a bicrystal impacts the boundary of the computational domain, which is rigid. The ratio of the initial velocity of the bicrystal to the fastest longitudinal wave speed is 0.001, so the insult is a very weak shock. The stiffer grain is on the left, the longitudinal wave speed ratio for the grains is $\sqrt{2}$. A gray line indicates the grain boundary which is oriented at 45° to the boundary. The top surface of the bicrystal is free. The spatial variation of the magnitude of the velocity of the bicrystal is depicted, with black corresponding to zero and light gray to the initial velocity of the bicrystal.

The left frame depicts the bicrystal as the first wave traverses from bottom to top, bringing material to rest. The wave propagates faster in the left grain. The right frame depicts the bicrystal after "breakout" (wave reflection at the free surface) has occurred in the left grain, and just as breakout is occurring in the right grain. The white region at the intersection of the free surface and the grain boundary indicates a higher rarefaction velocity there. The material property mismatch and grain boundary orientation result in energy localization at the free boundary. While more careful validation between codes is needed, the GIMP calculation qualitatively reproduces the original prediction.

CONCLUSIONS

The focus of this paper is on verification and validation calculations for finite deformation wave propagation through material interfaces. Verification calculations using several GIMP method algorithm variants reveal a minimum algorithmic complexity required for accurate solutions. The calculations demonstrate that particle methods can handle material interfaces well. The credibility of particle method computations of wave propagation through heterogeneous media is enhanced.

ACKNOWLEDGMENTS

This work is a collaboration between researchers in C, MST, P, T and X divisions at LANL (under the auspices of the United States Department of Energy), the University of Utah's C–SAFE, and the AFRL/MNMW, Eglin AFB.

REFERENCES

1. Sulsky, D., Zhou, S.-J., and Schreyer, H. L., *Comput. Phys. Commun.*, **87**, 236–252 (1995).
2. Bardenhagen, S. G., and Kober, E. M., *Comput. Model. Eng. & Sci.* (submitted).
3. Bardenhagen, S. G., Roessig, K. M., Byutner, O., Guilkey, J. E., Bedrov, D., and Smith, G. D., "Direct Numerical Simulation of Weak Shocks in Granular Material," in *The Proceedings of the 12th International Detonation Symposium*, edited by J. M. Short, in press.
4. Baer, M. L., *Thermochimica Acta*, **384**, 351–367 (2002).
5. Krautkrämer, J., and Krautkrämer, H., *Ultrasonic Testing of Materials*, Springer–Verlag, Berlin, 1983, pp. 23–25.
6. Roessig, K. M., Foster, J. C., and Bardenhagen, S. G., *Exp. Mech.*, **42**, 329–337 (2002).
7. Greenfield, D. L., S. R. Paisley, and Koskelo, A. C. (This volume).
8. Koskelo, A. C. (This volume).
9. Niemczura, J., Paisley, D. L., and Swift, D. (This volume).
10. Greening, D. R., and Koskelo, A. (This volume).
11. Peralta, P., Loomis, E., McClellan, K. J., and Swift, D. C. (This volume).
12. McClellan, K. J., Swift, D. C., Paisley, D. L., and Koskelo, A. C. (This volume).

CP706, *Shock Compression of Condensed Matter - 2003*
edited by M. D. Furnish, Y. M. Gupta, and J. W. Forbes
© 2004 American Institute of Physics 0-7354-0181-0/04/$22.00

THREE-DIMENSIONAL IMPACT SIMULATIONS BY CONVERSION OF FINITE ELEMENTS TO MESHFREE PARTICLES

S. R. Beissel, C. A. Gerlach and G. R. Johnson

Army HPC Research Center, NetworkCS, P.O. Box 581459, Minneapolis, MN 55415-1459

Abstract. The simulation of high-velocity impact and penetration is inhibited by complex material behavior and large deformations. Lagrangian formulations best model complex materials because history-dependent variables and material boundaries are not advected. However, Lagrangian finite elements are limited by large deformations. Recently, meshfree particle methods have been used to avoid such limitations, and have demonstrated greater accuracy than traditional erosion methods (wherein deformed elements are removed). Though the variable connectivity of particles enables them to model large deformations, it requires more computational effort than (fixed-connectivity) elements. Therefore, an algorithm was designed to convert deformed elements to particles, thus providing the ability to model large deformations where needed, while maintaining the efficiency of elements elsewhere. This combination is essential in three dimensions, where problem size demands efficiency. In this paper, the conversion algorithm is demonstrated for several three-dimensional simulations of high-velocity impact and penetration.

INTRODUCTION

Numerical simulations of high-velocity impact and penetration have been performed for years with both Eulerian and Lagrangian finite-element formulations. Eulerian formulations incur errors from the advection of boundary conditions (esp. contact) and the advection of history-dependent material variables (e.g., plastic strain and damage). Conversely, Lagrangian finite-element formulations do not easily accommodate the large material distortions typical of penetration events, and the traditional numerical approach of eroding distorted elements may result in unacceptably large errors.

The variable nodal connectivity of meshfree particle methods [1] allows them to simulate large distortions in a Lagrangian formulation without erosion errors. In effect, simplified adaptivity is performed each timestep. The cost of variable connectivity is greater computational effort.

Therefore, an algorithm to automatically convert distorted finite elements to particle nodes has been developed [2,3] so that finite elements efficiently model the material that undergoes mild and moderate distortions, and particle nodes only model the material that has undergone severe distortions. This combination of efficiency and accuracy is essential in three dimensions, where spatial resolutions of interest result in very large data sets. This paper demonstrates the conversion algorithm with three-dimensional computations of ballistic impact and penetration. The formulation of the particle algorithms is documented elsewhere [4,5].

ALGORITHM DESCRIPTION

Figure 1 compares the results of two computations of a tungsten rod impacting a steel target using erosion (left) and conversion (right).

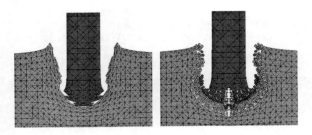

FIGURE 1. Comparison of erosion (left) and conversion (right) in penetration computation

The computations are in 2D for demonstration purposes only; the effects are the same in 3D. The erosion algorithm generates gaps along the interface between the penetrator and the target when elements are eroded. The sudden introduction of these gaps produces pressure drops that may significantly affect the strength of pressure-dependent (frictional) materials. In the conversion algorithm, distorted elements become particles and the gaps are avoided, resulting in greater continuity and accuracy along the material interface.

Figure 2 depicts three examples of conversion from a tetrahedral element to a spherical particle. The criterion for conversion is a critical value of the element equivalent plastic strain. The process of conversion begins with the removal of the element and the transfer of its variables (stress, strain, etc.) to the new particle. This is simply a one-to-one mapping for constant-strain tetrahedra. Although the process is completed during a single timestep, the algorithm provides a smooth transition in time by locating the new particle at the center of mass of the deleted element, and by setting the mass and velocity of the new particle to the total mass and net velocity of the deleted element.

The algorithm models a continuum across the interface between the elements and particles by attaching new particles to the finite-element mesh. This entails tracking the surfaces of the elements on the interface in a manner similar to the eroding master surface of a contact interface, then attaching new particles to these master surfaces as though they were slave nodes which cannot slide along or separate from the surface. Interparticle forces are evaluated by the GPA algorithm [4,5], and particle motion is determined by these forces, subject to the constraints of attachment (if any).

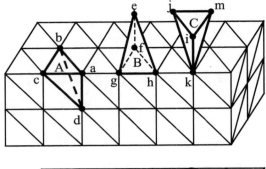

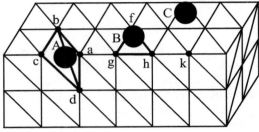

FIGURE 2. Conversion of tetrahedral elements (top) to particles (bottom)

During the conversion of element A to particle A in Fig. 2, surfaces *a-b-c* and *a-c-d* are removed from the list of interface surfaces, and surfaces *a-b-d* and *c-d-b* are added. For the conversion of element B, three surfaces are removed and one is added. For the conversion of element C, which is attached to the mesh only at node k, four surfaces are removed, none are added, and particle C is not attached to any surface.

EXAMPLES

The first example is a tungsten rod impacting spaced aluminum and steel plates. The impact velocity is 1500 m/s and the obliquity is 45 degrees. The rod is 1.27 cm in diameter and 12.7 cm in length. The plates are 1.27 cm and 2.54 cm thick. The initial mesh is 952,512 tetrahedral elements.

Figure 3 shows the computed results. The top image of Fig. 3 shows the result 50 μs after impact, when the rod has perforated the aluminum plate. Conversion of elements around the impact allows material in both the rod and target to accommodate large plastic flows. The constitutive models include continuum damage, which causes loss of strength, fracture, and eventually the spray of particles.

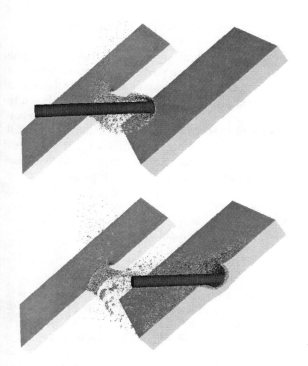

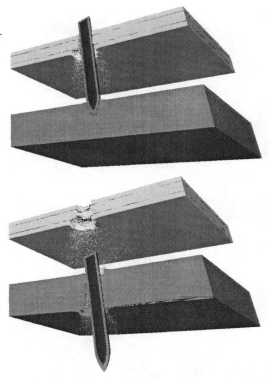

FIGURE 3. Rod and spaced plates 50 μs (top) and 117.5 μs (bottom) after impact

FIGURE 4. Projectile and concrete structure 6 ms (top) and 12 ms (bottom) after impact

The bottom image in Fig. 3 shows the results 117.5 μs after impact, when the rod has penetrated deep into the steel plate, and 74,199 elements (7.8 %) have converted to particles. Much of the debris (in the form of particles) from the backside of the aluminum plate has impacted the front side of the steel plate.

The second example is the computation of a steel projectile impacting a concrete structure. The projectile includes explosive fill and an ogive nose. The impact velocity is 305 m/s and the obliquity is 15 degrees. The projectile length is 2.0 m and its diameter 0.254 m. The structure consists of two slabs: a top slab 0.381 m thick, including rebar at 7.62 cm cover; and a bottom slab 1.016 m thick. The initial mesh is 458,934 tetrahedral elements.

Figure 4 shows the results of this computation at 6 ms (top) and 12 ms (bottom) after impact. At 6 ms, the projectile has perforated the first slab and begun penetrating the second slab. At 12 ms, the projectile has perforated both slabs. Because steel is much stronger than concrete, the strains in the projectile are small compared to the concrete target, and all 12,413 particles (2.7 %) are concrete.

The final example is a projectile impacting a layered SiC/Al target. The projectile is a hard steel cylinder 0.8 cm in diameter and 2.4 cm in length. The impact velocity is 850 m/s and the obliquity is 8 degrees. The layers are each 0.7 cm thick. The initial mesh contains 635,040 tetrahedral elements.

Figure 5 shows the results of this computation. The top image of Fig. 5 shows the results at 25 μs after impact, when the projectile has penetrated most of the ceramic. The red particles on the top surface of the ceramic are the result of the projectile dwelling on the surface before penetration begins. The yellow particles under the projectile represent comminuted ceramic.

The middle image of Fig. 5 shows the results at 100 μs, when 27,381 (4.3 %) particles have been created. The projectile has perforated the target, carrying a layer of comminuted ceramic, and a plug of (unconverted) aluminum. In addition, a gap has opened up between the two target layers, and failed material from both the projectile and the ceramic has filled the gap near the hole.

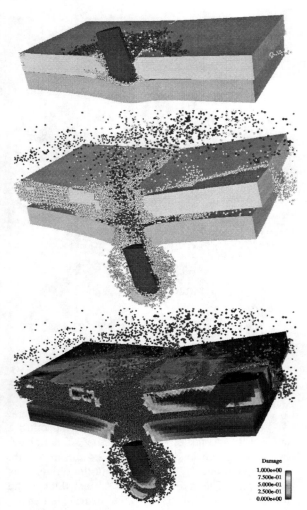

Damage
1.000e+00
7.500e-01
5.000e-01
2.500e-01
0.000e+00

FIGURE 5. Projectile and layered SiC/Al target 25 μs (top) and 100 μs (middle and bottom) after impact; color fills (bottom) indicate damage; red is full damage

The bottom image of Fig. 5 shows damage at 100 μs. Three cracks, characterized by full damage (red) and particles, appear in the ceramic. One lies along the symmetry plane to the left of the impact; one lies at 90 degrees to the symmetry plane; and one lies to the right, about 15 degrees from the symmetry plane.

The strengths of ceramic and concrete strongly depend on pressure. Thus, errors are introduced if the last two examples are computed with erosion (due to pressure drops) instead of conversion.

SUMMARY AND CONCLUSIONS

An algorithm to convert distorted finite elements to meshfree particles is demonstrated for three-dimensional computations of ballistic impact and penetration. The algorithm has been shown to provide greater accuracy than the traditional erosion technique. However, some refinements are still necessary to increase efficiency. In three dimensions, the large number of particles and their need to find neighbors each timestep (variable connectivity) place great importance on improved searching and sorting. The variable connectivity also increases the complexity and communication costs of parallel computations.

ACKNOWLEDGMENTS

The research reported in this document was performed in connection with contract DAAD19-03-D-0001 with the U.S. Army Research Laboratory. The views and conclusions contained in this document are those of the authors and should not be interpreted as presenting the official policies or positions, either expressed or implied, of the U.S. Army Research Laboratory or the U.S. Government unless so designated by other authorized documents. Citation of manufacturer's or trade names does not constitute an official endorsement or approval of the use thereof. The U.S. Government is authorized to reproduce and distribute reprints for Government purposes notwithstanding any copyright notation hereon.

REFERENCES

1. *Computer Methods in Applied Mechanics and Engineering* **139** (1996).
2. Johnson, G.R., Stryk, R.A., Beissel, S.R., and Holmquist, T.J., *Impact Engineering* **27**, 997-1013 (2002).
3. Johnson, G.R., Stryk, R.A., *Impact Engineering* **28**, 947-968 (2003).
4. Johnson, G.R., Beissel, S.R., and Stryk, R.A., *Computational Mechanics* **25**, 245-256 (2000).
5. Johnson, G.R., Beissel, S.R. and Stryk, R.A., *International Journal for Numerical Methods in Engineering* **53**, 875-904 (2002).

CP706, *Shock Compression of Condensed Matter - 2003*
edited by M. D. Furnish, Y. M. Gupta, and J. W. Forbes
© 2004 American Institute of Physics 0-7354-0181-0/04/$22.00

STRONGLY NONLINEAR WAVES IN 3D PHONONIC CRYSTALS

C. Daraio, V. Nesterenko, S. Jin

Materials Science and Engineering Program, University of California at San Diego, La Jolla CA 920, USA

Abstract. Three dimensional phononic crystal ("sonic vacuum" without prestress) was assembled from 137 vertical cavities arranged in hexagonal pattern in Silicone matrix filled with stainless steel spheres. This system has unique strongly nonlinear properties with respect to wave propagation inherited from nonlinear Hertz type elastic contact interaction. Trains of strongly nonlinear solitary waves excited by short duration impact were investigated. Solitary wave with speed below sound speed in the air and reflection from the boundary of two "sonic vacuums" were detected.

INTRODUCTION

Linear elastic phononic crystals are materials with a periodic structure causing acoustic band gap [1,2]. An approach for modeling of compressional waves in *weakly nonlinear* phononic materials "the phononic lattice solid with fluids (PLSF)" at the microscopic scale was proposed in [3]. This paper presents the results on wave dynamics in *strongly nonlinear* [4] phononic crystals based on granular chains in a Silicone elastomer or Teflon matrix.

STRONGLY NONLINEAR WAVES

Non-classical wave behavior appears if a chain of grains is "weakly" compressed [4]. The principal difference between this case and the "strongly" compressed chain is due to a lack of a small parameter with respect to a wave amplitude in the former case. Long wave equation for displacement u in this case is:

$$u_{tt} = -c^2 \left\{ (-u_x)^{\frac{3}{2}} + \frac{a^2}{10} \left[(-u_x)^{\frac{1}{4}} \left((-u_x)^{\frac{5}{4}} \right)_{xx} \right] \right\}_x , \qquad (1)$$

$$-u_x > 0 , \qquad c^2 = \frac{2E}{\pi \rho_0 (1-v^2)} , \qquad c_0 = \left(\frac{3}{2}\right)^{\frac{1}{2}} c \xi_0^{\frac{1}{4}} .$$

Here c is not a sound speed, instead c_0 is a sound speed corresponding to initial strain ξ_0.

This equation has no characteristic wave speed independent on amplitude (equation for general interaction law can be found in [4]). Despite its complex nature the equation has simple stationary solutions with unique properties. For example, supersonic solitary wave propagates with a speed V_s depending on the ratio ξ_r of initial ξ_0 and maximum ξ_m strains:

$$V_s = \frac{c_0}{(\xi_r - 1)} \left(\frac{4}{15} \left[3 + 2\xi_r^{\frac{5}{2}} - 5\xi_r \right] \right)^{\frac{1}{2}} . \qquad (2)$$

This strongly nonlinear solitary wave is of a fundamental interest because Eq. 1 is more general than weakly nonlineaar KdV equation. In a system moving with a speed V_p, its periodic solution is represented by a sequence of humps ($\xi_0=0$) [4]:

$$\xi = \left(\frac{5V_p^2}{4c^2} \right)^2 \cos^4 \left(\frac{\sqrt{10}}{5a} x \right) . \qquad (3)$$

Solitary shape can be taken as one hump of periodic solution (it has only two harmonics) with finite length equal five particle diameters. This unique wave was observed in numerical calculations and detected in experiments [4]. Solitary wave can be considered as a quasiparticle with mass equal about 1.4 mass of grain in the chain and its speed V_s has a nonlinear dependence on maximum strain ξ_m (or particle velocity v_m):

$$V_s = \left(\frac{4}{5}\right)^{\frac{1}{2}} c\, \xi_m^{\frac{1}{4}} = \left(\frac{16}{25}\right)^{\frac{1}{5}} c^{\frac{4}{5}} \upsilon_m^{\frac{1}{5}}. \qquad (4)$$

We may see that the speed of this wave can be infinitely small if the amplitude is small! It means that using this material as a matrix in NTPC (Nonlinear, Tunable Phononic Crystals) we can ensure infinite elastic contrast of components, important for monitoring of band gaps. At the same time speed of solitary waves can be considered as constant at any relatively narrow interval of amplitudes due to power law dependence with small exponent. These properties allow using NTPCs as effective delay lines with exceptionally low speed of signal. Simple estimation based on Eq. 4 shows that it is possible to create materials with impulse speed in the interval 10 – 100 m/s corresponding to the amplitude of audible signal.

EXPERIMENTAL RESULTS AND DISCUSSION

We processed a 3-D phononic crystal (Fig. 1) based on a Silicone elastomer matrix filled with one-dimensional chains of steel spheres. We will present a results describing how these chains support waves of different amplitude and duration.

In experiments we measured the force between the bottom plate and the last particle in the chain resting on this plate (Fig. 2). Piezoelectric gauges were placed under the plates of different diameters allowing support one or seven chains. They were connected with a wave guide - a long steel rod with a length about 20 cm embedded into the massive steel block. Typical time of the electric circuit of

Figure 1. 3-D phononic crystal

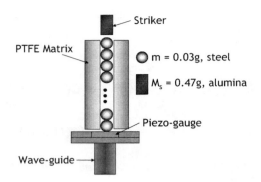

Figure 2. Set-up for testing of 1-D chain.

the gauge was RC=10^3 μs which was enough to ensure a good quality of signals with characteristic period up to 100 μs. The gauges were calibrated using impact with the parallel detection of acceleration.

One dimensional testing was performed using chains of balls placed in Teflon or Silicone elastomer matrixes to investigate how it may influence wave propagation in the chains (Fig. 3).

From Fig. 3 a remarkable feature of "sonic vacuum" is evident – very rapid decomposition of initial impulse on the distances comparable with the soliton width. In fact, the impulse is split after traveling only through 20 particles. This example also demonstrates that "short" duration impact on highly nonlinear ordered periodic systems (lattices) with weak dissipation may result in a chain of solitary waves instead of intuitively expected shock wave. Increase of the duration of impact results in shock wave with oscillatory structure where the leading pulse can be KdV-type for weakly nonlinear chain or compacton-like for strongly nonlinear case [4].

This property of *strongly nonlinear* phononic crystal can be used for controlled impulse transformation in relatively short transmission lines. If chains of grains are placed into a polymer matrix the nonlinear elastic behavior is accompanied by strong dependence of electrical resistivity on local pressure [5]. This behavior can result in a new phenomena like train of locally conductive solitary waves.

Single solitary wave can be generated in the strongly nonlinear chains under impact of particles (pistons) with a mass equal or smaller than mass of particle in the chain [4].

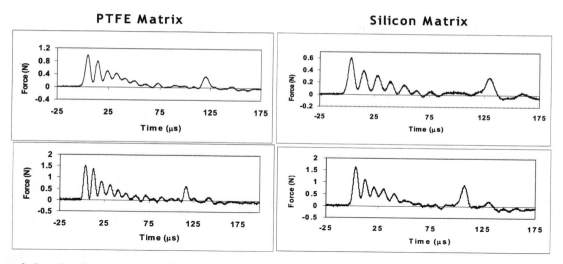

PTFE Matrix   **Silicon Matrix**

Figure 3. Impulses in a 1-D chains (20 steel particles with diameter 2 mm embedded into different matrixes) under identical loading using alumina striker (mass 0.47 g, velocity 0.4 m/s (top Figures) and 0.6 m/s (bottom Figures).

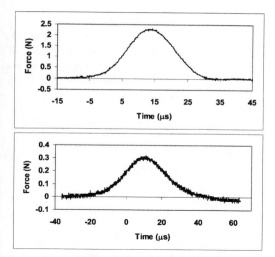

Figure 4. Solitary wave with amplitude significantly larger than gravitational prestress (top) and comparable to gravitational prestress.

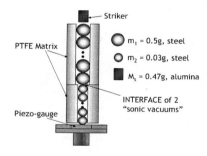

Figure 5. Set-up for the testing of reflection from interface of two "sonic vacuums".

The earlier experiments were conducted with amplitudes of solitary waves about 10 N (see references in [4]).

At the same time it is important to determine the possibility of these chains to support solitary waves with very small amplitude and small speed. For this reason we conducted experiments with loading of 5 mm diameter chains with 20 particles by impact of a single 2 mm diameter particle.

The solitary signals were detected with small amplitudes 2.5 N and 0.3 N (Fig. 4). Based on the amplitudes (Eq. 4) a speed of solitary waves are equal 350 m/s and 317 m/s correspondingly, the latter is below sound speed in air.

Another type of testing was performed on interface between two "sonic vacuums" (Fig. 5) From a comparison of the signals (Fig.6) it is apparent that impulses marked by arrows correspond to the reflection of wave, originated on the bottom of the system, from the interface of two strongly nonlinear chains. Because the system is strongly nonlinear this reflection may not be described in terms of linear acoustics impedances.

Preliminary experiments with 3-D materials

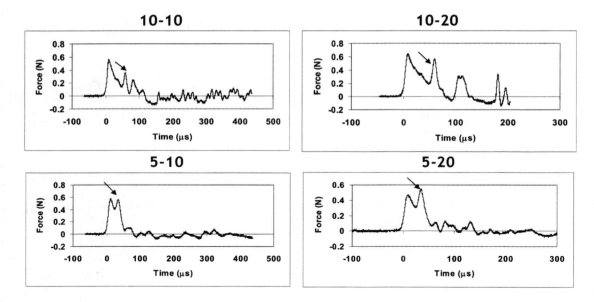

Figure 6. Wave reflection (shown by arrow) from the interface of strongly nonlinear chains.

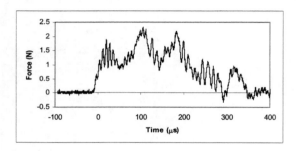

Figure 7. Wave in 7 central chains caused by impact of a steel ball (ø 10.5 mm, mass M_s = 5.3g and u_s = 0.4m/s) on a steel cover top plate with mass 7.65 g.

were conducted using the central part of our phononic crystal (Fig. 1.) Seven chains were supported by the single plate placed on the top of piezogauge. The recorded signal (Fig. 7) corresponds to their collective action. It has oscillatory front and waves reflected from the cover plate back to the gauge are evident.

CONCLUSIONS

1-D and 3-D strongly nonlinear phononic crystals were assembled and investigated for different conditions of loading. Chain of solitary waves, reflection from the interface of two strongly nonlinear chains and single solitary waves with amplitude two orders of magnitude smaller than previously reported for chain of stainless steel spheres were observed. Small amplitude solitons broke "sound barrier" having speed of propagation below sound speed in air.

ACKNOWLEDGEMENTS

Funding was provided by the National Science Foundation, NSF grant #0301322.

REFERENCES

1. Kushwaha, M.S., Classical band structure of periodic elastic composites. *Int. J. of Modern Physics B*, 1996. **10**, 977.
2. Liu, Z., Zhang, X., Mao, Y., Zhu, Y., Yang, Z., Chan, C., and Sheng, P., Locally resonant sonic materials. *Science*, 2000. **289**, 1734.
3. Lian-Jie, H. and Mora, P. The phononic lattice solid with fluids for modelling non-linear solid-fluid interactions. *Geophysical Journal International*, 1994. **117**, 529.
4. Nesterenko, V.F., *Dynamics of Heterogeneous Materials,* Springer-Verlag, New York, 2001.
5. Jin, S., Tiefel, T.,Wolfe, R., Sherwood, R., and Mottine, J., Optically Transparent, Electrically Conductive Composite Medium. *Science*, 1992. **255**, 446.

CP706, *Shock Compression of Condensed Matter - 2003*
edited by M. D. Furnish, Y. M. Gupta, and J. W. Forbes
© 2004 American Institute of Physics 0-7354-0181-0/04/$22.00

DYNAMICS OF THE LOAD TRANSFER IN A SINGLE STRAIGHT CHAIN OF DISKS: FEM SIMULATIONS

A. Goldenberg, A. Britan, G. Ben-Dor, O. Igra, I. Hariton, and B. Glam

Shock Tube Laboratory,
Department of Mechanical Engineering, Ben Gurion University of the Negev,
P.O. Box 653, Beer-Sheva 84105, Israel.
britan@bgumail.bgu.ac.il   avnergo@bgumail.bgu.ac.il

Abstract. The wave propagation phenomenon in a single straight chain of disks made of PSM9 was simulated numerically using the finite element code ABAQUS [1]. The results yield information on the stress wave propagation along the chain. Qualitative agreement with experimental data that was obtained using an optical method of dynamic photo-elasticity and strain gages was obtained. The results of the comparison clearly demonstrate that the stress propagation phenomena are largely governed by the quality of the contacts between the disks. Based on the obtained results justifications for further research efforts on this subject are presented.

INTRODUCTION

The effects of the local microstructure on the dynamic of the stress propagation in both one- and two-dimensional assemblies of particles were extensively discussed in the literature during the past three decades. By means of dynamic photo-elasticity, Rossmanith and Shukla [2] conducted dynamic experiments using the common concept of modeling the granular media by an array of elastic disks. The explosive loading of a single straight chain composed of polymeric disks (Homalite-100) yield information on the stress wave velocity, the amplitude attenuation and the wave spreading characteristics. These experiments were extended later by Shukla and Damania [3] by using a strain gage technique to examine the stress propagation history in longer chains of disks. Shukla *et al.* [4] and Sadd *et al.* [5] used the Distinct Element Method (DEM) introduced by Cundall, and Strack [6] to simulate the stress wave propagation inside 2-D assembles of disks. The DEM is capable of accounting for various types of packing and the interactions between the disks. By using the

information of the material response obtained from the single chain experiments DEM shows good correlation with the experiments (*eg.*, [4] and [5]).

More recently Roessig *et al.* [7] also used the photo-elasticity method to study the stress phenomena in a single chain and 2-D packing of disks, loaded via a split Hopkinson bar. The experimental data obtained were further used for validation of a numerical code based on the Material Point Method (MPM). The MPM uses the arbitrary Lagrangian/Eulerian formulation and shows qualitative agreement with experiments in a variety of contact conditions.

The main objective of this work is to achieve a better understanding of the phenomena of stress wave propagation in a single straight chain of disks. Using the commercial FEM code, ABAQUS, a dynamic loading of symmetrical model of 11 disks was numerically investigated. The simulated stress propagation history was compared with the experimental finding to clarify the phenomenon and obtain information for further experimental efforts.

EXPERIMENTAL SETUP

The experimental setup is described by Glam *et al.* [8], and is only briefly discussed here. A test section made of perspex holds a single straight chain of disks in plane with approximately 0.25 mm tolerance on each side. The disks, 20 mm in diameter and 9.5 mm in width are made of a photoelastic material PSM-9. The material properties are shown in table 1.

The test section is connected to the bottom of a vertical shock tube channel and closed on the top with a thin rubber film preventing air filtration. The shape, duration and amplitude of the impact pulse are determined by the incident conditions in the channel and the shock wave Mach number.

For all the studied cases a step-wise pressure profile with amplitude of 0.375 MPa and duration of about 5 ms was used. Both a dynamic photoelasticity method and strain gages were used to observe the stress propagation history and the contact stress distribution in the chain.

FINITE ELEMENT MODEL

A symmetric model assembly of 11 circular disks with contact interaction was employed for the analysis as is shown in Fig. 1a. The disks dimensions and properties are identical to the disks used in the experiment. A deformable puncher of the same material closes the top disk. The puncher was also used in the experiment to ensure normal reflection of the incident shock wave at the chain entrance. The bottom disk of the chain rests on a rigid end-wall.

Each half disk in the model is discretized into 1130 triangular elements CPS3 with significant refinement towards the contact surface (Fig. 1b). The size of the elements around the contact surface (Fig. 1c) is 0.1 mm, which is less, than the contact area defined from the experiments. The puncher is discretized with CPS4R elements.

The contact interaction constraints are enforced using the ABAQUS/Explicit kinematic contact algorithm and the 'hard' contact relationship. All the contacts in the simulation are similar.

The chain is instantly loaded by a step-wise pressure jump acting over the puncher area. according to experimental conditions the amplitude

TABLE 1. PSM9-Properties

Property	Disk width [mm]	Young's modulus [MPa]	Poisson's ratio	Density [kg/m^3]
Symbol	h	E	υ	ρ
Value	9.5	4600	0.378	1240

of the simulated pulse is 0.375 MPa and its duration is 5 ms.

RESULTS AND DISCUSSION

Simulations

Figure 2 shows typical results of the numerical simulations where the normal force (F_{cy}) is normalized by value of the static normal force (F_{cys}) calculated as the input pressure acting on the puncher (about 35 N). The letters on the graphs indicate marked points in evolution of the contact force. The arrival of the transmitted stress wave (TSW) in contact N1 increases the contact force at (A) and causes oscillations of the signal over a mean value. When the reflected stress wave (RSW) reaches the contact it results in a strong jump at (B). Since contact N2 positioned closer to the bottom, the rise in the contact force at (B) is observed there earlier. In contact N1 the transmitted expansion wave (TEW), resulted by reflection of RSW at the top puncher reduces the force at (C) and afterward oscillations of the signal over the same mean value of the static force are repeated. In contact N2, because the TEW arrives there later, the quasi-steady period in the force signal behind the RSW (Fig. 2b) is longer than that in contact N1 (Fig. 2a). Similar to the case shown in Fig. 2a the contact force here is reduced in (C) and is also followed by oscillations until the reflected expansion wave REW arrives at (D). At point (D) the reflected expansion wave REW arrives from the bottom and relieves the force almost completely. The cycle then starts again at point (E) and repeats itself. The oscillations tend to converge, especially at the early stage of the process while proceeding to later time, situation becomes more complex for analysis.

Comparison between the contact forces reached

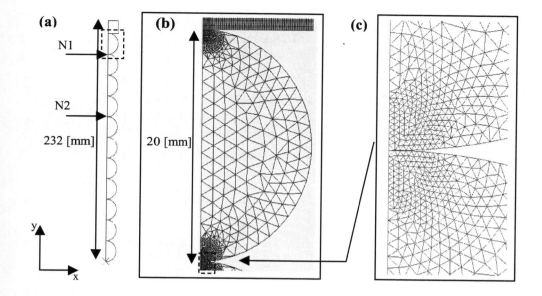

FIGURE 1. FEM model: (a) 11 symmetric disks with a punch on top and a rigid surface at the bottom. Arrow N1 shows position of the contact between disk #1 and disk #2, while arrow N2 - the contact between disk #4 and disk #5; (b) symmetric disk discretization; (c) magnification of the area around the contact.

at different points inside the chain clearly shows that the force drops after the REW is stronger in contact N2. Initial oscillations of the signals are stronger in contact N2. Arrival of the RSW and the REW in the observed contact between the disks complicate further comparison between the signals.

From the simulated x-t diagram in Fig. 3 it is evident that the average velocity of the TSW is $\overline{V}$ =1316 m/s. This is almost half of the stress wave velocity $\overline{V}$ =2300 m/s obtained in our simulations for dynamic impact of solid plate of material PSM-9. The next important result visible in this figure is that the average velocity simulated for RSW is about twice smaller than that for TSW.

Comparison with Experiment

In Fig. 4 the normal strain (ε_y) registered by the strain gauge bonded 1.6 mm downstream from contact N2 is compared with our numerical prediction and with the results obtained from the photo-elastic measurements. The strain is normalized by the quasi-static value (ε_{ys}) obtained

during the final stage of the test time period of each signal. Letters A to D in this figure indicate the similar marked points in the strain history as those introduced early in Fig. 2.

Whereas qualitatively, simulation correlates well with the experiment, idealization of the simulated strain history is evident. Judging from the positions of the points (a) on the curves it appears that the simulated velocity of the TSW is much faster than that registered in the experiment. This difference can be ascribed to the neglecting in the simulations of the initial imperfection of the contact between the disks. Although only qualitative agreement exists at this stage between the simulation and the experiment, comparison studies of the results shows the way for future research. It became evident, in particular, that a "stiff" contact law used in the numerical model is an over simplified approximation and further work is needed to simulate this feature more accurately. A new series of tests with pre-stressed disks might be a first step in the same direction in the experiments. These measurements are also bound to clarify the effect of

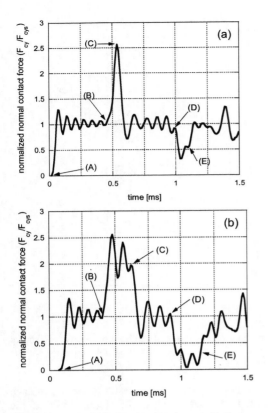

FIGURE 2. Normalized normal contact force versus time at contact: N1 (a) and contact N2 (b). (see Fig. 1)

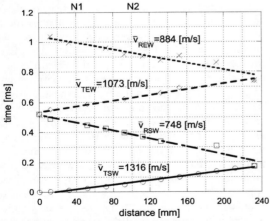

FIGURE 3. x-t diagram of the stress propagation in a single straight chain of disks. The positions of the contact points N1 and N2 are shown by dotted lines.

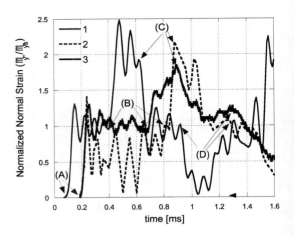

FIGURE 4. Normalized normal strain versus time at 1.6 mm under contact N2. 1-Simulations data, 2-Photo-elasticity data 3-Strain Gages data

the contact quality in the propagation history of the stress waves inside the chain. These experiments are now in preparation stage.

ACKNOWLEDGMENT

The authors would like to acknowledge the Israel Science Foundation under grant No. 190/01-1 and the United State–Israel Binational Science Foundation under grant No. 1999248 for their support, which made this research possible.

REFERENCES

1 ABAQUS, version 6.3, Hibbitt, Karlsson and Sorensen, Inc. (2002)
2 Rossmanith, H. P. and Shukla, A., *Acta Mech.*, **42**, 211-225, (1981).
3 Damania, C. and Shukla, A., *J., Exp. Mech.,* **27**, 268-281, (1987).
4 Shukla, A., Sadd, M. H., Singh, R., Tai, O. and Vishwanathan, S., *Optics and Lasers in Eng.*, **19**, 99-119, (1993).
5 Sadd, M. H., Tai, O. and Shukla, A., *J. Non-Linear Mech.,* **28**, 251-265 (1993).
6 Cundall, P. A. and Strack, O. D. L., *Geotechnique,* **29**, 47-65 (1979).
7 Roessig, K. M., Foster, J. C. and Bardenhagen, S. G., *J. Exp. Mech.* **42**, 329-337, (2002).
8 Glam, B., Goldenberg, A., Britan, A., Ben-Dor, G. and Igra, O., Presented in **Shock Compression of Condensed Matter-2003,** Portland, Oregon, USA.

CP706, *Shock Compression of Condensed Matter - 2003*
edited by M. D. Furnish, Y. M. Gupta, and J. W. Forbes
© 2004 American Institute of Physics 0-7354-0181-0/04/$22.00

ANALYTIC SOLUTIONS AND CONSTITUTIVE RELATIONS FOR SHOCK PROPAGATION IN POROUS MEDIA

Dennis Grady

Applied Research Associates, 4300 San Mateo Blvd., A-220, Albuquerque, New Mexico 87110

Abstract. The present effort examines shock-wave propagation in porous ceramic powders. Computational models of varying sophistication have been developed to treat the dynamic compaction of porous media. The preponderance of computational treatments in production codes, however, have used relatively straightforward engineering modeling approaches to the stress-wave induced compaction of porous matter such as the Herrmann p-alpha model, and a more recent method identified as the p-lambda model. Analytic solutions of shock propagation in porous media have also been fruitfully pursued as exemplified by the seminal solutions of Kompaneets outlined in the second volume of the Zeldovich and Raiser treatise on shock wave physics. Analytic solutions offer instructive insight into the phenomena of shock propagation in porous media and allow scaling of the governing equations to identify the prevailing material and boundary properties. Here the solution methods of Kompaneets have been extended to include specific compaction models. Relationships between compaction models and the resulting shock propagation are explored.

INTRODUCTION

The propagation of shocks in porous media is a fascination special case in the applications of the tools of shock wave physics. Observations in the present note have resulted from the efforts of this author to better understand the response of powders of ceramic materials when subjected to intense impact or explosive loading.

Porous solids are a field of shock wave science which has been pursued by many. Herrmann (1) formulated the p-alpha model of dynamic compaction, which has been implemented into numerous numerical shock physics codes. Several years later Carroll and Holt (2) put forth a theoretical study of dynamic compaction through extensive analysis of the dynamics of ductile hole compaction. A range of issues associated with shocks in porous media are addressed in the collection of articles edited by Davison et al. (3). The monogram by Nesterenko (4) explores a number of selected applications in dynamic compaction. Others, too numerous to mention, have pursued many facets of shock compaction in a plethora of organic and inorganic porous material in which the phenomena is relevant.

ANALYTIC SOLUTIONS AND SCALING RELATIONS

Rapid growth in computational methods has allowed the solution of complex shock compression problems not dreamed of a decade ago. Analytic solutions, however, still offer richness in physical insights and a depth of understanding not fully appreciated from a strictly numerical regimen.

An insightful analytic solution approach to the shock compaction of porous media was pursued by Kompaneets (5) and is summarized in the second volume of Zeldovich and Raiser (6). Kompaneets considered the deposition of a large amount of energy at a point source in a porous media and the subsequent propagation of outwardly directed spherical shock waves. The material was considered to have a porosity p fully compacted in the shock process (the snow-plow model) but was otherwise incompressible. Through an energy balance solution Kompaneets showed that the Hugoniot particle velocity at the shock front is given by,

$$u(X) \propto X^{\frac{3(1+\beta)}{2\beta}} , \qquad (1)$$

where X is the radial position of the shock front and where,

$$\beta = 3/(p + p^{2/3} + p^{1/3}) . \qquad (2)$$

Other properties, including the shock velocity and Hugoniot pressure, are calculated through the Hugoniot conservation relations. The Kompaneets solution is illustrated in Figure 1 where the post-shock incompressible flow is provided through the solution of Grady and Passman (7).

Attenuation coefficients for properties of a spherical shock wave in porous media are readily derived and reveal, in particular, that energy is conserved as porosity approaches zero whereas momentum is conserved in the limit of unit porosity. This Kompaneets result was offered by Hosapple (8) in the clarification of scaling properties in impact crater phenomena.

The Kompaneets solution method is readily extended to cylindrical and planar shock propagation geometries with the solution provided by Equation 1 with Equation 2 replaced by $\beta = -\ln p /(1-p)$ for cylindrical geometry and $\beta = 1$ for planar geometry.

COMPACTION SOLUTIONS IN PLANAR GEOMETRY

Working with the Kompaneets solution in planar geometry one readily discovers the opportunity for richer characterization of the problem, both in terms of the compaction properties of the porous material as well as in the description of the dynamic loading conditions. Consider, for example, planar loading on an infinite half space, $x \geq 0$, of porous material of initial density ρ_o which compacts to a density $\rho(P)$ dependent on the shock amplitude P but is otherwise incompressible. A material compaction relation which depends on the shock pressure compares, for example, to the material response relations used in the computation p-alpha model and allows for the prediction of densification with propagation distance.

The planar impact of an incompressible plate of mass M per unit area at velocity u_o on to a porous media at $x = 0$ will be considered. The analytic solution is equally applicable to the detonation of a plate of high explosive where M is equal to one-half of the charge mass per unit area and u_o is proportional to the explosive Gurney velocity. Either energy or momentum conservation yields the governing shock propagation relations. Momentum conservation readily yields,

$$(M + \rho_o X)\frac{du}{dX} + \rho_o u = 0 , \qquad (3)$$

with solution for the Hugoniot particle velocity as a function of distance,

$$u(x) = u(X) = \frac{u_o}{(1 + X / b)} , \qquad (4)$$

where $X(x)$ is the position of the shock front. The characteristic distance $b = M/\rho_o$ separates the near field solution, independently dependent on the loading parameters M and u_o, from the far field solution dependent only on the product Mu_o. The pressure, strain, and shock velocity at the position X are calculated from,

$$P(X)\varepsilon(X) = \rho_o u(X)^2 , \qquad (5)$$

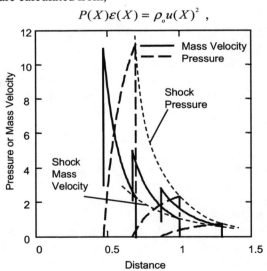

FIGURE 1. The Kompaneets solution for pressure and mass velocity spherical shock amplitude and profile propagation in 30% initial porosity medium.

$$U(X)\varepsilon(X) = u(X) \ . \qquad (6)$$

Closure requires a material dependent compaction relation of the functional form $\varepsilon(P)$.

ROBUST COMPUTATIONAL MODELS

A number of computational models have been developed to treat the dynamic compaction of porous media. Here we refer only to two models which have been actively used in the present efforts. Both are implemented into the Sandia CTH shock physics code (9). First is the p-alpha model originating from the seminal studies of Herrmann (1). Second is the more recently developed p-lambda model with somewhat wider applications to material mixtures but in the spirit of the p-alpha model (10). Briefly, in the p-alpha model an equation of state of the solid component of the porous material is identified. Compression of the porous media is captured through a pressure-dependent compaction relation for the distention $\alpha(P) = \rho_s/\rho(P)$. An average elasticity of the partially compacted material describes the decompression properties of the porous solid at intermediate states of compaction.

In the p-lambda model a two-state description allows the material to exist in either an uncompacted strain-equilibrated state or in a fully compacted pressure-equilibrated state. In the model's simplest form the two nonlinear compression relations correspond to the Ruess and Voigt bounds for the linear elasticity of mixtures. Intermediate states of compaction and decompression elasticity are captured with a mass average of the two compression states through a compaction parameter $0 \le \lambda \le 1$. A compaction relation is provided through a pressure-dependent $\lambda(P)$.

COMPACTION RELATIONS FOR POROUS SOLIDS

There is considerable uncertainty as to the appropriate compaction relation to use in either the analytic solutions or the computational models pursued here. The appropriate constitutive relation would certainly be expected to depend on the porous material of interest. Here interest is in powders of relatively high strength ceramics. The production version of the CTH p-alpha model currently supports only a quadratic form for the compaction distention $\alpha(P)$. This function has not been found to be satisfactory for the available compaction data on ceramic powders. A strength-of-materials approach has led to a compaction relation of the form,

$$\lambda(P) = 1 - \exp(-(P/\sigma)^n) \ , \qquad (7)$$

for the CTH p-lambda model (10). This functional form has the flexibility for reasonably describing the available compaction data. One of the few theoretical approaches to the compaction of porous solids is that of Carroll and Holt (2) which has provided the compaction relation,

$$P = \tfrac{2}{3} Y \ln((\alpha - 1)/\alpha) \ , \qquad (8)$$

based on a plastic medium with flow stress Y. Although this development, which was focused on ductile material, might appear inappropriate to brittle ceramic powders, there is some evidence that the model may be applicable over at least a portion of the compaction process. Compaction of HMX powder has been found to be sensibly described by a power law compaction relation in the distention,

$$P = a\alpha^{-n} \ , \qquad (9)$$

and there is evidence that this form also captures compaction of harder ceramics. Another empirical form for the compaction relation has been guided by shock wave data,

$$P = K\varepsilon /(1 - \varepsilon / p)^n \ , \qquad (10)$$

where here $\varepsilon = 1 - v/v_o$ is the compaction portion of the compression strain and p is the initial porosity.

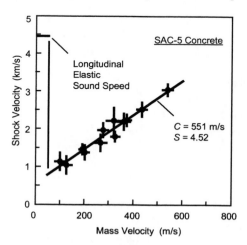

FIGURE 2. Shock wave data for concrete illustrating shock velocity versus mass velocity slope and intercept data consistent with pore compaction.

For $n = 2$ the functional form is that of the linear shock velocity versus particle velocity compression where evidence such as that for porous concrete in Figure 2 is suggestive of the appropriateness of Equation 10.

SHOCK ATTENUATION IN POROUS MEDIA

Any of the previously described compaction relations would adequately serve as a constitutive model in the analytic solution leading to the solutions for shock pressure, strain and particle velocity as a function of propagation distance provided in Equations 4, 5 and 6. Here we have used the relation developed within the p-lambda model provided in Equation 7. Parameters in this relation have been adjusted to describe the compaction response of a ceramic powder with initial porosity of 60%. The powder was impacted at 335 m/s by a one-cm thickness metal plate. The problem was simulated using the p-lambda model with the CTH shock physics code. Calculated pressure histories at increasing distance from the impact plane are shown in Figure 3. Attenuation calculations from the analytic solution are compared with the CTH simulation in Figure 4. Some disagreement in strain is noted in the intermediate field; however, solutions converge in the far field.

CLOSURE

Both analytic and computational models of shock compaction of porous media are explored. Powders of high strength ceramics are of particular interest in this effort. Several compaction relations in which a compaction dependence on peak shock pressure are pursued. This issue requires further study and supporting theoretical efforts.

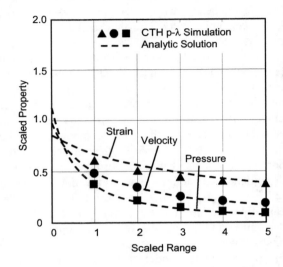

FIGURE 4. Comparison of CTH simulation and analytic solution of shock attenuation in porous media.

REFERENCES

1. Herrmann, W.J., Appl. Phys., **40**, 2490-2499 (1968).
2. Carroll, M.M., and A.C. Holt, J. Appl. Phys., **43**, 1626-1635 (1972).
3. Nesterenko, V.F., Dynamic of Heterogeneous Materials, Springer (2001).
4. Davison, L., Y. Horie, M. Shahinpoor, High Pressure Shock Compression of Solids IV, Springer (1997).
5. Kompaneets, A.S., Dokl. Akad Nauk SSSR **109**, 49-52 (1956).
6. Zeldovich, Y.B. and Y.P. Raiser, Physics of Shock Waves and High-Temperature Hydrodynamic Phenomena, Vol. II, Academic (1967).
7. Grady, D.E., and S.L. Passman, Int. J. Impact Engng. **10**, 197-212 (1990).
8. Hosapple, K.A., Int. J. Impact Engng. **5**, 243-356 (1987).
9. Bell, R.L., Baer, M.R., Brannen, R.M., Elrich, M.G., Hertel, E.S., Silling, S.A., and Taylor, P.A., Sandia National Laboratories Tech. Rept., April (1999).
10. Grady, D.E. and S.L. Passman, Int. J. Impact Engng **10**, 197-212 (1990).

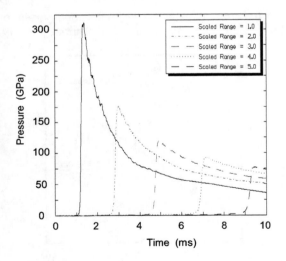

FIGURE 3. Shock wave attenuation in porous ceramic powder. Simulation with CTH code and p-lambda porous material model.

CP706, *Shock Compression of Condensed Matter - 2003*
edited by M. D. Furnish, Y. M. Gupta, and J. W. Forbes
© 2004 American Institute of Physics 0-7354-0181-0/04/$22.00

MODELING DYNAMIC PHASE TRANSITIONS IN TI AND ZR

C. W. Greeff*, **P. A. Rigg***, **M. D. Knudson**[†], **R. S. Hixson*** and **G. T. Gray, III***

*Los Alamos National Laboratory, Los Alamos, NM 87545
†Sandia National Laboratory, Albuquerque, NM 87185

Abstract. Both Ti and Zr exhibit phase transitions from the α (hcp) to the ω phase at pressures of a few GPa. In addition, the Hugoniot of Zr shows a second phase transition at 23 GPa. We have developed multi-phase equations of state for these metals based on ultrasonic, static compression, and shock data. The second transition in Zr is consistent with a phase diagram in which the high-temperature and high-pressure bcc phases are a single continuous phase. Time-resolved experiments using plate impact and continuous magnetic loading are compared to simulations to investigate the kinetics of these phase transitions. Strong kinetic effects are observed in the $\alpha - \omega$ transition in both metals, with the dynamic phase transition observed at pressures well above the equilibrium phase boundary. Data on Zr samples of varied purity are consistent with a strong reduction of the transformation rate by impurities.

INTRODUCTION

The elements of the Ti group, Ti, Zr and Hf, exhibit similarities in their phase diagrams. All three are hcp (α) under ambient and conditions and transform to the bcc (β) structure at high temperature [1], and the ω structure (a hexagonal structure with 3 atoms per cell [2]) under pressure. Under further compression at room temperature, Zr and Hf transform to the bcc phase at pressures of 35 and 71 GPa, respectively, while Ti is reported to remain in the ω phase to 116 GPa where it transforms to an orthorhombic structure [3]. There is substantial variation in the reported pressure of the $\alpha - \omega$ transition in these metals [4]. This is probably a result of the importance of kinetics, even under static loading [5], and the sensitivity of this transformation to impurities [6].

Earlier publications [7, 8] describe our analysis of the equation of state and the shock induced $\alpha - \omega$ transition in Ti. We have extended this analysis to Zr, which we discuss here. For Zr it is necessary to include three phases, since the sequence α, ω, and β is apparent along the Hugoniot. We also present our analysis of recent time-resolved experiments on

both Ti and Zr [9]. We compare direct numerical simulation with experimental measurements under loading by flyer impact and smooth magnetic drive to better understand the kinetics of the $\alpha - \omega$ transition in Ti and Zr.

ZR PHASE CHANGES AND EQUATION OF STATE

A previous analysis of the equation of state (EOS) of Ti [7] concluded that the measured Ti Hugoniot consists of a metastable α phase branch and an ω phase branch. There are no indications of further phase changes on the Hugoniot to at least 80 GPa. We have developed a similar multi-phase EOS for Zr. Our analysis is mostly empirical, and takes into account the measured properties at ambient pressure, and high pressure data from diamond anvil cell (DAC) and the Hugoniot. In cases where data were lacking or conflicting, electronic structure calculations have been used. This was done for the cold energies of the ω and bcc phases at high pressure.

Through this analysis, the free energies of the

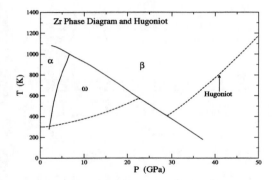

FIGURE 1. Phase Diagram and Hugoniot of Zr. Solid curves are phase boundaries and dashed line is calculated Hugoniot.

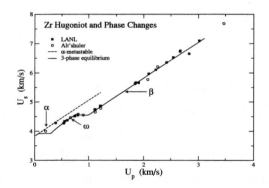

FIGURE 2. Hugoniot of Zr. Symbols are data points. Solid curve is equilibrium Hugoniot. Dashed curve is metastable α-phase Hugoniot.

three phases were obtained. By equating the Gibbs free energies and pressures, we obtain the phase boundaries shown in Figure 1. It is known from DAC measurements [10] that there is an ω-bcc transition at room temperature and $P \approx 35$ GPa, with $dP/dT < 0$ along the phase boundary. This is consistent with the Figure 1, where the high T bcc (β) and the high pressure bcc phases are a single continuous phase. The calculated Hugoniot is shown as the dashed line. It intersects the $\omega - \beta$ boundary at $P = 23$ GPa. This is consistent with shock measurements. We thus conclude that the high temperature and high pressure bcc phases are identical, and we use the the symbol β for this phase. The Hugoniot is shown in the $U_s(U_p)$ plane in Figure 2. The solid curve is the equilibrium Hugoniot, and the dashed curve is the metastable α-phase Hugoniot. As with Ti, the Zr Hugoniot stays in the α phase well above the equilibrium $\alpha - \omega$ transition point. The distinct appearance of the second $\omega - \beta$ transition on the Hugoniot contrasts with the case of Ti, where only the $\alpha - \omega$ transition is apparent [7].

DYNAMIC $\alpha - \omega$ TRANSITION

The $\alpha - \omega$ transition appears to be similar in Ti and Zr. In both cases the equilibrium transition pressure is ≈ 2 GPa [4]. The transition appears on the Hugoniot at ~ 12 GPa in Ti and ~ 10 GPa in Zr. This is due to kinetic effects. The transition is too slow to be observed at the equilibrium pressure under normal shock loading conditions.

To further investigate the kinetics of the $\alpha - \omega$ transition in Ti and Zr, we have done numerical simulations of experiments involving flyer impacts and magnetic loading. Temperature equilibrium is assumed between coexisting phases, and the specific volume and internal energy are taken to be additive [11],

$$
\begin{aligned}
V &= (1-\lambda)V_1 + \lambda V_2 \\
E &= (1-\lambda)E_1 + \lambda E_2,
\end{aligned} \tag{1}
$$

where the subscripts refer to the two phases, and λ is the mole fraction of phase 2.

To complete the description, an expression is needed for $\dot{\lambda} = d\lambda/dt$. Our basic idea is motivated by observations of Singh *et al.* [5], who found that under static pressure, the transformation rate in Ti is an exponential function of the pressure. The expression

$$\dot{\lambda} = (1 - \lambda)v\frac{(G_1 - G_2)}{B} \exp[(G_1 - G_2)/B]^2 \quad (2)$$

was proposed [8] as a phenomenological model for the transition rate in Ti, and is adopted here. Here v and B are material parameters. Eq. (2) is a simple model that gives $\dot{\lambda} = 0$ at the equilibrium phase boundary, and qualitatively describes the very strong variation of the transition rate with pressure.

In an idealized picture, a material undergoing a phase transition exhibits three waves, an elastic precursor, the P_1 wave, which takes the pressure up to the phase boundary, and the P_2 wave, which goes to the final stress. In this picture, the P_1 wave amplitude and speed are material properties related to the EOS and transition pressure. The scenario resulting from Eq. (2) (or any qualitatively similar model) departs from this ideal, because the P_1 wave is nonsteady. Its amplitude is expected to decay with propagation distance. The P_1 amplitude and velocity both increase with increasing peak stress [8]. This latter phenomenon is observed in our experiments.

Figure 3 shows calculated and measured velocity profiles for Ti impacted by a W flyer. The main impurity in the sample is Oxygen at 360 ppm. There is a sapphire window on the sample, and the velocity is measured at the Ti/sapphire interface. The sample thickness is 3.9 mm, and the two traces correspond to flyer velocities of 0.638 and 0.760 km/s. These generate peak stresses in the sample of 11.7 and 14.3 GPa, respectively. The simulations use Eq. (2) with $B = 5.42 \times 10^2$ J/mol and $v = 10^5$ s^{-1}. In the higher stress experiment, the P_1 wave has higher amplitude and moves faster. These phenomena are also shown by the simulations, although to a slightly lesser degree. This is a kinetic phenomenon in which the higher stress wave goes further up the metastable α-phase branch and has less time to relax.

In addition to flyer plate impact, we have carried out experiments on the $\alpha - \omega$ transition under magnetically generated ramp-wave loading (isentropic compression) at the Sandia Z-Machine. A resulting wave profile, along with corresponding simulations,

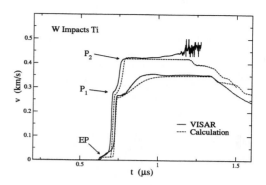

FIGURE 3. Velocity profile in high-purity Ti. Ti sample is impacted by W flyer with sapphire window. Lower traces are for for flyer velocity $v_f = 0.638$ km/s. Upper traces are $v_f = 0.76$ km/s. Solid curves are measured VISAR. Dashed curves are simulations. Labels show the elastic precursor (EP) and P_1 and P_2 waves.

is shown in Figure 4. Simulations of this type of experiment require the applied stress as a boundary condition. In this case this information was obtained by integrating backward in space from a VISAR measurement on an empty sample chamber to the surface of applied stress [12]. This procedure has a limited range of validity in time, due to wave reflections that are absent in the empty chamber. In practice this limits the calculation to times earlier than the point near the peak labeled "Interface Reflection" in the figure. The figure shows simulations including the $\alpha - \omega$ transition and with the α phase only. The experiment clearly shows the transition. The calculations use the same EOS and kinetic parameters as for the flyer experiments.

The observed $\alpha - \omega$ transition is strongly influenced by kinetics, which are sensitive to impurities [6]. Figure 5 shows velocity profiles for two Zr samples, one of very high purity, and the other of lower purity. The composition of the samples is shown in ref. [9]. The differences in purity between the two are at the level of a few parts in 10^4 of O and Hf. The samples are 3.0 mm thick, and are impacted by

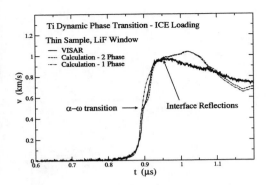

FIGURE 4. Velocity profile from isentropic compression experiment on Ti.

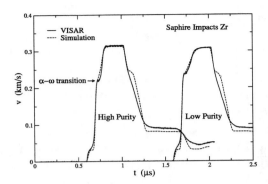

FIGURE 5. Impurity effects on $\alpha - \omega$ transition in Zr. Zr samples impacted by sapphire flyer with sapphire window. Solid curve - VISAR, dashed curve - simulation.

sapphire flyers. The flyer velocities are the same to within 2% in both cases, resulting in a peak stress of 11 GPa. There are sapphire windows, and the velocity is at the Zr/sapphire interface. The profile for the low purity specimen shows a higher P_1 amplitude and a broader P_2 wave. The simulations capture those trends by changing the parameter v in Eq. (2). Both simulations use $B = 5.42 \times 10^2$ J/mol. The high purity simulation uses $v = 10^6$ s^{-1}, and the low purity simulation uses $v = 5 \times 10^5$ s^{-1}. All other parameters are the same for both simulations. The match to experiment suggests that the main effect of impurities is an overall reduction in the transition rate.

CONCLUSIONS

Ti and Zr show similar $\alpha - \omega$ transitions under dynamic loading. The Hugoniot of Zr shows a further $\omega - \beta$ transition, which appears to be absent in Ti. We have compared numerical simulations to time-resolved data on the $\alpha - \omega$, and find that in both Ti and Zr there are strong kinetic effects. An exponential dependence of the transformation rate on pressure captures many features of the data. Small amounts of impurities substantially reduce the transformation rate in Zr.

REFERENCES

1. D.A. Young, *Phase Diagrams of the Elements*, (University of California Press, Berkeley, 1991)
2. J. Donohue, *The Structures of the Elements*, Wiley, New York, 1974.
3. Y. K. Vohra and P. T. Spencer, Phys. Rev. Lett. **86**, 3068 (2001).
4. V. A. Zilbershteyn, N. P. Chistotina, A. A. Zharov, N. S. Grishina, and E. I. Estrin, Fiz. Metal. Metalloved. **39**, 445 (1975).
5. A. K. Singh, M. Mohan, and C. Divkar, J. Appl. Phys. **53**, 1221 (1982).
6. Y. K. Vohra, S. K. Sikka, S. N. Vaidya, and R. Chidambaram, J. Phys. Chem. Solids, **38**, 1293 (1977).
7. C. W. Greeff, D. R. Trinkle, and R. C. Albers, J. Appl. Phys. **90** 2221 (2001).
8. C. W. Greeff, D. R. Trinkle, and R. C. Albers, in *Shock Compression of Condensed Matter - 2001*, edited by M. D. Furnish, N. N. Thadhani, and Y. Horie, AIP (Melville 2002).
9. P. A. Rigg, *et al.*, these proceedings.
10. H. Xia, A. L. Ruoff, and Y. K. Vohra, Phys. Rev. B **44**, 10374 (1991).
11. D. B. Hayes, J. App. Phys. **45**, 1208 (1974).
12. D. Hayes and C. Hall, in *Shock Compression of Condensed Matter - 2001*, edited by M. D. Furnish, N. N. Thadhani, and Y. Horie, AIP (Melville 2002).

CP706, *Shock Compression of Condensed Matter - 2003*
edited by M. D. Furnish, Y. M. Gupta, and J. W. Forbes
© 2004 American Institute of Physics 0-7354-0181-0/04/$22.00

CALCULATION OF GRAIN BOUNDARY SHOCK INTERACTIONS

D. R. Greening[1] **and A. Koskelo**[2]

[1]*Material Sciences Group, Applied Physics Division*
[2]*Advanced Diagnostics and Instrumentation Group, Chemistry Division*
Los Alamos National Laboratory, Los Alamos NM 87545

Abstract. As part of a larger project into the interactions of shock with grain boundaries, calculations that show a localization of energy in some configurations were performed. The verification and prediction of these localizations become important, due to their role in the initiation of a variety of significant material process. The prototype problem is a single grain boundary, inclined to the direction of shock propagation, separating regions of differing orientation. The calculations are made with a finite volume code using a continuum material model with explicit elastic, and plastic anisotropy. The response of NiAl is simulated using material property data from published sources, and from new experiments performed as part of the over-all project. The localization is seen internally as a small region of higher pressure at the intersection of the shock and the grain boundary. At the breakout surface the localization becomes manifest in velocity and displacement excursions at the grain boundary. This surface phenomenon provides an observable that can be used, with planned experiments, to validate the predictive behavior of the model.

INTRODUCTION

Improving our understanding of and predictive capacity for phenomena in shock-loaded materials requires an improved understanding of the complex interaction of the shock with material microstructure. The overall purpose of the project, of which this work is part [1-4], is the exploration of shock interaction with the most fundamental features of the microstructure of polycrystalline materials: differently oriented grains, and the grain boundary between them. Calculations indicate-- and initial experiments consistent with-- the formation of a small region of higher pressure occurring at the intersection of the shock with an inclined grain boundary in some configurations. The existence of such a localization of energy may have a significant role in a detailed understanding of many shock induced material processes.

In addition to the role of the grain boundary in inducing the localization, the spatial association of the localization with the grain boundary has implication for processes where a significant role may be being played. There is an observed association of ejecta with where some grain boundaries emerge along surfaces. A localization of energy along a grain boundary, with the resulting strain gradient along the free surface, offers a driving force for the ejecta. Further, the variation in the strength of the localization with grain geometry are suggests an explanation of why eject occurs at some grain boundaries and not others.

It is also well known that phase changes tend to nucleate at intersections of grain boundaries. The propagation of one or more regions of energy localization traveling along the grain boundaries can provide energy for nucleation, well above the mean field. Thus, understanding these localizations have implication for the science-based predictive capacity for these processes. Many damage mechanisms may be similarly coupled to the spatial variation of strain and strain rates associated with the localization.

Before these more complex interactions can be explored, the underlying phenomena of localization at grain boundaries must be verified, and understood quantitatively. The experimental portion of this program is measuring variations in surface motion at and near a grain boundary along a free surface when breakout occurs [1]. The current calculations are intended to simulate the experiment to support the experimental effort and to begin validation of the modeling approach.

COMPUTATIONAL APPROACH

The approach to both the experiments and the computation is to begin with the simplest geometry embodying the issues of grain boundary shock interaction. Thus we begin with a single grain boundary separating two crystalline regions--the bicrystal. The material in the initial investigation is stoichiometric nickel-aluminide (NiAl). The material was selected for a number of reasons. Of significance for this paper is that NiAl has significant elastic anisotropy, and being of crystal group (B2), has cubic symmetry. Cubic symmetry, allows the separation of the volumetric and deviatoric stresses and strain, greatly simplifying the development of the material model.

The model consists of an equation of state developed for NiAl during this project [2], coupled to an anisotropic elastic model with crystal plasticity that derives directly from the work of Rashid and Nemat-Nasser [5]. Apart from certain details in the numerical implementation, we part from the model as described in [5] in applying it solely to the portion of the velocity gradients associated with the deviatoric deformations and rotation, with the volumetric portion treated separately as alluded to above.

At this stage only the primary slip-systems are considered active. No special mechanisms are modeled for the grain boundary. These features may be revised as suggested by future experiment results. The values for the model parameters are from earlier experiments and published sources [4-6].

The model is used to describe the material response for each region of the simulation differing only by initial orientation. This is of course extensible to more and more complex regions as require.

The shock response of the bicrystal is then simulated for a given combination of crystal orientations and grain boundary angle with a conservative finite volume code [9]. Two sorts of boundary conditions are used along the bottom of the computational region: a constant velocity boundary condition; and a short duration flat-topped pressure boundary condition. These are intended to represent the two methods of shock generation, laser-driven flyer plate, and direct laser ablation respectively. The right and left margins have reflecting boundaries with the top free. In the images that follow, the direction of shock propagation is upwards. Meshing is uniformly 0.5μm vertically nominally 1μm horizontally, with the latter distorted in aligning a mesh line with the incline grain boundary.

Table 1. Parameter used in simulations of NiAl

Linear U_s-U_p Equation of State [4]:	
Density	5.86 (g/cm^3)
C_o	5.45715 (km/s)
S	1.23573 (--)
Γ	1.18738 (--)
Elastic Constants [6]:	
C_{11}	199 (GPa)
C_{12}	137 (GPa)
C_{44}	116 (GPa)
Crystal Plasticity on $\langle 100 \rangle, \{100\}$ System [7,8]:	
τ_{CRSS}	90.0 (GPa)
h (hardening coefficient)	900.0 (GPa)
M	40

RESULTS AND DISCUSSION

The computations results show a complex interaction of condition causing or failing to cause a localization of energy. A clear effect of the role that the orientations of the crystals play is seen in a pair of calculations that differ only in the orientation of the right hand crystal. In figures 1 and 2 the pressure profiles of the two computations are show for the same time. The color scales are the same and have been adjusted to highlight the variation within the shocked region, putting the unshocked region off-scale.

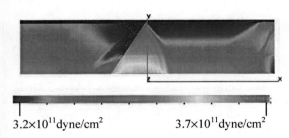

Figure 1. Pressure profile at 12.5 ns, left crystal (100) type direction up, right crystal (111) up, both with a (110) type direction out, 10^5 cm/s upward boundary condition on lower surface.

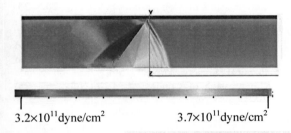

Figure 2. Pressure profile at 12.5 ns, left crystal (100) type direction up, right crystal (110) up, both with a (110) type direction out, 10^5 cm/s upward boundary condition on lower surface (note higher pressure, red, region).

While the mean fields in the two simulations are quite similar, the appearance of a small region of high pressure originating at the intersection of the shock and the grain boundary is seen in the second case. The observable effect for the experiments being conducted is shown in figure 3. The reflected shock seen in figure 1 helps to locally elevate the red curve in figure 3 resulting it a peak in each case. Much of the elevation difference in the red curve is due to the earlier arrival of the fast wave in the (111) region. The isolated peak in the green curve is a more significant variation from the surrounding displacements.

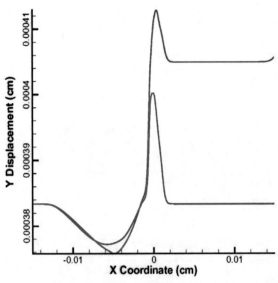

Figure 3. Surface displacement profile at 15 ns, left crystal (100) type direction up, right crystal green (110) up and red (111) up, both with a (110) type direction out, 10^5 cm/s upward boundary condition on lower surface.

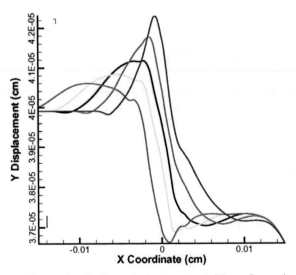

Figure 4. Surface displacements at 25 ns for grain boundary angles −30° (red) though 30° (blue) by 15° increments measured anti-clockwise from vertically downward for a 2.5 ns 15 kbar pulse.

The orientation of the grain boundary is also seen to produce a significant variation in the free surface displacements. Looking to results from short pulse calculations (figure 4), one sees a regular pattern of variation in the local surface motion. The peak displacement seen for the 30° grain boundary orientation is seen to be associated with another localization of pressure at the shock grain boundary intersection seen in figure 5.

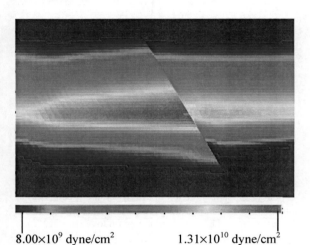

8.00×10^9 dyne/cm^2 1.31×10^{10} dyne/cm^2

Figure 5. Pressure localization in short pulse shock intersection with grain boundary left crystal (100) up, right crystal (111) up, both (110) out.

The patterns of localization differ between the calculations preformed with the two types of boundary conditions. This suggest that the wave shape has a role to play in the formation of energy localization along grain boundaries noting that in the case of the short pulse the shock calculated becomes much more dispersed as it transits the media. Whether the dispersive wave running ahead preconditions the material or has some other effect requires further investigation.

CONCLUSIONS

Thus far the ability to perform calculation predicting potential measurable experimental results has been exhibited [1,2]. The intent is to move forward with calculations in support of the experimental effort, and with the resulting feedback to improve the model. Already one sees prediction of localized events at the grain boundary, consistent with early experiment.

While a more complete parameter study is yet to be completed, the key sources of variation are clear. Crystal orientations, and grain boundary orientation are the key parameter for the control of localization. This work also indicates the time profile of the shock also plays a key role.

ACKNOWLEDGEMENTS

This work was funded through LANL LDRD #2002 0055 DR and performed under the auspices of the U.S. Department of Energy under contract number W-7405-ENG-36.

REFERENCES

1. Koskelo, A., "New windows into shock at the mesoscale", for presentation at 13th American Physical Society Topical Conference on Shock Compression of Condensed Matter July 20 - 25, 2003.
2. Greenfield, S. R., Paisley, D. L. and, Koskelo, A. C. "Transient interferometric studies of shocked bicrystals", for presentation at 13th American Physical Society Topical Conference on Shock Compression of Condensed Matter July 20 - 25, 2003.
3. McClellan, K., Swift, D., Paisley, D., and, Koskelo, A., "Dynamic properties of nickel-aluminum alloy", for presentation at 13th American Physical Society Topical Conference on Shock Compression of Condensed Matter July 20 - 25, 2003.
4. Swift, D. C., "An *ab initio* Grüneisen Equation of State for Nickel-Aluminum Alloy", Los Alamos National Laboratory report LA-UR-02-1757, 2002.
5. Rashid, M. M., and Nemat-Nasser, S., "A Constitutive Algorithm for Rate-Dependent Crystal Plasticity", Comp. Meth. In Appl. Mech. And Eng., vol 94, pp 201-228, 1992.
6. Rusovic, N., and, Warlimont, H., "The elastic behaviour of β_2-NiAl alloys", Phys. Status Solidi A, vol.44, no.2, p.609-19, 1977
7. Mielec, J., Novak, V., Zarubova, N., and, Gemperle, A., "Oreientation Dependence of plastic deformation in NiAl Single Crystals", Matl. Sci. and Eng. A234-236, pp. 410-413, 1997.
8. Maloy, S. A., Gray, G. T. III, and, Darolia R., "High Strain Rate Deformation of NiAl", Matl. Sci. and Eng. A192/193, pp. 249-254, 1995.
9. Cambell, J. and, Shaskov, M. Los Alamos National Laboratory report LA-CC-02-1757, 2002.

CP706, *Shock Compression of Condensed Matter - 2003*
edited by M. D. Furnish, Y. M. Gupta, and J. W. Forbes
© 2004 American Institute of Physics 0-7354-0181-0/04/$22.00

TWO DIMENSIONAL CONTINUUM PROPERTIES FROM MOLECULAR DYNAMICS SIMULATIONS

Robert J. Hardy[1], Seth Root[2], and David R. Swanson[3]

[1]*Department of Physics and Astronomy, University of Nebraska, Lincoln, Nebraska 68588-0111*
[2]*Institute for Shock Physics, Washington State University, Pullman, Washington 99164-2816*
[3]*Department of Computer Science, University of Nebraska, Lincoln, Nebraska 68588-0115*

Abstract. Techniques for obtaining continuously distributed local properties – such as density, velocity, pressure, and temperature – from atomistic simulations are discussed. The resulting local properties are averages over nanometer sized circular regions and are defined so that the continuum expressions for mass, momentum, and energy conservation are exactly satisfied. The techniques are illustrated by calculating the two dimensional spatial distribution of local temperature that results from a shock wave passing over a void.

INTRODUCTION

The behavior of matter can be described in terms of continuously distributed local properties – such as density, velocity, pressure, and temperature – or in terms of the motion of atoms and molecules. Experimental studies are often analyzed at the continuum level, while molecular dynamics simulations give an atomic level description. This paper discusses techniques for obtaining the continuum description from the results of molecular dynamics simulations.

The pioneering contribution to these techniques was made by Irving and Kirkwood,[1] who used Dirac delta functions. Hardy extended that work by introducing localization functions of finite width, which are practical in computer simulations.[2] At the 2001 SCCM conference variations of properties in one spatial dimension were considered and the basis for choosing the width of the localization function was discussed.[3] Investigations of one-dimensional spatial variations are informative, but are incapable of showing intrinsically multi-dimensional phenomena. For example, the turbulence found on a nanometer length scale can only be seen when variations in two dimensions are considered.[4] The study of two-dimensional spatial variations of both one and two particle properties is continued here.

ONE PARTICLE PROPERTIES

Since matter is made up of discrete entities, any description of matter in terms of continuously distributed properties necessarily involves some type of averaging. Presumably, the mass, momentum, and energy densities at point $\vec{R}$ in space represent weighted sums of the mass, momentum, and energy of the atoms in the vicinity of that point. In molecular dynamics simulations atom i is treated as an object of mass m_i at point $\vec{r}_i$, which moves as the system evolves. A function is needed to associate the behavior of atom i with the continuum properties at point $\vec{R}$. A convenient form for the function is $\Delta(\vec{r}_i - \vec{R})$. This *localization function* should be peaked at $\vec{r}_i = \vec{R}$ and go to zero as $|\vec{r}_i - \vec{R}| \to \infty$. Also, it should be normalized so that

$$\int dV_R \, \Delta(\vec{r}_i - \vec{R}) = 1, \qquad (1)$$

where dV_R is a volume element in $\vec{R}$ space. This is required so that the integral of a local property over the entire system equals the value of the property for the complete system. With the aid of a localization function, the mass density $\rho(\vec{R})$, the momentum density $\vec{p}(\vec{R})$, and kinetic energy density $E_K(\vec{R})$ at observation point $\vec{R}$ become

$$\rho(\vec{R}) = \sum_i m_i \, \Delta(\vec{r}_i - \vec{R}), \qquad (2)$$

$$\vec{p}(\vec{R}) = \sum_i \left(m_i \, \vec{v}_i\right) \Delta(\vec{r}_i - \vec{R}), \qquad (3)$$

and

$$E_K(\vec{R}) = \sum_i \left(\tfrac{1}{2} m_i \, v_i^2\right) \Delta(\vec{r}_i - \vec{R}), \qquad (4)$$

where the dependence on position $\vec{R}$ has been made explicit. The dependence on time t is left implicit and enters through the time dependence of the atomic positions $\vec{r}_i$ and velocities $\vec{v}_i$, as determined by molecular dynamics.

We use a localization function that depends on the magnitude $|\vec{r}_i - \vec{R}|$, so that the resulting densities are averages over circular regions centered on $\vec{R}$. The size of the region is characterized by its full width at half maximum (FWHM). Figure 1 shows the cross section of the function used. $\Delta(\vec{r}_i - \vec{R})$ can be interpreted as being associated with a fixed point in space, $\vec{R}$, and representing the region averaged over. Or, it can be interpreted as assigning a shape to atom i that move with the atom. From Eq. (1) it follows that the localization function has units of (volume)$^{-1}$, so that $\rho(\vec{R})$, $\vec{p}(\vec{R})$, and $E_K(\vec{R})$ give the mass, momentum, and kinetic energy per unit volume. For two-dimensional simulations (volume)$^{-1}$ is replaced by (area)$^{-1}$.

The local velocity $\vec{V}(\vec{R})$ and local temperature $T(\vec{R})$ are defined in terms of the above densities. The local velocity is related to the mass and momentum densities by

$$\vec{V}(\vec{R}) = \vec{p}(\vec{R}) / \rho(\vec{R}). \qquad (5)$$

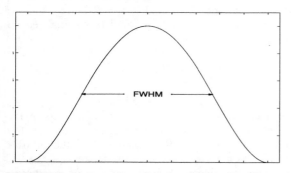

FIGURE 1. Cross section of the localization function.

The equipartition theorem assigns $\tfrac{1}{2} k_B T$ of energy to each kinetic degree of freedom, where k_B is Boltzmann's constant. It suggests that one estimate the local temperature from the local kinetic energy. Correct accounting of energy in the continuum description requires that one use the kinetic energy in the frame of reference moving with the local velocity. The resulting estimate is[4]

$$\tfrac{1}{2} f \, k_B T(\vec{R}) = \frac{E_K(\vec{R}) - \tfrac{1}{2}\rho(\vec{R}) V(\vec{R})^2}{\sum_i \Delta(\vec{r}_i - \vec{R})}, \qquad (6)$$

where f is the number of kinetic degrees of freedom per atom, which equals the number of dimensions of the space.

In continuum theory the conservation laws can be expressed by continuity equations that equate the time derivative of a density to the negative of the divergence of a flux. For example, the continuity equation for mass conservation is

$$\partial \rho / \partial t = -\nabla \cdot \left[\rho \vec{V}\right]. \qquad (7)$$

This relationship is an *exact* consequence of the definitions of ρ, $\vec{p}$, and $\vec{V}$ in Eqs. (2), (3), and (5) and the time dependence of the positions and velocities of the atoms as determined by the equations of motion for the atomic constituents.

The densities ρ, $\vec{p}$, and E_K involve sums of the properties of the individual atoms, and hence are conveniently referred to as one-particle properties. Since pressure is affected by the forces acting between atoms, it contains contributions that are two-particle properties.

TWO PARTICLE PROPERTIES

When the macroscopic flow of the material is accounted for, the continuity equation for momentum conservation becomes

$$\rho \left[\frac{\partial}{\partial t} + \vec{V} \cdot \nabla\right] \vec{V} = -\nabla \cdot \vec{P}, \qquad (8)$$

where $\vec{P}$ is the pressure tensor. In continuum mechanics this relationship is derived by applying Newton's laws to a small element of matter. The Navier-Stokes equation is obtained from Eq. (8) by arguing phenomenologically that $\vec{P}$ is the sum of an isotropic pressure term and a viscous contribution. The pressure tensor $\vec{P}$ is essentially the momentum flux in the frame of reference moving with the local velocity. The formula that relates the pressure tensor to the properties of the atoms is obtained by requiring that Eq. (8) is an exact consequence of the definitions of ρ, $\vec{p}$, $\vec{V}$, and $\vec{P}$. The resulting expression for the pressure tensor contains a kinetic contribution $\vec{P}_K$ and a potential contribution $\vec{P}_\Phi$,

$$\vec{P} = \vec{P}_K + \vec{P}_\Phi . \qquad (9)$$

The kinetic contribution is the one-particle property obtained by replacing the expression in parenthesis in Eq. (4) with $m_i \vec{v}_i \vec{v}_i$ with velocities calculated in the moving frame.

The potential contribution is a two-particle property. The formulas for the pressure developed in references 1 and 2 were obtained with the assumption that the potential energy of the system is a sum of two-body potentials $\phi(r_{ij})$, where $r_{ij} = |\vec{r}_i - \vec{r}_j|$ is the distance between atoms i and j. As illustrated by the reactive bond order (REBO) potential,[5] the accurate modeling of chemical reactions requires potentials more complicated than $\phi(r_{ij})$. Nevertheless, even when the potential energy contains complicated many-body contributions, the forces on the atoms can be expressed as sums of two-body forces. Typically, many-body potentials depend on the atomic positions $\vec{r}_i$ only through the inter-atomic distances r_{ij}. When this is so, forces occur in pairs, as is required by Newton's third law. Specifically, the force exerted on atom i by atom j, $\vec{F}_{ij}$, is equal in magnitude and opposite in direction to $\vec{F}_{ji}$, the force exerted on j by i. Even though the formulas for the pressure derived in reference 2 were obtained with the assumption of two-body potentials, they are valid whenever Newton's third law is satisfied. Thus, with many-body potentials, as well as with two-body potentials, the potential contribution to the pressure is

$$P_\Phi^{\alpha\gamma}(\vec{R}) = \frac{1}{2} \sum_{i,j} F_{ij}^\alpha \, r_{ij}^\gamma \, B_{ij}(\vec{R}) , \qquad (10)$$

where $\alpha, \gamma = x, y, z$ label Cartesian components and $\vec{r}_{ij} = \vec{r}_i - \vec{r}_j$. Evaluating Eq. (10) requires knowing what atom exerts each force, as well as the atom on which it is exerted. The computer programs used for simulations often keep track only of the atom on which a force is exerted, so that some re-programming may be needed before $\vec{P}_\Phi$ can be evaluated. We emphasize that the force $\vec{F}_{ij}$ exerted

on atom i by atom j may depend on distances r_{kl} between atoms other than i and j.

The size of the contribution to $\vec{P}_\Phi$ of atoms i and j is determined by $B(\vec{R})$, which is the integral along the straight line from atom j to atom i of the localization function centered at $\vec{R}$.[2] Consequently, the force $\vec{F}_{ij}$ between two atoms only contributes to the pressure at $\vec{R}$ if the atoms are in the immediate vicinity of $\vec{R}$. Although the appearance of $B(\vec{R})$ in Eq. (10) may not be intuitive, its appearance is plausible and intuitive arguments can be made for it. Nevertheless, its ultimate justification is: When the derived formulas for the pressure tensor are used, the continuum expression for momentum conservation in Eq. (8) is an *exact* consequence of the equations of motion for the system's atomic constituents.

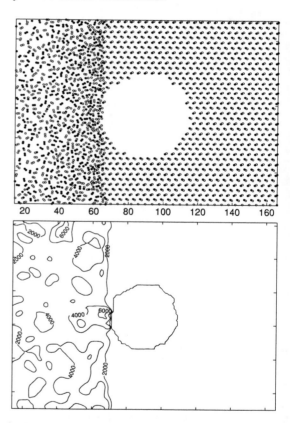

FIGURE 2. Views of a shock wave at $t = 1.0$ ps in a reactive system containing a 50Å diameter void.

ILLUSTRATION

We illustrate the techniques discussed by showing the two-dimensional variation of the local temperatures $T(\vec{R})$. A shock wave passing over a void in a two-dimensional system with periodic boundaries in the y-direction is considered. The model system contained two atomic species, A and B, and used the REBO interatomic potential.[5] The equilibrium spacing between bound pairs of atoms was 1.0Å and the minimum of the weak Van der Waals interaction between molecules was at 3.0Å. The initial loosely packed collection of AB molecules is shown at the right in the top frame of Fig. 2. Each AB molecules is bound in a 2.0 eV deep potential well, while the reacted AA and BB molecules at the left are in 5.0 eV deep potential wells. The initial diameter of the void is 50Å. At time $t = 0$ the system was impacted at $x = 0$ by a flyer plate moving at 6.0 km/s. The top frame in Fig. 2 shows the positions of atoms at $t = 1.0$ ps. The numbers on the x-axis are in Angstroms.

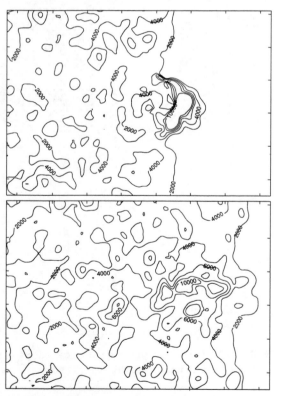

FIGURE 3. Contour plots of local temperature at 1.5 ps and 2.0 ps. FWHM = 7.5Å.

The bottom frame in Fig. 2 is a contour plot of the local temperature $T(\vec{R})$. The FWHM of the localization function was 7.5Å. The numbers on the contours are in kelvins. Figure 3 shows contour plots of the local temperature at times 1.5 ps and 2.0 ps. The maximum temperatures reached at 1.0 ps, 1.5 ps, and 2.0 ps were 7500 K, 22000 K, and 13600 K, respectively.

The hot spot resulting from the collapse of the void is easily seen in Fig. 3. Such a hot spot is expected. Perhaps more surprising are the inhomogeneities behind the shock front that exist even before the wave interacts with the void. In part, these are caused by statistical fluctuations, which can be large, since the number of atoms averaged over to obtain $T(\vec{R})$ was quite small (20 to 40).

SUMMARY

We have shown that calculations of continuum properties from atomic data with a localization function can be a useful tool for analyzing very anisotropic phenomena, such as those associated with shock waves. Although the model system considered was two-dimensional, the techniques are equally applicable to three-dimensional systems. Plots of two-dimensional spatial variations in planes at arbitrary positions and orientations can be very useful for analyzing atomic level simulations of realistic model systems.

REFERENCES

1. Irving, J. H. and Kirkwood, J. G., *J. Chem. Phys.* **18**, 817-829 (1950).
2. Hardy, R. J., *J. Chem. Phys.* **76**, 622-628 (1982).
3. Hardy, R. J., Root, S., and Swanson, D. R., in *Shock Compressions of Condensed Matter-2001*, edited by M. D. Furnish et al. AIP Conference Proceedings 620, Melville, New York, 2002, pp. 363-366.
4. Root, S., Hardy, R. J., and Swanson, D. R., J. Chem. Phys. **118**, 3161-3165 (2003).
5. White, C. T., Swanson, D. R., and Robertson, D. H. in *Molecular Dynamics Simulations of Detonations, Chemical Dynamics in Extreme Environments*, edited by R. A. Dressler, (World Scientific, Singapore, 2001) pp. 546-592.

CP706, *Shock Compression of Condensed Matter - 2003*
edited by M. D. Furnish, Y. M. Gupta, and J. W. Forbes
© 2004 American Institute of Physics 0-7354-0181-0/04/$22.00

TEMPERATURE DEPENDENCE OF SHOCK-INDUCED PLASTICITY: A MOLECULAR DYNAMICS APPROACH

Takahiro Hatano

Center for Promotion of Computational Science and Engineering, Japan Atomic Energy Research Institute, Ibaraki 319-1195, JAPAN

Abstract. Molecular dynamics simulation on a fcc perfect crystal with the Lennard-Jones potential is performed in order to investigate temperature dependence of shock-induced plasticity. It is found that the critical piston velocity above which stacking faults emerge shifts downwards once the temperature exceeds approximately half the melting temperature. Also Hugoniot elastic limit is found to be a decreasing function of temperature, whereas the corresponding critical strain is insensitive to temperature. The discrepancy between the simulation and the experiments where Hugoniot elastic limit is a increasing function of temperature is discussed.

INTRODUCTION

One of the important applications of shock wave is to study the properties of deformation processes in high strain rates which are qualitatively different from those of slow strain rates. In particular, temperature dependences of mechanical strength such as dynamic tensile strength (spall strength) or yield strength are important from the practical point of view. It is well known that, in quasistatic deformations, yield strength is very sensitive to the initial temperature; It decreases as the initial temperature increases. However, the situation seems to be very different for shock-loaded metals. Some of the recent experiments reveal anomalous temperature dependences of Hugoniot elastic limit (HEL), which are quite opposite to the case of quasisitatic deformation [1, 2]. A recent experiment on single crystals of aluminium and copper shows the rise of HEL for the increasing initial temperature, whereas the spall strength decreases [2].

In contrast to the arising of experiments, neither theories nor computer simulations have explained this anomalous temperature dependence. As to molecular dynamics simulations, the celebrated work of Holian and Lomdahl [3] which showed stacking faults initiated by shocks could not find any temperature dependence of the critical shock strength. However, since the initial temperature they adopted ranges from $0.001 T_M$ to $0.5 T_M$ where T_M denotes the melting temperature, it is possible that new feature may arise beyond that temperature range. In this paper I present a result on the temperature dependence of HEL via molecular dynamics simulations in a wide range of temperatures.

THE MODEL SYSTEM

The system considered here is an fcc perfect crystal where the intermolecular forces is calculated from the Lennard-Jones potential.

$$U(r) = 4\varepsilon \left[\left(\frac{\sigma}{r} \right)^{12} - \left(\frac{\sigma}{r} \right)^{6} \right] \quad (1)$$

Throughout this paper, σ and ε are set to unity (the LJ unit). Although there are various methods to mimick shock waves in molecular dynamics simulations, the most popular method is adopted here that a moving piston of infinite mass hits the still target. The velocity of the piston is denoted by u_p. A shock propagates along the $\langle 001 \rangle$ orientation. Periodic boundary conditions are applied to directions perpendicular to a shock. The whole system typically consists of $80 \times 80 \times 150$ unit cells. The melting temperature T_M of this system is about 0.7 [4], and the simulation is performed from $0.01 T_M$ to $0.99 T_M$, which covers a wider range than investigated in [3].

To detect stacking faults easily in molecular dynamics simulations, the centrosymmetry parameter has been introduced [5]. It is defined to represent the degree of symmetry of vectors from one atom to its nearest neighbors.

$$c_i = \frac{\sum_{(j,j')} |r_{ij} + r_{ij'}|^2}{2 \sum_j |r_{ij}|^2}, \quad (2)$$

where (j, j') is the 6 pairs of the opposite nearest neighbors in fcc lattices. This quantity is nonnegative, and 0 for an atoms in a perfect crystal. It increases as the coordination symmetry of neighbors is violated. For example, it is on the order of 10^{-1} for amorphous materials. Note that the centrosymmetry paramter is insensitive to thermal noise and hence suitable for the visualization of planer defects including stacking faults which can be hardly detected by other methods in high temperature region; e.g. Visualization by potential energy fails near the melting temperature due to strong thermal noise. In contrast, the centrosymmetry parameter for a perfect crystal in the presence of strong thermal noise is less than 0.01 which is well distinguishable from the value for stacking faults, 0.04. In order to visualize the atomistic configurations together with the centrosymmetry parameter presentation, "Atomeye" developed by J. Li is employed [6].

The procedure of simulation is as follows. The piston velocity u_p is gradually increased from 0, and the value of u_p above which the stacking faults can be observed is recorded as the critical piston velocity u_p^*. Then HEL is calculated by the following Hugoniot relation.

$$p_{zz} = p_0 + \rho_0 u_p u_s, \quad (3)$$

where p_0 and ρ_0 is the initial pressure and density, respectively. Since p_0 is extremely small compared to p_{zz} which denotes the normal (z direction) pressure behind the shock, it is neglected. The shock velocity u_s is measured from the elastic precursor. In this way, we can calculate HEL if the value of critical piston velocity is known.

$$p_{HEL} \simeq \rho_0 u_p^* u_s. \quad (4)$$

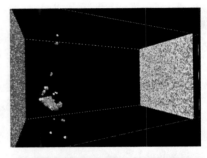
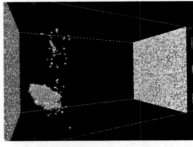
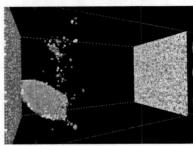

FIGURE 1. Snapshots of an emerging stacking fault. Atoms whose centrosymmetry parameter is less than 0.01 are not presented. The shock is propagating from left to right, where the piston velocity is slightly above the critical value. The initial temperature is $0.3 T_m$. The duration between the snapshots is 0.04. Note that growth of a stacking fault is very fast (faster than the shock velocity).

RESULTS

Figure 1 shows a growing stacking fault behind the shock front. We can see that one of the slip planes {111} begins to slide to form a stacking fault. The density of stacking faults becomes higher for stronger shocks which is visualized in Fig. 2. The temperature dependence of the critical piston velocity is shown in Fig. 3. We can see that the critical piston velocity begins to decrease once the initial temperature gets higher than a certain value T^*. The

FIGURE 2. Snapshot of intersecting stacking faults where the piston velocity is well above the critical value. ($u_p = 2.0$). Again, atoms whose centrosymmetry parameter is less than 0.01 are omitted.

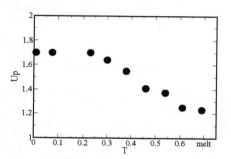

FIGURE 3. Temperature dependence of the critical piston velocity.

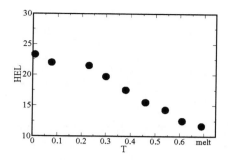

FIGURE 4. Temperature dependence of the Hugoniot elastic limit.

threshold temperature T^* is approximately $0.43T_m$, which complements the result of [3] where no temperature dependence was seen up to half the melting temperature. However, the existence of the threshold T^* is rather unexpected. To clearify the mechanism, further calculations are carried out in the following. Since the piston velocity is a parameter given from outside, we have to focus on more intrinsic quantities which can characterize the degree of deformation, such as stress or strain. Figure 4 shows the temperature dependence of HEL caluculated via Eq. 4, which constitutes the main result of this paper. We can see that HEL decreases gradually as the temperature increases. No plateau is seen but there seems to be two stages of decrease: The below and the above T^*. Although this two-fold temperature dependence is not investigated further, the result obtained here is quite plausible since the yield strength for slow deformations shows the same behavior with respect to the temperature. However, the result is opposite to the previous experiment [2] where the rise of HEL is observed as the initial temperature increases. The reason of this discrepancy is still unclear but it will be partially discussed at the last section of this paper.

Then we turn to the strain. The critical strain where the stacking faults emerge can be calculated via the following Hugoniot relation;

$$\frac{\rho - \rho_0}{\rho} = \frac{u_p}{u_s}. \qquad (5)$$

The obtained values have little temperature dependence; 0.135 ± 0.07 for a wide range of temperature $0.01T_M \leq T \leq 0.99T_M$. Note that the simulation of [3] also indicates that stacking fault emerges when the strain becomes 0.14, where the simulation was performed at $T = 0.001T_m$. Hence, it can be said that the critical strain is independent of, or at least, insensitive to the initial temperature. Also recalling that strain coincides with the stacking fault density [3], strain might be a characteristic measure of yielding rather than stress or piston velocity. In order to confirm the character of shock strain, temperature dependences of normal pressure and strain are shown in Fig. 5 with the piston velocity u_p fixed. The stress gradually decreases as the temperature rises, whereas the strain increases. No plateau is seen for the both graphs. It is quite intuitive that the strain with respect to the same piston velocity increases as the temperature rises, which means that the thermal softening takes place. These data show that the nature of shock-induced plasticity where ultrafast deformations are realized is not qualitatively different from slow plastic deformations realized in the conventional tensile tests, as long as perfect crystals concern.

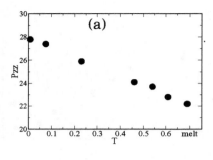

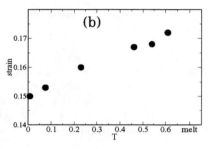

FIGURE 5. Temperature dependences of (a) the normal pressure P_{zz} and (b) the strain, where the piston velocity is fixed to be 2.0. Note that many stacking faults are seen at this piston velocity where the strain exceeds the critical value 0.135.

CONCLUSIONS AND DISCUSSIONS

In the present simulation, the decrease of HEL is found for the increasing temperature. The result is intuitive and complementary to the existing computational study [3]. Also strain increases and stress decreases as the temperature rises for the same piston velocity. These results indicate that the temperature dependence of shock-induced plasticity is quite similar to that of quasistatic deformations. However, the simulations cannot explain the experiments which shows the increase of HEL at the high temperature range. One of the reaons for this discrepancy is that specimens used in experiments contain various kinds of defects which deeply influence the nature of plastic deformation, whereas the simulations treat a perfect crystal. However, if dislocations preexist, the critical shock strength might further decrease since mobile dislocations increase as the temperature increases. This intuitive fact is also confirmed by the dynamic tensile test where high strain rates are realized up to the order of 10^4 /s [7]. It is hardly imaginable that some kinds of obstacles for moving dislocations might be activated when the temperature is increased. On the other hand, there are some experiments which report decreasing HEL which treat α-β titanium alloy [8], and stainless steel [9] as the specimens. At this point, the relation between the similar results for a perfect crystal and for the compounds is not clear. Further intensive efforts including simulations on defective crystals and theoretical analysis besed on dislocation dynamics are needed to understand the nature of dynamical yielding.

ACKNOWLEDGMENTS

The author would like to express his appreciation for the encouragements of H. Kaburaki and F. Shimizu.

REFERENCES

1. Rohde, R. W., Acta Metallugia **17**, 353-363 (1969).
2. Razorenov, S. V., Kanel, G. I., Baumung, K., and Bluhm, H. J., "Hugoniot Elastic Limit and Spall Strength of Aluminium and Copper Single Crystals over a Wide Range of Strain Rates and Temperatures," in *Shock Compression of Condensed Matter - 2001*, edited by M. D. Furnish et al., AIP Conference Proceedings 620, New York, 2002, pp. 503-506.
3. Holian, B. L. and Lomdahl, P. S., Science **280**, 2085-2088 (1998).
4. Agrawal, R. and Kofke, D. A., Mol. Phys. **85**, 43-59 (1995); van der Hoef, M. A., J. Chem. Phys. **113** 8142-8148 (2000).
5. Kelchner, C. L., Plimpton, S. J., and Hamilton, J. C., Phys. Rev. B 58, 11085-11088 (1998).
6. J. Li, Modelling Simul. Mater. Sci. Eng. **11**, 173-177 (2003).
7. Sakino, K., J. Phys. IV France **10**, Pr9 57-62 (2000).
8. Krüger, L., Kanel, G. I., Razorenov, S. V., Meyer, L., and Bezrouchko, G. S., "Yield and Strength Properties of the Ti-6-22-22S Alloy Over a Wide Strain Rate and Temperature Range," in *Shock Compression of Condensed Matter - 2001*, edited by M. D. Furnish et al., AIP Conference Proceedings 620, New York, 2002, pp. 1327-1330.
9. Gu, Z. and Jin, X, "Temperature Dependence on Shock Response of Stainless Steel," in *Shock Compression of Condensed Matter - 1997*, edited by S. C. Schmit et al., AIP Conference Proceedings 429, New York, 1998, pp. 467-470.

CP706, *Shock Compression of Condensed Matter - 2003*
edited by M. D. Furnish, Y. M. Gupta, and J. W. Forbes
© 2004 American Institute of Physics 0-7354-0181-0/04/$22.00

MOLECULAR DYNAMICS SIMULATIONS OF SHOCKS INCLUDING ELECTRONIC HEAT CONDUCTION AND ELECTRON-PHONON COUPLING

Dmitriy S. Ivanov[1], Leonid V. Zhigilei[1], Eduardo M. Bringa[2], Maurice De Koning[2], Bruce A. Remington[2], Maria Jose Caturla[2], and Stephen M. Pollaine[2]

[1] *Department of Materials Science & Engineering, University of Virginia, Charlottesville, VA 22904-4745*
[2] *Lawrence Livermore National Laboratory, Livermore CA 94550*

Abstract. Shocks are often simulated using the classical molecular dynamics (MD) method in which the electrons are not included explicitly and the interatomic interaction is described by an effective potential. As a result, the fast electronic heat conduction in metals and the coupling between the lattice vibrations and the electronic degrees of freedom can not be represented. Under conditions of steep temperature gradients that can form near the shock front, however, the electronic heat conduction can play an important part in redistribution of the thermal energy in the shocked target. We present the first atomistic simulation of a shock propagation including the electronic heat conduction and electron-phonon coupling. The computational model is based on the two-temperature model (TTM) that describes the time evolution of the lattice and electron temperatures by two coupled non-linear differential equations. In the combined TTM-MD method, MD substitutes the TTM equation for the lattice temperature. Simulations are performed with both MD and TTM-MD models for an EAM Al target shocked at 300 kbar. The target includes a tilt grain boundary, which provides a region where shock heating is more pronounced and, therefore, the effect of the electronic heat conduction is expected to be more important. We find that the differences between the predictions of the MD and TTM-MD simulations are significantly smaller as compared to the hydrodynamics calculations performed at similar conditions with and without electronic heat conduction.

INTRODUCTION

Molecular Dynamics (MD) has been intensively used to study ultrafast non-equilibrium phenomena in solids. In particular, MD has been proved to be an efficient tool for the microscopic analysis of shock waves [1]. Simulations up to date have dealt mostly with single crystal samples, which are in general far from the experimental "single crystals" and polycrystalline samples. Grain boundaries (GB) form one of the relevant groups of defects to be considered. MD simulations have suggested that GB can melt at lower temperature as compared to the bulk material [2], possibly leading to a decreased yield stress. This could explain recent experimental results on Rayleigh-Taylor (RT) instability growth in a polycrystalline Al [3]. MD simulations of shocks in fluids have been related to

predictions by hydrodynamic (HD) calculations [4]. Recent HD simulations of shock wave heating of a GB region, Fig. 1, show large differences (~2000 K at the GB) in the temperature profiles obtained with and without electronic heat conduction. In the HD simulation illustrated by Fig. 1, zones corresponding to the GB were described by the same equation of state (EOS) as the bulk Al with an initial/equilibrium density decreased to 0.75 of the bulk density. Note that the results of HD calculations may no longer be valid at small spatial scales, such as the thickness of a GB. Therefore, the thickness of the GB region in the HD simulation was chosen to be 10 nm, several times larger that an effective thickness of a typical GB. These rough approximations were intended to provide some initial feedback into possible GB evolution scenarios under shocks. In these simulations, fast electron conduction led to significant energy dissipation from the GB region.

MD simulations can be used to simulate a more realistic GB, without any pre-existing assumptions on the EOS, density, or thickness of the GB. The electronic heat conduction, however, cannot be reproduced within a classical MD model, which only includes the lattice contribution to the heat conduction. Since the electronic contribution to the thermal conductivity of a metal is orders of magnitude larger than the lattice contribution, conventional MD significantly underestimates the total thermal conductivity in metals. In order to overcome this limitation we combine MD with a continuum description of the electronic heat conduction, as briefly described below.

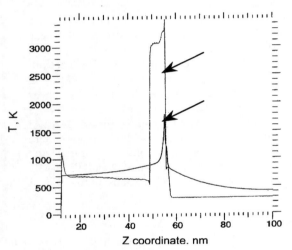

Figure 1. Temperature profiles obtained in HD simulations of a shock traversing a grain boundary, with and without electronic heat conduction.

THE COMBINED TTM-MD MODEL

A hybrid computational model that combines the classical MD method for simulation of fast non-equilibrium processes in the shocked material with a continuum description of the electronic heat conduction is used in the simulations. The model is based on the so-called two temperature model (TTM) [5], that describes the time evolution of the lattice and electron temperatures by two coupled non-linear differential equations. The TTM was originally developed for strong electron-phonon

non-equilibrium due to the fast electronic excitation. The model, however, is also appropriate for situations where the electron and lattice temperatures are close to each other and, in particular, is applicable for the description of the electronic heat conduction in a material undergoing shock wave heating. In the combined TTM-MD method [6-8], the MD completely substitutes the TTM equation for the lattice temperature. The diffusion equation for the electron temperature is solved by a finite difference method simultaneously with MD integration of the equations of motion of atoms. The electron temperature enters a coupling term that is added to the MD equations of motion to account for the energy exchange between the electrons and the lattice. A modified, as compared to earlier works [6,7], formulation of the coupling term is used in the model. The new formulation [8] distinguishes between the thermal velocities of the atoms and the velocities of their collective motion. It also does not require *a priori* knowledge of the lattice heat capacity of the model system, which is, in general, a function of temperature. The expansion and density variations, predicted in the MD part of the model, are accounted for in the continuum part of the model. A complete description of the combined TTM-MD model is given elsewhere [8]. In this paper, we report the results of MD and TTM-MD simulations of shock wave heating of a tilt GB in an Al bi-crystal. The effect of the electronic heat conduction on the temperature evolution in the GB region is discussed based on the simulation results.

The simulations reported in this article are performed for a 300 kbar shock wave propagating through an Al bi-crystal containing a (110) $\Sigma 5$ tilt boundary. The initial MD system is an FCC bi-crystal with dimensions $9.17 \times 12.3 \times 21.8$ nm, periodic boundary conditions in the directions parallel to the shock front, and a grain boundary located in the middle of the computational cell. Inter-atomic interaction is described by the embedded-atom method (EAM) in the form and parameterization suggested in [9]. Before applying the shock loading, the system is equilibrated at 300 K and zero pressure. The parameters used in the TTM equation for the electronic temperature are given in [10].

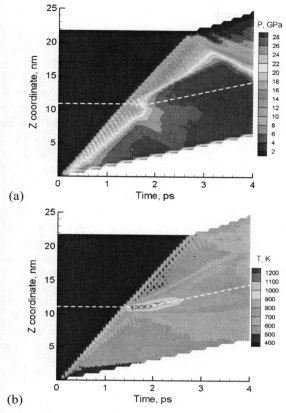

(a)

(b)

Figure 2. Pressure (a) and temperature (b) contour plots for a TTM-MD simulation of a 300 kbar shock in an Al sample with a tilt GB. The location of the GB is indicated by the dashed line.

In the combined TTM-MD model the cells in the finite difference discretization are related to the corresponding volumes of the MD system. The local lattice temperature and pressure are defined for each cell from the average kinetic energy of the thermal motion of atoms and through the virial equation, respectively.

The shock wave is generated by applying additional forces to a layer of atoms that make up a "piston" located on one side of the MD computational cell. The forces are chosen so that a constant pressure of 300 kbar is maintained in a region adjacent to the piston. In the following section, we present and compare results obtained in simulations performed with and without the electron conduction included.

RESULTS AND DISCUSSION

Both MD and TTM-MD simulations are run for four picoseconds so that the shock wave has enough time to reach the back surface but the reflected wave does not reach the GB. The temperature and pressure contour plots are shown in Fig. 2 for the TTM-MD simulation performed with the electron heat conduction included. In the pressure plot we can see a partial reflection of the shock (v_{shock}~7.5 km/s) from the GB at ~1.5 ps, and a complete reflection of the shock wave from the back surface of the computational cell at ~3 ps. After the shock reaches the GB, we can see that the GB starts to move with the piston velocity, as expected. In the temperature contour plot we can see that the interaction of the shock wave with the GB leads to a transient thermal spike in the GB region. A very similar temperature and pressure distributions are observed in MD simulations without the electron heat conduction. We observe only small differences in temperature distributions between the two simulations, whereas there is practically no difference in the pressure plots.

The difference in temperature observed in the GB region is ~50-200 K at t~1.6-2.6 ps ps. Therefore, the effect of the electronic heat conduction is relatively small. It can only play a significant role when the maximum local temperature reached at the GB is close to the melting point. In this case the electronic heat conduction may prevent melting in a system that would otherwise melt in a pure MD simulation.

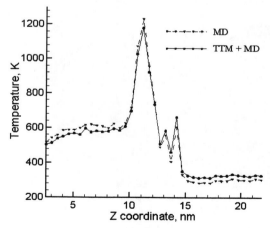

Figure 3. Temperature profiles for the MD and TTM-MD simulations at t = 1.8 ps.

The fast electronic heat conduction leads to the increasing temperature ahead of the shock wave and decreasing temperature behind the shock front in the TTM-MD simulation as compared to the MD simulation, Fig. 3.

The atomic-level picture of the structural changes induced by the shock wave can be seen in Fig. 4. The (110) Σ5 tilt boundary is initially located at Z~12 nm. At t=2.0 ps, a slab of material with a tilt several degrees smaller than the original one appears in a region adjacent to the grain boundary, Z~12-13 nm. By 2.8 ps this region has grown thicker, Z~13-16 nm, giving a velocity of growth of ~2.5 km/s ~ v_{shock}/3. The train of elastic precursors can be seen well ahead of the upper boundary for this region. No plastic front formation is observed at this shock pressure for a perfect crystal. This GB "splitting" appears in both MD and TTM-MD simulations. A more detailed study of this effect is in progress.

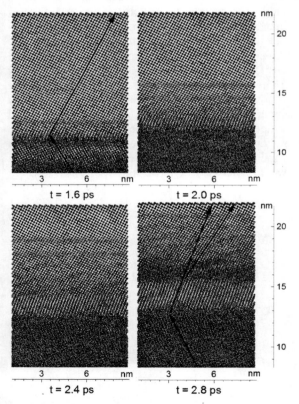

Figure 4. Grain boundary "splitting" for a 300 kbar shock. Crystal orientation is indicated by arrows.

SUMMARY

Simulations of a 300 kbar shock wave propagation in an Al target containing a single (110) Σ5 tilt grain boundary are performed with (TTM-MD) and without (MD) electronic heat conduction. We find that the differences between the predictions of the MD and TTM-MD simulations are relatively small. The difference in the lattice temperatures at the GB was ~50-200 K at t~1.6-2.6 ps. A significantly stronger effect of the electronic heat conduction is predicted in the HD calculations

A "splitting" of the initial GB is observed in both the MD and TTM-MD simulations. The mechanism of the formation of the second tilt GB is being further investigated. Additional studies including other ordered and disordered boundaries, at several shock pressures, are also needed to evaluate the change in the mechanical properties of the shocked grain boundary, which would be important to estimate its influence on the RT growth [3]. The TTM-MD model will be also used to investigate the role of the electronic heat conduction and the strength of the electron phonon coupling in plastic deformation and melting induced by strong shocks.

ACKNOWLEDGEMENTS

The work at LLNL was performed under the auspices of the U.S. Department of Energy and LLNL under contract No. W-7405-Eng-48.

REFERENCES

1. B. Holian and P. Lomdahl, *Science* **280**, 2085 (1998).
2. J. Lutsko et al., *Phys. Rev. B* **40**, 2841 (1989).
3. D. Kalantar et. al., *Phys. Plasmas* **7**, 1999 (2000).
4. B. Holian et al., *Phys. Rev. A* **22**, 2798 (1980).
5. M. I. Kaganov et al., *Sov. Phys. JETP* **4**, 173, (1957).
6. H. Häkkinen and U. Landman, *Phys. Rev. Lett.* **71**, 1023 (1993).
7. C. Schäfer et al., *Phys. Rev. B* **66**, 115404 (2002).
8. D. S. Ivanov and L. V. Zhigilei, *Phys. Rev. Lett.* **91**, 105701 (2003); *Phys. Rev. B* **68**, 064114 (2003).
9. W. Zhou et al., *Acta Mater.* **49**, 4005 (2001).
10. G. Tas and H. J. Maris, *Phys. Rev. B* **49**, 15046 (1994); V. Kostrykin et al., *SPIE Proc.* **3343**, 971 (1998).

CP706, *Shock Compression of Condensed Matter - 2003*
edited by M. D. Furnish, Y. M. Gupta, and J. W. Forbes
© 2004 American Institute of Physics 0-7354-0181-0/04/$22.00

ATOMISTIC SIMULATIONS OF SHOCK-INDUCED PHASE TRANSITIONS

K. Kadau, T. C. Germann, P. S. Lomdahl, B. L. Holian, and F. J. Cherne

Los Alamos National Laboratory, Los Alamos, NM 87545, U.S.A.

Abstract. We report on large scale non-equilibrium atomistic simulations of shock-induced solid-solid phase transformations. As an example the $\alpha \rightarrow \epsilon$ transformation in iron and the A11 (GaI)→cI12 (GaII) transformation in gallium are discussed. The use of semi-empirical descriptions of the inter-atomic forces and today's parallel computing resources allow for a quantitative comparison of the theoretically calculated data with the experimental results. The discussion will include the crystallographic orientation dependence on the transformation process in single crystals. Simulations containing several millions of iron atoms reveal that above a critical shock strength, many small close-packed grains nucleate in the shock-compressed bcc crystal. For shock waves in the [001] direction the initially small grains are growing on a picosecond time scale to form larger, energetically favored, grains. For the two other major crystallographic directions, here the annealing processes are slower and and have not finished within the time scales accessible with atomistic simulations (up to 50 ps). Furthermore, crystals shocked in [111] direction produce solitary waves ahead of the actual shock front.

INTRODUCTION

Atomistic computer simulations on shock-induced plasticity and phase transformations have been done by several authors within the past years. In 1991 the first split two-wave structure has been seen by non-equilibrium molecular-dynamics (NEMD) simulations in a polymorphic phase transition for a two dimensional material undergoing a dissociative transition [1]. Later, large-scale NEMD simulations have been used to study the plasticity of face-centered cubic materials [2, 3], the orientation dependence of shock-induced chemistry [4], and shock-induced structural transformations in iron [5, 6]. As a computationally cheaper alternative to the direct method of NEMD, ensembles of atoms can be driven by homogeneous uniaxial compression toward the final defective Hugoniot state [7, 8].

The use of the SPaSM ("Scalable Parallel Short-range Molecular dynamics") code allows one to simulate and analyze routinely millions of atoms on parallel machines [1] [9, 10], and has been tested for up to some 20 billion atoms on a fraction of Los Alamos'

Q-machine [11], paving the way for atomistic simulations on the μm scale [12, 13]. In this work we use the embedded atom method (EAM) [14, 15] for the description of the interaction of the iron atoms. EAM has been successfully applied for metals and has been extended to the modified EAM (MEAM) [16] in order to describe materials in which covalent character is important, like the element gallium [17].

SHOCK-INDUCED TRANSFORMATIONS IN IRON

Shock experiments found that iron undergoes a structural transition from the body-centered cubic (bcc) α ground state to the hexagonal close-packed (hcp) ϵ structure at a pressure of 13 GPa [18] [2]. Hydrostatic compression experiments at room temperature reveal about the same transition pressure for the $\alpha \rightarrow \epsilon$ transformation, but the reverse transformation takes place at 8 GPa [19]. This hysteresis decreases with

---

[1] See also http://bifrost.lanl.gov/MD/MD.html

[2] Here the transition pressure is the pressure component P_{zz} in the direction of the shock propagation.

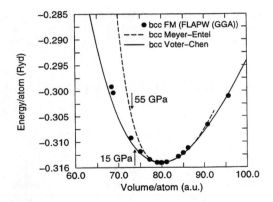

FIGURE 1. Cold curve for the Meyer-Entel potential (dashed line) and the Voter-Chen potential (full line). The circles correspond to FLAPW (Full Potential Augmented Plane Wave) with GGA (generalized gradient approximation) electronic structure calculations of the ferromagnetic bcc ground state of iron [20]. Arrows indicate where the shock-induced transformation occurs for each potential.

increasing temperature, and vanishes at the triple point of the phase diagram ($\approx$ 800 K, 10.5 GPa).

The atomistic modeling of iron has been done with different EAM potentials: a Voter-Chen potential [21] and the Meyer-Entel potential [22]. The latter has the advantages that it is also fit to zone-boundary phonon frequencies, and therefore can describe the vibrational spectrum well, which is, for example, important for the description of the temperature-induced martensitic phase transitions [23]. However, this potential is too stiff upon compression (Fig.1), resulting in an extremely high transition pressure (55 GPa) for the shock-induced $\alpha \rightarrow \epsilon$ transition. The Voter-Chen potential has the advantage of using the empirical Rose equation of state and compares well to *ab initio* calculations of the ferro-magnetic bcc ground state of iron [20], which is most important for a quantitative comparison to shock experiments. Therefore results presented here will be based upon the Voter-Chen description, where the transition pressure, as measured by NEMD in [001] direction, is 15 GPa, which compares well with experimental data. However, the qualitative results obtained with both potentials are quite similar [5]. Unless otherwise noted, simulations presented here initially start with T = 50 K. With increasing initial temperature, entropic contributions make the close-packed phase more favorable

and reduce the energy barrier for the transformation [23]. This leads to a slightly reduced transition pressure for samples simulated at room temperature.

Crystallographic orientation dependence

We have carried out a series of molecular-dynamics (MD) simulations for different shock strengths with shock waves propagating in the three major crystallographic directions [001], [011], and [111] of a bcc iron single crystal [6, 5, 24]. To initiate shock waves the momentum mirror method [2] has been employed. Simulation cells containing up to some eight million atoms (Fig.2) show a structural transition from the initial bcc structure into a close-packed structure (Fig.3) for pressures above 15 GPa in the [001] direction. For the two other directions the transition pressure is $\approx$ 20 GPa.

For shock strengths just above the transition pressure a split two-wave structure is observed, with a leading elastic wave followed by a slower transformation wave [3]. For high shock strengths, the shock is overdriven and has only one transformation front. This behavior is typical for a material that exhibits a phase transformation [25]. The Hugoniot for the different crystallographic shock directions compares favorably with experimental polycrystalline data [18] for the overdriven region (Fig.4). The fact that the shock speed for the [001] direction in that region is larger than for the other two directions reflects denser packing due to fewer grain boundaries. Different elastic properties for the various directions result in differences of the elastic part of the Hugoniot and therefore in in the split two-wave region. Yet it is not clear if the higher transition pressure in the [011] and [111] direction is an effect of the short time scales accessible in NEMD [8], or if it is an effect that can be measured experimentally. Preliminary simulations of polycrystalline samples show a reduction of the transition pressure by heterogeneous nucleation at grain boundaries, thus coming closer to the experimental 13 GPa threshold.

The different crystallographic shock directions favor different structural transformation pathways, so

---

[3] Due to the increased Hugoniot elastic limit in perfect single crystals, we see no evidence for bcc plasticity before the onset of the phase transformation.

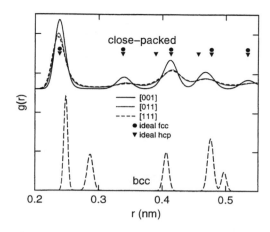

FIGURE 3. Radial distribution functions of the transformed material for crystals shocked in different directions with a piston velocity $u_p = 1087$ m/s. As a reference the ideal positions of the fcc and the hcp structure as well as the distribution function for the unshocked bcc are shown.

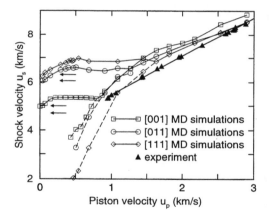

FIGURE 4. Hugoniot for the different crystallographic shock directions for iron single crystals as obtained by MD, with experimental polycrystalline data [18]. The arrows indicate the sound speeds of the potential for the bulk and the [001], [011], and [111] directions (bottom to top).

FIGURE 2. Samples containing $\approx$ eight million atoms ($\approx$ 40 nm×40 nm×57 nm) after 6.57 ps shocked in the bcc [001], [011], and [111] direction (top to bottom). Piston velocity $u_p = 1087$ m/s. Atoms are color-coded by the number of neighbors n within 2.75 Å: the unshocked bcc ($n = 8$) is gray, uniaxially compressed bcc ($n = 10$) is blue, and the transformed close-packed grains ($n = 12$) are red, separated by yellow ($n = 11$) grain boundaries.

different dynamics are observed for each direction (Figs. 5 and 6). For shocks in [001] only two equivalent twin variants of the product phase, separated by twin boundaries, are observed. Here, the initially small grains anneal on a picosecond time scale and form larger grains with most of the relaxation process completed within the accessible simulation time (up to 50 ps). However, in the two other directions, more variants with more complicated grain

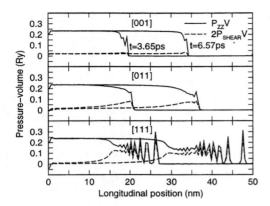

FIGURE 5. Profiles of the longitudinal ($P_{zz}V$) and twice the shear ($[P_{zz} - (P_{xx} + P_{yy})/2]V$) components of the pressure-volume tensor ($P_{ij}V$, $i,j = x,y,z$). The shock waves propagate from left to right. Profiles are shown for the three major cubic crystallographic orientations for a piston velocity $u_p = 1087$ m/s.

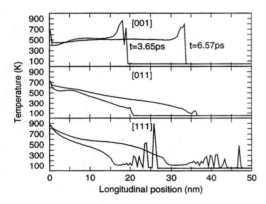

FIGURE 6. Temperature profiles for the major cubic crystallographic orientations for $u_p = 1087$ m/s. The shock waves propagate from left to right. For the [011] and [111] directions the annealing process of the transformed material takes much longer than for the [001] direction.

boundaries are formed. In these cases the relaxation processes are far from complete by the end of the simulation. The temperature (Fig.6) is rising, which indicates a structural relaxation to lower potential energies for the constant energy simulations presented here. The grain size remains small (Fig.2) and the number of grain boundary atoms is significantly larger than in the [001] case [this can also be seen in the broader peaks of the radial distribution functions (Fig.3)]. Hayes [26] has experimen-

tally observed different kinetics for the KCl shock-induced phase transformation for different orientations.

On the unloading process, by the reflection of the shock wave at the end of the samples, the transformation reverts almost perfectly for all three directions investigated (in the [001] case it is perfect). Since the transformation is done by a collective movement of the atoms over short distances (martensitic-like), rather than by reconstruction, the transformation process is likely to be reversible and this has been observed in experiments on iron and other materials [25]. However, for extremely high shock strength the system melts during the rarefaction wave (e.g., $u_p = 2899$ m/s).

Solitary waves

For shocks in the [111] direction solitary waves ahead of the actual shock front are observed (Figs.2,5, and 6). We observe these waves also for [001] shocks at very high shock strength, though there is also some evidence at early time for lower shock strength (Fig.5). These solitary waves have also been observed for shocks along [011] in fcc crystals [3]. The waves can be seen in properties like pressure, particle velocity, or density. However, they only affect the longitudinal temperature and can not be seen in the transverse velocity profiles.

These waves are relatively stable and do not significantly interfere with each other, as can be checked by inducing two shock waves with opposed propagation directions at each end of the sample. Yet, they appear not to be ideal solitons [27] on the simulation time scale. The peak height decays in time (the peak width does not significantly change), as their initial velocity decreases (Fig.7). This process gets more pronounced with increasing temperature and eventually the solitonic peak vanishes. However, the shock front can emit new solitary waves. As for the dependence on the shock strength, here the simulations show that the velocity of the solitons do not necessarily increase with shock strength. Rather, it looks like there might be an optimal shock strength for a given system. For example, solitons emitted at shocks with $u_p = 1087$ m/s are initially faster and their decay is slower than those created at fronts with $u_p = 1812$ m/s. For low shock strength

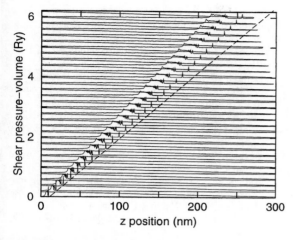

FIGURE 7. Time evolution (every 1.314 ps) of the shear stress profile ([111] bcc Fe, $u_p = 1812$ m/s, $T = 50$K). The decay in peak height of the leading solitary wave can be seen as can the decrease of the velocity of the solitary wave (the straight line is a guide to the eye to the initial velocity). The curves have been shifted for clarity.

($u_p = 471$ m/s), as well as for moderate temperatures (320 K), solitary waves have not been observed (some evidence of soliton-like behavior remains iin an oscillatory shock front).

Much more work on these interesting phenomena, in particular the connection to already existing theories and models of solitons, has to be done to understand this process in the framework of shock physics. However, a preliminary picture that emerges so far is that at the shock front, a soliton (as understood in the particle picture), emerges with an initial velocity and size. The initial velocity and size depends on the properties of the system (crystallographic direction, shock strength, temperature, ...). The soliton behaves nonideally (which gets more pronounced with increasing temperatures), as it interacts with its environment and loses energy, causing it to decay in size and velocity.

TRANSFORMATIONS IN GALLIUM

The ground state of gallium is an orthorhombic structure with a significant amount of dimerization (A11, GaI) (Fig.8). For T = 100 K, at P = 3 GPa a transformation into the bcc cI12 (GaII) structure occurs, the transition pressure dropping with increasing temper-

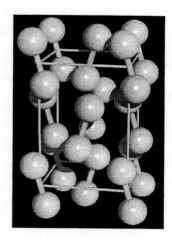

FIGURE 8. Unit cell of the Ga A11 ground-state structure. The structure is achieved by Gallium dimers occupying an orthorhombic cell.

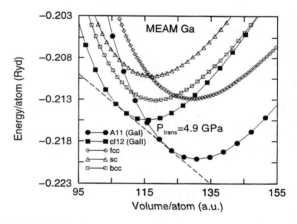

FIGURE 9. Cold curves for different structures as obtained by a MEAM potential [17]. The cold curve transition pressure between the A11 ground state (GaI) and the cI12 structure (GaII), as obtained by triangular construction, is 4.9 GPa. The experimental transition pressure at $T \approx 100$ K is ≈ 3 GPa.

ature [28]. Recently a MEAM model for gallium has been proposed by Baskes *et al.* [17] that describes the zero pressure properties for the various solid and liquid structures amazingly well. However, little is known about the quality of this potential under pressure. A first step is the calculation of the zero temperature transition pressures as obtained by the cold curves of the various structures (Fig.9). A tangent construction reveals A11→cI12 to have the lowest transition pressure, in agreement with the experimental phase diagram. The cold curve transition pressure is 4.9 GPa, which is a little bit larger than the experimental transition pressure at T = 100 K. However, the effect of temperature on the theoretical transition pressure is not clear at the moment. So far, dynamic simulations have just begun, yet no shock experiments on solid gallium have been performed.

OUTLOOK

Concerning the crystallographic orientation dependence of shock-induced transformations in iron, the authors look forward to future experiments, in order to see how they compare with our theoretical predictions. It would be interesting to see experimentally whether solitary waves can even be detected in single crystals at low temperatures. For gallium, we plan to do a series of NEMD shockwave simulations to study shock-induced structural transformations.

ACKNOWLEDGMENTS

We thank J. Roth, R. Ravelo, and M.I. Baskes for discussions and Heike Herper for her *ab initio* data on iron. Work has been supported by the DOE through LDRD-20020053DR and the ASCI program.

REFERENCES

1. Robertson, D. H., Brenner, D. W., and White, C. T., *Phys. Rev. Lett.*, **8**, 3132 (1991).
2. Holian, B. L., and Lomdahl, P. S., *Science*, **280**, 2085 (1998).
3. Germann, T. C., Holian, B. L., Lomdahl, P. S., and Ravelo, R., *Phys. Rev. Lett.*, **84**, 5351 (2000).
4. Zybin, S. V., Elert, M. L., and White, C. T., *Phys. Rev. B*, **66**, 220102 (2002).
5. Kadau, K., Germann, T. C., Lomdahl, P. S., and Holian, B. L., *Science*, **296**, 1681 (2002).
6. Kadau, K., Germann, T. C., Lomdahl, P. S., and Holian, B. L., "Shock-Induced Structural Phase Transformations Studied by Large-Scale Molecular-Dynamics Simulations," in *Shock Compression of Condensed Matter-2001*, edited by M. D. Furnish *et al.*, AIP Conference Proceedings 620, American Institute of Physics, New York, 2002, p. 351.
7. Maillet, J. B., Mareschal, M., Soulard, L., Ravelo, R., Lomdahl, P. S., Germann, T. C., and Holian, B. L., *Phys. Rev. E*, **63**, 016121 (2000).
8. Reed, E. J., Fried, L. E., and Joannopoulos, J. D., *Phys. Rev. Lett.*, **90**, 235503 (2003).
9. Lomdahl, P. S., Tamayo, P., Grønbech-Jensen, N., and Beazley, D. M., "50 GFlops Molecular Dynamics in the Connection Machine 5," in *Proceedings of Supercomputing 93*, edited by G. S. Ansell, IEEE Computer Society Press, Los Alamitos, CA, 1993, p. 520.
10. Beazley, D. M., and Lomdahl, P. S., *Computers in Physics*, **11**, 230 (1997).
11. Kadau, K., Germann, T. C., and Lomdahl, P. S., *to be submitted to Int. J. Mod. Phys. C* (2003).
12. Abraham, F. F., Walkup, R., Gao, H., Duchaineau, M., De La Rubia, T. D., and Seager, M., *PNAS*, **99**, 5777 (2002).
13. Abraham, F. F., Walkup, R., Gao, H., Duchaineau, M., De La Rubia, T. D., and Seager, M., *PNAS*, **99**, 5783 (2002).
14. Daw, M. S., and Baskes, M. I., *Phys. Rev. Lett.*, **50**, 1285 (1983).
15. Daw, M. S., and Baskes, M. I., *Phys. Rev. B*, **29**, 6443 (1984).
16. Baskes, M. I., *Phys. Rev. B*, **46**, 2727 (1992).
17. Baskes, M. I., Chen, S. P., and Cherne, F. J., *Phys. Rev. B*, **66**, 1041007 (2002).
18. Brown, J. M., Fritz, J. N., and Hixson, R. S., *J. Appl. Phys.*, **88**, 5496 (2000).
19. Wassermann, E. F., Acet, M., Entel, P., and Pepperhoff, W., *phys. stat. sol.*, **82**, 2911 (1999).
20. Herper, H. C., Hoffmann, E., and P., E., *Phys. Rev. B*, **60**, 3839 (1999).
21. Harrison, R. J., Voter, A., and Chen, S. P., "Embedded Atom Potential for bcc Iron," in *Atomistic Simulation of Materials Beyond Pair Potentials*, edited by V. Vitek and D. J. Srolovitz, American Institute of Physics, Plenum Press, New York and London, 2002, p. 219.
22. Meyer, R., and Entel, P., *Phys. Rev. B*, **57**, 5140 (1998).
23. Kadau, K., Entel, P., Germann, T. C., and Lomdahl, P. S., *J. Phys. IV (Paris)*, **11**, Pr8–17 (2001).
24. Kadau, K., Germann, T. C., Lomdahl, P. S., and Holian, B. L., *to be submitted to Phys. Rev. B* (2003).
25. Asay, J. R., *High-Pressure Shock Compression of Solids*, Springer, New York, 1993, p. 90.
26. Hayes, D. B., *J. Appl. Phys.*, **45**, 1208 (1974).
27. Remoissenet, M., *Waves Called Solitons*, Springer, Berlin, New York, 1999.
28. Young, D. A., *Phase Diagrams of the Elements*, University of California Press, Berkeley, Los Angeles, Oxford, 1991.

CP706, *Shock Compression of Condensed Matter - 2003*
edited by M. D. Furnish, Y. M. Gupta, and J. W. Forbes
© 2004 American Institute of Physics 0-7354-0181-0/04/$22.00

SHOCK WAVES PROPAGATION IN SCOPE OF THE NONLOCAL THEORY OF DYNAMICAL PLASTICITY

Tatyana A. Khantuleva

Department of Physical Mechanics, Faculty of Mathematics and Mechanics, S.Petersburg State University, S.Petersburg, 198904, Bibliotechnaya, 2, RUSSIA

Abstract. From the point of view of the modern statistical mechanics the problems on shock compression of solids require a reformulation in terms of highly nonequilibrium effects arising inside the wave front. The self-organization during the multiscale and multistage momentum and energy exchange are originated by the correlation function. The theory of dynamic plasticity has been developed by the author on the base of the self-consistent nonlocal hydrodynamic approach had been applied to the shock wave propagation in solids. Nonlocal balance equations describe both the reversible wave type transport at the initial stage and the diffusive (dissipative) one in the end. The involved inverse influence of the mesoeffects on the wave propagation makes the formulation of problems self-consistent and involves a concept of the cybernetic control close-loop.

INTRODUCTION

It is just time to formulate the problem on the shock compression of solids, unresolved in scope of the traditional continuum mechanics, in terms of nonequilibrium transport processes. For this purpous a critical examination and a change in the fundamental assumptions is necessary to realize predictive capabilities [1]. The correct relationships between the stress, strain and strain-rate through the loading regime should extend the well-known Maxwell relaxation model to include mesoscopic structure formation.

A first-principles theory has been developed on the base of the modern nonequilibrium statistical mechanics.. A new self-consistent nonlocal approach had been applied to the problem on planar shock propagation in semi-space full of a medium with relaxing internal structure [2-3]. The set of integro-differential balance equations for the mass, momentum and energy densities combines both

wave (elastic) and dissipative (plastic) transport properties in a non-additive way. Nonlinear integral conditions to the set determine the temporal evolution of the internal scale spectra (size effect regime) involving a cybernetic concept of a close-loop. The discrete spectrum of the internal structure scales becomes continuous both in elastic and in hydrodynamic limits pointing out the validity region of classical continuum mechanics. The fracture as a phase transition occurring in a threshold way on account of resonance effects is a result of the self-organization process occurring during the relaxation.

MESOSCALE DETERMINATION

The most profound result obtained in scope of the nonequilibrium statistical mechanics [4] consists in the general constitutive relationship between the conjugate thermodynamic flux J (stress tensor) and force G (strain-rate)

$$J(\mathbf{r},t) = J^0(\mathbf{r},t) + \int\limits_{-\infty}^{t} dt' \int\limits_{V} d\mathbf{r}' \mathfrak{R}(\mathbf{r},\mathbf{r}',t,t')G(\mathbf{r}',t'). \quad (1)$$

The first and second units in the relationship (1) correspond to the reversible and irreversible parts of the stress tensor respectively. The main meaning of the relationship (1) is that nonequilibrium statistical ensemble in the phase space at the microscale level gives rise to another nonequilibrium statistical distribution $\mathfrak{R}(\mathbf{r},\mathbf{r}',t,t')$ in space and time at the intermediate scale level between the macro- and microscale, called mesoscale. It means that a state of the system in a point $\mathbf{r}$ at the time instance t is determined by a history of the system evolution in the finite size space vicinity of the point.

In general the correlation function is an unknown nonlinear functional of the macroscopic variables G. However, a simple parametric modeling of the correlation function doesn't allow an adequate description of nonequilibrium processes and a satisfaction to the natural boundary conditions to continuum. Many decades this circumstance was an obstacle for the nonlocal models applied to nonequilibrium transport in real media.

At the initial instance $t=0$ the system had been subjected to an external loading during a time interval t_R, the typical stress relaxation time is t_r. The relaxation proceeds by stages. The initial stage of frozen relaxation $t_R \leq t << t_r$ corresponds to the case where the correlation is constant $\mathfrak{R}(\mathbf{r},\mathbf{r}',t,t') = K = const$. At this stage the system remembers the prehistory of its deformation $K \int\limits_{-\infty}^{0} dt' \int\limits_{V} d\mathbf{r}' G(\mathbf{r}',t')$. The final near-equilibrium hydrodynamic stage $t_r << t \leq t_R$ corresponds to the case where the memory and nonlocal effects can be neglected. Then the transport relaxation kernels determine the transport coefficients in scope of the linear thermodynamics of irreversible transport processes.

$$P(\mathbf{r},t) \approx k_0^{(0)}(\mathbf{r},t)G(\mathbf{r},t), \quad (2)$$

$$k_0^{(0)}(\mathbf{r},t) \equiv \int\limits_{-\infty}^{t} dt' \int\limits_{V} d\mathbf{r}' \mathfrak{R}(\mathbf{r},\mathbf{r}',t,t').$$

At this stage the system forgets its prehistory reducing to the boundary and initial conditio $\int\limits_{-\infty}^{0} dt' \int\limits_{V} d\mathbf{r}' \mathfrak{R}(\mathbf{r},\mathbf{r}',t,t')G(\mathbf{r}',t') = G(\mathbf{r}=\Gamma,0).$

The momentum transport coefficient $k_0^{(0)}$ at the relaxation stage near to equilibrium presents a medium viscosity.

In scope of the theory the nonlocal and memory effects prove to be connected due to the finite perturbation rate propagation in real medium: $\dot{\mathbf{r}} < \infty$, when there exists a trajectory $\mathbf{r} = \mathbf{r}(t)$. In the structureless medium the rate reaches the equilibrium sound velocity and $r = Ct$. Then one can choose the space nonlocal model to describe the internal structure evolution.

$$\int\limits_{-\infty}^{t} dt' \int\limits_{V} d\mathbf{r}' \mathfrak{R}(\mathbf{r},\mathbf{r}',t,t')G(\mathbf{r}',t') = \int\limits_{\Gamma}^{\Gamma+\Omega(t)} d\mathbf{r}' \mathfrak{I}(\mathbf{r},\mathbf{r}')G(\mathbf{r}',t')$$

$$(3)$$

The integration in Eqn (3) is going over the volume $\Omega(t)$ embraced by the perturbation that has already come from a source through out a boundary Γ. Resolution of the function G by Tailor near the point $\mathbf{r}' = \mathbf{r}$, and its substitution into Eqn (3) result an infinite order differential operator

$$\int\limits_{\Gamma}^{\Gamma+\Omega(t)} d\mathbf{r}' \mathfrak{I}(\mathbf{r},\mathbf{r}')G(\mathbf{r}',t') = \quad (4)$$

$$= k_0(\mathbf{r},t)G(\mathbf{r},t) + k_1(\mathbf{r},t)\frac{\partial G}{\partial \mathbf{r}}(\mathbf{r},t) + \frac{1}{2}k_2(\mathbf{r},t)\frac{\partial^2 G}{\partial \mathbf{r}^2}(\mathbf{r},t) + \dots$$

The first moments of nonequilibrium statistical distribution of the space correlation have definite physical sense related to the medium internal structure. The 0-order moment $k_0(\mathbf{r},t)=k_0^{(0)}(1+\alpha)$ defines effective values of the transport coefficients for the medium with internal structure. The 1-order moment $k_1(\mathbf{r},t)$ defines a mathematical expectation

236

of the vector $\langle \mathbf{r}' - \mathbf{r} \rangle_3 = \gamma$ that is not equal to zero by definition if the statistical distribution differs from normal. It means that in general nonequilibrium distribution gives rise to the medium polarization along the direction of the vector γ under an external loading. This vector determines an arm of a force acting on a finite size medium element in the inhomogeneous velocity and stress fields. The resulted structured medium becomes mesopolar and has asymmetrical stress tensor. The second moment $k_2(\mathbf{r},t) = \langle (\mathbf{r}' - \mathbf{r})^2 \rangle = \gamma^2 + \varepsilon^2$ determines the dispersion of the space correlation distribution. In the case where $\varepsilon \to 0$ the nonlocal effects can be neglected, the statistical distribution tends to the δ-function. This case corresponds to the fluid reaction to perturbations near equilibrium. In the opposite case where $\varepsilon \to \infty$ the correlation doesn't decay with a distance from the point $\mathbf{r}$. In the case of constant correlation the medium is embraced by correlation in a whole. In both limits the medium doesn't demonstrate the internal structure effects that is in scope of the continuum mechanics validity.

In the intermediate case $t_R \cong t \cong t_r$ under pulse loading the medium reaction is like one of a multiphase dispersed mixture of solid phase grains with rigid correlation and viscous liquid with weak correlation. The dynamic material properties depend on the correlation rate at a given stage of relaxation.

So, the nonlocal model for the correlation function depending only on the first moments of the space correlation statistical distribution results new nonequilibrium constitutive relationships between thermodynamical fluxes and forces with the mesoscale medium structure characteristics included [2-3]

$$J(\mathbf{r},t) = k_0 \frac{1+\alpha}{\varepsilon} \int\limits_{\Gamma}^{\Gamma + \Omega(t)} d\mathbf{r}' \, \omega \left\{ \frac{(|\mathbf{r}' - \mathbf{r} - \gamma|)}{\varepsilon} \right\} G(\mathbf{r}',t), \quad (5)$$

where the function ω determine a rate of the space correlation decaying with a distance from the point $\mathbf{r}$. All higher moments of the statistical distribution in Eqn (5) are not neglected. The structure parameters being functionally depending on G can be determined by using boundary and integral conditions imposed on the system. In general case the last conditions can be satisfied only on account of the parameters and therefore determine them.

$$\Phi_i \left[G(\mathbf{r},t), \alpha, \varepsilon, \gamma \right] \Big|_{\mathbf{r} = \mathbf{r}_\Gamma} = 0, \quad i = 1,2,...,5 \quad (6)$$

According to Eqn (6) the relationships for the space nonlocal models determine the structure parameters $\alpha, \varepsilon, \gamma(t)$ as time-depending functions.

The nonlinear relationships (6) complete the model in a self-consistent way and determine the time evolution of spectra of internal structure scales. In general nonequilibrium case the scale spectrum is discrete and in limiting cases it becomes continuous defining the validity region of continuous mechanics. The bifurcation points for the branching equations (6) define the structure transitions in the system and determine the threshold values of the loading parameters.

So, it means that nonequilibrium transport can be described in a correct way only involving a close-loop between the external loading propagating in a medium and the internal medium structure formed by the perturbation.

.

SHOCK WAVE PROPAGATION

For the plane wave the mass and momentum balance equations are

$$\frac{1}{\rho} \frac{\partial \rho}{\partial t} + \frac{\partial u}{\partial x} = 0; \quad \rho \frac{\partial u}{\partial t} + \frac{\partial}{\partial x}(P - S) = 0. \quad (7)$$

In scope of the new self-consistent nonlocal approach the reversible part of the stress P is determined by the pressure of the mesoscale fluctuations and the irreversible part S in general cannot be divided into elastic and plastic components. Instead of the classical Maxwell model for the viscous-elastic-plastic medium the nonlocal theory results [2]

$$P = \rho CAD, \qquad (8)$$

$$S = (\lambda + \frac{2}{3}\mu)(1+\alpha)^{-1} \int_{-\infty}^{\Omega(t)} \frac{dx'}{\varepsilon} \omega \left\{ -\frac{\pi}{\varepsilon^2}(x'-x-\gamma)^2 \right\} \frac{\partial u}{\partial x'}$$

The velocity dispersion $D(x,t)$ plays a role of the temperature for the mesoscale fluctuations.

In the elastic limit at $t << t_r$, when $\varepsilon \to \infty$, $\omega \to 1$, $\varepsilon = Ct_r$, and the mass velocity $v << C$, Eqn (8) results

$$P = 0, \ S = (\lambda + \frac{2}{3}\mu) \left[\int_{-\infty}^{0} \frac{dx'}{\varepsilon} \frac{\partial u}{\partial x'} + \int_{0}^{Ct} \frac{dx'}{\varepsilon} \frac{\partial u}{\partial x'} \right] =$$

$$= (\rho C^2 t_r + \frac{4}{3} G t_r) \left[\frac{0-C}{Ct_r} + \frac{0-v}{Ct_r} \right] =$$

$$= -(K + \frac{4}{3}G)(1 + e_{xx}) = -S^e = -\rho_0 C_l (C_l + v) \qquad (9)$$

Here $C_l^2 = C^2 + (4/3)G\rho^{-1}$

In the hydrodynamic limit at $t >> t_r$, when $\varepsilon \to 0$, $\omega \to \delta - function$

$$P = \rho AT, \ S = (\lambda + \frac{2}{3}\mu) \frac{\partial u}{\partial x}, \qquad (10)$$

The mass and momrntum balance equations (7) can be reduced to a one equation

$$\frac{\partial^2 u}{\partial t^2} - C_l^2 \frac{\partial^2 u}{\partial x^2} = -\frac{\partial^2}{\partial t \partial x}(P^m - S^m)\rho^{-1}. \qquad (11)$$

Eqn (11) reduces to the wave equation in the elastic limit and describes the wave-type reversible transport and in the hydrodynamic limit it converts into the momentum diffusion equation and describes dissipative irreversible momentum transport. In the intermediate case both transport types are present and the share of the wave transport grows with the nonlocality parameter ε.

The energy balance determines multi-stage and multi-scale energy exchange where the mesoscale play a role of an energy buffer between the macro- and micro- levels [2-3,5].

In the reference connected to the elastic precursor Eqn (8) on the condition $v << a$ results in dimensionless form of the relaxation of a weak pulse

$$v(x,t) = (1+\alpha(x)) \int_{0}^{\tau} d\tau' \omega \left\{ \frac{|\tau' - \tau + \theta(x)|}{t_r(x)} \right\} \frac{\partial v}{\partial \tau'} \qquad (12)$$

DISCUSSION

The theory of dynamic plasticity is developed on the base of nonequilibrium statistical mechanics where nonequilibrium phase ensemble naturally originates mesoscale. It becomes clear that high-rate and large-gradient process can be describe only involving concepts of the self-organization and self-regulation at mesoscale.

REFERENCES

1. Asay J.R. CP620 Shock Compression of Condensed Matter-2001 ed. M.D.Furnish, N.N.Thadhani, and Y.Horie, 26-35 (2002).

2. Khantuleva T.A. CP620 Shock Compression of Condensed Matter-2001 ed. M.D.Furnish, N.N.Thadhani, and Y.Horie,, 263-266 (2002).

3. Khantuleva T.A. High-Pressure Shock Compression of Solids VI. Old Paradigms and New Challenges. Ed. Y.Horie, L.Davison, N.N.Thadhani, Springer, 215-254 (2003).

4. Zubarev D.N. Doklady Acad. Nauk, SSSR, 92-95 (1961).

5. Mescharyakov Yu.I. High-Pressure Shock Compression of Solids VI. Old Paradigms and New Challenges. Ed. Y.Horie, L.Davison, N.N.Thadhani, Springer, 169-214 (2003).

CP706, *Shock Compression of Condensed Matter - 2003*
edited by M. D. Furnish, Y. M. Gupta, and J. W. Forbes
© 2004 American Institute of Physics 0-7354-0181-0/04/$22.00

NEW WINDOWS INTO SHOCKS AT THE MESOSCALE

A. Koskelo[1], Scott Greenfield[1], Doran Greening[2], Damian Swift[3]

[1]*Advanced Diagnostics and Instrumentation Group, Chemistry Division*
[2]*Material Sciences Group, Applied Physics Division*
[3]*Plasma Physics Group, Physics Division*
Los Alamos National Laboratory, Los Alamos NM 87545

Abstract. This paper presents experimental observation and modeling of time dependent energy localization occurring when a shock propagates along and through a grain boundary. This work is part of a larger program of investigation of the effect of grain boundaries on shock propagation in materials. The project is initially focused on the simplest of materials: a bicrystal. Our study covers the effects of grain orientation, the grain boundary angle, the boundary region, shock properties and the interplay between these in determining the characteristics of shock propagation. Ultimately, we will use this information as a basis for incorporation into models of polycrystalline materials. Some of the physics found in polycrystals is absent, but the use of simple, well-defined samples allows thorough measurements to be made. Laser-based experiments and diagnostics are used throughout, permitting us to perform the many experiments required in an economical way. NiAl was chosen as a suitable anisotropic material; single crystal and bicrystal samples were prepared. The EOS and single-crystal elasticity were estimated with *ab fere initio* quantum mechanics. Laser flyer impact and direct drive experiments, coupled with line-imaging VISAR, were used to test and refine the EOS, and to measure crystal plasticity. Initial models of bicrystals under shock loading have been developed, shock experiments have been conducted on bicrystals and recovered samples have been analyzed.

INTRODUCTION

Stockpile Stewardship places stringent demands on the understanding of materials and their behavior under extreme conditions. Many materials – e.g., metals, ceramics and explosives – are polycrystalline. In the past, largely empirical homogeneous isotropic models were developed to describe the response of a material to dynamic loading. These models include Equation of State (EOS), constitutive, and reactive properties models. Through macroscopic averaging over randomly oriented crystals, the models adequately describe bulk properties for the conditions under which they have been calibrated, but provide little connection to the underlying physics and cannot be extrapolated confidently beyond their calibrated regime. The inherently heterogeneous properties of the individual crystals and the effects of grain boundaries result in the localization of stresses and energy. Many important phenomena – shearing, fracturing, hot spot and ejecta formation – depend nonlinearly on the material state, and therefore are highly influenced by material heterogeneity. To develop accurate, quantitative and predictive models it is imperative to develop a sound physical understanding of the micro- and meso-scale material response in polycrystalline materials.

Our approach is the systematic study of the most basic structure: two grains and their common boundary. With the study of this simple system the effects of grain orientation, the grain boundary angle, the boundary region, shock properties and the interplay between these in determining the characteristics of shock propagation can be explored. Ultimately, we will use this information as a basis for incorporation into models of polycrystalline materials and the grain boundary region itself can be explored. The results will then become the basis of a better understanding of shock propagation in a polycrystalline material. While all of the physics found in shock propagation through bulk materials are not present in bicrystals, we believe starting with simple, well-defined samples with precisely known geometries is an appropriate scientific approach.

Our project incorporates a combination of capabilities in the measurement, analysis and modeling of the mechanical response of materials to dynamic loading on the microscopic to mesoscopic scale. We grow bicrystals in a controlled manner, characterize their structure, and use them in shock experiments with either laser direct drive or laser driven mini-flyers as the shock initiation mechanism. Using recovered samples, post-shot analyses are done. Stoichiometric NiAl is the material chosen to begin this study.

To date, *ab fere initio* calculations have been used to estimate the EOS of NiAl and elastic response of a single crystal of material [1]. Laser-driven mini-flyers coupled with line and point VISAR have been used to test and refine the EOS calculations [2]. The EOS has been incorporated into a bicrystal model [3]. NiAl bicrystals have been grown. A Transient Interferometric Microscope (TIM) has been integrated into the LANL TRIDENT laser facility and experiments on bicrystals have been conducted [4-6]. Post-shock analyses of bicrystals have revealed fracture mechanisms depending on the details of bicrystal structure and shock orientation [7,8]. Advances in the Material Point Method have been made necessary for applying to the bicrystal/shock problem [9].

This paper reports on some of the bicrystal calculations conducted to date using a finite volume code. Results obtained using the TIM to view the dynamics of the out-of-plane displacements of a bicrystal at the shock breakout surface will also be described. The details of the computational approach can be found in the paper by Greening and Koskelo elsewhere in these proceedings [3]. The details of the experimental approach can be found in the paper by Greenfield et al., also elsewhere in these proceedings [6]

RESULTS AND DISCUSSION

Calculations show that the propagation of a shock wave through a bicrystal involves the interaction of crystal anisotropy, relative subgrain orientation, grain boundary angle, and shock profile [3]. Depending on the values of all of these parameters, the out-of-plane displacement profile at any given snapshot in time is complex. Figure 1 shows calculations of the surface displacement variation as a function of grain boundary angle, 30ns after a shock is applied at the rear surface of a 100μm thick NiAl bicrystal. The insert of Figure 1 shows the geometry of the calculation. The coordinate system has Y in the out-of-plane direction from the breakout surface, X is in the plane of the breakout surface and perpendicular to the line of intersection of the grain boundary with the breakout surface. X=0 is the point where the grain boundary intersects the breakout surface of the bicrystal, opposite the side where the shock is applied. Positive angles are defined as those that subtend the green vertical line and the grain boundary.

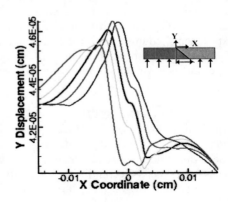

Figure 1. Calculated normal surface displacements at 30 ns after the initial shock for grain boundary angles −30° (red) though 30° (blue) by 15° increments measured anti-clockwise from vertically downward for a 2.5 ns 15 kbar pulse passing through a 100 μm thick NiAl bicrystal.

Each side of the bicrystal is shown in a shade of gray. The left side of the bicrystal has the (100) direction along the Y direction, the right side has the (111) direction along the Y direction. The two subcrystals both have the (110) direction pointing out of the plane of the paper.

Figure 1 shows that the grain boundary can lag or lead the motion of the adjoining material in the bicrystal depending on conditions. As shown be Greening and Koskelo this effect is also time dependent and in some cases the grain boundary lags in the first few nanoseconds after shock breakout, catches up and then may pass the neighboring material [3].

We have observed this motion of a grain boundary under shock loading using simultaneous TIM and line VISAR. Figure 2 is an image of a bicrystal taken with the TIM camera before the shock is created. The subgrain in the bottom part of the image has the (100) direction pointing out of the plane of the paper. The subgrain in the upper part of the image has the (111) direction pointing out of the plane of the paper. The bicrystal was grown from two subgrains sharing a common (110) direction. The grain boundary angle is at 55° in the coordinate system shown in the insert in Figure 1.

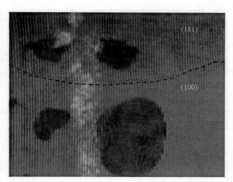

Figure 2. False color transient interferometric microscope (TIM) interferogram image of the grain boundary of a NiAl bicrystal. The dashed blue line is just below the grain boundary, the red region is the image of the line VISAR laser beam. The image is approximately 1.8mm square. The four dark regions are ink spots placed on the surface for alignment purposes.

Figure 3 through Figure 5 show three frames in a framing camera sequence taken of the bicrystal shown in Figure 2, taken during a shock produced by direct laser absorption on the side opposite that shown in the image. The data has been reduced from the raw interferograms to show the dynamic

out-of-plane displacement of the breakout surface in the region of the grain boundary. The color scale indicates the out-of-plane displacement and has been adjusted to emphasize the grain boundary movement. Figure 3 was taken just prior to shock breakout, Figure 4 was taken 13.3 ns after the image in Figure 3 and Figure 5 was taken another 6.6 ns later.

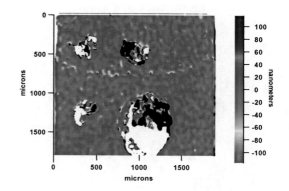

Figure 3. Transient Interferometric microscope image just prior to shock breakout in a NiAl bicrystal. The color scale is in nanometers of out-of-plane displacement. White and black are off-scale pixels.

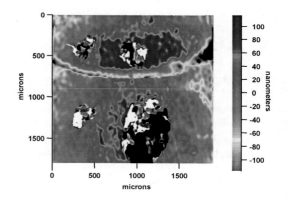

Figure 4. Transient Interferometric microscope image acquired 13.3 ns after the image in Figure 3. The color scale is the same absolute scale as in Figure 3

The blue areas in Figure 4 show that within 13.3 ns after the shock arrives at the breakout surface, the regions of the subcrystals near the grain boundary move approximately 100 nm out-of-plane. The grain boundary lags the bulk surface by approximately 100 nm compared to its pre-shock position.

After another 6.6 ns, the region near the grain boundary that was swollen by 100 nm has grown in extent, and the grain boundary has rebounded to about the same relative position it had before the shock, as shown in Figure 5. The region of displacement equal to the grain boundary can also be seen to have grown compared to that in Figure 4.

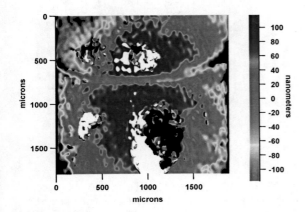

Figure 5. Transient Interferometric microscope image acquired 19.9 ns after the image in Figure 3. The color scale is the same absolute scale as in Figure 3

CONCLUSIONS

We have observed experimentally the dynamic motion of a grain boundary of a shocked bicrystal and the region around it. The out-of-plane displacement is on the order of 100 nm for the first tens of nanoseconds after shock breakout. The grain boundary first trails the bulk surface then rebounds as the shock dissipates at the surface. Calculations based on our experimentally derived equation-of-state show qualitatively similar behavior but depend greatly on the details of the shock event and bicrystal structure. A calculation for the specific conditions of the data presented is being conducted. Additional bicrystal experiments will be done to define quantitatively the complex interplay of grain boundary angle, relative subgrain orientation and shock profile.

ACKNOWLEDGEMENTS

The authors thank the rest of the project team for their hard work in making this project successful. We also thank similarly the LANL TRIDENT laser facility staff. This work was funded through LANL LDRD #2002 0055 DR and performed under the auspices of the U.S. Department of Energy under contract number W-7405-ENG-36.

REFERENCES

1. Koskelo, A., McClellan, K., Brooks, J., Paisley, D., Swift, D., Greenfield, S., and Greening, D., "Properties of Ni-Al under Shock Loading", Fourth Biennial International Workshop on New Models and Hydrocodes for Shock Wave Processes in Condensed Matter, Edinburgh, U.K., May 19-24 2002; LANL report LA-UR-02-2905.
2. McClellan, K., Swift, D., Paisley, D. L. and, Koskelo, A., "Dynamic properties of nickel-aluminum alloy", 13th American Physical Society Topical Conference on Shock Compression of Condensed Matter July 20 - 25, 2003.
3. Greening, D. and Koskelo, A. "Calculation of grain boundary shock interactions", *ibid*.
4. Greenfield, S.R. Greenfield, Casson, J. L., Koskelo, A. C., "Nanosecond interferometric studies of surface deformation induced by laser irradiation", Proceedings of the SPIE - The International Society for Optical Engineering; 2000; vol.3902, p.108-17.
5. Greenfield, S.R. Greenfield, Casson, J. L., Koskelo, A. C., "Nanosecond interferometric studies of surface deformations of dielectrics induced by laser irradiation", *ibid*; 2000; vol.4065, p.557-66.
6. Greenfield, S.R., Swift, D. C. and Koskelo, A.C. "Transient interferometric studies of shocked bicrystals", 13th American Physical Society Topical Conference on Shock Compression of Condensed Matter July 20 - 25, 2003.
7. Peralta, P., Swift, D. Loomis, E. Lim, C.-H. and McClellan, K.J., "Characterization of laser-driven shocked NiAl monocrystals and bicrystals", *ibid*.
8. Peralta, P., Swift, D., Loomis, E., McClellan, K. J., "Study of High Strain Rate Behavior of NiAl Single Crystals and Bicrystals using Laser driven Impacts", Submitted to Metallurgical and Materials Transactions A for review.
9. Bardenhagen, S. G., and Kober, E. M., *Comput. Model. Eng. & Sci.,* Under review.

CP706, *Shock Compression of Condensed Matter - 2003*
edited by M. D. Furnish, Y. M. Gupta, and J. W. Forbes
© 2004 American Institute of Physics 0-7354-0181-0/04/$22.00

MESODEFECTS COLLECTIVE PROPERTIES AND SELF-SIMILAR REGULARITY OF SHOCKED CONDENSED MATTER BEHAVIOR

O.B.Naimark

Institute of Continuous Media Mechanics of the Russian Academy of Sciences
1 Acad.Korolev str., 614013, Perm Russia

Abstract. The response of shocked materials in the pressure range 100 GPa reveals self-similar features that could be linked with collective properties of typical mesodefects (dislocation substructures, microcracks, microshears). The statistical approach was developed that allowed us to specify the Ginzburg-Landau theory for tensor order parameter of the mesodefect density and to establish the existence of self-similar solutions as specific collective modes in mesodefect ensemble related to the plastic strain and damage localization. The subjection of relaxation properties of dynamically loaded materials to mentioned collective modes provides the self-similar behavior of materials that experimentally were observed as the universality (fourth power law) of steady-state plastic front and delayed failure (failure wave) phenomenon as the resonance excitation of blow-up damage localization areas.

INTRODUCTION

Experimental study of material responses for a large range of loading rate reveals some specific features in failure and plasticity, and shows the linkage of solid behavior with evolution of the ensemble of typical mesodefects (dislocation substructures, microcracks, microshears). These features are particularly pronounced for dynamic and shock wave loading, when the internal times of the ensemble evolution for different structural levels are approaching to characteristic loading times. As a consequence, the widely used assumption in phenomenology of plasticity and failure concerning the subjective role of structural variables to stress-strain variables can not be generally applied. In this view the multifield theory of mesoscopic defects was developed in [1] and was applied for the explanation of the plastic front universality and failure wave effect in shocked materials [2].

DISLOCATION SUBSTRUCTURES

The dislocation density increases due to the plastic deformation and the consequent changes of dislocation substructures are observed. Despite the variety of the deformed materials the limited types of the dislocation substructures are observed. The transition between these substructures is the property of the dislocation interaction and leads to sharp changes in mechanical properties of metals and alloys. The main mechanisms of the dislocation friction (viscoplastic material responses), deformation hardening, damage localization have the relationship to the reconstruction of dislocation substructures.

Structural parameters associated with typical mesodefects (microcracks, microshears) were introduced in [1]. These defects are described by symmetric tensors of the form $s_{ik} = s v_i v_k$ in the case of microcracks and $s_{ik} = 1/2 s (v_i l_k + l_i v_k)$

for microshears. Here $\vec{v}$ is unit vector normal to the base of a microcrack or slip plane of a microscopic shear; $\vec{l}$ is a unit vector in the direction of shear; s is the volume of a microcrack or the shear intensity for a microscopic shear. The average of the "microscopic" tensor s_{ik} gives the macroscopic tensor of the microcrack or microshear density $p_{ik} = n\langle s_{ik} \rangle$ that coincides with the deformation caused by the defects, n is number of defects.

STATISTICAL AND PHENOMENOLOGICAL MODEL OF CONTINUUM WITH DEFECTS

Statistics of the microcrack (microshear) ensemble was developed in terms of the solution of the Fokker-Planck equation [3] in the phase space of the possible states of the microscopic variable s_{ik} concerning the size s and the orientation $\vec{v}, \vec{l}$ modes. According to statistical self-similarity hypothesis this solution, the distribution function of the defects, can be represented in the form $W = Z^{-1} \exp(-E/Q)$, where E is the mesodefect energy, Z is the normalization constant, Q is potential energy relief of the initial structure. The average procedure gives the self-consistency equation for the determination of defect density tensor

$$p_{ik} = n \int s_{ik} W(s, \vec{v}, \vec{l}) ds_{ik} . \tag{1}$$

The solution of equation (1) established qualitative different responses of materials depending on the value of structural δ-parameter, which represents the ratio of characteristic structural scales in material $\delta \approx (R/r_0)^3$, where R is the distance between defects, r_0 is the mean size of the defect nuclei (size of structural heterogeneity, for instance grain size). For uni-axial loading ($\sigma = \sigma_{xx}$) $p = p_{xx}$ consists of the volume part (induced by the microcracks) for a tension and only deviatoric part (induced by the microshears) for a compression or simple shear.

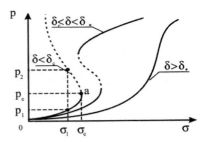

Figure 1. Characteristic solid responses on the defect growth.

It was shown [1] the ranges $\delta < \delta_c \approx 1$, $\delta_c < \delta < \delta_*$, $\delta > \delta_* \approx 1.3$ are characteristic for quasi-brittle, ductile and nanocrystalline responses of materials. The curves in Fig.1 represent the solution of the equation $\partial F / \partial p = 0$, where F is the part of the free energy caused by defects. The metastability for the stress $\sigma < \sigma_c$ is the consequence of the orientation interaction in the defect ensemble. The value $\sigma = \sigma_c$ determines the Hugoniot elastic limit for quasi-brittle materials. The stresses, which determine the metastability area for $\delta_c < \delta < \delta_*$, correspond to the range of HEL for materials with initially ductile response. The statistical description allowed us to propose the model of solid with defects based on the appropriate free energy form taking in view the characteristic non-linearity of the free energy on the defect density tensor in the corresponding ranges of δ. The simple phenomenological form of the part of the free energy caused by defects (for the uni-axial case $p = p_{zz}, \sigma = \sigma_{zz}, \varepsilon = \varepsilon_{zz}$) is given by sixth order expansion, which is similar to the well-known Ginzburg-Landau expansion in the phase transition theory [2].

$$F = \frac{1}{2} A(1 - \frac{\delta}{\delta_*}) p^2 - \frac{1}{4} B p^4 - \frac{1}{6} C(1 - \frac{\delta}{\delta_c}) p^6 - D\sigma p + \chi(\nabla_l p)^2 \tag{2}$$

The gradient term describes the non-local interaction in the defect ensemble; A, B, C, D are positive phenomenological material parameters, χ is the non-locality coefficient. According to the definition of s_{ik} the value of p consists of both deviatoric and volume parts in the rarefaction wave and only deviatoric in compressive wave. The damage kinetics is determined by the evolution inequality

$$\delta F/\delta t = (\partial F/\partial p)\dot{p} + (\partial F/\partial \delta)\dot{\delta} \leq 0,$$

that leads to the kinetic equation for the defect density p and scaling parameter δ

$$\dot{p} = -\Gamma_p(A(1-\delta/\delta_*)p - Bp^3 + C(1-\delta/\delta_c)p^5 - D\sigma) - \partial/\partial x_l(\chi \partial p/\partial x_l)),$$

$$\dot{\delta} = -\Gamma_\delta(-A/(2\delta_*)p^2 - C/(6\delta_c)p^6), \qquad (3)$$

where Γ_p, Γ_δ are the kinetic coefficients. Kinetic equations (3) and the equation for the total deformation $\varepsilon = \hat{C}\sigma + p$ ($\hat{C}$ is the component of the elastic compliance tensor) represent the system of constitutive equations of materials with mesodefects.

SELF-SIMILAR SOLUTIONS

In the region $\delta > \delta_*$ this equation is of the elliptic type with periodic solutions with spatial scale Λ and possesses p anisotropy determined mainly by the applied stress. This distribution of p gives rise to weak pulsations of the strain field. As $\delta \to \delta_*$ the solution of Eqns. 3 passes the separatrix, and the periodic solution transforms into a solitary-wave solution. This transition is accompanied by divergence of the inner scale $\Lambda : \Lambda \approx -\ln(\delta - \delta_*)$. In this case the solution has the form $p(\zeta) = p(x - Vt)$. The wave amplitude, velocity and the width of the wave front

are determined by the parameters of orientation transition:

$$p = \frac{1}{2}p_a[1 - tahn(\zeta l^{-1})], l = 4/p_a(2\chi/A)^{1/2}. \quad (4)$$

The velocity of solitary wave is $V = \chi A(p_a - p_m)/2\zeta^2$, where $(p_a - p_m)$ is the jump in p in the course of an orientation transition.

A transition through the bifurcation point δ_c is accompanied by the appearance of spatio-temporal structures of a qualitatively new type characterized by explosive accumulation of defects as $t \to t_c$ in the spectrum of spatial scales. The so-called «blow-up» self-similar solution appears under the pass of the critical point p_c

$$p = g(t)f(\xi), \xi = x/\varphi(t),$$
$$g(t) = G(1 - t/\tau_c)^{-m}, \qquad (5)$$

where τ_c is the so-called "peak time" ($p \to \infty$ at $t \to \tau_c$), $G > 0, m > 0$ are the parameters of non-linearity, which characterise the free energy release rate for $\delta < \delta_c$.

PLASTIC FRONT UNIVERSALITY. FAILURE WAVES

The unique feature of the large amplitude wave is the steady-wave plastic shock profile. The steady-wave profile propagates without change in form and, as pointed out in [4], is the consequence of stable balance between the competing processes of stress-strain non-linearity and dissipative material behavior. Eqns 3 describe the generation of shear localization areas with the boundaries having the solitary wave dynamics. The structural rescaling in term of the δ–decrease in the range $\delta_c < \delta < \delta_*$ and generation of finite-amplitude strain localization provide the anomalous dissipative ability similar to the energy absorption near the critical point in the phase transition and subjection of material relaxation property to the solitary wave dynamics of plastic strain localization. This is assumed to be the main reason for the universality of plastic strain rate dependence on the stress amplitude $\dot{p} \approx A\sigma_{amp}^4$, established in [4] for a wide class of materials for strain rates $\dot{\varepsilon} > 10^5 \ s^{-1}$. The fourth power law follows from the stable difference in the power in the $p(\sigma)$-dependence for upper and lower branches (Fig. 1) in the course of the δ–kinetics. The self-similar solution (5) can be excited for the appropriate energy density (stress amplitude and impulse duration) imposed by shock. Excitation of these blow-up dissipative structures leads to the phenomenon of delayed failure (failure waves [2]) in shocked glasses when the microshear interaction in the compressive wave provides the generation of blow-up damage kinetics behind the stress front with the delay time related to τ_c, Fig. 2. This result is in the correspondence with [5], where the failure wave front was considered as the propagating phase boundary.

Figure 2. Simulation of stress wave (S) and failure wave (F) propagation for different time.

CONCLUSIONS

The most important result of this work is the established linkage of collective modes in the mesodefect ensemble, having the nature of self-similar solutions, and specific nonlinear responses of shocked materials: steady-state profile of plastic wave and delayed failure phenomenon - failure waves. The excitation of these modes provides the subjection of the plastic wave dynamics to the slow

solitary wave dynamics of structural orientation transition in mesodefect ensemble (shear strain localization) and, as a consequence, the universality (four power law) of plastic front profile. The failure wave effect has the linkage with the excitation in the microshear ensemble under the shock loading the collective modes, having the "self-keeping blow-up" dynamics of damage localization. It is important that the blow-up damage localization kinetics is characterized by inherent spatial and temporal scales that assumes the existence of critical energy density of the shock pulse for the excitation of failure wave with delay related to the temporal scale τ_c.

ACKNOWLEDGEMENTS

The research was supported in part by the Russian Fund of Basic Research (project n. 02-01-0736), contract R&D 8936-AN-01S with the European Research Office of U.S. Army and the ISTC projects n. 1181 and 2146.

REFERENCES

1. Naimark, O.B. and Silbershmidt, V.V. "On the fracture of solids with microcracks", Eur.J.Mech., A/Solids., vol.10, pp. 607-619, 1991.
2. Naimark, O.B., "Defect induced transitions as mechanisms of plasticity and failure in multifield continua", in Advances in Multifield Theories of Continua with Substructure, 2003 (G.Capriz, P.Mariano, eds.), Birkhauser Boston Inc. ,pp.75-114.
3. Naimark, O.B., "Kinetic transitions in ensembles of defects (microcracks) and some nonlinear aspects of fracture ", in Proceedings of the IUTAM Symposium on Nonlinear Analysis of Fracture, 1997, (J.R.Willi, ed.), pp.285-298.
4. Swegle, J.W. and Grady, D.E., "Shock viscosity and the prediction of shock wave rise times", J.Appl.Phys. vol. 58, no. 2, pp. 692-701, 1985.
5. Clifton, R.J., "Analysis of failure waves in glasses", Appl. Mech. Rev., vol. 46, pp.540-546, 1993.

CP706, *Shock Compression of Condensed Matter - 2003*
edited by M. D. Furnish, Y. M. Gupta, and J. W. Forbes
© 2004 American Institute of Physics 0-7354-0181-0/04/$22.00

CALCULATING ELASTIC CONSTANTS OF MOLECULAR CRYSTALS USING CRYSTAL98

Troy Oxby[1], W. F. Perger[1], J. Zhao[2], J. M. Winey[2], and Y. M. Gupta[2]

[1]*Physics Dept., Michigan Tech. Univ., Houghton MI 49931*
[2]*Inst. For Shock Physics, Wash. State Univ., Pullman WA 99164*

Abstract. Calculations of the second-order elastic constants of pentaerythritol (PE), and urea were performed using the output of the modeling program CRYSTAL98. CRYSTAL98 performs first principles calculations of periodic systems using Gaussian type basis functions. The elastic constants were obtained by performing a polynomial fit to the calculated total energy of the crystal as a function of elastic strain. When the application of strains resulted in a reduction of crystal symmetry, the crystal unit cells often had to be redefined to use a space group of lower symmetry. To reduce computation times, the new space group was chosen to maximize remaining symmetry. A rigid molecule model was employed, where the internal geometry of the molecule was kept unchanged. An optimization technique was developed for lattice relaxation and tested for both urea and PE. Results obtained using different functionals in the Density Functional Theory (DFT) method were compared to Hartree-Fock (HF) calculations.

INTRODUCTION

The procedure for calculating second-order elastic constants is to calculate the total energy of the crystal for several values of strain. The second-order elastic constants are directly related to the 2nd derivative of the energy vs. strain curve. The strains must be small enough so that higher order effects are negligible. Because the intermolecular interactions are weak, the small deformations of the crystal result in relatively small changes in the total energy. Capturing the small changes in total energy requires a high quality calculation. For this reason, calculating elastic constants is a good test of the quality of a calculation method.

COMPUTATIONAL FRAMEWORK

Calculating the total energy of urea in CRYSTAL98 after applying strains of approximately 1% resulted in energy differences on the order of 10^{-3} hartrees. Therefore, to get accurate results for the elastic constants, the calculations should be at least that precise. There are many parameters that influence the accuracy of CRYSTAL calculations. The TOLINTEG keyword allows you to improve the quality of the calculation by specifying 5 values. These values influence the accuracy of a calculation by using an approximation when the overlap of two gaussians is less than 10^{-n}. The default values for TOLINTEG are "6 6 6 6 12".

The shrinking factor is another parameter that affects the accuracy of CRYSTAL98 calculations. It is specified using 3 values, but only the first and third are significant. The 1st value is the shrinking factor of the reciprocal lattice vectors and determines the number of k points used in diagonalizing the Fock matrix. The 3rd value is the shrinking factor of the reciprocal vectors in the Gilat net. For insulators the 1st and 3rd values should be equal. The values of shrinking factor used in the test case for urea included with CRYSTAL98 are "2 2 2".

To better understand the influence of these parameters on the total energy, a series of calculations were performed using different values for TOLINTEG and the shrinking factor. Figure 1 shows the total energy of urea as a function of TOLINTEG and the shrinking factor. TOLINTEG was varied by adding a constant integer to all 5 values. The number on the TOLINTEG axis of Fig. 1 is the value of the 1st four components.

In Fig. 1, it is clear that a shrinking factor of "2 2 2" is not sufficient and for higher values of TOLINTEG exhibits some peculiar behavior. For values of shrinking factor larger than 2, the energy difference found by increasing the shrinking factor is on the order of 10^{-6} hartrees. There is a discontinuity in the TOLINTEG direction at about 9. This discontinuity is approximately 3 millihartrees. From this chart it was decided that our calculations should use at least "11 11 11 11 17" for TOLINTEG and a shrinking factor of "4 4 4".

To further improve the accuracy, the NOBIPOLA keyword was used which forces CRYSTAL98 to make no approximations during the calculation of the bi-electronic integrals, and POLEORDR keyword was used to increase the order of multipole moment expansion to 6. The NOBIPOLA and POLEORDR options each decreased the total energy by approximately 1 millihartree.

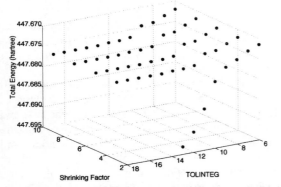

Figure 1. Plot of total energy of urea vs. TOLINTEG and shrinking factor

Rigid Molecule Approximation

With a reasonable precision in the total energy calculations, the next step was to find a method for applying strains to the materials in CRYSTAL98. CRYSTAL98 has an ELASTIC keyword that applies a strain matrix but was not used. The ELASTIC keyword strains the lattice, but because the atom locations are defined in fractional coordinates, it also strains the molecule. This artificial strain would greatly increase the total energy because the intramolecular forces are so much stronger than the intermolecular forces. A possible solution would be to relax the internal coordinates after using the ELASTIC keyword, but the non-orthogonal nature of linear combination of atomic orbitals (LCAO) calculations makes relaxation of internal coordinate difficult. The next version of the CRYSTAL software will include automatic internal spacing optimization[1]. Within the limitations of the current version, these calculations employ a rigid molecule approximation. During relaxation and straining of the crystals, the molecules maintain their size, shape, and orientation, but the distance between molecules is changed. It is expected that the intermolecular forces are the dominant component of the elastic constants so this model would be a reasonable approximation.

System relaxation

Both urea and pentaerythritol (PE) are tetragonal systems, so to find the minimum energy configuration in the rigid molecule approximation, it is necessary to perform a two-dimensional minimization of total energy over lattice spacings a, and c. The energy depends on lattice parameters a, and c in a non-linear way so the two-dimensional minimization cannot be replaced with two separate one-dimensional minimizations. The downhill simplex method[2] was tried but failed to consistently reach the minimum without requiring tuning of the various parameters for each system. The number of trials required to tune the parameters made it impractical.

The relaxation technique used was the modified gradient descent[3] technique illustrated in Fig. 2.

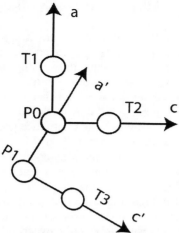

Figure 2. Diagram illustrating the modified gradient descent technique used in lattice relaxation.

The total energy was calculated at the initial point P0, as well as test points T1 and T2. Numeric derivatives were calculated at P0 and a linear combination of the derivatives points in the direction of the next point. The code then steps out in that direction and calculates the energy at point P1. Test point T3 is chosen so that it is orthogonal to P0 and P1. Numeric derivatives are taken in the a', c' coordinate system and projected back onto the a, c coordinate system. The linear combination of derivatives again points to the next point. This continues until the new point has a higher energy than the previous point. The step size was chosen to be 1% of the maximum dimension. This value is large enough to find the minimum in a reasonable number of terms, but small enough to give a good approximation to the derivatives.

Once the approximate minimum is found, the energy is calculated for a 4 x 4 grid centered at that point. A second order surface is then fit to the grid and a functional minimization is done to locate the minimum. Table 1 shows the relaxed lattice constants for urea using 2 different basis sets in Hartree-Fock calculations, DFT calculations using Becke exchange functional with Lee-Yang-Par correlation (B3LYP), and the generalized gradient approximation using the Perdew-Burke-Ernzerhof functional(GGAPBE). The relaxed lattice constants for urea match closely with experiment.

Table 1. Calculated relaxed lattice constants for urea using several basis sets and correlation potentials

urea	a (Å)	c (Å)
exp[4]	5.565	4.680
6-21g* HF	5.652	4.678
6-31g HF	5.837	4.735
6-21g* B3LYP	5.403	4.554
6-31g B3LYP	5.704	4.646
6-21g* GGAPBE	5.463	4.590
6-31g GGAPBE	5.632	4.643

Table 2 shows the results for the relaxed lattice constants for PE. The values for a are in good agreement but the values for c are over 20% too large. If one were to plot of the energy surface vs. a and c, they would see that the theoretical values are indeed minimums so the discrepency is probably a result of the inadequacy of the rigid molecule approximation or the inability for the basis set and method to capture the weaker intermolecular interactions.

Table 2. Calculated Relaxed Lattice Constants for PE using several basis sets and correlation potentials.

PE	a (Å)	c (Å)
exp[5]	6.083	8.726
6-21g* HF	6.148	10.696
6-31g HF	6.067	11.786
6-21g* B3LYP	6.059	10.460
6-31g B3LYP	5.990	11.428
6-21g* GGAPBE	6.040	10.471
6-31g GGAPBE	5.974	11.328

RESULTS AND DISCUSSION

All strains were applied relative to the relaxed configurations shown in Tables 1 and 2. The elastic constants were calculated using a 4th order polynomial fit of energy vs. strain.

Table 3 contains the calculated and experimental values for C_{11} and C_{33} for urea. To calculate C_{33} in urea, strains were applied in the c direction using the original space group.

Table 3. Calculated elastic constants of urea using different basis sets and correlation potentials.

urea	C_{11} (GPa)	C_{33} (GPa)
exp[6]	21.7	53.2
6-21g* HF	21.6	66.4
6-31g HF	20.5	63.3
621g* B3LYP	5.9	143.9
6-31g B3LYP	7.4	73.1
6-21g* GGAPBE	5.5	100.6
6-31g GGAPBE	8.7	78.8

To calculate C_{11} it was necessary to reduce the symmetry. The necessary calculations were done in space group $P2_12_12$ (no. 18). Hartree-Fock seems to produce reasonable values for the elastic constants using either basis set. Neither DFT method was able to produce reasonable values. The low values of C_{11} are due to noise in the energy vs. strain curve. The curve was smoother in the C_{33} calculations but results in values that are much too large.

With the relaxed lattice spacing of PE being so far from experimental, the elastic constants are not expected to be very accurate. For PE, the crystal was strained equally in the a and b directions. This preserves the symmetry, which decreases the calculation time, but fitting the energy vs. strain curve now yields $C_{11} + C_{12}$. As seen in Table 4, the 6-31g HF calculation is reasonably close to experiment but the DFT calculations all tend to overestimate the stiffness of the lattice. Again, the rigid molecule approximation is a likely cause for the inaccuracy of the elastic constant calculations. The PE molecule is much more flexible than urea and changes in the molecular geometry due to strain cannot be ignored.

Table 4. Calculated elastic constants of PE using different basis sets and correlation potentials.

PE	$C_{11}+C_{12}$ (GPa)	C_{33} (GPa)
exp[7]	67.1	13.9
6-21g* HF	95.5	55.0
6-31g HF	86.5	11.1
6-21g* B3LYP	126.1	26.0
6-31g B3LYP	126.3	8.2
6-21g* GGAPBE	126.8	89.2
6-31g GGAPBE	125.7	24.8

CONCLUSIONS/FUTURE WORK

From this preliminary work it appears that the rigid molecule model employed was not sufficient to accurately calculate the elastic constants. In urea, a more rigid system, Hartree-Fock produced reasonable results while neither of the two DFT methods tried were able to do so. None of the methods tried were able to produce good results for the more flexible crystal PE, and it is expected that the rigid molecule approximation must be removed in order to improve agreement with experiment. It is difficult to determine the influence of the basis set but in urea it appears that going from a 6-21g* basis set to a 6-31g basis set tended to push the calculated results toward the experimental values.

When the new version of CRYSTAL is released, these calculations will be reevaluated. Allowing for internal relaxation should do much to improve the results. Future refinements will also include optimizing the basis set for the lattice.

ACKNOWLEDGMENTS

The authors gratefully acknowledge the financial support of the Office of Naval Research and the MURI program.

REFERENCES

1. Civalleri, B., D'Arco, Ph., Orlando, R., Saunders, V. R., Dovesi, R. Chem. Phys. Lett. **348,** 131 (2001)
2. Nelder, J. A., and Mead, R., Computer Journal, **7,** 308 (1965)
3. Chong, E. K. P., Zak, S. H., *An Introduction to Optimization*, John Wiley and Sons, New York, 1996, pp. 102
4. Swaminathan, B. M., Craven, B. M., McMullan, R. K., Acta Crytallogr. **B40,** 300 (1984)
5. Eilermam, D. Rudman, R., Acta Crysta. **B35,** 2458 (1979)
6. Fischer, G., Zarembowitch, A. C. R., Seances Acad. Sci., Ser. B **270,** 852 (1970)
7. Nomura, H., Higuchi, K., Kato, S., Miyahara, W. Jpn. J. Appl. Phys. **11,** 304 (1972)

CP706, *Shock Compression of Condensed Matter - 2003*
edited by M. D. Furnish, Y. M. Gupta, and J. W. Forbes
© 2004 American Institute of Physics 0-7354-0181-0/04/$22.00

FIRST-PRINCIPLES INTERMOLECULAR BINDING ENERGIES IN MOLECULAR CRYSTALS WITH OPTIMIZED BASIS SETS

W. F. Perger*, Jijun Zhao†, Miguel Blanco** and Ravi Pandey*

Michigan Tech Univ
†*Institute for Shock Physics, Washington State Univ*
**Universidad de Oviedo*

Abstract. The intermolecular binding (lattice) energy is calculated for the molecular crystals RDX, PE,and PETN using the Crystal98 and Gaussian98 programs (with Gaussian basis sets) and the DMOL program (double-numerical basis set) and the CASTEP program (plane-wave basis set). These first-principles theoretical results are compared with each other and experiment. The importance of the basis sets used (e.g., 6-21G vs. 6-311G*) and choice of Hamiltonian (Hartree-Fock or density-functional theory) is illustrated by these comparisons. The relevance of calculating the theoretical intermolecular binding energy, being the difference of two nearly-equal numbers, as a tool for testing the intrinsic quality of a calculation is explained. The importance of optimization of the Gaussian basis sets for the Crystal98 program versus adding more terms to the Gaussian basis set is discussed. Work supported by the Office of Naval Research and the MURI program.

INTRODUCTION

The accurate calculation of organic molecular crystal (OMC) properties represents a formidable challenge of contemporary interest. Three structural levels of OMCs have been identified[1]: electronic, intramolecular and intermolecular. The relatively strong intramolecular binding is responsible for the fact that vibrational frequencies calculated with the Gaussian98 program compare reasonably well with experiment in, for example, PETN [2] (although, of course, the lattice modes cannot be calculated with this program). In this case, the agreement between theory and experiment for these high-frequency vibrational modes supports the idea that the properties of the molecule are more-or-less unaffected by the presence of the lattice. However, the binding of the molecules in the lattice, being due to the relatively weak van der Waals interaction, must be accurately described if other properties, such as the lattice vibrational modes (acoustic and optical) or the lattice energy itself, are to compare well with experiment. To address this need for an accurate intermolecular

potential, one can consider the molecular crystal at a perturbed isolated molecule level where the existence of adjacent molecules is simulated by introducing an external Coulomb potential produced by the charge distribution of these adjacent molecules into the molecular Hamiltonian [3]. This approach was used by Krijin, *et al.* [4] using a local-density approximation (LDA) for crystalline α-oxalic acid dihydrate.

If the molecular interaction energy is to be determined, then terms in addition to the electrostatic one must be included, such as with the two-body additivity approximation [3], which was used by Avoird *et al.* [5]. Neither of these approaches is fully periodic, in contrast to the all-electron method used in the Crystal98 [6] program, and the relatively-recent work of Dovesi *et al.* [7] on urea represents one of the first efforts using this first-principles Hartree-Fock linear combination of atomic orbitals (HF-LCAO) approach. As noted in that work, although the Crystal98 program is, in principle, superior to the aforementioned approaches, several practical issues arise, such as basis set completeness and numerical accuracy, which must be carefully considered prior to reaching a conclusion as to the quality of a specific calculated property.

Because the weak intermolecular interaction is responsible for the mechanical and elastic properties, which are in turn of importance for a detailed understanding of the shock initiation-to-detonation transition, one of the goals of the current work is to calculate the lattice energy for several OMCs, using different theoretical methods, and compare with experiment. Different basis sets are used and contrasted, and issues of basis set optimization, as well as a new script using a parallel version of Crystal98 for this purpose, are described.

METHODOLOGY

The intermolecular binding (lattice) energy is defined here as:

$$E_{lattice} = \frac{E_{crystal} - nE_{molecule}}{n}, \qquad (1)$$

where n is the number of molecules in a unit cell. Eq. (1) requires two, distinct, calculations for each value of the lattice energy. Because this results in taking the difference of two numbers which are identical to the 4th or 5th significant figure, on the order of tens of milliHartrees, basis set completeness and numerical error must be considered in order to achieve a reasonable value for the lattice energy. In addition to improving upon the basis sets currently used for OMCs, comparison with independent theoretical approaches offers further insight as to the overall quality of the calculation.

The first theoretical approach used in this work was to implement the Crystal98 [6] program, as it had been used with success for other studies with molecular crystals [8]. In that approach, a given crystalline orbital, $\psi_i(\mathbf{k}, \mathbf{r})$, is expressed as a linear combination of Bloch functions, $\phi_\mu(\mathbf{k}, \mathbf{r})$, which are defined in terms of a set of atomic orbitals, $\zeta_\mu(\mathbf{r} - \mathbf{A}_\mu - \mathbf{g})$ [6]:

$$\zeta_\mu(\mathbf{r} - \mathbf{A}_\mu - \mathbf{g}) = \sum_{j=1}^{n_G} d_j G(\alpha_j; \mathbf{r} - \mathbf{A}_\mu - \mathbf{g}). \qquad (2)$$

Gaussian-type orbitals, while having the advantage of being able to accurately describe the core and valence atomic orbitals with few functions relative to a plane-wave basis set, have the disadvantage of resulting in certain well-known characteristics, such as basis set superposition error (BSSE) [7]. Furthermore, computational demands have limited the use of basis sets to 6-21G in urea just a little more than a decade ago [7] and more recently to 6-21G for systems as large as RDX [8]. In the present work, we extend the quality of the basis sets used in two, distinct, ways: 1) expanding the GTFs to 6-311G*, and 2) performing basis set optimization on the exponents of the most diffuse Gaussians representing the valence electrons. The coefficients and exponents of the GTFs were all taken from the Pacific Northwest National Laboratory's website. Given the difficulty of optimizing all the exponents for all the electrons in a given basis set, the approach taken was to select the exponents of the most diffuse Gaussians describing the valence orbitals for optimization, then a modified version of an optimization script was used to call a parallel version of Crystal98 using parallel-virtual machine (PVM).

The second theoretical approach was to use the plane-wave basis set programs DMOL [9] and CASTEP [10], and then compare the results with the first theoretical approach and with experiment.

CRYSTAL98 AND GAUSSIAN BASIS SETS

The choice of the basis set is the most important step for obtaining accurate first-principles calculations in crystalline systems[6]. For OMCs, although the intramolecular bonding may be described well with basis sets borrowed from years of development by quantum chemists (standard Pople sets[11]), the intermolecular interaction must also be accurately described and it is not obvious that the basis sets developed for molecular calculations will, without adaptation, be suitable for many problems of interest for crystalline systems. Although in principle to obtain greater accuracy one could simply increase the number of GTFs in Eq. (2) to improve the representation of the atomic orbital, in practice this approach is intractable for many OMCs, owing to the rapid increase in the number of integrals necessary as a function of the number of GTFs. Furthermore, it is not obvious that simply increasing the size of the basis set by adding more Gaussians whose parameters were determined for a molecular calculation, will necessarily be more accurate than a smaller basis set which has been optimized for the crystalline system. To examine in detail the effects of the basis set on the lattice energy, the OMCs RDX, PE, and PETN, were studied. The first basis set chosen was the common 6-21G, which used $s(6)sp(2)sp(1)$ for carbon, nitrogen, and oxygen, and $s(2)s(1)$ for hydrogen. The second set was the 6-311G*, with $s(6)sp(3)sp(1)sp(1)d(1)$ for carbon, nitrogen, and oxygen, and $s(3)s(1)s(1)$ for hydrogen. In addition to the different basis sets, different potentials were also chosen: Hartree-Fock (HF), density-functional theory with B3LYP (DFT-B3LYP) and DFT local-density-approximation (DFT-LDA). Of the crystals studied, RDX represents the greatest computational challenge, with 1176 atomic orbitals required which results in an integrals file size of roughly 60GBytes for the 6-311G* basis set.

The optimization of the exponents of only the most diffuse Gaussians can be a computationally difficult task because it requires the simultaneous adjustment of several parameters in order to achieve a minimum of the total energy; each adjustment of a given parameter is another complete run of the Crystal98 program, which often means dozens or hundreds of runs for obtaining one set of optimized parameters. The procedure for optimization of the selected Gaussian exponents proceeds in the following manner. A given exponent is arbitrarily selected and a range of values around the nominal one is chosen, requiring one calculation of the Crystal98 program per value in this range. A determination is then made as to where in that range the minimum energy occurs; the value of that exponent which minimizes the total crystalline energy is then saved and the process repeated for the next exponent. Once every exponent is optimized using this procedure, another complete pass is made, searching for the value of each exponent in turn which optimizes the total energy. This process is slow and not guaranteed to get caught in local minima of the total energy surface. However, with the use of the parallel version of the Crystal98 program and PVM on a 15-node Linux cluster of 32-bit, 2.2GHz machines, the total execution time is now reasonable. As for the issue of unambiguously identifying the global minimum of the total energy, other techniques, such as gradient approaches, are being pursued for verification.

RESULTS

Table 1 shows a comparison of calculations of the lattice energy for the three OMCs selected, using the Crystal98, DMOL, and CASTEP programs, and compared with experiment. In addition to the different basis sets employed, different correlation potentials were also chosen. All calculations obtained using the Crystal98 program were corrected with the counterpoise method, as this was previously determined to be an important effect for molecular crystals [7]. The density-functional theory local-density approximation (DFT-LDA) was found to reproduce the lattice energy reasonably well, often slightly overestimating it. However, this agreement may be the consequence of a fortuitous cancellation of errors, because it has recently been shown that LDA overestimates the strength of the intermolecular potential in elastic constants [12]. It is interesting to note from Table 1 that LDA results obtained from the different programs with different basis sets agreed with each other for all three OMCs. Generally speaking, the 6-21G basis set with DFT-B3LYP results from the Crystal98 program produced results in better agreement with experiment but these results suggest that there remains a significant spread in the theoretical values, one that requires further work. Regarding the two Gaussian basis sets chosen, the results of Table 1 do not show an improvement in the agreement with experiment as the Crystal98 basis set was increased from 6-21G to 6-311G*, despite the dramatic increase in computational complexity.

To address this issue, the 6-21G basis set size dependence in PE was optimized by selecting the outer $s(1)$ Gaussian for hydrogen and the $sp(1)$ Gaussians for carbon and oxygen. Table 2 shows the relative improvement in the lattice energy compared with experiment, for both the HF and DFT-B3LYP correlation potentials. This finding, which must be examined for other OMCs, suggests that an optimized 6-21G basis set could be an accurate approach while also being much more computationally economical.

CONCLUSIONS AND FUTURE WORK

The lattice energy represents a test of the ability of a first-principles theoretical approach to model the relatively weak intermolecular potential found in

TABLE 1. Lattice energy (in units of kJ/mol, 1 eV=96.487 kJ/mol) for the chosen molecular crystals, with experimental values and calculations based on different programs and methods. All the structures for the crystal are taken from experiments. For CASTEP calculations, the ultrasoft pseudo-potential with 500 eV cut-off energy was used.

	Exper.	Crystal98(6-21G)		Crystal98(6-311G*)			DMOL		CASTEP	
		HF	B3LYP	HF	LDA	B3LYP	LDA	GGA	LDA	GGA
RDX	130.1[13]	65.0	72.7	.	123.7	44.1	143.77	63.30	135.59	61.47
PE	163.[14]	113.6	147.3	69.8	190.5	101.7	218.81	130.40	199.70	117.2
PETN	151.9[15, 16]	80.2	76.4	26.6	124.7	25.5	155.05	53.36	150.61	58.12

TABLE 2. Lattice energy (in units of kJ/mol, 1eV=96.487kJ/mol) for PE, with experimental value and using optimization of the outer valence functions, including basis-set superposition (BSSE) corrections.

	Crystal98 6-21G		Crystal98 6-21G-optimized	
Exper.	HF	B3LYP	HF	B3LYP
163.[14]	113.6	147.3	128.5	174.6

organic molecular crystals. Three crystal modeling programs were used to calculate the lattice energy of RDX, PE, and PETN, each employing a different type of basis set: the Crystal98 program, which uses Gaussian-type functions to represent the atomic orbitals, DMOL, which uses a double-numerical basis set, and CASTEP programs, which use plane waves. For the use of the Crystal98 program, it appears that the optimization of at least some of the most diffuse Gaussians may lead to a better estimate of the inter-molecular potential than a larger, but unoptimized, Gaussian basis set. Further examination of this idea is warranted and currently being pursued. Finally, in the present study, the optimized basis set affected both the intra- and inter-molecular potentials, but the relative amounts are unknown. It is therefore of interest to separate the effects of basis set optimization on these potentials, which suggests a series of optimizations performed on suitable molecular systems.

ACKNOWLEDGMENTS

The authors gratefully acknowledge the financial support of the Office of Naval Research and the MURI program.

REFERENCES

1. Silinsh, E. A., and Čápek, V., *Organic Molecular Crystals: Interaction, Localization, and Transport Phenomena*, AIP Press, New York, 1994.
2. Gruzdkov, Y. A., and Gupta, Y. M., *J. Phys. Chem. A*, **105**, 6197–6202 (2001).
3. Kitaigorodskii, A. I., *Molecular Crystals and Molecules*, Academic Press, Inc., New York, 1973.
4. Krijin, M. P. C. M., and Feil, D., *J. Chem. Phys.*, **89**, 4199 (1988).
5. van der Avoird, A., Wormer, P. E. S., Mulder, F., and Berns, R. M., *Top. Curr. Chem.*, **93**, 1 (1980).
6. Saunders, V. R., Dovesi, R., Roetti, C., Causà, M., Harrison, N. M., Orlando, R., and Zicovich-Wilson, C. M., *CRYSTAL98 User's Manual*, University of Torino, Torino, 1998.
7. Dovesi, R., Causà, M., Orlando, R., Roetti, C., and Saunders, V., *J. Chem. Phys.*, **92**, 7402–7411 (1990).
8. Perger, W. F., *Chem. Phys. Letters*, **368/3-4**, 319–323 (2003).
9. Delley, B., *J. Chem. Phys.*, **92**, 508 (1990).
10. Milman, V., Winkler, B., White, J. A., Packard, C. J., Payne, M. C., Akhmatskaya, E. V., and Nobes, R. H., *Int. J. Quant. Chem.*, **77**, 895 (2000).
11. Hehre, D. J., Radom, L., Schleyer, P. V. R., and Pople, J. A., *Ab initio molecular orbital theory*, John Wiley and Sons, New York, NY, 1986.
12. Zhao, J., Winey, J. M., Gupta, Y. M., and Perger, W. F., "Elastic properties of molecular crystals using density functional theory," in *Shock Compression of Condensed Matter-2003*, 13th APS Topical Group Meeting on Shock Compression of Condensed Matter, American Physical Society, Portland, OR, 2003.
13. Rosen, J., and Dickinson, C., *J. Chem. Eng. Data*, **14**, 120 (1969).
14. Semmingsen, D., *Acta Chem. Scand.*, **A42**, 279 (1988).
15. Chickos, J. S., "Heats of Sublimation," in *Molecular Structure and Energetics*, edited by J. F. Liebman and A. Greenberg, VCH Publishers, New York, 1987, chap. 2.
16. Edwards, G., *Tran. Faraday Soc.*, **49**, 152 (1953).

CP706, *Shock Compression of Condensed Matter - 2003*
edited by M. D. Furnish, Y. M. Gupta, and J. W. Forbes
© 2004 American Institute of Physics 0-7354-0181-0/04/$22.00

MODELING POLYMORPHIC TRANSFORMATIONS OF QUARTZITE IN DYNAMIC PROCESSES

A.V.Petrovtsev, Yu.N. Zhugin, G.V.Kovalenko

Russian Federal Nuclear Centre - Institute of Technical Physics, P.O. Box 245, Snezhinsk, Chelyabinsk Region 456770, Russia

Abstract. A model of dynamic behaviour of quartzite under loading and release is developed, which describes polymorphic transformations and strength properties of this matereal. The model is based on recently obtained data of static, laboratory and large-scale dynamic experiments. Main peculiarities of dynamic processes in quartzite are taken into account. They are related to considerable non-equilibrium of polymorphic transformations under loading characterized by large deviation of transformation beginning from equilibrium curve and large length of transformation area; large hysteresis of transformation under loading and release; kinetic nature of transformation. Appropriate quantitative characteristics of transformations are estimated on adiabatic curves of loading and release. Relaxation times of transformations are estimated as well. Conclusions are drawn on the value of residual shear strength. Simulation results are presented for a series of shock experiments used to calibrate the model.

INTRODUCTION

Quartz is the main rock-forming mineral, and this fact initiated comprehensive research into its properties. A great part of this research consists of dynamic studies stimulated by different applications related to the investigation of impact and shock induced phenomena in rocks.

One of the most interesting issues of fundamental importance is studying phase transitions in quartz and quartzite that is virtually pure polycrystalline quartz. These materials have a complicated phase diagram (Fig.1) that besides the low-pressure quartz phase (Q) shows two solid high-pressure phases of coesite (C) and stishovite (S). All these phases show a sophisticated crystalline structure and significantly differ in density [1]. Shock-induced polymorphous transformations in quartz are essentially non-equilibrium [2-11]: the states that correspond to coesite are almost absent and transformations into a stishovite-like phase are strongly metastable with an extended phase mixture region on the Hugoniot and

a relatively long transformation time. This is very similar to the graphite-diamond transformation in carbon [12,13], which is also rather delayed in contrast to the α-ε transformation in iron that is also metastable but to a lesser degree than in the above cases.

It should be noted that there are no concurrent views on the nature and mechanisms of shock-induced phase transformations in quartz, particularly mechanisms governing the formation of high-density amorphous quartz phases (see [14]).

The objective of this work is to study kinetics of the polymorphous Q-S transformation in quartzite through numerical simulations of experiments made at RFNC-VNIITF, using a phenomenological model of this transformation, based on all available data on quartzite properties.

KEY ELEMENTS OF THE MODEL

The description of thermodynamic properties of quartzite is based on the equation of state for quartz and stishovite similar to that proposed in [15]. The parameters of crystalline density, sound velocity and Gruneisen coefficient were adjusted according to data obtained for quartzite. Coesite was excluded from consideration and the line of Q-S equilibrium was taken identical with the line of coesite and stishovite equilibrium [15].

The kinetic Q-S transformation model is based on the kinetic relaxation equation [15-17] in which limiting concentrations in the phase mixture region account for the surfaces of metastable states on the adiabatic curves for loading and release. The approach to Q-S transformation modeling was in general close to that proposed in [18,19] except that we used a time dependent kinetics of transformation that was similar to [17]. A more detailed description of the model can be found in [20].

The detailed analysis presented in [6] shows that in the phase mixture region the Hugoniot for monolithic quartzite (Fig.2) consists of three sections for $\sigma^{max} \approx 13$-23, 23-30 and 30-60 GPa. The second section has the greater slope compared with the others and it is this section that corresponds to the most intensive production of the high-density phase ($\xi_S \approx 0.2$-0.8). Based on the comparison of different experimental and evaluated data, Ref. [6] concludes that the quartz Hugoniot has two branches on this section (Fig.2), which are related to laboratory (samples of thickness h≤10 mm) and large-scale (h~1 m) experiments that differ in the time dependence of the Q-S transformation kinetics. Following this, the surface of metastable states for the direct transformation in the model was accepted in accordance to the data from large-scale experiments [6].

Characteristics of the inverse transition were evaluated from the data obtained in the analysis of profiles describing the release of shock-compressed quartzite into buffer materials of different stiffness, with the registration of velocity profiles of the sample-buffer boundary by the technique of variable-induction pickup. Modeling of these experiments helped to relate the features of measured profiles to states on the P-T plane. It was found that the point where the inverse transition started was close to the phase equilibrium line. Following this, the surface of limiting states for the inverse transition in the model was taken to be close to the equilibrium one. Analysis of experimental data helped also to estimate the characteristic time

of the polymorphous Q-S transformation in quartz to be equal to ≈300 ns.

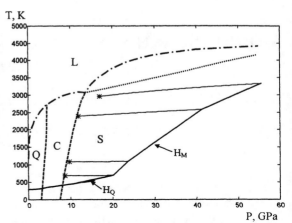

FIGURE 1. Quartz phase diagram [1] (the dashed lines show Q-C and C-S equilibria, dots and dashes show the melting line in static measurements, and the dotted one is for shock wave measurements), and calculated Hugoniot (the bold line, H_Q – quartz Hugoniot, H_M – phase mixture) and release adiabat (thin lines, * – the point where the inverse transition starts)

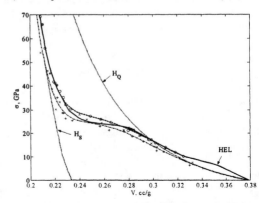

FIGURE 2. Quartzite shock compression: The dashed and solid lines are laboratory and large-scale experiment approximations of the data O, ⊕ [4,5,8,11] for monolithic quartzite. The dot-and-dash line is the approximation of data from mix experiments +, × [6,7]. The dotted line shows calculated hydrodynamic Hugoniots of quartzite and stishovite.

All quartzite data were analyzed in assumption that shear stresses were present. Associated estimates of the yield strength were made through the comparison of the quartzite Hugoniot with its "hydrodynamic" Hugoniot evaluated from the data obtained in the study on shock compressibility of disperse quartzite mixed with high plasticity materials (paraffin and fluoroplastic) [6,7] (see

Fig.2) and in static experiments on isothermal compressibility of quartz and stishovite [1]. According to the accepted model of quartzite elastic-plastic properties, the yield strength Y first grows achieving a maximum of ≈5 GPA for $\sigma^{max} \approx 6$ GPA on the Hugoniot elastic limit (HEL). Then quartzite softens and Y reduces to ≈2 GPa for $\sigma^{max} \approx 10$ GPa. In the phase mixture region, the yield strength begins growing as the concentration of stishovite increases, and achieves ≈8 GPa at $\sigma^{max} \approx 60$ GPa when the transformation is completed. Similarly, the data on elastic precursor velocities and sound velocities in shock-compressed quartz [10,21-23] (see Fig.3.) were used to obtain the relationship between shear modulus and pressure. It corresponds to the increase of the Poisson ratio ν of quartzite along Hugoniot from the initial value of 0.15 to ≈0.3 during quartz softening beyond the elastic limit and to the decrease of ν in the phase mixture region to $\nu=0.25$ for the state of the complete transformation into stishovite.

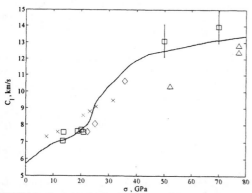

FIGURE 3. Sound velocity in shock-compressed quartz versus wave amplitude: ◊ - [21], × - [22], □ - [10], △ - [23]. The solid line shows data calculated with the quartzite model.

NUMERICAL MODELING OF EXPERIMENTS

The model was verified through numerical simulations of different experiments. Experiments of the first type included plane experiments [6,8] in which profiles of stress waves in quartzite samples were measured by a manganine transducer when the wave covers distances ≤10 mm. These experiments prove the two-wave structure of profiles with a strong elastic precursor. This structure does not change when the wave amplitude grows and the

polymorphous transformation in quartzite begins, until the shock strength achieves $\sigma^{max} \approx 23$ GPa. As the shock strength growing, the wave profiles show a feature evidencing that the front of the main plastic wave is not stable. In fact, data from [6,8] prove that laboratory experiments show the initial stage in the forming of the three-wave configuration of shocks in quartzite, with the elastic precursor and one more wave which is an analog of the phase precursor but with a significant ($\xi_S \approx 0.2$) quartz transformation in the front.

The results of our calculations and experimental data are shown in Fig.4. Because of the sharp increase in transformation completeness for $\sigma^{max} \geq 23$ GPa and its low rate, the main plastic wave where the most portion of the transformation occurs is seen to be virtually a compression wave. The presented data show the polymorphous transformation in quartzite to be a two-stage process. Its characteristic time ($\tau_{QS} \approx 300$ ns) is typical only for the second stage ($\sigma \geq 23$ GPa, $\xi_S \geq 0.2$). At the first stage ($\xi_S \leq 0.2$), the transformation rate is much higher and $\tau_S \approx 10$ns. This may evidence that the mechanism of the transformation changes at the second stage.

In experimental measurements [11], the time scale of the phenomenon changes by several orders of magnitude. This allows the front of the main plastic wave following the precursor to form more completely. Fig.5 shows calculated and experimental [11] velocities of the free surface of quartzite walls versus time. Note than since the elastic precursor is not registered in large-scale experiments, we introduced appropriate corrections into the description of shear strength (excluded the region of higher values of Y for initially strong quartzite at $\sigma^{max} \leq 10$ GPa). It is seen that calculated results are in a good qualitative and quantitative agreement with the data from [11]. A bit smaller separation of waves for the first distance from the explosion center may be a result of the underestimated amplitude of phase precursor in the model based on the one-stage transformation kinetics with $\tau_{QS}=300$ ns that is also shown in simulations of laboratory experiments.

Analysis of the two types of experiments proves conclusions in [6] about the evolution of the metastable state surface of the polymorphous Q-S transformation and the need in advanced kinetic equations for its description. Appropriate improvements of the model will be done in the further research.

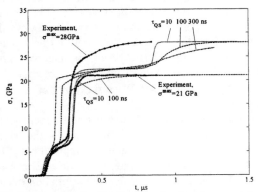

FIGURE 4. Stress profiles in quartzite for plane experiments [6,8] with $\sigma^{max} \approx 21$ and 28 GPa. The solid lines with points show experimental data. The dotted and dashed lines show calculated data with $\tau_{QS} \approx 10$, 100 and 300 ns.

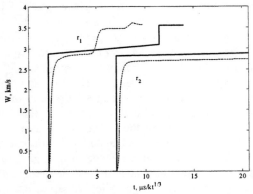

FIG. 5. Profile of velocity W(t) of the free surface of the quartzite wall. The solid lines show experimental data [11] and the dashed ones show calculated data, r is distance from the explosion center.

REFERENCES

1. Hemley, R.J., High-Pressure Behavior of Silica, in *Silica: Physical Behavior, Geochemistry and Materials Applications*, Reviews in Mineralogy **29**, 41-82, Mineralogical Society of America, Washington, D.C. (1994).
2. Wackerle, J., J.Appl.Phys. **33** (2), 922-937 (1962)
3. Fowles, R. J., Geoph. Res, **72** (22) 5729-5742 (1967)
4. Trunin R.F., Simakov G.V. et al., Trans. of the USSR Academy of Sciences, Geophysics, **1**, 13 (1971).
5. LASL Shock Hugonoit data, ed. by Marsh S.P., Berkeley, University of California Press, 1980, p.499.
6. Zhugin, Yu. N., The Behaviour of α-Quartz under High Dynamic and Static Pressures: New Results and Views, in *Shock Compression of Condensed Matter-1995*, Woodbury, New-York: AIP Press, 1996. Part I, pp.97-100.
7. Zhugin, Yu. N., Krupnikov, K.K. et al. Trans. of the USSR Academy of Sciences, Geophysics, **10**, 16-22 (1994).
8. Zhugin, Yu. N., Krupnikov, K.K., in *Problems of Non-linear Acoustics*, Proc. Of XI-th Int. (IUPAP-IUTAM) Symp. on non-linear acoust., Novosibirsk, Acad. Sci. of the USSR, Sib. Branch, 1987, Part II, p. 196-200.
9. Grady, D.E., Zhugin, Yu. N., *Bull. Amer. Phys. Soc.*, ser. 2, **39**, 1, 410-411 (1994).
10. Chhabildas, L.C., Shock Loading and Release Behaviour of X-Cut Quartz, in *Shock Waves in Condensed Matter -1985*, edited by Y.M. Gupta, Plenum Press, New-York, 1986, pp.601-605.
11. Vildamov, V.G., Gorshkov, M.M., Slobodenjukov, V.M., Senichev, P.N. Rus. J. Ch. Phys., **14**, 2-3, 122-125 (1995).
12. Gust, W.H., Phys. Rev. B **22** (10), 4744-4756 (1980).
13. Zhugin, Yu.N., Krupnikov, K.K., Tarzhanov, V.I., Investigation into Kinetics of Natural Ceylon Graphite Transformations in Shock Waves, in *Shock Wave in Condensed Matter – 1998*, edited by I.Yu. Klimenko et al., High Pressure SIC Press, St.-Petersburg, 1998, pp.163-166
14. Kuznetsov, N.M., in *Shock Waves and Extreme States of Matter*, edited by V.E. Fortov et al., Moscow, NAUKA Publishers, 2000, pp.199-218.
15. Swegle, J.W., J. Appl. Phys., **68**, 4, 1563-1579 (1990).
16. Andrews, D. J., J. Comp. Phys. **7**, 310-326 (1971).
17. Boettger, J. C., and Wallace, D. C., Phys. Rev. B **55**, 2840-2849 (1997).
18. Boettger, J. C., J. Appl. Phys. **71** (11), 5500-5508 (1992).
19. Boettger, J. C., Furnish M.D., Dey T.N., Grady D.E., J. Appl. Phys. **78** (8), 5155-5165 (1995).
20. Petrovtsev, A.V., Bychenkov, V.A., Kovalenko, G.V., in *Shock Compression of Condensed Matter - 2001*, AIP Conference Proceedings #620, editors M.D. Furnish, N.N. Thadhani, and Y. Horie, American Institute of Physics, 2002, pp.591-594.
21. Grady, D.E., Murri, W.J., DeCarli, P.S., J. Geophys. Res., **80** (35), 4857-4861 (1975)
22. Pavlovsky, M.N., Rus. J. Appl. Mech. Tech. Phys. **5**, 136-139 (1976).
23. McQueen, R.G., The Velocity of Sound behind Strong Shocks in SiO_2, in *Shock Waves in Condensed Matter - 1991*, edited by S.C.Schmidt, R.D.Dick, J.W.Forbes, D.G.Tasker, Elsevier Science Publishers B.V., 1992, pp.75-78.

CP706, *Shock Compression of Condensed Matter - 2003*
edited by M. D. Furnish, Y. M. Gupta, and J. W. Forbes
© 2004 American Institute of Physics 0-7354-0181-0/04/$22.00

A METHOD FOR TRACTABLE DYNAMICAL STUDIES OF SINGLE AND DOUBLE SHOCK COMPRESSION

Evan J. Reed[*], **Laurence E. Fried**[†], **M. Riad Manaa**[†] **J. D. Joannopoulos**[*]

[*]*Department of Physics, Massachusetts Institute of Technology, Cambridge, MA 02139*
[†]*Chemistry and Materials Science Directorate, Lawrence Livermore National Laboratory, Livermore, CA 94550*

Abstract. A new multi-scale simulation method is formulated for the study of shocked materials. The method combines molecular dynamics and the Euler equations for compressible flow. Treatment of the difficult problem of the spontaneous formation of multiple shock waves due to material instabilities is enabled with this approach. The method allows the molecular dynamics simulation of the system under dynamical shock conditions for orders of magnitude longer time periods than is possible using the popular non-equilibrium molecular dynamics (NEMD) approach. An example calculation is given for a model potential for silicon in which a computational speedup of 10^5 is demonstrated. Results of these simulations are consistent with the recent experimental observation of an anomalously large elastic precursor on the nanosecond timescale.

INTRODUCTION

Molecular dyanmics simulations have provided valuable insight into atomic scale dynamical processes in shock waves.[1, 2, 3, 4] However, existing methods of performing these simulations are generally limited to the 10 picosecond timescale. The popular non-equilibrium molecular dynamics (NEMD) approach to atomistic simulations of shock compression involves creating a shock on one edge of a large system and allowing it to propagate until it reaches the other side. The computational work required by NEMD scales at least quadratically in the evolution time because larger systems are needed for longer simulations. When quantum mechanical methods with poor scaling of computational effort with system size are employed, this approach to shock simulations rapidly becomes impossible. Another approach that utilizes a computational cell moving at the shock speed has the same drawbacks.[5] This paper presents a method which circumvents these difficulties by requiring simulation only of a small part of the entire system. The effects of the shock wave passing through this small piece of the system are simulated by dynamically regulating the applied stress which is obtained from a continuum theory description of the shock wave structure. Because the size of the molecular dynamics system is independent of the simulation time in this approach, the computational work required to simulate a shocked system is nearly linear in the simulation time. By circumventing the scaling problems of NEMD, molecular dyanmics simulation of shocked materials for orders of magnitude longer timescale becomes possible.

Molecular dynamics simulations have been performed that utilize a shock Hugoniot-based thermodynamic constraint for the temperature at fixed volume.[6] This approach is a thermodynamic one for a single shock wave that constrains the system to a path through pressure, density and energy space that is inconsistent with a steady shock wave, allowing the possibility of formation of unphysical states. The new method outlined in this paper is a method for the dynamical simulation of shock waves that ensures correct material pathways are sampled by constraining the system to the pressure, density and energy states of a steady shock. It enables simulation of shock waves in systems that have material instabilities which lead to the formation of multiple shock waves and requires no *a priori* knowledge of the system phase diagram, metastable states, or sound speeds.

SIMULATION OF A SINGLE SHOCK WAVE

We model the propagation of the shock wave using the 1D Euler equations for compressible flow, which neglect thermal transport. These equations represent the conservation of mass, momentum, and energy respectively everywhere in the wave. Neglecting thermal transport in high temperature shocks is valid in systems where electronic mechanisms of heat conduction are not important, i.e. usually less than a few thousand K in insulators.[7] While continuum theory is not rigorously applicable at elastic shock fronts, the correct dynamics will be approximated in these special regions. We seek solutions of these equations which are steady in the frame of the shock wave moving at speed v_s. This substitution, and integration over x yields a variation of the Hugoniot relations,

$$u = v_s \left(1 - \frac{\rho_0}{\rho}\right), \qquad (1)$$

$$p - p_0 = v_s^2 \rho_0 \left(1 - \frac{\rho_0}{\rho}\right), \qquad (2)$$

$$e - e_0 = p_0 \left(\frac{1}{\rho_0} - \frac{1}{\rho}\right) + \frac{v_s^2}{2} \left(1 - \frac{\rho_0}{\rho}\right)^2. \qquad (3)$$

Here u is the local speed of the material in the laboratory frame (particle velocity), v is the specific volume, $\rho = 1/v$ is the density, e is the energy per unit mass, and p is the negative component of the stress tensor in the direction of shock propagation, $-\sigma_{xx}$. Variables with subscripts 0 are the values before the shock wave, and we have chosen $u_0 = 0$, i.e. the material is initially at rest in the laboratory frame. In the language of shock physics, Eq. 2 for the pressure is the Rayleigh line and Eq. 3 for the internal energy is the Hugoniot at constant shock velocity. These equations apply to a system which has a time-independent steady-state in the reference frame moving at the shock speed v_s.

For the molecular dynamics simulation, we employ the Lagrangian,

$$L = T\left(\{\dot{\vec{r}}_i\}\right) - V\left(\{\vec{r}_i\}\right) + \frac{1}{2}Q\dot{v}^2 +$$
$$\frac{1}{2}\frac{v_s^2}{v_0^2}(v_0 - v)^2 + p_0(v_0 - v) \qquad (4)$$

where T and V are kinetic and potential energies per unit mass, and Q is a mass-like parameter for the simulation cell size. It can be seen that Eq. 4 implies Eq. 3 when $\dot{v} = 0$ because $T + V = e$. The equation of motion for the system volume is,

$$Q\ddot{v} = \frac{\partial T}{\partial v} - \frac{\partial V}{\partial v} - p_0 - \frac{v_s^2}{v_0^2}(v_0 - v) \qquad (5)$$

which reduces to Eq. 2 when $\ddot{v} = 0$. We use the scaled atomic coordinate scheme of Ref [8] to deal with the variable computational cell size. This scheme introduces a volume dependence for T and V. Strain is only allowed in the shock direction, i.e. $v_0 - v = -\varepsilon_{xx}v_0$ where ε_{xx} is the uniaxial strain. The pressures in Eq. 5, including the thermal contribution, are taken to be the uniaxial x component of stresses. Computational cell dimensions transverse to the shock direction are fixed, as in NEMD simulations. This approach allows the simulation of shocks propagating in any direction using the same simulation cell. By choosing a small representative sample of the shocked material, it is assumed that stress gradients and thermal gradients in the actual shock wave are negligible on the length scale of the sample size. While the thermal energy is assumed to be evenly distributed throughout the sample, thermal equilibrium is not required. It can be shown that the stable states of the constraint equations satisfy shock wave stability requirements. [9]

To simulate a shock to a given pressure, the initial state parameters which define the MD constraint in Eq. 4 are chosen (ρ_0, p_0, e_0) during an equilibration phase. A guess for v_s is made for the constraint to take the system to the desired final pressure and the molecular dynamics system is compressed slightly (typically -10^{-4} strain) to ensure the simulation proceeds along the compressive branch of the Rayleigh line. Upon simulation, if the final pressure is other than the desired one, improved guesses for v_s can be made and simulated again until the desired v_s is determined. The simulation of a shock to a given parti-

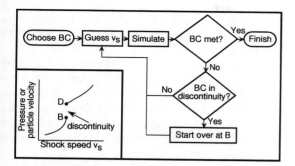

FIGURE 1. Flowchart for simulation of a shock to a chosen pressure or particle velocity boundary condition (BC). Instabilities due to regions where $\frac{d^2p}{dv^2} < 0$ along the Hugoniot can give rise to a discontinuity in the inset plot.

cle velocity using this approach is a straightforward extension.

TREATMENT OF MULTIPLE SHOCK WAVES

The above method describes the simulation of a single stable shock wave. However, it is not always possible to shock to a given pressure or particle velocity using this technique. For example, it may not be possible to connect a straight Rayleigh line to all final pressures when there is a region of negative curvature in the Hugoniot, $\frac{d^2p}{dv^2} < 0$. Such regions of negative curvature are common in condensed phase materials and may be a result of phase transformations or may be the shape of a single phase Hugoniot. While a single Rayleigh line is insufficient to meet the pressure boundary condition in this region, two Rayleigh lines are sufficient.

Figure 1 shows a flowchart that illustrates how to determine the set of Rayleigh lines that are stable and meet the boundary conditions without any *a priori* knowledge of the system. A shock wave instability exists when the boundary condition falls within a discontinuity in the set of final pressures as a function of shock speed, as in the inset figure in Figure 1. The existence of such a discontinuity can be determined when sufficient trial values of v_s have been simulated. If the boundary condition falls within the discontinuity, Equation 4 is modified to include a second wave with a transition between waves at point D. The entire process is then repeated by fixing the 1st shock speed at point D and varying the 2nd shock

speed to find the speed that meets the boundary condition. If further instabilities are discovered that prevent the boundary condition from being met with a single shock, the process is continued.

The formation and evolution of multiple waves becomes more complicated when chemical reactions or phase transitions occur, leading to the phenomenon of elastic precursor decay. Volume decreasing phase transformations cause the pressure at point B in Figure 1 to decrease with time. Parameterization of the *p-v* space path with Rayleigh lines is valid when the timescale of this pressure change is less than the time required for a material element to reach the final shocked state. This condition can be made rigorous though the so-called shock change equation and shown to hold for times longer than some timescale. [9]

APPLICATION TO SILICON

As an illustrative example, we apply the new method to an elastic-plastic transition in a model potential for silicon. Figure 2 shows shock speed as a function of particle velocity for shock waves propagating in the [011] direction in silicon described by the Stillinger-Weber potential.[10] This potential has been found to provide a qualitative representation of condensed properties of silicon. Data calculated using the NEMD method are compared with results of the new method presented in this paper. NEMD simulations were done with a computational cell of size $920\text{Å} \times 12\text{Å} \times 11\text{Å}$ unit cells (5760 atoms) for a duration of about 10-20ps. Simulations with the new method were done with a computational cell size of $19\text{Å} \times 12\text{Å} \times 11\text{Å}$ unit cells (120 atoms). Both simulations were started at 300K and zero stress. Since the NEMD simulations were limited to the 10ps timescale by computational cost, simulations with the new method were performed to calculate the Hugoniot on this 10ps timescale for comparison. The final particle velocity in these simulations was taken to be a point of steady state after a few ps.

Figure 2 indicates a single shock wave exists below 1.9 km/sec particle velocity. Above this particle velocity, an elastic shock wave precedes a slower moving shock characterized by plastic deformation. Agreement between the two methods is good for all regions except for the plastic wave speed for parti-

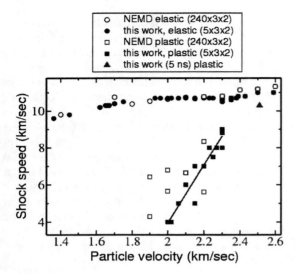

FIGURE 2. Comparison of calculated Hugoniots for the NEMD approach and the method presented in this paper for roughly 10 ps runs. Note the ability to utilize much smaller computational cell sizes with the new method. Also included is one data point for a 5 ns simulation using this work which would be prohibitive with NEMD requiring a factor of 10^5 increase in computational effort.

cle velocities less than 2.1 km/sec. The wide range of values for the plastic wave speeds in NEMD simulations in this regime is due to finite simulation cell size effects. Better agreement in this regime can be obtained by using simulation cells with larger cross sectional area.

One of the primary advantages of using the method outlined in this paper is the ability to simulate for much longer times than is possible with NEMD. As an example, Figure 2 shows the result of a 5 ns simulation performed along a Rayleigh line corresponding to a shock speed of 10.3 km/sec. The uniaxially compressed elastic state required 5 ns to undergo plastic deformation. The difference in particle velocity between the 10 ps and 5 ns simulations at this shock speed is 0.8 km/sec, suggesting that the elastically compressed state is metastable with an anomalously large lifetime. This is consistent with experimental observations of shocked silicon that indicate an anomalously high pressure elastic wave exists on the nanosecond timescale.[11] In addition to the simulations performed with the Stillinger-Weber potential, we have performed more accurate tight-binding[12] 120 atom simulations using the method

of this paper that also suggest an anomalously high pressure elastic wave precursor exists on the 10 ps timescale.

This simulation done with NEMD would require *more* than 5 ns simulation time due to the time required for the equilibration of the first and second wave speeds. For an $\mathcal{O}(\mathcal{N})$ method of force evaluation, the computational cost of this simulation with the NEMD method would be at least 10^5 times greater, and therefore not tractable.

ACKNOWLEDGMENTS

We thank J. Forbes, C. Tarver, and D. Hare for helpful discussions. We acknowledge support from the NDSEG Fellowship and the LLNL MRI/EMC Graduate Fellowship. This work was performed under the auspices of the US Department of Energy by the University of California, LLNL under contract number W-7405-Eng-48.

REFERENCES

1. Kadau, K., Germann, T. C., Lomdahl, P. S., and Holian, B. L., *Science*, **296**, 1681 (2002).
2. Germann, T. C., Holian, B. L., Lomdahl, P. S., and Ravelo, R., *Phys. Rev. Lett.*, **84**, 5351 (2000).
3. Kress, J. D., Bickham, S. R., Collins, L. A., Holian, B. L., and Goedecker, S., *Phys. Rev. Lett.*, **83**, 3896 (1999).
4. Holian, B. L., and Lohmdahl, P. S., *Science*, **280**, 2085 (1998).
5. Zhakhovskiĭ, V. V., Zybin, S. V., Nishihara, K., and Anisimov, S. I., *Phys. Rev. Lett.*, **83**, 1175 (1999).
6. Maillet, J. B., Mareschal, M., Soulard, L., Ravelo, R., Lomdahl, P. S., Germann, T. C., and Holian, B. L., *Phys. Rev. E*, **63**, 016121 (2001).
7. Zel'dovich, Y. B., and Raizer, Y. P., *Physics of shock waves and high-temperature hydrodynamic phenomena*, Academic Press, New York, NY, 1967.
8. Andersen, H. C., *J. Chem. Phys.*, **72**, 2384 (1980).
9. Reed, E. J., Fried, L. E., and Joannopoulos, J. D., *Phys. Rev. Lett.*, **90**, 235503 (2003).
10. Stillinger, F. H., and Weber, T. A., *Phys. Rev. B*, p. 5262 (1985).
11. Loveridge-Smith, A., Allen, A., Belak, J., Boehly, T., Hauer, A., Holian, B., Kalantar, D., Kyrala, G., Lee, R. W., Lohmdahl, P., Meyers, M. A., Paisley, D., Pollaine, S., Remington, B., Swift, D. C., Weber, S., , and Wark, J. S., *Phys. Rev. Lett.*, **86**, 2349 (2001).
12. Sawada, S., *Vacuum*, **41**, 612 (1990).

CP706, *Shock Compression of Condensed Matter - 2003*
edited by M. D. Furnish, Y. M. Gupta, and J. W. Forbes
2004 American Institute of Physics 0-7354-0181-0/04/$22.00

UNIVERSAL RELATIONS FOR ACCELERATION WAVE SPEEDS IN NONLINEAR VISCOELASTIC SOLIDS

Mike Scheidler

US Army Research Laboratory, APG, Maryland 21005-5069

Abstract. For finite deformations of nonlinear viscoelastic solids, the speed of propagation of acceleration waves (i.e., ramp waves) generally depends not only on the current state of strain at the wave front but also on the prior strain history. Consequently, explicit formulas for the wave speed can be quite complicated. Simple formulas for the wave speed do exist for special classes of materials and/or special deformation histories, and in this regard we consider one-dimensional motions of viscoelastic solids governed by single integral laws. Some of the relations obtained are universal in the sense that they hold for all materials in a given class and do not explicitly involve the relaxation kernel function in the hereditary integral defining these materials.

INTRODUCTION

We consider the speed of propagation of acceleration waves in viscoelastic solids undergoing uniaxial strain, as would occur in a normal plate impact experiment prior to the arrival of lateral release waves. The front of a ramp wave is an example of an acceleration wave.[1] Expansive (unloading) ramp waves can be generated by reflection of a shock wave from a free surface or a lower impedance material [1]. Compressive ramp waves can be generated by use of fused silica buffer plates or by graded density impactors [1, 3] and also by fast pulsed power techniques [3].

For nonlinear viscoelastic solids, the acceleration wave speed U is generally a nonlinear function of the current strain as well as the past strain history at the material point instantaneously situated on the wave front. In particular, for viscoelastic materials governed by single or multiple integral laws, the dependence of the wave speed on the strain history typically involves an explicit dependence on the relaxation kernel function(s) in the hereditary integral(s).[2]

However, for special classes of viscoelastic solids and/or special strain histories, simple explicit formu-

las for the wave speed exist in terms of quantities which have a direct physical interpretation. An example of such a relation was given by Nunziato et al. [1] for an acceleration wave propagating into a deformed region in equilibrium in a *finite linear* viscoelastic solid.[3] They showed that

$$\rho_0 U^2 = \frac{d\sigma_E}{d\varepsilon_1} + \frac{1-\varepsilon_1}{1-\frac{1}{2}\varepsilon_1} \frac{\sigma_I(\varepsilon_1) - \sigma_E(\varepsilon_1)}{\varepsilon_1}, \quad (1)$$

where σ_I and σ_E are the instantaneous and equilibrium elastic response functions, and ε_1 is the equilibrium uniaxial strain ahead of the wave.[4] This relation is *universal* in the sense that it holds for all materials in the indicated class and does not explicitly involve the relaxation kernel function G' in the hereditary integral (17) defining the materials in this class.

The paper begins with a discussion of the general, one-dimensional, nonlinear, single integral law for viscoelastic response. This is followed by some general results on acceleration wave speeds in such materials. These results are used to derive relations analogous to (1) but for more general classes of nonlinear viscoelastic solids and non-equilibrium conditions ahead of the wave.

---

[1] More precisely, an *acceleration wave* is a propagating singular surface across which the stress, strain and particle velocity are continuous but their spatial gradients and time derivatives suffer jump discontinuities [1, 2].
[2] See eq. (13) below for the general single integral case.

---

[3] This class of nonlinear viscoelastic materials is defined by equations (17) and (20) below.
[4] Precise definitions of all terms are given below.

SINGLE INTEGRAL LAWS

Let $\mathbf{F}$ denote the deformation gradient relative to a fixed, unstressed reference state, let $J = \det \mathbf{F}$, and let $\mathbf{T}$ denote the Cauchy stress tensor. Then $\boldsymbol{\Sigma} = J\mathbf{T}(\mathbf{F}^{-1})^T$ and $\mathbf{S} = J\mathbf{F}^{-1}\mathbf{T}(\mathbf{F}^{-1})^T$ are the 1st and 2nd Piola-Kirchhoff stress tensors. Introduce a Cartesian coordinate system with the 1-axis a symmetry axis of the material, and consider a time-dependent uniaxial strain along this axis. The normal component of stress, taken positive in compression, is

$$\sigma \equiv -T_{11} = -\Sigma_{11} = -\lambda_1 S_{11}, \qquad (2)$$

where $\lambda_1 = F_{11}$ is the (principal) stretch along the 1-axis. Let ε denote the nominal measure of uniaxial strain taken positive in compression: $\varepsilon = 1 - \lambda_1$.

We consider viscoelastic materials governed by a nonlinear single integral law, the most general one-dimensional form of which is

$$\sigma(t) = \sigma_I\big(\varepsilon(t)\big) + \int_0^\infty \mathscr{G}'\big(\varepsilon(t), \varepsilon(t-s), s\big)\,ds. \quad (3)$$

Here $\mathscr{G}'$ denotes the partial derivative of $\mathscr{G}$ with respect to its third (or temporal) argument s:

$$\mathscr{G}'(\varepsilon_1, \varepsilon_2, s) = \frac{\partial}{\partial s}\,\mathscr{G}(\varepsilon_1, \varepsilon_2, s). \qquad (4)$$

Fading memory requires that the relaxation kernel $\mathscr{G}'(\varepsilon_1, \varepsilon_2, s)$ decay to zero sufficiently rapidly as $s \to \infty$. There is some non-uniqueness in the functions σ_I and $\mathscr{G}$. This may be removed by the assumptions

$$\mathscr{G}'(\varepsilon_1, 0, s) = 0 \quad \text{and} \quad \mathscr{G}(\varepsilon_1, \varepsilon_2, 0) = \sigma_I(\varepsilon_1). \quad (5)$$

Indeed, since $\mathscr{G}'$ rather than $\mathscr{G}$ itself appears in the constitutive relation (3), we are free to choose the initial value $\mathscr{G}(\varepsilon_1, \varepsilon_2, 0)$. The choice $(5)_2$ simplifies the results below. By $(5)_1$ the upper limit ∞ in (3) may be replaced with t whenever $\varepsilon(\tau) = 0$ for $\tau < 0$. Condition $(5)_1$ implies that σ_I is the *instantaneous elastic response function*, i.e., $\sigma(t) = \sigma_I(\varepsilon_1)$ for the jump strain history

$$\varepsilon(\tau) = \begin{cases} \varepsilon_1, & \text{if } \tau = t; \\ 0, & \text{if } \tau < t; \end{cases} \qquad (6)$$

then $\sigma_I(0) = 0$. Also note that (5) implies

$$\mathscr{G}(\varepsilon_1, 0, s) = \mathscr{G}(\varepsilon_1, 0, 0) = \sigma_I(\varepsilon_1). \qquad (7)$$

Observe that (6) is the strain history experienced by a point which, at the instant t, lies on the front of a shock wave propagating into an unstrained region. Thus the instantaneous elastic response function may be inferred from measurements of the stress jump across shocks in undeformed materials [1].

For any material with fading memory, let $\sigma_R(\varepsilon_1, t)$ denote the stress at time $t \geq 0$ for the *stress relaxation test*

$$\varepsilon(\tau) = \begin{cases} \varepsilon_1, & \text{if } \tau \geq 0; \\ 0, & \text{if } \tau < 0. \end{cases} \qquad (8)$$

Then σ_R is called the *stress relaxation function* (cf. [4]), and

$$\sigma_R(\varepsilon_1, 0) = \sigma_I(\varepsilon_1), \quad \sigma_R(\varepsilon_1, \infty) = \sigma_E(\varepsilon_1), \quad (9)$$

where σ_E is the *equilibrium elastic response function*. From (3)–(5) it follows that for $t \geq 0$,

$$\mathscr{G}(\varepsilon_1, \varepsilon_1, t) = \sigma_R(\varepsilon_1, t), \qquad (10)$$

and that $\mathscr{G}(\varepsilon_1, \varepsilon_2, t)$ is the stress at time $t > 0$ for the strain history

$$\varepsilon(\tau) = \begin{cases} \varepsilon_1, & \text{if } \tau = t; \\ \varepsilon_2, & \text{if } 0 \leq \tau < t; \\ 0, & \text{if } \tau < 0. \end{cases} \qquad (11)$$

The 1-D linear theory of viscoelasticity, namely

$$\sigma(t) = G(0)\,\varepsilon(t) + \int_0^\infty G'(s)\,\varepsilon(t-s)\,ds, \quad (12)$$

is the special case of (3) with $\mathscr{G}(\varepsilon_1, \varepsilon_2, s) = G(s)\varepsilon_2$. Here $G'(s) = dG/ds$; $G(0)$ and $G(\infty)$ are the instantaneous and equilibrium elastic moduli; $\sigma_I\big(\varepsilon(t)\big) = G(0)\varepsilon(t)$ is the instantaneous elastic response; and $\sigma_R(\varepsilon_1, t) = G(t)\varepsilon_1$. In this case G rather than σ_R is usually referred to as the stress relaxation function.

ACCELERATION WAVE SPEED

Let ρ_0 and ρ be the densities in the undeformed and deformed states. Let $U(t)$ denote the referential or Lagrangean acceleration wave speed (measured with respect to distance in the undeformed reference state). For a general class of materials with fading memory, Coleman et al. [2] showed that $\rho_0 U^2(t)$ is given by the derivative of the stress response functional with respect to the current strain $\varepsilon(t)$, holding

264

the past strain history fixed. When applied to the single integral law (3), this yields the formula

$$\rho_0 U^2(t) = \sigma_I{}'(\varepsilon(t))$$
$$+ \int_0^\infty \partial_1 \mathcal{G}'(\varepsilon(t), \varepsilon(t-s), s)\, ds, \qquad (13)$$

where $\sigma_I{}'(\varepsilon) = \frac{d}{d\varepsilon}\sigma_I(\varepsilon)$ and $\partial_1 \mathcal{G}'$ denotes the partial derivative of $\mathcal{G}'$ with respect to its first argument. The Eulerian wave speed (measured with respect to distance in the deformed state) is given by $\mathcal{U} = \lambda_1 U$. And since $\rho_0 = \lambda_1 \rho$, we also have $\rho \mathcal{U}^2 = \lambda_1 \cdot \rho_0 U^2$.

For the linear theory (12), we see that (13) reduces to $\rho_0 U^2 = G(0)$, and hence we recover the well-known result that the wave speed is a constant, independent of the deformation ahead of the wave. However, for the nonlinear theory, (13) implies that the acceleration wave speed $U(t)$ is generally a complicated function not only of the current strain $\varepsilon(t)$ but also of the past strain history at the material point instantaneously situated on the wavefront.

If the material ahead of the wave was initially undeformed and subjected to a step in uniaxial strain of amount ε_1 at time zero (i.e., the stress relaxation test (8)), then (13) simplifies to

$$\rho_0 U^2(t) = \partial_1 \mathcal{G}(\varepsilon_1, \varepsilon_1, t)$$
$$= \partial_1 \sigma_R(\varepsilon_1, t) - \partial_2 \mathcal{G}(\varepsilon_1, \varepsilon_1, t) \qquad (14)$$

for $t > 0$, where $(14)_2$ follows from (10).

More generally, for the strain history (11) ahead of the wave, the acceleration wave speed at the instant t is given by

$$\rho_0 U^2(t) = \partial_1 \mathcal{G}(\varepsilon_1, \varepsilon_2, t). \qquad (15)$$

Actually, this statement requires some qualification since ε is not continuous at the instant t. Let $t > 0$ be fixed, and consider the strain history

$$\varepsilon(\tau) = \begin{cases} \hat{\varepsilon}(\tau), & \text{if } \tau \geq t; \\ \varepsilon_2, & \text{if } 0 \leq \tau < t; \\ 0, & \text{if } \tau < 0; \end{cases} \qquad (16)$$

where $\hat{\varepsilon}(\tau)$ is any continuous function of τ such that $\hat{\varepsilon}(t) = \varepsilon_1$. Let $U(\tau)$ denote the acceleration wave speed at time $\tau > t$ for the strain history (16) ahead of the wave. Then on taking the limit as τ approaches t from above, we obtain (15). That is, (15) gives the acceleration wave speed in the state immediately following the second strain jump (at time t) for the strain history (11).

FINITE LINEAR VISCOELASTICITY

Now we consider single integral laws of the form

$$\sigma(t) = \sigma_I(\varepsilon(t))$$
$$+ \int_0^\infty G'(\varepsilon(t), s) \cdot f(\varepsilon(t-s))\, ds, \qquad (17)$$

where $G'(\varepsilon_1, s) = \frac{\partial}{\partial s} G(\varepsilon_1, s)$. This is the special case of (3) with

$$\mathcal{G}'(\varepsilon_1, \varepsilon_2, s) = G'(\varepsilon_1, s) \cdot f(\varepsilon_2). \qquad (18)$$

To satisfy $(5)_2$ we require that

$$\mathcal{G}(\varepsilon_1, \varepsilon_2, s) = f(\varepsilon_2)\left[G(\varepsilon_1, s) - G(\varepsilon_1, 0)\right]$$
$$+ \sigma_I(\varepsilon_1). \qquad (19)$$

Here f is interpreted as a strain measure, so that $f(0) = 0$ and $f'(0) = 1$, in which case $(5)_1$ is satisfied. Nunziato et al. [1] considered the special case of (17) with f given by

$$f(\varepsilon) = \varepsilon - \tfrac{1}{2}\varepsilon^2. \qquad (20)$$

This case arises from the 3-D single integral law

$$\mathbf{S}(t) = \mathscr{S}_I(\mathbf{E}(t)) + \int_0^\infty \mathbb{G}'(\mathbf{E}(t), s)\left[\mathbf{E}(t-s)\right] ds, \qquad (21)$$

where $\mathbf{E} = \frac{1}{2}(\mathbf{F}^T\mathbf{F} - \mathbf{I})$ is the Green strain tensor and $\mathbb{G}'(\mathbf{E}(t), s)$ is a fourth order tensor. This class of viscoelastic materials was termed *finite linear* by Coleman and Noll [5]. The quadratic term in (20) results from conversion to the nominal strain measure ε. Replacing $\mathbf{E}(t-s)$ by other finite measures of past strain results in a different class of materials and in particular a different choice for f in (17).

On setting $\varepsilon_2 = \varepsilon_1$ in (19) and using (10), we see that

$$G(\varepsilon_1, t) - G(\varepsilon_1, 0) = \frac{\sigma_R(\varepsilon_1, t) - \sigma_I(\varepsilon_1)}{f(\varepsilon_1)}. \qquad (22)$$

Now consider an acceleration wave in a material governed by the integral law (17), with the material ahead of the wave undergoing the stress relaxation test (8). Then from $(14)_2$, (19) and (22), the wave speed at time $t > 0$ is given by

$$\rho_0 U^2(t) = \partial_1 \sigma_R(\varepsilon_1, t)$$
$$+ \frac{f'(\varepsilon_1)}{f(\varepsilon_1)}\left[\sigma_I(\varepsilon_1) - \sigma_R(\varepsilon_1, t)\right]. \qquad (23)$$

Note that under the given assumptions, $\sigma_R(\varepsilon_1, t)$ is the stress at the wave front. On taking the limit as $t \to \infty$ and using $(9)_2$, we obtain the speed of a wave propagating into a region which has been in equilibrium for all time at strain ε_1 and stress $\sigma_E(\varepsilon_1)$:

$$\rho_0 U^2 = \frac{d\sigma_E}{d\varepsilon_1} + \frac{f'(\varepsilon_1)}{f(\varepsilon_1)} \left[\sigma_I(\varepsilon_1) - \sigma_E(\varepsilon_1) \right]. \quad (24)$$

For the special case where f is given by (20), this reduces to the formula (1) of Nunziato et al. [1, §21]. They used this to calculate the speed of expansive acceleration waves propagating into deformed regions in equilibrium, the deformation having been induced by the passage of a steady shock wave. The strain history resulting from the passage of a steady shock is only approximately given by (8), but due to fading memory this approximation leads to small errors.

PIPKIN-ROGERS MATERIALS

Next we consider single integral laws of the form

$$\sigma(t) = \sigma_I(\varepsilon(t)) \\ + h(\varepsilon(t)) \cdot \int_0^\infty G'(\varepsilon(t-s), s) \, ds, \quad (25)$$

where $G'(\varepsilon_2, s) = \frac{\partial}{\partial s} G(\varepsilon_2, s)$, $G(0, s) = 0$, and $h(0) = 1$. This is the special case of (3) with

$$\mathscr{G}'(\varepsilon_1, \varepsilon_2, s) = h(\varepsilon_1) \cdot G'(\varepsilon_2, s). \quad (26)$$

Condition $(5)_1$ is satisfied, and $(5)_2$ holds if we take

$$\mathscr{G}(\varepsilon_1, \varepsilon_2, s) = h(\varepsilon_1) \left[G(\varepsilon_2, s) - G(\varepsilon_2, 0) \right] \\ + \sigma_I(\varepsilon_1). \quad (27)$$

Pipkin and Rogers [4] considered the 3-D single integral law

$$\mathbf{S}(t) = \mathbf{G}(\mathbf{E}(t), 0) + \int_0^\infty \mathbf{G}'(\mathbf{E}(t-s), s) \, ds, \quad (28)$$

where $\mathbf{G}(0, s) = 0$, so that the first term on the right represents the instantaneous elastic response. For the 1-D case considered here, (28) reduces to (25) with $h(\varepsilon) = 1 - \varepsilon = \lambda_1$; this term is a consequence of conversion from the second to the first Piola-Kirchhoff stress measure (see (2)). Replacement of $\mathbf{S}$ in (28)

with other (Lagrangean) stress tensors would result in different functional forms for h in (25).

Now consider an acceleration wave in a material governed by the integral law (25), with no restrictions on the (uniaxial) strain history ahead of the wave. From (13) and (26), we see that $\rho_0 U^2(t)$ is given by $\sigma_I'(\varepsilon(t)) + h'(\varepsilon(t)) \int_0^\infty G'(\varepsilon(t-s), s) \, ds$. Then on solving (25) for this integral and substituting the result into the above expression, we obtain

$$\rho_0 U^2(t) = \sigma_I'(\varepsilon(t)) \\ - \frac{h'(\varepsilon(t))}{h(\varepsilon(t))} \left[\sigma_I(\varepsilon(t)) - \sigma(t) \right], \quad (29)$$

with $\varepsilon(t)$ and $\sigma(t)$ the strain and stress at the wave front. When $h(\varepsilon) = 1 - \varepsilon$, the term $h'(\varepsilon)/h(\varepsilon)$ reduces to $1/(1-\varepsilon) = 1/\lambda_1$. For a stress relaxation test (8) ahead of the wave and $t > 0$, (29) reduces to

$$\rho_0 U^2(t) = \frac{d\sigma_I}{d\varepsilon_1} - \frac{h'(\varepsilon_1)}{h(\varepsilon_1)} \left[\sigma_I(\varepsilon_1) - \sigma_R(\varepsilon_1, t) \right]. \quad (30)$$

Finally, we note that (29) and (30) remain valid if (25) is replaced by the more general relation

$$\sigma(t) = \sigma_I(\varepsilon(t)) + h(\varepsilon(t)) \cdot \underset{s>0}{\psi}(\varepsilon(t-s)), \quad (31)$$

given appropriate restrictions on the functional ψ. In particular, it is assumed that ψ does not depend on the current strain $\varepsilon(t)$. This includes the case where ψ is given by multiple hereditary integrals of the type considered by Green and Rivlin [6].

REFERENCES

1. Nunziato, J. W., Walsh, E. K., Schuler, K. W., and Barker, L. M., "Wave Propagation in Nonlinear Viscoelastic Solids," in *Handbuch der Physik*, edited by C. Truesdell, Springer, New York, 1974, vol. VIa/4.
2. Coleman, B. D., Gurtin, M. E., and Herrera, I., *Arch. Rational Mech. Anal.*, **19**, 1–19 (1965).
3. Asay, J. R., "Isentropic compression experiments on the Z accelerator," in *Shock Compression of Condensed Matter—1999*, edited by M. D. Furnish et al., AIP Conference Proceedings 505, Melville, New York, 2000, pp. 261–266.
4. Pipkin, A. C., and Rogers, T. G., *J. Mech. Phys. Solids*, **16**, 59–72 (1968).
5. Coleman, B. D., and Noll, W., *Rev. Modern Phys.*, **33**, 239–249 (1961).
6. Green, A. E., and Rivlin, R. S., *Arch. Rational Mech. Anal.*, **1**, 1–21 (1957).

CP706, *Shock Compression of Condensed Matter - 2003*
edited by M. D. Furnish, Y. M. Gupta, and J. W. Forbes
© 2004 American Institute of Physics 0-7354-0181-0/04/$22.00

NUMERICAL SIMULATION OF DIFFUSION OF ELECTRONS AND HOLES IN SHOCKED SILICON

Yuri Skryl[a)] and Maija M. Kuklja[b)]

[a)] *Institute of Mathematics and Computer Science, University of Latvia, Riga, LV-1459, Latvia*
[b)] *Division of Materials Research, National Science Foundation, Arlington VA, 22230, USA*

Abstract. The numerical method for simulation of diffusion of electrons and holes in the inertial field of a crystal under shock loading is developed. To analyze this diffusion, a complete system of electro-diffusion equations for charge carriers has been solved by means of the Poisson equation. Inertial forces were taken into account also by the drift-diffusion and convection-diffusion approximations. The equation system has been solved numerically by difference methods. Inertial currents in shocked Si are calculated for the shock wave with an amplitude of 1 GPa and different concentrations of doped impurity. It is shown that the shock wave traveling across the crystal creates an effective region of the electrical space charge that is moving along with the shock wave. The convection-diffusion model, describing inertial effects, is able of explaining large electric currents in shocked metals while the drift-diffusion model is not.

INTRODUCTION

Studies of electromotive forces (EMF) induced by shock waves in condensed matter have quite a long history because of both the fundamental interest for behavior of materials under extreme conditions and technological applications. It was first discovered that deformation of ionic crystals results in an electric potential between the deformed surfaces[1]. Later, an appearance of so-called impact EMF in linear and nonlinear dielectrics[2,3], semiconductors[4], and metals[5] under shock loading was observed.

Physics of impact EMF is complex and multifaceted. It is mostly developed for dielectrics[6]. Phenomenological models often used here assume polarization of a dielectric in the front of a plane shock wave[7,8,9]. Advantages and problems of this approach are discussed in great detail in the review of Mineev and Ivanov[6]. An explanation of the shock induced EMF is not that simple for semiconductors. Attempts to describe EMF with polarization

mechanisms alone fail. Many authors conclude that in line with polarization there are other mechanisms such as Frenkel pair type point defect formation and migration[4]. The situation is even more complex for metals because polarization models used for dielectrics do not work here at all. Therefore, other dynamic models were suggested, for example, diffusion of charge carriers from the wave front[10] and the increase of the number of carriers by the deformation of the lattice[11].

Inertial mechanisms of EMF induction are complex dynamic phenomena that are largely neglected[6]. Only a few reports are available in the literature[6]. Some estimates for the current induced by inertial jumps of carriers in the wave front have been obtained[10,12]. However, they were not satisfactory for interpretation of large currents observed experimentally[5,11].

Although experiments on the electro-conductivity of condensed matter under high compressions have being performed for a long time, there are some difficulties in interpretation of the measurements[13]. From a theoretical point of view, there is a lack of computational and theoretical tools to describe in detail inertial currents created by shock waves[14].

In this work we study electric currents generated in crystals by inertial forces of the shock wave. The method for calculation of inertial diffusion of electrons and holes, which allows obtaining magnitudes of currents and voltages in shocked solids, is developed.

The general methodology developed for semiconductor devices[15] has been applied in this study for the simulation of conductivity induced by inertia. Many efficient numerical schemes are well known in this field, so some success was expected. It turned out however, that the problem of calculating inertial currents in a shock wave is very complicated and the solution is not always possible. The methods work fairly well only in the regions of modest concentrations of charge carriers. Therefore, they are not quite applicable to metals. In the meantime, the most interesting problem is the determination of currents in metallic samples because there is no clear picture in interpretation of a large EMF in shocked metals.

First, the numerical method developed was applied to silicon crystal with a low concentration of doped impurity. Then, an extrapolation of the results obtained was made to the metallic samples. The conclusion that the inertia may contribute significantly to the shock induced EMF in metals is hardly surprising.

METHOD OF CALCULATIONS

The methodology of calculation of inertial currents in doped silicon is based on numerical methods for semiconductors[16,17]. We will briefly show here only the main steps.

Let us consider the basic equations for the motion of charge carriers in semiconductors. The complete system of electro-diffusion equations can be written as follows:

$$\mathbf{J}_n = -D_n\left(\nabla n - \frac{q}{k_B T} n \nabla(\varphi + \varphi_T)\right) \quad (1)$$

$$\nabla \cdot \mathbf{J}_n = -\frac{\partial n}{\partial t} \quad (2)$$

$$\nabla(\varepsilon\nabla\varphi) = -\frac{q}{\varepsilon_0}(N_d - n) \quad (3)$$

$$q\nabla\varphi_T = -m\frac{\partial v}{\partial t} \quad (4)$$

where n, m, and D_n are the concentration, the effective mass, and the diffusion coefficient of electrons, respectively. N_d is the concentration of ionized donor impurity, v is the mass velocity, φ and φ_T are the electric and effective potentials, respectively, ε and ε_0 are dielectric constants, k_B is Boltzmann's constant, q is the electronic charge, and T is the temperature.

A similar system of equations (1)-(4) can be also written for holes. However, in our case it does not make any difference whether n- or p-type conductivity is considered. Therefore, we restrict ourselves to n-type. Equation (1) for the electron flow includes both diffusion and drift components. The drift flow is generated by the inner field $E = \nabla\varphi$ which can be found from Poisson's equation (3). The continuity equation (2) defines the relationship between currents and the change of electron concentration in time.

The effective potential φ_T, Eq. (4), is introduced here to take into account the inertial force applied to the electrons. This is essentially a common drift-diffusion approximation that works fairly well for large concentrations of scattering centers. In a similar manner, an inertial electrical current induced in a rapidly slowing down copper coil was calculated in the classical experiments of Tolman and Steward[18].

It should be noted here that, if the wave front is as narrow as an electron free mean path, the drift-diffusion approximation is not valid. In this case, charge carriers move freely across the wave front with acceleration as free electron gas. Then, Eq. (1) for the particle flow can be written as:

$$\mathbf{J}_n = -D_n\left(\nabla n - \frac{q}{k_B T} n \nabla\varphi\right) - nv. \quad (5)$$

Eq. (5) describes convection motion of electrons with no scattering. The convection-diffusion model is good in the initial stages when the wave enters the crystal. Thus, this crude approximation is useful because it permits the current to be estimated.

The boundary conditions are:
$$n(0,t) = n(L,t) = N_d \quad (6)$$

$$\varphi(0,t) = \varphi(L,t) = 0 \qquad (7)$$

where L is the length of the silicon sample. We consider the response of the silicon crystal uniformly doped with a donor impurity to the shock wave loading so the initial concentration of electrons and donor impurities was chosen as a constant.

A numerical scheme for solution of the system of equations (1)-(4) is described in detail elsewhere[19,20]. The structure of the shock wave was modeled using an analytical approach based on an elastic approximation[21]. Parameters of the model were corrected later to obtain a shock front width of the same value as is obtained by ab initio calculations of shock waves in solid silicon[22]. A shock wave amplitude of 1 GPa was chosen for the calculations, producing a corresponding moderate heating due to impact.

NUMERICAL RESULTS AND DISCUSSION

Electrical characteristics of n-type silicon for several different concentrations of the donor impurity are shown in Fig. 1. The calculated potential and electric current functions are plotted by solid lines at different moments in time. The shock wave travels across the crystal from left to right and it is shown by the dot-dashed line.

Time 0 corresponds to the moment when the wave has travelled a distance equal to two wave front widths from the crystal surface. It is seen from the figure that the shock wave progressing throughout the crystal within the front width, creates a space charge region (the dotted line), moving along with the shock wave. This space charge changes its sign from minus to plus while moving across the crystal. This happens because of the inertial delay of electrons under dynamic compression.

Thus, the wave front forms a double electrical layer, the existence of which was a priori assumed in phenomenological models discussed in the literature (see, for example[6]). In this work, formation of the double electric layer naturally follows from the simulation.

The potential induced by the space charge is spread out over the whole sample both in front of the wave and behind it as well. Thus, a nonzero electric current appears that has the same magnitude

at all times while the wave travels in the crystal. This current relaxes back to zero when the wave emerges from the crystal. This kind of behavior agrees well with experiments on shocked metals[6]. For semiconductors, the time dependence of the current is more complex because the polarization is induced not only by a redistribution of free carriers but also by the lattice deformation and defect interactions[4].

The results obtained lead us to the following interpretation. The shock wave front behaves like a moving voltage generator, which generates a non-zero electric current over the crystal structure. As is seen from Fig. 1, the currents are relatively small, 6.54×10^{-2} μa/cm^2 at $N_d = 10^{14}$ cm^{-3} and 6.54×10^{-1} μa/cm^2 at $N_d = 10^{15}$ cm^{-3}, probably because of the low conductivity of the silicon sample. The magnitude of the potential is not dependent on the impurity concentration and varies within the range from -0.1 to +0.1 μV.

An important observation here is that the inertial current increases by an order of magnitude with increase of the impurity concentration by an order of magnitude. Extending this to metals where the concentration of free electrons is 10^{23}/mole, yields a current of 6.54 A/cm^2. This is a relatively significant current yet insufficient to explain the large EMFs observed in experiments on shocked metals. It has to be concluded here that the diffusion approximation in Eqs (1)-(4) predicts small currents.

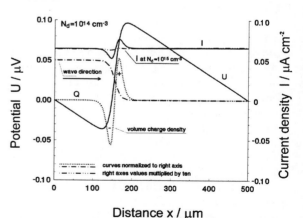

FIGURE 1. Electrical characteristics of n-type silicon for several different concentrations of the donor impurity. The shock wave and the charge density curves are scaled to show them in the same figure.

Once the amplitude of the shock wave is increased or plastic flow is allowed in the model, the width of the wave front decreases significantly. Hence, the diffusion approximation, Eqs. (1)-(4), does not work because carriers move through the wave front without scattering. This will result in an electrical current increase.

Estimates based on a convection approximation, Eq. (5), yield a 3-4 orders of magnitude increase in the current compared to the diffusion approximation. This estimate is in agreement with experimental values[6].

CONCLUSION

A computational scheme for calculation of inertial currents in shocked solids is developed. The analysis is first tested on semiconductors and then extended to metals. Numerical modelling of electron and hole diffusion due to inertial forces in shocked silicon crystals is performed by means of electro-diffusion equations. It is shown that the shock wave forms a region of significant charge in the crystal, which generates an electric current in the sample. This current relaxes back to zero when the shock wave leaves the crystal. The behavior observed is in qualitative agreement with experiments on shocked metals. The small magnitude of the currents is explained by the low conductivity of silicon.

Extension of the results obtained for silicon to metals also yields small currents, which shows the disadvantage of the drift-diffusion model. A convection-diffusion approximation applied here provides an increase of electric current of 3-4 orders of magnitude, which is, in fact, comparable with the results of experiments on shocked metals.

The essential conclusion here is that the convection-diffusion model is able to explain the EMF produced in metals under shock loading while the classic druft-diffusion model is insufficient.

The numerical method developed here is rigorous yet general and may be easily improved to incorporate more details of materials physics and chemistry. We believe the method is applicable for an accurate interpretation of experimental data on EMF in shocked materials.

ACKNOWLEDGEMENT

M.K. is grateful to the DMR of the US National Science Foundation for the support under the Individual Research and Development Program. Discussions with Frank Zerilli are greatly appreciated.

REFERENCES

[1] Stepanow A.W., *Zs. Phys.* **81**, 560 (1933).
[2] Linde R.K., Murri W.J., Doran D.G., *J. Appl. Phys.* **37**, 2527 (1966).
[3] Reynolds C.E., Seay G.E., *J. Appl. Phys.* **33**, 2234 (1962).
[4] Mineev V.N., Inanov A.G., Lisicyn Yu.V., Novickij E.Z., Tunjaev Yu.N., *JETP* **59**, 1091 (1970).
[5] Kanel G.I., Dremin A.N., *Doklady AN* **211**, 1314 (1973).
[6] Mineev V.N., Inanov A.G., *UspehiFN* **119**, 75 (1976).
[7] Allison F.E., *J. Appl. Phys.* **36**, 2111 (1965).
[8] Zel'dovich Ya. B., *JETP* **53**, 237 (1967).
[9] Inanov A.G., Lisicyn Yu.V., Novickij E.Z., *JETP* **54**, 285 (1968).
[10] Nesterenko V.F, *Fizika Gorenija I Vzryva* **10**, 752 (1974).
[11] Migault A., Jacguesson J., C.R.Ac. Sci. **B264,** 507 (1967).
[12] Mineev V.N., Inanov A.G., Tunjaev Yu.N. in *Combustion and Explosion* (in Russian), p. 597, Moscow, Nauka (1972).
[13] Keller R.N., *Physics of High Energy Density*, ed. by P.Caldirola and H.Knoepfel, Academic Press, New York and London (1971).
[14] Kanel G.I., Razorenov S.V., Utkin A.V., Fortov V.E., *Shock-Wave Phenomena in Condensed Matter*, (in Russian) (Yanusz -K, Moscow, 1996) pp. 407.
[15] Mock M., *Analysis of Mathematical Models of Semiconductor Devices*, Boole Press, Dublin (1983)
[16] Polsky B., *Numerical Modelling of Semiconductor Devices* (in Russian), Riga, Zinatne (1986).
[17] Polsky B., Rimshans J., Solid-St. Electron. 24, 1081 (1981).
[18] Tolman R.C, Steward T.D., Phys. Rev. **8**, 97 (1916).
[19] Martuzans B., Skryl Yu., J. Chem. Soc., Faraday Trans. **94**, 2411, (1998).
[20] Skryl Yu., Phys. Chem. Chem. Phys. **2**, 2969 (2000).
[21] Martuzans B., Skryl Yu., Kuklja M.M., Latv. J. Phys. Tech. Sci., **3**, 40 (2002).
[22] Swift D.S., Ackland G.J., Hauer A., Kyrala G.A., Phys. Rev. **64,** 214107 (2001)

CP706, *Shock Compression of Condensed Matter - 2003*
edited by M. D. Furnish, Y. M. Gupta, and J. W. Forbes
© 2004 American Institute of Physics 0-7354-0181-0/04/$22.00

A MODULAR MATERIAL MODELING ARCHITECTURE FOR NON-LINEAR DYNAMICS

O. Vorobiev, M. Cowler and N. Birnbaum

Century Dynamics Incorporated,
Concord, CA 94520, USA

Abstract. An architecture for material modeling in non-linear dynamic analysis programs is described that allows complex material models to be developed and rigorously examined independently of the codes or solvers in which they are to be used. Object-oriented architecture permits the package to be used independent of any specific code structure, allowing a separate driver program to apply a variety of load conditions, thereby offering a rigorous examination of model response prior to application in large simulations. The driver also allows model coefficients to be optimized (fitted) automatically to experimental data. Owing to its modular structure, the package can easily be interface to a variety of codes / solvers.

INTRODUCTION

Material modeling is an important part of the computer codes designed for fluid dynamics and structural mechanics simulations. Coefficients for material models are normally determined from the experiments where simple loading conditions are applied to the material, such as uniaxial stress or strain loading, shock compression and isentropic expansion. There is no need in sophisticated computer codes to interpret such experiments it is sufficient to use analytical methods or simple 1D solutions. Yet these material models are often applied in 2D/3D computer codes to model much more complex loading paths without examining the models. Therefore they may not produce results that are in good agreement with the measurements. If this is the case one needs to modify the original model so that it produces results that match experimentally tested loading conditions and then try it again in complex problems. That is why it is very convenient to have a tool at hand that can

analyze and compare material models for various loading conditions. We will later refer to this tool as Material Drive (MD). Since we need to modify material models many times, it makes sense to separate all material model routines from the rest of the code. The vector with the material state variables (state vector) and loading conditions is sent to material modeling routines anytime the state of the material needs to be updated (see Fig.1). When the update is done the same vector is sent back to the solver and material variables for the current element are saved in the data structure, which may have a solver specific architecture.

The MD accesses the modeling material response routines using the same state vector, so regardless if the current state has come from one of the solvers or from the Material Driver it is processed the same way.

This scheme has been implemented in AUTODYN [1], a non-linear dynamic analysis program for a number of solvers (Lagrange, ALE, Euler, Shell).

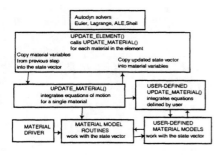

Figure 1. Scheme of communications between material modeling package and the solvers

MODELING ARCHITECTURE

There may be one (for Lagrange and Shell solver) or several (for Euler, ALE solvers) materials within one finite element. For each material in the element there is a vector of variables stored. Depending on the material model settings different sets of variables are activated for the material. These sets of variables are subset from the list of all possible variables.

In the case of multiple materials loading conditions known for the whole element need to be partitioned among the participating materials. Bulk modulus weighted velocity gradient partitioning [2] has proven to be a very robust method. It accounts for the difference in impedance between solids and gases within a single finite element.

More complex, "plain stress", loading conditions are used for SHELL solvers. Two principle strain rates and one fixed stress (zero stress normal to the shell plane) are sent to the material model routines. In the 2D case, when one of the strains is zero, the stress increments for elastic-plastic shell can be calculated analytically [3], whereas in 3D iterations are required.

MATERIAL DRIVER AND PARAMETRIC STUDY

Material Driver can be applied anytime to examine materials used in the problem run by AUTODYN. It includes uniaxial and triaxial strain loadings, hydrostatic compression and expansion, Hugoniot loading, plain stress loading and uniaxial

stress loading. Stress loading requires iterations to find the appropriate strain load that forces stresses to evolve in certain way. An alternative method to test material behavior in uniaxial stress loading would be a single cell model with the appropriate stress boundary conditions. But in this case the result will also depend on the cell size and numerical method used for that cell. Using MD offers superior way to test material models. Besides, some of the loadings, for example, shock Hugoniot, cannot be modeled with a single cell method.

To illustrate new approach let us, for example, consider material models for concrete. Figure 2 shows compressive strength calculated with MD at three different strain rates for two widely used material models for concrete, RHT [4] and HJC[5], and a new Generic material model [6], which has recently been added to the library of AUTODYN's material models. The Generic model allows a user to switch on and off different factors such as hardening, bulking, porous compaction, third invariant dependence etc. and choose various functions of forms for these factors. The function of form for the pressure hardening for the Generic model was the same as in the RHT and HJC models. Parameters were fitted to match the RHT model. MD is equipped with optimization routines that allow fitting parameters in the model so that the history profile will match data obtained from experiments or other models.

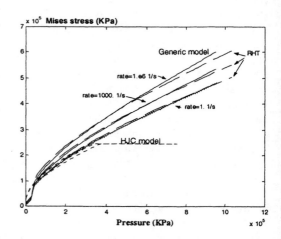

Figure 2. Rate dependence of the compressive strength of concrete for different models

The Generic strength model can combine advantages of HJC and RHT models. For example, it accounts for material damage due to porous compaction (or dilation) similar to HJC and includes dependence on the loading path (the third stress invariant) similar to RHT. In addition it can model effects of bulking and shear enhanced compaction as well as rate dependent porous compaction and dilation.

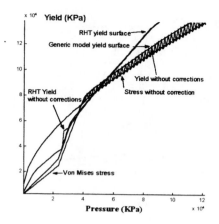

Figure 4. Yield strength and Von Mises stress calculated for RHT and the Generic model in uniaxial loading with and without plastic strain rate correction.

Material driver enables us also to detect multiple defects in a material model that otherwise may go unnoticed in numerical simulations. For example, Figure 4 shows, that material models with explicit plastic strain rate dependence may exhibit unstable behavior that can be fixed using strain rate corrections [7].

We have set up a simple problem with concrete to see how the models examined before will perform. The test problem we have chosen is an explosion of 41 Kg of TNT at 1.6 m thick concrete wall. 2D Lagrange solver was used with two grids to model the wall and the charge. For the Generic model the strength of the damaged material is a fraction $F(\Omega)$ of its virgin strength if the pressure is low. As pressure increases the strength of damaged material approaches to the strength of the virgin material as

$$Y = Y(P, \theta, \varepsilon_p)\left(1 + (F(\Omega) - 1)e^{-\left(\frac{P}{P_r}\right)}\right), \quad (1)$$

Where, P_r is the reheal pressure. Results of 2D calculations for different P_r values are shown in Fig.5. When P_r is big effect of damage is significant and the result looks close to one without strength.

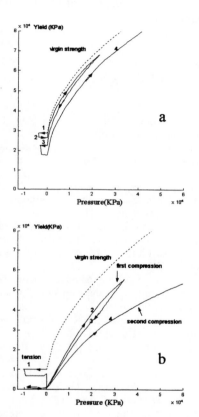

Figure 3. (a,b) Strength degradation in expansion-compression loading cycle: **a**- is HJC model, **b** - is the Generic strength model. Deformation strain rate is 1000 1/s.

Figure 3 shows how concrete strength evolves in a cyclic loading. The amount of strength degradation can be controlled by the measure of porosity damage (volumetric plastic strain in [5]) and the distortional damage related to the plastic strain.

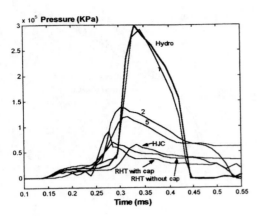

Figure 5. The pressure profiles calculated in the middle of the wall for different concrete models. The numbered curves correspond to the values of the reheal pressure shown in Fig.6.

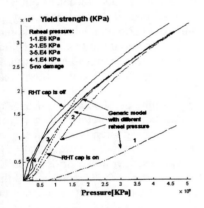

Figure 6. Yield strength for the Generic model calculated by MD for the different values of the reheal pressure parameter. Results for the RHT model calculated with (dashed line) and without cap effect are also shown.

CONCLUSIONS

A new architecture for material modeling has been developed for AUTODYN. It allows user to quickly and conveniently test material models before running large complex problems where these models are to be used. The same material model routines are accessed by the Material Driver and AUTODYN solvers. This approach removes possible differences in material model implementation for the solvers. Thus differences in simulation results from various solvers can be attributed only to the numerical methods used for these solvers. It is expected that the new architecture will facilitate material model development for AUTODYN and make simulation results obtained with the different solvers more consistent with each other. The package of material model routines can also be conveniently maintained and tested independent of AUTODYN solver development.

REFERENCES

1. Century Dynamics, Inc., "AUTODYN Users Manual", Version 4.3, 2003
2. Vorobiev, O.Yu., Lomov, I.N., "Numerical Simulation of gas-solid interfaces with large deformations", in *Advances in Computational Engineering & Sciences*, edited by S.N.Atluri, F.W.Brust, Tech Science Press, Palmdale, CA, USA,Vol.1, 2000, pp.922-927
3. Cowler, M. S., "The Numerical Solution of the Response of Elastic-Plastic Strain Rate Sensitive Shells of Revolution" UKAEA, TRG Report 6294(R/X), 1973
4. Riedel, W., "Beton unter dynamishen Lasten Meso- und makromechanishe Modelle und ihre Parameter" EMI Report, 6/00, July, 2000
5. Holmquist, T. J., Johnson, G.R. and Cook, W. H., "A Computational Constitutive Model for Concrete subjected to Large Strains, High Strain Rates, and High Pressure", in *Proc. 14th International Symposium on Ballistics*, Quebec, pp.591-600, 1993
6. Rubin, M.B., Vorobiev, O.Yu., Glenn,L.A., *Int.J.Solids and Struc.*, **37**, 1841-1871(2000)
7. Vorobiev, O., Yu., "Improved numerical integration of elastic-viscoplastic models with hardening and rate dependence in AUTODYN", in *Structures under shock and impact VII*, edited by N.Jones, C.A. Brebbia, A.M. Rajendran, WIT Press, Southampton, Boston,2002, pp.455-466

CP706, *Shock Compression of Condensed Matter - 2003*
edited by M. D. Furnish, Y. M. Gupta, and J. W. Forbes
© 2004 American Institute of Physics 0-7354-0181-0/04/$22.00

FIRST-PRINCIPLES VIBRATIONAL STUDIES OF PENTAERYTHRITOL

S. Vutukuri[1], W.F. Perger[1], Z.A. Dreger[2], and Y.M. Gupta[2]

[1]Department of Physics, Michigan Technological University, Houghton, MI-49931
[2]Institute for Shock Physics, Washington State University, Pullman, WA-99164

Abstract. The first-principles periodic Hartree-Fock calculations of the unit cell and vibrational frequencies of pentaerythritol crystal are presented. The basis set 6-31G was used to calculate the vibrational frequencies. The calculated OH stretching mode vibrational frequencies using the Crystal program are about 200 wave numbers lower than those calculated using the Gaussian98 program. This is to be expected for crystals of this type having an intermolecular hydrogen bonding interaction. The pressure effect on vibrational frequencies is calculated and compared with experimental data, up to 3 GPa.

INTRODUCTION

Pentaerythritol, $C(CH_2OH)_4$, (PE) is a molecular solid with an attractive combination of hydrogen bonding and van der Waals interactions. Because of its relatively simple structure, it is considered a model compound for studying hydrogen-

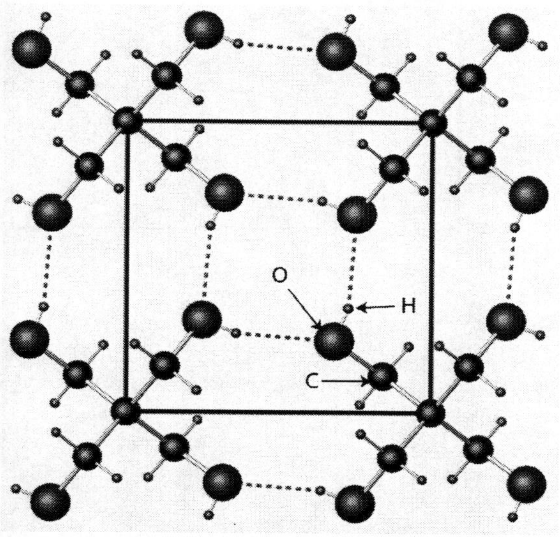

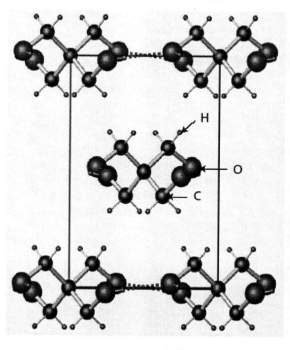

FIGURE 1a Crystal Structure of PE in ab plane. The locations of H, C, and O atoms are shown by arrows and hydrogen bonding by dashed lines.

FIGURE 1b Crystal Structure of PE in ac plane

bonded molecular crystals. The study of vibrational spectra combined with high pressure provides a tool to understand the effects of intermolecular interactions on the crystal stability. At ambient pressure and room temperature, PE crystallizes in the body-centred tetragonal structure [1, 2]. The primitive cell contains only one molecule and the molecule possesses S_4 symmetry. The molecules are arranged in parallel *ab* planes and hold by hydrogen bonds. Then the layers are bonded by van der Waals interactions (Fig. 1a & 1b).

COMPUTATIONAL METHODS

All-electron Hartree-Fock (HF) calculations were performed in the framework of the periodic linear combination of atomic orbitals (LCAO) approximation. In the LCAO-HF approximation, the localized atomic orbitals are constructed from a linear combination of normalized Gaussian type functions from which Bloch functions are constructed by a further linear combination with plane-wave phase factors. All these features are implemented in the CRYSTAL program [3].

Vibrational frequencies are obtained by diagonalization of the mass-weighted Hessian matrix, which is found by making a series of small displacements of atoms and calculating forces acting on the others.

Hydrostatic pressure calculations on tetragonal systems (with two independent variables *a, c* describing the lattice parameters) involve the optimization of the *c/a* ratio and internal coordinates. For a given unit-cell volume, the *c/a* ratio is varied until the energy is minimized. At that *c/a* ratio, the internal coordinates of the atoms are relaxed until a new minimum energy is found. This procedure is then repeated for different unit-cell volumes, resulting in a curve of energy vs. volume. Fitting the E(V) curve to a standard equation of state gives the pressure-volume relation.

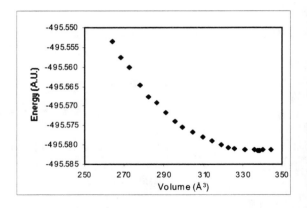

FIGURE 2 Energy change with unit-cell volume

RESULTS AND DISCUSSION

The potential energy surface (Fig. 2) of PE is found and fitted to the Murnaghan equation of state to obtain the pressure corresponding to different volumes. We applied the hydrostatic pressure up to 4 GPa and found the variation of unit cell parameters with respect to pressure (Fig. 3), using the procedure described above.

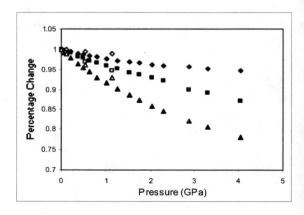

FIGURE 3 Unit-cell changes with pressure.
Calculated   : a/a_o (♦), c/c_o (■), v/v_o (▲);
Experimental [4] : a/a_o (◊), c/c_o (□), v/v_o (Δ).

Since the van der Waals interaction between the molecular layers (along c-axis) is weaker than that of the hydrogen-bonding within the layers (along a-axis), the variation of the unit-cell dimension along the c-axis is greater compared to the unit-cell dimension along the a-axis. The variation of volume (V/V_0) with respect to the pressure is also shown in the Fig. 3.

To facilitate a comparison of our calculations with experiment, we calculated the pressure at the unit-cell volume 299.3 $\mathring{A}^3$. The experimental pressure at that volume has been measured to be 1.15 GPa [4] and our calculated pressure is 1.5 GPa, which is in reasonable agreement with the experiment. When we used the rigid-molecule approximation (no relaxation of atoms), we got 12 GPa pressure at the same volume, which shows that the relaxation of atoms is necessary to obtain the correct pressure.

Since the PE molecule contains 21 atoms, there are 57 (3n-6) fundamental frequencies and, because of symmetry, many frequencies are degenerate. Using group-theoretical methods we found the irreducible representation as 14A+15B+14E. We calculated the vibrational frequencies using the 6-31G basis set and Hartree-Fock approximation (Fig. 4).

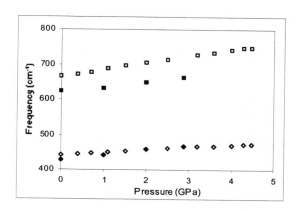

FIGURE 5. Variation of frequencies with pressure. OH torsion : experimental (□), calculated (■); CCC bend : experimental (◊), calculated (♦).

We scaled these frequencies with the appropriate scale factor [5] and compared with experiment [6, 7]. Except the range 500 - 900 cm^{-1} (OH torsion and CCC bend modes), all the modes are in reasonable agreement with experimental values. A shift in frequencies from gas phase to crystalline phase was observed because of intermolecular interaction [8]. For the gas phase molecular vibrational calculations, we used the

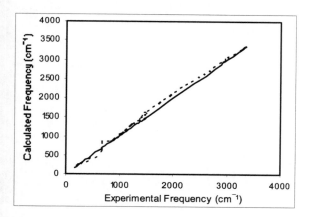

FIGURE 4. Comparison of calculated vibrational frequencies (dashed) using 6-31G basis set and Hartree-Fock approximation with experiment (solid) [6, 7].

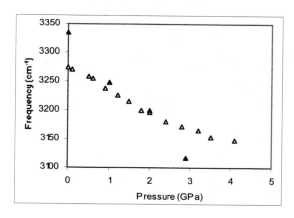

FIGURE 6. Variation of OH stretching mode under pressure: calculated (▲), experimental (△).

Gaussian98 (HF/6-31G) suite program. Particularly, for OH stretching modes, we observed a decrease of around 200 wave numbers since these are branch modes.

The effect of pressure on the vibrational frequencies is calculated and compared with experimental data [9]. Under pressure, all modes shifted gradually toward higher frequencies except the OH stretching modes. As an example of this general trend, Fig. 5 shows the CCC bend and OH torsion modes calculated using 6-31G basis set and compared with experiment.

Regarding the OH stretching, Fig. 6 shows the decrease in calculated vibrational frequency with increasing pressure, which agrees with experiment [9]. This increase is caused by an increase in the O-H...O intermolecular hydrogen bond strength under pressure and thus by increase of the O-H bond length. Note that the recent theoretical work of Ramamoorthy, et.al [10] showed an increase in frequency for increasing pressure for this OH mode, in contrast with our calculation and experiment.

CONCLUSIONS

The vibrational frequencies of Pentaerythritol were obtained using first-principle calculations. We were able to apply hydrostatic pressure theoretically and found that the relaxation of the atoms (internal coordinates) is necessary to get the correct hydrostatic pressure. We calculated the variation of frequencies with pressure. The results are in good agreement with the experiment.

ACKNOWLEDGEMENTS

The authors gratefully acknowledge the financial support of the Office of Naval Research and the MURI program.

REFERENCES

1. Eilerman, D., and Rudman, R., *Acta Cryst. B* **35**, 2458 (1979).
2. Ladd, M. F. C., *Acta Cryst. B* **35**, 2375 (1979).
3. Dovesi, R., Roetti, C., and Saunders, V. R., 1998 CRYSTAL Program
4. Katrusiak, A., *Acta Cryst. B* **51**, 873 (1995).
5. Scott, A. P., and Radom, L., *J. Phy. Chem.* **100**, 16502 (1996).
6. McLachlan, R. D., and Carter, V. B., *Spectrochim. Acta A* **27**, 853 (1971).
7. Marzocchi, M. P., and Castellucci, E., *J. Mol. Struct.* **9**, 129 (1971).
8. Decius, J. C., and Hexter, R. M., *Molecular Vibrations in Crystals*, McGraw-Hill Inc. N.Y., 1977.
9. Park, T-R., Dreger, Z. A., and Gupta, Y. M. submitted.
10. Ramamoorthy, P., Rajaram, R. K., and Krishnamurthy, N. *Cryst. Res. Technol.* **36**, 169 (2001).

CHAPTER VI

MOLECULAR DYNAMICS MODELING: NONREACTIVE MATERIALS

CP706, *Shock Compression of Condensed Matter - 2003*
edited by M. D. Furnish, Y. M. Gupta, and J. W. Forbes
© 2004 American Institute of Physics 0-7354-0181-0/04/$22.00

SHOCK HUGONIOT AND MELT CURVE FOR A MODIFIED EMBEDDED ATOM METHOD MODEL OF GALLIUM

F. J. Cherne*, M. I. Baskes*, T. C. Germann*, R. J. Ravelo*† and K. Kadau*

*Los Alamos National Laboratory, Los Alamos, NM 87545
†Department of Physics, University of Texas El Paso, El Paso, TX 79968

Abstract. Molecular dynamics (MD) simulations have been performed on the complex material gallium using a literature modified embedded atom method (MEAM) potentials which reproduces the unusual behavior of this element. Both liquid and solid properties will be examined using the equilibrium MD "Hugoniostat" method and molecular statics. The calculated pressure dependence of the melt curve is found to agree well with experiment. The calculated Hugoniot is in reasonable agreement with the experimental Hugoniot EOS even better agreement can be obtained through a slight modification of the MEAM parameters.

INTRODUCTION

Gallium is a complex element. It exhibits a negative Clapeyron slope between the low pressure solid phase (A11, oC8, GaI) and liquid (see Fig. 1). The equilibrium phase at low pressure is orthorhombic showing a significant amount of dimerization. The dimerization has been used to describe the existance of a slight shoulder in the structure factor for liquid Ga[1, 2]. The phase diagram shows three additional solid phases at high pressure: face centered tetragonal (fct) (cI12, GaII), A6 (tF4, tI2, GaIII), and fcc (GaIV, not shown in Fig. 1)[3]. Gallium also melts slightly above room temperature at 303 K.

Recently, a modified embedded atom method (MEAM) potential for gallium was developed[5]. Besides gallium, MEAM has been used to describe a variety of other complex materials, namely, tin[6] and plutonium[7]. In the gallium paper[5], the authors examined a variety of static and transport properties of the MEAM gallium potential. Here we will focus on the effect of pressure upon the melt curve as well as the shock behavior of liquid gallium. Using the equilibrium MD Hugoniostat technique [8], we calculate the liquid shock Hugoniot, which can be directly compared with recently available experimental data[9].

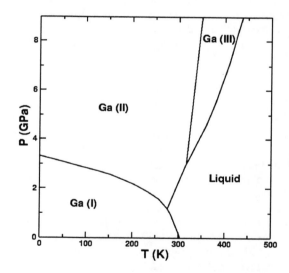

FIGURE 1. Phase diagram of gallium[4].

MODEL

For a system containing one type of atom using the embedded atom method (EAM) formalism[10, 11, 12] the total energy is given by an expression of the form,

$$E = \sum_i \left(F(\overline{\rho}_i) + \frac{1}{2} \sum_{j \neq i} \phi(R_{ij}) \right), \qquad (1)$$

where the indices i and j denote the atoms. The embedding function F is the energy to embed an atom into the background electron density, $\overline{\rho}_i$ at site i; and ϕ is a pairwise interaction between atoms i and j whose separation is given by R_{ij}. For EAM, $\overline{\rho}_i$ is given by a linear superposition of spherically averaged atomic electron densities, however in MEAM, $\overline{\rho}_i$ has an angular dependence[13, 14]. The embedding function F is equal to $A_i E_c \overline{\rho} \ln(\overline{\rho})$ where A is a constant about unity, and E_c is the cohesive energy.

By using a reference equation of state (EOS) $E^u(R)$, where R is the nearest neighbor distance, the pair potential $\phi(R)$ between 2 atoms is given by

$$\phi(R) = \frac{2}{Z} \{ E^u(R) - F(\overline{\rho}^0(R)) \}, \qquad (2)$$

where $\overline{\rho}^0(R)$ is the background electron density at an atom in what is termed the reference structure, and Z is the number of first neighbors in this structure. The reference EOS is taken to be the universal EOS of Rose et al. [15],

$$E^u(R) = -E_c \left(1 + a^* + \frac{r_e}{R} \delta a^{*3} \right) e^{-a^*} \qquad (3)$$

with

$$a^* = \alpha \left(\frac{R}{r_e} - 1 \right), \qquad (4)$$

and $\alpha^2 = 9\Omega B / E_c$, where r_e, Ω, and B are the nearest neighbor distance, atomic volume, and bulk modulus, respectively, all evaluated at equilibrium in the reference structure. The cubic anharmicity parameter δ has been added to better represent the pressure derivative of the bulk modulus and will be discussed further below. The specific parameters for Ga are given in Baskes et al. [5].

RESULTS AND DISCUSSION

In Fig. 2, we show the calculated energy as a function of volume for a variety of structures: A11 (GaI), cI12 (GaII), bcc, sc, and fcc. For these calculations, we calculated the energy of the structure using a three dimensional periodic cell with approximately

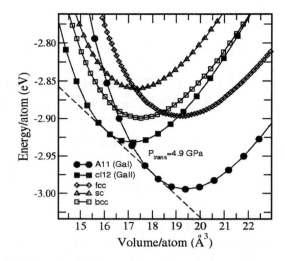

FIGURE 2. Energy diagram of various phases calculated with the MEAM gallium potential at 0 K.

250 atoms expanding the volume isotropically. Figure 2 shows that at zero pressure the most stable phase as predicted by the potential is the A11 phase which corresponds to the equilibrium phase observed experimentally. In addition to this we predict the A11 to cI12 transition pressure (triangular construction, $-dE/dV$) at 0K to be 4.9 GPa. We compare this pressure with approximately 3.2 GPa from the experimental phase diagram shown in Fig. 1[4] and the 4.0 GPa determined from Fig. 5 of reference [3]. We see that the pressure for the transition is predicted to be a little high yet very close to the experimental values. From this we are encouraged that the potential could potentially capture semi-quantitatively the phase transition behavior of gallium.

In addition to the transition pressure calculated above, we performed a series of molecular dynamics calculations to obtain the melting point at pressure. The method we used is the so called moving interface method. This method consists of a three step iterative process. The first step is to equilibrate a solid at isothermal-isobaric conditions maintaining crystal symmetries at the temperature one assumes the melting temperature to be. This step provides the necessary lattice parameters that can then be used in step two. Step two uses a periodic planar liquid-solid-liquid geometry. The simulation is performed maintaining a constant in-plane lattice spacing (plane

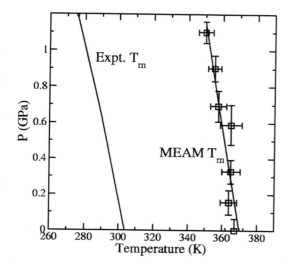

FIGURE 3. Low pressure melting temperature of MEAM gallium (A11) compared with the experimental melt curve. The line is a linear fit to the MEAM gallium calculations.

TABLE 1. Properties for A11 to liquid transition in gallium. Quantities presented are the heat of fusion ΔH, the relative volume change on melting $\Delta V/V$, the melting point (T_m) at zero pressure, and the pressure derivative of the melting point (dT_m/dP). The heat of fusion and volume change were calculated at 375 K.

	MEAM	**Experiment***
ΔH (eV/atom)	0.063[†]	0.058
$\Delta V/V$	-2.5%[†]	-3.2%
T_m (K)	367±5[†]	303
dT_m/dP (K/GPa)	17.3	23.0**

* Reference [16]
[†] Reference [5]
** Reference [4]

parallel to the liquid-solid interface) yet allowing the out of plane box dimension to expand and contract while maintaining the desired pressure and temperature. Once the system has equilibriated its pressure and temperature we begin step three by performing a simulation for 125 ps at constant volume and energy. If the average temperature or average pressure deviates significantly from the initial temperature and desired pressure, the above process is repeated. In Fig. 3, we show the calculated melting points as a function of pressure compared with the experimental melting points of gallium. As we previously reported [5], the calculated melting point of Ga is approximately 60 K higher than experiment. We see here that the change in melting point with pressure, namely the negative Clapeyron slope, is in excellent agreement with experiment. In Table 1, we reiterate the data suggesting that the potential captures the nature of the solid- liquid phase transition[5].

The liquid Hugoniot was calculated using the method of Maillet *et al.* [8]. A liquid consisting of 1372 atoms was equilibrated at T = 308 K and P = 0, which resulted in ρ = 5.983 g/cm^3 compared with the experimental ρ of 6.078 g/cm^3. The Hugoniostat was then implemented with target pressures of up to 200 GPa. The results are shown in Fig. 4. The

P versus u_p prediction agrees almost exactly with experiment[9], but the predicted P versus V curve is slightly below (softer than) experiment. By varying the parameter δ, we see that there is a much closer agreement could be obtained without affecting the P versus u_p curve. Furthermore, we determined that there was no size dependence by performing simulations with 10976 atoms. The u_s versus u_p curve is slightly below experiment because the predicted equilibrium bulk modulus of liquid Ga is below experiment. The sound speed calculated ($\sqrt{B/\rho}$) for MEAM Ga is 2.799 km/s whereas the experimental sound speed is 2.911 km/s. The slope of the $u_s - u_p$ curve agrees well with experiment. The slight change in the slope at the low u_p values could be attributed to structural changes due to undercooling because the simulations were begun about 60 K below the MEAM Ga T_m. Another possible explanation in the change of slope is that gallium has dimers present in its low temperature liquid. Finally, the "Hugoniostat" method may not be sensitive in the low pressure regime. The variation in the δ parameter does not appear to affect the low u_p portion of the curve, but improves the agreement with experiment at high u_p.

CONCLUSIONS

The literature MEAM gallium model is in excellent agreement with experimental phase stability, melting point, and liquid Hugoniot as a function of pressure.

A slight modification of the MEAM parameters can improve the agreement of the liquid Hugoniot

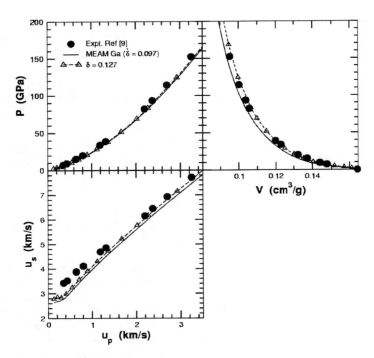

FIGURE 4. Predicted liquid Hugoniot using the Ga MEAM parameters from Baskes *et al.* [5] and a slightly modified Ga MEAM potential compared with experiment[9].

with experiment with no significant change to the solid properties.

ACKNOWLEDGMENTS

We would like to thank J. N. Fritz, B. L. Holian, J. E. Vorthmann, W. W. Anderson, R. S. Hixson, and P. A. Rigg for useful comments surrounding discussion of this work. This work was supported by the US DOE under contract W-7405-ENG-36.

REFERENCES

1. Bernasconi, M., Chiarotti, G. L., and Tosatti, E., *Phys. Rev. B*, **52**, 9988 (1995).
2. Rapeanu, S. N., and Padureanu, I., *Physica Scripta*, **T57**, 18 (1995).
3. Schulte, O., and Holzapfel, W. B., *Phys. Rev. B*, **55**, 8122 (1997).
4. Young, D. A., *Phase Diagrams of the Elements*, University of California Press, Berkeley, CA, 1991.
5. Baskes, M. I., Chen, S. P., and Cherne, F. J., *Phys. Rev. B*, **66**, 104107 (2002).
6. Ravelo, R., and Baskes, M., *Phys. Rev. Lett.*, **79**, 2482 (1997).
7. Baskes, M. I., *Phys. Rev. B*, **62**, 15532 (2000).
8. Maillet, J.-B., Mareschal, M., Soulard, L., Ravelo, R., Lomdahl, P., Germann, T. C., and Holian, B. L., *Phys. Rev. E*, **63**, 016121 (2000).
9. Fritz, J. N., and Carter, W. J., The hugoniot EOS for liquid Ga and Hg, Tech. Rep. LA-13844-MS, Los Alamos National Laboratory (2001).
10. Daw, M. S., and Baskes, M. I., *Phys. Rev. Lett.*, **50**, 1285 (1983).
11. Daw, M. S., and Baskes, M. I., *Phys. Rev. B*, **29**, 6443 (1984).
12. Daw, M. S., Foiles, S. M., and Baskes, M. I., *Mater. Sci. Rep.*, **9**, 251 (1993).
13. Baskes, M. I., *Phys. Rev. B*, **46**, 2727 (1992).
14. Baskes, M. I., and Johnson, R. A., *Modelling Simul. Mater. Sci. Eng.*, **2**, 147 (1994).
15. Rose, J. H., Smith, J. R., Guinea, F., and Ferrante, J., *Phys. Rev. B*, **29**, 2963 (1984).
16. Brandes, E. A., editor, *Smithells metals reference book*, Butterworths, London, 1983.

CP706, *Shock Compression of Condensed Matter - 2003*
edited by M. D. Furnish, Y. M. Gupta, and J. W. Forbes
© 2004 American Institute of Physics 0-7354-0181-0/04/$22.00

LARGE-SCALE MOLECULAR DYNAMICS SIMULATIONS OF EJECTA FORMATION IN COPPER

Timothy C. Germann[1], James E. Hammerberg[1], and Brad Lee Holian[2]

[1]*Applied Physics Division, Los Alamos National Laboratory, Los Alamos NM 87545*
[2]*Theoretical Division, Los Alamos National Laboratory, Los Alamos NM 87545*

Abstract. Non-equilibrium molecular dynamics simulations are used to investigate the ejection of matter which takes place when a planar shock wave encounters a free surface. We will focus on Cu fcc single crystals, using an empirical embedded-atom method interatomic potential, and present results for the ejecta mass dependence on shock strength, as well as size and velocity distributions.

INTRODUCTION

The ejection of material from shocked surfaces has been experimentally and theoretically studied for a number of years, beginning in the modern era with the pioneering work by Asay and colleagues in the 1970s [1-3]. Recently, ejecta formation, expansion, and recompaction has been suggested as an important mechanism for void collapse-induced "hot spots," which may serve as initiation sites in energetic materials with a significant void volume [4].

Prior work on roughened surfaces [2,3,5] has demonstrated that the total ejected mass is approximately equal to the "defect" mass in the solid state, but can be significantly enhanced upon surface melting. Measurements of particle size distributions [6,7] exhibit a power-law scaling with exponents in accord with two- or three-dimensional percolation theory for spall-dominated and jet-dominated breakup, respectively.

In this work we investigate the fundamental *lower limit* of ejecta mass, that for a defect-free perfect crystal with an atomically smooth free surface and impactor. We find that even for shock strengths in the melting regime, only on the order of one surface monolayer (corresponding to areal densities of less than 1 $\mu g/cm^2$) are ejected from the free surface. (By comparison, experimental densities typically range from 1 $\mu g/cm^2$ to 10 mg/cm^2.) This suggests that even in the melting regime, heterogeneous ejecta formation dominates homogeneous production by several orders of magnitude.

NEMD SIMULATIONS OF PERFECT FCC COPPER CRYSTALS

Large-scale molecular dynamics simulations, integrating the classical equations of motion for millions of atoms over picosecond to nanosecond timescales, have been used to investigate fundamental mechanisms of shock-induced plasticity [8,9] and polymorphic phase transformations [10], among other phenomena. Preliminary studies of ejecta indicate that for perfect fcc crystals shock loaded (below melting) in the <100> direction, atoms and small clusters of atoms are primarily ejected at the trijunctions where two stacking faults on different {111} slip planes intersect the (100) free surface [11]. Moreover, the amount of ejected material can be significantly increased at surface heterogeneities, such as machining grooves [11,12].

In order to eliminate the effects of surface roughness, inclusions, grain boundaries, etc., and focus on the fundamental lower limit of ejecta due to plasticity and melt, we consider homogeneous ejecta formation in perfect, defect-free crystals which are shocked in specific crystallographic directions. We employ an embedded-atom method (EAM) empirical potential for copper [13], and quasi-2D samples with a thin x dimension (periodic boundary conditions are applied in the transverse x and y directions). Results presented below are for a sample having dimensions of approximately 3 nm $\times$ 72 nm $\times$ 276 nm, a total of 3 million atoms. The sample travels at a specified particle (or "piston") velocity u_p in the $-z$ direction towards a perfectly reflecting "momentum mirror" as described in [8]. (Velocities presented below are corrected from this piston-centered computational frame to the lab reference frame.)

Instead of the standard experimental method of measuring the ejected mass arriving downstream at some stand-off distance, we have developed an efficient (parallel) analysis subroutine which identifies individual clusters of atoms based on a bond-length criterion. If all ejecta are generated at the same instant (as is generally assumed in interpreting the experimental signals), then this algorithm will give an ejecta mass independent of the time at which the measurement is made. This assumption is confirmed for shocks resulting in a solid phase, but becomes less valid for stronger shocks in the melting regime, as seen in Fig. 1. The increase in ejecta mass from 32 ps (open circles and dashed line) to 43 ps (solid circles and solid line) for <100> shocks is presumably due to a continued evaporation of atoms from the free liquid surface, which becomes more pronounced as the temperature increases.

We note that there is no abrupt increase in ejecta mass upon melting, but rather a gradual increase with areal densities remaining below 0.1 $\mu g/cm^2$, or one Cu(100) surface monolayer. This suggests that homogeneous melting alone cannot account for the mg/cm^2 densities measured in many experiments; surface roughness effects and heterogeneous production of ejecta are still dominant.

Orientation effects seem most important in the solid phase, as expected; the lower-energy close-packed (111) surface deforms primarily by buckling rather than by ejection of atoms, so <111> ejecta masses (triangles in Fig. 1) are smaller than <100> ones for shocks below or close to the melt boundary. On the other hand, the initial temperature (counterintuitively) influences mainly the ejecta production from strong shocks in the melting regime, where one might expect no effect because the shock temperature $T_1 \gg T_0$ (see Fig. 1 inset). Most simulations here were carried out with a very low initial temperature $T_0 = 7$ K, but as seen in Fig. 1, virtually identical results are obtained for $T_0 = 300$ K in the mixed-phase region, but a *lower* ejecta mass is found in the liquid regime. Finally, we mention that the expression (in 3D) for the critical shock strength at which ejection occurs [4], $u_p^* = (12\varepsilon/m)^{1/2}$ for bond energy ε and atomic mass m, gives a quite good estimate, $u_p^* = 3.3$ km/s.

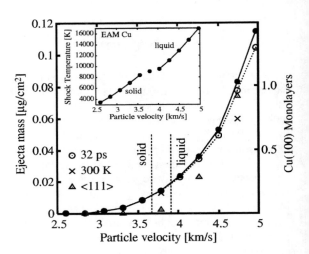

Figure 1. Areal mass density of ejecta from Cu perfect crystals. The solid circles and line are for the <100> orientation at an initial temperature $T_0 = 7$K, at 43 ps after impact. Other symbols are for earlier measurement times (32 ps, open circles and dashed line), an elevated initial temperature ($T_0 = 300$ K, crosses), or the <111> crystal direction (triangles). Vertical dashed lines indicate the approximate locations where the <100> shock Hugoniot enters and exits the melting curve for this EAM potential, as determined by the shock temperature (see inset). (We note that the Hugoniot melting is noticeably higher than experiment; roughly 9000 K and 300 GPa for EAM Cu versus 6000 K and 250 GPa from experiment [14].)

Data for a <100> perfect crystal shocked to a particle velocity u_p = 5.21 km/s (well into the shock-melting regime) are shown in Fig. 2. After the initial shock traverses the sample, a spreading rarefaction fan rebounds towards the impact plane, while leaving a small but well-defined surface region moving at 11.60 km/s, noticeably greater than the nominal $2u_p$ free surface velocity due to an additional thermal expansion component. At the free surface — ejecta interface, there is a distinct kink in particle velocity due to fast ejecta particles.

Using the cluster detection algorithm, we find that the ejecta cloud for this snapshot consists of 1823 atoms in 888 clusters, whose sizes and longitudinal velocities are shown in the middle panel of Fig. 2. Due to downstream collisions after being ejected, some of these clusters (predominantly with sizes of N = 1-3 atoms) are traveling more slowly than the free surface and will be recollected as the free surface advances. A least-squares fit of all 888 data points to an asymptotic (N $\rightarrow \infty$) velocity $\langle u_{ej} \rangle \rightarrow u_{surface}$ with a $N^{-2/3}$ falloff based on surface thermodynamic considerations is shown; allowing the exponent to vary gives an equally good fit, with a more gradual $N^{-0.40}$ falloff.

The velocity distributions for N = 1 and 2 atom ejecta are shown in the bottom panel of Fig. 2; neglecting the slight dependence of average ejecta velocity on mass, we show the results of a fitting both distributions to a 1D Maxwellian (Gaussian) distribution with a single mean velocity ($\langle u_{ej} \rangle$ = 12.78 km/s) and temperature (T_{ej} = 7653 K). Fitting each distribution separately produces only a slightly better fit, with 12.85 km/s and 7448 K for monomers and 12.57 km/s and 8653 K for dimers. In any event, this temperature is much less than that of the initial shock (T_1 = 19240 K).

The size distribution of ejecta particles (clusters of atoms) for this particular snapshot is shown in Fig. 3. Ignoring the three largest clusters (N = 33, 34, and 46 atoms), the remaining data closely fits a $N^{-2.3}$ power-law distribution, or a $d^{-6.9}$ relation to particle diameter d. Experimental d exponents range from -2.8 $\pm$ 0.4 for copper [6] to -5.8 for aluminum [7], although these are dominated by heterogeneous ejecta formation at surface imperfections, leading to significantly (3 orders of magnitude or more) greater areal densities and ejecta sizes.

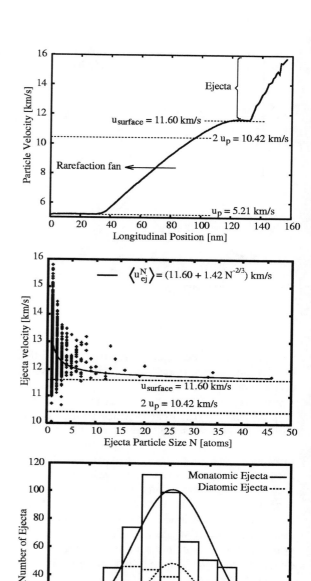

Figure 2. (Top) Longitudinal velocity profile 22 ps after shock loading of a Cu <100> perfect crystal to u_p = 5.21 km/s. (Middle) Velocity scatter of the 888 ejecta particles (with masses N = 1 to 46 atoms, totaling 1823 atoms); a linear least-squares fit suggests a slight decrease in the average velocity with increasing particle size N. (Bottom) Observed velocity distributions (bars) of N = 1 and 2 ejecta, along with a fit of each to the same mean velocity and temperature.

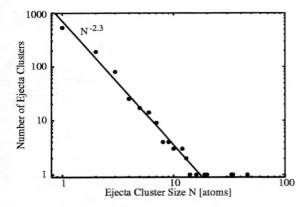

Figure 3. Ejecta particle size distribution for a shocked Cu <100> perfect crystal with u_p = 5.21 km/s (see Fig. 2). The solid line represents a fit to the N < 30 clusters.

SUMMARY

Even in the shock melting regime, we find that only about one surface monolayer of material (corresponding to an areal density around 0.1 $\mu g/cm^2$) is ejected from the free surface of a perfect crystal. This suggests that even for strong shocks, surface roughness is the determining factor in Cu ejecta production. Tin appears to behave similarly; in experimental measurements of strong (50 GPa) shock loading of samples with different surface preparations, the ejected mass rises from less than 1 mg/cm^2 for a polished surface up to 25 mg/cm^2 for a roughly machined one [15]. On the other hand, it has been suggested that lead ejection is dominated by a melt contribution solely dependent on shock pressure [2]. Further investigations of nanojet formation and breakup at surface grooves using molecular dynamics are underway.

ACKNOWLEDGEMENTS

This work was performed under the auspices of the U.S. Dept. of Energy at Los Alamos National Laboratory, operated by the Univ. of California under Contract No. W-7405-ENG-36.

REFERENCES

1. Asay, J. R., Mix, L. P., and Perry, F. P., *Appl. Phys. Lett.* **29**, 284-287 (1976).

2. Asay, J. R., "Material Ejection from Shock-Loaded Free Surfaces of Aluminum and Lead," Sandia Report, SAND76-0542, October 1976.

3. Asay, J. R., and Bertholf, L. D., "A Model for Estimating the Effects of Surface Roughness on Mass Ejection from Shocked Materials," Sandia Report, SAND78-1256, October 1978.

4. Holian, B. L., Germann, T. C., Maillet, J.-B., and White, C. T., *Phys. Rev. Lett.* **89**, 285501 (2002).

5. Cheret, R., Chapron, P., Elias, P., and Martineau, J., "Mass Ejection from the Free Surface of Shock-Loaded Metallic Samples," in *Shock Waves in Condensed Matter*, edited by Y. M. Gupta, Plenum Press, New York, 1985, pp. 651-654.

6. Werdiger, M., et al., *Laser and Particle Beams* **14**, 133-147 (1996).

7. Sorenson, D. S., Minich, R. W., Romero, J. L., Tunnell, T. W., and Malone, R. M., "Ejecta Particle Size Distributions for Shock-Loaded Sn and Al Targets", in *Shock Compression of Condensed Matter – 2001*, edited by M. D. Furnish, N. N. Thadhani, and Y. Horie, AIP Conference Proceedings 620, Melville, New York, 2002, pp. 531-534; *J. Appl. Phys.* **92**, 5830-5836 (2002).

8. Holian, B. L., and Lomdahl, P. S., *Science* **280**, 2085-2088 (1998).

9. Germann, T. C., Holian, B. L., Lomdahl, P. S., and Ravelo, R., *Phys. Rev. Lett.* **84**, 5351-5354 (2000).

10. Kadau, K., Germann, T. C., Lomdahl, P. S., and Holian, B. L., *Science* **296**, 1681-1684 (2002).

11. Holian, B. L., et al., "Shock waves and their aftermath: a view from the atomic scale," in *Shock Compression of Condensed Matter – 1999*, edited by M. D. Furnish, L. C. Chhabildas, and R. S. Hixson, AIP Conference Proceedings 505, Melville, New York, 2002, pp. 35-41.

12. Chen, J., Fu-qian, J., Jing-Lin, Z., Dong-quan, C., and Ju-hai, W., *J. Phys.: Cond. Mat.* **14**, 10833-10837 (2002).

13. Voter, A.F., *Phys. Rev. B* **57**, 13985-13988 (1998).

14. Hayes, D., Hixson, R. S., and McQueen, R. G., "High pressure elastic properties, solid-liquid phase boundary and liquid equation of state from release wave measurements in shock-loaded copper," in *Shock Compression of Condensed Matter – 1999*, edited by M. D. Furnish, L. C. Chhabildas, and R. S. Hixson, AIP Conference Proceedings 505, Melville, New York, 2002, pp. 483-488.

15. Andriot, P., Chapron, P., and Olive, F., "Ejection of Material from Shocked Surfaces of Tin, Tantalum, and Lead-Alloys," in *Shock Waves in Condensed Matter – 1981*, edited by W. J. Nellis, L. Seaman, and R. A. Graham, AIP Conference Proceedings 78, New York, 1982, pp. 505-509.

CP706, *Shock Compression of Condensed Matter - 2003*
edited by M. D. Furnish, Y. M. Gupta, and J. W. Forbes
© 2004 American Institute of Physics 0-7354-0181-0/04/$22.00

QUANTUM MOLECULAR DYNAMICS SIMULATIONS OF SHOCKED MOLECULAR LIQUIDS

J. D. Kress[1], S. Mazevet[1], L. A. Collins[1], P. Blottiau[2]

[1]*Theoretical Division, Los Alamos National Laboratory, Los Alamos, NM 87545*
[2]*CEA, BP12 F91680, Bruyeres Le Chatel, France*

Abstract.   Using Quantum Molecular Dynamics, we study the dissociation of nitrogen oxide (NO) and carbon monoxide (CO) along both the principal and reshocked Hugoniots. We obtain good agreement with experimental data in terms of pressure and density. As the molecules dissociate at high pressure and temperatures, we characterize the myriad of species that form. As NO dissociates along both the principal and reshocked Hugoniot, a significant amount of molecular nitrogen forms. As CO dissociates along the principal Hugoniot, first at low pressures, CO_2 forms and large particles form (both polymer chains and rings) that contain both carbon and oxygen. At higher pressures (above 30 GPa), the CO_2 dissociates and the particles breakup and form a mixture of CO, atomic carbon, and small transient clusters with lifetimes less than a typical molecular vibrational period.

The study of materials under extreme conditions of temperature and pressure ("Warm, Dense Matter") has made significant progress in the past few years, due ot noticeable advances in both the experimental and theoretical techniques. On the experimental side, gas-gun, Z-pinch, and laser-driven experimental setups have pushed principal and reshock Hugoniot pressure measurements up to the Mbar (100 GPa) range. On the theoretical side, Quantum Molecular Dynamics (QMD), where the electrons receive a fully quantum mechanical treatment using density functional theory (DFT), is particularly suited for the study of such chemical processes as ionization, recombination, dissociation, and association of the various atomic species present in the warm, dense media. In many circumstances, a quantum ("*ab initio*") treatment of the interaction between electrons and ions provides the only reliable procedure for gaining information and understanding on the state and interaction of matter in these extreme conditions for which experiments remain difficult. Up to now, applications of QMD primarily focussed on one-component systems, such as molecular hydrogen(1), nitrogen(2) and oxygen(3). In the present work, we study the equation of state (EOS) and dissociation of two-component systems

(NO and CO) along the principal and reshock Hugoniots using QMD. NO serves as a prototype for the study of explosive compounds and the associated reactive chemistry. Both CO and NO and their reaction products can be present in the detonation products of explosives(4). The two-component systems provide for a much richer set of dissociation products as compared to the one-component systems. Particular attention is paid to the constituency of the fluid along each Hugoniot as the rich chemistry, induced by such an increase of pressure and temperature, remains among the challenging aspects in the simulation of such systems.

In DFT methods, the total energy is written as a functional of the electron density, which is obtained by summing the probability density over the occupied orbitals. Further, for the Generalized Gradient Approximation (GGA), the electronic exchange and correlation energy are approximated using a functional which depends only on the electron density and its spatial derivatives. GGA methods provide a highly accurate means of studying the thermochemistry of chemical bonding by representing the inhomogeneities inherent in the electron charge density relative to the homogeneous electron gas. We employed the VASP plane–wave pseudopotential code, which was developed at the Technical University of Vienna (5). This code implements the Vanderbilt ultrasoft pseudopotential scheme in a form supplied by G. Kresse and J. Hafner and the Perdew-Wang 91 parameterization of GGA. A finite-temperature density functional procedure (Mermin functional) is employed by setting the electron and ion temperatures equal (local thermodynamic equilibrium) using a Fermi-Dirac distribution. Finite-temperature, fixed–volume molecular dynamics simulations were performed for density and temperature points selected to span the range of the princial and reshock Hugoniot experiments. A simulation cell with periodic boundary conditions with a fixed number of atoms and an isokinetic ensemble was employed. 27 nitrogen (carbon) and 27 oxygen atoms were used in the NO (CO) simulation cell. Typically, MD trajectories (with time step=2 fs) were equilibrated for a few ps after which the results were sampled from another 1 to 2 ps. Further details on the QMD approach can be found in Ref. (6).

The Rankine-Hugoniot equation

$$(U_0 - U_1) + \frac{1}{2}(V_0 - V_1)(P_0 + P_1) = 0, \qquad (1)$$

describes the shock adiabat through a relation between initial ($V_0 = 1/\rho_0$, U_0, P_0) and final($V_1 = 1/\rho_1$, U_1, P_1) volume, internal energy, and pressure. The initial densities for the principal Hugoniot were set to ρ_0=1.22 and 0.807 g/cm^3, respectively, for NO and CO. We have chosen $P_0 = 0$, and U_0 so that the energy of the isolated molecule is zero. To find the Hugoniot point for a given V_0, P, and U derived from the QMD results were fit by a least-squares prescription to a quadratic function in T. P_1 and T_1 were found by substituting these functions and solving Eq. 1.

In Fig. 1, we compare the results from the QMD calculations with measurements for the principal Hugoniot and the reshock Hugoniot from explosively-driven shock experiments(7). For the highest density reached on the experimental principal Hugoniot (ρ_1=2.38 g/cm^3), we find remarkable agreement between the calculation and the measurement. As we go to lower densities, the agreement along the principal Hugoniot deteriorates; the QMD pressure is a factor of two lower at the lowest density measured. For the reshock Hugoniot, we plot the

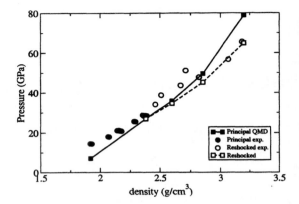

FIGURE 1. Principal and reshock Hugoniot for NO. Squares=QMD. Circles=experimental data, Ref. (7).

experimental results corresponding to initial reshock densities between ρ_0=2.28 and 2.38 g/cm^3. For the theoretical reshock, we concentrated on the initial condition corresponding to the highest density experimental principal Hugoniot ρ_0=2.38 g/cm^3, with P_0=25.8 GPa and U_0=-5.38 eV/atom as given by the QMD principal Hugoniot. Such a small variation in the initial conditions does not affect substantially the final reshock pressures. This variation does not, for example, explain the consistently higher experimental pressures for densities between 2.5 and 2.75 g/cm^3. However, a closer inspection of Fig. 1 also reveals that this disagreement is less as the density increases; there is also significant scatter in experimental data for densities below 2.75 g/cm^3. Taking these two points into consideration along with the good description of the principal Hugoniot in the region around 2.38 g/cm^3 suggests that the QMD provides a satisfactory description of the state of the fluid as the increase in pressure and temperature breaks the molecular bonds.

To quantify the state of the fluid as the density increases, we calculate the pair corrletion functions (PCFs) representative of each possible combination (N-O, O-O, and N-N). Fig. 2 shows the PCFs along the principal Hugoniot and extending to higher densities. At the lowest density, ρ=1.92 g/cm^3, the N-O PCF (no-shaded region) peaks around 2.0 a_B (1 a_B = one atomic unit), which corresponds to the intermolecular bond distance of a NO molecule. The

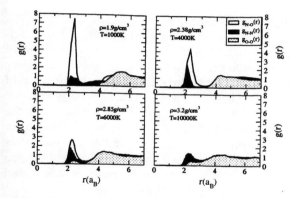

FIGURE 2. Pair correlation functions along the principal Hugoniot for NO.

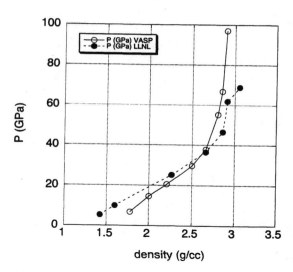

FIGURE 3. Principal Hugoniot for CO. Open circles=QMD. Closed circles=experimental data, Ref. (8).

lack of structure in the PCFs for the other two pairs indicates that the fluid is mostly composed of NO molecules at T=1000K and ρ=1.92 g/cm^3. As the density increases to ρ=2.38 g/cm^3, there is a sizeable reduction and broadening of the first peak in the N-O PCF at 2.0 a_B, indicating that the fluid has partially dissociated. At this density, a sharp peak is now seen in the N-N PCF (black-shaded region) centered near 2.0 a_B as well. We find that at ρ=2.38 g/cm^3 and T=4000K, the NO fluid has partially dissociated into a mixture of NO and N_2 molecules and some atomic oxygen. This behavior has also been confirmed by performing a standard cluster analysis (defined below). Note that at this density there is a lack of structure in the O-O PCF (light-shaded region), which indicates, in contrast to nitrogen, oxygen atoms do not recombine upon dissociation of NO. As inferred from Fig. 2, this basic behavior is preserved for T=6000K and ρ=2.85 g/cm^3 and T=10000K and ρ=3.20 g/cm^3 as well.

In Fig. 3, we compare the results of the QMD calculations with measurements of the principal Hugoniot for CO gas-gun shock experiments(8). For pressures below 20 GPa, the QMD EOS is softer than the experimental measurements. A chemical equilbrium analysis(8) of the experimental data suggested that the temperature is too low (2000 K and less) to allow the CO to dissociate on a timescale less than a typical shock-passage time ($\sim$200 ns) and therefore the fluid is undissociated. In contrast, for the lowest QMD density of ρ_1=1.78 g/cm^3, where T=1470

K and P=6.8 GPa, the fluid is already dissociated and has formed some CO_2, a 12-atom ring (C_6O_6), and small linear clusters with carbon backbones. A dissociated fluid will lead to a softer EOS relative to an undissociated fluid. For densities of 2.2 g/cm^3 and larger in Fig. 2, the QMD results are in good agreement with experiment. For pressures greater than 30 GPa, the experimental results were consistent with chemical equilbrium calculations(8), where the carbon is allowed to form both graphite and diamond.

To quantify the constituency of the warm, dense fluid, we have performed standard cluster and nearest-neighbor (NN) analyses of the QMD trajectories. For the cluster analysis, an effective radius is selected (for C-C bonds r_{eff}=3.2 a_B), and all atoms within this distance are considered bound to a reference atom. This produces, at a selected time step in the QMD trajectory, a distribution of monomers, dimers, and larger clusters. For the NN analysis, at each timestep each carbon atom is examined and the number of oxygen atoms within a cutoff distance r_{bond}=3.0 a_B is tabulated. We then average the distributions over the extent of the QMD trajectory.

In Fig. 4 we plot the probability of the occurence of a carbon cluster of size n for four different densities along the CO principal Hugoniot. For $\rho \leq 2.2$ g/cm^3 and T$\leq$4000K, a bimodal distribution is ob-

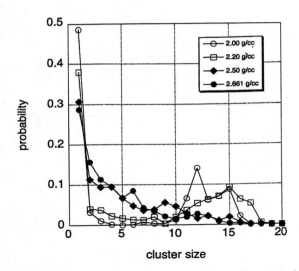

FIGURE 4. Carbon cluster distribution along the principal Hugoniot for CO.

served which has a first sharp peak decaying rapidly from n=1 and a second broad peak centered around n=13 to 15. The second peak indicates there are large clusters forming with carbon backbones. Based on a lifetime analysis, these particles tend to be long-lived (~20 fs) relative to the period of a molecular vibration. The small finite sample of 27 carbon atoms is trying to mimic graphite-like and diamond-like species by forming these large clusters. The NN analysis indicates that the C-O distribution is about 1/3 C-O and 2/3 CO_2. The C-O bonds exist both as diatomic molecules and in small clusters with carbon backbones. For $\rho \geq 2.5$ g/cm^3 and T$\geq$6000 K, the second peak disappears and the first peak decays monotonically from n=1 with a decay width broader than the lower density cases. The large clusters break up into smaller clusters that are transient (lifetimes of only a few fs) and the NN analysis indicates that the CO_2 dissociates. For example, the C-O bond distribution at ρ=2.9 g/cm^3 and T=20000 K is 17% CO_2, 38% C-O, and 44% zero C-O. The density region where the bimodal distribution begins to disappear in Fig. 4 corresponds to the density (between ρ =2.2 and 2.5 g/cm^3) where the slope softens slightly in the QMD results in Fig. 3. The P-ρ curve first softens slightly as the large clusters dissociate into smaller transient clusters and then eventually stiffens at high

densities as the CO_2 and C-O bonds dissociate. The origin of the softening of the slope in the principal Hugoniot is different than that in simulations of one-component systems(1, 2, 3), where the change in slope for the latter was attributed to the onset of the dissociation of the molecules. Finally, based on a lack of a peak in the O-O PCF (not shown) at the molecular oxygen bond distance (as was the case for the NO principal Hugoniot), we find that no molecular oxygen is formed from recombining O atoms for any of the densities and temperatures considered along the principal Hugoniot in Fig. 3.

Finally, the constituency analyses are preliminary since we have not systematically studied finite-size effects. We plan on running simulations with 108 total atoms (as compared to the present work with 54 atoms). For example, there may be some O_2 formation, especially at the lowest densities and temperatures for NO.

This work was supported under the auspices of the U.S. Dept. of Energy at Los Alamos National Laboratory under Contract W-7405-ENG-36.

REFERENCES

1. T. J. Lenosky, S. R. Bickham, J. D. Kress, and L. A. Collins, Phys. Rev. B **61**, 1 (2000).

2. J. D. Kress, S. Mazevet, L. A. Collins, and W. W. Wood, Phys. Rev. B **63**, 024203(1-5), (2001).

3. J. D. Kress, S. Mazevet, and, L. A. Collins, AIP Conference Proceedings No. 620, eds. M. D. Furnish, L. C. Chhabildas, and R. S. Hixson, 91 (2002).

4. M. S. Shaw, AIP Conference Proceedings No. 620, eds. M. D. Furnish, L. C. Chhabildas, and R. S. Hixson, 157 (2002).

5. G. Kresse and J. Hafner, Phys. Rev. B **47**, 558 (1993); *ibid.* **49**, 14 251 (1994); G. Kresse and J. Furthmüller, Comput. Mat. Sci. **6**, 15 (1996); Phys. Rev. B **55**, 11 169 (1996).

6. L. A. Collins, S. R. Bickham, S. Mazevet, J. D. Kress, T. J. Lenosky, N. J. Troullier, and W. Windl, Phys. Rev. B **63**, 184110 (2001).

7. G. L. Schott, M. S. Shaw, and J. D. Johnson, J. Chem. Phys. **82**, 4264 (1985).

8. W. J. Nellis, F. H. Ree, M. van Thiel, and A. C. Mitchell, J. Chem. Phys. **75**, 3055 (1981).

CP706, *Shock Compression of Condensed Matter - 2003*
edited by M. D. Furnish, Y. M. Gupta, and J. W. Forbes
© 2004 American Institute of Physics 0-7354-0181-0/04/$22.00

QUANTUM MOLECULAR DYNAMICS CALCULATIONS OF ROSSELAND MEAN OPACITIES

S. Mazevet, J. D. Kress, N. H. Magee, J. J. Keady, and L. A. Collins

Theoretical Division, Los Alamos National Laboratory, Los Alamos, NM 87545

Abstract. We show that Quantum Molecular Dynamics provides a powerful tool to extend and benchmark current opacity libraries into the complex regime of warm dense matter. In this regime, the medium can be constituted of electrons, protons, atoms and molecules, while plasma and many body effects can not be treated as perturbations. Among the most notable features of this new approach for calculating Rosseland mean opacities is the ability to obtain a consistent set of material, optical and electrical properties for various mixtures from the same simulation.

Many opacity libraries commonly in use for standard macroscopic modeling programs such as in hydrodynamic codes use absorption cross sections derived for isolated atoms or ions while many-body and plasma effects are modeled either as perturbation to the population distribution over these atomic states and/or reintroduced as adjustments to the final results[1]. In the mean time, developments in a wide variety of fields, including dense plasma[2], ICF, and astrophysics[3], require an extension of the opacity libraries into new and complex regimes where such a simple approach to the many-body effects quickly becomes questionable.

The relatively low-temperature, high-density regime (a few eV and few g/cm^3), often labeled as "warm dense matter", is an example of such a situation where the intricate nature of the medium, partially dissociated and ionized, intermediate coupling, $\Gamma \sim 1$, and partially degenerate $\eta \sim 1$, requires a careful validation, either from experiments or more sophisticated theoretical methods, of the physical models that produce the opacity data. In this regime, usually reached experimentally by shock compression in the Mbar range, the accurate knowledge of opacities drives the modeling of such diverse systems as white dwarf atmospheres[3], which play a key role in latest attempts to date various astrophysical objects[4], to strongly coupled plasmas

produced by exploding aluminum wires[5, 6].

Since experiments have proven difficult within these new realms, as witnessed by the continuing controversy over the equation-of-state(EOS) of compressed hydrogen[2], we show in the present work that Quantum Molecular Dynamics (QMD) simulations provides an effective venue to calculate and benchmark Rosseland mean opacities. This *ab-initio* approach produces a consistent set of material, electrical, and optical properties from the same simulation. This contrasts with the opacity libraries, which consist of an integrated collection of approximate models leading sometimes to inconsistencies on the overall nature of the medium. While the method presented is not restricted to the type of medium considered and can be applied without restriction to various mixtures of atomic, ionic or molecular species, we focus here on warm dense hydrogen. We choose warm dense hydrogen [7] to first validate our approach by directly comparing our result with one of the standard opacity libraries available in this regime, the Los Alamos Opacities Library LEDCOP [8], but also to illustrate its flexibility at describing the rapid change in the constituency of the fluid in a region where the increase in temperature and density breaks molecular bonds and ionizes the media.

THEORETICAL METHOD

We briefly review some of the main points of our simulations approach; more details appear in earlier publications[9]. In the present work, the molecular dynamics trajectories were calculated using the VASP plane–wave pseudopotential code, which was developed at the Technical University of Vienna [10]. This code implements the Vanderbilt ultrasoft pseudopotential scheme[11] in a form supplied by G. Kresse and J. Hafner[12] and the Perdew-Wang 91 parameterization of the Generalized Gradient Approximation (GGA)[13].

The main ingredient of the opacity calculation is the frequency-dependent conductivity, $\sigma(\omega)$, which has both real and imaginary parts:

$$\sigma(\omega) = \sigma_1(\omega) + i\sigma_2(\omega). \qquad (1)$$

The real part is derived from the Kubo-Greenwood (KG) formulation[15, 14]

$$\sigma_1(\omega) = \frac{2\pi}{\Omega\omega} \sum_{ij} F_{ij} |D_{ij}|^2 \delta(\varepsilon_j - \varepsilon_i - \omega), \qquad (2)$$

where Ω is the atomic volume and ω the frequency. F_{ij} stands for the difference between the Fermi-Dirac distributions at temperature T. In (2), the velocity dipole matrix element is given as

$$|D_{ij}|^2 = \frac{1}{3} \sum_{\alpha} | < \psi_i |\nabla_\alpha| \psi_j > |^2, \qquad (3)$$

where α represents the spatial directions x, y, or z. The quantities ε_i and ψ_i represent the energy and wavefunctions of the i-th orbital found from the diagonalization of the Kohn-Sham equations. For each snapshot in a given trajectory, the dipole matrix elements (3) are evaluated using the Projector Augmented (PAW) method as supplied by Kresse et $al.$ [16, 17]. Eq.(3) applies to a spatial configuration of N atoms at a single time step within an MD trajectory. We calculate a trajectory-averaged conductivity by averaging the conductivity obtained for a set of configurations spaced at time steps separated by at least the correlation time, the e-folding time of the velocity autocorrelation function[18].

Various optical properties follow directly from the knowledge of the frequency-dependent real-part of the conductivity[14, 19]. We focus here on the two key ingredients for the calculations of Rosseland mean opacities, the absorption coefficient and index of refraction, respectively, defined as

$$\alpha(\omega) = \frac{4\pi}{n(\omega)c} \sigma_1(\omega),$$

$$n(\omega) = \sqrt{\frac{1}{2}[|\varepsilon(\omega)| + \varepsilon_1(\omega)]}. \qquad (4)$$

The dielectric functions are, in turn, immediately obtained from the two parts of the conductivity:

$$\varepsilon_1(\omega) = 1 - \frac{4\pi}{\omega}\sigma_2(\omega), \qquad (5)$$

$$\varepsilon_2(\omega) = \frac{4\pi}{\omega}\sigma_1(\omega), \qquad (6)$$

while the imaginary part, $\sigma_2(\omega)$ arises from the application of a Kramers-Kronig relation as

$$\sigma_2(\omega) = -\frac{2}{\pi}P \int \frac{\sigma_1(v)\omega}{(v^2 - \omega^2)} dv. \qquad (7)$$

In Eq.(7), P stands for the principal value of the integral.

Finally, the Rosseland mean opacity is defined as[20]

$$1/\kappa_R(\rho, T) = \frac{\int_0^\infty dv n^2(v) \partial B(v, T)/\partial T /\alpha(\rho, T, v)}{\int_0^\infty dv n^2(v) B(v, T)}. \qquad (8)$$

where the derivative of the normalized Planck function is

$$\partial B(v, T)/\partial T = 15/(4\pi^4 T)u^4 e^u/(e^u - 1)^2. \qquad (9)$$

The dimensionless variable $u = hv/T$, with hv the photon energy and T the temperature of the media in energy units. $\partial B(v, T)/\partial T$ is a slowly varying function that peaks around $4k_B T$.

RESULTS

An understanding of the differences in the opacities between the LEDCOP library and the QMD approach requires at first a detailed examination of the underlying material properties such as EOS. The QMD approach, while extremely expensive computationally, includes all transient effects such as dissociation/association of chemical bonds, and ionization/recombination. Consequently, the total pressure

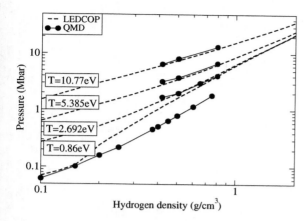

FIGURE 1. Comparison between the LEDCOP and QMD for hydrogen at four different temperatures[7]. Reproduced by permission of Astronomy and Astrophysics.

obtained from the QMD method reflects not only the constituency of the fluid at a given temperature and density but also various density effects, such as pressure ionization, which are only introduced in a phenomenological fashion in the LEDCOP code.

In LEDCOP, the EOS model is based on the Saha equation, where the bound Rydberg sequences are cutoff by plasma corrections. The modified Saha equation is solved iteratively to obtain a consistent set of ion abundances, bound state occupancies, and free electrons. In contrast to the QMD simulations, LEDCOP does not account at this point for the possibility of molecular formation.

Fig.1 shows a comparison of pressure as a function of densities for four hydrogen isotherms as obtained with LEDCOP and the QMD simulations. At the highest temperature shown, T=10.77eV (125000K), a reassuring concurrence exists in the predictions of the two methods. At this temperature and for the densities shown, hydrogen is largely a fully ionized plasma, consisting of non-degenerate electrons and protons. The good agreement between the model calculations and the QMD results demonstrates the adequacy of the approximation used in LEDCOP to describe hydrogen in this regime. As the temperature is lowered [e.g., 2.692eV (31250K)], the nature of the fluid becomes more complex, and a significant departure between the LEDCOP calculations and the simulations becomes noticeable, with significant discrepancy at the lowest temperature investigated, T=0.86eV (10000K).

Further analysis shows that the disagreement at the lowest temperature comes, at densities less than $0.1g/cm^3$, from the appearance of molecules in the QMD calculations which are not accounted for in the LEDCOP model. When the density is further increased, the LEDCOP model shows a rapid rise in the total pressure indicating that the atomic fluid rapidly ionizes within a very small density range. This appears as the consequence of the pressure ionization model used which places bound states of radius greater than the ion sphere radius into the continuum. The inadequacy of the model is directly highlighted by the QMD simulation which suggests a smoother increase in the total pressure and thus indicates that the fluid is still not fully ionized up to densities of $1g/cm^3$. Before investigating the repercussion of such a discrepancies in the EOS for the the Rosseland mean average, it is instructive to first make a direct comparison of the absorption coefficients and index of refraction as obtained by each approach.

We show in fig.(2), a comparison of these two quantities for the highest density-temperature point investigated, where we find that the respective EOSs are in good agreement. In fig.(2a) one sees that the LEDCOP and QMD results are in good agreement for photon energies around 20eV while the two methods diverge below and above this energy region. At high photon energies, the disagreement is due to the the limited number of bands used in the QMD method [upper limit of the sum in Eq.(2)]. Rather than increasing the number of bands in the QMD method which is rather expensive computationally, we instead extrapolated the high photon energy behavior of the absorption coefficient as an inverse power of the square of the photon energy. We have found that such an extrapolation suffices for the calculation of the Rosseland mean opacity and gives good agreement for high photon energies when compared with calculations using a larger number of bands.

For low photon energies, the rapid increase in the LEDCOP absorption coefficient can be traced back to the free-free contribution (inverse Bremschtralung) and the $1/\nu^3$ behavior of the Kramer's relation. It should be stressed also that in hand with the semiclassical Kramer's relation, various quantal, relativistic, and many body corrections are usually included in the form of a product of Gaunt factors which, in the present situation are extrapolated in

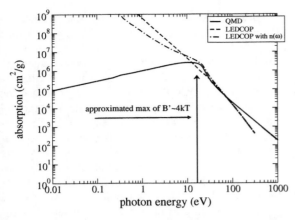
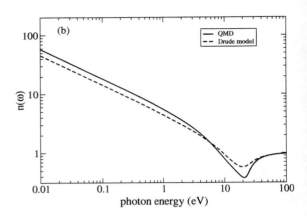

FIGURE 2. (a) Variation of the LEDCOP and QMD hydrogen absorption coefficients at a density of $\rho = 0.65 \text{g/cm}^3$ and T=48000K. Comparison of the QMD and Drude index of refraction for the same conditions as in (a). The vertical bar in (a) shows the position of the maximum of the derivative of the Planck function. Reproduced by permission of Astronomy and Astrophysics.

regmes outside of the intended conditions[1]. The comparison between the LEDCOP and QMD results shows that in the low photon energy range(<10eV), the absorption coefficient is dominated by the free-free contribution whose evaluation in this regime is tied to the calculation of the electrical conductivity of a poor metal. For the conditions shown in fig.(2), the DC electrical conductivity found using the QMD method is on the order of a few thousands (ohm cm)$^{-1}$. We also point out that, within the QMD method, the absorption coefficient, the conductivity, and the index of refraction are consistent by definition and recent conductivity measurements indicate a reasonable agreement between the QMD calculations and the available experimental data [19]. In light of the large discrepancies at low photon energy between the LEDCOP and QMD calculations shown in fig.2(a), it is useful to recall that the Kubo-Greenwood relation embodies the behavior of a Drude model for a simple metal[15]. We compare in fig.(2b) the index of refraction obtained with the outcome of a Drude model to further convince ourselves of the legitimacy of the QMD calculations. In fig(2b), starting at 100eV, the QMD index of refraction is close to unity down to photon energies corresponding to the plasma frequency where it becomes temporarily less than one before rapidly increasing to a value of about sixty at 0.01eV. This behavior is followed rather closely by the Drude model. As the LEDCOP model approximates the index of refrac-

tion as a step function equal to unity at the plasma frequency and higher energies, fig.(2b) shows that this approximation, also valid for a low-density high-temperature regime needs to be improved for a fair comparison between the two methods. We choose in the present case to directly use the QMD index of refraction to correct the LEDCOP opacity as described in [20]. Fig.(2b) shows the effect of such a correction on the LEDCOP absorption coefficient.

We finally turn to a direct comparison of the Rosseland mean opacities obtained using the two methods. Fig.(3) shows the variation of the Rosseland mean opacity for hydrogen as obtained for fixed density and for increasing temperature, fig.(3a), and for fixed temperature but increasing density, fig.(3b). For the LEDCOP results, we show both the original calculations and the ones corrected using the QMD index of refraction. Fig(3a) shows that as the temperature increases and the media, initially in a quasi-molecular state, continuously dissociates and ionizes, the QMD and LEDCOP calculations converge to similar values of the Rosseland mean opacities. Considering that the absorption coefficient, as displayed in fig.(2a), is only marginally affected by the temperature variation, we see that this result can be anticipated by the position of the maximum of the derivative of the Planck function. We see in fig.(2a) that at the highest temperature considered, the Rosseland average probes the photon energy region where the absorption coefficients are in the best agree-

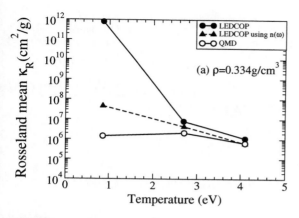 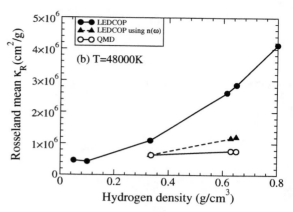

FIGURE 3. Variation of the LEDCOP and QMD Rosseland mean opacities (a) as a function of temperature and for a fixed density of $\rho = 0.334 \text{g/cm}^3$, (b) as a function of density and for a fixed temperature of T=48000K[7]. Reproduced by permission of Astronomy and Astrophysics.

ment. As the temperature is lowered, the difference in the opacities obtained can be traced back to the low photon energy behavior discussed above. For a fixed temperature of 48000K, one sees in fig.(3b) that when the density is increased, discrepancies arise as the many-body effects become more important. We further see in fig.(3b), that introducing collective effects in the LEDCOP results via the use of the QMD index of refraction improves the agreement between the two methods.

In conclusion, we showed, by making a direct comparison with an atomic opacity code, LEDCOP, that QMD simulations can be used effectively to calculate Rosseland mean opacities. While the QMD method is computationally expensive and can only be applied to a limited density-temperature region compared to atomic modeling codes, it provides however a powerful tool to validate plasma models used in atomic physics approaches in the warm dense matter regime by providing a consistent set of material, electrical and optical properties from the same simulation.

ACKNOWLEDGMENTS

Work supported under the auspices of the U.S. Department of Energy at Los Alamos National Laboratory under Contract W-7405-ENG-36.

REFERENCES

1. J.P. Cox, and R. T. Giuli, *Principles of Stellar Structure*, Gordon & Breach London (1968).
2. G. W. Collins *et al.*, Science **281**(5380) 1178 (1998) and references therein.
3. C. A. Iglesias *et al.*, APJL **569**, L111 (2002).
4. G. Fontaine *et al.*, PASP **113**, 409 (2001).
5. J.F. Benage *et al.*, Phys. Rev. Lett. **83**, 2953 (1999).
6. J. P. Chittenden *et al.*, Phys. Rev. E **61**, 4370 (2000).
7. S. Mazevet *et al.*, Astronomy. and astrophysics, **405**, L5 (2003).
8. C. Neuforge-Verheecke *et al.*, APJ **561**, 450 (2001).
9. T.J. Lenosky *et al.*, Phys. Rev. B **61**, 1 (2000).
10. G. Kresse and J. Hafner, Phys. Rev. B **47**, RC558 (1993); G. Kresse and J. Furthmüller, Comput. Mat. Sci. **6**, 15–50 (1996); G. Kresse and J. Furthmüller, Phys. Rev. B **54**, 111.
11. D. Vanderbilt, Phys. Rev. B **41** 7892 (1990). 69 (1996).
12. G. Kresse and J. Hafner, J. Phys. Condens. Matter **6**, 8245 (1994).
13. J. P. Perdew, in *Electronic Structure of Solids*, ed. by F. Ziesche and H. Eschrig (Akademie Verlag, Berlin, 1991).
14. J. Callaway, *Quantum Theory of the solid state*, Academic Press New York, (1974).
15. W. A. Harrison, *Solid State Theory*, Mc Graw-Hill, (1970).
16. G. Kresse, and J. Joubert, Phys. Rev. B **59**, 1758 (1999).
17. P.E. Blöchl, Phys. Rev. B **50**, 17953 (1994).
18. M. Desjourlais, *et al.*, Phys. Rev. E **66**, 025401 (2002).
19. L. Collins, *et al.*, Phys. Rev. B **63**, 184110 (2001).
20. F. Perrot, Laser and particle Beams **44**, 731 (1996).

CP706, *Shock Compression of Condensed Matter - 2003*
edited by M. D. Furnish, Y. M. Gupta, and J. W. Forbes
© 2004 American Institute of Physics 0-7354-0181-0/04/$22.00

SIMULATIONS OF RAPID PRESSURE-INDUCED SOLIDIFICATION IN MOLTEN METALS

Mehul V. Patel* and **Frederick H. Streitz***

*Lawrence Livermore National Laboratory, L-045, 7000 East Ave, Livermore CA 94550

Abstract. The process of interest in this study is the solidification of a molten metal subjected to rapid pressurization. Most details about solidification occurring when the liquid-solid coexistence line is suddenly transversed along the pressure axis remain unknown. We present preliminary results from an ongoing study of this process for both simple models of metals (Cu) and more sophisticated material models (MGPT potentials for Ta). Atomistic (molecular dynamics) simulations are used to extract details such as the time and length scales that govern these processes. Starting with relatively simple potential models, we demonstrate how molecular dynamics can be used to study solidification. Local and global order parameters that aid in characterizing the phase have been identified, and the dependence of the solidification time on the phase space distance between the final (P,T) state and the coexistence line has been characterized.

INTRODUCTION

The nature of the solidification process has long been an area of scientific and technical interest. The confluence of peaked experimental interest in this problem and recent theoretical and computational advances that may allow for a breakthrough in understanding, have resulted in many recent studies [1]. While solidification is most commonly studied by varying temperature at fixed (ambient) pressure, *i.e.* freezing, the goal of this research is to study the other axis, namely rapid, pressure-induced solidification. The questions to be answered are numerous and include the determination of the final state (crystalline or glassy?), the transition path, and the time scales for the relevant processes. Atomistic simulations show great potential for answering many of the detailed questions describing how a disorder to order phase transition proceeds.

Figure 1 shows the phase diagram of tantalum, with the horizontal arrow depicting the isothermal compression path being studied in this work. This path lies close to an isentrope, a path being investigated in the experiments that are also part of this effort [2]. (Also shown is the principle hugoniot, which

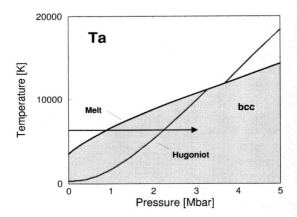

FIGURE 1. Phase diagram of tantalum. The arrow demonstrates an isothermal compression, which is the solidification path being studied. Also shown is the principle hugoniot.

is the locus of all possible states attainable by a shock from ambient temperature and pressure). In this paper, we will present some preliminary results studying the dependence of solidification time on the degree of over-pressurization. This is analogous to the

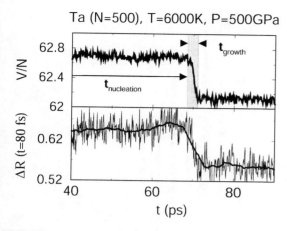

Ta (N=500), T=6000K, P=500GPa

FIGURE 2. Typical simulation output for Ta at 6000 K, showing phase transition signatures in the volume and average displacement. The nucleation time and growth time are noted on the volume trace.

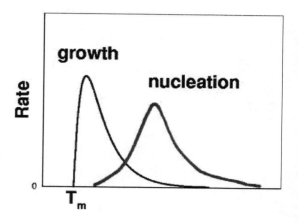

FIGURE 3. Sketch showing the dependence of the nucleation and growth rates on the degree of under-cooling below the melting temperature, T_m. The solidification time is minimized in the region where the nucleation rate and growth rate overlap.

time-temperature-transition studies that look at the dependence of the solidification rate on the degree of under-cooling.

METHODOLOGY

Constant NPT molecular dynamics simulations using a stochastic Langevin thermostat are employed for this study. Copper is modeled using the EAM potential developed by Johnson and Oh [3]. For Ta, we utilize the more sophisticated Model Generalized Pseudopotential Theory (MGPT) potentials [4]. These potentials were developed for metals having more complex electronic structure, and have been used with success in a number of other high pressure studies [5].

An equilibrated molten sample is prepared and subjected to a sudden compression, reaching final pressure in 1-2 ps. After pressurization, the system is observed until a phase transition, if any, occurs. Evidence for the transition can be seen in several global observables such as the energy, density (atomic volume), diffusion constant, and pair correlation function. Local order parameters [6] were also used to study the transition and are especially useful when a metastable phase is present.

For example, Figure 2 shows output from a typical tantalum simulation. One can see signatures of the

transition in the atomic volume as well as the average displacement (in this case plotted for an interval of 80 fs). Figure 2 also shows how we define the solidification timescales: $t_{\text{solidification}} = t_{\text{nucleation}} + t_{\text{growth}}$.

Using our machine-precision restart capability (even for this stochastic, parallel code), we can study the simulation starting from just before the transition using a variety of analysis tools. With so many parameters to contend with (final state, loading rates, etc), we choose to focus on final pressure (analagous to under-cooling) as the independent variable and fix the others.

TIME-PRESSURE-TRANSITION ANALYSIS FOR EAM COPPER

TTT Plots

In classical nucleation theory (developed to describe the freezing process), nucleation and growth rates are described by Arrhenius equations that depend on quantities such as the nucleation energy and the diffusive energy barrier. They have been sketched in Figure 3 which highlights the non-monotonic dependence of the nucleation and growth rates on the amount of under-cooling. The overlap region in which both nucleation and growth are favored de-

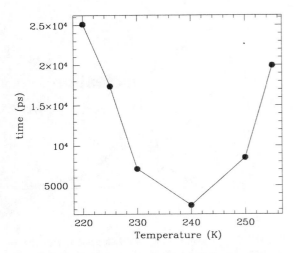

FIGURE 4. An example of a Time-Temperature-Transition (TTT) plot obtained from MD simulations. Average crystallization time for water as a function of temperature shown for a density of 1.15 g/cm³. Plot adapted from Ref. [7].

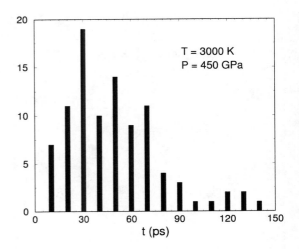

FIGURE 5. Distribution of solidification times in copper (N=500 atoms) for a fixed final temperature and pressure.

fines the domain of the Time-Temperature-Transition plots, in which the transition time is plotted against the final temperature as shown in Figure 4.

The intersection of the nucleation rate and growth rate in Figure 3 defines the optimum under-cooling temperature (corresponding to the "nose" in Figure 4) where the transition time is minimized. As is clear from the figure, changing the final temperature by only 10% can lead to an order of magnitude change in the crystallization time. Given the limited time scales accessible to MD simulation, such an order of magnitude shift in expected crystallization time can be the difference between observing a transition during the simulation or not. In the present study, we seek to determine if similar behavior is present along the pressure axis, namely, is there an optimal degree of over-pressurization that minimizes the liquid-solid transition time?

Preliminary Results

The solidification times, as defined in Figure 2, were obtained for a variety of different final pressures and temperatures for a large set of independent samples having N=500 atoms. We obtain an average solidification time with error estimates for

each temperature and final pressure by analyzing the (rather broad) distribution of independent solidification times. Figure 5 displays one such distribution. We plot the resulting average solidification time against the final pressure to create a time-pressure-transition (TPT) plot, shown in in Figure 6 for EAM copper at T = 3500K. The transition time initially decreases with increasing over-pressure, but eventually reaches a plateau, which defines a minimum solidification time. The initial decrease is expected, since the transition time must diverge to infinity at the melt line. However, the lack of a transition time minimum stands in stark contrast to the behaviour demonstrated in a TTT plot, where there is a clear minimum in the transition time as a function of under-cooling. The viscosity is a sharply increasing function of pressure, which should serve to hinder the nucleation process and drive the solidification times up. The lack of a clear minimum may indicate that the relative loss of mobility with over-pressure is compensated by the increased thermodynamic driving force for solidification, and suggests that the formation of a metastable state by rapid over-pressurization is unlikely. A detailed discussion of these issues will be presented elsewhere [8].

Similar behavior is observed in TPT plots over a range of different temperatures. If we define the optimal pressure to be that pressure at which the transition time plateaus to its minimum value, we can plot these optimal phase space points on the phase

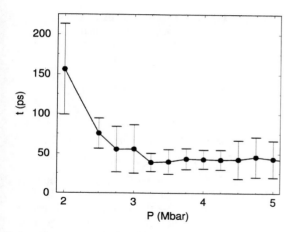

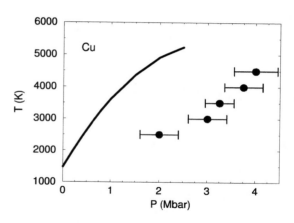

FIGURE 6. Time-Pressure-Transition (TPT) plot for EAM copper at T = 3500 K (N=500) showing the dependence of the solidification time on the final pressure.

FIGURE 7. Locus of phase space points where solidification appears optimal plotted along side the copper melt line from Ref. [9].

diagram as is shown in Figure 7. At the level of the error bars shown, these points seem to follow the melt curve. Further analysis using larger simulation cells is needed to make better quantitative comparisons and to determine the role of finite size effects in these preliminary calculations.

CONCLUSION

Atomistic simulations of pressure induced solidification pose many challenges but also show potential for answering basic questions about this process, and for bridging the gap to the next length and time scales. Some preliminary results have been presented here to illustrate the kinds of questions that we hope to answer. Simulations on larger systems are underway to assess the magnitude of finite size effects and to study any size scaling that can be used advantageously.

ACKNOWLEDGMENTS

Work performed under the auspices of the U.S. DOE at the University of California/Lawrence Livermore National Laboratory under contract W-7405-ENG-48.

REFERENCES

1. Day, C., *Physics Today*, **7**, 24 (2003).
2. Nguyen, J. H., Orlikowski, D., Streitz, F. H., Minich, R., Holmes, N. C., and Moriarty, J. A., *in preparation* (2004).
3. Oh, D. J., and Johnson, R. A., *J. Mater. Res.*, **3**, 471 (1988).
4. Yang, L. H., Söderlind, P., and Moriarty, J. A., *Phil. Mag. A*, **81**, 1355 (2001).
5. Moriarty, J., Belak, J., Rudd, R., Söderlind, P., Streitz, F., and Yang, L., *J. Phys. Cond. Matt.*, **14**, 2825 (2002).
6. Steinhardt, P., Nelson, D. R., and Ronchetti, M., *Phys. Rev. B*, **28**, 784 (1983).
7. Yamada, M., Mossa, S., Stanley, H. E., and Sciortino, F., *Phys. Rev. Lett.*, **88**, 35501 (2002).
8. Patel, M. V., and Streitz, F. H., in preparation (2004).
9. Jeong, J., and Chang, K. J., *J. Phys. Cond. Matt.*, **11**, 3799 (1999).

CP706, *Shock Compression of Condensed Matter - 2003*
edited by M. D. Furnish, Y. M. Gupta, and J. W. Forbes
© 2004 American Institute of Physics 0-7354-0181-0/04/$22.00

SHOCK WAVES AND SOLITARY WAVES IN BCC CRYSTALS

J.ROTH

*Institut für Theoretische und Angewandte Physik, Universität Stuttgart, Pfaffenwaldring 57, 70550
Stuttgart, Germany*

Abstract. The effect of shock waves in bcc crystals have been studied by means of molecular dynamics simulations. The interaction between the atoms is modelled by Dzugutov potentials [M. Dzugutov, J. Non-Cryst. Sol. **131** (1991) 62]. Depending on the strength of the shock wave and the propagation direction different phase transition processes have been observed, leading to perfect and defective close-packed phases.

Elastically-dispersive and plastic wave fronts accompany the expanding shock waves. If the propagation is along the three-fold direction additional solitary waves are observed. At elevated temperatures and increased cross-sections the solitary waves get damped. The conditions for the occurance of solitary waves will be discussed.

INTRODUCTION

The purpose of this paper is two-fold. First we will present a selection of our results on shock waves in media with Dzugutov potential interactions[1]. A full account will be published elsewhere[2]. Second we will discuss the phenomenon of solitary waves which has been observed on several occasions together with the shock waves[3, 4, 5, 6]. The paper is organized as follows: in the first part we will present the interaction, its phase diagram and the general setup for the simulations. The second part contains the shock wave effects and the third part the properties of the solitary waves and our solitary wave model.

INTERACTION AND SIMULATION SETUP

In 1991 Dzugutov[1] introduced a new potential, designed to avoid crystallization in cooling simulations of monatomic fluids. The attractive part of the potential is similiar to the well-known Lennard-Jones potential. The special feature of the potential is a maximum at about $\sqrt{2}$ times the nearest neighbor distance a, designed to disfavor close-packed phases. The phase diagram of the Dzugutov potential has

been described in [7]. At $T^* = 0^1$ and $P^* = 0$ the stable phase is bcc. With increasing pressure a σ phase (which will not be considered in the following) becomes more stable in a short pressure interval, and at $P^* > 5.5$ close-packed (cp) phases are the most stable structures. It has not been possible to distinguish between the stability of fcc, hcp or more complex stackings.

The shock wave simulations were carried out by the standard molecular dynamics method in the microcanonical ensemble. Samples with up to 5 million atoms have been studied, with transverse dimensions of up to 120 nearest neighbor distances a and length up to 600 a. After preparation, the samples were equilibrated for a time interval of $T = 10a\sqrt{m/\varepsilon}$ using the Nosé-Hoover thermostat to simulate the canonical ensemble. Along the transverse direction we applied periodic boundary conditions, and open boundaries along the along the longitudinal direction. The shock waves were generated by the momentum mirror method[8].

---

[1] All results are given in units of the nearest neighbor distance a, the particle mass m, and the depth of the potential minimum ε.

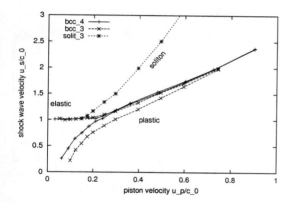

FIGURE 1. Hugoniot diagram of some selected waves. The indices indicate the 3. and 4-fold propagation direction.

SHOCK WAVES

The Hugoniot diagram

The u_s-u_p-diagram for the bcc initial structure has been determined for several crystal orientations (See Fig. 1). A feature common to all directions is the constant velocity of the elastic wave which is the longitudinal sound velocity c_0. Already at very low piston velocities u_p a plastic wave is observed which indicates a phase transition to a cp structure. If we compare the pressures we find that the plastic wave starts at the point where the longitudinal stress reaches the transition pressure from bcc to close packed. The plastic and the elastic wave flow into a common wave at about $u_p = 0.2c_0$. The differences at this point will be explained in the following subsections.

In addition to these wave phenomena a solitray wave is observed along the three-fold direction. The velocity of this wave is larger than the elastic and plastic wave and grows rapidly. Solitons in fcc-crystals[3, 6], on the other hand, have the same velocitiy as the plastic wave. There is also an unstable branch at lower piston velocities (not shown here, see [4]).

The four-fold direction

If the shock direction is parallel to a four-fold symmetry axis we get a sharp phase transition. The crystal structure changes from bcc to cp within two lattice planes. This is the reason why we could determine only one wave velocity if $u_p > 0.25$.

At the beginning of the simulation, close to the momentum mirror, we find two single cp crystals which are separated by a grain boundary parallel to the shock direction. Later, close to the position of the plastic wave front, we find only a single crystal. A dislocation or triple junction must terminate the grain boundary. A three-fold direction of the new cp phase is always parallel to the shock direction, and a two-fold direction parallel to the face diagonal of the sample cross section. The stacking along the three-fold direction frequently changes between fcc and hcp. This is the common picture at moderate piston velocities up to about $u_p = 0.4$. If the shock waves are stronger, and the plastic wave gets faster, there is no longer a perfect single crystal but an increasing number of differently oriented crystallites.

Other directions

Along other crystal directions we find similar phenomena. The details, however, are different. There is no sharp, clean transition as in the four-fold direction, and also no single cp crystal. Therefore it is possible to distinguish between the time when the transition starts and when it is completed. This is indicated by two branches of the plastic wave in the three-fold direction shown in Fig. 1. If we look at pictures of the samples after shocking, we observe that all atom rows parallel to the shock direction are bent or broken in the same way. This indicates that lattice planes perpendicular to the shock direction have been shifted more or less undistorted in the transverse direction. Alternatively, it happens that the lattice planes are not shifted but rotated around an axis parallel to the shock direction. A single crystal is only observed close to the momentum mirror, with stackings switching between fcc and hcp. At increasing piston velocities we find again that the correlated motion is broken up and many crystallites form.

In the four-fold direction it was possible to determine the transformed crystal structure directly at moderate piston velocities. Here and at high piston velocities in the four-fold direction we have determined the structure of the crystallites by means of the radial and angular distribution functions, which yielded definitely that we have a defective cp structure.

SOLITARY WAVES

Solitary waves have been observed on several occasions in the history of shock wave simulations[3, 4, 5]. First they have been found and explained in one-dimensional simulations, but the solitary waves there are not of the same type as the ones discussed here[9, 10]. The solitary waves are qualitatively independent of the interaction. In bcc they have been observed with the Dzugutov potential and EAM potentials for iron[5]. They are dependent on the crystal structure and orientation. In fcc they occur along a two-fold direction[3], in bcc along the three-fold direction, and in a specially prepared sc crystal they have been observed along the four-fold direction[4].

Properties of the solitary waves

Here we describe the solitary waves in the bcc-Dzugutov model. Solitary waves in other structures and with different interactions might differ in detail. At moderate piston velocities $u_p \approx 0.4$, there are several solitary wave trains. If the shock intensity increases, their number shrinks, and finally we are left with only one wave maximum at $u_p \approx 0.8$. For simplicity, we will concentrate on this situation. The full width of the solitary wave maxima at have maxmium is always about 6 a. No dispersion is observed.

Fig. 2 displays the relative displacement of atoms along three arbitrarily selected atom rows. The solitary wave is at about $x = 167a$. The digram clearly shows that there is one distance which is shortened by about 25% of the interatomic distance. The preceding and the following distances are shortened by a small amount only. The shortening is strongly correlated between different atom rows. If there are several solitary wave maxima, then there are also several of displacement jumps. The crystal structure is not altered by the solitary wave, but the displacement modulation after the wave shows that the longitudinal temperature of the sample has increased from $T^* = 0.001$ to about 0.02.

The sharpness of the displacement jump clearly indicates that a continuum model soliton description is not possible, a discrete model is required.

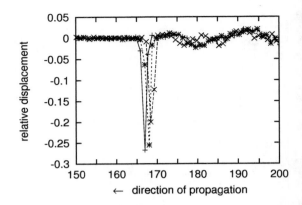

FIGURE 2. Relative displacement of neighboring atoms along three atom rows parallel to the shock direction

The decay of the solitary waves

Solitary waves have been studied in samples with transverse dimension from 5 to 120 interatomic distances a. Even in samples with open transverse boundaries they have been observed. If the transverse dimension gets larger than about 40 a, it seems as if the solitary waves get damped, but there is still no dispersion (See Fig. 3). The reason for the apparent damping has been derived from the surface formed by the maxima of the local pressure. The surface is perfectly flat at the beginning of the simulation. After some time the surface starts to fluctuate randomly with increasing amplitude. Then certain modes start to grow whereas the other modes decay, and a surfaces with a periodic modulation is created. The periodic modulation is commensurate with the periodic boundary conditions, and there are always one or two maxima along a transverse directions. In thin samples the solitary waves also decay if the temperature is increased. Due to the lack of space this damping will not be discussed here (See [4]).

The solitary waves decay if the transverse direction is large enough since the correlation between the atom rows parallel to the shock direction is lost. The decay further indicates that the solitary waves are three-dimensional phenomena and a modelling would require linear differential equations for the transverse directions.

To make sure that the solitary waves are not only artefacts of the simulation setup and the initial conditions we have tried to suppress them. At the beginning of the simulation a singularity occurs when the

304

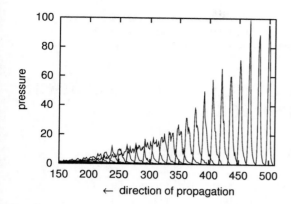

← direction of propagation

FIGURE 3. Time sequence of the decay of the local pressure at the position of the soliton

sample and its mirror image collide at full speed. To avoid this we have started the simulation at low velocity and accelerated the sample up to the final piston velocity. The solitary waves are modified indeed, but the phenomenology only shifts to that at lower piston velocities. If the acceleration is slow enough such that no solitary waves are generated at the beginning then they show up later! Thus it is clear, that the solitary waves are no artefact of the simulation setup. Due to the damping, however, it may be hard to detect them in experiments.

A model for solitary waves

Instead of setting up three-dimensional non-linear difference-differential equations for the solitary waves which would require a number of approximations and could still not be solved analytically we have chosen another way to explain the occurance or avoidance of solitary waves along the major symmetry directions.

First we mention some further observations. Solitary waves have always been observed along directions where atoms form uninterrupted rows of nearest neighbor atoms (with distance a). The shortening of the interatomic distance leads to an increase in potential energy of about 70ε. In bcc we never find solitary waves along directions perpendicular to a four-fold direction, this means especially that there are no solitary waves along the two- and four-fold direction.

The model is the following: solitary waves can be described as the propagation of a direct kick between atoms along a row. The atoms behave more or less like hard spheres due to the large energy increase. The neighboring rows are also largely independent due to this large energy. The solitary waves cannot propagate if the atom rows are interrupted, and they get damped if the continuous rows are thinned out by different atom arrangements in between them. No solitary waves occur especially perpendicularly to the four-fold direction in bcc since in this case the kick goes directly into a gap between two atoms.

The model can explain solitary waves along the major symmetry directions in the monatomic structures studied so far. It cannot explain the complicated pattern of occurance and avoidance of solitary waves along arbitrary directions in bcc crystals[4]. For example along the [321] direction solitary waves have been observed which are generated by very complicated diagonal kick sequences.

But it is also doubtful that the non-linear difference-differential equation description could help, since the only parameters which enter these equations are the elastic constants and they vary smoothly with the crystal orientation.

ACKNOWLEDGMENTS

The author wants to acknowledge helpful discussions with H.-R. Trebin and especially with K. Kadau. Part of this research was supported by the DFG through the SFB 382, and by the Los Alamos National Lab.

REFERENCES

1. Dzugutov, M., *J. Non-Cryst. Sol.*, **131**, 62 (1991).
2. Roth, J., *Shock waves in Dzugutov crystals, to be published* (2003).
3. Germann, T., Holian, B., Lohmdahl, P., and Ravelo, R., *Phys. Rev. Lett.*, **84**, 5351 (2000).
4. Roth, J., *to be published* (2003).
5. Kadau, K., *these proceedings* (2003).
6. Zvbin, S., Zhakhovskii, V., Oleynik, I., Elert, M., and White, C., *these proceedings* (2003).
7. Roth, J., and Denton, A., *Phys. Rev. E*, **61**, 6845 (2000).
8. Holian, B., and Lomdahl, P., *Science*, **280**, 2085 (1998).
9. Holian, B., and Straub, G., *Phys. Rev. B*, **18**, 1593 (1978).
10. Holian, B., Flaschka, H., and McLaughlin, D., *Phys. Rev. A*, **24**, 2595 (1981).

CP706, *Shock Compression of Condensed Matter - 2003*
edited by M. D. Furnish, Y. M. Gupta, and J. W. Forbes
© 2004 American Institute of Physics 0-7354-0181-0/04/$22.00

MOLECULAR DYNAMICS STUDY OF NON-REACTING SHOCK WAVES IN ANTHRACENE

S. V. Zybin[1], M. L. Elert[2], and C. T. White[3]

[1]*Department of Chemistry, The George Washington University, Washington, DC 20052*
[2]*Chemistry Department, U. S. Naval Academy, Annapolis, MD 21402*
[3]*Naval Research Laboratory, Washington, DC 20375*

Abstract. We use nonequilibrium molecular dynamics (MD) simulations to study the behavior of solid anthracene under shock compression in different crystallographic directions in the absence of chemical dissociations. The interatomic forces were modeled using a recently modified reactive empirical bond order (REBO) potential with intermolecular interactions, termed the adaptive intermolecular REBO potential (AIREBO). These simulations were undertaken to get insight into the microscopic mechanisms of shock compression and energy dissipation in molecular crystals, and in particular how the shape of anthracene molecules affects the properties of the shock layer.

INTRODUCTION

An important development in the study of shock-induced phenomena in solids is the observation of orientation dependence of shock-induced chemistry both in experiments on single-crystal explosives [1] and simulation of covalently-bonded solids [2], which has so far been explained by an anisotropy of plastic flow under shock compression. Another possible mechanism for non-equilibrium rate phenomena is an anisotropy of kinetics (i.e. energy transfer) within a shock layer. In particular, MD simulations in rare gas solids [3] have shown that a momentum flow along the shock direction has highly non-Maxwellian character at the shock front.

However, there have been as yet no considerable efforts on MD simulations of such phenomena in organic molecular crystals, which can have quite anisotropic potential energy surfaces [4]. They are characterized by two distinct interactions: strong covalent bonding between atoms within a single molecule and, on the other hand, weak London dispersion forces between the molecules.

In this paper, we describe the first MD simulations of shock propagation in different directions of anthracene single crystals, which is regarded as a model system for molecular crystal studies [4]. We use the AIREBO potential [5], which includes both covalent bonding and van der Waals interactions. This potential has been recently developed for hydrocarbons on a basis of REBO potential [6,7] and applied in studies of tribology of hydrocarbon-coated diamond surfaces [8] and shock-induced chemistry in hydrocarbon solids [9]. The simulations could help in understanding shock-induced phenomena in molecular crystals such as energy transfer between translational and internal degrees of freedom, which is important for the initiation of chemical reactions.

SIMULATION MODEL

The model system is an anthracene molecular crystal, represented by a nanosample of about 2000 anthracene molecules arranged into a supercell with periodic boundaries applied in the transverse (*x*,*y*) directions. The supercell length in the shock direction *z* varied from 20 nm to 30 nm with a corresponding cross-section area of 4×4 nm^2 to 7×4 nm^2.

The anthracene crystal structure obtained from X-ray diffraction data [10-12] is the monoclinic space group P2$_1$/a (see Fig.1), which is characterized by the lattice parameters shown in Table 1. The unit cell parameters as well as positions of the anthracene molecules within it were determined for the AIREBO potential (see Table 1) by allowing the crystal to change its shape and sizes at 0 K in order to minimize the total potential energy of interactions using the steepest descent algorithm. The anthracene crystal is quite anisotropic: the *ab* planes consist of pairs of anthracene molecules arranged in a herringbone pattern while *ac* planes displays layered structure along the crystallographic *c* direction.

In the piston-driven simulations reported here, shock waves are produced by pushing the crystal supercell toward a repulsive potential wall at the particle velocity −*u$_p$*.

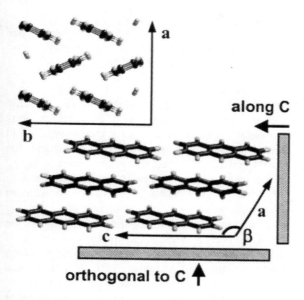

FIGURE 1. Structure of the anthracene crystal, illustrating the anisotropy of shock compression in different directions in our piston-driven MD simulations.

TABLE 1. Experimental (T=295 K) and AIREBO potential (T=0 K) unit-cell parameters of anthracene

Unit-cell parameters	Experiment	AIREBO
a (Å)	8.553	8.481
b (Å)	6.016	6.036
c (Å)	11.172	11.574
β (degrees)	124.60	123.04

The supercell axis z of shock compression is oriented along c direction or, otherwise, orthogonal to it, while the transverse axis y is always oriented along the b direction.

RESULTS AND DISCUSSION

In a series of simulations, we have shocked an anthracene crystal at particle velocities *u$_p$* below the threshold for initiation of chemical reactions.

The experimental studies have provided an evidence of possible dimerization of the anthracene molecules at *u$_p$* > 2.22 km/s [13,14]. However, in our MD simulations of shock-induced chemistry in anthracene [15] no dissociation or dimerization was observed up to *u$_p$* = 4 *km/s*. This difference between the simulation and the experiment may be explained by too stiff intermolecular repulsion in the AIREBO potential which results in relatively higher pressure required to compress an anthracene sufficiently (V/V$_0$ ~ 0.65) that dimerization can occur. Besides, a typical time of shock passage through the anthracene supercell is a few picoseconds, which seems too short for dimerization rate at particle velocities *u$_p$* < 4 km/s.

MD simulations are capable of providing unmatched insight into the nanoscale structure of the shock layer. Figure 2 shows snapshots from the simulations in different directions at the same particle velocity. The shape of anthracene molecules makes the process of their deformation and molecular bond distortion greatly dependent on the shock orientation. In particular, the shock wave "along *c*" direction exhibits significant bending deformation of molecules in contrast to the "orthogonal to *c*" direction.

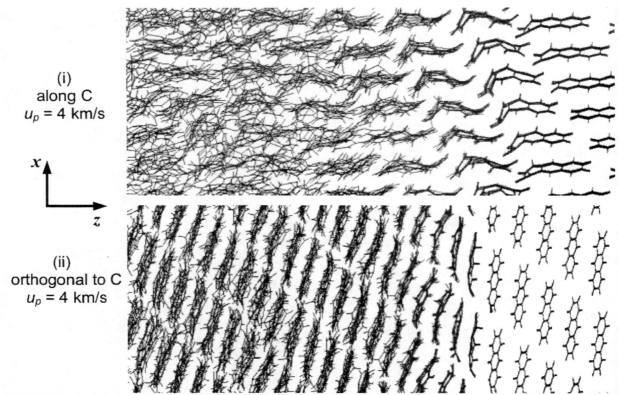

FIGURE 2. Snapshots of shock simulations in the anthracene supercell (only small part of the crystal is shown). Shock wave is propagating from left to right along the supercell axis z in two different directions: (i) along c, (ii) orthogonal to c. Shock wave velocity U_s=10.8 km/s for the particle velocity u_p=4 km/s and initial density ρ_0=1.192 g/cm^3.

To find out how the kinetics of material response to shock loading is affected by orientation we have calculated the local flow densities using Hardy's approach [16]. Shock profiles are shown in Figs. 3 and 4, where the components of mass velocity and kinetic temperature are defined as

$$u_a = \frac{1}{M}\sum_i m_i v_{ia}, \quad M = \sum_i m_i, \quad a = x, y, z$$

$$T_{aa} = \frac{1}{N}\sum_i m_i (v_{ia} - u_a)^2,$$

$$T_{aa}^{mol} = \frac{1}{N^{mol}}\sum_k \frac{1}{n}\sum_{i\in(k)} m_i (v_{ia} - u_a^{(k)})^2, \quad k = 1..N^{mol}$$

where m_i and v_i are the mass and velocity of the i-th particle, u^k is the center of mass velocity of the k-th molecule, N and N^{mol} are the numbers of particles and molecules in the averaging slab, respectively. Here T_{xx}, T_{yy} and T_{zz} represent the dispersion of non-equilibrium momentum distributions inside the whole slab, while the T^{mol} components describe *intramolecular* kinetic fluctuations. The shock wave "along c" exhibits overshoot in the T_{xx} temperature due to bending in the transverse direction x. On the other hand, the "orthogonal to c" shock shows overshoot in the T_{zz} component (also typical for atomic solids) because of the highly non-Maxwellian momentum distribution along the shock direction z.

The intramolecular temperature T^{mol} follows the corresponding components of the total kinetic temperature without any significant delay, which suggests quite fast excitation of deformations and vibrations of anthracene molecules which agrees well with Fig. 2, which shows molecules bending and hydrogen bonds flapping near to the front. Of course, this preliminary analysis should be complemented by calculations of molecular internal modes.

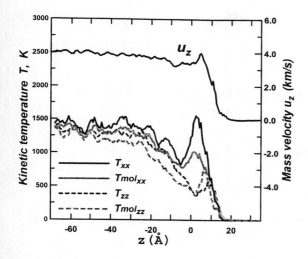

FIGURE 3. Profiles of mass velocity v_z and components of kinetic temperature for a shock wave propagating in "along c" direction.

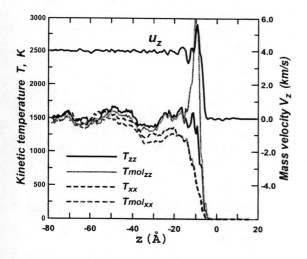

FIGURE 4. Profiles of mass velocity v_z and components of kinetic temperature for a shock wave propagating in "orthogonal to c" direction.

In summary, we have performed the first MD study of orientation effects in non-reacting shock waves in an anthracene crystal using the AIREBO potential, which incorporates intra- and intermolecular interactions.

ACKNOWLEDGEMENTS

Funding was provided by the Office of Naval Research. The authors thank J.A. Harrison for useful discussions.

REFERENCES

1. J.J. Dick, *J.Phys.Chem.* **97**, 6193-6196 (1993).
2. S.V. Zybin, M.L. Elert, and C.T. White, *Phys. Rev. B,* **66**, 220102(R) (2002).
3. V.V. Zhakhovskii, S.V. Zybin, K. Nishihara, and S.I.Anisimov, *Phys.Rev.Lett.* **83**, 1175-1179 (1999).
4. A.I. Kitaigorodsky, *Molecular Crystals and Molecules (Acad. Press, NY, 1973).*
5. S.J. Stuart, A.B. Tutein, and J.A. Harrison, *J. Chem. Phys.* **112**, 6472-6486 (2000).
6. D.W. Brenner, *Phys. Rev. B,* **42**, 9458-9471 (1990).
7. D.W. Brenner, O.A. Shenderova, J.A. Harrison, S.J. Stuart, B. Ni, and S.B. Sinnott, *J. Phys.: Condens. Matter*, **14**, 783-802 (2002).
8. A.B. Tutein, S.J. Stuart, and J.A. Harrison, *J. Phys. Chem. B,* **103**, 11357-11365 (1999); Langmuir, **16**, 291-296 (2000).
9. M.L. Elert, S.V. Zybin, and C.T. White, *J. Chem. Phys.*, **118**, 9795-9801 (2003).
10. R. Mason, *Acta Crystallogr.* **17**, 547-555 (1964).
11. C.P. Brock and J.D. Dunitz, *Acta Crystallogr. Sect. B: Struct. Sci.* **46**, 795-806 (1990).
12. G.R. Desiraju and A. Gavezzotti, *Acta Crystallogr. Sect. B: Struct. Sci.* **45**, 473-482 (1989).
13. R.H. Warnes, *J. Chem. Phys.* **53**, 1088-1094 (1970).
14. R. Engelke and N.C. Blais, *J. Chem. Phys.* **101**, 10961-10972 (1994).
15. M.L. Elert, S.V. Zybin, and C.T. White, this proceedings.
16. S. Root, R.J. Hardy, and D.R. Swanson, *J. Chem. Phys.* **118**, 3161-3165 (2003).

CP706, *Shock Compression of Condensed Matter - 2003*
edited by M. D. Furnish, Y. M. Gupta, and J. W. Forbes
© 2004 American Institute of Physics 0-7354-0181-0/04/$22.00

MOLECULAR DYNAMICS STUDIES OF ORIENTATION DEPENDENCE OF SHOCK STRUCTURE IN SOLIDS

S. V. Zybin[1], V.V. Zhakhovskii[2], M. L. Elert[3], and C. T. White[4]

[1]*Department of Chemistry, The George Washington University, Washington, DC 20052*
[2]*Institute of Laser Engineering, Osaka University, Osaka 565-0871, Japan*
[3]*Chemistry Department, U. S. Naval Academy, Annapolis, MD 21402*
[4]*Naval Research Laboratory, Washington, DC 20375*

Abstract. Molecular dynamics (MD) simulations using empirical potentials have proven to be an efficient tool for study at the lattice level of non-equilibrium phenomena in solids under shock loading. Such anisotropic properties of a single crystal as elastic constants, slip directions and especially shear stresses might significantly affect the internal structure of a shock wave in different directions of high strain rate uniaxial shock compression. Specifically, the mechanisms of lattice deformation, plasticity, and relaxation of shear stresses, structure of the elastic precursor, as well as shock-induced chemistry are found to depend crucially on the direction of shock propagation.

INTRODUCTION

Classical MD simulations provide unmatched insight into non-equilibrium processes of shock compression in solids at the atomistic level. Such simulations have been successfully applied to complex shock-induced phenomena such as elastic [1] and plastic [2,3] deformations, shock splitting caused by elastic-plastic [4-7] and phase [8,9] transformations, melting [10,11], and simple detonations [12]. However, until recently their potential has remained largely untapped, not only because sample sizes and times are still limited to the nanoscale, but also because of only a gradually emerging appreciation of the ability of MD simulations to make useful predictions both for shock wave theory and experiment.

This paper will briefly review some of the recent achievements of classical MD simulations on orientation dependence of shock-induced plasticity, melting and chemistry in solids. Three prototypical potential models will be discussed here, representative of three classes of crystalline solids, namely: (1) Lennard-Jones potentials (LJ) for simple atomic

solids, (2) reactive empirical bond order (REBO) potentials [13-15] for covalently bonded solids, and (3) adaptive intermolecular REBO potentials (AIREBO) [16], which include both covalent bonding and intermolecular van der Waals interactions, intended for hydrocarbon molecular crystals.

OSCILLATORY ELASTIC PRECURSOR

Even early small-scale MD simulations of shock waves in single crystals [17-19] revealed a quite complex structure of the shock layer in solids, including temperature overshoot and non-monotonic character of shock profiles. Holian and Lomdahl [2] studied stacking fault production and its dependence on the volumetric strain $V/V_0 = 1 - u_p/U_s$ in a shock wave propagating along the [100] direction in a LJ crystal. At about the same time, the fine-grained structure of shock profiles was revealed using the "moving window" technique of time-averaging in the shock reference frame [1,3] which allowed for highly accurate measurements within the non-equilibrium layer in steady shock waves. It was found that a weak elastic shock wave in the [100] direction of a 3D LJ crystal (at non-zero initial temperature) has an oscillatory profile (see Fig. 1) which becomes steady after a transient period [1]. Note that the shear stress remains unrelaxed, indicating that the compression remains purely uniaxial. A dissipative mechanism of oscillatory train decay is thought to be related to the transverse distortions and collisions in the crystal lattice [21].

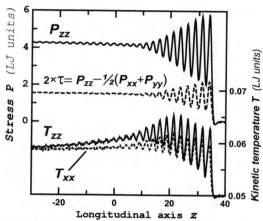

FIGURE 1. Profiles of pure elastic shock wave [1] in [100] direction of LJ crystal exhibit a stationary oscillatory train within the shock layer. Shock and piston velocities are U_s=1.46 and u_p=0.055 (in LJ units [20]).

Further simulations have studied the crystallographic orientation dependence in both the elastic-plastic and melting regimes. The structure of the elastic precursor as well as plastic deformation mechanism were found to be essentially different in the low-index directions in the LJ solid. The [110] shock wave exhibits a quite long oscillatory train ahead of the plastic front in comparison to the other two directions. Remarkably, even overdriven plastic waves in all directions exhibit an oscillatory elastic precursor which, however, remains stationary and propagates at the same plastic shock velocity (see Fig. 2), implying that no actual shock splitting has occurred. Using piston-driven simulations, elastic-plastic splitting was detected [5,6] for weaker shocks in both the [110] and [111] directions, though not in the [100]. This peculiarity of [100] shock waves might be related to the different shear stress behavior under uniaxial compression.

In addition, an unsteady wave train (which the authors refer to as solitary train) ahead of the elastic compression wave was observed in the [110] direction of LJ crystals [6] and later in fcc metals [7]. Recent ramp-up loading simulations by Germann and Kadau [22] and shock waves in the [111] direction in bcc metals [23] also exhibit similar waves. Still, their nature and relationship with the oscillatory elastic precursor remain obscure, and might depend on solitary wave solutions in the zero-temperature limit of the 3D lattice [21].

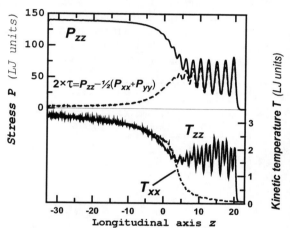

FIGURE 2. Structure of plastic shock wave in [110] direction of LJ crystal showing the stationary oscillatory elastic precursor ahead of plastic front. Shock and piston velocities are U_s=3.13 and u_p=0.89 (in LJ units [20]).

SHOCK-INDUCED MELTING

In contrast to melting of a uniformly heated crystal which is always initiated on its surface, the melting of a shock loaded solid begins inside its body. As a result, an overheated state might be reached in the shocked perfect crystal without preexisting defects. Yet in his MD study Belonoshko [10] stated an absence of any overheating for shock melting in the [100] direction in argon. However, his calculations might be affected by large statistical fluctuations typical for shock simulations in a small system.

In a series of shock-induced melting simulations for different directions in a single LJ crystal [11], a "moving window" technique was used to reduce the uncertainty of measurements and to get smooth profiles of flow variables and their distributions. This study has shown that a metastable overheated state exists in the [100] shock wave but not in the [110] and [111] cases. Figure 3 displays the simulation results for solid argon. Corresponding shock wave parameters are listed in Table 1. For analysis of the structure of shock compressed material, the radial distribution function was calculated in a few thin slabs within the thermally equilibrated region far behind the shock front. It showed that the [100] shocked crystal at compression $V/V_0 = 0.661$ is not melted within the simulation cell in contrast to the other two directions. The metastable overheated state at $U_s = 4.4$ km/s has a temperature beyond the melting line in Fig. 3. An increase of the transverse size of the system did not change the results.

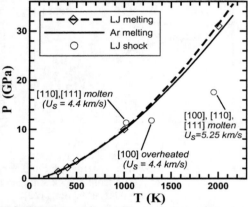

FIGURE 3. Data from MD simulations of shock melting in LJ solid (circles), shown in relation to the melting curves of argon. The dashed curve represents a Simon law for the theoretical melting points of LJ solid: $P=(1.219\times10^{-4})T^{1.639}$. The solid curve is fitted to the argon experimental data [24]: $P=(2.172\times10^{-4})T^{1.556}-0.21$.

Our [100] simulations are in good agreement with the observations by Maillet et al. [25] of a new, highly distorted but still ordered, structure at similar compression in their MD studies.

One of the important factors which can be used to explain the discrepancy between static and shock wave experiments is shear stress produced by high strain rate uniaxial shock compression. It was recently discussed how shear stress might affect the melting transition under shock loading in iron [26]. Melting in a perfect crystal could be facilitated by emission of dislocations during a plastic defor-

TABLE 1. Parameters of shock waves in Fig.3 (data in parentheses are for shock waves at U_s=5.25 km/s)

Parameters in LJ units [20]	[100]	[110]	[111]
Shock vel. U_s	4.0 (4.8)	4.0 (4.8)	4.0 (4.8)
Pressure P_{zz}	282 (419)	272 (419)	272 (419)
Temperature T	10.8 (16.3)	8.5 (16.3)	8.5 (16.3)
Compression V/V_0	0.661 (0.645)	0.671 (0.645)	0.671 (0.645)

mation. The driving force for creation of extended defects is the shear stress, which consists of two parts: kinetic and virial (potential). The former results from the anisotropy of the momentum distribution at the shock front, while the latter is associated with the lattice distortion under uniaxial strain and can be estimated from the strain-stress curve of cold compression.

Figure 4 sheds light on how a buildup of shear stress in an elastic compression wave is affected by shock orientation in the LJ crystal. The virial part of the [100] shear stress increases more slowly with uniaxial strain in comparison to the other two directions. It exhibits non-monotonic behavior when reaching a maximum at strain value $\varepsilon=0.82$ and then dropping to negative minimum at $\varepsilon=0.67$. There is also a special symmetry point of zero shear stress at $\varepsilon=1/\sqrt{2} \approx0.707$, where the initial *fcc* cell is turned into *bcc* under uniaxial compression. Note that for the strain $\varepsilon = 0.661$, corresponding to the compression V/V_0 in the overheated [100]

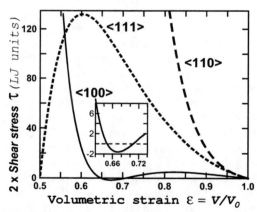

FIGURE 4. Twice the shear stress $2\times\tau=P_{zz} - \frac{1}{2}(P_{xx}+P_{yy})$ as a function of uniaxial compression along axis z in different directions of LJ crystal at $T=0$.

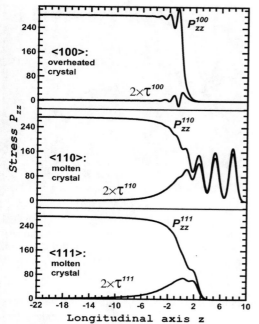

FIGURE 5. Longitudinal stress P_{zz} (in LJ units [20]) and twice the shear stress $2\times\tau$ at shock velocity U_s=4.4 km/s.

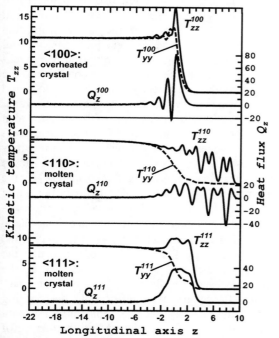

FIGURE 6. Longitudinal T_{zz} and transverse T_{yy} components of kinetic temperature with longitudinal heat flux Q_z for shock waves in Fig. 5. LJ units [20] are used.

shock wave, the virial part of the shear stress becomes negative. It might have a significant effect on plastic flow by hampering the generation and motion of dislocations. Thus, the melting transition in the [100] shocked perfect LJ crystal might be delayed, leading to the overheating phenomenon.

Because shock compression involves considerable kinetic energy fluctuations at the shock front due to the anisotropy of momentum distribution, their contribution to the shear stress should also be evaluated. Dynamics of the kinetic temperature, stress and heat flux within the shock layer are illustrated in Figs. 5 and 6. The shear stress profiles are consistent with a preliminary analysis based on the cold compression calculations. While the kinetic part of shear stress might compare to its virial part in [100] shock waves [27], it cannot diminish the substantial difference between [100] and the other two directions highlighted in Fig. 4. In our view, momentum fluctuations, at least in [110] and [111] shocks, would rather play the role of the trigger than of the driving force for dislocation production.

Even in such strong shock waves there still exists a stationary oscillatory region of elastic compression ahead of the plastic deformation wave. Again, the shear stress relaxation begins when its oscillations in the elastic wave are damped out. Note that the heat flux is also oscillating during elastic shock compression. When thermal equilibrium is reached ($T_{xx}=T_{yy}=T_{zz}$), the heat flux becomes equal to zero indicating that the energy flow (coming from both kinetic and virial parts of the heat flux) plays the role of the driving force for all non-equilibrium processes within the shock layer.

SHOCK-INDUCED CHEMISTRY

Recently we have carried out a series of MD shock simulations in hydrocarbon molecular crystals [28] and covalently-bonded solids [29], using both REBO and AIREBO potentials. One important question which can be addressed by such simulations is what role the shock orientation plays in non-equilibrium rate phenomena including various exothermic and endothermic chemical reactions. In particular, experiments and theory have suggested that the sensitivity of single crystal explosives to shock initiation is strongly dependent

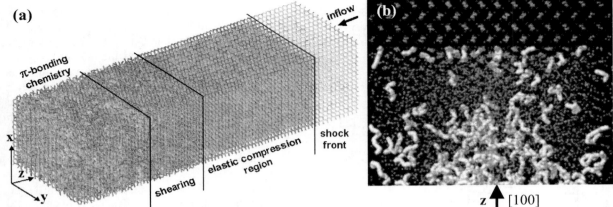

FIGURE 7. (a) Snapshots of shock wave in REBO diamond propagating along axis z in [110] direction. Atoms are shaded according to their potential energy. Shock velocity U_s=25 km/s, piston velocity u_p=6.1 km/s. **(b)** Snapshot of non-stationary shock wave in solid acetylene (C_2H_2) after a "flyer plate" impact. Shock wave is propagating along axis z in [100] direction. At this time, longitudinal stress, temperature and mass velocity at shock front were $P_{zz}\approx40$ GPa, $T_{zz}\approx3500$ K, and $u_z\approx5$ km/s, correspondingly. C-C bonds for product chains of three or more carbons are shown in bold white.

on the crystal orientation and related plasticity mechanisms [30].

Our simulations, employing the REBO model of diamond, indicate that anisotropy of shock-induced plastic flow might have a significant effect on the breaking and making of covalent bonds [29]. Figure 7(a) displays the structure of a [110] strong shock wave in REBO diamond, where a shear deformation wave (preceded by a region of elastic compression) causes tetrahedral sp^3-bond breaking and formation of amorphous carbon inclusions (π-bonding chemistry). Because of anisotropy of the deformation mechanisms, sp^2-bonded atoms dominate the chemistry region in the [110] shock wave while the [111]-oriented shock wave retains a relatively large number of sp^3-bonded atoms (see Table 2). Recent experimental and density-functional studies have also shown that graphitization of diamond due to applied shear stress is expected to be highly anisotropic [31].

Shock compression of molecular crystals has considerably more complex character because of

TABLE 2. Orientation dependence of bond hybridization for shock-induced amorphization in REBO diamond.

Bond hybridization	sp (%)	sp^2 (%)	sp^3 (%)
[110] U_s=30 km/s	4.20	61.55	33.81
[111] U_s=31 km/s	2.52	28.31	68.81

the existence of both strong intramolecular forces that bind atoms into molecules and weak intermolecular dispersion forces that bind molecules into solids. In addition to anisotropy of the plasticity mechanisms, there might also be anisotropy of kinetic energy transfer into molecular internal modes. In the vicinity of the shock front, the molecules are subjected to very high rate uniaxial strain as well as highly non-equilibrium momentum flow. Under these extreme conditions, the energy might be directly channeled into bond breaking modes rather then dissipated into transverse directions by relaxation of the crystal to a state of thermal equilibrium and hydrostatic stress.

We have performed a series of MD simulations to investigate shock-induced chemistry in single crystal acetylene, ethylene and methane [28] by the impact of a "flyer plate" of a hydrocarbon solid against a larger crystal. Figure 7(b) shows a snapshot of a shock wave in solid acetylene taken 1.3 ps after flyer impact (the crystal is rotated 45° around the vertical propagation axis z so that the density of material appears to increase in the center). Because the flyer thickness is much smaller than the length of target crystal, the resulting shock wave quickly becomes non-stationary with gradual decay of its profiles. At the beginning of the impact the shock velocity is $U_s \approx 15.5$ km/s for particle velocity u_p=8 km/s and initial density ρ_0=0.76 g/cm^3. Note that a number of the dissociation and polymer production (bold white chains) reactions are initiated right at the shock front. Reactions continue until the impact energy is dissipated in the target crystal and peak pressure and temperature at the front are decreased below the reaction threshold. Thus, quite fast molecular fragmentation and polymerization is possible in relatively strong shock waves in molecular solids.

Our preliminary simulations in a single crystal of anthracene [32] show that the shape of anthracene molecules makes the process of their bending and molecular bond distortion greatly dependent on the shock orientation. We also find that the excitation of deformations and vibrations of anthracene molecules begins in the vicinity of the shock front. Such simulations might help in understanding how energy transfer between translation and internal molecular modes within the non-equilibrium shock layer depends on shock orientation in the molecular crystals, especially when complex transient phenomena, like oscillatory elastic precursors, could significantly affect the relaxation and thermalization processes.

ACKNOWLEDGEMENTS

Funding was provided by the Office of Naval Research. The authors thank Sergey Anisimov, Vladimir Klimenko, Jim Belak, Brad Holian, Tim Germann, Ramon Ravelo, Jim Hammerberg, Kai Kadau, and Jean-Bernard Maillet for discussions.

REFERENCES

1. V.V. Zhakhovskii, S.V. Zybin, K. Nishihara, and S.I.Anisimov, in *Proceedings of Symposium on Shock Waves'98*, Aoyama Gakuin Univ., Tokyo, 1999, pp. 241–244.
2. B.L. Holian and P.S. Lomdahl, *Science*, **280**, 2085-2088 (1998).
3. V.V. Zhakhovskii, S.V. Zybin, K. Nishihara, and S.I.Anisimov, *Phys.Rev.Lett.* **83**, 1175-1179 (1999).
4. D.H. Robertson, D.W. Brenner, and C.T. White, in *High Pressure Shock Compression of Solids III*, edited by L. Davison and M. Shahinpoor (Springer, New York, 1998), p.37.
5. V.V. Zhakhovskii and S.V. Zybin, in *Seminário Brasileiro de Análise 49*, edited by S.A. Tozoni et al., Campinas, SP, Brasil, 1999, pp. 635-649.
6. T.C. Germann, B.L. Holian, P.S. Lomdahl, and R. Ravelo, *Phys. Rev. Lett.* **84**, 5351-5354 (2000).
7. O. Kum, *J. Appl. Phys.* **93**, 3239-3247 (2003).
8. D.H. Robertson, D.W. Brenner, and C.T. White, *Phys. Rev. Lett.* **67**, 3132-3135 (1991).
9. K. Kadau, T.C. Germann, P.S. Lomdahl, and B.L. Holian, *Science*, **296**, 1681-1684 (2002).
10. A.B. Belonoshko, *Science*, **275**, 955-957 (1997); see also a reply in *Science*, **278**, 1474-1476 (1997).
11. V.V. Zhakhovskii, S.V. Zybin, K. Nishihara, and S.I. Anisimov, *Prog. Theor. Phys. Suppl.* **138**, 223-228 (2000).
12. D.W. Brenner, D.H. Robertson, M.L. Elert, and C.T. White, *Phys. Rev. Lett.* **70**, 2174-2177 (1993).
13. D.W. Brenner, *Phys. Rev. B*, **42**, 9458-9471 (1990).
14. D.W. Brenner, O.A. Shenderova, J.A. Harrison, S.J. Stuart, B. Ni, and S.B. Sinnott, *J. Phys.: Condens. Matter*, **14**, 783-802 (2002).
15. D.W. Brenner, J.A. Harrison, C.T. White, and R.J. Colton, *Thin Solid Films*, **206**, 220 (1991).
16. S.J. Stuart, A.B. Tutein, and J.A. Harrison, *J. Chem. Phys.* **112**, 6472-6486 (2000).
17. V.Y. Klimenko and A.N. Dremin, *Sov. Phys. Dokl.* **25**, 288-289 (1980).
18. B.L.Holian, *Phys. Rev. A*, **37**, 2562-2568 (1988).
19. J.Belak, LLNL Rep.# UCRL-JC-109989 (1992).
20. LJ units for argon: $[z]$=3.405 Å, $[u]$=1.093 km/s, $[T]$=120 K, $[P]$=41.9 MPa, $[Q]$=45.8 GW/m^2.
21. S.V. Zybin, V.V. Zhakhovskii, T.C. Germann, K. Kadau, and B.L. Holian (unpublished).
22. T.G.Germann, K.Kadau (private communication).
23. J. Roth, these proceedings.
24. F. Datchi, P. Loubeyre, and R. LeToullec, *Phys. Rev. B*, **61**, 6535-6546 (2000).
25. J.-B. Maillet, M. Mareschal, L. Soulard, R. Ravelo, P. S. Lomdahl, T. C. Germann, and B. L. Holian, *Phys. Rev. E*, **63**, 016121 (2001).
26. M.A. Podurets, *High Temp.* **38**, 860-866 (2000).
27. The kinetic part of shear stress due to anisotropy of the kinetic temperature, say, $\Delta T = T_{zz} - T_{xx}$=500K, is $2 \times \tau = n_1 k \Delta T \approx 280$ MPa for shocks in Figs.5 and 6.
28. M.L. Elert, S.V. Zybin, and C.T. White, *J. Chem. Phys.*, **118**, 9795-9801 (2003).
29. S.V. Zybin, M.L. Elert, and C.T. White, *Phys. Rev. B*, **66**, 220102(R) (2002).
30. J.J. Dick, *J. Phys. Chem.* **97**, 6193-6196 (1993); C.S. Yoo, N.C. Holmes, P.C. Soures, C.J. Wu, F.H. Ree, and J. J. Dick, *J. Appl. Phys.* **88**, 70-75 (2000); C.S. Coffey and J. Sharma, *J. Appl. Phys.* **89**, 4797-4802 (2001).
31. Y.G. Gogotsi, A. Kailer, and K.G. Nickel, *J. Appl. Phys.* **84**, 1299-1304 (1998); H. Chacham and L. Kleinman, *Phys. Rev. Lett.* **85**, 4904-4907 (2000).
32. S.V. Zybin, M.L. Elert, and C.T. White, these proceedings.

CHAPTER VII

CHAPTER VII

MODELING AND SIMULATION: REACTIVE MATERIALS

CP706, *Shock Compression of Condensed Matter - 2003*
edited by M. D. Furnish, Y. M. Gupta, and J. W. Forbes
© 2004 American Institute of Physics 0-7354-0181-0/04/$22.00

HOTSPOT MECHANISMS IN SHOCK-MELTED EXPLOSIVES

Shirish M. Chitanvis

Theoretical Division, Los Alamos National Laboratory, Los Alamos, New Mexico 87545

Abstract. This report is a review of various initiation mechanisms in shock-melted explosives. This paper focuses on pre-ignition phenomena which cause a local temperature rise when a single void in HMX collapses under the action of a shock wave. Working in the melting regime, a timeline can be associated with the collapse of a single void, through a consideration of the time scales on which these mechanisms are activated. We have studied the *hydrodynamic mechanism*, in which the shock driven incident side of the void impinges on the shadow side of the void, and is brought to rest, causing a considerable temperature rise in the HMX. Clearly this mechanism comes into consideration after the void collapses completely. Another mechanism we studied is that of *shear heating*. It is important for extremely small voids, or for large voids after the void has been compressed to a sufficiently small scale. This mechanism comes into play after the void has collapsed, and a remnants of the void have been spun off into a vortex. The phenomenon of *gas compression* as the gas-filled void collapses is difficult to ignore, in view of the fact that large temperatures are generated by this mechanism. We speculate that this mechanism could be important if the initial, endothermic induction step in a reaction scheme is shorter than the time of collapse.

INTRODUCTION

The complexity of multiple phenomena involved in the ignition mechanisms of shocked HMX has forced upon modelers the necessity of treatments at various length scales. Many formalisms begin at a macroscopic scale where experiments are utilized to determine the parameters of the models[1]. The Johnson-Tang-Forest model[2], while still tied to experiments, acknowledges a sub-grid hot-spot mechanism, described via a progress variable. Hot-spot mechanisms associated with the collapse of microscopic voids in the sample are generally acknowledged to be the primary cause of initiation in HMX[3]. In a generalized sense this mechanism encompasses shear heating phenomena and viscoplastic effects in solid HMX. There are theoretical treatments which attempt to come to grips with microscopic mechanisms responsible for the initiation of HMX. Some of these papers remain focused pre-dominantly on a single mechanism. Others like that by Field[4] discuss how a variety of ignitions mechanisms might come into dominance under differing conditions. The discussion in this paper falls into this latter category.

Frey argues in his paper that dissipative mechanisms can be important when the viscosity of the material under consideration is large. Intuitively, one sees that this mechanism must dominate for solid samples. In this case, one uses the full set of Navier-Stokes equations, with the assumption that viscous terms can be used effectively in the solid state. This set is a mixed hyperbolic-parabolic type. The convective terms describing the progress of shocked conditions provide the hyperbolic character, while the dissipative mechanisms, of the diffusive sort, give it a parabolic flavor. The numerical treatment is thus made greatly difficult. The numerical tools available to us use the conventional method of treating the fast, convective mechanisms accurately, while accounting for the dissipative mechanisms perturbatively. Such a treatment is useful for indicating the onset of dissipative heating mechanisms, but become unreliable soon thereafter.

Given this situation, it becomes reasonable to ask if and when such complications can be ignored. It turns out that if we shock the HMX with a shock wave so as to melt it, and estimate the viscosity of the melt, a consideration of the Reynolds number shows that dissipative mechanisms can be ignored at the outset of the collapse of large voids. Towards the end of the void collapse, the Reynolds number becomes sufficiently small and dissipative mechanisms can no longer be ignored. As pointed out in the abstract, other hydrodynamic mechanisms come into dominance and can lead to temperatures large enough to initiate ignition.

THE TIME-LINE

In this paper we shall consider the case when a shock is sent through a sample of HMX with sufficient energy to cause melting immediately behind it. Consider the case that a single void is present in this sample. The figure below gives an overview of the ideas developed in this paper.

FIGURE 1. A Time-line to resolve heating mechanisms.

Void Gas Heating: As the shock wave goes over the void, the gas in the void begins to get compressed, sending out a rarefaction wave into the surrounding medium. Adiabatic Compression of the gas in the void causes the temperature to go up. A simple example of a spherical void being compressed gives an estimate of the temperature rise:

$$T(t) = T(0) \frac{\rho(t)^{\gamma-1}}{\rho(0)^{\gamma-1}} = T(0) \left(\frac{R(0)}{R(t)} \right)^{3/2} \quad (1)$$

$R(t)/R(0) = 1/4 \rightarrow T(t)/T(0) = 8$, so that gas cavity temperatures will reach $2400K$ quite easily. These estimates for spherical cavities are expected to be larger than the corresponding void collapse in higher dimensions.

Some of this internal energy in the gas will be transferred to the inside surface of the void. According to the model developed experimentally by Henson and co-workers[5], this could cause evaporation of HMX molecules into the void and lead to a self-sustained reaction. However, the initial step in this reaction scheme is endothermic in which a phase change occurs from one form of HMX to another. If the void is sufficiently large such that the the endothermic reaction step is concluded and the exothermic surface reaction begins before complete collapse

occurs, then a self-sustaining burning reaction could take place.

Quantitative estimates of the critical size of the void cannot be made as the experimental model was developed, for obvious reasons, in a low-pressure, cook-off-like regime. Extrapolation of the model to the high pressure regime we are interested in lead to unreliable estimates for the temporal length of the reaction steps.

Such mechanisms have been discussed earlier in the literature, but not in the context of the recent model of Henson et al[5].

Hydrodynamic mechanism: If the gas void mechanism does not initiate burning strongly, then the next logical phenomenon that needs to be considered is the hydrodynamic mechanism first considered a long time ago by Mader[6]. In this model, the gas in the void plays a minimal role, and for expository purposes will be ignored in this sub-section. The shock wave emerges into the void. The surface velocity of the front entering the void is approximately twice the particle velocity, as determined empirically in many experiments. A rarefaction wave will proceed backward into the medium. The *ejecta* will impact the shadow side of the void. In this impact, kinetic energy is converted into internal energy as the ejecta slow down. In one-dimensional simulations of void collapse, we have observed a fairly large temperature rise due to this mechanism ($\sim 200°K$).

Figure 2 is a pictorial depiction of this mechanism.

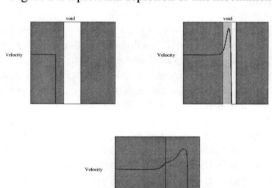

FIGURE 2. A physical picture of the hydrodynamic mechanism, showing the initial void configuration, the void during collapse and the impacted stage, when some of the kinetic energy has been converted into internal energy.

A description of this mechanism involves the use of the Euler equations:

$$\partial_t \rho + \vec{\nabla} \cdot (\rho \vec{u}) = 0$$
$$\partial_t (\rho \vec{u}) + \vec{\nabla} \cdot (\rho \vec{u} \otimes \vec{u}) + \vec{\nabla} P = 0$$
$$\partial_t (\rho (\vec{u}^2/2 + \varepsilon)) + \vec{\nabla} \cdot (\rho \vec{u} (\vec{u}^2/2 + \varepsilon)) + \vec{\nabla} \cdot (P\vec{u}) = 0$$

$$(2)$$

Shear heating: The physical notion is that when there is shear present in a flow, viscosity effects between neighboring layers of fluid will lead, through friction, to shear heating. Note that we are considering purposely the case that the HMX melts behind the incident shock wave. This reduces the viscosity from the solid state. Now the value of the viscosity of HMX is the subject of current research, both in the solid state as well as the melted state. Estimates of its magnitude may be made using available data[7].

The hydrodynamic formulation of this phenomenon reflects a positive semi-definite source term in the energy equation. One begins with the following set of conservation equations:

$$\partial_t \rho + \vec{\nabla} \cdot (\rho \vec{u}) = 0$$
$$\partial_t (\rho \vec{u}) + \vec{\nabla} \cdot (\rho \vec{u} \otimes \vec{u}) - \vec{\nabla} \cdot \overleftrightarrow{\sigma} = 0$$
$$\partial_t (\rho E) + \vec{\nabla} \cdot (\rho E \vec{u}) - \vec{\nabla} \cdot (\overleftrightarrow{\sigma} \cdot \vec{u}) = 0 \quad (3)$$

where

$$E = \vec{u}^2/2 + \varepsilon$$
$$(\overleftrightarrow{\sigma})_{ij} = -P\delta_{ij} + \Delta\sigma_{ij}$$
$$\Delta\sigma_{ij} = 2\mu e_{ij}$$
$$(\overleftrightarrow{e})_{ij} = 1/2(\partial_i u_j + \partial_j u_i) \quad (4)$$

The energy equation can be conveniently recast in terms of the internal energy:

$$\partial_t (\rho\varepsilon) + \vec{\nabla} \cdot (\rho\varepsilon\vec{u}) + P\vec{\nabla} \cdot \vec{u} = \underline{2\mu \overleftrightarrow{e} : \overleftrightarrow{e}} \quad (5)$$

The positive semi-definite form of the source term is manifest in this formulation.

We can write down the above equations in dimensionless form by using a characteristic velocity u^*, a characteristic length L^*, a characteristic time L^*/u^*, a characteristic density ρ^*, and a characteristic energy $\rho^* u^{*2}$. In particular, the energy equation can be recast as:

$$\partial_{t'} (\rho'\varepsilon') + \vec{\nabla}' \cdot (\rho'\varepsilon'\vec{u}') + P' \vec{\nabla}' \cdot \vec{u}' = \frac{2}{\mathscr{R}} \mathbf{e'} : \mathbf{e'}$$
$$\mathscr{R} = \frac{\rho^* u^* L^*}{\mu}$$

$$(6)$$

where the primed quantities indicate dimensionless variables.

If we put in characteristic values: $\mu \sim 10^{-6}$ (in cgs units), $\rho^* \sim 2$ (g/cc), $u^* \leq 1$ (cm/μs), $L^* \sim 0.001 - 0.01$(cm) (*typical dimension of voids used in our simulations*), we get:

$$\mathscr{R} \sim 10^2 - 10^3 \quad (7)$$

It follows that for voids in this range, the viscous heating term is negligible, in the initial stages of collapse. Of course, as the void continues to collapse, the Reynolds number gets smaller, and viscous heating effects will eventually begin dominating.

In highly resolved numerical simulations performed for such voids, it was found that there was a noticeable time lag between the hydrodynamic mechanisms considered above, and shear heating effects, which set in later. The hydrodynamic mechanisms provide a considerable temperature rise before shear heating effects set in.

Figures 3-5 show stages in the two-dimensional simulation of the collapse of a cylindrical void in an HMX sample characterized by a Mie-Gruneisen equation of state with parameters available standardly. The CFD code *Comadreja* was used to perform these simulations using a mesh about 1000 × 1500. The code was developed by Prof. Benson at UCSD. The void was filled by an ideal gas. For the purposes of simulation, shear viscosity was turned off, in order to obtain a base-line. We have performed other simulations where this term was retained. It was observed to become important when void collapse has occurred, as indicated in Fig. 5.

ACKNOWLEDGMENTS

I wish to acknowledge useful discussions with John Bdzil, Brad Holian and Ralph Menikoff. The work presented here was performed under the auspices of the DOE.

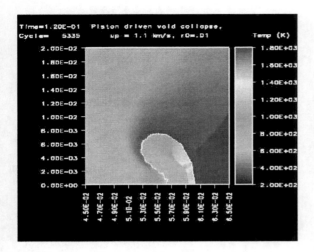

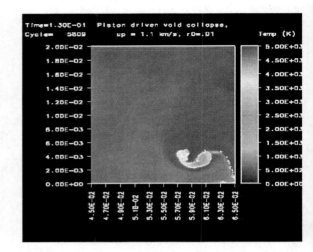

FIGURE 3. Two-dimensional void collapse simulation at about $0.12\mu s$. The incident shock has sent a shock wave into the gaseous void, which has reached the other side. The incident wave has curled around the void and is seen to interact with the shocked gas in the void.

FIGURE 5. The two-dimensional void collapse simulation approximately $0.145\mu s$ into the simulation. The void has collapsed. The temperature scale again differs from that in the previous two figures. The void has split off into several smaller voids, and the temperature rise is rather high. It is suspected that the these two features at this stage of the collapse might have a numerical origin.

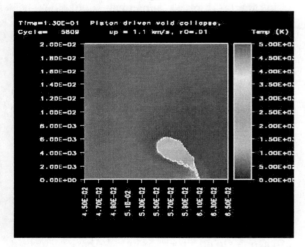

FIGURE 4. Simulation results at about $0.13\mu s$. The incident side of the void has just begun impacting the shadow side of the void. Note that the temperature scale in this figure is different from that in the previous figure. Notice also the hydrodynamic mechanism disucssed in the text becomes evident in the vicinity of the center of the void.

REFERENCES

1. Y. Horie, Y. H., and Greening, D., Mechanistic Reactive Burn Modeling of Solid Explosives, Tech. Rep. LA-14008, Los Alamos National Laboratory (2003).
2. J.N. Johnson, P. T., and Forest, C., *J. Appl. Phys.*, **57**, 4323–4334 (1985).
3. Bdzil, J. (2002), private communication.
4. Field, J., *Acc. Chem. Res.*, **25**, 489–496 (1992).
5. B.F. Henson, B. A., L. Smilowitz, and Dickson, P., *J. Chem. Phys.*, **117**, 3780–3788 (2002).
6. Mader, C., and Kershner, J., 3-D Hydrodynamic Hotspot Model Applied to PETN, HMX, TATB and NQ, Tech. Rep. LA-10203-MS, Los Alamos National Laboratory (1984).
7. Shepherd, J. (2001), cIT ASCI Alliance Program Review, Los Alamos, unpublished.

CP706, *Shock Compression of Condensed Matter - 2003*
edited by M. D. Furnish, Y. M. Gupta, and J. W. Forbes
© 2004 American Institute of Physics 0-7354-0181-0/04/$22.00

MOLECULAR DYNAMICS STUDY OF SHOCK-INDUCED CHEMISTRY IN ANTHRACENE

M. L. Elert[1], S. V. Zybin[2], and C. T. White[3]

[1]*Chemistry Department, U. S. Naval Academy, Annapolis, MD 21402*
[2]*Department of Chemistry, The George Washington University, Washington, DC 20052*
[3]*Naval Research Laboratory, Washington, DC 20375*

Abstract. Molecular dynamics simulations employing a reactive empirical bond-order (REBO) potential are used to investigate shock-induced chemical reactions in anthracene. Previous studies have shown that the dominant shock-induced reaction for smaller unsaturated hydrocarbons is polymerization, but fragmentation and pyrolysis are expected to be more prevalent for larger molecules. In agreement with recent experimental results, it is found that dimerization is the dominant chemical reaction in anthracene subjected to shock above a threshold strength. In addition, anthracene exhibits significant anisotropy in the solid phase, leading to orientation dependence of shock-induced chemistry in this material.

INTRODUCTION

Recently, molecular dynamics simulations have shown [1] that shock impact can induce significant polymerization reactions in acetylene and ethylene. For larger unsaturated hydrocarbon molecules, one expects that fragmentation should become an increasingly probable reaction channel. In light of recent experimental evidence [2] that amino acids can survive and even form polypeptide oligomers under shock conditions consistent with shallow-angle cometary impacts on earth's atmosphere, it is of considerable interest to examine shock-induced chemical reaction processes for molecules in a size range comparable to that of naturally occurring amino acids. Energetic shock waves in larger, anisotropic molecular systems also offer the possibility of observing significant orientation effects in shock-induced chemical reactions in such materials.

In the present study, molecular dynamics simulations of shock impact in anthracene crystals are reported. Anthracene was chosen for this investigation because it is a relatively large, highly anisotropic, unsaturated hydrocarbon system which has been experimentally shown [3] to undergo shock-induced dimerization. In fact, reference 3 reports the surprising result that anthracene dimers are the *only* observed reaction product upon shock impact at a pressure of 18 GPa.

THE MODEL

Molecular dynamics simulations of shock wave propagation in solid anthracene were carried out using the AIREBO empirical bond order potential for hydrocarbons.[4] The potential was designed to reproduce the enegetics of reactive hydrocarbon systems, including hybridization and conjugation effects for carbon, and also includes torsional and van der Waals forces. It has been employed in tribological simulations of hydrocarbons on diamond surfaces[5,6] as well as our recent molecular dynamics study [1] of shock-induced chemistry in acetylene, ethylene, and methane.

Anthracene crystallizes in a monoclinic lattice with the long axis of the molecule nearly aligned with the crystallographic *c* axis. A projection of the molecules onto the *ab* plane forms a herringbone pattern. The crystal structure of anthracene is illustrated in Figure 1. For the simulations reported here, the experimental crystallographic parameters [7] were used as a starting point for a steepest descent minimization of the energy with respect to unit cell parameters and molecular positions to obtain the minimum-energy configuration for crystalline anthracene within the AIREBO potential model. The final unit cell parameters were within a few percent of the experimental values.[8]

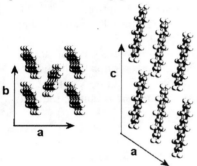

FIGURE 1. Structure of the anthracene unit cell.

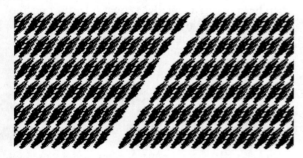

FIGURE 2. Initial configuration for simulation with shock direction along the *a* axis. Crystal segments are launched towards each other horizontally, closing the gap and generating shock waves propagating to the left and right.

Shock impact simulations were carried out in a "flyer plate" configuration by launching two zero-temperature anthracene crystal segments towards each other at a defined relative velocity. Orientation dependence was investigated by carrying out simulations with the shock propagation direction oriented along both the *a* and *c* crystallographic axes. Figure 2 shows the initial configuration for a simulation with the shock direction along the *a* axis. Periodic boundary conditions were imposed along the other two unit cell axes. Note that the shock front forms an oblique angle with the shock propagation direction in this configuration. All simulations were performed with 40320 atoms or 1680 anthracene molecules for a time period of at least 5 ps.

RESULTS

At relative impact speeds below 8 km/s, no chemical reactions were observed at either crystal orientation. Above this threshold, significant fragmentation and polymerization occurred in the vicinity of the impact region between the two crystal segments. A snapshot of the impact region for an impact velocity of 12 km/s is shown in Figure 3. Crystal orientation is the same as in Figure 2. The picture corresponds to a time of 5 ps after impact, when carbon backbone reactions are largely complete. For clarity, all unreacted anthracene molecules have been deleted, and only carbon atoms are shown.

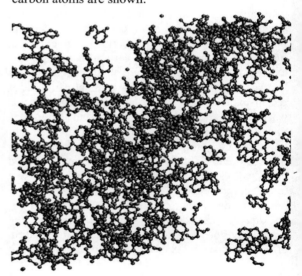

FIGURE 3. Impact region after collision of the two crystal segments shown in Figure 2 at a relative velocity of 12 km/s. Unreacted anthracene molecules are not shown.

Near the reaction threshold at 8 km/s, the peak density and temperature in the impact region are approximately 1.8 g/cm^3 and 3000 K, respectively, as shown in Figure 4. At an impact speed of 12 km/s those values increase to 2.0 g/cm^3 and 7500 K, resulting in substantially more chemical reactivity in the impact region.

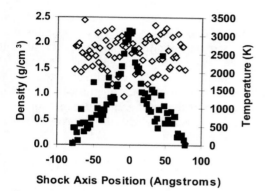

FIGURE 4. Density and temperature profiles for simulation with shock direction along c axis and impact speed of 8 km/s, at 1 ps after start of simulation. Solid squares represent temperature (right axis) and open diamonds are average densities (left axis) for bins of width 2 Ångstroms.

The extent of chemical reaction for each simulation was determined by calculating the number of carbon atoms in each connected cluster (molecule), from which a simulated "mass spectrum" could be produced. However, in the interest of simplicity the number of hydrogen atoms in each molecule was not considered. In what follows, therefore, "unreacted" anthracene molecules are those with fourteen carbon atoms, although they may actually not contain exactly ten hydrogen atoms. In fact about ten percent of "unreacted" anthracene undergo single hydrogen abstraction reactions in 10 km/s simulations and about 20% at 12 km/s, with multiple hydrogen abstractions being much more rare.

Figure 5 shows a comparison between simulations at ten and twelve km/s impact speeds for shock direction along the a axis. At 10 km/s, about 1400 of the 1680 anthracene molecules in the simulation

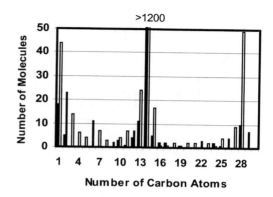

FIGURE 5. The number of molecules containing a given number of carbon atoms for simulations with shock direction along the a axis at impact velocities of ten (solid bars) and twelve (open bars) km/s. The peaks for unreacted anthracene molecules (14 carbons) are far off scale with more than 1200 molecules.

remain unreacted. Of those that do react, most gain or lose a single carbon atom or form dimers with 28 carbon atoms. At 12 km/s the extent of reaction is much more pronounced, but the major reaction channels remain the same. At the higher impact speed a few higher oligomers (not shown in the figure) are also produced.

Possible orientation dependence of shock-induced chemistry in the anthracene system was investigated by comparing simulations with the impact direction along the a axis, as shown in Figure 2, with those in which the impact direction was along the c axis. In the latter case, the shock velocity is oriented along the long axis of the molecules and collisions occur "end on." Figure 6 shows a comparison of the product distributions for the two shock orientations at an impact speed of 12 km/s. Both simulations produce essentially the same total reactivity, with about 25% of reactant anthracene molecules exhibiting some change in carbon backbone structure in each case. Dimerization is the dominant reaction pathway in both simulations. However, the c axis simulation shows significantly more fragmentation and fewer simple dimerization reactions than the a axis simulation. In fact, the latter simulation produces 49 anthracene dimers (28 carbons) compared to 23 dimers for the c axis case.

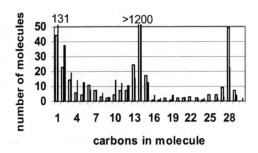

FIGURE 6. Product distribution for 12 km/s impact along *a* axis (unfilled bars) and *c* axis (filled bars). Heights of off-scale peaks are shown at top of figure.

DISCUSSION

The threshold for shock-induced chemical reactivity found in these simulations is 8 km/s, corresponding to a density of 1.8 g/cm³ as shown in Figure 4. These conditions closely match the discontinuity in shock velocity vs. particle velocity reported by Warnes [9] which he attributed to an intermolecular coupling reaction. The pressure-volume Hugoniot constructed from Warnes' data shows that this discontinuity corresponds to a pressure of 18 GPa in agreement with the dimerization threshold observed by Engelke and Blais [3]. It should be noted, however, that the AIREBO potential is known to overestimate the stiffness of repulsive intermolecular interactions at elevated pressure, and therefore the actual pressure in the reaction zone in our simulations is likely to be significantly higher than 18 GPa.

In an observation that they characterize as "remarkable," Engelke and Blais [3] report that the *only* new chemical species observed in the mass spectral analysis of the products of their experimental shock wave study near the 18 GPa reaction threshold is the anthracene dimer. Dimerization is also found to be the dominant reaction channel in our simulations, although we also observe small numbers of higher oligomers and small hydrocarbon fragments. Of course, while primary chemical reactions are essentially complete by the end of our simulations at 6 ps, small reactive fragments may not persist on the timescale of the time-of-flight mass spectral

analysis of Engelke and Blais [3]. Our results are therefore in substantial agreement with that experimental study.

CONCLUSION

Dimerization is found to be the dominant process for shock-induced chemistry in anthracene at impact speeds between 8 and 14 km/s. The robustness of this large aromatic ring structure, and its tendency to form even larger molecules under these extreme conditions, is consistent with the recently observed shock-induced behavior of amino acids [2] under conditions expected for terrestrial cometary impact. Significant orientation dependence of the reaction products is found, with "end-on" collisions along the *c* axis producing greater fragmentation.

ACKNOWLEDGEMENT

This work was supported by the Office of Naval Research. MLE received additional funding from ONR through the Naval Academy Research Council.

REFERENCES

1. Elert, M. L., Zybin, S. V., and White, C. T., *J. Chem. Phys.* **118**, 9795-9801 (2003).
2. Blank, J. G., Miller, G. H., Ahrens, M. J., and Winans, R. E., *Origins Life Evol. Biosphere* **31**, 15-51 (2001).
3. Engelke, R. and Blais, N. C., *J. Chem. Phys.* **101**, 10961-10972 (1994).
4. Stuart, S. J., Tutein, A. B., and Harrison, J. A., *J. Chem. Phys.* **112**, 6472-6486 (2000).
5. Mikulski, P. T. and Harrison, J. A., *J. Am. Chem. Soc.* **123**, 6873-6881 (2001).
6. Gao, G. T., Mikulski, P. T., and Harrison, J. A., *J. Am. Chem. Soc.* **124**, 7202-7209 (2002).
7. Kitaigorodskii, A. I., *Organic Chemical Crystallography*, Consultants Bureau, New York, 1961, pp. 420-422.
8. See S. V. Zybin, V.V. Zhakhovskii, M. L. Elert, and C. T. White, these proceedings.
9. Warnes, R. H., *J. Chem. Phys.* **53**, 1088-1094 (1970).

CP706, *Shock Compression of Condensed Matter - 2003*
edited by M. D. Furnish, Y. M. Gupta, and J. W. Forbes
© 2004 American Institute of Physics 0-7354-0181-0/04/$22.00

SIMULATIONS OF FLUID NITROMETHANE UNDER EXTREME CONDITIONS

Laurence E. Fried[1], Evan J. Reed[2], and M. Riad Manaa[1]

[1]*L-282, Chemistry and Materials Science Directorate, Lawrence Livermore National Laboratory, Livermore, California 94550*
[2]*Department of Physics, Massachusetts Institute of Technology, Cambridge, Massachusetts 02139*

Abstract. We report density functional molecular dynamics simulations to determine the early chemical events of hot (T = 3000 K) and dense (1.97 g/cm^3, V/V$_0$ = 0.68) nitromethane (CH$_3$NO$_2$). The first step in the decomposition process is an intermolecular proton abstraction mechanism that leads to the formation of CH$_3$NO$_2$H and the aci ion H$_2$CNO$_2{}^-$, in support of evidence from static high-pressure and shock experiments. An intramolecular hydrogen transfer that transforms nitromethane into the aci acid form, CH$_2$NO$_2$H, accompanies this event. This is the first confirmation of chemical reactivity with bond selectivity for an energetic material near the condition of fully reacted specimen. We also report the decomposition mechanism followed up to the formation of H$_2$O as the first stable product.

INTRODUCTION

The reaction chemistry of energetic materials at high pressure and temperature is of considerable importance in understanding processes that these materials experience under impact and detonation conditions. Basic questions such as: (a) which bond in a given energetic molecule breaks first, and (b) what type of chemical reactions (unimolecular versus bimolecular, etc.) that dominate early in the decomposition process, are still largely unknown. The most widely studied, and archetypical example of such materials is nitromethane (CH$_3$NO$_2$), a clear liquid with mass density 1.13 g/cm^3 at 298°K. Static high-pressure experiments[1] showed that the time of explosion for deuterated nitromethane is approximately ten times longer than that for protonated materials, suggesting that a proton or hydrogen atom abstraction is involved in the rate determining step. Isotope-exchange experiments, using diamond cells methods, also gave evidence[2] that the aci ion concentration (H$_2$CNO$_2{}^-$) increases with increased pressure. Other studies[3] also suggested that reactions occur more rapidly and are pressure enhanced when small amount of bases are present, giving further support to the aci ion production. Shock wave studies of the reaction chemistry are still inconclusive and at odds: mass spectroscopic studies suggesting condensation reactions[4], time-resolved Raman spectroscopy suggesting a bimolecular mechanism[5], UV-visible absorption spectroscopy indicating no sign of chemical reaction[6], or the production of H$_3$CNO$_2{}^-$ intermediate for amine-sensitized nitromethane[7]. It was noted, however, that part of the discrepancy is due to the fact that the ring-up experiments are mapping lower temperature regimes ($\approx$1000 °K) than experienced under detonation conditions (T$\approx$2500-5000 °K)[4].

In this work, we use spin-polarized, gradient corrected density-functional calculations to determine the interatomic forces, and simulate the initial decomposition steps of hot (T = 3000K) dense (1.97 g/cm^3, V/V$_0$ = 0.68) nitromethane at constant-volume and temperature conditions. The studied state is in the neighborhood of the Chapman-Jouget state, which is achieved behind a steady detonation front when the material has fully reacted. This state could be achieved through a sudden heating of nitromethane in a diamond anvil cell under constant volume conditions. Our results emphatically show that the first chemical event is a proton extraction to form CH_3NO_2H, the aci ion $H_2CNO_2^{-}$, and the aci acid $H_2CNO_2\,H$. These results are uniquely associated with the condensed-phase rather than the energetically favored C-N decomposition expected in the gas-phase.

COMPUTATIONAL APPROACH

The electronic structure calculations of the molecular forces were performed using density functional theory (DFT).[8] For the exchange-correlation potential, we used the spin-polarized generalized gradient corrected approximation of Perdew -Wang (PW91)[9]. Electron-ion interactions were described by Vanderbilt-type ultrasoft pseudopotentials,[10] and orbitals were expanded in a plane wave basis set with kinetic energy cutoff of 340 eV. We used two k-point spacing in the Brillouin zone, each with a total number of 2921 plane waves. Minimization of the total density functional from DFT utilized the charge density mixing scheme[11]. Calculations on a single unit cell were performed using the CASTEP program[12], while those on larger cells employed the VASP program.[11,13]

Molecular dynamics simulations were carried out under constant volume and temperature using a Nose thermostat. For each MD run, random initial velocities were chosen, and a first-order Verlet extrapolation of the wave functions was used. Periodic boundary conditions, whereby a particle exiting the cell on one side is reintroduced on the opposing side with the same velocity were imposed. A dynamical time step of 0.25 fs was employed for all runs, the longest of which was 4.5 ps. Simulations were performed at a constant temperature of 3000 K using either one unit cell of nitromethane crystal (4 molecules, 28 atoms), a supercell with 8 molecules, and a supercell with 16

molecules. The unit cell was fully optimized at the reduced (compressed) volume, V = 205.36 Å^3.

RESULTS AND DISCUSSION

The initial configuration at density 1.974 g/cm^3 was determined by compressing the simulation cell and performing full relaxation of all atomic coordinates. From this initial structure, molecular dynamics were performed using the Nose-Hoover thermostat at 3000K.

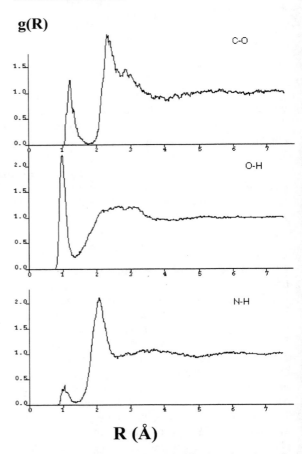

Figure 1. Calculated radial distribution functions for various intra and intermolecular bonds.

Here, we report the results obtained from the simulation using the largest supercell, consisting of four unit cells of nitromethane molecules with a repetition of 2x2x1 of the unit cell and corresponding to a volume of 821.5 Å^3. The total time of this simulation was 1.16 ps.

Figure 1 shows the C-O, O-H, and N-H radial distribution functions obtained over the total time of the simulation. It is evident from the distributions for the C-O, O-H and N-H that significant rearrangement of the bonds have occurred and chemistry has ensued. For C-O, the dominant population around 1.2 Å is due mainly to the formation of CO_2, while for N-H, the small population around 1.0 Å is due to the formation of radical intermediates of CH_2NHO. Most interestingly is the significant population growth of O-H at 1.0 Å, which encompasses the formation of H_2O, and in the early stages, to inter and intramolecular hydrogen bonding that leads to proton transfer.

To examine the early steps of the simulation, Figure 2 displays the variation of the N-O, C-H, and O-H bond distances with time. As shown, the C-H bond clearly undergoes a significant stretch that eventually leads to a hydrogen ejection and subsequent capture by the oxygen of a nearby nitromethane molecule, leading to the formation of CH_3NO_2H and CH_2NO_2 species.

A snapshot of the MD simulation at 59 fs where the formation of CH_3NO_2H and CH_2NO_2 takes place is shown in Figure 3. This process of proton transfer is initially facilitated by enhancement in the C-N double bond character, and an accelerated rotation of the methyl groups (CH_3), rotations that are omnipresent even at ambient temperatures.[14]

The proton transfer process described above is uniquely associated with the condensed fluid phase of nitromethane. This bond specificity is remarkable, since in the gas phase the C-N bond is the weakest in the molecule (D_0 = 60.1 kcal/mol)[15], and is therefore expected to be the dominant dissociation channel and the initial step in the decomposition of nitromethane even at high temperature. In contrast, the C-H bond is the strongest in the nitromethane molecule. In the condensed phase, however, vibrational energy is the highest in the C-H mode. Due in part to a caging effect, this vibrational motion eventually leads to a proton extraction.

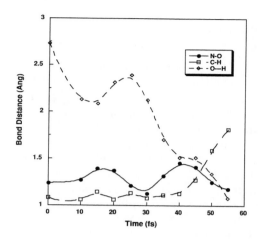

Figure 2. Time variation of intramolecular C-H and N-O, and the intermolecular O···H bonds.

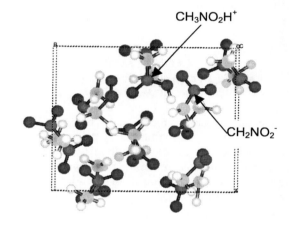

Figure 3. A snapshot of the MD simulation at 59 fs. The formation of CH_3NO_2H and CH_2NO_2 due to intermolecular hydrogen abstraction is shown.

The formation of CH_3NO_2H and CH_2NO_2 via proton extraction was observed in all three simulations of different supercell sizes. In the simulation on a single unit cell, the event occurs at 785 fs of the simulation time. We performed Mulliken charge analysis and determined the net charges on the two moieties CH_3NO_2H and CH_2NO_2. We notice that the negative charge on the carbon atom of CH_2NO_2 is larger than in

CH_3NO_2H, while the opposite trend is exhibited for the positive charge on nitrogen. This is a manifestation of electronic charge redistribution in the region between the C and N atoms.

It is noteworthy that all three simulations (one, two and four unit cells) have yielded the same results in the formation of CH_3NO_2H, $H_2CNO_2^-$, and CH_2NO_2H. Experimental concurrence for the production of the aci ion in highly pressurized and detonating nitromethane abound. Shaw et al.[1] observed that the time to explosion for deuterated nitromethane is about ten times longer than that for the protonated materials, suggesting that a proton or hydrogen atom) abstraction is the rate-determining step. Isotope-exchange experiments provided evidence that the aci ion concentration is increased upon increasing pressure[2], and UV sensitization of nitromethane to detonation was shown to correlate with the aci ion presence[16]. Finally, we note that a recent electronic structure study of solid nitromethane determined a significant C-H stretch upon compression, which eventually lead to proton dissociation[17].

CONCLUSION

We studied the early chemical events of hot (T = 3000 K) and dense (1.97 g/cm^3, V/V_0 = 0.68) nitromethane using density functional molecular dynamic simulations. Three simulations on one, two, and four unit cells of crystal nitromethane have shown that the first step event in the decomposition process is an intermolecular proton abstraction mechanism that leads to the formation of CH_3NO_2H and the aci ion $H_2CNO_2^-$, which lends support to experimental results from static high-pressure and shock experiments. An intramolecular hydrogen transfer that transforms nitromethane into the aci acid form, CH_2NO_2H, accompanies this event. This is the first confirmation of chemical reactivity with bond selectivity for an energetic material near the condition of fully reacted specimen.

ACKNOWLEDGMENTS

This work was performed under the auspices of the U.S. Department of Energy by the Lawrence Livermore National Laboratory under contract number W-7405-Eng-48.

REFERENCES

[1] R. Shaw, P. S. Decarli, D. S. Ross, E. L. Lee, and H. D. Stromberg, Combust. Flame 35, 237 (1979); R. Shaw, P. S. Decarli, D. S. Ross, E. L. Lee, and H. D. Stromberg, Combust. Flame 50, 123 (1983).

[2] R. Engelke, D. Schiferl, C. B. Storm, and W. L. Earl, J. Phys. Chem. 92, 6815 (1988).

[3] J. W. Brasch, Journal of Physical Chemistry 84, 2084 (1980); D. L. Naud and K. R. Brower, High-Pressure Research 11, 65 (1992).

[4] N. C. Blais, R. Engelke, and S. A. Sheffield, Journal of Physical Chemistry A 101, 8285 (1997).

[5] J. M. Winey and Y. M. Gupta, Journal of Physical Chemistry B 101, 10733 (1997).

[6] J. M. Winey and Y. M. Gupta, Journal of Physicall Chemistry A 101, 9333 (1997).

[7] Y. A. Gruzdkov and Y. M. Gupta, Journal of Physical Chemistry A 102, 2322 (1998).

[8] P. Hohenberg and W. Kohn, Phys. Rev. 136, B864 (1964).

[9] J. P. Perdew and Y. Wang, Phys. Rev. B 46, 6671 (1992).

[10] D. Vanderbilt, Phys. Rev. B 41, 7892 (1990).

[11] G. Kresse and J. Furthmuller, Phys. Rev. B 54, 11169 (1996).

[12] A. Inc., Cerius 2 Modling Environment (Accelrys Inc., San Diego, 1999).

[13] G. Kresse and J. Hafner, Phys. Rev. B 47, RC558 (1993).

[14] M. E. Tuckerman and M. L. Klein, Chem. Phys. Lett. 283, 147 (1998); D. C. Sorescu, B. M. Rice, and D. L. Thompson, J. Phys. Chem. B 104, 8406 (2001).

[15] J. B. Pedley, R. D. Naylor, and S. P. Kirby, *Thermochemical Data of Organic Compounds*, 2nd ed. ed. (Chapman, New York, 1986).

[16] R. Engelke, W. L. Earl, and C. M. Rohlfing, J. Phys. Chem. 90, 545 (1986).

[17] D. Margetis, E. Kaxiras, M. Elstner, T. Frauenheim, and M. R. Manaa, J. Chem. Phys. 117, 788 (2002).

CP706, *Shock Compression of Condensed Matter - 2003*
edited by M. D. Furnish, Y. M. Gupta, and J. W. Forbes
© 2004 American Institute of Physics 0-7354-0181-0/04/$22.00

DETERMINATION OF JWL PARAMETERS FOR NON-IDEAL EXPLOSIVE

H. Hamashima[1], Y. Kato[2], and S. Itoh[1]

*[1]Shock Wave and Condensed Matter Research Center, Kumamoto University,
2-39-1 Kurokami, Kumamoto 860-8555, Japan*
[2]NOF CORPORATION, 61-6 Kitakomatsudani, Taketoyo-cho, Chita-gun, Aichi 470-2398, Japan

Abstract. JWL equation of state is widely used in numerical simulation of detonation phenomena. JWL parameters are determined by cylinder test. Detonation characteristics of non-ideal explosive depend strongly on confinement, and JWL parameters determined by cylinder test do not represent the state of detonation products in many applications. We developed a method to determine JWL parameters from the underwater explosion test. JWL parameters were determined through a method of characteristics applied to the configuration of the underwater shock waves of cylindrical explosives. The numerical results obtained using JWL parameters determined by the underwater explosion test and those obtained using JWL parameters determined by cylinder test were compared with experimental results for typical non-ideal explosive; emulsion explosive. Good agreement was confirmed between the results obtained using JWL parameters determined by the underwater explosion test and experimental results.

INTRODUCTION

The ammonium nitrate (AN) -based explosives are well known to show non-ideal detonation behavior. Their detonation waves steadily propagate. However their characteristics are significantly affected by the conditions such as charge diameter or confinement. The emulsion explosives show non-ideal detonation behavior, and their detonation velocities are easily controlled by selecting the void size and adjusting the quantity of voids included.

JWL EOS[1-4] is widely used because of its simplicity in hydrodynamic calculations. JWL EOS contains parameters which describe the relationship among the volume, energy and pressure of detonation products. These parameters may be determined by the metal cylinder expansion test. In the cylinder test, the expansion of detonation products is estimated by the metal cylinder

expansion, but not by the real expansion of the products for non-ideal explosive. Therefore, the expansion of products used to be underestimated by the effect of metal cylinder confinement. We have confirmed both in experiments and in numerical simulations that there is a strong correlation between underwater shock waves and expansion waves produced by the expansion of detonation products.[5]

In order to study the expanding process of detonation products of non-ideal explosives, the optical observation of the underwater explosion of cylindrical non-ideal explosive detonation was carried out. Using a method of characteristics applied to the configurations of underwater shock waves and applying one-dimensional hydro-dynamic analysis for the axis symmetric flow, the expanding process of detonation products is made clear in all stages. Therefore the expanding process of detonation products is predicted. The pressure and density of detonation products can be determined by making the underwater expanding process of the detonation products clear. The parameters of JWL EOS are obtained by using this technique.

EXPERIMENTAL PROCEDURE

Sample emulsion explosive (EMX) was composed of emulsion matrix and polystyrene resin balloon of multi-cell structure (average diameter; 2.2mm, bulk density; 43kg/m^3). Composition of the emulsion matrix was AN/sodium nitrate (SN)/ Hydrazine nitrate (HN)/ Water/ EDTA/ Wax and emulsifier = 72.29/ 6.22/ 5.52/ 11.04/ 0.10/ 4.82 (wt.%). Detonation velocity of sample EMX was 2520m/s, and its initial density was 900kg/m^3. Sample explosive was set in the aquarium made of Polymethylmethacrylate (PMMA). Sample explosive was initiated by No. 6 electric detonator. Streak photographs and framing photographs are taken by a high-speed camera (IMACON468, HADLAND PHOTONICS, Framing rates; 200000 fps, Streak window; 30μs) using a conventional shadowgraph system. The configurations of underwater shock waves were obtained from the streak photographs. The experimental device for the cylindrical non-ideal explosive EMX is shown in Figure 1. In this figure, the slit shows the optical slit for taking the streak photograph.

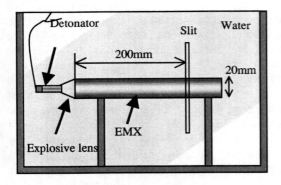

FIGURE 1. Experimental configuration. The non-ideal explosive EMX is 250mm in length, 20mm in diameter.

METHOD OF CHARACTERISTICS

Nomenclature

r	Radius
P	Pressure
ρ	Density
e	Internal energy
D	Detonation velocity
U_s	Shock velocity
u_p	Particle velocity
υ	Particle velocity in stationary coordinate
C	Sound velocity
M	Mach number
μ	Mach angle
ν	Prandtl-Meyer function
θ	Shock front angle
Γ	Coefficient of Grüneisen
δ	Deflection angle

Method of characteristics

The theory to use a method of characteristics is easily described as follows. The underwater shock wave system described in stationary coordinate system fixed to detonation front is shown in Figure 2. The properties of detonation and the propagating process of underwater shock wave are assumed as follows. Detonation wave propagates into explosive with a constant velocity D and has the steady detonation behavior. Underwater shock wave keeps it similar shape and moves toward x at a constant velocity of D with detonation wave. Consequently, detonation wave and underwater shock wave can be stopped by adding the reverse velocity of $-D$, which has the reverse direction of x, to the whole stream field. Boundary between detonation products and water is shown by curve AB at stationary coordinate system. Curve of Characteristics S_1B_1 is described between this boundary and underwater shock wave AS. If the equation $U_s = C_0 + su_p$ is applied to the relation of the oblique shock by using a method of characteristics and the change of δ along streamline S in the direction of streamline and the change of υ among streamlines, the equations for underwater shock wave are obtained as follows.

$$P = \rho_0 U_S u_P \qquad (1)$$

$$\rho = \rho_0 U_S / (U_S - u_P) \qquad (2)$$

$$\upsilon^2 = \frac{[U_S(s-1)+C_0]^2}{s^2} + D^2 - U_S^2 \qquad (3)$$

$$\tan\delta = \frac{(U_S - C_0)\sqrt{D^2 - U_S^2}}{sD^2 - U_S(U_S - C_0)} \qquad (4)$$

$$\frac{d\upsilon}{dU_S} = \frac{\sqrt{M^2 - 1}\,[U_S(1-2s)+C_0(s-1)]}{U_S^2(1-2s)+2U_S C_0(s-1)+C_0^2+s^2 D^2} \qquad (5)$$

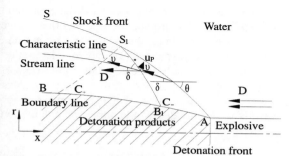

FIGURE 2. Stationary Coordinate System.

If the configuration of underwater shock wave is given, physical quantities of a range between AS and AB are obtained by using the above equations in calculations. Using one-dimensional hydrodynamic analysis for the axis symmetric flow, the pressure and density of products are found by making the underwater expanding process of the detonation products clear. Thus, if the configurations of underwater shock waves are known, the expanding process of the detonation products is made clear, even if the composition of explosive is unknown.

NUMERICAL SIMULATION

The numerical simulation of the underwater explosion of cylindrical explosive was conducted by Arbitrary-Lagrangian-Eulerian (ALE) method[6], by using C-J Volume Burn Technique[7] and by using the laws of conservation of mass, momentum, energy and EOS.

Mie-Grüneisen EOS is used for water.[8]

$$P = \frac{\rho_0 C_0^2 \eta}{(1-s\eta)^2}\left(1 - \frac{\Gamma\eta}{2}\right) + \Gamma\rho_0 e \qquad (6)$$

where

$$\eta = 1 - \frac{\rho_0}{\rho} \qquad (7)$$

The constants of Mie-Grüneisen EOS for water are shown in Table 1.

TABLE 1. Constants of Mie-Grüneisen EOS.

Material	r_0 (s/m^3)	C_0(m/s)	s	Γ
WATER	1000	1489	1.79	1.65

JWL EOS using new parameters was obtained by the proposed method used for the detonation products.

$$P = A\left(1 - \frac{\omega}{R_1 V}\right)\exp(-R_1 V)$$
$$+ B\left(1 - \frac{\omega}{R_2 V}\right)\exp(-R_2 V) + \frac{\omega\rho_e e}{V} \qquad (8)$$

where A, B, R_1, R_2, ω are JWL parameters. V is ρ_e (density of explosive)/ ρ (density of detonation products). JWL parameters of EMX are shown in Table 2.

TABLE 2. JWL Parameters of EMX. UWS is JWL parameters obtained by the underwater explosion test. CYL is JWL parameters obtained by the metal cylinder expansion test.

	A (GPa)	B (GPa)	R_1	R_2	ω
UWS	1.51	4.43	5.80	1.7	0.24
CYL	209.3	11.6	23.4	2.6	0.34

RESULTS AND DISCUSSION

The configurations of underwater shock wave for cylindrical non-ideal explosive EMX obtained from the numerical results using JWL parameters obtained by the underwater explosion test (UWS) and experimental results are shown in Figure 3. The cylindrical explosive has 20 mm in diameter and 250 mm long. Good agreement is obtained between numerical and experimental results. The numerical results using JWL parameters obtained by UWS and by the metal cylinder expansion test (CYL), and experimental results (EXP) obtained for cylindrical explosive are compared in Figure 4. The vertical axis is the distance in the direction of radius. The horizontal axis is the distance measured from the detonation front. Open circles indicate the configuration of underwater shock wave obtained experimentally and Open triangles indicate the boundary between detonation products and water, obtained experimentally. A solid line shows the configuration of underwater shock wave obtained in numerical simulation with UWS and a broken line shows the boundary between detonation products and water obtained in numerical simulation with UWS. A dot-dashed line shows the configuration of underwater shock wave obtained in numerical simulation with CYL and a dot-dot-dashed line shows the boundary between detonation products and water obtained in numerical simulation with CYL. A good agreement between the numerical results with UWS and experimental results is confirmed in the cases of both underwater shock wave and a boundary between detonation products and water.

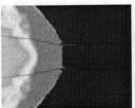 

(a) Numerical result (UWS)   (b) Experimental result

FIGURE 3. Configurations of underwater shockwave by cylindrical non-ideal explosive.

CONCLUSIONS

A new technique in determining the JWL parameters of detonation products for non-ideal explosive is proposed in this paper. The strong correlation between the underwater shock wave and the expansion wave produced by the expansion of detonation products was confirmed in our underwater experiments. This technique developed the method of characteristics in the relation

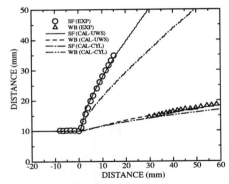

FIGURE 4. Configurations of the underwater shock wave and the boundary between detonation products and water.

between underwater shock wave and the expansion wave of detonation products. Using this theory, we can estimate the relation between the pressure and volume in the expanded region of detonation products. Then finally we can get the parameters of JWL EOS. It is concluded that for cylindrical charge, the configuration of the underwater shock wave is well estimated by the numerical calculation.

REFERENCES

1. Lee, E., Finger, M., Collins, W., "JWL Equation of State Coefficients for High Explosives", Lawrence Livermore Laboratory, Rept-UCID-16189, (1973)
2. Lan, I. F., Hung, S. C., Lin, W. M., Chen, C. Y., Niu, Y. M. and Shiuan, J. H., *Kogyo Kayaku*, Vol.**53**, No.3, 149-157, (1992)
3. H. Hornberg, *Propellants, Explosives, Pyrotechnics* **11**, 23-31, (1986)
4. Kury, J. W., Hornig, H. C., Lee, E. L., McDonnel, J. L., Ornellas, D. L., Finger, M., Strange, F. M. and Wilkins, M. L., "Metal Acceleration by Chemical Explosive", *4th Symposium on Detonation*, 1965, pp.3-13
5. Itoh, S., Liu, Z., Nadamitsu, Y., *Transactions of the ASME*, Vol.**119**, 498 (1997)
6. Amsden, A. A., Ruppel, H. M., Hirt, C. W., "SALE: A Simplified ALE Computer Program for Fluid Flow at All Speeds", LA-8095, UC-32 (1980)
7. Mader, C. L., *Numerical Modeling of Detonations*, University of California Press Berkeley and Los Angeles, California, 1979
8. Takahashi, K., Murata, K., Torii, A., Kato, Y., "Enhancement of Underwater Shock Wave by Metal Confinement", *11th Symposium on Detonation*, 1998, p.466

CP706, *Shock Compression of Condensed Matter - 2003*
edited by M. D. Furnish, Y. M. Gupta, and J. W. Forbes

A STATISTICAL APPROACH ON MECHANISTIC MODELING OF HIGH-EXPLOSIVE IGNITION

Y. Hamate and Y. Horie

Applied Physics Division, Los Alamos National Laboratory

Abstract. This paper presents the development of a mechanistic reactive burn model for solid explosives through use of a size distribution function for hotspots. The model couples the unifying hot-spot model developed earlier and the Lagrangian hydrodynamic flow equations. The hot-spot model incorporates key features of energy localization without introducing the mechanism-specific traits and is applicable to the three primary mechanisms of energy localization: pore collapse, friction. and shear banding. The coupling of the model to the hydrodynamic flow equations include models for energy localization, the growth of hot spots, and a two-phase aggregation of distributed hot spots, and a mixing rule for a product gas and a reacting solid. Proof-of-concept calculations for shock initiation are carried out in one spatial dimension, using RDX as a model material. Results include (1) shock ignition and growth-to-detonation, and (2) quenching.

INTRODUCTION

The concept of hot spots has been used to describe the ignition of solid explosives by mechanical insults such as impact and shock loadings over the last fifty years. Primary mechanisms responsible for the creation of hot spots (energy localization) are thought to be void collapse, friction, and shear [1, 2, 3]. These mechanisms transform overall mechanical energy into localized heating. But none of the models that are focused on a single physical mechanism has a universal appeal, because it is doubtful that any one mechanism dominates in vastly different ranges of mechanical loading [4].

The unifying model was developed [5, 6] to consider various regimes of loading on a common framework by focusing attention on a mathematical abstraction of the fundamental features of three primary mechanisms of energy localization: void collapse, shear banding, and friction. These features were, however, described without introducing the mechanism-specific traits. In this way, the issue of how to consider the localization as dependent on the state of the material was set aside temporarily. Instead, attention was focused on the parameteric representation of the energy localization that is consistent with detailed mechanistic models.

In the unifying model the geometry of energy localization was simplified to a plane structure consisting of the following key features: a region of localized heating surrounded by material with lower or zero heating rate; the existence or creation of a space next to the heated zone that is occupied by gases; energy transfer between the regions; and reaction chemistry within the gas, at the gas-solid surface, or both. What is important in this model is that attention is shifted from "hot spots" to "gas producing, hot interfaces."

In order to simulate shock-to-detonation phenomena, a linkage model, which incorporates the above mentioned unifying hot spot model into global hydrodynamic calculations, needs to be developed. We introduce two key ideas for the linkage. One describes how the global energy is deposited locally into the hot spot and another is a mixture EOS which couples the two-phase (gas/solid) hot-spot with single phase (phase-averaged) hydrodynamic calculations. The former is developed using a statistical treatment of hot-spot evolution and energy localization.

MODEL DESCRIPTION

Figure 1 shows how the linkage model couples gloabal hydro calculations and the unifying hot-spot

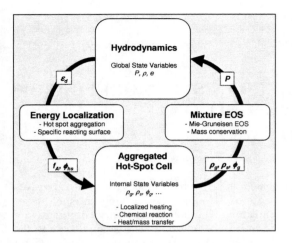

FIGURE 1. Schematic diagram of the linkage module

model. Each linkage module will be reviewed in the following sections.

Energy localization

Energy localization is modeled by the specific surface area of locally heated sites contributing to the reaction (hot-spots), and the amount and rate of energy deposited into the hot spots. They are assumed to be a function of global state variables, which in this report is the dissipative energy related to artificial viscosity. The specific surface area of hot spots is derived from a hypothesized exponential area distribution.

Statistical Treatment of Hot Spots

A statistical treatment of hot-spot evolution is based on two main ideas. The first is a modified idea of Cochran [7] and assumes that shock (or impact) loading creates a number of locally heated sites. For our initial attempt, we hypothesize that the size distribution for heated sites is exponential,

$$f(r) = \frac{N_0}{r_0} exp\left(\frac{r}{r_0}\right) \quad (1)$$

where $f(r)$ is the number of sites available for localized heating having size r, N_0 is the total number of the sites and r_0 is the average size. And the rate of production is propotional to the overall specific dissipated mechanical energy (ε_d),

$$\dot{S} = \frac{\dot{\varepsilon}_d}{\varepsilon_d^0} f(r) \left(1 - \xi \frac{4\pi \int_0^\infty Sr^2 dr}{a_m}\right) \quad (2)$$

where S is the number of locally heated sites per unit volume having size r at time t, ε_d^0 is a normalized constant, ξ is a parametric constant that controls growth. The parameter a_m is the maximum area available for the reaction.

$$a_m = \int_0^\infty 4\pi f(r) r^2 dr = 8\pi N_0 r_0 \quad (3)$$

Equation (2) is solved by the following assumption.

$$\frac{d}{dt} \int_0^\infty Sr^2 dr = \int_0^\infty \dot{S}r^2 dr \quad (4)$$

The result is an analytic expression for the specific surface area of locally heated sites (f_A) where solids transform into gases at the appropriate conditions of temperature and pressure.

$$f_A = 4\pi \int_0^\infty Sr^2 dr = \frac{a_m}{\xi}\left[1 - \alpha_m exp\left(-\xi \frac{\varepsilon_d}{\varepsilon_d^0}\right)\right] \quad (5)$$

where the parameters ξ and α_m are interrelated and depend on the conditions such as initial porosity, the saturation fraction of hot spots.

Assuming locally heated sites has a constant depth of δ, total volume of locally heated sites per unit volume is then,

$$\lambda_{hs} = \delta f_A = \frac{\delta a_m}{\xi}\left[1 - \alpha_m exp\left(-\xi \frac{\varepsilon_d}{\varepsilon_d^0}\right)\right] \quad (6)$$

Boundary conditions, $\lambda_{hs} = \lambda_0$ at $\varepsilon_d = 0$, and $\lambda_{hs} = \lambda_m$ at $\varepsilon_d = \varepsilon_d^0$ yield,

$$\lambda_0 = \frac{\lambda_m}{\xi}(1 - \alpha_m) \quad (7)$$

$$\xi = 1 - \alpha_m exp(-\xi) \quad (8)$$

The major differences between our model and Cochran's are (1) the source rate is a function of dissipated energy at the macrolevel in place of relative compression, and (2) the source rate is only concerned with the production of heated sites, and all the other mechanisms are treated mechanistically by the hot-spot model. So, for example, the ignition in our model occurs when self-sustained reactions are initiated in the reactant gases. The process is path-dependent. In the Cochran model, however, the ignition is preset to happen continuously at a specified rate as a function of pressure.

The second idea is that a collection of hot spots may be aggregated into a single, super hot spot through use of f_A for the heat and mass transport across the reacting gas-solid interface. The result is a scaled, localized heat flux $\hat{\Phi}_{hs}$ for the super hot spot in terms of overall dissipated energy ε_d.

$$\hat{\Phi}_{hs} = \rho_s \frac{d}{dt}\left(\frac{\varepsilon_d}{\lambda_{hs}}\right) \quad (9)$$

where ρ_s is the explosive solid density. In our model, $\hat{\Phi}_{hs}$ is the expression that transforms the mechanical energy into localized heating and governs the ignition behavior. Once ignited, the chemistry drives the surface burning wave.

Mixture EOS

We follow the formulation by Massoni et al. [8]. This approach assumes common particle velocities among constituents, and the mass-based averaging for mass, momentum, and internal energy. Additionally, the equation of state for each constituent is written in a Mie-Grüneisen form. The resulting overall mixture EOS is,

$$P = \Gamma \rho e + P^* \quad (10)$$

$$\frac{1}{\Gamma} = \frac{\phi_g}{\Gamma_g} + \frac{1 - \phi_g}{\Gamma_s} \quad (11)$$

$$P^* = \Gamma\left[\phi_g \frac{P_g^*}{\Gamma_g} + (1 - \phi_g)\frac{P_s^*}{\Gamma_s}\right] \quad (12)$$

where subscript g and s represent gas and solid, respectively and ϕ_g the volume fraction of gas. For more details about the EOS for each phase, see Ref. [9].

MODEL CALCULATIONS

One-dimensional hydrodynamic equations are solved for shock ignition and detonation using a standard leap-frog method and a two-term artificial viscosity. Materials parameters including chemical kinetics are those for RDX [1]. This choice is historical in that the hot-spot model is first tested for RDX

as a model material [5]. The material is assumed to have an initial porosity of 1%.

Shock loading is initiated by a near-symmetric impact where the impactor is assumed to be an inert solid RDX. A thin buffer is added between the impactor and the reactive RDX to avoid a false ignition due to oscillations near the impact plane. Figure 2 shows a typical run-to-detonation behavior observed in simulation where the computational cell size was 10 μm. Impact velocity of 1.0 km/s generates the initial shock pressure of 3.5 GPa, which grows to a steady wave propagating at 7.2 km/s. Pressure history shows that the steady wave has a peak pressure of about 30 GPa and a kink around 20 GPa. The latter corresponds to the end of the reaction zone. Listed values of von Neuman spike and CJ pressure are 47 GPa and 33.7 GPa, respectively. This suggests that the EOSs used in the model need improvement.

Figure 3 is a demonstration of quenching as a result of rarefaction waves emanating from the back free surface of the flyer. An incident shock of 2.5 GPa indtroduces a reaction as shown in Fig 3 (b). However, as shock decays due to rarefaction wave, the reaction quenchs. Consequently the chemical reaction is limited to the region near the impact.

CONCLUSION

Global hydrodynamics and the unifying hot-spot model are coupled by the two linkage models. The energy localization model incorporates a statistical treatment of hot-spots, that leads to an analytical expression for the specific surface area of hot-spots. Proof-of-concept calculations are performed using a RDX as a material model. Results show that the model is capable of simulating shock ignition, growth-to-datonation, and quenching.

ACKNOWLEDGMENTS

This work is supported by the US Department of Energy under contract W-7405-ENG-36.

REFERENCES

1. Kang, J., Butler, P., and Baer, M., *Combustion and Flame*, **89**, 117–139 (1992).
2. Dienes, J., "A unified theory of flow, hot spot,

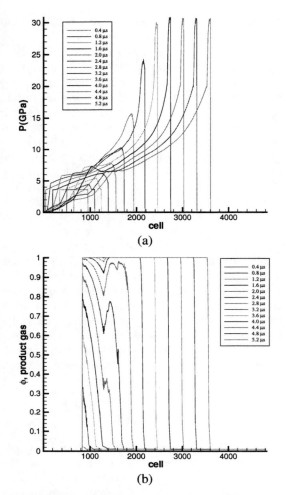

(a)

(b)

FIGURE 2. Growth-to-detonation behavior is shown in (a) where the steady detonation propagates at the velocity of 7.2 km/s. The value in a LASL report is 8.6 km/s at 33.7 GPa (CJ state). Fractions of the product gas shown in (b) indicate an heterogeneous ignition.

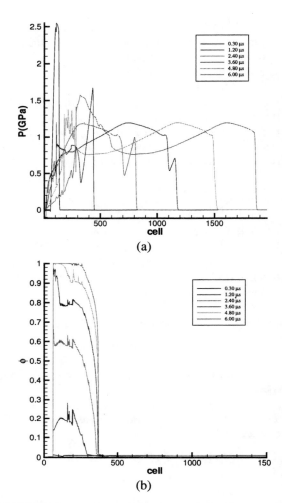

(a)

(b)

FIGURE 3. An illustration of quenching behavior for the statistical model.

and fragmentation with an application to explosive sensitivity," in *High-Pressure Shock Compression of Solids II*, edited by L. W. Davison, D. E. Grady, and M. Shahinpoor, Springer, New York, 1995, p. 366.

3. Frey, R., "The Initiation of Explosive Charges by Rapid Shear," in *Proceedings of 7th Symposium (International) on Detonation*, 1982.

4. Khasainov, B., Attetkov, A., and Borisov, A., *Chemical Physics Reports*, **15**, 987–1062 (1996).

5. Yano, K., Horie, Y., and Greening, D., A unifying framework for hot spots and the ignition of energetic materials, Tech. Rep. LA-13794-MS, LANL (2001).

6. Yano, K., Horie, Y., and Greening, D., "Mechanistic Model of Hot-Spot: A Unifying Framework," in *AIP Conference Proceedings*, 620, 2001, pp. 983–986.

7. Cochran, S., A statistical treatment of heterogeneous chemical reaction in shock-initiated explosives, Tech. Rep. UCID-18548, LLNL (1980).

8. Massoni, J., Saurel, R., Baudin, G., and Demol, G., *Physics of Fluids*, **11**, 710–736 (1999).

9. Horie, Y., Hamate, Y., and Greening, D., Mechanistic reactive burn modeling of solid explosives, Tech. Rep. LA-14008, LANL (2003).

10. Kipp, M., "Modeling Granular Explosive Detonations with Shear Band Concepts," in *Proceedings of 8th Symposium (International) on Detonation*, 1985.

CP706, *Shock Compression of Condensed Matter - 2003*
edited by M. D. Furnish, Y. M. Gupta, and J. W. Forbes
© 2004 American Institute of Physics 0-7354-0181-0/04/$22.00

REACTIVE FLOW IN NITROMETHANE USING THE CW2 EQUATION OF STATE

O. Heuzé[1], E. Martinez[1], S. Szarzynski[1], R. Mulford[2], D. C. Swift[2]

[1] CEA/DIF, B.P.12, 91680 Bruyères-le-Châtel, France
[2] Los Alamos National Laboratory, Los Alamos NM 87545

Abstract. The CW2 Equation of State (EoS) has been found to reproduce equilibrium chemistry predictions of thermodynamic states in the reaction products of explosives, and is also capable of representing unreacted material. Partially-reacted mixtures can then be represented by interpolating parameters in the EoS. Here we present CW2 EoS for nitromethane, unreacted and products. Reaction rate parameters in a two-step Arrhenius model were calibrated to reproduce single-shock initiation properties. Predictions were then made of initiation from a reflected shock, and compared with recent experimental data.

INTRODUCTION

Although numerous experiments have been performed to investigate the initiation of detonation in nitromethane (NM) by shock waves, this process is still not well-understood. In recent experiments [1], pure nitromethane was initiated from a reflected shock, shedding a new light on this process.

In 1980, Presles et al. [2] investigated initiation of NM from a shock reflected from aluminum. Although the reflected shock pressure was higher than the pressure required for a single shock initiation, no detonation was observed. The authors argued that the reflected shock temperature was lower than the ignition temperature and thus accounted for this phenomenon.

More recently, Higgins et al. [3,4,5] found it necessary to use sensitized NM to obtain its initiation following a shock reflected from steel. In that case, they found that there was little correlation between the pressure threshold upon incident and reflected shock but a sound link between the temperature thresholds, despite the large discrepancy in temperature determination with available EoS.

In 2002, Heuzé et al. [6] proposed a new experimental setup to obtain initiation of pure NM from a reflected shock, using a tungsten reflecting wall. The high impedance of tungsten leads to higher temperature behind the reflected shock. This experiment was undertaken by Szarzynski et al. [1]. The results confirmed that the temperature is probably the dominant parameter describing the initiation of nitromethane, as was also observed by Partom [7] for high explosives.

Although Heuzé et al's predictions led to this successful experiment, the velocity measurements were not in agreement with his *a priori* calculations fitted to single shock experiments. This suggests that the initiation process cannot be explored adequately by single shock initiation data alone, and that the modeling of NM initiation has to be improved in the light of the recent reflected shock experiments.

In this paper, we evaluate the phenomena and parameters leading to improved simulations of this experiment. Three main phenomena have been investigated : the equation of state of NM, the chemical kinetics scheme based on the Arrhenius law, and the effect of strength in the inert components.

The CW2 equation of state was used in the present study [8,9,10]. It is a thermodynamically complete equation of state which was designed for detonation products used in hydrocodes, but is also valid for the unreacted explosive.

The kinetics scheme was based on a two step Arrhenius law.

Plasticity was taken into account and showed a significant influence.

CW2 EQUATION OF STATE

This equation of state is defined from the free energy, as function of volume and temperature, and 9 parameters (Γ_o, a, λ, μ, C_v, δ, ρ_r, θ_r, S_r):

$$A(V,T) = E_K(V) + C_V.T[\ln(x) - \delta.x] - S_r T \quad (1)$$

with :

$$E_K(V) = y\left[\lambda\frac{\rho}{\rho_r} + \mu\ln\left(\frac{\rho}{\rho_r}\right)\right] \quad (2)$$

$$x = \frac{\theta_r.y}{T} \quad (3)$$

$$y = \left(\frac{\rho}{\rho_r}\right)^{\Gamma_0} \exp\left[\frac{a}{2}\left(\left(\frac{\rho}{\rho_r}\right)^2 - 1\right)\right] \quad (4)$$

$$\Gamma = \Gamma_0 + a\left(\frac{\rho}{\rho_r}\right)^2 \quad (5)$$

The first derivatives of A(V,T) give pressure and entropy:

$$P(V,T) = P_K(V) + \frac{\Gamma(V)}{V}C_V.T(1 - \delta.x) \quad (6)$$

$$S(V,T) = C_V[1 - \ln(x)] + S_r \quad (7)$$

with:

$$P_K(V) = -\frac{dE_K(V)}{dV} \quad (8)$$

then:

$$E(V,T) = E_K(V) + C_V.T(1 - \delta.x) \quad (9)$$

Γ_o is the Grüneisen coefficient at the null density limit. Cv is the specific heat capacity at constant volume.

ρ_r, θ_r, S_r are respectively reference density, temperature and entropy. They are arbitrary parameters.

It is suited to fast calculations in hydrocodes because it can be written in the P(V,E) form:

$$P(V,E) = P_K(V) + \frac{\Gamma(V)}{V}[E(V,T) - E_K(V)] \quad (10)$$

For detonation products, the form of this EoS has been designed to be in agreement with thermochemical calculations [8]. More recently [10], calculations with different thermochemical codes have validated its main assumptions : the Grüneisen coefficient G depends only on density (i.e. not on temperature or internal energy), the specific heat capacity is nearly constant. Moreover, in the present case, the assumption that the Grüneisen is a function of the square of the density is valid, although it was not the case for the highest densities reached by HMX detonation products.

For the unreacted explosive, the parameter are deduced from the shock polar $U_S = c + s\ U_P$ [9]. Their values are taken from [11]: $U_S = 1650 + 1.64\ U_P$.

CHEMICAL KINETIC SCHEME

It is known from [2] and [4] that the single shock pressure threshold is not valid for multiple shocks in NM. Then we chose to use an Arrhenius scheme for the present study. It was stated in [11] that a single Arrhenius rate could not reproduce the pressure/run-to-detonation relationship. Then we used a two steps Arrhenius scheme:

Explosive → "Radicals" → Products

In a former study [6], we showed that such a modeling could lead to a backward superdetonation (BSD). However it predicted initiation from the reflected shock, the existence of the BSD is not yet confirmed by the experiments. Further experiments with more diagnostics are required.

ELASTIC-PLASTIC EFFECTS

We found it necessary to include the constitutive behaviour of the metal components in order to reproduce details of the LDI (Laser Doppler Interferometry) records. The EoS were of the cubic Grüneisen form. The constitutive model used was elastic-perfectly plastic. Spall was modelled by the minimum pressure method. Material parameters were taken from the Steinberg compendium [12].

REFLECTED SHOCK EXPERIMENTS

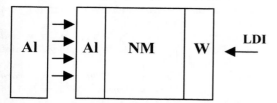

FIGURE 1. Reflected shock configuration.

An aluminum flyer impacts the aluminum cover plate (Fig. 1), creating a shock wave at the interface. This wave crosses the aluminum and the nitromethane and reflects from the tungsten slab. The experiment was designed so that the temperature behind the reflected shock in NM was expected to lead to its initiation.

The thickness of the different layers for Al/Al/NM/W were respectively 10/2/12/2 mm. Two experiments were fired, with impact speeds of 1792 and 1662 m/s respectively.

The free surface velocity of the tungsten plate was measured by LDI.

RESULTS AND DISCUSSION

The experiments were simulated using a 1D Lagrangian continuum mechanics program, with finite difference discretisation and a predictor-corrector numerical scheme. Shocks were stabilised using artificial viscosity, and reaction was operator-split from the hydrodynamics and subcycled. We found it necessary to use a cell size of 0.2 mm or smaller to reproduce reverberations in the tungsten plate at all accurately, and 0.05 mm cells were needed for the reactive behaviour to approach convergence.

Neglecting spall, the simulated velocity histories exhibited too much pull-back between reverberations. Neglecting strength, the shape and period of the reverberations was significantly less accurate.

With the modelling prescription described above, there was reasonable agreement with the general trend of the experimental records. The appearance of the elastic precursor at the surface of the tungsten did not match perfectly well, and the amplitude of the first few jumps in velocity were significantly lower than the experiment, suggesting inaccuracy in the unreacted or metal EoS. The agreement at later times was well within uncertainties in EoS, constitutive behaviour and numerical convergence, and clearly followed the data much better than did simulations in which the NM was inert or allowed to react more promptly. (Figs 2 and 3.)

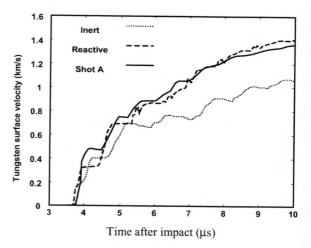

FIGURE 2. Velocity of the tungsten free surface versus time. Shot A : Vp = 1662 m/s.

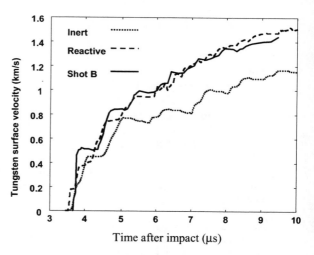

FIGURE 3. Velocity of the tungsten free surface versus time. Shot B : Vp = 1792 m/s.

341

CONCLUSIONS

The aim of the present study was to develop an improved model for the initiation of NM from shock waves. Simulations were compared with single shock and reflected shock data. This approach shed light on the results of the recent reflected shock experiments, and helped in the design of further ones.

We focused here on three aspects: the equation of state, chemical kinetics and strength effects.

The CW2 EoS was used for both unreacted NM and its reaction products. Parameters were adjusted to fit the linear U_S-U_P relationship for unreacted NM and thermochemical calculations for the products. These EoS seem adequate on the basis of available data.

Our chemical kinetics was based on a two steps Arrhenius scheme. The parameters used successfully to design the reflected shock experiments did not reproduce these experiments as well as had been hoped; a modified set of parameters will be developed. At this stage, we can not yet confirm the existence of the backward superdetonation.

This modeling was required to understood the main phenomena occuring in these recent experiments and to design further ones.

ACKNOWLEDGEMENTS

We would like to thank F. Chaissé for valuable comments.

This work was performed under the auspices of an agreement between CEA/DAM and NNSA/DP on cooperation on fundamental science.

REFERENCES

1. Szarzynski S., Heuzé O., Martinez E. *"Initiation of pure nitromethane by a reflected shock"*, this conference.
2. Presles H. N., Fisson F., Brochet C. *"Sur des conditions d'amorçage par onde de choc de la détonation du nitrométhane"*, Acta Astronautica, Vol. 7, pp 1361-1371,1980.
3. Higgins A.J., Jetté F.X., Yoshinaka A.C., Zhang F. *"Detonation Initiation in Preshocked Liquid Explosives"* 12th APS Conference on SCCM, Atlanta, 2001.
4. Higgins A.J., Jetté F.X., Yoshinaka A.C., Zhang F. *"Initiation of Detonation in Liquid Explosives by a*

4. Higgins A.J., Jetté F.X., Yoshinaka A.C., Zhang F. *"Initiation of Detonation in Liquid Explosives by a Reflected Shock Wave"*, 13th Symposium (Int.) on Detonation, San Diego, 2002.
5. Yoshinaka A.C., Zhang F. *"Shock Initiation and Detonability of Liquid Nitromethane"*, this conference.
6. Heuzé O., Chaissé F. *"On the backward superdetonation in homogeneous explosives"*, 13th Symposium (Int.) on Detonation, San Diego (CA), 2002.
7. Partom Y. *"Hydro-Reactive Computations with a Temperature dependent Reaction Rate"*, 12th APS Conference on SCCM, Atlanta, 2001 pp. 460-463.
8. Heuzé O. *"An equation of state of detonation products for hydrocode calculations"* 27th International Pyrotechnics, pp.15-19, Grand Junction (CO), 2000.
9. Heuzé O. *"A complete equation of state for detonation products in hydrocodes"* 12th APS Conference on SCCM, pp.450-453, Atlanta, 2001.
10. Heuzé O. ,Braithwaite M. , Swift D.C., Tanaka K. , Victorov S. , *"Towards the equation of state for detonation products. A benchmark for hydrocodes"* 5th HDP Conference, pp.291-299, Saint-Malo (France), June 23-27 2003.
11. Leal-Crouzet B., Baudin G., Presles H.N. *"Shock Initiation of Detonation in Nitromethane"*, Comb. and Flame 122:463-473, 2000.
12 Steinberg, D.J., *Equation of state and strength properties of selected materials*, Lawrence Livermore National Laboratory report UCRL-MA-106439 change 1 (1996).

CP706, *Shock Compression of Condensed Matter - 2003*
edited by M. D. Furnish, Y. M. Gupta, and J. W. Forbes
© 2004 American Institute of Physics 0-7354-0181-0/04/$22.00

PROGRESS IN THE DEVELOPMENT OF A SHOCK INITIATION MODEL

Philip M. Howe[1] and David J. Benson[2]

[1]*Los Alamos National Laboratory, Los Alamos, NM 87545*
[2]*Department of Mechanical and Aerospace Engineering, Univ. of Calif. San Diego, 9500 Gilman Drive, La Jolla, CA 92093*

Abstract. We used an Eulerian hydrocode to guide the development of an engineering model of shock initiation. The model in its current form has two types of hotspots– one from void collapse, and one from interactions at grain boundaries. The dependence of hotspot and bulk temperatures upon shock strength is estimated using a Gruneisen equation of state for the bulk solid, calibrated against measurements of reaction times for steady state detonation. Arrhenius kinetics are used to predict ignition times associated with hotspot temperatures. The hotspots contribute a small amount of energy to the shock front, thereby causing some shock front acceleration, and also serve to initiate erosive burning. The two erosive burn reactions that result from the two different types of hotspots compete to consume the material. The energy release rate resulting from the competition of these reactions was used as input to a method of characteristics code. This in turn was used to calculate particle velocity – time profiles at various simulated gauge locations. These calculated profiles were compared with experiment.

INTRODUCTION

In this effort, we used the calculated shock response of two- dimensional microstructures to help develop the framework of an initiation model. We used the calculations to describe the peak temperatures for hotspots generated at grain boundaries and voids as functions of shock strength. We used an experimentally determined rate equation for PBX 9501 decomposition (1), and an experimentally determined Hugoniot (2). We used a well-known function ($r = aP^n$), where r is the rate, P is the pressure, and n is a burning rate exponent, to describe the linear erosion rate and its dependence upon pressure. This was extrapolated well beyond the regime where there is data. In addition, we had to create a topology function for erosive burning that was without any theoretical or experimental support. We had to treat the representative hotspot temperatures as fitting

parameters, although the functional dependence upon shock strength was guided by the microstructure calculations. The main focus of this paper will be the effect of void collapse upon relatively weak shock initiation, although some results will be presented for stronger shocks.

Two types of hotspots, with different temperature dependences were included to account for the fact that introduction of porosity significantly increases the shock sensitivity of explosives. Yet, when explosives are very close to theoretical maximum density, explosives still initiate, at shock strengths for which the calculated bulk temperatures are still far too low to cause initiation. Thus, hotspots based upon pore collapse, and hotspots based upon grain boundary interactions are included. Hydrocode calculations have shown the pore collapse hotspots to be more efficient initiators (in the sense of generating higher temperatures at a given shock strength) than hotspots resulting from interactions at grain boundaries. The hotspot size was not taken into account explicitly – i.e., there was no heat transfer included in the hotspot temperature calculations. A hotspot size effect was implicitly treated by allowing the ratio of surface area to mass fraction of hotspots to be treated as a parameter. That is, the hotspots are modeled as affecting initiation in two ways: The explosion of a hotspot causes the release of some energy that can contribute directly to the flow field. Thus, some hotspots react, contribute their energy and –if too small – extinguish. Secondly, some fraction of the hotspots is assumed large enough to initiate erosive burning, which then spreads throughout the unreacted explosive.

Hydrocode calculations for the temperatures of the two types of hotspots have shown that hotspots formed from grain burning depend linearly upon particle velocity at the shock front. Those formed from void collapse have a linear dependence upon particle velocity at high shock strengths, but have a sharp drop in temperature at low shock strengths.

The chemical reaction leading to hotspot reaction is modeled by one step Arrhenius kinetics, with inclusion of the effects of adiabatic explosion. An attempt was made to calibrate the various temperatures to an equation of state developed by Davis (3) by fitting the equation of state to a reaction zone thickness estimated from measurements at the detonation state. This was only partially successful, due to the large uncertainty in the reaction zone length at the detonation state. In addition, it is may not be a correct approach, as there is evidence that erosive burning still is very influential at the detonation state. The equations for the model are summarized in reference (4).

HOTSPOT FORMATION

Hotspot formation was modeled using a two dimensional Eulerian hydrocode (Raven) developed by the second author. A representative microstructure is shown in Figure 1, where the

polygons represent grains of HMX, the interstices between the polygons are treated as binder, and the red spots represent voids. The HMX was modeled with an elastic viscoplastic material model and the binder was modeled as an elastoplastic material. The two models are described in references (5,6) respectively.

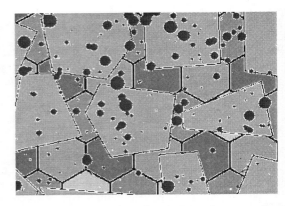

Figure 1. Representative microstructure for a porosity of 10%. The large central grain is nominally 100 μm across. Red spots represent voids voids.

We have used microstructures like these to construct hotspot temperature distributions as functions of shock strength and to examine the behavior of single voids. Both single voids and distributions of voids follow the general behavior shown in Figure 2, where temperature vs. time is plotted for a 10 micron void as a function of shock particle velocity.

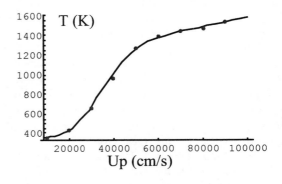

Figure 2. Void induced hotspot temperature vs. shock strength. Nonlinearity arises from failure of voids to close fully at low shock strengths.

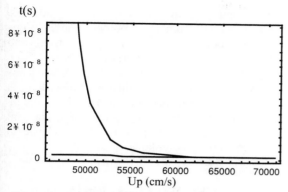

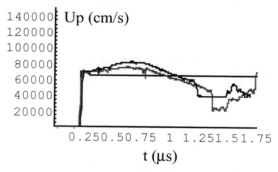

Figure 3. Hotspot ignition times. Upper curve includes effects of void partial closure. The nonlinearity of the lower curve is suppressed on this time scale but actually exhibits the standard Arrhenius dependence.

For low shock strengths, total void collapse did not occur for voids of any size considered. This increases the nonlinearity of the dependence of hotspot ignition time upon shock strength greatly and provides a very sharp threshold for initiation as a function of shock strength. A comparison of ignition times calculated with this effect included and ignition times where the linear behavior observed at high shock strengths is extended down to zero shock strength is shown in Figure 3.

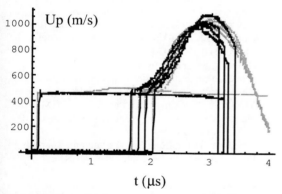

Figure 4. Slow shock acceleration for PBX 9501. Gray curve is calculated. Data are from (7,8). See Lambourn, B, this proceeding for thoughts on why the first experimental gauge record is flat.

This rapid drop in hotspot temperatures with reduced shock strength appears to be important in the initiation behavior at low shock inputs. Inclusion of this behavior in the model allows us to

capture the very slow shock acceleration and following flow at early times. Comparisons with data for a shock input of Up = 46400 cm/s are shown in Figure 4 (7,8).

Figure 5. Effect of hotspot ignition on wave formation (black is calculated.) Evidently, only a small fraction of the void volume leads to hotspot formation. PBX 9501 has about 2% void volume.

The percentage of material that actually contributes to form hotspots was estimated by turning off all reaction except the ignition reaction resulting from void collapse. A calculation with a hotspot mass fraction of 0.005 is shown in Figure 5. (Note that the input shock strength was of the order of 670000 cm/s). Even this mass fraction leads to an excursion in the calculated shock profile that exceeds that of the experimental record.

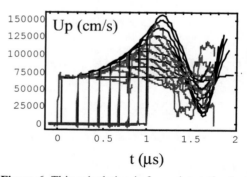

Figure 6. This calculation is for an input shock strength of up=670000 cm/s. The shock strength and time of arrival are well matched, but not the following flow.

We were curious to see how sensitive the buildup process is to the following flow. The set of shock profiles shown in Figure 6, as an example, does a poor job of matching the following flow, but captures shock growth rather well and the buildup process is modeled within experimental error (see Figure 7, which shows reaction progress in the t-h plane, where h is the Lagrange coordinate and the color contours vary from no reaction (orange) to 80 % completion (deep red)). Evidently, over most of the buildup process, reaction occurring very close to the front controls the shock evolution. For most of these calculations, reaction is not reaching completion during the buildup phase.

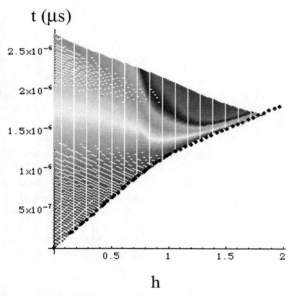

Figure 7. Time vs. Lagrange position plot for reaction progress corresponding to the waveforms shown in Figure 6. Also shown are experimental data (black points) for shock buildup converted to t-h plane.

CONCLUSIONS

In this model, hotspots caused by void collapse control ignition for low shock velocities (Up <600 m/s). These hotspots add a little energy to the shock, determine the time for ignition and initiate erosive burning. The very strong dependence of hotspot temperature upon shock strength at low velocities makes the ignition times extremely nonlinear in shock strength. The combination of this hotspot dependence and erosive burning allow a reasonable fit to the low strength behavior. The midrange is not adequately captured. For shocks whose input strength is above about 600 m/s, hotspots formed from grain boundary interactions also contribute energy to the shock front and initiate a second erosive burning process, which competes to consume the material. PBX 9501 has approximately 2% void volume. We found that fits to the data required that hotspot mass fractions be below 0.005, indicating that the volume of material that participates as hotspots in void collapse is smaller than the void volume.

ACKNOWLEDGEMENTS

Funding was provided by the U.S. Department of Energy. The authors are pleased to acknowledge the help of J.N. Johnson (Wash. State Univ., Institute for Shock Physics), and the use of his CHARADE code, and D. Shevitz, (and the use of his MOC code), L. Luck, and D. Zerkle, all of Los Alamos National Laboratory. The first author has benefited from discussions with B. Lambourn, AWE.

REFERENCES

1. Henson, B., private communication.
2. Gustavsen, R. private communication
3. Davis, W.C., "Complete Equation of State for Unreacted Solid Explosive", Combustion and Flame, 120:399-403 (2000).
4. Howe, P. M. and Benson, D. J., "An Engineering Model of Shock Initiation", 12th Symposium (International) on Detonation, (in press).
5. Menikoff, R. "Constituent Properties of HMX Needed for Meso-Scale Simulations,", LA-UR-00-3804-rev (2001).
6. Steinberg, D.J., "Equation of State and Strength Properties of Selected Materials", UCRL-MA106439 (1996).
7. Gustavsen, R., private communication.
8. Sheffield, S., et al, "Progress in Measuring Detonation Wave Profiles in PBX 9501", *Eleventh Symposium (International) on Detonation*, Snowmass, CO, Office of Naval Research, 821-827, (1998). Snowmass, CO, Office of Naval Research, 821-827, (1998).

CP706, *Shock Compression of Condensed Matter - 2003*
edited by M. D. Furnish, Y. M. Gupta, and J. W. Forbes
© 2004 American Institute of Physics 0-7354-0181-0/04/$22.00

EOS OF MIXTURES: PHASE TRANSFORMATION AND EXPLOSIVE REACTION

J. N. Johnson[1], D. H. Dolan[1], and P. M. Howe[2]

[1] *Institute for Shock Physics, Washington State University*
[2] *Los Alamos National Laboratory*

Abstract. Mixture equations of state are developed to model the freezing of water to ice VII, and the transformation of solid explosive reactants to gaseous reaction products. In each case there is a single reaction variable, w, the mass fraction of the initial phase, that proceeds from unity (initial phase, $i=1$) to zero (second phase, $i=2$). Common temperature and pressure between the two phases is assumed. Each phase is also assumed to be represented in terms of $p_i(V_i,T)$ and $e_i(V_i,T)$, where p_i, V_i, T, and e_i are the pressure, specific volume, temperature, and specific internal energy, respectively. Three independent equations are formed: (1) $p_1 = p_2$; (2) $wV_1 + (1-w)V_2 = V$; and (3) $we_1 + (1-w)e_2 = e$. These equations determine V_1, V_2, and T given the total specific volume V, the total specific internal energy e, and w. Three specific forms for $p_i(V_i,T)$ and $e_i(V_i,T)$ are used.

INTRODUCTION

The need to develop time evolving equations of state (EsOS) of mixtures arises in several areas of shock-wave research. One application is in a description of dynamic high-pressure phase transformation and another is in explosive reaction. A specific example of the former is shock-induced solidification of ordinary water. At high pressure a stable form of water is ice VII. The transformation of water from its liquid state to ice VII has been the subject of recent extensive study(1), and a significant part of the data analysis has been the development of an EOS of a mixture of liquid water and solid ice VII for arbitrary mass fraction w of the second (solid) phase.

A second application requiring a mixture EOS is that of explosive reaction. In this case only one of the components (the initial solid) is a condensed phase; the second phase is the gaseous detonation products, whose EOS takes on a much different form than normal solids or liquids. In this paper we propose two mixture EsOS; one for transformation between two condensed phases, each of which has well-known isotherms, and the other between a Mie-Grüneisen solid phase with linear shock velocity/particle velocity relationship and an ideal gas with known adiabat passing through the Chapman-Jouguet state.

GENERAL RULES FOR THE MIXTURE

There are many ways of specifying rules for the EOS of a mixture in the fluid approximation. Common pressure is often invoked. Common temperature is sometimes assumed, although it is less universal. In this work both assumptions are used, with the knowledge that the latter assumption may sometimes be violated, especially for very short times and coarse mixtures.

A binary mixture can be described mathematically in terms of the mass fraction of the initial component, w. In terms of w, the specific volume and specific internal energy is given by

$$wV_1 + (1-w)V_2 = V \ , \qquad (1)$$

$$we_1(V_1,T)+(1-w)e_2(V_2,T)=e \ . \qquad (2)$$

Common pressure and temperature then requires

$$p_1(V_1,T)-p_2(V_2,T)=0 \ . \qquad (3)$$

We want to solve for V_1, V_2, and T given V, e, and w. To do this we use Newton's method(2) and form the vector function $\mathbf{F}$ according to

$$F_1 = wV_1 + (1-w)V_2 - V \ , \qquad (4)$$
$$F_2 = we_1(V_1,T)+(1-w)e_2(V_2,T)-e \ , \qquad (5)$$
$$F_3 = p_1(V_1,T)-p_2(V_2,T) \ . \qquad (6)$$

We then search for solutions $\mathbf{x}=(V_1,V_2,T)$ such that $\mathbf{F}=0$.

By-products of the numerical calculation are derivatives such as

$$\left(\frac{\partial p}{\partial V}\right)_{e,w}=\left(\frac{\partial p}{\partial V_1}\right)_T\left(\frac{\partial V_1}{\partial V}\right)_{e,w} \\ +\left(\frac{\partial p}{\partial T}\right)_{V_1}\left(\frac{\partial T}{\partial V}\right)_{e,w} \ . \qquad (7)$$

This allows us to obtain the rate-dependent reaction rate law from $p=P(V,e,w)$ as

$$\dot{p}-(P_V-pP_e)\dot{V}=P_w\dot{w} \ . \qquad (8)$$

LIQUID WATER AND ICE-VII

Equations of state for liquid water and ice-VII are based on isothermal p,V reference curves and constant c_V. Thermodynamic consistency then allows $p(V,T)$ to be written as

$$p(V,T)=p(V,T_0)+b(V)(T-T_0) \ . \qquad (9)$$

The reference isotherm is assumed to be of the Murnaghan form

$$p(V,T)=p_0+(K/g)[(V_0/V)^g-1] \ . \qquad (10)$$

The function $b(V)=\rho\Gamma c_V$ (where Γ is the Grüneisen coefficient) is assumed to be expressible as

$$b(V)=\sum_{i=1}^{N+1}B_i\left(\frac{V}{V_0}-1\right)^{i-1} \ . \qquad (11)$$

as determined by a combination of Hugoniot and isothermal data in the case of liquid water. For ice-VII there is insufficient data to express $b(V)$ as anything more than a constant ($N=0$).

The specific internal energy and entropy can be written as

$$e(T,V)=e_0+\int_{V_0}^V[T_0 b(V')-p(T_0,V')]\,dV' \\ +c_V(T-T_0) \ , \qquad (12)$$

$$s(T,V)=s_0+\int_{V_0}^V b(V')\,dV'+c_V\ln(T/T_0) \ . \qquad (13)$$

Equation of state parameters are summarized in Table 1. The constant value of $b(V)$ and s_o for ice-VII are found from the equilibrium melting curve(1).

Equations (9) and (12) then provide sufficient information to carry out mixture calculation presented in the previous section.

TABLE 1. EOS parameters: Liquid water/Ice-VII.

EOS constant		Liq. water $i=1$	Ice-VII $i=2$
V_0 cm³/g		1.00296	0.6795
p_0 GPa		0.0001	0
T_0 K		298.15	300
e_0 J/g		0	-160.2
s_0 J/g K		0	-1.514
c_V J/g K		3	2
K GPa		2.210	25.04
g		6.029	3.660
B_1	10^{-4} GPa/K	+4.83640	+68.29
B_2	"	-195.275	0
B_3	"	-154.322	0
B_4	"	-2355.40	0
B_5	"	-11759.0	0

B_6	"	-17466.9	0
B_7	"	-8658.02	0

EXPLOSIVE REACTION

As the second application of EOS of mixtures we examine condensed reactants and gaseous reaction products of high explosives.

Solid HE (i = 1)

The solid reactants EOS is assumed to be in the Mie-Grüneisen form, with constant c_v, constant $\rho\Gamma$ and a linear shock-speed/particle-velocity relationship:

$$p(V,T) = p_H(V) + \frac{\Gamma_0 c_v}{V_0}[T - T_H(V)], \quad (14)$$

$$e(V,T) = e_H(V) + c_v[T - T_H(V)], \quad (15)$$

$$p_H(V) = p_0 + \frac{c_0^2(V_0 - V)}{[V_0 - s(V_0 - V)]^2}, \quad (16)$$

$$e_H(V) = \frac{1}{2}[p_0 + p_H(V)](V_0 - V), \quad (17)$$

for the shock speed given by $U = c_0 + su$. The subscript H refers to conditions on the principal Hugoniot.

The Hugoniot temperature is obtained by the method of C. Forest[3]

$$T_H(V) = T_0 \exp(\Gamma_0 \varepsilon) +$$
$$e^{\rho_0 \Gamma_0 V} s \int_0^u \frac{u^2}{c_0 + su} \exp\left[\frac{c_0 + (s-1)u}{c_0 + su}\right] du, \quad (18)$$

where $\varepsilon = 1 - V/V_0$ and u is parameter representing the particle velocity in a single shock corresponding to the specific volume V:

$$u = \frac{c_0(V_0 - V)}{V_0 - s(V_0 - V)}, \quad (19)$$

$$\rho_0 V = \frac{c_0 + (s-1)u}{c_0 + su}, \quad (20)$$

$$\frac{du}{dV} = -\frac{U^2}{c_0 V_0}. \quad (21)$$

The second term in (18) is evaluated numerically and fitted to a power law in u (omitting constant and linear terms):

$$T_H(V) = T_0 \exp(\Gamma_0 \varepsilon) + Au^2 + Bu^3. \quad (22)$$

Various total and partial derivatives needed for numerical solution of the mixture equations (below) are now listed:
For phase 1 (solid), $p = p_1$, $V = V_1$, $e = e_1$

$$\left(\frac{\partial p}{\partial V}\right)_T = p'_H(V) - \frac{\Gamma_0 c_v}{V_0} T'_H(V),$$

$$\left(\frac{\partial p}{\partial T}\right)_V = \frac{\Gamma_0 c_v}{V_0}, \qquad \left(\frac{\partial e}{\partial T}\right)_V = c_v,$$

$$\left(\frac{\partial e}{\partial V}\right)_T = e'_H(V) - c_v T'_H(V).$$

EOS constants for the solid (PBX 9502) are as follows:
Units (g, cm/μs, Mbar, Mbar·cm³/g, K)
$V_0 = 0.544$, $c_0 = 0.24$, $s = 1.883$, $p_0 = 10^{-6}$,
$c_v = 1.05 \times 10^{-5}$, $\Gamma_0 = 0.87$, $A = 12,604$, $B = 59462$.

Gas HE (i = 2)

The EOS of the gaseous products is assumed to be that of the gamma-law ideal gas (isentrope) with constant specific heat, here called c'_v. Thus,

$$p(V,T) = p_I(V) + \frac{\gamma - 1}{V} c'_v [T - T_I(V)], \quad (23)$$

$$e(V,T) = e_I(V) + c'_v [T - T_I(V)], \quad (24)$$

$$p_I(V) = p_{cj}(V/V_{cj})^{-\gamma}, \quad (25)$$

349

$$e_I(V) = e_{cj} + \frac{p_I(V)V - p_{cj}V_{cj}}{\gamma - 1} , \qquad (26)$$

$$T_I(V) = T_{cj}\left(V / V_{cj}\right)^{-\gamma + 1} . \qquad (27)$$

Again, as in the case of the solid component, various total and partial derivatives that are needed for numerical solution of the mixture equations (below) are now listed:

For phase 2 (gas), $p = p_2$, $V = V_2$, $e = e_2$

$$\left(\frac{\partial p}{\partial V}\right)_T = p_I'(V) - \frac{(\gamma - 1)c_v'}{V^2}\left[T - T_I(V) + VT_I'(V)\right],$$

$$\left(\frac{\partial p}{\partial T}\right)_V = \frac{(\gamma - 1)c_v'}{V} ,$$

$$\left(\frac{\partial e}{\partial V}\right)_T = e_I'(V) - c_v'T_I'(V) ,$$

$$\left(\frac{\partial e}{\partial T}\right)_V = c_v' .$$

EOS constants for the gas (PBX 9502) are as follows:
Units (g, cm/μs, Mbar, Mbar·cm³/g, K)
p_{cj}=0.358, V_{cj}=0.4083, T_{cj}=2455, e_{cj}=0.024295, c_v'=2.043×10⁻⁵, γ=3.0.

Figure 1 shows the reference curves (Hugoniot and adiabat) for the solid and gaseous components of the high explosive.

SUMMARY

In this paper we have presented equations of state for (1) liquid water, ice-VII and (2) solid explosive (PBX 9502, solid reactants and gaseous products) in a form suitable for mixture calculations. FORTRAN subroutines are available for liquid water and Ice-VII (WSUmix) and explosive reaction (HExmix, ref. 4). These routines have V, e, w input and respond with p, T, c output (along with other thermodynamic derivatives). These routines find use in finite-difference and characteristic code calculations.

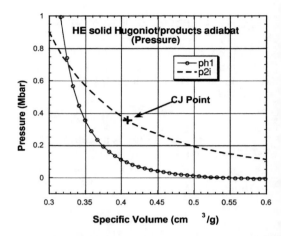

FIGURE 1. Pressure reference curves for solid and gas.

ACKNOWLEDGMENTS

This work was supported at the Institute for Shock Physics at Washington State University through DOE Grant DE-FG03-97SF21388.

REFERENCES

1. Dolan, D. H., "Time Dependent Freezing of Water Under Shock Wave Reverberation Compression," *Ph.D. Thesis*, Dept. of Physics, Washington State University (2003).

2. Johnson, J. N., Tang, P. K., and Forest, C. A., *J. Appl. Phys.* **57**, 4323 - 4334 (1985).

3. Forest, C. A., "Isentrope Energy, Hugoniot Temperature, and the Mie-Grüneisen Equation of State," *in Shock-Compression of Condensed Matter - 1995*, ed. by S.C. Schmidt and W.C. Tao, AIP Conference Proceedings 370, part I, pp. 31 - 34 (1996)]

4. Johnson, J. N., "Tabular CHARADE: Planar, Cylindrical and Spherical Flow," Los Alamos Report, in preparation (2003).

CP706, *Shock Compression of Condensed Matter - 2003*
edited by M. D. Furnish, Y. M. Gupta, and J. W. Forbes
© 2004 American Institute of Physics 0-7354-0181-0/04/$22.00

DEVELOPMENT OF SCALABLE COOK-OFF MODELS USING REAL-TIME IN SITU MEASUREMENTS

M. J. Kaneshige, A. M. Renlund, R. G. Schmitt and W. W. Erikson

Sandia National Laboratories, Albuquerque, NM 87185

Abstract. Scalable thermal runaway models for cook-off of energetic materials (EMs) require realistic temperature- and pressure-dependent chemical reaction rates. The Sandia Instrumented Thermal Ignition apparatus was developed to provide in situ small-scale test data that address this model requirement. Spatially and temporally resolved internal temperature measurements have provided new insight into the energetic reactions occurring in PBX 9501, LX-10-2, and PBXN-109. The data have shown previously postulated reaction steps to be incorrect and suggest previously unknown reaction steps. Model adjustments based on these data have resulted in better predictions at a range of scales.

INTRODUCTION

Prediction of time to ignition in cook-off of energetic materials (EMs) is frequently based on several-step chemical reaction mechanisms. Detailed mechanisms are too computationally expensive to use routinely in realistic problems, and simple mechanisms are generally able to reproduce observed ignition time in a given apparatus over a large temperature range. The steps in these simple mechanisms are only loosely associated with real processes and mostly serve as degrees of freedom for non-linear parameter fits. For ignition time data from a single apparatus, 3 or 4 steps typically provide extraneous degrees of freedom, resulting in non-unique fits. Extrapolation to other geometries or scales often results in poorer agreement with experimental data.

One approach to constraining the extra degrees of freedom and improving the range of validity of a model is by simultaneously fitting data from experiments at different scales and/or geometries. However, this brute force method is expensive and generally results in only marginal improvement since the model details do not correspond to real properties. To improve model scalability fundamentally, the details of the model should be individually optimized to describe real processes and parameters. In situ measurements allow these processes to be observed directly, providing physical insight not available from ignition time alone, and a more rigorous and complete constraint for models. Sufficiently complete data from small-scale experiments can be more useful than global data from much larger ones.

Beyond improving ignition time predictions, there are other reasons that understanding the details of ignition is important. Gas generation is relevant to confinement response and mechanical damage of the EM, and its prediction requires that the model include gas evolution. The chemical (extent of decomposition) and mechanical (porosity) states are important for subsequent prediction of reaction violence.

The present work concerns the HMX-based plastic bonded explosives PBX 9501 (95% HMX, 2.5% Estane®, 2.5% BDNPA/F nitroplasticizer) and LX-10-2 (94.7% HMX, 5.3% Viton), and the RDX-based PBXN-109 (67% RDX, 20% aluminum, 13% HTPB/DOA binder).

EXPERIMENTAL ARRANGEMENT

Typical of cook-off experiments used to develop ignition models is the One-Dimensional Time to eXplosion (ODTX) [1,2] experiment. ODTX has been used to generate a large volume of ignition time data over large temperature ranges for a number of EMs, and multi-step reaction mechanisms have been fit to these data. This is a common starting point for cook-off modeling.

The Sandia Instrumented Thermal Ignition (SITI) apparatus [3], was developed to provide more detail than ignition time in a small-scale system. It is similar in many respects to the Los Alamos Radial Cookoff Test [4,5]. The internal temperature field is observed as the energetic material approaches ignition to gain direct insight into the thermal and chemical processes controlling thermal runaway. The system provides nearly isothermal boundary conditions, gas-tight sealing, and rigid confinement.

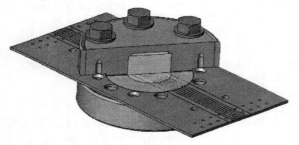

FIGURE 1. Cutaway view of the SITI apparatus.

As illustrated in Figure 1, a grid of nine thermocouple wires mounted on a fiberglass holder is sandwiched between two EM pellets and the halves of an aluminum cell. A gasket insulates the wires from the cell and provides a gas-tight seal. The thermocouple junctions are normally positioned to provide temperature measurements at the following radial positions: 0.0, 1.7, 2.6, 3.4, 4.3, 5.1, 6.0, 8.8, and 11.7 mm.

Figure 2 shows a cross-sectional view of the apparatus. Expansion gaps allow for volumetric expansion of the EM while maintaining compression between the pellets and the thermocouple grid. A rope heater wrapped around the aluminum cylinders supplies heat. Unlike ODTX, which is preheated and assembled quickly, SITI is assembled and then heated, so the temperature boundary condition cannot change instantaneously. Therefore, it cannot achieve as high temperatures or as fast ignition times as ODTX. The aluminum creates a nearly isothermal boundary condition on the outside of the EM. In a preliminary experiment with the thermocouple junctions arranged in a symmetric square pattern, the internal temperature field was found to be very symmetric.

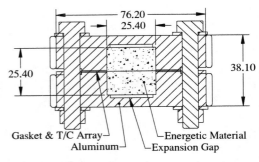

FIGURE 2. Cross-sectional view of the SITI apparatus. Dimensions are in millimeters.

In addition to the internal and control thermocouples, two external thermocouples monitor the top and bottom surfaces. Heavy insulation around the apparatus minimizes heat losses and results in good vertical symmetry. The final assembly, without the top insulation, is shown in Figure 3. The standard heating profile consists of a 10-minute ramp to the target temperature followed by an indefinite hold.

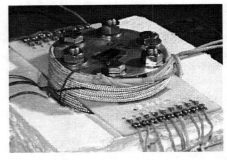

FIGURE 3. Photograph of the SITI apparatus assembled for a test.

For some experiments, a thin tube connects the internal volume to a pressure transducer for measuring the pressure of the gaseous decomposition products.

RESULTS

Figure 4 shows an example of temperature history data from a SITI experiment with LX-10-2. For clarity, only a few temperature signals are shown. The data show several distinct stages. During the initial heat ramp, the outer temperatures are higher than the inner temperatures. As the temperatures reach about 170°C, the endothermic HMX β–δ phase transition occurs, resulting in a brief lag in the temperatures and even a drop at the center. With this value of boundary temperature, exothermic reactions begin as the internal temperatures approach the boundary causing the internal temperatures to exceed the boundary and heat to flow outward. Heat losses prevent thermal runaway for a "plateau" period, the duration of which depends on the boundary temperature. Finally, the heat released by decomposition exceeds the heat losses and the internal temperatures run away, leading to ignition and violent explosion.

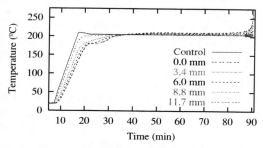

FIGURE 4. Example SITI temperature history data from LX-10-2 with 205°C boundary temperature.

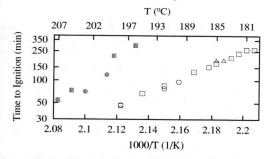

FIGURE 5. SITI ignition time data for PBX 9501 and LX-10-2, represented by open and solid symbols, respectively. Squares, circles, and triangles represent 9.6%, 13.8%, and 14.1-15.3% expansion volume.

While the temperature history data provide an enormous amount of physical information, ignition time is a simple basis for comparing a large number of experiments. An interesting case study is the comparison of ignition time data for PBX 9501 and LX-10-2, shown in Figure 5. These materials contain about the same amount of the same high explosive, HMX. The remainder in PBX 9501 is a polyurethane binder and energetic plasticizer, while LX-10-2 contains a non-energetic binder. ODTX data show the materials to behave the same at higher temperatures, but at the lower temperatures presented here, the difference is marked.

Based on these data, different reaction models may be fit for the two materials. However, these global data alone do not suggest what steps or parameters of the reaction mechanism account for the difference. Figure 6 shows a drop in the internal temperatures during the plateau period of the experiment shown in Figure 4, after exothermic reactions have begun (indicated by the radial temperature inversion). This occurs in almost all LX-10-2 experiments but never with PBX 9501. While these data do not directly indicate the processes responsible for this difference, useful observations can be made. The temperature drop suggests an endothermic process, perhaps a chemical reaction or phase transition. Since it occurs after exothermic reactions have begun, the underlying process is not an initial endothermic step postulated in the Tarver-McGuire mechanisms [1,2]. It is likely related to the binder. Perhaps the Viton has an endothermic reaction, or it is an HMX process that is masked in PBX 9501 by the exothermic effects of the energetic binder.

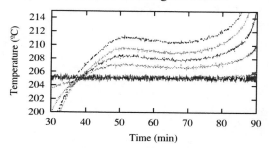

FIGURE 6. Internal temperatures in LX-10-2 during the "plateau" stage, showing a drop in temperature following onset of exothermic reactions.

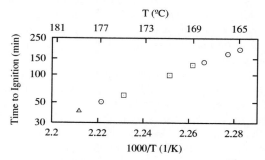

FIGURE 7. PBXN-109 ignition time data from SITI. Squares, circles, and triangles represent 5.1%, 9.6%, and 13.8% expansion volume.

Figure 7 shows PBXN-109 ignition time data. Insight from these and the related temperature histories has led to reaction model adjustments and improved large-scale ignition time predictions [3]. The adjustment involved eliminating the endothermicity of the first Tarver-McGuire step for RDX while maintaining the total reaction energy and re-fitting the reaction rate parameters to ODTX data [1,2]. While the resulting mechanism still does not necessarily represent real chemistry, it is significantly closer, demonstrating the usefulness of this type of data.

Figure 8 shows static pressure measured simultaneously with internal temperature in PBX 9501. Interestingly, the gas pressure does not continue to accelerate as the temperatures run away. Instead, the pressurization slows during this stage. This suggests that thermal runaway is associated with the completion of a gas-generating process.

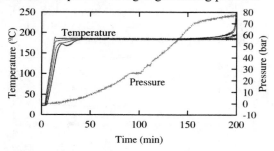

FIGURE 8. Static pressure and internal temperature data from PBX 9501.

CONCLUSIONS

Development of practical models for truly predictive analysis of cook-off is an ongoing challenge. However, small-scale real-time in situ measurements are able to provide unique physical insight into the governing processes, leading to scalable models.

Marked differences have been observed between PBX 9501 and LX-10-2. LX-10-2 exhibits an endothermic process after exothermic reactions have started that slows its ignition relative to PBX 9501. This may be associated with the Viton binder, or could be present in PBX 9501 but masked by its energetic binder.

As shown elsewhere [3], SITI data indicate that the strongly endothermic first reaction postulated in the Tarver-McGuire RDX mechanism is incorrect for PBXN-109. A revised mechanism has shown better agreement with larger-scale experimental results.

We have not observed a significant effect of expansion volume on ignition time for any of these materials. While we do not have enough data to say for certain that there is no such effect, it is clearly not strong.

ACKNOWLEDGEMENTS

Michael Oliver designed and built SITI and performed the experiments presented here. Thanks to Deanne Idar of LANL for providing PBX 9501, and Matt McClelland of LLNL for providing LX-10-2 specimens. Sandia is a multiprogram laboratory operated by Sandia Corporation, a Lockheed Martin Company, for the United States Department of Energy under Contract DE-AC04-94AL85000. This work was also supported by the DoD/DOE Memorandum of Understanding for conventional munitions technology.

REFERENCES

1. Tarver, C.M., McGuire, R.R., Lee, E.L., Wrenn, E.W., and Brein, K.R., "The Thermal Decomposition of Explosives with Full Containment in One-dimensional Geometries," in *Seventeenth Symposium (International) on Combustion*, p. 1407, 1978.
2. McGuire, R.R. and Tarver, C.M., "Chemical Decomposition Models for the Thermal Explosion of Confined HMX, TATB, RDX, and TNT Explosives," in *Seventh Symposium (International) on Detonation*, p. 56, Annapolis, MD, 1981.
3. Kaneshige, M.J., Renlund, A.M., Schmitt, R.G., and Erikson, W.W., "Cook-off Experiments for Model Validation at Sandia National Laboratories," in *Twelfth Symposium (International) on Detonation*, San Diego, CA, 2002.
4. Asay, B., Dickson, P., Henson, B., Smilowitz, L., and Tellier, L., "Effect of Temperature Profile on Reaction Violence in Heated, Self-ignited, PBX-9501," in *Shock Compression of Condensed Matter-2001*, AIP Conference Proceedings 620, Melville, New York, 2002, p. 1065. Also LANL report LA-UR-01-3364.
5. Dickson, P.M., Asay, B.W., Henson, B.F., Fugard, C.S., and Wong, J., "Measurement of Phase Change and Thermal Decomposition Kinetics During Cookoff of PBX 9501," in *Shock Compression of Condensed Matter*, p. 837, Snowbird, UT, 1999. Also LANL report LA-UR-99-3272.

CP706, *Shock Compression of Condensed Matter - 2003*
edited by M. D. Furnish, Y. M. Gupta, and J. W. Forbes
© 2004 American Institute of Physics 0-7354-0181-0/04/$22.00

HOMOGENEOUS MECHANISMS FOR DETONATION OF HETEROGENEOUS HE

Vladimir Yu. Klimenko

High Pressure Center,
Institute of Chemical Physics, Moscow, Russia

Abstract. To prepare a precise numerical model of detonation it is necessary to take into account all mechanisms of the detonation process. Hot spot mechanism is a conventional one, but there are classical experiments, which indicate that at high pressure the decomposition is homogeneous in nature. Recent experiments also show that the width of the chemical spike depends greatly on the porosity of the HE. The analysis of these results leads to the conclusion that explosive decomposition is different for low (<0.05) and high (>0.05) porosities. Two-dimensional numerical simulations of shock wave propagation in porous HMX were carried out, which showed that at low porosity the fast frontal mechanism dominates in explosive decomposition. At high porosity a standard thermal decomposition takes place.

INTRODUCTION

Now we have very powerful computers (1-10 Teraflops), which allow us to perform numerical simulations of real explosive devices. Many excellent hydrocodes have been created recently for these simulations. But, accuracy of modeling is limited by quality of physical models, which describe behavior of materials under shock compression. Current models for inert materials have good accuracy – about 3-5 %. However, accuracy of the models for energetic materials is only about 30%.

In our applied calculations we use numerical models, which were developed 15-20 years ago - the model of C.Tarver (1980-85) [1,2] and the model of J.Johnson, P.Tang, C.Forest (1985) [3]. These models are based on the hot spots mechanism of explosive decomposition.

Analysis of old and recent experiments shows that explosive decomposition represents a mixture of basic decomposition mechanisms, with each of them having own region of domination. The hot spots mechanism (heterogeneous mechanism) dominates in the pressure range of 30 to 200 kbar. Homogeneous mechanism dominates at pressures P > 300 kbar. In the intermediate region of 200-300

kbar we have mixture of both heterogeneous and homogeneous mechanisms. At very low pressures P < 20 kbar decomposition is governed by the so-called dislocation mechanism. It is necessary to emphasize that these values of pressures – 30, 200 and 300 kbar are only qualitative and have concrete values for each specific explosive.

The hot spots mechanism was studied very well and many numerical versions (numerical models) have been developed for it. The same cannot be said about the homogeneous mechanism. The present paper will discuss two homogeneous mechanisms: the standard thermal mechanism for HE with large porosity and the frontal mechanism for HE with small porosity.

CLASSICAL EXPERIMENTS IN SUPPORT OF HOMOGENEOUS DECOMPOSITION

In 1974-77 A.Krivchenko and K.Shvedov [4] investigated initiation of detonation in pressed powdered explosives. They studied TNT and RDX with grain size of 100-300 µm and porosity 10-40 %. They filled pores with different liquids. Their main results are presented in Fig.1. While the authors had studied about 20 liquids, for our discussion we have selected only four characteristic fillers (air, pentane, water and bromoform).

The most important result is the existence of a characteristic pressure P* [5]. At pressure P > P* decomposition of explosive does not depend on the quality of the pore filler (air, pentane, water or bromoform). It means that the decomposition is governed not by the processes in the pores, but those on the outside of the pores, i.e. in the grains of the explosive. HE grains compose about 95 % in volume of a standard explosive. Krivchenko and Shvedov had also shown that in the pressure region P > P* decomposition of liquid and pressed solid TNT has a similar homogeneous nature.

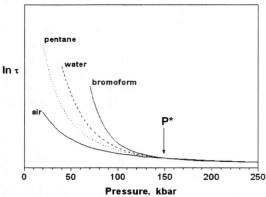

FIGURE 1. Detonation reaction time at initiation by different pressure for explosives with various fillers (air, pentane, water, bromoform) of pores [4].

MECHANISM OF HOT SPOTS IGNITION

Presented classical experiments give us an opportunity to establish details of the hot spots mechanism. Considering shock Hugoniots of the filler liquids (Fig. 2), one can see that there is a clear correlation between the dynamic compressibility of the filler and decomposition rate – more compressive filler gives more fast reaction.

To understand the physical nature of this correlation we have performed 2D numerical simulations of compression of pore in HMX filled by air, pentane, water and bromoform. In all cases we observe large plastic deformation of explosive around the pore. Clearly that this deformation gives rise of temperature due to viscous or plastic work. This rise is the main reason of explosive ignition.

The addition of temperature has two components: viscous and plastic. At pressure P > 20

kbar the most large is temperature increase due to viscous work:

$$\Delta T_{vis} = \frac{1}{C_v} \int \left(\frac{dQ}{dt} \right)_{vis} dt = \frac{2\mu}{C_v} \int \dot{e}_{ij} \dot{e}_{ij} dt$$

where μ is viscosity, C_v is heat capacity and $\dot{e}_{ij}$ are deviatoric strain rates. We do not know exactly the values of viscosity and heat capacity for HMX at high pressure and therefore we cannot calculate exactly "viscous" temperature. But, we can calculate the following value

$$\int \dot{e}_{ij} \dot{e}_{ij} dt = \frac{C_v}{2\mu} \Delta T_{vis} = \Delta \widetilde{T}_{vis}$$

which is temperature with precision to coefficient. In our numerical simulations we have calculated these "temperatures" for three cases of pores filled with : (1) pentane, (2) water and (3) bromoform. Fig. 3 shows calculated temperature in the vicinity of the pores after passing of shock wave front over it.

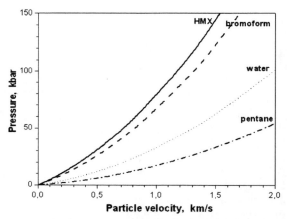

FIGURE 2. Shock Hugoniots of liquids filling pores in explosive.

One can see that pentane gives the largest rise of temperature and bromoform gives the smallest. At the same time pentane gives the largest reaction rate while bromoform gives the smallest (see Fig. 1). It means that the decomposition is governed by the temperature generated at deformation of the explosive in the vicinity of the pore due to viscous work.

Thus, viscous heating is governing mechanism of the hot spots decomposition.

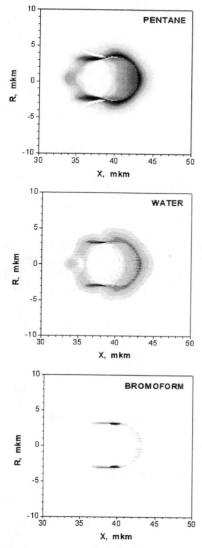

FIGURE 3. Pores filled with pentane, water and bromoform after passage of a shock wave. Addition of temperature due to viscous work is illustrated by the amount of darkness in the figure.

HOMOGENEOUS MECHANISMS

As we concluded before, at pressure P < P* decomposition of heterogeneous HE has a nature of hot spots. At pressure P > P* decomposition has a homogeneous nature. The question arises why is there a transition from a heterogeneous to a homogeneous decomposition in a heterogeneous

HE?

The direct 2D numerical simulations give a clear answer to this question. Figure 4 shows qualitative temperature of HMX in the vicinity of the pore filled with air after passing a shock wave through it. The upper and lower figures show the results of a weak shock wave (P ~ 50 kbar) and a strong shock wave (P ~ 400 kbar) respectively. At high pressure (when the pore temperature is close to the bulk temperature) the decomposition rates in both pore and grain become comparable and the total process becomes homogeneous.

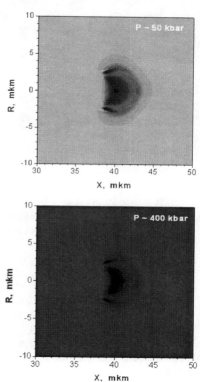

FIGURE 4. Temperature in HMX at low and high pressure. Shade of darkness is proportional to temperature.

We know two homogeneous mechanism – the standard thermal mechanism and the so-called frontal mechanism. The frontal mechanism was developed in our earlier publications [6,7]. It works only at high pressures P > 250-300 kbar.

Our 2D numerical simulations show that this mechanism works effectively if porosity of HE is small (see Fig. 5a). If porosity is large, only small part of explosive is decomposed by the frontal mechanism (see Fig. 5b). In this case the thermal mechanism becomes dominant.

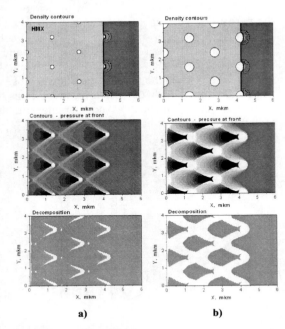

FIGURE 5. Density contours (upper); pressures at shock front (middle); decomposition by the frontal mechanism (black color) and by the thermal mechanism (white color). a) porosity ~ 1%, b) porosity ~ 8%.

DISCUSSION

Now we can explain recent results [8] of S.Lubyatinsky, who studied experimentally the width of the chemical spike in HE with different porosities. We have analyzed his data and the dependence of decomposition rate on porosity is shown in Fig. 6.

At very small porosity the decomposition proceeds by the fast frontal mechanism in the timeframe of 0.1 – 1 ns. At higher porosity a portion of explosive decomposed by the fast frontal mechanism decreases and the remaining part is decomposed by the slow thermal mechanism. Both the frontal and the thermal mechanisms work here. As a result total decomposition lasts about 1-5 ns.

At some critical porosity (4-6 %) decomposition becomes mainly thermal and the characteristic decomposition time equals 10-50 ns. Further rise of the decomposition rate is caused by the increase of the bulk temperature of explosive, because increase of porosity gives rise of temperature.

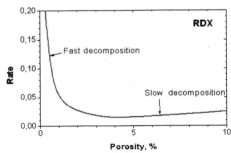

FIGURE 6. Decomposition rate vs. porosity [8].

CONCLUSIONS

At low pressure the HE decomposition proceeds by the hot spots mechanism. At high pressure the decomposition proceeds by the frontal mechanism (if porosity is small) or by the thermal mechanism (if porosity is large).

ACKNOWLEDGEMENTS

The author would like to thank Dr. A.L.Krivchenko (Samara State Technical University, Russia), Dr. G.F.Tereschenko (Saint-Petersburg Division of the Russian Academy of Science, Russia) and Dr. P.A.Urtiew (Lawrence Livermore National Laboratory, USA) for useful discussions.

REFERENCES

1. E.L. Lee, C.M. Tarver, *Phys. Fluids,* **23**, p. 2362 (1980).
2. C.M. Tarver, J.O. Hallquist, *VII Symposium (Int.) on Detonation,* USA, 1981, pp. 488-497.
3. J.N. Johnson, P.K. Tang, C.A. Forest, *J. Appl. Phys.,* **57**, pp. 4322-4334 (1985).
4. A.L. Krivchenko, *Private Communication,* Institute of Chemical Physics, Chernogolovka, 1977.
5. A.N. Dremin, K.K. Shvedov, *VI Symposium (Int.) on Detonation,* USA, 1976, pp. 29-35.
6. V.Yu. Klimenko, in *Shock Compression of Condensed Matter-1995,* Seattle, USA, pp. 361-364 (1996).
7. V.Yu. Klimenko, *Chem. Phys. Reports,* **17** (1), pp. 13-30 (1998).
8. S.N. Lubyatinsky, B.G. Loboiko, in *Shock Compression of Condensed Matter-1997,* Amherst, USA, pp. 743-746 (1998).

CP706, *Shock Compression of Condensed Matter - 2003*
edited by M. D. Furnish, Y. M. Gupta, and J. W. Forbes
© 2004 American Institute of Physics 0-7354-0181-0/04/$22.00

PRESSURE CALCULATION OF REACTING EXPLOSIVE BY MIXTURE RULE

Shiro Kubota[1], Kunihito Nagayama[2], Hideki Shimada[3] and Kikuo Matsui[3]

[1]*Research Center for Explosion and Safety, National Institute of Advanced Industrial Science and Technology, Tsukuba, 305-8569, Japan*
[2]*Department of Aeronautics and Astronautics, Faculty of Engineering, Kyushu University, Fukuoka, 812-8581, Japan*
[3]*Department of Earth Resources Engineering, Faculty of Engineering, Kyushu University, Fukuoka, 812-8581, Japan*

Abstract   In order to discuss on the assumptions of simple mixture rule for reacting explosive, shock to detonation transition processes were calculated by finite difference method for high explosive. The reacting explosive was regarded as the simple mixture phase of the reactants and products components. Four kinds of assumptions, such as thermal equilibrium, isentropic reactants, etc. were adopted as one of the assumptions to solve two components variables and to determine the pressure. The numerical results that were obtained under the each assumption had a good agreement with each other. In order to answer the reason why the difference of the assumption is insensitive to the numerical results of shock initiation, the relations of reactants and products components were considered in the specific internal energy and specific volume plane with constant pressure lines for both components.

INTRODUCTION

The simple mixture theory, in which the reacting explosive is regarded to be a simple mixture phase of reactant and product components, has been widely used for the pressure calculation in reacting explosive. There are many reports and discussions on issues important to the numerical simulation of the initiation process, the reaction rate law and equations of state for both reactant and product components. Although pressure calculation for the reacting explosive is unavoidable in the numerical simulation of the initiation process, there are few reports and discussions on the calculation methods. In this paper, we shall discuss the calculation methods of the pressure for the reacting explosive. In this simple theory, to calculate the pressure of the reacting explosive, we must obtain the solutions for four unknown variables: specific volumes V_s, V_g and internal energies E_s, E_g. Subscripts s and g indi-

cate the solid reactant and the gaseous product components, respectively. The internal energy and specific volume of the reacting explosive have been represented by a linear combination of individual internal energies and specific volumes, respectively. These are the first and second conditions for the pressure calculation. The third condition used to obtain these solutions is the mechanical equilibrium, $P = P_s (V_s, E_s) = P_g (V_g, E_g)$. Since we have the above three conditions, another physical assumption is needed to obtain the four solutions. The temperature equilibrium assumption has often been used as the fourth condition, but due to a lack of accurate information on the temperature for reactant and product components, another assumption of isentropic reactants, has been adopted instead. It can be thought that the pressure for the reacting explosive is insensitive to the fourth assumption. However a clear reason has not been shown why the fourth assumption hardly affects the numerical results of the shock initiation process. In this paper, to examine the effect of the fourth assumption on the numerical results of shock initiation, numerical results obtained under each assumption are compared. Four equations were adopted as the fourth assumption to solve two component variables and to calculate the pressure. The samples used for this calculation are PBX 9404, Composition B and PETN. The ignition and growth model and the JWL equation of state for reactant and detonation products are employed in these hydrodynamic calculations. The state quantities for both components are calculated using the mixture equation of state. We shall discuss the relationship between the reactant and product components and clarify why the fourth assumption has little effect on the numerical results of the shock initiation process.

MIXTURE RULE FOR DETONATING EXPLOSIVE AND EQUATIONS OF STATE

Many researchers have adopted the thermal equilibrium condition,

$$T = Ts (Vs, Es) = Tg (Vg, Eg), \qquad (1)$$

as the fourth assumption. However, it is sometimes difficult to obtain actual information on the temperature by either the theoretical or experimental approach. The relation of the isentropic reactants has often been adopted instead, which is

$$Es = Esi, \qquad (2)$$

where subscript i indicates the isentropic state. Under the assumption of isentropic reactants, the two phases are thermally isolated. In addition to the above assumptions, we can adopt other ones. Liang et al[1]. adopted the following relation instead of the thermal equilibrium condition.

$$Eg = Es \qquad (3)$$

One of the greatest works on the modeling of detonation was by Mader.[2] He showed various equations of state for a mixture cell. HOMSG is the mixture EOS for condensed and gaseous components, and it can probably be used in the pressure calculation in reacting explosive. The assumption in HOMSG is expressed by

$$Es/Eg = Esh/Egi, \qquad (4)$$

where Esh is the specific internal energy along the Hugoniot of reactants and Egi is that along the isentrope of products. Using equations (1)–(4) as the fourth assumption, we will discuss the relations for the reactant and product components obtained from the mixture EOS, and the effect of the mixture rule on the numerical results of shock initiation process.

JWL EOS is one of the most useful equations of state

for detonation products. In this calculation, for both of the reactant and product components, JWL EOS is employed. In order to examine the assumption of temperature equilibrium, temperature-dependent JWL EOS is also employed. The isentropic relation for the reactant is estimated by

$$Esi = Vs/\omega s(Psi - Psh) + Esh. \qquad (5)$$

Psi is obtained by solving the following differential equation.

$$\frac{\partial Psi}{\partial Vs} + \frac{\omega s}{Vs} Psi = \frac{\partial Psh}{\partial T} - \frac{\omega s}{Vs} Esh \qquad (6)$$

NUMERICAL SIMULATION OF SHOCK INITIATION PROCESS

In order to obtain the paths of the state quantities along the particle during the SDT process, numerical simulations of the impact problem were carried out for various impact velocities. The one-dimensional Lagrangian code was used in this simulation. The governing equations are mass, momentum and energy conservation laws, which are solved by the finite difference method. In addition, the reaction rate model is necessary to estimate the degree of decomposition of explosive. In this study, the ignition and growth model was used.

The reaction rate parameters were calibrated by a simulation in which the thermal equilibrium condition for pressure determination was adopted. Figure 1 shows the paths along the particle during the SDT process on the pressure-specific volume plane obtained by the simulation with various impact velocities. Because the particle near the impact surface traces a typical SDT path, the profiles of state quantities near the impact surface are very important. In this figure, the thermal equilibrium condition has been adopted, and the observation point was initially 3 mm from the initial impact surface. Path A corresponds to the case of 0.6 km/s impact velocity, and Paths B and C correspond to those of 0.8 and 1.0 km/s, respectively. Using the same reaction rate parameters, the same impact problem was solved, varying the mixture rule: isentropic reactants, HOMSG and Es=Eg. Similar calculations of impact problems were carried out for Composition B and PETN with a different projectile.

Figures 2 indicate the numerical results obtained using each of the mixture rules. Good agreements for each pressure distribution are confirmed.

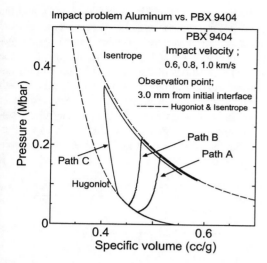

Figure 1. The paths along the particle during SDT process on pressure-specific volume plane obtained by numerical simulation of impact problem for Aluminum vs. PBX 9404.

DISCUSSION

Since the specific volumes and specific internal energies for reactant (Vs, Es) and product (Vg, Eg) components are estimated by the linear combinations under the assumption of mechanical equilibrium, V and E for each component and for the reacting explosive are located on the same straight line in the specific internal energy and specific volume, E-V, plane. The conceptual diagram of the E-V plane for estimating the mixture rule is shown in Figure 3. Let us assume that the three conditions, thermal equilibrium, isentropic reactants and Es = Eg, give the same results in the numerical simulation of the shock initiation process. Then all components calculated using specific (V,E,λ) values are located on exactly the same constant-pressure lines as shown Figure 3. The two constant-pressure lines are obtained using the equations of state for the reactant and product component, respectively. V for reacting explosive satisfies

$$Vg-V=(1-\lambda)(Vg-Vs) \tag{7}$$
$$V-Vs=\lambda(Vg-Vs), \tag{8}$$

and those relations hold independent of remaining assumptions for mixture rules. Of course, for specific internal energy, similar relations can be introduced. The above concept is convenient for investigating why the influence

for the pressure calculation of the reacting explosive, due to the difference of the adopted the fourth assumptions for mixture rule is very small.

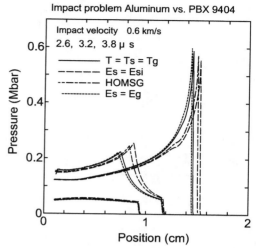

(a)   Impact velocity 0.6 km/s

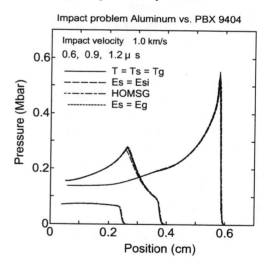

(b) Impact velocity 1.0 km/s

Figure 2. Comparison of the numerical results from different four assumptions for mixture rules; such as thermal equilibrium, isentropic solid, HOMSG and Es=Eg. These figures show the Shock propagation process in PBX 9404 obtained by the impact problem for Aluminum vs. PBX 9404.

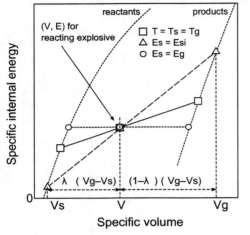

Figure 3. The conceptual diagram of specific internal energy and specific volume plane for estimating the mixture rule

The condition under which all assumptions yield the exact same results is that, in the E-V plane, the slope of the constant pressure line for the reactant component is equal to that for the product component at any pressure level:

$$\left(\frac{\partial Es}{\partial Vs} \right)_P = \left(\frac{\partial Eg}{\partial Vg} \right)_P \qquad (9)$$

When the subscripts 1 and 2 indicate the assumptions of thermal equilibrium and isentropic reactants, respectively, the condition under which the two assumptions yield similar results is that $(Es_1-Es_2)/(Vs_1-Vs_2)$ is almost equal to $(Eg_2-Eg_1)/(Vg_2-Vg_1)$ at any pressure level. It can be considered that the assumptions of thermal equilibrium and isentropic reactants correspond to the extreme cases, and that a physical nonequilibrium state for reacting explosive exists between the two assumptions. The component Vs must be between Vs_1 and Vs_2, and Vg must be between Vg_1 and Vg_2. Similar relations are also required for Es and Eg. In the case of assuming $Es = Eg$, Vs, Vg, Es and Eg do not exist in the above ranges. However, in the range that includes the predicted state quantities of each component, when the slopes of the constant pressure lines on the E-V plane for each component have almost the same values, all the assumptions yield similar results.

Figures 4 shows the E-V planes for the reactant and product components of reacting PBX 9404 with constant pressure lines. Each component was calculated using the

four mixture rules defined by equations (1)-(4), and the input data for mixture variables were specific internal energies, specific volumes and mass fractions of detonation products for the reacting explosive obtained from the numerical simulations of the shock initiation process under each assumption. In Figure 4, Path A is similar to path A in Figure 1. Although the results were obtained under the four different assumptions for the mixture rule, excellent agreement is confirmed for the set of (E,V) for Figures 4. Similar relations on the E-V plane were also confirmed for Composition B and PETN.

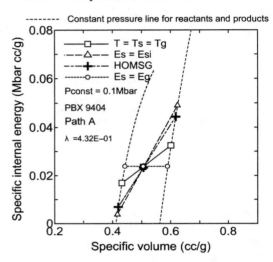

Figure 4. Specific internal energy and specific volume plane for reactants and products components of reacting PBX 9404 with constant pressure line. Each components were calculated by using four mixture rules; thermal equilibrium, isentropic solid, HOMSG and Es=Eg.

CONCLUSION

Because the slopes of the constant-pressure lines for both components on the E-V plane are almost the same, the different assumptions are insensitive to the numerical results.

REFERENCES

1. D. Liang, W. J. Flis and P. C. Chou, thenth symposium (international) on detonation (Office of Naval Research ONR 33395-12, Boston, 1993), p.1003

2. C. L. Mader, "Numerical modeling of detonations.", Univ. of Calif. Press, (1979)

CP706, *Shock Compression of Condensed Matter - 2003*
edited by M. D. Furnish, Y. M. Gupta, and J. W. Forbes
© 2004 American Institute of Physics 0-7354-0181-0/04/$22.00

AB INITIO CALCULATIONS OF THE ELECTRONIC STRUCTURE OF 1,1-DIAMINO-2,2-DINITROETHYLENE

M. M. Kuklja[1], S. N. Rashkeev[2], and F. J. Zerilli[3]

[1]*Division of Materials Research, National Science Foundation, Arlington, VA 22230*
[2]*Department of Physics and Astronomy, Vanderbilt University, Nashville TN 37235*
[3]*Research and Technology Department, Naval Surface Warfare Center, Indian Head, MD 20640*

Abstract. The atomic and electronic structure of the molecular crystal 1,1-diamino-2,2-dinitroethylene (FOX-7) is studied by means of first-principles Hartree-Fock and density-functional methods. The unimolecular decomposition pathway of the molecule and the crystal has been also investigated. It was found in both cases that the $C-NO_2$ dissociation energy has the lowest value indicating that the decomposition will start with breaking of this bond. Also, the decomposition energy in the solid state is found to be lower than that in the gas phase. This fact probably indicates that the mechanisms for decomposition in these two materials states are different. It is shown that the presence of "reversed-orientation-molecule" defects narrows the band gap and lowers the decomposition barrier of the material in the solid phase. The appearance of additional defect-related electronic states in the band gap is consistent with our previous results for dislocation simulation in other high explosive solids and with the experimentally observed pre-explosion conductivity and luminescence in some energetic materials.

INTRODUCTION

Recent experimental developments in ultrafast optical techniques provided a unique opportunity not only to understand the mechanisms and dynamics of photo-induced processes in reactive organic molecular crystals but also to control chemical reactions by laser by inducing some targeted chemical reactions[1]. Because of that, ab initio calculations of the electronic structure and decomposition pathways of these materials became a subject of a particular interest for practical applications.

Results of our previous investigations of the electronic structure of ideal and defective crystals devoted to RDX ($C_3H_6N_6O_6$), PETN ($C_5H_8N_4O_{12}$), and metal azides were summarized in the recent reviews[1,2,3]. In this article we present and discuss the results obtained[4,5] for 1,1-diamino-2,2-dinitroethylene (FOX-7) that has attracted

substantial interest because it is expected that its sensitivity could be as low as that of TATB ($C_6H_6N_6O_6$)[6], while its performance is comparable with the performance of RDX and HMX ($C_4H_8N_8O_8$)[7].

The unimolecular decomposition pathway of this molecule has been recently investigated using density functional theory (DFT). Politzer, et al.[8] showed that $C-NO_2$ dissociation energy is lower than $C-NH_2$ dissociation energy and is equal to 70 kcal/mol. Gindulyté, et al.[9] proposed that the first step in the decomposition is the transformation of the NO_2 group attached to the carbon atom into an O-N-O chain-like structure, with one oxygen atom attached to the carbon. According to their calculations, such a reorganization of the NO_2 group costs 59.7 kcal/mol, which they noted agrees with the experimental activation energy of 58 kcal/mol for temperatures between 210-250°C [7]. One needs to note here that the calculated values[8,9] are related to the dissociation mechanism in the gas phase, while all the experimental studies have been performed for the solid phase. Recently, it has been shown for RDX that the activation barrier is different for the solid state and the gas phase molecule and that the decomposition barrier near a defect is lower than in a perfect crystal[10]. Defects formed by change of orientation of molecules in the solid phase may play a significant role in activation of decomposition process[11].

Another interesting fact is the observation of pre-explosion conductivity and pre-explosion luminescence in some high-energy materials[1]. All of these materials are dielectrics with a wide band gap. The appearance of conductivity means that some electronic states may appear in the gap. What is the nature of these states and how are they related to the dissociation mechanism? If one could answer these questions, it may become possible to control certain desirable properties of these materials based on electrical and optical manipulations.

In this work, we performed DFT and HF calculations to illustrate how the changes of mutual orientation of molecules in a molecular crystal and internal stresses can affect the electronic structure and decomposition for crystalline FOX-7. For this purpose we generated a "reversed-orientation-molecule" defect to mimic the structure deformation in the vicinity of a dislocation core, stacking fault, or grain boundary. In a perfect crystal, hydrostatic or uniaxial pressure cannot produce any additional states in the band gap[12] while in a crystal with defects additional electronic bands appear in the gap[3,13,14]. We suggest that in crystal with reversed molecules, the decomposition barrier is lower than in a perfect crystal and in excellent agreement with experiment.

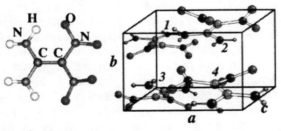

Figure 1. The molecular and crystalline structure of FOX-7. The crystal has space group P2$_1$/n with four molecules (56 atoms) per unit cell.

DETAILS OF CALCULATIONS

The self-consistent field calculations for both molecular and crystalline FOX-7 were done using the VASP[15] codes. DFT in the generalized gradient approximation (GGA) for exchange-correlation, and plane waves with ultrasoft pseudopotentials for C, N, O, and H, were used[16]. For the crystalline material we used the experimental crystal structure provided by Gilardi[17]. The structure was determined by Bemm and Östmark[18] to have space group P2$_1$/n with four FOX-7 molecules (56 atoms) per unit cell (Fig.1). The relaxed structure and electronic density of states (DOS) of the defect-free molecular crystal agrees well with previously published results based on force field and DFT[6], and with our recent investigation based on the Hartree-Fock method[5] using CRYSTAL98[19] codes.

RESULTS AND DISCUSSION

The geometry configurations obtained by the full optimization of the internal molecular parameters in both methods agree well with each other and with experimental[7,17] and other theoretical[2,8] studies. The lattice parameters obtained with HF methods are much closer to the experimental values at their equilibrium configuration than those by DFT.[5]

The highest occupied molecular orbital – lowest unoccupied molecular orbital (HOMO-LUMO) gap of the molecule is about 2.3 eV. Both the HOMO and LUMO states are mostly related to the p functions for both oxygen and nitrogen atoms from the NO$_2$ groups and p functions of the carbon atom attached to the NO$_2$ groups. The lowest of the bond-breaking energies is the energy for the C-NO$_2$ bond (72 kcal/mol), what is in excellent agreement with previous studies[8,9].

The structure of the ideal FOX-7 crystal looks like a set of parallel corrugated, zigzag like planes. In each of the layers, molecules are arranged in a checkerboard order, and the NO$_2$ (NH$_2$) groups of molecules in a given layer are neighboring with the NH$_2$ (NO$_2$) groups of molecules in the next (previous) layer[18]. The positions of the DOS peaks for the crystal have a one-to-one correspondence with the positions of the energy levels of the isolated molecule. Also, the electronic states of the crystal have the same nature as the molecular orbitals of the isolated molecule. The band gap is 2.2 eV for the crystal, very close to the HOMO-LUMO gap of 2.3 eV for the molecule. This means that the electronic states of a perfect crystal are mostly defined by intramolecular bonds. The intermolecular interactions, however, are responsible for the cohesion in the solid. This is typical for molecular crystals and has been found in a study of RDX[20].

When a hydrostatic pressure is applied to the perfect crystal, the band gap does not change significantly even for large volume changes (the change is only 0.1 eV for $V/V_0 = 0.92$). A similar conclusion was obtained for perfect RDX crystals[12] and perfect nitromethane crystals[11]. In addition, it was found that the band gap is sensitive to the presence of local strains due to defects in these materials[1,3,11].

Further, we simulated a simple structural defect, "reversed-orientation" molecule, where one of the molecules is rotated by 180° about an axis perpendicular to the C-C axis. It corresponds to an interchange between the NO_2 and NH_2 groups. In the unit cell that consists of four molecules there are now two FOX-7 molecules with normal orientation (the NO_2 groups of the molecule are facing the NH_2 groups of the adjacent molecule) and two molecules with anomalous orientation (the NO_2 groups of the adjacent molecules are facing each other). In other words, the system consists of alternating zigzag like molecular planes each of which consists of layers with normal and anomalous mutual orientation. The total energy of this system is 2.5 eV per cell higher in comparison with perfect material. This structure corresponds to a local energy minimum, and the barrier to invert the molecule back to its normal orientation by rotation is high.

The band gap of the crystal with the reversed-orientation-molecule defect (1.3 eV) is 0.9 eV narrower than the gap of the perfect structure. Also, the additional states that appear in the band gap of the ideal crystal have a dominant contribution from the oxygen atoms associated with the anomalous intermolecular NO_2-NO_2 bonds.

In order to simulate the local strain in the vicinity of defects such as grain boundaries,

stacking faults, and dislocations, two different series of calculations in which intermolecular distances were decreased, were performed. The total energy and DOS were compared to those of the original undisturbed structure.

First, we shorten the normal intermolecular distance. For this purpose, molecule 1 in the unit cell was translated parallel to its original orientation towards the neighboring molecule 2 in the same layer (see Fig. 1). The positions of atoms C_1 and C_2 were fixed and the remaining atoms in the unit cell were relaxed. The density of states then was analyzed as a function of the parameter α, which represents the changed distance d between the atoms C_1 and C_2 so that $d=d_e-\alpha$, where d_e is the equilibrium distance between these atoms in the fully relaxed system. The position of the bottom of the conduction band gradually decreases in energy as the parameter α increases, that is, the gap narrows. One of the two NO_2 groups splits away from molecule 1 when $\alpha=0.85$ Å. The decomposition energy for this case is about 4 eV (92 kcal/mol), which is higher than the experimental value[7]. The size of the band gap is very sensitive to the magnitude of the displacement α and the value of the gap is decreased by about 2 eV at $\alpha=0.75$ Å in comparison with the perfect crystal.

The same procedure, that is, shortening of the intermolecular distance was applied to the anomalous molecular arrangement (Fig. 1). We shifted molecule 3 towards molecule 4, fixed the positions of the carbon atoms C_3 and C_4, and relaxed the rest of the system. The distance between the two carbon atoms was reduced until one of the NO_2 groups becomes detached from the molecule. The gap is nearly unchanged with α up to the critical value of $\alpha=0.65$ Å corresponding to the breaking of the molecule when the gap narrows down by about 2 eV. The decomposition barrier in this case is 2.6 eV (59 kcal/mol), which is in excellent agreement with the experimental value.

In both the normal and anomalous intermolecular bonding we suggest that the decomposition starts with detachment of one of the NO_2 groups. We did not observe the formation of the O-N-O chains attached to oxygen predicted for a single molecule[9]. Reduction of the band gap supports the experimental observations of pre-explosion conductivity and luminescence in metal

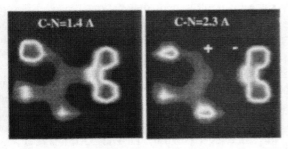

Figure 2. C-NO$_2$ bond dissociation in FOX-7: electronic density.

azides[1]. The appearance of new local electronic levels in the gap predicts new absorption peaks in optical spectra and new luminescence lines. This is also consistent with previous experimental[21] and theoretical results for RDX materials[3,12,14].

CONCLUSIONS

In conclusion, we investigated the electronic structure and decomposition mechanisms in crystalline FOX-7. We also simulated a reversed-orientation-molecule defect in the crystal, which should be a very representative element of such common defects in molecular crystals as dislocations, stacking faults, and grain boundaries. The decomposition mechanisms in the solid phase and gas phase are different. In spite of the fact that in the FOX-7 crystal the electronic structure is mostly defined by intramolecular interactions, it is the intermolecular interactions that play a crucial role in the decomposition process. The perfect crystal is relatively stable, the decomposition barrier which is associated with C-NO$_2$ bond breaking being about 92 kcal/mol (4 eV). This decomposition barrier is much higher than the experimentally measured one. The presence of reversed-orientation-molecule defects produces broken C-NO$_2$ bonds at 59 kcal/mol that is in excellent agreement with experiment. These defects reduce the band gap by changing the hybridization between the N-p and O-p molecular orbitals thereby introducing new local electronic states. Further work on dynamic modeling of decomposition mechanisms and real defects is currently underway.

ACKNOWLEDGEMENTS

M.K. is grateful to the DMR of NSF for the support under the Individual Research and Development Program. F.Z. was supported by the NSWC Core Research Program.

REFERENCES

[1] M.M.Kuklja, Appl. Phys. A **76**, 359 (2003).

[2] M.M.Kuklja, B.P.Aduev, E.D.Aluker, V.I.Krasheninin, A.G.Krechetov, and A.Yu.Mitrofanov, J. Appl. Phys. **89**, 4156 (2001).

[3] M.M.Kuklja, A.B.Kunz, J.Appl.Phys. **87**, 2215 (2000).

[4] S.N.Rashkeev, M.M.Kuklja, and F.J.Zerilli, Appl. Phys. Lett. **82**, 1371 (2003).

[5] M.M.Kuklja, F.J.Zerilli, and S.M.Peiris, J. Chem. Phys. **118**, 11073 (2003).

[6] D. C. Sorescu, J. A. Boatz, and D. L. Thompson, J. Phys. Chem. A **105**, 5010 (2001).

[7] H. Östmark, A. Langlet, H. Bergman, N. Wingborg, U. Wellmar, and U. Bemm, Proceedings of the 11th Int. Detonation Symposium, Snowmass, Colorado, 1998, ONR 33300-5 (ONR, Arlington, Virginia, 2000), p. 807.

[8] P.Politzer, M.C.Concha, M.E.Grice, J.S.Murray, P.Lane, D.Habibollazadeh, J.Mol. Struct. (THEOCHEM), **452**, 72 (1998)

[9] A. Gindulyté, L. Massa, L. Huang, and J. Karle, J. Phys. Chem. A **103**, 11045 (1999).

[10] M.M.Kuklja, J. Phys. Chem. B, **105**, 10 159 (2001).

[11] E.J.Reed, J.D.Joannopoulos, L.E.Fried, Phys. Rev. B., **62**, 16500 (2000).

[12] M.M.Kuklja and A.B.Kunz, J. of Appl. Phys., **86**, 4428 (1999).

[13] M.M.Kuklja, E.V.Stefanovich and A.B.Kunz, J. Chem. Phys., **112**, 3417 (2000).

[14] M.M.Kuklja, A.B.Kunz, J.Appl.Phys. **89**, 4962 (2001).

[15] G. Kresse and J. Hafner, Phys. Rev. B **48**, 13 115 (1993); G. Kresse and J. Furthmüller, Comput. Mater. Sci. **6**, 15 (1996).

[16] M.C. Payne, M. P. Teter, D. C. Allan, T. A. Arias, and J. D. Joannopoulos, Rev. Mod. Phys. **64**, 1045 (1992).

[17] R.Gilardi, private communication, 2001.

[18] U.Bemm, H. Östmark, Acta Cryst. **C54**, 1997 (1998).

[19] R.Dovesi, V.R.Saunders, C.Roetti, M.Causà, N.M. Harrison, R.Orlando, and C.M.Zicovich-Wilson, CRYSTAL 98 User's Manual, University of Torino, Torino, 1998.

[20] M.M.Kuklja and A.B.Kunz, J. of Phys. and Chem. of Solids, **61**, 35 (2000).

[21] A.B.Kunz, M.M.Kuklja, T.R.Botcher, and T.P.Russel, *Thermochimica Acta* **384**, 279-284 (2002).

CP706, *Shock Compression of Condensed Matter - 2003*
edited by M. D. Furnish, Y. M. Gupta, and J. W. Forbes
2004 American Institute of Physics 0-7354-0181-0/04/$22.00

AN INTERPRETATION OF PARTICLE VELOCITY HISTORIES DURING GROWTH TO DETONATION

B D Lambourn

AWE Aldermaston, READING, Berkshire RG7 4PR, England

Abstract. The most important advance in shock initiation experimentation has been the development and application of in-material gauges, measuring pressure or particle-velocity histories. However, the gauge records show a number of puzzling features. These include the near linear growth of particle velocity with depth into the HE and the changing trajectory of the peak relative to the shock front.

By simplifying the equation of state, an analytic model for a reacting fluid has been developed that is an exact solution of the equations of motion. The model qualitatively explains the phenomena in the early stages of the growth to detonation, and by extrapolation, leads to a complete picture of the hydrodynamics of the growth-to-detonation process. Amongst other conclusions is that the peak particle velocity is more associated with the peak reaction rate than with completion of reaction.

1. INTRODUCTION

For many years, only the trajectory of the shock was measured in shock initiation experiments, e.g. [1]. Initially the shock velocity increases very slowly, but after a depth dependent on the initial shock strength, the shock velocity increases rapidly to full detonation.

The most important advance in recent years has been the development of in-material gauges. Multiple gauges, either measuring pressure (p) histories or particle velocity (u) histories along (Lagrangian) particle paths are mounted at regular intervals through the charge. Figure 1 shows a typical set of particle velocity histories, obtained by Gustavsen et al [2], for a 5.12GPa sustained shock into PBX9501. The shock was generated by the plane 1D impact of a thick Vistal impactor.

The particle velocity at the impacted interface is initially nearly constant, but gradually u falls, indicating that the higher pressure generated by exothermic reaction is decelerating the interface. For subsequent gauges, u rises to a widish maximum. The deeper the gauge from the interface

the higher the maximum, and the narrower the width of the peak.

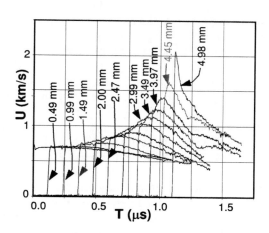

FIGURE 1. Shot 1133 - Particle Velocity Histories [2]

The times of the peaks of the particle velocity histories form a locus on a plot of time (t) versus depth (h) into the charge (Fig 2), on which are marked stages in the growth to detonation.

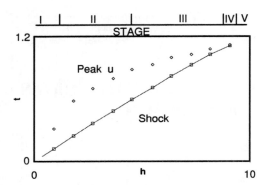

FIGURE 2. Trajectory of the Shock and Peak u

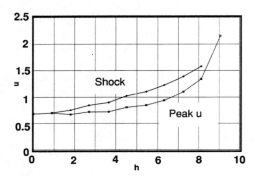

FIGURE 3. Particle Velocity at the Shock and Peak u

The velocity of the locus is slower than that of the shock in Stage I. Gradually the locus speeds up, runs parallel with the shock (Stage II), accelerates further (Stage III) and eventually catches up with the shock.

Peak u, plotted against gauge depth in Fig 3 grows nearly linearly with depth until the peak catches up with the shock.

After the peak catches up, the shock strength grows rapidly (Stage IV) with u falling immediately behind the shock front, rather like in a von Neumann spike.

Many of these features described above are puzzling. The aim of the model described below is to explain the features and to build up a qualitative description of the reactive-hydrodynamics of the growth-to-detonation for a sustained shock.

2. CIM - A CONSTANT IMPEDANCE MODEL FOR REACTIVE-HYDRODYNAMIC FLOW

The positive and negative characteristic relations for 1D Lagrangian reactive flow are

$$\frac{dp}{dt} \pm \rho c \frac{du}{dt} = \rho c^2 \sigma \dot{\lambda} , \qquad (1)$$

along the directions of the characteristics

$$\frac{dh}{dt} = \pm \rho c , \qquad (2)$$

where h is areal mass, ρ is density, c is the frozen velocity of sound, σ is the thermicity coefficient [3], and $\dot{\lambda}$ is the reaction rate.

By assuming that the material impedance, $Z = \rho c$ is constant, it follows that the characteristics are straight lines in h – t space. In the simplest form of CIM, it is also assumed that the shock impedance is the same as the material impedance and that $\rho c^2\sigma=(\partial p/\partial\lambda)_{v,e}$, is a constant. The characteristic relations (1) are then easily integrable.

In Stage II of the growth to detonation, the trajectory of the peak particle velocities is running parallel to the shock trajectory. This suggests that the reaction rate history on any particle path is independent of its neighbours. The fourth assumption of CIM is that the reaction rate is only dependent on the time since the shock passed. Then integrating along any positive or negative characteristic JK:

$$p_K \pm Zu_K = p_J \pm Zu_J + \rho c^2\sigma \int_{t_J}^{t_K} \dot{\lambda}\langle t - t_{shk} \rangle dt \qquad (3)$$

where the integral is called the reaction integral, in which t_{shk} varies with h along the characteristic.

The problem considered is the impact of a thick inert plate of constant impedance Z_1 travelling at velocity u_1, with an explosive of impedance Z. Immediately after impact the state is p_2, u_2.

The analytic solution is found at the intersection of a positive and negative characteristic successively along the reflected and transmitted shocks, the interface, and for any point within the shocked explosive. The general solution is:

$$p = p_2 + \left(\frac{Z_1}{Z_1 + Z}\right)\frac{\rho c^2\sigma}{2}\lambda + \frac{\rho c^2\sigma}{2}\dot{\lambda}\cdot\frac{h}{Z} \qquad (4)$$

$$u = u_2 - \frac{\rho c^2\sigma}{2(Z_1 + Z)}\lambda + \frac{\rho c^2\sigma}{2Z}\dot{\lambda}\cdot\frac{h}{Z} \qquad (5)$$

Equations (4) and (5) encompass the whole of the solution. At the interface (h = 0), only the first two terms apply. The decrease in particle velocity and the increase in pressure along the interface is proportional to the variation of the extent of reaction, λ, with time. Thus there is no peak in u at the interface. The third term in (4) and (5) will be discussed later

As an example consider a very simple reaction rate increasing linearly with time, but limited by a depletion term $(1 - \lambda)$, i.e.

$$\dot{\lambda} = (1 - \lambda)b(t - h/Z) \qquad (6)$$

where b is a constant, chosen so that the reaction rate peaks at $\sim 0.4\mu s$.

Figure 4 shows how particle velocity histories grow nearly linearly with depth for the first five gauge positions in the explosive. The trajectory of the peak particle velocity, shown in Fig 5, starts at a velocity less than that of the shock, gradually accelerates and then runs parallel to the shock, delayed by $\sim 0.4\mu s$. At least qualitatively, the particle velocity histories behave in a similar manner to the early gauge results in Figure 1 to 3. It follows that the simple model can be used as a basis for describing the reactive-hydrodynamics of the SDT process.

3. THE FIVE STAGES OF THE SHOCK TO DETONATION PROCESS

3.1. Stage I: Velocity of Peak Significantly Less than Shock Velocity

In Stage I, the shock strength is effectively constant and the reaction rate is growing, but has not yet reached its peak rate.

The additional pressure, generated by the release of energy in the explosive, pushes on the interface and starts to slow it down, so that the explosive near the interface expands. Effectively a compression wave is propagated into the inert impactor and a rarefaction 'reflected' into the explosive. There is a peak in u between the shock and the interface.

For a reaction rate that grows initially linearly with time, the initial velocity of the trajectory of the peak particle velocity is:

$$\dot{h}_0 = \frac{z^2}{z_1 + 2z} \qquad (7)$$

Hence $\dot{h}_0$ is lower for a higher impedance impactor

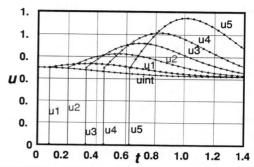

FIGURE 4. CIM – Particle velocity Histories

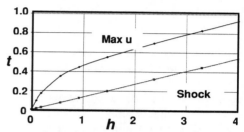

FIGURE 5. CIM Trajectory of Shock and Peak

3.2. Stage II – Peak Particle Velocity Growing and Running Parallel to the Shock

Stage II includes particles further away from the interface and reaction occurring at later times. The shock strength is still virtually constant. Each particle has the same reaction rate history, i.e. the reaction rate is constant on h-t lines parallel to the shock trajectory.

The dominant term in equations (3) and (4) is the final term. Along each positive characteristic, p and u grow linearly with h, with a gradient proportional to the reaction rate.

The rate of growth in p and u is greatest at the maximum reaction rate. Hence peak p and u grow linearly with depth. Because u is reducing and p is increasing along the interface, in Phase II, peak u occurs slightly before and peak p after peak reaction rate.

The trajectory of the peak particle velocity accelerates from its initial velocity (7) and becomes nearly parallel with the shock trajectory close to peak reaction rate. The trajectory is subsonic.

The fall in u behind the peak is not due to the rarefaction 'reflected' from the interface, but is due to the fall in reaction rate behind its peak.

369

The above does not explain why the peak grows. The two essential factors are that the generation of pressure by reaction pushes adjacent particles away and that the reaction rate history for each particle is delayed until its path is crossed by the shock. Each release of energy by the next particle to be shocked reinforces, at exactly the right time, all the release of energy that has gone before. The 'reaction wave' accumulates the energy release and accelerates the next particle slightly more than the previous particle. It might be called an 'Accumulation Wave', with a strength that grows nearly linearly with the mass of explosive consumed.

3.3. Stage III – Peak Particle Velocity Trajectory Supersonic Relative to the Shock

The simple CIM model would allow the peak u and p to grow without limit. This is where the assumption of constant impedance in CIM breaks down. In reality, the trajectory becomes supersonic relative to the shock for two reasons:

1. Peak p and u grow sufficiently that a compression wave forms, which travels supersonically relative to the shock wave.
2. Energy released behind the shock feeds forward and accelerates the shock. As the shock strength increases, the time to peak reaction rate reduces.

3.4. Stage IV and V – Pressure and Particle Velocity Decreasing Behind the Growing Shock

After the locus of the peak particle-velocity catches up the shock, p and u both fall behind the shock, akin to a von Neumann spike. Despite the fall in u, shock evolution theory shows that in a reacting fluid, the shock strength grows if the reaction rate is more vigorous than the decay. Full detonation (Stage V) occurs when the reaction rate just balances the decay.

4. CONCLUSIONS AND DISCUSSION

A simple reactive-hydrodynamic model has been developed that adequately explains the behaviour of particle-velocity gauge records in the early stages of the growth-to-detonation for a sustained shock in high explosive.

It has been shown that particle velocity histories are a reflection of the reaction rate, where the reaction lasts ~1µs for a 5GPa shock into PBX9501. However, the trajectory and growth of the peak particle velocity is not due to increasing extent of reaction, but is a consequence of the reactive-hydrodynamics of the SDT process.

The assumptions in the model are in the properties of the explosive, not in the equations of motion. Hence, the analytical form of the model is an exact solution of the 1D Lagrangian equations of motion.

The assumptions in CIM are of course challengeable. However, it is not felt that increasing the complexity of the model will substantially change the conclusions. The assumption that the shock impedance is the same as the impedance of the shocked material is the weakest of the approximations in CIM1. The theory has been extended to allow for different shock impedance. It then shows some of the features of Stage III.

ACKNOWLEDGEMENTS

The author is indebted Rick Gustavsen (LANL) for providing the particle velocity histories on Shot 1133 and to Hugh James for many interesting and useful discussions.

REFERENCES

[1] Campbell, A. W., Davis, W. C., Ramsay, J. B. and Travis, J. R., *Phys. Fluids*, **4**, 511-521, (1961)
[2] Gustavsen, R. L., Sheffield, S. A., Alcon, R. R. and Hill, L. G., *LA-13634-MS, (1999)*.
[3] Fickett, W., and Davis, W. C., *Detonation*, University of California Press, Berkeley and Los Angeles 1979, p78.

CP706, *Shock Compression of Condensed Matter - 2003*
edited by M. D. Furnish, Y. M. Gupta, and J. W. Forbes
© 2004 American Institute of Physics 0-7354-0181-0/04/$22.00

SIMULATION OF VOID COLLAPSE IN AMMONIUM NITRATE USING A MESHFREE LAGRANGIAN PARTICLE METHOD

L.D. Libersky[1], P.W. Randles[2], Neil Bourne[3], Rade Vignjevic[4]

[1]Los Alamos National Laboratory, Los Alamos, NM 87545
[2]Defense Threat Reduction Agency, Kirtland AFB, NM 87117
[3]Royal Military College of Science, Cranfield University, Shrivenham, Swindon, SN6 8LA, UK
[4]Cranfield University, Cranfield, MK43 OAL UK

Abstract. A meshfree Lagrangian particle code is used to simulate void collapse in Ammonium Nitrate. A 4.3 GPa shock is introduced into the emulsion through impact with a PMMA flyer traveling at 2 mm/μs. The jet created by the shock-induced void collapse is examined, and the temperature in the region where the jet impacts the opposite side of the void is estimated.

INTRODUCTION

Void collapse has been simulated previously using hydrodynamic Euler codes [1,2,3]. Recently, shock-induced collapse of a cylindrical air cavity in water has been simulated using a Free-Lagrange method [4]. This paper presents a simulation of void collapse in inert ammonium nitrate using the meshfree Lagrange particle method Dual Particle Dynamics [5]. The Lagrange nature of the method should provide improved temperature estimates in the hot spot and enhanced resolution of the jet boundary evolution. We envision that the meshfree nature of DPD can extend Lagrangian computing to larger deformations than traditional grid-based techniques. Void collapse is an interesting and challenging problem that will test this hypothesis.

We consider a spherical void 6 mm in diameter inside inert Ammonium Nitrate (ρ=1.25 g/cc). The virtual sample is impacted with a PMMA (ρ=1.18 g/cc) flyer at 2 mm/μs. The initial particle setup is shown in Fig. 1.

Parameters used for the Ammonium Nitrate are C=1.6 mm/us, S=1.8, γ=2.6, C_v=1500 J/kg K. For the PMMA we use C=2.6 mm/us, S=1.5, γ=1.0. Here C and S are coefficients in the linear shock speed – particle speed relationship, γ is the Grünei-

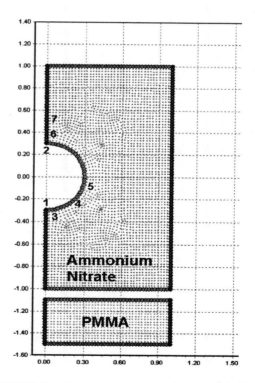

FIGURE 1. Setup with history particles labeled.

sen coefficient and C_v is the specific heat.

In order to examine the free surface motion of the jet tip and subsequent impact we tag two boundary particles on the poles of the void (labeled 1,2 in Fig. 1). To study the behavior of material in the jet we tag three particles below the void (labeled 3,4,5 in Fig. 1). Also, to inspect the hot spot we tag two additional particles in the region where the jet impacts the opposite side of the void (labeled 6,7 in Fig. 1).

Fig. 2 shows stress-free boundary particles defining the surface of the void at subsequent stages during the collapse.

Figures 3 and 4 show history plots (vertical coordinate and vertical velocity respectively) for opposing velocity particles on the poles of the void. Figures 5, 6 and 7 show history plots (density, temperature and number of neighbors) for three particles initially below the void and which become part of the jet as the collapse ensues. Figures 8 and 9 show history plots (pressure and vertical velocity) of two particles in the region just downstream of the impact point.

We estimate the peak temperature in the jet to be 1800 °K and at the impact point 1900 °K. It is important to note that the number of neighbors is changing rather rapidly in the jet (Fig. 6) but the fields at those points (Figs 5, 7) are remaining smooth. Success of meshfree particle methods will require that the connectivity change in such a way as to maintain stability while not introducing significant noise into the simulations.

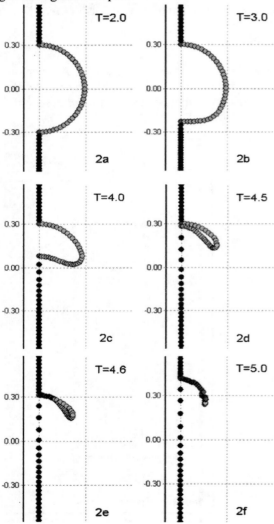

FIGURE 2. Void collapse sequence. Times are μs.

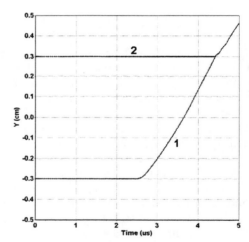

FIGURE 3. Vertical coordinate for particles (1,2) in Fig. 1.

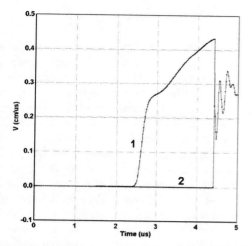

FIGURE 4. Vertical velocity for particles (1,2) in Fig. 1.

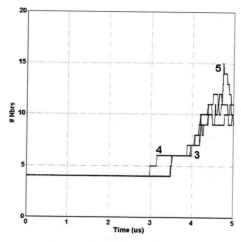

FIGURE 6. Number of neighbors for particles (3,4,5) in Fig. 1.

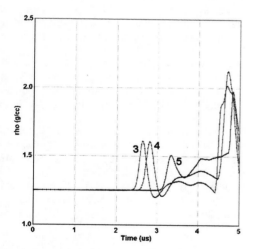

FIGURE 5. Density for particles (3,4,5) in Fig. 1.

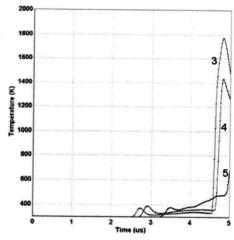

FIGURE 7. Temperature for particles (3,4,5) in Fig. 1.

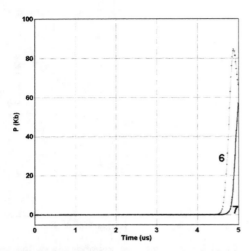

FIGURE 8. Pressure for particles labeled (6,7) in Fig. 1.

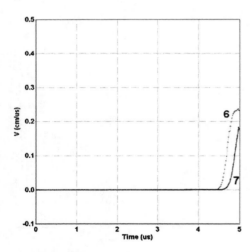

FIGURE 9. Vertical velocity for particles (6,7) in Fig. 1.

CONCLUSIONS

Void collapse in Ammonium Nitrate has been simulated with the meshfree particle code Dual Particle Dynamics. The meshfree nature of the method has allowed for simulation of the highly distorted flow in a Lagrange frame, thereby gaining increased accuracy in the constitutive modeling leading to temperature estimates as well as interface tracking. Our results agree qualitatively with the simulations and experiments reported in [5]. For the collapse of a 6 mm void by a 4.3 GPa shock we estimate the peak temperature in the region impacted by the jet (hot spot) to be 1900 °K. Future work will incorporate reaction physics so that conditions necessary for detonation can be investigated.

REFERENCES

1. Mader, C. L., and Kershner, J.D., "The three-dimensional hydrodynamic hot spot model", in Proc. 8th Int. Symp. Detonation, 1985, Albuquerque, NM.
2. Ding, Z., and Gracewski, S.M., "Behavior of a gas cavity impacted by a weak or strong shock wave", J. Fluid Mech., 309, 183, 1995.
3. Milne, A.M., and Bourne, N.K., "Experimental and Numerical Study of Temperatures in Cavity Collapse", in Shock Compression of Condensed Matter, 2001, (Y. Horie, N. Thadani, M. Furnish, eds.), pp. 914-917.
4. Ball, G.J., Howell, B.P., Leighton, T.G., Schofield, M.J., "Shock-Induced Collapse of a Cylindrical Air Cavity in Water: A Free-Lagrange Simulation", Shock Waves 10, 265, 2000.
5. Randles, P.W. and Libersky, L.D., "Normalized SPH with Stress Points", Int. J. Num. Meth. Engng., 48, 1445, 2000.

CP706, *Shock Compression of Condensed Matter - 2003*
edited by M. D. Furnish, Y. M. Gupta, and J. W. Forbes
© 2004 American Institute of Physics 0-7354-0181-0/04/$22.00

IMPLEMENTATION OF A HIGH EXPLOSIVE EQUATION OF STATE INTO AN EULERIAN HYDROCODE

David L. Littlefield[1], Ernest L. Baker[2]

[1]*Institute for Computational Engineering and Sciences, The University of Texas at Austin, Austin TX 78712*
[2]*Armament Research Development and Engineering Center, Picatinny, NJ 07806*

Abstract. The implementation of a high explosive equation of state into the Eulerian hydrocode CTH [1] is described. The equation of state is an extension to JWL [2] referred to as JWLB [3], and is intended to model the thermodynamic state of detonation products from a high explosive reaction. The EOS was originally cast in a form $p = p(\rho, e)$, where p is the pressure, ρ is the density and e is the internal energy. However, the target application code requires an EOS of the form $p = p(\rho, T)$, where T is the temperature, so it was necessary to reformulate the EOS in a thermodynamically consistent manner. A Helmholtz potential, developed from the original EOS, insures this consistency. Example calculations are shown that illustrate the veracity of this implementation.

INTRODUCTION

Accurate and realistic modeling of detonations is important to a number of civilian and military applications. Among the requirements for modeling detonation phenomena is an accurate equation of state (EOS) for the explosive reactants and products. In 1968, Lee *etal.* [2] proposed an algebraic form for the EOS of explosive products referred to as the Jones-Wilkins-Lee, or JWL, EOS. This equation of state was an extension of the earlier work by Wilkins [4] to extend the applicability to geometries involving large expansions of detonation products. The EOS is still in widespread use.

The JWL equation of state is based on a first order expansion of the principle isentrope, assumed to be of the form

$$p_s = Ae^{-R_1\rho_0/\rho} + Be^{-R_1\rho_0/\rho} + C\left(\frac{\rho}{\rho_0}\right)^{\omega+1} \quad (1)$$

where p is the pressure, ρ the density, ω the Grüneisen coefficient, and A, B, C, R_1 and R_2 are constants. The subscripts s and 0 imply a value along the adiabat and reference condition, respectively. Expansion of the adiabat is accomplished by using the isentropic identity and Grüneisen coefficient definition, given by

$$p = \left.\frac{\partial e}{\partial \rho}\right|_s ; \quad \omega = \frac{1}{\rho}\left.\frac{\partial p}{\partial e}\right|_\rho \quad (2)$$

Use of these identities yields the well-known relationship

$$p = A\left(1 - \frac{\omega\rho}{R_1\rho_0}\right)e^{-R_1\rho_0/\rho}$$
$$+ B\left(1 - \frac{\omega\rho}{R_2\rho_0}\right)e^{-R_2\rho_0/\rho} + \omega\rho(e - e_0), \quad (3)$$

where e is the specific internal energy and e_0 is a reference energy. Since the most common use of

JWL is to model states of detonation products, the adiabat passing through the Chapman-Jouget (C-J) point, also called the C-J adiabat, is usually the one selected.

THE MODEL

Even though it was developed 35 years ago, the JWL equation of state is still widely used today to model the state of reaction products in detonation calculations. Not surprisingly, since JWL is constrained to lie on the C-J isentrope, the states most reliably represented using the EOS are states near this isentrope (for example, the state of reaction products behind a planar detonation wave propagating in a semi-infinite slab of reacting material).

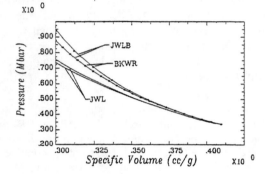

FIGURE 1. Reactive Hugoniot and principle isentrope for Octol 75/25 above C-J state (reprinted from [3]).

However, for states deviating substantially from the isentrope, the errors can be considerable. This is illustrated by the results shown in Fig. 1, where Hugoniots determined from JWL are compared to calculations from BKWR (a thermochemical potential code) and JWLB (the extension to JWL described later in this paper) for states at pressures above the C-J state. As is evident, errors in the pressure can be 20% or more when using JWL in this regime.

To correct these deficiencies, Baker [3] proposed a modified form for the principle isentrope, given by

$$p_s = \sum_{i=1}^{n} A_i e^{-R_i \rho_0 / \rho} + C \left(\frac{\rho}{\rho_0} \right)^{\omega+1} \quad (4)$$

where the Grüneisen coefficient λ is a function of ρ given by

$$\lambda = \sum_{i=1}^{n_\lambda} \left(\frac{A_{\lambda i} \rho_0}{\rho} + B_{\lambda i} \right) e^{-R_{\lambda i} \rho_0 / \rho} + \omega \quad (5)$$

Expansion of the isentrope is again achieved using the isentropic identity and definition of the Grüneisen parameter given in Eq. (2) (with ω replaced with λ). This expansion yields

$$p = \sum_{i=1}^{n} A_i \left(1 - \frac{\lambda \rho}{R_i \rho_0} \right) e^{-R_i \rho_0 / \rho} +$$
$$\lambda \rho (e - e_0) + C \left(1 - \frac{\lambda}{\omega} \right) \left(\frac{\rho}{\rho_0} \right)^{\omega+1}. \quad (6)$$

This form of JWL is often referred to as JWLB. Note that when $\omega = \lambda$ and $n = 2$ the original form of JWL is recovered. A reference condition for energy must be specified to determine the constant e_0.

The form of JWLB given in Eq. (6) interfaces naturally with codes requiring an EOS of the form $p = p(\rho, e)$. The target application code in this case, however, required an EOS of the form $p = p(\rho, T)$, where T is the temperature, so it was necessary to reformulate the original EOS.

It is essential to maintain thermodynamic consistency when recasting an equation of state to a different form. This is assured if the EOS is rewritten in terms of a thermodynamic potential. The natural choice of potential for an EOS of the form $p = p(\rho, T)$ is the Helmholtz potential a, defined as $a = e - Ts$, where s is the specific entropy. If a is known, expressions for the internal energy and pressure are easily determined as:

$$p = \rho^2 \frac{\partial a}{\partial \rho} \bigg|_T ; \quad e = a - T \frac{\partial a}{\partial T} \bigg|_\rho . \quad (7)$$

Likewise, if e or p are known then Eq. (7) can be integrated to determine a. To carry out the integration, a functional form for the dependence of energy on temperature has to be assumed. In this study, the specific heat c_v is assumed constant so that $e - e_0 = c_v T + f(\rho)$, where $f(\rho)$ is some function of density (a similar procedure was also used to derive the temperature form for JWL used today in many codes [5]). The function f is not arbitrary; it must satisfy compatibility conditions that are derived from Eq. (7). Using these conditions, a general expression for f is given by:

$$f(\rho) = \sum_{i=1}^{n} \frac{A_i}{R_i \rho_0} e^{-R_i \rho_0 / \rho} + \frac{C}{\rho_0 \omega} \left(\frac{\rho}{\rho_0} \right)^{\omega+1} \qquad (8)$$
$$+ C_1 \exp\left(\int \frac{\lambda}{\rho} d\rho \right),$$

where C_1 is an arbitrary constant. A further restriction that can be placed on f is that the original temperature form for JWL be recovered when $\omega = \lambda$ and $n = 2$. Applying this result yields the final form for f:

$$f(\rho) = \sum_{i=1}^{n} \frac{A_i}{R_i \rho_0} e^{-R_i \rho_0 / \rho}$$
$$+ \frac{C}{\omega \rho_0^{\omega+1}} \left[\rho^\omega - \exp\left(\int \frac{\lambda}{\rho} d\rho \right) \right]. \qquad (9)$$

Using this equation we get the final temperature form for JWLB given by:

$$p = \sum_{i=1}^{n} A_i e^{-R_i \rho_0 / \rho} + \lambda \rho c_v T$$
$$+ C \left(1 - \frac{\lambda}{\omega} q \right) \left(\frac{\rho}{\rho_0} \right)^{\omega+1} \qquad (10)$$

where q is a function of density. Evaluation of q requires integration of the Grüneisen coefficient which cannot be accomplished in closed form using the expression for λ given in Eq. (5). However, a Taylor series approximation for q can be determined in closed form and was found to be sufficient in the case studies examined thus far. Using four terms in this expansion yields an approximate expression for q given by

$$q \cong \exp\left(\sum_{i=1}^{n_\lambda} \left\{ \frac{A_{\lambda i}}{R_{\lambda i}} e^{-R_{\lambda i} \rho_0 / \rho} + B_{\lambda i} \mu e^{-R_{\lambda i}} X \right. \right.$$
$$\left[1 + \frac{\mu}{2} (R_{\lambda i} - 1) + \frac{\mu^2}{6} (R_{\lambda i}^2 - 4R_{\lambda i} + 2) \qquad (11) \right.$$
$$+ \frac{\mu^3}{24} (R_{\lambda i}^3 - 9R_{\lambda i}^2 + 18R_{\lambda i} - 6)$$
$$\left. \left. \left. + \frac{\mu^4}{120} (R_{\lambda i}^4 - 16R_{\lambda i}^3 + 72R_{\lambda i}^2 - 96R_{\lambda i} + 24) \right] \right\} \right)$$

where $\mu = (\rho - \rho_0)/\rho_0$.

RESULTS

This equation of state was implemented into the hydrocode CTH [2]. To verify the veracity of the implementation, a large number of calculations were performed simulating a cylinder test with a variety of explosives. The results shown herein

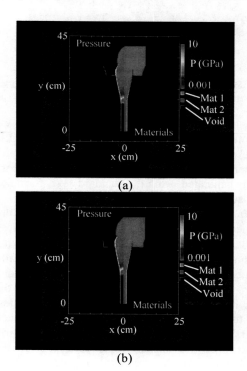

FIGURE 2. Pressure profiles at 20 μs for a cylinder test simulation using the (a) JWL EOS and (b) JWLB EOS.

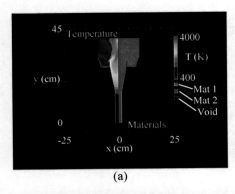

(a)

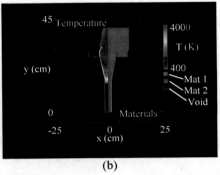

(b)

FIGURE 3. Temperature profiles at 20 μs for a cylinder test simulation using the (a) JWL EOS and (b) JWLB EOS.

were taken from one of those runs using a JWLB model for PETN, with $n = 5$ and $n_\lambda = 2$, and constants taken from Ref. [6]. For comparison, results generated using the standard JWL formulas are also given.

In Figs. 2 and 3, pressure and temperature profiles for the JWL and JWLB runs are compared at 20 μs after detonation. Figure 2 shows that pressures from the two calculations are nominally identical, with only slight differences in expanded states. The temperatures shown in Figure 3, on the other hand, are very different in expanded states; with the temperatures predicted from JWLB much lower. Temperatures and pressures are nearly identical at the shock fronts for each of the calculations, which is a result of matching C-J

adiabats to the C-J conditions for each EOS. However, away from the C-J points the adiabats can be different, which accounts for the differences seen in expanded states.

CONCLUSION

The implementation of the JWLB high explosive equation of state into a hydrocode has been described in this work. A thermodynamically consistent temperature form of the EOS was derived as a necessary step of the implementation. Example calculations illustrate the veracity of the implementation. The tests were performed using programmed burn, but the EOS should work equally well with reactive burn models.

ACKNOWLEDGEMENTS

This work was supported by the DoD High Performance Computing Modernization Office (HPCMO) under the Programming, Environment and Training (PET) Program.

REFERENCES

1. McGlaun, J. M., Thompson, S. L., Elrick, M.G., "CTH: A three dimensional shock wave physics code", *Int. J. Impact Engng.*, **10**, 351–360 (1990).
2. Lee, E. L., Hornig, H. C. and Kury, J. W., *Adiabatic Expansion of High Explosive Detonation Products*, Lawrence Radiation Laboratory Report No. UCRL-50422, Livermore CA (1968).
3. Baker, E.L., *An Explosives Products Thermodynamic Equation of State Appropriate for Material Acceleration and Overdriven Detonation: Theoretical Background and Formulation*, ARAED-TR-91013, Picatinny Arsenal, NJ (1991).
4. Wilkins, M. L., *The Equation of State for PBX-9404 and LX04-01*, Lawrence Radiation Laboratory Report No. UCRL-7797 (1964).
5. Tarver, C., private communication (2003).
6. Baker, E. L. and Stiel, L. I., *Improved Quantitative Explosive Performance Prediction using JAGUAR*, Proceedings of the Insensitive Munitions and Energetic Materials Technology Symposium, Tampa, FL (1997).

CP706, *Shock Compression of Condensed Matter - 2003*
edited by M. D. Furnish, Y. M. Gupta, and J. W. Forbes
© 2004 American Institute of Physics 0-7354-0181-0/04/$22.00

MD SIMULATIONS OF HOT SPOTS

Jean-Bernard Maillet

CEA, Département de Physique Théorique et Appliquée, BP12, 91680 Bruyères-le-Châtel, France

Abstract. The microscopic approach applied to shock-induced phenomena has often brought new understanding of the relevant physical and chemical processes. The interaction of a shock wave with pre-existing structural or chemical defects in a material can lead to the formation of hot spots, i.e. local regions of significant over heating. These hot spots play a key role in thermally activated processes such as detonation or phase transitions. We propose a microscopic mechanism for the interaction of a shock wave with defects leading to the formation of hot spots. This mechanism involves the vaporization of the material into the pore, the stagnation of a low-density gas at the far side of the void and the recompression of the gas back to the original shocked density during the collapse of the pore. The subsequent increase of temperature is analysed in terms of pressure-volume work associated with the compression. In case of reactive materials, molecular dynamics simulations show that the detonation threshold is significantly lowered in the presence of voids.

I. INTRODUCTION

The concept of hot spot is of fundamental importance in the area of high explosive materials. Indeed, the ignition under shock conditions of a heterogeneous reactive material is known to proceed in two steps: first, the propagation of the shock wave through the sample leaves some localized regions of high energy deposit, or hot spots. Secondly, these hot spots grow through chemical decomposition mechanism, leading finally to the formation of a self-sustained detonation wave. This paper is concerned with the description and understanding of the first step, i.e. how can the interaction between a shock wave and a local defect lead to the formation of hot spots ?

Little phenomenological information is available to help defining the concept of hot spot. This is mainly due to experimental difficulties to observe this phenomenon in such extreme conditions (high pressure, high temperature, small induction and response time, small spatial dimensions of defects, destructive trial, opacity). Several models have been developed in the literature, based on different origins for the formation of hot spots; they include the visco-plastic work occuring during the void compaction or associated with the appearance of shear bands, purely hydrodynamic phenomena when the pore collapses [1], and the heating related to the compression of the gas contained in the pore.

The continuous increase of computer power has made possible the investigation of hot spot formation at the atomic scale using Non Equilibrium Molecular Dynamics calculations. Several defects have been considered as potential candidates for initiation sites, ranging from chemical impurities, vacancies, elongated and spherical voids, and "nanocracks". Similarly to other thermally activated processes [2], spatially extended defects are required to activate chemical reactions. Indeed, if the defect is small like a vacancy or a chemical impurity, the shock wave passes over it without any major change. If the defect is big enough, it interacts more strongly with the shock wave, and energy will be deposited in the defect. In the first section, the properties of the energetic material are examined, followed by a description of the microscopic model for hot spot formation.

II. PROPERTIES OF THE EXPLOSIVE MATERIAL

Molecular dynamics simulations of reactive systems using the REBO (Reactive Empirical Bond Order) potential introduced by Brenner [3,4] have been performed [5]. This potential allows for covalent bond breaking and forming. The system is initially composed of diatomic molecules AB, that can evolve following the globally exothermic reaction $2AB \rightarrow A_2 + B_2$. This model has been found to reproduce many of the properties of condensed phase explosives [6], including the critical flyer plate velocity for initiation, the independency of detonation velocity versus initiation conditions (for unsustained shock) and failure diameter for detonation of 2D ribbons [7]. For perfect single crystals the detonation wave develops around $1.6 km.s^{-1}$ in two dimensions and between 1.56 and $1.72 km.s^{-1}$ in three dimensions. Simulations of critical diameter in three dimensions have been performed, as represented in figure 1. Samples usually contain between one and two millions particles in a crystalline lattice and are run for about 15 ps (about 1,245000 and 1,940000 atoms for cylinders of 80 and 150 nm respectively). Free boundary conditions are used.

A flyer plate is thrown into the cylinder, initiating a shock wave. For a strong enough shock, chemical reactions occur, leading to the formation of a reactive wave. The free surface at the boundary of the cylinder is a source of rarefaction waves travelling towards the centre of the cylinder. For small cylinders, they can quench the reactive wave - i.e. the detonation will not propagate. For cylinders large enough i.e. above the failure diameter, their effect is to curve the wave front. In the case reported in figure 1, a stationary detonation wave is obtained. The stationarity of the reactive wave is deduced from the analysis of the velocity of the front. In figure 2 the time evolution of the position of the wave

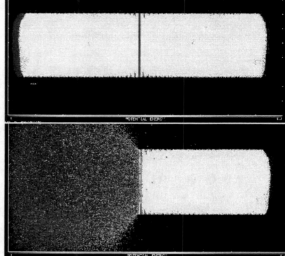

Figure 1. Potential energy of particle in a 3D cylinder geometry of radius 150 Å before and during the shock.

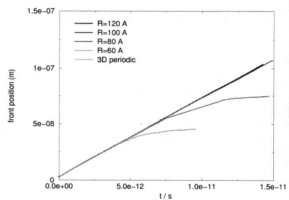

Figure 2. position of the reactive wave as a function of time for cylinders of different size. The position of the front is not determined as an average of particles positions over the entire curved front but rather by the position of the first particle in the shock front (given by an arbitrary criterium on the particle velocity) - this particle is usually located closed to the centre of the cylinder, where the wave velocity is maximum.

front for different cylinder sizes is shown.

The reactive wave is being quenched (transforming to an unreactive wave and vanishing) when the front position tends to a finite value. On the other hand, the linear regime corresponds

380

to a stationary propagating wave, the velocity of which is given by the slope of the line. The failure diameter is immediately deduced from this figure and is equal to $100\mathring{A}$. As expected, the velocity of the detonation wave depends on the cylinder radius. This is consistent with continuum theory. For 100 and $120\mathring{A}$ cylinder radii, velocities of 6990 and $7095m.s^{-1}$ are obtained, to be compared to the velocity of $7140m.s^{-1}$ for a detonation using periodic boundary conditions. A detailed analysis of the properties of the detonation wave is currently under treatment and will be reported in a near future [8].

III. DEFECTIVE MATERIAL

For a defective material, this detonation threshold can be considerably lowered. A typical example is the spherical void, which is used in many models as a simple and representative defect. The study of the interaction of an unreactive shock wave with a spherical void displays several effects: a jet of particles is ejected when the wave collides the upstream side of the pore. This jet is convergent and it impacts the back side of the pore without immediately causing any chemical reaction [9]. The fact that chemical reactions are not observed immediately after the first particle impact (i.e. the most energetic collision) reveals that the initiation process is not a direct mechanical one. At the same time the wave travels around the closed void, but the collapse of the void occurs before the wave could initiate any reactions on the far side. Chemical reactions appear nearly at the end of the closure of the pore, causing the appearance of a circular expanding pressure wave. These reactions may ultimately propagate through the entire system, causing the appearance of a reactive wave. It has been found that the use of periodic boundary conditions in the direction perpendicular to the shock enhances the capability of the first decomposition reactions to develop. This is related to the interaction between defects, and their influence on the sensitivity of the material (the detonation threshold is lowered as the concentration of defects increase).

Geometrical effects have been studied in [5]; it was shown that the probability of a defect to initiate reactions not only depends on its surface (in 2 dimensions) but also on its shape. Indeed, elliptical voids of aspect ratio 2:1 are more efficient to initiate reaction when the shock compression occurs along the long axis. These results may provide a good criterium in order to differentiate models of pore compaction.

More intriguing are the results of White et al [6] on the interaction of a shock wave with a planar gap (nanocrack). A crack perpendicular to the shock direction is introduced in the sample (see figure 3); periodic boundary conditions are employed in the direction transverse to the shock, making this crack of infinite dimension. It was found that chemical reactions can be initiated for shocks weaker than the shock-to-detonation transition in a perfect single crystal, provided the crack is large enough. Narrow cracks do not cause any sufficient changes to initiate chemical reactions. On the other hand, cracks larger than a threshold value lead to detonation. As in the case of the spherical pore, chemical reactions were not induced immediately after the collision of the ejected particle with the far side of the crack but later, during the collapse of the crack. As in the case of spherical or elliptical voids, detonation induced by hot spot formation requires the presence of extended defects. Studies on spherical and elliptical voids demonstrated the existence of geometrical effects; however, initiation of detonation on nanocracks shows that hot spot formation finds its origin elsewhere.

In the following part of the paper, attention will be given to 1-dimensional gap geometry, which retains the essential properties of the problem without including additional complexity from geometrical effect. The crack is made by removing slices of particles in the initial sam-

Figure 3. Configuration of the system with a nanocrack. Particles in green and red represent the piston and the explosive respectively.

ple. The system of interest is made of Lennard-Jones particles located on a fcc lattice (this unreactive potential has been selected in order to quantify the over heating only due to the defects, eliminating any contribution from chemical reactions). Particles are given an initial velocity of $-u_p$, moving towards an infinitively massive wall standing at $x = 0$. A shock wave is then created, propagating at velocity $u_s - u_p$. The particle flow velocity behind the shock wave in this frame is zero. When the shock wave emerges on the surface of the crack, it is reflected and propagates back in the shocked material as a rarefaction wave. At the same time, particles are ejected inside the crack, moving toward the far side. A collision occurs, creating a shock wave which propagates forwards in the unshocked material and backwards in the expanded material (it can eventually catch up with the rarefaction wave, ultimately leaving the material in the first shocked state). The different processes are illustrated in figure 4 where a density profile during the closing of the fracture is shown and in figure 5 where a temperature profile is shown after the closing of the fracture, providing evidence of the hot spot effect. Indeed, this profile exhibits a peak well above the shocked temperature, associated to a less dense material (the recompressed

vapor-ejecta). The first step in the process of hot spot formation is the emergence of the shock wave at the surface of the gap. Velocity profiles have been computed during the collapse of the nanocrack and are reported in figure 6, for simulations at low and high impact velocity. The differences in the two profiles reported in figure 6 are analysed in terms of the melting of the material. Indeed, for a weak shock, particle velocities range from zero (in the shocked material) up to approximately u_p (in this reference frame) and the profile tends to become linear. The fracture is progressively filled with the dense solid material. On the other hand, when melting occurs, particles are ejected from the surface forming a spray, and they appear to travel faster than those of the bulk liquid. This is similar to the transition from a smooth and symetric collapse to a turbulent collapse of the pore that has been observed by Mintmire et al [10]. Two distinct linear regimes are found in the velocity profile, corresponding to the particle velocities in the gas and liquid state respectively - the density profile, evidencing the presence of the two phases, is shown in figure 4. In both cases, the velocity of the condensed phase is equal to u_p (or $2u_p$ in the laboratory frame). The production of a gas inside the pore is of importance with respect to the mecha-

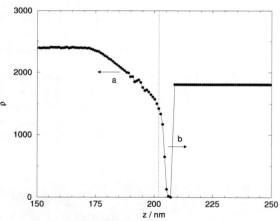

Figure 4. density profile just before the closing of the fracture. a: rarefaction wave moving backward in the shocked material and b: vapor-ejecta of average density ρ_{00} moving at center-of-mass velocity u_ℓ. The left and right parts correspond respectively to the shocked and unshocked material.

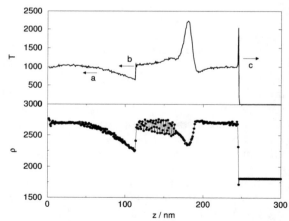

Figure 5. Temperature and density profile after the collapse of the fracture, showing a: the rarefaction wave propagating backward, b: a shock wave travelling backward in the expanded material and c: a shock wave travelling forward in the unshocked material. The fracture were located at $z \sim 180nm$

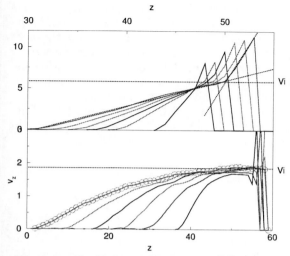

Figure 6. velocity profile during the closing of the gap for piston velocities of 5 and 2.

nism of hot spot formation. Indeed, it was found that there is no substantial increase of temperature upon void collapse in the absence of melting. However, if melting occurs, gas is ejected which is shock-recompressed during the collapse of the pore. The PV (pressure-volume) work associated with the recompression of the stagnating gas is seen as the source of the temperature increase.

In order to quantify the temperature increase due to the presence of the pore, temperature profiles were computed during the simulation (taking into account only the z-component of the particle velocity). Considering that the system is far from equilibrium, T. Hatano [11] uses the number of energetic collisions as a criterium for enhanced chemical reactivity. It has to be noted at this point that, provided chemical reactions are exothermic, the onset of enhanced chemical reactivity is ultimately given by the first chemical reaction, which has an occurence probability depending on the temperature of the system (since it is a thermally activated process). Each temperature profile has a maximum, which arrives at the vapor velocity. For small ℓ (the gap width), the temperature maximum corresponds to the shocked temperature and the vapor velocity to the expected value of $2u_p$. As ℓ increases, the vapor velocity increases, so as the temperature maximum. It was found that the evolution

of the vapor velocity results were correctly fitted with an exponential law. Similarly, the temperature maxima follow the following analytical expression:

$$T = T_1 + \Delta T_m(1 - e^{-\frac{\ell}{\ell_0}}) \tag{1}$$

where T_1 is the shocked temperature, ΔT_m is the difference between the maximum temperature (for large gaps) and the shock temperature and ℓ_0 is a phenomenological length. This rise in temperature explains the existence of a minimum size of the defect in order to induce chemical reaction. A similar expression can be deduced from the recompression of an expanding spray of ejecta [12].

The PV work W done upon recompression can be approximated by the area of the triangle in a PV diagram of the two Rayleigh lines associated with the first shock and the recompression respectively (this approximation leads to an over estimation of the work). An average density of the vapor ejecta is considered, moving at velocity u_ℓ toward the wall. From the expression of the mass of the vapor-ejecta, an approximate expression of the PV work is obtained :

$$W = \frac{1}{2}u_s u_p(1 - e^{\frac{\ell}{\ell_0}}) \tag{2}$$

In the ideal gas approximation, the maximum overheat that can be obtained is then given by:

$$k\Delta T_{max} = \frac{mu_s u_p}{d} \tag{3}$$

where d is the dimensionality. Simulations in 2 dimensions have been realized for different values of ℓ and u_p in order to check the validity of the above expression. As expected, the predicted values of ΔT are found to be greater than the results of the simulations. Nevertheless, this expression gives a good estimate of the overheat due to the hot spot effect.

In figure 7 is shown an example of a simulation in three dimensions where a defective energetic material is shocked (the shock strength is weaker than the detonation threshold). The REBO potential has been used and the sample contains 2,200000 atoms. From the pictures shown in figure 7, it can be seen that chemical reactions are initiated on the far side of the pore where

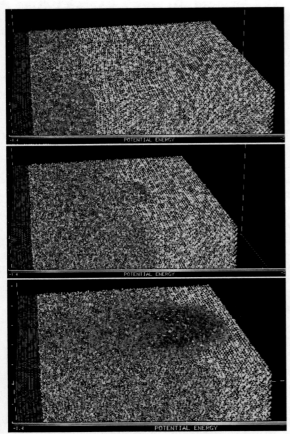

Figure 7. Potential energy of particles during the three dimensional shock of a defective material. The radius of the spherical pore is 77Å.

ejecta molecules converge, before the shock wave completed its travel around the void. The shock strength in this simulation is 10% below the initiation threshold, providing evidence of the important role of defects on the sensitivity of an explosive material.

IV. CONCLUSION

It has been shown that the hot spot effect finds its origin in the recompression of an expanding vapor-ejecta. The work achieved during the recompression is converted into heat, leading to a local increase of temperature. In the case of a nanocrack, the maximum temperature that can be produced is predicted via a simple model. However, geometrical effects (i.e. focusing of the ejected particles) can locally enhance the heat production and should be taken into account in a more precise model. Also the interaction between defects have been deliberately omitted but should be considered at some point.

T.C. Germann, B.L. Holian and C.T. White are gratefully acknowledged for their collaboration in this work.

[1] C.L. Mader, *Numerical Modeling of Detonations* (University of California Press, Berkeley, CA, 1979).

[2] B.L. Holian and P.S. Lomdahl, Science 280, 2085 (1998).

[3] D.H. Robertson, D.W. Brenner and C.T. White, Phys. Rev. Let. 67, 3132 (1991).

[4] D.W. Brenner, D.H. Robertson, M.L. Elert and C.T. White, Phys. Rev. Let. 70, 2174 (1993), (E) 76, 2202 (1996).

[5] T.C. Germann, B.L. Holian, P.S. Lomdahl, A.J. Heim, N. Gronbech-Jensen and J.-B. Maillet, Proceedings of the Twelfth Symposium (International) on Detonation, San Diego (2002).

[6] C.T. White, D.R. Swanson and R.H. Robertson, "Chemical Dynamics in Extreme environments", R.A. Dressler (Ed.); World Scientific, Singapore, 2001, p547.

[7] C.T. White, R.H. Robertson, D.R. Swanson and M.L. Elert, Shock Compression of Condensed Matter - 1999. Ed by M.D. Furnish, L.C. Chhabildas and R.S. Hixon.

[8] J.-B. Maillet and C. Matignon, in preparation.

[9] T.C. Germann in *Shock Compression of Condensed Matter*-2001, edited by M.D. Furnish *et al.*, AIP Conference Proceedings 620, New York 2002, p333

[10] J.W. Mintmire, D.H. Robertson, and C.T. White, Phys. Rev. B 49, 14859 (1994).

[11] T. Hatano, cond-mat/0306629 (2003).

[12] B.L. Holian, T.C. Germann, J.-B. Maillet and C.T. White, Phys. Rev. Lett. 89, 285501 (2002), (E) 90, 069902 (2003).

CP706, *Shock Compression of Condensed Matter - 2003*
edited by M. D. Furnish, Y. M. Gupta, and J. W. Forbes
© 2004 American Institute of Physics 0-7354-0181-0/04/$22.00

Multiscale simulation of detonation

J.-B. Maillet, B. Crouzet, C. Matignon, L. Mondelain and L. Soulard

CEA-DAM Ile-de-France, BP12, 91680 Bruyères-le-Châtel, France

Abstract. A multiscale approach has been developed and is currently applied to detonation related processes. Modeling the explosive material behavior at a macroscopic level requires several data that can be extracted from a lower level of description. The first one is the equation of state of reactants and products, which can be built from an interatomic potential using the method of Lysnes and Hardesty[3]. Secondly, the system chemical behavior (rate of reaction) can be deduced from MD simulations using a reactive potential which allows for chemical bond breaking and creation and which is particularly well suited for the study of detonation. Finally, mixing laws between reactants and products in the detonation reaction layer can be obtained via non equilibrium simulations of reactive wave where partial thermodynamics variables can be directly computed. A test *a posteriori* of this global method is performed: the Dremin-Trofimov geometry for the measure of the self-confined failure diameter. This setup can be modeled with either the direct (microscopic) simulation or the macroscopic model.

I. INTRODUCTION

Ignition and propagation of a detonation wave involve several coupled processes associated with different length and time scales. In the particular case of an homogeneous liquid explosive (i.e. without any structural and chemical defects or impurities), only two scales could be considered in a first approximation. The smallest should describe the chemical decomposition mechanisms of the organic material. The largest is associated with the propagation of reactive and mechanic waves resulting from the combustion of the explosive. For example, chemical reaction rates should be treated at the lower scale while the determination of a failure diameter has to be envisaged at the larger scale. The links between these two scales have to be built and validated. Our approach relies on three points: (a) the choice of a simple reactive system (a diatomic molecular material AB allowed to form A_2 and B_2 molecules through an exothermic reaction),

(b) a numerical scheme designed to link the two scales and (c) a selection of theoretical tools transforming the numerical results into hydrodynamic suitable input data (EOS of reactants, products and mixture, and the reaction kinetics). This approach is validated by performing a direct comparison between MD simulations, theoretical predictions, and hydrodynamic results on representative setup: ignition, failure diameter and propagation of detonation. The MD method has been used to compute the properties of the diatomic AB material, given the interatomic potential. A distinction is made between equilibrium and non equilibrium simulations. The former allows the calculation of properties in an homogeneous system evolving slowly in time (adequate for EOS calculation), while the latter is characterized by large gradients in space (suitable for shock wave propagation). The REBO potential is selected to describe the interatomic interaction. In this potential, the chemical composition of the system could evolve through the creation and break of covalent bonds. It has been shown that this potential allows the simulation of a self-sustained reactive wave [1]. In the next section, the characteristics of the microscopic reactive wave are extracted, giving the consistence of the global multiscale approach. The following section describes this method and preliminary results are shown in the last section.

II. NATURE OF THE REACTIVE WAVE

Non equilibrium Molecular Dynamics has been employed in order to propagate a self-sustained reactive wave. The structure of the reaction zone has then been analysed. Density profiles taken at different times during the simulation are displayed in figure 1 (for convenience, these profiles are shown in the leading shock attached reference frame). The steady character of the reac-

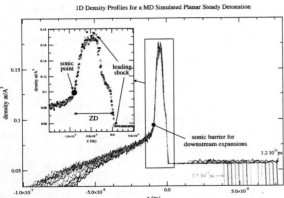

Figure 1. Density profiles of a 2D detonation wave simulated by MD (using periodic boundary conditions in the direction transverse to the shock and free conditions at both sides of the system) with the REBO potential, presented in the shock reference frame.

tive wave is evidenced by the superposition of the different curves. One may easily identify a sonic point from where downstream expansions are blocked. This structure is conform to the ZND scheme (a leading shock followed by a zone of exothermic chemical transformations ending on a sonic surface). Similar studies have been performed on steady curved detonations propagating through a cylindrical confined charge.

The results on the charge axis confirm this structure. This analysis validates the microscopic concept of detonation. In the following, a set of theoretical tools for multiscale modeling is described.

III. GLOBAL FRAMEWORK

The EOS of the explosive is generally represented by a mixture of the EOS of the unreacted material and the EOS of detonation products *via* arbitrary combination rules [2]. The EOS of either the unreacted material or the detonation products is obtained using the method of Lysnes and Hardesty [3]. The MD method is used to construct a set of hugoniot curves starting from different points on an isobare. The integration of the differential equations associated with the variation of the entropy and temperature along the hugoniot curves allows to compute the tabulated EOS in the range defined by the hugoniot curves. This method is currently applied to the determination of the EOS of the unreacted REBO potential. The Hugoniostat technique [4,5] has been used in order to calculate hugoniot curve continuum (see figure 2). This quasi equilibrium method drastically reduces the CPU time of such calculations. The choice of

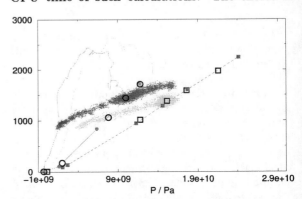

Figure 2. Hugoniot curve and release isentrope starting from two initial states at different densities. Full and open symbols correspond to equilibrium hugoniostat and NEMD results respectively.

mixing rules for the construction of the EOS of the reaction zone can be probed by MD: given the thermodynamic local equilibrium, ideal mixing rules are usually employed; the thermodynamic conditions of the reaction zone are then required. In the case of liquid explosives, the hypothesis of thermal and mechanical equilibrium (i.e. different constituents at the same temperature and same partial pressure) is intuitive. Simulations results show that the isothermal hypothesis is valid. Work is in progress to determine the validity of the mechanical equilibrium.

The chemical decomposition kinetics is now the only missing data to fulfill the hydrodynamic modeling. The thermal explosion model has been used to compute the chemical reaction rates under the arrhenius assumption. This model describes the energy balance between the chemical reaction and the dissipative processes such as conduction, convection and radiation. The concept of critical temperature, i.e. the temperature above which the heat coming from chemical reactions is greater than the dissipated heat, was introduced by J.H. Van't Hoof [6]. Above this temperature, the system explodes after a time τ defined as:

$$ ln\tau = n\left(\frac{C_v R T_i^2}{Q E_{act} Z f(\{c_k\})}\right) + \frac{E_{act}}{RT_i} \qquad (1) $$

where usual notations are used. Given the function $f(\{c_k\})$ (which is related to the chemical reaction mechanism), the activation energy can be deduced from the curve $ln\tau = f(\frac{1}{T_i})$. MD simulations in the microcanonical ensemble have been performed in two dimensions starting from a set of independent configurations at a temperature T_i above the ignition temperature. Several spots of chemical reactions appear in the system after some delay, and spread over the entire system. For this particular choice of potential parameters, the reaction is exothermic enough so the temperature could be used as a criterium for the reaction growth. The time evolution of the temperature of the system is shown in figure 3. The induction time can be inferred from this curve. Repeating the same operation for different initial temperatures T_i allows us to compute the variation of the induction time as a function of the inverse of the initial temperature. According to the thermal explosion model, a linear fit of $ln\tau = f(\frac{1}{T_i})$ represents correctly the simulation results (the slope gives the activation energy).

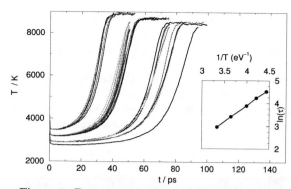

Figure 3. Decomposition kinetics as a function of time for different initial temperatures. The figure inside represents the variation of $ln\tau$ with the inverse of the initial temperature.

IV. GLOBAL VALIDATION: THE DREMIN-TROFIMOV GEOMETRY

An ensemble of links between microscopic data and macroscopic modeling has been constructed. A global test of the method, involving several complex phenomena is required to fully validate the multiscale method. The extinction or re-ignition mechanism of a self-confined detonation provides a good example. The explosive material is confined by a rigid cylinder of diameter ϕ_1. A steady detonation wave is obtained with a planar piston impact. This detonation emerges in a cylinder of diameter ϕ_2 greater than ϕ_1. Experimental results show that a failure wave appears at the borders of the detonation wave and propagates toward the center of the cylinder. Below a diameter $\phi_1 = \phi_c$ the detonation wave vanishes. Above ϕ_c, a secondary detonation wave appears in the compressed unreacted explosive and is transmitted to the large cylinder. ϕ_c defines the seld-confined detonation failure diameter. In order to model these phenomena, three descriptions have been employed. The first one is a direct microscopic simulation using MD. The second relies on hydrodynamic approach within the ZND framework. This requires the knowledge of the complete EOS and of the chemical kinetics, extracted from microscopic simulations as discussed above. The last description is a theoretical model developed by Dremin and Trofimov [7] and formulated by Enig and Petrone [8].

The qualitative agreement between these three approaches constitutes a first validation test (the macroscopic models being fed with microscopic data).

MD simulations using the REBO potential have been performed in the Dremin-Trofimov geometry in three dimensions using about 10 millions atoms. Different cylinder diameters have been used in order to calculate the value of the self-confined failure diameter (see figure 4). In fig-

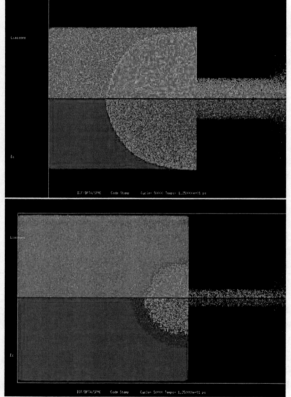

Figure 4. Simulations in the Dremin-Trofimov geometry. The upper picture shows a system where $\phi_1 > \phi_c$ so the detonation is transmitted (see text). In the lower picture, $\phi_1 < \phi_c$ and the detonation wave is quenched.

ure 4-a the detonation is transmitted to the large cylinder. In figure 4-b, a rarefaction wave coming from the confinement discontinuities is observed; it quenches the emerging detonation wave. These results are in agreement with Dremin and Trofimov's description. In the same way, bidimen-

sional ZND simulations using the Esione eulerian code lead to similar results (not shown here).

V. CONCLUSION

These first results show that Molecular Dynamics brings new insight in the understanding of detonation phenomena for condensed phase explosive. Indeed, this method succesfully accounts for (i) a propagation of a reactive wave consistent with a ZND scheme (ii) the construction of reactive EOS (probing the reaction zone) in order to feed the macroscopic model and (iii) a good agreement with the experimental phenomenology with no more additional hypothesis about the structure of the detonation wave but from the interatomic potential. A coherent, multiscale approach of detonation has then been constituted. The next step resides in the quantitative agreement between microscopic simulations and macroscopic predictions.

[1] C.T. White, D.R. Swanson and R.H. Robertson, "Chemical Dynamics in Extreme environments", R.A. Dressler (Ed.); World Scientific, Singapore, 2001, p547.

[2] C.L. Mader, *Numerical Modeling of Detonations* (University of California Press, Berkeley, CA, 1979).

[3] P.C. Lysne and R.H. Hardesty, J. Chem. Phys., 59(12), 6512 (1973).

[4] L. soulard, *Shock Waves in Condensed Matter*, p185 (1999).

[5] J.B. Maillet, M. Mareshal, L. soulard, R. Ravelo, P.S. Lombdahl, T. Germann and B. Holian, Phys. Rev. E, 63, n16121 (2001).

[6] J.H. Van't Hoff, "Etudes de dynamique chimique". Amersterdam, 1884.

[7] A.N. Dremin and V.S. Trofimov, Zhur. Prikl. i Tekh. Fiz (1964), n 1, p126.

[8] J.W. Enig and F.J. Petrone, Proceedings of the fifth symp. Int on Detonation, Pasadena, office of Naval Research, Dpt of the Navy, 1970, p99.

CP706, *Shock Compression of Condensed Matter - 2003*
edited by M. D. Furnish, Y. M. Gupta, and J. W. Forbes
© 2004 American Institute of Physics 0-7354-0181-0/04/$22.00

DIRECT NUMERICAL SIMULATIONS OF PBX 9501

E. M. Mas, B. E. Clements, and D. C. George

(T-1) Theoretical Division, Los Alamos National Laboratory, Los Alamos, NM 87545

Abstract. We have explicitly gridded HMX crystals in PBX 9501 from 25 μm in diameter up to $\sim$.5 mm. We used HMX particle size distributions found in the literature to determine the relative numbers of different sized particulates. We applied our modified Mori-Tanaka theory to model the smaller crystals embedded in the plasticized estane binder (the dirty-binder). This model was modified to accommodate the large amount of HMX in the dirty binder. We then subjected the $\sim$1 million element PBX 9501 realization to boundary conditions commensurate with a Split Hopkinson Pressure Bar experiment. We compare results to experiment and a micro-mechanical model we have reported on earlier. We also discuss the information which can be extracted from these direct numerical simulations.

INTRODUCTION

Modeling composites with complex microstructures has been a continuing problem in mechanics of material modeling. The desire to create computationally fast and efficient models often necessitates homogenization of the microstructure. This leads to models which cannot predict localization phenomena, or any non-uniformity in stress fields when an applied load is simple, *i.e.* uni-axial stress. A complex microstructure, however, can give rise to highly fluctuating fields even under simple loading. Elastically stiffer grains will carry more of the stress load than softer material; the microstucture of the stiff grains could isolate other stiffer grains by bridging the stress around the isolated grain, thus changing the stress distribution. A soft matrix will act as a shock moderator around some grains while other grains might experience direct contact with another grain, again, modifying the stress distributions. The goal in this work is to determine the feasibility of doing a direct numerical simulation (DNS) on PBX 9501 wherein as much of the microstructure is explicitly gridded into a finite element analysis (FEA) simulation as is practicable. After establishing a prescription for conducting a DNS, the fidelity of the simulation to experiment validates the simulation.

Finally, we analyze some of the resulting data to give us insight into how the microstructure affects the local fields. This insight can be used to improve meso- and macro-mechanical models of the composite.

DIRECT NUMERICAL SIMULATIONS

In order to conduct a DNS on PBX 9501 two key factors must be addressed; a grid must be created which captures as much of the microstructure as possible without being too complex to simulate in a reasonable amount of time, and second, the constitutive models must capture the necessary physics and also be fast and efficient. We will address both of these issues in the next sections.

In 1997 Skidmore *et al.* reported a study of HMX particle sizes in PBX 9501 [1]. They removed the binder with a solvent and measured the particle size distribution by laser diffraction. We utilized the size distribution of the "Pressed Piece" (Fig. 2 of Ref. [1]). We approximated the distribution with seven crystal sizes (450, 350, 250, 200, 150, 100, and 50 micron diameters) effectively accounting for grains down to a 25 μm diameter. Assuming that we have one largest grain we used the distribution to calculate the approximate number of other grains leading us to a total of approximately 1600 grains. We can also infer from the distribution that ~80 volume percent of the HMX resides in grains with a diameter of 25 μm or larger which implies that the explicitly gridded up HMX grains should account for 75 volume percent of the PBX 9501. The remaining 25% therefore, must be filled with dirty binder which contains ~75% HMX and 25% binder matrix (note that it is a coincidence that both of these mixtures are 75/25).

The grid was generated using LaGriT, the Los Alamos Grid Toolbox [2], on a regular tetrahedral mesh of ~1 million elements. The center of the largest grain was randomly positioned within the mesh, restricted to be at least 90% of the grain radius from the sides of the mesh. The grain was then expanded about this center by specifying all of the elements which share a node with the initial element as HMX. Successive shells are specified as HMX in this way until the appropriate volume is attained. The center of the next largest grain is then randomly placed in the mesh, this time with the restriction that its center be 81% (.9x.9) of the grain radius from the sides of the mesh and the sides of the other grain. The remaining grains are similarly alloted within the mesh. A picture of the microstructure is provided in Fig. 1. The front surface is approximately 20% into the mesh. Individual grains can be observed, as can percolating paths of dirty binder. This appears to be a much better representation of PBX 9501 than has previously been seen in the literature.

Two constitutive models are necessary to conduct the DNS, one for the dirty binder and one for the explicitly gridded HMX grains. The dirty binder model is extensively discussed earlier [3] and we will only give a cursory explanation here as well as a description of how the model was modified to accommodate the higher HMX concentrations. The larger HMX grains are modeled as elastic-plastic with brittle fracture. The model is similar to the ones we have used for the large HMX grains in our Method of Cells (MOC) formulation [4, 5, 6]. The bulk and shear moduli are 12.5 GPa and 5.4 GPa [7] and the yield stress applied by a von Mises formulation is 66 MPa. The micro-crack growth model was captured using the model of Johnson, Addessio, and Dienes [8]. The crack parameters were the same as those used before [5] with the exception of the K_0 parameter which was changed to 2.9MPa$\sqrt{\text{m}}$. This was necessary because the former parameters were fit to plate impact experiments and this parameter has been observed to be rate dependent.

In order to model a dirty binder containing 75% HMX and 25% plasticized estane binder (bulk modulus taken to be 3.65 GPa) our original dirty binder which contains 40% HMX had to be modified. We have discussed [9] how to accommodate high filler concentrations in our model. Here we only quote that the final result is that the effective shear relaxation function can be written as

$$\mu^* = \mu_0(t)\left(1 + 2.5\frac{c_I}{1 - f_1 c_I}\right), \qquad (1)$$

where c_I is the concentration of the filler and f_1 is an empirically determined parameter associated with filler-filler correlation effects. A value of $f_1 = 0$ leads to the Einstein result, $f_1 = 1$ is the Mori-Tanaka effective medium theory, and $f_1 > 1$ accounts for further particle-correlation effects. A value of $f_1 = 1.2$ was found to sufficiently stiffen the dirty binder.

The value for f_1 and the new K_0 parameter were fit to an experimental SHPB stress-strain curve. Since doing a fit using the DNS would be prohibitive (a run takes about 4 days on a single processor alpha workstation) we used our MOC-dirty binder hybrid model [6]. In this model a representative volume element (RVE) is created which mimics the composites microstructure. In our case we use a 2x2x2 RVE in order to create a computationally fast model. Seven of the subcells in the RVE are filled with the 40% HMX, 60% binder dirty binder and the other subcell is pure HMX which encompasses about 90% of the RVE. In order to attain values for our DNS we changed these values to a 75, 25% mixture for the dirty binder and the pure HMX crystal occupying 75% of the RVE. The resulting fit is shown in Fig. 2 and the fit parameters are given in the above text.

The model was ported into the FEA code EPIC [10]. We modeled a load similar to a SHPB experiment, specifically uniaxial stress with an applied strain rate of 2000 s^{-1} by applying velocity boundary condition on the top and bottom nodes. A run to a strain of 8 percent took about 4 days on a single processor alpha workstation. We are hoping that the arrival of a parallel version of EPIC will allow us to simulate larger, more representative samples of PBX 9501 under more complicated loading conditions.

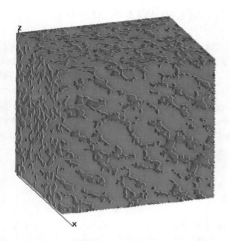

FIGURE 1. The microstructure for the DNS. Green denotes HMX and pink is the dirty binder

RESULTS

The purpose of the results section is two-fold; first to show agreement with experiment to validate the procedure, and second to analyze the data for information which might tell us qualitative or quantitative information about what is happening in the composite. This information can then be used as a tool for general understanding, and as a guide for meso- and macro-models which might be missing some important properties of the composite.

The agreement with the experimental stress-strain curve is good as seen in Fig. 2. The agreement is also close to the MOC simulation described above which produced the parameters.

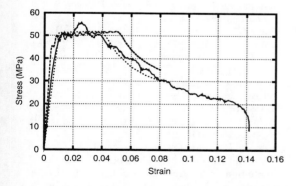

FIGURE 2. Stress-strain curves. Experiment is the solid line, the DNS is the dashed line, and the MOC simulation is the dots.

Other information gleaned from the simulation is more interesting. In the course of a simulation we observed stress bridging between HMX crystals, as one might expect. We also observed that the distribution of most quantities, including stress, pressure, and crack size varied considerably from grain to grain and within grains. Figure 3 shows the distribution of pressures at the same surface as shown in Fig. 1 at a strain of 4%. As can be seen, the distribu-

FIGURE 3. Pressure distribution in the simulation at a 17 μs or 4% volume averaged strain. Blue, lt. blue, green, yellow and red denote zero, 10, 25, 35 and 45 MPa respectively.

tion is far from uniform. In fact, the volume average of the pressure is about 17 MPa (light blue-green in the figure) and the pressures are seen to reach easily three times that in some grains. Further, there are regions which are predicted to be in tension (very dark blue). None of these results could be observed with a homogenized macro-model of the composite. As mentioned before, similar distributions are observed for other observables, but for now we will continue to focus on the pressure data. Figure 4 shows a histogram (here the tops of the histogram are connected with a line for ease of reading) of pressure versus the number of elements at that pressure for different times in the simulation. We were surprised that the negative pressures became so high, but for each negative pressure there are an order of magnitude more elements at the positive value of that pressure. As can readily be seen by the distribution all values of pressure are present at any given time out to a value of approximately three times the volumetric average. This is interesting from the point of view that no matter what the volumetric average of the pressure is there are always grains at low pressure. This might be important when considering chemistry of detonation. Many processes in the chemistry of decomposition are believed to be affected by high pressure. Despite this, there is no evidence present in the time to ignition data from thermal explosion to detonation indicating a kinetic affect due to pressure build up [11]. This could be explained by the results of these calculations, which show the persistence of low pressure conditions in the heterogeneous sample.

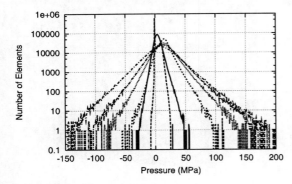

FIGURE 4. Histogram of pressures (from inside out) at 0.1, 1, 2, 5, 10, and 25 μs. (0.1 means zero elements.)

CONCLUSIONS

A DNS was successfully conducted on PBX 9501 where HMX particulates down to a diameter of 25μm were explicitly gridded up. The remainder of the HMX was accounted for by using a dirty binder model. The grid of $\sim$1 million tetrahedral elements was used to simulate a SHPB experiment. The run took $\sim$4 days on a single processor alpha workstation. The resulting stress-stain curve agreed well with experiment. Analysis of the simulations showed much variation in the pressure field in the sample, a result which would not have been seen in a homogenized model. The authors speculated that the fluctuations in the pressure field could explain why the expected pressure dependence of the HMX ignition kinetics can be ignored in plate impact, run to detonation experiments.

ACKNOWLEDGMENTS

The authors would like to thank William Blumenthal for all of his work and advice on experimental matters. The authors would also like to thank Laura Smilowitz and Bryan Henson for their discussions about chemistry in PBX 9501. The authors gratefully acknowledge financial support for this work from the Joint DoD/DOE Munitions Technology Development Program.

REFERENCES

1. Skidmore, C. B., Phillips, D. S., Son, S. F., and Asay, B. W., *Shock Compression of Condensed Matter-1997*, edited by S. C. S. et al., AIP Conference Proceedings 429, AIP, New York, 1997, pp. 579–582.
2. George, D., LaGriT User's Manual, http://www.t12.lanl.gov/home/lagrit.
3. Clements, B. E. and Mas, E. M., *J. Appl. Phys.*, **90**, 5522–5534 (2002).
4. Mas, E. M., Clements, B. E., Blumenthal, B., Cady, C., Gray III, G. T., *Shock Compression of Condensed Matter-2001* edited by M. D. F. et al., AIP Conference Proceedings 620, AIP, New York, 2001, pp. 539–542.
5. Mas, E. M., Clements, B. E., and Schlei, B., "Applying Micro-Mechanicals to Finite Element Simulations of Plate Impact Experiments on High Explosives," in *New Models and Hydrocodes for Shock Wave Processes in Condensed Matter - 2002*, Edinburgh, to be published.
6. Clements, B. E. and Mas, E. M., *Shock Compression of Condensed Matter-2001* edited by M. D. F. et al., AIP Conference Proceedings 620, AIP, New York, 2001, pp. 427–430.
7. Zaug, J. M., "Elastic Constants of β-HMX and Tantalum, Equation of State of Supercritical Fluids and Fluid Mixtures and Thermal Transport Determinations" in *Proc. Eleventh Detonation Symposium* 1998.
8. Bennett, J., Haberman, K., Johnson, J., Asay, B., and Henson, B., *J. Mech. Phys. Solids*, **46**, 2303, (1998).
9. Mas, E. M. and Clements, B. E, *J. Appl. Phys.*, **90**, 5535–5541 (2002).
10. Johnson, G. R., Stryk, R. A., Holmquist, T. J., and Beissel, S. R., User Instructions of the 1997 Version of the APIC Code, Wright Laboratory, Armament Directorate, Eglin Air Force Base report, WL-TR-1997-7037, 1997.
11. Henson, B. F., Smilowitz, L., Asay, B. W., Dickson, P. M., and Howe, P. M., "Evidence for Thermal Equilibrium in the Detonation of HMX" in *Twelfth International Detonation Symposium*, Wyndham, San Diego, 2002.

CP706, *Shock Compression of Condensed Matter - 2003*
edited by M. D. Furnish, Y. M. Gupta, and J. W. Forbes
© 2004 American Institute of Physics 0-7354-0181-0/04/$22.00

PORE COLLAPSE AND HOT SPOTS IN HMX

Ralph Menikoff

Theoretical Division, MS-B214, Los Alamos National Laboratory, Los Alamos, NM 87545

Abstract. Hot spots are critical for initation of explosives because reaction rates are very temperature sensitive. For a plastic-bonded explosive, shock desensitization experiments imply that hot spots generated by pore collapse dominate shock initiation. Here, for the collapse of a single pore driven by a shock, the dependence of the temperature distribution on numerical resolution and dissipative mechanism is investigated. An inert material (with the constitutive properties of HMX) is used to better focus on the mechanics of pore collapse. Two important findings result from this study. First, insufficient resolution can significantly overpredict the hot-spot mass. Second, up to moderate piston velocities (< 1 km/s), shock dissipation alone does not generate sufficient hot-spot mass for initiation. Two other dissipative mechanisms investigated are plastic work and viscous heating. In the cases studied, the integrated temperature distribution has a power-law tail with exponent related to a parameter with dimensions of viscosity. The parameter of either dissipative mechanism can be fit to obtain the hot-spot mass needed for initiation of any single experiment. However, the dissipative mechanisms scale differently with shock strength and pore size. Consequently, to predict initiation behavior over a range of stimuli and as the micro-structure properties of a PBX are varied, sufficient numerical resolution and the correct physical dissipative mechanism are essential.

INTRODUCTION

It has been known since the 1950s that initiation in a plastic-bonded explosive (PBX) is due to thermal reactions but requires hot spots [1]. Hot spots reconcile the large discrepancy between the time to detonation from Pop-plot data and the adiabatic induction time based on the bulk shock temperature and an Arrhenius reaction rate, see figure 1. For a strong shock ($P = 10$ GPa) at the high end of the measured Pop plot in HMX-based PBX-9501, we note that the time to detonation is $\simeq 200$ ns. Hot spots must react fast on this time scale, say within 20 ns, and would require a temperature of from 800 K based on the liquid phase reaction kinetics of Rogers [2] to 1500 K based on the "global reaction rate" of Henson *et el.* [3], see figure 2.

Shock desensitization experiments [4] and the increased sensitivity of a PBX with increasing porosity, as displayed in Pop-plot data, imply that hot spots generated by pore collapse dominate a shock-to-detonation transition. Early hydrodynamic simulations of heterogeneous initiation by Mader [5, sec. 3.3] utilized artificial viscosity for shock waves as the only dissipative mechanism. They showed that, when a strong shock impinges on a pore, a micro-jet is formed and subsequently produces a hot spot on impact with the downstream side of the pore. Furthermore, Mader's simulations with arrays of pores in an explosive showed a shock-to-detonation transition. When the simulations were performed (2-D flow in the 1960s and 3-D in the 1980s), the available computer power limited the resolution. In addition, the equation of state for the explosive had a constant specific heat. The specific heat for HMX varies by a factor of 2 between room temperature and 1000 K, and can have a large effect on the reaction rate because of its sensitivity to temperture.

With only shock dissipation, the peak pore collapse temperature can be estimated based solely on the equation of state (EOS) of the explosive and simple Riemann problems. For β-HMX a complete EOS derived from data currently available and estimates of the hot-spot temperature have been presented [7]. Since the EOS neglects solid-liquid phase transition, which has a latent heat equivalent to $\Delta T \simeq 200\,\mathrm{K}$, we take T=1000 K as the critical hot-spot temperature for fast reaction. The estimates in [7] indicate that shock dissipation alone does not generate sufficiently high hot-spot temperatures, even for a strong shock at the high end of the measured Pop plot.

Other dissipative mechanism applicable to pore collapse are viscous heating and plastic work. In the past, hot-spot temperatures were estimated based on simplified models due to the limited computer power available at the time, see for example [8, 9]. Here, for the collapse of a single pore driven by a shock, the dependence of the temperature distribution on numerical resolution and dissipative mechanism is investigated.

We consider both shear viscosity and rate-dependent plasticity. These dissipative mechanisms introduce a parameter η with dimensions of dynamic viscosity. For plasticity, the parameter determines the relaxation rate to the yield surface. The viscous parameter gives rise to two dimensionless parameters: Reynolds number, $Ry = \frac{\rho u R}{\eta}$ and (shock width)/(pore radius). Consequently, scaling of hot-spot temperature with pore radius and particle velocity will depend on the dissipative mechanism. To study this dependence we choose to hold the pore radius fixed and vary the viscosity parameter.

SIMULATIONS

Initial conditions for two-dimensional simulations are a gas-filled pore of radius 0.1 mm centered at (0.4,0) mm and surrounded by an inert material at 300 K with the EOS properties of HMX. A piston at the left boundary with a velocity of 1.3 km/s is used to drive a shock wave with a pressure of 13 GPa and temperature of 630 K.

For hydrodynamic pore collapse, in which the only dissipation is at shock fronts, figure 3A shows the temperature field after the shock front has passed over the pore. Pore collapse gives rise to an outgoing rarefaction wave in the material compressed by the lead shock, followed by an outgoing shock wave. These secondary waves give rise to the main features seen in the temperature field. We note that the sec-

ondary shock has caught up to the lead shock resulting in a Mach wave pattern. The temperature discontinuity corresponds to the contact emanating from the Mach triple point at (0.82,0.14). The gas pore has been highly compressed and distorted by the vortex set up from the impact of the micro-jet, formed when the lead shock overtakes the pore, on the downstream side of the pore. Since the vortex and the gas interface are expected to be unstable, the shape of the pore is presumably inaccurate in detail.

The temperature distribution is shown in figure 3B. The first peak at 300 K corresponds to the ambient state ahead of the lead shock front. The second peak centered at 575 K corresponds to the material heated by the lead shock and then cooled by the rarefaction from the pore implosion. The third peak centered about 675 K corresponds to the material heated by the lead shock and backward expanding portion of the secondary shock from the explosion of the pore. The low broad peak between 700 and 850 K corresponds to the region between the Mach stem and the material directly impacted by the micro-jet. The highest HMX temperatures, above 850 K, occur only near the pore and are numerical artifacts of the material interface treatment. The peak temperature is consistent with the estimates based on Riemann problems [7].

We note that the third peak is similar to what Hayes [10, Fig. 5] used to model initiation in HNS. The temperature of this peak depends on the explosive through its equation of state and specific heat. Since 1000 K is needed for fast reactions in HMX, we conclude that shock dissipation alone is not sufficient for initiation.

With additional dissipative mechanisms, plastic work or viscous heating, the tail of the temperature distribution can be greatly enhanced. Consequently, the reaction from hot spots is associated with the extreme tail of the temperature distribution. The tail of the temperature distribution is best described by the integrated temperature distribution, mass(T_1) at $T_1 > T$. It is convenient to normalize the mass relative to the equivalent mass in the pore volume at the initial explosive density.

We next examine the effect of mesh resolution on the temperature distribution. Figure 4 shows the integrated temperature distribution, with shock dissipation only, as the resolution is varied from 5 to 100 cells in the initial pore radius. We note that the distributions are nearly the same up to 800 K, but differ substantially in the high temperature tail. The differences are largely due to truncation errors from discretizing the flow equations in the underresolved region of high vorticity around the pore.

The low resolution case has sufficient hot-spot mass above 1000 K that, if the simulation included chemical reaction, a substantial amount of burn would occur, roughly a burn mass equivalent to 25% of the volume of the pore. Thus, errors from too low a resolution can substantially affect simulations of initiation. This is an important concern for meso-scale simulations of hot-spot initiation and is a determining factor in the size or number of grains in a PBX that can be included in the computational domain.

Simulations with additional dissipative mechanisms are summarized by the integrated temperature distribution shown in figure 5. A striking feature of the integrated distributions is that the tail of the distributions, above 600 K, are approximately linear on a log-log scale. This implies that the tail of the distribution has a power-law behavior. Moreover, the exponent of the power-law is related to the viscous parameter such that the effective hot-spot mass increases as the viscous parameter increases.

The viscous parameter has not been directly measured. Instead it is usually fit to reproduce integral data in a limited class of experiments. The fact that both the viscous heating and plastic work have similar distributions implies that either dissipative mechanism can be used in a fit. To discriminate between these mechanism requires a range of experiments that are sensitive to differences in the scaling behavior of each mechanism.

CONCLUSIONS

Two important findings resulted from this study. First, too low a resolution can significantly enhance the hot-spot mass. Second, up to modest piston velocities (< 1 km/s), shock dissipation alone does not generate sufficient hot-spot mass. Two other dissipative mechanism investigated are plastic work and viscous heating. In the cases studied, the integrated temperature distribution has a power-law tail with exponent related to a parameter with dimensions of viscosity. The dissipative mechanisms scale differently with shock strength and pore size. Consequently, to predict initiation behavior over a range of stimuli and

as the micro-structure properties of a PBX are varied, sufficient numerical resolution and the correct physical dissipative mechanism are essential.

ACKNOWLEDGMENTS

This work was carried out under the auspices of the U. S. Dept. of Energy at LANL under contract W-7405-ENG-36. The author thanks Prof. David Benson, Univ. of Calif. at San Diego, for providing the code used for the simulations.

REFERENCES

1. Bowden, F. P., and Yoffe, Y. D., *Initiation and Growth of Explosion in Liquids and Solids*, Cambridge Univ. Press, Cambridge, UK, 1952.
2. Rogers, R. N., *Thermochimica Acta*, **11**, 131–139 (1975).
3. Henson, B. F., Asay, B. W., Smilowitz, L. B., and Dickson, P. M., Ignition chemistry in HMX from thermal explosion to detonation, Tech. Rep. LA-UR-01-3499, Los Alamos National Lab. (2001).
4. Campbell, A. W., and Travis, J. R., "The Shock Desensitization of PBX-9404 and Composition B-3," in *Proceedings of Eighth Symposium (International) on Detonation, Albuquerque, NM, July 15–19, 1985*, Naval Surface Weapons Center, White Oak, Silver Spring, Maryland 20903–5000, 1986, pp. 1057–1068.
5. Mader, C. L., *Numerical Modeling of Explosives and Propellants*, CRC Press, Baca Raton, FL, 1998, second edn.
6. Gibbs, T. R., and Popalato, A., editors, *LASL Explosive Property Data*, University of California Press, 1980.
7. Sewell, T. D., and Menikoff, R., "Complete Equation of State for beta-HMX and Implications for Initiation," in *Shock Compression of Condensed Matter*, 2003, these proceedings.
8. Khasainov, B. A., Attetkov, A. V., Borisov, A. A., Ermolaev, B. S., and Soloviev, V. S., "Critical Conditions for Hot Spot Evolution in Porous Explosives," in *Progress in Astronautics and Aeronautics*, 1988, vol. 114, pp. 303–321.
9. Frey, R. B., "Cavity Collapse in Energetic Materials," in *Eighth Symposium (International) on Detonation*, 1985, pp. 68–80.
10. Hayes, D. B., "Shock Induced Hot-Spot Formation and Subsequent Decomposition in Granular, Porous HNS Explosive," in *Progress in Astronautics and Aeronautics*, 1983, vol. 87, pp. 445–467.

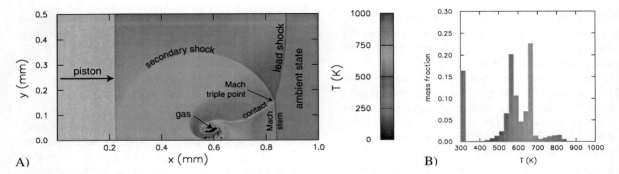

FIGURE 3. Temperature after pore collapse. Simulation with piston velocity of 1.3 km/s, shock dissipation only and resolution of 100 cells in the initial pore radius (0.1 mm). A) 2-D temperature field. Bottom boundary is symmetry plane. B) Temperature distribution.

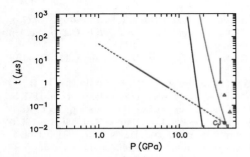

FIGURE 1. Time to detonation as function of shock pressure for HMX. Red line is Pop plot for PBX-9501 [6]. Blue and green lines are adiabatic induction time for Arrhenius reaction rate based on liquid HMX [2] and "global reaction rate" [3], respectively. Symbols are from wedge experiments of single crystal HMX by Craig [4].

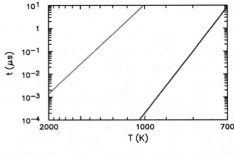

FIGURE 2. Adiabatic induction time for HMX. Blue and green lines are based on Arrhenius reaction rate for liquid HMX [2] and "global reaction rate" [3], respectively.

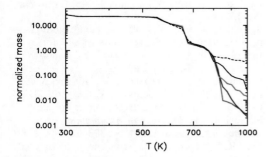

FIGURE 4. Variation of temperature distribution with resolution for shock dissipation only. Cells in pore radius: red, 100; blue, 50; green, 20; black, solid and dashed, 10 and 5, respectively.

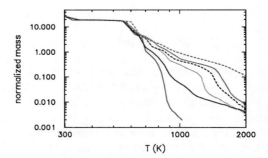

FIGURE 5. Variation of temperature distribution with dissipative mechanism. Dissipative mechanisms: red, shock heating; blue, shear viscosity, solid and dashed $\eta = 10$ and 100 Poise, respectively; green, rate dependent plasticity, solid, dashed and dotted $\eta = 800$, 8000 and 80 Poise, respectively.

CP706, *Shock Compression of Condensed Matter - 2003*
edited by M. D. Furnish, Y. M. Gupta, and J. W. Forbes
© 2004 American Institute of Physics 0-7354-0181-0/04/$22.00

ALE3D STATISTICAL HOT SPOT MODEL RESULTS FOR LX-17

Albert L. Nichols III, Craig M. Tarver, and Estella M. McGuire

Lawrence Livermore National Laboratory
P.O. Box 808, L-282, Livermore, CA 94551

The Statistical Hot Spot shock initiation and detonation reactive flow model for solid explosives in the ALE3D hydrodynamic computer code provides physically realistic descriptions of: hot spot formation; ignition (or failure to ignite); growth of reaction (or failure to grow) into surrounding particles; coalescence of reacting hot spots; transition to detonation; and self-sustaining detonation. The model has already successfully modeled several processes in HMX-based explosives, such as shock desensitization, that can not predicted by other reactive flow models. In this paper, the Statistical Hot Spot model is applied to experimental shock initiation data on the insensitive triaminotrintrobenzene (TATB) based explosive LX-17.

INTRODUCTION

The Statistical Hot Spot reactive flow model (1) was formulated in the ALE3D hydrodynamic code (2) to provide physically realistic descriptions of "hot spot" ignition, growth and coalescence. The details of its mathematics and material models were discussed previously (1), along with applications to HMX-based explosives. Besides shock initiation and detonation properties that the Ignition and Growth model (3) predicts, the Statistical Hot Spot model successfully calculated shock desensitization results (4) on PBX 9404 (94% HMX, 3% nitrocellulose, 3% CEF) that no other model has simulated. In this paper, the Statistical Hot Spot model is applied to the insensitive high explosive LX-17 (92.5% TATB and 7.5% KelF binder).

The alternating amino and nitro groups on TATB's benzene ring lead to strong intra- and inter-molecular hydrogen bonding. This hydrogen bonding is a major factor in TATB's high activation energy for thermal decomposition of 60 kcal/mol, compared to 40 – 50 kcal/mol for other explosives (5). Hydrogen bonding also contributes to the high thermal diffusivity of TATB, which is 1.6 to 3 times those of other explosives (6). This high thermal diffusivity causes shock induced "hot spots" to cool before they can react and grow (7). LX-17 requires sustained shock pressures of 6.5 GPa to cause any reaction. An 8.4 GPa sustained shock pressure in LX-17 causes slow growth of hot spots (8). Once ignited, LX-17 deflagrates an order of magnitude more slowly than other solid explosives. Under extremely high pressures in a diamond anvil cell, TATB deflagrates at rates less than 22 m/s (9), whereas HMX deflagrates at rates approaching 1000 m/s (10). Due to the slow growth rates of spreading hot spots, shock to detonation transitions in LX-17 take longer than those of other explosives (11). The 3 mm reaction zone length of fully detonating LX-17 causes wave curvature (12). TATB reactive flow models must predict the experimental observations.

NEW MODEL IMPROVEMENTS

In the previous version of the model (1), a second order accurate pressure, temperature equilibration scheme was used to the close the explosive mixture equation of state. In this version, a self-consistent second order accurate pressure equilibration with non-equilibrium temperatures is used. The temperature is determined by tracking the flow of energy from reactant to product states. The pressure dependent change in composition at each time step is determined self-consistently using an average of the initial and final pressures. The final state volumes are adjusted by iteration until the errors in

Table 1. Reaction Rate Parameters

Parameter	HMX Value	LX-17 Value
P_0	0.1 GPa	0.1 GPa
P^*	10 GPa	15 GPa
A	2000 cm-µs/g	1000 cm-µs/g
μ	5 µs^{-1}	10 µs^{-1}
D	11.3	30
ρ_P^{0}	1.4 x 10^{10} cm^{-3}	1.4 x 10^{10} cm^{-3}
ε	1.5 x 10^{-4} cm	1.5 x 10^{-4} cm

pressure and composition reach acceptable levels.

LX-17 MODEL PARAMETERS

The Statistical Hot Spot model requires an unreacted equation of state, a reaction product equation of state, and a set of 8 reaction rate parameters: P_0, P^*, A, μ, v, D, ρ_P^{0}, and ε (1). The unreacted equation of state is the Jones-Wilkins-Lee (JWL) fit used in the Ignition and Growth model (12). The reaction products are described by LEOS tables calculated by the CHEETAH code (13). P_0 is the ignition rate threshold pressure at which pores collapse and is related to the yield strength. P^* is the saturation pressure at which all potential hot spots have been created. A is the ignition pre-factor, which is related to the reacting surface area. μ and D are the hot spot death rate and constant death rate parameter, respectively. D is related to the shock pressure at which the explosive begins to ignite. The deflagration velocity v can be experimentally determined in diamond anvil experiments (9,10). ρ_P^{0} is initial number of hot spot sites, and ε is the initial hot spot diameter. The seven constant reaction rate parameters for LX-17 are listed in Table 1, along with those used for HMX. Table 2 lists the assumed dependence of v as a function of pressure for LX-17 and HMX. Tables 1 and 2 show that the reaction rates for LX-17 are much slower and the LX-17 hot spots die out more completely than those of HMX. The

Table 2. Deflagration Speed v vs Pressure

Pressure(GPa)	HMX v(cm/µs)	LX17(cm/µs)
0.0001	2.35x10^{-7}	2.35x10^{-7}
0.1	5x10^{-5}	5x10^{-5}
3	7x10^{-4}	4x10^{-4}
12	3x10^{-2}	2x10^{-3}
50	9x10^{-2}	4x10^{-3}

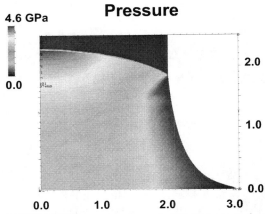

FIGURE 1. Pressure contours in LX-17 impacted into a stone wall at 0.7 km/s after 7 µs of propagation

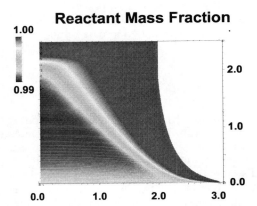

FIGURE 2. Fraction unreacted contours for LX-17 7 µs after collision with a stone wall at 0.7 km/s.

saturation pressure P_0 for LX-17 is 15 GPa, while 10 GPa was used for HMX.

LX-17 MODELING RESULTS

The first test of the LX-17 model is shock initiation caused by driving the explosive into a stone wall at various initial velocities. Figure 1 shows the pressure contours after 7 µs of propagation following a 0.7 km/s impact. This initial shock pressure is approximately 6.5 GPa, which induces very little reaction in LX-17. The maximum shock pressure has decreased to 4.6 GPa in Fig. 1. Figure 2 shows the fraction unreacted contours for the same time. The maximum fraction is 0.6%, which agrees with experiments at 6.5 GPa. Figure 3 shows the pressure contours in LX-17 3.9 µs after impact with a stone wall at 1.8 km/s. The

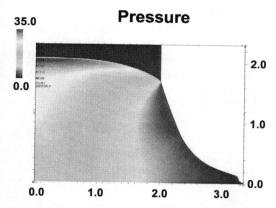

Pressure

FIGURE 3. Pressure contours in LX-17 3.9 μs after impact with a stone wall at 1.8 mm/μs.

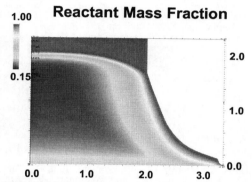

Reactant Mass Fraction

FIGURE 4. Fraction unreacted contours for LX-17 3.9 μs after impact with a stone wall at 1.8 mm/μs.

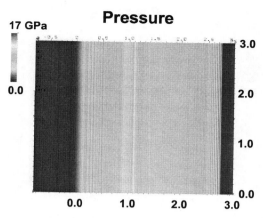

Pressure

FIGURE 5. Pressure contours in LX-17 6.4 μs after steel flyer impact at 1 km/s and reflection by copper

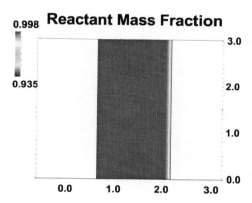

Reactant Mass Fraction

FIGURE 6. Fraction unreacted contours in LX-17 6.4 μs after steel impact at 1 km/s and reflection at a copper disc.

pressure has reached 34 GPa, the von Neumann spike value for LX-17 detonation. Figure 4 shows the corresponding fraction unreacted contours, which show a maximum of over 90% reacted as this wave builds to steady detonation. At steady state detonation, the calculated detonation velocity agrees closely with the experimental value of 7.6 mm/μs. Thus the LX-17 model predictions agree well with single shock initiation and detonation experiments.

Another test for the LX-17 model is the series of 12 reflected shock experiments reported by Tarver et al. (14). LX-17 was subjected to impact shock pressures of 4.4 GPa to 8.6 GPa, and then these shocks were reflected back into the LX-17 using aluminum, copper, or tantalum discs. The lowest pressure shocks caused desensitization, or "dead pressing," while higher pressures caused partial reactions behind the front shock and faster reactions

or detonation behind the reflected shocks. All 12 experiments were modeled with the Statistical Hot Spot model with good results. Two examples are shown in Figs. 5 – 8. Figure 5 shows pressure contours 6.4 μs after steel flyer impact at 1 km/s and reflection off a copper disc. The impact pressure was 6.8 GPa and the reflected shock pressure was 14 GPa. This shot exhibited shock desensitization on the embedded pressure gauges shown in Fig. 2 of Tarver et al. (14). Figure 6 shows that the LX-17 model predicts this desensitization, because its maximum fraction reacted is only 6% at 14 GPa. Figure 7 shows LX-17 pressure contours 5 μs after steel flyer impact at 1.19 km/s and reflection off a tantalum disc. The impact pressure is 8.1 GPa, which causes

Pressure

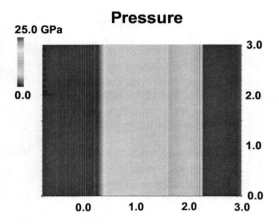

Figure 7. Pressure contours in LX-17 5 μs after steel impact at 1.19 km/s followed by tantalum reflection.

Reactant Mass Fraction

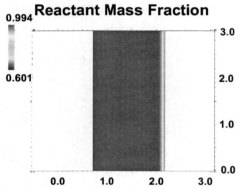

Figure 8. Fraction unreacted contours in LX-17 5 μs after steel impact and reflection off a tantalum disc.

reaction. The reflected shock pressures in the reacting LX-17 are over 20 GPa, resulting in rapid reaction and detonation. Figure 8 shows the fraction unreacted contours at 5 μs. The maximum fraction reacted has reached 40% at the LX-17-Ta boundary and later grows to 100% and produces detonation. The LX-17 parameters simulate the main features of the 12 reflected shock experiments.

SUMMARY

The LX-17 Statistical Hot Spot model accurately simulates single and reflected shock initiation data, because it is based on physical descriptions of hot spot ignition, death, growth, and coalescence.

ACKNOWLEDGMENTS

This work was performed under the auspices of the U.S. Department of Energy by Lawrence Livermore National Laboratory (contract no.W-7405-ENG-48).

REFERENCES

1. Nichols III, A. L. and Tarver, C. M., "A Statistical Hot Spot Model for Shock Initiation and Detonation of Solid Explosives," *Twelfth International Detonation Symposium*, Office of Naval Research, San Diego, CA, August 2002, in press.
2. Nichols III, A. L., Couch, R., McCallen, R. C., Otero, I., and Sharp, R., *Eleventh International Detonation Symposium*, Office of Naval Research ONR 33300-5, Snowmass, CO, 1998, pp. 862-871.
3. Tarver, C. M., Hallquist, J. O. and Erickson, L. M., *Eighth Symposium (International) on Detonation*, Naval Surface Weapons Center, NSWC MP 86-194, Albuquerque, NM, 1985, pp. 951-961.
4. Campbell, A. W. and Travis, J. R., *Eighth Symposium (International) on Detonation*, Naval Surface Weapons Center, NSWC MP 86-194, Albuquerque, NM, 1985, pp. 1057-1068.
5. Chidester, S. K., Tarver, C. M., Green, L. G., and Urtiew, P. A., *Combustion and Flame* **110**, 264-280 (1997).
6. Cornell, R. H. and Johnson, G. L., "*Measuring Thermal Diffusivities of High Explosives by the Flash Method,*" Lawrence Livermore Laboratory Report UCRL-52565, October 1978.
7. Tarver, C. M., Chidester, S. K., and Nichols III, A. L., *J. Phys. Chem.* **100**, 5794-5799 (1996).
8. Bahl, K., Bloom, G., Erickson. L., Lee, R., Tarver, C., Von Holle, W., and Weingart, R., *Eighth Symposium (International) on Detonation*, Naval Surface Weapons Center, NSWC MP 86-194, Albuquerque, NM, 1985, pp. 1045-1056.
9. Foltz, M. F., *Propellants, Explosives, Pyrotechnics* **18**, 210-216 (1993).
10. Esposito, A. P., Farber, D. L., Reaugh, J. E., and Zaug, J. M., *Propellants, Explosives, Pyrotechnics* **28**, 83-88 (2003).
11. Forbes, J. W., Tarver, C. M., Urtiew, P. A., and Garcia, F., *Eleventh International Detonation Symposium*, Office of Naval Research ONR 33300-5, Snowmass, CO, 1998, pp. 145-152.
12. Tarver, C. M. and McGuire, E. M., "Reactive Flow Modeling of the Interaction of TATB Detonation Waves with Inert Materials," *Twelfth International Detonation Symposium*, Office of Naval Research, San Diego, CA, August 2002, in press.
13. Fried, L., Howard, W. M., and Souers, P. C., "EXP6: A New Equation of State Library for High Pressure Thermochemistry," *Twelfth International Detonation Symposium*, Office of Naval Research, San Diego, CA, August 2002, in press.
14. Tarver, C. M., Cook, T. M., Urtiew, P. A., and Tao, W. C., *Tenth International Detonation Symposium*, Office of Naval Research ONR 33395-12, Boston, MA, 1993, pp. 697-703.

CP706, *Shock Compression of Condensed Matter - 2003*
edited by M. D. Furnish, Y. M. Gupta, and J. W. Forbes
© 2004 American Institute of Physics 0-7354-0181-0/04/$22.00

COMPUTER SIMULATIONS TO STUDY THE HIGH-PRESSURE DEFLAGRATION OF HMX

John E. Reaugh

HE and Organics Group, Physics and Applied Technologies, Lawrence Livermore National Laboratory, Livermore CA 94551

Abstract. The accepted micro-mechanical picture of the build-up of detonation in solid explosives from a shock is that imperfections are a source of hot spots. The hot spots ignite and link up in the reaction zone by high-pressure deflagration. Although the deflagration is subsonic, there are so many ignition sites that the pressure build-up is rapid enough to strengthen the initial shock. Quantitative advances in this research require a detailed understanding of deflagration at the high pressure, 1 to 50 GPa, which is present in the reaction zone. We performed direct numerical simulations of high-pressure deflagrations using a simplified global (3-reaction) chemical kinetics scheme. We used ALE-3D to calculate coupled chemical reactions, heat transfer, and hydrodynamic flow for finite-difference zones comprising a mixture of reactants and products at pressure and temperature equilibrium. The speed of isobaric deflagrations depends on the pressure and initial temperature. We show how this dependence changes with kinetic parameters, including the order of the last reaction step and the heat of formation of the species formed, relative to the reactant.

INTRODUCTION

Starting in the 1940's, explosives researchers became aware of the safety hazards that resulted from the presence of 0.1 to 10 µm gas bubbles or voids in solid and liquid explosives [1,2]. Such defects made the explosive more sensitive to shock. In the design of slurry explosives, defects are added to make the mixture detonable in the desired borehole diameter, but in solid and liquid explosives, they are reduced to minimize the sensitivity to handling accidents. In the reaction zone of a detonation, these voids become hot spots by their asymmetric collapse in the shock front. A deflagration starts at the hot spot and grows spherically until it begins to interact with the outward growing deflagrations from neighboring hot spots. Behind the deflagration front, the solid explosive crystal has transformed into gas products at high pressure and temperature.

We have used the computer hardware and software at our laboratory to study the details of the deflagration of high pressure and high-temperature explosives, as part of our Grain-Scale Dynamics in Explosives project for the Accelerated Strategic Computing Initiative. In the work reported here, we are using our computer simulations of deflagration to examine the dependence of the calculated deflagration speed on the chemical reaction kinetics and the species energetics.

NUMERICAL METHODS

We report the results of direct numerical simulations of deflagrations at high pressure. The computer simulation program, ALE-3D, which is under development at this laboratory, includes numerical methods to solve the heat flow equations either explicitly or implicitly. At present, the only solution method available for compressible hydrodynamic flow in the absence of shear strength is explicit. The resulting time step, as determined by the Courant condition for hydrodynamics [3] is larger than the explicit time step for heat transfer [3] for all mesh sizes Δ larger than the critical size Δ_c, where

$$\Delta_c = 2\kappa/c. \qquad (1)$$

Here κ is the thermal diffusivity, and c is the sound speed. Typical values for explosives and explosive products are $\kappa \sim 10^{-7}$ m^2/s, $c \sim 5 \cdot 10^3$ m/s, so that $\Delta_c \sim 0.04$ nm.

In standard solution methods for heat transfer, temperature is defined on the nodes of a finite element mesh. This is not directly compatible with standard solution methods for hydrodynamic flow in which all thermodynamic variables are defined within the volume of the finite element. Although methods have been developed to treat coupled, explicit hydrodynamics and heat transfer on the same mesh [4] the methods were not readily extended to include implicit heat transfer solutions. Instead, Nichols [5] developed a method that forces communication between the node-centered and element-centered temperature fields to have the appropriate time delay. This methodology was used for all the computer simulations reported here.

A challenging part of our numerical simulations is the incorporation of multiple chemical reactions using Arrhenius kinetics. Since the net result of these reactions is typically a 2500 K increase in temperature, the coupled equations become mathematically stiff. Nichols [6] developed a solution method that enforces simultaneous solution of pressure and temperature equilibrium for all species present in a finite element that is consistent with the chemical reaction rates. That method of coupling changes to species concentrations with temperature and hydrodynamics is used in all the simulations reported here.

In all of these simulations, we have ignored the contribution of species diffusion in smearing out the concentration of the intermediate products. The transport properties as calculated by Bastea [7] show that in the pressure range of interest to detonation, between 1 and 50 GPa, the ratio of species

to temperature diffusivity changes from the low-pressure, ideal gas value of 1.0 to the high-pressure value of 0.01. We can estimate the effect that change has on the calculated deflagration speed. Zel'dovich and Barenblatt [8] report a simplified set of equations for one-dimensional flame propagation where the ratio of species to thermal diffusivities is a parameter. For that case, the change in ratio from 1 to 0.01 results in a change of calculated flame speed of about a factor of four. From the solution of their equation set, our results will be qualitatively correct, but quantitatively in error. Some quantitative comparisons of our simulations with Diamond Anvil Cell (DAC) experiments in this pressure range have already shown that to be the case [9].

MATERIAL PROPERTIES

The review of Menikoff and Sewell [10] describes some of the properties needed for mesoscale simulations of explosives. They reported mainly the mechanical properties of the unreacted HMX explosive crystal. In addition, we need to describe the chemical reaction rates, and the mechanical, thermal, and transport properties of the various species present. We begin with the simplified global kinetics scheme reported by McGuire and Tarver [11] for HMX. That scheme includes a first, endothermic step, followed by a second, mildly exothermic step to form relatively high-molecular weight gas mixture, and a final exothermic step to form the final products. We identify the product of the first endothermic step as δ-HMX, although the endothermicity of Tarver's reaction scheme is significantly larger than the measured latent heat of the β-δ transition. We assumed that the intermediate gas products of the second step are those identified by Tarver: CHOH, N_2O, HCN, and HNO_2.

We have taken the solid equation of state for the β- and δ-phase HMX to be simple Mie-Gruneissen forms with constant specific heat. The compressibility of HMX is based on Yoo's DAC measurements [12]. We used Bridgman's correlation [13] for high-density fluids to estimate the pressure-dependent thermal conductivity. We constructed complete equation of state tables for the intermediate and final products using CHEQ [14].

Since exponential-6 potential parameters are not available for the intermediate products, they were estimated by Bastea based on results of classical MD simulations performed by Wu [15]. The transport properties for the two gas species are calculated by a method developed by Bastea [7] and incorporated as additional table entries.

SIMULATION GEOMETRY AND BOUNDARY CONDITIONS

The underlying simulation geometry is one-dimensional slab symmetry, which is treated as a stack of cubes in ALE-3D. All four lateral faces are adiabatic symmetry planes. The material occupying the cubes is β-HMX at the desired pressure, $p0$, and at room temperature. The input boundary is held at the constant pressure $p0$, and is given a temperature ramp from room temperature to the approximate flame temperature over a time of five (or sometimes ten) psec. The flame is propagated a distance of approximately one μm, which is about half the initial length of the stack of cubes. At that time, typically 200 nsec, the propagation velocity is observed to be constant. Most of the simulations were performed with a cube dimension of four nm, which results in an average time step of 0.5 psec, for a total of 0.6 M calculational cycles. A few simulations were performed with smaller mesh sizes (2 and 1 nm) to confirm that convergence with zone size had been achieved. The simulations with nominal resolution take 15 to 20 hours each on a 400 MHz SG workstation.

RESULTS AND ANALYSIS

Computer simulations of deflagration for the nominal set of properties are well-represented by the functional form

$$v = Ap^n \qquad (2)$$

where A takes the value 2.6 and n takes the value 0.38 for pressure in GPa and propagation velocity measured in m/s.

We performed variations on the nominal properties to determine the effects of changes in them to the calculated deflagration speed. We altered the heat of formation for the product of the first, endothermic step from its nominal value of 420 J/g relative to the heat of formation of the reactant, β-HMX. For the cases where the heat of formation was taken to be 0 or -420 J/g, the values of the burn rates were identical to within 0.2%, which is the precision that we can extract the flame speed from the simulations. Changing the activation energy of the first reaction to 90% of the nominal value, and changing the frequency factor to 110% of the nominal value similarly had no affect on the calculated flame speed.

Changes to the energy of formation of the intermediate products had a noticeable effect. Increasing (or decreasing) the heat of formation by 420 J/g from the nominal value of -540 J/g relative to β-HMX had the effect of decreasing (or increasing) the burn velocity at all pressures by 10%. The power-law dependence of the flame speed was not changed. Changing the kinetic parameters of the second reaction in a similar way to the way we changed them for the first reaction resulted in no change to the calculated flame speed.

In contrast, the heats of formation of the final product and the kinetic parameters to form the final product have a substantial effect on the calculated burn rate. Reducing the heat of formation of the final product from -6000 J/g relative to β-HMX to -6400 increases the flame speed by 18% at all pressures. Increasing the heat of formation to -5600 J/g reduces the velocity to 85% of the nominal value at all pressures. Reducing the activation energy of the last step by 10% increases the velocity to 145% of nominal at all pressures. Increasing the frequency factor by 10% increases the burn velocity at all pressures to 195% of nominal. We changed the order of the third reaction from second to first order. The pressure dependence of the flame speed changed from n = 0.38 to n = 0.20. Figure 1 shows the effect of changes to the final reaction on the flame speed.

CONCLUSIONS

The results of our simulations show that only those parameters associated with the substantially exothermic step have a significant influence on the propagation speed of high-pressure deflagrations. Reducing the activation energy or increasing the frequency factor of the exothermic reaction makes the temperature gradient steeper, which is associated with faster flame propagation. Decreasing the

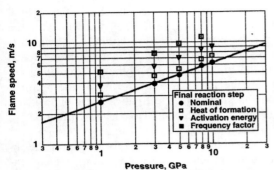

Figure 1. Calculated isobaric flame sped as a function of pressure for various perturbations to the nominal properties of the exothermic reaction step. The activation energy and frequency factor were changed 10%, and the head of formation of the final products was reduced 7%.

heat of formation of the final product results in an increase of the final temperature. At 10 GPa, for example, the flame temperature increases from the nominal value of 2820 to 2960 K. This temperature rise, in turn, affects the kinetic rates as well.

ACKNOWLEDGEMENTS

The author has benefited greatly from the ALE 3D models for chemistry and heat transfer devised and implemented by Albert L. Nichols, and for his advice on their effective use. In addition, it is a pleasure to acknowledge the contribution of Sorin Bastea in developing and implementing consistent models of the transport properties of explosive products. The ALE 3D computer program team, lead by Richard Sharp, has earned our gratitude for having been so responsive to the needs of our project, and for implementing the changes to that program that we have requested. This work was performed under the auspices of the U.S. Department of Energy by University of California, Lawrence Livermore National Laboratory under Contract W-7405-Eng-48.

REFERENCES

1. Bowden, F. P., and Yoffe, Y. D., Initiation and growth of explosion in liquids and solids, Cambridge University Press, Cambridge, 1952.

2. Campbell, A. W. et al., "Shock Initiation in Solid Explosives," The Physics of Fluids, vol. 4, no. 4, pp. 511-521, 1961

3. Richtmyer, R. D., and Morton, K. W., Difference Methods for Initial-Value Problems, second ed., John Wiley & Sons, New York, 1967.

4. Reaugh, J. E., "Calculation of Quasistatic Elastic-Plastic Deformation with Heat Conduction," Transactions, American Nuclear Society, vol. 18, p135-136, 1974

5. Nichols, A. L., private communication on secondary zonal temperature method, Lawrence Livermore National Laboratory, 1999.

6. Nichols, A. L., private communication on the closely coupled hydro-thermal scheme, Lawrence Livermore National Laboratory, 2001.

7. Bastea, S., "Transport properties of fluid mixtures at high pressures and temperatures, Application to the detonation products of HMX," 12th International Detonation Symposium, San Diego, CA, August 2002.

8. Zel'dovich, Ya. B. and Barenblatt, G. I., "Theory of Flame Propagation," Combustion and Flame, vol. 3, p61, 1959.

9. Esposito, A. P. et al., "Reaction propagation rates in HMX at high pressure," Propellants, Explosives and Pyrotechnics, vol. 28, no. 2, pp83-88, 2003.

10. Menikoff, R. and Sewell, T. D., "Constituent properties of HMX needed for mesoscale simulations," Combustion Theory and Modelling, vol. 6, no. 1, pp 103-125, 2002.

11. McGuire, R. R., and Tarver, C. M., "Chemical decomposition models for thermal explosion of confined HMX, TATB, RDX, and TNT explosives," 7th Symposium (International) on Detonation, White Oak, MD, August 1976.

12. Yoo, C. -S. and Cynn, H., "Equation of state, phase transition, decomposition of β-HMX (octahydro-1,3,5,7-tetranitro-1,3,5,7-tetrazocine) at high pressures," Journal of Chemical Physics, vol. 111, no. 22, pp10229, 1999.

13. Bird, R. B. et al., Transport Phenomena, John Wiley, New York, 1960.

14. Ree, F. H. "A statistical mechanical theory of chemically reacting multiphase mixtures: application to the detonation properties of PETN," Journal of Chemical Physics, vol. 81, pp. 1251, 1984.

15. Wu, C. J., private communication, Lawrence Livermore National Laboratory, 2001.

CP706, *Shock Compression of Condensed Matter - 2003*
edited by M. D. Furnish, Y. M. Gupta, and J. W. Forbes
© 2004 American Institute of Physics 0-7354-0181-0/04/$22.00

MOLECULAR DYNAMICS STUDY OF DETONATION

Z.A. Rycerz [1], P.W.M. Jacobs [1] and I.E. Hooton [2]

[1] *Department of Chemistry, University of Western Ontario, London, ON, Canada, N6A 5B7*
[2] *Defence R & D Canada – Suffield, PO Box 4000, Medicine Hat, AB, Canada, T1A 8K6*

Abstract. An advanced O(N) molecular dynamics (MD) computer program has been used to study the detonation of a model explosive, which simulates ammonium perchlorate (AP). The ten pair potentials and two bond-bending three-body potentials describe the cubic structure of AP accurately at or above 520 K, but the program does not include the complex reactions associated with the chemical decomposition of AP. Detonation is marked by large particle displacements and velocities, accompanied by an extremely rapid (that is, within a few fs) rise in the particle velocities and in the temperature, pressure and potential energy of the system. The calculated average velocity of all the atoms in the system at detonation exceeds 4 km/s and the pressure in the simulated system is 40-60 GPa. The MD results reported here are for 3-D geometry and were performed on PC computers.

INTRODUCTION

Much of the scientific and technological interest in ammonium perchlorate (NH_4ClO_4 or AP) is due to its extensive use as the oxidizer in solid-fuel rocket propellants. The stable form at room temperature has an orthorhombic structure (space group 62) but at 513 K AP transforms into a face-centered cubic (fcc) structure (space group 216). Only the high-temperature fcc structure was simulated in this work. The crystal consists of NH_4^+ and ClO_4^- molecular ions both of which have a tetrahedral structure. The onset of free rotation of ClO_4^- accompanies the phase transition but the NH_4^+ ions rotate freely down to low temperatures. A specified number of atoms N (in this work $N = 320$, 2560 or 20,480) are placed at their appropriate lattice sites and given random velocities that satisfy the conditions that the whole system has components of linear momentum that are zero and kinetic energy that corresponds to the desired temperature. The system is made pseudo-infinite by the application of periodic boundary conditions (PBC) [1,2]. The classical equations of motion are then solved at successive time steps of length $\Delta t = 0.5$ fs, using the *leap-frog* algorithm. The temperature, pressure, potential energy and total energy of the system are calculated at run-time and the positions and velocities of the atoms are stored for later calculation of other properties.

THE MD CALCULATION

In the absence of external fields the potential energy of the system of N atoms is

$$\Phi = \Phi^c + \Phi_2^{sr} + \Phi_3^{sr} = \sum_{i<}\sum_j \frac{Z_iZ_j}{r_{ij}} + \sum_{i<}\sum_j \phi_2(r_{ij}) + \\ \sum_{i<}\sum_{j<}\sum_k \phi_3(\mathbf{r}_i,\mathbf{r}_j,\mathbf{r}_k) \quad (1)$$

where: Φ^c denotes Coulombic energy, Φ_2^{sr} and Φ_3^{sr} are the short-range (*sr*) two-body (2-b) and three-body (3-b) energies; Z_i, Z_j are the electrical charges of ions i and j; ϕ_2 is the 2-b potential ($r_{ij}=|\mathbf{r}_i-\mathbf{r}_j|$) and ϕ_3 is the 3-b potential.

Long-range Coulomb forces are evaluated using a tabulated version of the Ewald summation [1]. The crucial step in any MD code is the location of all pairs of atoms within the chosen cutoff distance R_c.

For this task we used the *pyramid algorithm* [3,4] which is the most efficient order of O(N) algorithm known to us. The cutoff radius R_c we used was 9 Å and a time step $\Delta t = 0.5$ fs.

The short-range potentials used were:

a) Two-body (non-bonded – pair potentials), either Buckingham (B):

$$V_{nb}(r_{ij}) = A \exp(-Br_{ij}) - Cr_{ij}^{-6} \qquad (2)$$

or Morse (M):

$$V_{nb}(r_{ij}) = A\left[\left\{1 - e^{-B(r_{ij}-r_m)}\right\}^2 - 1\right] \qquad (3)$$

b) Three-body (bond-stretching and bond-bending; harmonic):

$$V_b(r,\Theta) = \tfrac{1}{2}k_s(r - r_o)^2 + \tfrac{1}{2}k_b(\Theta - \Theta_o)^2 \qquad (4)$$

Potential parameters are given in Tables 1 and 2.

TABLE 1. Two-body interaction potentials (eqs. 2 and 3)

Nr	Interaction	Type	A/eV	B/Å^{-1}	C/eVÅ^{-6}	r_m/Å
1	N-N	B	0.0	0.0	0.0	
2	N-H	M	1.00	2.20	0.0	1.01
3	N-Cl	M	3.00	6.00	0.0	3.815
4	N-O	B	5050.4	8.8778	0.0	
5	H-H	B	2000.0	8.00	0.0	
6	H-Cl	B	7000.0	3.979	20.00	
7	H-O	B	2756.0	4.00	40.17	
8	Cl-Cl	B	0.0	0.0	0.0	
9	Cl-O	M	0.80	2.00	0.0	1.50
10	O-O	B	3485.8	4.07	6.98	

TABLE 2. Three-body interaction potentials (eq. 4)

species	k_s /eV Å^{-2}	k_b /eV rad^{-2}
H-N-H	0.0	20.0
O-Cl-O	0.0	20.0

A well-equilibrated system was subjected to a thermal or a mechanical shock by either increasing the velocities of all atoms in the same ratio at a constant rate, or by decreasing the length of the system in the z direction at a linear rate, for a fixed period of time t_{exp} (usually 10 or 100 fs). The fractional increase in velocity or the fractional decrease in length are defined by a parameter F_{exp}. For sufficiently large departures of F_{exp} from 1, detonation occurred, marked by an extremely rapid increase in particle velocities, temperature and pressure, which resulted in termination of the run. A mechanical shock was delivered in either of two ways: in the *hard piston* case the walls of the box

normal to the z direction are moved inwards at a constant rate for time t_{exp} so that the PBC in the z direction are changed. In the *leaking piston* case the contents of the box are compressed in the z direction but PBC are not changed, so that the system is free to expand after the shock to fill the original box. Both these cases resulted in detonations (*cf*. [2]).

RESULTS AND DISCUSSION

Projections of the contents of the simulation box on planes normal to the x, y and z axes in an unshocked run confirm the cubic structure of AP.

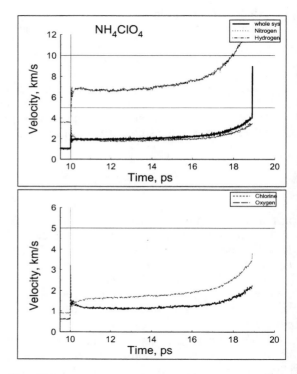

FIGURE 1. Average velocities of atoms (2.8% compression – *leaking piston*)

Similar projections of the positions of individual atoms show the rotation of NH_4^+ and the much slower rotation of ClO_4^-. A well equilibrated system is allowed to evolve further for $10 - t_{exp}$ ps and then subjected to a thermal or mechanical shock of magnitude controlled by F_{exp}. The total energy remains constant after the shock until it rises rapidly at the time of the explosion because of the breakup of the system, which therefore no longer has the structural properties of AP. Projections of the contents of the simulation box in runs that end in detonation, confirm the high mobility of H and the break-up of the crystal structure before the runs terminate.

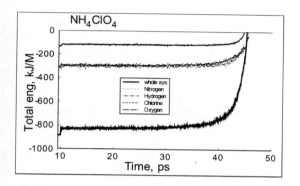

FIGURE 2. Total energy of the system vs. time (2% compression – *hard piston*)

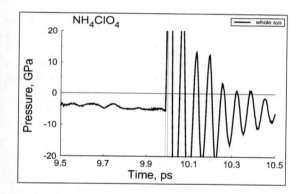

FIGURE 3. Pressure vs. time (2% compression – *hard piston*)

Fig. 1 for run Ta972 (a 2.8% compression by a *leaking piston* over 10 fs) shows how the average velocities of the individual species and of the whole system are affected by the shock. Sharp spikes in the velocity of N and Cl settle down rapidly to new steady values which are higher than before the shock. Slower growth in the velocities of H and O confirms that z compression sets the NH_4^+ and ClO_4^- molecular ions vibrating and that part of this lattice vibrational energy is then transformed into vibrational energy of the N-H and Cl-O bonds. The average velocities of all species then increase,

slowly at first and then at a rapidly accelerating rate until explosion occurs 8.9 ps after the shock. The average velocity increases so rapidly at the point of detonation that it is difficult to give a precise value for the detonation velocity. In this run it is about 4 km/s, which is close to the experimental values of Seely *et al.* [5], but in several others it appears to be more like 6 or 6.5 km/s. Our data are also consistent with those reported in [6].

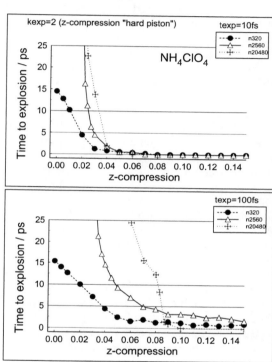

FIGURE 4. Time to explosion for *mechanically* induced shock in system of different size

Figs. 2 and 3 show the total energy and pressure as functions of time for a 2% compression by a *hard piston* over 10 fs. In this case detonation occurs 35.6 ps after the shock. The apparently negative pressure before the shock is an artifact caused by the fact that P is calculated as the difference between the kinetic energy and the *virial*, divided by the volume. The *virial* is not calculated exactly since it involves a long-range correction that is based on the assumption of a uniform distribution of atoms beyond the cutoff distance R_c, an approximation which makes the calculated *virial* too large. The fluctuations in the pressure are due to the kinetic energy contribution. The $P(t)$ plot shows that the effect of the shock is to set the NH_4^+ and ClO_4^- vibrating at a frequency of about 15 Thz. Large-scale potential energy plots show that this is an optic mode lattice vibration of the NH_4^+ and ClO_4^- molecular ions. Fig. 3 shows that the amplitude of these pressure oscillations decreases rapidly as lattice vibrational energy is transferred into vibrational energy of N-H and Cl-O bonds. At the end of the run a rapid increase in P occurs in which the pressure in the simulated system rises to about 50 GPa at detonation.

Figure 4 summarizes the effects of the size of the MD system and the duration of the shock (t_{exp}). As the magnitude of the shock increases the time to explosion decreases until it reaches a constant value, at about 5% compression for a 10 fs shock. With a shock of longer duration, it takes larger compressions to reach this saturation region because of the longer time available for the system to lose kinetic energy (which is transformed into configurational PE) during the shock. The initial post-shock steady temperatures are therefore lower (as observed) and the time to explosion longer for the same compression. Systems with $N = 320$ are more sensitive than corresponding systems with larger N, mainly because R_c is too small. (The maximum size for R_c is half the length of the box.) For small enough t_{exp}, the time to explosion does not change much for $N > 2560$, though the size of the system becomes more important for small compressions when the duration of the shock is longer.

CONCLUSIONS

It has been shown that by the use of advanced O(N) MD code detonation in relatively large ($N = 20,480$) and complex ionic systems (AP) can be simulated effectively on PC computers. The calculated pressure, temperature and particle velocity all rise suddenly (that is within a few fs) at detonation to very large values; that for the average particle velocity lies in the range 4-6 km/s while the pressure reaches 40-60 GPa. Detonations are induced by subjecting the system to compressive shocks either 'instantaneously' (within one time step of 0.5 fs) or at a constant rate over a fixed time (usually 10 or 100 fs). Studies of the dependence of the time to explosion on the size of the system and the magnitude, duration and type of the shock have led to the formulation of this model. A compressive shock excites an optic mode lattice vibration and the kinetic energy of the atoms is manifested as an increase in temperature. This kinetic energy is partially transformed into intramolecular vibrational energy of N-H and Cl-O bonds and lattice configurational potential energy. The post-shock system is not in an equilibrium configuration and so potential energy is transformed into kinetic energy, accompanied by a rise in temperature and the accumulation of sufficient vibrational energy in the N-H and Cl-O bonds to cause bond fission (shown by plots of projections of particle positions on planes normal to the x, y and z axes) and eventually, very high particle velocities and sufficiently large particle displacements that cause the disruption of the crystal lattice and the program to terminate.

ACKNOWLEDGEMENTS

The funding for this research was provided by Defence R&D Suffield under Contract PWGS W7702-1-R873/001/EDM. Presentation at this Conference was supported in part by a grant from the Swedish Foundation for International Cooperation in Research and Education (STINT).

REFERENCES

1. Sangster M.J.L. and Dixon M., "Interionic potentials in alkali halides and their use in simulation of the molten salts", *Adv.Phys* **25**, 247, 1976.
2. Holian B.L. "Modeling shockwave deformation via molecular dynamics", in *Shock Waves in Condensed Matter 1987. Proc.of the APS conference*, 185, 1988
3. Rycerz Z.A., "Molecular dynamics simulation program of order N for condensed matter systems", *Computer Phys. Commun.* **61**, 361, 1990
4. Rycerz Z.A. and Jacobs P.W.M., "A molecular dynamics algorithm of order N", *Grand Challenges in Supercomuting,Proc.SupercomutingSymposium'92*, Atmospheric Environment Science, Environment Canada, 241, 1992
5. Seely L.B., Blackburn J.H., Reese B.O. and Evans M.W., "Shock front reaction and initiation of detonation in low density ammonium perchlorate", *Comb. and Flame* **11**, 375, 1969.
6. Walker F.E., "A new kinetics and the simplicity of detonation", *Propellants, Explosives, Pyrotechnics*, **19**, 315, 1994.

CP706, *Shock Compression of Condensed Matter - 2003*
edited by M. D. Furnish, Y. M. Gupta, and J. W. Forbes
© 2004 American Institute of Physics 0-7354-0181-0/04/$22.00

DIRECT SIMULATION OF DETONATION PRODUCTS EQUATION OF STATE

M. Sam Shaw

Group T-14, MS B214, Los Alamos National Laboratory, Los Alamos NM 87545

Abstract. Theoretical calculations are made for the equilibrium detonation products equation of state using a Composite Monte Carlo method. In this study the method is extended to include small (~2nm) diamond clusters and to allow for surface chemistry of the groups that cap dangling bonds. An analytic representation of the Gibbs free energy of the cluster as a function of surface composition is incorporated in a fashion similar to that used for internal degrees of freedom for molecules. Bulk carbon phases are incorporated as analytic terms while the molecular fluid mixture is explicitly included in the Monte Carlo simulation. Starting from a very general partition function, equilibrium chemical composition results from a correlated interchange of atoms between species, whether fluid, bulk, cluster, or cluster surface group. Also allowed is fluid-fluid phase separation from terms that reduce to the Gibbs ensemble as a special case. Hugoniots and detonation velocities are determined from interpolation of tabulated equation of state results. Quantum calculations are in progress to better characterize the diamond clusters, surface groups, and their interactions with the molecular fluid.

INTRODUCTION

The products of a high explosive detonation form a very complicated mixture at extreme conditions of high pressure and temperature. In addition to a mix of molecular fluids, the solid carbon is known to form diamond clusters of ~ 1000 atoms of which around 30% are on the surface. Important issues in the calculation of the Equation of State (EOS) are non-ideal mixing, fluid-fluid phase separation, solid phase transitions, and especially chemical equilibrium composition, including the composition of the surface of the diamond clusters.

The Monte Carlo method [1] was developed 50 years ago to numerically simulate thermodynamic properties essentially exactly (except for statistical uncertainty) for a given potential interaction. Extensions of the method have been developed to treat special cases such as fluid-fluid [2] phase separation and chemical equilibrium composition

of fluids [3]. My recently developed Composite Monte Carlo Method [4] was designed to incorporate multiple special cases in a single simulation method. In addition, allowance was made for solid carbon phases, in chemical equilibrium with the fluid mixture, through a model Gibbs free energy. In this study the method is extended to include small (~2nm) diamond clusters and to allow for surface chemistry of the groups that cap dangling bonds.

THEORETICAL METHOD

The Composite Monte Carlo method [4] is derived in the NPT ensemble where the total number of atoms of each type remains constant and the independent variables are P and T. However, the partition function is written in a very general form that incorporates important options. For example, the ways of correlating atomic positions is included as a sum over allowed molecular compositions. Then chemical equilibrium enters the Monte Carlo simulation by discrete steps inside the summation leading to changes in molecular composition. Inclusion of solid phases was accomplished by separation of the partition function into a fluid part times a solid part. Since the symmetry constraints in a solid make it very difficult to change the number of particles by one inside a simulation, the solid term is collected into a factor inside the distribution of a form $e^{-\beta G_c(N_c,P,T)}$ where G_C is the Gibbs free energy of the solid carbon phase with N_C particles from an accurate analytic model.

An analytic model for the Gibbs free energy of a diamond cluster with surface groups acting as caps on the dangling bonds has previously been developed [5]. Characterization of the cluster begins with a diamond lattice with all atoms removed beyond a given radius. In the original method, the number of dangling bonds on each surface atom was counted. In this study, a surface group at an empty lattice site may also have multiple single bonds to all the nearest neighbor surface carbon atoms.

The cluster is divided into three parts, core carbon atoms with only carbon neighbor, surface carbon atoms, and surface groups such as H, OH, and O (including groups such as N that replace a surface carbon atom). The core is modeled as bulk diamond and the two surface parts are treated as additive units with an associated effective volume, estimated bond energies, and internal degrees of freedom from similar small molecules. When more than one type of surface group occupies a given type of site, there is a counting term contribution to the entropy similar to that found in ideal mixing. Further details are found in the references [3,4,5].

In order to avoid simulations involving thousands of particles, the clusters have to be included in the Monte Carlo method in a fashion requiring a relatively small number of particles. The key here is that the cluster model is additive in its contribution to the Gibbs free energy (and hence additive terms in the exponent of the probability distribution). For a fixed size cluster, this means setting the relative numbers of core carbon atoms, surface carbon atoms, and surface group sites. In the simulation, all of the total number of each type

have to have integer values since we are dealing with atoms. In principle, a set of rational ratios could be preserved, but the number of particles involved would be too large to maintain reasonable acceptance probabilities for such moves.

Instead, the ratios are constrained to be near the desired value, but with enough room for reactions involving a small number of particles to occur. In Table 1, a set of reactions are illustrated, where C* indicates a core carbon atom, CH* indicates a surface C and an H bound by a single bond, and CO* indicates a surface C and an O bound by a double bond. For this case, no dangling bonds are allowed, so a carbon atom has to be added to the surface along with a new surface group. The notation for the table is that of the change in number of molecules or group of a given type as

TABLE 1. Change in numbers of particles with reaction type.

	N₂	CO	CO₂	H₂O	C*	CH*	CO
1		-2	+1		+1		
2		-1	+1	-1	-2	+2	
3		-1					+1
4			+1	-1	-2	+2	-1
5		-1			-1		+2
6	+1		-1		-1		+1

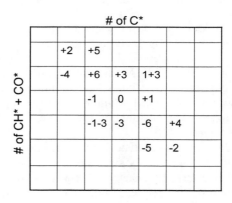

FIGURE 1. Net changes in the number surface groups and core carbon atoms with reaction steps in the Monte Carlo simulation.

the result of a given reaction. For example, the first reaction is given by

$$2CO \rightarrow CO_2 + C* \qquad (1)$$

and the reverse reaction has the opposite signs on each entry.

The steps in chemical reaction are illustrated in Fig. 2, where 0 stands for the starting point and the result of a particular reaction is denoted by the reaction number (with a minus sign for the reverse reaction). The ratio of surface groups to core carbon atoms is preserved by a line of constant slope. Allowing moves to states within one or two of the line will keep the ratio nearly constant, but give enough room for sampling all allowed states by a series of reaction steps.

The volume occupied by the surface groups has a large effect on the PV term in the Gibbs free energy because the pressures in detonation products are very high. Some of the important issues are illustrated in Fig. 3. The carbon volume can be estimated to have the same volume as in bulk diamond at the given P and T.

total volume of the cluster is determined by the effective surface. (Note that a 1 A shift in a 10 A radius cluster makes over 30% difference in the total volume of the cluster.)

Monte Carlo calculations were made on a model system to estimate the location of the surface. A system of N_2 molecules with an exponential-six spherical potential forms the background fluid and one N_2 interacts with the rest through just the exponential part. It mocks up a cluster by setting the interaction distance between the "cluster" and molecules to be the distance from the surface of a sphere of radius R_c (centered on the cluster) and the molecule. The change in volume with $(V+V*)$ and without (V) the cluster at P=30GPa and T=3000 K determines the radius R_c+R* that gives a sphere equal in volume to the change with and without the cluster. Then $R*$ is the effective offset of the radius from the geometric radius R_c.

The value of $R*$ is nearly constant for various values of R_C and is shown in Fig. 4, along with the exponential-six potential (chain dash) and exponential only (dash). Note that $R*$ is well separated from the potential at 3000 K.

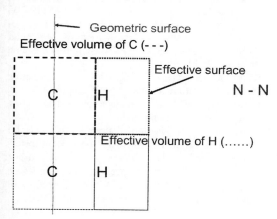

Geometric surface
Effective volume of C (- - -)

Effective surface

N - N

Effective volume of H (......)

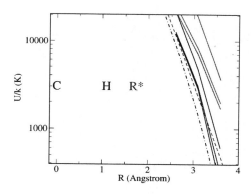

Figure 2. Illustration of the volume occupied by a H atom on the surface.

Figure 3. Location of $R*$ relative to potentials, see text.

The effective surface of the cluster depends on its interaction with the molecular fluid, denoted by the N_2 molecule at approximately the location of its closest approach to the surface. In the additive term treatment, the volume in between is associated with the surface H atom such that the

The solid lines are quantum calculations of the interaction of a N_2 molecule with a 111 layer of diamond for several configurations. The MondoSCF code[6] was used with a 6-31G** basis set and the BLYPxc density functional. For these preliminary calculations, only two layers in the diamond are included and H atoms cap the dangling bonds on both sides. Periodic boundary conditions are applied in two directions. The blue lines are for the N_2 parallel to the surface above three different positions on the surface. The green lines are the same except the N_2 is perpendicular to the surface and the radius is that to the center of mass. Further calculations are in progress to characterize other parts of the carbon cluster model with surface groups. This will allow a more reasonable choice of parameters.

Preliminary Monte Carlo calculations were made to illustrate the effect of diamond clusters with surface chemistry. A large set of EOS points were calculated for detonation products of HMX. By interpolation, the Hugoniot and CJ state were found as a function of initial density. The size of the carbon cluster has an effect similar in magnitude to that found from variations in cross-potentials. Removing the constraint of a fixed ratio of surface groups to carbon atoms, the equilibrium ratio (for a given set of potentials and parameters) that corresponds to a given cluster size typically tends to the large cluster limit. In some cases, overdriven data does not show a diffusion-limited transient. This could possibly be due to a shift in equilibrium to the small cluster limit.

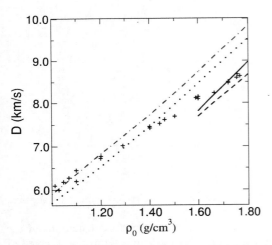

Figure 4. Detonation velocity versus initial density in HMX and RDX, data (+). Monte Carlo results for a typical diamond cluster ~2 nm (line) and for a much smaller cluster (dash). Previous bulk graphite results (chain dash, dot) for two different sets of cross-potentials.

CONCLUSIONS

With this extension of the Monte Carlo method to include clusters with surface chemistry, we have a thermodynamics tool that can accurately handle all of the major cases needed for detonation products in a single method. The focus is now on improving the potential interactions. A much more thorough set of quantum calculations needs to be made to guide the choice of parameters in the carbon cluster model.

ACKNOWLEDGEMENTS

The author thanks Dr. C. J. Tymczak for valuable discussions on quantum methods and how to use the MondoSCF code.

REFERENCES

1. Metropolis, N., Rosenbluth, A. W., Teller, E. H., and Teller, E., J. Chem. Phys., 21, 1087-1092 (1953).
2. Panagiotopoulos, A. Z., Quirke, N., Stapleton, M., and Tildesley, D. J., Mol. Phys.63, 527-545 (1988).
3. Shaw, M. S., "Monte Carlo Simulation of Equilibrium Chemical Composition of Molecular Fluid Mixtures in the $N_{atoms}PT$ Ensemble," J. Chem. Phys. 94, 7550 (1991).
4. Shaw, M. S., "Direct Simulation of Detonation Products Equation of State by a Composite Monte Carlo Method", 12th International Detonation Symposium, to be published.
5. Shaw, M. S., "A Theoretical Equation of State for Detonation Products with Chemical Equilibrium Composition of the Surface of Small Carbon Clusters," in Shock Compression of Condensed Matter - 1999, edited by M. D. Furnish et al., AIP Conference Proceedings 505, New York, 2000, pp. 235-238.
6. Matt Challacombe, Eric Schwegler, C.J. Tymczak, Chee Kwan Gan, Karoly Nemeth, Anders M.N. Niklasson, Hugh Nymeyer, and Graeme Henkleman, "MondoSCF, A Program suite for massively parallel, linear scaling SCF theory and ab initio molecular dynamics." Los Alamos National Laboratory (LA-CC 01-2)

CP706, *Shock Compression of Condensed Matter - 2003*
edited by M. D. Furnish, Y. M. Gupta, and J. W. Forbes
© 2004 American Institute of Physics 0-7354-0181-0/04/$22.00

MOLECULAR DYNAMICS SIMULATION OF SHOCK INDUCED DETONATION

Vikas Tomar and Min Zhou

The George W. Woodruff School of Mechanical Engineering
Georgia Institute of Technology, Atlanta, GA 30332-0405

Abstract. This research focuses on molecular dynamics (MD) simulation of shock induced detonation in Fe_2O_3+Al thermite mixtures. A MD model is developed to simulate non-equilibrium stress-induced reactions. The focus is on establishing a criterion for reaction initiation, energy content and rate of energy release as functions of mixture and reinforcement characteristics. A cluster functional potential is proposed for this purpose. The potential uses the electronegativity equalization to account for changes in the charge of different species according to local environment. Parameters in the potential are derived to fit to the properties of Fe, Al, Fe_2O_3, and Al_2O_3. NPT MD simulations are carried out to qualitatively check the energetics of the forward (Fe_2O_3+Al) as well as backward (Al_2O_3+Fe) thermite reactions. The results show that the potential can account for the energetics of thermite reactions.

INTRODUCTION

Stress-induced initiation of chemical reactions in reactive metal/oxide powders (RMPs) are dominated by effects that have a multi-scale character. However, it is important to characterize the influence of nanoscale effects, which determine the energy release rate and extent of reaction in RMPs. In this context, molecular dynamics (MD) can be used as an important tool for framing local initiation criteria that are useful in performing meso-scopic and macro-scopic calculations at length scales ranging from hundreds of nanometers to millimeters. Accordingly, a MD model for nanoscale analyses of stress-induced reactions in metal/oxide thermite mixtures is presented. The system under study is a Fe_2O_3+Al thermite mixture. The model is able to describe the multi-component behavior of the system as well as the qualitative energetics of the association and the dissociation during thermite reactions. The atomic level behavior is described by a cluster functional potential with variable charge electrostatic interactions. The charges of the species are calculated using the principle of electronegativity equalization[1]. Additionally, a screening length is introduced to delineate the charged and the neutral species. The parameters of potential are calculated by fitting it to the properties of Fe, Al, Fe_2O_3, and Al_2O_3. Since the focus is on the development, the modeling here involves the NPT MD simulations at 4 different temperatures for two different values of external pressure. The results show that the model is able to describe pressure-induced reaction of thermite mixtures.

MD MODEL

The required potential must not only be able to describe the behavior of individual components of the system, it must also be able to characterize the *intra-component* interactions with a good description of intermediate structures.

This entails the ability of the potential model to respond well to the environment an ion or an atom is in. With this in mind, we use a variable charge cluster functional potential for the description of our system. The total energy of the system is expressed as

$$E(\mathbf{r}, q) = E_{glue}(q) + E_{es}(\mathbf{r}, q), \qquad (1)$$

where E_{es} is the electrostatic energy in the form of

$$E_{es}(\mathbf{r}, q) = \sum_i q_i \chi_i + \frac{1}{2} \sum_{i(\neq j)} q_i q_j V_{ij}. \qquad (2)$$

with q_i being the charge of ith component, χ_i being the instantaneous electronegativity and V_{ij} being electrostatic pair interaction[1]. E_{glue} is the cluster energy and is given as

$$E_{glue}(\mathbf{r}) = \sum_i F_i(\rho_i) + \frac{1}{2} \sum_{i(\neq j)} \phi_{ij}. \qquad (3)$$

Here, F is the embedding energy, ρ_i is the atomic density at the ith site, and ϕ_{ij} is the pair interaction. The atomic density ρ has angular dependence besides the usual radial dependence and follows the functional form in [2]. Functional forms of the pair potentials are based on those in [3]. The calculation of the fluctuating charges is based on the approach in [1] using the principle of electronegativity equalization, see [4]. The final functional forms of the terms in potential are taken as;

$$\chi_i = \chi_i^0 + \sum_{j \neq i} Z_j \left(\int d^3 r \frac{f_j(r)}{|r - r_i|} - \int d^3 r_1 \int d^3 r_2 \frac{f_i(r_1) f_j(r_2)}{r_{12}} \right), \qquad (4)$$

where Z_i is the effective nuclear charge which must satisfy the condition $0 < Z_i < Z_i$, with Z_i being the total nuclear charge of the atom and

$$f_i(r) = \frac{\zeta_i^3}{\pi} \exp(-2\zeta_i r). \qquad (5)$$

Also in (2),

$$V_{ij} = \begin{cases} J_i^0 & i = j; \\ \int d^3 r_1 \int d^3 r_2 \dfrac{f_i(r_1) f_j(r_2)}{r_{12}} & i \neq j. \end{cases} \qquad (6)$$

It should be noted that $0 \leq q^{Al} \leq +3$, $0 \leq q^{Fe} \leq +3$, and $-2 \leq q^O \leq 0$. The form of the atomic density distribution is given as (see [2])

$$\rho = \sum_{h=0}^{3} t^{(h)} \left(\rho^{(h)} \right)^2 \qquad (7)$$

such that, $\rho^{(0)}$, $\rho^{(1)}$, $\rho^{(2)}$, and $\rho^{(3)}$ correspond to radial, cos, $\cos^2$ and $\cos^3$ dependent atomic densities. However, unlike [2] we avoid using a reference equilibrium structure which results in an equivalent crystal type model. The pair interaction functions are based on the work in [3], i.e.

$$\phi_{ij}(r) = \psi(r) - \psi(D_p) + \frac{D_p}{20} \left[1 - \left(\frac{r}{D_p} \right)^{20} \right] \psi'(D_p) \qquad (8)$$

so that interaction disappears at the cutoff value of D_p. The form for $\psi(r)$ is

$$\psi(r) = \psi_0 \left[\exp(-2\gamma(r-t)) \right] - 2\psi_0 \left[\exp(-\gamma(r-t)) \right]. \qquad (9)$$

The embedding function is given as

$$F(\rho) = E_0 A (\rho \ln(\rho))^B. \qquad (10)$$

Parameters in the potential are determined after fitting to the cohesive energy and elastic constants of Fe, Al, Fe_2O_3, and Al_2O_3. The determination involves the use of genetic algorithm based fitting to find the global optima followed by the Newton-Raphson method based minimization of the least

TABLE 1. Values of Monatomic Potential Parameters

	A	B	E_0 (eV)	r_0 (Å)	β_0 (Å⁻¹)	t_0	β_1 (Å⁻¹)	t_1	β_2 (Å⁻¹)	t_2	β_3 (Å⁻¹)	t_3	χ (eV)	J (eV)	ζ (Å⁻¹)	Z (eVÅ)
Al	2	1	3.3	-1.5	1.03	5	0.62	7.5	1.7	8.6	1.55	9.6	0	10.3	0.96	0.75
Fe	1.6	1	4.3	0.07	1.499	1	2.4	1.9	1.26	1.1	0.19	4.1	1.83	12.1	1.00	0.8
O	-0.7	1	2.6	-0.6	1.63	1.0	0.16	2.1	0.014	2.1	0.32	3.1	5.48	14.0	2.14	0.0

square residuals. Since Al and Fe are charge neutral, the glue potential parameters for them are determined first. This is followed by the determination of Fe-Al pair potential parameters. The glue and the electrostatic parameters for Oxygen and the electrostatic parameters for Al and Fe are then determined by fitting to Fe_2O_3, and Al_2O_3 properties. Table 1 shows the monatomic potential parameters for Al, Fe and O. The pair potential parameters for Al-Al, Al-Fe, Fe-Fe, Al-O, Fe-O and O-O are shown in Table 2.

NPT MD ANALYSES

In order to study the energetics of the Fe_2O_3+Al thermite reaction

$$2Al(s) + Fe_2O_3(s) \rightarrow Al_2O_3(l) + 2Fe(l)$$
$$\Delta H_R = -849 \text{ KJ/mol,} \quad (11)$$

NPT MD simulations are carried out at four different temperatures (300K, 600K, 900K, 1200K) for two different values of external pressure (10GPa, 20GPa). The Fe_2O_3+Al system consists of 8 primitive unit cells (80 atoms) of Fe_2O_3 surrounded by 960 Al atoms which form a confinement. Similarly, to study the possibility of the reverse reaction, the Al_2O_3+Fe system is analyzed under the same conditions. The Al_2O_3+Fe system consists of 560 atoms with 8 primitive unit cells (80 atoms) of Al_2O_3 surrounded by 480 Fe atoms. Long range electrostatics calculations based on Ewald summation method are carried out each time step. The NPT thermostat and barostat parameters, see [5], are chosen so as to have proper equilibration of energy and minimum box size oscillations. Figure 1 shows the morphology of both systems. A screening length is introduced to delineate the charged and the neutral species. This length is used to form a sphere around every oxygen atom. Any Fe or Al atom that falls within this sphere is considered charged and the charge is determined during the electronegativity equalization calculations. The atoms not counted during this

TABLE 2. Values of Pair Potential Parameters.

	Ψ (eV)	t (Å)	γ (Å⁻¹)	Cutoff (Å)
Al-Al	11.6	1.04	0.03	5
Al-O	243.18	6.01	-0.02	5
Al-Fe	-0.78	5.04	-3.59	5
Fe-Fe	1.53	0.94	0.087	5
O-O	68.8	-14.9	0.177	5
Fe-O	45.7	0.77	-0.04036	5

process are considered neutral. This parameter is useful since the system has both charged and neutral atoms. During the calculation, the screening length is kept at 4 Å for the Fe_2O_3+Al system and at 3 Å for the Al_2O_3+Fe system.

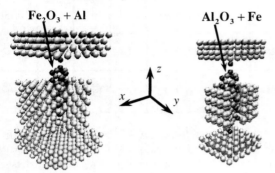

Fe$_2$O$_3$ + Al   **Al$_2$O$_3$ + Fe**

FIGURE 1. Molecular Dynamics Systems

These values correspond approximately to the equilibrium lattice constant of FCC Al and BCC Fe respectively. However, this value needs careful calibration with regard to the energetics of the reactions. During the NPT MD simulations, the total energy of the system just before the explosion is recorded and used as a qualitative indication of the energetic input needed to initiate the thermite reaction under given temperature and pressure. Figure 2 shows the energy per atom at different temperatures and the pressure of 10 GPa for both systems studied. As expected, the energy before explosion increases with temperature. Additionally,

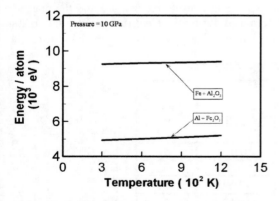

FIGURE 2. System Energetics Comparison

the energy before explosion for the reverse reaction is considerably more than that for the forward reaction. Similar result is obtained at the pressure of 20 GPa. This indicates that the current potential yields the desired attribute of the forward (Fe_2O_3+Al) reaction being energetically more favorable than the reverse (Al_2O_3+Fe) reaction. However, proper calibrations of the screening length and the energetics of the reactions are needed for accurate description of the thermite reactions. Additionally, the values of potential parameters also need to be improved through the use of better fitting procedures.

CONCLUSION

A MD model is established to simulate pressure-induced thermite reactions. A cluster functional potential is developed for this purpose. Initial results show that the model is capable of accounting for the reaction energetics. However, careful calculation and calibration of potential parameters is still needed.

ACKNOWLEDGEMENTS

Funding through an AFOSR MURI grant at Georgia Tech and a NaST fellowship from Georgia Tech is greatly appreciated. The authors thank Dr. Julian D. Gale for providing the GULP 1.3 code used in the calculations. Computations are carried out at ERDC and NAVO MSRCs

REFERENCES

Strietz, F.H., and Mintmire, J.W., *Phys Rev B* **50**(16), 11996-12003 1994.

Baskes, M. I., *Phys Rev B* **46**(5), 2727-2742 1992.

Besson, R., and Morillo, J., *Phys Rev B* **55**(1), 193-204 1997.

Rappe, A.K., and Goddard III, W.A., *J. Phys. Chem.* **95**, 3358-3363 1991.

Melchionna, S., Ciccotti, G., and Holian, B.L., *Mol. Phys.* **78**(3), 533-544 1993.

CP706, *Shock Compression of Condensed Matter - 2003*
edited by M. D. Furnish, Y. M. Gupta, and J. W. Forbes
© 2004 American Institute of Physics 0-7354-0181-0/04/$22.00

PARTICLE-VELOCITY DEPENDENT RATE CONSTANTS FROM TRANSITION-STATE THEORY

S. M. Valone[1]

[1] *Materials Science and Technology Division, Los Alamos National Laboratory, Los Alamos NM 87545*

Abstract. Shock loading activates complex kinetic processes, including detonations, deformations, fracture, spall, and phase transformations. Many practical rate models have succeeded as fitting forms; they have limited predictive capabilities. One route to greater predictability may be to revise contemporary theories of activated rate processes to depend explicitly on properties of the shock environment. The rate model presented here depends on the projection of the particle velocity along a reaction coordinate (or order parameter or slip system depending on the context) describing the transformation process. The model is able to fit both thermal- and shock-activated PETN and nitromethane (NM) barriers in a physically sensible way. The model behaves qualitatively correctly to account for the difference in two NM shock experiments. The model predicts hypersensitivity to dispersion in the particle velocity distribution as it projects onto a reaction coordinate.

INTRODUCTION

Greater predictive capabilities in modeling shock activated kinetics is needed a number of areas, including detonations, hot spots, spall, and phase transformations. Shock-activated kinetics has been treated less rigorously that other aspects of shock dynamics in terms of starting from based physical principles and deriving the consequences. The most rigorous treatment to date is the starvation kinetics (SK) model of Eyring [1], which has been used and extended by Tarver and coworkers [2]. The SK model recognizes that there are other degrees of freedom at the transition state that can siphon energy from reactants, thereby reducing the reaction rate. Starvation kinetics is discussed most frequently in the context of the nonequilibrium Zel'dovich-von Neumann-Doring (NEZND) kinetics model [3]. NEZND recognizes that the shock environment introduces an essential element of nonequilibrium into the kinetics environment. The nonequilibrium element has not been formally introduced into the derivation of the kinetics models.

Moreover, there are a number of curious observations that are modeled less than satisfactorily. Two examples are the experimental rate constants for pentaerythritol (PETN) [4] and nitromethane (NM) [5] under thermal and shock activation (Fig. 1). The shock-activated values of both prefactor and activation energy are lower than their respective thermally-activated values. The activation energies appear to be dependent on properties of the shock environment. Such dependencies do not appear in existing rate models. In general, it appears that rarely can thermally derived rate constants be used in a shock environment. A third example, numerical in nature, appears in the work of Yano and Horie [6]. They model the shock-activated mesoscale α-ε phase transformation of polycrystalline iron. The transformation is modeled as a first-order process:

$$d\Lambda/dt = (\Lambda_{eq}-\Lambda)/\tau_0 , \qquad (1)$$

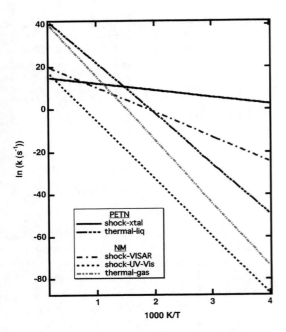

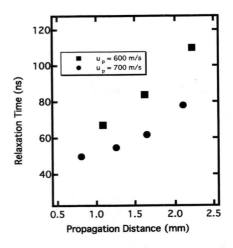

Figure 2. Particle-velocity dependence in the phase transformation relaxation time (Redacted from Figure 11 of Ref. 6 with permission from Elsevier).

Figure. 1. Arrhenius plots for PETN [4] and NM [5].

where Λ and Λ_{eq} are the dynamic and equilibrium transformation progress variables, respectively. The variable τ_0 is a characteristic time for the transformation and the inverse of the rate constant. Their Fig. 11 is reproduced in Fig. 2. It very clearly shows a particle-velocity dependence in τ_0 without explicitly assuming such a dependence in their simulation technique. Other examples appear in the work of Germann and coworkers [7] in the simulating anisotropy in the phase transformation of single crystals. Finally, the differences in kinetic rates of liquid and solid RDX are poorly understood at this time.

In this paper, the implicit nonequilibrium assumption of NEZND is made explicitly in the nonequilibrium Hamiltonian used in the derivation of transition-state theory (TST) rate constants [8,9]. TST rates are among the simplest to derive. Yet, the simple nonequilibrium effect explored here exhibits anisotropy, saturation, and extreme sensitivity to particle velocity and its dispersion. The presence of these behaviors in the aforementioned examples is discussed after a derivation of the rate model.

TRANSITION-STATE THEORY UNDER SHOCK CONDITIONS

The concept discussed here applies to first-order rate processes as expressed in Eq. (1). In the context of explosives, the focus is most naturally on the first stage of the four-stage NEZND model of Tarver [10]. The first stage is immediately behind the shock front. The notation that the system is most strongly out of equilibrium at this point is adopted for the general case.

Starting from the phenomenological Eq. (1), one can transform this relation into one for $k_0 = 1/\tau_0$, the rate constant. A clear example of how this is done can be found in Berne [11]. A series of approximations leads to the TST approximation, $k_0 \sim k_{TST}$, which is a short-time, low-friction approximation to the rate. The classical result in an arbitrary number of spatial dimensions is

$$k_{TST} = \frac{\int du\,dq\,\delta\big(f\big(\mathbf{q},\mathbf{q}^{\ddagger}\big)\big)\nabla f \quad \mathbf{u}\theta(\nabla f \quad \mathbf{u})\,exp(-\beta H(\mathbf{u},\mathbf{q}))}{\int du\,dq\,\theta(-f)\,exp(-\beta H(\mathbf{u},\mathbf{q}))},$$

(2)

where $\mathbf{u}$ and $\mathbf{q}$ are the vectors of velocity and space coordinates, $\beta = 1/(k_B T)$, T is the temperature, k_B is the Boltzmann constant, H is the Hamiltonian, and $f(\mathbf{q},\mathbf{q}^{\ddagger})$ is the flux dividing surface at the transition state (TS). This expression obviates the concept of projection onto the reaction coordinate, which is $\mathbf{q}^{\ddagger}$ at TS.

In an equilibrium setting the velocity distribution is Maxwellian, so that H depends quadratically on $\mathbf{u}$ and is centered on zero. If the potential surface is also quadratic at the reactant well, all of the integrals can be evaluated and one derives a multi-dimensional form of TST. The discussion here is restricted to a 1D form of Eq. (2), whereby

$$k_{TST} \sim \omega/2\pi \exp(-\beta E_a) ,\tag{3}$$

where ω is the frequency in the reactant well along the reaction coordinate, q, and E_a is the free energy of activation.

Under shock loading conditions and immediately behind the shock front, assuming a Maxwellian velocity distribution seems inappropriate. In the reactant well, the center of the velocity distribution is $u_q = u_p \cos \theta$, where u_p is the particle velocity, and θ is the angle between the reaction coordinate q and the direction of shock propagation. In this simple 1D picture, only the component of u_p aligned with q can contribute to the shock activation of the rate process.

TST assumes that there is no friction operating. The reactants are free to propagate toward the TS as a result of the shock loading. Effectively then, the barrier is reduced by the projected shock kinetic energy, $1/2\ m\ u_q^2$, where m is some effective mass of the reactant entities. If $1/2\ m\ u_q^2 > E_a$, then, at the transition state, there will be excess velocity to contribute to the flux at the TS dividing surface. It is worth noting that in several of his early works, Eyring accounted for mechanical loading.

Evaluating Eq. (2) under these nonequilibrium conditions one finds

$$k_{STST} = k_{TST}\exp\left(1/2\,\beta m u_q^2\right)\tag{4}$$

for $1/2\ m\ u_q^2 < E_a$, and

$$k_{STST} = k_{TST}\exp\left(\beta E_a\right)$$
$$\left(\exp\left(-1/2\beta m u_q^{\ddagger 2}\right) + u_q^{\ddagger}(1/2)\sqrt{2\pi\beta m}\ \text{erfc}\left(-\sqrt{\beta m u_q^{\ddagger 2}/2}\right)\right),\tag{5}$$

where

$$u_q^{\ddagger} = \sqrt{u_q^2 - 2E_a/m}\ .\tag{6}$$

When there is excess kinetic energy, the rate constant has a thermal and a convective contribution. At high excess kinetics energies, the thermal contribution damps out and the convective contribution continues to increase, but only linearly in the excess projected particle velocity.

Testing the assumptions of this derivation is done through dynamical simulations in a 1D system with the trajectories starting in the reactant well and a barrier separating the reactant well from a product region. The product region has no repulsive wall, which eliminates any reverse reactions. The distribution of initial velocities is sampled from a Maxwell distribution centered on u_q. The reaction rate is proportional to the flux of reactants crossing a dividing surface placed at the top of the barrier, divided by the number of trajectories. By starting the trajectories in the reactant well, we are testing the assumption of the initial transient being that of moving the reactants toward the barrier, as well as testing the TST assumption that the dynamics at the TS determines the rate. As in conventional TST, no frictional effects are considered.

For this purpose, a simple 1D potential modeled after the Hénon-Heiles potential is used. The barrier E_a is set to 0.7 eV and m is chosen to be 12 AMU. Simulations are performed at both 300 K and 3000 K. Figure 3 shows that the model tracks the simulations qualitatively correctly. There is a rapid rise in the rate as a function of u_q, but then that rise saturates because all of the reactants get promoted to the TS by shock activation. Finally, the isotropic average of Eq. (4) shows an effective rate drop of several orders of magnitude at its extremes.

DISCUSSION

The model represented by Eq. (4) can account for many of the observations discussed in the Introduction. First, the reduction in activation energy in both PETN and NM under shock versus the thermal activation conditions makes sense. The

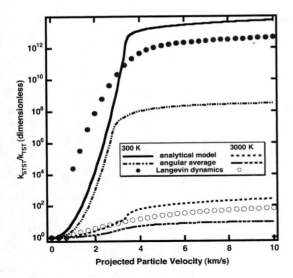

Figure 3. Numerical simulation of shock activated rate, relative to thermal values. The lines correspond to Eq. (4). The marker-curves correspond to numerical simulations of Newton's equations using the model potential described in the text. The isotropic average of Eq. (4) over θ is also shown.

effective u_q for PETN is 1.0 km/s. The NM data consists of two sets of shock experiments. The VISAR experiments employ single shock conditions and indicate strong shock activation. The UV-VIS experiments employ multiple shocks and give an activation energy nearly identical to the thermal value.

The prefactors are more problematic. Shock activation lowers the prefactor in PETN and NM. Eq. (4) has no mechanism for lowering the prefactor. There are three possibilities. One is that friction plays a role. A version of Eq. (4) with friction can be found in Ref. (7). If that model applies, it implies that the frictional effects are increasing as the result of the shock compression. The second possibility is that recrossing events are reducing the prefactor. TST ignores recrossing, but a whole literature on the corrections exists. The third is that TST is inappropriate, and some other kinetics model, such as the low-friction Kramers limit [12], is more appropriate.

The model dependence on particle velocity is consistent with the findings of Yano and Horie [6]. The dependence of phase transformation rates on the alignment of the shock loading direction with

the transformation reaction coordinate, as seen in molecular dynamics simulations [7], is captured in the model.

CONCLUSIONS

A rigorously derived shock activation kinetics model is derived from transition state theory under the assumption of a nonequilibrium Hamiltonian. It exhibits a number of interesting features. The model is anisotropic because it captures the alignment of the shock loading direction with the reaction coordinate. Subjecting the model to an orientational averaging process reduces the rates relative to their rate at optimum alignment.

ACKNOWLEDGEMENTS

This work was performed under the auspices of the US Department of Energy at the Los Alamos National Laboratory under contract W-7405-ENG-36 to the University of California.

REFERENCES

1. Stearn, A. E., and Eyring, H., *Chem. Rev.* **29**, 509 (1941); Eyring, H., Powell, R. E., Duffey, G. H., and Parlin, R. B., *Chem. Rev.* **45**, 69 (1948).
2. Tarver, C. M., Breithaupt, R. D., and Kury, J. W., *J. Appl. Phys.* **81**, 7193 (1997).
3. Zel'dovich, Y. B., *J. Exptl Theor. Phys. (U.S.S.R.),* 10, 524 (1940); von Neumann, J., O.S.R.D. Report No. 549, 1942; Doring, W., *Am. Physik,* **43**, 421 (1943).
4. Dick, J. J., *J. Appl. Phys.* **81**, 601 (1997); Yoo, C. S., Holmes, N. C., Souers, P. C., Wu, J., Ree, F. H., and Dick, J. J., *J. App. Phys.* **88**, 70 (2000).
5. Dick, J. J., in *Shock Compression of Matter - 1985,* edited by Y. M. Gupta, (Plenum, 1986), p. 903; Winey, J. M., and Gupta, Y. M., *J. Phys. Chem.* **101**, 9333 (1997).
6. Yano K. and Horie Y., *Inter. J. Plasticity* **18**, 1427 (2002).
7. Kadau, K., Germann, T. C., Lomdahl, P. S., and Holian, B. L., *Science* **296**, 1681 (2002).
8. Eyring, H., *J. Chem. Phys.* **3**, 107 (1935).
9. Valone, S. M., *J. Chem. Phys.* **118**, 6289 (2003).
10. Tarver, C. M., *Combust. Flame* **46**, 111 (1982)
11. Berne, B. J., in *Activated Barrier Crossing* edited by P. J. Hänggi and G. R. Fleming, World Scientific, 1993.
12. Kramers, H. A., *Physica* **7**, 284 (1940).

CP706, *Shock Compression of Condensed Matter - 2003*
edited by M. D. Furnish, Y. M. Gupta, and J. W. Forbes
© 2004 American Institute of Physics 0-7354-0181-0/04/$22.00

A FAST THREE-DIMENSIONAL LIGHTING TIME ALGORITHM

Jin Yao

Scientific B-Division, Lawrence Livermore National Laboratory, Livermore, CA 94550

Abstract. A narrow band level-set method to calculate the physical lighting time in three dimensions has been implemented with arbitrary hexahedral element systems. This method does not involve finite difference calculation of spatial derivatives. Mesh regularity, and the local topologic equivalence to a regular mesh are not required. The lighting surface is represented by a collection of curved facets contained in partially burnt cells. Level-set functions are calculated by direct measurement of distance to the lighting surface, and are carried only for nodes of a set of elements in a narrow band that covers the lighting surface. In the case of a concave boundary, the nodal distances are calculated with geodesics. A polynomial fitting of nodal level-set values across elements provides third-order spatial accuracy where the surface is sufficiently smooth. The DSD (Detonation Shock Dynamics) level-set equation is integrated directly in the normal direction. The new algorithm is self-initialized and allows easy boundary treatment with body fitting meshes.

INTRODUCTION

Existing level-set lighting time algorithms for detonation propagation [1] [2] usually require mesh regularity, external initialization, and integration on all data points at every time step. These constraints make these algorithms inefficient with nontrivial boundary treatment, and of lower accuracy. The method presented here is efficient, self-starting, of higher accuracy, and easily capable of treating general boundaries.

NARROWBAND ALGORITHM

The new method integrates the DSD level-set equation only for data points in a narrowband at a time step. The cost of numerical integration is one dimension lower than a full method. A narrowband method in two-dimensions can be easily implemented [3] [4]. In comparison, a three-dimensional narrowband level-set method seems to require a significantly simplified approach because of the nontrivial connectivity of surfaces. The work described in this paper is the first complete description of a simple three-dimensional narrowband level-set method and its numerical implementation.

Body-fitting Mesh

There is no requirement on mesh regularity with the new method. Without loss of generality, a finite element system is assumed to represent the numerical problem. The external boundaries and the material interfaces are assumed represented with cell faces. This is not required for the new method to work but by taking advantage of body-fitting meshes, the boundary treatment becomes trivial.

Elementary Approach

The new method is essentially an elementary approach based on straightforward distance calculations and least squares fitting. An outline of the procedures is as follows

1) Determine for each partially burnt cell a 'facet' using nodal level-set values, thereby constructing the lighting surface.

2) Calculate the signed normal distances to the surface for the nodes in a narrowband.

3) Integrate the DSD level-set equation in the normal direction for each narrowband node.

Steps 1), 2), and 3) are repeated until all HE regions are burnt.

Self-Initialization

To start a full level-set calculation requires having the signed minimum distances from the entire set of data points to the initial burn surface. This is logically contradictory in the sense that initialization of such a problem is essentially equivalent to solving the entire problem. In the case of complex boundaries, initialization with signed minimum distances is nontrivial. In contrast, the new method only requires the calculation of signed minimum distances for points in a narrowband around the lighting surface at any given time. Since the narrowband data is sufficient to derive the solution, the new method is self-initialized, once the initial lighting surface is specified. The initial burn surface is assumed specified by the signed distances to the surface from the nodes of partially burnt cells.

Surface Representation

With the new method, the burn surfaces are represented with a collection of 'facets', which are contained in partially burnt cells. A facet is the portion of lighting surface cut off by a partially lighted cell. This concept can efficiently be used for general front capturing problems. Complexity of surface connectivity in three dimensions can then be handled with ease.

Level of Neighbors

Level of neighbors [5] is a powerful way to loop over a set of items that has connectivity of neighbor relations. It is an efficient approach for managing the facets on the lighting surface, as demonstrated in Figure 1.

TABLE 1. The level of neighbors.

3	3	3	3	3	3	3
3	2	2	2	2	2	3
3	2	1	1	1	2	3
3	2	1	0	1	2	3
3	2	1	1	1	2	3
3	2	2	2	2	2	3
3	3	3	3	3	3	3

Narrowband

The narrowband is defined as the set of all the surface nodes (which define all the partially lighted cells) and the nodes of the first level of unburned neighbor cells of partially burnt cells. Among these nodes, the surface nodes are used to determine the burn surface at the current time with their level-set values. The rest will have their level-set values calculated at the current time step, and are going to define the burn surface at later time steps as they have become new surface nodes. The definition of narrowband here makes it the thinnest possible thus the method is optimized.

Characteristic length

Any node in the narrowband cannot possibly have a distance to the region of partially burnt cells greater than the maximum cell dimension (the diameter of a smallest sphere that may contain any cell). The maximum cell dimension is the characteristic length with this method.

Facet

The vertices of a facet are first determined with a linear interpolation of nodal level-set values on the edges of a given partially burnt cell. Such an approximation can exactly propagate a planar wave. We fit the facet with a plane to define a local Cartesian coordinate for an initial nodal distance calculation. This facet-coordinate has its origin at the vertical projection of a node along an axis that coincides with the facet normal.

Nodal Distance

The nodal distance is defined as the shortest path from a node to a facet. The distance from a given narrow band node to the surface is the minimum among the distances measured from the node to all facets. The facet corresponding to the shortest path will be used to refine the initial nodal distance calculation in its facet-coordinate.

Region of Influence

The region that contains all points with a distance less or equal to the characteristic length (the maximum cell dimension) to a given cell is defined as the "region of influence" of this cell. For a given facet, we only need to calculate for the nodes inside the region of influence and inside the narrowband. Since the information for these nodes is sufficient to propagate the surface, all of the other nodes can be eliminated from the calculations of the nodal distance to this *facet*.

Polynomial Surface Fitting

To more accurately calculate the *nodal distance*, on the smooth portion of the burn surface, one can fit the surface with a quadratic polynomial of two variables in the *facet-coordinate* when curvature is small, using known nodal level-set values. This gives the third order of accuracy.

Least Squares Fitting and Finite Difference

The polynomial fitting is done by selecting a set of surface nodes in the region of influence of a facet and by fitting their level-set values with a least squares method. It is worth noting that general finite difference methods are strictly equivalent to special cases of the least squares method as used here[5]. Compared to finite difference methods, the new algorithm is much more flexible.

Curvature

The integration of the DSD level-set equation requires an explicit curvature calculation. This calculation is tedious, even on regular meshes with finite difference methods. A second order method does not provide the required Taylor expansion terms to obtain curvature. Thus a second order method that claims to have curvature effect is not self-consistent. The third order of accuracy of the new method is necessary and sufficient for the curvature term to be included in the calculation. The calculation of curvature is trivial with the quadratic surface fitting polynomial.

Integrating on the Normal

The DSD level-set equation is expressed as

$$\frac{\partial \varphi}{\partial t} + D_n(\kappa) \,|\, \nabla \varphi \,|= 0,$$

for a quasi-steady detonation, here φ is the level-set function, D_n is the normal detonation speed and κ is the curvature. With a finite difference method, the curvature and the norm of gradient cannot be calculated in simple ways. However, on the surface normal, the DSD level-set equation reduces to

$$\frac{d\varphi}{dt} + D_n(\kappa) = 0.$$

This is much easier to solve using the new method since the surface normal vector and curvature of the burn surface are naturally obtained. The integration of the DSD level-set equation can be done trivially in the normal direction. If the detonation velocity is a constant, one simply has $\phi(t + dt) = \phi(t) - D_n(dt)$, which is the same as a simple Huygens construction.

Time Step

The time step is determined by the minimum time it takes for the current surface to reach a non-surface node in the narrowband. This means the lighting surface at the next time step is completely determined by information in the current narrowband. A factor less than the unity can be used to further limit the time step.

Aspect Ratio

Naturally, the new algorithm can be performed on a regular mesh. It is most effective when the cells are cubic. For non-cubic cells, the time step of this method may be reduced when the detonation wave is traveling parallel to the short cell dimension. In addition, a curved burn front can light the center of a given cell without any of its nodes being lit. This situation may cause a few facets to be missing from the lighting surface. Although polynomial fitting of the surface tends to have the missing portion reconstructed, particularly thin cells should be avoided if possible.

Multiple Detonation Fronts

On the facet level, the case of multiple lighting surfaces and the case of single lighting surface are not treated differently with the new method. Let us consider the case of interactions between multiple lighting fronts in some detail. In theory, it is possible for a partially burnt cell to contain several pieces of lighting surfaces. It looks like a complicated situation. In practice, the solution is simple with the narrowband approach. We argue that a) a cell in which such a complex situation occurs is not going to be used to generate a facet for determining nodal distances of other narrowband nodes, and b) the nodal distances in this cell are determined in some previous time steps. The accuracy of the new method is therefore not affected by multiple front interactions.

Material Interfaces

A detonation front is assumed to intersect material interfaces with determined angles. With the assumption of body-fitting mesh, a facet in a boundary cell can be determined with the angle between the facet and the cell walls. This makes the boundary condition trivial to apply.

Geodesics

Since a given facet needs only to be considered within its region of influence, the new method can treat complex boundary geometry in an easy fashion.

When the regions of HE material have concave boundary portions, the path of minimum lighting time may not be determined by a straight line. To compute the shortest path between two points, one does a geodesic calculation on the plane that contains the principal surface normal and the points where the boundary is smooth. It can be easily shown that such an approximation is 3rd order accurate. If sharp concave edges are present in the region of influence, one may use a *plumb line* to determine the nodal distance.

Lighting Time on Cell Centers

Nodal average can give at most 2nd order of accuracy for the lighting time at cell centers. If computing cost is not a burden, one should treat cell centers as nodes and carry level-set values on them, as well. An alternative is to use high order MLS interpolation of nodal values.

NUMERICAL EXPERIMENT

A cubic of size *40* by *40* by *40* is ignited at corner (*20, 20, 20*) and (*-20, -20, -20*) at time zero. The lighting velocity is set to *1*. The initial burn surfaces are two 1/8 spheres with a radius of *10* thus about *1.64%* of the total volume is burnt when the calculation starts. *4* different cell sizes are tested. The non-dimensional nodal errors of lighting time compared to theoretical solution are shown in table 1 (the amounts inside parentheses are for narrowband nodal distances measured at the first time step).

It is easy to see the high accuracy and the nonlinear convergence as the cell size decreases.

TABLE 2. The nonlinear convergence.

CELL SIZE	4 x 4 x 4	2 x 2 x 2	1 x 1 x 1	2 x 2 x 1/2
L2 ERROR	.002158 (.002078)	.000118 (.000065)	.000028 (.000008)	.000198 (.000175)
MAX ERROR	.009588 (.009588)	.000410 (.000385)	.000088 (.000053)	.012225 (.002008)
TIME STEPS	9	18	35	64

CONCLUSIONS

The narrowband nature of the new three-dimensional lighting algorithm makes it a fast one. The concept of region of influence has simplified the full problem to a collection of simple geometry problems. Direct distance calculations and least squares fitting are basically all that are needed for the method to work. The high order of accuracy and the low order of computing cost by the new method can probably make it a preferred method to solve general surface propagation problems.

ACKNOWLEDGMENT

This work was performed under the auspices of the U.S. Department of Energy by University of California Lawrence Livermore National Laboratory under contract No. W-7405-Eng-48. The author wishes to thank Tom Adams and Richard Sharp of Lawrence Livermore National Laboratory for their encouragement and guidance.

REFERENCES

1. Aslam, T.D., J. B. Bdzil, and D. S. Stewart, Level Set Methods Applied to Modeling Detonation Shock Dynamics, JCP. 126, pp. 390-409, 1996.
2. Wen, S., C. Sun, F. Zhao, and J. Chen, The Level Set Method Applied to Three-dimensional Detonation Wave Propagation, The Twelfth International Det. Symp., San Diego, CA, 2002.
3. Adalsteinsson, D. and J. A. Sethian, A Fast Level Set Approach for Propagating Interfaces, JCP. 118, pp. 269-277, 1995.
4. Sethian, J. A., Level Set Methods and Fast Marching Methods, Cambridge University Press, pp. 80-85, 1999.
5. Yao, J., M. E. Gunger, and D. A. Matuska, Simulation of Detonation Problems with the MLS Grid-Free Methodology, The Twelfth International Det. Symp., San Diego, CA, 2000.

CP706, *Shock Compression of Condensed Matter - 2003*
edited by M. D. Furnish, Y. M. Gupta, and J. W. Forbes
© 2004 American Institute of Physics 0-7354-0181-0/04/$22.00

SIMULATING THE THERMAL RESPONSE OF HIGH EXPLOSIVES ON TIME SCALES OF DAYS TO MICROSECONDS

Jack J. Yoh* and Matthew A. McClelland*

Energetic Materials Center, Lawrence Livermore National Laboratory, Livermore CA 94551

Abstract. We present an overview of computational techniques for simulating the thermal cookoff of high explosives using a multi-physics hydrodynamics code, ALE3D. Recent improvements to the code have aided our computational capability in modeling the response of energetic materials systems exposed to extreme thermal environments, such as fires. We consider an idealized model process for a confined explosive involving the transition from slow heating to rapid deflagration in which the time scale changes from days to hundreds of microseconds. The heating stage involves thermal expansion and decomposition according to an Arrhenius kinetics model while a pressure-dependent burn model is employed during the explosive phase. We describe and demonstrate the numerical strategies employed to make the transition from slow to fast dynamics.

INTRODUCTION

In the DoD and DOE communities, computational models are being employed to an increasing extent to analyze the performance and safety of weapons systems. Hydrocodes have been used successfully to model the high frequency behavior in shocks and detonations. The multi-time scale behavior encountered in off-normal thermal (cookoff) events such as fires provides additional modeling challenges. For example, the Navy is interested in the behavior of munitions in shipboard fires to help with the design of storage systems and the development of fire fighting procedures. In these fires, time scales for behavior can range from days to microseconds. During the relatively slow heating phase, the response of an energetic materials system is paced by thermal diffusion and chemical decomposition while the mechanical response is essentially a quasi-static process. As the decomposition reactions accelerate, heat is generated faster than it can be removed. Product gases are formed and the resulting pressure rises accelerate the energetic and containment materials. The resulting violence can range from a pressure rupture to a detonation.

The accurate modeling and simulation of cookoff requires an understanding of the mechanical, thermal, and chemical behavior during slow heating and the subsequent explosion [1, 2, 3]. The explosion time and temperature have been successfully predicted using relatively simple thermal analysis codes [4]. However, the prediction of reaction violence requires detailed mechanical models throughout the cookoff event. In particular, dynamic gaps and the generation of thermal damage (porosity and cracks) during slow heating are likely to be important factors. These processes are typically orders of magnitude slower than those associated with the burn phase. Thus, to model a slow cookoff event will require computational tools and models that can handle a wide variety of physical processes and time scales.

In this paper, we describe how the cookoff of high explosives is simulated with the arbitrarily Lagrangian-Eulerian code [5], ALE3D. In particular, we discuss the numerical methodology to transition from slow to fast time scales. We apply our modeling capability to a Scaled Thermal Explosion Experiment (STEX) [6] and compare calculated and measured curves for the wall strain during both the heating and explosive phases of the test.

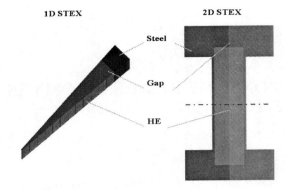

Steel

Gap

HE

FIGURE 1. Schematics of 1D and 2D STEX domains.

NUMERICAL MODELING OF THE STEX EXPERIMENT

Fig. 1 shows one and two-dimensional modeling domains for a STEX test involving PBXN-109 (64% RDX, 20% Al, 16% DOA/HTPB). A wedge slice is taken from the center line of the STEX system shown in the right image. This one-dimensional wedge represents an axisymmetric section of the STEX system in which variations occur only in the radial direction. The boundaries at planes of constant θ are rigid slip surfaces. In the actual experiment [6], the HE, nominally 5.08 cm diameter is encased in a 0.4 cm thick steel cylinder. The 5% ullage by volume is located at the outside radius of the HE in the 1D model. The gaps are treated in two different ways in the 2D models. In Model 2Da, a 4% gap by volume is included at the top end of the cylindrical charge, and a 1% gap is used at the outside radius of the HE. In Model 2Db, a 5% gap is included at the top of the cylinder, and there is no gap on the side. The cookoff simulation starts with an increase of the set-point temperature at the steel surface at 5 °C/h to 130 °C, followed by a hold for 5 h, and then an increase at a rate of 1 °C/h until cookoff. As the PBXN-109 is heated, it thermally expands to fill in the gap. At a temperature above 130 °C, exothermic decomposition begins and eventually ignition occurs near the midplane of the system. On a time scale of microseconds, the propagation of flame through the PBXN-109 causes the temperature and pressure to rise, and ultimately causes a break in confinement. In the simulations, three different mesh resolutions (1X, 2X, 4X) are considered. In the base case (1X), there are

12 elements across the HE in the radial direction, and in the fine mesh case there are 48 elements in this direction. The air/HE interface is not tracked explicitly, and zones with both air and HE have properties determined by mixing rules.

The equations of mass, momentum, energy, and chemistry are solved on the long time scale of the heating phase and on the short time scale of the thermal runaway phase in a single simulation. The momentum equation is integrated explicitly during both the slow and fast phases. In order to provide computationally feasible step sizes , the method of variable mass scaling [7] is applied during the slow heating phase. The density is increased in the momentum equation to reduce the sound speed and allow larger step sizes consistent with the Courant condition. However, if the time step size and material density are too large, spurious fluctuations, characteristic of a simple harmonic oscillator, appear. Thus, a tradeoff is required between numerical efficiency and accuracy. In practice, the time step size is fixed during the slow heating phase with the density calculated from the Courant condition.

During the transition phase in which the decomposition reactions are accelerating, the time step size is reduced to meet error specifications for the calculation of thermal and composition fields. At the same time, the artificial density is reduced following the Courant condition until the physical value is obtained. When the HE reaches a user-specified temperature, the Arrhenius kinetics expression are replaced by a burn model. A level-set method is used in the modeling of the advancing burn front.

We use the Backward Euler method for the integration of the thermal equations and reaction kinetics during the heating, and transition phases. During the slow heating phase, the time step size is the value selected for the integration of the hydrodynamic equations. A switch is made to an explicit method when the time step size is a user-specified multiple of the Courant time step size calculated with no mass scaling.

ONE-DIMENSIONAL RESULTS

Calculated results for the wall hoop strain are shown in Fig. 2 for the one-dimensional STEX test. For comparison, the theoretical thermal expansion of an

TABLE 1. Material properties used in the cookoff simulation. Shown are the density (ρ), coef. of thermal expansion (CTE), shear modulus (μ), and yield strength (Y_0).

	ρ (g/cm^3)	CTE (^{o}C^{-1})	μ (GPa)	Y_0 (GPa)
PBXN-109	1.67	1.21e-4	4.628e-3	0.06
Steel 4340	7.83	1.20e-5	77.0	1.03

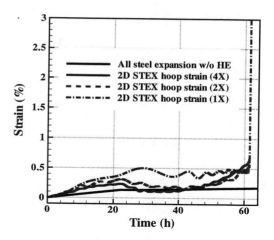

FIGURE 3. Calculated mechanical response of confined HE in a 2D STEX experiment.

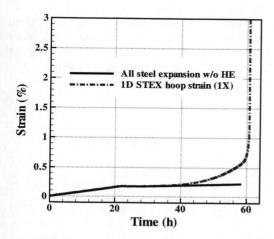

FIGURE 2. Simulated mechanical response of confined explosive in a 1D STEX experiment.

empty steel vessel is shown in the same plot. Note that these expansions are constrained to occur in the r-direction alone and cannot be calculated from the CTEs alone. As the temperature increases, the HE thermally expands inside the steel vessel at a rate approximately 10X greater than the steel vessel itself (see Table 1). For $t < 38$ hours, the 1D STEX results are in excellent agreement with the theoretical results for the empty vessel. After this point, decomposition gases pressurize the vessel, generating strain values larger than the empty vessel values.

The reaction rate accelerates to the point where the expansion of the STEX system is taking place on the scale of microseconds. During this acceleration, the three-step kinetics [4] is replaced by the burn model as discussed earlier and the calculation is continued to completion.

TWO-DIMENSIONAL RESULTS

The two dimensional simulations add the influence of variations in the z direction. This system is assumed to be axisymmetric, and a cylindrical wedge was selected for the calculation domain. In Fig. 3, calculated wall hoop strains for the STEX system are presented. As was done for the one-dimensional case, the theoretical expansion of the empty steel vessel is plotted together with the calculated strain results. Shown are the hoop strains for Model 2Da on meshes of three different resolutions. These results are generally higher than the results for pure steel, and show spurious oscillations from the method of mass scaling used in the integration of the momentum equation as described above. The results for the three meshes appear to be approaching a converged solution, but convergence has not yet been achieved. The results for Model 2Da, should match the empty vessel results until the 1% gap at the side of the HE cylinder closes at $t = 9.4$ h and a strain of 0.056%. It is seen that the Model 2Da curves are higher than the empty vessel curve during this period, indicating that numerical errors of the scale 0.1% remain. Although these absolute errors are small, the associated changes in HE pressure are large given the large elastic modulus for steel.

A rapid expansion of the vessel wall follows the slowing heating and ignition phases. In Fig. 4, wall

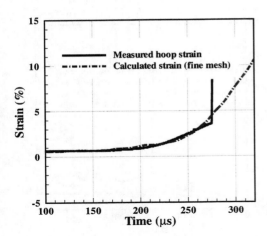

FIGURE 4. Comparison of measured and calculated hoop strains during a thermal runaway.

strain results for Model 2Db (no side gap) are compared with the measured hoop strain results for the STEX test. The model results compare favorably with the measured results until $t = 275\,\mu s$. At this time, the measured strain rate changes dramatically. It is likely, that the gauge failed at this point.

Although the model provides a good representation of the measurements, mesh refinement results need to be completed to establish the numerical accuracy of the calculations. In addition, the gap at the side of the HE needs to be added to provide a more complete model.

CONCLUSIONS

Progress has been made towards the modeling of violence in slow cookoff using the ALE3D code. Numerical procedures have been developed to model the thermal, mechanical, and chemical behavior during the slow heating, transition, and explosive phases in a single simulation. In this paper attention is focused on the accuracy of mechanical results for the simulation of a STEX test with PBXN-109. For the heating phase, an explicit hydro scheme with mass scaling provides numerically accurate results for wall strains in one-dimension and approximate results in two dimensions. For the rapid expansion, the model provides a good representation of mea-

sured wall strains. However, a better treatment of gaps is needed during the heating phase to confirm the numerical accuracy of these model results. An implicit hydro scheme with slide surfaces is being developed to provide improved accuracy for cookoff systems with gaps.

ACKNOWLEDGMENTS

Support for this work was provided by the DoD Office of Munitions through DoD/DOE Memorandum of Understanding. The work was performed under the auspices of the U.S. Department of Energy by the University of California, Lawrence Livermore National Laboratory under Contract No. W-7405-Eng-48.

REFERENCES

1. McClelland, M. A., Maienschein, J. L., Nichols, A. L., Wardell, J. F., Atwood, A. I., and Curran, P. O., "ALE3D Model Predictions and Materials Characterization for the Cookoff Response of PBXN-109," in *Proceedings of JANNAF 38th Combustion and 20th Propulsion Systems Hazards Subcommittee Meetings*, Destin, FL, 2002.
2. McClelland, M. A., Maienschein, J. L., Nichols, A. L., and Yoh, J. J., "Ignition and Initiation Phenomena: Cookoff Violence Prediction", UCRL-ID-103482-02, 2002.
3. Williams, F. A., *Combustion Theory*, Addison-Wesley, Redwood City, CA, 1985.
4. McGuire, R. R., and Tarver, C. M., "Chemical Decomposition Models for the Thermal Explosion of Confined HMX, TATB, RDX, and TNT Explosives," in *Proceedings of 7th International Detonation Symposium*, Annapolis, MD, Naval Surface Weapons Center, 1981, pp. 56–64.
5. Belytschko, T., "An Overview of Semidiscretization and Time integration Procedures," in *Computational Methods for Transient Analysis*, edited by T. Belytschko and T. J. R. Hughes, North-Holland, 1983, pp. 1–65.
6. Wardell, J. F., and Maienschein, J. L., "The Scaled Thermal Explosion Experiment," in *Proceedings of 12th International Detonation Symposium*, San Diego, CA, Office of Naval Research, 2002.
7. Prior, A. M., "Applications of Implicit and Explicit Finite Element Techniques to Metal Forming", *J. of Material Processing Technology*, **45**, pp. 649–656, 1994.

CP706, *Shock Compression of Condensed Matter - 2003*
edited by M. D. Furnish, Y. M. Gupta, and J. W. Forbes
© 2004 American Institute of Physics 0-7354-0181-0/04/$22.00

ELASTIC PROPERTIES OF MOLECULAR CRYSTALS USING DENSITY FUNCTIONAL CALCULATIONS

Jijun Zhao[1], J. M. Winey[1], Y. M. Gupta[1], and Warren Perger[2]

[1]*Institute for Shock Physics, Washington State University, Pullman, WA 99164*
[2]*Dept. of Physics and Dept. of Elect. Eng., Michigan Tech University, Houghton, MI 49931*

Abstract. The elastic properties of several molecular crystals (PE, PETN, urea) have been investigated using plane-wave pseudopotential methods based on density functional theory (DFT). The lattice constants, elastic constants and bulk modulus of the molecular crystals were calculated and compared with experiments. Two prevalent density functional methods, LDA and GGA-PBE were tested and compared with each other. We find that LDA typically overestimates the stiffness of these crystals at least by factor of two, while the GGA calculations with the PBE exchange-correlation functional are more reasonable. A large cutoff of the plane-wave basis and a modest sampling of **k** space are required to describe the molecular crystals. The elastic behavior of urea and PE crystals under uniaxial compressions show interesting anisotropic effects.

INTRODUCTION

The elastic properties of molecular crystals are important for understanding the intermolecular interaction and the high-pressure behavior of the crystals. The second-order elastic constants have been measured experimentally [1,2] for many molecular crystals and calculated theoretically using empirical potentials [3]. So far, there are only limited first principle studies on the ambient properties and high-pressure behavior of molecular crystals like urea [4,5,13]. In particular, the nature of hydrogen bonds and van der Waals forces in the intermolecular interactions of molecular crystals make the theoretical computations based on density functional methods more difficult. Moreover, shock compressions of molecular crystals along different orientations show anisotropic effects [6]. Thus, it is important to examine the elastic properties of molecular crystals under uniaxial compression using first principles methods. Meanwhile, the elastic constants of these crystals can be calculated as a benchmark. In this paper, we

report density functional calculations of molecular crystals at ambient conditions as well as under uniaxial and hydrostatic compression. Several molecular crystals with tetragonal lattice structures such as urea, pentaerythritol (PE), and pentaerythritol tetranitrate (PETN), are studied.

COMPUTATIONAL DETAILS

In this work, we perform density functional calculations of the molecular crystals based on the self-consistent field plane-wave technique [7] and a Troullier-Martin norm-conserving pseudopotential [8]. For the electron exchange-correlation interaction, both the local density approximation (LDA) in PW92 form [9] and the generalized gradient approximation in PBE form [10] have been used for a comparative study.

Within the constraint of the known space group, we simultaneously optimize the atomic coordinates and lattice constants of the molecular crystals. The crystals at both ambient conditions and under hydrostatic compression are optimized in this way. After the lattice constants at ambient conditions are determined, it is easy to apply uniaxial strain along the specific crystal orientation and relax the atomic coordinates. It is noteworthy that the molecular crystals with an originally tetragonal lattice may lose part of their symmetry under uniaxial strain. For example, the space groups of urea and PETN crystals reduce from P$\bar{4}2_1$m or P$\bar{4}2_1$c to the same P2_12_12 group with (100) strain. The space group for PE crystal under (100) strain reduce from I$\bar{4}$ to C2. On the other hand, (001) strain will not change the space group for these tetragonal crystals. For the crystal with a given uniaxial strain η, the elastic energy can be expressed by the following strain-energy relation

$$E = E_0 + \frac{1}{2}C_{\alpha\alpha}\eta^2 + O(\eta^3)$$

from which one can fit the second order elastic constant C_{11} or C_{33}.

RESULTS AND DISCUSSION

In the plane-wave pseudopotential calculations, there are several critical parameters such as cutoff energy of the plane-wave basis and the number of k-points for sampling the reciprocal space of solids. Because of the computational complexity, many test calculations have been done to find a practical choice of the parameters so that the computation time is minimized without downgrading the accuracy. First, we found that an unusually large cutoff of the plane-wave basis is needed to achieve desirable accuracy. Figure 1 (lower plot) is the total energy of urea and PE crystals versus cutoff energy. For both urea and PE, a fairly large cutoff, about 80 Ry (~1100 eV) is needed to obtain the convergence of total energy to less than 10 meV per atom. As shown in the upper plot of Figure 1, the total energy of both urea and PE crystals converge at a modest sampling of k-space (around 4×4×4). Therefore, in the remaining calculations, we choose the cutoff energy as 80 Ry and use the k-points sets of (4,4,2) and (2,2,4) for the molecular crystals PE and PETN, respectively. All these benchmark results in the Figure 1 are

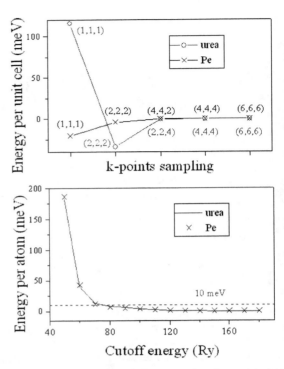

FIGURE 1. Total energy of PE (crosses) and urea (circles) molecular crystals as function of k-points sampling (upper plot) and cutoff energy (lower plot) by the DFT-GGA plane-wave pseudopotential calculations.

based on GGA, while the comparison between GGA and LDA is made in the following.

In density functional methods, the choice of exchange-correlation function of electron density is critical. In general, it is known that LDA may overestimate the binding of weakly bonded systems like van der Waals interactions or hydrogen bonds. The GGA is believed to improve on LDA to a certain extent but might overcorrect sometimes. Previous studies reveal that GGA-PBE functional can be used to describe weakly bonded systems like noble gas dimers [11]. Thus, it is critical to test the performance of both LDA and GGA in the molecular crystals. We choose urea as a representative because of its relative simplicity. In Table 1, we compare the lattice constant (**a** and **c** for tetragonal crystal), bulk modulus **B**, and elastic constants C_{11} and C_{33} of urea from LDA and GGA calculation with experiments. It is clearly shown

Table 1. Lattice constants at ambient conditions, bulk modulus and elastic constants C_{11} and C_{33} of urea, PE, and PETN crystals from GGA calculations, compared with experiments [1,2,12].

Crystal	Space group	Method	a (Å)	c (Å)	C_{11} (GPa)	C_{33} (GPa)	B (GPa)
Urea	P$\bar{4}2_1$m	LDA	5.32	4.53	34.88	106.1	27.2
		GGA	5.75	4.70	13.72	54.44	12.88
		Exper.	5.661	4.712	11.73	53.98	11.13
PE	I$\bar{4}$	GGA	6.101	9.508	33.64	5.54	8.79
		Exper.	6.067	8.799	38.98	13.43	12.2
PETN	P$\bar{4}2_1$c	GGA	9.785	6.899	10.10	6.90	--
		Exper.	9.378	6.708	17.24	12.17	9.85

that LDA overestimates the intermolecular interaction of urea. The lattice constants from LDA are much smaller than experiments, while the bulk modulus and elastic constant are overestimated by a factor of two with LDA. On the contrary, the overall performance of GGA is much better than LDA and the agreement with experiments is good. Hence, in the remaining *ab initio* calculations, the GGA functional in PBE form will be used. It is worthy to note that the current LDA and GGA lattice constants compare well with previous DFT results on urea (**a**=5.299 Å, **c**=4.532 Å by LDA[4]; **a**=5.604 Å, **c**=4.689 Å by GGA-PBE [5]).

In Table 1, we also include the ambient lattice constants, bulk modulus **B**, and elastic constants C_{11} and C_{33} of PE and PETN crystals from GGA calculations, compared with experiments. For urea, as we discussed above, the agreement between GGA calculations and experiment is very good. However, the GGA calculations obviously underestimate the intermolecular interaction along (001) direction in case of PE. The GGA lattice constant **c**=9.508 Å is much larger than experimental value 8.799 Å, while the lattice constant **a** for PE crystal at (100) direction is well described by GGA. Such a remarkable difference can be understood by considering the different types of intermolecular interaction. There are hydrogen bonds between PE molecules along the (100) direction of the crystal, while the intermolecular interaction is almost pure van der Waals-like for (001) direction. Our current results imply that DFT calculations at GGA level can model the intermolecular interaction with hydrogen bonding fairly well but fails in describing the van der Waals interaction in the molecular crystals. This argument also helps to explain why the GGA

lattice constants of urea agree well with experiments, because urea is hydrogen bonded [13]. In the case of PETN crystals, there is weak bonding in both (100) and (001) directions. Accordingly, our GGA calculations overestimate the lattice constants by about 0.2 to 0.4 Å and underestimate the elastic constants by about 40%.

As for the elastic constants, as shown in Table 1, when GGA gives a reasonable description of the ambient lattice constants, the corresponding elastic constants in this direction will also be consistent

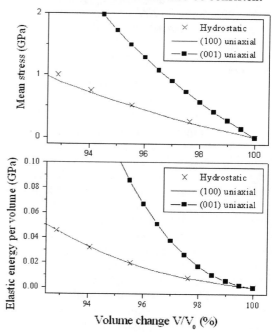

FIGURE 2. Mean stress (upper plot) and elastic energy (lower plot) of urea crystal as function of volume change under different compression: hydrostatic (crosses), (100) uniaxial (line), and (001) uniaxial (dots connected with line).

with experiments to an acceptable accuracy. Based on the benchmark calculations on both lattice constants and elastic constants, we conclude that the DFT method at the GGA level can describe molecular crystals reasonably well if there are hydrogen bonds between the molecules.

We further discuss the elastic properties of the molecular crystals under different compressions. Figure 2 shows the mean stress and elastic energy versus the volume change under hydrostatic pressure and uniaxial compression for urea crystal. It is interesting to find that uniaxial compression along (100) direction is close to hydrostatic case, while the curve of (001) uniaxial compression is very different. This anisotropic effect can be understood by the large difference between the elastic constants C_{11} and C_{33} of urea. Both DFT calculations and experiments show that C_{33} is about five times larger than C_{11}. Thus, the stiffness along (100) is much less than that for (001) direction. Under hydrostatic pressure, the urea crystal will be primarily compressed along the softer (001) direction. A similar anisotropic effect is found for PE, for which C_{11} is much larger than C_{33}.

CONCLUSIONS

In summary, we have performed DFT plane-wave pseudopotential calculations on the structural and elastic properties of molecular crystals including urea, PE, and PETN. Benchmark calculations show that a large cutoff of the plane-wave basis (up to 80 Ry) and a modest **k** point set are needed to describe the molecular crystals. We compared results from LDA and GGA and found that LDA typically overestimates the stiffness of these crystals at least by a factor of two, while the GGA results are more reasonable. The lattice constants and elastic constants of these three molecular crystals were then calculated within GGA and compared with experiments. The DFT calculations at the GGA level can describe intermolecular hydrogen bonding in the molecular crystals reasonably well, but they underestimate the intermolecular van der Waals interaction. The elastic behavior under both hydrostatic compression and uniaxial strain for urea and PE show interesting anisotropic effects. If the difference between C_{11} and C_{33} is large, then the hydrostatic compression is very close to the uniaxial compression along the orientation with the smaller elastic constant.

ACKNOWLEDGEMENTS

This work is supported by the Office of Naval Research under Grant N00014-01-1-0802 and the Department of Energy under Grant DEFG0397SF21388.

REFERENCES

1. Haussuhl, S., *Z. Kristal.* 216, 339-353(2001).
2. Winey, J. M., and Gupta, Y. M., *J. Appl. Phys.* 90, 1669-1671 (2001); Morris, C. E., in *Sixth Symposium (International) on Detonation*, edited by J. Short (Office of Naval Research, Arlington, VA 1976), pp. 396.
3. Day, G. M., Price, S. L., and Leslie, M., *Crystal Growth & Design* 1, 13-27(2001).
4. Miao, M. S., Doren, V. E. Van, Keuleers, R., Desseyn, H. O., Alsenoy, C. Van, Martins, J. L., *Chem. Phys. Lett.* 316, 297-302(2000).
5. Morrison, A. C., and Siddick, M. M., *Chem. Eur. J.* 9, 628-634(2003).
6. Dick, J. J., and Ritchie, J. P., *J. Appl. Phys.* 76, 2726-2737(1994); Dick, J. J., *J. Appl. Phys.* 81, 601-612(1997).
7. Payne, M. C., Teter, M. P., Allen, D. C., Arias, T. A., and Joannopoulos, J. D., *Rev. Mod. Phys.* 64, 1045-1097 (1992).
8. Troullier, N., and Martins, J. L., *Phys. Rev. B* 43, 1993-2006 (1991).
9. Perdew, J. P., and Wang Y., *Phys. Rev. B* 45, 13244 (1992).
10. Perdew, J. P., Burke, K., Ernzerhof, M., *Phys. Rev. Lett.* 77, 3865-3868 (1996).
11. Zhang, Y. K., Pan, W., Yang, W. T., *J. Chem. Phys.* 107, 7921-7925(1997).
12. Conant, J. W., Cady, H. H., Ryan, R. R., Yarnell, J. L., Newsam, J. M., LANL Report, #LA-7756-MS, (1979).
13. R. Dovesi, M. Causa', R. Orlando, C. Roetti, V. R. Saunders, *J. Chem. Phys.* 92, 7402-7411(1997).

CHAPTER VIII

SPALL, FRACTURE, AND FRAGMENTATION

CP706, *Shock Compression of Condensed Matter - 2003*
edited by M. D. Furnish, Y. M. Gupta, and J. W. Forbes
© 2004 American Institute of Physics 0-7354-0181-0/04/$22.00

MORPHOLOGICAL STUDIES OF AN INERT POLYMER BINDER SUBJECTED TO SHOCK LOADING

J. Akhavan, J.C.F. Millett, N.K. Bourne and R. Longjohn

Royal Military College of Science, Cranfield University, Shrivenham, Swindon, SN6 8LA, United Kingdom.

Abstract. To improve the dimensional stability of a solid rocket propellant or a plastic bonded explosive (PBX) and to allow it to deform under stress a rubbery polymeric network is incorporated in the composition. The rubbery binder used in this investigation was hydroxy terminated polybutadiene (HTPB) crosslinked with a polyisocyanate. The cured polymer was then subjected to shock loading and its resulting morphology investigated. The results indicate that there is no significant change in the morphology of the shocked polymer which agrees with the results from the mechanical response.

INTRODUCTION

Polymers are widely used in explosive compositions and are known as plastic bonded explosives (PBXs) [1]. A PBX can be regarded as a highly filled polymer, consisting of *ca.* 88 % wt of the explosive crystals embedded in *ca.* 12 % wt binder. The binder is a polymer, which is used to bind two different phases together, providing good processing and the ability to tailor properties, while maintaining the role of the material [2]. The binder surrounds the explosive crystals (oxidiser) enabling processing, as well as being the fuel for the reaction [3]. PBXs are good examples of systems where the polymeric binder is used to improve the processing and safety of an explosive, while retaining its efficiency. Since the development of PBXs in the early 1950's many polymers have been employed, but to date the most widely used is hydroxy terminated polybutadiene (HTPB). The function of HTPB as a rubbery binder is to give the traditional nitramine compositions (with their brittleness, poor mechanical properties and high vulnerability), better mechanical properties, processability and lower sensitivity than the nitramine alone. HTPB's low viscosity allows a high solids loading which is established in its cure with an isocyanate to form a polyurethane linkage.

The shock response of the elastomer binder to shock loading is becoming increasingly important and information relating to polyurethane and fluoroelastomeric polymers has already been reported [4,5]. Wingborg [6] has shown that the mechanical properties of hydroxy terminated polybutadiene (HTPB) are influenced by the choice of curing agent, but gives no information regarding the changes in the chemistry and morphology of the polymer. Many other authors [7-10] have carried out experimentation into the relationship between the degree of crosslinking, reactivity ratios and the mechanical properties of the polymer, but have not studied the effect of high rate loading on this polymer.

This paper investigates the changes in the chemistry and morphology of HTPB crosslinked with an isophorone diisocyanate (IPDI) that has been subjected to high-rate loading.

EXPERIMENTAL

Hydroxy terminated polybutadiene (HTPB) was supplied by Krahn Chemie GmbH under the trade name of Liquiflex P. Before any experimental work was conducted on this prepolymer, it was degassed for several hours by applying a high vacuum in order to remove any residual dissolved gases which may have been present due to natural ageing, as well as any reactants from the synthesis process. The crosslinking agent was an isophorone diisocyanate (IPDI) and was supplied by Bayer (UK). The catalyst dibutyl tin dilaurearte was supplied by Aldrich.

88 parts of HTPB was mixed with 12 parts of IPDI and 0.05 parts of dibutyl tin dilaureate to give a reactivity ratio of 1 [11]. This mixture was cast into acetal cups of diameter 20 mm and depth 4 mm. These were included in specimen assemblies based upon the soft recovery techniques developed at Los Alamos National Laboratories [12,13].

Impact stresses were generated with 3 mm copper flyer plates accelerated using a 6 m, 50 mm bore single stage gas gun at the Royal Military College of Science. Impact velocities were 380 and 544 m s^{-1}, generating stresses of 0.8 and 1.5 GPa.

The cured polymeric material, and the material which had been subjected to the high-rate loading, was then analysed to see if any structural changes had occurred.

In order to determine the glass and degradation temperatures, and to detect the presence of a crystalline melting temperature, a Mettler DSC 30 with RT11 Processor was used.

The degree of crosslinking in the network polymer was determined from swelling experiments using toluene as the swelling agent. The samples were immersed in excess toluene and the degree of swelling was measured at regular intervals. The molecular weight per crosslinked unit (M_x) and crosslink density (v_e) were calculated using the Flory-Rhener relation [14], as shown in equation 1.

$$v_e = \frac{-\left[\ln\left(1 - V_2\right) + V_2 + \chi V_2^2\right]}{V_s \left(V_2^{1/3} - V_2/2\right)} = \frac{\rho_2}{M_c} \quad (1)$$

where V_2 = volume fraction of the polymer,

χ = polymer-solvent interaction parameter,
V_s = molar volume of the solvent, and
ρ_2 = density of the polymer.

RESULTS AND DISCUSSION

The results from the thermal analysis, shown in Fig. 1, suggest that there is no change in the glass transition temperature and thermal degradation temperature when HTPB is subjected to high shock loading. This further suggests that the degree of crosslinking has not altered even though the polymer has been subjected to high shock loading. There was no detection of a crystalline melting peak in the cured HTPB and the shocked HTPB which therefore suggests that the shock loading has not caused the polymeric chains to become more ordered.

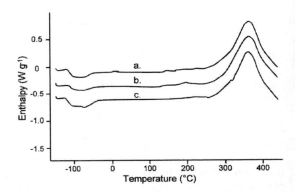

Figure 1. Differential scanning calorimetry traces for HTPB. a) – as received, b) – 0.8 GPa, c) – 1.5 GPa.

In Fig. 2, we present the variation of weight gain with time of HTPB immersed in toluene as a function of time. Note that in both unshocked and shocked samples, the weight gain is near constant, at around 400%, suggesting that no significant changes in chemistry have taken place. Observe, however, that the uptake in the sample shocked to 1.5 GPa is much faster than in the other samples. We point out that this higher shocked sample was heavily damaged by the shock loading process, and thus the faster take up may simply be due to the greater surface area present due to the presence of tears in the recovered sample.

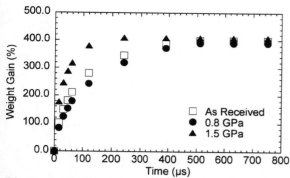

Figure 2. Swelling tests for as received and shocked HTPB.

The calculated values for the molecular weight per crosslinked unit (M_x) and the crosslink density (ν_e) using the Flory-Rhener relation [14], are presented in table 1.

TABLE 1. Reactivity ratio, molecular weight per cross linked unit (M_s) and the crosslink density (ν_e) of as received and shocked HTPB.

State	Reactivity	M_x	ν_e, mol m^{-3}
As received	1	5866	157
0.5 GPa	1	5646	163
1.5 GPa	1	6155	150

It can be seen from table 1 that there is no significant difference between the results for the cured HTPB and the shocked loaded HTPB.

These results therefore indicate that the molecular chains of the HTPB and the crosslinked chains of IPDI, when subjected to shock loading do not undergo any type of deterioration.

Therefore, to summarise the effects of shock loading on the morphology of HTPB, it would appear that no effects have been noticed from the tests performed in this study. However, the results presented here are consistent with the much earlier work of Kargin *et al.* [15] who shocked and recovered thin films of thermoplastics and observed them through cross polarising filters. They showed that the original structure was reformed in a new spheritic structure, with the new spherulite size increasing with shock amplitude. However, viscosity measurements both pre and post shock were unchanged, indicating that the molecular

weight was also unchanged. Therefore, our own results in HTPB are in agreement with these observations.

CONCLUSIONS

HTPB crosslinked with an isophorone diisocyanate (IPDI) has been subjected to shock loading. The results reported in this paper indicate that no significant change in the chemistry and morphology of the polymer has occurred for shocked HTPB. The results from Fourier transformed infra-red spectroscopy confirm that the chemistry of the polymer chains has not altered. The results from the thermal studies show that the glass transition temperature remains unaltered, therefore suggesting that the degree of crosslinking has remained the same. The presence of a crystal structure has not been detected in either the as received or shocked samples. Lastly, the crosslink density and the number of units between each crosslink did not change for the shocked HTPB.

Therefore it can be concluded that the chemistry and morphology of cured HTPB when subjected shock loading does not change. These results are in agreement with the previous work of others. Concomitantly the mechanical properties should also remain the same. The results for the mechanical response for the polymer invested in this paper is reported in a sister paper [16].

ACKNOWLEDGEMENTS

We are grateful to Gary Cooper, Matt Eatwell and Gary Stevens of RMCS for technical assistance.

REFERENCES

1. Akhavan, J., *Chemistry of Explosives*, Royal Society of Chemistry, Cambridge 1998, Chapter 1.
2. Meyer, R., *Explosives,* 3rd Ed, VCH Publisher Inc, New York, 1987, pp. 285.
3. Conkling, J.A., *Chemistry of Pyrotechnics: Basic Principles and Theory*, Marcel Dekker Inc, New York, 1985.
4. Anderson, M.U., *Shock Compression of Condense Matter –1991*, S.C. Schmidt et al. Editors, , Elsevier: Amsterdam, 1992, pp.875.

5. Bourne, N.K., Gray III, G.T., *Plasticity 99: Constitutive and Damage Modeling of Inelastic Deformation and Phase Transformation*, A.S. Khan Editor, Neat Press: Fulton, Maryland, 1998, pp.619.

6. Wingborg, N., *Polymer Testing*, **21**, (2002), 283-287.

7. Sekkar, V., Bhagawan, S.S., Prabhakaran, N., Rama Rao, M., and Ninan, K.N., *Polymer*, **41**, (2000) p6773-6786.

8. Jain, S.R., Sekkar, V., and Krishnamurthy, V.N., *J. Appl. Polymer Science*, **48**, (1993), p1515-1523.

9. Ramarao, M., Scariah, K.J., Ravindran, P.V., Chandrasekharan, G., Alwan, S., and Sastri, K.S., *J. Appl. Polymer Science*, **49**, (1993), p435-444.

10. Desai, S., Thakore, I.M., Sarawade, B.D., and Devi, S., *European Polymer Journal*, **36**, (2000), p711-725.

11. Mehmet, S. Eroğlu., *J. Appl. Poly. Sci.*, **70**, 1129 (1998).

12. Gray III, G.T. in Shock Compression of Condensed Matter – 1989, S.C. Schmidt, J.N. Johnson and L.W. Davison, Editors. 1990, North Holland: Amsterdam. P407.

13. Gray III, G.T. in High Pressure Shock Compression of Solids, J.R. Assay and M. Shahinpoor, Editors. 1991 Springer-Verlag: new York. P.187

14. Flory, P.J, *Principles of Polymer Chemistry*, Cornell University Press, Ithaca, New York, 1953, pp. 579.

15. Kargin, V.A., Andrianova, G.P., Tsarevskaya, I. Yu., Goldanskii, V.I., Yampolskii, P.A., J. Polym. Sci. 9 (1971) 1061

16. Meziere. Y., Akhavan, J., Stevens G.S., Millett J.C.F., Bourne N.K., *These proceedings*.

CP706, *Shock Compression of Condensed Matter - 2003*
edited by M. D. Furnish, Y. M. Gupta, and J. W. Forbes
© 2004 American Institute of Physics 0-7354-0181-0/04/$22.00

FRACTURE ENERGY EFFECT ON SPALL SIGNAL

N. Bonora[1], A. Ruggiero, and P.P. Milella[2]

[1]*DiMSAT- University of Cassino, Cassino, Italy, I-03043*
[2]*APAT,Env.Protection Agency, Rome, Italy I-00144*

Abstract. Numerical simulations of flying plate impact test usually show discrepancies between the calculated and observed velocity vs time plot under spalling conditions. Very often these differences are ascribed either to the constitutive model or to the numerical scheme. In this paper it is shown that, at least in the case of ductile metals, these differences can be the understood as the presence of a dissipation process during fracturing due to the viscous separation of spall fracture plane surfaces. An advanced CDM model for ductile metals has been used in order to simulate soft spall in metals and standard fracture mechanics concepts have been used to estimate the fracture energy dissipated during the separation of the fracture surfaces.

INTRODUCTION

Today, the flying pate impact test is a consolidated experimental technique, mainly used to determine the material equation of state (EOS), which can also be employed to investigate material spall fracture resistance and to check constitutive and damage model performance and its effective predicting capabilities. Usually, predicted spall signals show differences with respect to the experimental measurements that are usually ascribed either to the model or to the numerical scheme. On the contrary, as it will be discussed here, these discrepancies can reveal much more information on the dynamics of the spall fracture process. In this paper the continuum damage mechanics (CDM) model developed by Bonora [1] and later extended to incorporate strain rate and temperature effects by Bonora and Milella [2] has been used to predict soft spall in ductile metals. Numerical simulations have been performed using both standard implicit fem code (MSC/MARC) and lagrangian hydrocode (AUTODYN), in order to highlight possible differences due to the different numerical formulations. The predicted spall signal for OFHC copper, compared with experimental data available in the literature, shows dissimilarities that, at least for some classes of metals, are a clear indicator of the existence of an additional dissipation process associated with the material fracturing. Here, these differences have been explained by accounting for the fracture energy associated with the separation of the spall plane surfaces using standard fracture mechanics concepts.

CDM DAMAGE MODEL

The CDM model used in this work has been initially developed to predict and simulate ductile failure in metals under very general loading conditions and geometries. In the following the basic features of the model are briefly summarized, more detailed discussion can be found elsewhere [1-3]. The model is derived in the framework of CDM initially developed by Lemaitre [4]. With respect to the standard formulation, the damage model proposed in [1] differs in the following: a) damage affects the material stiffness only and not the material yield function which is already the result of the work hardening and damage softening competing actions; b) damage does not accumulate under a compressive state of stress where the damage effects are temporarily recovered ("compression healing"); c) the dissipation potential depends on the accumulated plastic strain ε^p.

Consequently, this model shows the following features: a limited number (4 only) of material parameters needed; all the parameters have a physical meaning and can be identified with simple uniaxial tests [3], parameter transferability to different stress triaxiality geometry conditions; the possibility to account in the damage parameter for strain rate and temperature effect, [2]; lack of localization effects. The damage evolution law derived from the damage potential is given below:

$$\dot{D} = \alpha \cdot \frac{(D_{cr} - D_0)^{\frac{1}{\alpha}}}{\ln(\varepsilon_f / \varepsilon_{th})} \cdot f\left(\frac{\sigma_H}{\sigma_{eq}}\right) \cdot (D_\sigma - D)^{\frac{\alpha-1}{\alpha}} \cdot \frac{\dot{\varepsilon}^p}{\varepsilon^p} \quad (1)$$

The material parameters needed are: ε_{th}, the damage threshold strain at which damage processes are activated, ε_f, the theoretical failure strain under constant uniaxial loading (TF:0.333); D_{cr} the critical damage at which failure occurs and α, the damage exponent, which determines the shape of damage evolution law. D_0 is the initial damage usually neglected. Stress triaxiality effects are accounted by the following function of the stress triaxiality factor (TF) σ_H/σ_{eq}:

$$f\left(\frac{\sigma_H}{\sigma_{eq}}\right) = \frac{2}{3}(1+\nu) + 3 \cdot (1-2\nu) \cdot \left(\frac{\sigma_H}{\sigma_{eq}}\right)^2 \quad (2)$$

where σ_H is the hydrostatic pressure.

NUMERICAL SIMULATIONS OF FLYING PLATE IMPACT TEST

In the FEM code MSC/MARC, the flying plate impact has been modeled using a simple single axisymmetric strip mesh for both the flyer and the target plate. Appropriate tying has been applied in order to assure pure uniaxial strain conditions. Calculations have been performed using direct integration procedure, large displacement, finite plasticity and lagrangian updating. Heat generation due to the plastic work also has been considered as well as temperature effect on material strength and damage parameters.

TABLE 1- Material/damage parameters for copper

Material Properties	
Elastic modulus	124 GPa
Poisson ratio	0.34
Shear modulus	46 GPa
Bulk modulus	129 GPa
Density	8960 kg/m^3
Conductivity	389 W/mK
Specific heat, c_p	383 J/kgK
Expansion coef. α	0.00005 K^{-1}
Melting temperature	1356 K

Damage parameters			
ε_{th}	ε_f	D_{cr}	α
0.0095	3.2	0.85	0.63

Johnson & Cook parameters	
A	90 MPa
B	292 MPa
C	0.025
m	1.09
n	0.31

Material strength was modeled using Johnson and Cook model, Table 1. Half of the entire disk geometry, for symmetry reasons, was modeled in the simulations performed with hydrocode AUTODYN. A number of impact cases for which experimental data are available in the literature have been simulated (i.e., OFHC copper, tungsten, ARMCO iron). The results presented here are limited to the case of OFHC copper. The geometric configuration is a flyer 2 mm thick and a target plate 9 mm thick with a reference impact velocity of 185 m/s, [5]. In order to comply with CDM reference volume element requirements a mesh size of 100 μm has been used. Since the impact pressure in all investigated cases is well below the reference limit of 100 times the material dynamic yield strength, a linear EOS has been used. In Fig. 1 the comparison of the calculated velocity plot vs time is given. The predictions are in a very good agreement with experimental measurements up to the spall time. When spall occurs the calculated spall signal has almost the same frequency as the experimental

440

one while the velocity amplitude is higher than measured. This raises the question about the effective dissipation mechanism during spall plane formation and fracture surface separation.

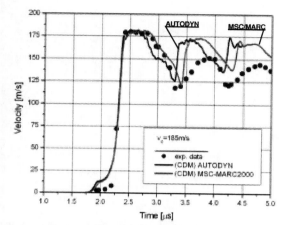

FIGURE 1. Damping effect on the target rear surface velocity plot calculated with FEM and hydrocode.

THE ROLE OF FRACTURE ENERGY

The comparison given in Fig. 1 reveals the following important features:

a) both codes calculations are in a very good agreement with the experimental data up to the spall time first arrival. The maximum velocity plateau is very well predicted both as far as concern the rise time and the signal amplitude (i.e., velocity). Similar good agreement is found also for the portion of the signal relative to the arrival of the release wave up to the time of the arrival of the first spall stress wave. This agreement confirms that the choice of the numerical damping and the material constitutive modeling (i.e. strength model, damage and EOS) is appropriate.

b) Calculated spall signal, with both codes, shows that velocity pull-up after spall takes place with higher acceleration (steeper slope) than experimentally measured. This difference cannot be ascribed to the numerical damping since there are no reason why, if it has been calibrated to match the first part of the curve, this parameter should change after spall. Since the spall signal slope is related to the intensity of the generated release stress waves, the only reason why the acceleration should

decrease with respect to the first compressive wave is the existence of a dissipative process which occurs during the separation of the spall surface just after spall fracture plane formation.

An other clear indicator of this dissipation is the amplitude of the first velocity peak in the spall signal (~150 m/s) which is clearly much lower than the maximum velocity reached with the arrival of the first compressive wave (~180 m/s) (i.e, spall strength). What can cause this effect is probably the work needed to separate the two fracture surfaces immediately after the spall. This work is spent by the material in order to fail the ligaments between smaller nucleated voids along the spall plane resulting in a viscous action which damps the stress wave travel, dissipating some of the elastic strain energy available. Since the formation of the spall plane is analogous to the formation of a ductile crack, using fracture mechanics concepts it is possible to estimate the work required to generate two free surfaces and to compare it with the amount of surplus energy in the calculated spall signal, responsible of higher velocity in the spall signal peak. From the Griffith theory, the energy required to generate two free surfaces in a material is given by:

$$G = \frac{dU}{da} = 2\Gamma \qquad (3)$$

where Γ is the total free surface energy, including both elastic and plastic contributions. The strain energy release rate G can be directly related to the material fracture toughness, K_{Ic}, which even though it is a linear-elastic concept, can still be taken as a reference value since plastic deformation along the spall plane is usually very contained in its absolute value and confined to a thin layer across the separation, plane. Thus it follows:

$$\Gamma = \frac{1}{2}\alpha \frac{K_{Ic}^2}{E} \qquad (4)$$

Since $K_{Ic} = 60$ MPa$\sqrt{m}$ is a reasonable value for copper, assuming pure plane strain conditions $\alpha = (1-v^2)$ and recalling that there is a factor of 2π due to axial-symmetry, we finally get $\Gamma \simeq 2000$ J/m^2. This is approximately the energy dissipated due to the formation of new surface energy in copper. The comparison of calculated and experimental spall signal as a function of time can be used to estimate the excess in kinetic energy that in the numerical simulation is not spent in other dissipative processes, such as new surface formation. As a matter of fact, it is possible to subtract, at each time instant, to the calculated velocity, the corresponding experimental value. The resulting plot, as a function of time, gives the evolution of the velocity difference for which an effective value can be defined as follows:

$$\Delta v_{eff} = \sqrt{\frac{1}{T} \int_0^T \left[v_{fem}(t) - v_{exp}(t) \right]^2 dt} \qquad (5)$$

The surplus of kinetic work per unit surface can be then calculated as follows:

$$\frac{\Delta W}{\Delta S} = \frac{1}{2} \rho \Delta v_{eff}^2 h_f \left[1 - \frac{h_f}{h_t} \right] = 3842 \frac{J}{m^2} \qquad (6)$$

where h_f and h_t are the flyer and target thickness respectively. Dividing by a factor of 2 (2 surfaces!) Eqn. (6) we get the calculated value for Γ, $i.e$ 1921 J/m^2, which is in a very good agreement with the value estimated with fracture mechanics concepts.

It has to be noted that similar behavior would also be expected for Armco iron that is substantially pure ductile material. On the contrary, differences between the first and the second peak velocity are much less pronounced than for copper. This can be explained in a stronger strain rate material sensitivity that under dynamic loading fails in a brittle manner at least for pressure lower than 13 GPa at which transformation phase occurs. Finally if this dissipation is incorporated in the numerical simulation in form of viscous forces, the new predicted spall signal matches with high accuracy the experimental data, Fig. 2.

CONCLUSIONS

The investigation performed on the possibility to accurately predict velocity evolution with time in a flying plate impact test has shown that, in the case of ductile metals with contained strain rate sensitivity, spall fracture surface separation may occur in a viscous manner with a direct effect on the spall signal shape and amplitude. Using fracture mechanics concepts it has been demonstrated that the energy loss is equal to the work required to generate new fracture surface in the material.

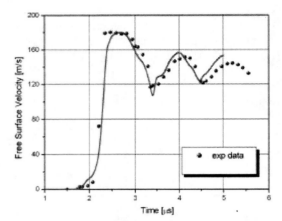

FIGURE 2. Predicted spall signal incorporating the additional dissipation due to fracture work dissipation.

ACKNOWLEDGEMENTS

This work was funded by the U.S. Air Force contract n° F61775-01-C0003.

REFERENCES

1. Bonora, N., *Eng. Frac. Mech.* **58**, 11-28 (1997)

2. Bonora, N. and Milella, P.P., *Int. J. of Impact Eng.*, **26**, 53-64 (2001)

3. Bonora, N., *J. Strain Anal.* **36**, 463-478 (1999)

4. Lemaitre, J., *J. Eng. Mat. and Tech.*, 83-89 (1985).

5. Rajendran, A. M., Dietenberger, M.A., Grove D.J., *J. Appl. Phys.* **67**, 3275 (1990)

CP706, *Shock Compression of Condensed Matter - 2003*
edited by M. D. Furnish, Y. M. Gupta, and J. W. Forbes
© 2004 American Institute of Physics 0-7354-0181-0/04/$22.00

DEVELOPMENTS TOWARD A CONTINUUM-LEVEL NON-SPHERICAL VOID GROWTH MODEL USING A MICRO-MECHANICS BASIS

B. E. Clements, E. M. Mas, and P. J. Maudlin

Theoretical Division, Los Alamos National Laboratory, Los Alamos, NM 87545

Abstract. A phenomenological damage model, based on a successful theory of Continuum Damage Mechanics (CDM), has been investigated for the goal of developing a predictive non-spherical void growth model. Using Green's functions, validated against standard Eshelby analysis, which provides reliable solutions for the growth of non-spherical voids, we attempt to verify several common assumptions of CDM. Non-spherical void growth has been observed in various steels for which we use HY-100 steel as our test system.

INTRODUCTION

Metallographic fractography shows that low concentrations of manganese sulfide (MnS) impurities have a significant role in the fracture of 1080 and HY-100 steels[1]. Examination of spalled 1080 and HY-100 samples reveal that the MnS impurities act as microvoid initiation sites necessary for ductile fracture to occur. The MnS form into high aspect ratio aligned inclusions. The alignment is the origin of the orientation dependence observed in fracture. Consequently, to model the fracture occurring in these steels it is important to include the non-spherical shape and orientation of the MnS impurities.

While the scientific literature contains many articles on spherical void growth, substantially fewer attempts have been made to model ductile failure in materials characterized by non-spherical void growth. It is of considerable interest to inquire if a predictive, *albeit*, purely continuum model can be constructed that has the advantage over any micro-mechanics-based model for being of greater numerical efficiency. There are several candidate theories in the scientific literature but in this work we have focused on the Continuum Damage Mechanics (CDM) that appears to have been originated by Kachanov[2] and Rabotnov[3]. Since the work of these authors CDM has burgeoned –

background literature can be found in Lemaitre and Chaboche[4], for example. In spite of successes demonstrated by CDM very little literature is available that directly tests the conjectures of the theory. Given that the conjectures are not transparent, in the present work we have undertaken the task of applying a rigorous method, namely Green's Functions (GF), to test the reliability of CDM theory. This rigorous micro-mechanics theory can also be used to steer future directions for CDM. In the next section we summarize several conjectures of CDM theory. This is followed by a discussion of Green's functions applied to the void problem and then a test of CDM conjectures.

CONJECTURES OF CDM

Because of the vastness of the literature on CDM, no attempt here is made for a complete survey. Kachanov[2] and Rabotnov[3] were among the first to conjecture the existence of an effective stress $\tilde{\sigma}$ defined such that the complete continuum mechanics of a damaged material can be described by invoking a corresponding set of constitutive equations, given for an undamaged material, but with the simple replacement, $\sigma \to \tilde{\sigma}$. As discussed below, this replacement includes, for example, the stresses appearing in inelastic constitutive equations. Sidoroff and Cordebois[5] conjectured that $\tilde{\sigma}$ could be uniquely determined by making an assumption of elastic energy equivalence. Chow and coworkers[6] have extended this idea and assume, in the presence of inelastic work (for example, plasticity), one should determine $\tilde{\sigma}$ and an effective strain $\tilde{\varepsilon}$ by equating the total stored energies. Letting D denote a measure of the damage state, called the damage field, the conjecture states that

$$W(\sigma, D) = W(\tilde{\sigma}, D = 0), \quad (1)$$

where the left (right) side of the equality is the energy stored in the damaged (undamaged, with $\sigma \to \tilde{\sigma}$, $\varepsilon \to \tilde{\varepsilon}$).

Kachanov and Rabotnov gave a physical interpretation for $\tilde{\sigma}$ by proposing that for systems with isotropic damage, $\tilde{\sigma}$, the Cauchy stress σ, and D should obey the simple relationship:

$$\tilde{\sigma} = \frac{\sigma}{1 - D}, \quad (2)$$

where

$$D \equiv \frac{S - \tilde{S}}{S}. \quad (3)$$

In Eq. (3), for any cross-section of the damaged material, with cross-sectional area S, the quantity $\tilde{S}$ represents the cross-sectional area that is capable of supporting a load (the undamaged area). Thus D is interpreted as the fraction of damaged to the total cross-sectional areas. These concepts have been extended to anisotropic damage, for example, a system containing aligned non-spherical voids. In the principal axis system of the voids there are three orthogonal principal directions and Eq. (2) becomes

$$\tilde{\sigma} = M(D)\sigma. \quad (4)$$

$M(D)$ is called the damage tensor. Chow and coworkers assume a simple form for $M(D)$[6],

$$M_{ij}(D) = \delta_{ij} / (1 - D_i). \quad (5)$$

A large body of work has followed from these initial ideas. We refer to the book by Lemaitre and Chaboche[4] for further discussions. Interestingly, we find little theoretical work has been done to justify these conjectures.

GREEN'S FUNCTION ANALYSIS

The elastic-plastic stress and strains surrounding a void in an isotropic matrix can be calculated by Green's function techniques[7]. Let υ and μ denote the Poisson's ratio and shear modulus of the material. Next introduce a void into the system. The displacement rate in and in the vicinity of the void, contained in volume V, is given by the integral-differential equation

$$\dot{u}_m(x) = \dot{u}_m^\infty(x) - \int_V dx' \partial_j g_{mi}(x - x') C_{ijkl}^0 \dot{\varepsilon}_{kl}(x')$$

$$- \int_M dx' \partial_j g_{mi}(x - x') C_{ijkl}^0 \dot{\varepsilon}_{kl}^P(x'), \quad (6)$$

where M is the region bounding the matrix, $\dot{u}^\infty$ is the applied displacement rate at the boundary and is related to the applied strain rate through $\dot{u}_i^\infty = \dot{\varepsilon}_{ij}^\infty x_j$, C_{ijkl}^0 is the elastic stiffness tensor of the elastic homogeneous material, $\dot{\varepsilon}(x)$ is the local total strain rate tensor, $\dot{\varepsilon}^P(x)$ is the local plastic strain rate tensor, and $g_{ij}(r)$ is the Green's function given by

$$g_{ij}(r) = \frac{1}{4\pi\mu r} \delta_{ij} - \frac{1}{16\pi\mu(1 - \upsilon)} \partial_i \partial_j r. \quad (7)$$

In these expressions we use $\partial_k \equiv \partial / \partial x_k$. Thus, under the assumption of small strains the elements of the total strain rate are given by

$$\dot{\varepsilon}_{ij}(x) = \frac{1}{2} \left(\partial_i \dot{u}_j(x) + \partial_j \dot{u}_i(x) \right). \quad (8)$$

Given a constitutive relation for the plastic strain rate, Eqs. (6) and (8) are sufficient to solve, by iteration, for the displacement, strain, and stresses in the entire system.

There a several tests we can perform to check the accuracy of our solutions of Eq. (6). Let Ω be the volume of the entire system and let it be bounded by an external surface S. We have solved Eq. (6) on rectangular grids where we commonly use Ω up to (60x60x60). The first check is that our boundary value problem is satisfied, namely, $\varepsilon(x) = \varepsilon^\infty$ when $x \in S$. By making Ω large enough we can satisfy the condition always for elastic fields, but for plastic fields, the restricted domain of integration (the second integral in Eq. (6)) at the corners of the box will always cause a small error. This error is tolerable since it is not linked directly to the distortion in the fields caused by the void. We also checked that the average strain theorem (AST) was always satisfied for all elements of the strain tensor:

$$\varepsilon_{ij}^\infty = \frac{1}{\Omega} \int_\Omega dx \varepsilon_{ij}(x) \equiv \langle \varepsilon_{ij} \rangle_\Omega = c_v \langle \varepsilon_{ij} \rangle_V + c_m \langle \varepsilon_{ij} \rangle_M. \quad (9)$$

The first equality is the AST and c_v and c_m are the volume concentrations of the void and matrix. That c_v calculated from Eq. (9) is in good agreement with that obtained directly from the system, indicates that our numerical solutions are accurate. Second, Eshelby determined an analytic solution for the case of an ellipsoidal void in an elastic matrix, when the system is subjected to remote loading[8]. The integrals in Eq.(6) can be computed analytically the resulting strain in the void is given by

$$\varepsilon = [I - S]^{-1} \varepsilon_{ij}^\infty. \quad (10)$$

where S is the Eshelby tensor[8]. The solutions of Eq. (10) can be compared to our numerical solutions of Eq. (6). For an example consider an ellipsoidal void with aspect ratio 1.2. Let x_3 be the

long axis of the void and $\dot{\varepsilon}_{33}^* = 1000\,s^{-1}$, and $\dot{\varepsilon}_{11}^* = \dot{\varepsilon}_{22}^* = -309\,s^{-1}$.

Figure 1 shows the case where the loading strains have reached $\varepsilon_{33}^\infty = 0.070$, $\varepsilon_{11}^\infty = \varepsilon_{22}^\infty = -0.022$. The match with Eshelby's solution is extremely good except for the lip in the numerical solution caused by approximating the ellipse with rectangular grid elements. HY-100 bulk (1.6×10^5 MPa) and shear (1.6×10^4 MPa) moduli are used.

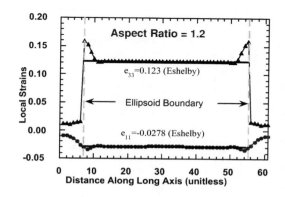

FIGURE 1. Comparison of solutions of Eq. (6) and analytic solutions of Eq. (7).

GF and CDM COMPARISONS

Consider a elastic-perfectly plastic (yield stress = 0.76 GPa) matrix surrounding a single void. A tensile uniaxial strain ($\dot{\varepsilon}_{33}^\infty = 30\,s^{-1}$) is applied. The initial void concentration and aspect ratio is 0.01 and 3, respectively. Figure 2 shows the resulting GF stress strain curves ($\langle \sigma_{33} \rangle_\Omega$ vs. $\langle \varepsilon_{33} \rangle_\Omega$) for the undamaged state, and for two orientations of damage. The most softened stress-strain behavior occurs for when the void's long axis is perpendicular (transverse) to the 33-direction. Less softening occurs when the void is aligned (longitudinal) in the 33-direction.

Next, we calculate, using CDM, the matrix $M(D)$ using Eq. (5). $M(D)$ is substituted into a standard set of equations for a homogeneous elastic perfectly plastic material with the simple substitutions $\sigma \rightarrow \tilde{\sigma}$ and $\varepsilon \rightarrow \tilde{\varepsilon}$ where to maintain Eq. (1), we take $\tilde{\varepsilon} = M^{-1}(D)\varepsilon$.

By comparing to the GF solution we found that the prescription of using the substitutions $\sigma \rightarrow \tilde{\sigma}$ and $\varepsilon \rightarrow \tilde{\varepsilon}$, derived from the conjectures of CDM, and the simple form of Eq. (5), worked remarkably well for general loads (see Figure 3). However, seldom could we use the exact area ratios given by the GF solution to achieve agreement. For example, the agreement in Figure (3) came about by using CDM area ratios of $D_1 \approx 0.01$, and $D_2 = D_3 \approx 0.03$, about half the GF values, indicating that effective area ratios must be used in the CDM theory. Using the actual GF areas in the CDM gave results typically about 5% of the GF solutions.

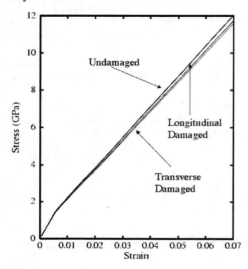

FIGURE 2. Stress-strain GF solutions of Eq. (6).

CONCLUSIONS

We have used rigorous Green's functions solutions for the problem of a void in an elastic perfectly plastic material to test the validity of several conjectures of CDM. We invoked the conjecture that effective stresses and strains can be defined such that by solving the corresponding homogeneous system using them in place of the strain and Cauchy stress we can accurately calculate the mechanical behavior of the damaged material. Using effective area ratios rather than the Green's function calculated ones, CDM predictions could be made to match the GF solutions.

ACKNOWLEDGMENTS

This work was supported by the Joint DoD/DOE Munitions Technology Development Program and the (LANL) Laboratory. Research supported by the USDOE under contract W-7405-ENG-36.

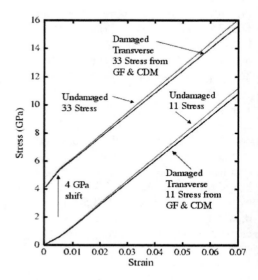

FIGURE 3. Comparison of CDM and GF solutions.

REFERENCES

1. Jablokov, V., Goto, D., Koss, D., and McKirgan, J., Mat. Sci. Eng., A302, 197-205 (2001). Clements, B. E., Mas, E. M., and Gray III, G. T., Shock Compression in Condensed Matter–2001, ed. Furnish, M. D., Thadhani, N. N., and Horie, Y., AIP Conference Proceedings 620, New York, 2002, pp. 535-538.
2. Kachanov, L. M., Izv. Akad. Nauk USSR Otd. Tekh. 8, 26-31 (1958)
3. Rabotnov, Y. N., *Creep Problems in Structural Members*, North Holland, Amsterdam (1969).
4. Lemaitre, J. and Chaboche, J. L., *Mechanics of Solid Materials*, University Press, Cambridge (1990).
5. Cordebois, J. P., and Sidoroff, F., Euromech, 115 ed. Villard de lans (1979).
6. Lu, T. J., and Chow, C. L., Theoretical and Appl. Fract. Mech. 14, 187-218 (1990) and references therein.
7. Mura, T., *Micromechanics of Defects in Solids*, 2nd Ed., Martinus Nijhoff, Dordrecht, (1987).
8. Eshelby, J. D., Proc. Roy. Soc., A241, 376-396 (1957).

CP706, *Shock Compression of Condensed Matter - 2003*
edited by M. D. Furnish, Y. M. Gupta, and J. W. Forbes
© 2004 American Institute of Physics 0-7354-0181-0/04/$22.00

ON THE ROLE OF CRACK ORIENTATION IN BRITTLE FAILURE

J.K. Dienes, J. Middleditch, Q.H. Zuo, and J.D. Kershner

Los Alamos National Laboratory, Los Alamos, NM 87545

Abstract. Many materials contain a large number of microcracks that can propagate under sufficiently high stress, but their stability is sensitive to crack orientation. We have explored this sensitivity using classical fracture mechanics with the added feature that interfacial friction is accounted for in the behavior of compression cracks. Our analysis shows that four types of unstable crack growth are possible for a penny-shaped crack under a general state of stress, depending on crack orientation: opening without shear, mixed opening and shear, pure shear without friction, and shear with interfacial friction. In addition, interfacial friction prevents crack growth at all stress intensities in a certain range of compressive stress. It will be shown that these analytic results are captured by the SCRAM brittle-failure algorithm, and that friction strongly affects the orientation of the most unstable shear crack as well as the range of unstable orientations. A second study examines the variations in material response as a function of the number of orientations represented. This is done by computing the dynamic response of an axisymmetric thick ring to internal pressure. With the traditional 9 crack orientations the fluctuation in porosity is about 28%, while with 480 orientations the fluctuation drops to just over 2%.

INTRODUCTION

At sufficiently low temperatures and high strain rates most materials exhibit brittle behavior. In Statistical CRAck Mechanics (SCRAM) we have been attempting to characterize the behavior of brittle materials by superimposing the effects of an ensemble of penny-shaped cracks of various orientations and sizes[1]. Since the theory of penny-shaped cracks has been well established, we can draw on the literature for many results. Until recently, however, published results generally ignored interfacial friction in closed cracks, though this is an important feature of compression failure. The frictional behavior of closed cracks and the propagation of rupture is now a subject of active research[2]. The failure and strain softening of brittle materials can be represented by the instability of penny-shaped cracks. An analysis of this instability has been carried through for a general state of stress[3], showing that just 4 kinds of instability are possible: opening without shear, mixed opening and shear, pure shear without friction, and shear with interfacial friction. Growth of shear cracks in a failure plane, both with and without friction, is clearly observed in various impact experiments[4,5]. When the compressive stress is sufficiently large, interfacial friction prevents crack growth.

We review briefly the effects of crack orientation under a general state of stress, noting that there is a regime wherein the most unstable closed cracks are oriented so as to propagate without friction[3]. The effect of orientation on stability and the range of unstable orientations near the most critical orientation are illustrated with a parameter study. In a second investigation the effect of the number of crack orientations on the dynamic response of a thick ring to a ramped internal pressure is shown. This serves as a basis for testing a new coalescence algorithm that accounts for the formation of macroscopic cracks by coalescence of microcracks, resulting in spall, and/or radial and circumferential cracks near an impact or explosion.

PENNY-SHAPED CRACKS

Theoretical expressions for the strain energy of a penny-shaped crack subject to axisymmetric tension[6] and pure shear[7] have been determined, and extended to combined shear and tension[8]. However, the effect of interfacial friction when the crack is under compression is crucial. It can be obtained by superimposing the stress in the far field and the frictional stress[9], so long as the crack is thin. (This superposition does not hold, for example, for a spherical cavity.) The effect of friction explains the fact that brittle materials are much stronger in compression than in tension, typically by a factor of about ten[10]. The failure surface that follows from an analysis of open and closed cracks is shown in Fig. 1.

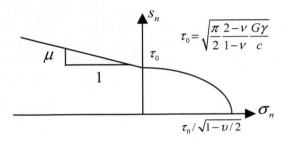

$$\tau_0 = \sqrt{\frac{\pi}{2}\frac{2-\nu}{1-\nu}\frac{G\gamma}{c}}$$

FIGURE 1. The failure (defined by the onset of crack instability) surface for combined shear (s_n) and normal (σ_n) stresses on a penny-shaped crack with interfacial friction. Note that the result is similar to that for Mohr-Coulomb failure, but here the strength depends on crack dimensions. G denotes shear modulus; ν, Poisson's ratio; c, crack radius; γ, surface energy; and μ, friction coefficient.

The idealization of brittle materials by representing the defects as an ensemble of penny-shaped cracks is to be taken in the same spirit as idealizing molecules by spheres in the kinetic theory of gases, a surprisingly good approximation. It has been shown that an ensemble of elliptic cracks can be very well approximated with an ensemble of circular cracks[11]. Furthermore, it can be argued that an unstable tensile crack that is nearly circular will tend toward a circular shape as

it grows[12]. In shear cracks, the onset of instability in Mode III precedes that in Mode II, (though only by a small amount). Thus, we are hopeful that assuming a penny-shape geometry is not an excessively idealized approximation.

In many cases real materials can be considered to contain a large number of cracks to account for their brittle behavior. This is the underlying concept in SCRAM, which considers the behavior of materials to be accounted for by the opening, shear, growth and coalescence of an ensemble of penny-shaped cracks[1]. The validity of super-imposing the effects of an ensemble of defects has been shown theoretically[13], suggesting that superposition of strain rates has much wider applicability than the well-established super-position of elastic and plastic strain rates. This superposition makes it possible to synthesize constitutive laws for materials containing many kinds of defects as well as nonlinear continuum behavior, so long as the strain rate for the appropriate physics can be represented in a computer algorithm.

THE CRITICAL CRACK ORIENTATION AND THE ANGULAR RANGE OF UNSTABLE ORIENTATIONS

A stand-alone version of SCRAM has been run specifying a constant strain rate and incrementing the stresses at regular intervals using the resulting stress rate. The strain rate corresponds to triaxial compression with a small radial compressive stress. Results of four such calculations with varying friction coefficient μ are shown in Figure 2. Nine crack orientations are accounted for in the nine adjacent columns. The stability of cracks with each orientation is noted in the figure with an s if stable, or with a dark region if not. Realistic values of the parameters for PBX 9501, which we have studied extensively, were selected so that we use the entire SCRAM algorithm; the effects of nonlinearities in the equation of state and crack interactions are accounted for, but are small.

time	σ_{11}	σ_{22}	μ = 0.1	μ = 0.5	μ = 1.0	μ = 1.2
0.00	-0.29	-0.02	ssssssss	ssssssss	ssssssss	ssssssss
0.01	-0.57	-0.04	ssssssss	ssssssss	ssssssss	ssssssss
0.01	-0.86	-0.06	ssssssss	ssssssss	ssssssss	ssssssss
0.02	-1.15	-0.08	sssss	ssssssss	ssssssss	ssssssss
0.02	-1.43	-0.10	ssss	ssssssss	ssssssss	ssssssss
0.03	-1.72	-0.12	sss	ss sssss	ssssssss	ssssssss
0.03	-2.01	-0.14	sss	ssss	ssssssss	ssssssss
0.04	-2.30	-0.16	ss	sss	ssssssss	ssssssss
0.04	-2.58	-0.18	ss	sss	ssssssss	ssssssss
0.05	-2.87	-0.21	ss	sss	ssssssss	ssssssss
0.05	-3.16	-0.23	ss	ss	ssssssss	ssssssss
0.06	-3.45	-0.25	ss	ss	ssssssss	ssssssss
0.06	-3.74	-0.27	ss	ss	sssss sss	ssssssss
0.07	-4.03	-0.30	s	ss	sss sss	ssssssss
0.07	-4.32	-0.32	s	ss	sss sss	ssssssss
0.08	-4.61	-0.34	s	ss	sss sss	ssssssss
0.08	-4.90	-0.37	s	ss	sss sss	ssssssss
0.09	-5.19	-0.39	s	ss	sss sss	ssssssss
0.09	-5.48	-0.42	s	ss	sss sss	ssssssss
0.10	-5.77	-0.44	s	s	ss ss	sssss sss
0.10	-6.07	-0.47	s	s	ss ss	ssss sss
0.10	-6.36	-0.49	s	s	ss ss	ssss sss
0.11	-6.65	-0.52	s	s	ss ss	ssss sss
0.11	-6.94	-0.54	s	s	ss ss	ssss sss
0.12	-7.24	-0.57	s	s	ss ss	ssss sss
0.12	-7.53	-0.60	s	s	ss ss	ssss sss
0.13	-7.82	-0.62	s	s	ss ss	ssss sss
0.13	-8.12	-0.65	s	s	ss ss	ssss sss
0.14	-8.41	-0.68	s	s	ss ss	ssss sss
0.14	-8.70	-0.70	s	s	ss ss	ssss sss
0.15	-9.00	-0.70	s	s	ss ss	ssss sss
0.15	-9.29	-0.76	s	s	ss ss	ssss sss
0.16	-9.59	-0.79	s	s	ss ss	ssss sss
0.16	-9.89	-0.82	s	s	ss ss	ssss ss
0.17	-10.18	-0.85	s	s	ss ss	ssss ss
0.17	-10.48	-0.88	s	s	ss ss	sss ss
0.18	-10.77	-0.91	s	s	ss ss	sss ss
0.18	-11.07	-0.94	s	s	ss ss	sss ss
0.19	-11.37	-0.97	s	s	ss ss	sss ss

FIGURE 2. The history of crack behavior for 4 values of μ, the coefficient of friction. The stress increases as the on the left increases. The 9 columns for each μ represent the stability (s) or instability (shaded) for the 9 orientation bins, ranging from 47.5° in the first column to 87.5° in the ninth. These are the angles of the crack normal with the axial loading direction. A bin is a range of 5°. For low friction, the range of instability is rather wide, narrowing as μ increases. The stress levels need to be multiplied by 1e8 to get cgs units, and the time is in microseconds. The strain rate is -6e4/s in the axial direction and -2e4/s in the radial direction.

THE EFFECT OF DISCRETIZING CRACK ORIENTATION

In the earliest SCRAM calculations[14,15], which were considered part of a feasibility study, 9 crack orientations were accounted for. The associated caps on the unit sphere could be considered to

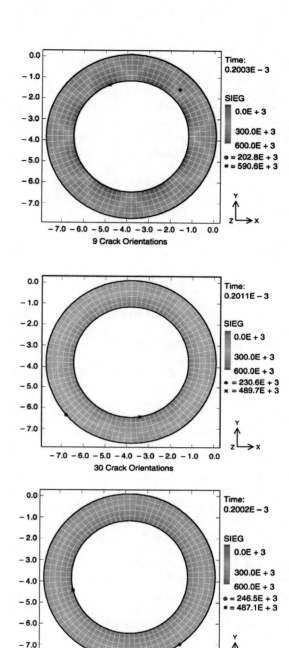

FIGURE 3. The response of a ring to internal pressure for three levels of resolution of crack orientation. With 9 orientations the anisotropy is quite large, while with 480 the uniformity is good. The internal pressure ramps to 25 bars in 30 μs.

449

divide it into 18 equal parts, since reversal of the crack normal has no effect, and the caps were taken to have equal areas. It was convenient in original studies to take 3 of the 9 orientations parallel to the coordinate axes. However, this does not lend itself to highly accurate calculations. To verify the coding and examine the importance of angular resolution we have introduced orientations based on the Platonic solids. Thus, we can compare calculations with 9, 30, and 480 orientations. When half the unit sphere is divided into 480 caps, each defines a cone with a half angle of 3.7^0. Figure 3 shows the effect of increasing resolution of crack orientation. The colors specify SIEG, the energy per gram consumed in brittle fracture.

CONCLUSIONS

The orientation of a crack strongly influences its behavior, especially when the components of stress have different signs or friction is important[3]. The earliest SCRAM work assumed that 9 crack orientations are sufficient to represent failure, but recent resolution studies suggest that this may not be enough in many applications. In one example using the stand-alone SCRAM we find that the range of unstable crack orientations covers only 20^0. In a second series of calculations, simulating an internally pressurized ring, we find that the fluctuation in porosity due to discretizing crack orientation is 28% with 9 orientations, while with 480 orientations the fluctuations drop to 2.3%. Still, as the resolution changes, the average varies only from 1.8% to 2.2%, so if average values are sought, high resolution may not be necessary; the required resolution depends on the application. When concerned with gross failure in a large mass, spurious fluctuations in the calculation may be important, leading to the formation of unphysical macroscopic cracks. Thus, in detailed calculations with many defects, it may be important to account for 30 or more crack orientations.

ACKNOWLEDGEMENTS

This work was jointly supported by the MoU, ASCI high explosive, and weapons programs at Los Alamos National Laboratory.

REFERENCES

1. Dienes, J.K., "A Unified Theory of Flow, Hot Spots, and Fragmentation with an Application to Explosive Sensitivity," in *High-Pressure Shock Compression of Solids II: dynamic fracture and fragmentation,* edited by L. Davison, D.E. Grady and M. Shahinpoor, Springer-Verlag, New York, 1996, pp. 366-398.
2. Rice, J.R., "New Perspectives on Cracks and Fault Dynamics," in *Mechanics for a New Millennium,* Proceedings of the International Union of Theoretical and Applied Mechanics, edited by H. Aref and J.W. Phillips, Chicago, Kluwer Academic Publishers, 2000, pp. 1-24.
3. Zuo, Q.H., and. Dienes, J.K, On the Types of Brittle Failure, Los Alamos Report LA-13962, July, 2002.
4. Howe, P.M., Gibbons, G.G., and Webber, P.E., "An Experimental Investigation of the Role of Shearing Initiation of Detonation," in *Proceedings of the Eighth International Symposium on Detonation,* edited by J.M. Short, Albuquerque, NM, 1985, pp. 294-306.
5. Rosakis, A.J., *Advances in Physics* **51**, 1189-1257 (2002).
6. Sack, R.A., *Proc. Phys. Soc.* **58**, 729-736 (1946).
7. Segedin, C.M., *Proc. Camb. Phil. Soc.* **47**, 396-400 (1950).
8. Keer, L.M., *J. Mech. Phys. Solids,* **14**, 1-6 (1966).
9. Dienes, J.K., "Theory of Deformation, Part II, Physical Theory," Los Alamos Report LA-11063-MS, 1989.
10. Dienes, J. K., *J. Geophys. Res.* **88**, 1173-1179 (1983).
11. Sevostianov, I. and Kachanov, M., *Int. J Fracture,* **114**, 245-257 (2002).
12. Kanninen, M.F., and Popelar, C.H., *Advanced Fracture Mechanics,* Oxford University Press, New York, 1985, p. 153.
13. Dienes, J.K., Finite Deformation of Materials with an Ensemble of Defects, Los Alamos Report LA-13994-MS, 2003.
14. Meyer, H.W. Jr, Abeln T., Bingert, S. Bruchey, W.J., Brannon, R.M, Chhabildas, L.C., Dienes, J.K., and Middleditch, J. "Crack Behavior of Ballistically Impacted Ceramic", *Shock Compression of Condensed Matter,* edited by M.D. Furnish, L.C. Chhabildas and R.S. Hixson, 1109-1112, 1999.
15. Dienes, J.K., Middleditch, J., Kershner, J.D., Zuo, Q., and A. Starobin, "Progress in Statistical Crack Mechanics, An Approach to Initiation," Proceedings of the Twelfth International Detonation Symposium, San Diego, CA, 2002, to appear.

CP706, *Shock Compression of Condensed Matter - 2003*
edited by M. D. Furnish, Y. M. Gupta, and J. W. Forbes
© 2004 American Institute of Physics 0-7354-0181-0/04/$22.00

FRAGMENTATION OF REACTIVE METALLIC PARTICLES DURING IMPACT WITH A PLATE

David L. Frost[1], Samuel Goroshin[1], Stephen Janidlo[1], Jason Pryszlak[1], Jeff Levine[1],
and
Fan Zhang[2]

[1]*McGill University, Department of Mechanical Engineering, 817 Sherbrooke St. W.,*
Montreal, Quebec, Canada H3A 2K6
[2]*DRDC - Suffield, PO Box 4000, Stn Main, Medicine Hat, Alberta, Canada T1A 8K6*

Abstract. The effect of the detonation of a spherical heterogeneous charge on the loading applied to a nearby structure has been investigated experimentally. The charge consists of a packed bed of solid reactive particles saturated with a liquid explosive. When the charge is detonated, the particles ignite while rapidly accelerating to high speeds, then impact either a rigid plate or a cantilever gauge consisting of a plate attached to a rod clamped at one end. The positive phase reflected pressure impulse at the rigid plate surface and the cantilever bend angle is measured at various distances from the charge. Comparison of the results for charges containing large (millimeter-sized) zirconium particles with small (sub-micron) aluminum particles suggests that the fragmentation of the large particles upon impact contributes to the loading. To investigate this possibility further, single zirconium spheres are accelerated with a tubular explosive. When the burning sphere impacts a rigid plate, fine zirconium fragments are produced in all cases. The impact fragmentation of burning particles is an efficient mechanism for increasing the overall rate of energy release from the particle combustion process.

INTRODUCTION

When solid particles are added to high explosives, the acceleration and impact of the particles when the charge is detonated can generate significant loading on nearby structures. Earlier work with a heterogeneous charge consisting of a packed bed of inert particles saturated with a liquid explosive has shown that the particles are accelerated to speeds of 1-2 km/s by the detonation of the charge[1]. Later experiments with the same charge showed that the particle momentum flux provides the primary contribution of the multiphase flow to the near-field impulse applied to a nearby small structure[2]. In these experiments the inert particle velocities were not sufficient to lead to

particle fragmentation upon impact. However, if relatively large reactive particles are explosively dispersed, the impact of a burning (and hence partially molten) particle with a structure may generate a large increase in particle surface area (by fragmentation of the liquid layer or solid kernel) which will enhance the overall rate of energy release from the particle combustion. If this energy release is sufficiently rapid to equilibrate with the local thermodynamic conditions, this will result in increased loading on the structure.

Over the past 50 years many researchers have investigated the breakup of metallic particles during hypervelocity impact with either a semi-infinite obstacle or a thin bumper plate[3-6]. However, relatively little work has been done on the impact fragmentation of burning metallic particles. In the present paper, the impact of explosively dispersed particles with a plate is investigated. The effect of dispersing both large and small reactive particles is compared with the earlier work with inert particles. To help interpret the results, additional experiments are carried out in which a single reactive sphere is accelerated to elucidate the impact fragmentation process.

PARTICLE DISPERSAL EXPERIMENTS

The dispersal experiments were carried out by detonating spherical glass-cased charges (12.3 cm dia) containing a packed bed of particles saturated with sensitized nitromethane adjacent to a rigid steel plate (1.83 m square x 5.1 cm). To determine the loading on the plate, the reflected pressure was measured with a tourmaline pressure bar (PCB 134A22) flush-mounted on the plate. The surface of this transducer was protected from direct particle impact with a screen containing holes located at a radius just larger than the transducer face. Calibration tests carried out with NM charges showed that whereas the screen attenuated the peak pressure by up to 50%, the positive phase pressure impulse varied by less than 5% with or without the presence of the screen. Further details of the experiments can be found in earlier publications[1,2].

Two different reactive particles were used to illustrate the range of possible fragmentation behavior: i) large zirconium particles (725±125 μm dia from Atlantic Equipment Eng.) which are rapidly accelerated and maintain their speed due to their inertia, and hence may fragment upon impact, and ii) small aluminum particles (100-200 nm dia, denoted Alex™ from Argonide Corp) which rapidly burn and accommodate to the local flow velocity, and are not expected to undergo impact fragmentation.

The results for reactive particles are compared with the earlier results with inert iron particles (463±38 μm dia) as well as pure NM charges, with all charges having the same volume (about 1 *l*). The solid/liquid masses in the four cases were: Zr/NM:2130g/655g, Alex/NM:300g/900g, Fe/NM:4450g/430g, and NM only:1000g.

The variation with distance of the side-on pressures and side-on and reflected positive-phase

pressure impulses obtained for the four different charge configurations are shown in Figs. 1 and 2.

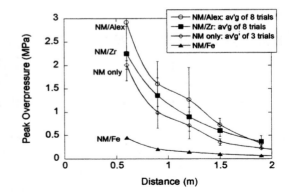

FIGURE 1. Peak incident (side-on) blast wave overpressure for four different charges configurations.

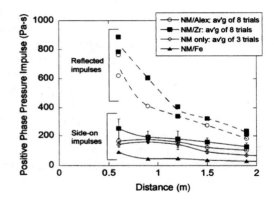

FIGURE 2. Side-on and reflected positive phase pressure impulse for four different charges.

Several observations can be made from the above figures. The shape of the decay with distance of the side-on pressures and impulses for the NM/Zr and NM/Fe charges are similar due to a similar interaction between the particles and the flow field. Also the (*p, I*) profiles for the NM/Alex and NM only charges are qualitatively alike. Apparently due to the small size of the Alex particles, the NM/Alex charges have a blast wave profile similar to that of a homogeneous explosive. Also note from Figs. 1 and 2 that while the NM/Alex charges consistently produce higher peak blast overpressures, the NM/Zr charges generate larger impulses. Particularly striking from Fig. 2 is the augmentation in the reflected pressure impulse for NM/Zr charges in the near-field (0.6-1.2 m), which based on previous work[1,2] corresponds to the distance over which the majority of the particles arrive at the plate on a timescale on the same order as the arrival of the blast wave at the plate. A credible explanation for the increased reflected impulse is that the Zr particles fragment upon impact and the combustion of the fragments contributes to the overall pressure loading.

Further indirect evidence for the possibility of Zr particle fragmentation is given in Fig. 3, which shows the cantilever gauge bending for the different charges. Although a complete analysis of the cantilever loading and motion is complex, the large bending for NM/Zr charges suggests that combustion of the impact-induced Zr liquid/solid fragments may play a role. This is supported by an examination of the cantilever plate surface which is covered with a gray layer (presumably zirconia) and exhibits few impact craters which are probably due to the impact of the solid particle kernels (Fig.4).

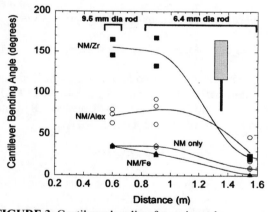

FIGURE 3. Cantilever bending for various charges.

SINGLE PARTICLE EXPERIMENT

To investigate the impact fragmentation of burning Zr particles directly, a second apparatus was used, as shown in Fig. 5. The length of the hollow detaprime explosive driver was varied from 0.4-2.3 m (354 g/m) and connected to a 1 m steel launch tube. Spherical 6.4 mm Zr particles were numerically machined from a high purity Zr rod (Al,

Mg, and W spheres were also tested but they did not ignite and hence the results are not relevant here).

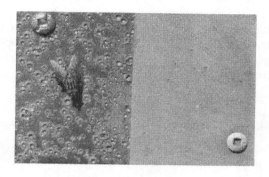

FIGURE 4. Cantilever impact plates showing impact of 1 mm iron (left) and 725 μm Zr particles (right) at 1 m.

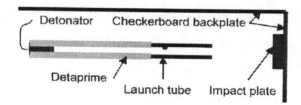

FIGURE 5. Top view of apparatus for accelerating single 6.4 mm dia spherical particle.

The zirconium spheres were accelerated to speeds between 150 m/s and 2.5 km/s. In all cases, the sphere ignited within the launch tube and partially melted during the acceleration phase. For a Zr sphere, the characteristic thermal boundary layer thickness, δ, is on the order of 0.1 mm ($\delta \sim (\kappa t)^{1/2}$) after a time of 1 ms. Hence, the burning Zr particles will have only a thin reacting liquid layer around the solid particle kernel just prior to impact (about 5 ms after launch). As the particle moves through the launch tube, fine burning zirconium droplets were stripped off the particle and exited the tube continuing to burn, as shown in Fig. 6. In all cases, a spray of fine burning droplets was generated upon impact with the plate. As the impact velocity increased, the fragmentation behavior was qualitatively similar, but at higher velocities, the solid particle kernel sometimes fragmented into several smaller fragments upon impact which then generate secondary fragmentation events when they impact the checkerboard backplate (e.g., see Fig. 7 with an impact velocity of 650 m/s). At higher velocities still, the backscatter angle of the ejecta increases, as shown in Fig. 8, where one of the secondary fragments is visible in the lower photographs, moving to the left at about 150 m/s, shedding fragments like a micrometerorite as it travels.

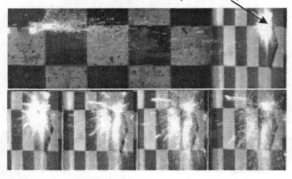

FIGURE 6. Impact of a burning Zr particle moving at 160 m/s with a plate. Small burning fragments stripped from the particle during acceleration within the launch tube are visible in the top frame on the left. Background squares are 30 cm; time between frames is 0.5 ms.

FIGURE 7. Impact of a burning zirconium particle moving at 650 m/s with a plate. Time between upper 4 frames is 0.5 ms. Frame at bottom, taken 17 ms after impact, shows impact fragmentation of secondary fragments on checkerboard backplate.

In conclusion, the single sphere acceleration experiments have demonstrated that even millimeter-sized Zr spheres can be ignited during

acceleration by an explosive. Since zirconium is known to burn according to a heterogeneous surface reaction, the particle surface is partially melted prior to plate impact. The impact of the reacting particles is an efficient means by which to increase the total burning surface of the particle, and hence the overall energy release rate. Further work is required to clarify the mechanism of fragmentation of the surface layer and particle kernel and determine how the increased burning feeds back to influence the thermodynamic flowfield.

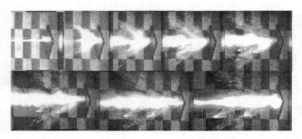

FIGURE 8. Impact of a burning zirconium particle moving at about 2.5 km/s with a plate. Time between frames is 0.5 ms. Particle appears as a streak in the first frame due to the long duration of the frame (278 µs).

ACKNOWLEDGMENTS

The authors wish to acknowledge the assistance of students C. Orthanalai, S. Jacobs, and DRDC technical staff, T. Storrie, D. Boechler, R. Linde, and K. Gerrard, during the field trials.

REFERENCES

1) Zhang, F., Frost, D. L., Thibault, P. A. and Murray, S. B., *Shock Waves* **10**, 431-443 (2001).
2) Frost, D. L. et al., "Near-Field Impulse Effects from Detonation of Heterogeneous Explosive," in *Shock Compression of Condensed Matter*, 2001.
3) Merzhievshii, L. A. and Titov, V. M., *Combustion, Explosion, and Shock Waves* **23(5)**, 92-108 (1987).
4) Grady, D. E. and Kipp, M. E., *Int. J. Solids Structures* **32(17/18)**, 2779-2791 (1995).
5) Piekutowski, A. J., *Int. J. Impact Engng.* **20**, 639-650 (1997).
6) Lavrukhov, P. V., et al. V. V., *Combustion, Explosion, and Shock Waves* **37(6)**, 707-716 (2001).

CP706, *Shock Compression of Condensed Matter - 2003*
edited by M. D. Furnish, Y. M. Gupta, and J. W. Forbes
© 2004 American Institute of Physics 0-7354-0181-0/04/$22.00

THE STATISTICAL FRAGMENTATION THEORY OF N. F. MOTT

Dennis Grady

Applied Research Associates, 4300 San Mateo Blvd., A-220, Albuquerque, New Mexico 87110

Abstract. For a brief period during the height of World War II, Neville F. Mott left his position at the University of Bristol and headed up a concerted theoretical effort at Fort Halstead, UK, to investigate the operational science of weapons and armor technology. The seminal achievements resulting from the efforts of the participating scientists are extraordinary and have provided the basis for much of the continuing research in this field over the intervening six decades. N. F. Mott chose to study the phenomenon of the explosive-driven fragmentation of exploding shell cases. The approaches pursued by Mott are documented in several interim reports and open literature publications and offer a fascinating look into the insightful thinking and scientific methods of one of the preeminent physicists of the last century. This presentation offers a perspective into the several theoretical approaches pursued by Mott. In particular, the hallmark relation for the representation of exploding munitions fragmentation data to the present day is the Mott distribution. The efforts of Mott leading to this distribution are explored and a judgment is offered as to whether Mott himself would use this distribution today.

INTRODUCTION

The fragment size distribution developed in the seminal investigation of Mott (1) to describe the mass distribution consequences of exploding metal shells has seen continued application over the intervening decades. Minor efforts have been undertaken to expand on or modify his original relation. It has remained relatively unscathed, however, and is actively used in its original form to the present day.

Efforts to document the original developments of Mott by other authors have been confusing and often wrong. Mott's original development was put forth in rather terse internal reports and this eminent physicist took rather large steps in explaining the progression of his thoughts. His later open literature report (2) did not cover pertinent issues on the proposed Mott statistical fragment size distribution.

The present paper documents the interpretations of this author on Mott's theoretical developments and

Mott's justification of his own (the Mott) distribution from the original internal reports.

THE MOTT DISTRIBUTION AND GEOMETRIC JUSTIFICATIONS

Mott proposed his statistical fragment size distribution within the first few pages of his first internal report. Much of his subsequent efforts were focused on justification of his original proposition. As an aside it is worth noting that this intuitive approach pursued by Mott is illustrative of much of the original physics developed by this premier scientist over a long and successful career. His incisive physical intuition frequently allowed him to leap to, or close to, the fundamental outcome. His strong mathematical talents then provided him the means to build a rigorous theory around his original ideas.

The fragment size distribution put forth by Mott can be written,

$$p(m) = \frac{1}{\mu\sqrt{2m/\mu}}e^{-\sqrt{2m/\mu}} \quad , \qquad (1)$$

where m is the fragment mass, $p(m)$ the probability density function and μ the average fragment mass. This proposition was apparently guided by two inputs. First, Lineau (3) published a paper a few years earlier demonstrating that the distribution of random lengths l along the length of a line is a Poisson process and gives the distribution,

$$p(l) = \frac{1}{\mu}e^{-l/\mu} \quad . \qquad (2)$$

Second, Mott had access to exploding munitions fragment size distribution data which he found to plot sensibly like Equation (1).

It then remained for him to provide some theoretical basis for Equation (1). In the spirit of the study of Lineau on the statistical fragmentation of a line he chose to compare the proposed fragment size distribution with the random geometric fragmentation of an area. He first pursued the random fragmentation algorithm illustrated in Figure 1a in which the random spacing of vertical and horizontal lines was determined through the Lineau relation in Equation (2). Through an elegant analysis he arrived at,

$$p(a) = \frac{2}{\mu}K_o\left(2\sqrt{a/\mu}\right) , \qquad (3)$$

where K_o is a modified Bessel function. Their comparisons of Equation (3) with the proposed distribution in Equation (1) were not fully satisfying with the latter geometric distribution providing a somewhat tighter distribution (a variance of 3 as opposed to 5). They also considered the problem of partitioning the area with randomly oriented lines but this result was not fully analytic and did not provide additional enlightenment.

One further geometric algorithm for randomly partitioning an area was explored and is illustrated in Figure 1b. It was suggested by Mott that this algorithm would more closely replicate fragmentation behavior of exploding shells. This tacit recognition that the statistical partitioning of an area is dependent on geometric algorithm selected, as hinted at by Mott, weakens this random geometric approach as theoretical justification for the proposed Mott distribution. Even more

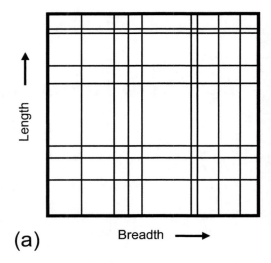

(a)

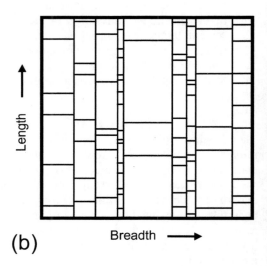

(b)

FIGURE 1. Geometric fragmentation constructions pursued in justification of the Mott distribution.

fundamental in this regard, however, is that Mott also selected a different algorithm for distributing the spacing of vertical lines and horizontal line segments. Rather than the relation of Lineau from Equation (2) he chose to use the distribution,

$$p(l) = \frac{4}{\mu}\frac{l}{\mu}e^{-2l/\mu} \quad . \qquad (4)$$

Equation (4) results from the construction in which points are placed at random on a line as in the

456

Lineau development but then fractures are assumed to occur at the bisectors of each point pair. This algorithm is the one-dimensional analog of the two- and three-dimensional Voronoi tessellation of a body. Mott's reasoning for selecting this alternative statistical distribution of the partitioning lines is not clear. This algorithm, however, results in statistically less closely spaced fractures (Equation (4)) and the present author suspects that there was an emerging recognition by Mott that the physics of the fracture process precludes arbitrarily close fractures. In any case, analysis of the area fragmentation illustrated in Figure 1b resulted in the integral expression,

$$p(a) = \frac{1}{\mu\sqrt{a/\mu}} \int_0^\infty \left(\sqrt{4a/\mu x} - 1\right) \times \\ \left(1 + \frac{1}{x^2}\right) e^{-\left(\sqrt{4a/\mu x} + 1/x^2\right)} dx \; . \quad (5)$$

Equation (5) indeed compares quite well with the Mott distribution in Equation (1) with the geometric distribution this time slightly broader than the Mott distribution.

At this point in his study Mott dropped the geometric fragmentation line of pursuit in attempting to justify Equation (1) and undertook a uniquely original physics-based approach to the statistical fragmentation problem.

- Mott I (Lineau)

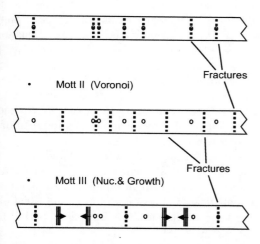

FIGURE 2. Algorithms pursued by Mott in one-dimensional statistical fragmentation.

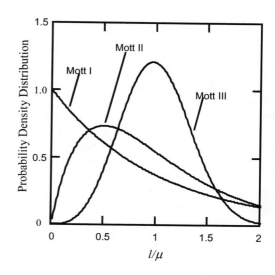

FIGURE 3. Probability density distributions in fragment length for statistical fragmentation algorithms shown in Figure 2.

THE MOTT NUCLEATION AND GROWTH THEORY OF STATISTICAL FRAGMENTATION

The decidedly different theoretical approach to the statistical fragmentation problem subsequently pursued by Mott is best understood through further reflection on the one-dimensional fragmentation problem or the statistical fracturing of a line. In Figure 2 the random placement of points on a line and the identification of these points as fractures is the straightforward logic followed by Lineau (3) in deriving the exponential fragment length distribution shown in Figure 3. The same random points but with the pair-wise bisectors identified as fractures leads to the markedly different one-dimensional Voronoi fragment length distribution. These same random points can yet again be considered in the Mott development to be described, but the physics-based algorithm which associates these points with the statistical placement of fractures and fragments is markedly different. The algorithm developed by Mott can be identified as a process of statistical fracture nucleation and growth, or, perhaps more accurately, a process of fracture activation and interaction. The latter two terms succinctly identify the two physical issues pursued by Mott in his theoretical development.

First fracture activation: In the Mott theory fractures occur on the same points placed randomly on the line but only after some time interval from the start of the event. As a physical model Mott envisioned a uniformly expanding and stretching metal ring or cylinder (the Mott cylinder). Hardening properties of the plastically stretching metal had saturated and the cylinder is expanding under a constant tensile force Y and at a constant stretching rate identified as $\dot{\varepsilon} = V / R$. The radial velocity and radius of the Mott cylinder are V and R, respectively, and do not vary significantly over the problem duration. Thus the occurrence (activation) of a fracture at a point on the line is assumed to depend on the strain (or the time) interval, $\varepsilon = \dot{\varepsilon}t$, after onset of fracture.

Mott then proposed a function of the strain $\lambda(\varepsilon)d\varepsilon$ which determined the likelihood that fracture would occur at the specified point on the line (actually any point on the line) within $d\varepsilon$. $\lambda(\varepsilon)$ is known as the hazard function, or conditional mortality function, in the statistical theory of reliability. Mott explored both the power law and the exponential law hazard functions,

$$\lambda(\varepsilon) = \frac{n}{\sigma}\left(\frac{\varepsilon}{\sigma}\right)^{n-1} , \quad \lambda(\varepsilon) = \frac{1}{\sigma}e^{-(\varepsilon-\mu)/\sigma} , \quad (6)$$

and was more enamored with the latter. The cumulative probability of fracture at a point is then,

$$P(\varepsilon) = 1 - e^{-\int \lambda(\varepsilon)d\varepsilon} , \quad (7)$$

which is an extreme value distribution of the Weibull form for a power law and of the Gumbel form for an exponential law.

With the occurrence of fracture at any point on the line statistically characterized, Mott moved on to the second issue of fracture interaction. When fracture occurs at a point the tensile force drops rapidly to zero from the value Y prior to fracture. Through a deceptively simple analysis Mott was able to show that the release of the tensile force propagated a distance $d(t)$ from the point of fracture according to,

$$d(t) = \sqrt{2Yt / \rho\dot{\varepsilon}} , \quad (8)$$

where t is the time after fracture. It is readily shown that the stress release from the point of fracture into the plastically flowing medium is a diffusion wave

(the Mott wave) rather than a constant velocity elastic wave.

With the physics composed, the algorithm for statistical fracture of the line (or the Mott cylinder) is as follows: At onset of the problem the time (or strain) at which fracture occurs at a point is determined by the fracture hazard and probability functions in Equations (6) and (7). When fracture occurs at a point Mott release waves emanate from the fracture and encompass regions of the line according to Equation (8). If points on the line are subsumed by Mott waves before fracture occurs, fracture at these points is no longer allowed. Fracture and subsequent stress release proceeds until release waves encompass the entire line at which time the fracture process is complete. The fracture problem is quantified through the following relations: The number of fractures per unit length of the line at any time (or strain) is given by,

$$N(\varepsilon) = \int_0^\varepsilon (1 - F(\eta))\lambda(\eta)d\eta , \quad (9)$$

where $F(\eta)$ is the fraction of the line encompassed by stress release waves at a strain η. $F(\eta)$ is in turn provided by,

$$F(\eta) = \int_0^\eta 2d(\eta - \xi)\lambda(\xi)d\xi . \quad (10)$$

Working with Equations (9) and (10) one can calculate the number of fractures and the distribution in fragment lengths and this was done by Mott. There is a problem, however. Equation (10) does not account for the overlapping of Mott release waves (impingement) nor does it account for the activation of fractures in already stress relieved regions (exclusion). Thus, Mott used dimensional arguments and graphical methods to extract fragment number and fragment distribution relations from Equations (9) and (10).

It is possible to analyze Equations (9) and (10) in a statistically rigorous manner. A method for accounting for exclusion and impingement is found in the statistical treatment of Johnson and Mehl (4) and has been applied to Mott fragmentation by the present author (5,6). Briefly $F(\eta)$ is calculated through Equation (10) but $F(\eta)$ in Equation (9) is replaced by $1\text{-}\exp(-F(\eta))$.

The solution to Equations (9) and (10) provide a fragment number after fracture completion for a power-law hazard function in Equation (6) of,

$$N = \beta_n \left(\frac{\rho \dot{\varepsilon}^2}{2\pi Y} \frac{n}{\sigma} \right)^{n/(2n+1)} , \quad (11)$$

where β_n is a numerical constant approaching unity for large n. Strain rate dependence ranges between two-thirds and one. For the exponential hazard function, preferred by Mott,

$$N = \sqrt{\frac{\rho \dot{\varepsilon}^2}{2\pi Y} \frac{1}{\sigma}} , \quad (12)$$

and a linear dependence on strain rate is found. The Mott γ parameter in his original development is $\gamma = 1/\sigma$ in the present representation.

Mott, working with a deck of cards and graph paper, extracted the one-dimensional fragment length distribution from Equations (9) and (10). An analytic solution was obtained by this author assuming a constant hazard function ($n=1$ in the power-law hazard function). This distribution is,

$$p(l) = \frac{\beta}{\mu} \left(\frac{l}{\mu} \right)^3 e^{-\frac{1}{4}(l/\mu)^3} \int_0^1 \left(1 - y^2\right) e^{-\frac{3}{4}(l/\mu)^3 y^2} \, dy \quad (13)$$

where β is a numerical constant. The Mott graphical and the later analytic fragment length distributions are compared in Figure 4. The two

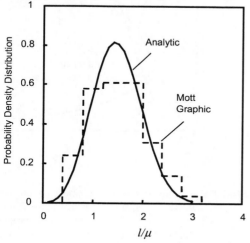

FIGURE 4. Comparison of graphical and analytic solution to the Mott one-dimensional fragment length distribution.

distributions compare well and suggest as Mott pointed out that the theoretical predictions are not overly sensitive to the fracture hazard function selected.

How does the physics-based one-dimensional theory compare with experimental data? Where data exist which are sensibly statistically homogeneous to satisfy the theoretical criteria, the theory seems to do quite well. Expanding metal ring fragmentation data reported by Grady and Benson (7) and by Grady and Olsen (8) show strain rate dependencies of fragment number in the range of two-thirds to one. Also, fragment distribution data from these studies compare well with the theoretical distributions shown in Figure 4 as due the extensive aluminum cylinder fragmentation data of Wesenberg and Sagartz (9) compared with the Mott theory in Grady (6). The latter data are shown in Figure 5 and compared with both the graphic distribution of Mott and the later analytic distribution. The temporal history of axial fractures in the expanding metal tube experiments of Winter (10) also support the Mott predictions.

What then of the two-dimensional distribution in Equation (1) initially posed by Mott to describe the exploding tubular munitions data? This closure was never fully completed. The one-dimensional fracture nucleation and growth statistical distribution he considered as descriptive of the spacing of axial fractures along the axis of the exploding cylinder. He was apparently reasonably pleased with this development as it was the portion that he emphasized in his openly published 1947 paper. The length of fragments in exploding munitions was addressed in a separate analysis within his final internal report. His conclusion was that the length to width ratio of fragments was constant (or nearly constant) at approximately 3.5. Nevertheless Mott did not reject his original distribution (Equation (1)). His thoughts are probably summarized in his quote, "apart from the theoretical significance of formula 3 (the present Equation (1)) it provides a convenient practical method of comparing the fragmentation of different projectiles". Mott had strong faith in his own physical intuition and, unlike many theoreticians, was strongly guided by experimental observation.

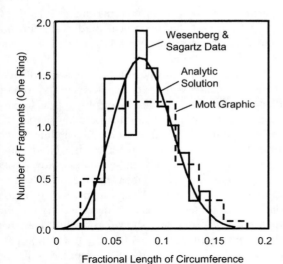

FIGURE 5. Aluminum cylinder fragment size data of Wesenberg and Sagartz compared with the Mott nucleation and growth theory.

REFERENCES

1. Mott, N.F., Ministry Supply Repts., AC3348 (with E.H. Linfoot), January; AC3642, March; AC4035, May (1943).
2. Mott, N.F. (1947), Proc. Royal Soc., **A189**, 300-308, (1947).
3. Lineau, C.C., J. Franklin Inst., **221**, 485-494, 674-686, 769-787 (1936).
4. Johnson, W.A. and R.F. Mehl, Trans. AIMME, **135**, 414-458 (1939).
5. Grady, D.E., *J. Geophys. Res.* **86**, 1047-1054 (1981).
6. Grady, D. E., Shock Waves and High-Strain-Rate Phenomena in Metals, M. A. Meyers and L. E. Murr, Eds., Plenum, New York, 181-191 (1981).
7. Grady, D.E. and D.A. Benson , Exp. Mech., **23**, 393-400 (1983).
8. Grady, D.E. and M.L. Olsen, Int. J. Impact Engng., to be published (2003).
9. Wesenberg, D.L. and M.J. Sagartz, J. Appl. Mech., **44**, 643-646 (1977).
10. Winter, R.E., *Inst. Physics, Conf. Series 47,* Bristol, 81-89 (1979).

A CLOSING CALCULATION

As a closing exercise it is intriguing to go full circle with Mott's analysis by extending his latter one-dimensional distribution to a two-dimensional fragment distribution through his earlier geometric methods. The two-dimensional distribution of Mott in Equation (1) is arrived at directly by assuming the Lineau distribution in Equation (2) for the distribution in the breadth of the fragments and by assuming that the fragment aspect ratio is constant. Alternatively, Mott assumed that the Lineau distribution described the distribution of horizontal and vertical lines as illustrated in Figure 1 and through the analysis described earlier arrived at the Bessel distribution provided in Equation 3. The same assumptions, utilizing the later one-dimensional distribution provided in Equation (13), results in the corresponding distributions shown in Figure 6. These distributions are compared with the original Mott distribution from Equation 1. Either of the two-dimensional fragment distributions based on the one-dimensional Mott nucleation and growth algorithm differ markedly from Mott's initial distribution.

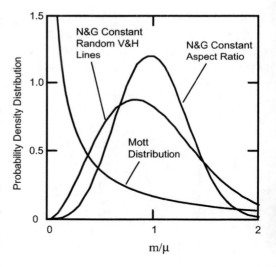

FIGURE 6. Comparison of Mott distribution with one-dimensional nucleation and growth fragment algorithm and assuming, (1) constant aspect ratio fragments and, (2) random vertical and horizontal line construction.

CP706, *Shock Compression of Condensed Matter - 2003*
edited by M. D. Furnish, Y. M. Gupta, and J. W. Forbes
© 2004 American Institute of Physics 0-7354-0181-0/04/$22.00

INFLUENCE OF SHOCK-WAVE PROFILE SHAPE ("TAYLOR-WAVE" VERSUS SQUARE-TOPPED) ON THE SHOCK-HARDENING AND SPALLATION RESPONSE OF 316L STAINLESS STEEL

G.T. Gray III, N.K. Bourne*, J.C.F. Millett*, and M.F. Lopez

Los Alamos National Laboratory, Los Alamos, NM 87545
* Defense Academy of the United Kingdom, Cranfield University, Shrivenham, Swindon, SN6 8LA, UK.

Abstract. While much has been learned over the past five decades concerning shock hardening and the spallation response of materials shock-loaded using "square-topped" shock profiles, achieved via flyer plate loading, considerably less quantitative information is known concerning direct in-contact HE-driven or triangular-wave loading profile shock prestraining on metals and alloys. In this paper the influence of shock-wave profile, using both "square-topped" and triangular-wave pulses, on the shock hardening and spallation response of 316L stainless steel is presented. The shock hardening in 316L SS, using a triangular-shaped pulse and square-topped pulse (pulse duration of 0.75 μsec) to a peak shock pressure of 6.6 GPa was found to be reasonably similar. Square-wave loading at 6.6 GPa is observed to result in incipient spallation in 316L SS while triangular-wave loading to an equivalent peak stress is quantified to exhibit no wave-profile "pull-back" nor damage evolution.

INTRODUCTION

A great deal has been learned over the past five decades concerning shock hardening and the spallation response of materials shock-loaded using "square-topped" profiles. These are achieved via flyer-plate loading from either a gas launcher or energetic drive[1,2]. Considerably less quantitative information is known concerning the effect of direct, in-contact, high explosive (HE)-driven "Taylor-wave" or triangular-wave loading profile shock loading[3]. The influence of shock prestraining, using both triangular-wave loading, via both direct HE and triangular-wave pulses on a gas launcher, as well as "square-topped" shock prestaining via conventional flyer-plate impact, is crucial to understanding shock hardening and spallation of materials. Development of predictive capability for HE-driven shocks therefore requires the formulation, verification, and validation of advanced constitutive models that include the

quantitative effect of shockwave profile shape on the mechanical behavior of materials.

Studying the physical properties of materials during the very rapid loading rate and short time interval during the actual passage of a shock wave through a material remains difficult. Experimental programs which couple both real-time and post-mortem aspects have demonstrated the greatest progress towards achieving an understanding of shock processes in ductile materials and to support the development of physically-based models describing shock loading[1,4]. In this paper the influence of shock-wave loading pulse shape on the process of shock hardening and damage evolution during spallation in 316L stainless steel (316L SS) is illustrated via "soft" recovery techniques and mechanical behavior.

EXPERIMENTAL TECHNIQUES

The material used for this investigation was a 316L stainless steel in 12.5 mm-thick plate form. The analyzed chemical composition of this 316L SS (wt. pct.) was: 0.022 carbon, 0.40 silicon, 10.03 nickel, 16.16 chromium, 2.08 molybdenum, 0.19 cobalt, 0.39 copper, 1.70 manganese, 0.0004 sulfur, 0.063 nitrogen, and 0.029 phosphorus. The texture of the 316L SS investigated was quantified using X-ray diffraction to be nearly purely random in nature. The wave speeds in the 316L SS samples were measured to be: C_L = 5.71 km/sec and C_S = 3.09 km/sec.

To examine the influence of shock-wave profile shape (triangular-shaped - "Taylor-wave" versus square-topped shock) on the spallation and shock hardening response of 316L SS, 50-mm x 50-mm square (spall) and 38-mm diameter (shock recovery) by 5 mm-thick samples were sectioned from the 316L SS plate. To assess the influence of shock-pulse shape on the shock hardening in 316L SS, samples were shock prestrained to 6.6 GPa (same peak stress) for both pulse shapes using samples fully momentum trapped within "soft" shock recovery assemblies as described previously[2]. Triangular-shaped "Taylor-wave" loading of spall and recovery samples in the current study was achieved on a 50-mm gas launcher using composite flyer plates comprised of 0.5-mm of tungsten backed by 5-mm of polymethylmethacrylate (PMMA).

Square-topped shock profiles with a pulse duration of ~0.9 µsec were produced utilizing 2.5-mm-thick stainless steel impactors to introduce a central spall plane in the sample. The 316L SS samples were impacted at 350 m s^{-1} for a nominal shock pressure of 6.6 GPa using a square-topped impactor. The spallation and shock recovery assemblies were also subjected to a "Taylor-wave" triangular-shaped pulse at 383 m/s to match the 6.6 GPa peak pressure. The square and "Taylor-wave" triangular-shaped shock loading pulses achieved are seen in Figure 1.

The longitudinal stress profiles in the spalled 316L SS samples were measured with commercial manganin stress gauges placed on the rear face of the specimens and supported with thick PMMA blocks. These gauges (Micromeasurements type LM-SS-125CH-048) have been calibrated previously by Rosenberg[5]. The impact velocity was measured to an accuracy of 0.5% using a sequential pin-shorting method and tilt was fixed to be less than 1 mrad by means of an adjustable specimen mount. The spalled samples and shock-recovery assemblies were "soft-recovered" by deceleration in layers of rags. Specimens for optical metallography were sectioned from the shock-recovered samples and prepared for optical metallography.

RESULTS AND DISCUSSION

Manganin-Gauge Spall Data

The present study was focused on quantifying the influence of triangular-wave and square-topped shock loading to a peak pressure of 6.6 GPa on the spallation and shock hardening behavior of 316L SS. The manganin-gauge traces for the 316L SS loaded using both pulse shapes shocked to 6.6 GPa are shown in Figure 1. The gauge profiles for the 316L SS loaded with both pulse shapes are seen to display a number of essentially identical features. The Hugoniot-elastic limits (HEL's) in Fig. 1 are seen to be similar for both loading pulse shapes. The HEL measured in the PMMA (seen at 0.15 GPa) converts to approximately 0.9 GPa in the 316L SS sample using elastic impedance corrections for 316L SS. Both pulse shapes are seen to achieve the same peak pressure and similar loading rise times.

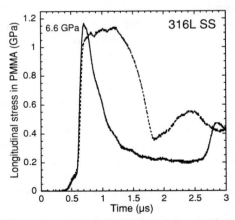

FIGURE 1. Gauge trace for 316L stainless steel spalled as a function of pulse shape (triangular and square-topped) for a peak pressure of 6.6 GPa.

The 316L SS sample loaded using a square-topped pulse with a pulse duration of ~0.9 µsec displays a "classic" pull-back signal consistent with spallation damage within the sample while the sample loaded to an identical peak shock pressure, but with a triangular-shaped loading pulse, immediately unloads the sample after the peak Hugoniot stress is achieved. Metallographic examination of the recovered samples loaded for the two-pulse shapes displayed incipient spallation damage in the square-topped 6.6. GPa sample while the triangular-wave loaded sample loaded to 6.6 GPa peak pressure displayed *NO* evidence of damage. This observation is consistent in concept with the strong influence of pulse duration on spallation documented previously[4,6]. The timing of the reverberation "ringing" in the spallation signal for the square-topped loaded sample after the "pull-back" is consistent with the traditional trapping of momentum in the scab rear portion of the spalled sample following the generation of a "free-surface" (although complete separation of neither target occurred at this stress level; i.e the spall was incipient).

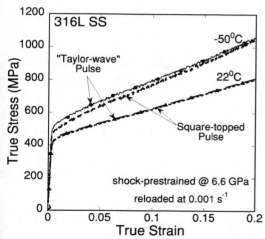

FIGURE 2. Reload stress-strain response of 316L SS as a function of pulse shape (triangular and square-topped) shock prestrained at 6.6 GPa.

Shock Hardening

The influence of shock-pulse shape on the shock

hardening in 316L SS shocked to a peak shock pressure of 6.6 GPa is presented in Figure 2. The quasi-static stress-strain response at 22°C of the 316L SS shocked prestrained via either shock-wave pulse shape is observed to be identical while the flow stress response measured at –55°C displays a slightly higher yield strength for the triangular-wave loading pulse. The total elastic and plastic strain magnitudes during the shock cycle are determined by the Hugoniot of that material. Accordingly, the observation of similarity in the total shock hardening between the two pulse shapes is understandable. The total amount of stored defects due to shock prestraining is principally controlled by the magnitude of the peak shock pressure applied and not related to the time interval at peak pressure. The one complexity to this view is the fact that the square and triangular-shaped shock loading pulses, while possessing similar loading strain rates, exhibit radically different unloading strain rates. The triangular-wave shape shock pulse exhibits a steeper unloading strain rate which is consistent with the slightly higher yielding behavior displayed during the –50°C reload sample. Future studies should focus on the quantification of the influence of shock-wave loading pulse shape on the shock hardening in a spectrum of crystal structures.

Damage Evolution

Figure 3 shows the damage evolved in the incipiently spalled 316L SS samples as a function of pulse shape for the 6.6 GPa shock peak pressure. The 316L SS sample loaded to 6.6 GPa using a triangular-wave "Taylor-Wave" pulse shape was found to exhibit no evidence of damage evolution. The square-topped pulse samples loaded to peak pressures of 6.6 GPa was seen to display void coalescence and in isolated areas evidence of the initiation of strain localization / shear-crack linkage connecting some voids.

The placement of the spall-damage region immediately adjacent to the rear of the sample is consistent with the fact that "Taylor-Wave" / triangular-wave spall drives the sample into tension immediately upon release[3].

The results of the current study illustrate the

complex and evolving nature of the effects of shock-wave loading studies of materials. Variation of the shock-wave loading pulse shape from triangular to square-topped exhibits substantially different effects on the shock hardening and damage evolution of 316L SS. The morphological changes in damage evolution for the triangular-shaped "Taylor-wave" loading support the critical importance of pulse duration on the volume-additive, and therefore time-dependent damage processes of void nucleation, void growth, void coalescence, and thereafter strain / shear localization and linkage. Conversely, the phenomenological micro-mechanisms controlling shock hardening in 316L SS appear to be similar with the exception of the influence of the higher strain rate during rarefaction release afforded by the triangular pulse shape. Future experimental and modeling studies should quantify the separate and synergistic effects of shock pulse length, pulse duration, and rarefaction rate on defect generation, storage, and damage evolution.

SUMMARY AND CONCLUSIONS

The current study of the effect of shockwave profile shape and peak shock pressure on the spallation response of 316L SS reveals that: 1) Shock-wave pulse shape (square-topped vs. triangular-wave "Taylor-wave") has a strong influence on the spallation response of 316L SS. This is reflected in both the "real-time" wave profiles and the micromechanisms of damage evolution, 2) Square-wave loading at 6.6 GPa is observed to result in incipient spallation in 316L SS while triangular-wave loading to an equivalent peak stress is quantified to exhibit no wave-profile "pull-back" nor damage evolution, and 3) The shock hardening in 316L SS was essentially identical for both "Taylor-wave" and square-topped shock loading at 6.6 GPa reloaded quasi-statically at 22°C, although the triangular-shaped shock-wave shock exhibits somewhat higher reload yield response at −50°C consistent with the higher inherent strain rates imparted by the triangular-shaped shock-wave loading pulse.

ACKNOWLEDGMENTS

This work was supported under the auspices of the US Department of Energy, specifically in part by the Joint DoD/DOE Munitions Technology Development Program. The authors acknowledge the assistance of G. Cooper for assistance with the spallation testing at the Defense Academy of the U.K.

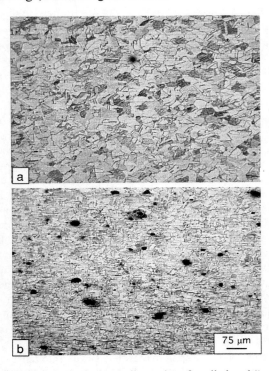

FIGURE 3. Optical metallography of spalled and "soft" recovered 316L SS samples exhibiting the differences in damage evolution between samples spalled using: a) a triangular "Taylor-wave" shaped pulse versus using b) a square-topped pulse each at a peak stress of 6.6 GPa.

REFERENCES

1. G. T. Gray III, in *High-Pressure Shock Compression of Solids*, edited by J. R. Asay and M. Shahinpoor (Springer-Verlag, New York, 1993), p. 187-216.
2. G. T. Gray III, in *ASM Handbook. Vol. 8: Mechanical Testing and Evaluation*, edited by H. Kuhn and D. Medlin (ASM International, Materials Park, Ohio, 2000), p. 530-538.
3. J. N. Johnson, J. Appl. Phys. 52, 2812-2825 (1981).
4. J. N. Johnson, G. T. Gray III, and N. K. Bourne, J. Appl. Phys. 86, 4892-4901 (1999).
5. Z. Rosenberg, D. Yaziv, and Y. Partom, J. Appl. Phys. 51, 3702-3705 (1980).
6. A. K. Zurek and M. A. Meyers, in *High Pressure Shock Compression of Solids II: Dynamic Fracture and Fragmentation*, edited by L. Davison, D. E. Grady, and M. Shahinpoor (Springer-Verlag, New York, 1996), p. 25-70.

CP706, *Shock Compression of Condensed Matter - 2003*
edited by M. D. Furnish, Y. M. Gupta, and J. W. Forbes
© 2004 American Institute of Physics 0-7354-0181-0/04/$22.00

RAPID EXPANSION AND FRACTURE OF METALLIC CYLINDERS DRIVEN BY EXPLOSIVE LOADS

T. Hiroe[1], K. Fujiwara[1], T. Abe[2], and M. Yoshida[2]

[1]*Kumamoto University, Kumamoto 860-8555, Japan*
[2]*National Institute of Advanced Industrial Science and Technology, Ibaraki 305-8565, Japan*

Abstract. Smooth walled tubular specimens of stainless steel and low-carbon steels were explosively expanded to fragmentation. The driver was a column of the high explosive PETN inserted into the central bore and initiated by exploding a fine copper wire using a discharge current from a high-voltage capacitor bank. The variation of wall thickness and the effect of different explosive driver diameters are reported. A fully charged casing model was also exploded with initiation at the end surface for comparison. Streak and framing photos show both radially and axially symmetric expansion of cylinders at average strain rates of above $10^4 \, \text{s}^{-1}$ and a wall velocity of 417-1550 m/s. Some framing photos indicate the initiation and spacing of fractures during the bursting of the cylinders. Hydro codes have been applied to simulate the experimental behavior of the cylinders, examining numerical stresses, deformation and fracture criteria. Most of the fragments were successfully recovered inside a cushion-filled chamber, and the circumferential fracture spacing of measured fragments is investigated using a fragmentation model

INTRODUCTION

The understanding of high speed deformation and fracture has been of great concern in fracture control and safety evaluation for high energy storage containers. Previously, the authors developed explosive loading devices producing planar, imploding and diverging detonation waves in powder pentaerithritoltetranitrate (PETN) with the use of exploding wire initiation. The cylindrically expanding detonation has been applied to the dynamic response of metallic cylinders at high strain-rates [1]. In other studies, Forrestal et. al. [2] also achieved uniform explosively driven expansion of cylinders at strain rates of the order 10^3s^{-1}. Grady [3] proposed a fragmentation model for explosive-filled steel cylinders, detonated at one end. In this paper, deformation and fragmentation for exploding various steel cylinders are studied both experimentally and numerically, based on uniform expansion and comparing this with axially phased expansion.

EXPERIMENTAL PROCEDURE

Experiments are performed utilizing the explosion test facilities at the Shock Wave and Condensed Matter Research Center, Kumamoto University. The cylindrical diverging detonation wave generator [1] consists of a PETN column (charged density: 0.90-0.95 g/cc) and a bundle of three copper wires (diameter 175μm) set along the central axis of the column. Axially expanding detonation was generated [1] by exploding the wires using a discharge from a high-voltage capacitor bank, (40kV, 12.5μF). Test specimens were machined from steel tubes of 18Cr-8Ni stainless steel (JIS SUS 304, dynamic proof stress: 340 MPa) or a carbon steel A (carbon 0.14%, JIS SGP-E-G, dynamic proof stress: 450 MPa). The test specimens were cylinders of 100 mm length (*L*), outer diameter (*D*$_o$) 34 mm and wall thickness (*t*) 3 mm. PETN columns of 16 mm diameter were placed inside the metallic tubes so leaving an air-layer 6 mm thick, between the explosive and the

TABLE 1. Summary of experimental data

Mat.	Cylinder sizes, mm			PETN dia, mm	$*\dot{R}$, m/s	$*\dot{\varepsilon}$, $10^4 s^{-1}$	Ave. data of fragments		
	D_o	t	L				t_f, mm	S, mm	Γ, kJ/m²
SUS304	34	3	100	28/full	1023	3.80.	1.79	6.50	140
SUS304	ditto (standard cylinder)			16	667	2.38	1.81	9.47	170
SUS304	34	1.65	100	30.7/full	1556	5.56	0.97	3.92	73
SUS304	40	6	100	28/full	759	2.57	4.14	10.39	262
C. S./ A	34	3	100	28/full	1180	3.75	1.39	6.94	177
C. S./ A	ditto (standard cylinder)			16	417	1.50	1.83	(16.22)	**(420)
C. S./ B	38	3.5	100	31	950	2.79	2.27	8.16	155
C. S./ B	ditto (casing model)			31	1035	2.51	1.49	10.18	283

$*\dot{R}$, $\dot{\varepsilon}$ are the data at the estimated fracture initiation period. **Too few fragments for statistics.

cylinder wall. When the experiment required the cylinder to be fully filled the charge diameter was 28 mm. Specimens with varying wall thickness, 1.65, 6 mm, were fabricated from SUS304, for uniform expansion experiments. A model casing cylinder (D_o-t-L: 38-3.5-100 mm) of carbon steel B (carbon 0.25%, JIS: G3445 STKM13A) with a welded end plate (D_o 48 mm, t 6 mm) of a carbon steel (JIS S55C) was fitted copper wire rows for the initiation at the open end. These samples were completely filled with PETN. A uniform expansion test of the cylinder of carbon steel B was conducted for comparison.

The deformation behavior of expanding cylinders is observed with a high speed camera: IMACON 468. Lighting was provided by a combination of a xenon lamp as a back light and front lit by the use of a mirror to reflect the flash of exploding wires at the both edges of the cylinder. A steel chamber filled with waste cloth was used to recover the fragments of exploded specimens. Table 1 shows the test conditions.

EXPERIMENTAL AND NUMERICAL EXPLODING BEHAVIOR

Figure 1a and b-1 to b-3 show a typical streak record at the midlength and framing records for a cylinder of carbon steel A (PETN: full charged), showing symmetric radial and axially uniform expansion. Crack initiation and venting of detonation gas are seen at 34 and 39 μs, these features were also seen in other cases. A circumferential corrugation in b-3 seems to be a local reduction of area due to the axial tensile stress. Time-histories of expanding wall radii obtained from the streak records are shown for all the standard cylinders in Fig. 2. The carbon steel cylinders expand later than those of SUS304, however, the average expansion velocities are very similar for both metals except in the final stages of acceleration for SUS304. These phenomena seem to be related with lower proof stress and higher work-hardening of SUS304 compared with carbon steel. The wall velocities $\dot{R}$ and the circumferential logarithmic strain rates $\dot{\varepsilon}$ ($=\dot{R}/R$) of above $10^4 s^{-1}$ estimated at fracture initiation are shown in Table 1, where R is outer radius of the cylinder and the superscript dots denote differentiation with respect to time. These velocity values are 0.55-0.93% of those on the Gurney equation [4], which ignores wave propagation effects and energy consumption in deformation and fracture.

Numerical simulations are performed for all the experiments using a hydro code, Autodyn 2D, where the Johnson-Cook constitutive model (1006 steel) and the Steinberg model are adopted for the carbon steels and SUS304 respectively. The numerical results reproduce the experiments successfully, including the observed growth of outer wall radii and the small axial deformation. Figure 3 represents the comparison of experimental and numerical time-histories of wall radii of SUS304 cylinders (PETN: full charged) with variation of wall thicknesses (1.65, 3, 6 mm). Both results coincide well, and the disjunction radii R_f suggest the critical fracture strains (ε_f) of 39-50% ($=\ln(2R_f/D_o)$). These values are almost the same as in other experiments for SUS304 and carbon steels (49-62%). On the other hand, numerical time-histories of circumferential stresses in the wall, mid-way along the column length, of SUS304 suggest that the time when the stress reaches the value of 1.2 GPa corresponds with such disjunction points. But in case of carbon steel, the stress increase due to expansion is very small at the final stage and a critical value cannot be determined.

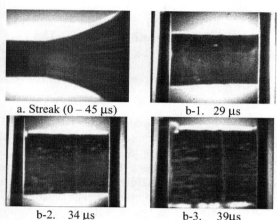

a. Streak (0 – 45 μs) b-1. 29 μs

b-2. 34 μs b-3. 39μs

Figure 1. A typical streak record (a) and framing records (b) of uniformly expanding standard cylinders (carbon steel A/ outer dia.: 34 mm, thickness: 3mm, PETN/ dia,: 28mm). Initial diameter in the streak record is 34mm.. Time 0 is a current discharged time.

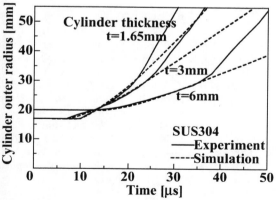

Figure 3. Experimental (solid line) and numerical (dotted line) time-histories of wall radii for uniformly expanding SUS304 cylinders with variations of wall thicknesses (1.65, 3, 6 mm).

Figure 4 show the agreement of experimental and numerical deformation behaviour for the casing model. The radial wall velocity is slightly smaller than that of the uniformly expanded cylinder (c. steel B), the initial fracture occurs at the welded joint of a cylinder and an end plate.

FRAGMENTATION AND DISCUSSION

Around 90 weight percent of the fragments were recovered without secondary damage for all tests.

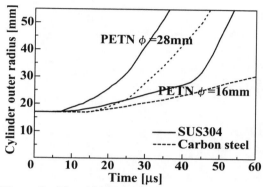

Figure 2. Time-histories of wall radii at mid-length of uniformly expanding standard cylinders (SUS304: solid line, carbon steel A: dotted line) with variations of PETN dia. (28, 16 mm).

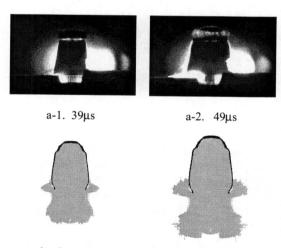

a-1. 39μs a-2. 49μs

b. Corresponding numerical simulations

Figure 4. Typical framing records (a) and numerical simulations (b) for an axially phased expanding casing model initiated at the end surface. The diameter of end plate in photo a-1 is around 50mm.

a. SUS304 cylinder

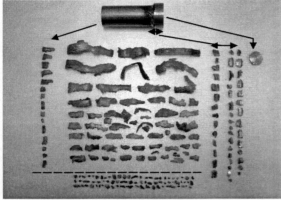

b. Casing model

Figure 5. Typical recovered fragments of a uniformly expanded cylinder (SUS304, wall thickness: 1.65mm, PETN dia.: 28mm) and an axially-phased expanded casing model (carbon steel B, t =3.5mm, PETN dia.: 31mm).

Typical recovered fragments are shown in Fig. 5a, b for a uniformly expanded cylinder and an axially-phased expanded casing model. It is seen that fracture of the cylinder portion is predominantly along elongated strips, with the fracture parallel to the axis. Most of the fragments are 3-6 times longer than they are wide, and shear fracture appears to be the dominant mechanism. Fragment size statistics of width S, thickness t_f at the central location for every fragment are determined except those of cylinder edges and very small ones, and furthermore the experimental fragmentation energy $\Gamma (= \rho \dot{\varepsilon}^2 S^3/24$, ρ: density) [3] is calculated using such data. Typical statistics of Γ for standard cylinders of SUS304 and carbon steel is expressed in Fig. 6. The average values of S, t_f and Γ are summarized in Table 1. The critical circumferential strains calculated using t_f values on the volume constant hypothesis almost correspond with those

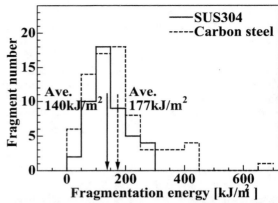

Figure 6. Typical statistics of fragmentation energy based on Grady's model for standard cylinders (PETN dia.: 28mm) of SUS304 (solid line) and carbon steel (dotted line).

estimated so far. The cylinders with thinner wall thickness expand faster and fracture with narrower spacing as shown for SUS304 steel, and the axially-phased expanded cylinder of casing model elongates to thinner fragments in comparison with the uniformly expanded cylinder. The Γ values for SUS304 and carbon steels differ only slightly except for carbon steel A (PETN dia.: 16 mm) where there is a more limited data set. However, the thickness effect in Γ is demonstrated and to account for this a more sophisticated fragmentation model is required.

REFERENCES

1. Hiroe, T., Matsuo, H., Fujiwara, K., Abe. T., and Kusumegi, K., "Uniform Expansion of Cylinders at High Strain Rates Using an Explosive Loading", Proc. Int. Conf. on *Condensed Matter under High Pressures,* National Institute of Science Communication, New Delhi, 1998, pp.458-465.

2. Forrestal, M. J., Duggin, B. W., and Butler, R. I., "An Explosive Loading Technique for the Uniform Expansion of 304 Stainless Steel Cylinders at high Strain Rates", *Trans. ASME J. of Applied Mechanic*s, Vol. 47, 1980, pp.17-20.

3. Grady, D. E., and Hightower, M. M., "Natural Fragmentation of Exploding Cylinders", in *Shock-wave and high-strain-rate phenomena in material*, Marcel Dekker, Inc., 1992, pp.713-721.

4. Gurney, R., "The Initial Velocities of Fragments from Bombs, Shells and Grenades", Report. No. 405, Ballistic Research Laboratory, 1943.

CP706, *Shock Compression of Condensed Matter - 2003*
edited by M. D. Furnish, Y. M. Gupta, and J. W. Forbes
© 2004 American Institute of Physics 0-7354-0181-0/04/$22.00

DYNAMIC DAMAGE INVESTIGATIONS USING TRIANGULAR WAVES[*]

R. S. Hixson, G. T. Gray, P. A. Rigg, L. B. Addessio and C. A. Yablinsky

Los Alamos National Laboratory, Los Alamos, NM 87545

Abstract. Many experimental investigations of dynamic damage (spall) have been done using gun-driven flat top shock waves to generate release waves that interact and cause tension. Such localized tensile pulses can cause damage to occur, and may cause the target to separate into two pieces. Metals that are subjected to shock loading as a result of being in contact with a detonating high explosive are well known to exhibit a triangular ('Taylor-wave') loading/unloading profile. When such a triangular wave reaches a free surface the lead shock is reflected as a release, interacts with the Taylor wave, and causes a tensile wave of increasing negative amplitude to propagate back into the sample. We describe here new experiments done to investigate the damage process for both flat top and triangular wave dynamic damage experiments. Both time-resolved free surface velocity results and post-experiment metallurgical examination of copper samples are described.

INTRODUCTION

Much experimental work on dynamic tensile damage has been done using shock wave techniques. (1,2,3) Most of these investigations have used supported shock waves to generate release fans; interaction of two release fans leads to dynamic tension. If the tension is of sufficient amplitude this can lead to enough damage to create a new interface in the material being studied. Even if the amount of damage created is not enough to cause complete separation, a region of substantial damage can alter material properties enough to cause different dynamic response. For example, it is known that a low impedence region in a material can cause wave reflections that look very much like complete spall separation. (4) Time-resolved experimental techniques such as VISAR (5) interferometry can provide much information on the damage process, but there still can remain ambiguities such as mentioned above. The addition

of sample recovery and post-experiment metallography can extend our understanding of dynamic damage processes considerably. We have applied these experimental techniques to investigate both flat top and triangular wave damages processes in copper, in a way similar to that done by Gray et. al in 316L stainless steel. (6)

EXPERIMENTAL TECHNIQUES

The material used in this study was Oxygen Free High Conductivity (OFHC) copper in two states; half-hard and annealed. Copper in the half-hard condition was used for the initial experiments, and the copper for the remaining experiments was studied following annealing at 600°C for 1 hour to yield equiaxed grains 40 μm in size. Samples were disk shaped (approximately 15 mm diameter by 4.5 mm thick) with three momentum-trapping rings

---
[*] This work supported by the US Department of Energy

pressed around them to minimize perturbations from edge releases. Combined VISAR and recovery experiments were performed on a 50mm gas gun facility using a technique described previously. (7) Projectile velocities and tilt were measured using electrical shorting pins. Samples were soft recovered using felt to decelerate them slowly to avoid re-shocking.

Two techniques were used to produce triangular waves in the sample. The first is similar to that used by Gray et. al (6) which uses a composite flyer. The composite flyers used in this work consisted of layers of Fe/Al/PMMA of nominal 0.25/0.40/0.50 mm. The second technique consisted of using a thin sapphire flyer combined with a tin buffer layer on front of the copper target assembly. These two techniques were used in preliminary experiments on half-hard copper; both produced triangular waves as shown in Figure 1.

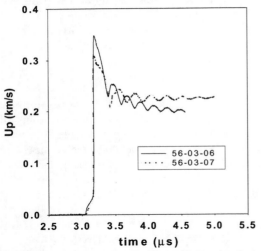

Figure 1. Comparison of two experimental techniques used to produce triangular-wave loading/unloading conditions.

The sapphire impactor technique was used for subsequent experiments because of the slightly smoother release response as well as simpler component fabrication. It should be noted that the release side of our triangular wave differs from that of a high-explosively driven metal in that it takes much less time to reduce stress.

RESULTS AND DISCUSSION

Shown in Figure 2 are data from experiments 15 and 27, as well as data from experiments 11 and 14. It appears from the VISAR data that samples for all experiments except 15 spalled completely or sustained considerable damage. Experiment 14 is noisy because VISAR light was significantly

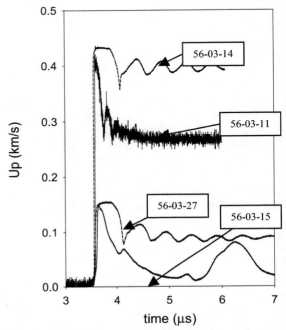

Figure 2. Experiments 15 and 27 done at a peak stress of about 30 kbar; experiments 11 and 14 done at a peak stress of about 80 kbar.

reduced at impact. As mentioned previously, the first two experiments (56-03-06 and 56-03-07) were done to prove the techniques used to generate triangular wave loading/unloading paths. The other four experiments were done in pairs of triangular/flat-top experiments at two peak stresses; roughly 80 kbar and 30 kbar. At 30 kbar peak stress the recovered samples had not been split into two pieces, but at 80 kbar peak stress complete spall was observed. Approximate spall strengths were calculated using:

$$\sigma_{spall} = \rho_0 C_b \Delta U_p \qquad (1)$$

As pointed out by Antoun et. al (8) this is an ap-

TABLE 1. Experimental details

Experiment	Flyer	Flyer Velocity (km/s)	Target Thickness (mm)	Tin Buffer Thickness (mm)	Peak Stress [a] (kbar)	Spall Stress (kbar)	Measured Spall Scab Thickness (calculated) (mm)
56-03-06	Sapphire (1mm)	0.499	4.545 (half hard)	2.469	65.2	-	-
56-03-07	Layered	0.464	4.465 (half hard)	none	58.0	-	-
56-03-11	Sapphire (1.043mm)	0.521	4.329 (annealed)	2.427	72.5	22.70	0.63 (0.429)
56-03-14	Sapphire (3.053mm)	0.401	4.246 (annealed)	none	83.0	12.82	1.08 (1.120)
56-03-15	Sapphire (1.025mm)	0.205	4.553 (annealed)	2.575	27.5	No separation	none
56-03-27	Sapphire (2.979mm)	0.142	4.563 (annealed)	none	28.5	13.2	1.147 (1.16)

[a] Peak stress for triangle waves changes as the wave moves through the sample. This stress is at shock arrival at the free surface.

proximate way to calculate tensile strength in a brittle material. A small correction (9) needs to be applied to take wave interactions into effect. In addition, it is not strictly correct to use bulk sound speed in this expression for ductile materials. It is done here for the purpose of comparison between spall done with flat top and triangular waves. From Table 1 we observe that if one calculates spall strength from a triangular wave experiment in the same way (using equation 1) as for a flat top experiment, the value of tensile strength obtained is much higher. For our copper experiments the two are almost a factor of two different; this cannot be correct. The work of Johnson (3) shows comparable spall strengths for flat top and triangle wave experiments in copper. Also shown in Table 1 are the measured spall scab thickness, where possible, and the thickness as calculated from the VISAR data. Agreement is generally good.

DAMAGE EVOLUTION

The damage evolution in OFE copper samples, subjected to flat-top-wave and triangular-wave shaped wave loading to examine spallation response, was seen to depend on both the shock pulse shape as well as the peak shock stress. Quantification of the damage evolution as a function of shock-wave profile shape and peak pressure was evaluated through metallographic examination of the "soft-recovered" damaged samples. Figure 3 shows the damage evolved in the incipiently spalled Cu samples as a function of pulse shape and peak shock stress. The Cu sample loaded to 30 kbar using a triangular-wave "Taylor-Wave" pulse shape was found to exhibit incipient ductile void formation with minimal coalescence (Fig. 4a). Conversely loading to 30 kbar using a flat-top-topped shock pulse as seen in Fig. 4b displays fully coalesced ductile void damage with final scab separation only being truncated by lateral release. This comparison graphically demonstrates the importance of pulse shape (both duration and unloading strain rate) on spallation damage in Cu. Increasing the peak shock stress was seen to lead to more extensive damage evolution. Flat-top-pulse shock spallation loading to 80 kbar is seen in Fig. 4c to result in a "clean" scab to be ejected from the sample with minimal secondary damage residual behind the main spall plane. Triangular-pulse loading to 80 kbar also resulted in a scab being ejected from the Cu sample (Fig. 4d). However in contrast to the flat-top pulse, the triangular-wave loading shock pulse at 80 kbar is seen to result in the additional formation of a fully coalesced secondary spall plane as well as tertiary void sheet planes oriented at 45 degrees to the shock propagation plane. These tertiary planes are consistent with late time damage due to lateral-release induced spallation. The damage evolution in the 80 kbar triangular-shaped loading sample (Fig.4d) illustrates the critical importance of post-mortem sample analysis in connection with "real-time" diagnostics as well as illustrating a limitation of VISAR. Once the scab has separated from the rear of the sample, the details of the time-dependent further damage evolution within the sample is cut-off from the VISAR that images ONLY the scab surface and its particle velocity.

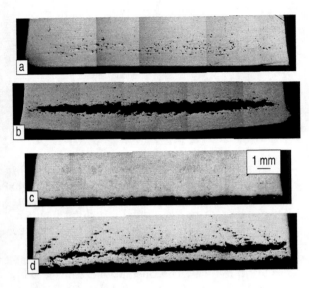

Figure 3: Optical metallography (unetched) of the spalled and "soft" recovered Cu samples illustrating the differences in damage evolution between samples spalled using triangular "Taylor-wave" versus flat-top-topped pulses for the peak pressures of 30 kbar – (a) triangular and (b) flat-top, and peak pressure of 80 kbar - (c) flat-top and (d) a triangular-pulse shock wave.

SUMMARY AND CONCLUSIONS

Experiments have been done at two peak stresses, 38 and 102 kbar, to investigate differences between flat top and triangular wave dynamic damage. We observed two spall planes for the high stress triangular wave experiments, and see clear differences in the data between triangular and flat top spall processes. This highlights the fact that in flat top spall experiments the region of maximum tension and damage is well controlled, but for triangular wave spall plane location is a variable, and there can be several regions of relatively high tension. More work is required to more fully understand these processes, but this is a good beginning. We also observed considerable differences in the amount of damage done between flat top and triangular experiments as observed from the soft-recovered samples from experiments done at 38 kbar peak stress. Work is continuing to understand root causes of these differences.

ACKNOWLEDGEMENTS

We gratefully acknowledge the work of M Byers and R. Martinez in constructing test assemblies. We are also grateful to J. Vorthman and W. Anderson for helpful comments and suggestions.

REFERENCES

1. D. E. Grady, J. Mech. Phys. Solids **36**(3), 353 (1988).
2. D. R. Curran, R. L. Seaman, and D. A. Shockey, Physics Reports **146**(5&6), 253 (1987).
3. J. N. Johnson, J. Appl. Phys. **52** (4), 2812, (1981).
4. W. R. Thissell, A. K. Zurek, D. L. Tonks, and R. S. Hixson, in Shock Compression of Condensed Matter, 1999, edited by M. D. Furnish, L, C. Chhabildas, and R. S. Hixson, number 505 in AIP Conference Proceedings, pages 451-454, Melville, New York, 2000, American Institute of Physics, AIP Press.
5. L. M. Barker and R. E. Hollenbach, J. A. P., **43**, 4669 (1972)
6. G.T. Gray III, N.K. Bourne, B.L. Henrie, and J.C.F. Millett: J. de Physique, (2003) in press.
7. R. S. Hixson, J. E. Vorthman, A. K. Zurek, W. W. Anderson, and D. L. Tonks , in Shock Compression of Condensed Matter, 1999, edited by M. D. Furnish, L, C. Chhabildas, and R. S. Hixson, number 505 in AIP Conference Proceedings, pages 489-492, Melville, New York, 2000, American Institute of Physics, AIP Press.
8. T. H. Antoun, R. L. Seaman, and D. R. Curran, Report DSWA-TR-96-77-V1, (1998).
9. V. I. Romanchenko and G. V. Stepanov, Zhurnal Prikladnoi Mekhaniki i Teknicheskoi Fiziki, **4**, 141 (1980)

CP706, *Shock Compression of Condensed Matter - 2003*
edited by M. D. Furnish, Y. M. Gupta, and J. W. Forbes
© 2004 American Institute of Physics 0-7354-0181-0/04/$22.00

DEVELOPMENT OF A NON-RADIOGRAPHIC SPALL AND DAMAGE DIAGNOSTIC

D. B. Holtkamp[1], D. A. Clark[1], M. D. Crain[2], M. D. Furnish[3], C. H. Gallegos[2], I. A. Garcia[1], D. L. Hammon[1], W. F. Hemsing[1], M. A. Shinas[1], and K. A. Thomas[1]

[1]*Los Alamos National Laboratory, Los Alamos NM 87545*
[2]*Bechtel Nevada, Los Alamos NM 87544*
[3]*Sandia National Laboratories, Albuquerque NM 87185*

Abstract. A new, non-radiographic diagnostic has been developed that appears to provide information on multiple spall and damage layers in metals. The velocities of multiple layers (up to 5 in copper) can be determined using this method, with additional information possible on damaged material between layers at densities less than the bulk metal value. Metals that are melted on release (tin) also seem to exhibit a distinctive signature that is quite different from conventional multi-layer spall. Experimental results on these metals and proton radiographs confirming these results are also presented.

INTRODUCTION

When a pressure wave produced by a high-explosive (HE) detonation reaches the free surface of most metals, different phenomena can occur: (a) one or more layers of solid material is produced from the fracture of the accelerated metal ("spall") or (b) the metal is melted on release and simultaneously accelerated. The detailed understanding of damage and spall phenomena in metals is an active area of research in shock physics but also in materials science and microstructural modeling and is of significant interest to both applied and basic science. Penetrating radiography using X-rays [1] or protons [2] is often employed to explore these phenomena. But, experimental requirements sometimes dictate that radiography may be too expensive or unwieldy to apply effectively. This led us to explore whether a non-radiographic approach might have value. Spall layer formation has historically been investigated with surface VISAR and Fabry-Perot optical probes [3]. Under some conditions, however, the surface velocity signature of damage *when considered alone* can be mistaken for an indication of complete spall and may lead to erroneous conclusions as to the extent of damage in the sample [4].

EXPERIMENTAL APPROACH

We undertook to develop a non-radiographic spall diagnostic that could be used in shock experiments where radiography was impractical. In an early paper on foil-based ejecta diagnostics, Asay suggested that ejecta masses from shocked surfaces could be measured using a thick witness foil in contact with a window [5]. Generally, investigations over the years have used thin (but robust) foils instead of coated windows to measure ejecta mass distributions since thin foils offered greater sensitivity. We named the current method after Jim Asay in recognition of this early suggestion.

The approach we adopted is shown schematically in Fig. 1. A VISAR views a metal surface through a window that is separated from the metal surface by a few mm. The concept is to use the VISAR to detect the free surface motion before impact and register the time of arrival of this surface at the window. Our hope was that after window impact there would be sufficient signal in the VISAR to measure the subsequent velocity history of the metal/window interface to obtain information on the first spall/damage layer and subsequent layers (if there are more than one). One pleasant surprise from these experiments is that the window seems to remain transparent after being impacted by the fast-moving metal and a velocity history of the interface can be reliably measured for several microseconds after impact.

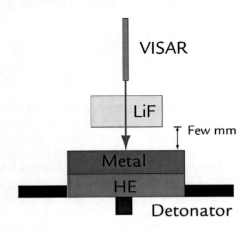

FIGURE 1. Configuration for Asay Window experiments.

A high explosive (HE) cylinder (PBX-9501, 2 inches in diameter and ½" thick) is point initiated with a single SE-1 detonator. A metal coupon (also always 2 inches diameter, of various thicknesses and materials) is glued with a thin layer of epoxy directly to the HE. A Lithium Fluoride (LiF) window is placed in an axially symmetric position a known distance (typically a few mm) above the metal coupon. The LiF window is typically 1-2" in diameter and ½" to 1" thick. A dual delay leg VISAR views the metal coupon free surface through the window near the coupon center (~1 mm spot). We exploited probe spatial filtering to be able to look through a window without seeing window surface reflections. Reconciling two different delay legs in each experiment (and the customary shocked-LiF corrections) produce quantitative, reliable data. Because the HE is point initiated, there is significant curvature to the shock wave (and the metal coupon). Modeling indicates that these 2D effects do not generally affect the

motion of the metal near the center of the coupon since the damage/spall occurs long before the edge releases play a significant role. Because the HE is unconfined, if multiple spall/damage layers are formed, they move on ballistic velocity trajectories without subsequent acceleration after the initial shock.

RESULTS AND DISCUSSION

Fig. 2 shows VISAR data for two experiments using a ½" thick OFHC copper coupon. The Taylor wave "pullback" after shock breakout near 7 μs indicates the spall strength (25.1 ± 2.9 kbars) from the change in velocity and the strain rate (~3.9 ± 0.5 x 10^4 s^{-1}) from the slope. The ring time in the velocity trace (415 ± 35 ns) can be used to calculate the thickness of the first damage layer (0.99 ± 0.08 mm) using the longitudinal sound speed of copper (4.76 km/s). The time axis is relative to detonator initiation. The downward spike in velocity (near 10 and 12 μs) at the LiF impact time is an artifact of the VISAR velocity analysis.

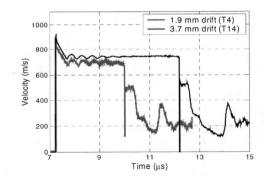

FIGURE 2. VISAR velocity data for ½" thick Cu at two metal-LiF gap distances – 1.9 mm (red) and 3.7 mm (black).

If one plots the data from Fig. 2 with the second set (3.7 mm gap) shifted earlier in time to align the LiF impact times, some interesting features are evident (Fig. 3). At LiF impact time (10 μs), the 1st spall layer undergoes a shock/release/reshock series of velocity changes after impact. Note that for the larger gap (black) the 1st layer has a longer time to undergo this successive deceleration against the window. A feature in each data set (near 11.5 and 12 μs in Fig. 3, respectively) is the velocity change at the metal/LiF interface resulting from the 2nd spall/damage layer having impacted the 1st, transferring its momentum to the interface measured by the VISAR.

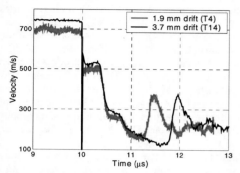

Figure 3. Plots of the same data as Fig. 2, but with the second trace (black) shifted earlier in time (by 2.19 μs) to align the LiF impact times (near 10 μs).

Using a "back-of-the-envelope" approach, one can calculate the velocity of the 2nd layer from the difference in metal/LiF gap distances (Δx) and the difference in the arrival time of the 2nd layer relative to shock breakout (Δt) for the two experiments shown in Figs. 2 and 3. This simple approach yields a 2nd layer velocity (Δx/Δt ≈ 663 m/s) as compared to a 1st layer velocity of 700-740 m/s (for the two experiments shown here).

A series of proton radiography experiments were also performed on several metals using an identical HE system and coupon geometry to that used here but without a LiF window [6]. Fig. 4 shows a magnified view of one frame (of 21) at 63.32 μs after detonator initiation. The 1st layer is clearly visible (near the top of the image) and has a radiographic thickness (vertically) of 1.03 ± 0.10 mm (FWHM). A remnant of the 2nd layer is just behind the first and has a radiographic thickness of 1.2 ± 0.1 mm and appears to have a somewhat lower radiographic density (~60% of nominal) than the 1st layer. By fitting the centroid position of the 2nd layer as a function of time in the proton radiograph series (6 were used), one obtains a velocity for the 2nd layer of 671 ± 4 m/s, in good agreement with the approximate value of 660 m/s estimated above.

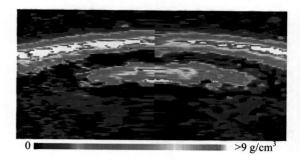

Figure 4. Magnified view of Abel inverted volume density derived from a proton radiograph of ½" Cu (ref. 6). The scale is 20 mm (H) by 10 mm (V). The HE drive is identical to the present experiments.

One may also perform a 1D calculation using a wave code such as MACRAME [7] to interpret the data. If one uses the measured thickness of the 1st layer (0.99 mm from the VISAR ring time) and the measured asymptotic VISAR velocity of 740 m/s (black trace **T14** in Figs. 2 & 3), then the parameters of the 1st damage layer are fixed by the observables in the VISAR trace before LiF impact A reasonable match between calculation and data (Fig. 5) is obtained using these parameters for the 1st layer and a 2nd layer 1.2 mm thick, having a density 60% of nominal (5.4 g/cm^3), moving at 630 m/s. Better agreement would likely result from using a more elaborate calculation with density profiles for the layers (instead of uniform density). This would also likely match the more gradual shock/release velocity profiles better, reflecting the effects of density gradients in the damaged layers.

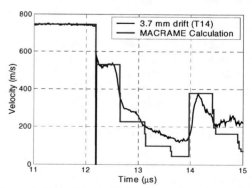

Figure 5. Comparison of 3.7 mm gap experiment (black) with a MACRAME calculation (blue).

When shocked metal is melted on release [6,8], distinctive profiles are obtained with the Asay Window technique. Fig. 6 shows two experiments with a high purity tin coupon (¼" thick) shocked with the same HE system with LiF windows at 2.4 and 3.55 mm. The velocity profile after shock breakout has very little evidence of pullback and no ringing. This indicates loss of strength, but perhaps not necessarily melt. The absence of any shock/reshock features after LiF impact though is unique for material that is melted on release, since cavitated liquid cannot sustain such wave interactions.

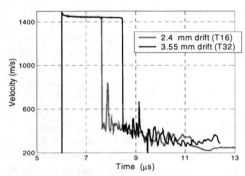

Figure 6. VISAR velocity data for ¼" thick Sn at two metal-LiF gap distances – 2.4 mm (red) and 3.55 mm (black).

Other experiments that we have conducted in this series on thinner Cu coupons (¼") show comparable features but for a much larger number of layers (up to 5). Similar experiments on tantalum, 6061-T6 aluminum, and unmelted tin also show layered features consistent with proton radiography [6]. Gas gun experiments with specially fabricated calibration targets also confirm the accuracy of the method [9]. Future work will include detailed comparisons with modern material strength and damage models in 2D geometries. Preliminary results are encouraging [10].

ACKNOWLEDGEMENTS

This work was supported in part by the Department of Energy under contract W-7405-ENG-36 to the University of California and we gratefully acknowledge their support.

REFERENCES

1. Breed, B. R., Mader, Charles L., and Venable, Douglas, "Technique for the Determination of Dynamic-Tensile-Strength Characteristics," *J. Appl. Phys.* **39**, No. 7 (1967) p. 3222; Thurston, Rodney S. and Mudd, William L., "Spallation Criteria for Numerical Computations," Los Alamos Scientific Laboratory report LA-4013 (1964); Mader, Charles L., Neal, Timothy R., and Dick, Richard D. eds., "LASL Phermex Data," Los Alamos Scientific Laboratory, University of California Press (1980).

2. King, N.S.P., *et al.*, "An 800-MeV Proton Radiography Facility for Dynamic Experiments," Nucl. Instr. & Meth. A424 (1999) pp84-91; and Kwiatkowski, K., *et al.*, "Development of Multiframe Detectors for Ultrafast Radiography with 800 MeV Protons," *IEEE Trans. Nuc. Sci.* **49**, Issue 1 (2002) pp 293-296.

3. Grady, D. E., and Kipp, M. E., "Dynamic Fracture and Fragmentation," in *High-Pressure Shock Compression of Solids*, J. R. Asay and M. Shahinpoor, eds., Springer-Verlag (1993), pp 265-322 and references therein.

4. Zurek, A.K., Thissell, W.R., Johnson, J.N., Tonks, D.L., and Hixson, R.," Micromechanics of Spall and Damage in Tantalum," *Journal of Materials Processing Technology* **60**, Issue: 1-4 June 15, 1996, pp. 261-267.

5. Asay, J. R., "Thick-Plate Technique for Measuring Ejecta from Shocked Surfaces," *J. of App. Phys.*, **49**, no.12 (1978) pp.6173-6175.

6. Holtkamp, D. B., *et al.*, "A Survey of High Explosive-Induced Damage and Spall in Selected Metals using Proton Radiography," *Shock Compression of Condensed Matter*, Portland, Oregon (2003) these proceedings.

7. Fritz, J. N., and Kennedy, J. E., "Air Cushion Effect in the Short-Pulse Initiation of Explosives," *Meeting of the Topical Group on Shock Compression of Condensed Matter of the APS* (27 Jul - 1 Aug 1997), Amherst, MA, AIP Conf. Proc., July 1998, **429**, no. 1, pp 393-396; and references therein.

8. Chapron, P., Elias, P., and Laurent, B., "Experimental Determination of the Pressure Inducing Melting in Release for Shock-Loaded Metallic Samples," *Shock Waves in Condensed Matter*, 1987 (S. C. Schmidt, N. C. Holmes eds.) pp. 171-173.

9. McCluskey, Craig W., et al (these proceedings) 2003.

10. Mason, T. A., Prime, M. B., (to be published).

CP706, *Shock Compression of Condensed Matter - 2003*
edited by M. D. Furnish, Y. M. Gupta, and J. W. Forbes
© 2004 American Institute of Physics 0-7354-0181-0/04/$22.00

A SURVEY OF HIGH EXPLOSIVE-INDUCED DAMAGE AND SPALL IN SELECTED METALS USING PROTON RADIOGRAPHY

D. B. Holtkamp[1]**, D. A. Clark**[1]**, E. N. Ferm**[1]**, R. A Gallegos**[1]**, D. Hammon**[1]**, W. F. Hemsing**[1]**, G. E. Hogan**[1]**, V. H. Holmes**[1]**, N. S. P. King**[1]**, R. Liljestrand**[1,2]**, R. P. Lopez**[1]**, F. E. Merrill**[1]**, C. L. Morris**[1]**, K. B. Morley**[1]**, M. M. Murray**[1]**, P. D. Pazuchanics**[1]**, K. P. Prestridge**[1]**, J. P. Quintana**[1]**, A. Saunders**[1]**, T. Schafer**[2]**, M. A. Shinas**[1]**, H. L. Stacy**[1]

[1]*Los Alamos National Laboratory, Los Alamos NM 87545*
[2]*Bechtel Nevada, Los Alamos NM 87544*

Abstract. Multiple spall and damage layers can be created in metal when the free surface reflects a Taylor wave generated by high explosives. These phenomena have been explored in different thicknesses of several metals (tantalum, copper, 6061 T6-aluminum, and tin) using high-energy proton radiography. Multiple images (up to 21) can be produced of the dynamic evolution of damaged material on the microsecond time scale with a <50 ns "shutter" time. Movies and multiframe still images of areal and (Abel inverted) volume densities are presented. An example of material that is likely melted on release (tin) is also presented.

INTRODUCTION

When a pressure wave produced by a high-explosive (HE) detonation reaches the free surface of most metals, different phenomena can occur: (a) one or more layers of solid material is produced from the fracture of the accelerated metal ("spall") or (b) the metal is melted on release and simultaneously accelerated. The detailed understanding of damage and spall phenomena in metals is an active area of research in shock physics but also in materials science and microstructural modeling and is of significant interest to both applied and basic science. We undertook a series of proton radiography (PRad) experiments to study HE-induced spall in several metals, including tantalum, copper, 6061-T6 aluminum, and tin. These experiments used the PRad facility in Area C at the Los Alamos Neutron Scattering Center (LANSCE) [1]. Analysis of the imagery obtained continues, but several important observations can be inferred from the data: (a) it appears that the PRad technique can produce data

on the thickness and velocity of multiple layers of spall (up to ~10 layers in tin for example); (b) although the first one or two layers appear to be quite smooth and reproducible, as one goes deeper into the damaged metal the phenomenology becomes "noisier" with a more statistical nature to the layering, at least in this HE/target geometry; and (c) when material is melted on release from the initial shock, this material is readily visible in the proton radiographs (e.g., tin) and information can be obtained on the density and velocity distribution of the melted material.

EXPERIMENTAL CONFIGURATION

The basic configuration of the experiment is shown in Fig. 1. A 2-in. diameter, 0.5-in.-thick cylinder of HE (PBX 9501) is initiated with an SE-1 detonator centered on the charge. Because the HE is point initiated, the shock wave has significant curvature. This curved geometry may be advantageous in PRad experiments since the integral of the proton path length is often shorter, and resolution and contrast may be improved, as compared to a pure planar geometry. The axial symmetry of the experiment is retained, however, which makes other advanced image processing possible (e.g., Abel inversion [2] used to produce "volume" density images). A velocity laser interferometer (VISAR) is also used to measure the time history of the surface during the experiment. Excellent agreement between VISAR and radiography results for the free surface velocity was obtained in all experiments to date. The "shutter time" of these proton radiographs is determined by the pulse width of protons that are used to produce each image frame. In these cases, the pulse width was typically less than 50 ns, a short enough time to produce minimal motion blur ($\approx$ 100 μm) even for the highest material velocity (aluminum at ~ 2 km/s).

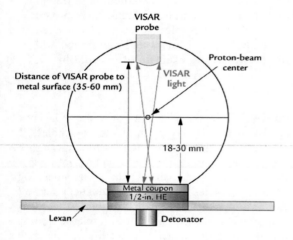

FIGURE 1. Schematic rendering of HE spall/damage experiment. The metal coupon thickness varies for different material experiments.

RESULTS AND DISCUSSION

Fig. 2 shows VISAR data for the materials studied with proton radiography. Since the HE drive is identical for all experiments, the various peak and asymptotic velocities reflect the different material properties (spall strength, shock impedance, etc). From the velocity pullback (ΔU_{fs})

in the VISAR trace, one can calculate the spall strength (σ)

$$\sigma = \tfrac{1}{2}\, \rho_0 \cdot c_0 \cdot \Delta U_{fs} \qquad (1)$$

using the density (ρ_0) and the sound speed (c_0).

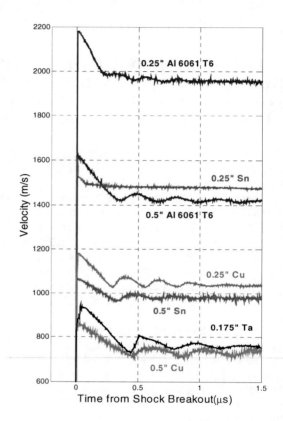

FIGURE 2. VISAR traces from several proton radiography experiments for differing thicknesses of 6061-T6 aluminum (blue), tin (red), copper (green) and tantalum (black). Zero time is adjusted to show breakout at the same time for each experiment.

There is also distinctive ringing period (Δt) in the VISAR trace for all the materials tested (except the ¼" Sn sample discussed below) that can be used to infer the thickness of the first spall/damage layer

$$\Delta x = \tfrac{1}{2}\, c_\ell \cdot \Delta t \qquad (2)$$

where c_ℓ is the longitudinal sound speed of the material.

TABLE 1. Peak pressures [3], free surface pullback velocity changes, spall strengths, ring times, and layer thicknesses obtained from the VISAR and PRad data.

PRad Shot No.	Material	P_{peak}, kbar	ΔU_{fs}, m/s	σ, kbar	Δt, ns	Δx, mm	PRad Δx, mm
#90	¼" Sn	~225	N/A	N/A	N/A	N/A	N/A
#146	½" Sn	~128	84.3 ± 9.9	8.0 ± 0.9	313 ± 14	0.536 ± 0.024	0.65 ± 0.05
#117	¼" Al 6061 T6	~197	187 ± 10	13.6 ± 0.7	209 ± 5	0.672 ± 0.017	0.62 ± 0.04
#118							0.73 ± 0.05
#119	½" Al 6061 T6	~137	186 ± 4	13.5 ± 0.3	303 ± 28	0.97 ± 0.09	0.99 ± 0.07
#120							0.93 ± 0.05
#121	¼" Cu	~250	148 ± 7	25.8 ± 1.2	268 ± 6	0.638 ± 0.014	0.60 ± 0.06
#122	½" Cu	~170	144 ± 16	25.1 ± 2.9	415 ± 35	0.99 ± 0.08	1.03 ± 0.10
#89	0.175" Ta	~310	217 ± 12	61.9 ± 3.5	364 ± 11	0.757 ± 0.022	0.87 ± 0.08

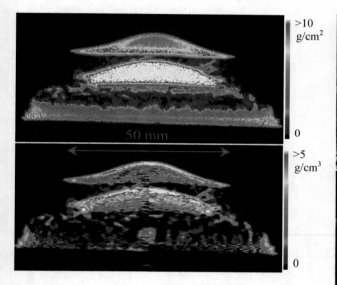

>10 g/cm²

0

>5 g/cm³

0

FIGURE 3. Proton radiograph of HE damaged Tantalum as areal densities (top) and Abel inverted volume densities (bottom) at 52.3 µs after detonator initiation. Each image is 80 mm (Horiz) by 40 mm (Vert).

Figure 3 shows a frame of the radiograph of HE damaged Tantalum (0.175") at 52.3 µs after SE-1 initiation. The colors were selected to highlight different density ranges in the areal (0-10 g/cm²) and volume density (0-5 g/cm³) images. The first damage layer is clearly visible above an undamaged portion of the Ta coupon. Below these is a region of release wave fragmented material. By analyzing 7 images (of 16) where the spall layer is clearly resolved from the rest of the Ta coupon,

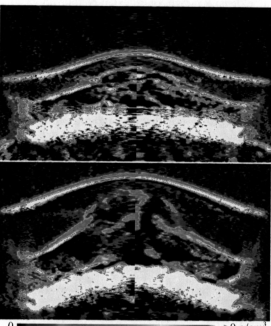

0   >9 g/cm³

FIGURE 4. Abel inverted volume densities of ¼" Copper at 32.3 µs (top) and 60.9 µs (bottom) after detonator initiation. Each image is 50 mm (Horiz) by 30 mm (Vert).

one obtains a radiographic thickness of the first layer of 0.87 ± 0.08 mm. This compares well with what one calculates using Eq. 2 (see Table 1) from the VISAR data for this experiment. It is interesting to note that immediately behind the first layer is a lower density (~1-3 g/cm³) region of damaged material. The strain rate in this experiment was ~8.4 ± 0.5 x 10^4 s^{-1} [4].

Fig. 4 shows two (of 21) Abel inverted volume densities for a ¼" Cu coupon shocked by the same HE drive at 32.3 and 60.9 µs after detonator initiation. Since the spall strength of Cu (~25 kbar) is lower than for Ta (~60 kbar), a larger number of damage layers are produced. It is evident that only the first layer remains intact, with deeper layers breaking up into 3D structures. From comparing the upper and lower images of Fig. 4, it appears that the center ~16 mm of the second layer separates from the rest of that layer and breaks up. Similarly, Fig. 5 shows one (of 21) Abel inverted volume density image for a ½" Cu coupon shocked with the same HE system. In this case the first and the partial second layer (near the top of the image) remain intact with more complicated structures arising deeper into the coupon. In the other images of the series it is evident that the second layer separates into several pieces with the central ~12 mm of the second layer following close behind the first. It is worth noting that although the strain rates

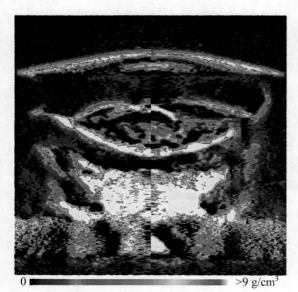

FIGURE 5. Abel inverted volume densities of ½" Copper at 63.3 µs after detonator initiation. The image is 50 mm (Horiz) by 50 mm (Vert).

of the two Cu experiments shown here differ by almost a factor of two [5], the spall strengths were observed to be the same. And yet, the evolution of

the inner layer phenomena differs significantly between the two thickness coupons (perhaps due to strain rate and/or 3D effects).

A lower spall strength material that we explored (Aluminum 6061-T6) also showed interesting results. Fig. 6 shows a single (of 21 images from PRad #120) Abel inverted volume density image of ½" Al-6061 T6 at 32.8 µs after HE initiation. The first layer is intact and well defined while the second layer is somewhat less well defined, showing some evidence of breakup near the axis of symmetry. Evidence of residual deeper layers is present but they appear to be increasingly fractured. Images from a duplicate shot (#119) look almost identical indicating good reproducibility. The strain rate in this experiment was ~5.8 ± 0.5 x 10^4 s^{-1}.

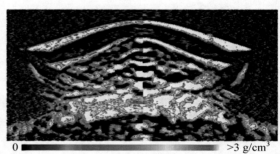

FIGURE 6. Abel inverted volume densities of ½" Aluminum 6061-T6 coupon at 32.8 µs after detonator initiation. The image is 60 mm (Horiz) by 30 mm (Vert).

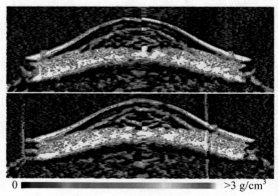

FIGURE 7. Abel inverted volume densities of ¼" Aluminum 6061-T6 coupons (#117 top and #118 bottom) at 23.2 µs after detonator initiation. Each image is 60 mm (Horiz) by 20 mm (Vert). The vertical bars (left and right side of the lower image) are artifacts of the tiled radiography.

At somewhat higher strain rate (~9 ± 1 x 10⁴ s⁻¹), thinner (¼") Al 6061-T6 coupons are shown in Fig. 7. The two images (#117 and #118) are taken at the same time with the same image processing applied. Even the first layer is beginning to break up in these images, and the residual layers deeper into the coupon have already fractured into small pieces. As before, the spall strengths (Table 1) are very reproducible over several shots but the damaged material densities inside the coupons have unique and complex structures.

The lowest spall strength material we explored was high purity (99.98%) Sn. At low pressures, it exhibited the lowest spall strength (~8 kbar) and a correspondingly large number (~10) of damage layers. Fig. 8 shows one (of 21) areal density image

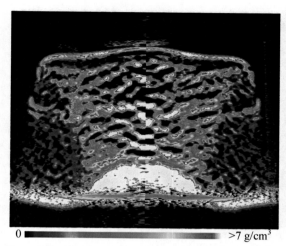

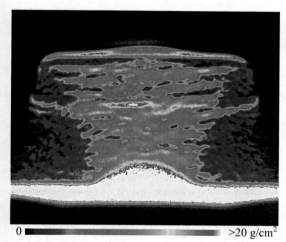

0 ▬▬▬▬▬▬▬▬▬▬ >20 g/cm²

FIGURE 8. Areal densities of ½" Tin at 43.4 μs after detonator initiation. The image is 60 mm (Horiz) by 50 mm (Vert). The saturated region near the bottom of the image is melted with the upper region still solid on release. The blue "bar" just above the first spall layer is an artifact of the radiography scintillator tile.

of a ½" thick Sn coupon at 43.4 μs after HE initiation. From this image and from the Abel inverted volume image (Fig. 9) processed from the image at the same time, it appears that much of these multiple layers, particularly near the axis of symmetry, are at near full (7.3 g/cm³) density. From this image (and others in the series), it is evident that the complex structure of the inner layers is the result of breakup of more intact layers as a function of time.

0 ▬▬▬▬▬▬▬▬▬▬ >7 g/cm³

FIGURE 9. Abel inverted volume densities of ½" Tin at 43.4 μs after detonator initiation. The image is 60 mm (Horiz) by 50 mm (Vert). The scintillator tile artifact appears just above the first layer.

When one increases the peak pressure from ~128 to ~225 kbar, the free surface of the tin is melted on release [6]. This results in a surface VISAR trace with little or no pullback and ringing (see trace labeled **0.25" Sn** in Fig. 1). The proton radiography vividly illustrates the dynamic nature of metal that is melted on release (Fig. 10).

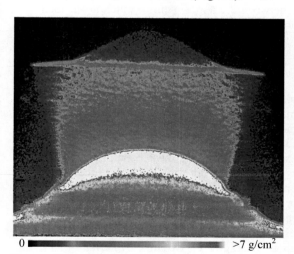

0 ▬▬▬▬▬▬▬▬▬▬ >7 g/cm²

FIGURE 10. Areal densities of ¼" Tin at 42.3 μs after detonator initiation. The image is 60 mm (Horiz) by 50 mm (Vert). Note the areal density color scale is much lower than for Fig. 8.

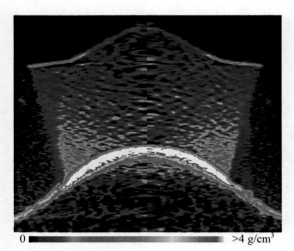

0 ███████████████████████ >4 g/cm³

FIGURE 11. Abel inverted volume densities of ¼" Tin at 42.3 μs after detonator initiation. The image is 60 mm (Horiz) by 50 mm (Vert).

When the Taylor wave reflects from the free surface of the liquid material, it generates a velocity distribution in the liquid that results in cavitation. This cavitation appears to generate "waves" in the density that are apparent over several frames in the radiography. Note that the density in the material above the uncavitated remnant (white curved disc in Fig. 11) is quite low – between 0.1 and 1 g/cm³ – much lower than the nominal density of the liquid (~7 g/cm³).

CONCLUSIONS

High explosive induced spall and damage in metals are complex phenomena with relevance to many areas of materials science. Proton radiography is becoming a useful quantitative tool for exploring fast phenomena in shock experiments. To date, these experiments have used high explosives but soon a powder breach gas gun and a pulsed power implosion facility are expected to become operational in combination with proton radiography at LANSCE. We expect that future experiments will shed light on several interesting areas relating to the dynamic evolution of shocked compressed matter.

ACKNOWLEDGEMENTS

This work was supported by the Department of Energy under contract W-7405-ENG-36 to the University of California and we gratefully acknowledge their support.

REFERENCES

1. King, N.S.P., *et al.*, "An 800-MeV Proton Radiography Facility for Dynamic Experiments," *Nucl. Instr. & Meth.* **A424** (1999) pp84-91; and Kwiatkowski, K., *et al.*, "Development of Multiframe Detectors for Ultrafast Radiography with 800 MeV Protons," *IEEE Trans. Nuc. Sci.* **49**, Issue 1 (2002) pp 293-296.
2. Abel, N. H., "Résolution d'un Problème de Mecanique," *J. Reine u. Angew. Math.* **1** (1826) p. 153; and Morris, C. L., private communication (2001).
3. Peak pressure is shown as approximate in the Table since it is difficult with VISAR to resolve the true peak free surface velocity from a HE wave profile.
4. Here the strain rate is calculated from the slope of the VISAR pullback and the bulk sound speed $\dot{\varepsilon} = -\dot{V}/(2 \cdot c_b)$.
5. PRad #121 used a ¼" Cu coupon and had a strain rate of ~7.1 ± 0.5 x 10^4 s⁻¹ while #122 used a ½" coupon and exhibited a strain rate of ~3.9 ± 0.5 x 10^4 s⁻¹.
6. Chapron, P., Elias, P., and Laurent, B., "Experimental Determination of the Pressure Inducing Melting in Release for Shock-Loaded Metallic Samples," *Shock Waves in Condensed Matter,* 1987 (S. C. Schmidt, N. C. Holmes eds.) pp. 171-173.

CP706, *Shock Compression of Condensed Matter - 2003*
edited by M. D. Furnish, Y. M. Gupta, and J. W. Forbes
© 2004 American Institute of Physics 0-7354-0181-0/04/$22.00

DISCRETE LAYER VERIFICATION OF THE LiF WINDOW SPALL DIAGNOSTIC

Craig W. McCluskey*, Mark D. Wilke*, William W. Anderson*, Mark E. Byers*, David. B. Holtkamp*, Paulo A. Rigg*, Michael D. Furnish[†] and Vincent T. Romero**

Los Alamos National Laboratory, Los Alamos, NM 87545
[†]*Sandia National Laboratory, Albuquerque, NM 87185*
***Bechtel Nevada, Los Alamos, NM 87544*

Abstract.
Recently, LiF windows suspended close to the surface have been employed as a non-radiographic spall diagnostic. Calibration has typically used HE to shock metals to produce spall layers. Because the exact characteristics of these layers cannot be pre-determined, we are using a gas gun to test the accuracy and repeatability of the diagnostic. We impact a LiF or PMMA window in front of a VISAR probe with a projectile consisting of four thin stainless steel disks spaced apart 200 microns with either vacuum or polyethylene. The measured signature from the VISAR probe is compared with what is expected from the layered assembly traveling at the projectile's velocity.

INTRODUCTION

The natural consequence of many situations of high shock loading is damage to the material shocked, frequently with layers of spall produced and occasionally with a lower density "rubble" between the spall layers. The existing methods of viewing spall layers include x-ray, VISAR[1] and proton radiography [2]. While it is possible to determine the number of spall layers, their density, and the density of material between them radiographically, there are many situations (*e.g.*, spherical or cylindrical geometries) where edge structures obscure the image making radiography impractical. This led to the development of a non-radiographic spall and damage diagnostic at the Los Alamos National Laboratory.[1] The series of experiments described in this paper, impacting a window in front of a VISAR probe with a known set of metallic layers, was conducted to help establish the validity of this new diagnostic, which, in honor of the inventor of the VISAR, is named the Asay Window.

---

[1] Holtkamp, D. B., Clark, D. A., Crain, M. D., Furnish, M. D., Gallegos, C.H., Garcia, I. A., Hammon, D. L., Hemsing, W. F., Shinas, M. A., Thomas, K. A., "Development of a non-radiographic spall and damage diagnostic," this volume.

EXPERIMENTAL PROCEDURE

For each shot, a standard Bechtel-Nevada VISAR probe was attached to a fixture that held a cylindrical window made of either LiF or PMMA in front of the probe. The fixture was then mounted on the end of the LANL Ancho Canyon single stage gas gun in a target assembly with electrical shorting pins for velocity and tilt determination and a piezoelectric pin for generating a trigger signal. Special projectiles fired from the gun carried four disks of 304 stainless steel shim stock (0.004", 0.004", 0.010", and 0.004" thick in order of impact) spaced apart by annular rings of 0.008" stainless shim stock (I.D. 25.4 mm). The volumes between the disks and between the last disk and the magnesium sabot were either left empty (vacuum) or filled with polyethylene.

The VISAR probe pointing was adjusted so the back reflections from the window surfaces went back into the source fiber, minimizing the unshifted light reflection into the collection fibers. The LiF windows were anti-reflection (AR) coated to further minimize reflections; the reflections from the PMMA windows were sufficiently low without AR coating that they were used uncoated. Two sizes of PMMA windows were used, 20 mm diameter x 20 mm long and

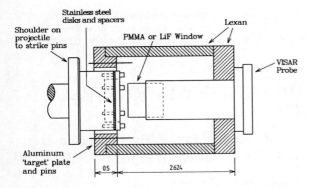

FIGURE 1. Fired projectile about to strike VISAR probe.

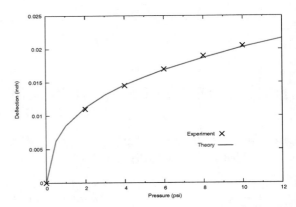

FIGURE 2. Theoretical versus measured deflection of 0.004" stainless disk.

10 mm diameter x 20 mm long; only one size of LiF window (20 mm diameter x 20 mm long) was used.

Figure 1 shows the front end of the projectile after it has been fired and is about to strike the target assembly, consisting of the shorting and piezo pins and the VISAR probe/window assembly. Typical muzzle velocity was ≈ 500 m/s.

RESULTS AND DISCUSSION

When fired from the gas gun, the projectile is still accelerating as it strikes the target plate, causing the obvious difficulty of bowing of the stainless disks. Initial attempts at modeling diaphragm deflection according to Eaton, *et al.*[3], gave non-physical results, so representative disks, larger than those used on the projectile, were mounted in a special test fixture and their deflections versus air pressure in the fixture were measured. These measurements indicated that bowing of the projectile's disks would be on the order of 0.012". The magnitude of bowing was confirmed by a gas gun test shot of a single 0.004" disk on a sabot fired into an array of piezoelectric pins, where the differences in arrival times of the pins' signals were translated into differences in position. Later discussions with William Eaton revealed the formula error in [3]. With the correct formula, as Figure 2 shows, theory does match experiment.

The effects of bowing (non-planar shock and early edge releases) were modeled using the Autodyn code[4] and found to be not significant for the on-axis light emitted and received by the LiF window of

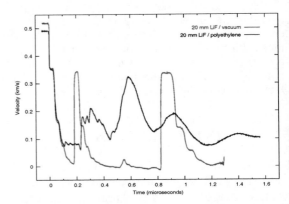

FIGURE 3. VISAR traces of vacuum and polyethylene spaced projectiles impacting 20 mm LiF windows.

the Asay Window diagnostic.[2]

Figure 3 shows the VISAR traces of vacuum and polyethylene spaced projectiles impacting 20 mm LiF windows while Figure 4 shows the VISAR traces

---

[2] The experiment was modeled in two dimensions using the cylindrical Lagrangian mesh option of the commercial hydrodynamics code Autodyne. Shock wave equations of state (EOS) were used for all materials. In addition, a Steinberg-Guinan strength model was used for the 304 stainless (SS-304). The Autodyn model included the disks, ring spacers, poly spacers when present, part of the sabot, and either the LiF or PMMA window. Each disk or spacer was modeled with 10 cells in the Z-direction. The r-direction was modeled with 430 um cells. It is likely that the large aspect ratio caused the disks to be stiffer in the model than in actuality. The LiF/PMMA was modeled with graded cells in r, size matched with the disk on the impact side.

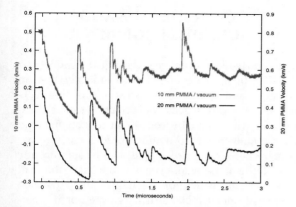

FIGURE 4. VISAR traces of vacuum spaced projectiles impacting 10 mm and 20 mm PMMA windows.

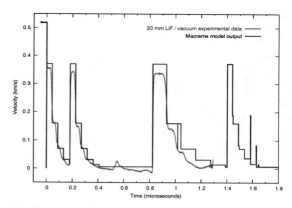

FIGURE 5. Comparison of theory and experiment for vacuum spaced projectile impacting 20 mm LiF window.

of vacuum spaced projectiles impacting 20 mm and 10 mm PMMA windows, plotted with offset double y-axes for clarity.

Most surprising in a comparison of Figures 3 and 4 is the early loss of light through the LiF windows, allowing the impacts of only three disks to be visible in the VISAR traces. The PMMA cases, in contrast, have four disks visible even with the smaller diameter window, though there is a lot going on in that trace between disks three and four, including missed VISAR fringes causing a raised baseline. The LiF windows are loaded to shock pressures of roughly 60 kilobars, well below LiF's opaque limit of ≈ 2 megabars, so the light loss must be due to the edge releases causing the stressed crystalline LiF to crumble and scatter the VISAR light. PMMA, being amorphous, is less brittle than LiF and so transmits the light longer.

Figure 4 also shows how repeatable the diagnostic is, even with different window diameters. Neglecting the timing differences due to different bowing of the disks and the baseline shift of the 10 mm PMMA trace, the ordering and shapes of the peaks are almost identical to after the impact of the fourth disk.

Figure 5 shows a comparison between the experimental data and a Macrame[5] simulation of a vacuum spaced projectile impacting a 20 mm LiF window. For this simulation, the projectile spacings were adjusted from the nominal 203.2 μm to 56 μm between the first and second disks and 290 μm between the second and third. The mechanical impulse of firing the projectile evidently excites the drumhead res-

onances of the disks, which makes it difficult to know exactly where the disks are. The Macrame modeling does not show the rebound (negative velocities) or the small bump between the impacts of the second and third disks. It also overestimates the peak velocities of the disk impacts on the LiF. These effects, as well as softening of the shock/reshock of the disks may be due to residual air in the volumes between the disks that did not have time to leak out of the ventilation holes into the evacuated barrel before firing. Air would not only cushion the impact of the second and following disks, but could also cause the second disk to rebound off the first disk and then re-impact as the third disk approaches.

Figure 3 also shows how polyethylene spacing of the disks significantly complicates the interpretation of the experiment. Nevertheless, as Figure 6 shows, it is possible to model this case in addition to that of the vacuum spaced projectile. Figure 6 also shows the difference in model results produced by adding a 35 μm (0.0014") gap between the back of the first polyethylene spacer and the front of the second disk. Polyethylene was chosen to simulate "rubble" between spall layers, but the data show it is too dense because it rapidly transmits significant shocks and releases. This complicates the analysis more than is desirable. We hope to perform shots with disks spaced by 100 mg/cm^3 open-cell foam to see if this more accurately simulates inter-layer "rubble".

Figure 7 shows a comparison between the experimental data and Autodyn[4] simulations of a vacuum spaced projectile striking a 20 mm PMMA win-

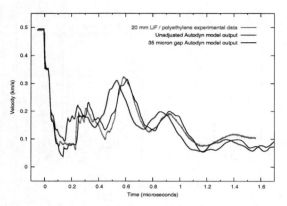

FIGURE 6. Comparison of theory and experiment for polyethylene spaced projectile impacting 20 mm LiF window.

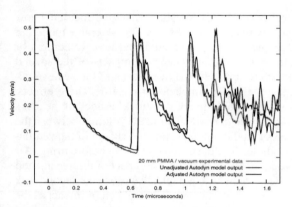

FIGURE 7. Comparison of theory and experiment for vacuum spaced projectile impacting 20 mm PMMA window.

dow. The unadjusted model used flat disks spaced 203.2 μm, while the adjusted model used flat disks with spacings of 222 μm, 99 μm, and 258 μm. There is some difference between the calculation and data during deceleration after the third disk impact, indicating that some of the disks must have been deflected and that, in contrast to our prior modeling, bowed disks do make a difference in this case. The difference may result from inadequate zoning in the r-direction, the fact the disks are not bowing equal amounts in the same direction as our earlier modeling assumed, or the difference between LiF and PMMA windows.

CONCLUSIONS AND ACKNOWLEDGEMENTS

The good agreement between the adjusted model and the data in the case of Figure 6 indicates that, in principle and given enough prior information, it is possible to model a density distribution from spalled material with simple hydrodynamic models and only simple adjustments to nominal predictions. We expect the case of HE driven spall will be significantly more difficult. Depending on how stochastic the spall process is, much iteration might be required to match the density and velocity distribution measured experimentally. This is the case in Figure 7, where several iterations were needed to match the data, and detailed features did not match probably because our model had mesh or curvature inadequacies. In addition, two dimensional effects such as uneven fractures with irregularities fine enough to not be predicted by a coarse mesh model would also present problems because of blurred impacts. Despite these limitations, the experiments in this paper demonstrate the utility of the Asay window diagnostic for measuring spall layers in shocked materials.

This work was supported by the United States Department of Energy under contract number W-7405-ENG-36; their support is gratefully acknowledged.

REFERENCES

1. Barker, L. M., "THE DEVELOPMENT OF THE VISAR, AND ITS USE IN SHOCK COMPRESSION SCIENCE," in *Shock Compression of Condensed Matter – 1999*, edited by M. D. Furnish, L. C. Chhabildas, and R. S. Hixson, 2003, pp. 11 – 17.
2. Amann, J. F. , *et al.*, High-Energy Test of Proton Radiography Concepts, LA-UR-97-1520, Los Alamos National Laboratory, Los Alamos, NM 87545 (1997).
3. Eaton, W. P., Bitsie, F., Smith, J. H., and Plummer, D. W., "A New Analytical Solution for Diaphragm Deflection and its Application to a Surface-Micromachined Pressure Sensor," in *Modeling and Simulation of Microsystems, MSM 99 (19 Apr 1999 - 21 Apr 1999: San Juan (Puerto Rico)*, SAND99-0573C Sandia National Laboratories, P. O. Box 5800, Albuquerque, NM 87185-1081, 1999.
4. Century Dynamics, Incorporated, Autodyn™, interactive non-linear dynamic analysis software, Revision 4.3, 1001 Galaxy Way, Suite 325, Concord, CA 94520, USA (2003).
5. Fritz, J. N., Los Alamos National Laboratory (1985).

CP706, *Shock Compression of Condensed Matter - 2003*
edited by M. D. Furnish, Y. M. Gupta, and J. W. Forbes
© 2004 American Institute of Physics 0-7354-0181-0/04/$22.00

SHOCK – INDUCED MESOSCOPIC PROCESSES AND DYNAMIC STRENGTH OF MATERIALS

Yuri I. Mescheryakov and A.K. Divakov

Institute of Problems of Mechanical Engineering RAS, V.O. Bol'shoi 61, St.Petersburg 199178, Russia

Abstract. Dynamic failure during spallation is found to be correlated with the instability threshold of a material under compression at the front of loading pulse. The instability occurs in the form of structural transition when the shock self-consistently establishes the scale features at the mesoscopic scale. The threshold of structural instability of material under shock compression is shown to have a concrete mechanical meaning - it determines a strength component of resistance of material to high-velocity penetration of rods into target in Alekseevskii-Tate model.

INTRODUCTION

Spall fracture of material is known to be a result of interference of release waves propagating from rear surfaces of impactor and target. A common thread is that spall-strength is a sufficiently objective characteristic of tensile strength of material at the microsecond region of dynamic loading. In reality, however, a preliminary compression of material takes place during passing the front of compressive pulse. If dynamic compression achieves a critical value, the irreversible structural changes of solid occur before the tensile stresses generate within spall zone of target. Thus, dynamic failure during spallation sensitively depends on plastic instability of a material under compression at the front of loading pulse. The plastic instability can be considered as a strain-rate dependent structural phase transition by means of which the shock self-consistently establishes the scale features at the mesoscopic scale. The structural re-arrangements of dynamically deformed material reflect the energy exchange between mesoscopic and macroscopic scale levels whereas the particle velocity dispersion, is a quantitative characteristic of this process[1,2]. In the paper presented, the effect of processes at the front of compressive pulse on dynamic strength of materials during spallation and penetration is explored.

EXPERIMENTS AND RESUTS

Experiments on shock loading of different solids under uniaxial strain conditions within impact velocity range of $50 \div 600$ m/s were performed by using a one stage light gas gun facility of 37 mm bore diameter. The thickness of target and impactor were adjusted to provide spallation. Free surface velocity profiles $u_{fs}(t)$, were recorded with two-channel velocity interferometer [1]. The experimental technique provides also recording a time history of distribution (dispersion) in particle velocity during the shock deformation. The shock wave is considered to be an averaged motion of medium at the background of which the velocity fluctuations take place at one or more scale levels

Shape of the front of compressive pulse is known to reflect a dynamics of processes flowing at different scale levels. The quantitative characteristics of macroscopic response of medium on impact is a time-resolved profile $u_{fs}(t)$. Just that profile is commonly registered in the uniaxial strain shock tests. Characteristic of fluctuative processes at the microlevel is a temperature as a measure of chaotic motions of atoms. At higher scale levels, such as mesoscopic scale level, the quantitative characteristic of fluctuative features of medium is a velocity dispersion of mesoparticles.

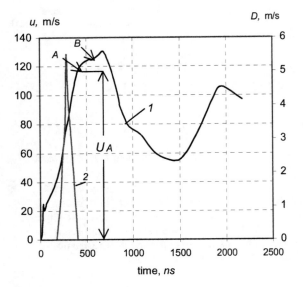

FIGURE 1. Free surface velocity profile (1) and velocity dispersion (2) for beryllium.

Behavior of velocity dispersion during the shock deformation depends on whether or not the shock wave is steady or non-steady [2].

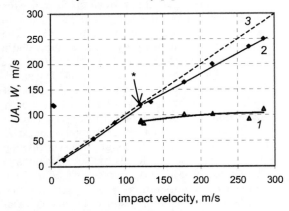

FIGURE 2. Threshold of structural transition U_A (1) and pull-back velocity W (2) versus impact velocity for the beryllium. Point of instability is indicated by symbol, line (3) is $U_{fs} = U_{imp}$.

Presented in Fig.1 free surface velocity profile shows that inclination of plastic front changes when the velocity dispersion becomes equaled to zero. Beginning from this moment, the stress relaxes by means of other mechanism.

Initiation of this mechanism happens at point A and continues for a transient stage AB, after that the particle velocity grows again, although much slowly. As a matter of fact, transition to new regime of dynamic deformation can be considered as a structural phase transformation initiated by shock loading. Value of mean particle velocity corresponding to the onset of transient stage, U_A, determines a threshold of structural transition.

In Fig. 2 a dependence of the free surface velocity corresponding to the instability threshold on the impactor velocity, $U_A = \mathrm{f}(U_{imp})$, for polycrystalline beryllium is presented. One can see that break of dependence occurs at the impact velocity of 120 m/s, after that a slope of the dependence decreases. In this figure, a dependence of pull-back velocity, $W = \mathrm{f}(U_{imp})$, is also plotted. Spallation begins just at the impact velocity where a break of the dependence $U_A = \mathrm{f}(U_{imp})$ happens.

The structural transitions under dynamic compression are identified for armco-iron, armor and maraging steels, aluminum alloys and other metals.

There exists a direct coupling of instability threshold and spall strength - the higher instability threshold, the higher spall-strength of the material. To check this assertion, two sets of 38KhN3MFA constructional steel targets have been tested under identical shock conditions. The first set of steel targets was as supplied while the second set was subjected to standard for this steel thermal treatment. Results of tests are presented in Fig. 4. For the first set of steel, the structural instability occurs at the impact velocity of 200 m/s while for the second set – at the velocity of 273 m/s. Accordingly, pull-back velocity equals 170 m/s for the first set of steel and 200 m/s for the second set. Thus, spall-strength of material is sensitively dependent on the threshold of dynamic instability under compression and spall-tests don't provide objective information about tensile strength of solid at the microsecond range of dynamic loading.

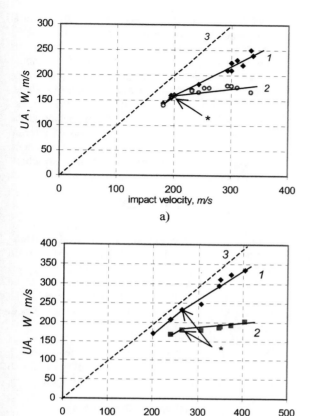

a)

b)

FIGURE 3. Velocity of structural transition U_A (1) and pull-back velocity W (2) versus impact velocity for the first set of 38KhN3MFA steel (*a*) and for the second set (*b*). Point of instability is indicated by symbol *, line (3) is $U_{fs} = U_{imp}$.

STRUCTURAL INSTABILITY UNDER COMPRESSION AND RESISTANCE TO HIGH-VELOCITY PENETRATION

Normal stress at which irreversible structural transition at the front of compression pulse occurs must be considered as independent strength-characteristic of material determining the threshold instability of material under dynamic compression. This strength characteristic is thought to be

important not only in developing the theoretical models of dynamic fracture but in calculating the parameters of high-velocity penetration processes. The penetration depth is known to be determined on the basis of modified Bernoulli equation [3]

$$Y + \frac{1}{2}\rho_s(v-u)^2 = \frac{1}{2}\rho_t u^2 + R . \qquad (1)$$

Here v *is* the impactor velocity, u is the particle velocity in the material of target, and Y and R are the empirical constants defining a dynamic strength for the material of penetrator and target, respectively. In the well-known reviews on high-velocity penetration it is claimed that physical meaning of parameters Y and R remains to be unclear [3,4]. Value R takes into account a deviation in behavior of the material of target from the hydrodynamic model of penetration. Micromechanisms of dynamic deformation responsible for physical nature and value of R are the subject of investigations in microplasticity. The parameter R is often identified with the dynamic hardness HD which connected to dynamic yielding limit, Y_D, by the following correlation dependence [3,4]

$$H_D = (3 \div 3.5)\, Y_D. \qquad (2)$$

In turn, dynamic yielding limit is determined by Hugoniot elastic limit, σ_{HEL}:

$$Y_D = \frac{1-2\nu}{1-\nu}\sigma_{HEL}, \qquad (3)$$

where ν is the Poisson coefficient.

The main conclusion following from the analysis of high-velocity penetration is that the strength-component of resistance of solids to penetration, as a complementary factor for the inertial forces, is determined by the resistance to plastic deformation. This means that if the character of plastic deformation changes, for example, because of change of structural mechanism of deformation, the strength-component of resistance to penetration changes as well. Analysis of shock-wave processes during penetration shows that inside the target, in vicinity of the head of penetrator, at the so-calles stagnation point (critical point of flows within target near the penetrator), the iniaxial strain conditions are realized [2].

The high-velocity penetration tests don't provide a quantitative information about change of macroscopic response of target on impact due to change of the mechanism of plastic deformation. As distinct from the high-velocity penetration, shock-wave tests under uniaxial strain conditions are sufficiently informative. This permits to apply the results obtained under planar collision experiments as a strength-characteristic of target for penetration tests. In this case, as strength-components of resistance of target to penetration R, the stress corresponding to instability threshold of material under compression may be taken.

$$R = \sigma_{\mathrm{i}} = \frac{1}{2} \rho_{\scriptscriptstyle M} C_{pl} U_{\scriptscriptstyle A} \qquad (4)$$

Here C_{pl} is the velocity of plastic front in uniaxial strain tests and U_{A} is the particle velocity corresponding to loss of structural stability under compression at the front of pulse.

In Table 1 the results of calculation of value R by using formulae (2)-(3), on one hand side, and formula (4), on another hand side, are presented for the set of construtional materials. It's seen that R values determined according Tate's technique, and dynamic instability threshold values determined under planar tests are practically coincide.

TABLE 1. Comparison of strength-component of resistance to high-velocity penetration R and instability threshold σ_i for different materials.

Material of target	U_{HEL} m/s	R. GPa	C_p km/s	U_A m/s	σ_i GPa
D-16 Al alloy	40	0,6	5,35	80	0,58
M-2 copper	5	2,44	4,38	132	2,54
38KhN3MFA st	75	4,01	5,0	210	3,96

This means that strength-component of material resistance to high-velocity penetration has a concrete physical meaning – its value is determined by the structural instability threshold of material under uniaxial strain conditions. In the case of steady plastic waves, instability under shock compression is initiated at the impact velocity when the particle velocity dispesion at the mesoscale, as a means for relaxation of internal stresses, is entirely exhausted. At that moment, relaxation of internal stresses at higher structural scale is initiated.

SUMMARY

Analysis of the processes at the front of compression pulse shows that spall-tests don't provide an objective information on the tension strength of material in microsecond region of dynamic loading. The spall-strength proves to be sensitively depends on whether or not the irreversible processes of structural stability loss at the front of pulse occur.

The threshold of structural instability of material under shock compression determines a strength component of resistance of material to high-velocity penetration of rods into target.

ACKNOWLEDGMENTS

The work was performed in the frame of Grant FPC of Ministry of Industry and Science of Russian Federation № 40.010.11.1195 and VW Research Project 1/74 645.

REFERENCES

1. Mescheryakov Yu.I. and Divakov A.K. Multiscale kinetics of microstructure and strain-rate dependence of materials. *Dymat Journal* № 1, 271-28(1994).
2. Khantuleva T.A. and Mescheryakov Yu.I. Nonlocal theory of the high-strain-rate processes in structured media. *Int. J. Solids and Structures*. **36**, 3105-3129 (1999).
4. Hohler V. and Stilp A.J. Long-rod penetration mechanics. High velocity impact dynamics. Edited by Jonas A. Zukas. ISBN 0-471-51444-6. John Wiley & Sons, Inc. 1990.
5. Tate A. *J. Mech Sci.* **19**, 121-122 (1977).

CP706, *Shock Compression of Condensed Matter - 2003*
edited by M. D. Furnish, Y. M. Gupta, and J. W. Forbes
© 2004 American Institute of Physics 0-7354-0181-0/04/$22.00

SUB-MICROSECOND YIELD AND TENSILE STRENGTHS OF METALS AND ALLOYS AT ELEVATED TEMPERATURES

S.V. Razorenov [1], A.S. Savinykh [1], G.I. Kanel [2], and S.N. Skakun [1]

[1]*Institute of Problems of Chemical Physics, Chernogolovka, Moscow region, 142432 Russia*
[2] *Institute for High Energy Densities,IVTAN, Izhorskaya 13/19, Moscow, 125412 Russia*

Abstract. The shock-wave tests have been carried out for Armco-iron, aluminum and Al-6%Mg alloy, and titanium of different purity at temperatures up to 600°C. The dynamic yield stresses are practically athermal for Al-6%Mg alloy and titanium of commercial purity, decrease for Armco-iron and anomalously increase for titanium of high purity with increasing the temperature. The different behaviors of metals and alloys are treated in terms of relationship between phonon drag of motion of dislocations and the drag forces created by obstacles. The spall strength slightly decreases with increasing the temperature and, for polycrystalline aluminum and aluminum alloy, precipitously drops as temperature approaches the melting point. The waveforms for pure titanium indicate irreversible polymorphic transformations at normal and elevated temperatures.

INTRODUCTION

Non-trivial behavior of the strength properties of polycrystalline metals and metal single crystals has been revealed[1-4] when temperature was introduced as a variable parameter in shock-wave experiments. It was found that the resistance to high-rate fracture of metals does not vary much when increasing the temperature at least up to 85-90% of absolute melting temperature, T_m. With further temperature increase polycrystalline metals exhibit a precipitous decrease of strength down to practically zero as soon as temperature approaches T_m, whereas the dynamic tensile strength of single crystals remains high even in a close vicinity of T_m. The dynamic yield stress of some metals anomalously increases with heating.

In order to understand what governs the resistance to high-rate inelastic deformation and fracture at elevated temperatures, it would be desirable to compare behaviors of different metals and alloys. In this paper we present new experimental data for Armco-iron, aluminum and Al-6%Mg alloy, and titanium of different purity at temperatures up to 600°C.

EXPERIMENTAL PROCEDURE

In the experiments, one-dimensional shock loads were created in the samples by impacts of aluminum flyer plates 0.4 to 2 mm in thickness. The impactors were launched by explosive facilities with velocities of 0.7 to 2 km/s. The test temperature was varied using resistive heaters as it was described earlier.[1] The free surface velocity histories of impacted samples were recorded with the VISAR laser Doppler velocimeter. All experiments were done in air without evacuation.

RESULTS OF MEASUREMENTS

Figure 1 compares the free surface velocity histories of aluminum single crystals[3] and Al-6%Mg alloy[5] at normal and elevated temperatures. The wave profiles for single crystals show significant increase in the precursor wave amplitude with increasing the temperature. Whereas at room temperature the peak stress in the elastic precursor wave for single crystals is much lower than that for the alloy, they gradually become equal as the temperature increases. At highest temperatures near the melting point single crystals maintain their spall strength whereas the alloy gradually loses it.

Figure 2 presents the yield stress data for aluminum single crystals and Al–6% Mg alloy. Since Poisson's ratio increases with temperature, the yield stress is less sensitive to the temperature than is the Hugoniot elastic limit. At room temperature the alloy has the shock yield stress which approximately equals to that of aluminum crystal at highest temperatures; at 600°C both materials show practically the same yield stresses.

Figure 3 summarizes the results of measurements of the dynamic tensile strength of Al–6% Mg alloy in comparison with similar data for high-purity polycrystalline aluminum, commercial aluminum AD1,[1] and aluminum single crystals. The point for alloy at 600°C indicates rather upper estimation of the strength; in fact this temperature is above the alloy solidus at which the alloy begins to melt. Independently on the purity, polycrystalline aluminum loses its strength with approaching the melting temperature. Whereas the alloy and commercial aluminum have practically the same strength at normal and moderate temperatures, partial melting of alloy at the solidus temperature sharply decreases its strength relatively aluminum near 600°C.

Different shock behavior of pure metals and

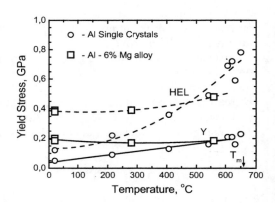

FIGURE 2. Hugoniot elastic limits and the yield stresses of aluminum single crystals and Al-6% Mg alloy as functions of the temperature. T_m is the aluminum melting temperature.

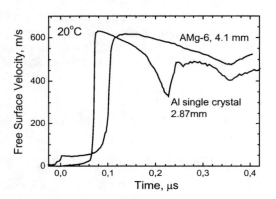

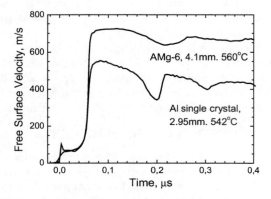

FIGURE 1. Free surface velocity histories of samples of Al-6%Mg alloy shocked at normal and elevated temperatures in comparison with the data for aluminum single crystals.

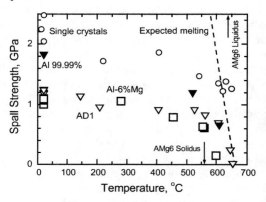

FIGURE 3. Spall strength of different aluminums and aluminum alloy as functions of the temperature. The data for aluminum of 99.99% purity have been obtained at higher strain rate than the rest data.

metal alloys at elevated temperatures was confirmed by experiments with titanium. Figure 4 presents examples of free surface velocity histories for commercial titanium VT1-0 of 99.28% purity and for titanium of 99.99% purity. The VT1-0 titanium contained 0.18% Fe, 0.1% Si, 0.07% C, 0.01% H, 0.04% N, and 0.12% O. As it could be expected, impurities increase the HEL of titanium. For both materials the data demonstrate growth of the Hugoniot elastic limits with heating. Note that at lower peak stress and, correspondingly, lower strain rate similar commercial titanium demonstrated decrease of HEL at elevated temperature.[6]

The plastic compression wave in pure titanium has a unique feature: the rate of increase of the surface velocity drops abruptly after the velocity reaches about 200 m/s at 20°C or ~270 m/s at 465°C. Obviously, we detected known α→ω

polymorphic transformation during shock compression in these experiments. The values of the transformation pressures are 2.37 GPa at 20°C and 3.05 GPa at 465°C. No features possibly to be related to reverse ω→α transformation were detected at the unloading parts of the velocity profiles. Commercial titanium did not exhibit the transformation at peak shock pressures up to 10 GPa.

Figures 5 and 6 compare the yield stress and spall strength data for the Ti-6-22-22S alloy,[7] commercial grade titanium, and titanium of 99.99% purity. Whereas the HEL of commercial titanium increases with temperature, its yield stress (which was calculated accounting for the

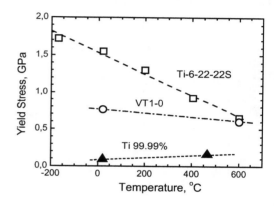

FIGURE 5. Temperature dependencies of dynamic yield strengths of Ti-6-22-22S alloy, commercial grade titanium, and titanium of 99.99% purity.

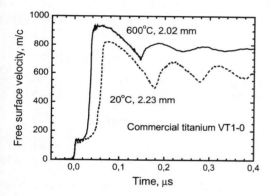

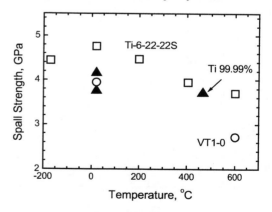

FIGURE 6. Temperature dependencies of the spall strength of Ti-6-22-22S alloy, commercial grade titanium, and titanium of 99.99% purity.

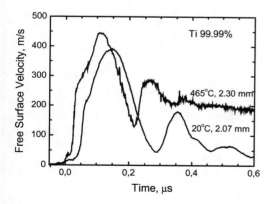

FIGURE 4. Free surface velocity histories of commercial titanium VT1-0 and high-purity titanium samples at the room and elevated temperatures.

temperature dependencies of elastic modules) slightly decreases with heating. As in the case of aluminum, pure titanium shows an anomalous increase of the yield strength with increasing temperature.

Figures 7 and 8 present the results of experiments with Armco-iron. The yield stress decreases with heating whereas the spall strength is maintained approximately constant. The wave profiles demonstrate decrease of the transformation pressure with increasing temperature.

DISCUSSION

Experiments confirmed different shock behavior of pure metals and metal alloys at elevated temperatures. In general, shock response of alloys does not contradict to the hypothesis[4] that phonon friction is the dominant mechanism at high strain rates and elevated temperatures. The flow stress in pure crystals is small and comparable with the phonon friction forces, therefore growth of the latter makes an essential contribution to the drag of the dislocations. In contrast to pure metals, the alloys contains numerous obstacles such as inter-phase boundaries, inclusions, etc., that have been created specifically to increase the yield strength. As a result, the stress needed to overcome these obstacles far exceeds the forces of phonon drag which are, therefore, unable to make a significant contribution into the resistance to plastic flow. The transition from "thermal softening" to "thermal strengthening" depends on the strain rate and occurs between 0.2 and 0.5 GPa of the yield stress for the strain rates of order of 10^5 s^{-1} to 10^6 s^{-1}.

Behavior of the spall strength of aluminum and aluminum alloy near the melting temperature is in agreement with the hypothesis of pre-melting phenomena in polycrystalline metals.

ACKNOWLEDGEMENT

The work was supported by Russian Foundation for Basic Research under grant number 03-02-16379.

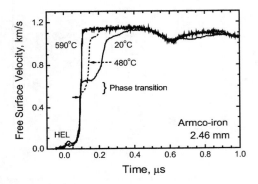

FIGURE 7. Free surface velocity histories of Armco iron plates impacted by aluminum flyer plates 2 mm thick at the impact velocity of 1.9±0.05 km/s.

REFERENCES

1. G.I. Kanel, S.V. Razorenov, A.A. Bogatch, A.V. Utkin, V.E. Fortov, and D.E. Grady. *J.Appl.Phys.* **79**(11), 8310–8317 (1996).
2. A.A. Bogatch, G.I. Kanel, S.V. Razorenov, A.V. Utkin, S.G. Protasova, and V.G. Sursaeva. *Physics of the Solid State* **40**(10), 1676–1680 (1998).
3. G.I. Kanel, S.V. Razorenov, K. Baumung, and J. Singer. *J. Appl. Phys.* **90**(1), 136–143 (2001).
4. G.I. Kanel, S.V. Razorenov, K. Baumung, and H. Bluhm. *In: Shock compression of condensed matter – 2001,* eds. M. D. Furnish, Y. Horie, and N. N. Thadhani, *AIP CP* **620**, 603–606 (2002)
5. S.V. Razorenov, G.I. Kanel, and V.E. Fortov. *Phys. of Metals and Metallography,* **95**(1), 86–91 (2003).
6. G. I. Kanel, S. V. Razorenov, E. B. Zaretsky, B. Herrman, and L. Meyer. *Physics of the Solid State,* 45(4), 656–661 (2003).
7. L. Krüger, G. I. Kanel, S. V. Razorenov, L. Meyer, and G. S. Bezrouchko. *In: Shock compression of condensed matter – 2001,* eds. M. D. Furnish, Y. Horie, and N. N. Thadhani, *AIP CP* **620**, 1327–1330 (2002)

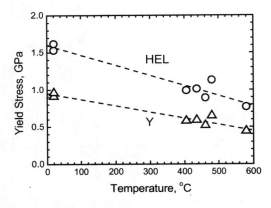

FIGURE 8. Temperature dependencies of the Hugoniot elastic limit and the yield stress for Armco iron.

CP706, *Shock Compression of Condensed Matter - 2003*
edited by M. D. Furnish, Y. M. Gupta, and J. W. Forbes
© 2004 American Institute of Physics 0-7354-0181-0/04/$22.00

DYNAMIC FAILURE RESISTANCE OF TWO TANTALUM MATERIALS WITH DIFFERENT MELT PRACTICE SEQUENCES

W. Richards Thissell[1], Davis L Tonks[2], Dan Schwartz[3], and Joel House[4]

[1]MST-8: Structure-Property Relationships, MS: G755, [2] X-7: Materials Modeling, MS: F699, [3] NMT-16: Plutonium Metallurgy, MS: G721, Los Alamos National Laboratory, Los Alamos NM 87545
[4] United States Air Force, AFRL/MNMW, 101 West Eglin Blvd, Suite 135, Eglin AFB, FL 32542-6810

The dynamic failure resistance of a Cabot Ta is compared to that of a Starck Ta under nearly identical loading conditions. The two materials have nominally very similar grain sizes, texture, and bulk impurity contents. The two materials do differ in the melt practice used, the Cabot material underwent triple e-beam re-melting, while the Starck material underwent a double e-beam re-melting followed by a vacuum arc re-melt (VAR). Melt practice strongly influences the material cleanliness in most materials and hence greatly influences fracture properties such as fatigue resistance and fracture toughness. The samples were tested in a flyer plate experiment with momentum trapping and soft recovery. A VISAR recorded the free surface velocity profile of the samples. The resulting damage in the microstructures was quantified, statistically reduced and used in developing separate parameters for a damage model. Comparisons between simulation predictions and experimental measurements of free surface velocity, porosity distributions, and volumetric number density distributions of voids are presented.

INTRODUCTION

Fracture properties such as fatigue resistance, fracture toughness, and dynamic failure resistance are strongly influenced by a material's microstructure. Secondary melt practice (SMP) is also known to strongly influence fatigue resistance in many materials. SMP is the sequence of melting stages a material undergoes with the goal of improving its cleanliness by the removal of second phase particles and possible selective evaporative of undesirable tramp elements. The solidification texture of certain SMPs can sometimes be resilient towards subsequent thermo-mechanical processing steps and can therefore still be manifested in the finished microstructure. [1] Hence, it can be challenging to create identical microstructures from material that underwent different SMP.

This paper describes the dynamic incipient failure response of two different tantalum materials whose microstructures are somewhat similar but were processed via different SMPs.

DESCRIPTION OF MATERIAL

The Cabot material has a secondary SMP that consisted of triple e-beam re-melting. The subsequent thermomechanical processing has been described. [2, 3] The Starck material had a secondary SMP that consisted of double e-beam re-melting followed by a vacuum arc re-melting (VAR). The subsequent thermo-mechanical processing included swaging in 12 mm increments from the starting diameter of 150mm to 63mm followed by upset forging. Each material received a one hour vacuum recrystallization anneal heat treatment. Starck was heat treated at 1150°C, and Cabot was annealed at 1100°C for one hour. Both materials have a mid-plane <111> texture component of about 8 times random. The Cabot material had some grain size banding but the Starck material had a relatively homogenous grain size distribution. Table 1 lists the grain size using the 1.125 times the mean intercept length method.

Material	Lab	Grain Size, μm	C	N	O	H	Fe	Ni	Cr	Mo	W	Nb
Cabot	Cabot	48	10	<10	<50	<5	<5	<5	<5	11	<25	<25
Starck	Starck	30	14	25	71	6	12	16	n/a	99	55	49

Table 1: Tantalum Impurity Levels, in ppm wt %

RESULTS

Mechanical tests

Momentum trapped incipient fracture gas gun flyer plate experiments were performed on both materials using nearly identical sample dimensions and impact velocities. Details and results of the Cabot material have been reported [4-7]. This paper will compare and contrast the material response of the Cabot and Starck materials to one set of experimental conditions, 2 mm flyer on 4 mm target, symmetric impact, 254.9 m/s for the Cabot, 243.2 m/s for the Starck. Figure 1 compares the back free surface velocities for these experiments, measured using a velocity interferometer for any reflector (VISAR). The Stepanov corrected [8, 9]spall strength for this Cabot experiment is 4.7 GPa, whilst for this Starck experiment it is 5.4 GPa.

Microstructures

The soft captured specimens were sectioned and metallographically prepared using Kelly's chemical polish and etch. [10] A computer controlled optical microscope acquired digital images of the entire sample section plane at 1.40 and 1.71 μm/pixel for the Cabot and 0.706 and 0.859 μm/pixel for the Starck , x and y directions respectively. Figure 2 shows details of the incipient damage in the two experiments. The shock wave propagation direction is from top to bottom.

Damage quantification

The damaged microstructures from one section plane (~ 19 mm wide) of each specimen were quantified using image analysis and optical profilometry. The resulting damage was statistically reduced to volumetric spatially resolved size distributions, number densities and porosity

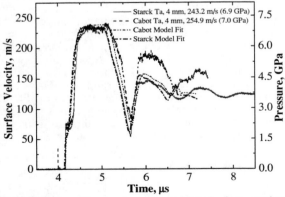

FIGURE 1. Measured and simulated free surface velocities of the two materials.

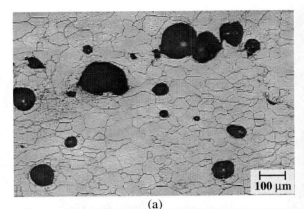

(a)

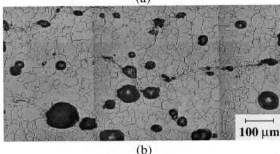

(b)

FIGURE 2. Optical micrograph details of (a) Cabot and (b) Starck experiments.

distributions. The spall plane is defined as the plane of maximum porosity in each specimen. Figure 4 shows the porosity distributions, Figure 3 shows the void number density spatial distributions, and Figure 5 shows the void expected size spatial distributions, assuming that a log-normal distribution describes the statistically reduced data.

Damage model fitting

The Tonks damage model [11]has been calibrated for the Cabot material and validated under uniaxial strain loading conditions. [12] Parameters for the Starck material have been determined using this experiment. Table 2 summarizes the model parameters for these two materials. Three model parameters were changed from the Cabot fit to achieve a good fit to the Starck free surface velocity profile and quantified damage. P_0 was increased 12.5 % to match the VISAR pull back signal. γ was increased considerably to agree with the greater void number density in the Starck material compared to the Cabot material. σ_0 was increased by 50 % to match the observed porosity distribution.

The agreement between the simulations of these experiments with the measurements is shown in Figure 1, 4-5. The standard error, e, of the integral of the model predictions to the experimental data is shown in the graphs.

DISCUSSION AND CONCLUSIONS

The Starck experiment exhibited about five times the void number density than the Cabot experiment. The Cabot maximum volumetric expected size was over twice the Starck value, but the Cabot average expected size across the damage regions is only about 25 % larger than the Starck experimental value. The maximum porosity value of the Starck experiment is about 13 % higher than the Cabot experiment, with a somewhat broader

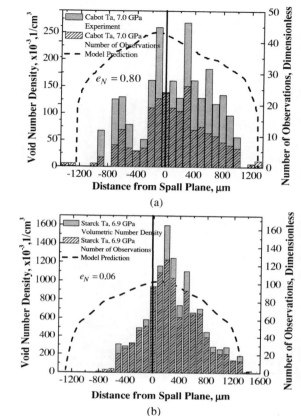

FIGURE 3. Experimental statistically reduced and simulated volumetric void number density for (a) Cabot and (b) Starck experiments.

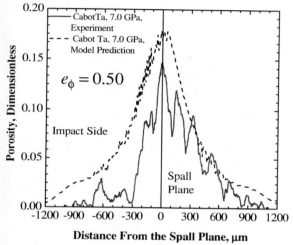

(a)

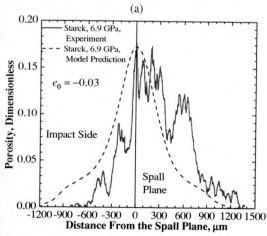

(b)

Figure 4. Porosity distribution measurements and simulation predictions for (a) Cabot, and (b) Starck experiments.

distribution of porosity. The Starck experiment exhibited a significantly larger pull-back signal than the Cabot, indicating that nucleation occurred at a higher tensile stress in the former material.

Void clustering and long range plastic instabilities between voids are two manifestations of coalescence. The Cabot experiment also exhibited a significantly lower amount of void clustering and localized plastic linking between voids than did the Starck experiment. 24 % of the voids in the Cabot experiment were cluster members, whilst over 53 % of the voids in the Starck experiment were cluster members. The Starck experiment exhibited over three times the number of plastic instabilities between voids (133 vs. 40) than observed in the Cabot experiment.

The lack of a strong spatial distribution of void volumetric expected size in the Starck experiment indicates that most of the porosity distribution may be explained by the void number density distribution. In contrast, the sharp increase in the

497

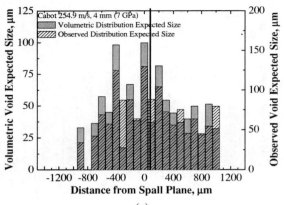

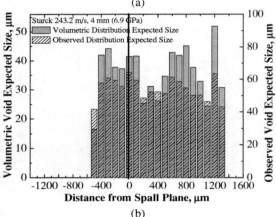

FIGURE 5. Experimental statistically reduced observed and volumetric void expected size for (a) Cabot and (b) Starck experiments.

Cabot volumetric expected size near the spall plane indicates that void growth and possibly coalescence contribute significantly to the observed porosity distribution. The significant role of void coalescence in the Cabot experiment is substantiated by the decrease in void number density at the spall plane.

ACKNOWLEDGMENTS

This work was sponsored by the DoD/DOE Joint Munitions Technology Program. Carl Trujillo and Rey Chavez are thanked for performing the gas gun experiments.

REFERENCES

1. J. B. Clark, R. K. Garrett, T. L. Jungling *et al.*, *Metallurgical Transactions A Physical Metallurgy And Materials Science*, 2959-2968 (1991).

Table 2: Tonks Damage Model Parameters			
Nucleation and Cavitation Growth Parameters			
Parameter	Definition and Units	Cabot	Starck
P_o	Nucleation Threshold Pressure, GPa	4	4.7
r_o	Void Nucleation Radius, μm	1	
γ	Factor in void number density, GPa/cm^3	2×10^8	18×10^8
α	Void velocity coefficient, g/cm^3	16	
P_c	Cavitation Pressure, GPa	4	
$C_{plas\ wave}$	Release Wave Velocity, cm/μs	0.34	
β	Factor in cavitation transition criteria, dimensionless	2	
Porosity Growth Component			
y	Flow Stress, GPa	0.33	
σ_o	Stress Scaling Factor, GPa	0.4	0.4
$\dot{\epsilon}_o$	Strain Rate Scaling Factor, μs^{-1}	1	1.5
n	Exponent, dimensionless	1	

2. J. B. Clark, R. K. Garrett, T. L. Jungling *et al.*, *Metallurgical Transactions A Physical Metallurgy And Materials Science*, 2039-2048 (1991).
3. J. B. Clark, R. K. Garrett, T. L. Jungling *et al.*, *Metallurgical Transactions A Physical Metallurgy And Materials Science*, 2183-2191 (1992).
4. W. Richards Thissell, Anna K. Zurek, Duncan A. S. Macdougall *et al.*, *Journal de Physique Colloque*, **C3**, 769-774 (2000).
5. W. R. Thissell, A. K. Zurek, J. M. Rivas *et al.*, "Damage Evolution and Clustering in Shock Loaded Tantalum," in *Proceedings of 31st Annual Convention of the International Metallographic Society*, edited by: Eugen Abramovici, Derek O. Northwood, Mahmoud T. Shehata *et al.*, ASM, Microstructural Science, 1998, pp. 497-505.
6. W. R. Thissell, A. K. Zurek, D. L. Tonks *et al.*, "Quantitative Damage Evolution and Void Growth Model Predictions in Tantalum Under Spallation Loading," in *Proceedings of 22nd International Symposium on Shock Waves*, edited by: R. Hillier G. J. Ball, G. T. Roberts, 1999, ISBN: 085432 706 1, 085432 711 8.
7. J. M. Rivas, A. K. Zurek, W. R. Thissell *et al.*, *Metallurgical Transactions*, **31** (3A), 845-851 (2000).
8. G.V. Stepanov, *Problems of Strength (USSR)*, (8), 66-70 (1976).
9. V. I. Romanchenko and G. V. Stepanov, *Zhurnal Prikladnoi Mekhaniki i Teknicheskoi Fiziki*, **4**, 141-6 (1980).
10. A. M. Kelly, S. R. Bingert, and R. D. Reiswig, *Microstructural Science*, **23**, 185-195 (1996).
11. D. L. Tonks, A. K. Zurek, and W. R. Thissell, "The Tonks Ductile Damage Model (U)," in *Proceedings of NECDC 2002*, edited by, 2002.
12. D. L. Tonks, A. K. Zurek, and W. R. Thissell, "Spallation Modeling in Pure Tantalum," in *Proceedings of Explomet 2000*, edited by: K. Staudhammer, 2000.

CP706, *Shock Compression of Condensed Matter - 2003*
edited by M. D. Furnish, Y. M. Gupta, and J. W. Forbes
© 2004 American Institute of Physics 0-7354-0181-0/04/$22.00

METALLURGICAL CHARACTERIZATION OF ATLAS CYLINDRICALLY CONVERGENT SPALLATION EXPERIMENTS

[1]W. R. Thissell, [1]B. L. Henrie, [1]E. K. Cerreta, [2]W.A. Anderson, [3]W. L. Atchison, [4]J. C. Cochrane, [3]A. M. Kaul, [5]R. K. Keinigs, [6]J. S. Ladish, [3]I. R. Lindemuth, [6]D. M. Oro, [7]D. Paisley, [8]R. Reinovsky, [9]G. Rodriguez, [2]M. A. Salazar, [6]J. L. Stokes, [9]A. J. Taylor, [10]D. L. Tonks, [1]A.K. Zurek

[1]*MST-8: Structure-Property Relationships, MS: G755,* [2]*MST-7: Polymers and Coatings,* [3]*X-1: Plasma Physics, MS: B259,* [4]*Bechtal-Nevada,* [5]*X-4, MS: T086,* [6]*P-22: Hydrodynamics and X-Ray Physics , MS: D410,* [7]*P-24: Plasma Physics, MS: E526,* [8]*DX-DO: Dynamic Experimentation, MS: D420,* [9]*MST-10: Condensed Matter Physics, MS: K764,* [10]*X-7: Materials Modeling, MS: F699*
Los Alamos National Laboratory, Los Alamos NM 87545

The microstructural distribution and nature of damage from three different cylindrically convergent spallation experiments performed on the pulsed power machine named Atlas are presented. Longitudinal momentum trapping was used to minimize the influence of release waves and thereby decrease the dimensionality of the experiments. Two of the experiments involved soft capture of the spalled piece. The material used is a proprietary directionally cast Al alloy with a mostly equiaxed grain morphology and essentially random texture in the region of spallation. The damage was most distributed in the lowest impact velocity shot and became progressively more narrow with increasing impact velocity. The effectiveness of the momentum trap design increased with increasing impact velocity.

INTRODUCTION

The Atlas pulsed power machine is capable of delivering up to 30 MA of current at voltages of up to about 240 KV in about 5-6 µs. This machine has been employed to perform a series of experiments investigating cylindrically convergent spallation, friction, strength at high strain rates, and hydrodynamic instabilities.

Traditional Cartesian uniaxial strain experiments, such as plate impact, involve very high stress triaxialities, $-P/\tau > 10$. The Atlas spallation experiments employ a magnetic field accelerated cylindrically convergent flyer impacting a cylindrical annulus target that results in cylindrically convergent shock waves. This shock wave, although uniaxial in the cylindrical coordinate system, results in lower stress triaxialities than what occurs in plate impact experiments, $5 < -P/\tau < 10$. This paper describes three cylindrically convergent spallation experiments performed on Atlas. Complete details of these experiments will be published [1].

DESCRIPTION OF MATERIAL

The alloy selection criteria for these experiments required a low Z number material for compatibility with the radiography equipment and a material with either isotropic texture and grain morphology or a texture in the cylindrical longitudinal axis. The material used was a high purity Al-0.2 wt. % Mn solid solution alloy. It was DC cast as a 0.179 m diameter billet at the Alcan Kinston Research and Development Centre in Kingston Ontario, Canada; chemical analysis is given in Table 1.

Grain size measurements were performed in each of the three cylindrical orthogonal directions at seven radial and theta locations of the casting from one section, using the 1.125 times the mean intercept length. The location that corresponded to where spallation occurred in the experiments had a radial section normal plane grain size of 156 µm and equiaxed grains; the theta section normal plane had a grain size of 223 µm with elongated equiaxed

Table 1: Al alloy chemistry, weight percent			
Al	Mn	Si	Fe
Balance	0.2	0.08	0.08

grains; and the longitudinal section normal plane had a grain size of 91.8 μm, consisting of a mixture of equiaxed and fine columnar grains. The latter indicates incomplete recrystallization.

Electron-backscattered diffraction measurements were performed on two of the sample locations and indicated a weak (3x random) <110> texture where the material re-crystallized and a very strong (25x random) <110> texture where the material had not re-crystallized.

MECHANICAL TESTS

A cross-section of the experimental load cassette is shown in **Error! Reference source not found.**. The current passes over the surface of the Al liner and the resulting **B** field accelerates it toward the target. The target had two longitudinal momentum traps on each of the two longitudinal sides. These traps are bonded to each other and the target using epoxy. The purpose of the momentum traps is to greatly reduce the magnitude of the longitudinal release waves which would otherwise greatly complicate the sample loading. Dynamic experimental diagnostics included both longitudinal and radial radiography and velocity interferometry system for any reflector (VISAR) measurements.

The initial liner thickness is 2 mm, but this increased at impact due to convergence effects.

The target outer radius is 37 mm and the inner radius is 15 mm. The ratio of the liner to target thickness is therefore about 1:10, so that the release wave would overtake the shock wave and result in

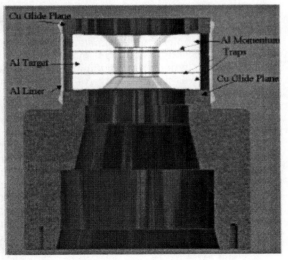

FIGURE 1. Cross section of the load cassette used

Taylor wave like loading in the spallation region.

Experiments were performed using one third of the capacitor bank at three different voltages, 136.7 kV, 148 kV, and 173.4 kV, which result in different impact velocities. The impact velocities were not measured, but are estimated using simulations of the experiments using RAVEN [2], a magneto-hydrodynamic explicit finite element code. RAVEN estimated impact velocity for the 136.7 kV experiment is 448.7 m/s; 535.2 m/s for the 148 kV experiment, and 728.9 m/s for the 173.4 kV experiment. The 136.7 kV and 173.4 kV experiments had VISAR turning mirrors that were 0.254 mm thick quartz that permitted soft recovery of the spalled pieces.

RESULTS

Free Surface Velocities

A comparison of the VISAR measurements at the longitudinal center of the target is made in Figure 2. The 173.4 kV experiment missed one fringe that was manually inserted into the analysis. The listed stress on the right side is calculated assuming 6061 properties because such parameters

do not exist for this alloy. The graph shows the data with the experiments Hugoniot elastic limits (HEL) aligned. The material strength causes the spalled layer to decelerate in the two lower shock amplitude experiments, but not in the highest shock amplitude experiment. The effect of release catch-up is also clearly observed by the decay in the shock plateau duration with decreasing shock amplitude and the differences in shock plateau shape in each of the three experiments.

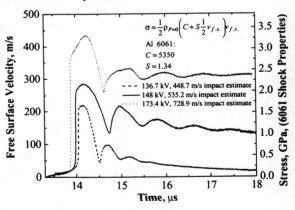

FIGURE 2. VISAR free surface velocity profiles.

Microstructures

Figure 3-5 show cross-sections of the recovered sample from the 136.7, 148, and 173.4 kV experiments, respectively. The surface normal is the theta direction, and the shock direction is from bottom to top in each case. Ductile damage is observed on both the spallation region at the top in each case. Incipient damage is also observed in the center of the specimen, where the two longitudinal release waves intersected in the 136.7 kV experiment, but not the other two experiments.

Figure 6 is a scanning electron micrograph (SEM) of the fracture surface of the 173.4 kV experiment.

Soft Capture

Computed tomography was performed on the soft recovered spalled annular layers in the 136.7 and 173.4 kV experiments and are shown in Figure 7.

DISCUSSION AND CONCLUSIONS

The anomalous incipient damage in the

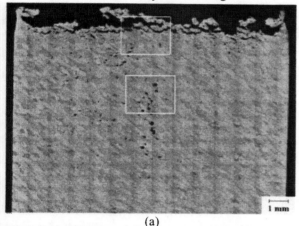

(a)

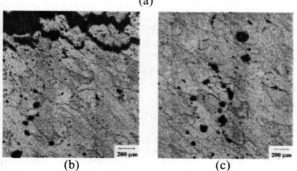

(b)   (c)

FIGURE 3. Optical micrographs of the sample from the 136.7 kV experiment. (a) overview, (b) top detail, (c) bottom detail.

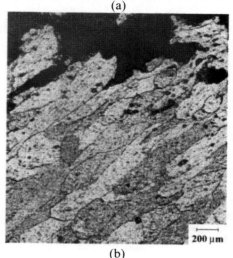

(a)

(b)

FIGURE 4. Optical micrographs of the 148 kV experiment. (a) overview, (b) detail.

longitudinal center of the lowest shock amplitude experiment, illustrated in Figure 3(c), may be explained by residual tensile strength of the epoxy bonding the momentum traps to the sample. The initial passage of the compressive shock wave in this experiment was insufficient to completely fail the bond. The lack of this damage in the other experiments indicate that the residual tensile strength of the bond was low enough to prevent sufficient release wave passage to result in exceeding the nucleation threshold upon rarefaction intersection.

Ductile void nucleation was observed to preferentially occur in all three experiments at grain boundaries, followed by a ductile tearing through the shorter grain length. The elongated equiaxed grain morphology results in a scalloped fracture surface most clearly observed in Figure 3(a), but present in all three experiments.

The radial spatial extent of the damage decreases with increasing shock pressure. Assuming symmetry of damage across the fracture surface, the spatial extent of damage is about 5-6 mm in the 136.7 kV experiment, 3-4 mm in the 148 kV experiment, and about 2 mm in the 173.4 kV

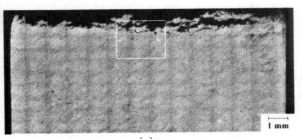

(a)

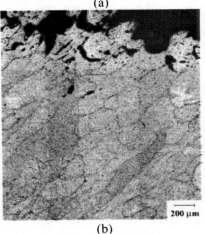

(b)

FIGURE 5. Optical micrographs of the 173.4 kV experiment. (a) overview, (b) detail.

experiment. The microstructural observations of the spatial extent of damage correlates with the decreasing post-spall damping of the free surface velocities with increasing shock pressure, shown in

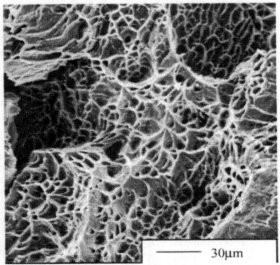

FIGURE 6. (SEM) of the fracture surface of the 173.4 kV experiment.

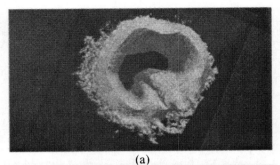

(a)

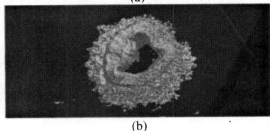

(b)

FIGURE 7. Computed tomography renderings of the spalled and soft-recovered piece from (a) 136.7 kV and (b) 173.4 kV experiments, at the same scale.

Figure 2.

SEM analysis of the fracture surfaces indicates that some void nucleation occurred at super micrometer sized second phase particles, an example is shown in Figure 6 near the white arrow.

The soft recovered pieces showed that convergence instabilities occurred that resulted in buckling. The strength of the material was sufficient to prevent complete convergence in both experiments. A comparison between Figure 7 and Figure 2 indicates that the strength induced deceleration of the spalled annulus in the 173.4 kV experiment must have occurred at late times, after the end of the VISAR signal.

ACKNOWLEDGMENTS

Stuart MacEwen, of Alcan International Limited, is greatly appreciated for providing technical expertise in material selection and the material used. David Philips and Anthony Davis of Hytec, Inc. (www.hytec.com) are thanked for performing the computed tomography.

REFERENCES

1. W.A. Anderson, W.A. Atchison, E. K. Cerreta *et al.*, *to be submitted to The Journal of Applied Physics*, (2003).
2. Thomas A. Oliphant and Kathleen Witte, Los Alamos National Laboratory, Los Alamos, Report No. LA-10826 (UC-32), January 1987.

CP706, *Shock Compression of Condensed Matter - 2003*
edited by M. D. Furnish, Y. M. Gupta, and J. W. Forbes
© 2004 American Institute of Physics 0-7354-0181-0/04/$22.00

MECHANICAL PROPERTIES FROM PBX 9501 PRESSING STUDY

Darla Graff Thompson and Walter J. Wright

DX-2, Materials Dynamics, Los Alamos National Laboratory, Los Alamos, NM, 87545

Abstract. A PBX 9501 pressing study was conducted by researchers in ESA-WMM, LANL, to identify the hydrostatic pressing parameters most important in fabricating high-density parts with uniform density. In this study, 31 charges were pressed using a full permutation of six pressing parameters. Five charges from the set of 31 were selected for an evaluation of their mechanical properties, specifically uniaxial compression and tension. Charges were selected to 1) span the density range of the study, and 2) allow two direct comparisons of pressing parameters independent of bulk density (density has a well-established affect on some material properties). Three PBX 9501 charges pressed isostatically at Pantex Plant in Amarillo, TX were also included in the study. The tensile properties of the 8 charges varied significantly. Careful evaluation of the results suggests that an increase in pressing temperature may correlate with an increase in tensile stress (strength) and a decrease in strain (ductility). Trends in compression exist but are less pronounced. In an effort to explore the relationship between pressing temperature and tensile strength, four sheets of Estane polymer (a component of the PBX 9501 binder) were compression molded at 70, 90, 110 and 130°C. The tensile strength of Estane was observed to increase by a factor of nearly 20 when the molding temperature was increased from 70 to 90°C (strength increase was negligible beyond 90°C). We present an outline of ongoing work that will irrefutably quantify the mechanical property affects of both pressing temperature and dwell time on PBX 9501.(*LA-UR 03-4842*)

INTRODUCTION

Uniaxial compression and tension measurements are used to characterize plastic bonded explosives (PBX) for nuclear stockpile surveillance, new lot certifications, and PBX model development. In the charge fabrication process, PBX molding powders are compacted in large hydraulic presses. A recent PBX 9501 pressing study was conducted by ESA-WMM [1] to identify the pressing parameters most important in manufacturing high- and uniform-density parts. A total of thirty-one (31) hemispherical charges were fabricated using a permutation of six pressing parameters: number of cycles, pressure, dwell time, rest time, temperature, and sack thickness. Statistical analysis indicated that of the six parameters, the most important for achieving high density was first, number of cycles and second, pressure. Four of the resultant charges

were cross-sectioned for an evaluation of bulk density distribution (1 cm³ resolution). Density gradients were shown to be nearly identical in each of the charges studied, only the bulk density varied. From the set of 31, five charges were selected for a study on mechanical properties. The highest and lowest density charges were selected. Also selected were two pairs of charges with identical densities that resulted from two different parameter sets (Table 1); in this way, the affect of parameter sets could be directly compared independent of density. Three additional charges included in the mechanical testing study were from the same lot of PBX 9501, pressed at BWXT Pantex Plant (PX) in Amarillo, TX. These charges were pressed isostatically with heated pressing fluid (70 °C), whereas LANL charges were pressed hydrostatically with unheated pressing fluid (25°C). The three PX charges were pressed using

Table 1: Pressing Parameters for Charges in Mechanical Properties Study.

charge	part#	density (g/cm^3)	pressure (kpsi)	# of cycles	powder temp (°C)	dwell time (min)	sack thickness (in)	rest time (min)
PX1[a]	30015	1.8369	25	3	85	5	(c)	2
PX2[a]	90606	1.8368	"	"	"	"	(d)	"
PX3[a]	30012	1.8351	"	"	"	"	(e)	"
LA4[b]	009-0003	1.8347	19.54	5	73.1	10	3	1
LA5[b]	029-0001	1.8346	20.1	5	95.2	2	1	1
LA6[b]	024-0001	1.8297	10.05	5	66.5	10	1	5
LA7[b]	038-0001	1.8297	20.22	2	73.7	10	1	5
LA8[b]	014-0002	1.8232	10.04	2	73.5	10	3	1

(a) Pantex pressings are isostatic with press temperature around 70°C.
(b) LANL (ESA-WMM) pressings are hydrostatic with press temperature around 23°C.
(c) Sack configuration C.
(d) Sack configuration D.
(e) Sack configuration E.

three different sack configurations (Table 1). Correlations of mechanical properties with bulk density have been previously observed [2,3]. The objective of this study was to identify pressing parameters that have an affect on PBX mechanical properties independent of bulk density.

EXPERIMENTAL PROCEDURE

Tension and compression specimens were cored from the same locations/orientations in each of the hemispherical charges. Mechanical properties data were analyzed for effects of orientation (i.e. density gradient relative to uniaxial stress), but significant differences were not observed. Specimen densities were obtained using immersion methods. All specimens were dessicated for 14 days prior to testing. Quasi-static tensile and compressive tests were conducted in a temperature-controlled chamber on an Instron 5567. Tests were conducted at 23°C with a strain rate of 0.0001 s^{-1}. In compression, tests were also conducted at −20 and 50°C (data not shown). For each charge at each test condition, 8 specimens were tested in tension, 6 specimens in compression.

Stress-strain curves were characterized using three parameters, σ_m, ε_m, and E_{25}. The ultimate stress, σ_m, is the highest stress value achieved in the test. E_{25} is the modulus or slope of the line at 25% of σ_m. Because the low-strain region of the curves can contain loading artifacts, a toe-correction is performed to shift the strain data such that the E_{25} line passes through the strain origin. Ultimate strain, ε_m, is the toe-corrected strain value corresponding to σ_m.

RESULTS AND DISCUSSION

All stress-strain curves at 23°C are shown for tension and compression in Figures 1 and 2. Characterization parameters for tension and compression are plotted in Figure 3 along with specimen densities (points are averages with standard deviations shown in error bars).

Variations in tensile and compressive properties are significant with trends that correlate somewhat

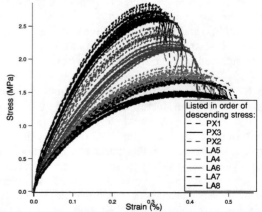

Figure 1: Tension Stress-Strain Curves for 8 Differently Pressed Charges, 23°C at 0.0001 s^{-1}.

with specimen density (Figure 3). Of particular interest are the charges where the mechanical properties do not correlate with density. Specifically, the density of PX2 is very similar to the densities of PX1 and PX3, however, its σ_m and E_{25} values are significantly lower and ε_m is significantly higher. This observed difference is assumed to be an affect of the particular sack configuration used for the PX2 charge; the results must be repeated to assure that the observed difference is statistically significant. Similarly, data for specimens LA4 and LA5 show that bulk densities are nearly equal but LA5 has σ_m and E_{25} values significantly higher than those of LA4 and ε_m values significantly lower. From Table 1, the differences in LA4 and LA5 pressing parameters included temperature, dwell time, and sack thickness. Comparing these two charges, LA4 density was obtained using a longer dwell time, and an equivalent LA5 density was obtained using a higher temperature (in the ESA-WMM pressing study, sack thickness was not observed to have a detectable effect on density). Careful examination of Figure 3, particularly the tensile E_{25} data, suggests to us that the pressing temperature may be responsible for tensile property differences that are independent of specimen density. The tensile E_{25} values for charges LA4, LA6, LA7 and LA8 are all much lower than those of the PX* charges and they do not track directly with density; these four LA* charges were all pressed at the lowest temperature, with LA5 being pressed at a higher temperature

and showing a higher tensile E_{25} value. A similar analysis applies to the tensile ε_m values, suggesting to us that the effect of pressing temperature on the tensile properties of PBX 9501 should be further explored. Reference made [4] to an earlier study at Pantex Plant, Amarillo, TX, indicates potentially-related observations on LX-04.

To explore a hypothesis that increased pressing temperature may promote better binder flow and melding of PBX prills, four sheets of pure Estane polymer (a binder component of PBX 9501) were compression-molded at temperatures of 70, 90, 110 and 130°C. Tensile dogbones were cut from these sheets, and representative tensile stress-strain curves are shown in Figure 4. The σ_m value for the specimen molded at 70°C is on the order of 30 times smaller than the σ_m values for specimens molded at or above 90°C. The ε_m value is reduced by a factor of 40 when the specimen is molded at 70°C as opposed to the higher temperatures. By-eye observation of the specimens molded at 70°C showed clear "meld lines" between the Estane pellets, and tensile failure was observed to occur along those meld lines. For specimens molded at

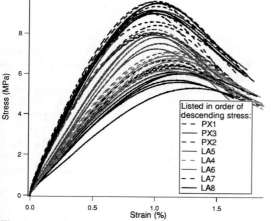

Figure 2: Compression Stress-Strain Curves for 8 Differently Pressed Charges, 23°C at 0.0001 s⁻¹.

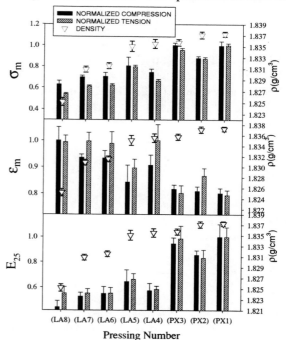

Figure 3: Stress-Strain Parameters and Densities for Compression and Tension, 23°C at 0.0001 s⁻¹.

90°C and above, no meld lines were observed and tensile failure occurred in a line perpendicular to the uniaxial stress. These data support the hypothesis that PBX 9501 mechanical properties may be significantly affected by the pressing temperature.

Polarized microscopy is being applied to the post-test PBX 9501 specimens. Initial images show meld lines between prills for specimens pressed at low temperature. The work is ongoing.

To independently evaluate the affects of temperature and dwell time on the material properties, independent of density, a specimen matrix has been created for an upcoming series of measurements (Figure 5). The three lines on the

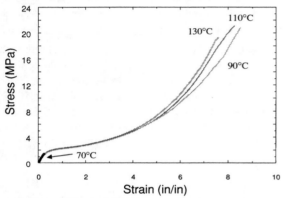

Figure 4: Tension Stress-Strain Curves of Pure Estane Compression-Molded at Different Temperatures; Tested at 23°C and 10 in/min.

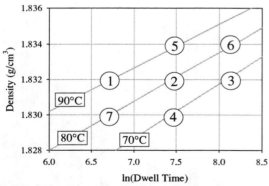

Figure 5: Specimen Preparation Matrix to Explore the Independent Affects of Temperature and Dwell Time on Material Properties.

plot were experimentally determined and correlate specimen density with dwell time at three pressingtemperatures: 70, 80 and 90°C. Specimens 1 thru 7 will be pressed according to the figure, and densities and material properties measured. A comparison of results for specimens 7, 2 and 6 with those for specimens 4, 2 and 5 will provide two sets of material properties differences over an identical specimen density range. The affects of pressing temperature and dwell time will thus be isolated and quantified. This study is underway.

CONCLUSIONS

From a set of 31 PBX 9501 charges produced by a permutation of pressing parameters, we have identified temperature as being a potentially important parameter for determining the mechanical properties, particularly in tension. Estane tensile data also suggest an important role of pressing temperature. Work is underway to further isolate and quantify the affect of temperature on PBX 9501 tensile properties. The quantitative results may have significant implications in PBX mechanical properties measurements for surveillance and certification.

ACKNOWLEDGEMENTS

Thanks to Bart Olinger, Sandra Powell, Bruce Orler, Paul Peterson, Mike Oldenborg. Funded by W76 SLEP, ESC and Non-Shock Initiation MOU.

REFERENCES

1. Olinger, Bart "Densities of Hollow Hemispheres of PBX 9501 Pressed Under Different Conditions," ESA-WMM-03-001, Los Alamos National Laboratory Memo, January 2003.
2. Meyer, Thomas et al., "Compression Properties of PBX 9501 at Densities from 1.60 to 1.83 g/cc," Pantex Plant report, DOE/AL/65030-00-04, February 2000.
3. Thompson, Darla Graff et al., "Quasi-Static Compression Study of PBX 9501, PBXN-9 and PBXN-110," Los Alamos National Laboratory Report, LA-13940-MS, May 2002.
4. Wilson, A.L. and Neff, G.W. "Physical Properties of Explosives," engineering order# 816-00-002, Pantex Plant quarterly report, April, May, June 1966.

CP706, *Shock Compression of Condensed Matter - 2003*
edited by M. D. Furnish, Y. M. Gupta, and J. W. Forbes
© 2004 American Institute of Physics 0-7354-0181-0/04/$22.00

MODELING INCIPIENT COPPER DAMAGE DATA FROM THE TENSILE HOPKINSON BAR AND GAS GUN

D. L. Tonks, W. R. Thissell, D. S. Schwartz

Los Alamos National Laboratory, Los Alamos, NM 87545.

Abstract. Ductile damage in copper has been created using a split tensile Hopkinson pressure bar. Precise momentum trapping has made it possible to arrest the damage after a short tensile pulse before complete fracture. This process has been modeled with a void nucleation and growth law. 2D calculations have been performed to compare with final porosity and void number density data. The tensile bar damage modeling has been supplemented with modeling of incipient spallation of copper in a plate impact gas gun experiment. These two experiments differ widely in the (negative) pressure level and modeling both their porosity results will permit the creation of a nucleation model that spans the resulting wide pressure range.

INTRODUCTION

The split tensile Hopkinson pressure bar presents the opportunity of creating damage at fairly high strain rates (10^4/s) with large plastic strains (100%). Careful momentum trapping permits incipient damage states to be arrested and recovered for metallurgical examination. The use of notched samples permits the pressure – flow stress, or triaxiality, to be varied from 1/3 to about 1.2 to study the interplay of pressure and deviatoric stress.

In this paper, we will concentrate on modeling the ductile damage process in pure copper (Hitachi). We also study the damage evolution in a gas gun experiment with a sample of the same material to obtain a contrasting behavior at a much higher triaxiality of about 4. The goal of the modeling is to obtain a unified ductile damage model that will work for both.

In the operation of the bar, a projectile and sabot sleeve are launched to collide with a flange at the end of the tensile bar to produce a tensile pulse. Gaps between the bars and the momentum trap bars serve to contain unwanted energy within the trap bars to prevent late time loading of the specimen.

DAMAGE MODELING

The ductile damage modeling involved a void nucleation component, an early cavitation void growth component, and a stress equilibrium growth component for later times. Our samples did not fracture so no fracture modeling was employed.

The first component is a stress based void nucleation model. In spallation conditions, where the negative pressures are large, void nucleation tends to be determined more by stress than strain. On the other hand, in low pressure conditions, like the tensile test, strain is very available and significant for nucleation, in contrast to the spallation condition. The nucleation modeling assumes heterogenous instantaneous nucleation around inclusions with a distribution of nucleation strengths, σ_I. The local pressure, σ_L, taken positive in tension, must exceed σ_I to open up a void. Earlier work has shown that a combination of ambient pressure and strain produces the local stress, σ_L (1). The plastic strain produces a local

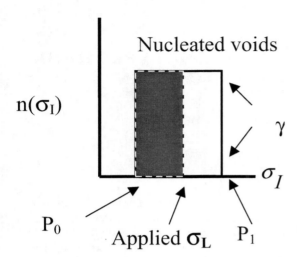

FIGURE 1. Distribution of nucleated void density versus nucleation strength.

dislocation build-up around a void with an accompanying local pressure contribution to which the ambient pressure adds or subtracts to produce a total local tension. We model these two contributions

$$\sigma_L = \sigma_H + e\sqrt{\psi}, \qquad (1)$$

where σ_H is the ambient pressure taken positive in tension and e is the coefficient giving the contribution from the plastic strain ψ. Figure 1 shows that, for our modeling, $n(\sigma_I)$, the differential density of nucleated voids, is taken to be a rectangle. Adjustable parameters are P_0 and P_1, the local stresses at which nucleation starts and stops, and γ ', the height of the distribution. The void number density produced by a negative local pressure of magnitude σ_L is $\gamma(\sigma_L - P_o)$, if $P_0 < \sigma_L < P_1$.

The copper studied was pure enough that void nucleation occurred on too small a scale to be observed in the optical microscope. For this reason, the nucleation distribution parameters were treated as adjustable, with the constraint that no void nucleation was observed in tensile bar samples below a certain plastic strain.

An early "cavitation" void growth law is used just after void nucleation. It involves a void that is expanding behind elastic and plastic release waves started from the nucleation event. This growth law is appropriate just after the high stresses suddenly break out a nucleated void. During this early time, more conventional porosity growth models that assume no wave or inertia effects are not appropriate. It is assumed that the void surfaces expand with the velocity, V(P), given by:

$$|P| = P_c + \alpha V^2, \qquad (2)$$

where P is the applied pressure, α is a constant roughly equal to the mass density, and P_c is a cavitation pressure roughly equal to four times the yield stress. This formula for V(P) was obtained from a modification of the work of Hunter and Crozier (2) who studied the expansion of the cavity ahead of a penetrating warhead. In an earlier porosity growth law (3), the creation times of individual voids are kept track of but here a simplification was made to obtain a simpler growth law: an average radius, r_{ave}, was used for all the voids The resulting expression for the total porosity, ϕ, is:

$$\phi = N(4/3)\pi(r_{ave})^3, \qquad (3)$$

where N is the volume void number density. Using this form for ϕ, the following simplified porosity growth law for the cavitation void growth can be obtained:

$$\dot{\phi} = (4/3)\pi R_o^3 \dot{N} + 3[(4\pi/3)N]^{1/3}\phi^{2/3}V(P), \qquad (4)$$

where R_o is the void radius at nucleation. This law involves both the porosity and the volume void number density.

At some point, after release waves have had time to reflect among the damage structures, the cavitation void process must transition to a more conventional "macroscopic" one in which, for example, no large waves run and stress equilibrium is present. In our model, the transition criteria is taken to be a small transition porosity threshold, typically of order 0.0005. In earlier calculations of a gas gun experiment with a tantalum sample, the result of this procedure was compared with that of using a more physically based transition criteria based on wave transit times. There was little difference in the calculated final porosity profile or

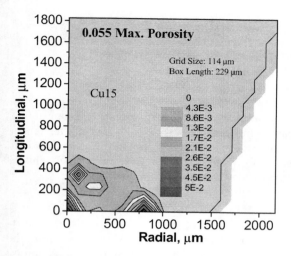

FIGURE. 2. Measured porosity, Cu 15notch quadrant.

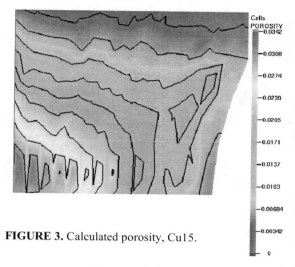

FIGURE 3. Calculated porosity, Cu15.

in the calculated free surface velocity between the two procedures. No further void nucleation is provided after the transition since, by then, the main negative pressure excursion is over and no new voids are nucleated.

The macroscopic porosity growth law used has the form of a stress potential whose derivative with respect to pressure gives the porosity growth rate. This potential was obtained by an upper bound procedure using a two component trial plastic flow field with a cell consisting of a spherical void in a spherical cell with a pressure and deviatoric stress boundary condition. An overstress power law was assumed for the plastic behavior of the matrix. The contour of zero porosity growth in pressure – deviatoric stress space is the Gurson surface for plastic yielding in a porous plastic material. Reference 4 contains a brief description with formulas.

COMPARISON OF CALCULATION WITH EXPERIMENT

Figures 2 and 3 show the calculated and experimental final porosities for tensile bar shot Cu15. The data was averaged over four quadrants to obtain the quadrant shown. The calculation was axially symmetric. The calculated porosity shows one peak in the center of the notched sample while the data shows two peaks. This could be due to statistical variation in nucleation. The experimental

and calculated final void number densities (not shown) exhibited a similar behavior.

The sample notch of Cu15 produced an initial triaxiality of 0.6, which increased during the deformation. The void nucleation parameters used were the following: P_o and P_1 were 16.04 and 16.61 kbar. The values of e and γ' were 24 kbar and about 6×10^7/kbar cm^3, respectively.

The same nucleation and growth modeling was applied to a gas gun shot of the same copper, but annealed whereas the tensile bar sample was half hard. The data and calculational fits to the final porosity profile, the final void number density profile, and the VISAR trace of the free surface particle velocity are shown in Fig. 4.

The nucleation strength distribution used here had a starting stress of 1.22 GPa whereas that for the tensile bar modeling was 1.66 GPa. These preliminary results suggest that the same nucleation modeling could serve for both.

The void growth modeling found necessary differed between the tensile bar and the gas gun. The cavitation stress was not reached for the tensile bar so the early void growth component was not used. The matrix flow stress of about 0.4 GPa produced by the Johnson – Cook plasticity model was used for the yield stress parameter in the macroscopic porosity growth law. For the gas gun modeling, a constant flow stress of 0.22 GPa produced a good fit. Also a higher value of the porosity rate scaling parameter was used to give a

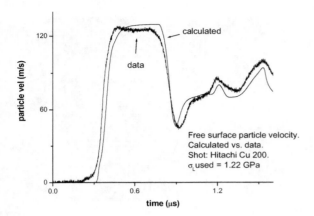

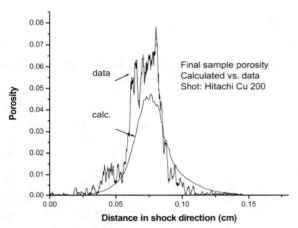

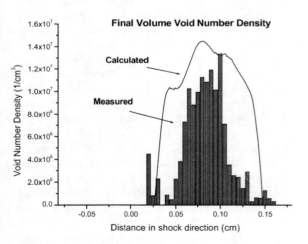

FIGURE 4. Calculated and measured free surface particle velocity, final porosity profile, and final void number density for gas gun shot Cu200.

higher overall porosity growth rate for the gas gun. Hence, the existing modeling cannot seamlessly span the two experimental regimes.

CONCLUSION

It appears that the same void nucleation model can be made to work for copper for both the tensile Hopkinson bar and the gas gun. The void growth modeling requires further generalization for this to be possible.

ACKNOWLEDGMENTS

We are pleased to acknowledge support from the US Joint DoD/DOE Munitions Technology Development Program and the US DOE.

REFERENCES

1. Goods, S.H. and Brown, L.M., "The Nucleation of Cavities by Plastic Deformation," Acta Met. **27**, 1-15 (1979).
2. Hunter, S.C. and Crozier, R.J.M., "Similarity Solution for the Rapid Uniform Expansion of a Spherical Cavity in a Compressible Elastic-Plastic Solid," Quart. Journ. Mech. and Applied Math. **21**, 467 – 486 (1968).
3. Tonks, D.L., Zurek, A.K. and Thissell, W.R. "Modeling Incipient Spallation in Commercially Pure Tantalum". in *Fundamental Issues and Applications of Shock-Wave and High-Strain-Rate Phenomena, EXPLOMET 2000*, edited by K.P. Staudhammer et al, Elsevier Science Ltd., Amsterdam, 2001, pp. 517 - 523.
4. Tonks, D. L., Zurek, A. K., and Thissell, W. R., in *Metallurgical and Materials Applications of Shock-Wave High-Strain-Rate Phenomena*, edited by L. E. Murr et al, Elsevier Science BV, Amsterdam, 1995, pp. 171 – 178.

CP706, *Shock Compression of Condensed Matter - 2003*
edited by M. D. Furnish, Y. M. Gupta, and J. W. Forbes
© 2004 American Institute of Physics 0-7354-0181-0/04/$22.00

STUDY OF SPALLING FOR HIGH PURITY IRON BELOW AND ABOVE SHOCK INDUCED α ⇔ ε PHASE TRANSITION

Christophe Voltz and Gilles Roy

Commissariat à l'Energie Atomique – Centre de Valduc – 21120 Is sur Tille France

Abstract. The study of the dynamic fracture of iron has been undertaken below and above phase transition. Light gas launcher and powder gun combined with Doppler Laser Interferometry (DLI) measurements have been used to evaluate the loading and spall fracture values. Symmetrical impacts have been conducted without windows below and above phase transformation to study the effects of both phase change kinetics and release transition on spallation response of iron. Shots performed below and above phase transition pressure show noticeable differences on pull back signals. Samples were soft recovered for SEM examination. Spall plane surfaces reveal several aspects linked to changes in material behavior. Simulations are conducted with a 2D hydrocode to compare experimental data with numerical results.

INTRODUCTION

Iron is known to exhibit an α ⇔ ε [body centered cubic to hexagonal close packed] phase transition when shock loaded above ≈ 13 GPa. We are interested in spall studies of phase-transformed iron by planar impact techniques. In symmetrical conditions (without backing window) within the range of our experiments (impact velocities greater than to 660 m.s^{-1}) iron phase transition takes place both in impactor and target. It leads to a two wave rarefaction from flyer plate rear side which interacts with another two waves rarefaction from the target free surface. Since we don't use any backing window, DLI signals allow us to record the free surface velocity, without any more details on shock wave release and reverted phase transition after spalling occurrence.

EXPERIMENTAL PROCEDURE

Iron characterization

The polycrystalline iron used for this study is a high purity Armco type IV grade. This material was provided by AKSTEEL from as wrought cast ingot. The bar was made by high temperature open die forging followed by thermal treatment (under inert gas - 4 hours at 920 °C). Chemical analysis of this iron (in wt.%) is C 0.0125, O 0.0245, Si < 0.005, Cr 0.031, Ni 0.0185, Cu 0.0064, Mn 0.0025, and balance Fe. The grain morphology is equiax with no preferential elongation, and no twins. The grain size is between 200 and 300 µm. The Vickers hardness is 66 Hv (under 10 N) indicating a fully relaxed structure. The as-received iron is characterized in compression at temperatures between 77 K and 741 K and at strain rates ranging from 10^{-4} to 2000 s^{-1}. The stress-strain response (figure 1) of this iron is similar to the other body cubic centered metals with low symmetry (Ta, Nb, Mo and W), and has been presented elsewhere [1].

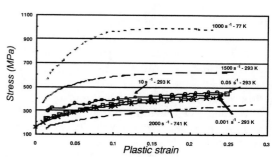

Figure 1. Stress-strain response of iron as a function of strain rate and temperature.

Spall experiments

The spall tests were conducted in symmetric-impact geometry, with a 2-mm thick iron impactor supported by polyethylene sabot and a 4-mm thick iron target. For low impact velocity (< 500 m.s^{-1}) we used a single stage light gas gun (80 mm in diameter) and for higher impact velocity a 50 mm diameter powder gun was employed. Three shots were conducted at 248, 367 and 437 m.s^{-1} in α phase and three others at 779, 1079 and 1457 m.s^{-1} in ε phase. Free surface velocity is recorded by DLI. Impactor velocities and tilt are measured by two sets of three piezoelectric pins. The tilt is always less than 1 mrad, the impactor velocity is measured with an accuracy of 0.5 %. The targets were soft recovered to conduct SEM spall surfaces examinations.

RESULTS

Typical DLI signals are presented in figure 2.

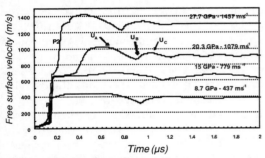

Figure 2. Free surface velocity curves. Zero time is adjusted for each experiment so that all first plastic waves (P1) would all appear at the same time.

Several points can be outlined.

- First, for the impact in α phase (437 m.s^{-1}) the elastic precursor is followed by the main plastic wave (P1). After a plateau there are a dip and a pull-back signal indicative of spalling,
- Second, for the impact slightly above phase transition a weak surface velocity acceleration appears before the dip,
- Third, for the two impacts conducted far above phase transition, the plastic wave P1 is

followed by the main plastic wave P2 corresponding to α to ε phase change. The decreasing delay before P2 occurrence (≈ 0.2 µs for 1079 m.s^{-1} shot, null for 1457 m.s^{-1} test) is related to increasing shock wave velocity in ε phase. Spall occurs in both cases.

These observations are consistent with results reported in [2]: for similar shock pressure and time duration.

The spall strength (σ_{spall}) is determined from the pull-back velocity (ΔU_{pb}) i.e. the free surface velocity difference between plateau (U_A) and dip (U_B) that is (figure 2) :

$$\Delta U_{pb} = U_A - U_B \tag{1}$$

$$\sigma_{spall} = \frac{1}{2} \rho_0 C_l \Delta U_{pb} \tag{2}$$

where ρ_0 is the initial density and C_l the longitudinal elastic wave velocity. Moreover, according to [3] damage can be quantified by a ratio R defined by :

$$R = \frac{U_A}{U_C} \tag{3}$$

where U_C is the post pull back maximum free surface velocity, due to recompression waves from spall plane. Figure 3 presents σ_{spall} and R values.

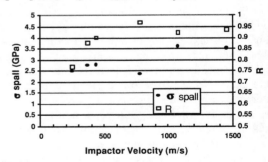

Figure 3. σ_{spall} and R values.

The spall strength ranges from 2.5 to 2.75 GPa below the phase transition and 3.5 GPa for tests conducted above phase transition. In the case of shot at 779 m.s^{-1}, slightly above polymorphic transition the spall is calculated at 2.5 GPa. This value is similar to those observed in the α phase.

The R value shows another discontinuity between tests done far above phase transition and those done below or near phase transition. The values are ranging from 0.75 to 0.97 for impact from 248 to 779 m.s^{-1} and ≈ 0.93 for tests at 1079 and 1457 m.s^{-1}.

We see for the shot just above the phase transition (779 m.s^{-1} impactor velocity) that the σ_{spall} and R values as consistent with those found for shocks below the polymorphic transition.

SEM observations

For impact velocity below up to 779 m.s^{-1}, we observe a single spall plane near the target mid thickness. Brittle fracture is observed through the radius [4] [5].

At higher impact velocity (1079 and 1457 m.s^{-1}), spalling occurs in two planes (figure 4). Neverless we have recovered only two layers, because the third one near the free surface has completely spread out in fine material particles. So the third surface layer should have scattered during free flying stage.

Flyer surface contact

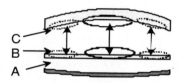

Free surface

Figure 4. Schematic description of multi spalled samples

The three recovered fracture surfaces present the following patterns:

- A near target free surface, the material has smooth surface with little parts of iron glued on it.

- B and C, near plate impact face, a central smooth area of ductile fracture behavior is surrounded by an area showing brittle fracture. This type of observations have been reported elsewhere [5].
Iron is known to exhibit ductile fracture with increasing temperature [4], and brittle fracture with increasing strain rate. The unloading is roughly the same through the sample (instead of lateral released waves which affects very small part of the target), in our case the ductile behavior near the center is due to temperature (shock induced heating).

SIMULATIONS

We have performed calculations with the two-dimensional lagrangian hydrocode HESIONE. Three tests were simulated: 437 m.s^{-1} in the α phase ; 779 m.s^{-1} just above phase transition and at 1457 m.s^{-1} in ε phase.

We used a spall strength criterion with values taken from experiments. Figure 5 shows density evolution in space-time lagrangian diagram.

3 main points can be outlined:
- shot a) - we are in "classical" configuration of spalling by interacting of two single release waves both issued from target free surface and impactor rear face reflections.

- shot b) - spall fracture occurs in material which didn't undergo phase transition. The fracture takes place in locus which is submitted to these two types of waves: -1) the P1 compression wave interacts with target free surface and travels back in target as a release wave, so this fan meets P2 wave and lowers pressure -2) in impactor the rarefaction wave issued from backside of the flyer presents a sudden release due to P1 to P2 interaction.

- shot c) - simulation shows multi spalling like we observed in experiments. Fracture occurs when the two release waves interact. After the maximum pressure there is a plateau followed by the rarefaction shock from ε to α phase reversion. For spall plane near free surface (right part of the figure 5 c) the fracture is due to interaction of rarefaction shock from free surface and release waves travelling from flyer plate back face. This sudden release leads to a material fragmentation.

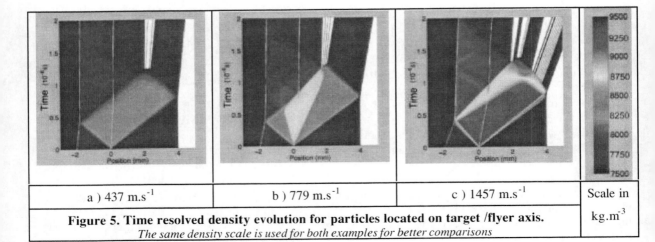

a) 437 m.s⁻¹	b) 779 m.s⁻¹	c) 1457 m.s⁻¹	Scale in

Figure 5. Time resolved density evolution for particles located on target /flyer axis. kg.m⁻³
The same density scale is used for both examples for better comparisons

The spall plane located at ≈ 2 mm inside the target is the consequence of interaction the rarefaction shock from back flyer face and rarefaction waves from free surface. Before crossing, this wave meets the main plastic wave followed by its plateau, so the release rate is affected and lowered.

In case of 1079 and 1457 m.s⁻¹ tests the spall strength is 3.5 GPa, simulations show that in these cases material submitted to fracture have done an excursion in ε domain. Previous work [1] shows that iron which undergone α to ε transition has a deeply affected post-shocked mechanical response (increased strength, modified crystallographic texture). In consequence it leads to a higher spall strength than for not transformed iron

CONCLUSIONS

We have conducted iron symmetrical impacts below and above polymorphic phase transition.
We have found that for shocks up to 13 GPa and slightly above, the fracture is brittle and spall strength is ≈ 2.5 GPa. For tests done far above α ⇔ ε transition multi-spall occurs. The spall strength is evaluated at 3.5 GPa for spalling near target free surface.
Numerical simulations show, for 1079 and 1457 ms⁻¹ shots, that the whole material is ε transformed and the two front release wave

structures lead to 2 types of fracture. These results are confirmed by recovered sample observations.

ACKNOWLEDGEMENTS

The authors wish to thank P.Antoine and P.Martinuzzi for conducting the impact tests and F.Buy for his helpful for code simulation.

REFERENCES

1. Buy F. and Voltz C. – Shock loading influence on mechanical behavior of high purity iron – Proceedings of HDP-V – pp.415-423 - (2003).
2. Veeser L.R., Gray G.T., Vorthmann J.E., Rodriguez P.J., Hixson R.S. and Hayes D.B. - High-pressure response of a high purity iron – Shock compression of condensed matter – pp 73-76. (1999).
3. Cochran S., Banner D. – J. Appl. Phys. 48, p. 2729. (1997).
4. Curran D.R., Seaman l., Shockey D.A. – Dynamic failure of solids. Physics reports vol. 147 n°5 & 6 – pp. 253-388. (1987).
5. Trunin I.R., Koristskaya S.V.Palenova T.S. and Arnold W. – Numerical-experimental study of Armco-80 iron spall study fracture – J. Phys. IV France 10 Pr9 – pp. 421-426. (2000).
6. Nahme H., Hitl M. – Dynamic properties and microstructural behavior of Armco iron shock-loaded at high temperatures – Structures under shock and impact – pp. 427-437.

CHAPTER IX

CONSTITUTIVE AND MICROSTRUCTURAL PROPERTIES OF METALS

CP706, *Shock Compression of Condensed Matter - 2003*
edited by M. D. Furnish, Y. M. Gupta, and J. W. Forbes
© 2004 American Institute of Physics 0-7354-0181-0/04/$22.00

MESOSCALE STUDIES OF SHOCK LOADED
TIN SPHERE LATTICES

M. R. Baer and W. M. Trott

Engineering Sciences Center 9100, Sandia National Laboratories[*], *Albuquerque, New Mexico 87185*

Abstract. The shock response of heterogeneous materials involves highly fluctuating states and localization effects that are produced by mesostructure. Prior studies have examined this shock behavior in randomized inert and reactive media. In this work, we investigate the shock behavior in a porous lattice consisting of hexagonally packed layers of 500 μm tin spheres impacted at 0.5 km/s. This ordered geometry provides a well-defined configuration to validate mesoscale material modeling based on three-dimensional CTH calculations. Detailed wave fields are experimentally probed using a line-imaging interferometer and transmitted particle velocities are compared to numerical mesoscale calculations. Multiple shock fronts traverse the porous layers whereby particle-to-particle interactions cause stress bridging effects and the evolution of organized wave structures.

INTRODUCTION

It is now known that the shock behavior of heterogeneous materials, at the mesoscale, involves multiple waves that interact with material heterogeneities or internal boundaries[1]. In a randomized media, the stochastic nature of microstructure produces highly fluctuating states and localization effects with multiple length and time scales.

Prior mesoscale modeling has revealed that shocked porous materials produce wave fields that arise at contact points and coalesce to produce a variety of thermal and mechanical states. Statistical interrogation techniques have been devised to probe the vast quantity of four-dimensional data in detailed numerical simulations[2]. Results from these studies suggest that the spatial variations of thermal and stress states persist for a long time behind a shock and are not uniform but instead are distributions in space and time.

Time-resolved measurements using line-imaging interferometry have investigated transmitted wave characteristics in impact flyer plate experiments and have verified the existence of fluctuation states during shock loading[3]. Although details of wave behavior in the heterogeneous material cannot be directly probed using this diagnostic, the interactions at physical boundaries of the mesostructure can be observed. The integration of mesoscale modeling and time-resolved experimental measurement technique offers much promise to improve our understanding of shock behavior in heterogeneous materials by providing useful information for validation and calibration of models.

Although the actual mesostructure for most real materials often has stochastic geometry, this study examines the shock behavior in a well-defined porous lattice of hexagonally packed tin spheres. The choice of a simplified geometry and metal spheres eliminates much of the uncertainty in material orientation and properties, such as those encountered in polycrystalline composites. In complimentary experiments, porous layers of spheres are impacted in gas gun experiments and transmitted wave profiles are measured using line-imaging interferometry.

[*] Sandia is a multiprogram laboratory operated by Sandia Corporation, a Lockheed Martin Company, for the US Department of Energy under contract DE-AC04-94AL8500.

EXPERIMENTAL OVERVIEW

The gas-gun studies of waves transmitted by a well-ordered bed of tin spheres were motivated by earlier work on low-density (65% TMD) pressings of sieved, coarse-grain (212-300μm) HMX[4]. The wave profiles generated by this material exhibit distinct ordered wave structures that are both distributed over multiple grain dimensions and largely coherent temporally. It is likely that these features are driven in part by the tendency of nearly monodisperse particles to pack in a somewhat layered structure. Investigation of close-packed samples of spherical particles in a narrow size range provides a favorable opportunity to relate measured wave structures to a well-characterized sample microstructure.

Accordingly, we have adapted the gas gun target design used in earlier tests to probe close-packed samples of tin spheres. A schematic diagram of this design is shown in Figure 1. The experimental assembly consists of a Kel-F impactor and a Kel-F sample cup containing the bed of spherical particles. A 0.225-mm-thick buffer layer of Kapton (not shown) and an aluminized PMMA interferometer window confine the porous bed. The target fixture is designed to accommodate simultaneous measurements using a fiber-coupled, single-point VISAR and a line-imaging optically recording velocity interferometer system (ORVIS). The Kapton buffer is used to mitigate the loss in reflected light intensity that can occur upon shock arrival at the window.

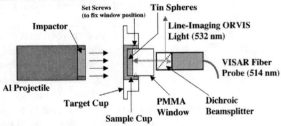

FIGURE 1. Schematic diagram of experimental design for gas-gun studies on tin spheres.

Samples were prepared from 500-μm-diameter spheres obtained from Alpha Metals Inc. The spheres are >96.5% tin with a small fraction of silver and copper and only trace quantities of other materials. The spheres are very precisely controlled

in diameter with a 3σ variation of only 8 μm. With careful layer-by-layer assembly of the sample bed, it was possible to achieve a uniform close-packed configuration containing five layers or more. In these initial tests, we did not attempt to generate a strict hexagonal closest packed (hcp) or face centered cubic (fcc) symmetry. A hybrid assembly is the most likely configuration. Uniform packing was achieved over nearly the entire 1.6" diameter of the sample cup. (A photograph illustrating the uniformity of the top layer in one sample is shown at the top of Figure 6.)

Consistent with earlier results on sieved HMX, transmitted wave profiles with distinct structures are observed. Figure 2 shows a line-imaging ORVIS record of wave behavior obtained with a 5-layer sample under impact at 0.5 km/s.

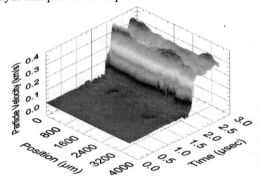

FIGURE 2. Spatially resolved velocity profile of wave transmitted through close packed tin spheres.

COMPUTATIONAL MODEL

Three-dimensional numerical simulations of impact on a lattice of tin spheres were conducted using the Eulerian CTH shock physics code[5] including the projectile and buffer/window materials. A porous lattice is constructed with five layers of 500 μm diameter tin spheres packed in a hexagonal configuration within a representative volume element (RVE) having a transverse and lateral width of 2000μm x 1800μm, respectively. This volume of material is sufficient to capture interparticle interactions that are representative of the response of the mesostructure. Periodic boundary conditions are imposed on the transverse boundaries of the RVE. The numerical resolution is fixed at a cell size of 10 μm. Each tin sphere is assigned a distinct material number and material interfaces are allowed to slip in mixed cells. Figure 3 displays the 3D material geometry prior to impact on the lower surface.

FIGURE 3. Initial material configuration prior to impact at 0.5 km/s.

The hexagonal packing of tin spheres forms an initial configuration with a density of ~74% TMD. Implicit in this direct numerical simulation is the assumption of constitutive relationships that may be only approximate at the mesoscale. Each sphere is modeled using a Mie-Grüneisen EOS and a Steinberg-Guinan-Lund viscoplastic material strength description using CTH EOS database parameters[6].

A shock load is imparted to the porous lattice by incorporating a symmetric impact of Kel-F onto a Kel-F target layer. Included in the simulations are the Kapton/PMMA materials of the buffer layer and optical window in the gauge package. All of these polymers are modeled using a Maxwell viscoelastic description with the CTH database parameters[7]. Tracers at the Kapton/PMMA interface monitor the conditions at the location of the line-imaging ORVIS.

RESULTS AND DISCUSSION

Figure 4 displays a time sequence of material deformation following a 0.5 km/s impact at the lower surface. Painted over the tin surfaces are stress contours. After initial impact, elastic waves

traverse the porous layer concentrated at the contact points in the lattice of spheres. Stress bridging is clearly evident as diagonal "fringes" that span the contact points of the tin spheres. As the stress wave propagation continues, material deformation fills the interstitial pores to consolidate the lattice. Material jet effects are evident after 3 μs when the deformed tin spheres interact with the buffer layer interface.

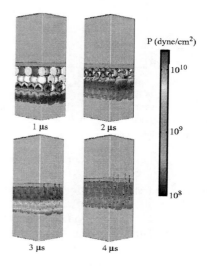

FIGURE 4. Time sequence of deformation and stress contours following impact at 0.5 km/s.

Figure 5 shows a comparison of the spatially averaged particle velocity measurements to the numerical CTH calculations at the tracer locations on the Kapton/PMMA interface. (The time scale is arbitrarily shifted.) As evident in Figure 2, the spatially resolved velocity profile exhibits an extended precursor region that precedes the main wave by up to 1μs over parts of the line segment. This phenomenon reflects the facile pathways for elastic stress propagation that are available in a well-ordered sample. Also consistent with the experimental measurements are the pronounced ramp wave of ~350 ns duration and the prominent fluctuations due to interactions of the particle mesostructure. Similar temporal variations are seen in the VISAR measurements.

As the consolidated lattice interacts with the optical window wave reflections are produced by a change in shock impedance. Additional reflections between the consolidated tin layers and the window

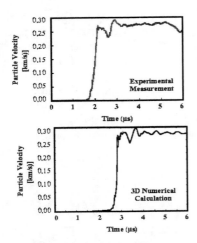

FIGURE 5. Comparison of experimental and CTH numerical calculations of particle velocity at the buffer/PMMA interface.

materials are consistent with the experimental observations.

At the measurement location, two sets of tracers are included in the numerical calculation corresponding to the orthogonal orientations of the spheres in the lattice packing. Figure 6 shows that the calculated particle velocities for each orientation are distinctly different. At the interstitial locations higher velocities are due to material jet effects. This implies that the packing of the spheres in the target must be well controlled to preserve consistent hexagonal packing and the line-imaging probe orientation carefully aligned.

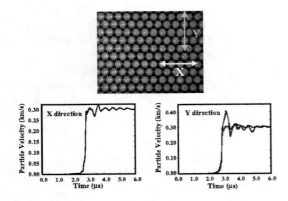

FIGURE 6. Particle velocities along orthogonal directions at the buffer/PMMA interface location.

CONCLUSIONS

The integration of mesoscale modeling and time-resolved experimental measurements offers much promise to greatly improve our understanding of the effects of mesostructure in the shock response of heterogeneous materials. In this work, detailed wave fields are experimentally probed using line-imaging interferometry and transmitted particle velocities are compared to numerical mesoscale calculations representing a heterogeneous media with a well-defined geometry. Our preliminary results indicate good agreement between modeling and experiment. To refine this comparison, a well-controlled and optimized experimental design requires precise packing symmetry and probe alignment along specified particle orientations.

REFERENCES

1. Baer, M. R., et al., "Computational Modeling of Heterogeneous Reactive Materials at the Mesoscale," in Shock Compression of Condensed Matter, 1999 (M.D. Furnish, L.C. Chhabildas, R.S. Hixson, eds.), part I, pp. 27-33.
2. Baer, M. R. and Trott, W. M., et al., "Mesoscale Descriptions of Shock-Loaded Heterogeneous Porous Materials," in Shock Compression of Condensed Matter, 2001 (M.D. Furnish, N. N. Thadhani, Y. Horie, eds.), part I, pp. 713-716.
3. Trott, W. M., et al., "Measurements of Spatially Resolved Velocity Variations in Shock Compressed Heterogeneous Materials using a Line Imaging Velocity Interferometer," in Shock Compression of Condensed Matter, 1999 (M.D. Furnish, L.C. Chhabildas, R.S. Hixson, eds.), part II, pp. 993-998.
4. Trott, W. M., et al., "Investigation of Dispersive Waves in Low-Density Sugar and HMX Using Line-Imaging Velocity Interferometry," in Shock Compression of Condensed Matter, 2001 (M.D. Furnish, N. N. Thadhani, Y. Horie, eds.), part II, pp. 845-848.
5. McGlaun, J.M., Thompson, S.L., and Elrick, M. G., "CTH: A Three-Dimensional Shock Wave Physics Code," *Int. J. Impact Eng.*, Vol. 10, 351-360, 1990.
6. Hertel, E. S., and Kerley, G. I, "CTH EOS Package," Sandia National Laboratories, Albuquerque, NM, SAND98-0945, 1998.
7. Hertel, E. S., and Kerley, G. I, "Recent Improvements to CTH EOS Package," Sandia National Laboratories, Albuquerque, NM, SAND99-2340, 1999.

CP706, *Shock Compression of Condensed Matter - 2003*
edited by M. D. Furnish, Y. M. Gupta, and J. W. Forbes
2004 American Institute of Physics 0-7354-0181-0/04/$22.00

HYDROCODE MODELLING AND ANALYSIS OF A DYNAMIC FRICTION EXPERIMENT.

Andrew J. Barlow, Ronald E. Winter, Derek Carley and Peter Taylor

AWE, Aldermaston, Reading, Berkshire RG74PR, UK.

Abstract. An experiment has been performed to measure dynamic friction between two metal plates after the passage of an explosively generated shock wave. In this experiment friction was measured indirectly from the curvature of fiducial markers placed inside the plates. Hydrocode simulations were then performed to determine how much friction must be present in the experiment to match the observed fiducial shapes. The experimental data is also being used to validate friction models and has motivated the development of a new friction model. The new model defines the magnitude of the frictional force as a function of the difference in tangential velocity across the interface, but then constrains the frictional shear stress to not exceed the von Mises yield condition for the weaker material. This new friction model enables the experimental data to be matched more accurately than was possible with a normal force model.

INTRODUCTION

Dynamic friction refers to the physics that governs the tangential force across a material interface between to strong materials under shock loading. It is termed dynamic to indicate that the physics involved may be different from conventional friction due to the presence of strong shock waves.

Fiducial [1] and recovery [2] experiments are currently being performed at AWE to study dynamic friction. The fiducial experiments allow a direct comparison between code and experiment to validate friction models, whilst the recovery experiments provide insight into the physics of dynamic friction.

In [3] a method was developed for extending the slide algorithm in CORVUS [4] AWEs 2D Arbitrary Lagrangian (ALE) code to include a frictional force. This paper focuses on the modelling of one of the fiducial experiments and how the experimental data has been used to validate and improve the friction models in CORVUS.

FIDUCIAL EXPERIMENTS

The first fiducial based dynamic friction experiment at high pressures and sliding velocities to the author's knowledge was performed by Hammerberg et al at LANL [5]. In these experiments the PEGASUS pulsed power facility at LANL was used to implode a cylindrical aluminium Liner which then impacted with the outside of a cylindrical target. This produces fairly constant pressures and sliding velocities for several μs's. However, it also places a limitation on the size of the target (~3 cm diameter) and the duration of the experiment (~2 μs), which limits the resolution. The AWE fiducial experiments employ a high explosive (HE) drive [1], which allows larger targets (~10 cm) and longer problem times (~20-40 μs).

The experiment considered here (FN1/2) consisted of a 10 cm long target made from a sandwich of three metal plates. Aluminium the higher shock speed material was placed in the middle with steel plates on top and bottom. An EDC37 block was used to provide the drive and was initiated with a central lineator as shown in Fig. 1.

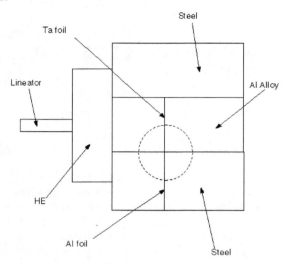

FIGURE 1. Initial geometry for fiducial experiment.

A fiducial marker was placed inside each metal perpendicular to the interface. A 0.5 mm aluminium foil was used in the steel and a 0.5 mm tantalum foil was used in the aluminium as discussed in [1]. The idea being that the higher the tangential force or friction across the interface the more fiducial bending should be observed. The relative curvature should then provide both an indirect measure of the amount of friction present and the distribution of plastic work near the interface. AWE's Mogul D flash X-ray machine was used to radiograph the fiducial at 25.0 μs.

COMPARISON OF CALCULATED AND EXPERIMENTAL FIDUCIALS

The FN1/2 experiment was initially calculated with CORVUS for the two limiting cases of true slip and a locked interface as described in [6]. Figure 2 shows a comparison of the calculated fiducial shapes obtained compared against the experimental edge locations.

Simply comparing the slope of the experimental fiducial against those for the two calculations suggests that there was some friction present in the experiment as the experimental slope lies between the two calculated slopes. However, there was a significant timing difference between the calculations and the experiment, which needed to be resolved.

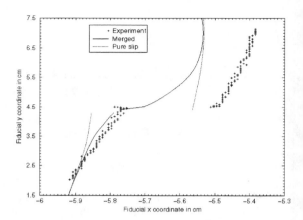

FIGURE 2. Initial comparison of calculated and experimental fiducial shapes.

A sensitivity study was then performed to try to determine what was responsible for the timing difference. This suggested that the shock wave in the aluminium was being attenuated too rapidly by the release wave from the back of the explosive. In order to overcome this problem the mass confinement around the explosive was artificially increased until a reasonable match was achieved to the position of fiducial in the middle of the aluminium plate as shown in Fig. 3.

A normal force based friction model was then applied where the frictional force is simply a function of the normal force acting across the interface. The friction coefficient for this model was then varied to obtain the best match to the experimental fiducial shapes. This was achieved with a friction coefficient of 0.01 as shown in Fig. 3. However, the calculated fiducial in the aluminium close to the interface shows catastrophic failure and the overall match to the slope of the fiducial in the steel is not as good as that in the aluminium. If the friction coefficient is adjusted to improve the match to either of the two fiducials it can only be achieved at the cost of degrading the match to the other fiducial.

The catastrophic failure calculated in the aluminium suggested that close to the interface the aluminium is yielding too rapidly because the frictional force is too high. So the frictional force in the aluminium at the fiducial for this best fit was compared against the von Mises yield limit for a 1D shear layer as shown in Fig. 4. This confirmed that the frictional force acting on the interface was exceeding the maximum shear stress that the aluminium can support when the shock wave first arrives. The calculated frictional force then oscillates in noisy manner, which probably results from the master slave slide algorithm rather than physics.

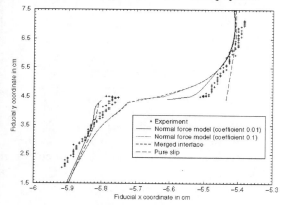

FIGURE 3. Comparison of calculated and experimental fiducial positions with increased mass confinement.

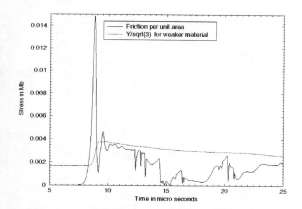

FIGURE 4. Frictional force per unit area at fiducial in the aluminium with normal force model and a friction coefficient of 0.01.

YIELD LIMITED FRICTION MODEL

A new friction model has been implemented in CORVUS to address these deficiencies. The new model initially defines the magnitude of the frictional force using the velocity dependent model described in [3], where the frictional force is defined as a function of the difference in tangential velocity across the interface. The magnitude of the frictional

force obtained is then constrained, so that the frictional shear stress σ_{xy} cannot exceed von Mises yield limit for one dimensional shear flow in the weaker material,

$$\text{if } \sigma_{xy} \geq \frac{Y}{\sqrt{3}} \text{ then}$$

$$\sigma_{xy} = \frac{Y}{\sqrt{3}} \tag{1}$$

where Y is the yield strength of the weaker material.

COMPARISON OF NEW MODEL AGAINST EXPERIMENT

The new model was then used to calculate FN1/2 with the friction coefficient again being adjusted until the best fit was obtained to the experimental fiducial data. This was achieved with a friction coefficient of 0.1 as shown in Fig. 5. The variation in the frictional shear stress is also given in Fig. 6.

A significant improvement in the match to the experimental data was obtained over that using the normal force model. The slopes of both fiducials are well matched and there was no sign of the catastrophic failure observed close to the interface in the aluminium when the normal force form was used.

The frictional shear stress profile was also more believable in that it initially follows the yield surface, but then falls away from it smoothly at lower velocities. The areas under the two curves in Fig. 6 after the first shock arrives relate to impulses applied across the interface. The area under the yield strength curve is the maximum impulse that the weaker material can support. The area under the applied frictional shear stress is the impulse that was actually required to match the experimental data. The ratio of the two areas gives a figure of merit indicating how significant friction was in the experiment. This suggests friction was quite significant in the experiment since about 50% of the maximum impulse that could be supported by the weaker material was required to match the experimental data.

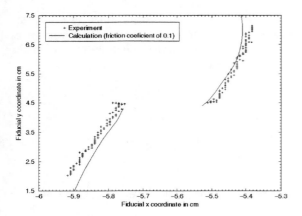

FIGURE 5. Fiducial shape calculated with yield limited velocity dependent friction model with friction coefficient of 0.1.

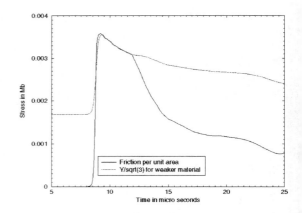

FIGURE 6. Frictional force per unit area at fiducial in the aluminium with yield limited velocity dependent model with friction coefficient of 0.1.

FUTURE WORK

The new model now needs to be applied to a wider range of problems to test its general applicability and improve it as required. The first step will be to apply it to the fiducial experiments that have already been performed at AWE [2]. These experiments are interesting as they appear to exhibit two different types of interface behaviour in one experiment. At one point on the interface the behaviour is more like that of a locked interface and at another it is more like slip. This provides a particularly good test for the model to see whether it alone can discriminate between these two cases.

DISCUSSION AND CONCLUSIONS

In attempting to simulate the fiducial data obtained from the dynamic friction experiment a number of deficiencies have been highlighted with the normal force friction model. This has lead to a new yield limited velocity dependent model being developed. This model has been shown to enable a significantly improved fit to be achieved to the experimental fiducial shapes. It has also been shown that the amount of friction required to match the experimental data is significant compared to the maximum amount of friction that the weaker material can support without exceeding the von Mises yield condition.

REFERENCES

1. Winter, R. E., Taylor, P., Carley, D. J. , Barlow, A. J., Pragnell, H., "Strain Distribution at High Pressure, High Velocity Sliding Interfaces.", *Proceedings of the American Physical Society meeting on materials under high strain rate*, Atlanta, USA, 2001.
2. Winter, R. E., Smeeton, V. S., De'ath, J., Taylor, P., Markland, L., Barlow, A. J., "Metallography of Sub-Surface Flows Generated by Shock-Induced Friction.", *Proceedings of the American Physical Society meeting on Shock Compression of Condensed Matter*, Portland, USA, 2003.
3. Barlow, A. J., "Friction in CORVUS a 2D ALE code", *Proceedings of the 22nd International Symposium on Shock Waves*, Imperial College, London, 1999, 653-657.
4. Barlow, A. J., "An Adaptive Multi-material Arbitrary Lagrangian Eulerian Algorithm for Computational Shock Hydrodynamics", PhD thesis, University of Wales Swansea, 2002.
5. Hammerberg, J. E., Kyrala, G. A. , Oro, D. M., Fulton, R. D., Anderson, W. E., Obst, A. W., Oona, H., Stokes, J., "A PEGASUS Dynamic Liner friction Experiment.", *Shock Compression of Condensed Matter*, 1999, pp. 1217.
6. Barlow, A. J., "Hydrocode Modelling of Dynamic Friction Experiments", *Proceedings of New Models and Hydrocodes for Shock Wave Processes in Condensed Matter, International Workshop held in Edinburgh*, Scotland, 2002.

CP706, *Shock Compression of Condensed Matter - 2003*
edited by M. D. Furnish, Y. M. Gupta, and J. W. Forbes
© 2004 American Institute of Physics 0-7354-0181-0/04/$22.00

EVOLUTION OF CRYSTALLOGRAPHIC TEXTURE AND STRENGTH IN BERYLLIUM

W.R. Blumenthal, D.W. Brown, and C.N. Tomé

Materials Science & Technology Division, Los Alamos National Laboratory, Los Alamos, NM 87545

Abstract. The evolution of the dynamic mechanical behavior and crystallographic texture in polycrystalline beryllium with different initial textures was measured and compared to a polycrystalline plasticity model. The split-Hopkinson pressure bar compression behavior and the activity of deformation mechanisms were found to be highly dependent on the initial texture and the loading orientation. Neutron diffraction measurements of the bulk texture as a function of strain were made at the Manuel Lujan Jr. Neutron Scattering Center. The activation of deformation twinning at high strain rates in beryllium was observed to cause both anisotropy in the mechanical behavior and rapid evolution of the texture compared to slip deformation alone. A visco-plastic self-consistent (VPSC) polycrystalline plasticity model was used to closely simulate the texture and flow strength evolution by accounting for contributions from both slip deformation and twinning mechanisms.

INTRODUCTION

Polycrystalline beryllium was deformed under high-strain-rate ($\sim 10^3$ s^{-1}) uniaxial compression to strains of over 20% at temperatures down to 77°K [1]. High-strain-rate loading has also been shown to activate $\{10\bar{1}2\}\langle10\bar{1}1\rangle$ twinning in beryllium [2].

Twinning is a significant deformation mechanism because it causes both anisotropy in the constitutive strength behavior and rapid changes in the of the crystallographic orientation of the microstructure or "texture" compared to slip deformation alone [3]. The rapid texture change is a manifestation of the extreme 84.4° rotation of basal poles toward the loading direction upon twinning.

In this paper, the evolution of the dynamic mechanical behavior and crystallographic texture in polycrystalline beryllium was measured and compared to a polycrystalline plasticity model. The split-Hopkinson pressure bar (SHPB) was used to dynamically deform specimens with varying initial textures to target strain levels in uniaxial compression. Neutron diffraction measurements of the crystallographic texture were made as a function of strain. A visco-plastic self-consistent (VPSC) polycrystalline plasticity model was then used to simulate the texture and strain-hardening evolution.

Polycrystalline beryllium with three distinct initial textures was studied. The first material was hot-pressed (HP) grade S200F beryllium (Brush Wellman, Elmore, OH) with near-random crystallographic texture. The second beryllium tested was a heavily rolled (~16:1) sheet beryllium which developed a strong basal (c-axis) fiber texture in the through-thickness (TT) direction. Dynamic compression testing was conducted along both the TT direction and the in-plane (IP) direction of the rolled sheet, which is nearly free of basal poles. Pre-existing twins were not observed in either of the starting materials by metallography.

The purpose of this study was to gain a fundamental understanding of the mechanisms of deformation, their interaction, and the role of texture in the mechanical response of polycrystalline beryllium.

EXPERIMENTAL TECHNIQUES

Incremental-strain SHPB compression tests were designed and conducted at room temperature at strain rates of ~4500 s^{-1} using precision-machined strain-limiting fixtures. IP specimens were compressed to strains of 0, 1.7, 4.8, 8.2, and 20%. HP specimens were compressed to 0, 5.6 and 14% strain. Incremental strain tests were not conducted for the TT specimens because the texture did not evolve with up to 20% strain. The SHPB method is described in detail elsewhere [1, 2].

Bulk crystallographic texture was measured with time-of-flight neutron diffraction techniques using the High Intensity Powder Diffractometer (HIPD) as described elsewhere[4, 5]. The volume fraction within the sample that has re-oriented during deformation was calculated from the integrated difference in the initial and final orientation distribution functions (ODFs).

The visco-plastic self-consistent (VPSC) model simulates a population of grains assigned orientations and weights based on the initial texture. Each grain is modeled as an ellipsoidal visco-plastic inclusion embedded in a homogenous effective medium (HEM). The deformation of an individual grain depends on its strength with respect to the HEM which is a function of its orientation. The mechanical response of the HEM is the average of all grains and is solved iteratively. A four-parameter Voce hardening law was used to describe the yield and hardening behavior of each slip mode (basal, prismatic, and pyramidal) and the tensile twinning mode. Grain reorientation due to twinning is accounted for by two extra parameters [3]. The VPSC calculation monitors the development of both the macroscopic stress and the texture at each imposed strain step. The results are then compared to experimental measurements and final parameters are determined by iteration. Note that the same parameters are used to describe the behavior of *all* three initial texture conditions. The activity of individual deformation modes was also calculated.

RESULTS AND DISCUSSION

Figure 1 shows the observed and calculated dynamic compressive stress-strain curves for the TT, IP, and HP specimens at 25°C. The behavior of the TT specimen is typical of a ductile metal. Twinning does not occur in the TT specimen as corroborated by microstructural observations and texture measurements. Therefore, the applied deformation is accommodated by dislocation slip *including* the very hard pyramidal mode. The calculated activity level of each deformation mode is shown in Fig. 2 and indicates a relatively constant combination of basal and pyramidal slip is responsible for the deformation. The TT case provides an important bound for the model and the calculated flow curve and pole figures simulate the experimental behavior very well.

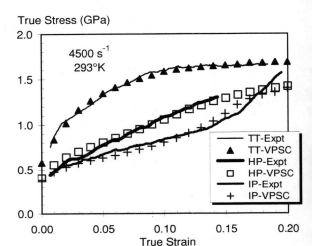

FIGURE 1. Measured (solid lines) and VPSC-calculated (symbols) compressive stress-strain curves for the rolled-TT, rolled-IP, and hot-pressed (HP) beryllium.

The IP specimen displays a yield and flow strength about half that of the TT orientation up to about 15% strain. After 15% strain a dramatic increase in hardening occurs until the flow stress is almost comparable to the TT specimen at 20% strain. The calculated IP strength satisfactorily reproduces the observed mechanical response, including a rapid increase in hardening above 15% strain. The activity of various deformation modes varies greatly with strain in the IP specimen as shown in Fig. 2. Initial deformation is dominated by prismatic slip but is slowly replaced by twinning and basal slip until twinning peaks at about 15% strain. After 15% strain basal slip dominates and pyramidal slip begins to increase while twinning and prismatic slip activity decrease.

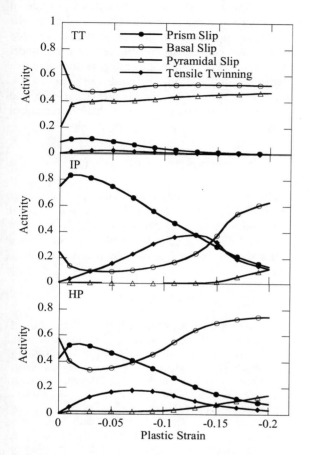

FIGURE 2. Calculated activity of deformation modes during dynamic compression at 25°C for TT, IP, and HP orientations.

Due to its random texture, the HP specimen lies between the TT and IP behaviors and displays a hardening rate that is nearly constant up to about 15% strain. The calculated strength curve is also in very good agreement with experiment. The activity of various deformation modes also varies with strain in the HP specimen, but less dramatically than the IP specimen (Fig. 2). Initial deformation is controlled by a combination of prismatic and basal slip, but twinning slowly increases and peaks at about 10% strain while prismatic slip steadily

declines from nearly the outset. After 10% strain basal slip dominates and pyramidal slip begins to increase.

Figure 3 shows the observed and calculated basal (0002) pole figures of the IP specimen compressed to strains of 0%, 1.7%, 4.8%, 8.2%, and 20% at 25°C. Initially the strong basal fiber texture, 8 multiples of random distribution (mrd), is perpendicular to the loading direction. After 2% deformation, the texture has changed little, but by 5% deformation the basal fiber is beginning to weaken slightly from a maximum of 8 to 7 mrd. After 8% deformation, a significant textural reorientation is apparent, with basal poles aligning parallel to the compression direction. Finally, after 20% deformation, the IP texture has been almost completely reoriented and a strong basal fiber (7.3 mrd) has developed parallel to the applied load. Since the new texture is nearly equivalent to the TT specimen, its strength also approaches that of the TT specimen (Fig. 1). The two lobes observed in the pole figure are part of the original texture and represent orientations unfavorable for twinning. The calculated texture development during IP compression at various strain intervals reproduces the experimental data well, including the two remnant lobes.

HP specimens were also compressed iteratively to 0%, 5.6% and 14% strain. For brevity the measured and calculated pole figures are not shown. Initially the texture is nearly random, but after 5.6% strain a basal fiber parallel to the applied load is apparent and after 14% deformation it is relatively strong (3.4 mrd). Calculated pole figures reproduced the texture development adequately.

Twins first appear in the IP specimens between 2% and 5% strain and the twin volume fraction subsequently increases linearly to 0.67 at 20% strain. The rate of twin formation between 5% and 20% strain is 0.041/percent strain. Twinned develops more slowly in the HP specimens because fewer grains are favorably oriented. After 14% strain the twin volume fraction is 0.29 compared to roughly 0.42 in the IP material. Between 5% and 20% strain, the rate of twin formation is 0.019/percent strain or about half the rate in the IP material. The calculated rates of twin formation from the VPSC model were in good agreement with experimentally measured rates.

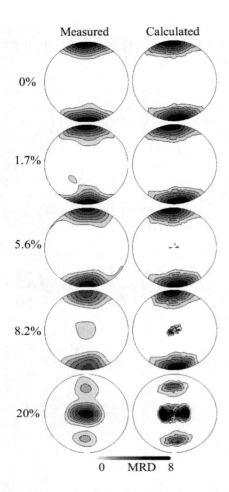

Measured   Calculated

0%

1.7%

5.6%

8.2%

20%

0   MRD   8

FIGURE 3. Measured and calculated (0002) pole figures of rolled beryllium after 0, 1.7, 5.6, 8.2, and 20% in-plane deformation. Compressive direction is into the page.

SUMMARY AND CONCLUSIONS

The dynamic compression behavior of beryllium was successfully simulated for vastly different initial textures using a polycrystalline plasticity model (VPSC) model that accounted for three anisotropic slip systems (basal, prismatic, and pyramidal), plus the $\{10\bar{1}2\}\langle10\bar{1}1\rangle$ tensile twinning mode.

Accounting for twinning deformation in the VPSC model is extremely important because it causes rapid and significant evolution of the texture compared to slip deformation alone. Neutron diffraction measurements after small and large strain increments accurately defined the texture evolution and the activity of the twinning mode. The different starting texture conditions provided a high degree of constraint and validation for the self-consistent model. The ultimate goal is to develop an accurate continuum constitutive strength model that accounts for both the initial texture and the texture evolution from the strain history. Future work will attempt to incorporate strain-rate and temperature dependence into the mechanical behavior and texture evolution modeling.

ACKNOWLEDGMENTS

This work was supported under the auspices of the U.S. Dept. of Energy at Los Alamos National Laboratory under contract W-7405-ENG-36. The Manuel Lujan Jr. Neutron Scattering Center is a national user facility funded in part by the U.S. DOE.

REFERENCES

1. Blumenthal, W.R., Carpenter, R.W., Gray, G.T. III, Cannon, D.D., and Abeln, S.P., "Influence of strain rate and temperature on the mechanical behavior of beryllium," in *Shock Compression of Condensed Matter -1997*, edited by S.C. Schmidt, Dandekar, D.D., Forbes, J.W., AIP Conf. Proc, Woodbury, NY., 1998, pp. 411-414.
2. Blumenthal, W.R., Abeln, S.P., Mataya, M.C., Gray, G.T. III, and Cannon, D.D., "Dynamic behavior of beryllium as a function of texture," in *Proc. of Plasticity '99: 7th Int. Symp. on Plasticity and its Current Applications*, edited by A.S. Khan, Neat Press, Fulton, MD, 1999, pp. 615-618.
3. Tomé, C.N., Agnew, S.R., Blumenthal, W.R., Bourke, M.A.M., Brown, D.W., Kaschner, G.C., and Rangaswamy, P., *Mater. Sci. Forum: Textures of Materials* **408-412**, 263-268 (2002).
4. Von Dreele, R.B., *J. Appl. Crystall.* **30**, 517-525 (1997).
5. Bourke, M.A.M., Goldstone, J.A., and Holden, T.M., *Measurement of residual and applied stress using neutron diffraction*, Kluwer Academic, Amsterdam, The Netherlands, 1992, p. 369.
6. Kocks, U.F., Tomé, C.N., and Wenk, H.-R., *Texture and Anisotropy*, Cambridge University Press, Cambridge, U.K., 1998, p. 676.

CP706, *Shock Compression of Condensed Matter - 2003*
edited by M. D. Furnish, Y. M. Gupta, and J. W. Forbes
© 2004 American Institute of Physics 0-7354-0181-0/04/$22.00

COUPLED PLASTICITY AND DAMAGE MODELING AND THEIR APPLICATIONS IN A THREE-DIMENSIONAL EULERIAN HYDROCODE

Michael W. Burkett[1], Sean P. Clancy[2], Paul J. Maudlin[3], and Kathleen S. Holian[4]

[1]*Applied Physics Div., Primary Design and Assessment Grp. (X-4), MS T086*
[2]*Applied Physics Div., Integrated Physics Methods Grp. (X-3), MS F644*
[3]*Theoretical Div., Fluid Dynamics Grp. (T-3), MS B216*
[4]*Computing Communications & Networking Div., Scientific Software Engineering Grp. (CCN-12), MS B295*
Los Alamos National Laboratory, Los Alamos, NM 87545

Abstract. Previously developed constitutive models and solution algorithms for continuum-level anisotropic elastoplastic material strength and an isotropic damage model TEPLA have been implemented in the three-dimensional Eulerian hydrodynamics code known as CONEJO. The anisotropic constitutive modeling is posed in an unrotated material frame of reference using the theorem of polar decomposition to compute rigid-body rotation. TEPLA is based upon the Gurson flow surface (a potential function used in conjunction with the associated flow law). The original TEPLA equation set has been extended to include anisotropic elastoplasticity and has been recast into a new implicit solution algorithm based upon an eigenvalue scheme to accommodate the anisotropy. This algorithm solves a two-by-two system of nonlinear equations using a Newton-Raphson iteration scheme. Simulations of a shaped-charge jet formation, a Taylor cylinder impact, and an explosively loaded hemishell were selected to demonstrate the utility of this modeling capability. The predicted deformation topology, plastic strain, and porosity distributions are shown for the three simulations.

INTRODUCTION

Accurate constitutive descriptions are required for high-strain-rate deformation processes involving metals whose mechanical responses show significant directional dependence. The viability of utilizing anisotropic elastoplastic constitutive modeling to predict the large rigid-body rotation and plastic deformation was shown previously (1). Those calculations showed a sensitivity to the yield function description for a hexagonal close-packed material with high-yield anisotropy. Constitutive models and solution algorithms for anisotropic elastoplastic material strength (2) and damage evolution (3) developed for a Lagrangian continuum mechanics code (EPIC) (4) have been implemented in the three-dimensional CONEJO

codes (5). CONEJO was developed under the Accelerated Strategic Computing Initiative (ASCI) Blanca code project.

HYDROCODE ARCHITECTURE

CONEJO is an explicit Eulerian continuum-mechanics code that is used to predict formation processes associated with large material deformation at elevated strain rates. Some special features of CONEJO include a high-order advection algorithm, a material interface tracking scheme, and van Leer monotonic advection limiting. CONEJO utilizes a Parallel Object-Oriented Methods and Applications (POOMA) framework developed for advanced parallel computing.

CONSTITUTIVE MODEL

The TEPLA (3) constitutive module addresses a ductile, porous, inelastic solid subjected to a total rate of straining $\underline{\underline{D}}$ and computes the Cauchy stress $\underline{\underline{\sigma}}$ while evolving porosity ϕ and von Mises plastic strain ε_{vM}^p (dilatational and shear damage modes, respectively). TEPLA is based on the Gurson flow surface but contains phenomenological extensions beyond the original modeling work of Johnson and Addessio (3), such as elastoplastic anisotropy, a rate-dependent dislocation-mechanics-based flow stress model, and decohesion for the material failure process. The TEPLA equation set, which is solved implicitly over each time increment in the above-discussed CONEJO code, is described briefly here.

Consistent with experimentally observed metallic behavior, the Cauchy stress rate $\dot{\underline{\underline{\sigma}}}$ is assumed to be constitutively related to linear elasticity (hypoelasticity) for its deviatoric response and further related to nonlinear elasticity for the spherical response (i.e., an equation of state [EOS]):

$$\dot{\underline{\underline{\sigma}}} = \mathbf{M} : \widetilde{\mathbf{E}}' : \left(\underline{\underline{D}}' - \underline{\underline{D}}^{p'} \right) - \mathbf{M} : \left(\frac{\partial \widetilde{P}}{\partial \varepsilon_v^e} \dot{\varepsilon}_v^e + \frac{\partial \widetilde{P}}{\partial S} \dot{S} \right) \underline{\underline{1}} + \dot{\mathbf{M}} : \mathbf{M}^{-1} : \underline{\underline{\sigma}} \tag{1}$$

Here an additive decomposition of the total rate of straining $\underline{\underline{D}}$ into elastic, isotropic damage and incompressible plastic parts is assumed:

$$\underline{\underline{D}} = \underline{\underline{D}}^e + \underline{\underline{D}}^d + \underline{\underline{D}}^p \tag{2}$$

The last term appearing in Eq. 1 represents a softening effect due to the rate of increase of damage. The EOS partial derivatives appearing in Eq. 1 are easily relatable to the isentropic bulk modulus and the Grüneisen derivative. The quantity $\widetilde{\mathbf{E}}'$ is the deviatoric (denoted by primes) elastic stiffness tensor for the pristine or matrix material (the overscore tilde designates a pristine material variable), S is the entropy, and ε_v^e is a volumetric elastic strain that can be related to the change in density. The tensors of Eq. 1 are material frame tensors (unrotated tensors with respect to the original configuration, but related to the current configuration via a rigid rotation), and the damage effect tensor $\mathbf{M}$ is isotropic:

$$\mathbf{M} = (1 - \phi)\boldsymbol{\delta} \tag{3}$$

Here ϕ is defined as the ratio of void volume over the total volume, and $\boldsymbol{\delta}$ is the fourth-order identity tensor.

In the absence of void nucleation, the evolution of porosity due to the growth of an initial amount of void ϕ_0 can be kinematically derived from the deformation gradient:

$$\dot{\phi} = (1 - \phi)\,\mathrm{trace}\left(\underline{\underline{D}}^d \right) \tag{4}$$

A Mie-Grüneisen analytical EOS form is typically used to obtain the partial derivatives appearing in Eq. 1, although a more accurate tabular EOS option (SESAME [6]) is also available, which has been used in the problem simulations described below.

The rate equations for stress and porosity, Eqs. 1 and 4, respectively, are closed once the equilibrium state $\underline{\underline{\sigma}}$ is known. The equilibrium stress $\underline{\underline{\sigma}}$ and porosity ϕ are constrained to be consistent with a Gurson flow surface:

$$G = \frac{3}{2} \underline{\underline{\sigma}} : \boldsymbol{\alpha} : \underline{\underline{\sigma}} - \sigma^2 \left\{ 1 + q_3 \phi^2 - 2q_1 \phi \cosh \beta \right\} = 0 \tag{5}$$

where σ is the flow stress, $\boldsymbol{\alpha}$ is a yield surface shape tensor typically expressed in a five-dimensional orthonormal basis (7), $\beta \equiv -3q_2 P / 2\sigma_s$ is a modified triaxiality, and the q_i are the Tvergaard (8) void-growth parameters.

A local over-stress formulation for a rate-dependent material is postulated in order to maintain a well-posed numerical solution:

$$\underline{\underline{\sigma}} = \underline{\underline{\sigma}}^\infty + \tau_r \underline{\underline{D}}^p \tag{6}$$

where $\underline{\underline{\sigma}}^\infty$ is the equilibrium stress solution (from Eqs. 1 through 5), and τ_r is a viscosity parameter related to a stress-relaxation time for the material. The implied length scale in this overstress model can be approximated from τ_r using the elastic properties of the material for an isotropic infinite medium (7).

The accumulation of damage within the TEPLA equation set is represented dilatationally by the porosity, and deviatorically by the von Mises equivalent plastic strain, as discussed above. These two strain invariants are combined with the stress triaxiality P/σ_{vM} in an isotropic empirical failure surface (3) of the form

$$F = 1 - \left(\frac{\phi}{\phi_f}\right)^2 - \left(\frac{\varepsilon_{vM}^p}{\varepsilon_f^p}\right)^2 = 0 \qquad (7)$$

where ϕ_f and ε_f^p are failure values for the porosity and equivalent strains, respectively, as obtained from plate-impact and tensile-bar testing driven to specimen fracture. The failure strain is determined from the classic Hancock-MacKenzie work (9).

MODEL APPLICATION AND DISCUSSION

Simulations of a shaped-charge jet formation, a Taylor cylinder impact, and an explosively loaded hemishell were selected to demonstrate the utility of the above modeling capability. The Mechanical Threshold Stress (MTS) flow-stress model (10) was used in the simulations. CONEJO simulations were performed for a jet formation with a full, three-dimensional anisotropic plasticity material description without damage as well as for two fully coupled simulations with three-dimensional anisotropic plasticity and isotropic damage: a Taylor cylinder impact and an explosively loaded hemishell.

The jet formation problem for a hemispherical liner material with directional yield dependency involves elevated strain rates ($\sim 10^4$ s^{-1}), large strains (>500%), and significant rigid-body rotations in some regions of the jetting material. The predicted jet topology for a uniform 1 mm resolution and planar views of the jet are provided in Fig. 1. Figure 2 shows the Euler-Rodrigues vectors and angles representing the rigid-body

rotation that the material is undergoing during formation. The vectors indicate the axis of rotation (right-hand rule applies), and the vector's color quantifies the magnitude of the rotation in radians. For the hemispherical liner geometry, the jet tip region experiences small rigid-body rotations while the tail section experiences large rotations, as indicated in Fig. 2. Also, as the tip region of the jet begins to form, small rigid-body rotations on the jet surface and radial gradients in the rotation angle exist. These kinematical features could be nucleation sites for necking to develop and subsequently lead to jet particulation.

A tantalum Taylor cylinder impact problem (length of 3.81 cm, diameter of 0.381 cm and impact velocity of 175 m/s) was selected to demonstrate the coupling of the anisotropic plasticity and isotropic damage at moderate strain rates. The impact of the cylinder with the rigid boundary induces stress-wave interactions that result in the formation of center-line tension and porosity growth. Figure 3 shows the porosity distribution at the center of the cylinder at 2 μs after impact. The void growth is short-lived and eventually crushes out (voids collapse) as the cylinder continues to deform in a compressive mode against the rigid anvil. The nonuniform deformation characteristics of the event (cylinder topology and deformation footprint [end of cylinder striking boundary]) are shown in Fig. 3 and are the results of the assumed initial yield surface orientation and rigid-body rotation and the direction of the applied load. The plastic flow is larger along the Y-axis because the flow stress is softer in the Y-direction, as a consequence of the yield surface shape for this material. In this calculation, the strong axis of the yield surface was initially oriented along the Z-axis, perpendicular to the

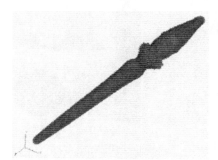

FIGURE 1. Jet topology at 60 μs.

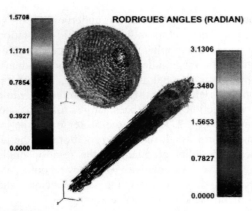

FIGURE 2. Euler-Rodrigues vectors and angle magnitudes, initially and at 60 µs.

loading direction (X-axis). This configuration produces an elliptical footprint with some flattening along 45° lines.

A high strain rate and large strain deformation with anisotropic plasticity and isotropic damage is realized in a highly resolved (30 computational cells through the shell thickness), explosively loaded hemishell simulation shown in Fig. 4. The angular dependence of the loading to the hemishell by the detonation wave can induce significant temporal and spatial gradients in plastic strains, strain rates, and porosity for any isotropic material. The effect of applying these loading conditions to an anisotropic material while evolving damage can provide an additional mechanism to enhance instabilities in the plastic flow. Figure 4 shows the predicted nonhomogenous distribution of plastic strain and porosity after 2 µs of loading by the detonation wave.

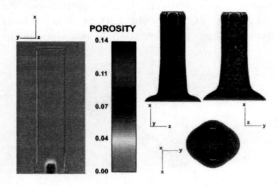

FIGURE 3. Ta Taylor cylinder porosity distribution, major and minor side profiles, and footprint.

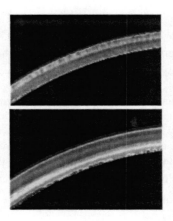

FIGURE 4. Porosity (top) and plastic strain distribution (bottom) in the hemispherical shell after 2 µs of expansion. Porosity and plastic strain distributions vary linearly (red:highest value; blue: lowest value) from 0-20% and 0-58%, respectively.

ACKNOWLEDGMENTS

The authors acknowledge the effort of Lynne Atencio in the preparation of this paper. The financial support of the Joint DoD/DOE Munitions Technology Development Program, HE/Metal Interactions Project Leader L. Hull and ASCI was appreciated. This work was performed under the auspices of the US Department of Energy.

REFERENCES

1. Clancy, S. P., Burkett, M. W., and Maudlin, P. J., *J. Phys. IV France* **7**, Colloq. C3 (DYMAT 97), 735–740 (1997).
2. Maudlin, P. J., and Schiferl, S. K., *Comput. Methods Appl. Mech. Engrg.* **131**, 1–30 (1996).
3. Johnson, J. N., and Addessio, F. L., "Tensile Plasticity and Ductile Fracture," *J. Appl. Phys.* **64**, 6699–6712 (1988).
4. Johnson, G. R., and Stryk R. A., Air Force Armament Laboratory, Eglin Air Force Base report AFATL-TR-86-51 (August 1986).
5. Holian, K. S., et al., *Proceedings of the Conference on Numerical Simulations and Physical Processes related to Shock Waves in Condensed Media*, Oxford, England, September 1997.
6. Holian, K. S., (ed.), *T-4 Handbook of Material Data Bases, Vol. 1c: Equation of State*, Los Alamos National Laboratory report LA-10160-MS (November 1984).
7. Maudlin, P. J., Harstad, E. N., Mason, T. A., Zuo, Q. H., and Addessio, F. L., "TEPLA: An Elastoplasticity, Damage and Failure Approach for Ductile Materials," Los Alamos technical document in progress, to be published in 2003.
8. Tvergaard, V., "Influence of Voids on Shear Band Instability under Plane Strain Conditions," *Int. J. Fracture* **17**, 389–407 (1981).
9. Hancock, J. W., and MacKenzie, A. C., *J. Mech. Phys. Solids* **24**, 174 (1976).
10. Follansbee, P. S., and Kochs, U. F., *Acta Metall.* **36**, No 1, 81–93 (1988).

CP706, *Shock Compression of Condensed Matter - 2003*
edited by M. D. Furnish, Y. M. Gupta, and J. W. Forbes
© 2004 American Institute of Physics 0-7354-0181-0/04/$22.00

SHOCK LOADING INFLUENCE ON MECHANICAL BEHAVIOR OF HIGH PURITY IRON

François Buy and Christophe Voltz

Commissariat à l'Energie Atomique – Centre de Valduc – 21120 Is Sur Tille

Abstract. This paper proposes the analysis of shock wave effects for high purity iron. The method developed is based on the characterization of the mechanical behavior of as received and shocked material. Shock effect is generated through plate impact tests performed in the range of 4 GPa to 39 GPa on a single stage light gas gun or a powder gun. Therefore, as-received and impacted materials are characterized. A formalism proposed by J.R.Klepaczko [1] and based on physical relations has been adopted to describe stress strain curves.

INTRODUCTION

Pure iron is known to exhibit a α to ε (body-centered cubic to hexagonal close packed) phase transition when shocked to pressure above 13 GPa. This pressure induced solid-solid transformation is completely reversible. Post shock recovered iron, which undergoes α to ε and return transition, cumulates strain-hardening, defects caused by imposed pressure and shear. The mechanical response to shock of pure iron is reviewed in this paper regarding to shock pressure: below or above 13 GPa.

EXPERIMENTAL PROCEDURE

Iron characterization

The polycrystalline iron used for this study is a high purity Armco type IV grade. The bar was made by high temperature open die forging followed by thermal treatment (under inert gas - 4 hours at 920 °C). Chemical analysis of this iron (in wt.%) is C 0.0125, O 0.0245, Si < 0.005, Cr 0.031, Ni 0.0185, Cu 0.0064, Mn 0.0025, and balance Fe. The grain morphology is equiax with no preferential elongation. The grain size is between 200 and 300 μm. The Vickers hardness is 66 Hv (under 10 N) indicating a fully relaxed structure. The as-received iron is characterized in compression at temperatures between 77 K and 741 K and at strain rates ranging from 10^{-4} to

$2000\,s^{-1}$. The stress-strain response (figure 1) of this iron is similar to the other body cubic centered metals with low symmetry (Ta, Nb, Mo and W).

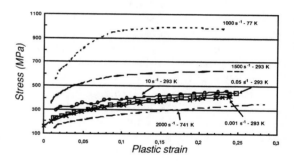

Figure 1. Stress-strain response of iron as a function of strain rate and temperature.

For this class of materials strain-hardening responses after yielding are nearly invariant as a function of strain rate, the stress-strain curves are parallel although offset in their initial yields [2]. Recent work allows us to define the two main mechanical behaviors: athermal plateau and thermally activated domain. The limit is found at $2\,s^{-1}$ – 293K, so further mechanical tests will be done at $0.05\,s^{-1}$ – 293K (athermal plateau domain) and $1500\,s^{-1}$ – 293K (thermally activated domain). Tests done at $1000\,s^{-1}$ – 77K exhibit particular behavior, a high strength level and important strain-hardening response due to twinning deformation mode.

Shock-wave recovery tests

The shock recovery experiments are performed using a 80-mm single-stage light gas gun for shock pressure below 13 GPa and a 60-mm powder gun for upper pressure. The samples, 50-mm-diameter, 7 or 5-mm-thick (low pressure or high-pressure test) are fitted in a precisely bored recess inside momentum trap. Backing the target assembly prevents sample spallation by 20-mm thick spall plate, which also serves as longitudinal momentum trap. All apparatus components were fabricated from A33 low carbon steel to ensure good impedance matching during shock loading with high purity iron. The samples are soft recovered. For shock pressure under phase transition, flyer is 3.1-mm-thick, 70-mm diameter austenitic steel (EN 1.4404) and for higher shock pressure flyer is 2.5-mm-thick, 58-mm diameter tantalum. The pulse duration is ranging from 1.22 to 1.32 µs. Sample were shock loaded at 4, 6.1, 8.6, 19.7, 31.6, and 38.3 GPa.

RESULTS

The Vickers hardness is measured to the target center through the thickness, for all samples the values are constant. For shock pressure below phase transition the hardness increases gradually with increasing peak pressure, from 66 H_v (as received material) to 119 H_v (8.6 GPa shock). For tests above 13 GPa, results are ranging from 235 to 256 H_v. It seems that a limit is reached. These results are consistent with other works [3,4].

Mechanical behavior

Compression specimens were machined from the recovered targets. The axis of the cylinders corresponds to the direction of shock wave propagation. Mechanical characterizations are presented figure 2 and 3.

- First, the elastic limit exhibits a high sensitivity to strain rate. This trend is typical of b.c.c. structure. The flow stress can be considered as the sum of two components. The first one depends on long range obstacle effects and is not very sensitive to temperature and strain rate. The second one denotes the effects of short-range obstacles that can be overcome by thermal fluctuations. This sensitivity reveals to be weakly influenced by the shock pressure (200 MPa) which

is consistent with results observed on tantalum, another b.c.c. structure metal.

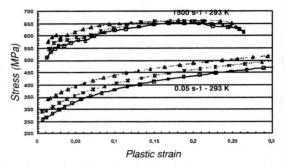

Figure 2. Shocks below 13 GPa - ■ 4 GPa ; * 6.1 GPa ; Δ 8.6 GPa.

- Second, the strain hardening diminishes both with strain rate and shock pressure level. The extreme case is the dynamic compression on the 8.6 GPa shocked specimens. The flow stress shows a slight hardening before decreasing. This can be accounted for by a microstructure reorganization (mainly of dislocations) which tends to annihilate during the deformation.

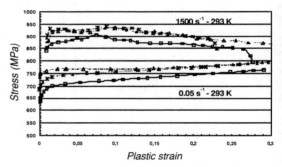

Figure 3. Shocks above 13 GPa - ■ 19.7 GPa ; * 31.6 GPa ; Δ 38.3 GPa.

- Finally the flow stress tends to a saturation value, which depends on the solicitation conditions. This illustrates the fact that microstructure reaches a maximum default concentration. We can notice a great influence of the shock on the ulterior mechanical behavior of the sample, which corroborates the microstructure observations.

DISCUSSION

Plastic deformation evaluation

However, in order to take into account the effective influence of the shock on the mechanical behavior of iron, we have to evaluate the level of pre-straining due to this loading condition. The plastic strain is the sum of plastic strain evaluated by simulation (shock and release stages) and residual one. So, the total plastic strain (called γ further in this paper) reaches the levels: 0.08–0.107–0.116 for shocks below phase transition and 0.52–0.8–0.87 for shocks above 13 GPa.

Constitutive law

We consider the following stress evolution with dislocation density [5]. The shear stress is decomposed in two stresses: internal stress τ_a and effective stress τ^* (1) (2).

$$\tau_a = M_\sigma \alpha G b \sqrt{\rho} \qquad (1)$$

$$\tau^* = \tau_0^* \left[1 - \left(\frac{kT}{\Delta G_0} \ln \frac{\nu(\rho_m)}{\dot{\gamma}} \right)^{\frac{1}{q}} \right]^{\frac{1}{p}} \qquad (2)$$

with :

$$\frac{d\rho}{d\gamma} = M_{II}(\dot{\gamma}) - k_a(\dot{\gamma}, T)(\rho - \rho_0) \qquad (3)$$

For monotonic test in the region of athermal plateau for iron, it can be deduced that the dislocation density is related to plastic strain through the expression (4):

$$\rho = \rho_0 + \frac{M_{II}}{k_a}(1 - e^{-k_a \gamma}) \qquad (4)$$

Parameters identification for as-received iron

To evaluate M_{II} and k_a we make some assumptions: τ^* remains constant during the deformation, and $M_\sigma \alpha = 1.5$ ($M_\sigma = 3$ and $\alpha = 0.5$). We have measured by TEM the dislocation density to $3\ 10^{+13}$ m^{-2} for initial state. With (4) we calculate this quantity at $2.55\ 10^{+13}$ m^{-2}. This is in good accordance and validates the hypothesis for $M_\sigma \alpha = 1.5$. For further tests the M_{II} and k_a calculated will be kept and introduced in following development. Results are listed below:

TABLE 1. Constitutive law parameters.

		0.05 s^{-1}	1500 s^{-1}
M_{II}	m^{-2}	$7{,}7\ 10^{+14}$	$2{,}6\ 10^{+15}$
k_a	---	3.2	11.9
Y_0	MPa	155	156
τ^*	MPa	----	152
σ_{sat}	MPa	504	636

Parameter identification for shocked iron

Considering that the dislocations density at the end of shock prestraining (strain jump) is identical to those at the beginning of the reloading, the dislocation density law integration leads to:

$$\rho = \rho_0 + \frac{M_{II}}{k_a}(1 - e^{-k_a(\gamma - \gamma_j)}) + (\rho_j - \rho_0)e^{-k_a(\gamma - \gamma_j)} \qquad (5)$$

with : γ_j the shear strain at the jump (estimated by code simulation and sample measurements) and ρ_j the dislocation density at the jump.

Below phase transition

The best fit is achieved by taking into account a new athermal stress process: twinning. So the total stress is the sum of three terms:

$$\tau = \tau_a + \tau^* + \tau_i \quad (6) \qquad \text{and} \qquad \tau_i = KG\sqrt{\frac{b}{\Delta}} \quad (6)$$

with: K a constant, b burgers vector, G shear modulus and Δ twinning spacing. The best fit for the three tests gives results below (table 2).

TABLE 2. Optimization results

		4 GPa	6.1 Gpa	8.6 Gpa
τ_i	MPa	0	15	30
τ^*	MPa	158	160	158
ρ_j	m^{-2}	$6{,}9\ 10^{+13}$	$8{,}3\ 10^{+13}$	$10{,}6\ 10^{+13}$

Several observations can be done:
- The effective stress is not affected by shock loading below phase transition. The average value of 158 MPa is quite the same for as-received material.

- Twinning has an increasing effect on internal stress for increasing shocks pressure. For 4 GPa test this effect is negligible.

- Dislocations density is multiplied by ≈ 4 for 8.6 GPa shock.

According to (6) we calculate the twinning inter spacing values. These are listed table 3. For 6.1 GPa sample metallography we find 70 µm for Δ, this in quite in accordance with calculated value, which can be taken as realistic values.

TABLE 3. Parameter identification for τ_i.

	Δ (µm)	K
8,6 GPa	12	0.0803
6,1 GPa	50	0.0814

Above phase transition

It is impossible to find a parameter setup that fits the data with the formalism used. In fact we observe that shock wave prestraining above 13 GPa deeply alters iron mechanical response (figure 4) :

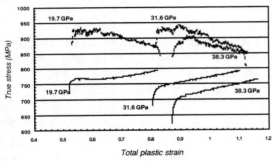

Figure 4. Stress strain curves for above phase transition shocks (solid line quasi-static reloading – dashed line dynamic reloading).

- *yield stress* – upper and lower yield stress disappear whereas this phenomenon is present for below phase transition shocked material. For quasi static tests yield stress diminishes with increasing pressure shock (100 MPa).

- *strain hardening* – for quasi static tests strain hardening is linear and the same for all three shock pressure. This type of stress strain curve can be noted for quenched low carbon steel from upper part of α domain.

- *saturation stress* – for quasi-static loading there is no evidence of saturation stress for the applied plastic strain. For dynamic testing the three

curves have the same behavior after saturation level (930-900 MPa) the stress decreases.

CONCLUSION

We have shown that the model proposed by Klepaczko for b.c.c. is able to reproduce stress strain curves for shock wave prestraining iron below phase transition and to take into account of twinning effect on stress.

Even recovered material (after shock wave prestraining) is in α iron state, the mechanical behavior is deeply affected by prestraining for shock pressure above 13 GPa. Under phase transition there is a slight increase in stress due mainly by twinning (athermal process).

For shock above phase transition in case of quasi-static reloading the upper and lower yielding is suppressed, this is showing that dislocations are not yet impinging on interstitial impurities.

Common dislocation evolution law is not able to reproduce the mechanical behavior for material which undergone α ⇔ ε phase transition. We need better knowledge of dislocation nature and post-shock crystallography for recovered iron.

ACKNOWLEDGEMENTS

Many thanks to C.Valot from Bourgogne University (UMR n°5613) for X-rays analysis and comments on texture. The authors gratefully acknowledge P.Antoine for conducting impact test, S.Clement, P.Martinuzzi and C.Mathieu for their contribution to material characterization.

REFERENCES

1. J.R. Klepaczko – Mat. Sc. & Eng 18. p 121.
2. High pressure response of high-purity iron – L.R.Veeser, G.T.Gray, J.E.Vorthman, P.J.Rodrigez, R.S. Hixon and D.B. Hayes. Shock compression of condensed matter - pp. 73-76-1999.
3. Effect of shock waves on armco iron and copper at different temperatures – Z.M. Gelunova, V.P. Lemyakin and P.O.Pashkov. Physics metal. Metalloved 28, N°6. pp. 1064-1069-1969.
4. Dieter G.E. - Response of metals to high velocity deformation. pp. 409-445 – Shewmon editors-1960.
5. Evaluation of the parameters of constitutive model for b.c.c. metals based on thermal activation – F.Buy, J.Farre, J.R. Klepaczko and G.Talabart – J.Phys IV France 7 C3 – pp. 631-636-1997.

CP706, *Shock Compression of Condensed Matter - 2003*
edited by M. D. Furnish, Y. M. Gupta, and J. W. Forbes
© 2004 American Institute of Physics 0-7354-0181-0/04/$22.00

SHEAR LOCALIZATION-MARTENSITIC TRANSFORMATION INTERACTIONS IN Fe-Cr-Ni MONOCRYSTAL

B. Y. Cao[1], M. A. Meyers[1], V. F. Nesterenko[1], D. Benson[1], and Y.B. Xu[2]

[1] *University of California, San Diego, La Jolla, CA 92093,*
[2] *Institute of Metal Research, Chinese Academy of Sciences, Shenyang 110016, P. R. China*

Abstract. A Fe-15wt%Cr-15wt%Ni alloy monocrystal was deformed dynamically (strain rate~$10^4 s^{-1}$) by the collapse of an explosively driven thick-walled cylinder under prescribed temperature and strain. The experiments were carried out under the following conditions: (a) Alloy in austenitic state; (b) Alloy in transformed state; (c) Alloy at temperature slightly above M_s. The last case is characterized by precipitating concurrent shear-band propagation and martensitic transformation. The alloy exhibited profuse shear-band formation depending on the deformation condition. Stress-assisted and strain-induced martensitic transformation competed with shear localization. The alloy that was deformed at a temperature slightly above M_s showed a significantly reduced number of shear bands. The anisotropy of plastic deformation determined the evolution of strains and distribution of shear bands. Calculated shear-band spacings based on the Grady-Kipp and Wright-Ockendon theories are compared with observed values. The microstructure within shear bands was characterized by transmission electron microscopy. Regions of sub-micron grain sizes exhibiting evidence of recrystallization were observed, as well as amorphous regions possibly resulting from melting and rapid resolidification of heavily deformed material inside shear band.

INTRODUCTION

Shear localization is an extreme case of inhomogeneous deformation where plastic deformation localizes in a thin region of a specimen. In most cases, shear localization is associated with a local softening of the material. This softening can be due to thermal or geometrical reasons, or due to microfracture.

In thermal softening, the local increase in temperature leads to localization. In the extreme case when the strain rate is so high that the generated local heat cannot escape from the deformation area, shear bands are formed, which are called "adiabatic". Regardless of the initial softening mechanism, the localized deformation leads to an accelerated strain rate. Eventually, there are heat concentration and thermal effects in most situations [1,2].

Nesterenko, Meyers and co-workers [3-5] have established that shear bands in metals organize themselves with a characteristic spacing that is a function of material parameters; it was found that this spacing evolves with the development of the shear bands, so that smaller shear bands have a smaller spacing.

The goal of this research was to determine the interaction between shear localization and martensitic transformation.

EXPERIMENTAL METHODS

The FCC alloy Fe-15wt%Cr-15wt%Ni was chosen for this investigation because it has a very well characterized deformation and transformation response. A monocrystalline cylindrical specimen with 32 mm diameter was grown at the Advanced Research and Development Laboratory at Pratt and

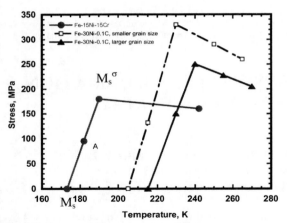

FIGURE 1. Temperature dependence of the material yield strength for Fe-15Cr-15Ni: Fe-15Cr-15Ni, M_s=173 K (data from Stone &Thomas [6] and Guimaraes et al. [7], D = 0.211mm; D = 0.101mm.

Whitney Aircraft. This alloy was homogenized at 1,500K for 72 hours. Figure 1 shows the estimated M_s temperature for this alloy, obtained by using point A (182 K, 80 MPa [6]) and an assumed slope of the stress-assisted range. For comparison, the measured results for Fe-30Ni-1C [7] at two grain sizes (D) are shown in the same plot. The reduction of grain size affects the M_s temperature.

The dynamic method of plastic deformation used in this investigation, the collapsing thick-walled cylinder, has been used and thoroughly characterized for multiple shear band generation [8]. The thick-walled cylinder (TWC) implosion technique was introduced by Nesterenko and Bondar [9] and represents an improvement from the confined exploding cylinder technique developed by Shockey [10].

In TWC test, the specimen is sandwiched between a copper driver tube and a copper stopper tube and is collapsed inward during the test. Figure 2(a) shows the initial and imploded configurations. OFHC copper was used to make these tubes. The internal diameters of the inner copper tube were selected to produce prescribed and controlled final strains. In some special cases, a central steel rod was also used. The explosive is axi-symmetrically placed around the specimen (Fig. 2(b)). The detonation is initiated on the top. The expansion of the detonation products exerts a uniform pressure on the cylindrical specimen and drives the

specimen to collapse inward. The detonation velocity of the selected explosive is approximately 4000 m/s and its density is 1 g/cm³. The velocity of the inner wall of the tube was determined by an electromagnetic gage [9].

The conditions for the three experiments were:

1. $T < M_s$: alloy in martensitic state; test was conducted at 190 K. The sample was first cooled down to the 77 K. After that, temperature was raised to 190 K, when explosive charge was detonated.

2. $M_s < T < M_s^\sigma$: concurrent deformation and transformation. Experiment conducted at 188 K (-850 C).

3. $T > M_s^\sigma$: no transformation. Experiment was conducted at 273 K.

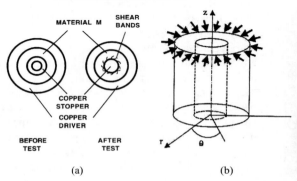

FIGURE 2. Thick-Walled Cylinder Test.

RESULTS AND DISCUSSION

Figure 3 shows the differences in collapse geometry between monocrystalline Fe-15%Cr-15%Ni (Fig. 3(a)) and polycrystalline AISI 304 stainless steel (Fe-18%Cr-8%Ni) (Fig. 3(c)). The crystallographic anisotropy creates a global mechanical anisotropy in the monocrystal. This results in a difference in resistance to the collapse. Stone and Thomas [6] show the stress strain response obtained for the [001] and [011] directions.

The Fe-Cr-Ni monocrystal has a response characteristic of FCC monocrystals: the [001] orientation has four slip systems with identical Schmid factors and therefore a higher yield stress. Thus, the [001] direction shows a greater resistance to collapse. This symmetry is revealed by the tendency of the cylinder to become square.

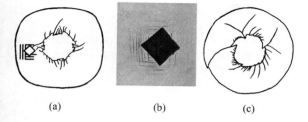

(a)   (b)   (c)

FIGURE 3. Configuration of shear bands in specimen subjected to implosion with $\varepsilon_{eff} = 0.92$: (a) in Fe-15Cr-15Ni monocrystal: cylinder axis: [100]; (b) Microhardness indentation and traces of [111] slip planes (the traces are [110] orientation); (c) Polycrystalline Fe-18Cr-8 Ni.

The number of shear bands formed was also affected by the conditions of deformation. The untransformed and pre-transformed conditions showed a large number of shear bands (N_s equal to 41 and 56, respectively). If transformation was active during deformation, then a marked decrease in the number of shear bands: $N_s=10$, was observed. This drastic reduction is the direct result of the competition between shear localization and martensitic transformation in the process of strain accommodation. The decrease in the number of shear bands and the associated increase in their spacing can be explained, partially, by the reduced yield stress. Both Grady-Kipp [11] and Wright-Ockendon-Molinari [12,13] analyses predict an inverse relation between yield stress and shear-band spacing. The corresponding equations can be expressed as:

$$\text{Grady} - \text{Kipp} \quad L = 2\pi \left[\frac{kC}{\dot{\gamma}^3 a^2 \tau_0} \right]^{1/4} \cdot \frac{9^{1/4}}{\pi}, \quad (1)$$

$$\text{Wright} - \text{Ockendon} \quad L = 2\pi \left[\frac{kC}{\dot{\gamma}^3 a^2 \tau_0} \right]^{1/4} \cdot m^{3/4}, \quad (2)$$

$$\text{Molinari} \quad L = 2\pi \left[\frac{kC}{\dot{\gamma}^3 \tau_0 a^2} \right]^{1/4} \cdot \left[\frac{m^3 (1-aT_0)^2}{(1+m)} \right]^{1/4}, \quad (3)$$

where the heat capacity is C = 500 J/KgK, thermal conductivity K = 14.7 W/mK, thermal softening factor a = 7.2 x 10^{-4} K^{-1}, strain rate hardening index

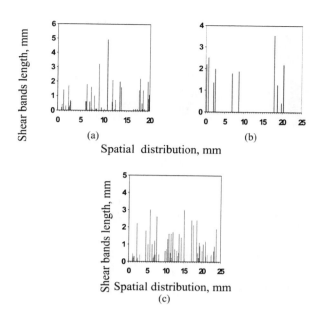

FIGURE 4. Shear bands distributions in three specimens deformed at: (a) Ambient temperature-no transformation; (b) 188 K, simultaneous transformation; (c) Pre-transformed structure.

m = 0.026, and shear strain rate $\dot{\gamma} = 6$ x 10^4[8].

These models predict an increase in the spacing by 50%, if yield stress is reduced by 80%. The difference in the experimental results is much more dramatic, as shown by the quantitative measurements in Figure 4. The spacing of shear bands was calculated assuming that they initiated at a strain of 0.1. The spacing in the condition undergoing simultaneous transformation and deformation (L=4.03 mm) is about four times the spacings for the other conditions (L=0.98 mm and L=0.72 mm). This test was carried out at 188 K; the shear yield stress is reduced from 90 MPa to 50 MPa.

The analysis of the microstructure reveals important features. The austenitic monocrystal had a pattern of gray crisscrossing regions. These regions are diffuse and indicate that the composition of the alloy is not entirely homogeneous. The pre-transformed material shows shear bands cutting through the pre-existing martensite. Figure 5 shows the microstructure of the material imploded at 188 K. We can see the existence of stress-assisted/strain-induced martensite as well as shear localization. Fig.

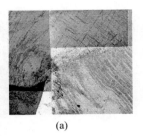 

(a)   (b)

FIGURE 5. Martensite transformation during deformation (Fe-15Cr-15Ni collapsed cylinder at T=188 K); (a) Interaction of shear bands and stress-assisted martensite; (b) Region along internal surface where deformation was accommodated by martensite transformation.

5(a) shows one principal shear band and several martensite laths forming at its tip, in a fan-like pattern. Fig. 5(b) shows a region along the surface in which the plastic deformation was accommodated by transformation. Several laths can be seen, emanating from the surface. No shear bands were formed in this region.

Transmission electron microscopy was conducted inside the shear band. The features revealed are identical to the ones seen for a polycrystalline AISI 304 stainless steel. In general, an equiaxed structure with grains in the range of 100-300 nm was observed. This is shown in Figure 6. The formation of sub-micrometer grains has been attributed to a rotational dynamic recrystallization process [14]. Fig. 6 also shows an amorphous region. The formation of the amorphous region is thought to be due to melting and very high rate resolidification. A similar

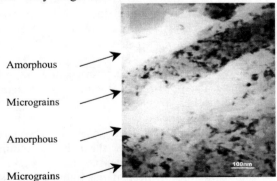

Amorphous

Micrograins

Amorphous

Micrograins

FIGURE 6. TEM of shear band shows microcrystalline and amorphous regions in Fe-15wt%Cr-15wt%Ni.

microstructure has been seen earlier by Meyers et al. [14] for AISI stainless steel. However, this was only possible for higher carbon concentrations. The cooling rate inside the shear band is extremely high and could possibly lead to melting and resolidification in the glassy state. We can estimate the cooling rate $\dot{T} = \dfrac{T_m}{\tau}$, where $\tau \sim \dfrac{\delta^2}{D}$, thickness of shear bands δ, thermo-diffusivity D. In our case, $\dot{T} \sim 4 \times 10^9 \, \mathrm{K/s}$.

ACKNOWLEDGMENT

We would like to thank Dr. S. Usherenko, Dr. D. Lassila, Dr. Q. Xue and Professor G. Stone.

REFERENCES

1. Rogers, H. C., *Annu. Rev. Mater. Sci.*, **2**, 283 (1979).
2. Nesterenko, V. F., *Dynamics of Heterogeneous Materials,* Spinger-Verlag, NY, 2001, pp. 307-384.
3. Nesterenko, V. F., Meyers, M. A., and Chen, H. C., *Acta Metall.* **44**, 2017 (1996).
4. Nesterenko, V. F., Meyers, M. A., LaSalvia, J. C., Bondar, M. P., Chen, Y. J., and Lukyanov, Y. L., *Mater. Sci. Eng. A*, **229**, 23 (1997).
5. Nesterenko, V. F., Meyers, M. A., and Wright, T. W., *Acta Mater.* **46**, 327 (1998).
6. Stone, G., and Thomas, G., *Metall. Trans.* **5**, 2095 (1974).
7. Guimaraes, J.R.C., Gomes, J.C., and Meyers, M.A., *Suppl. Trans. J. I. M.,* **17**, 411 (1976)
8. Xue, Q., Nesterenko, V.F., and Meyers, M.A., *International Journal of Impact Engineering,* **28**, 257 (2003).
9. Nesterenko, V.F., and Bondar, M. P., *DYMAT Journal,* **1**, 245 (1994).
10. Shockey, D.A., In *Metallurgical Applications of Shock-Wave and High-Strain-Rate Phenomena,* edited by Murr, L.E., Staudhammer, K.P., and Meyers, M.A., Dekker, NY, 1986, pp. 633-656.
11. Grady, D.E., and Kipp, M.E., *J. Mech. Phys. Solids,* **35**, 95 (1987).
12. Wright, T. W., and Ockendon, H., *Int. Journal of Plasticity,* **12**, 927 (1996).
13. Molinari, A., *J. Mech. Phys. Solids*, **45**, 1551 (1997).
14. Meyers, M. A., Xu, Y. B., Xue, Q., Perez-Prado, M. T., and McNelley, T. R., *Acta Mat.*, **51**, 1307 (2003).

CP706, *Shock Compression of Condensed Matter - 2003*
edited by M. D. Furnish, Y. M. Gupta, and J. W. Forbes
© 2004 American Institute of Physics 0-7354-0181-0/04/$22.00

THE INFLUENCE OF PEAK STRESS ON THE MECHANICAL BEHAVIOR AND THE SUBSTRUCTURAL EVOLUTION IN SHOCK-PRESTRAINED ZIRCONIUM

E. Cerreta, G.T. Gray III, B.L. Henrie, D.W. Brown, R.S. Hixson and P.A. Rigg

Los Alamos National Laboratory, Los Alamos NM 87545

Abstract: The post shock mechanical behavior and substructure evolution of zirconium (Zr) under shock prestrained at 5.8 and 8 GPa, above and below the pressure induced α-ω phase transition, has been quantified. The reload yield stress of Zr shock prestrained to 8 GPa was found to exhibit enhanced shock hardening when compared to the flow stress measured quasi-statically at an equivalent strain. In contrast, the reload yield behavior of Zr specimens shocked to 5.8 GPa did not exhibit enhanced shock hardening. The microstructure of the as-annealed and shock prestrained materials were examined. The presence of a reduced available glide distance due to a relatively more well developed dislocation substructure and increased twinning over quasi-static specimens deformed to comparable strains correlates with the increased yield stresses after shock prestraining at 8 GPa. Additionally, the retention of ~ 40% by volume metastable high-pressure ω-phase in specimens shocked to 8 GPa and its absence in the 5.8 GPa specimen, is thought to contribute to the increased yield stress in the 8 GPa specimens.

INTRODUCTION

The effect of shock wave loading on the structure property response of metals has been the subject of numerous investigations. While many of these investigations have focused on the influence of peak stress on mechanical properties and microstructure in cubic metals and alloys, only a few have examined these phenomena in hexagonal metals [1-5]. In the case of pure face-centered-cubic (fcc) metals, like Cu and Ni, it is well documented that shock loaded specimens exhibit increased hardening behavior in reload tests compared to the same metal deformed quasi-statically to an equivalent strain [1]. Studies on fcc metals have shown that mechanisms for shock strengthening can be examined in terms of the work hardening behavior observed in subsequent lower strain rate testing. Previous work on shock prestrained fcc metals such as nickel and copper has shown enhanced hardening compared to quasi-

static deformation, but some body-centered-cubic (bcc) metals like tantalum have exhibited minimal or no improvement in yield stress compared to quasi-static deformation. In the case of Zr, previous work has focused on the pressure induced α-ω phase transition, specifically temperature dependence, the crystallography of transformation, and its effect on spall fracture [6-12].

Typically, the quasi-static and dynamic work hardening behavior of hcp metals differs from fcc metals in that hexagonal-closed-packed (hcp) metals are substantially more sensitive to strain rate and temperature of deformation than pure fcc metals. Pressure induced phase transitions have also been shown to play a role in the shock-induced mechanical properties and substructural evolution of metals and alloys. Work on iron has shown that peak shock pressure increases the volume fraction of transformed second phase material within shocked specimens [13]. Differences in strength and defect storage between the parent and product phases lead to differences in mechanical properties. Recently the effects of shock hardening in hcp metals and alloys, such as Zr, have received increased interest. The purpose of the present study is to present results quantifying the influence of peak pressure on the shock loading response and the substructural evolution in Zr.

EXPERIMENTAL

The Zr used was high-purity crystal-bar grade, (0.14 wt% O, 0.003 wt% H, 0.008 wt% N, 0.027 wt% C) which was upset forged from remelted bar and clock rolled to form a 6-mm thick plate. It was annealed at 550°C for 1 hour and had an average grain size of 25μm. This processing produced a strongly textured Zr plate, whose normal is nearly aligned with the c-axis. Shock recovery experiments were performed utilizing an 80-mm single stage launcher and recovery techniques as described previously [14]. Zr specimens were shock loaded in Ti-6Al-4V shock recovery assemblies to 5.8 and 8 GPa for 1μs pulse durations using Ti-6Al-4V impactors. Compression samples were electro-discharged machined (EDM) from the as-annealed plate and the shock prestrained samples and reloaded at room temperature and at strain rates of 10^{-3} and 10^{-1}/s. Textural evolution as well as detection and quantification of retained ω-phase within the shocked specimens were performed through neutron diffraction at the Manuel Lujan Jr. Scattering Center at LANSCE with a pulsed neutron source and through electron back-scattered diffraction (EBSD) on a Phillips XL30 scanning electron microscope (SEM). All substructural analysis of the shocked specimens was performed on a Phillips CM30 transmission electron microscope (TEM) operating at 300kV.

RESULTS

The reload mechanical response of shock prestrained Zr was found to depend on the peak shock pressure. Figure 1 presents a plot of the quasi-static reload stress-strain behavior of as-annealed Zr as well as the samples shock prestrained to 5.8 and 8 GPa. The plot contains

data at strain rates of 10^{-3}/s and 10^{-1}/s. The reload shock curves have been offset with respect to the annealed Zr by the strain accumulated by shock prestraining. This strain is defined as $4/3\ln(V/Vo)$ where V and Vo are the final and initial volumes of Zr during the shock cycle.

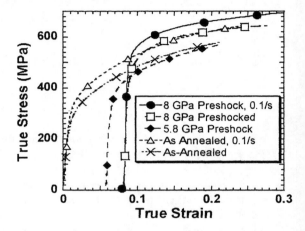

Figure 1. Quasi-static stress-strain response at a strain rate of 10^{-3}/s and 25°C of annealed and shock recovered Zr after 5.8 and 8 GPa shocks.

The quasi-static yield behavior of the annealed Zr is seen to be sensitive to strain rate, with higher flow stresses observed at higher strain rates. The reload mechanical response of the specimen shock prestrained to 5.8 GPa exhibits a slightly reduced flow stress level compared to the quasi-static data. However, the 8 GPa stress-strain curve exhibits an increased shock hardening response compared to annealed Zr deformed to the same strain level. The case of the 8 GPa reload stress strain curve tested at a strain rate of 10^{-3}/s is coincidently nearly identical to the annealed specimens tested at a strain rate of 10^{-1}/s, which exhibit a higher yield and flow stress than the annealed specimens tested at 10^{-3}/s. This is similar to behavior observed in many fcc metals [4]. As in the case of the annealed Zr, the hardening of shock prestrained specimens is sensitive to the rate of deformation upon reloading. Zr preshocked at 8 GPa and reloaded at 10^{-1}/s exhibited a higher flow stress for a given strain than did the 8 GPa preshocked Zr reloaded at 10^{-3}/s.

An examination of the annealed and preshocked microstructures revealed that the microstructure of the Zr was sensitive to the shock peak stress. SEM images indicate that while little twinning existed in the as-annealed microstructure, twinning increased with peak shock pressure. In the Zr shock prestrained at 5.8 GPa, approximately 7% by volume of the specimen twinned, but twinning in the specimen shock prestrained at 8 GPa was so extensive that quantification of the volume fraction of twins within the microstructure using EBSD was too extensive to quantify. Neutron diffraction indicated that within the α-phase there was significant reorientation of the texture with respect to the basal plane texture in the 8 GPa shock prestrained specimens. The reorientation due to shock loading within the specimen shocked at 5.8 GPa was not as significant, indicating that the volume fraction of twinning was far less than in the 8 GPa case. Neutron diffraction as well as EBSD data indicated that while no ω-phase was present in the annealed or in the 5.8 GPa shock prestrained Zr, almost 40% by volume of the high-pressure ω-phase was retained in the 8 GPa sample.

The microstructures of the as-annealed, the 5.8, and the 8 GPa shock prestrained Zr specimens were examined using TEM. The as-annealed microstructure contained few dislocations and most grain interiors were free from defects. The microstructure of the 5.8 GPa shock prestrained material displayed a high dislocation density. The dislocations, $<c+a>$ type, were tangled in some cases and others were freely gliding through the interior of the grains. Figure 2 depicts the dislocation substructure of the 5.8 GPa specimen. The substructure of the α-phase of the 8 GPa shock prestrained material consisted of a very high density of both $<a>$ and $<c+a>$ type dislocations that were highly tangled. Few dislocations were observed to be gliding in the interior of the grains. Many grains contained groups of thin twins with a high density of $<a>$ type dislocations within the twins as depicted in Figure 3a. These dislocations were also highly tangled. In other grains, a high contrast substructure associated with high strains within the metal was observed. Selected area diffraction analysis, in areas such as the one shown in Figure 3b, was used to identify the phase and orientation of the material. In this case, the material was identified as the retained pressure induced ω-phase.

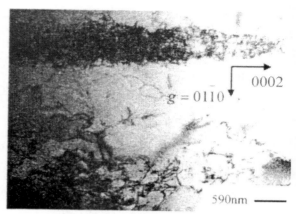

Figure 2: Bright field image of the substructure of the 5.8 GPa shock prestrained substructure.

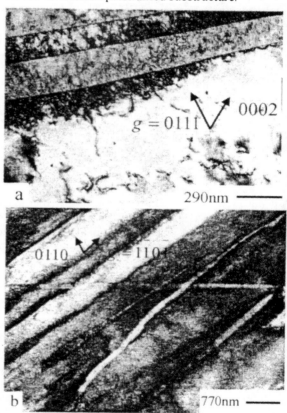

Figure 3: Bright field images of 8 GPa shock prestrained (a) α-phase and (b) ω-phase Zr.

DISCUSSION

The differences in the shock hardening behavior between the 5.8 and 8 GPa specimens may be attributed to substructural differences present within these specimens. In the 5.8 GPa case, dislocations and twins were stored in the substructure due to shock prestraining. Some of these dislocations were not tangled and therefore able to glide upon reloading. Seven percent of the microstructure by volume was twinned prior to reload and may have created a back stress within the matrix due to piled up dislocations at twin boundaries without significantly reducing available glide distances. Most two phase metals and some single phase metals that deform by twinning exhibit a reduced yield stress upon reversing the direction of stressing quasi-statically. This is termed the Bauschinger Effect [15]. This effect was observed in Si-bronze and in two –phase Al-Cu alloys [16]. This along with the availability of dislocations may account for the reduced flow stresses observed in the 5.8 GPa specimens.

Specimens shock prestrained at 8 GPa, displayed an enhanced hardening. Twinning had increased measurably over the as annealed and the 5.8 GPa shocked microstructure and dislocation glide occurred on two different slip planes resulting in a highly tangled or dislocation cell substructure. The substructure of the pressure-induced ω-phase contains high strains and commensurate high dislocation density due to the phase transition, making it difficult for easy glide of dislocations. With 40% of the substructure metastable retained ω-phase, flow stresses are also increased within the specimen, consistent with that observed in Figure 1.

SUMMARY AND CONCLUSIONS

Based upon a study on the effects of peak shock pressure on the mechanical response of Zr the following conclusions can be drawn: 1.) the reload yield behavior of shock prestrained Zr has a hardening behavior that is dependent upon the peak shock pressure; this is directly related to the microstructure induced by the shock prestraining and 2.) the development of dislocation cells, the increased twinning, and the formation and the retention of ω-phase at increased peak shock pressures contribute to enhanced hardening upon reload.

ACKNOWLEDGEMENTS

The authors are grateful for the technical assistance of M.F. Lopez in performing the experiments. This work was performed under the auspices of the U.S. Department of Energy.

REFERENCES

1. Gray, G.T. & K. Vecchio, Metal. Mat. Trans. A, 1995. **26A**: p. 2555-2563.
2. Mahajan, S., Phys. Stat. Sol. A, 1970. **2**: p. 187-201.
3. Meyers, M., U. Andrade, & A. Chokshi, Metal. Mat. Trans. A, 1995. **26**(11): p. 2881-2893.
4. Murr, L.E., *Shock Waves and High Strain Rate Phenomena in Metals*, M. Meyers & L.E. Murr, Eds. 1981, Plenum Press: NY. p. 607-673.
5. Vecchio, K. & G. Gray, Metal. Mat. Trans A, 1995. **26**(10): p. 2545-2553.
6. Xia, H., A. Ruoff, & Y. Vohra, Phys. Rev. B, 1991. **44**(18): p. 10374-10376.
7. Kozlov, E., J. de. Phys. III, 1991. **1**(8): p. 675-679.
8. Song, S. & G. Gray, Phil. Mag. A, 1995. **71**(2): p. 275-290.
9. Jyoti, G., et al., Bulletin Mat. Sci., 1997. **20**(5): p. 623-635.
10. Jyoti, G., et al., Phil. Mag. Lett., 1997. **75**(5): p. 291-300.
11. Podurets, A., V. Dorokhin, & R. Trunin, High Temp., 2003. **41**(2): p. 216-220.
12. Kozlov, Y., et al., Fizika Metallov I Metallovedenie, 1995. **79**(6): p. 113-127.
13. Murr, L.E., *Shock Waves and High Strain-Rate Phenomena in Metals*, M. Meyers & L.E. Murr, Eds. 1981, Plenum Press: NY. p. 753-777.
14. Gray, G.T., *High Pressure Shock Compression of Solids*, J.R. Asay & M. Shahinpoor, Eds. 1993, Springer-Verlag: NY. ch.6.
15. Gray, G.T., R.S. Hixson, & C.E. Morris, *Shock Compression of Condensed Matter*, S.C. Schimdt, et al., Eds. 1991, Elsevier Science Publishers.
16. Gray, G.T., *Shock Wave And High Strain Rate Phenomena in Materials*, M. Meyers, L.E. Murr, & K.P. Staudhammer, Eds. 1992, Marcel Dekker Inc.: NY. p. 899-911.

CP706, *Shock Compression of Condensed Matter - 2003*
edited by M. D. Furnish, Y. M. Gupta, and J. W. Forbes
© 2004 American Institute of Physics 0-7354-0181-0/04/$22.00

A NOVEL EXPERIMENTAL TECHNIQUE FOR THE STUDY OF HIGH-SPEED FRICTION UNDER ELASTIC LOADING CONDITIONS

Paula Crawford*, Kevin Rainey*, Paul Rightley*and J.E. Hammerberg[†]

Dynamic Experimentation Division, Los Alamos National Laboratory, Los Alamos, NM 87545
[†]*Applied Physics Division, Los Alamos National Laboratory, Los Alamos, NM 87545*

Abstract. The role of friction in high strain-rate events is not well understood despite being an important constitutive relationship in modern modeling and simulation studies of explosive events. There is a lack of experimental data available for the validation ofmodels of dynamic sliding. The Rotating Barrel Gas Gun (RBGG) is a novel, small-scale experimental facility designed to investigate interfacial dynamics at high loads and sliding speeds. The RBGG utilizes a low-pressure gas gun to propel a rotating annular projectile towards an annular target rod. Upon striking the target, the projectile imparts both an axial and a torsional impulse into the target at a timescale relevant to explosively-driven events. Resulting elastic waves are measured using strain gages attached to the target rod. The coefficient of friction is obtained through an analysis of the resulting strain wave data. Initial experiments have been performed using dry copper/copper interfaces. We find that the measured coefficient of friction can evolve significantly over a 30 μs event.

INTRODUCTION

The study of high-speed sliding events is essential to understanding the nature of friction for dynamic events. For example, at high sliding speeds the heat created during sliding can be enough to produce melt lubrication, but on what time-scale can this occur? Previous studies [1–4] have investigated friction during high-speed (>100 m/s) sliding events using various experimental configurations, including pin-on-disk, modified Hopkinson bars, and pressure-shear (planar) impacts. These experimental methods provided insight into the nature of contact events at high sliding speeds and relatively long time-scales (>1 ms) [1–3] or at shorter time-scales and lower sliding speeds [4]. For explosively-driven tests, the event time-scales are on the order of tens of microseconds. This study aims to investigate the frictional contact events under high-speed, high-load conditions at these time-scales.

The Rotating Barrel Gas Gun (RBGG) experimental facility was designed to eliminate some of the experimental limitations present in previous experi-

mental investigations of dynamic sliding events. The RBGG propels a spinning projectile, with an annular cross-section, towards an annular target rod. Unlike using a rifled gun to produce both spin and translation, the spinning barrel of the RBGG can produce spin rates independent of the translational speed of the projectile (unlike the fixed angle of rifling). The current RBGG arrangement allows investigation of sliding speeds up to 10 m/s. The impact load is limited to the elastic regime of the target rod and projectile material (150 MPa to 2 GPa).

The objectives of the present study were to evaluate the quality of data produced by the RBGG and the effect of gluing samples onto the ends of the projectile and target rod on the measurement accuracy. A collaboration with Dave Rigney's group at the Ohio State University aims to investigate the microstructural changes occurring during these dynamic events, but those results will not be presented here.

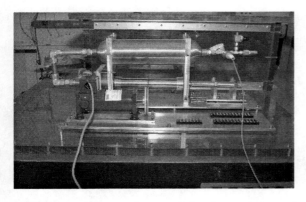

FIGURE 1. The Rotating Barrel Gas Gun experimental set-up.

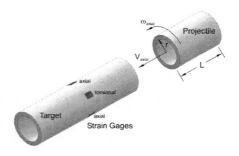

FIGURE 2. A schematic showing the geometry of the target and projectile used in the RBGG experiments.

EXPERIMENTAL METHODS

The Rotating Barrel Gas Gun, shown in Fig. 1, consists of a stainless steel barrel (1.695 in. OD, 1.350 in. ID, 12.5 in. long) mounted on rotational bearings. The barrel is connected to a variable speed electric motor via a belt/pulley system to provide barrel rotational speeds up to 10,000 rpm. A tank pressurized with nitrogen gas, connected to the end of the barrel through a solenoid valve, is used to propel the projectile. The projectile is a copper tube (1 in. OD, 0.75 in. ID, 3 in. long) placed inside a tight fitting polyethylene sleeve that slides on the inner surface of the barrel as the projectile is traveling along the length of the barrel. Located in-line with the rotating barrel is an 8 in. long copper target rod, with the same annular dimensions as the projectile, mounted on aluminum supports. Fig. 2 shows the alignment geometry of the target rod and the projectile. The target rods and projectiles used in the present experiments were machined from OFHC 10100 copper (99.99% pure). The impact surfaces were prepared by sanding with 400-grit sandpaper in circular motions and then cleaned with isopropyl alcohol.

The target rod was outfitted with strain gages at three locations (0.25, 1.5 and 3.0 in. from the impact surface) to measure the dynamic axial (normal) and shear strains. For the axial strain measurement, 0.032 in. long strain gages in a "2 gages in opposite arms" Wheatstone bridge configuration was used. For the shear strain measurement, 0.062 in. long shear pattern strain gage rosettes in a "full torsional" Wheatstone bridge configuration was used. A strain am-

plifier was used to amplify the bridge outputs prior to being read by the data acquisition system. The barrel rotational speed was measured with an optical tachometer and the pressure in the tank was measured with a pressure gage.

After the projectile was placed into the barrel, the target rod was aligned with the projectile by placing a stainless steel rod through the center of both and adjusting the target rod support. The plenum was then filled with nitrogen gas to the desired pressure (20 psi for the present data). At this point the entire setup was secured inside a lexan box for safety reasons. After setting the rotation to the desired speed, the solenoid valve is triggered by the data acquisition system. The rotating projectile is accelerated toward the target rod over a distance of about 6 in. The projectile impacts the target rod while most of the projectile is still inside the barrel. Upon impact, the rotating projectile imparts a substantial axial load, as well as a rotational shear onto the impact surface of the target rod.

For future testing, it would be advantageous to use removable samples at the contacting ends of the projectile and target rod. Therefore, a second projectile and target rod set was tested with identical configuration as above, except that a 0.50 in. long, annular copper sample (same material and annular dimensions) was glued onto the ends of both the projectile and target rod using a high strength epoxy. This increases the projectile and target rod lengths to 3.5 and 8.5 in., respectively. The sample on the target rod also contained a fourth set of axial and torsional strain gages located 0.25 in. from the impact surface.

DATA ANALYSIS

For the present analysis, it is assumed that the copper material used behaves according to Hooke's Law and that any plastic deformation due to the impact, as opposed to sliding, at the contacting surfaces can be neglected. When the projectile impacts the target rod, compressive and torsional waves of duration $2L/c_o$ are imparted to the target rod. L is the length of the projectile and c_o is the longitudinal wave speed given by

$$c_o = \sqrt{\frac{E}{\rho}}, \quad (1)$$

where E is the elastic modulus and ρ is the material density. After this impact duration, the projectile and target rod separate due to the arrival of the tensile wave reflected from the distal end of the projectile at the contact surface. The attached strain gages measure the axial, ε, and maximum shear, γ_{max}, strains. From ε, the axial stress, σ, is calculated as

$$\sigma = E\varepsilon, \quad (2)$$

where E is the bulk modulus of the impacted material. From γ_{max}, the maximum torsional stress, τ_{max}, is calculated as

$$\tau_{max} = G\gamma_{max}, \quad (3)$$

where G is the shear modulus of the impacted material. However, in order to ultimately calculate the kinetic friction coefficient, μ_k, the shear stress averaged over the contact surface area, τ_{avg}, is needed and is expressed by

$$\tau_{avg} = \frac{\int_r^R \rho\tau\,d\rho}{\int_r^R \rho\,d\rho}, \quad (4)$$

where

$$\tau = \frac{G\gamma_{max}}{R}. \quad (5)$$

Solving for τ_{avg} gives

$$\tau_{avg} = \frac{G\gamma_{max}}{R}\frac{2}{3}\frac{(R^2 + Rr + r^2)}{(R + r)}. \quad (6)$$

Finally, the kinetic friction coefficient, μ_k, for steady-state sliding conditions is given as

$$\mu_k = \frac{\tau_{avg}}{\sigma}. \quad (7)$$

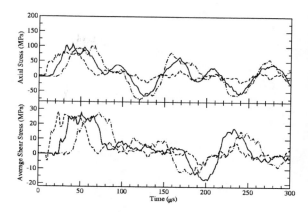

FIGURE 3. Axial and Shear stress versus time during an impact. The three curves for each plot correspond to the three strain gages located along the length of the target rod.

RESULTS

Fig. 3 shows the σ and γ_{max} versus time results for an impact using the target rod and projectile without a sample. A tank pressure of 20 psi and a barrel rotational speed of 3000 rpm were used for the experiments discussed here. This produced a sliding speed of approximately 3.5 m/s (calculated from barrel rotational speed). From Fig. 3, the σ traces show a fast rise time of approximately 15 μs. The trace then peaks at a relatively constant average value of 80 MPa for 30 to 35 μs before dropping off, which compares well with the estimated contact duration of 42 μs. The apparent oscillation of the peak value during the contact duration may be due to a stick-slip phenomena. The γ_{max} traces show a behavior similar to the σ traces, peaking out at a relatively constant average value of 24 MPa. In addition, there is a local peak in the trace during the rise time, which is repeatable in nature. The source of this local peak is unknown.

Comparing the σ traces at the three different gage locations shows some variability in the waveform that could be due to small differences in strain gage location or alignment. However, there appears to be no appreciable attenuation as the wave propagates down the target rod. The average longitudinal wave speed was measured from the three different traces as 3455 m/s which compares well with the estimated value of 3633 m/s (using Eq. 1). It is interesting to note that the reflected tensile waves for both

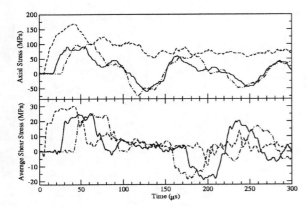

FIGURE 4. Axial and shear stress versus time during an impact event with samples attached to the target rod and projectile. The three curves correspond to the three locations of the strain gages.

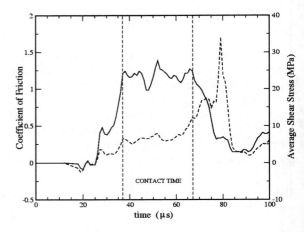

FIGURE 5. Average shear stress and friction coefficient.

the σ and γ_{max} traces show significant smearing of the waveform, indicating that the aluminum support used to hold the target rod is causing spurious reflections.

Fig. 4 shows the σ and γ_{max} results for an impact with a 0.50 in sample attached to the impact end of the target rod and projectile. The tank pressure and barrel speed conditions of the impact were the same as the previous data in Fig. 3 which gives the same sliding speed as in the previous experiment. Comparing the data in Fig. 4 with the previous data in Fig. 3 reveals some significant problems. The first set of strain gages (which are attached to the sample itself) show a significantly higher peak σ than the two sets of gages further down the target rod. Additionally, there is a residual axial stress/strain in the sample. This indicates that the glue joint is reflecting a significant portion of the axial wave and causing plastic deformation within the sample. Interestingly, the peak value of both σ and γ_{max}! seen at the other two strain gage locations are approximately equivalent to the peak values for the previous data in Fig. 3 without a sample attached.

Ultimately, the use of the RBGG is to study the behavior of the coefficient of kinetic friction, μ_k during the impact event. Fig. 7 shows the μ_k versus time for the no-sample data of Fig. 3. As can be seen, it appears that during the impact duration, μ_k is continually increasing. This result has been found to be repeatable, as well. If proved correct, a non-constant coefficient of kinetic friction with time at these time-

scales presents a significant impact on interfacial dynamics modeling and simulation efforts.

CONCLUSIONS

Following are the major conclusions of this analysis of the use of the Rotating Barrel Gas Gun (RBGG) in studying interfacial friction at impact conditions.

- The RBGG appears to provide good, repeatable stress data of high speed frictional impacts.
- Having a sample (and requisite glue joint) on the end of a target rod appears to cause plastic deformation in the sample, but, the significance on the data taken from the target rod strain gages is unclear.
- The coefficient of kinetic friction appears to be non-constant during the contact duration, however, this result needs to be more fully investigated.

REFERENCES

1. Bowden, F. P., and Freitag, E. H., *Proc. Roy. Soc. A,* **248**, 350–367 (1958).
2. Bowden, F. P., and Persson, P. A., *Proc. Roy. Soc. A,* **260**, 433–458 (1961).
3. Montgomery, R. S., *Wear*, pp. 275–298 (1976).
4. Ogawa, K., *Experimental Mechanics*, **37**, 398–402 (1997).

CP706, *Shock Compression of Condensed Matter - 2003*
edited by M. D. Furnish, Y. M. Gupta, and J. W. Forbes
© 2004 American Institute of Physics 0-7354-0181-0/04/$22.00

DYNAMIC COMPRESSION OF ALUMINUM FOAM PROCESSED BY A FREEFORM FABRICATION TECHNIQUE

Kathryn A. Dannemann[1], James Lankford, Jr. [1], Arthur E. Nicholls[1], Ranji Vaidyanathan[2], Catherine Green[2]

[1]*Southwest Research Institute, Mechanical and Materials Engineering Division
P.O. Drawer 28510, San Antonio, TX, 78228*
[2]*Advanced Ceramics Research, 3292 East Hemisphere Loop, Tucson, AZ 85706*

Abstract. The compressive deformation behavior of a new type of aluminum foam was assessed under static and dynamic loading conditions. The aluminum foam investigated was processed by Advanced Ceramics Research using an extrusion freeform fabrication technique. The foam contained approximately 50 to 60 % porosity. The dynamic compression response was evaluated in air using a split Hopkinson pressure bar (SHPB) system with aluminum bars, and strain rates ranging from 600 s^{-1} to 2000 s^{-1}. Compression tests were also conducted at lower strain rates (10^{-3} s^{-1} to 4 s^{-1}) to determine the extent of strain rate strengthening. The low strain rate tests were performed with a servo-controlled hydraulic test machine. The results were analyzed as a function of foam density, structure, and process conditions.

INTRODUCTION

Processing developments and advancements have resulted in the introduction of new and improved porous metallic materials over the last few years [1-3]. These include the implementation of new foaming agents, as well as improved powder processes. An extrusion freeform fabrication (EFF) technique is presented as yet another method of processing metal foams [4]. The benefit of this in-situ fabrication technique over other foam processing methods is the ability to produce complex foam shapes and structures with greater control over pore size and orientation at relatively low cost.

The advantage of ultra-lightweight metallic foam materials for high strength and stiffness properties is well documented [5]. Numerous studies have enhanced current understanding of the low strain-rate deformation behavior of both open and closed-cell metal foams. The deformation capacity of metal foams may also be beneficial for energy absorption applications. Thus, the high strain rate and ballistic performance of metal foams have been the subject of recent investigations [6-10]. Some closed-cell aluminum (Al) foams have been reported to exhibit a strain rate effect [6-8]; others appear relatively strain rate insensitive [9,10]. The objective of the present study is to determine if strain rate strengthening occurs for aluminum foams processed by EFF.

MATERIALS

The Al foams of interest were processed by Advanced Ceramics Research (ACR) using an extrusion freeform fabrication process, developed and patented by ACR. The EFF process is a versatile rapid prototyping (RP) process that provides for the fabrication of near net-shape metallic foam components from highly loaded polymer binders filled with metallic powder. Metal foam parts are fabricated by the sequential deposition of multiple discrete raw material layers.

The aluminum foam samples for this study were processed by blending metal powders (86.9 w/o

Al, 6.7 Mg, 6.4 Sn) powders in a high-shear mixer with a molten thermoplastic binder. The compounded powder/binder material blend (50% binder, 50% metal powder) was pressed into feedrods, and placed in the high-pressure extrusion head barrel of the RP machine, a Stratasys 3D Modeler. Following heating of the specially designed extrusion tip (0.635 mm dia), the material was extruded into individual cylinders for testing. Each cylinder was slightly larger than the desired size of the compression test samples (i.e., 2.54 cm dia, 1.5-3.0 cm long). The green aluminum cylinders were then subjected to a binder burnout process; the hold temperature controlled the pore size. Sintering and a homogenization anneal followed.

Aluminum foams with two different layups were evaluated: $[0/90]_n$ and $[0/45/90/135]_n$. The different layups were obtained by adjusting the orientation sequence of the material layers. Representative longitudinal and transverse microstructures for the two different orientations are illustrated in Fig. 1. The transverse micros show the open-cell structure of the EFF foams on a macroscale. On a microscale, however, individual struts comprising the structure contain pores with a closed-cell morphology as shown in Fig. 1 for the longitudinal sections. Individual sample densities ranged from 1.43-1.54 g/cc for the 0/90 layup, and 1.43-1.58 g/cc for the 0/45/90/135 foams. The porosity level was approximately 55% (ρ/ρ_s = 0.53-0.58) with the 0/90 samples generally having slighter higher porosity levels.

EXPERIMENTAL PROCEDURE

Compression test samples were electro-discharge machined (EDM'd) from slightly oversized aluminum foam cylinders. The test samples measured approximately 1.27 cm long by 2.36 cm in diameter. The sample diameter was chosen to maximize the number of open cells across the diameter within the limitations imposed by the SHPB dimensions. There were at least 12 cells across the diameter of each compression test sample. A length to diameter (L/D) ratio of approximately 0.5 was selected to minimize dispersion effects in the high strain rate tests.

Compression tests were conducted at strain rates ranging from 10^{-3} s^{-1} to 2000 s^{-1}. The dynamic compression response was evaluated at room temperature in air using a split Hopkinson pressure bar (SHPB) system with 2.54 cm diameter aluminum bars; strain rates ranged from 600 s^{-1} to 2000 s^{-1}. Aluminum bars were utilized to obtain a higher signal to noise ratio than possible with a higher modulus bar material. A long (50.8 cm) aluminum striker bar was utilized to achieve sufficient pulse duration in the lower velocity range. Strain gages were attached to the pressure bars. Low strain rate (10^{-3} s^{-1} to 4 s^{-1}) tests were also performed to determine the extent of strain rate strengthening. The low strain rate tests were conducted with a servo-controlled hydraulic test machine.

The ACR aluminum foams were fully compressed during SHPB tests. Several

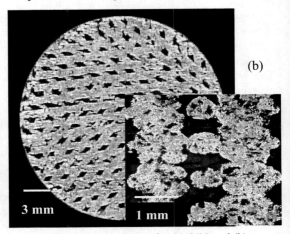

(a)

(b)

FIGURE 1. Representative transverse and longitudinal microstructures of EFF aluminum foams for (a) 0/90 and (b) 0/45/90/135 layups. The longitudinal microstructures in the foreground show closed-cell porosity.

interrupted, high strain rate compression tests were also conducted to allow evaluation of deformed microstructures at a series of increasing strains. Steel spacers were used to limit the maximum strain response. Following testing, interrupted and fully loaded test samples were sectioned and evaluated with optical and scanning electron microscopy.

RESULTS AND DISCUSSION

Compression test results for the foams with 0/90 and 0/45/90/135 layups are shown in Figs. 2a and 2b, respectively. Each curve represents a single test. For ease of comparison, the test results shown in each figure are for samples with similar densities. Lower density quasistatic data are also included for comparison. The EFF aluminum foams evaluated had higher relative densities (ρ/ρ_s ~0.55) than for previously published aluminum foam data [6-10]. This accounts for the higher measured strengths (>50 MPa) of the EFF foams versus Al foams processed by other techniques.

The general shape of the compression stess-strain curves obtained differ from the typical three-stage curve, characteristic of other foams [11]. The measured curves for the EFF foams show a more gradual elastic-plastic transition. The curves did not exhibit the characteristic plateau region, nor a marked densification region with sharply rising stress. The absence of a distinct plateau for the EFF-foams can be linked to the higher relative density. Since there are fewer cells for collapse, the foam is behaving more like a solid, which is characterized by continuous hardening. In essence, the EFF material is a macroscopically open-cell, high-density foam, with "thick" pillar-like walls that neither buckle nor collapse. The closed cells in the walls/columns are basically widely distributed voids, and, therefore, do not significantly degrade the material strength via foam collapse mechanisms.

The SHPB test results for the EFF foams were very consistent, especially given the 10% density variation for the range of samples tested. The reproducibility of the data is evident in the plots in Fig. 2 for both foam layups. Many of the high rate curves, including some not shown, are almost identical.

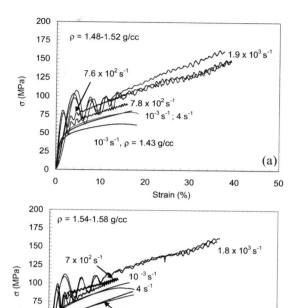

FIGURE 2. Engineering stress-strain response of EFF aluminum foam for (a) 0/90 and (b) 0/45/90/135 layups.

Both foam layups exhibited a weak dependence of compression strength on strain rate, for strain rates less than 100 s^{-1}. Some strengthening is evident for the SHPB tests relative to the lower strain rate tests. The extent of the strain rate strengthening effect is particularly evident in the flow stress versus strain rate plot in Fig. 3. Each point represents one test. The stress at 10% strain was selected for comparison to overcome the inherent variability in the SHPB data at low strain values. Since the compressive response was very sensitive to foam density, as predicted by the Gibson-Ashby [11] relationship, the stress values were normalized by relative density to allow direct comparisons over the range of densities tested. The density sensitivity is illustrated in Fig. 2 for the low strain rate results.

Post-test sample evaluation revealed some crushing of the voids within the walls/columns; the extent of void collapse is more severe at higher strains and strain rates. The surrounding solid

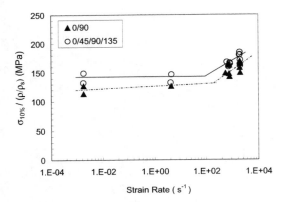

FIGURE 3. Normalized flow stress (at 10% strain) vs strain rate plot showing slight strain rate dependence.

provides continuum constraint until bending of the columns becomes unavoidable, as shown in Fig. 4. The primary damage mechanism, however, is due to plastic flow of the highly-voided aluminum. The continuum constraint on the 0/45/90/135 foam was somewhat greater owing to more substantial cell walls versus thinner struts with the 0/90 layups. Buckling and rupture of individual cell walls, characteristic of other metal foams was not observed. The EFF foams did not exhibit any localized flow or interactions.

FIGURE 4. Deformation of a 0/90 EFF foam sample following SHPB testing at strain rate = 1950 s^{-1}, 36% strain.

CONCLUSIONS

An extrusion freeform fabrication technique was employed to fabricate metallic foams with controlled pore size and orientation. Aluminum foams with different layups are possible by varying the deposition sequence. Compression test results indicate that EFF aluminum foams are stronger than aluminum foams processed by alternative methods; this is related to the higher density of the EFF foams and intrinsic structural differences. Strain rate strengthening was

observed and is attributed to plastic flow of the EFF foam (i.e., void containing solid).

ACKNOWLEDGMENTS

The authors acknowledge the support of the U.S. Army Research Office (SBIR Phase II Contract No. DAAD19-00-C-0025), and Dr. David Stepp, ARO technical monitor. Mr. Ron Cipriani of ACR is acknowledged for his assistance in processing the foam materials. The technical contributions of the following SwRI staff are also gratefully acknowledged: Dr. James Walker, Dr. Dan Nicollela, and Mr. Jim Spencer.

REFERENCES

1. C.J. Yu, H. Eifert, J. Banhart and J. Baumeister, *Mater. Res. Innovations*, **2**, p. 181-188, 1998.
2. T. Miyoshi, M. Itoh, S. Akiyama, A. Kitahara, *Advanced Engineering Materials*, **2**, p. 179-183, 2000.
3. S.J. Yu, H.H. Eifert, M. Knuewer, M. Weber, J. Baumeister, in *Porous and Cellular Materials for Structural Applications*, D.S. Schwartz, D.S. Shih, A.G. Evans, H.N.G. Wadley (eds.), Materials Research Society, Warrendale, PA, **521**, p. 145-150, 1998.
4. R. Vaidyanathan, J. Walish, J.L. Lombardi, S. Kasichainula, P. Calvert and K.C. Cooper, *J. of Metals*, p. 34-37, Dec. 2000.
5. M.F. Ashby, A. Evans, N.A. Fleck, L. Gibson, J.W. Hutchinson, H.N.G. Wadley, *Metal Foams: A Design Guide*, Butterworth-Heinemann, Woburn, MA, 2000.
6. T. Mukai, H. Kanahashi, T. Miyoshi, M. Mabushi, T.G. Nieh and K. Higashi, *Scripta Mater.*, **40** (8), p. 921-927, 1999.
7. A. Paul and U. Ramamurty, *Mater. Sci. Eng.A*, **281**, p. 1-7, 2000.
8. K.A. Dannemann and J. Lankford, Jr., *Mater. Sci. Eng. A*, **293**, p. 157-164, 2000.
9. V. S. Deshpande and N. A. Fleck, *Intl. J. of Impact Engrg.*, **24**, p. 277-298, 2000.
10. I.W. Hall, M. Guden and C.-J. Yu, *Scripta Mater.*, **43** (6), p. 515-521, 2000.
11. L.J. Gibson and M.F. Ashby, *Cellular Solids: Structure and Properties*, 2nd edition, Pergamon Press Oxford, 1997.

CP706, *Shock Compression of Condensed Matter - 2003*
edited by M. D. Furnish, Y. M. Gupta, and J. W. Forbes
© 2004 American Institute of Physics 0-7354-0181-0/04/$22.00

KINETICS OF MESOSTRUCTURE AND RELOADING BEHAVIOR OF DYNAMICALLY COMPRESSED SOLIDS

A.K. Divakov *, T.A. Khantuleva**, Yu.I. Mescheryakov*

* Institute of Problems of Mechanical Engineering RAS
V.O. Bol'shoi 61, 199178 Saint-Petersburg, Russia
** Saint-Petersburg St. Univ., Petrodvoretz, Biblioyechhaya, 2,

Abstract. Experimental study of shock reloading processes during the uniaxial strain loading of aluminum alloy targets reveals a dependence of reloading response of material on the particle velocity dispersion behavior. This response is found to be elastic-plastic in case the velocity dispersion at the plateau of compressive pulse equals zero. In the presence of the velocity dispersion the reloading response is pure plastic. This phenomenon is considered from the position of non-local hydrodynamic theory developed for description of high-nonequilibrium processes in heterogeneous media.

INTRODUCTION

There are sufficiently reliable experimental facts in dynamic plasticity discovered last 10-15 years that cannot be explained in the frame of traditional continuum mechanics:

1. Evidence of heterogeneity of dynamic plastic deformation quantitatively characterized by the particle velocity distribution (dispersion), measured in real time together with mean particle velocity [1].
2. Velocity loss (fluctuative decay of mean velocity) resulting from the energy exchange between fluctuative modes of dynamic plasticity and mass velocity of plastic flow.
3. Shock-induced structure formation at the mesoscopic scale.
4. Elastic-plastic behavior of a medium during the dynamic reloading of solid. [2]. Response of a solid during dynamic reloading is found to be elastic-plastic though amplitude of the first loading step corresponds to the pure plasticity.

EXPERIMENTS

In the present work, a series of experiments on dynamic loading of D-16 aluminum alloy under uniaxial strain conditions were conducted by using a light gas gun facility and special interferometer to provide the real-time measurements both the free surface velocity profile and time history of distribution (dispersion) in the particle velocity during the shock deformation. In all the experiments, impactors were made in the form of two-layer constructions to provide reloading scheme of shock-wave initiation. The first layer is a steel disk that initiates the first step of shock front and the second layer is an aluminum alloy disk connected through thin epoxy gluing. The thickness of aluminum layer is varied, which allows to change a delay of reloading step position relatively the first plastic front.

The reloading shock front appears to be as well two-wave shape (elastic precursor plus plastic front) as one-wave shape (only plastic front) depending on the particle velocity dispersion behavior. Specifically, when the dispersion at the plateau of compressive pulse equals zero, the reloading shock has the evident two-wave shape. This situation is illustrated in Fig.1a. Reloading begins just at the moment when the velocity dispersion equals zero. One can see that the reloading velocity profile has elastic precursor and following plastic front.

u, m/s   D, m/s

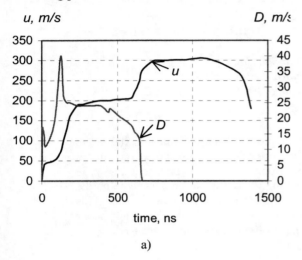

time, ns

a)

u, m/s   D, m/s

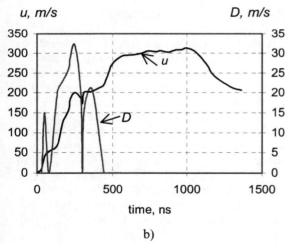

time, ns

b)

FIGURE 1. Free surface velocity profile, $u(t)$, and time history of the velocity dispersion, $D(t)$, for two reloading experiments: a) elastic-plastic reloading response; b) plastic reloading response.

The opposite situation is shown in Fig.1b. The reloading step has the evident one-wave structure

corresponding to pure plastic response of material during the reloading.

MULTISCALE MOMEMTUM AND ENERGY EXCHANGE DURING HIGH-RATE STRAINING OF SOLIDS

Explanation of the obtained results can be done only in scope of the theory connecting macroscopic behavior of a medium with its internal structure effects at mesolevel during the dynamic loading. At present it is already understood that all the processes going inside the wave front are highly nonequilibrium. From the point of view of the modern nonequilibrium statistical mechanics the main feature of highly nonequilibrium processes is strong correlation between all the field characteristics in a medium both in space and in time. The correlation effects can be considered as the effects of the collective or cooperative effects that leads to the new internal structure formation in a medium under loading and found experimentally [4]. Often the new structure scales generated in the medium during dynamical loading have no connection to the initial medium structure but strongly depend on the size and geometry factors of a specimen and on the loading regimes. This universal behavior of different type media can be a result of a selforganization and is generated by the macroscopic fields in the media under the similar loading conditions. The theoretical description of the synergetic processes supposes to have a large-scale inverse coupling (closeloop) between the newstructure formation and macroscopic field in a medoum. structure formation and macroscopic fields in a medium.

We begin with the mass and momentum balance equations in Lagrangian coordinates in the one-dimensional case [3,4]:

$$\frac{\rho_0}{\rho^2}\frac{\partial \rho}{\partial t} + \frac{\partial u}{\partial x} = 0;$$

$$\rho_0 \frac{\partial u}{\partial t} + \frac{\partial}{\partial x}(S - P) = 0; \qquad (1)$$

The deviator S and spherical part P of the stress tensor should be formally divided in turn into two parts: $S=S^e+S^m$, $P=P^e+P^m$. The first items S^e, P^e correspond to the elastic shear and compression respectively: $S^e=\frac{4}{3}G\frac{\rho-\rho_0}{\rho}$, $P^e=(\rho-\rho_0)C^2$. The second ones S^m, P^m assume being connected with the mesoscopic effects. Herewith, S can be determined by the nonlocal relaxation model [3,4] with the bulk viscosity included and P^m determined further. Then the mass and momentum balance equations (1) can be reduced to a wave-type equation with a nonlinear intergro-differential source depending on u:

$$\frac{\partial^2 u}{\partial t^2} - a_l^2 \frac{\partial^2 u}{\partial x^2} = -\frac{\partial^2}{\partial t \partial x}(S^m - P^m)\rho_0^{-1}. \qquad (2)$$

Here $a = C\rho/\rho_0$ is the Lagrangian velocity of the wave propagation and $a_l^2 = a^2 + (4/3)G\rho_0^{-1}$ is the longitudinal sound velocity in Lagrangian coordinates. Equation (2) is hyperbolic differential equation in the solid-like limit ($\varepsilon \to \infty$) and describes the wave type reversible momentum transport at the velocity. a_l. In the opposite limit ($\varepsilon \to 0$) it becomes parabolic corresponding the diffusive momentum transport followed by the viscous dissipation.

By using the Green function for the wave operator, Eqn.(5) can be reduced to the D'alamber form

$$u(x,t) = u_0(x-a_l t) + \frac{1}{a_l \rho_0} \int_0^t d\tau \int_{x-a_l(t-\tau)}^{x+a(t-\tau)} d\xi \left[- \frac{\partial^2}{\partial \varpi \xi}(S^m - P^m) \right]. \qquad (3)$$

Here the following initial and boundary conditions are taken: $u(x,0) = \frac{\partial u}{\partial t}(x,0) = 0$, $u(0,t) = U_0\Omega(t)$ and a zero order solution has a form of a simple wave $u_0(x-a_l t) = U_0\Omega(x-a_l t)$ propagating without changing its form while the second term defines the form changing of the wave front. In the reference connected to the front the form changing of the wave front profile due to the influence of the mesoscopic effects during non-steady shock wave propagation can be calculated explicitly:

$$\Delta u(x,t) = -\frac{1}{a_l \rho_0} \Delta(S^m - P^m). \qquad (4)$$

So, the form of the plastic front is entirely characterized by the processes at the mesolevel while in all limiting cases the last ones don't arise.

The spherical part of the stress tensor connected with mesofluctuations P^m can be called mesoscopic fluctuative pressure and determined like pressure due to heat fluctuations

$$P^m = \rho_0 a D. \qquad (5)$$

The internal energy density E should also includes two components: the energy of the elastic compression and shear E^e and the energy of mesoscopic degrees of freedom E^m respectively $E = E^e + E^m$. Then the general balance equation can be split into two parts related to the different scales and stages of the relaxation

$$\rho_0 \frac{\partial E^e}{\partial t} + (S^e - P^e)\frac{\partial u}{\partial x} +$$

$$\rho_0 \frac{\partial E^m}{\partial t} + (S^m - P^m)\frac{\partial u}{\partial x} = 0; \qquad (6)$$

The elastic component of internal energy is determined by the potential of a lattice and the mesoscopic component should be determined by the following equation of state

$$E^m = c_m a D + E^{ms}, \quad c_m = \left(\frac{\partial E}{\partial D}\right), \qquad (7)$$

where c_m is an energy capacity of the mesoscopic fluctuations. The full energy at the mesoscopic scale level E^m consists of two parts: kinetic energy of the mesoscopic fluctuations $c_m D$ and potential energy E^{ms} stored in the mesoscopic structures.

At the initial stage of relaxation $t \ll \tau_r$ when the mesoscopic fluctuations have not yet excited $c_m \to \infty$, the mesoscopic scale level should consider being frozen and the nonlocal correlation embracing the entire specimen becomes large ($\varepsilon \to \infty$). Eqn. (9) is reduced to the two first items while the second ones are still zero. As the mesoscopic level has already relaxed $c_m \to 0$, velocity fluctuations become the heat ones and Eqn. (9) would describe the last stage of the relaxation for large typical time $t>>t_r$. This is the irreversible dissipation stage of relaxation corresponding to the fluid-like behavior of a medium without internal structure and space correlation ($\varepsilon \to 0$), when the compression and high-rate form changing converts into a heat. At the intermediate stage when the material properties are characterized by finite values c_m, the macro-meso-energy exchange has both reversible and irreversible features depending on whether $\partial E^{ms}/\partial t=0$ or not, respectively. It means that the irreversible macro-meso-energy exchange corresponds to a structure transition. Substitution from the Eqn. (7) and Eqn (4) into the balance Eqn. (6) results in the front following reference

Integration of the Eqn. (8) with respect to time inside the wave front till results

$$c_m D(t_R) - \Delta U = -\frac{\Delta E^{ms}(t_R)}{\rho_0 a_l};$$

$$\Delta U = \int_0^{t_R} \Delta u \, \frac{dt}{t_R}; \quad t_R = \left(\frac{\partial u}{\partial x}\right)_{max}$$

$$(9)$$

The velocity loss at the pulse plateau ΔU on account of the macro-mesolevel momentum and energy exchange not finished inside the wave front had been found in experiments on the shock loading of metals [3-4]. In the case of the reversible macro-meso-energy exchange without structure transitions $\partial E^{ms}/\partial t=0$ where all the processes involving mesolevel had completed inside the front and at the pulse plateau the dispersion D and the velocity loss ΔU are zero: $D(t_r)=\Delta U=0$, the medium behavior is entirely solid-like ($\varepsilon \to \infty$). Just this situation takes place at Fig.1(a): the medium response to the shock loading corresponds to the case where the second reloading goes like the first one as the medium after the front doesn't conserve the memory about the initial loading. New reloading should begin from the elastic stage followed by the plastic flow initiating the mesoscopic degrees of freedom. If any irreversible process occurs inside the front and at any rate one of the values in the Eqn. (12) is not zero at the pulse plateau, the nonlocal radius ε is finite, and the situation should correspond to the intermediate stage, where the initial state for the second loading is plastic determined by the macro-meso-energy exchange that hasn't yet finished at the pulse plateau. The state of the medium

remembers the history of the first loading through the velocity dispersion. This situation can be seen in Fig. 1(b).

CONCLUSIONS

In scope of the results, the physical nature of plasticity consists in the intermediate stage of the macro-meso-energy exchange, when a part of macroscopic kinetic energy goes to the mesolevel generating fluctuations and forming new structures. The reversible macro-meso-level momentum and energy exchange without the amplitude loss can form quasi-stationary wave fronts. The velocity dispersion profiles inside the front are symmetrical: first the dispersion grows then falls down. Due to anysotropic parts of fluctuations arising on account of the interaction between mesoscopic volumes along the x-direction becomes different from the transversal one, the mean velocity fluctuation velocity is not zero and can be referred to the macrolevel velocity. It means that the energy of mesofluctuations comes back to macrolevel. There is no contradiction with the second principle of thermodynamics as the energy transition from macro-level to mesofluctuations can't be called the dissipation into a heat. Mesofluctuations don't bring up the temperature growing. The irreversible macro-meso-energy exchange transforms the internal structure of a medium making the medium response more fluid-like or plastic. The behavior of a medium during dynamic loading is entirely defined by the variety of processes following the momentum and energy exchange between macro- and mesolevel degrees of freedom.

REFERENCES

1. Mescheryakov Yu. I. and Divakov A.K.. *Dymat Journal*, **1**, № 4, 271-287 (1994).
2. Lipkin J., Asay J.R. *J.Appl.Phys*.**48**.182-195 (1977)
3. Khantuleva, T.A and Mescheryakov Yu. I. *Int. J. Phys. Mesomechanics* , **2**, No 5,. 5-17 (1999).
4. Khantuleva T.A., Mescheryakov Yu.I. *Intern. Journ. Sols and Sts*, **36**, 3105-3129 (1999).
5. Khantuleva T.A. "High-pressure compression of solids VI: old paradigms and new challenges" Y.Y Horie, L.Davison, N.N.Thadhani, Eds., Springer, Berlin, 121-161 (2002).

CP706, *Shock Compression of Condensed Matter - 2003*
edited by M. D. Furnish, Y. M. Gupta, and J. W. Forbes

QUASI-STATIC AND SHOCK INDUCED MECHANICAL RESPONSE OF AN ALUMINIUM-ZINC-MAGNESIUM ALLOY AS A FUNCTION OF HEAT TREATMENT

M.R. Edwards, J.C.F. Millett, N.K. Bourne

Royal Military College of Science, Cranfield University, Shrivenham, Swindon, SN6 8LA. United Kingdom.

Abstract. Samples of an aluminium-zinc-magnesium alloy, typical of high strength weldable aluminium alloys, have been heat treated to produce two different microstructural conditions, these being peak-aged and under-aged. Mechanical tests have been performed, both at quasi-static strain rates and under shock loading conditions to determine how the mechanical properties change with heat treatment. Results indicate that the material has its highest strength when peak aged. Properties are discussed in relation to observed features within the microstructure, as recorded by optical and scanning electron microscopy.

INTRODUCTION

Aluminium alloys have long been seen as useful armour materials due to their combination of low densities and (in some alloys at least) high strengths. These higher strength materials are the result of their alloy additions resulting in a heat treatable microstructure, which through careful control confers a useful balance of properties. These microstructures range from the solution treated state, where all alloy elements exist in solid solution to where the microstructure consists of an equilibrium mixture of phases as dictated by the phase diagram. Engineering aluminium alloys generally have a balance of elements that can be entirely dissolved into solid solution at high temperatures. Due to the nature of the phase diagram, this is a non-equilibrium state, and thus the elements will precipitate out as intermetallic phases, which will effect the properties of the alloy. The way this occurs differs from composition to composition, but a number of features are common to all and are briefly discussed here. In the first instance, alloy elements will begin to concentrate into clusters or platelets of the order of a few nanometers, but still maintaining the structure of the original microstructure (*i.e.* forming coherent precipitates). This creates elastic strains around the precipitates due to mismatches in the lattice spacings, and thus strengthening occurs. As heat treatment progresses, they begin to order, losing coherency with the parent matrix until the equilibrium phase is reached and the precipitate has no coherency with the parent matrix. Consequently, there are no strains around the precipitates, and thus any strengthening effects are less effective. A more complete discussion of age hardening can be found in Smallman [1]. The way the microstructure affects the (quasi-static) mechanical properties has been extensively studied, but its effects upon the dynamic and shock induced mechanical response has been less studied. One of the few investigations was by Rosenberg *et al* [2] who showed in the aluminium – copper based 2024, the Hugoniot

Elastic Limit (HEL) and dynamic-tensile (spall) strengths displaying the same trends as the quasi-static properties with the solution treated state having the lowest strengths under all testing regimes.

EXPERIMENTAL

Two rolled plates of aluminium alloy 7017 (150 mm x 150 mm x 75 mm) were solution treated at 475 °C, followed by an aging treatment at 120°C. One block was aged for 2 hours to produce an under aged microstructure (UA) and the other aged for 24 hours to produce an optimized balance of properties in the peak aged (PA) condition. Samples for shock loading were machined such that the impact axis was in the short transverse direction of the original plates. Manganin stress gauges (Micromeasurements type LM-SS-124CH-048) were supported on the back of 5 mm thick target plates with 12 mm thick plates of polymethylmethacrylate [3]. In material stresses (σ_x) were calculated from those measured in the PMMA (σ_P), using the relation,

$$\sigma_x = \frac{Z_x + Z_P}{2Z_P} \sigma_P, \qquad 1.$$

where Z is the shock impedance and the subscripts x and P refer to the aluminium alloy and PMMA respectively.

Shock stresses were induced using 2.5 mm flyer plates of aluminium alloy 6082-T6, such that the impacts were as symmetrical as possible. Impact velocities were in the range 200 to 865 m s^{-1}. The thickness of flyer was such that the spall plane, due to interactions of release waves from the rear of the flyer and target occurred in the center of the target. Quasi-static tests were performed on a 150 kN servo-hydraulic test rig, with samples of 9.06 mm diameter, and a gauge length of 25 mm, to a strain-rate of 4.76 x 10^{-4} s^{-1}.

The acoustic properties of both heat treatments were identical; density (ρ_0) = 2.74±0.01 g cm^{-3} and longitudinal sound speed (c_L) = 6.25±0.03 mm μs^{-1}.

RESULTS AND DISCUSSION

In Figure 1, the undeformed microstructure of 7017 in the peak aged condition is shown. It can be seen to consist of elongated "pancake" grains that are typical of hot worked aluminium alloys that do not undergo dynamic recrystallization. Also (Fe, Mn)Al$_6$ inclusions can be seen along the grain boundaries.

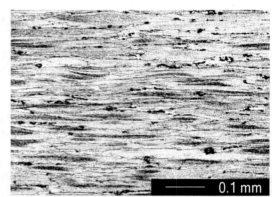

Figure 1. Undeformed microstructure of the peak-aged material. The under-aged microstructure was similar in form, and is not presented here.

In Table 1, we present quasi-static mechanical properties data for 7017.

Table 1. Quasi-static mechanical properties of 7017

Condition	0.2% Proof Stress (MPa)	Tensile Strength (MPa)	% Elongation
UA	275	435	13
PA	402	469	5

As would be expected, the peak aged material displays a higher 0.2% proof (yield) strength than the under aged material. However, it is interesting to note that the tensile strengths of both conditions are similar.

In Figure 2, gauge traces from impacts of 458 m s^{-1} are presented. Observe that the HELs of both conditions are different (calculated from the traces using equation 1), with the underaged material at 463 MPa and the peakaged at 839 MPa, in common with the trends displayed by the quasi-static tests.

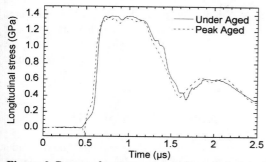

Figure 2. Rear surface gauge traces from under aged and peak aged 7017. Impact stress is 3.6 GPa

However, when the spall signals are examined, it can be seen that both are near identical at *ca.* 200 MPa, corresponding to a strength of the material of 626 MPa (also determined using equation 1). Whilst this is higher than the quasi-static results, indicating a degree of strain-rate sensitivity, the fact that both spall strengths are the same is again in agreement with the trends observed in the quasi-static results. A possible cause can be observed from SEM fractographs of the quasi-static tensile tests. A representative overview of fracture on 7017 (from the under aged material in this case) is shown in Figure 3. Here it can be seen that ductile failure has been initiated around the (Fe, Mn)Al$_6$ inclusions. The fractograph from the peak aged material is near identical, and thus it would seem that the tensile (and spall) strength in 7017 is thus controlled by these features, whilst the yield strength and HEL is controlled by the degree of age hardening due to the heat treatment.

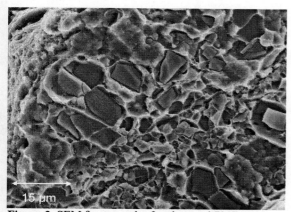

Figure 3. SEM fractograph of under aged 7017.

In Figure 4, we present an optical micrograph of the spall region in the peak-aged sample, shocked to 3.6 GPa. The under-aged specimen shows similar features. We also point out that although this sample has been 'soft recovered', no attempt was made to prevent lateral release. However, the essential features of the spalled regions are preserved.

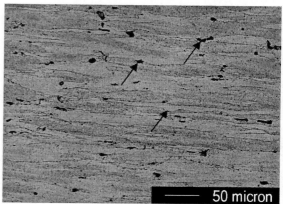

Figure 4. Optical micrograph of spall region in peak aged specimen. Impact stress is 3.6 GPa. Impact is in the vertical direction.

The spall features in this figure are subtle but noticeable. Here it can be seen that in contrast to most grain boundaries, which are rather diffuse in nature, some are clearly delineated (as indicated by the arrows). Through focusing shows relief across grains where these features occur, suggesting a small amount of grain movement, indicating cracking. Note that these would thus be intergranular cracks. Similar cracking, but to a greater extent was noted by Letian [4] in the aluminium alloy 2219-T6. We have also observed that some of these cracks are also associated with the (Fe, Mn)Al$_6$ inclusions (also indicated with the arrows). This would appear to occur with the fractograph of the quasi-statically deformed specimen shown in Figure 3. Again, this would appear to confirm the hypothesis made from the quasi-static mechanical properties and the shock properties, where the yield strength and HEL are controlled by the heat treated microstructure, whilst the tensile strength and spall strength controlled by the existing intermetallic inclusions.

Finally, we note that cracks shown in Figure 4 are not connected, and thus do not form a continuous spall plane, despite the fact that clear spall responses are seen in both heat treated states. A possible answer comes from the work of Chhabildas *et al* [5]. They placed tantalum into a condition of incipient spall, whereby the spall plane consisted of a series of unconnected voids. Observation of the rear surface velocity by a line-imaging velocity interferometer (ORVIS), over 2 mm with a spatial resolution of 50 µm, showed that the spall response was dependent upon precise location, and in fact showed close agreement to microstructures produced from similar experiments [6]. However, in our own experiments, we have used a stress gauge to determine the spall response. This has an active area of 20 mm^2, thus it samples over a large area. Therefore, even though the spall response in this alloy is incipient, the measurement technique we have used averages over a large area compared to the spall features themselves, and thus the results are repeatable.

CONCLUSIONS

The quasi-static and shock induced tensile response of the aluminium alloy 7017 has been investigated as a function of heat treatment. Results show that the yield strength and HEL are affected as would be expected, with the peak-aged material being the stronger. In contrast the quasi-static tensile strengths are very close and the spall strengths are identical. Microstructural examination has shown that in both heat-treated states, failure is initiated around intermetallic inclusions in both loading regimes. Finally, we note that despite the incipient nature of the spall, clear spall signals were produced in both material states. We believe that this be due to stress gauges averaging the spall response of a large area.

ACKNOWLEDGMENTS

We would like to thank Matt Eatwell, Gary Cooper and Paul Dicker for valuable technical assistance. We are also grateful to Adrian Mustey for performing the quasi-static work and micrographic preparation.

REFERENCES

1. R.E. Smallman, *Modern Physical Metallurgy*. 4th ed. 1985, London: Butterworths.
2. Z. Rosenberg, G. Luttwak, Y. Yeshurun, Y. Partom. J. Appl. Phys., 1983, **54**, 2147.
3. Z. Rosenberg, D. Yaziv, Y. Partom, J. Appl. Phys., 1980, **51**, 3702.
4. S. Letian, Z. Shida, B. Yilong, L. Limin. Int. J. Impact Engng., 1992, **12**, 9.
5. L.C. Chhabildas, W.M. Trott, W.D. Reinhart, J.R. Cogar, G.A. Mann .in *Shock Compression of Condensed Matter - 2001*, M.D. Furnish, N.N. Thadhani, and Y. Horie, Editors. 2002, AIP Press: Melville, NY. p. 483.
6. A.K. Zurek, W.R. Thissel, D.L. Tonks, R. Hixson, F. Addessio. J. Phys. IV, 1997, **7**, 903.

CP706, *Shock Compression of Condensed Matter - 2003*
edited by M. D. Furnish, Y. M. Gupta, and J. W. Forbes
© 2004 American Institute of Physics 0-7354-0181-0/04/$22.00

DYNAMIC PROPERTIES OF NICKEL-TITANIUM ALLOYS

Robert Hackenberg[1a], Damian Swift[1b], Neil Bourne[2], George (Rusty) Gray III[1c], Dennis Paisley[1b], Dan Thoma[1a], Jason Cooley[1a], and Allan Hauer[1d]

[1a]*MST-6,* [1b]*P-24,* [1c]*MST-8, and* [1d]*P-DO, Los Alamos National Laboratory, Los Alamos, NM 87545*
[2]*Royal Military College of Science, Cranfield University, Shrivenham, Swindon SN6 6LA, UK*

Abstract. The shock response of near-equiatomic Ni-Ti alloys have been investigated to support studies of shock-induced martensitic transitions. The equation of state (EOS) and elasticity were predicted using ab initio quantum mechanics. Polycrystalline NiTi samples were prepared with a range of compositions, and thicknesses between about 100 and $400\,\mu$m. Laser-driven flyer impact experiments were used to verify the EOS and to measure the flow stress from the amplitude of the elastic precursor; the spall strength was also obtained from these experiments. The laser flyer EOS data were consistent with Hugoniot points deduced from gas gun experiments. Decaying shocks were induced in samples, by direct laser irradiation with a variety of pressures and durations, to investigate the threshold for martensite formation.

INTRODUCTION

Although shape memory alloys are in widespread use, the process of martensitic transformations is not understood well, particularly under dynamic loading. We have investigated the dynamic response of near-equiatomic NiTi, which is the most common shape memory alloy.(1,2) Laser ablation was used to induce decaying shock waves in samples of NiTi of several compositions, to explore the stress, temperature, and duration of loads needed to induce phase transformations. Here we report the results of theoretical and experimental work undertaken in support of the phase transition experiments. Ab initio quantum mechanics was used to predict the equation of state (EOS). Flyer impact experiments allowed the EOS to be tested and a basic constitutive model developed. A computational model was developed to simulate the laser ablation experiments using radiation hydrodynamics, and thus to predict the loading history experienced at each point in each sample, to correlate with microstructural changes observed in recovered specimens.

QUANTUM MECHANICAL PREDICTIONS

A thermodynamically complete EOS is desirable to predict the detailed loading experienced by each sample. An ab initio EOS was calculated using quantum mechanics, using a simplified version of the method applied previously to Si,(3) among other materials. Ab initio non-local pseudopotentials were used to model the nucleus and inner electrons of each atom. The outer electrons were represented with a plane-wave basis set, and exchange and correlation effects were included through the local density approximation (LDA).(4) The frozen-ion cold curve was calculated by finding the ground state of the outer electrons with respect to the atoms, for different values of the lattice parameters. The thermal part of the EOS was predicted using the Grüneisen approximation, by estimating Grüneisen's Γ from the cold curve and applying the Dugdale-Macdonald formula.

This procedure was applied to NiTi in the CsCl (B2) structure. As normal when using the LDA, the lattice spacing predicted at STP varied from the observed value by ~1%. An ab fere initio EOS was constructed by adding a constant pressure offset to bring the EOS into agreement with observation at STP. The EOS was constructed in tabular form following the format and units employed in the SESAME database.(5)

A common first approximation for the EOS of alloys is to average the EOS of the components. For NiTi, the average cold curve varied from the CsCl calculation by a small amount at low pressures, rapidly increasing to tens of percent above 60 GPa. This indicates that the effect on the electrons of forming an intermetallic compound may influence the binding energy and hence pressure greatly for this material. Performing ab initio calculations of the actual alloy structure – at least to generate the cold curve, which is relatively quick – is the preferable approach when considering an untested material. (Fig. 1.)

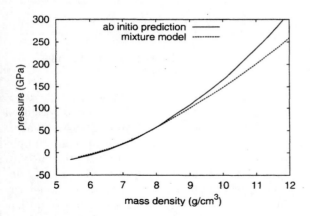

FIGURE 1. Comparison between ab initio cold curve for NiTi in the CsCl structure and a simple average of the cold curve of Ni and Ti separately.

Calculations were also performed of elastic strains applied to the CsCl lattice cell. Given the configuration of the electrons in the ground state, the stress tensor on the lattice cell could be deduced. The variation of stress with strain was used to estimate some of the elastic constants c_{ijkl} in deviatoric form, e.g. 73 GPa for c_{1111}.

FLYER IMPACT EXPERIMENTS

States on the principal Hugoniot of NiTi were measured using flyer impact data from gas gun or laser launch experiments. Gas gun experiments with stress gauges have been reported previously.(6) The laser flyer experiments used pulses ~600 ns long from the TRIDENT facility, absorbing the energy in a layer of C sandwiched between a transparent PMMA substrate and the flyer.(7) Additional layers of Al and alumina were included to contain the resulting plasma and reduce heat conduction. Experiments have been performed in which EOS points were deduced for Cu(8); these demonstrated that gross pathological problems associated with the use of a laser drive did not occur.

The TRIDENT flyer experiments included two compositions: "5B" with 52.0 at % Ni, and "6B" with 54.2%, both prepared by arc melting high purity Ni and Ti under Ar. Three main types of experiments were performed: NiTi flyers impacting NiTi targets which released into PMMA windows; Cu flyers impacting NiTi targets which released into vacuum; and NiTi flyers impacting LiF windows. In each case, point and line VISARs were used to monitor the velocity history of the flyer and the surface of the sample. The flyers were 5 mm in diameter, the targets semicircles of the same diameter and positioned so that roughly half of the flyer could be seen, to allow access for the VISAR beams. NiTi flyers ranged from 100 to 200 μm thick, Cu flyers 50 to 100 μm, and NiTi targets 100 to 400 μm. Cu flyers were punched from rolled foils of high purity material from Goodfellow Metals. The NiTi parts were prepared from ingots by electro-discharge machining, grinding and polishing. LiF windows were (100) crystals obtained from Saint-Gobain (Bicron) Corp.

In the laser flyer experiments, the line VISAR did not use quadrature encoding and the velocity deduced had an uncertainty of up to ~50 m/s, compared with up to ~5 m/s for the point VISAR, which did use quadrature encoding. The most accurate stress states were therefore deduced from experiments in which the accuracy of the point VISAR dominated: impact of the sample against a window. The EOS data deduced were consistent with the gas gun data, and filled in states around 3 to 4 GPa, where multiple waves were observed in records from stress gauges. According to the laser flyer data for alloy 6B, the inflection in the response curve was smoothly curved with an onset closer to 2.5 than the 3.5 GPa reported previously for material of 50.4 wt % Ni.(6) (Fig. 2.)

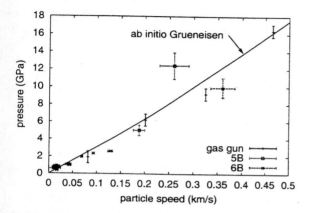

FIGURE 2. Comparison between theoretical Hugoniot and flyer impact data for NiTi.

VISAR records for alloy 5B exhibited precursor waves of around 25 or 70 m/s at the interface with the PMMA window, running 9 ± 3 ns ahead of the main shock per $100 \mu m$ of sample thickness. X-ray diffraction indicated some texture in both alloys; different precursors may reflect the textures. These results imply a shear modulus of 175 ± 80 GPa and a flow stress of 0.35 ± 0.05 or 0.95 ± 0.05 GPa respectively for the different interface speeds behind the precursor. The shear modulus cannot be compared directly with the ab initio predictions because the predictions have not so far included enough crystal orientations to enable the polycrystal properties to be estimated. The value deduced from the impact experiments is however consistent with the calculations in that compression along the $\langle 111 \rangle$ directions should produce a higher restoring force, hence the predictions based on compression along $\langle 100 \rangle$ should give elastic constants lower than the polycrystal average. Interestingly, the flow stresses deduced bracket values for quasistatic (0.5 GPa) and gas gun (0.8 GPa; projectile 3 mm thick) experiments.(6) The higher value observed here may reflect strain rate dependence in the plastic response; the lower value

is more of a mystery and again may have been caused by variability between samples.

There was clear evidence of spall and subsequent reverberations; the corresponding spall strength was estimated from the magnitude of the pull-back to be 1.8 ± 0.2 GPa for alloy 5B, and 2.4 ± 0.2 for 6B.

LASER IRRADIATION EXPERIMENTS

For parametric studies of the loading conditions necessary to induce martensitic phase change – results reported elsewhere(2) – the TRIDENT laser was used in nanosecond mode to induce dynamic loading by ablation of one surface of a sample several millimeters in diameter and $\sim 100 \mu m$ thick; the opposite surface was monitored with VISAR velocimetry.(7) The drive pulse was between 0.2 and 3.6 ns long and applied over a spot 5 mm in diameter using a Fresnel zone plate to smooth spatial variations in the beam. The drive irradiance was roughly constant in time. For samples over $\sim 50 \mu m$ thick, shock waves induced by laser pulses of this duration decay as they pass through the sample. Thus a single experiment can explore a range of loading conditions, if changes to the microstructure are investigated in cross-section. Relatively thick samples are easier to recover after the shot with relatively little perturbation to the microstructure, particularly if a release window is used. The ablative drive is benign in that the sample deceleration system need not cope with the energy and momentum of a sabot or detonation products, and the TRIDENT laser was capable of firing shots of this type at a rate up to 14 per day.

One concern was that direct laser drive is more likely than laser flyers to suffer from pathological laser-related problems such as preheating of the sample by x-rays or hot electrons. Samples recovered and sectioned indicated that, for NiTi at the irradiance levels needed to induce shocks of ~ 20 GPa or less, the region affected was at most $\sim 1 \mu m$ wide at the drive side.

In the direct drive experiments, the main diagnostic for applied load was the record of laser power history. Line VISAR records were obtained on almost every experiment, but because of the decaying shock, the velocity history at the opposite surface was used principally to verify that simulations of the loading conditions were accurate. The simulations were based on a prescription developed for laser-induced shocks of up to ~100 GPa or so driven by green light.(9) The EOS and strength model developed above were used in radiation hydrodynamics simulations of the effect of the laser drive. However, the EOS was not valid for the plasma region generated by laser deposition, so the generic "QEOS" model(10) was used for the first 1 μm of material at the drive side.

The QEOS model underpredicted the ablation pressure somewhat in this regime. Scaling the pressure history to improve the match to the peak surface velocity, the overall agreement was reasonable though with some smearing by the time of the reverberation (Fig. 3).

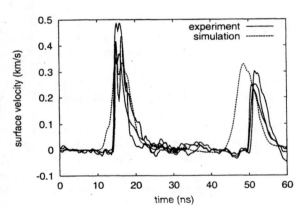

FIGURE 3. Surface velocity history for TRIDENT shot 12129 (80 μm NiTi, no window, 25 J drive in 3.6 ns).

CONCLUSIONS

The Hugoniot EOS and dynamic strength of NiTi alloys of different compositions were measured using laser-launched flyer impact experiments. The EOS data matched and extended previous gas gun results using stress gauges, and the flow stresses bracketed a value reported from static tests. Ab initio quantum mechanics was used to predict isotropic and uniaxial compression curves for NiTi in the CsCl structure, and

hence a simple – but thermodynamically complete – EOS and also a subset of the elastic constants. The principal Hugoniot deduced from the theoretical EOS matched the overall trend of the experimental Hugoniot states quite well. The theoretical elastic constants were consistent with the shear modulus inferred from precursors in the impact experiments on polycrystal samples. The amplitude of precursor waves confirms the previous identification of the inflection near 3 GPa as the onset of plastic flow. A prescription was developed for simulating direct laser drive using radiation hydrodynamics; reasonable agreement was obtained with laser ablation experiments.

ACKNOWLEDGMENTS

The electron code CASTEP was made available courtesy of the UK Car-Parrinello Consortium. TRIDENT staff laser provided substantial experimental help. This work was performed under the auspices of the US Department of Energy, contract W-7405-ENG-36.

REFERENCES

1. Saburi, T., in *Shape Memory Materials*, edited by K. Otsuka and C.M. Wayman, Cambridge University Press, London, 1998.
2. Hackenberg, R.E., Swift, D.C., Cooley, J.C., Chen, K.C., Thoma, D.J., Paisley, D.L., and Hauer, A., "Phase changes in Ni-Ti under shock loading," in *New Models and Hydrocodes for Shock Compression of Condensed Matter – Edinburgh 2002*, edited by V. Klimenko et al (to appear).
3. Swift, D.C., Ackland, G.J., Hauer, A., and Kyrala, G.A., Phys. Rev. B **64**, 21, 214107 (2001).
4. Kohn, W. and Sham, L.J., Phys. Rev. **140** 4A (1965).
5. Holian, K.S. (Ed.) *T-4 Handbook of Material Property Data Bases, Vol 1c: Equations of State*, Los Alamos National Laboratory report LA-10160-MS (1984).
6. Millett, J.C.F, Bourne, N.K., and Gray III, G.T., J. Appl. Phys. **92**, 6, pp 3107-10 (2002).
7. Paisley, D.L., Swift, D.C., Johnson, R.P., Kopp, R.A., and Kyrala, G.A., "Laser-launched flyer plates and direct laser shocks for dynamic material property measurements," ibid, pp 1343-6.
8. Niemczura, J., Paisley, D.L., and Swift, D.C., "Accuracy of laser-launched flyer experiments for measuring equations of state," this conference.
9. Gammel, J.T., Swift, D.C., and Tierney IV, T.E., "Shock response of iron on sub-nanosecond time scales," this conference.
10. More, R.M., Warren, K.H., Young, D.A., and Zimmerman, G.B., Phys. Fluids **31**, 10 (1988).

CP706, *Shock Compression of Condensed Matter - 2003*
edited by M. D. Furnish, Y. M. Gupta, and J. W. Forbes
© 2004 American Institute of Physics 0-7354-0181-0/04/$22.00

SLIDING FRICTION AT COMPRESSED Ta/Al INTERFACES

J.E. Hammerberg*, R. Ravelo†*, T.C. Germann*, J.D. Kress** and B.L. Holian**

Applied Physics Division, Los Alamos National Laboratory, Los Alamos NM 87545
†*Physics Department and Materials Research Institute, University of Texas, El Paso TX 79968-0515*
**Theoretical Division, Los Alamos National Laboratory, Los Alamos, NM 87545*

Abstract. The physics of sliding at compressed Ta/Al interfaces is discussed based upon the results of large scale 3D NEMD simulations. A new set of Embedded Atom Method potentials has been constructed to treat the Ta-Al interaction. Pressures of order 15 GPa are studied and the velocity dependence of the frictional force is studied for several interfacial configurations including Al(100)/Ta(100) and Al(111)/Ta(110). Generic behavior is observed, characterized by a linear increase at low velocities followed by a power law decrease at high velocities associated with near interface structural transformation in Al.

INTRODUCTION

The properties of sliding interfaces between ductile metals at high sliding speeds ($v \geq 1$ km/s) and under large compressions are largely unknown. Early measurements of Bowden et al. [1] showed a decreasing velocity dependence in the frictional force at high velocities ($v > 100$ m/s) for many metal tribo- pairs, which they attributed to melting. Recent large-scale Non Equilibrium Molecular Dynamics (NEMD) simulations [2, 3] have shown similar behavior for Cu/Cu, Lennard-Jones/Lennard-Jones, and Cu/Ag systems, with the high velocity weakening in the frictional force related to the onset of a dynamically induced microstructural transformation commonly seen in ductile metal dry sliding at substantially lower velocities but similar local interfacial strain states [4]. For incommensurate interfaces theoretical arguments and simulations [3] show an increasing velocity dependence at low velocities.

The Ta/Al interface is interesting in the above context since it represents a bcc/fcc system with a generally incommensurate interface for which the two components differ dramatically in hardness and shock wave properties. It is also the subject of recent high energy density radiography (ATLAS) experiments [5] as well as Proton Radiography (PRad) experiments at Los Alamos National Laboratory. We present here a series of NEMD simulations for (111) and (100) single crystal interfaces sliding in the <100> direction for a wide range of velocities, using a new Ta-Al interaction, in the high pressure (15 GPa) regime of interest in current ATLAS and PRad experiments.

TABLE 1. Zero temperature, zero pressure elastic constants, cohesive energy and elastic constants of pure Al and Ta EAM potentials and corresponding experimental values (in parentheses).

	Al	**Ta**
a_0 (Å)	4.05 *	3.304 *
E_{coh} (eV)	3.36 *	8.1 *
C_{11} (GPa)	113.9 (117.9)†	263.85 (263.9) **
C_{12} (GPa)	62.0 (62.2)†	160.0 (160.0)**
C_{44} (GPa)	31.5 (32.5)†	82.8 (82.0) **

* Values equal to experiment by construction
† Ref. [11]
** Ref. [12]

INTERATOMIC POTENTIALS

Pure Al, pure Ta and their binary alloys are described within the embedded atom method (EAM) [6]. The pure Al EAM potential functional forms and parameters used in this work were taken from [7]. The pure Ta EAM potential was obtained by fitting to the zero pressure elastic constants, vacancy formation energy, unrelaxed (100) surface free energy, experimental equation of state and first-principle calculations of the energy difference between fcc and bcc phases at

TABLE 2. Cohesive energies (from heats of formation) and lattice constants at zero pressure, zero temperature of binary alloys Ta$_3$Al and Al$_3$Ta and experimental values from Ref [13] (in parentheses).

	Ta$_3$Al	Al$_3$Ta
Structure	bcc	DO$_{22}$
a_0 (Å)	6.54 (6.57)	3.83 (3.840)
c_0 (Å)	-	8.57 (8.56)
E_{coh} (eV/atom)	6.90 (7.01)	4.71 (4.79)

fcc Al(100) R45 / bcc Al(100)

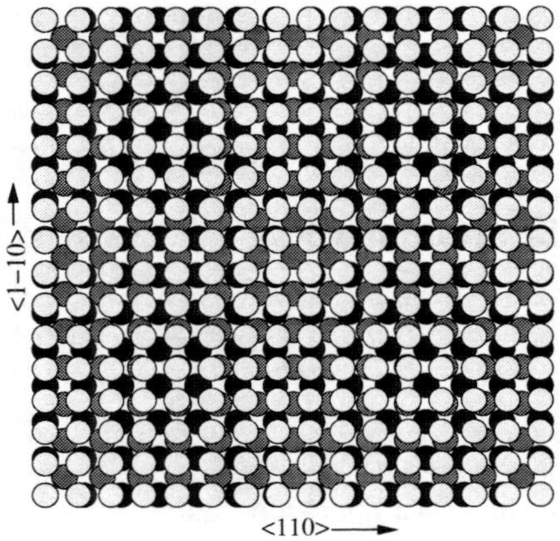

FIGURE 1. Coincidence site lattice (CSL) of fcc Al(100)R45° /bcc Al(100). Atoms are shaded according to depth: white circles represent topmost layer.

different volumes [8]. Functional forms and parameters are provided in [9]. Elastic constants of pure Al and Ta are given in Table 1. The fitting procedure used in developing the Ta-Al cross-potential is similar to that found in [10]. The cross pair potential parameters were fitted to the heats of formation and lattice constants at zero pressure of Al$_3$Ta (DO$_{22}$ structure) and Ta$_3$Al (BCC structure). Results of this fit are provided in Table 2.

RESULTS AND DISCUSSION

We investigated the minimum energy interface structures by performing kinetic annealing of Al/Ta slabs containing up to 200,000 atoms. Two low index configurations were studied: Al(100)/Ta(100) and Al(111)/Ta(110).

The lattice misfit between Al(100)/Ta(100) is about 13.3% at zero pressure. We constructed slabs of various sizes matching the surface misfit as closely as possible. Periodic boundary conditions were employed in the in-plane directions while free boundary conditions were used in the normal directions. The energy was minimized employing kinetic annealing. This process revealed that Al grows epitaxial on Ta(100) up to a thickness of about 3 monolayers and then transforms to fcc.

This surprising result prompted us to study the Al/Ta(100) interface employing *ab initio* calculations. We did this using density functional theory (DFT) as implemented in the VASP code [14]. Ultrasoft pseudopotentials [15] were used to represent the core electrons and the valence electronic wavefunctions were expanded in plane waves. Details of these calculations are provided in [9]. The simulation cell consisted of 3 (5×5) or (7×7) layers of Ta atoms and 1 layer of Al atoms. One Ta layer was

held fixed at a lattice constant $a_0 = 3.30$ Å. Periodic boundary conditions were applied in the directions parallel to the surface of the slab. Al coverages from (2×2) up to (6×6) were used but always less than the Ta surface periodicity. The optimal structure was obtained by energy minimization. In all coverages investigated, it was found that the Al layer relaxed to near epitaxial coverage with the Ta substrate, thus confirming the EAM results.

Energy minimization of bigger slabs employing the EAM potentials found that above 3 monolayers, the structure of the Al layers on Ta(100) changes from epitaxial bcc(100) to fcc(100) R45°. The transition from bcc(100) $\rightarrow$ fcc(100) stacking is accomplished by alignment of the $[110]_{fcc}/[100]_{bcc}$ axes. The misfit along this direction is $[110]_{fcc}/[100]_{bcc} \approx$ 13/15 or 13.3%. The transition layer creates as a result of this misfit, a super cell with a (15×15) coincidence site lattice (CSL) on the bcc Al substrate. The CSL is shown in Fig. 1. The 4 corner atoms are coherent with the substrate atoms while the inner atoms are incoherent. It is noted that similar epitaxial growth and structures have been observed on ultra thin layers of Ni and Cu grown on Fe(100) using molecular beam epitaxy (MBE) [16].

The epitaxial growth of Al/Ta(100) was not found in minimum energy interface studies of Al(111)/Ta(110). The crystal orientation of many fcc metals deposited on bcc(110) metals can have a

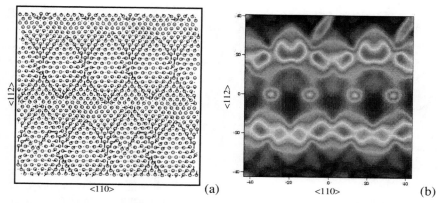

FIGURE 2. (a) Top view of Al(111)/Ta(110) interface showing Al atoms and their corresponding displacement (arrows) after minimization from their initial lattice positions. (b) Grayscale potential energy map of structure in (a). White corresponds to higher potential energy (-3.45 eV/atom); black to the lowest (-3.90 eV/atom).

specific orientation of the Nishiyama-Wassermann (N-W) or Kurdjumov-Sachs (K-S) relation [17]. Calculations of the interface energy as a function of the ratio of nearest-neighbor separation d_{bcc}/d_{fcc} within a rigid lattice model find two optimal values of this ratio for the N-W orientation: $d_{bcc}/d_{fcc} = 0.866$ and 1.061; and only one for the K-S case: $d_{bcc}/d_{fcc} = 0.919$ [18]. For Al/Ta(110), the nearest-neighbor ratio $d_{Ta}/d_{Al} \approx 1.0$. Since the K-S orientation requires a higher strain at the interface, we expect the N-W orientation to have the lowest interface energy. Our energy minimization studies are consistent with this picture: Al(111)/Ta(110) interface prefers the N-W orientation relation: $[1\bar{1}0]_{fcc}/[001]_{bcc}$, $[11\bar{2}]_{fcc}/[1\bar{1}0]_{bcc}$. In this configuration, the first few layers near the interface show in-plane distortions as depicted in Fig. 2(a). The strain is localized within the first few layers, decaying quickly as a function of distance from the interface. Fig. 2(b) shows the surface energy map of the first Al layer at the interface. Even though this layer is tightly bound to the Ta substrate, high energy regions can be seen where Al atoms sit on top of Ta atoms in the substrate.

Simulations of Al/Ta sliding interfaces

Large-scale non-equilibrium molecular dynamics (NEMD) simulations were carried out to investigate the microstructure evolution and dissipation mechanisms for dry slip along two interfacial orientations in the Al-Ta system: Al(100)/Ta(100) and Al(111)/Ta(110). The simulations were performed employing the SPaSM code [19]. The simulation methodology is as follows: the computational cell consists of two workpieces of nearly equal size. The interface is located at y=0 and periodic boundary conditions are employed in the x- and z-directions. Two reservoir regions (about 20 planes) are defined in the upper and lower work pieces. A normal force F_{norm} is applied to the reservoir regions equivalent to a pressure of 15 GPa and the temperature in the reservoirs is thermostated to 232K. A tangential force F_{tang} is applied to all atoms in the reservoir and adjusted at each time step so that the velocity of the reservoirs (averaged over all particles) is $\pm u_p$. The relative velocity of the moving work pieces is then $v_{rel} = 2u_p$. The dimensions of the computational cells used in the simulations were about (142 (X), 523 (Y), 132 (Z))Å containing approximately 600,000 atoms. For Al(111)/Ta(110), the sliding < direction was along the $[1\bar{1}0]_{fcc}/[001]_{bcc}$ orientation.

The tangential force F_{tang} was monitored until it reached steady state (typically about 200 psec). Fig. 3 shows the tangential force as a function of v_{rel} for the two interfaces. While the magnitude of F_{tang} is different, it exhibits the same behavior in both cases. There is a linear regime, where little plastic deformation is observed, and a plastic zone where the tangential force decays as v_{rel}^{β} with $\beta \approx 3/4$. The interface structures associated with the high velocities all exhibit strong disorder in the tangential directions and density correlations in the normal directions akin to a confined fluid or amorphous solid. In all cases the slip surface (zero tangential velocity) occurs in the

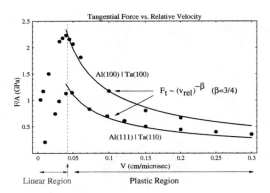

FIGURE 3. Tangential force vs. relative sliding velocity for the two interfacial orientations investigated.

weaker medium, Al, with strong bonding at the Al/Ta interface.

CONCLUSIONS

We have investigated the minimum energy structures of Al/Ta interfaces employing EAM potentials. It was found that Al grows epitaxially on Ta(100). This result was confirmed by *ab initio* simulations of Al/Ta(100). This can be explained by nearly equal lattice constants of bcc Ta and bcc Al (metastable minimum) and the large heats of formation of Al-Ta alloys. Energy minimization of Al(100)/Ta(100) reveal that the interface structure consists of about 3 monolayers of bcc(100) Al followed by fcc(100) R45°. The morphology of this interface is similar to what is observed on ultra thin films of Ni/Fe(100).

Large-scale NEMD simulations were performed for two interface orientations: Al(100)/Ta(100) and Al(111)/Ta(110). It was found that the frictional force as a function of relative sliding velocity exhibits a linear regime at low velocities, consisten with the behavior expected from an incommensurate interface [4], and a highly localized defective region at high velocities where the frictional force decreases with velocity as $v_{rel}^{-3/4}$. The structures associated with the high velocities all exhibit strong disorder in the tangential directions and density correlations in the normal directions akin to a confined fluid or amorphous solid.

ACKNOWLEDGMENTS

This work was supported by the U.S. Deparment of Energy under contract W-7405-ENG-36.

REFERENCES

1. Bowden F.P. and Freitag E.H., *Proc. Roy. Soc.* **248A**, 350 (1958); Bowden F.P. and Persson P.A., *Proc. Roy. Soc.* **260A**, 433 (1960).
2. Hammerberg J.E., Holian B.L., Roeder J., Bishop A.R. and Zhou S.J., *Physica D* **123**, 330 (1998).
3. Hammerberg J.E., Germann T.C., and Holian B.L., Los Alamos Nat. Lab. Rept. LA-UR-02-4654.
4. Rigney D.A. and Hammerberg J.E., *MRS Bull.* **23**, 32 (1998).
5. Faehl R.J. and Hammerberg J.E, Los Alamos Nat. Lab. Rept., LA-UR-02-1142 (2002).
6. Daw M.S. and Baskes M.I., *Phys. Rev. Lett.* **50**, 1285 (1993); Daw M.S. and Baskes M.I., *Phys. Rev.B*, **29**, 6443 (1984).
7. Angelo J.E., Moody N.R. and Baskes M.I., *Modelling Simul. Mater. Sci. Eng.* **3**, 289 (1995).
8. Söderlind P.S. and Moriarty J.A., *Phys. Rev B* **57**, 10340 (1998).
9. Ravelo R., Hammerberg J.E., Germann, T.C., Kress J.D. and Holian B.L., to be published.
10. Voter A.F., "The Embedded Atom Method", in *Intermetallic Compounds: Principles and Practice*, edited by J.H. Westbrook and R.L. Fleischer, John Wiley and Sons, 1993.
11. Kamm G.N. and Alers G.A., *J. Appl. Phys.* **35**, 327 (1964).
12. Cynn H. and Yoo C.S., *Phys. Rev. B*, **59**, 8526 (1999).
13. Subramanian P.R., Miracle D.B and Mazdiyasni S., *Metall. Trans.* **21A**, 539 (1990)
14. Kresse G. and Hafner J., *Phys. Rev. B* **47**, 558 (1993); **49**, 14251 (1994); Kresse G. and Furthmüller J., *Comput. Mater. Sci.* **6**, 16 (1996); *Phys. Rev. B* **55**, 11169 (1996).
15. Vanderbilt D., *Phys. Rev. B* **41**, 7892 (1990); Kresse G. and Hafner J., *J. Phys. Condens. Matt.* **6**, 8245 (1994).
16. Kamada Y. and Matsui M. *J. Phys, Soc. Jap.* **66**, 658 (1997) and references therein.
17. Bruce L.A. and Jaeger H., *Phil. Mag.* **38**, 223 (1978).
18. Gotoh Y., Entani S. and Kawanowa H., *Surf. Sci.* **507-510**, 401 (2000).
19. Lomdahl P.S., Tamayo P., Gronbech-Jensen N. and Beazley D.M.,in *Proceedings of Supercomputing 93*, IEEE Computer Society Press, Los Alamitos, CA 1993, pp. 520-527.

CP706, *Shock Compression of Condensed Matter - 2003*
edited by M. D. Furnish, Y. M. Gupta, and J. W. Forbes
© 2004 American Institute of Physics 0-7354-0181-0/04/$22.00

ANISOTROPIC FAILURE MODELING FOR HY-100 STEEL

E. N. Harstad[1], P. J. Maudlin[1], and J. B. McKirgan[2]

[1]*Theoretical Fluid Dynamics, Los Alamos National Laboratory, Los Alamos NM 87545*
[2]*Code 614, NSWC, Carderock Division, 9500 MacArthur Blvd., West Bethesda, MD 20817*

Abstract. HY-100 steel is a material that behaves isotropically in the elastic and plastic region and acts anisotropically in failure. Since HY-100 is a ductile metal, a more gradual failure process is observed as opposed to the nearly instantaneous failure in brittle materials. We extend our elasto-plastic-damage constitutive model by including of a decohesion model to describe material behavior between the onset of failure and fracture. We also develop an anisotropic failure surface to account for directionality in material failure. Both the anisotropic failure and decohesion models have been implemented into a finite element code, where the effects of these models are studied in a uniaxial stress simulations, a plate impact simulations, and a quasistatic notched round bar tensile test simulations.

INTRODUCTION

Based on experimental observation the fracture behavior of HY-100 steel is anisotropic and also contains a decohesive region between the fracture point and the complete failure of the material [1].

Quasi-static tension tests were conducted on A, B, D, and E shaped Navy HY-100 steel round-bar specimens by NSWC [1]. A typical example of these results in shown in Figure 1. These specimens were cut from plate stock in the rolling and transverse directions. They show a directional dependence in the failure point, with the rolling direction being tougher than either transverse direction. This directionality is thought to be attributed to Magnesium Sulfide inclusions[2], which are the the long ellipsoidal-type inclusions aligned with the rolling direction of the plate. When the loading is perpendicular to the long side of the inclusions, we see enhanced damage and earlier failure in the specimens. The point of failure initiation is characterized as the point where enough damage has accumulated that the structural integrity of the material changes dramatically and complete fracture is inevitable [2]. The stress state at the point of failure initiation is called the critical stress.

In the past, elasto-plastic-damage constitutive models have been developed and used successfully in describing metal behavior up to the onset of material failure. In our computer simulations of tension tests using the TEPLA model[3], once the stress-strain path has evolved to the insect with a failure surface, the stress was zeroed out in the material. This discontinuous drop in stress is not physical, and can cause numerical issues in continuum mechanics codes.

In order to account for a more complicated failure process, we introduce a general anisotropic failure surface, which includes the isotropic Hancock-MacKensize failure curve[4]. We extend the modeled mechanical response of the deformation from the onset of failure to zero stress state using a discrete constitutive equation, which prescribes the manner in which the stress state relaxes. In this softening regime, the material undergoes decohesion. The model relaxes the critical stress by softening the flow stress by a decohesion parameter, which has a value between one and zero. The evolution of this parameter is

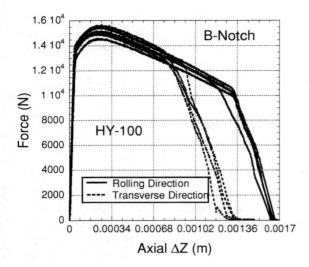

FIGURE 1. HY-100 steel quasi-static tension test results of B-notch geometry round bar.

governed by kinematical and stress conditions at the failure onset point.

ANISOTROPIC FAILURE SURFACE

The failure strain is a prediction of the failure point of a material (in shear mode),

$$\varepsilon_f^{HM} = \gamma_0 + \gamma_1 e^{\gamma_2 P/Y} , \qquad (1)$$

in which γ_0, γ_1, and γ_2 are material constants, and (P/Y) is the stress triaxiality defined to be the ratio of the Pressure, P, to the second invarient of the Cauchy stress, Y[3]. When this shear mode is combined with volumetric strain (porosity), ϕ, and a failure porosity, ϕ_f, a mixed-mode failure surface can be formulated, such as the isotropic surface used by Johnson and Addesio[2]. We consider a generalized failure surface of the form,

$$F = \underline{\varepsilon} \cdot \alpha_f \cdot \underline{\varepsilon} - r^2 = 0 , \qquad (2)$$

in which $\underline{\varepsilon}$ is the logarithmic strain, α_f is the fourth order failure surface shape tensor, and $r^2 = f\left(P/Y, \phi_f, \varepsilon_f^{HM}\right)$ is the average radius of the surface. The failure shape tensor can be split into an spherical and deviatoric parts

$$\alpha_f = aP^{sph} + \alpha^{dev} , \qquad (3)$$

where a is a material constant. In the separation of the shape tensor, we associate the spherical part of the failure surface with the porosity, and the deviatoric portion with incompressible plasticity. Note that we could further separate the deviatoric tensor into an isotropic part (associated with Hancock-MacKensize) and an anisotropic part. If we use the principle of superposition, we can postulate the existance of a form of r^2 that is seperable into

$$F = \left(\frac{\phi}{\phi_f}\right)^2 + \left(\frac{\underline{\varepsilon} \cdot \alpha^{dev} \cdot \underline{\varepsilon}}{b\left(\varepsilon_f^{HM}\right)^2}\right) - 1 , \qquad (4)$$

in which b is a material constant. To find the material constants which appear in Eqs. (1) and (4), quasi-static tensile tests are pulled to failure on notched round bar specimens. Using either a Bridgeman analysis or a finite element simulation for the triaxiality, the strain-to-failure for a given state can be determined and used to fit Eq. (1). The normalization constant b is then determined as the value that provides equivalent failure under uniaxial stress between the isotropic and anisotropic failure surfaces.

DECOHESION MODEL

We define the criterion for failure initiation as that point the material has evolved to a stress/porosity state that lies on the failure surface. Once the failure surface is reached, the stress state is relaxed to zero in an indirect manner by degrading the flow stress by the parameter

$$\sigma^* = \eta\sigma \qquad (5)$$

This softens the flow surface in a prescribed manner without changing its shape, leading to damage induced softening of the stress-strain path. A simple form is chosen for the evolution of η, which is consistent with to Needleman's ideas for decohesion[4],

$$\eta = 1 - \frac{[u]^d}{\delta} , \qquad (6)$$

in which δ is a material parameter and $[u]^d$ is the instantaneous distance between the two fracture surfaces. The distance between the fracture planes, $[u]^d$, is determined by

$$[u]^d = \int_{\text{failure}} (\psi \Delta x)^{\text{failure}} dt, \qquad (7)$$

where $\psi = \underline{n} \cdot \overline{\underline{D}} \cdot \underline{n}$, $\overline{\underline{D}}$ is the material frame rate of deformation tensor, and Δx is the length scale associated with the region of failure. The effect of Eq. (7) is that once failure begins the two surfaces move apart from each other at a prescribed velocity determined by the rate of deformation tensor's value in the direction normal to the failure plane. There is no mechanism for crush-up or healing once the decohesion process has begun. The direction normal to the plane of failure is,

$$\underline{n} = \frac{\lambda_1}{\lambda_s}\underline{v}_1 + \frac{\lambda_2}{\lambda_s}\underline{v}_2 + \frac{\lambda_3}{\lambda_s}\underline{v}_3$$

$$\lambda_s = \sqrt{\lambda_1{}^2 + \lambda_2{}^2 + \lambda_3{}^2}, \qquad (8)$$

where $(\lambda_i, \underline{v}_i)$ are the eigenvalues and eigenvectors of the material frame stress at failure.

RESULTS AND DISCUSSION

The models were implemented into the 3D finite element code EPIC[6]. We used the HY-100 material properties as determined by Goto et al. [6]. The anisotropic failure surface was characterized for HY-100 using a uniaxial stress path. The material constants for the failure model are as follows:

$$\alpha^{dev} = \begin{pmatrix} 1.6 & 0 & 0 & 0 & 0 \\ 0 & 1 & 0 & 0 & 0 \\ 0 & 0 & 1 & 0 & 0 \\ 0 & 0 & 0 & 1 & 0 \\ 0 & 0 & 0 & 0 & 1 \end{pmatrix}$$

$$b = 1.51229 \qquad (9)$$

$$\gamma_0 = -0.61; \quad \gamma_1 = 2.07; \quad \gamma_2 = 0.50$$

In (10), the shape tensor is expressed in a 5D basis, which allows us to represent the fourth-order tensor conveniently [7].

Figure 2 shows a uniaxial stress loading for the rolling and transverse plate directions. We see that the anisotropic failure surface is accounting for the smaller failure strain observed in the transverse

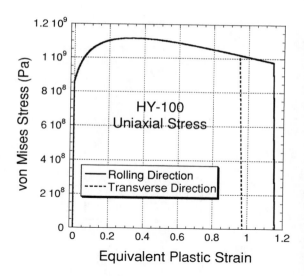

FIGURE 2. von Mises stress as a function of equivalent plastic strain for a uniaxial stress path. For HY-100 Steel, we see the anisotropy in the failure strain for the rolling and transverse directions of the plate.

direction. After the failure point, the decohesion model provides some slope to the failure.

Simulations of the Navy HY-100 D-Notch tension test were done using a three-dimensional quarter symmetry mesh. An isotropic failure surface was used in conjunction with the decohesion model. Figure 3 compares the calculation to the experimental data for a rolling direction D-notch geometry. The yield point and stress magnitude are too high because we are running EPIC at a strain rate of $\sim 100\ 1/s$ to facilitate a reasonable computational time. We see good agreement with the prediction of the failure point and good replication of the decohesion behavior. Further comparison with the transverse direction using the anisotropic failure surface is in progress.

HY-100 plate impact simulations were conducted for two material orientations. The impact velocity was 200 m/s and the flyer was one half the thickness of the target. The free surface velocity of the target is shown in Fig. 4 for the cases with and without decohesion. Additionally, we show the velocity profile for an isotropic failure surface.

The addition of the decohesion model further depresses the velocity past the pull-back point

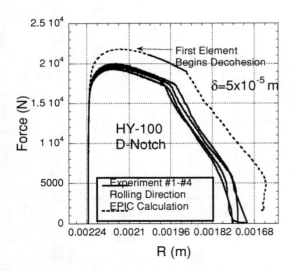

FIGURE 3. Comparison between experiment and EPIC simulation of a D-notch round bar tensile test for HY-100 Steel.

leading to an increased spall strength as defined in Baumung et. al. [8]. Since the flyer plate test is dominated by volumetric damage, we see little difference in the transverse and rolling directions, since the anisotropic surface primarily affects the shear mode.

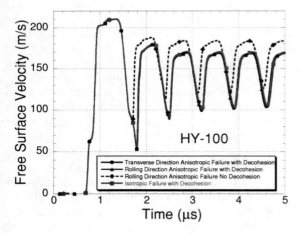

FIGURE 4. Free surface velocity as a function of time for a HY-100 steel plate impact simulation. Impact velocity was 200 m/s. The inclusion of the decohesion process shows a predicted increase in spall strength.

CONCLUSIONS

We have expanded our material models by the addition of an anisotropic failure surface and decohesion model. This furthers our design capabilities for more complex materials such as HY-100. The difference in the failure strain predicted by the anisotropic failure surface agrees well with notched round experiments. The decohesion model predicts an increase in the spall strength of the material. Flyer plate experiments for HY-100 are needed to investigate this.

REFERENCES

1. P.J. Maudlin , E.M. Mas, B.E. Clements, and J.B. McKirgan, "Anisotropic Damage Analysis of HY100 Steel Under Quasistatic Loading Conditions" , Plasticity, Damage, and Fracture at Macro, Micro and Nano Scales, Editors A.S. Khan and O. Lopez-Pamies, Neat Press (2002).
2. V. Jablokov, D.M. Goto, D.A. Koss, and J.B. McKirgan, Mater. Sci. Eng. A-Struct. Mater. Prop. Microstruct. Process. , v. 302, no. 2, p197-205, (Apr 2001)
3. J.N. Johnson and F.L. Addessio, J. Appl. Phys., vol 64., number 12, Dec. 1988
4. J.W. Hancock, and A.C. MacKenzie, J. Mech. Phys. Solids, 1983, vol 31. pp 1-24.
5. Needleman, J. Appl. Mech., 54, 525-531, (1987).
6. G.R. Johnson, R.A. Stryk, T.J. Holmquist, S.R. Beissel, Wright Laboratory, Armament Directorate, Eglin Air Force Base Report, WL-TR-1997-7037
7. D.M. Goto, J.F. Bingert, S.R. Chen, G.T. Gray, and R.K. Garrett, Metall. Mater. Trans. A-Phys. Metall. Mater. Sci., v. 21, no. 8, p. 1985-1996 , (Aug. 2000).
8. P.J. Maudlin, J.F. Bingert, G.T. Gray III, Int. J. Plast, vol 19, pp 483-515, (2003).
9. K. Baumung, G.I. Kanel, S.V. Razorenov, D. Rusch, J. Singer, and A.V. Utkin, Journal de Physique IV (Colloque) ; Aug. 1997; vol.7, no.C3, p.927-32

CP706, *Shock Compression of Condensed Matter - 2003*
edited by M. D. Furnish, Y. M. Gupta, and J. W. Forbes
© 2004 American Institute of Physics 0-7354-0181-0/04/$22.00

PRECURSOR SUPPRESSION BY SHEAR STRESS RELAXATION IN U-Nb(6-WT%)

D. B. Hayes, G. T. Gray III, R. S. Hixson, A. K. Zurek, J. E. Vorthman, W. W. Anderson

Los Alamos National Laboratory, Los Alamos NM 87545

Abstract. U-Nb(6-wt%) exhibits plastic yield strength of a few-tenths of a GPa that can vary depending upon the starting microstructure and heat-treatment. However, when a several-mm thick specimen of U-Nb(6-wt%) is shock loaded in the range between 1.5 and 10 GPa, no elastic precursor is observed in interferometer measurements at the rear free surface. The absence of the elastic precursor and other features of the compression and release measurements are explained by assuming shear stress-relaxation rate is dependent upon the shear stress. The resulting stress waves are unsteady and broaden so that shear stress relaxation can occur in the front preventing the plastic yield point from being reached. U-Nb(6-wt%) is known to twin in quasi-static compression and shear-induced, rate-dependent twinning is likely the underlying cause for shear stress relaxation in our experiments. Recent experiments in which U-Nb(6-wt%) was heavily cold-rolled (work-hardening to ~25% strain) display no evidence of a precursor, admitting the possibility of a pressure or temperature induced stress-relaxation process.

EXPERIMENTS

During the past few years several experimenters have noted that an elastic precursor is not present during dynamic compression of the alloy U-Nb(6-wt%)[1,2]. This alloy does have a small shear modulus that would cause the elastic precursor to move at a velocity only slightly exceeding that of the plastic wave. However, an assessment of the experimental resolution based on wave speeds showed that an elastic precursor should be resolvable in most of these experiments – yet none was present.

Six experiments on planar U-Nb(6-wt%) samples 3-5-mm thick produced stresses from 1.6 - 10.7 GPa. Plate impact experiments were done on a 50-mm bore gas gun. Impact velocities were measured with electrical pins and free surface velocity histories were measured at the back surface of the alloy using a VISAR velocity interferometer. One experiment had a Z-cut sapphire window bonded to the VISAR measurement surface. The others were unconfined (free surface). VISAR results resolve motion to ±1 ns and are accurate to about ±1% of peak free surface velocity. Experiments are summarized in Table 1.

Table 1. Experimental Details. Experiments 1, 2 and 6 have not been previously reported.

#	Experiment	Flyer	(mm)	Target	(mm)	U_d (km/s)	σ_X (GPa)	VISAR plane
1	5600-05	Z-Quartz	2.994	U-Nb(6-wt%)	5.11	0.1279	1.58	Z-sapphire
2	5600-01	Z-Quartz	5.074	U-Nb(6-wt%)	3.809	0.108	1.35	free surface
3	5699-19	Z-Quartz	4.075	U-Nb(6-wt%)	2.55	0.223	2.84	free surface
4	5699-27	Z-Quartz	4.081	U-Nb(6-wt%)	2.52	0.328	4.24	free surface
5	5699-21	Z-Quartz	4.078	U-Nb(6-wt%)	2.53	0.419	5.49	free surface
6	5601-48	Z-Sapphire	5.102	U-Nb(6-wt%)	5.12	0.446	10.73	free surface

MATERIAL MODEL

Pressure-strain for U-Nb(6-wt%) was described by the common Mie-Grüneisen equation of state with $\rho_0 = 17.34$ g/cm^3, $C = 2.65$ km/s, $S = 1.6$, $\rho\gamma =$ constant $= 17.34 \times 1.7$ g/cm^3.

The constitutive behavior is that for a stress-relaxing material with constant plastic yield strength, $Y = 0.5$ GPa and Poisson's ratio $= 0.42$. The shear stress at fixed Lagrangian position follows the evolutionary equation:

$$\dot{\tau} = \mu\frac{\dot{\rho}}{\rho} - \frac{\tau - \tau_{crit}}{t_0} \quad , (1)$$

subject to the von-Mises yield condition $|\tau| < Y/2$. Shear modulus is denoted μ and τ_{crit} is the critical shear stress for the onset of shear stress relaxation. If the second term on the right hand side is not present, Eq. (1) is the classical elastic result and by imposing the maximum shear stress von-Mises criterion it becomes the description of an elastic/perfectly plastic material; the second term on the right-hand side provides for relaxation of positive shear stress. If $-\tau_{crit} < \tau < \tau_{crit}$ then the second term in the Eq. (1) is set zero. Also when shear stress is negative the minus signs in the second term must be changed in an obvious way.

SIMULATIONS OF THE EXPERIMENTS

All experiments were simulated by setting the parameters in Eq. (1): $\tau_{crit} = 0.02$ GPa and $t_0 = 62$ ns whenever $p < 2$ GPa, and $t_0 = 124$ ns above that pressure; these were determined by trial and error when simulating the leading ramp up, roll-over, flat top and release arrival of the experiments 3 through 6 with a finite difference wave-propagation code.[3]

DISCUSSION

We have analyzed six shock compression experiments and have determined that shear stress relaxation can prevent generation of sufficient shear stress to cause the material to plastically yield in a far-field bulk sense. This happens because the wave immediately begins to spread as it propagates into the sample and the second term on the right hand side of Eq. (1) reduces the shear stress rapidly enough so the plastic yield shear

stress is not reached. Furthermore, the constitutive relation is time dependent and compression waves are unsteady and continue to spread.

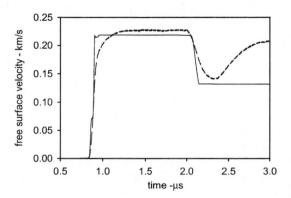

Figure 1. Experiment 5699-21 simulated as elastic, perfectly plastic. Note the elastic precursor is just barely discernable during the rise initial rise of the simulation and that the sharper rise and fall of velocity compare poorly with experiment.

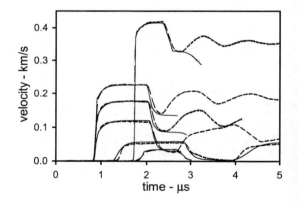

Figure 2. Simulations using the time-dependent shear stress relaxation model (solid) compared with experiments (dashed). Simulations are carried only slightly beyond the first pullback. Experiments "ring" at late times because of the compression wave trapped in the spall scab.

Interpreting the constitutive relation Eq.(1) in terms of the specific microstructural processes is complex at best. The most obvious explanation is that shear stress relaxation occurs from twinning[4] and is driven by excess shear stress. Field, et. al.[5] published a detailed study of the structural and metallurgical processes that occur when U-Nb(6-wt%) is deformed. The ambient metastable α'' alloy has a monoclinic structure. This material has several different deformation twin systems with different activation stresses. Because of the symmetry of the α'' phase, each twin has 12 variants.

Static *one-dimensional stress* measurements on U-Nb (6-wt%) show that the stress-strain curves exhibit two separate yield points as the compressive strain is increased. At small strains the material is elastic and remains so until the threshold for twin shear stress generation is reached. The shear stress remains relatively constant until most of the available twin sites have been exhausted at about 8% strain. When the maximum twinning has occurred, the stress again rises pseudo-elastically with increased strain until the von-Mises yield point is reached. At this and larger strains, the shear stress again remains relatively constant. In our finite difference simulations of compression in *one-dimensional strain*, maximum plastic strain was ~3%, far lower that the 8% that twinning in this material can accommodate. Therefore, maximum "plastic strain" from twinning is not probed in these experiments. Furthermore, threshold shear stress for twin generation is almost zero and setting it to zero has essentially no effect on the numerical simulations. We conclude that for the response of U-Nb (6-wt%) we could have set the threshold shear stress, τ_{crit}, to zero in our simulations with no adverse effect on the goodness-of fit and therefore the details of the effect of deformation twinning on the static shear stress are of no importance in interpreting the dynamic experiments. What is important, in fact dominant in these experiment, is the time-dependence of the twin formation and concomitant stress-relaxation.

However, there are some recent experimental observations that do not support the hypothesis that twin formation occurs solely from excess shear stress. A recent experiment (not shown) probed pre-strained material by extensive cold rolling in an attempt to pre-twin and or sufficiently work harden the substructure with dislocations enough to suppress the stress-relaxation twin process. However, when this material was shock loaded, again, no elastic precursor was observed. This result suggests the possibility that twinning and the corresponding shear stress relaxation is not caused by excess shear stress but some other environmental factor such as pressure. That would produce the following conundrum. Precursor suppression requires twin formation along directions of maximum shear stress. These are the very directions in which the pre-rolling presumably has already formed all possible twins and/or hardened the material via dislocation accumulation. Therefore if pressure is responsible, the twins formed must be on twin systems different than previously exercised by the pre-rolling.

There is evidence that pressure is not responsible for the stress relaxation we observe. The detailed structure of the release waves measured with the VISAR can only be replicated in simulations whenever significant stress relaxation of the opposite sign occurs upon initial release that occurs while the material is still at significant pressure, obviating the pressure-induced hypothesis. This negates the possibility that increased pressure or temperature is solely responsible for the shock observations and suggests that their influences may be negligible here.

The characteristic shear stress relaxation time is ultimately determined from several microstructural parameters: time for twins to form; relation between volume fraction of twins formed to reduction in shear stress; size of twinned regions, etc. However our intent here is to present experimental results and therefore results are expressed as observed relaxation time and not in terms of these detailed microstructural properties. These experimental results can ultimately be compared with some parts of various microstructural theories that exist or may be developed.

Finally it is important to note that if twinning were instantaneous upon application of sufficient shear stress, it is unlikely an elastic precursor would be seen. In *uniaxial stress*, $\sigma_T = 0$ and the pressure (mean stress) stays at ~Y/3. The bulk compression of the sample is therefore negligible, even for very large plastic strains. Under *uniaxial strain* in shock-compression experiments this is not the case. There $\sigma_L \sim \sigma_T$ and the generation of large plastic strain only occurs when the total elastic strain and hence the pressure rises. If eight-percent plastic strain corresponds to twin saturation in uniaxial stress then twin saturation will occur at 12% strain in uniaxialstrain, or about $\sigma_L > 23$GPa, a stress larger than achieved in any of our experiments. At this stress it is likely that any elastic precursor would be overdriven, a non-linear phenomenon that occurs from the tendency of the plastic wave speed to increase with stress until it exceeds the elastic precursor velocity.

SUMMARY

We have accounted for the absence of an elastic precursor in shock compression experiments by adding a shear stress relaxation term to classical elastic/perfectly plastic strength response. This modification to the constitutive response resolves all experimental data reported to date. Our conclusion is that the shear stress relaxation process is rapid enough to prevent U-Nb (6-wt%) from plastically yielding so that no elastic precursor is expected. The working hypotheses are that stress relaxation stems from twinning and that the rate of stress relaxation is dependent on the applied shear stress. However, it remains to be demonstrated that twinning is the cause and that other influences do not contribute to the lack of an HEL.

These shock-compression experiments reveal the time-dependence of twin formation but do are remarkably insensitive to details of the equilibrium dependence of shear stress on plastic strain at low strain rate.

REFERENCES

1. R. S. Hixson, J. E. Vorthman, A. K. Zurek, W. W. Anderson, and D. L. Tonks, 'Proceedings of the 1999 Conference on Shock Compression of Condensed Matter, M.D. Furnish, L. C. Chhabildas, and R. S. Hixson editors, (1999) p. 489.
2. A. K. Zurek, R. S. Hixson, W. W. Anderson, J. E. Vorthman, G. T. Gray III and D. L. Tonks, Journal de Physique IV, **10** (#9) p677-682
3. M.E. Kipp and R.J. Lawrence,Sandia National Laboratories Report SAND81-0930, June 1982
4. J. N. Johnson and R. W. Rohde, J. Appl. Phys., **42**, 4171 (1971)
5. R. D. Field, D. J. Thoma, P.S. Dunn, D. W. Brown and C. M. Cady, Phil. Mag. A 81#7, pp1691-1724 (2001)
6. R. M. Brannon, Sandia National Laboratories, (private communication).

CP706, *Shock Compression of Condensed Matter - 2003*
edited by M. D. Furnish, Y. M. Gupta, and J. W. Forbes
© 2004 American Institute of Physics 0-7354-0181-0/04/$22.00

U-0.75Ti AND Ti-6Al-4V IN PLANAR AND BALLISTIC IMPACT EXPERIMENTS

B. Herrmann[1], A. Venkert[1], V. Favorsky[2], D. Shvarts[1, 2], E. Zaretsky[2]

[1] *Nuclear Research Center - Negev, P.O.Box 9001, Beer-Sheva 84106, Israel.*
[2] *Mechanical Engineering Dept., Ben Gurion University, P.O. Box 653, Beer-Sheva 84105, Israel.*

Abstract. The response of U and Ti alloys has been studied in planar and ballistic impact experiments performed with a 25 mm light-gas gun. Free surface velocities were monitored by VISAR. The velocity profiles and the damage maps were simulated using 2D AUTODYN™ finite differences code. A modified Steinberg-Cochran-Guinan constitutive model was calibrated by simulating planar impact experiments. Bauschinger effect and a single-parameter spall model were added to describe the unloading and the tensile paths. The ballistic experiments were simulated by using the calibrated model. Softly recovered samples revealed different degrees of spall fracture (planar impact) and of adiabatic shear bands (ballistic experiments). The results demonstrate a possibility to combine experimental and numerical techniques, VISAR and AUTODYN, to calibrate constitutive models of solids in a wide range of shock-induced strain.

INTRODUCTION

Uranium and titanium alloys are susceptible to extensive deformation localization when subjected to high rate loading. It is attributed primarily to local thermal softening, produced by strain-induced heat generation, and to strain and strain-rate hardening[1]. Dislocations and dislocation-related structures participate in plastic flow of uranium[2] and titanium[3] alloys. Deformation by twinning can effectively strengthen or weaken the alloy, in contrast to the strengthening in dislocation-governed plastic flow[4]. Numerical simulations of reverse ballistic experiments of a tungsten heavy metal[5], performed with Steinberg-Cochran-Guinan (SCG) constitutive model[6], revealed a correlation between strain localization and strain hardening. Simulations of planar and reverse ballistic impact experiments of U-Ti[7] and titanium[8] have been performed previously. In the present work an additional type of loading, symmetrical ballistic impact, was performed. The

purpose here is to find a set of constitutive parameters, enabling the simulation of sample free surface (rear or lateral) velocity profiles and failure events in three types of impact experiments.

EXPERIMENTAL

A solution heat-treated, water-quenched and aged U-0.75Ti (U-Ti) and an annealed Ti-6Al-4V (Ti64) were impacted in a 25mm light-gas gun. The impact velocity (100-500m/sec) and the impactor-sample misalignment (< 0.4mrad) were defined by charged pins. The free surface velocity of the samples was monitored by VISAR[9] and the samples and impactors were softly recovered for further examination. The impactors and samples in the 1D planar impact (PI) experiments were disks of 24 mm diameter. The thickness of all impactors was 2mm while the U-Ti and the Ti64 samples were 6 and 5mm thick, respectively. The VISAR beam was focused on the center of the free surface of the sample. The 2D experiments were performed in two configurations employing cylindrical samples of 8mm diameter and 20mm length. In the symmetrical ballistic, SB, experiments the impactors were made of the same material and had the same geometry as the samples. In the reverse ballistic, RB, experiments the impactors were discs of 24mm diameter and 5mm thickness. The U-Ti samples were impacted with disks of the same material while tungsten heavy alloy disks were used in the case of the Ti64. The VISAR beam in both 2D configurations was focused on the sample's cylindrical surface, about 3mm from the impacted edge.

All types of the experiments were simulated by using an AUTODYN 2D commercial code. The 1D experiments, simulated with a 20μm mesh, were used to calibrate the AUTODYN library constitutive models of the studied materials by tuning strain-hardening parameters and adding Bauschinger effect[10] and a bulk failure model[11]. Modeling 1D experiments required minor modification of the AUTODYN library parameters of the studied materials. The constitutive parameters, obtained by calibrating the PI experiments, served in the simulations of the 2D experiments performed with a 100×100 μm mesh. For these simulations the heat generation caused by shear deformation, the thermo-mechanical coupling[1], was added to the AUTODYN calculations. Any modification of the constitutive

parameters required for reliable 2D simulations, was retested in 1D simulation and the values that fit both 1D and 2D experiments were chosen.

RESULTS AND DISCUSSION

The simulated velocity profiles and damage maps, calculated with a single set of hardening parameters for each material (Table 1), were compared with the experimental profiles and with the metallographic observations.

Table 1. Hardening parameters.

Alloy	Y_0 (GPa)	Y_{max} (GPa)	β	n
U-Ti	0.90	2.20	9,000	0.095
Ti64	1.45	1.67	1,200	0.080

Planar Impact Experiments

Bauschinger effect[10] was added to the calculations in order to improve the match between the simulated and experimental velocity profile (Fig. 1) in unloading. A bulk AUTODYN library failure model was used to describe the dynamic tension failure. Spall strength values of -2.25, -2.9 and -4.1GPa for 131, 272 and 500m/sec impact velocities in U-Ti and of -3.7 and -4.0GPa for 420 and 538m/sec impact velocities in Ti64 were

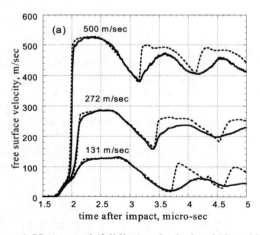 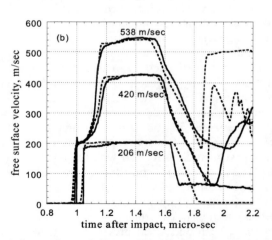

Figure 1. PI measured (full line) and calculated (dotted line) free surface velocity profiles of (a) U-Ti and (b) Ti64. Impact Velocities are shown in m/sec.

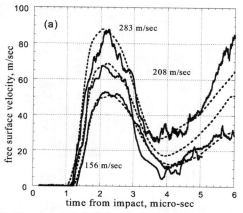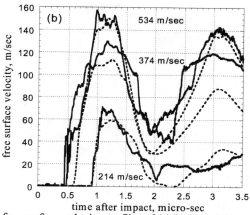

Figure 2. SB measured (full line) and calculated (dotted line) free surface velocity profiles of (a) U-Ti and (b) Ti64. Impact Velocities are shown in m/sec.

chosen to fit the experimental pull-back velocities.

The effective strain in the first compressive phase of PI experiments is less than 0.08, depending on the material and impact velocity. Metallographic observation show that the spall in U-Ti and in Ti64 occurred by micro-cracks, parallel to the impact plane, that are joined by normal shearing.

Ballistic Impact Experiments

An AUTODYN built-in erosion model described the failure with an onset criterion of 0.6 equivalent plastic strain (Fig. 2 and 3). Samples impacted in high velocity experiments were damaged along a conical surface (Fig. 4). The dominant deformation mode is shearing and failure occurred by creating the separated (crack-like) surfaces within the adiabatic shear bands. The Ti64 failed along a single surface while in the case of U-Ti a damage zone, about 1mm wide, containing a net of shear bands, has been observed. The simulated damage maps (Fig. 4) show states of the failure, corresponding to the instant about 5μsec after the impact events. A single path for Ti64 and several narrow damage paths for U-Ti reproduce the

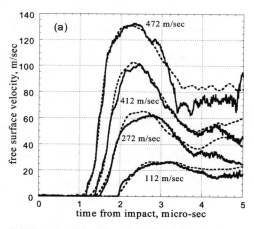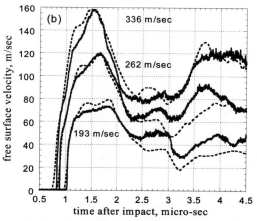

Figure 3. RB measured (full line) and calculated (dotted line) free surface velocity profiles of (a) U-Ti and (b) Ti64. Impact Velocities are shown in m/sec.

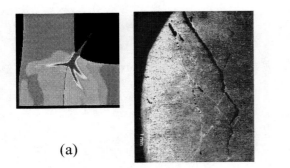
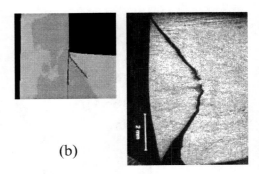

(a)   (b)

Figure 4. Metallograhic cross-sections and simulated damage maps of (a) U-Ti and (b) Ti64 RB samples.

metallography features. The distribution of the equivalent strain in the damaged zone, developed in the samples, was analyzed. The effective strain reached in the vicinity of the shear band in the U-Ti and Ti64 samples, 6μsec after SB and RB impacts, was found to be 0.16 and 0.35, respectively. These values were calculated for points located at a distance of two numerical cells from the erosion boundary. Obviously, the deformations taking place inside the eroded strip are much higher. In the case of SB experiments the absence of restricting the radial material flow results in a failure started by random geometry perturbation and propagating along the corresponding random plane, inclined to the sample axis. The microstructure features of the shear band were found to be similar for both SB and RB experiments, although the latter are characterized by conical failure surfaces. The simulation of the U-Ti experiment also revealed plugging, as observed in recovered impactors from RB experiments.

CONCLUSIONS

A numerical/experimental method for calibrating a material model was demonstrated for two alloys. Introducing minor modifications and employing a single set of hardening parameters, for each material, resulted in satisfactory simulations of PI, SB and RB experiments over a wide range of impact velocities.

The simulations reproduced the recorded VISAR signal as well as the main features of the damage,

for both materials. The shear failure modes in both materials were found to be comparable, exhibiting characteristic adiabatic shear banding with more profound localization in the Ti64 alloy.

Although the damage/failure models used in the present study reproduced strain localization, a more comprehensive constitutive description of the damage/failure process is still required.

REFERENCES

1. Estrin Y., Molinari A., Mercier S.; *J. Eng. Mater. Tech.* **119**, 322-331 (1997).
2. Armstrong R.E., Follansbee P. S., Zocco T, in: IPCS-102; ed. Harding J., 1989, pp. 237-244.
3. Timothy S. P., Hutchings I. M., *Acta Metall.* **33(4)**, 667-676 (1985).
4. Yoo M. H., *Metall. Trans.* **12A**, 409-418 (1981).
5. Zaretsky E., Levi-Hevroni D., Ofer D., Shvarts D., in: SCCM-1999; ed. Furnish M. D., Chhabildas L. C., Hixson R. S., 2000, pp. 593-596.
6. Steinberg D.J., Cochran S.J. and Guinan M.W. *J. Appl. Phys.* **51(3)**, 1498-1504 (1980).
7. Herrmann B., Landau A., Shvarts D., Favorsky V., Zaretsky E., in: SCCM-2001; ed. Furnish M. D., Thadhani N. N., Horie Y., 2001, pp. 1306-1309.
8. Herrmann B., Venkert A., Kimmel G., Landau A., Shvarts D., Zaretsky E.,; in: SCCM-2001, ed. Furnish M. D., Thadhani N. N., Horie Y., 2001, pp. 623-626.
9. Cochran S. G., Guinan M. W., UCID-17105; 1976.
10. AUTODYN, Theory Manual. Rev. 4.0, 1998.
11. Barker L.M., Hollenbach R.E., *J. Appl. Phys.* **43(11)**, 4669-4675 (1972).

CP706, *Shock Compression of Condensed Matter - 2003*
edited by M. D. Furnish, Y. M. Gupta, and J. W. Forbes
© 2004 American Institute of Physics 0-7354-0181-0/04/$22.00

DYNAMIC CHARACTERIZATION OF SHAPE MEMORY TITANIUM ALLOYS

V. S. Joshi* and M. A. Imam**

*Code 920L, Research and Technology Department, Naval Surface Warfare Center, Indian Head Division, Indian Head, Maryland 20640
**Code 6320, Materials Science and Technology Division, Naval Research Laboratory, Washington, D.C. 20375

Abstract: Evaluation of high strain rate behavior of materials at pre-fracture strains is very important where the materials are considered for ballistic applications. High compression strain rate response of shape memory titanium alloy including a typical titanium alloy are determined using the split Hopkinson pressure bar (SHPB). The conventional SHPB technique has been routinely used for measuring high strain rate properties of high strength materials. A split Hopkinson bar consisting of 10-mm diameter Maraging 350 alloy incident, transmitter, and striker bars was used to determine the compressive response of these alloys. Attempts are underway to use this technique to extract useful information required to design a material for improving its impact resistance. Initial test results performed on these different titanium alloys show an interesting trend with change of composition. Attempts were made to compare the stress-strain data of these alloys with the published data for titanium alloys. Stress-strain data and changes resulting in the microstructure from strain rates in the regime 1800-4000/s are presented.

INTRODUCTION

Titanium and its alloys are becoming materials of choice over conventional steels due to their strength –to-weight ratios. Several new titanium alloys have been a topic of considerable interest for investigators in aerospace industry and government research laboratories. These alloys offer higher strain capabilities without significantly sacrificing other mechanical property, especially tensile and yield. Some of these alloys also offer increased resistance to deformation under dynamic loading, which could make them suitable for applications demanding impact resistance.

Stress-strain data for these materials at low as well as high strain rates are very crucial for generating improved material models required to determine application of the material. While standard titanium wrought products have been studied at length, new alloys with specific heat treatments must be evaluated for their performance and compared to the existing standard grades of wrought titanium.

One such readily available new material is the nickel based titanium alloy, known as shape memory alloy. This alloy has already found numerous applications in semiconductor industry as thin film actuator, by making use of its low temperature phase transformation behavior. This property can also be applied to obtain a limited increase in resistance to deformation under dynamic loading. This mechanism is in sharp contrast to classical route of developing slip systems to alter mechanical property of material under specific conditions. In order to achieve increase in strain hardening, this material is manufactured and heat- treated for evaluating mechanical properties.

The objective of this paper is to explore the behavior of shape memory alloy at strain rates in the range of 1800-4000/s and compare the stress-strain data with a conventional titanium alloy.

EXPERIMENTAL CONFIGURATION

The Split Hopkinson Pressure Bar (SHPB) system used in this work consists of a 10-mm diameter steel pressure bars made of hardened M350 maraging steel capable of handling high loads generated by sample under high strain and strain rates. Mechanical impedance of maraging steel bar is suitable for testing the titanium alloy under investigation. The arrangement is shown schematically in Figure 1.

The length of the incident, transmitter and the striker bars were 1.22-m, 1.22-m, and 0.30-m, respectively. The bars are aligned in specially configured bar supports to achieve impact planarity. Pairs of 350-Ω strain gages were mounted on the incident and transmitter pressure bars 0.61-m away from the specimen/bar interfaces. The striker bar is launched at velocities of 15 to 45 m/s using a compressed helium gas gun, equipped with a fast valve.

Striker bar impact on the incident bar produces a compressive stress pulse that propagates along the length of the incident bar. The magnitude of the stress pulse, σ, is given by

$$\sigma = \rho\ C_0\ V_s$$

where ρ (7800 kg/m^3) is the density, C_o (4850 m/s) is the speed of sound in bar material (M350 steel), and V_s (m/s) is the striker bar velocity. The stress pulse width for a 0.30-m long steel striker bar is 120 μs. This compressive pulse in the incident bar subjects the specimen to a compressive load. A fraction of the incident compressive pulse, ε_i, is transmitted through the specimen ε_t and the remainder is reflected back in the incident bar ε_r. The amplitude of the incident, reflected, and transmitted pulses are recorded by the strain gages mounted on the pressure bars. Using the recorded strains ($\varepsilon_i, \varepsilon_r, \varepsilon_t$) in the bars, the following equations are applied to find the time dependent stress (σ), strain (ε) and strain rate ($\dot{\varepsilon}$) in the specimen:

$$\sigma(t) = E\ \frac{A_b}{A_s}\ \varepsilon_t(t) \qquad (1)$$

$$\dot{\varepsilon}(t) = \frac{2 \cdot C_o}{L}\ \varepsilon_r(t) \qquad (2)$$

$$\varepsilon(t) = \frac{2 \cdot C_o}{L} \int_0^t \varepsilon_r(t)\,dt \qquad (3)$$

where A_b and A_s are the cross-sectional area of the pressure bar and the specimen in the gauge section, respectively. L represents the gauge length of the specimen. The values of stress and strain are instantaneous, but the strain rate is given as an average value over a specified time. These values are determined by assuming a uniform stress-state condition [1].

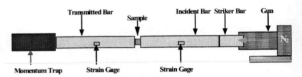

FIGURE 1. Schematic of the SHPB configuration.

RESULTS AND DISCUSSION

Compression tests were performed on two materials: (1) Ti-15-3 alloy, and (2) Ti-Ni shape memory alloy. The Ti-15-3 was furnished from Timet Corporation in the form of 38 mm (1,5 in) thick hot rolled plates. The chemical composition of the alloy in weight percent was V=14.79, Al=2.91, Cr=3.06, Sn=3.03, Fe=0.128, O=0.122 and N=0.009. The as-received material was annealed at 550°C for 5 hours after machining to samples. This material (Ti-15-3) like titanium alloy Ti-6Al-4V, has been extensively studied. Ti-Ni alloy was obtained from Metaltex International Corp. in the form of 12.7mm (0.5in) thick plate which went through forging and cross rolling at 900°C from 200mm diameter castings. It contained less than 500 PPM of oxygen and very low carbon with Ti being 50%. The transformation temperature was determined to be ~90°C. After machining, the alloy was annealed at 550°C for 5 hours.

Measured flow stress on 3.5-mm thick and 3.5-mm diameter Ti-15-3 specimens, determined from plots were about 1450 MPa. Experiments were targeted at two strain rates of 1800/s and 4,000/s. Velocity of striker determined the actual strain rate.

Stress values at strains of 10% are similar to the published values for Ti-6Al-4V, reported by Weerasuraiya et al [2]. Such good agreement convinced us that our SHPB configuration and stress-strain data analysis are reliable. Stress-strain data on Ti-15-3 at strain rates of 4,100/s and 2,050/s are shown in Figure 2 and 3 respectively.

the samples could not be loaded up to failure conditions, and therefore limited strain values are available. Experiments using longer striker bar is required to obtain actual failure strain capability of the material.

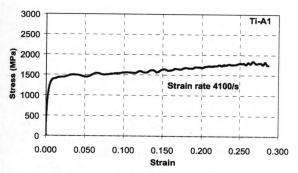

FIGURE 2. Stress-strain curve for Ti-15-3

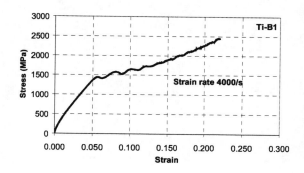

FIGURE 4. Stress-strain curves for Ti-Ni.

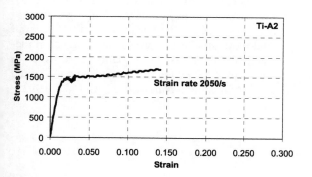

FIGURE 3. Stress-strain curve for Ti-15-3

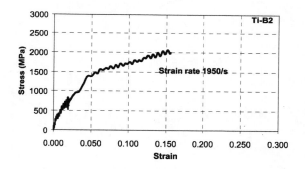

FIGURE 5. Stress-strain curves for Ti-Ni.

Microstructural Analysis

Test specimens of both types of alloys Ti-15-3 and Ti-Ni measured 3.5 -mm in diameter and 3.5 - mm in thickness. Stress-strain data on Ti-Ni at a strain rate of 1,950/s and 4,000/s are shown in Figure 4 and 5 respectively.

Magnitudes of compressive stresses of Ti-Ni alloy at strains of 10% are similar to Ti-15-3, but at 20% are significantly higher, even compared to those reported for Ti-6Al-4V, using similar bar and sample sizes.

Both samples were able to withstand 20 % deformation. Since the strike bar was only 0.3 m long,

Microstructures of samples were taken before the experiment after polishing the samples (manual) on 400 and 600 grit paper, followed by 0.1-micron alumina on a wet cloth. The samples were etched by a solution consisting of 3.2% HF, 14.1 % HNO3, and 82.7 % water. The microstructures before the test were compared to the ones after to determine the type of changes during deformations and look for any significant alterations in the materials, based on earlier work [3]. Both these alloys show significant deformation, but transformations are not readily detectable. Further analysis of the sample is in progress. Typical microstructures of two titanium alloys are shown below.

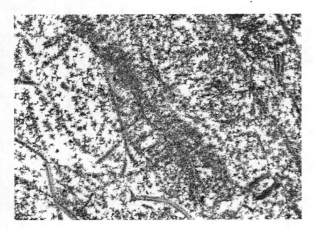

FIGURE 6. Microstructure of Ti-15-3 alloy showing fine precipitate due to aging.

FIGURE 7. Microstructure of Ti-Ni Alloy showing large equiaxed grains with fine precipitates.

CONCLUSIONS

Compression stress-strain curves at strain rates of ~2000-4000/s for two types of titanium alloys are generated using SHPB technique. Preliminary high strain rate data on these samples are presented. These results indicate effect of strain rate sensitivity of material due to composition changes and phase change mechanisms operating during dynamic deformation. More experiments are currently being pursued to evaluate the finer levels of microstructural changes and their influence on material flow under high strain rates.

ACKNOWLEDGEMENTS

This work was conducted under collaborative research between NSWC, Indian Head, MD, and Naval Research Laboratories, Washington D.C. for development of materials. The authors wish to acknowledge useful discussions with Dr. N. S. Brar of UDRI, Dayton.

REFERENCES

1. Gray, G. T., "Classical split Hopkinson bar techniques" in ASM Handbook, 10th edition, 8, pp. 462-476, 2000.
2. Weerasooriya T., Magness L., and Burkins M., " High strain-rate behavior of two Ti-6Al-4V alloys with different microstructures," in Fundamental Issues and Applications of Shock-Wave and High Strain Rate Phenomena, K.P. Staudhammer, L.E. Murr and M.A. Meyers eds., Elsavier Science Ltd., 2001, pp 33-37.
3. M.A. Imam, P.K. Poulose, B.B. Rath, "Effect of cold work and heat treatment in a region on mechanical property of Ti 15-3 alloy", published in the Proc. of Ti'92, ed. F. H. Froes and I. Caplan, TMS Publication, Warrendale PA, p 177. 1993

CP706, *Shock Compression of Condensed Matter - 2003*
edited by M. D. Furnish, Y. M. Gupta, and J. W. Forbes
© 2004 American Institute of Physics 0-7354-0181-0/04/$22.00

EFFECT OF PROCESSING INDUCED MICROSTRUCTURAL BIAS OF PHASE DISTRIBUTION ON SPALL STRENGTH OF TWO-PHASE TIB₂-AL₂O₃ CERAMICS

G. Kennedy[1], L. Ferranti[2], N.N. Thadhani[2]

[1]*Earth and Planetary Science, Harvard University, Cambridge, MA 02138*
[2]*Materials Science and Engineering, Georgia Institute of Technology, Atlanta, GA, 30332-0245*

Abstract. The influence of microstructural bias on the spall strength of two-phase $TiB_2+Al_2O_3$ ceramics was investigated in this study. The microstructural bias generated by varying the extent of ball-milling time prior to hot pressing of combustion synthesized or mechanically-mixed pre-alloyed powders, includes differences in connectivity of TiB_2 around Al_2O_3 and grain size of respective phases. Tensile spall properties were measured using plate-impact gas-gun experiments, with velocity interferometry. The results of these tests correlated with quantitative microscopy analysis of the microstructural bias illustrate that increase in connectivity of TiB_2 around Al_2O_3 lowers the spall strength due to tensile residual stresses developing during processing of the two-phase ceramics.

INTRODUCTION

Dynamic behavior studies have typically focused on single-phase ceramics, although those with glassy (impurity) phases, e.g., AD85 Al_2O_3, have also been investigated. In such ceramics, the intergranular oxide glass significantly lowers the tensile (spall) strength, and therefore, the fracture resistance [1]. Dynamic behavior of ductile metal-matrix composites, e.g., Al-alloys consisting of embedded ceramic particles, has [2] shown that both, the dynamic yield and tensile spall strengths are reduced, in comparison to strength increases in particle-reinforced composites under quasi-static loading. The composite acts as a mechanical energy trap due to scattering of waves from incoherent boundaries and interfaces between the matrix and reinforcement phases of dissimilar shock impedance.

The high-strain-rate mechanical behavior of $Al_2O_3+TiB_2$ (70/30 mass ratio) two-phase ceramic shows different strain-rate dependence with 80% increase in compressive strength with increasing strain rate (3.5 GPa at 10^{-4} s^{-1} to 5.8 GPa at 10^{3} s^{-1}) [3-7]. The $Al_2O_3+TiB_2$ ceramics also show better penetration resistance than the monolithic constituents [8]. The objective of the present work was to investigate the dependence of spall strength on microstructural bias generated vial ball-milling hot-pressed and combustion synthesized (SHS) or mechanically-mixed (MM) precursors of two-phase $TiB_2+Al_2O_3$ ceramics. The microstructures involved either a continuous (interconnected) TiB_2 network surrounding Al_2O_3, or where TiB_2 and Al_2O_3 are interdispersed and uniformly inter-twined [3,9].

EXPERIMENTAL PROCEDURE

Details of approaches used for fabricating these two-phase $TiB_2+Al_2O_3$ ceramics are described elsewhere [10]. The impact experiments used an 80-mm diameter, single-stage gas gun. Fig. 1 shows the experimental setup, illustrating the target sample (without any backer) being impacted by a projectile consisting of flyer of same material mounted on an aluminum sabot with an air gap to allow full unloading required for the spall strength measurements. All sample surfaces were lapped for flatness and parallelism. The ceramic targets were polished with 5-μm diamond paste to ensure reflectivity required by the VISAR beam.

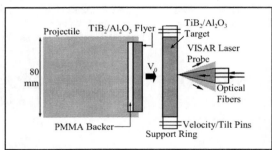

Figure 1. Setup used for measurements of free-surface velocity traces with VISAR interferometry.

RESULTS AND DISCUSSIONS

Optical micrographs of the various microstructurally-biased samples studied, are shown in Figure 2 (a-g).

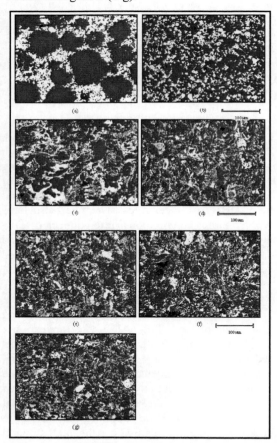

Figure 2. Optical micrographs of samples ball-milled for various times: (a) MM-0, (b) MM-2, (c) SHS-4, (d) SHS-8, (e) SHS-15, (f) SHS-23, (g) SHS-30

It can be seen that the MM samples show two extremes with (a) bright-contrast continuous TiB_2 phase surrounding Al2O3 and (b) the two phases being interdispersed. The SHS samples (c) to (g) show varying degree of TiB_2 phase connectivity as a function of ball-milling time.

Tensile spall strength experiments were performed on each of these samples at impact stress less than the Hugoniot Elastic Limit (~4 GPa) [10]. The output from the VISAR interferometer was captured on the oscilloscopes as interference fringes from the photomultiplier tubes. The raw data was combined in the VISAR data reduction program to provide free surface velocity versus time profiles. Figures 2 shows free surface velocity traces for each sample tested, illustrating a distinct "Spall signal" providing a clear measurement of ΔU_{fs}, the difference of maximum velocity before unloading and minimum velocity before increase in velocity associated with loading in spalled material. Table-I below lists the measured free surface velocities and spall strengths for each of the ceramic samples tested.

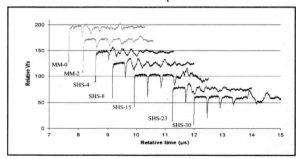

Figure 3. Spall data traces from all experiments. Time and velocity of each are shifted for ease of presentation.

Table I: Measured parameters of spall experiments

Sample (Ball mill time)	Peak Free Surface Velocity	ΔU_{fs} (m/s)	Spall Strength (MPa)	Bulk Wave Speed (m/s)	Density (g/cm^2)
MM-0	265.8 m/s	17	260	7.54	4.068
MM-2	266.3	20.2	309	7.51	4.096
SHS-4	258.8	18	276	7.65	4.037
SHS-8	273.7	28.1	427	7.59	4.015
SHS-15	270.8	40.2	610	7.58	4.024
SHS-23	267	29.5	447	7.56	4.015
SHS-30	264.8	41.5	631	7.60	4.007

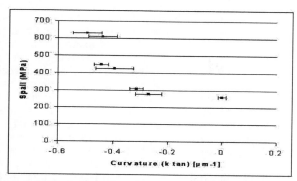

Figure 4. Variation of spall strength versus curvature representing connectivity of TiB_2

It can be seen from Table I that the spall strength of the hot-pressed samples increases with increasing time of prior ball-milling, due to its influence on final microstructure. Using a parameter, "average integral curvature", calculated from measured microstructural features, a measure of phase connectivity was determined [9]. The average "phase size" was also likewise measured. Samples with visually identified, connectivity of TiB_2 were measured to have less negative (approaching zero) values of curvature, and correspondingly lower spall strength, as shown in Figure 4.

Increasing ball-milling time also reduces the size of both the TiB_2 and the Al_2O_3 phases. The effect of decreasing phase size (with increasing ball mill time) on the Spall strength is shown in Figure 5, which reveals a clear trend illustrating drastic decrease in spall strength with increase in size of both TiB_2 and Al_2O_3 phases.

Although the variation in spall strength with phase size is clearly revealed in the plots in Figures 5, it is uncertain what would be the reason for such drastic reduction in strength with only minor increases in average phase size. Examination of the microstructures and the conditions responsible for their creation implies that the average phase size is perhaps not a good determinator of the spall strength. Larger average particle sizes exist in samples processed with shorter ball milling times. These shorter ball milling times have higher (less negative) curvature values indicating more connected TiB_2 regions with longer length dimensions than accounted for in the average particle size measurement. Hence, while the correlation of strength with phase size is shown to follow a certain trend, the change by a factor of two in Spall strength with an increase in particle size of 15 percent is too large. It is thus possible

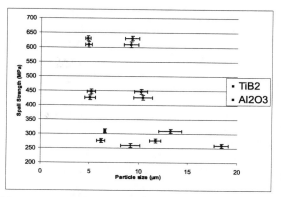

Figure 5. Variation of spall strength versus size of TiB_2 and Al_2O_3 phases

that some other property may be causing the Spall strength to change so drastically.

EFFECT OF RESIDUAL STRESSES

In the preparation of $TiB_2+Al_2O_3$ samples by SHS processing, it was noted that the highly interconnected TiB_2 structure arises from the physical geometry of the chemical reaction that takes place during the SHS reaction. Ball milling breaks down the TiB_2 phase connectivity of the SHS material. The curvature values determined from microstructures of samples ball milled for longer times clearly show this trend of decrease in connectivity with increasing ball-milling time. The break up of TiB_2 phase connectivity can be explained due to a residual stress phenomenon. During the high temperature SHS reaction the material reacts and forms the interconnected structure with the TiB_2 phase surrounding the Al_2O_3 regions [3]. The zero stress point for the reacted morphology is near the temperature of reaction, ~3000°C. As the reacted products cool, the TiB_2 and Al_2O_3 particles contract differentially due to their different thermal expansion coefficients. This causes residual stresses to arise at the interfaces and within the particles. Hence, the TiB_2/Al_2O_3 interfaces are highly stressed regions serving as stress concentrators that cause break-up of TiB_2 from Al_2O_3 during subsequent ball milling.

Residual stresses should also arise from the hot pressing process. The zero stress point will be below the hot pressing temperature of 1600°C, with residual stress increasing during cooling of the hot-pressed compacts. Thermally induced residual stresses were calculated to determine the magnitude and state of the residual stress. The residual stresses in individual phase can be calculated from the experimentally determined composite bulk modulus and the theoretically calculated coefficient of thermal expansion (CTE) of the composite given by:

$$\alpha_r = \frac{\alpha_1 K_1 F_1 / \rho_1 + \alpha_2 K_2 F_2 / \rho_2}{K_1 F_1 / \rho_1 + K_2 F_2 / \rho_2}$$

Where, α_1, K_1, F_1 and ρ_1 are CTE, bulk modulus, volume fraction, and density of one phase and α_2, K_2, F_2 and ρ_2 follow for the other phase. After α_r is obtained, the residual stress in an individual component (σ_i) can be estimated using:

$$\sigma_i = K(\alpha_r - \alpha_i)\Delta T$$

Where, K is the bulk modulus, α_i is the CTE of the particular phase of interest, and ΔT is the temperature change from the zero stress point. The residual stress in the TiB_2 was thus, calculated to be ~260 MPa in tension, while the Al_2O_3 was determined to be ~100MPa in compression. The stress is introduced due to the differences in thermal expansion coefficients of TiB_2 and Al_2O_3. Upon cooling from the hot pressing temperatures, the TiB_2 phase contracts more than the Al_2O_3. For the calculation, $1000^{\circ}C$ was assumed as the temperature range for the effect of thermal stress. Hence, cooling through this temperature range induces thermal residual stresses, such that the TiB_2 phase has a residual tension (~260 MPa) while the Al_2O_3 phase is under compression (~100MPa). The interface between the two phases is therefore a highly stressed region that can serve as a crack nucleation site upon further load application.

The above method provides an order of magnitude approximation for global residual stresses based on the assumption of spherical particles in a matrix. No effect of actual microstructural properties, curvature or phase size, is taken into consideration.

The elongated morphology of the TiB_2 connected network seen within samples of more less negative curvature (with residual tension) provides interconnected pathways for crack generation and propagation that can lead to lowering of the Spall strength. In contrast, samples with discrete, or less connected, TiB_2 phases, have higher Spall strength because the cracks must propagate through more of the alumina phase. The unconnected structures are quantified with values of more negative curvature. This alumina phase is also under residual compression stress, thereby increasing the stress necessary to reach the inherent Spall strength. It should be noted that the method employed for determining the residual stresses, assumes a homogeneous distribution of spherical particles, providing an order of magnitude value of the residual stresses. It does not account for the inhomogeneous distribution of TiB_2 networks present in the samples with high connectivity. Residual stress in local areas could be higher or slightly lower than those predicted. If the residual stresses in local areas rise above the material strength, cracks will form and thus be present in the material following processing. Presence of cracks would also serve to lower the Spall strength of the material quite significantly. Not only will the connectivity of the TiB_2 phase affect residual stresses, but phase size also will play a role. Larger phase sizes lead to larger residual stress values. The MM samples have a larger phase size and therefore have larger residual stresses. Hence, the Spall strengths of the two MM materials are lower than all but the most connected SHS material made with 4 hours of ball milling time.

ACKNOWLEDGEMENTS

Funding provided by U.S. Army Research Office, under Grant DAAG55-98-1-0454 (Dr. David Stepp program monitor). The authors thank Dr. K.V. Logan for providing the samples and valuable comments.

REFERENCES

1. S. Sundaram and R.J. Clifton, Mechanics of Materials, Vol. 29, (1997), pp. 233-251.
2. J.N. Johnson, R.S. Hixon, and G.T. Gray, J. Appl. Phys., 76 (10) (1994) 5706-5718.
3. K.V. Logan, U.S. Patent, 5,141,900, Aug 25,'92.
4. A. Keller, "An Experimental Analysis of the Dynamic Failure Resistance of TiB_2/Al_2O_3 Composites," Georgia Tech M.S. Thesis, 2000.
5. Andrew Keller, Greg Kennedy, Louis Ferranti, Min Zhou, and Naresh Thadhani, "Correlation of Dynamic Behavior With Microstructural-Bias In $TiB_2 + Al_2O_3$ Ceramics," in Proc. of Fourth Int. Symp. on Impact Engineering, Japan, July 2001.
6. G. Gilde, J.W. Adams, M. Burkins, M. Motyka, P.J. Patel, E. Chin M. Sutaria, M. Rigali, and L. Prokurat Franks, in Proc. of 25th Acers Cocoa Beach Conference, January 23-28, 2001.
7. J. Zhai and M. Zhou, Int. J. of Fracture, special issue on *Failure Mode Transition in Solids*, R. C. Batra, Y. D. S. Rajapakse, and A. J. Rosakis, eds., **101**, pp. 161-180, 2000.
8. K.V. Logan, Ph.D. dissertation, Georgia Institute of Technology, 1993.
9. Louis Ferranti, M.S. thesis, Georgia Institute of Technology, 2001.
10. Greg Kennedy, M.S. thesis, Georgia Institute of Technology, 2003.

CP706, *Shock Compression of Condensed Matter - 2003*
edited by M. D. Furnish, Y. M. Gupta, and J. W. Forbes
© 2004 American Institute of Physics 0-7354-0181-0/04/$22.00

METALLURGICAL INVESTIGATION OF DYNAMIC DAMAGE IN TANTALUM

Fabrice Llorca, Gilles Roy

CEA/Valduc, DRMN, 21120 Is/Tille, France.

Abstract. In this paper, we investigate the question of dynamic damage in tantalum. A wide range of shock pressure and pulse duration loadings is investigated. The aim of this work is to propose physical statements for the evaluation of multiscale damage models. From an experimental point of view, classical plate impact tests are performed to generate damage and fracture in tantalum. The recovery of shocked samples allows to make a complete investigation of the development of damage mechanisms. We investigate more particularly the problems of nucleation and growth of microvoids. The use of various apparatus (optical microscopy, SEM ...) give us the opportunity for understanding and quantifying the evolution of the microstructure. We discuss the three main stages of damage and fracture, giving at each time a physical understanding of the involved mechanisms and quantified data as well.

INTRODUCTION

Dynamic damage and fracture phenomena in metals have been studied for a long time. Because of the actual increase of hardware and software capabilities, there is a great need for developing physically based models. The building of these new multiscale approaches requires the qualification and the quantification of all damage sequences. It appears that a new orientation for Material Science investigations has to be developed in order to bring a logical and physical scheme of the damage events generated during shock wave experiments. In this work, we investigate the problem of the characterization of damage of tantalum under shock wave loading. The aim is to understand the damage mechanisms using observation devices and quantification tools in order to propose validated hypotheses which will be used for the elaboration of new multiscale models.

MATERIAL

The material used in this investigation is a high purity tantalum grade provided by Cabot Corporation. Chemical composition is listed in table 1. The elaborating process furnishes an equiaxed grain structure with a 90 μm average grain size.

Table 1. Chemical composition (in weight ppm)

H	O	C	N	Nb	Fe	W	Si
<5	15	15	<10	70	<5	30	20

SHOCK WAVE EXPERIMENTS

Plate impact tests were performed using a 80 mm single stage gas gun. Back free surface velocity of targets was measured using Doppler Laser Interferometry and recovery of a number of specimens has also been achieved. The range of loading is 5 GPa $< \sigma_{shock} <$ 30 GPa, 0.5 μs $< \Delta t_{shock}$ $<$ 1.5 μs. Free surface velocity measurement has not been realized above 15 GPa. The influences of shock pressure and pulse duration on free surface velocity profiles are reported in figures 1 and 2. These velocity profiles are very similar to those reported in [1,2]. The interruption of the release signal due to the creation of the spall plane is observed for all experiments even if fracture is not completely achieved. We concluded in a previous paper [3] that pullback signal was not systematically the result of fracture but in some cases the effect of an advanced coalescence stage.

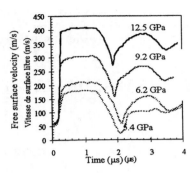

Figure 1. Effect of shock pressure levels on free surface velocity at given pulse duration (1.4 µs).

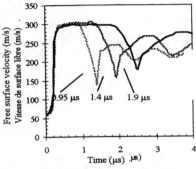

Figure 2. Influence of pulse duration on free surface velocity at given shock pressure (9.2 GPa).

Spall stress values are comprised between 4.5 GPa and 6 GPa and critical time at fracture between 500 ns and 900 ns for a [4 GPa.µs^{-1} - 12 GPa.µs^{-1}] stress rate range. Moreover, a sudden deceleration occurs in the pullback signal. This phenomenon is not always recorded during spall tests on metals and has already been observed in tantalum [2]. We agree with the conclusions of Zurek et al. [2] who have underlined the presence of "a restoring force" which "decelerates the spalled piece". The origin of this particular event is related by the previous authors to a two stage sequence of fracture: the partial coalescence first creates cracks parallel to the spall plane, then further coalescence develops which undergoes additional linking of these cracks to form the separated spall plane. Plastic instabilities (shear bands) are clearly observed between the basic coalescence in experimental results reported by Zurek et al [1]. This particular point is investigated in next paragraphs.

GENERAL DESCRIPTION OF DAMAGED AND FRACTURED SAMPLES

A clear understanding of the physical mechanisms involved in spall damage and fracture can be obtained from optical metallography analyses on etched sections of soft recovered specimens. Figure 3 reports some of the typical damaged and fractured sections at a given magnification level.

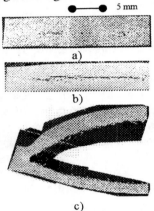

Figure 3. Typical recovered samples shocked at a) 5.2 GPa - b) 6.2 GPa - c) 9.2 GPa with 1.4 µs pulse duration.

The two halves of the specimens are not separated in tests a) and b): damage is called incipient. In test c), the two surfaces are completely separated: spall is complete. Each specimen exhibits ductile mechanisms of damage, mainly based on nucleation and growth of isolated microvoids. Nucleation seems to be random on both sides of the theoretical spall plane location (see figure 3.a) while isolated growth of quasi-spherical cavities is mainly observed (figures 3a and 3b). The cross sections show narrow damage bands 1 mm wide and normal to the shock wave propagation axis. The localized matrix relaxation attached to the primary stage of coalescence interacts with the propagation of complex compression and release waves. So, the lack of coalescence during primary stage leads to the formation of secondary planes (see figure 4). The link of the different blocks of coalesced microvoids is achieved if new cavities are activated (on the major and secondary coalescence planes): this is typical of a direct impingement mechanism (see next to last paragraph). This could be one explanation of some restoring force effect observed on free surface velocity profiles.

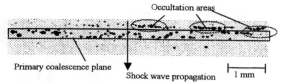

Figure 4. Primary and secondary coalescence planes.

NUCLEATION AND GROWTH

In very high purity bcc polycrystalline metals, the lack of inclusions and second phase particles leads to strong difficulties to track the origin of microvoid nucleation. Secondary Ions Mass Spectrometry technique (SIMS) has been applied to investigate element concentrations [4]: carbon and oxygen contents are uniformly spread over the microstructure. Looking for nucleation sites, SEM was used for the investigation of microcavity bottoms. No evidence was found of particle presence but some interesting "wrinkles" were observed (see figure 5). In our opinion, this residual aspect can be related to the origin of some intergranular nucleation at triple points. In the case of the high purity grade of tantalum, nucleation is classified as homogeneous [4] in the sense that it is activated on submicroscopic heterogeneities (such as low-angle grain boundaries...).

Figure 5. SEM observation of some cavity bottoms.

The process may be either intra or extra granular. The observations done in this work are not sufficient to confirm its origin. Nevertheless, we will see in next part that informations about the activation of nucleation can be drawn through indirect evaluations. Magnification of local areas of limited coalescence shows some isolated microvoids of spherical shapes ranging from 15 μm up to 400 μm in diameter, that is five times the mean grain size. A statistical evaluation of 2D damage statements has been performed using a technique based on dynamic focussing (using a UBM system) [4] closed to the one presented in [1,2]. Damaged areas of selected samples are recorded that way. Data is then computed to give a 3D reconstruction of the intercepted cavities. The final composite

image is made of 30 individual digitized views and the real surface resolution is 39.7 μm²/pixel. Each picture is analyzed using a specific toolbox developed at CEA/Valduc on Visilog software. In figure 6, we demonstrate that most observations lead to aspect ratios less than 1.2 : this result confirms the quasi sphericity of microcavities. Sphericity is assumed to be related to high triaxiality factor imposed by the uniaxial deformation loading. Moreover some numerical simulations have shown that micro inertia effects can affect the growth interactions between adjacent cavities [5]. In figure 7, 3D assessment of radius distribution puts in evidence the underestimated microvoid size with the 2D technique while the position of the maximum radius locations remains correctly confirmed. The important dispersion of the data in the vicinity of the spall plane proves that nucleation is activated gradually during all the damage process in the target. This strong coupling between nucleation and growth is an interesting track for modeling.

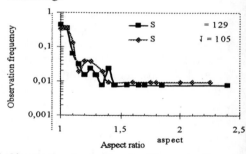

Figure 6. Observation frequency of void aspect ratio.

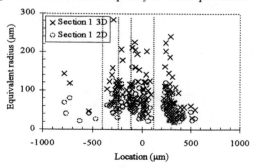

Figure 7. Evolution of radius as a function of location (x=0 mm: theoretical spall plane)

COALESCENCE AND FRACTURE

Coalescence occurs in a geometric process of intercavity link fracture (see figure 8a): this mechanism named "direct impingement" has been observed in all low and medium shock pressure tests. The viscoplastic instabilities reported in [1,2] have not been observed in this domain. But some similar localization (shear band or thin cracks) has been shown at 30 GPa (see figure 8b). The origin of these microcracks connecting cavity clusters could be connected to some brittle fracture mechanisms (induced by the very high strain rate) or viscoplastic shear bands (amplified by the increase of temperature) like those observed in [1,2].

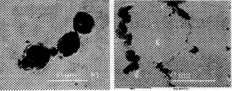

Figure 8. Coalescence mechanisms – direct impingement and cracks link.

Figure 9 shows the porosity with respect to distance from the spall plane (of maximum porosity). This calculation includes isolated microvoids and coalescence aspects. Maximum value of porosity (0.4) is in good agreement with classical observation or predicted values estimated by growth models. The spatial resolution gives access to a complex structure of primary and secondary damaged planes which have already been discussed. One interesting parameter is the mean dimple size observed on the fracture surface as a function of the applied shock pressure (see figure 10). This evolution shows that to reach fracture, nucleation rate increases with shock pressure. It can be explained by the effect of micro inertia, which would limit growth (and consequently final cavity size) and increase nucleation (through a relaxation process).

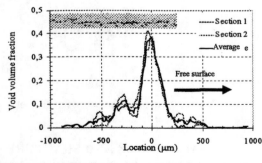

Figure 9. Porosity as a function of distance from the spall plane (for 6.2 GPa, 1μs shock)

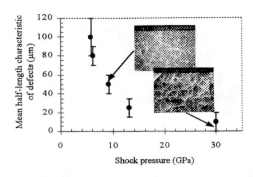

Figure 10. Mean void/dimple size at spall plane (for ~ 1 μs pulse duration)

CONCLUSIONS

Damage in tantalum under shock wave loading has been metallurgically investigated. Complete sequences of damage events in tantalum have been examined in a large range of shock pressure (5 GPa – 30 GPa). The collected informations constitute an interesting database for the proposition of a logical scheme of modeling. This work is in progress.

ACKNOWLEDGEMENTS

The authors are grateful to A. Dragon (ENSMA Poitiers) and H. Trumel (CEA) for helpful discussions on theoretical issues.

REFERENCES

[1] W.R. Thyssel, A.K. Zurek, D.A.S. Macdougall, D. Tonks, J Phys IV France, 10, 2000, pp 769-774.
[2] A. Zurek, W.R. Thyssel, J.N. Johnson, D.L. Tonks, R. Hixson, J. Materials Research Technology, 60, 1996, pp. 261-267.
[3] F. Llorca, G. Roy, P. Antoine, J Phys IV France, 10, 2000, pp. 775-780.
[4] G. Roy, PhD Thesis, Université de Poitiers, France, 2003.
[5] F. Buy, G. Roy, F. Llorca, Proceedings HDP, Saint Malo, 2003 (to be published).

CP706, *Shock Compression of Condensed Matter - 2003*
edited by M. D. Furnish, Y. M. Gupta, and J. W. Forbes
© 2004 American Institute of Physics 0-7354-0181-0/04/$22.00

DYNAMIC PROPERTIES OF NICKEL-ALUMINUM ALLOY

Kenneth J. McClellan[a], Damian C. Swift[b], Dennis L. Paisley[b], and Aaron C. Koskelo[c]

[a]MST-8, [b]P-24, and [c]C-ADI, Los Alamos National Laboratory, Los Alamos, NM 87545

Abstract. We are investigating interactions between shocks and grain boundaries in the anisotropic alloy NiAl; as part of this work, we need to know the shock response of single crystals. The equation of state (EOS) and elasticity were predicted using ab initio quantum mechanics. NiAl crystals were obtained from GE, and also prepared locally, and cut to thicknesses between about 100 and 500 microns. Laser-driven flyer impact experiments were used to verify the EOS and to measure the elastic precursor wave as a function of crystal orientation. Shocks induced by direct laser irradiation were used to investigate the elastic precursor and to demonstrate that the imaging VISAR system had a temporal resolution adequate to distinguish between different orientations. A single-crystal plasticity model is under development to design and interpret experiments on bicrystals.

INTRODUCTION

We are investigating the effect of grain boundaries on shock propagation in anisotropic materials, focusing on the localization of energy and deformation at a boundary between two crystals. Impact by laser-launched flyers was selected as the primary type of experiment, to allow a greater number of bicrystal configurations to be explored than would be the case with most other forms of dynamic loading. The intermetallic compound NiAl was chosen as a prototype, being relatively straightforward to prepare in the form of single crystals and bicrystals, and being strong enough to cut and polish down to the relatively thin (few hundred micron) samples necessary for a fully-supported shock to be induced by flyers ~100 μm thick. No equation of state (EOS) for NiAl was found in the literature, and it is highly adviseable to measure the dynamic strength for the exact composition and time scales (or strain rates) for our experiments, so the supporting work described here was undertaken to determine these properties.

Ab initio quantum mechanics was used to predict a thermodynamically complete EOS and also the variation of elastic constants with compression. Laser flyer experiments were used to verify the predicted EOS and to measure the flow stress under shock loading conditions.

QUANTUM MECHANICAL PREDICTIONS

The EOS was to be used in the simulation and interpretation of shock wave experiments exploring up to a few tens of gigapascals, and with physically-based models of flow stress in which temperature would be a parameter. Thus the EOS was required to be valid on the principal Hugoniot and for release states, and thermodynamic completeness was desirable. Quantum mechanics was used to calculate an ab initio EOS, by predicting the frozen-ion cold curve and estimating thermal effects by the Grüneisen approximation. This is a simplified variant of the technique applied previously to several elements.(1)

Ab initio nonlocal pseudopotentials were used to model the nucleus and inner electrons of each atom. The outer electrons were represented with a plane-wave basis set, and exchange and correlation effects were included through the local density approximation (LDA).(2) The frozen-ion cold curve was calculated by finding the ground state of the outer electrons with respect to the atoms, for different values of the lattice parameters. The thermal part of the EOS was predicted using the Grüneisen approximation, by estimating Grüneisen's Γ from the cold curve and applying the Dugdale-Macdonald formula.

This procedure was applied to NiAl in the CsCl structure. The pseudopotentials for Ni and Al had been exercised previously by calculating frozen-ion cold curves for the elements; the accuracy was typical of LDA predictions, and adequate for calculating EOS. As normal when using the LDA, the lattice spacing predicted at STP varied from the observed value by ~1%. An ab fere initio EOS was constructed by adding a constant pressure offset to bring the EOS into agreement with observation at STP. The EOS was constructed in tabular form following the format and units employed in the SESAME database.(3)

Calculations were also performed of elastic strains applied to the CsCl lattice cell. Given the configuration of the electrons in the ground state, the stress tensor on the lattice cell could be deduced. The variation of stress with strain was used to estimate some of the elastic constants c_{ijkl} in deviatoric form: 73 GPa for c_{1111}.

FLYER IMPACT EXPERIMENTS

States on the principal Hugoniot of NiAl were measured using flyer impact data from gas gun or laser launch experiments. The laser flyer experiments used pulses ~600 ns long from the TRIDENT facility, absorbing the energy in a layer of C sandwiched between a transparent PMMA substrate and the flyer.(4) Additional layers of Al and alumina were included to contain the resulting plasma and reduce heat conduction. Experiments had been performed previously with the same experimental setup in which EOS points were deduced for Cu(5) and the intermetallic compound NiTi(6); the results agreed well with data or EOS deduced from gas gun experiments, giving confidence in the validity of data from the laser flyer apparatus.

The accuracy of EOS and strength data varied with the details of the experimental config-uration used, depending on surface finish and the tolerance with which different dimensions of the components could be determined. Experiments were performed in several different configurations as the accuracy was optimized, and on different orientations of single crystals of NiAl. Experimental configurations included a Cu flyer impacting a NiAl target which released into a PMMA window or vacuum; a Cu flyer impacting a Cu base on which the NiAl sample was mounted, releasing into vacuum; and a NiAl flyer impacting a LiF window. Point and line VISARs were used to monitor the velocity history of the flyer and the surface of the sample. The flyers were 5 mm in diameter, the targets approximate semicircles of the same diameter and positioned so that roughly half of the flyer could be seen, to allow access for the VISAR beams. NiAl flyers ranged from 100 to 200 μm thick, Cu flyers 50 to 100 μm, and NiAl targets 100 to 400 μm. Cu flyers were punched from rolled foils of high purity material from Goodfellow Metals. The NiAl parts were prepared from oriented crystals by cutting with a diamond saw and polishing. LiF windows were (100) crystals obtained from Saint-Gobain (Bicron) Corp.

In the laser flyer experiments, the line VISAR did not use quadrature encoding and the velocity deduced had an uncertainty of up to ~50 m/s, compared with up to ~5 m/s for the point VISAR, which did use quadrature encoding. The most accurate stress states were therefore deduced from experiments in which the accuracy of the point VISAR dominated: impact of the sample against a window. As one would expect, no difference was observed in the Hugoniot states deduced for samples of different crystal orientation. (Figs 1 and 2.)

There was clear evidence of spall and subsequent reverberations; the corresponding spall strength was estimated from the magnitude of the pull-back to be 0.5 ± 0.1 GPa for crystals cut parallel to (100) planes.

LASER IRRADIATION EXPERIMENTS

The main interest of this project is to study interactions between shocks and grain bound-

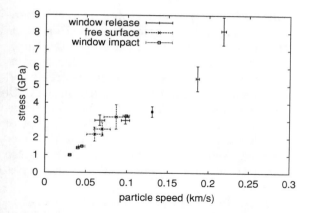

FIGURE 1. Hugoniot states deduced from different types of flyer impact experiment – (100) crystals.

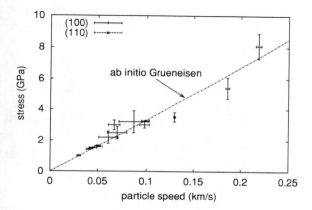

FIGURE 2. Comparison between theoretical Hugoniot and flyer impact data for NiAl.

aries in NiAl, by line VISAR and surface displacement maps. Diagnostics with high time resolution but finite data storage are difficult to synchronize with shock breakout in a flyer impact experiment, so experiments were performed using direct laser irradiation to verify that variations in arrival time could be detected in crystals of different orientation, and to develop techniques for aligning targets with adequate precision.

The TRIDENT laser was used in nanosecond mode to induce dynamic loading by ablation of one surface of the sample.(4) The drive pulse was 2.5 ns long and applied over a spot 5 mm in diameter using a Fresnel zone plate to smooth

long-wavelength spatial variations in the beam. The drive irradiance was roughly constant in time, here in the range 0.02 to 0.5 PW/m² (1 to 30 J per pulse) in order to generate a shock only slightly stronger than the elastic precursor. The ablative drive is benign in that the sample deceleration system need not cope with the energy and momentum of a sabot or detonation products, and the TRIDENT laser was capable of firing shots of this type at a rate up to 14 per day. As was found for NiTi,(6) for laser pulses inducing shock pressures ~20 GPa or less, the use of laser ablation affected a layer of the sample perhaps 1 μm thick; the remainder of the sample was apparently unaffected by any non-hydrodynamic effects such as x-rays or hot electrons.

Experiments were performed using pairs of samples of the same thickness and different crystal orientation, mounted side-by-side. Initially, the samples were glued to a release window – BK7 glass 12.7 mm thick – to allow part of the deceleration to be 1D, hence introducing less perturbation to the microstructure of recovered samples (Fig. 3). It was also anticipated that the glue layer would reduce the amount of drive light transmitted to the VISAR optics, which might otherwise disrupt the signal. Problems were experienced in obtaining adequate VISAR signals through the glue layer and window, so for subsequent experiments the samples were held side-by-side by clamping their edges (Fig. 4). There was generally a small gap between the samples, but at the low irradiances used here the drive light collected by the VISAR optics was found to be insignificant with the use of a notch filter in front of the optical streak cameras.

Records of the laser irradiance history and the surface velocity history – using a line-imaging VISAR – were obtained on every experiment. The line VISAR records clearly showed the difference in arrival time through the samples of different crystal orientation (Fig. 5). The amplitude of the elastic precursor also varied with crystal orientation.

CONCLUSIONS

The Hugoniot EOS and dynamic strength of approximately equiatomic NiAl alloys was mea-

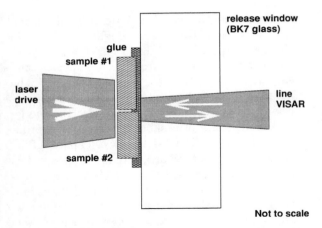

FIGURE 3. Schematic of direct drive experiments with pair of samples mounted on a release window.

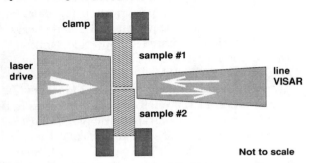

FIGURE 4. Schematic of direct drive experiments with pair of samples clamped by the edges.

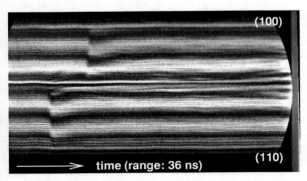

FIGURE 5. Example line VISAR interferogram demonstrating difference in shock wave arrival through different orientations of NiAl crystal. Both samples were $266\,\mu m$ thick. The VISAR sensivivity was $800\,m/s.$fringe, with a field of view of $1.5\,mm$. The VISAR record was delayed such that the center of the record was $55\,ns$ after the start of the laser pulse. The drive energy was $5 \pm 1\,J$.

sured using laser-launched flyer impact experiments. Experiments were performed on single crystals of two different orientations; the Hugoniots were not distinguishable. Ab initio quantum mechanics was used to predict isotropic and uniaxial compression curves for NiAl in the CsCl structure, and hence a simple – but thermodynamically complete – EOS and also a subset of the elastic constants. The principal Hugoniot deduced from the theoretical EOS matched the overall trend of the experimental Hugoniot states quite well.

Experiments were performed using laser ablation to induce shock waves in samples of different orientation simultaneously. Using a line-imaging VISAR, it was possible to observe the difference in arrival time of the weak shocks – dominated by the elastic precursor – and also the difference in amplitude of the elastic precursor.

ACKNOWLEDGMENTS

The electron code CASTEP was made available courtesy of the UK Car-Parrinello Consortium. TRIDENT staff laser provided substantial experimental help. This work was performed under the auspices of the US Department of Energy, contract W-7405-ENG-36.

REFERENCES

1. Swift, D.C., Ackland, G.J., Hauer, A., and Kyrala, G.A., Phys. Rev. B **64**, 21, 214107 (2001).
2. Kohn, W. and Sham, L.J., Phys. Rev. **140** 4A (1965).
3. Holian, K.S. (Ed.) *T-4 Handbook of Material Property Data Bases, Vol 1c: Equations of State*, Los Alamos National Laboratory report LA-10160-MS (1984).
4. Paisley, D.L., Swift, D.C., Johnson, R.P., Kopp, R.A., and Kyrala, G.A., "Laser-launched flyer plates and direct laser shocks for dynamic material property measurements," in *Shock Compression of Condensed Matter-2001*, edited by M.D. Furnish et al, AIP Conference Proceedings 620, Melville, New York, 2002, pp 1343-6.
5. Niemczura, J., Paisley, D.L., and Swift, D.C., "Accuracy of laser-launched flyer experiments for measuring equations of state," this conference.
6. Hackenberg, R, Swift, D, Bourne, N., Gray III, G, Paisley, D., Thoma, D., Cooley, J, and Hauer, A., "Dynamic properties of nickel-titanium alloys," this conference.

CP706, *Shock Compression of Condensed Matter - 2003*
edited by M. D. Furnish, Y. M. Gupta, and J. W. Forbes
© 2004 American Institute of Physics 0-7354-0181-0/04/$22.00

COMPARATIVE ANALYSIS OF UNIAXIAL STRAIN SHOCK TESTS AND TAYLOR TESTS FOR ARMOR AND MARAGING STEELS

Yu. I. Mescheryakov, N.I. Zhigacheva, Yu. A. Petrov, A.K. Divakov, C.F.Cline *[)]

Institute of Problems of Mechanical Engineering RAS, V.O. Bolshoi 61, Saint-Petersburg, 199178, Russia
[)] Lawrence Livermore National Laboratory, Livermore 94506-4528 CA , USA

Abstract. High-strength constructional 38KhN3MFA steel and 02H18K9M5-ВИ maraging steel were tested to determine the yield stress under dynamic loading. The 38KhN3MFA steel was used as central test material to work out the experimental technique. For both kinds of steel the results obtained in the plane shock tests under uniaxial strain condition show approximately the identical yield stress values as those obtained in Taylor tests. Cracking of maraging steel occurs along the shock-induced austenite bands where microhardness is much smaller than that for the rest of the matrix.

INTRODUCTION

One of the most important characteristic of material behavior under dynamic load is the *dynamic yield stress*. In the present study, the dynamic yield stress for two kinds of steel is found by using two different experimental techniques. The first technique is based on the collision of rod and rigid anvil. This technique has been developed by Taylor [1] and experimentally checked and confirmed by Wiffin [2]. Intensive studies of dynamic characteristics of material have shown that techniques based on wave propagation in thin rod have a limitation on strain rate. Therefore, in the middle of 70-th, instead of dynamic tests under *uniaxial stress conditions*, plane-wave tests under *uniaxial strain conditions* have been developed. This permits to study the high-velocity processes up to strain rates of the order of $10^7 - 10^8 \, \text{s}^{-1}$.

Nevertheless, to date a lot of materials are already tested under much simpler conditions, such as provided by Taylor technique. So, the problem is how these experimental data correlate with those obtained in plane shock tests. One of objectives of present study is to perform both Taylor tests and plane shock load tests by using materials with well-known dynamic properties. Additional objective is to explore maraging steel which behavior under dynamic load is very complex.

EXPERIMENTAL TECHNIQUE

Taylor Test

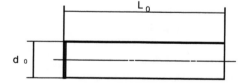

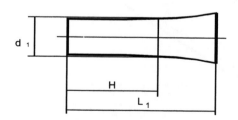

FIGURE 1. Shape and dimensions of rod before and after collision.

Value of dynamic yield stress in the Taylor tests Y_D is determined by using the following formula [2]:

$$\frac{Y_D}{\rho U_{imp}^2} = \frac{L_0 - H}{2(L_0 - L_1)} \frac{1}{\ln(L_0/H)} \qquad (1)$$

Here U_{imp} is the impactor velocity and ρ is the material density. Dimensions of the rod before and after collision are indicated in Fig. 1. In accordance with [1,2], we define the plasticity boundary inside the rod head as corresponding to the rod diameter increase up to 0.2 %. This position of boundary is determined by using the precise optical technique.

Diameter of the rod in all the tests equals 7 mm. As a material for anvil, we use a high- strength KhVG instrumental steel (Russian mark) of 62-64 HRC hardness. In order to avoid an influence of lateral effects of the anvil on rod reloading, dimensions of anvil in three directions are much greater than twice length of rod.

Plane Shock Experiments

The main quantitative characteristic of material response here is a time-resolved free surface velocity profile u_{fs} (t) which is recorded by using velocity interferometry. To infer dynamic yield $Y_{0.2}^{strain}$ from the uniaxial strain tests a series of tests at different impact velocities must be performed for the dynamic stress-strain diagram to

be built. Values of normal stress and strain for the dynamic stress-strain diagram are calculated by using the elastic-plastic model:

$$\sigma = \frac{1}{2}\rho\left(U_{fs}^{max}c_p + U_{HEL}(C_l - C_p)\right), \qquad (2)$$

$$\varepsilon = \frac{1}{2}\left(\frac{U_{HEL}}{C_l} + \frac{U_{fs}^{max} - U_{HEL}}{C_p}\right), \qquad (3)$$

where σ and ε are the normal stress and strain, U_{fs}^{max} is the peak value of the free surface velocity, U_{HEL} is the free surface velocity corresponding to Hugoniot Elastic limit, C_l is the longitudinal sound velocity and C_p is the plastic wave velocity.

The sought dynamic yield stress value corresponds to the residual normal strain of 0.2% after unloading from Hugoniot curve. An example for 02H18K9M5-VI maraging steel is shown in Fig.2. Dynamic yield stress under uniaxial stress conditions $Y_{0.2}^{stress}$ can be calculated from the following relationship between uniaxial strain and uniaxial stress values of yield stress through the Poisson coefficient:

$$Y_{0.2}^{stress} = \frac{1-2\upsilon}{1-\upsilon} Y_{0.2}^{strain} \qquad (4)$$

RESULTS OF EXPERIMENT

38KhN3MFA Steel

The yield stress under uniaxial strain conditions obtained from stress-strain diagram for the 38KhN3MFA steel equals $Y_{0.2}^{strain} = 2.81$ GPa.

Then $Y_{0.2}^{stress} = \frac{1-2\upsilon}{1-\upsilon} Y_{0.2}^{strain} = 1.59$ GPa.

Taylor tests were performed within the range of impact velocities from 90 m/s to 450 m/s. Within that range there exist four stages of the rod head deformation and fracture:

1). Uniform deformation with gradual increase of the rod head;
2). Deformation and fracture of rod head resulting in splitting of rod head by two pieces at 45 ^{0};
3). Fracture of rod head, which results in splitting of rod head by one large piece and small fragments;
4). Fracture by numerous fragments .

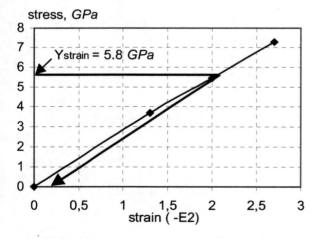

FIGURE 2. Scheme to infer the yield stress from dynamic stress-strain diagram for maraging steel.

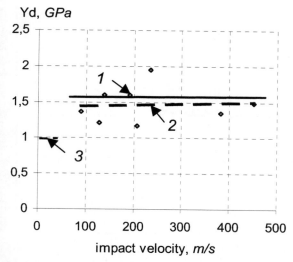

Yd, GPa

impact velocity, *m/s*

FIGURE 3. Yield stress versus impact velocity for 38KhN3MFA steel: 1 – plane tests; 2 – Taylor tests; 3 – quasistatic yield stress.

In Figure 3 the experimental data obtained in Taylor tests and plane tests are presented together. One can see that dynamic yield stress values for 38KhN3MFA steel obtained in Taylor tests and plane tests lie closely (~1.6 GPa for plane tests and mean value of ~1.5 GPa for Taylor tests).

02H18K9M5-VI Maraging Steel

Taylor tests were performed with two kinds of rods. This was done to check the influence of rod length on dynamic yield stress of material: (i) diameter of rod d_r = 7 mm, length L_r = 40 mm; (ii) diameter of rod d_r = 7 mm , length L_r = 20 mm. The value of yield stress under uniaxial strain conditions corresponding to the strain value of 0.2 % at the dynamic stress-strain diagram equals $Y_{0.2}^{strain}$ = 5.8 GPa. Then the value of yield stress under uniaxial stress condition equal $Y_{0.2}^{stress}$ = 3.31 GPa.

In Figure 4 we present the experimental data obtained in quasistatic tests, Taylor tests and plane tests. One can see that plane test data give a yield stress value 1.66 times greater than quasistatic tests. As for Taylor tests, the overall range of impact velocities can be subdivided by four regions

1). The first set of experimental data corresponds to homogeneous dynamic straining of the rod head occurring at the impact velocities below 130 m/s. The average value of yield stress for that set of experiments equals approximately 3.41 GPa, which is very close to the value of 3.31 GPa obtained in plane experiments.

2). The second set of experimental data corresponds to the specimens the head of which splits into two parts in a plane inclined under 45^0 relatively rod axis. This occurs within velocity range from 130 m/s through 230 m/s. The value of yield stress for that set approximately equals to quasistatic data .

3). The third set of data correspond to full fragmentation of the rod head which occurs from 230 m/s to 350 m/s. Average value of yield stress here is $Y_{0,2}$ = 1.3 GPa

4). The fourth set of specimens gives the yield stress of the order of $Y_{0,2}$ = 2.1 GPa, which corresponds to impact velocities between 350 m/s and 470 m/s.

Thus, for 02H18K9M5-ВИ maraging steel, the Taylor tests can be used only within narrow range of impact velocities below 130 m/s. The value of yield stress obtained in plane shock experiment coincides only with the Taylor tests data corresponding to undamaged specimens.

Structural investigations of maraging steel

Initial martensite-austenite structure of 02H18K9M5-ВИ maraging steel exhibits three types of parallel zones oriented along the specimen axis:

1. The first zone is a white austenite bands of microhardness value of 260 kgs/mm^2.

2. The second zone is a globular structure having the microhardness of 485 kgs/mm^2.

3. The third zone is lamellar austenite structure having the microhardness of 501 kgs/mm^2.

Whereas austenite structure of zone 1 (microhardness ~ 260 kgs/mm^2) remains unchanged in Taylor tests, the structures of zones 2 and 3 suffer a notable change. Although the initial structures of zone 2 and zone 3 are of different morphology - zone 2 has a globular structure of ~ 485 kgs/mm^2 microhardness whereas zone 3 has a needle structure of ~ 500

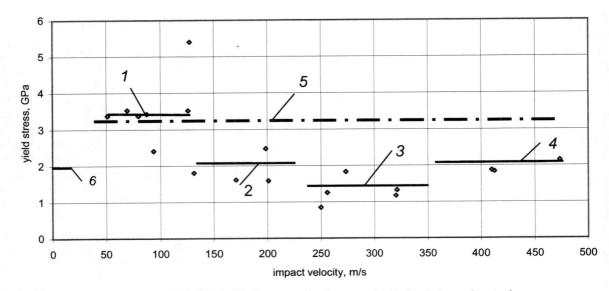

FIGURE 4. Yield stress versus impact velocity for the maraging steel
(1 – 4) – Taylor tests; (5) Plane shock tests; (6) Quasistatic yield stress

kgs/mm^2 microhardness - the initial difference in microhardness is not too much (~ 15 kgs/mm^2).

At the impact velocity of 87,4 m/s the microstructure state of the steel doesn't change. The difference in microhardness between zone 2 and zone 3 remains sufficiently small (~52 kgs/mm^2). This region of impact velocities corresponds to the maximum value of yield stress of maraging steel. With increasing the impact velocity to 256.1 m/s a valuable change in the steel structure occurs. So, the difference in microhardness increases (~ 160 kgs/mm^2) Just that region of impact velocities corresponds to minimum value of yield stress for the 02H18K9M5-ВИ maraging steel.

As microscopy evidence shows, further increase of impact velocity up to 321,6 m/s results in large increase of grain size. Microhardness for zone 2 equals 396 kgs/mm^2 and for zone 3 equals 486 kgs/mm^2, so the difference in microhardness between zone 2 and zone 3 remains sufficiently large (~ 90 kgs/mm^2).

At higher impact velocities, in particular at the velocity of 473 m/s, the difference in hardness of zone 2 and 3 decreases. It becomes equal to the value corresponding maximum yield stress (~ 44 kgs/mm^2). The yield stress determined in Taylor tests for that velocity equals 2.2 GPa, which is close to its average value obtained for small impact velocities.

Thus, microstructure heterogeneity for 02H18K9M5-ВИ maraging steel after Taylor tests correlates with the yield stress behavior. The maximum yield stress corresponds to the minimum difference in microhardness between neighbor structural zones. Cracking happens along the austenite band which microhardness is much smaller than that for the rest of the matrix. Thus, on the basis of studying the microstructure of specimens it can be concluded that the main reason for dynamic fracture of 02H18K9M5-ВИ maraging steel is a dynamic fragmentation of material along the austenite bands initiated by shock loading .

REFERENCES

1. 1. G.I. Taylor. *Proc. Roy. Soc.* London. A. **194,** p. 2899 (1948).
2. A.C. Wiffin. *Proc. Roy. Soc.* London A. **194,** p. 300 (1948).

CP706, *Shock Compression of Condensed Matter - 2003*
edited by M. D. Furnish, Y. M. Gupta, and J. W. Forbes
© 2004 American Institute of Physics 0-7354-0181-0/04/$22.00

CHARACTERIZATION OF LASER-DRIVEN SHOCKED NiAl MONOCRYSTALS AND BICRYSTALS

P. Peralta[1], D. Swift[2], E. Loomis[1], C.-H. Lim[1] and K. J. McClellan[3]

1. Arizona State University. Department of Mechanical Engineering. Tempe, AZ 85287-6106
2. Los Alamos National Laboratory. Physics Division. Los Alamos, NM 87545
3. Los Alamos National Laboratory. Materials Science and Technology Division. Los Alamos, NM 8754

Abstract. Disks of oriented single crystals and bicrystals of selected misorientations of NiAl were tested under direct laser-driven shock conditions. Shocked specimens were recovered and characterized to study cracking and slip behavior. In addition, the crystallographic orientation of the tested samples was studied using Orientation Imaging Microscopy (OIM). Results indicate that direct laser-driven shocks in monocrystals induce cracking on {110} planes, with a high crack density for <100> samples and a low crack density for <110> and <111> specimens. The crack density was much higher on the impact side. In one bicrystal, a Grain Boundary Affected Zone (GBAZ) was observed close to the boundary in one grain, where both cracking and slip were present, whereas no cracking or slip traces were observed in the other grain. OIM revealed that specimens developed gradients of orientation due to bowing of the foil caused by the impact. The changes in the speed of sound across the inclined interface correlated with the cracking mode, i.e., a shock propagating from a "slow" to a "fast" grain resulted in intergranular cracks, whereas the reverse resulted in transgranular cracks.

INTRODUCTION

Predictions of failure in polycrystalline materials require the consideration of failure modes at Grain Boundaries (GBs) as well as inside the crystals. The GBs may induced failure initiation or act as barriers to failure modes originating inside the grains. These modes can in turn be affected by the heterogeneous strain around GBs due to material anisotropy. The failure mechanisms under dynamic conditions at the grain level have been examined for steel by Mescheryakov *et al* [1] using velocity interferometry and fractography. They found that the dispersion of the local velocity distribution is of the same order as the mean flow and that the deformation mechanisms at this level can exert a strong influence over fracture and dynamic deformation. The study of this local behavior in polycrystals is complicated due to variations between samples. Hence, the use of bicrystals as

models of individual interfaces can facilitate the study of failure associated to GBs under dynamic loads, where relatively few experiments with single crystals and bicrystals have been carried out.

EXPERIMENTAL PROCEDURE

The dynamic behavior of the materials was studied with small-scale samples (3-5 mm in diameter, 100-400 μm thick) using a laser-driven testing setup located within the TRIDENT facility at Los Alamos National Laboratory (LANL). The Laser-driven flyer technique is a well-known technique [2] that provides well-defined shock pressures and wave profiles similar to those generated by traditional gas gun experiments. Laser-driven flyers, however, use smaller samples, which makes it more practical to study valuable single crystal and bicrystals. In addition, it is easier to recover the specimen and more experiments can

Table 1. Sample and laser drive characteristics for all samples tested.

Sample			Drive					
Label	Orientation	Thickness [μm]	Duration [ns]	Energy [J]	Irradiance [PW/m^2]	Pressure [GPa]	Mount	Type
P3	(110) and (111)	149	1	35	1.52	3.78	window	Pair
P9	(001) and (111)	185	1	27	1.17	3.26	window	Pair
100-02-6	(001)	228	2.4	8	0.14	0.73	clamp	Single
B1	(3 4 55) and (5 7 17)	346	2.4	11	0.20	0.90	clamp	Bicrystal
B3	(3 4 55) and (5 7 17)	186	2.4	8	0.14	0.73	clamp	Bicrystal
B2a1	(123) and (111)	200	2.5	16	0.34	1.81	clamp	Bicrystal
B2a2	(123) and (111)	200	2.5	17	0.35	1.86	clamp	Bicrystal

be carried out in a given amount of time. In this study, direct drive was used, whereby the beam is shined directly on the back of the sample forming hot plasma that supports the shock. Laser pulses were 1 to 2.5 nanoseconds long, with a wavelength of 527 nm (green) and had energies ranging from 8 to 35 J, with 85% of this energy distributed uniformly over a 5 mm circle. NiAl was chosen for this study, since it has well-known slip systems, {110}<001> [3], and its dynamic behavior has been documented for single crystals [4]. NiAl also has significant strength at room temperature, so samples can be prepared with minimal artifacts, and it has a high elastic anisotropy [3], which enhances grain boundary interactions [5]. Monocrystalline samples were obtained from stoichiometric NiAl single crystals provided by General Electric. Slices were cut parallel to {100}, {110} and {111} planes and used to prepare disks with thicknesses from 400 to 100 μm. The bicrystals were grown using a modified Czochralski technique [6]. Slices ≈ 400 μm thick were cut using electrodischarge machining. All samples were then mechanically polished down to the desired thickness and finished with 0.05 μm colloidal silica. Some single crystals with different orientations were mounted side-by-side and tested in pairs. Samples were either attached to a glass window with epoxy or clamped in position with a metallic ring holder and a damping flotation pad. The testing conditions are shown in Table 1. The shock pressure listed on this table was obtained from an approximate equation of state for NiAl that does not include strength. Post-shock characterization was carried out using optical and scanning electron microscopy. The surfaces of the specimens on the opposite side of the shock were examined using OIM.

RESULTS AND DISCUSSION

Optical microscopy pictures of the impact side of the (001) and (111) single crystals tested in pair P3 are shown in Fig. 1.

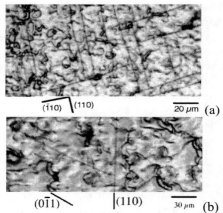

FIGURE 1. Optical micrographs of the shock side for monocrystalline samples. (a) (001); (b) (111).

Note the array of crystallographic cracks on the <100> oriented foil (Fig. 1a). These cracks are parallel to traces of {110} planes, which are cleavage planes for NiAl [3]. Similar behavior was found at lower irradiance values, except that crack spacing was slightly larger. Given the expected 1-D nature of the compressive shock, it is unlikely that cracks are produced during the shock itself, but rather during the propagation of the tensile release wave. However, the compressive wave probably plays a role on subsequent fracture by producing defects, e.g., dislocations, which could seed the cracking process through internal stresses produced

602

via pile-ups. This also suggests that cracking is a dissipation mechanism for shocks propagating along the <001> direction for NiAl. This is quite consistent with the well-known "hard" nature of this direction [3]. The behavior of the samples oriented along <111> is shown in Fig. 1b. The crack density was much lower than for <001> samples and cracks were parallel to traces of {110} planes. (110) foils behaved similarly. A low crack density was expected since <110> and <111> are "soft" directions for NiAl [3]. All samples showed surface roughness on the impact side, probably caused by small-scale spatial variations in the beam. This roughness may have contributed to cracking in the samples with soft orientations.

Regarding the bicrystal, a picture of the side opposite to the shock for sample B1 is shown in Fig. 2. Note that the grain boundary has been drawn on the micrograph, since it was difficult to observe in the specimen. The boundary in this specimen was inclined to the left through the thickness by about 30°, i.e., grain 2 is underneath grain 1, which implies that a shock wave initiating in the back of the sample went from grain 2 to grain 1 after crossing the boundary. This means that the trace of the boundary in the opposite side is located about 170 μm to the left of the boundary trace shown in Fig. 2.

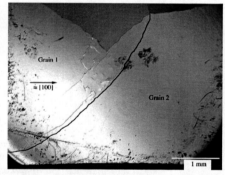

FIGURE 2. Cracking and slip in bicrystal B1 after testing. The load axis in Grain 1 is [3 4 55] and [5 7 17] for Grain 2. The shock is perpendicular to the page.

Note from Fig. 2 that cracking was mostly transgranular and took place in Grain 1, which is close to [001]. This is a "hard" orientation and cracking is expected. Nonetheless, it is interesting that fracture occurred in a region close to the grain

boundary and there were only a few cracks extending towards the bulk of the grain, in contrast to the <001> oriented single crystals, which had a uniform distribution of cracks within the impact region. This indicates that the boundary acted as a site for damage nucleation forming a clear GBAZ. It is likely that defects were produced by the shock at the grain boundary, which made it a nucleation site for cracks on the driven side of the sample. Cracks then nucleated and propagated through the thickness due to the tensile release wave. An area of this sample was examined with OIM and a misorientation map is shown in Fig. 3.

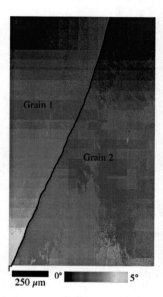

FIGURE 3. Misorientation map for sample B1. This map shows the angle of lattice rotation of each point in the grain with respect to the orientation before the test.

Note that strong vertical misorientation gradients can be observed. The lattice rotation increases towards the bottom, which is close to the clamped edge; however, there is also a horizontal component to the gradient, which was not the case for the <001> single crystal. This horizontal component is different for the two grains, and it is qualitatively stronger in the <001> grain, since the region with low misorientation angles (dark) gives way rather quickly to higher rotations along the horizontal axis as compared to the other grain. Larger misorientation gradients in grain 1 are consistent with the presence of the GBAZ in this grain, since the cracking and crystallographic slip present there would certainly lead to changes in lattice orientation.

Finally, the crack paths were examined in samples B2a1 and B2a2. These samples had the boundary inclined about 35° through the thickness. In sample B2a1 the shock crossed the boundary it from a grain with slow speed of sound, (123), to one with faster speed of sound, (111). The situation was reversed in sample B2a2. The crack paths for the two specimens are shown in Fig. 4.

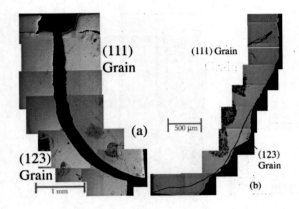

FIGURE 4. Crack paths in (a) bicrystal B2a1 and (b) sample B2a2. The boundary in (b) is marked to show the transgranular crack path.

The samples shown in Fig. 4 did not present a clear GBAZ as the one seen in sample B1. The main crack, which ran intergranularly in sample B2a1 and mostly transgranularly in sample B2a2, was enough to release the energy due to the shock for the two specimens. The cracking mode shown in (b) is consistent with that observed in sample B1, since they were both transgranular and occurred for a test configuration were the shock went from a plastically soft grain, (5 7 17) or (111), with faster speed of sound to a plastically hard grain, (001), with a slower speed of sound. Further studies are being carried out to investigate the defect structure leading to this result.

CONCLUSIONS

Laser driven shocks in NiAl single crystals lead to surface roughness and crystallographic cracking on {110} cleavage planes. Samples with <100> orientations had the highest crack density.

Cracking was mostly transgranular and more pronounced in the GBAZ of the hard (≈<001>) grain for bicrystal B1. The crack and slip lines were parallel to traces of {110} planes.

Misorientation maps showed that misorientation gradients were present in the bicrystals along vertical and horizontal directions and were affected by the presence of the boundary.

Cracking behavior in bicrystals depended on the way the grain boundary was inclined with respect to the shock and the change in sound speed across the interface. Shocks propagating from a slow orientation into a fast one induced intergranular cracking, whereas shocks going from a fast to a slow orientation promoted transgranular cracking.

ACKNOWLEDGEMENTS

This research was supported by LANL. under LDRD-DR #20020055DR.

REFERENCES

1. Y.I. Mescheryakov, N.A. Mahutov, and S.A. Atroshenko, "Micromechanisms of Dynamic Fracture of Ductile High-Strength Steel." *J. Mech. Phys. Solids*, **42**, (1994), 1435-1457.
2. D.L. Paisley, R.H. Warnes, and R.A. Kopp. In *Proceedings of the American Physical Society Topical Conference (7th)*. 1991. Williamsburg, VA, USA. North-Holland.
3. R.D. Noebe, R.R. Bowman, and M.V. Nathal, "Physical and Mechanical Properties of the B2 compound NiAl." *Int. Mater. Reviews*, **38**, (1993), 193-232.
4. S.A. Maloy, G.T. Gray, and R. Darolia, "High Strain Rate Deformation of NiAl." *Mat. Sci. Eng.*, **A192/193**, (1995), 249-254.
5. P. Peralta and C. Laird, "Compatibility Stresses in Fatigued Bicrystals: Dependence on Misorientation and Small Plastic Deformations." *Acta materialia*, **45**, (1997), 5129-5143.
6. J. Garrett, P. Peralta, J.R. Michael, F. Chu, K.J. McClellan, and T.E. Mitchell, "Growth of Oriented $C11_b$ $MoSi_2$ Bicrystals Using a Modified Czochralski Technique." *Journal of Crystal Growth*, **205**, (1999), 515-522.

CP706, *Shock Compression of Condensed Matter - 2003*
edited by M. D. Furnish, Y. M. Gupta, and J. W. Forbes
© 2004 American Institute of Physics 0-7354-0181-0/04/$22.00

LASER-INDUCED SHOCK COMPRESSION OF COPPER AND COPPER ALUMINUM ALLOYS

M. S. Schneider[1], F. Gregori[2], B. K. Kad[1], D. H. Kalantar[3], B. A. Remington[3], M. A. Meyers[1]

[1] *University of California, San Diego, La Jolla, CA 92093-0411,* [2] *University of Paris, 13, Paris, France*
[3] *Lawrence Livermore National Laboratory, Livermore, CA 94551*

Abstract. Single crystal copper and copper 2-wt% aluminum alloy with $[\bar{1}34]$ and [001] orientations are compressed by means of a high energy short pulse laser. Pressures ranging from 20 GPa to 60 GPa are achieved. The shocked samples are recovered and the residual defect substructure is analyzed by transmission electron microscopy. Results show systematic differences depending on orientation and stacking fault energy. Samples with orientations [001] are symmetrical with simultaneous activation of eight slip systems. This leads to a higher work hardening rate. The $[\bar{1}34]$ orientation is asymmetrical with one dominating slip system, and thus a reduced work hardening rate due to a prolonged easy glide region for dislocations. These differences in work hardening response affect the stresses required to achieve the twinning threshold pressure. The effects of stacking fault energy on the defect substructure and threshold twinning are also characterized. Experimental results are rationalized in terms of a constitutive description of the slip-twinning transition using a modified MTS equation. Differences in the mechanical response of the orientations and the chemical compositions are responsible for differences in the shear stress in the specimens at the imposed pressures and associated strains.

INTRODUCTION

The effects of shock waves on metals have been studied for over fifty years [1]. Most experiments have used explosives and flyer plates as the means of creating the compression pulse. The short duration of the shock pulse (0.1 – 2 μm) renders direct measurements of deformation mechanisms nearly impossible, and therefore have to be inferred from post-shock examination of the residual defect substructure. It is only recently [2,3] that pulsed x-ray diffraction has been used to obtain quantitative information of the lattice distortions at the shock front. Recent experiments by Meyers et al. [4] using copper single crystal specimens showed that dislocation configurations and the twinning threshold pressure using laser-induced shock waves are nearly identical to those obtained at durations 10 – 100 times longer as in explosively driven flyer plate studies. The early

experiments by Johari and Thomas [5] showed decreasing cell sizes with increasing shock pressures in copper aluminum. More recently, results by Murr [6] confirm that cell sizes decrease and dislocation densities increase with increasing shock pressures. These results are a clear confirmation that defects are generated at the shock front.

EXPERIMENTAL TECHNIQUES

The shock experiments were carried out at the OMEGA Laser Facility at University of Rochester's Laboratory for Laser Energetics (LLE). Single crystals of copper and copper 2-wt% aluminum were obtained for two orientations as shown in Figure 1: [001] (symmetrical) and $[\bar{1}34]$ (asymmetrical) and prepared into 3 mm diameter cylinders for the laser shock experiments.

The input laser energies used in the presented experiments are 70 J and 205 J. The laser spot size

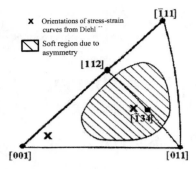

FIGURE 1 Stereographic projection showing [001] and [$\bar{1}$34] orientations as used in the laser-shock experiments.

was on the order of 2.5 mm to 3.0 mm depending on the size of the sample. The pulse duration was 2.5 ns. The energies can be translated into pressures using Lindl's equation [7]:

$$P = 4000\left(\frac{I_{15}}{\lambda}\right)^{\frac{2}{3}} \qquad (1)$$

where P is pressure (GPa), I_{15} is laser intensity (10^{15} W/cm^2), and λ is wavelength in micrometers. Thus, 70 J is equivalent to 20 GPa and 205 J is equivalent to 40 GPa. The results obtained by TEM of the shock recovery conditions are reported herein.

EXPERIMENTAL RESULTS AND

Laser Energy of 70J (Pressure of 20 GPa)

The low energy shock in pure copper creates a cellular organization of a low density of 1/2<110> dislocations. The cells are homogeneous, Figure 2(a), with an average diameter of 0.25 μm for the [001] orientation. The results obtained in general confirm previous observations, albeit at a pulse duration that is lower by a factor of 10-100, than those achieved by Murr [6]. The predicted cell size from the plot in [4] at a pressure of 20 GPa, is 0.22 μm.

The [$\bar{1}$34] orientation shocked at 20 GPa contains a similar well-defined cellular network with a slightly larger (0.3-0.4 μm) average cell size, Figure 2(b). The dislocation density is on the order of 10^{13} m^{-2}. The cells are comprised

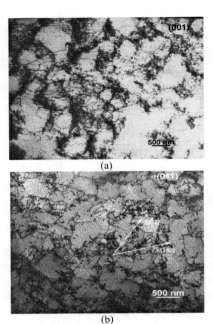

FIGURE 2. Dislocation cells in pure copper shocked with laser energies of 70 J; (a) [001] orientation is, B = [001]; g = [020]; (b) [$\bar{1}$34], B = [011], g = [$\bar{2}\bar{2}$2].

primarily of three dislocation systems: (111)[$\bar{1}$01], (111) [1$\bar{1}$0], and ($\bar{1}$11)[101].

The dislocation substructure observed in Cu-2wt% Al oriented to [001] shows a transitional structure between dislocation cells and stacking faults when shocked at 20 GPa. The dislocations are arranged into square cells as precursors to stacking faults and twins. The average cell sizes are found to be slightly larger (0.4 μm as in Figure 3(a)). It is also important to note the nature of the dislocations. In the alloyed materials, the dislocations have a greater energy barrier to overcome for cross-slip to occur. Instead, dislocations tend to pile-up and form planar arrays.

In the Cu-2wt% Al single crystal oriented to [$\bar{1}$34], the dislocation substructure is less organized. The dislocations have a greater line length compared to the pure Cu samples (Fig. 3(b)). Large densities of dislocation loops are observed. The dislocations are found to be from three predominant slip systems found as expected by Schmid factor calculations. Dislocations densities are lower than expected, but this could be a result of sample thickness.

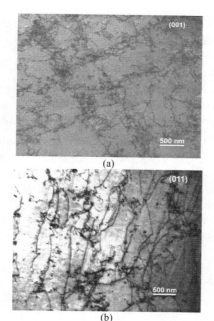

(a)

(b)

FIGURE 3. Cu-2wt%Al shocked with laser energy of 70 J (20 GPa); (a) [001] orientation, B = [001]; (b) [$\bar{1}$34], B = [011].

Laser Energy of 205J (Pressure of 40 GPa)

Impacting pure copper oriented to [001] with 205 J energy input creates dense dislocation tangles, stacking faults and micro-twins. There are no readily discernible dislocation cells. These traces (Figure 4(a)) are characteristic of stacking-fault bundles and twins which are analogous to previous observations by Murr [6], especially, Figs. 20, 21, and 23 of [6]. These features are significantly different than the ones at the lower energy. Traces of planar features are seen when the beam direction is <101>.

For the [$\bar{1}$34] orientation, the deformation sub-structure is cellular, albeit finer at a 0.15 μm average cell size and a significantly higher dislocation density, 10^{14} m^{-2}, Figure 4(b). This is in direct contrast to the mechanism change observed in [001]. Again, the three slip systems previously described dominate the deformation sub-structure. A large number of loops were also visible. These were found to contribute to the cell walls and were often commonly found in the cells at lower concentrations.

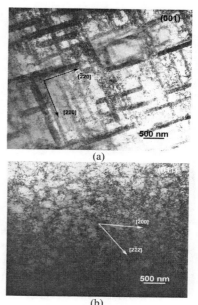

(a)

(b)

FIGURE 4. Pure Cu shocked with laser energy of 200 J (40 GPa); (a) [001] orientation, B = [001]; (b) [$\bar{1}$34], B = [011].

As expected, in the Cu-2w% Al [001] copper shocked at 200 J, the material twins readily as shown in Figure 5(a). At least two variants are observed and they are both well defined. When imaged at B = [001], they appear at exactly 90° to each other aligned along [220] and [$\bar{2}$20] directions. They are also present roughly in the same proportion. The density of twins is quite high. The twins vary in size and length with an average width of 20-30 nm and length on the order 1 μm. A large number of dislocations are also readily observed between the twins, which are not observed in pure copper when shocked at pressures above the twinning threshold [4].

The results obtained for Cu-2%wt Al oriented to [$\bar{1}$34] showed two twinning variants activated where the domain of one twin may be the nucleation site for the second activated twinning system (Figure 5(b)). The twins are in lower proportions than one would expect. The twins vary in size and proportion with the primary variant, (111)[$\bar{2}$11], having the largest size and the secondary variant, (1$\bar{1}$1)[$\bar{1}$$\bar{1}$2], being greater in number.

(a)

(b)

FIGURE 5. Cu-2wt%Al shocked with laser energy of 200 J (40 GPa); (a) [001] orientation, B = [001]; (b) [$\bar{1}$34], B = [011].

ANALYSIS
Prediction of Threshold Amplitude for Twinning

FCC metals have a stacking-fault energy dependent threshold pressures for the initiation of twinning [8]. If one assumes that slip and twinning are competing mechanisms, where plastic deformation by slip has a strain rate and temperature dependence well described by the theory of thermally-activated obstacles, then it is reasonable to assume that slip is highly favored at most conditions. This methodology to predict the threshold shock amplitude was delineated by Murr et al. [9] and Meyers et al. [10] The application of this criterion to the shock front necessitates the knowledge of the strain rate. The strain rate at the shock front has been established by Swegle and Grady [11] to be:

$$P = k_{SG}\,\dot{\varepsilon}^{1/4} \qquad (2)$$

Two separate aspects have also to be considered in the analysis: (a) shock heating, and (b) plastic strain at the shock front. By applying the Rankine-Hugoniot and Grüneisen relationships, equations relating these terms can be developed (See [12]). The response of the copper monocrystal is represented by the modified mechanical threshold stress (MTS) expression below:

$$\sigma=\sigma_0 f(\varepsilon)\left[1-\left(\frac{kT}{Gb^3 g_0}\ln\left(\frac{\dot{\varepsilon}_0}{\dot{\varepsilon}}\right)\right)^{2/3}\right]^2 +kd^{-1/2} \qquad (3)$$

For twinning, one assumes a strain rate and temperature independent σ_T. Setting $\sigma_T = \sigma_s$, one can obtain the critical twinning stress as a function of ε, $\dot{\varepsilon}$, and T. The work hardening term, $f(\varepsilon)$, incorporates orientation dependence. Figure 6 shows the predicted threshold for different initial temperatures and orientations in pure copper. It is clear that the [001] orientation has a lower twinning threshold pressure, in agreement with earlier results by De Angelis and Cohen [13].

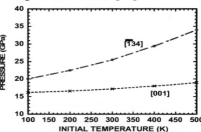

Figure 6. Calculated threshold twinning pressure for [001] and [$\bar{1}$34] orientation as a function of initial temperature.

Research supported by the DOE (Grant DE-FG03-98DP00212) and by Lawrence Livermore Nat. Lab.

REFERENCES

1. Smith, C.S., *Trans. AIME* **212**, 574 (1958).
2. Johnson, Q., et al, *Phys. Rev. Let.* **25**, 109 (1970).
3. Wark, J.S., et al. *Phys. Rev. B* **40**, 5705 (1989).
4. Meyers, M.A. et al., *Acta Mat.* **51,** 1211 (2003).
5. Johari, O. and Thomas, G., *Acta Met.* **12**, 1153 (1964).
6. Murr, L. E., *Scripta Met.* **12**, 201 (1978).
7. Lindl, J., *Phys Plasmas* **2**, 3933 (1995).
8. Murr, L.E, in *Shock Waves in Condensed Matter*, eds. S.C. Schmidt and N.C. Holmes, Elsevier, Amsterdam, 1988, pp. 315-320.
9. Murr, L.E., et al. *Acta Mater.* **45**, 157 (1997)
10. Meyers, M.A, et al.: *Matls. Sci. and Eng* **A322**, 2002 194.
11. W. Swegle and D. E. Grady, *J. Appl. Phys.* **58**, 941 (1983).
12. Meyers, M.A.: *Dynamic Behavior of Materials*, J. Wiley, NY, 1994.
13. De Angelis, R.J. and Cohen, J.B., *J. of Metals* **15**, 681 (1963).

CP706, *Shock Compression of Condensed Matter - 2003*
edited by M. D. Furnish, Y. M. Gupta, and J. W. Forbes
© 2004 American Institute of Physics 0-7354-0181-0/04/$22.00

FORMATION of ROTATION in TITANIUM ALLOYS at SHOCK LOADING

M.A Skotnikova[1], T.I. Strokina[1], N.A. Krylov[1], Yu. I. Mescheryakov[2], A. K. Divakov[2]

[1]*St. Petersburg Machine Building State Institute. Russia, Skotnikova@mail.ru*
[2]*Institute of Mechanical Engineering Problems, RAS, Russia, Ymesch@impact.ru*

Abstract. TEM and X-ray diffraction analyses are used to investigate the structural and phase changes occurring in a material of planar target samples of VT-6 two-phase titanium alloy, tested over the impact velocity range of 400 to 600 m/s. It is shown that the compressional plastic deformation wave modulates the material structure, breaking it up into micro-blocks of 4 to 40 μm in size. Along the boundaries of these blocks, the unloading wave produces a coordinated displacement of blocks relative to each other in such a way that the smaller the blocks and the closer they are to the center of rotation, the greater the mutual displacements of the blocks.

INTRODUCTION

According to modern scientific approaches, fracture of solid is considered as a multistage and multiscale kinetic process of gradual accumulation of damages in solids, exhausting a resource of plasticity, formation of a critical structure and, lastly, nucleation of a critical microcrack. Probably therefore, plastic deformation, destruction, and partitioning of energy, no less than physical and chemical processes, proceed discretely, that is, accompanied by change of waves of overhead tensions, and results in periodic localization of plastic deformation, as a main mode of stress concentrator motion in a sample [1].

EXPERIMENTAL PROCEDURE

TEM and X-ray diffraction analyses are used to investigate the structural and phase changes occurring in plane targets - samples of VT6 two-phase $(\alpha+\beta)$ titanium alloy, tested at shock conditions within the impact velocity range of 400 to 600 m/s, which corresponds to a strain rate range of the order of $4 \cdot 10^4$ to $1 \cdot 10^5$ sec^{-1}. Tests have been performed with planar targets with a thickness of 7 mm and diameter of 52 mm, shock loaded with a pneumatic gas gun facility 30 mm bore diameter [2]. To record the velocity of the free surface of the targets, a two-channel high-velocity interferometer was used, which recorded a time history not only of the average velocity of a free surface of a target, but also velocity distribution of particles (ΔV). The time resolution of the interferometer is about two nanoseconds, and the spatial resolution is 100 μm.

RESULTS AND DISCUSSION

In Fig. 1 the free surface time-velocity (V) profiles for targets № 1 and № 2 of VT6 titanium alloy are presented. The impact velocities ($V_{imp.}$) for tests equal 568 m/s and 446 m/s, respectively. The pull-back velocity ($V_{max.}-V_{min.}$), [2] for the tests is 346 m/s and 416 m/s, and distribution of particle velocities (ΔV) for the tests equals 40 and 28 m/s, respectively.

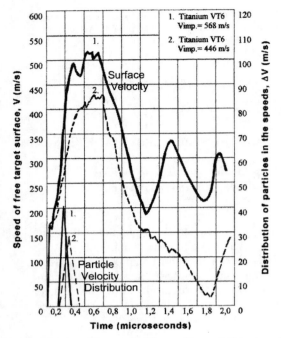

FIGURE 1. Time velocity profiles of free targets surface № 1 and № 2 from VT6 titanium alloy, loaded at impact velocities 568 and 446 m/s, respectively.

The target №1 has been damaged with full separation of spall plate. The target №2 after shock loading has been cut in halves along one of planes perpendicular to the free surface, and then investigated by a method of X-ray diffraction and TEM.

X-ray Diffraction Researches

In Fig. 2 the results of analysis of a phase composition for target material №2 before and after shock loading are provided. As it is seen from obtained diffractograms, after deformation the diffraction peaks from β- phase have decreased and peaks from oxide of titanium Ti_3O_5 have practically disappeared, while the intensity diffraction a peak from rutile TiO_2 has considerably increased. The obtained results specify a possible redistribution of oxygen in VT6 alloy during the high-velocity loading. The change of a stoichiometric relationship of oxygen and titanium flows according to formula $Ti_3O_5 + O \rightarrow 3(TiO_2)$ [3]. It is possible to believe, that at shock stressing in a target material, in places of

localization of plastic deformation process of internal oxidation develops.

Electron Microscopy Researches

32 samples (foils for transmission electronic microscopy) were cut out at different levels on thickness of a sample - target, lengthways and crosswise directions of movement of a shock wave, Fig. 3.

As the panoramas from electron microscope photos and scheme shows, Fig. 4, in a field of action of internal stresses, rotations of material are seen. The rotations are formed in a volume of the material, which was preliminary broken into rectangular blocks of 40, 27 and 4 μm width. Boundaries of such blocks represented the shear bands of 0,7 μm width, probably, of lower density, than metal of a basis. Along such bands, the coordinated mutual displacements of blocks around a common direction of shock are clearly seen. The displacements are proceed in such a way that the smaller block and the closer it is to the center of rotation, the greater the displacement. Thus, near the rotation axis the radial velocity of material is estimated to be sufficiently high to cause redistribution of atoms of oxygen.

As seen in Fig. 4, the coordinated displacement of micro- blocks of material takes place in a plane perpendicular to direction of the shock wave propagation. The prismatic boundaries of two colonies of plates α- phase are oriented at angles of 90° to each other. As it has been shown previously [4], along such boundaries α- grains, in colonies of plates α- phase parallel to boundaries, where the local heating occurs, stretch due to internal stresses, which work as sinks of vacancies and impurity oxygen atoms. The results of shock loading show that near the rotation axis, a localization of plastic deformation happens, where a decomposition of non-equilibrium β- phase takes place, Fig. 5. Under the action of a shock wave the interlayers of β- phase were subjected to an intensive bending, granulating, fractional or complete decomposition (dissolution), which results in β→α - transformation.

In such places as near the micro-cracks, the superfine structure is formed. Analysis shows a

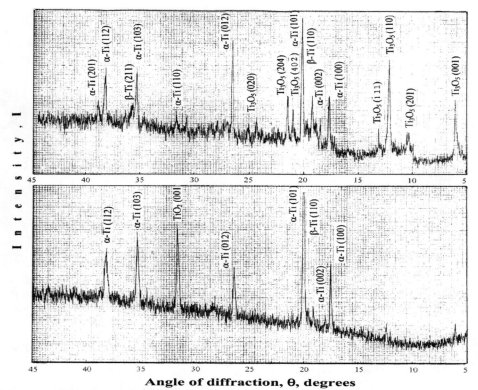

FIGURE 2. Results of the phase analysis of target material №2 of VT6 titanium alloy in an initial condition (a) and after shock stressing at impact velocities 446 m/s (b).

mode of microdiffraction identified as "ring" [5] or even "continuous ring" electronograms, as shown in Fig. 5, with interplane distances which are well conterminous with an oxide of titanium Ti_3O_5.

CONCLUSIONS

As a result of heterogenization of plastic deformation, instead of convective (translational) mode of uniform plastic deformation, a rotation mode of localized deformation is initiated.

During the shock deformation, the microvolumes of material can implement the coordinated rotational displacements relative each other and around of a direction of action of maximum macro-stress. The rotational mechanism of plastic deformation results in fracture of a sample; the damage has a localized character.

In places of localization of deformation the compression wave of plastic deformation

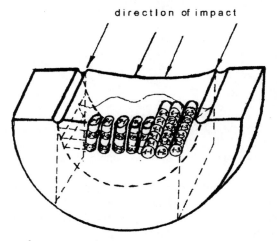

FIGURE 3. The scheme of a transmission electronic microscopy samples cutting.

611

displacement of the material. So, probably, rotations in titanium alloys under shock loading are formed. At the center of rotations the functional structure effects a convective current within the material.

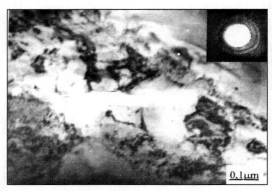

FIGURE 5. A structure and "continuous ring" an electronograme of VT6 titanium alloy near to microcrack in a plane perpendicular to movement of a shock wave. A sample № 4-3. x 82000.

ACKNOWLEDGEMENTS

The authors thank S.V. Shtelymakh for the help in decryptions of diffractograms

REFERENCES

1. Panin V.E. et. al., "Structural Levels of Plastic Deformation and Destruction", Novosibirsk, Russia, Science, 1990.
2. Mescheryakov Yu.I., Divakov A.K., "Interference Method for Registration of a High-Velocity Inhomogeneity of Particles in it Elastic - Plastic Waves of Loading in Solids", RAS, Saint - Petersburg, Russia, 1989, a pre-print no. 25, 36.
3. Luchinskij G.P., "Chemistry of Titanium", Moscow, Russia, Science, 1971.
4. Skotnikova M.A. et. al., "About Nature of Dissipative Processes at Cutting Treatment of Titanium Blanks", in *Titanium - 99, Science and Technology*, International IX proceedings, Saint Petersburg, Russia, 1999, vol. 3, pp. 1668 - 1674.
5. Skotnikova M.A. et. al., "Structural - Phase Transformation in Two-Phase Alloys at Pulse Loading", in *Actual problems of durability-2002*, International XL proceedings, Novgorod, Russia, 2002, pp. 12 - 15.

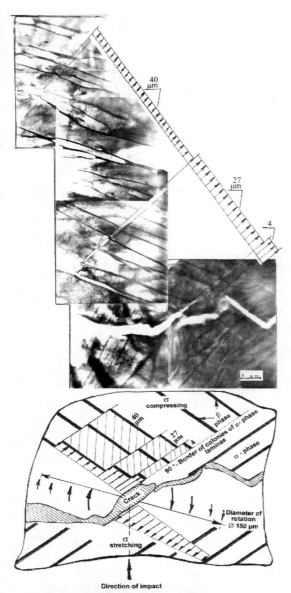

FIGURE 4. Electron microscope pictures and scheme of VT6 titanium alloy structure in a plane perpendicular to movement of shock wave. A sample № 1-2. x 3000.

modulates the material structure, breaking it into rectangular micro-blocks of the size 4 to 40 µm, along the boundaries of which the unloading wave produces a coordinated displacement. The displacements are realized in such a way that the smaller the distance from the center the greater the

CP706, *Shock Compression of Condensed Matter - 2003*
edited by M. D. Furnish, Y. M. Gupta, and J. W. Forbes
© 2004 American Institute of Physics 0-7354-0181-0/04/$22.00

HIGH STRAIN RATE COMPRESSION AND TENSION CHARACTERIZATION OF HIGH STRENGTH (AUTOMOTIVE) SHEET STEELS

I. H. Syed and N. S. Brar

Mechanical and Aerospace Engineering and Research Institute, University of Dayton, Dayton, OH 45469-0182

Abstract. Compression and tension split Hopkinson bar techniques are used to characterize high strength automotive sheet steels at strain rates of 500/s and 1000/s. Compression and tension data on two representative steel grades are given. There are differences in the measured compression and tension flow stresses for both the steels. The effect of strain rate and steel strength on these differences is discussed.

INTRODUCTION

Sheet steel products are the principal materials used to fabricate automobile body structure and chassis. Over the last decade, the sheet steel suppliers and automotive companies have undertaken a number of studies to determine the crashworthiness of the motor vehicle body structure and chassis. These studies involve numerical simulation of a motor vehicle structure to adequately protect the occupants from injury in the event of a crash against a rigid structure or another automobile. These simulations require high strain rate constitutive material models (e.g., Johnson Cook strength model) of the sheet steel, which are based on the material characterization at high strain rates and high/low temperature. Recently, the University of Dayton Research Institute (UDRI) performed high strain rate measurements for a small number of steels as part of the AISI/DOE Technology Roadmap Program, project 0038. Comprehensive high strain rate tension data, along with quasi-static data, and complete analysis are presented in Reference 1. The objective of this paper is to present the high strain rate compression and tension characterization techniques for thin steel sheet materials along with high strain rate data for two representative steel grades, a conventional High Strength Low Alloy (HSLA) steel, -HSLA350 and a high strength Dual Phase (DP) steel grade, -DP800. DP800 has higher yield strength and tensile strength compared to that of HSLA350.

EXPERIMENTAL METHOD

Materials

As part of the AISI/DOE project, Ispat Inland Inc. East Chicago, IN, supplied to UDRI a number of steel grades produced by several steel companies in the form of ~1.4-mm thick and 12"x12" sheets. Compression specimens in the shape of 3.17-mm diameter buttons through plate thickness within specified tolerances were machined. Tension specimens in the dog bone configuration (ASTM D1822 Type L) were fabricated from each steel sheet along the marked rolling direction using water jet technique.

Compression Split Hopkinson Technique

The Split Hopkinson Bar (SHB) is routinely used to characterize different materials (alloys, polymers, and ceramics) at high strain rates in the range of 100 to 2,000 sec^{-1} and at low/high temperatures. The schematic of the Compression Split Hopkinson Bar at the University of Dayton Research Institute is shown in Figure 1 [2]. It is configured with a striker bar and two pressure bars mounted on rigid supports on top of a steel beam aligned longitudinally. The 12.7-mm diameter pressure bars are made of Inconel 718. This pressure bar was chosen as the bar material due to very small degradation of its modulus at high temperatures to 1000°F. Two (1000 Ω) strain gauges are mounted on the incident and transmitter bars 0.91-meters away from the specimen to monitor strains in the pressure bars. The striker bar (0.76-m long) is accelerated with a torsional spring arrangement to a desired impact velocity. The striker bar impacts the incident bar (3.7-m long) and generates a compressive stress pulse in incident bar.

A 3.17-mm diameter disc shaped compression specimen is placed between the two pressure bars. The specimen is subjected to a compressive stress pulse generated by the impact of the striker bar with the incident bar. A portion of this incident compressive pulse, ε_i, is transmitted through the specimen ε_t and the remainder is reflected back in the incident bar ε_r. The strain gauges mounted on the pressure bars record the amplitude of incident, reflected, and transmitted pulses. Using the recorded strains, the stress (σ), strain (ε) and strain rate ($\dot{\varepsilon}$) in the specimen are determined using Equations (1) to (3).

$$\sigma(t) = E \frac{A_b}{A_s} \varepsilon_t(t) \qquad (1)$$

$$\varepsilon(t) = \frac{2 \cdot C_o}{L} \int_0^t \varepsilon_r(t) dt \qquad (2)$$

$$\dot{\varepsilon}(t) = \frac{2 \cdot C_o}{L} \varepsilon_r(t) \qquad (3)$$

where A_b and A_s are the cross-sectional area of the pressure bar and the specimen in the gauge section, respectively, and L is the gauge length of the specimen. The stress, strain, and strain rate are the average values and are determined by assuming a uniform uniaxial stress-state condition which implies that

$$\varepsilon_r = \varepsilon_i - \varepsilon_t$$

Direct Tension Split Hopkinson Bar Technique

Incident and transmitter pressure bars of the DTSHB are made of 25.4-mm diameter 7075 Aluminum. Two strain gages (1000 Ω) are mounted on each bar 1.22-m away from the specimen to monitor strains in the pressure bars. Tension specimens are placed in specially designed grips screwed into the threaded incident and transmitting aluminum bars. A 0.76-m long aluminum tube is launched around the incident bar and the impact of the aluminum tube against the aluminum anvil (rigidly attached to the end of the incident bar) generates a tensile stress pulse (Figure 1). This tensile stress pulse subjects the specimen to tensile loading. A portion of this incident tensile pulse, ε_i, is transmitted through the specimen ε_t and the remainder is reflected back in the incident bar ε_r. The amplitude of incident, reflected, and transmitted pulses are recorded by the strain gauges mounted on the pressure bars. Incident, reflected, and transmitted stress pulse data are analyzed following the procedure outlined above for compression split Hopkinson bar to obtain stress (σ), strain (ε), and strain rate ($\dot{\varepsilon}$) in the specimen.

RESULTS AND DISCUSSION

As stated, the aim of this paper is to present the compression and tension high strain rate techniques. Comprehensive high strain rate tension data, along with quasi-static tension data on a number of automotive steels are given in Reference 1. We present compression and tension data on two representative automotive sheet steels: HSLA350 and DP800 in order to point out the differences between their compressive and tension flow stress at strain rates of 500/s and 1000/s.

Three compression and three tension tests were performed on specimens fabricated from the two steel grades at the strain rates of ~500/s and ~1000/s. Elastic modulus (E) and bar wave velocity (C_O) for Inconel (compression bar material) used in the strain gauge data analysis are: $E = 1.95 \times 10^5$ MPa, and $C_O = 4.83 \times 10^3$ m/sec, respectively.

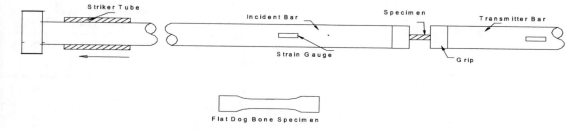

FIGURE 1. Schematic of the Direct Tension Hopkinson Bar

Elastic modulus (E) and bar wave velocity (C_O) for Aluminum (tension bar material) used in the strain gage data analysis are: $E = 6.968 \times 10^4$ MPa and $C_O = 5.08 \times 10^3$ m/sec, respectively. Compression and tension true stress versus true strain curves for the two steel grades are presented in Figures 2-3.

Tension stress-strain data for the two steel grades have large fluctuations, especially for strains below 0.05, compared to the relatively smooth compression data. This is likely due to the geometrical (flat dog bone) configuration. Tension specimens are configured in the flat (6.35-cm) dogbone shape with a relatively long gage length of 9.5-mm, where as the compression specimens are short 3.17-mm diameter button shaped cylinders of length equal to plate thickness (~1.4-mm). Stress equilibration between the specimen/bar interfaces in a tension test happens over a relatively long time giving rise to stress fluctuations at strains below 0.05. Once the stress equilibrium is attained, oscillation amplitude decreases.

Compression flow stresses for HSLA350 steel grade at a strain rate of 500/s are significantly higher (~15%) compared to the tension flow stresses for strains above 5% (Figure 2). The difference between compression and tension flow stresses for this steel notably decreases at a strain rate of 1000/s. In the case of high strength steel, DP800, compression and tension flow stresses agree within 5% at a strain rate of 500/s. At a strain rate of 1000/s, compression and tension data agree within experimental error. We conjecture that the surface layer of sheet steel may be harder as a result of the temper rolling or galvanization processes. These processes may have a significant effect in enhancing the compressive strength of steel, especially in case of thin (~1.4-mm thick) plates. Compression tests are performed on thin specimens punched out through the thickness dimension of the plates, whereas tension tests are performed on specimens fabricated in the rolling direction. Thus, in a tension test the interior of the specimen (bulk material) dominates the tensile properties with an insignificant effect of the surface layer. But surface layer hardness is likely to dominate the compressive strength of sheet steel. In addition, differences in the strain condition in the compression and tension testing modes also contribute to the differences in compressive and tensile flow stresses.

Disclaimer: The analysis and conclusions drawn in this paper are those of the authors and no claim is made as to the views and interpretations of the U. S. Department of Energy or those AISI companies sponsoring the Technical Roadmap Project #0038.

ACKNOWLEDGMENTS

The research reported in this paper was performed for the AISI/DOE TRP Project #0038 and was coordinated by Dr. Benda Yan of Ispat Inland Research. The authors thank the US DOE and ISI/Automotive Applications Committee for permission to use a part of the testing results for presentation and publication.

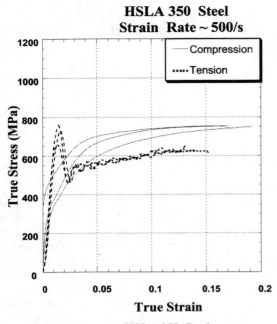

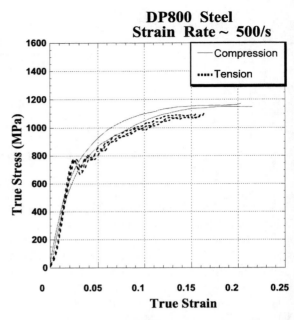

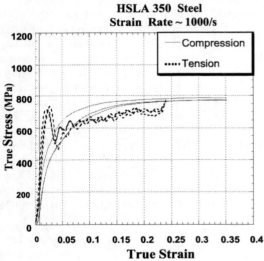

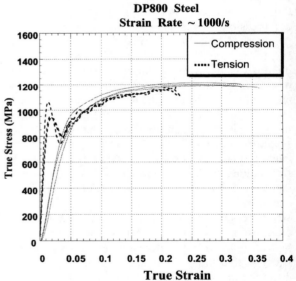

FIGURE 2. Stress-strain data for HSLA 350 steel at strain rates 500/s (top) and 1000/s (bottom).

FIGURE 3. Stress-strain data for DP800 steel at strain rates 500/s (top) and 1000/s (bottom).

REFERENCES

1. Yan, B. and Xu, K., "High Strain Rate Behavior of Advanced High Strength for Automotive Applications", Proceedings, The 44th Metal Working and Steel Processing Conferences, Vol. XL, pp. 493- 507, 2002.

2. Nicholas, T. Impact Dynamics. Eds. J.A. Zukas, T. Nicholas, H.L. Swift, L.B. Greszczuk, & D.R. Curran (Krieger Publishing Company, Malabar, FL), pp. 277-332, 1992.

CP706, *Shock Compression of Condensed Matter - 2003*
edited by M. D. Furnish, Y. M. Gupta, and J. W. Forbes
© 2004 American Institute of Physics 0-7354-0181-0/04/$22.00

A DISTRIBUTIONAL MODEL FOR ELASTIC-PLASTIC BEHAVIOR OF SHOCK-LOADED MATERIALS*

T. J. Vogler[1] and J.R. Asay[2]

[1]*Solid Dynamics & Energetic Materials Dept., Sandia National Laboratories, Albuquerque NM 87185*
[2]*Institute for Shock Physics, Washington State University, Pullman WA 99164*

Abstract. To address known shortcomings of classical metal plasticity for describing material behavior under shock loading, a model which incorporates a distribution in the deviatoric stress state is developed. This distribution will translate in stress space under loading, and growth of the distribution can be included in the model as well. This proposed model is capable of duplicating the key features of a set of reshock and release experiments on 6061-T6 aluminum, many of which are not captured by classical plasticity. The model is relatively simple, is only moderately more computationally intensive, and requires few additional material parameters.

* This work was supported by the U. S. Department of Energy under contract DE-AC04-94AL85000. Sandia is a multiprogram laboratory operated by Sandia Corporation a Lockheed Martin Company, for the United States Department of Energy

INTRODUCTION

Classical elastic-plastic theory has proven quite useful in simulating the behavior of metals in plate impact experiments. However, limitations of continuum rate-independent plasticity have been found over the years. For example, shocks are generally not homogeneous due to grain boundaries and crystal anisotropies, and elastic precursor decay is common in single crystal experiments. Discrete-element models [1] have illustrated fascinating local behavior such as vortex formation in polycrystals. Experimentally, the line-VISAR [2] has allowed researchers to probe spatial variations in particle velocities.

Other limitations of classical elasto-plastic theory are revealed by conventional plate impact experiments. For example, many materials display a gradual rise in particle velocity after the initial elastic precursor. This may be partly due to hardening, but there is probably some rate-dependence and variations in strength (due to orientation, presence of flaws, etc.) among material grains. Another anomalous behavior is the presence of an elastic precursor observed upon reshock [3]. This indicates that the shocked material is not on the current yield surface. Thus, it appears that the shocked material has either hardened after the initial shock has passed or "relaxed" away from the yield surface. Thermal trapping in regions of localized shear deformation has also been proposed [4].

The direct simulation of a polycrystalline material accounting for individual slip systems and grain orientations would be prohibitive for full-scale problems. Therefore, simplified models that allow for variations of the state of the material, especially its deviatoric state, are necessary. In this paper, the deviatoric stress model of Lipkin & Asay is extended to allow for evolution of the distribution of states. It is then used to simulate experimental data for 6061-T6 aluminum.[5]

MODEL DESCRIPTION

The present model is implemented as the strength component of a constitutive model in the one-dimensional explicit finite-difference code WONDY [6]. The hydrostatic response is determined independently using a variety of equation-of-state models.

In the center of a planar impact experiment, the deformation field is one of uniaxial strain. If projectile motion is in the x-direction and the material is isotropic, all components of shear stress and strain are zero, transverse strains are zero, and $\sigma_y = \sigma_z$. For a material capable of supporting shear stresses (*i.e.* it has strength), $\sigma_x \neq \sigma_y$. The deviatoric stresse is defined as

$$\sigma'_x = \sigma_x - \tfrac{1}{3}\left(\sigma_x + \sigma_y + \sigma_z\right) = \sigma_x - \bar{\sigma}. \qquad (1)$$

where $\bar{\sigma}$ is the mean stress. The maximal value of shear stress is thus $\tau = 0.75\,\sigma'_x$. As in Eq. 1, the deviatoric rate of deformation is defined as

$$d'_x = d_x - \tfrac{1}{3}\left(d_x + d_y + d_z\right) = \tfrac{2}{3} d_x \qquad (2)$$

where d_x etc. are the rates of deformation.

Following Lipkin & Asay [3], a Gaussian distribution of shear stresses is assumed for each material point as given by

$$N\left(\tau - \tau_m\right) = \frac{1}{\beta\sqrt{2\pi}} \exp\left[-\frac{\left(\tau - \tau_m\right)^2}{2\beta^2}\right], \qquad (3)$$

where N is the fraction of a material point that has a certain shear stress, β describes the width of the distribution, and τ_m is the center value of shear stress as illustrated in Fig. 1. In [3], the initial shock loading was ignored and the distribution of Eq. 3 was assumed present in the shocked state. In the current model, a distribution is assumed to be present initially and centered at $\tau_m = 0$; this may be regarded as residual stresses in the material or initial differences in grain orientations. As the material undergoes an increment in uniaxial strain, the shear stress is changed. The fraction of the distribution, λ, that currently lies within the yield surface is given by

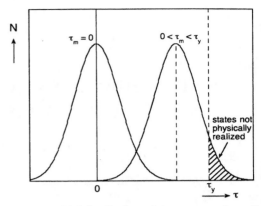

Figure 1. Initial distribution of shear stresses ($\tau_m = 0$) and distribution after loading and partial yielding.

$$\lambda = 0.5\left[erf\left(\frac{\tau_y - \bar{\tau}_m}{\sqrt{2}\beta}\right) + erf\left(\frac{\tau_y + \bar{\tau}_m}{\sqrt{2}\beta}\right)\right] \qquad (4)$$

where τ_y is the current value of the yield stress. For the variable $\bar{\tau}_m$, the value of τ_m from the previous step is used, unless the shear stress changes rapidly. In that case, the new shear stress is first estimated and then averaged with the previous step's value. Initially, the deviatoric stress increment, $\Delta\sigma'_x$, is proportional to the strain increment. Eventually, part of the stress distribution reaches the yield stress of the material as shown in Fig. 1. Then the stress increment includes only the fraction λ of the distribution which is less than the yield stress

$$\Delta\sigma'_x = 2\Delta t\, G\, d'_x\, \lambda \qquad (5)$$

where Δt is the time increment and G is the shear modulus. Thus, only the material which has not yielded contributes to the deviatoric stress. In the event of unload, the increment in deviatoric strain is entirely elastic and λ is omitted from Eq. 5. During the same strain increment, the value of τ_m changes according to

$$\Delta\tau_m = \tfrac{3}{2}\Delta t\, G\, d'_x. \qquad (6)$$

Note that τ_m will exceed τ if yielding has occurred and can become greater than τ_y. To prevent large overshoots in τ_m, its value is constrained to not increase if the value of λ is small, say less than

0.01. This constraint also prevents τ from exceeding τ_y by more than a modest amount. Since the mean stress of the distribution, τ_m, should be closely related to the actual shear stress in the material, τ, the value of the mean stress is relaxed toward the shear stress by

$$\Delta\tau_m = \frac{(\tau - \tau_m)\Delta t}{\zeta_1 h/C} \qquad (7)$$

where h is a characteristic dimension of the material (here the grain size), C is the current bulk sound speed, and ζ_1 is a constant that controls the relaxation time.

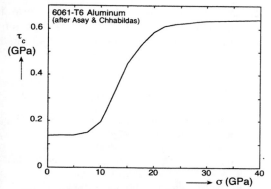

Figure 2. Shear strength of 6061-T6 aluminum.

The remaining inputs to the model are the yield stress τ_y and the distribution width β. For the simulations of aluminum, data for the critical shear stress [5] are used for τ_c, the limiting value of the yield stress as a function of the normal stress σ_x. The value of τ_c is assumed to increase but never to decrease. The current value of τ_y grows toward τ_c in a time-dependent manner as

$$\Delta\tau_y = \frac{(\tau_c - \tau_y)\Delta t}{\zeta_2 h/C} \qquad (9)$$

where ζ_2 is again a constant to control the rate of change. Finally, the distribution width, β, should depend upon the current stress, loading history, and micromechanical details of the material. Lacking data for a more accurate choice, it is chosen to be a function of the current yield stress $\beta = \tau_y / m$ where m is a constant. For large m, the model

reduces to an elastic-perfectly plastic model with hardening which occurs over time.

Table 1. Equation of state parameters for 6061-T6 aluminum alloy.

C_o (m/s)	s	ρ_o (kg/m^3)	Γ_o
5250	1.35	2703	2.14

Table 2. Parameters used in distributional model.

ν	$\tau_y^{initial}$ (GPa)	m	ζ_1	ζ_2	h (μm)
0.335	0.137	5	20	100	15

RESULTS

Six experiments [5] involving an initial shock and subsequent release or reshock were simulated with the current model. The equation-of-state parameters are listed in Table 1, while those for the strength model are shown in Table 2. As can be seen in Fig. 3, agreement with the experimental results is very good for all but the two lowest pressure shots. The model captures the precursor observed on reloading, as well as the finite rise time of reshock, very well. During release and at the lowest impact stress, though, it exhibits a more sharply defined yield point than is actually observed. The results do not strongly dependent on the choice of time parameters ζ_1 and ζ_2.

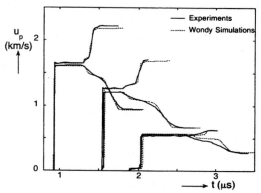

Figure 3. Measured and predicted particle velocities for reshock and release plate experiments on 6061-T6 aluminum.

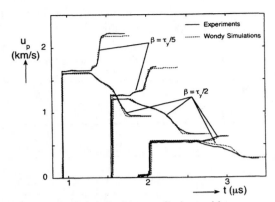

Figure 4. Particle velocities predictions with distribution parameter β varied to improve agreement.

To improve agreement with experiments, a wider distribution (m smaller) was used in simulating four of the experiments. When $m = 2$, the results for all three release experiments as well as the lowest pressure reshock experiment improve significantly as seen in Fig. 4. The yield during release is much less sharply defined, resulting in better agreement with experiment. It is unclear whether this better agreement reflects a physical process which is being captured by the model.

DISCUSSION AND CONCLUSIONS

Motivated by the experimental data on reshock and release of Asay & Chhabildas, a model has been developed that includes a distribution of deviatoric stresses. This seems plausible as individual grains will have different slip system orientations than their neighbors. While the choice of a Gaussian distribution may be somewhat incorrect, it is a reasonable first approximation. The presence of the distribution makes the precursor on reshock less sharp and the release path more smooth. A significant unknown is how the distribution evolves as a function of time and loading. The distributions probably evolve as a result of loading after the initial shock. There may also be a time-dependent evolution; for example, the distribution may grow or translate over time after the passage of the initial shock. This could be investigated by conducting one experiment in which the reshock arrives soon after the initial shock and a second in which it arrives later.

However, the time scale of evolution may be too small to be captured by such experiments.

The second interesting aspect of the problem is the presence of an elastic precursor upon reshock. There are at least two possible explanations: the deviatoric stress is "relaxing" so the material moves toward hydrostatic compression or the yield surface moves away from the current stress state. The first seems plausible, but it would lead to decreasing particle velocities after the initial shock, which does not typically occur. The second is just a variation on the concept of hardening. Again, this issue could be probed experimentally.

While the model provides good results for aluminum, it is also expected to be useful for high strength materials such as ceramics. Ceramic materials often have strong crystallographic anisotropies and very high yield/failure strengths that may be captured by a model of this type.

REFERENCES

1. K. Yano and Y. Horie, "Discrete-element modeling of shock compression of polycrystalline copper," Phys. Rev. B **59**, pp. 13672, 1999.
2. W. M. Trott, M. D. Knudson, L. C. Chhabildas, and J. R. Asay, "Measurements of spatially resolved velocity variations in shock compressed heterogeneous materials using a line-imaging velocity interferometer," in *Shock Compression of Condensed Matter*, 1999 (M. D. Furnish et al., eds.), pp. 993-998.
3. J. Lipkin and J. R. Asay, "Reshock and release of shock-compressed 6061-T6 aluminum," J. Appl. Phys. **48**, pp. 182, 1977.
4. J. W. Swegle and D. E. Grady, "Calculation of thermal trapping in shear bands," in *Metallurgical applications of shock-wave and high-strain-rate phenomena*, edited by L. E. Murr et al. (New York, 1986).
5. J. R. Asay and L. C. Chhabildas, "Determination of the shear strength of shock compressed 6061-T6 aluminum," in *Shock Waves and High-Strain-Rate Phenomena in Metals*, edited by M. A. Meyers and L. E. Murr (Plenum, New York, 1981), pp. 417-431.
6. M. E. Kipp and R. J. Lawrence, Sandia National Laboratories report SAND81-0930 (1982).

CP706, *Shock Compression of Condensed Matter - 2003*
edited by M. D. Furnish, Y. M. Gupta, and J. W. Forbes
2004 American Institute of Physics 0-7354-0181-0/04/$22.00

DIRECT COMPARISON OF A RANGE OF STRAIN-RATE DEPENDENT STRENGTH MODELS ON TAYLOR IMPACT TESTS

Stephen J White

Design Physics Department, AWE, Aldermaston, Berkshire, RG7 4PR, United Kingdom

Abstract. A number of strain rate dependent material strength models have been implemented in AWE hydrocodes. This paper serves to highlight the current capability to model material deformation, the advances made with the implementation of rate dependent models, and the basic differences between them in modelling a simple dynamic experiment. The comparison was carried out on tantalum - a material for which every rate-dependent model has published data.

INTRODUCTION

In recent years, it has become clear that the deviatoric stress state of a condensed material under shock loading does not follow a simple stress-strain relationship. Instead, the strain *rate* must be included in all calculations in order to accurately recover material response. Although other state parameters may well have an influence (such as pressure), the effect is not as strong as with strain rate. It took quite some time for the materials community to realise that the constitutive models formulated for low strain rates ($\sim 10^1 \mathrm{s}^{-1}$) were not suitable for application at higher rates.

At present, the only material strength options available to the user are based around simple stress-strain loci evaluations:

1. Constant Y,
2. the Steinberg-Guinan model, and
3. the Modified Steinberg-Guinan model.

Of the models available in the open press, the author has chosen to implement the following:

- Johnson-Cook (JC)[4],
- Steinberg-Lund (SL)[5],
- Armstrong-Zerilli (AZ)[1], and
- Mechanical Threshold Stress (MTS)[2,3].

Although there are only four items in the above list, in reality there are a total of seven new forms implemented in the production code. This is because

AZ has a different form for body centre cubic (bcc) and face centre cubic (fcc) materials, and MTS has a similar split, plus another for hexagonal close packed (hcp) structures.

Johnson-Cook

$$\sigma = [A + B\varepsilon^n][1 + C\ln\dot{\varepsilon}^*][1 - T^{*m}] \qquad (1)$$

where ε is the equivalent plastic strain, $\dot{\varepsilon}^* = \dot{\varepsilon}/\dot{\varepsilon}_0$ is the dimensionless plastic strain rate for $\dot{\varepsilon}_0 = 1.0\mathrm{s}^{-1}$ and $T^* = (T - T_{room})/(T_{melt} - T_{room})$ the homologous temperature. The five material constants are A, B, C, n and m.

Steinburg-Lund

$$Y = \left[Y_T\left(\dot{\varepsilon}_p, T\right) + Y_A f_{whd}\left(\varepsilon_p\right)\right]\frac{\mu(P, T)}{\mu_0} \qquad (2)$$

This is an extension to Steinberg's original method, with the addition of a thermal term, Y_T, which allows for rate dependence whilst maintaining the ability to be solved via a Prandtl-Reuss treatment in the hydrocodes. Strictly, Y_T accounts for the plastic strain rate dependent over-stress encountered during plastic deformation thus:

621

$$\dot{\varepsilon}_p = \left\{ \frac{1}{C_1} \left[\frac{2U_K}{k_b T} \left(1 - \frac{Y_T}{Y_P} \right)^2 \right] + \frac{C_2}{Y_T} \right\}^{-1} \quad (Y_T \le Y_P) \tag{3}$$

Armstrong-Zerilli

Face-centered-cubic.

$$\sigma = \Delta \sigma'_G + kl^{-1/2} + c_2 \varepsilon^{1/2} e^{(-c_3 T + c_4 T \ln \dot{\varepsilon})} \tag{4}$$

Body-centered-cubic.

$$\sigma = \Delta \sigma'_G + kl^{-1/2} + c_5 \varepsilon^n + c_1 e^{(-c_3 T + c_4 T \ln \dot{\varepsilon})} \tag{5}$$

Mechanical Threshold Stress (MTS)

$$\sigma = \sigma_a + \sum_{i=1}^{n} \hat{\sigma}_i s_i (\dot{\varepsilon}, T, P) \tag{6}$$

where s_i is a constant-structure deformation term, which modifies the mechanical threshold stress to account for the standard deformation state variables and reduces to unity at 0K. There is a complementary equation to 6 which governs the evolution of the dislocation population:

$$\frac{\partial \hat{\sigma}_i}{\partial \varepsilon} = \Theta_0 [1 - F(X)] \qquad \text{for interaction i} \tag{7}$$

where $X = \hat{\sigma}_i / \hat{\sigma}_{i_s}$. The parameter Θ_0 represents the material hardening from dislocation generation, and the product $\Theta_0 \cdot F$ gives the softening. The variable $\hat{\sigma}_{i_s}$ is called the saturation threshold stress, which is the threshold stress at zero strain hardening.

Face Centred Cubic.

$$\sigma = \sigma_a + \hat{\sigma}_d s_d \tag{8}$$

where the d subscript represents dislocation interactions only. σ_a is a constant.

Body Centred Cubic.

$$\sigma = \sigma_a + \hat{\sigma}_d s_d + \hat{\sigma}_p s_p \tag{9}$$

The extra term is designed to handle these Peierls barriers, subscripted p. Although the dislocation term maintains the same form as in the fcc case, the Peierls term has a constant value of $\hat{\sigma}_p$ instead of a strain-dependent differential. This is because it does not evolve after yielding.

Hexagonal Close Packed.

$$\sigma = \sigma_a + \hat{\sigma}_d s_d + \hat{\sigma}_i s_i + \hat{\sigma}_s s_s \tag{10}$$

Once again, only the dislocation interaction has an evolution term, meaning $\hat{\sigma}_i$ and $\hat{\sigma}_s$ are constants.

PERFORMANCE OF ALL STRENGTH MODELS

As a means of comparing the performance and accuracy of each model, a standard test problem on a standard test material was required. The vehicle used to compare all trials is the G I Taylor test. This is a simple but effective material strength test, whereby a rod of test material is fired end-on at a witness surface and the resulting shape of the rod analysed. The final shape is sensitive to the material strain history and is a good gauge of a model's capability. All the calculations below are carried out on a 10cm long, 1.5cm diameter rod, impacting a rigid wall at 0.019cm/μs. All runs were on a Ta rod with a 1/2mm mesh, half-space axisymmetric geometry, and ran until 99.99% of the initial energy had been converted into plastic work.

FIGURE 1. Constant Y Final G I Taylor Shape

The AWE default form for tantalum has a Balmoral EoS form with a constant strength treatment, i.e. Y and μ are held constant at 10.6kb and 688kb respectively. The final rod shape after 99.99% energy

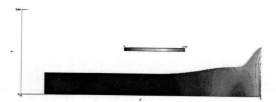

FIGURE 2. Weakened Y Final G I Taylor Shape

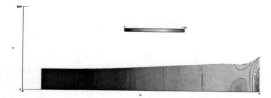

FIGURE 3. Johnson-Cook Final G I Taylor Shape

FIGURE 5. Armstrong-Zerilli Final G I Taylor Shape

FIGURE 6. Mechanical Threshold Stress Final G I Taylor Shape (with Plastic Strain Contours)

conversion can be seen as Figure 1, with plastic work contours overlayed. This result shows a final rod profile which divides naturally into two parts: the plastically deformed portion nearest the impact surface, and the remaining undeformed part.

Before contemplating the effects the various strength models have on the calculation of the problem, it is worthwhile noting the effect that simply weakening the material (ie by reducing Y) has on the trial. Figure 2 shows the resulting profile and plastic work contours of a G I Taylor test with Y lowered to 7.0kb (μ held at 677kb).

It is clear from the comparison between the two figures above that weakening the strength causes more of the rod to undergo some form of plastic deformation, as would be expected. There is also a reduction in the overall length of the rod at the end of the run, which would naturally follow from the previous statement. Also the thickness of the deformed

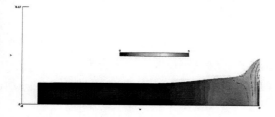

FIGURE 4. Steinberg-Lund Final G I Taylor Shape

portion is greater, both at the impact surface and further back along the deformed length.

By comparing these figures with Figure 1, the immediate difference can be seen in the apparent weakening of the rod when using Steinberg-Guinan. Looking at the figures for the Steinberg-Guinan tantalum model, at zero pressure, room temperature and zero plastic strain, the yield stress is 7.7kb. Although the model allows the stress to increase with both pressure and strain, the G I Taylor test here does not bring the yield surface to the 10.6kb figure used in the first constant strength calculation (the most it reached in this run was approximately 9.5kb). Therefore although the model gives a roughly 20% increase in flow stress due to hardening, the effect is short-lived before the material relaxes and softens due to rarefactions from the rod's surface.

The Johnson-Cook model is quite distinct from all the other runs in the magnitude of plastic work that is done; it is almost half the value of others. Reading around, it is clear that Johnson-Cook fails to adequately model most high strain rate situations, including plate impact.

The run predicts yield strengths as high as 17kb, suggesting the lack of any limiting yield strength (as in Steinberg-Lund) allows unrealistic results to be calculated in high strain rate regimes.

Comparing the Steinberg-Lund result against the Steinberg-Guinan shape, there are differences, although none simply attributable to one model predicting greater strength than the other. For instance, although the length of the undeformed section suggests that Steinberg-Lund predicts a stronger response, the degree of deformation at the impact face suggests the contrary.

The expected result of a direct comparison between the models would be to find Armstrong-Zerilli predicting a weaker response. It is thought that this is related to the fact that Steinberg-Guinan predicts a yield surface which (for the deformations seen in this problem at least) increases with strain, so the rod's impact face (where strains are highest) becomes stronger than the remainder of the rod. Thus deformation at the impact end is inhibited and, instead, a slightly greater share of the plastic work required to stop the rod occurs in its hitherto undeformed section. In the Armstrong-Zerilli case, for the strains seen at the impact face, the yield strength actually falls with increasing strain, causing the extreme sideways movement of material along the witness plane. This effect is similar to the one seen in the Steinberg-Lund case, where the yield surface at the impact face decreased at late time, allowing more deformation.

Finally, for the MTS run, the most striking difference that can be seen with this run up to the previous ones in the dramatic reduction in the overall length of the final rod. This length difference is seen between Figures 1 and 2, suggesting that MTS is indeed a generally weaker model for Ta. Notwithstanding, the final topology of the impact face is no wider than for the other rate dependent runs. The only answer to this riddle is that the bulk of the rod remains at a lower yield point for longer in the MTS model, compared to its rate dependent counterparts. Starting out at 4kb, the athermal yield stress is very low, but the strength soon increases due to work (dislocation) hardening.

CONCLUSION

By running challenging tests such as G I Taylor, the implementation or $\dot{\varepsilon}$ models is tested, and comparisons drawn between the various methods now available. There are still other models that have not been implemented yet, including the LANL model entitled PTW (Preston, Tonks & Wallace)[6], and an elastic viscoplastic model developed at AWE some years ago by Ian Gray[7].

ACKNOWLEDGMENTS

The author would like to acknowledge the effort of Michael Jeffery of the AWE Design Physics Department for his help in the gathering of data.

REFERENCES

1. F J Zerilli & R W Armstrong. *Description of Tantalum Deformation Behaviour by Dislocation Mechanics Based Constitutive Relations.* J. Appl. Phys, **68** (4), Aug 1990. Pp 1580-1591.

2. S R Chen & G T Gray III. *Constitutive Behaviour of Tantalum and Tantalum-Tungsten Alloys.* 2nd International Conference on Tungsten and Refactory Metals, 1995.

3. P J Maudlin, R F Davidson & R J Henninger. *Implementation and Assessment of the Mechanical-Threshold-Stress Model Using the EPIC2 and Pinon Computer Codes.* Los Alamos National Laboratory Report, LA-11895-MS, Issued September 1990.

4. G R Johnson & W H Cook. *A Constitutive Model and Data for Metals Subject to Large Strains, High Strain Rates and High Temperatures.* 7th International Symposium of Ballistics, April 1983, The Hague. Pp 541-547.

5. D J Steinberg & C M Lund. *A Constitutive Model for Strain Rates from 10^{-4} to 10^6.* J Appl Phys, 65, Pp 1528-1533. 1989.

6. D L Preston, D L Tonks & D C Wallace. *Model of Plastic Deformation for Extreme Loading Conditions.* J Appl Phys, 93, Pp 211-220. 2003.

7. I N Gray. *A Rate Sensitive Constitutive Model for Metals.* Internal AWE Technical Report. 1994.

CP706, *Shock Compression of Condensed Matter - 2003*
edited by M. D. Furnish, Y. M. Gupta, and J. W. Forbes
2004 American Institute of Physics 0-7354-0181-0/04/$22.00

METALLOGRAPHY OF SUB-SURFACE FLOWS GENERATED BY SHOCK-INDUCED FRICTION

R E Winter, V S Smeeton, J De'Ath, P Taylor, L Markland and A J Barlow

AWE, Aldermaston, Reading, Berks, RG7 4PR, UK

A configuration is described in which an explosively-generated shock is driven obliquely across the interface between aluminium and steel anvils causing them to slide against each other. The interface pressures and the time at which the surfaces separate, are estimated using hydro-code calculations. The recovered aluminium samples are then sectioned, etched, and photographed allowing sub-surface flows at the time of separation to be estimated from the grain distortions. The results are compared with predictions from the hydrocode CORVUS.

INTRODUCTION

One method of determining the sub-surface flow in samples recovered from shock-induced friction experiments is to make use of the fact that the grains in extruded metal bars are aligned along the axis. An aluminum alloy bar was machined on its end faces and sectioned and etched. It was seen that the fiducials provided by the grains appeared accurately straight but that there were slight variations, of order of a degree or two, in the alignment of the grains along the axis. Further, there was no evidence of local deformation at the end faces (caused for example by the machining process). Based on this evidence it was judged that observations of the bending of the grains could indeed provide useful information on sub-surface flows generated by shock-induced friction.

A configuration, designated FN6, has been designed using the AWE Lagrangian hydrocode CORVUS. A general account of the work was described in Ref. 1. This paper records the results of recent further analysis of the FN6 experiments.

THE FN6 CONFIGURATION

An "infinite slab" version of FN6 set-up using the AWE Lagrangian code CORVUS is shown in Fig. 1. A shock generated by a 12 mm slab of explosive is driven obliquely across the interface between aluminium and steel anvils causing them to slide against each other. The angle at which the shock wave intersects the interface, and the stress generated, vary along the length of the interface. It was anticipated that the great majority of any sub-surface flow would be found in the aluminium since this is the softer material. A 5 mm region adjacent to the interface was meshed more finely than the rest of the problem. In the example shown the fine meshes measured 0.125 mm perpendicular to the interface. The three sites marked L, M and R (Left, Middle and Right) are positioned 15, 30 and 45 mm from the left-hand side of the inner components.

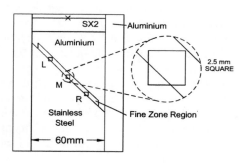

FIGURE 1 Set-up used in CORVUS

The problem was run (a) with a merged interface representing infinite friction and (b) with a sliding, or free, interface representing zero friction.

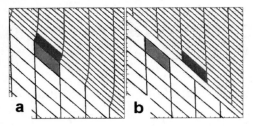

FIGURE 2. CORVUS Calculations run with merged and free interfaces. At the time of this frame, 16 μs, the Aluminium and steel components have just separated in the free calculation. The frames shown are 2.5 mm square.

Figure 2 (a) shows output at location "R" from the merged CORVUS calculation. Note that the merged interface treatment does not allow the surfaces to separate. Figure 2 (b) shows output from the same location and time but using an interface treatment that allows the metals to both slide and separate. It can be seen that, at 16 μs, the anvils are just separating. We conclude from these calculations that, in an experiment based on the FN6 configuration, the aluminium and steel anvils would separate at ~16 μs and that after this time no further sub-surface flow would occur. Therefore examination of the flows in the recovered samples provides a picture of the flow patterns at all times after 16 μs.

Figure 2 (b) also illustrates that, as expected, the relative motion at and near the interface is much more with the slip than with the merged treatment. Further the figure shows the differences between the distortion of the originally vertical mesh lines for merged and free treatments. In the merged calculation the axial mesh lines, which were straight and vertical in the time-zero set–up, are now curved. In the slide calculation the lines have remained nominally straight but their angle has changed very slightly.

The mesh lines at the three chosen locations are depicted together in Fig. 3. On the basis of these comparisons it was decided that there were good prospects that experimental observation could distinguish between these extremes and therefore that it was worth proceeding with an experimental investigation.

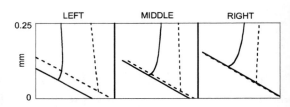

FIGURE 3. Grid lines generated at locations L, M and R by CORVUS run with merged and free interfaces.

EXPERIMENTS

60 mm diameter cylinders of aluminium alloy (EN AW5083 to BS EN 573-4) and stainless steel (321512 to BS EN 10088-3 (X2CrNi 18-9)) with one end machined at 45° were assembled with their angled faces in contact. The mating surfaces were machined to a surface roughness of 3.2 μm and were flat such that a gap of > 25 μm at any position was unlikely. The drive was provided by a centrally initiated disc of the RDX-based sheet explosive SX2.

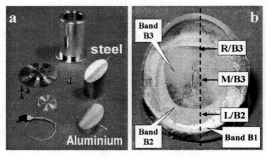

FIGURE 4. (a): FN6 Components. (b): the rubbed surface of the aluminium component from the experiment with 12 mm of SX2 showing the different surface bands and the points from which sections were taken.

A photograph of the components for a typical experiment is seen in Fig. 4 (a). The sliding elements were surrounded by a tubular aluminium momentum trap (shown as the outer cylinder in Fig. 1), which was itself surrounded by sand contained in a 1 m³ wooden box. Four experiments, designated FN6/1-4, were fired with different thicknesses of SX2. The sliding elements from all four experiments were recovered intact. The aluminium element from the most severely loaded experiment (12 mm of SX2) is shown in Fig. 4 (b).

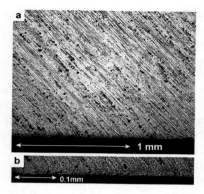

FIGURE 5. Micro-graphs from location M/B3

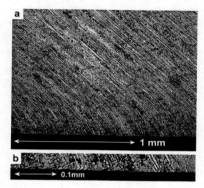

FIGURE 6. Micro-graph from location L/B2.

The sliding surfaces of the aluminium showed three sharply delineated, roughly concentric regions or bands. The bands, here designated B1, B2 and B3, can be seen in Fig. 4 (b). The outermost band, B1, ~ 10mm wide, shows a surface pattern similar to that seen on the as-machined component. The next band in from the outside, B2, again averaging ~ 10 mm wide, showed fine striations in the expected sliding directions suggesting that significant sub-surface deformation had occurred. The border between B1 and B2 was formed by a series of splash-like features, strongly suggesting that the aluminium in this region had melted. Finally in the elliptical area which formed the central region, B3, the surface appeared to be lightly imprinted by the machining pattern of the stainless-steel anvil.

The recovered aluminium cylinders were sectioned along planes close to their centres. Three samples were machined from each section at positions 1/4, 1/2 and 3/4 of the distance from the

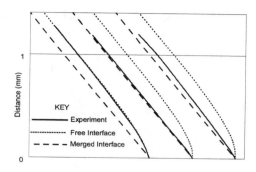

FIGURE 7. Comparison between experimental and CORVUS run with merged and free interfaces. Locations L, M & R are shown from left to right.

left-hand edge of the system as depicted by the arrows in Fig. 4 (b). The samples were then polished, etched and photographed. A set of micrographs for the experiments fired with 3, 9 and 12 mm charges were published in Ref. 1. Photomicrographs from the experiment with 12 mm of SX2 are reproduced in Figs. 5 and 6 which shows sub-surface structure from sites M/B3 and L/B2 respectively. Tracing paper was over-laid on each micrograph and a number of fiducial lines were drawn by eye. The set of fiducials obtained from each micrograph was then digitised and a representative average profile for each micrograph deduced.

ANALYSIS AND DISCUSSION

It was noted earlier that in pristine samples from the aluminium stock the angle of the grain fiducials could vary from axial by a degree or two. Because of this uncertainty we chose to focus on the curvature of the fiducials rather than the angle. In Fig. 7, the x-axis represents the interface. All of the curves have been rotated such that their slopes at a depth of 1 mm into the sample are exactly 45°. This enables the curvature of the fiducials in the 1 mm band next to the interface to be compared. The diagram shows the measured fiducials compared with the CORVUS predictions obtained assuming merged and free interfaces.

This discussion focuses mainly on the regions L/B2 and R/B3, (as defined in Fig. 4 (b)), since the results indicate that different mechanisms are at play in these two regions. The pressures and incident angles of the shocks at the two locations of interest are as depicted in figure 8.

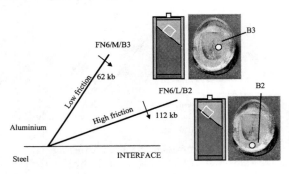

FIGURE 8. Showing the different shock conditions at locations L/B2 and M/B3.

Near the centre of the aluminium sample, the wave front makes a relatively large angle to the interface which would be expected to generate a relatively high relative velocity between the aluminium and steel. Further the pressure is relatively low which will tend to give a relatively low frictional force. We postulate, therefore, that the stress field at M/B3 could produce early localised flow at the interface, leading to an increase in temperature of the material near the interface. The increase in temperature leads to thermal softening which counteracts the effect of the work hardening. The splashes observed at the boundary between Bands 1 and 2 support the suggestion that thermal softening played a role here. The situation envisaged appears similar to that responsible for the formation of adiabatic shear bands in which deformation remains confined to a thin localised band within which the flow stress is much lower than the normal material strength. The result of thermal softening in the FN6 configuration is that the frictional forces appear low. This hypothesis is supported by the observation, shown in figures 5 (a) and (b), that the sub-surface deformation was mainly confined to a narrow band a few tens of microns thick.

As shown in figure 8, at location L/B2, the shock wave front makes an angle of ~ 23° to the interface and has a peak pressure of ~112 kb. Sections taken at this location, seen in figure 6, show that sub-surface flow extends to a depth of over 1 mm but, in this case, there is little evidence of localised surface flow. The curvature of the flow lines, as determined by examining the grain structure, are similar to those which hydrocode calculations suggest would occur if the aluminium and steel were bonded together. This bonding may result from the higher normal applied stress and the fact that the initial differential velocity of the metal components is less. Once bonded behaviour has started, flow in the aluminium layer near the interface leads to work hardening which causes the shear stresses to be transmitted into the bulk of the material.

The regions B1, B2 and B3 seen in Fig. 4 (a) are remarkably sharply delineated, an observation which suggests a sharp transition from one mechanism to another. The occurrence of B2 or B3-type behaviour depends whether thermal softening or work-hardening win the contest to control the distribution of deformation. Intermediate flow patterns in which thermal softening and work hardening mechanism co-exist may not be stable.

CONCLUSIONS

This study suggests that a phenomenon akin to catastrophic (or adiabatic) shear plays a role in shock-induced friction. Depending on the direction and magnitude of the incident shock the interface can act as a nucleus for the formation of a thin, thermally softened, layer. Therefore any physics-based approach to modelling shock-induced friction should incorporate a treatment of those parameters, such as thermal softening and heat conduction, which also control catastrophic shear. (See, for example, Ref. 2).

REFERENCES

1. Winter, R E., Smeeton, V. S., Taylor, P., Carley, D. J. and Markland L., Proceedings of 5th Int. Symp. on Behaviour of Dense Media under High Dynamic pressures, St-Malo, France, June 2003.
2. Recht, R. F., J Appl. Mech. **31**, 189, 1964.

CP706, *Shock Compression of Condensed Matter - 2003*
edited by M. D. Furnish, Y. M. Gupta, and J. W. Forbes
2004 American Institute of Physics 0-7354-0181-0/04/$22.00

PROGRESS IN COMPUTATIONAL MODELS FOR DAMAGE FROM SHEAR BANDS AND VOIDS

T.W. Wright[1,2], S.E. Schoenfeld[1], K.T. Ramesh[2], and X.Y. Wu[3]

[1]*U.S. Army Research Laboratory, Aberdeen Proving Ground, MD 21005*
[2]*Mechanical Engineering, Johns Hopkins University, Baltimore, MD 21218*
[3] *Department of Surgery and Biomedical Engineering, Emory University, Atlanta, GA 30329*

Abstract. Computational solid mechanics (CSM) for ballistics, as it exists today, does not have predictive capabilities that are comparable to those of computational fluid dynamics (CFD). The most important reason for this is the lack of high quality physical models for the damage and failure processes that occur during high-speed impact. An approach for modeling impact damage in metals, based on physical modeling of adiabatic shear bands and voids, will be presented. Scaling laws for damage from shear bands have been developed over the past decade from detailed analysis of the thermomechanical processes, and are being adapted for efficient use in large-scale computations. A similar approach for ductile void growth and spall has also been initiated. This approach is expected to lead to models that are based on the essence of the physics, rather than on fitting of phenomenological models to large databases.

INTRODUCTION

In the fields of ballistics and impact protection failure is expected as part of proper functioning, and therefore, the kinetics and dynamics of all major damage mechanisms should be part of the design. However, because of the localized nature of most damage mechanisms, and because of the strong coupling between mechanical and thermal effects in their formation, it is usually impractical to calculate their occurrence entirely from first principals in large-scale numerical simulations. Alternate means of predicting their behavior must, therefore, be sought. This paper will discuss an attempt to develop this fundamental idea for the particular examples of the formation and growth of adiabatic shear bands and voids.

The approach advocated here is to make a thorough study of the physics and mechanics of the formation of each damage mechanism, using analysis of simple, yet physically representative,

problems in continuum mechanics, and by extracting from those problems algebraic (or equally simple) scaling laws that capture the essential physics of band formation and of void initiation and growth. The resulting scaling laws must then be injected into large-scale calculations in a realistic way so that the macroscopic effect is adequately captured. The necessity and difficulty of this last step should not be underestimated, and the application to realistic design problems will always require computational mechanics at a high level, accompanied by experimental verification.

Much can be learned from simple dynamical problems with only one spatial dimension. Because of its high aspect ratio an adiabatic shear band may often be analyzed as a one-dimensional canonical problem from which one may deduce scaling laws. Similarly much can be learned about void formation and growth from simple spherically symmetric problems. These too are canonical problems with only one spatial dimension.

NUCLEATION OF SHEAR BANDS

Many scaling laws for adiabatic shear bands now exist in the literature so that today simple representations may be given for the timing of formation, for the cross sectional structure of a fully formed band, for the most probable spacing between shear bands both at initiation and in full development, for the structure around the tip of a propagating shear band in either mode-II or mode-III type motion, and for the speed of propagation. Furthermore, experimental evidence, which tends to validate the theoretical predictions qualitatively and to some extent quantitatively, is slowly accumulating. These developments have been summarized by Wright [1]. More recent developments, [2], will be briefly sketched below.

The timing or critical strain when an adiabatic shear band first nucleates is not a fixed material property, as is sometimes supposed. Although timing does depend on material properties, it also depends on material or geometric defects and on the process of deformation itself, as influenced for example by loading, boundary conditions, and the geometry of the material being deformed. Molinari and Clifton [3] predicted and Duffy and Chi [4] verified experimentally that the nominal strain at localization in a torsional Kolsky bar experiment would depend logarithmically on the strength of a particular geometric defect. Heat conduction appears to explain certain discrepancies between the results in [3] and [4], see Fig. 6.11 in [1].

A simple model for shearing of a short strip illustrates these features.

Momentum:	$s_y = \rho v_t$	,	$v(\pm h,t) = \pm v_0$
Energy:	$\rho c \theta_t = k\theta_{yy} + sv_y$	,	$\theta_y(\pm h,t) = 0$
Flow Law:	$s = F(\kappa, \theta, v_y)$		
Work Hardening:	$\dot{\kappa} = M(\kappa, \theta) sv_y$		

$$(1)$$

These equations represent an idealized material that deforms as a rigid plastic solid (with work hardening, but no elasticity effects). In addition the material softens as it heats up due to plastic work and hardens with increasing strain rate.

Subscripts y and t indicate partial differentiation with respect to space and time. The dependent variables are the driving shear stress, s, the particle velocity parallel to the band, v, the temperature difference from the initial average temperature, θ, and the work hardening parameter, κ. The flow stress, F, and the hardening rule, M, must be prescribed. Various rules are possible, but power law work hardening, linear or exponential or power law thermal softening, and power law or logarithmic rate hardening are most often used.

Physical constants are density, ρ, heat capacity, c, thermal conductivity, k, and strain rate sensitivity, m. The Taylor-Quinney coefficient, which designates the fraction of plastic work, sv_y, converted to heat, has been set equal to 1.0, but all other cases are easily recovered. The strip is subjected to a nominal strain rate of $\dot{\gamma}_0 = v_0/h$.

The basic procedure in developing scaling laws is to analyse (1) for slightly nonuniform initial conditions by a straightforward perturbation procedure, which is described in detail in [5]. The homogeneous or adiabatic case, which corresponds to uniform initial conditions, shows an increasing temperature, whereas the stress increases at first, but then passes through a maximum followed by a decreasing stress. Initial perturbations grow slowly until maximum stress; then grow far more rapidly. Typical numerical solutions confirm these results.

The main result from [5] to be used here is that the nominal plastic shear strain, γ_{crit}^p, at which band formation might be expected should be related approximately to the initial defects by a simple scaling law as shown in (2).

$$\eta = f(\beta) \text{ , where}$$

$$\eta = \sqrt{\left(-S_{\gamma^p\gamma^p} / 2mS\right)_{max}} \left(\gamma_{crit}^p - \gamma_{max}^p\right), \quad (2)$$

$$\beta = -\left(\frac{F_\kappa M}{mSM_0}\right)_{max} \quad \lambda_0(0) = \frac{(v_y)_{max}}{\dot{\gamma}_0} - 1.$$

The subscript 'max' indicates that the terms in parentheses should be evaluated at maximum homogeneous stress, and the subscript '0' indicates evaluation at the initial value. S indicates the stress for adiabatic response. Note the dependence on the curvature of the adiabatic response, which is to be evaluated at peak stress where the slope of the adiabatic response vanishes. F and M are the response functions in (1), and $\lambda_0(0)$ is a composite initial perturbation in the center of the band, which includes a linear combination of both strength and temperature.

Later in [6], the linear perturbation results were compared with numerical results, and it appeared that a good upper bound approximation for the strain at localization could be written

$$\eta = \frac{1}{\sqrt{\pi}} \ln \frac{1}{\beta} \qquad (3)$$

It was recognized only recently, however, that equation (3) could serve as the basis of a damage model if the second form for β in equation (2) is used rather than the first, which is based on initial perturbations that are largely unknown in realistic problems. But the second form, also derived from the perturbation analysis, refers to the strength of the perturbation in the velocity gradients at the moment of peak adiabatic stress. As interpreted in [2] equation (3) becomes

$$\left(t_{cr} - t_{max}\right)\dot{\gamma}_0 = \sqrt{\left(\frac{2mS}{-\pi S_{\gamma^p\gamma^p}}\right)_{max}} \ln \frac{\dot{\gamma}_0}{\dot{\gamma}_{max}^p - \dot{\gamma}_0}$$

where $\dot{\gamma}_0 = \left(\sqrt{2\mathbf{D}^p : \mathbf{D}^p}\right)_{ave}\Big)_{max}$ $\qquad (4)$

and $\dot{\gamma}_{max}^p = \sqrt{\left(2\mathbf{D}^p : \mathbf{D}^p\right)_{max}}$

where $\mathbf{D}^p$ is the finite plastic stretching tensor. In the second of (4) the subscript *ave* refers to the average of the quantity over a selected subset of the surrounding elements, and the subscript *max* indicates that evaluation should be made at each point in the material that reaches peak adiabatic stress. After the moment of localization, the stress is reduced in an appropriate manner. Equation (4) has been used to predict depth of penetration and plugging with some success, although the algorithm is still under development. Validation experiments, such as rapid punching of a slab of material, the shearing of a hat shaped specimen in a Kolsky bar, or rapid compression of slightly tilted cylinders, are also being developed.

VOID GROWTH AND SPALL

Wu, Ramesh, and Wright [7,8] considered the growth of a spherical void in an infinite isotropic material. In their studies they accounted for elasticity, work hardening, rate hardening, thermal softening, heat conduction, and inertia. Their intent was to examine one of the fundamental processes thought to lead to spallation in a ductile material. (Other processes that sometimes dominate, such as cleavage, or secondary effects, such as shear between damaged zones, were not examined.) Previous studies of dynamic void growth beginning with Carroll and Holt [9] have considered various aspects of this problem, but a comprehensive study that included all the constitutive and thermal effects, as well as inertia, has been lacking.

The basic configuration was assumed to be spherically symmetric with a spherical void of arbitrary initial size located at the origin. The surface of the void was assumed to be stress free, and an all around tensile pressure, applied at infinity, was taken as the driving force. This pressure was assumed to have a linear rise from zero for a finite time, and then subsequently to be held constant. The resulting profile is a rough approximation that corresponds to a typical tensile loading pulse following a shock reflection.

In a spherically symmetric problem only the radial balance of momentum is required

$$\frac{\partial \sigma_{rr}}{\partial r} - 2\frac{\sigma_e}{r} = -\rho \frac{\partial \psi}{\partial r}, \qquad (5)$$

where σ_e and ψ (the usual effective stress and the acceleration potential from [9]), are given by

$$\sigma_e = \sigma_{\theta\theta} - \sigma_{rr} = f(\varepsilon, T)$$
$$\psi = \frac{a^2\ddot{a}}{r} + \frac{2a\dot{a}^2}{r} - \frac{a^4\dot{a}^2}{2r^4} \qquad (6)$$

The current size of the spherical void, the radius, and the temperature are a, r, and T respectively and the effective strain is ε. The effective stress can also be modified to include rate dependence, but qualitatively the results do not change. In [7,8] power law work hardening and linear thermal softening were used.

In the static case integration of (5) shows that the relation between the void size and the applied stress at infinity is

$$p = \frac{2}{3}\sigma_y + \int_0^{2\ln a/a_0} \frac{\sigma_e d\varepsilon_p}{\exp\left(3\varepsilon_p/2 + \sigma_e/2G\right) - 1} \quad (7)$$

where σ_y is the tensile yield stress, ε_p is effective plastic strain, and G is the elastic shear modulus. In the limit that $a \to \infty$ the integral converges to give a value for the critical stress (or cavitation stress), $p_c = \lim_{a \to \infty} p$. Parametric calculations show that p_c increases with the usual work hardening exponent, n. Two limiting curves may be distinguished: an upper limit corresponding to the isothermal case and a lower limit corresponding to the adiabatic case, the upper curve increasing somewhat more rapidly with n than the lower one, [7]. Calculated values of p_c seem to approximate a lower limit for observed spall strengths in many materials, [8].

Because the integral in (7) has the same value for $a_0 \to 0$ as for $a \to \infty$, p_c is also a bifurcation value for the nucleation of voids. Therefore any small variation in the initial microstructure may be the source of a growing void.

The effect of heat conduction is to delay nucleation of a new void until the applied stress reaches the upper or isothermal curve, [8]. But if a void grows large enough, the critical stress will lie close to the lower or adiabatic limiting curve because the characteristic length scale for thermal conductivity will be much less than the size of the heated plastic zone surrounding the void, [8].

With inertia included integration of (5) leads to

$$\rho\left(a\ddot{a} + \frac{3}{2}\dot{a}^2\right) = p - p_c \quad (8)$$

Integration of (8) with a variable driving stress must be done numerically, but two limiting cases

may be distinguished. With a linear pressure rise it may easily be shown that

$$2a \approx \sqrt{\dot{p}/\rho}\left(t - t_c\right)^{3/2} \quad (9)$$

for the growth of a freshly nucleated void. For a nearly steady driving stress all voids eventually approach a limiting velocity

$$\dot{a} \to \sqrt{2\left(p - p_c\right)/3\rho} \quad (10)$$

This last equation holds true for rate dependent materials, as well. In all cases it has the effect of greatly narrowing the distribution of void sizes for a given applied stress pulse if its duration is greater than several hundred nanoseconds.

CONCLUSION

The approach taken in this paper is leading to damage and failure models that are based on the essence of the physics. An added virtue is that it tends to place a premium on accurate measurement and modeling of homogeneous processes, rather than on failure as a separate phenomenon.

REFERENCES

1. Wright TW, *The Physics and Mathematics of Adiabatic Shear Bands*. Cambridge University Press, Cambridge, 2002.
2. Schoenfeld SE and Wright TW. *Int. J. Solids and Structures*, **40**, 3021-3037, 2003.
3. Molinari A and and Clifton RJ. *J. Appl. Mech.*, **54**, 806-812, 1987.
4. Duffy J and Chi YC. *Mat. Sci. and Eng.*, **A157**, 195-210, 1992.
5. Wright TW. *Int. J. Plas.*, **8**, 583-602, 1992.
6. Wright TW. *Mech. of Mater.*, **17**, 1994.
7. Wu XY, Ramesh KT, and Wright TW. *J. Mech. Phys. Solids*, **51**, 1-26, 2003.
8. Wu XY, Ramesh KT, and Wright TW. *Int. J. Solids and Structures*, **40**, 4461-4478, 2003.
9. Carroll MM and Holt AC. *J. Appl. Phys.*, **43**, 1626-1636, 1972.

CP706, *Shock Compression of Condensed Matter - 2003*
edited by M. D. Furnish, Y. M. Gupta, and J. W. Forbes
© 2004 American Institute of Physics 0-7354-0181-0/04/$22.00

INFLUENCE OF SHOCK PRESTRAINING ON SHEAR LOCALIZATION IN 316L STAINLESS STEEL

Qing Xue, George T. Gray III, Shuh Rong Chen

*Materials Science and Technology Division, Los Alamos National Laboratory,
MST-8, MS G755, Los Alamos, NM 87545, USA*

Abstract: The effect of "Taylor-Wave" shock prestraining on the adiabatic shear localization response of annealed and shock pre-shocked 316L stainless steel (316L SS) was investigated. A forced shear technique using "hat-shaped" specimens on a compression split-Hopkinson bar was utilized. The mechanical responses at two different strain rates showed that the shock pre-strained specimens exhibit much higher yield stresses but much lower strain hardening effects than those in the annealed steel. The dynamic shear responses indicate that the shear stress in the shock pre-strained steel arrived at the instable point at much smaller plastic strain than the annealed steel. The initial microstructures in these two materials exhibit significant differences of textures and defects that may dominate the initiation of shear bands. The pre-shocked steel, which contains more deformation twins, was found to display a higher propensity to trigger localized deformation than the annealed steel.

INTRODUCTION

Shock wave pre-straining has been found to substantially influence the subsequent behavior of materials[1-3]. The instability of materials during large plastic deformation is very sensitive to the initial microstructure. Adiabatic shear localization (Shear band) is a typical unstable deformation and failure mode under high strain rate loading[4]. Their formation and development substantially depend on material microstructure. There is limited work on the influence of shock prestaining effects on shear banding[5].

This paper presents an experimental study of the influence of shock prestraining on the formation and development of shear localization in the AISI 316L stainless steel (hereafter 316L SS). The forced shear technique was utilized on a compressive split-Hopkinson bar using hat-shaped specimens. This technique was first introduced by Hartman et al.[6] and has been modified by several investigators [7-8]. The evolution of shear bands in this study was examined by using interrupted tests. The mechanical responses of the 316L SS during shear localization were correlated to the microstructural evolution of localized deformation.

EXPERIMENTAL

AISI 316L SS was studied in two microstructural / stress history states: the as-received (annealed) and a prestrained state that resulted from the as-received steel being subjected to an explosive "Taylor wave" shock loading pulse of ~40 GPa for ~15 μs. The transient strain during the shock (defined as $4/3 \ln V/V_0$, using the initial and shock compressed volumes)[2] is calculated as 0.214. The nominal chemical composition of the 316L SS is (in wt. %): 0.0084 O, 0.053 C, 1.75 Mn, 0.023 S, 0.51 Si, 8.42 Ni, 18.69 Cr, 0.30 Mo, 0.33 Cu, with the remainder Fe.

The as-received 316L SS exhibits a nominally equiaxed microstructure with an average grain size of 30 μm. Some annealing twins and MnS precipitates are present in the annealed 316L SS. The microstructure of the shock prestrained 316L SS exhibits a high density of slip bands, microbands, and deformation twins.

The quasi-static and dynamic constitutive response of the as-received and shock prestrained 316L SS of was examined. Cylindrical specimens 5 mm in diameter and 5 mm in height were utilized. The dynamic tests were conducted on the compression split-Hopkinson pressure bar at a strain rate of ~2×10^3 s^{-1}.

Shear localization was investigated using forced shear tests conducted on the Hopkinson bar with hat-shaped specimens. The schematic profile of the hat-shaped specimens is shown in Figure 1. The shock loading direction is parallel to the axis of the hat-shaped sample. The true shear stress was calculated as:

$$\tau = \frac{2 E_b hd_b^2 \varepsilon_t}{(d_1 + d_2) \cdot \left[(d_1 - d_2)^2 + 4h^2\right]} \quad (1)$$

$$h = h_0 - \delta \quad (2)$$

where E_b and d_b are the Young's modulus and the diameter of the Hopkinson bar. d_1 and d_2 are the diameters of the hat and the hole and h_0 is the length of shear section. ε_t is the elastic strain calculated from the transmitted Hopkinson bar signal. If the axial displacement of the hat shaped specimens is δ, "h" represents the residual length of shear section.

A series of forced shear tests were carried out under different loading durations. Four lengths of striker bars, 61.8mm, 76.4 mm, 89.1 mm and 101.6 mm, were used to create loading pulses of 25.1 µs, 31.0 µs, 36.1 µs, and 41.2 µs in length, respectively. Evolution of shear localization in these materials was examined in the samples after testing and the microstructural variation was correlated with the transient shear responses.

RESULTS AND DISCUSSION

The quasi-static and dynamic responses of the as-received and shock prestrained 316L SS is shown in

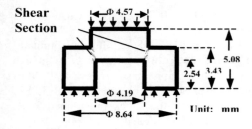

Shear Section

Φ 4.57

5.08
2.54 3.43

Φ 4.19

Φ 8.64   Unit: mm

FIGURE 1. Configuration of the hat-shaped specimens

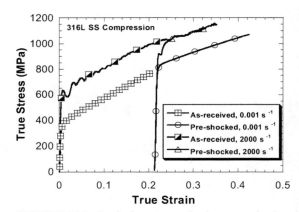

FIGURE 2 Mechanical responses in the as-received and the pre-shocked 316L SS. A strain shift for the shock transient strain was made in the pre-shocked curves.

Figure 2. Each material state displays a pronounced work hardening rate, although the strain hardening rate is lower for the prestrained 316L SS. The yield stresses measured at two different loading rates in the pre-shocked 316L SS are seen to be roughly double those for the as-received 316L SS. The calculated shock transient strain due to shock prestraining, although a triangular "Taylor wave" is different from a square-topped wave, is seen to correlate well with the as-received 316L SS flow stress response when shifted to account for the shock prestraining. That suggests no "net" enhanced shock hardening is displayed by the 316L SS shock prestrained to 40 GPa.

The dynamic forced shear experiments were used to characterize the evolution of localized deformation. The forced shear responses for the as-received 316L SS are plotted in Figure 3. A nearly linear strain hardening rate was seen to result from the loading pulses of 25.1 µs to 36.1 µs. The shear stress at 36.1 µs reached a peak and after that point the hardening process saturated. Additional loading for a duration of 41.2 µs initially displayed the same stress level and then the stress started to decrease. The peak stress has been frequently correlated to reflect the initiation of unstable localized shear deformation [4].

Microstructural examination of the recovered specimens displays the evolving process of localized deformation. The post-mortem observation of deformed samples at different stages was used to correlate the microstructural characteristics to the macroscopic mechanical

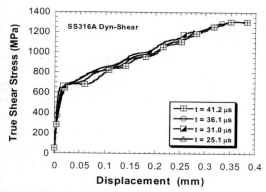

FIGURE 3 Forced shear responses in the as-received 316L SS. Each curve corresponds to a certain loading duration.

responses. Figure 4 shows the last two stages for the loading durations of 36.1 µs and 41.2 µs. Extensive shear plastic deformation is observed within the shear section loaded to 36.1 µs (Figure 4(a)). It is seen that no distinct shear localization has developed to this stage. This result is in agreement with the estimation of the stress responses. Loaded to 41.2 µs, the sample in Figure 4(b) reveals initiation of shear localization. A heavily deformed band-like region distinguishes itself from the surrounding area. However, this type of shear localization is still distinct from the typical shear bands that have sharp edges separating the strong shear localized flow inside the band from the heavy plastic deformation outside. A relatively continuous plastic deformation pattern can be traced across the band from one side of the intact matrix to the other. Therefore, shear localization in the as-received samples loaded for a duration of 41.2 µs, if any, should remain at an starting initiation stage.

The process of shear localization in the pre-shocked 316L SS is seen to be quite different from the as- received steel. The forced shear responses of the pre-shocked specimens reveal a strong softening effect as seen in Figure 5. The shear stress curves of the pre-shocked samples initially display hardening to a maximum strength following yielding and then decrease rapidly. Even the sample with the shortest loading duration of 25.1µs displays hardening, then stress saturation, finally substantial softening. Further tests using longer loading durations followed the same pattern. This response represents

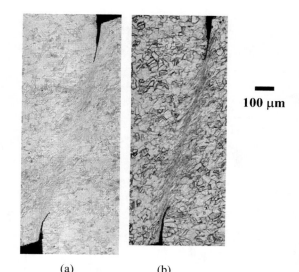

100 µm

(a)   (b)

FIGURE 4 Localized deformation in the as-received 316L SS at the loading duration of (a) 36.1 µs and (b) 41.2 µs.

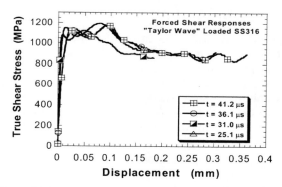

FIGURE 5 Forced shear responses in the "Taylor Wave" pre-shocked 316L SS

a localized deformation softening within the shear section. Each of these stress curves achieved their peak strength at a displacement of about 0.05~0.1mm. Generally, the stress drop is an indication of unstable deformation that signifies the on-set of shear localization. The evolution of shear localization in the pre- shocked 316L SS is characterized in Figure 6. Two principal stages of localization, loaded to the shortest and the longest pulse durations (25.1 µs and 41.2 µs), respectively, are selected to exhibit the development of shear bands. Comparison of the these two samples indicates that once the shear band initiated in the

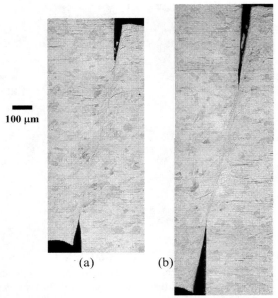

100 µm

(a)   (b)

FIGURE 6 Shear localization in the re-shocked 316L SS loaded at (a) 25.1 µs and at (b) 41.2 µs.

preshocked 316L SS, the subsequent deformation almost completely localized in the band. The shear deformation outside the bands in these two samples looks almost identical except for a shorter residual length of the shear section in the sample loaded to 41.2 µs. The details of shear localization in these samples are illustrated in Figure 7. The width of the shear band in the sample loaded to 25.1 µs is about 7 µm. The continued loading from 25.1µs to 41.2 µs results in an apparent increase in the shear band width to 24 µm. The expansion of the shear band width can be explained as due to thermal diffusion and thermal softening effects during continuous shear deformation. The shock prestrain accumulates a high density of defects and leads to a decreased

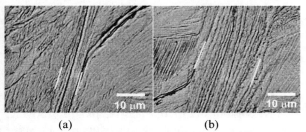

(a)   (b)

FIGURE 7 Shear band evolution with the increasing loading duration at (a) 25.1 µs and at (b) 41.2 µs. Markers show the width of the shear bands.

strain hardening capacity which is critical to the early development of shear localization in pre-shocked 316L SS.

The forced shear technique was previously applied to examine the microstructure of shear bands by using hat-shaped specimens and stop rings [8]. However, stop rings disturb the mechanical response and the recorded stress-strain curves do not represent the complete localization behavior. In the present study, the loading pulses were explicitly controlled for shear localization. The transient shear behavior could be directly correlated to the microstructure of the localized deformation.

CONCLUSIONS

The influence of "Taylor-Wave" shock prestraining on the adiabatic shear localization in annealed and shock pre-shocked 316L stainless steel was studied. The shock pre-strained 316L SS exhibited much higher yield stresses but lower strain hardening effects than the annealed steel. The dynamic shear responses indicate that the shear stress in the shock pre-strained steel arrived at an instable point at much smaller shear displacements than the annealed steel. The pre-shocked steel, which contains a high density of dislocations, slip bands, and deformation twins, was found to display a higher propensity to trigger localized deformation than the annealed 316L SS.

REFERENCES

1. Mahajan, S, *Phys. Status Solidi (A)*, **2**, 187 (1970)
2. Gray, G.T. III, in *High-Pressure Shock Compresssion of Solids,* Ed. by J.R. Asay and M. Shahinpoor, Springer-Verlag, 1993, pp.187-215.
3. Murr, L.E. In *Shock Waves and and High Strain Rate Phenomena in Metals,* Ed. by M.A. Meyers and L.E. Murr, Plunum, 1981, p.607-673.
4. Bai Y. and Dodd B. *Adiabatic Shear Localization*, Pergamon, Oxford, 1992, p.24-.
5. Hines JA, Vecchio KS. *Acta Mater*, **45**(1997), 635-49.
6. Hartman KH, Kunze HD, Meyer LW., In., Shock Waves and High Strain Rate Phenomena in metals, ed. by MA Meyers, LE Murr, Plenum Press, New York, 1981, p. 325-337.
7. Beatty, J., Meyer, L.W., Meyers, M.A., and Nemat-Nasser, S., in *Shock Waves and High-Strain-Rate Phenomena in Materials,* eds. by M. A., Meyers et al., Plenum Press, New York, (1992), pp. 645-656.
8. Meyers MA, Subhash G, Kad BK, Prasad L. *Mech. Mater.* **17**(1994), 175-193.

CP706, *Shock Compression of Condensed Matter - 2003*
edited by M. D. Furnish, Y. M. Gupta, and J. W. Forbes
© 2004 American Institute of Physics 0-7354-0181-0/04/$22.00

THE INTERACTION OF DISLOCATIONS AND RADIATION-INDUCED OBSTACLES AT HIGH-STRAIN RATE

J. A. Young[a] , and B. D. Wirth[b]

[a]*Chemistry and Materials Science, Lawrence Livermore National Laboratory, Livermore, CA 94550*
[b]*Nuclear Engineering Department, University of California Berkeley, Berkeley, CA 94703*

Abstract. Improved understanding of the plastic deformation of metals during high strain rate shock loading is key to predicting their resulting material properties. This paper presents the results of molecular dynamics simulations that identify the deformation modes of aluminum over a range of applied shear stresses and examines the interaction between dislocations and irradiation induced obstacles. These simulations show that while super-sonic dislocation motion can occur during impact loading, the finite dimensions of the materials render this motion transient. Larger applied loads do not stabilize supersonic dislocations, but instead lead an alternate deformation mode, namely twinning. Finally, the atomistic mechanisms that underlie the observed changes in the mechanical properties of metals as a function of irradiation are examined. Specifically, simulations of the interactions between moving edge dislocations and nanometer-sized helium bubbles provide insight into increases of the critical shear stresses but also reveal the effect of internal gas pressure on the deformation mode. The information gained in these studies provides fundamental insight into materials behavior, as well as important inputs for multi-scale models of materials deformation.

INTRODUCTION

Computer simulations are currently capable of providing new insights into fundamental mechanisms of materials deformation at the atomic and nano-scale, as well as predictions of material properties. For example, the study of plastic deformation of metals during shock loading yields insights into the fundamental behavior of dislocation and atomic deformation mechanisms controlling plastic deformation. Traditionally, it has been thought that dislocations could not move faster than the speed of sound, yet recent simulations have shown the existence of supersonic dislocations under high-applied strains[1]. One of the questions to be addressed in this paper is the subsequent stability of these dislocations.

The stress regime required to nucleate supersonic dislocations also lends itself to the activation of deformation modes otherwise inaccessible under quasi-static deformation. For example, high stacking fault energy materials, such as aluminum, have high critical stress for deformation twinning. These high stresses render experimental examination of the deformation details difficult at best, therefore, this paper presents results which identify alternate modes of deformation and their atomistic origins.

Materials, especially those exposed to radiation or utilized in aggressive corrosive or thermally-cycled environments, are not defect free. While many of the basic physical mechanisms governing irradiation induced hardening or environmentally-assisted degradation are understood, it is still unclear what mechanisms are dominant under specific conditions. Likewise many multi-scale models require accurate estimates of key material parameters. For example, the interaction between dislocations and nanometer voids and/or helium bubbles that form during irradiation, are known responsible for increased strength and decreased ductility; the manner in which factors such as the number density, size and internal pressure, these defects determine the final material properties remain to be determined. It is for these reasons that we will examine the interactions between edge dislocations in aluminum and He bubbles over a range of internal gas pressure.

PROCEDURE

Molecular dynamics (MD) simulations are an ideal tool to investigate deformation phenomena at high-strain rates, since they directly account for core and non-linear effects ignored by elastic theory. The motion of edge dislocations and their interactions with obstacles in aluminum has been studied using the molecular dynamics code MDCASK[2]. The simulation system used in our simulations has basis vectors along the X = [$\bar{1}$11] Y=[110] and Z = [1$\bar{1}$2] directions. Periodic boundary conditions are used in the Y and Z directions while the X = [$\bar{1}$11] faces are free surfaces. Each system contains approximately one million atoms. The Ercolessi and Adams force-matching embedded atom method potential[3] is used to model the aluminum/aluminum interactions while pair potentials are used for the aluminum/helium interactions[4] and helium/helium interactions.

An edge dislocation is introduced into the cell by removing two (220) half planes. For studying the dislocation interactions with radiation-induced obstacles, helium bubbles are introduced into the center of the cell by removing a 2.6 nm diameter sphere of atoms and introducing the desired number of helium atoms. Helium/lattice site ratios of 0.0 (i.e. a void), 0.5, 1.0 and 2.0 are considered. Dislocation motion is then studied as a function of applied shear stress by applying a constant surface traction in the [110] direction to the atoms in the two ($\bar{1}$11) surfaces. The range of applied stresses from 25 MPa to 1000 MPa, corresponding to 0.15% to 3.0% of the shear modulus is covered in this work.

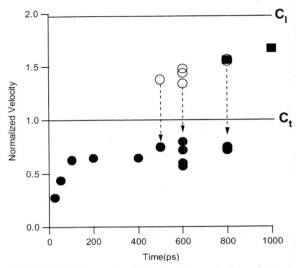

FIGURE 1 The velocity and deformation mode as a function of applied shear stress. Glide deformation is denoted by circles, open circle indicate transient velocities and closed circles indicate stable steady-state velocities, and squares indicate twinning.

RESULTS/DISCUSSION

Both the rate of plastic deformation and the mode of deformation are important in determining the mechanical properties of a metal. This is particularly true at high strain rates were deformation modes not accessible to a material under normal quasi-static deformation may be activated. Figure 1 summarizes the steady-state velocities of edge dislocations over a range of applied shear stresses. For applied shear stresses of up to 500 MPa, classic velocity/stress relationships are seen. These include the linear region (25 MPa-100 MPa) where frictional forces limit the velocity of the dislocation as well as the plateau region (100 MPa-400 MPa) where the velocity is limited by the transverse sonic wall. The saturation velocity of 2350 m/s corresponds to approximately 68% of the transverse sound speed, C_t, yielding a strain rate of ~1x10^8 s^{-1}. For the parameters of this interatomic potential for aluminum, the transonic sound speed is $C_t \approx 3475$ m/s.

Higher applied stresses produce trans-sonic dislocation motion. Figure 1 plots the velocity of dislocations as a function of applied stress and shows not only that transonic dislocations are transient but that twin deformation is also activated at high stress. To investigate the origin of the transonic instability we begin by examining the case in which the trans-sonic dislocation slows to a steady subsonic velocity while maintaining dislocation glide.

Examination of the position of the dislocation and the stacking fault width as a function of time for different size cells (X=50 and X=80 lattice units), is plotted in Figure 2 and shows that the existence of the free surfaces are responsible for the slowing of the dislocation. Specifically, it takes 0.5 X/C_t ps, traveling at the speed of sound, for the applied shear stress to be transmitted to the dislocation that is located in the center of the cell and initiate transonic motion. It takes an additional 1.0 X/C_t ps for the wave composed of the applied stress and the shock wave excited by the transonic dislocation to reach the free surface, excite surface waves and arrive back at the dislocation. Thus, an interaction between the transonic dislocation and the surface waves occurs 1.5 X/C_t ps after application of the applied stress and causes the slowing of the dislocation to a steady-state subsonic saturation velocity. As seen in figure 2, the breaking effect of the surface wave is manifested through an increase in the width of the stacking fault at the time 1.5X/Ct. Shortly thereafter, the partial dislocations recover their equilibrium spacing at a stable subsonic velocity. These simulations provide direct evidence that while dislocations can be accelerated to transonic speeds as a result of high-strain rate impulse loading, the excitation of Rayleigh waves from internal surfaces and interfaces will rapidly result in steady-state dislocation motion at a subsonic saturation velocity[5].

It is also important to note that at these high strain rates another deformation mode becomes accessible, namely deformation twinning. These simulations reveal that nano-twin formation results from the homogenous nucleation of dislocation dipoles along the stacking fault. While this is not an anticipated deformation mode for aluminum due to a high stacking fault energy, 110-130 mJ/m^2, similar results have been published for nano-crystalline aluminum[6].

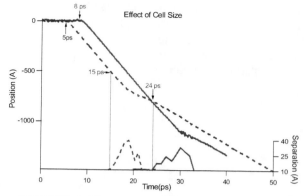

FIGURE 2. The average position of the dislocation and the separation of the partial dislocations (stacking fault width) as a function of time following an applied shear stress of 600 MPa. The broken lines are for X=50 cell and the solid lines are for the X=80 cell The time for the applied shear wave to reach the dislocation and the time at which the dislocation starts to widen are shown.

Atomistic simulations of the interaction between moving edge dislocations and 2.6nm helium bubbles as a function of internal gas pressure provide information key material behavior parameters. Estimates of critical angles and stresses are central to models such as Orowan's theory of dispersed obstacle hardening[7]. Helium bubbles having He/lattice ratios of 0-1 shear as a result of interaction with the dislocation. The bubbles are sheared by one Burger's vector with the passage of each dislocation. A critical stress of less than 25 MPa has been determined for the void and 35 MPa for the He/lattice=0.5 bubble. This value is in reasonable agreement with critical stresses obtained from TEM observations of irradiated copper[8].

At higher pressures corresponding to He/lattice= 2.0 a different interaction is seen. In this case the internal pressure of the bubble creates a significant disruption of the aluminum lattice. As the dislocation interacts with the bubble it not only shears the bubble, but also adsorbs vacancies, resulting in dislocation climb with the formation of a pair of less mobile super jogs. These super jogs then act to reduce the glide mobility of the edge dislocation. These results indicate that in irradiated metals the internal pressure of the bubbles not only determines the critical stress but also the mechanism by which dislocations interact with the obstacles and the rate of plastic deformation.

FIGURE 3. The jogged dislocation which results from interaction with a bubble having He/lattice=2.0.

SUMMARY AND FUTURE WORK

The results presented in this paper address various aspects of materials deformation at high strain rates. Specifically, the motion of supersonic dislocations has been shown to be transient resulting from the finite dimensions of the material. While twinning in aluminum is difficult to observe experimentally, these simulations reveal the onset of twin deformation by homogenous partial dislocation dipole nucleation at shear stresses of 800 MPa. Finally, both the critical stress and the interaction between a dislocation and He bubbles in irradiated metals is shown to depend of the internal gas pressure. Future work will involve additional simulations to examine the effect of dislocation morphology (i.e. screw dislocations) as well as the geometry of the dislocation/obstacle interactions.

ACKNOWLEDGEMENTS

The authors would like to acknowledge many fruitful discussions with Drs. J.S. Robach, I.M. Robertson, A. Arsenlis, V.V. Bulatov and E.M. Bringa, and funding for this research from the LLNL Dynamics of Metals program. This work was performed under the auspices of the U. S. Department of Energy by the University of California, Lawrence Livermore National Laboratory under Contract No. W-7405-Eng-48.

REFERENCES

1. Gumbsch, P. & Gao H. *Science* **283**, 965 (1999).
2. Diaz de la Rubia, T. & Guinan, M.W. *Journal of Nuclear Materials*, **174**, 151 (1990).
3. Ercolessi, F. & Adams, J. *Europhysics Letters* **26,** 583 (1994).
4. Private communication: Chris Mundy, Lawrence Livermore National Laboratory.
5. Hirth, J. & Lothe, J. *Theory of Dislocations* (McGraw-Hill Book Company, New York, 1968).
6. Yamakov, V., Wolf, D., Phillpot, S.R., Mukherjee, A.K. & Gleiter, H. *Nature Materials*, **1**, 1 (2002).
7. Orowan, E. *Fracture* (Wiley, New York, 1959).
8. Robertson, I.M., Robach, J., Wirth, B., & Arsenlis, A., Dislocation Behavior During Deformation —Combining Experiments, Simulation and Modeling. MRS April 21-25, 2003, San Francisco CA.

CP706, *Shock Compression of Condensed Matter - 2003*
edited by M. D. Furnish, Y. M. Gupta, and J. W. Forbes
© 2004 American Institute of Physics 0-7354-0181-0/04/$22.00

DEFECT ANALYSIS IN Ti-BASED ALLOYS DEFORMED AT DIFFERENT STRAIN-RATES

M. Zakaria, X. Wu, M. H. Loretto, J.C.F. Millett*, N.K. Bourne*

IRC in Materials, The University of Birmingham, Edgbaston B15 2TT, United Kingdom.
**Royal Military College of Science, Cranfield University, Shrivenham, Swindon,*
SN6 8LA. United Kingdom.

Abstract. Transmission electron microscopy (TEM) has been used in an attempt to assess the nature and density of dislocations present in samples of a number of Ti alloys subjected to low and high strain rate deformation. It has been found that conventional defect analysis is straightforward in samples strained at low strain rates of $10^{-1} s^{-1}$ but it is not possible to carry out defect analysis in samples strained at $5 s^{-1}$. In samples strained at higher strain rates, such as $10^3 s^{-1}$ or as high as $10^8 s^{-1}$, similar difficulties can be encountered in some grains but in others defect analysis is possible. These results are interpreted in terms of the elastic anisotropy of the hexagonal phase in these alloys and in terms of the level of internal stress introduced by deformation in these two-phase alloys.

INTRODUCTION

To gain a full understanding of the response of a material to external loading, it is necessary to investigate both its mechanical and its microstructural response. Whilst this is comparatively simple at quasi-static and intermediate (*e.g.* Hopkinson bar) strain-rates, it becomes significantly more difficult as the strain-rate increases (for example during shock loading). In these regimes, it is essential to load and unload the sample under known conditions of stress and strain (for example one dimensional strain during plate impact shock loading) if quantitative mechanical and microstructural information is to be gained. To determine the one-dimensional strain deformation structure, shock recovery is necessary. For this to be successful, samples must be flat and planar to tolerances of less than 5 µm, but also require the machining of complex momentum traps to prevent lateral releases and spall (tensile) interactions from affecting the region of interest. Much of the work in this area has been performed at Los Alamos National Laboratories [1, 2].

Over the past few years, the aerospace industry has become increasingly aware of the importance of high-strain-rate loading of materials. In particular, the alloy Ti-6Al-4V has received attention due its use as a fan blade material in jet turbines. This makes it vulnerable to impact events such as foreign object damage (FOD). As such, a body of work discussing the impact [3, 4] and shock response [5-7] is available. Very recently, alternative alloys have been developed, and it is the microstructural response from quasi-static to shock induced strain-rates that this paper aims to discuss.

EXPERIMENTAL

Shock loading experiments were performed on a 75 mm bore, 1 m long single stage gas gun at the Royal Military College of Science. Target assemblies were made from Ti-6Al-4V, with 10 mm diameter by 3 mm inserts of the alloys of interest. These were impacted at a velocity of 300 m s^{-1} with a 3 mm Ti-6Al-4V flyer plate, to induce a shock stress of *ca.* 3.4 GPa. Past experience has shown that alloys from the same base element tend to have similar Hugoniots [8, 9] (in stress – particle velocity space), and thus we have confidence that one-dimensional strain conditions will have been maintained in the alloy inserts. Samples deformed at 10^{-1} and 5 s^{-1}, were done using an MTS servo-hydraulic machine, with cylindrical specimens of 6.4 mm diameter, by 25.4 mm in length. Specimens deformed at 10^3 s^{-1} were of dimensions 4.3 mm diameter by 17.8 mm, and tested using a split tensile Hopkinson bar.

Once recovered, samples were prepared for electron microscopy by spark machining 3 mm foils from the shocked samples, which were then thinned using a Gatan PIPS ion beam thinner until perforation. Foil normals were selected such that they were parallel to the axis of deformation. The foils were examined in a Philips CM20 or JEOL 4000FX transmission electron microscope (TEM) operating at 200 kV, in either bright field, weak beam, dark field and selected area diffraction modes.

MATERIALS

The materials under investigation were Ti-6Al-4V, and an α-β alloy, TIMET 550. The Ti-6Al-4V was heat treated for 2 hours at 950°C and furnace cooled, followed by 2 hours at 900°C with air-cooling. The TIMET 550 was solution treated at 910°C for 1 hour, helium cooled, and then aged at 500°C for 24 hours, and finally air-cooled. Microstructural examination showed that the Ti-6Al-4V had an equiaxed primary α structure with an average grain size of 10-20 μm. The TIMET 550 had a primary α size of 5-10 μm. In both alloys, the microstructural balance was *ca.* 40% α and 60% transformed β (the body-centred cubic β phase with a high proportion of transformed α lathes). The alloy chemistries are shown in table 1 –

TABLE 1. Alloy Chemistries

Element	Ti-6Al-4V	Timet 550
Al	6.25	3.70
V	4.62	
Mo		4.83
Sn		2.10
Si		0.54
Ti	balance	balance

RESULTS

In figure 1, we present micrographs of Timet 550, deformed to 10^{-1} s^{-1} with the foil normal close to $1\bar{2}10$. In the examples shown here, we have imaged along g=0002 and g=$1\bar{2}10$, and the c-component dislocations with b = 1/3 <$1\bar{1}23$> visible in one are clearly visible in the other.

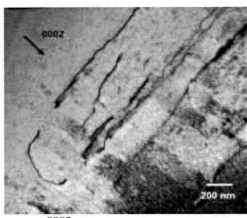

a.   g=0002

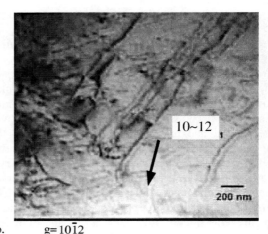

b.   g=$10\bar{1}2$

Figure 1. TIMET 550 deformed to 10^{-1} s^{-1}.

In figure 2, we now show micrographs deformed to a strain-rate of 5 s^{-1}.

a.　g=0002

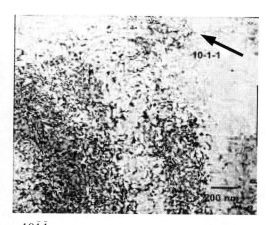

b. g= $10\bar{1}\bar{1}$

Figure 2. Timet 550 deformed to 5 s^{-1}.

In contrast to the previous figure, where it can be seen that dislocations can be imaged under different imaging conditions, only when g=0002 are they visible when strained at 5 s^{-1}. The visible dislocations have b=1/3<11$\bar{2}$3>, whilst those with b=1/3<11$\bar{2}$0> will be out of contrast with this diffracting vector. The dislocations of b = 1/3<11$\bar{2}$3> should also be visible with both or either g=$10\bar{1}\bar{1}$ and $\bar{1}01\bar{1}$. Similarly, the dislocations of b = 1/3<11$\bar{2}$0> should be visible with both diffracting vectors. However, it is clear that no useful dislocation contrast was observed and attempts to use other imaging conditions or even weak beam microscopy were not successful.

Finally, in figure 3, we present images from TIMET 550, deformed under conditions of one-dimensional shock loading. In this case, the specimen was subjected to a loading of 3.4 GPa for a duration of *ca.* 1 µs. Unlike the specimen deformed at 5 s^{-1}, reasonable images could be obtained when using diffracting vectors other than g = 0002 in common with specimens deformed at *ca.* 10^3 s^{-1} [10]. One feature that does stand out is the lack of any evidence of twinning. This shows similarities to the shock and recovery work of Gray and Morris [11], who in examining the microstructural response of Ti-6Al-4V, also showed little evidence of twinning when shocking to 3 GPa, although twins did become more prevalent as shock amplitude increased.

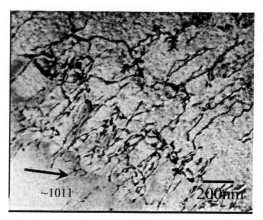

a. g= $10\bar{1}1$

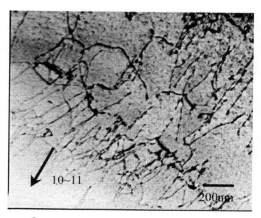

b. g= $\bar{1}011$

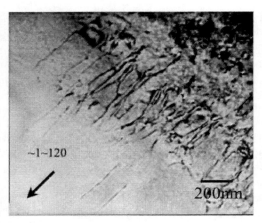

c. g=11$\bar{2}$0

Figure 3. Timet 550 shocked to 3.4 GPa. Imaging conditions are indicated.

DISCUSSION AND CONCLUSIONS

The dislocations images obtained at 0.1 s^{-1} show image contrast which allows Burgers vector analysis to be carried out. This was not possible in samples strained at 5 s^{-1} since no useful dislocation contrast can be obtained. The difficulty in imaging dislocations at 5 s^{-1} is associated with the high density of dislocations of b = 1/3<11$\bar{2}$0> and 1/3 <11$\bar{2}$3> and any defect debris they produced. The major displacements produced by dislocations of b = 1/3<11$\bar{2}$0> will distorts all planes other than the (0001) planes. Likewise, the secondary displacements associated with dislocations of b = 1/3<11$\bar{2}$0> and 1/3 <11$\bar{2}$3> will not distort the (0001) planes to the same extent as any other planes since the elastic modulus is higher along the [0001] direction than the <11$\bar{2}$0>. Thus, contrast from individual dislocations will be obscured when imaged with all diffracting vectors other than the 0002. The fact that dislocations can be imaged with a number of diffracting vectors, including the 10$\bar{1}$0 type, in some area of samples strained at 10^{8}s^{-1} samples suggests that some localized recovery through adiabatic heating is taking place where the strain associated with the defects is relieved at this high strain rate.

ACKNOWLEDGMENTS

This work is partly supported through Rolls Royce via the University Technology Partnership and by EPSRC. The help of Dr. S. Fox of Timet and I. Wallis of QinetiQ for supply of samples strained at 10^3 s^{-1} is gratefully acknowledged. We would also like to thank Gary Cooper and Paul Dicker of RMCS for their assistance in the shock loading part of this paper. Support through the ORS scheme is also acknowledged.

REFERENCES

1. Gray III, G.T., in *Shock Compression of Condensed Matter - 1989*, S.C. Schmidt, J.N. Johnson, and L.W. Davison, Editors. 1990, North-Holland: Amsterdam. p. 407-414.
2. Gray III, G.T., in *High-Pressure Shock Compression of Solids*, J.R. Asay and M. Shahinpoor, Editors. 1991, Springer-Verlag: New York. p. 187-215.
3. Peters, J.O., Ritchie, R.O. Int. J. Fatigue, 2001. **23** S413.
4. Martinez, C.M., Eylon, D., Nicholas, T., Thompson, S.R., Ruschau, J., Birkbeck, J., Porter, W.J. Mater. Sci. Engng. A, 2002. **325** 465.
5. Dandekar, D.P., Spletzer, S.V. in *Shock Compression of Condensed Matter 1999*, M.D. Furnish, L.C. Chhabildas, and R.S. Hixson, Editors. 2000, American Institute of Physics: Melville, New York. p. 427.
6. Hopkins, A., Brar, N.S. in *Shock Compression of Condensed Matter 1999*, M.D. Furnish, L.C. Chhabildas, and R.S. Hixson, Editors. 2000, American Institute of Physics: Melville, New York. p. 423.
7. Rosenberg, Z., Meybar, Y., Yaziv, D. J. Phys. D: Appl. Phys., 1981. **14** 261.
8. Bourne, N. Millett, J. Scripta Mat., 2000. **43** 541.
9. Millett, J.C.F., Bourne, N.K. Rosenberg, Z., Field, J.E. J. Appl. Phys., 1999. **86** 6707.
10. Zakaria, M., Voice, W., Wilson, A., Loretto, M.H., Wu, Xinhua. Phil. Mag. A, 2003 In press.
11. Gray III, G.T. Morris, C.E, in *Sixth World Conference on Titanium*. 1988: France. p. 269.

CP706, *Shock Compression of Condensed Matter - 2003*
edited by M. D. Furnish, Y. M. Gupta, and J. W. Forbes
© 2004 American Institute of Physics 0-7354-0181-0/04/$22.00

ANOMALOUS HIGH-TEMPERATURE SHOCK-INDUCED STRENGTHENING OF TWO SUPERALLOYS

E. B. Zaretsky[1], G. I. Kanel[2], S. V. Razorenov[3], and K. Baumung[4]

[1]*Department of Mechanical Engineering, Ben-Gurion University, P.O.Box 653, Beer-Sheva 84105, Israel*
[2]*Institute for High Energy Densities of R A S, IVTAN, Izhorskaya 13/19, Moscow, 125412 Russia*
[3]*Institute of Problems of Chemical Physics of R A S, Chernogolovka, 142432 Russia*
[4]*Forschungszentrum Karlsruhe, P.O. Box 3640, 76021 Karlsruhe, Germany*

Abstract. The dynamic yield and tensile strength of two high-strength high-temperature superalloys Inconel IN738LC and PWA 1483 have been studied in planar impact experiments at test temperatures ranged from 25°C to 680°C. Both the alloys exhibit abrupt increase of the dynamic yield strength within the temperature interval from about 500°C to 600°C. The dynamic tensile strength of the alloys is found less sensitive to the temperature and shows less pronounced anomaly. Most probably the observed non-monotonous variations of dynamic mechanical properties are caused by the structural rearrangement starting in the alloys in the vicinity of 550°C.

INTRODUCTION

The decrease of the yield strength with the increase of temperature of quasi-static test is common for most of polycrystalline low-alloyed metals [1]. A similar, monotonous with increasing temperature, decrease of both the dynamic yield and spall strength was also observed in several high-strength alloys loaded by impact [2]. The increase of the alloying complexity may result in departure of the material behavior from this trend. The dynamic response of two high-temperature turbine alloys (superalloys [3]) studied in this work provides an example of such deviation.

The high strength of superalloys derives from high, typically 60 to 75 %, volume fraction of fine precipitates of the long-range ordered γ'-phase embedded into disordered FCC γ-matrix. Measured under quasi-static loading conditions the yield strength of superalloy [4] is virtually independent of the deformation temperature: a weak minimum at 400 - 450°C is followed by a modest maximum at 700 - 750°C. Further heating results in the precipitous decrease of the superalloy strength.

MATERIALS AND EXPERIMENTAL

The materials tested were Inconel IN738LC (in wt. %: 16 Cr, 8.5 Co, 1.75 Mo, 3.5 Ti, 3.5 Al, 3 W with small additions of Ta, Nb, Zr, C, and B) and a new Ni-Co-Fe alloy whose Pratt & Whitney Co abbreviation is PWA 1483 (in wt. %: 28 Ni, 35 Co and 25 Fe with small additions of Cr, Nb, Ti, Al, Mn, and Si). The latter was directionally grown poly-crystal of preferable [200]-orientation parallel to the sample axis coinciding with the impact direction. The measured density of IN738LC is 8.3 g/cm^3, longitudinal and shear sound speeds are c_l = 6.06±0.04 km/s and c_s = 3.15±0.04 km/s, respectively. The density of the Ni-Co-Fe alloy is ρ_0 = 8.37 g/cm^3, longitudinal sound speed is c_l = 5.61±0.04 km/s and the shear sound speed measured in the sample plane is c_s = 3.74±0.04 km/s.

All the samples were 12-mm diameter disks of 1.7 to 1.9-mm in thickness cut from one rod of each material and grinded to 0.06-degree parallelism. No additional thermal treatment was performed with the samples.

The planar impact loading of the samples was produced by copper impactors of 0.5±0.01-mm thickness supported by thick PMMA backing. The impactors were accelerated up to velocities of 400 to 500 m/s by 58-mm gas gun. High-temperature tests were done using resistive heaters glued directly on the sample surface with alumina cement. The test temperature was controlled with an accuracy of ± 5°C by two thermocouples welded on the rear surfaces of the samples at the distance of about 3 mm from the sample axis. The free surface velocity histories $w(t)$ of the impacted samples were monitored by VISAR [5].

RESULTS OF THE MEASUREMENTS

The free surface velocity histories obtained after impact experiments with heated IN 738 LC and PWA 1483 samples are shown in Fig. 1. Being qualitatively similar, the wave profiles at the room and elevated temperatures differ quantitatively.

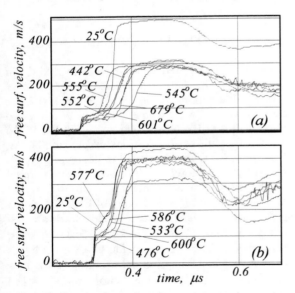

FIGURE 1. Free surface velocity histories of IN 738 LC (a) and PWA 1453 (b) samples at normal and elevated temperatures.

Within a narrow temperature interval between ~500°C and 600°C the amplitudes of elastic precursor waves (HEL) of both the alloys were found varying strongly. Similar behavior display

the time intervals Δt_{ep} between the elastic and the plastic waves and, to a lesser extent, the values of the velocity pullbacks Δw_{pb}.

DATA EVALUATION AND ANALYSIS

In the following analysis we used for both alloys an approximate relationship between the plastic shock velocity U_s and the particle velocity u_p,

$$U_s = 4600 + 1.4 u_p \text{ (m/s)}, \tag{1}$$

which has been obtained by averaging the Hugoniots of nickel, cobalt, stainless steel [6] and Co-Ni alloy [7]. Closeness of the shock compressibilities of these materials justifies the approximation. Since the bulk modulus of Inconel 718 alloy, a superalloy similar to the studied materials, remains practically unchanged with temperature [8] we also assumed invariability of the $U_s(u_p)$ dependence within given temperature range.

The temperature correction of the longitudinal sound speeds c_l of the alloys was based on known temperature dependence of the shear modulus of similar superalloy [8]. The dynamic yield strength $Y(T) = 3/4 \left(1 - c_b^2/c_l^2\right) \rho_0 c_l w_{HEL}$ was estimated on the base of the w_{HEL} values of the free surface velocity profiles at corresponding temperature. The dependencies $Y(T)$ for both the superalloys are shown in Fig. 3. It should be noted here that the study of the elastic moduli of a similar superalloy [8] does not indicate any their anomaly between 500°C and 600°C. This means that the observed variations of the Δt_{ep} values cannot be caused by anomalies the longitudinal and bulk compressibilities of the alloys in this temperature interval. Most probably the observed unstable response is related with the hardening-softening effects accompanying the impact loading.

The above-HEL part of the free surface velocity profile contains some relevant information concerning the strain-hardening behavior of the material. This information was extracted from the stress-strain diagrams of the alloys received from the measured velocity profiles $w(t)$. The procedure is based on the assumptions that the free surface velocity is equal to the doubled particle velocity, $u_p = w/2$, and the compressive parts of the velocity profile may be considered as a centered simple wave with characteristics [9] $h/t = a(\sigma)$. The Lagrangian phase velocities $a(\sigma)$ are equal to $a(\sigma) = c_l h / [h + c_l t(\sigma)]$, where h is the Lagrangian coordinate of the point at which the stress history $\sigma_x(t)$ of the compressive wave is analyzed, and $t(\sigma)$ is the time interval after the elastic precursor front. Integrating the conservation equations for mass, $d\varepsilon_x = -dw(\sigma)/2a(\sigma)$, and for momentum, $d\sigma_x = \rho_0 dw(\sigma)a(\sigma)/2$, provides the parametric form of the stress-strain diagram $\sigma = \sigma(\varepsilon)$. The material Hugoniot (1) allows obtaining the stress deviator $s(\varepsilon) = \sigma_x - p = \sigma_x - \rho_0 c_b^2 \varepsilon_x / (1 - s\varepsilon_x)^2$ important for analysis of the material plasticity. The $s(\varepsilon)$ diagrams derived from the free surface velocity profiles of Fig. 1 are shown in Fig. 2.

The effect of heating on the $s(\varepsilon)$-diagrams of the studied superalloys is apparently similar: in the temperature range from ~500°C to 600°C the steep increase of both the yield strength and the hardening modulus $\theta = ds/d\varepsilon$ is immediately followed by their abrupt drop. Since the hardening moduli for all the compressive curves were found decreasing continuously with the deformation, it is reasonable to compare the θ-values at the same level of plastic strain, e.g. $\varepsilon_{pl} = 0.002$ (0.2%).

The spall strength σ_{sp} of the tested samples was derived from the pullbacks and free surface velocity derivatives of the corresponding velocity profiles using analysis described elsewhere [10]. The spall strength of the superalloys was found less sensitive to the temperature variations than the yield stress.

Figure 3 summarizes the results of the study of the high strain rate response of IN738LC and PWA1483 superalloys. The available temperature dependence[3] of quasi-static ($\dot{\varepsilon} = 10^{-3} s^{-1}$) strength data is shown for comparison.

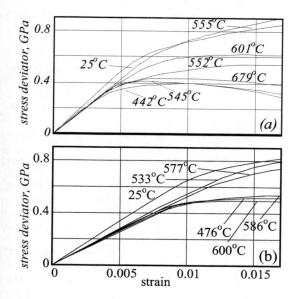

FIGURE 2. Stress deviators of IN 738 LC (a) and PWA 1483 (b) alloys derived from the free surface velocity profiles.

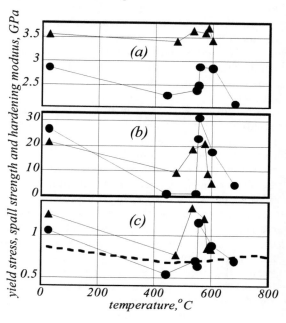

FIGURE 3. Spall strength (a), hardening modulus (b) and yield stress (c) of IN738LC (circles) and PWA1483 (triangles) alloys at different temperatures.

CONCLUDING REMARKS

The abrupt changes of the yield stress and of the hardening moduli that were observed in both tested materials within the temperature interval of 500°C to 600°C are the striking result of this study. It is apparent from Fig. 3c that the strong surge of the yield stress evidences the effect of testing at a high strain rate; no sign of such behavior is displayed under quasi-static ($\dot{\varepsilon} = 10^{-3}\,\mathrm{s}^{-1}$) conditions.

The increase of the sub-microsecond yield strength is accompanied by the increase of the strain hardening and, to a lesser extent, of the spall strength. The presence of similar features on the temperature dependence of the mechanical properties of Inconel 738 and of PWA 1483 suggests that the observed phenomena are common for two-phase Ni-based superalloys. Heating the alloys shifts the equilibrium between hardening (strain- and strain rate-related) and softening (dynamic recovery caused mainly by evolution of the dislocation substructure and the dislocations annihilation) processes that accompany impact loading. The hardening that prevails at room temperature is overcome by dynamic softening that occurs at test temperature above 400°C. In the 500 to 600°C temperature interval hardening again dominates: the duration of the loading cycle is apparently insufficient for complete dynamic recovery.

It is hardly possible to identify the dislocation-based process responsible for the property anomaly in the temperature interval of 500°C to 600°C. All likely processes differ only by their speed but all of them lead to the simplification and stabilization of the dislocation structure as is an inevitable outcome of any dynamic recovery. In Ni-based superalloys we, however, are able to point at some atomic rearrangements which may induce the shift in the delicate hardening-softening equilibrium. One of them is the change of the short range order. The strong deviation of specific heat of Ni-based Inconel 718 from the rule of mixtures at about 550°C has been attributed to the short order variation [11] which, in turn, may effect on the alloy strength [12]. In addition, the coherency stress resulting from the temperature dependent lattice misfit $\eta = \left(a_{\gamma'} - a_{\gamma}\right)/a_{\gamma}$ between ordered γ'- and disordered γ-phase matrix may play essential role in superalloy hardening and in its softening. In Ni-based superalloys [13] the misfit is constant up to 500°C and above this temperature it starts to decrease rapidly.

In conclusion, the dynamic properties in the complex alloy systems may display non-monotonous temperature dependences which is not observed in quasi-static experiments but can be revealed by impact testing.

REFERENCES

1. Shtremel., M.A., Strength of Allots, v.II "Deformation", MISIS, Moscow, 1997.
2. Zhuowei GU and Xiaogang JIN, In: *Shock compression of condensed matter – 1997*, eds. D. Schmidt et al., AIP CP 429, 467-470 (1998)
   Krüger, L., Kanel, G.I., Razorenov, S.V., Meyer, L.,and Bezruchko., G.S., *In: Shock compression of condensed matter – 2001,* eds. M. D. Furnish et al., *AIP CP* **620**, 1327 (2002)
3. Superalloys 1996 (Eds: R. D. Kissinger et al.), TMS, Warrendale, PA 1996.
4. Bettge, D; Osterle, W; Ziebs, J., *Z. f. Metallkd.* (Germany), **86**, 190 (1995)
5. Barker, L. M., and Hollenbach, R. E., *J.Appl.Phys.*, **45**, 4872, (1974).
6. *LASL Shock Hugoniot Data*, edited by. S.P. Marsh. (Univ. of California Press, Berkeley, 1980).
7. Trunin, R.F., Belyakova, M.Yu., Zhernokletov, M.V., and Sutulov, Yu.N., *Izv. Akad. Nauk SSSR. Fiz. Zemli* **2**, 99 (1991)
8. Fukuhara, M., and Sanpei, A., *Journal of Materials Science Letters.* **12**, 1122 (1993)
9. Fowles R. and Williams R.F.,*J. Appl. Phys.*, **41**, 360 (1970).
10. G.I. Kanel. Dynamic strength of materials. *Fatigue & Fracture of Engineering Materials and Structures*, **22**, 1011 (1999)
11. Brooks, C.R., Cash, M., and Garcia, A., *Journal of Nucl. Mate.* **78**, 419 (1978)
12. Cohen, J.B., and Fine, M.E., *Journ. Phys. Rad.,* **23**, 749 (1962)
13 T. Yokokawa, K. Ohno, H. Murakami, T. Kobayashi, T. Yamagata and H. Harada, *Advances in X-ray Analysis,.***39**, 449 (1997).

CHAPTER X

MECHANICAL PROPERTIES OF POLYMERS

CP706, *Shock Compression of Condensed Matter - 2003*
edited by M. D. Furnish, Y. M. Gupta, and J. W. Forbes
© 2004 American Institute of Physics 0-7354-0181-0/04/$22.00

SHOCK COMPRESSION OF SILICON POLYMER FOAMS WITH A RANGE OF INITIAL DENSITIES[†]

R. R. Alcon, D. L. Robbins, S. A. Sheffield, D. B. Stahl, and J. N. Fritz

Los Alamos National Laboratory, Los Alamos, NM 87545

Abstract. We report here on a collection of shock compression experiments on a silicon polymer foam with varying degrees of distension from low (~0.4 g/cm^3) to the near fully dense material (~ 1 g/cm^3). These experiments are being carried out on a two-stage gas-gun (50 mm bore) with a Kel-F 81 impactor at velocities between 1.5 and 3.1 km/s. Particle and shock velocity measurements are made with magnetic gauges by inserting the gauge package (0.001 inches thick) between layers of 2.3 mm thick foam. Special attention is required for assembly of these targets due to the foam's low strength. To minimize compression and gaps at interfaces, the foams are positioned between support rings, which are machined to match the foam's thickness. The Hugoniot data from these experiments is compared to unpublished data obtained with explosively driven flyers at Los Alamos National Laboratory in the early 1980's.

INTRODUCTION

Shock compression experiments require meticulous assembly and attention to detail. Target materials typically require machining to tolerances of only a few ten thousandths of an inch. The smallest (*a-priori* insignificant) mistake will frequently result in a lost shot or, minimally, an increase in the uncertainty of the data. Perhaps these unfortunate facts are part of the reason that little shock compression data exists for foams. Foams are in general difficult, if not impossible, to machine and are sometimes even more difficult to make geometrical measurements upon—especially for easily compressible, highly distended foams.

This paper focuses on some of the experimental intricacies of shock compression gas-gun experiments on silicon foams comprised of foamed polydimethylsiloxanes (PDMS). The goal of this work is to generate Hugoniot data and particle velocity-time wave profiles over a range of pressures for these silicon foams, which have varying degrees of distension, from ~ 0.4 to 1 g/cc. An obvious prerequisite to this work is to develop the best procedures for assembling the foam targets.

These foams are not machinable, are very easily compressed, and exhibit significant heterogeneity with respect to void size for the lowest density foams. Initial experiments have been successful, but demonstrate the subtle difficulties associated with avoiding gaps, while at the same time minimizing initial compression. We chose to do the initial experiments on the highest density foams to minimize these effects.

Embedded (sandwiched) "stirrup" type magnetic gauges are used here to obtain particle velocity/time profiles at three depths in the target. Analysis of these wave profiles yields Hugoniot data and also aids in assessing the quality of the target assembly. The Hugoniot data are also compared to unpublished data generated by Fritz [1] in the early '80s (Fig. 1). Fritz' high-pressure data were acquired with explosively driven flyer plates and explosively driven shocks. The pressures attainable with the two-stage gas gun used in this study will extend the existing Hugoniot data to lower pressures and allow for some overlap. In addition to the Hugoniot data the present study yields dynamic particle velocity wave profiles

---

[†] Work performed under the auspices of the U.S. Dept. of Energy.

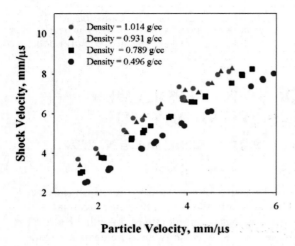

FIGURE 1. Unpublished Hugoniot data on a silicon polymer foams obtained in the early 1980's by Joseph Fritz at Los Alamos National Laboratory.[1]

offering the possibility to learn more about the foams crush-up process and reactions.

EXPERIMENTS

Material Description – The silicon foam mixture supplied from Dow Corning contains a proprietary mixture of PDMSs, cross-linkers, and diatomaceous earth. Dow Corning's trade name for this foam precursor is S5370. The mixture is foamed by addition of a tin catalyst. The foaming and compression to final density was completed at Los Alamos National Laboratory specifically for these shock compression experiments. The foams were made to four (nominal) densities, 0.4, 0.8, 0.9 and 1.0 g/cc. These densities were chosen to mimic the foam densities used by Fritz [1] for comparison. Each pressing was made to a thickness of ~ 2.5 mm and diameter of 250 mm. Gas-gun target coupons were punched from these pressings (Fig. 2). These 50 mm diameter by 2.5 mm thick disk were used as cut in the target assembly, since no machining or lapping was possible with the foam. The coupons have a "skin" or smooth film on both sides, which becomes more apparent for the highly distended foams.

Technique -- Experiments were done on a 50-mm-bore two-stage gas gun capable of launching projectiles to velocities of 3.1 km/s. Kel-F 81

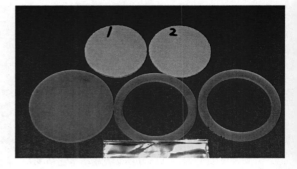

FIGURE 2. Silicon foam target discs were prepared with initial densities ranging from 0.4 to 1.0 g/cc. These foam discs are very flexible and compressible and were not amenable to machining. Kel-F front, PMMA rings, and silicon polymer foam discs are shown.

impactors were used on each of these experiments. Stirrup type magnetic gauges were used to acquire dynamic particle velocity-time profiles. The implementation of magnetic gauges to shock compression research has been recently reviewed by Sheffield [2] and is not discussed here. The stirrup gauge used for these measurements is shown in Fig. 3 and consists of a 5 µm thick aluminum conductor sandwiched between two thin Teflon sheets for a total thickness of 30 µm. There are actually two gauges shown in Fig. 3. Since they lie in the same plane they record similar velocities and are both used on occasions for redundancy.

Target Assembly – A schematic of the sandwich type target assembly is shown in Fig. 4. The assembly consists of two foam targets sandwiched between two target plates. The target plates were used to add rigidity to the assembly, since the foams are very flexible. The target plate at the impact side is made of Kel-F 81 and is an impedance match to the Kel-F 81 impactor. The target plate on the opposite side is made of PMMA. Clear PMMA windows were chosen to allow for velocity interferometry measurements at the foam/PMMA interface in subsequent experiments. Also shown in Fig. 4 are two spacer rings. Rings were machined and lapped to final thickness, which were unique to each foam's thickness. This is necessary to prevent compression of the foam or gap formation at the interfaces. Foam thickness

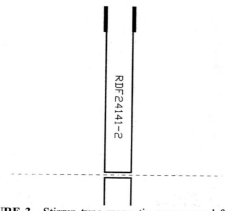

FIGURE 3. Stirrup type magnetic gauges used for these experiments consist of a thin 5 μm aluminum foil sandwiched between two Teflon sheets for a total thickness of ~ 30 μm.

was measured at multiple locations on the foams 50 mm diameter with a height gauge. Two different "foot" diameters were used with the gauge setup to assess measurement error associated with compression. It appears from work with foams near 1 g/cc that an optical based technique will be required for the fully distended (0.4 g/cm³) foams. Gauges were positioned at the three locations shown in Fig. 4. To prevent adhesive from filling the foam's voids, the gauges were pulled taunt and glued to the target plate and/or to the spacer rings, depending on the gauges position.

DATA

Three shots have been completed on the two-stage gas-gun each at a nominal projectile velocity of 2.1 km/s. Three foam densities ranging from 0.82 to 1.0 g/cm³) were used in these initial shots. Recorded particle velocity–time profiles for the 0.93 g/cm³ silicon foam target is shown in Fig. 5. The first gauge shows an abrupt jump in velocity and a fairly fast drop before stabilizing to a particle velocity of 1.31 mm/μs. The early spike is evidence of a gap between the first foam and the Kel-F 81 target plate. The spike arises because of the free surface associated with the gap. Gaps are not surprising due to the variation in the foams thickness across the disks diameter. The gauge appears to return the correct velocity, since it agrees well with the particle velocity recorded with the second gauge. The particle velocity recorded with

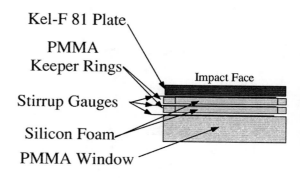

FIGURE 4. Schematic of the sandwich type target assembly used in these experiments. Rigid target plates (Kel-F 81 and PMMA) were used to hold the foams in place. Spacer rings machined from PMMA were tailored to each foams thickness to help prevent compression or gaps in the assembly. Three magnetic gauges positioned as shown, allow the evolution of the shock front to be viewed as a function of time.

the third gauge is lower due to the higher impedance of the PMMA target plate.

Particle velocity for these shots was calculated from the average of the first 0.5 μs of the first gauge for each shot. The shock velocity was calculated by measuring the shock's arrival time at each gauge and linearly fitting to the gauge's position. Results for these first three shots are given in Table 1. The only variable in these three shots was the foam's density. These data are plotted along with a subset of Fritz' data in Fig. 6. The new data (circled in Fig. 6) fall just below the lowest pressures achieved in Fritz' experiments, but show relatively good agreement. The two Hugoniot points for the higher densities (0.93 and 1.01 g/cm³) overlap and are in line with Fritz higher density Hugoniot data. The Hugoniot data point for the 0.82 g/cc foam lies just above Fritz' Hugoniot data for the 0.789 g/cm³ foam.

CONCLUSIONS

Initial shock compression experiments have been completed on silicon foams with densities of 0.82, 0.93, and 1.01 g/cm³. Special care was taken in the design and construction of the silicon foam target assemblies due to the flexible and easily compressible nature of the foams. Hugoniot data from these first experiments agrees with existing

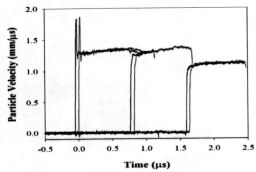

FIGURE 5. Dynamic particle velocity wave profiles from shot 2S-121. Data from the second set of gauges have been shifted by 50 ns to distinguish it from the first and show the agreement. The first waveform represents the particle velocity at the Kel-F 81/foam interface. The spike at the beginning indicates a gap was present at the interface. The second set of profiles beginning just before 1 μs represents the particle velocity between the two silicon foam discs. The third set shows the particle velocity at the foam/PMMA interface.

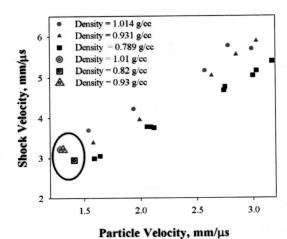

FIGURE 6. Hugoniot data from the current work is shown within the ellipse and agrees with earlier work by Fritz on silicon foams.

explosively driven (high pressure) data [1]. In addition to generating Hugoniot data, embedded magnetic gauges used in this work are yielding dynamic particle velocity waveforms. No evidence of reaction is seen on these initial low-pressure shots, but the profiles do show rounding of the shock front as it propagates through the first and second foam discs. The waveforms also show spikes in some of the data, which are evidence of gaps in the target assembly. Analysis and modeling efforts are ongoing and should help in understanding the shock compression of foams.

ACKNOWLEDGMENTS

Seth Gleiman and Tom Stephens supplied the Dow Corning silicon foam mixtures and foamed the samples to the desired densities.

REFERENCES

1. Unpublished data from experiments completed by Joseph Fritz in the early 1980s at Los Alamos National Laboratory.
2. Sheffield, S. A., Gustavsen, R. L., and Alcon, R. R., In-Situ Magnetic Gauging Technique used at LANL-Method and Shock Information Obtained," in Shock Compression of Condensed Matter-1999, Eds. M. D. Furnish, L. C. Chhabildas, and R. S. Hixon, AIP Conference Proceedings No 505 p 1043 (2000).

TABLE 1. Hugoniot data for initial silicon foam shock compression experiments

Shot #	Impactor	Impactor Velocity (mm/μs)	Silicon Foam Density (g/cm³)	Shock Velocity (mm/μs)	Particle Velocity (mm/μs)	Pressure (GPa)
2S-113	Kel-F 81	2.02	1.01	3.22	1.29	4.19
2S-120	Kel-F 81	2.02	0.82	2.95	1.41	3.41
2S-121	Kel-F 81	2.01	0.93	3.21	1.31	3.91

CP706, *Shock Compression of Condensed Matter - 2003*
edited by M. D. Furnish, Y. M. Gupta, and J. W. Forbes
© 2004 American Institute of Physics 0-7354-0181-0/04/$22.00

SPALL BEHAVIOUR OF POLYMETHYLMETHACRYLATE

D.L.A. Cross[1], D.P. Dandekar[2], and W.G. Proud[1]

[1]*Physics and Chemistry of Solids Group, Cavendish Laboratory, Madingley Road, Cambridge, CB3 0HE, UK*
[2]*Army Research Laboratory, Aberdeen Proving Ground, Aberdeen, Maryland, USA*

Abstract Polymeric materials are noted for their strong change in behaviour with strain-rate. Amongst these materials polymethylmethacrylate has been widely used in the shock physics community as a window material for velocity interferometry (VISAR) measurements. The Hugoniots measurements of PMMA from various commercial suppliers has been performed and reported. The spall behaviour of this widely used material, is however, not so well-documented. In this paper, VISAR is used to study the spall process. A series of plate impacts with PMMA fliers striking PMMA targets were performed. The process is shown to be strongly dependent on the impact geometry, the thickness of the impactor plate and the target plate. Overall, the VISAR signals show that the spall process does not simply scale with geometry .

INTRODUCTION

The spall strength of a material is the measure of its high-rate tensile strength. Spall occurs when the release fans from the impactor and the target free surfaces overlap. The accurate prediction and the modelling of spall requires an accurate knowledge of both the shock properties, such as the Hugoniot, and the release process, the isentrope, of the system.

Simple wave diagrams of the interaction of the releases and the fracture limit, are shown in figure 1. The spall strength indicated in figure 1 should be recognised as being highly dependent on the loading cycle and is not simply a fixed value of the material

In metals, the variation in value as a function of the initial shock input and the incipient formation of the spall plane have been studied. Recent interest has been directed into the studies of the mesoscale properties where individual grains affect the signal.

Polymers have been subject to shock investigation and, in general, the situation is more complex, the fracture does not form in an intra or trans-granular fashion, rather it passes through amorphous or semi-crystalline regions. Some polymers such as polycarbonate show very large strain to failure values.

The studies of spallation of polymers has been limited to a few materials, Lexan, polycarbonate[1], estane[2], rubber[3] and polymethylmethacrylate (PMMA)[4].

Overall PMMA has been relatively well characterised in shock compression as it is extensively used as a window in VISAR[5] experiments. The spallation experiments on PMMA[4] used material produced by Rohm and Haas and was their ultra-violet absorbing (UVA) Type II Plexiglass. The target and impactor thickness were maintained throughout at 6 and 3 mm respectively and the research showed a complex dependence on velocity.

This study seeks to extend this work by using a greater variety of target geometries, three groups of experiments were performed each with a different impactor: target thickness.

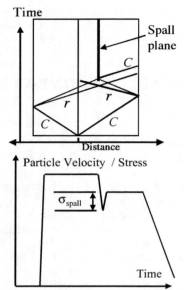

Figure 1. Top – Simple wave propagation diagram, c = compression wave, r = release fan. Bottom - velocity of rear of target plate.

EXPERIMENTAL

Plate impact experiments were carried out on the 50 mm bore gun at the University of Cambridge[6]. Impact velocity was measured to an accuracy of 0.5% using a sequential pin-shorting method and tilt was fixed to be less than 1 mrad by means of an adjustable specimen mount.

Targets and impactors were made from PMMA (Perspex, ICI). The impactor took the form of a 48 mm disc mounted on the front of a polycarbonate sabot. The thicknesses of the discs were varied from 2 - 6 mm. The targets were blocks 60 mm square with thicknesses of 6, 8 or 12 mm. The rear surface of each target was made specularly reflective by sputter coating < 5 μm with aluminium.

The aluminium surface was observed using VISAR developed by Barker and Hollenbach[7]. These data were reduced using the Valyn analysis programme.

In a series of experiments, not reported here, a high-speed camera was used to record the spallation process.

RESULTS

Figure 2 shows the velocity profile of an aluminium layer 12 mm inside a 24 mm thick target. This target was prodiced by aluminium sputtering a 12 mm thick plate and then gluing a second 12mm plate onto the aluminised surface using a slow setting epoxy resin. This was struck by a 4 mm flier plate at 300 m s[-1].

The data record shows the acceleration of the aluminium layer, followed by deceleration caused by the release wave from the rear of the impactor. The trace is truncated as the layer is again accelerated, this time by a release from the rear of the target.

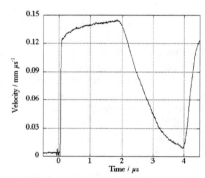

Figure 2. Velocity history of 5 μm aluminium layer, 12 mm inside PMMA target struck by 4 mm PMMA flier at 300 m s[-1]

Figure 3 shows the rear surface velocity of 12 mm thick targets when struck by 4 mm impactors at velocities between 300 – 750 m s[-1]. It can be seen that the width of the initial shock impulse decreases with increasing impact velocity.

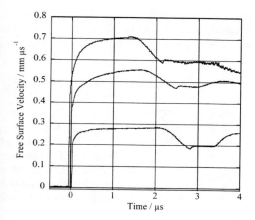

Figure 3 Spall traces of (a) 4 mm impactor against 12 mm target at 300, 600 and 750 m s⁻¹

At the lower velocity, the release profile shows a fall to a minimum followed by a reload signal of approximately 20 m s⁻¹ after 50 ns, A plateau follows for 500 ns and finally a second reload step. The variation in velocity does not affect the small reload signal nor the plateau. As the impact velocity increases, the main effect is to reduce the size of the second reload step. At impacts in the region of 750 m s⁻¹, this step no longer occurs.

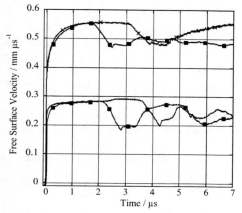

Figure 4. 4 mm impactor and 6 mm impactors against 12 mm targets at 300 and 600 m s⁻¹

In figure 4, spall traces for impacts between 6 mm thick impactors as well as 4 mm impactors on 12

mm thick targets are shown. Overall the same form as seen in figure 3. The effect of the 6 mm impactor plate is to produce a longer initial shock state. This also produces broader release fans given the larger distances over which the release moves prior to the spall. The overall effect is to reduce the size of the second reload signal, effectively degrading the spall strength of the PMMA.

Figure 5 shows the results from 2 mm fliers against 6 mm targets. The spall signature again shows a three part form. A simple analysis would suggest that this should scale by a factor of 2 on the time axis with impacts using a 4 mm flier / 12 mm target combination. This does *not* occur. Instead the small 20 m s⁻¹ reload and the 500 ns plateau is seen. The reload signal height is significantly smaller as well.

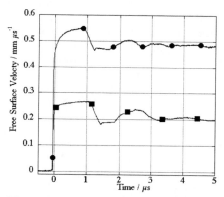

Figure 5. 2 mm impactors / 6 mm targets at 300 and 600 m s⁻¹

DISCUSSION

The shock response of the polymer is a complex phenomenon. The idea of a single spall strength is not applicable in these systems.

During the initial shock loading it can be seen that the loading moves rapidly to a steady value at low impact velocities while the loading becomes significantly more ramped at the higher velocities. The embedded aluminium plane shows a steady increase in velocity after the initial shock.

The maximum free surface velocity reached is always significantly below that of the impact velocity. To verify if this was an effect of the particular diagnostic system used, impacts on metals and ceramics sputter-coated with aluminium were performed. In these cases, it was seen that the velocity was as expected from simple conservation laws within a few percent. Overall there appears to be another effect occurring in the polymer system.

Discussion and some initial calculation[8] suggest that this velocity deficit a potential result of energy absorption in the polymer. Experiments with polystyrene will be performed as this material is known to have a smaller loss tangent than PMMA materials under quasi-static conditions.

This difference in ultimate velocity was not seen in the Rohm and Haas material[4]. Attention, therefore, must also be paid to the industrial processing of the polymer prior to acquisition as this is known to have a significant effect on the ultimate material properties. Other factors could include the mean molecular weight and any additives. Experiments on PMMA from a variety of suppliers will also be performed.

The release always seems to be accompanied by a sharp reload of 20 m s^{-1} or so and a 500 ns plateau. The height of the sharp reload seems to increase slightly with shock width as seen in figure 4. The plateau remains relatively unaffected. This region is probably associated with the nucleation of voids and microcracks followed by the growth of the spall layer.

Finally the main reload signal is shown to be dependent on shock duration and height. Generally it decreases with both, however, figure 5 shows small reload signals from thin 2 mm impactors, where the duration of the shock is much smaller and a higher spall strength would be expected.

Overall these results emphasise that the spall signature can be seen to vary with the impact velocity and the geometry in a non-trivial fashion.

Future publications will bring together high-speed photographic studies and the experiments outlined above to present a more complete data set for this complex spall process.

ACKNOWLEDGEMENTS

Philip Church of QinetiQ, Fort Halstead gave support to this reasearch. The European Office of the US Army provided support for D.P. Dandekar. EPSRC supported the high-speed imaging equipment though a series of grants. D. Porter of QinetiQ provided valuable insight into polymer systems. Prof J.E. Field gave his support to this research area.

REFERENCES

1. D.R. Curran, L. Seaman and D.A. Shockey, Phys. Today 30, 46-55 (1977)
2. J.N. Johnson and J.J. Dick, *Spallation Studies in Estane*, CP505, Shock Compression of Condensed Matter – 1999, 543-437
3. G.I. Kanel, S.V. Razorenov and A.V. Utkin, "*Spallation in Solids under Shock-wave loading, analysis and dynamic flow, Methodology of Measurements and Consituitive factors*" in High-Pressure Shock Compression of Solids II, edited by L. Davison, D.E. Grady and M. Shanipoor, Springer, New York, (1989), 1-24
4. P.T. Bartkowski and D.P. Dandekar, *Recompression of PMMA following shock induced tension*, CP505 Shock Compression of Condensed Matter, 1999, 539-542, (2000)
5. L.M. Barker, R.E. Hollenbach, J. Appl. Phy., 41, 4208 – 4226, (1970)
6. N.K. Bourne, Z. Rosenberg, D.J. Johnson, J.E. Field, A.E. Timbs and R.P Flaxman, Meas. Sci. Technol.,6, 1462 – 1470, (1995)
7. L.M. Barker and R.E. Hollenbach, J. Appl. Phy., 43, 4669-4675, (1972).
8. D. Porter, QinetiQ Ltd, Private Communication.

CP706, *Shock Compression of Condensed Matter - 2003*
edited by M. D. Furnish, Y. M. Gupta, and J. W. Forbes
2004 American Institute of Physics 0-7354-0181-0/04/$22.00

THE RESPONSE OF POLYURETHANE FOAM TO EXPLOSIVELY GENERATED SHOCKS

E J Harris, P Taylor and R E Winter

Hydrodynamics Department, AWE, Aldermaston, Reading, Berkshire, RG7 4PR, UK

A Plane Wave Lens (PWL) was characterised using a control experiment in which the surface of a copper plate driven by the explosive was measured by laser interferometry. A 1D computational model that closely matched the observed output of this experiment was developed. The PWL was then used to shock a polyurethane foam sample sandwiched between two copper plates. Again the surface velocity of the final copper plate was measured. Initial shock compressions were ~4 GPa rising to ~40 GPa after reflections between the metal plates. Hydrocode calculations using the previously published Equation of State (EoS) matched the shape of the measured velocity profile reasonably well but did not calculate the shock transit time through the foam correctly. A modified foam EoS which matched both the shape and timing of the observed experimental response has been developed.

INTRODUCTION

AWE has a requirement to determine the Equation of State (EoS) of low-density foams. A series of experiments in which discs of explosive initiated at the centre of one end face, delivered shocks to foam samples was reported by Maw, Whitworth and Holland in 1995 (1). The experiments provided the basis for the improvement of our foam Equations of State but suffered from the disadvantage that they were two dimensional (2D).

The purpose of this paper is to present the results of new, one-dimensional (1D), experiments and to describe how these experiments have been used to further optimise the Equation of State of polyurethane foam.

An explosive Plane Wave Lens (PWL) was used to drive a planar shock into the foam sample. The foam was subject to multiple shock loadings between the two copper plates. It was not possible to measure the input shock directly. Instead a simplified (1D) computational model was developed which accurately matched a calibration experiment. This computational model was then used to develop an improved EoS.

EXPERIMENTS

The experimental configuration, termed SI10, is shown in Fig 1. It consists of a 150 mm diameter bi-explosive lens constructed from EDC1S and a TNT/HNS explosive mixture, which detonates a 25.4 mm thick slab of EDC32 explosive. The EDC32 drives a planar shock into a 1.5 mm thick copper plate. The free surface of the copper is monitored by velocity interferometry. Velocity records obtained both on and off axis demonstrated that the experiment was acceptably 1D.

In a second experiment a 20 mm thick sample of 0.3 g/cc polyurethane foam and a second 1.5 mm thick copper plate were added to the front of the first copper plate. Velocity laser interferometry was again used to monitor the free surface of the second copper plate and hence the shocks transmitted by the foam sample.

MODELLING THE PWL

A 1D computational model of the SI10 geometry was developed using the AWE Eulerian code PETRA. The model consisted of a 70.8 mm length of explosive to represent the PWL, a 25.4

mm length of EDC32 and 1.5 mm of copper. Initial calculations used a TNT EoS from the LLNL Explosives Handbook (2) to model the PWL. However, it was necessary to adjust this EoS to obtain a good fit to experiment. An AWE EoS which switches between two forms depending on whether the density is less or greater than the Chapman-Jouget density was found to give a good fit. An EoS of similar form was used for the EDC32. The Copper was modelled using a constitutive model which had constant yield strength, included the Johnson spall model and used an EoS developed by R. K. Osborne of LANL.

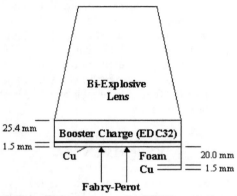

FIGURE 1. SI10 geometry. In SI10/2 a single copper disc was replaced with three discs of copper, foam and copper.

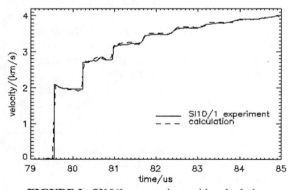

FIGURE 2. SI10/1: comparison with calculation.

Figure 2 shows the velocity trace obtained from SI10/1 experiment compared with calculation. The calculation is in excellent agreement with experiment. The early reverberations in the Copper

and the late time explosive drive are both well matched.

FOAM EXPERIMENT

The 1D model described above was used to calculate the foam experiment SI10/2. The foam was modelled with a modified Puff EoS as described by Maw et al (1). A snowplough model was used to simulate porosity.

The x-t plot shown in Fig. 3 illustrates the sources of the velocity jumps in the experimental data. The first jump is caused by the initial transmitted shock and the second jump results from a reverberation in the front copper plate. The third and largest jump results from a reflected shock in the foam, which is marked as R1 to C in Fig. 3.

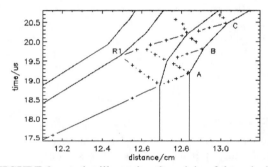

FIGURE 3. x-t plot illustrating the origin of the velocity jumps.

Figure 4 shows the velocity trace obtained from the experiment compared with calculation. It is seen that although the magnitudes of the first and second velocity jumps are calculated well with the Puff EoS, the initial shock transit time is too slow compared to experiment. In contrast, the transit time of the reflected shock, R1, is calculated well but the magnitude of the calculated velocity jump is larger than experiment. It is also apparent that the fourth velocity jump is calculated poorly.

Maw and Whitworth (3) reviewed data on 0.3 g/cc polyurethane foam and found that if they used a Gruneisen EoS then gas gun data (4) could be fitted with a value for Gamma (Γ) of ~ 1.0 which decreased to 0.75 at a reflected shock pressure of 9 GPa.

SI10/2 was calculated using a linear Gruneisen EoS and various values for Γ. If Γ is taken to be

1.0 up to solid density and then decreased linearly with density, the velocity profile of the copper is similar to that obtained with the Puff EoS. The velocity magnitudes are slightly lower and the transit time of the reflected shock, R1, is slightly shorter but otherwise the profile is very similar to that shown in Fig. 4.

It was found that if Γ was taken to be 0.8 at solid density, slowly decreasing to 0.775 at 2.1 g/cc, then the magnitude of the velocity jumps could be matched well. A comparison with experiment is shown in Fig. 5. The fourth velocity jump is better matched but the transit time of the initial shock is still too slow and the reflected shock, R1, is now too fast.

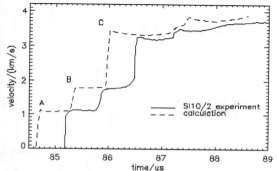

FIGURE 4. SI10/2: Comparison with a calculation using the Maw modified Puff EoS.

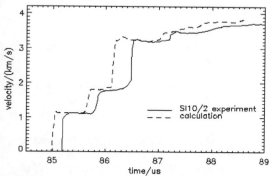

FIGURE 5. SI10/2: Comparison with a calculation using a Linear Gruneisen EoS.

In an attempt to match shock transit times *and* the initial velocity jumps, a cubic Gruneisen EoS was used for the foam. The bulk sound speed and the slope of the U_s-u_p Hugoniot were kept unchanged at 2486 m/s and 1.577 respectively but Γ

and the quadratic and cubic coefficients of the Shock Velocity Vs. Particle Velocity (U_s-u_p) relationship were varied. Figure 6 shows the agreement obtained with the SI10/2 experiment using a cubic Gruneisen EoS. It is seen that the shock transit times and the first two velocity jumps are well matched but the calculation overestimates the velocity jump caused by the reflected shock, R1.

The difference between the reflected shock pressures from the linear Gruneisen EoS of Fig. 5 and the cubic Gruneisen EoS of Fig. 6 can be seen in Fig. 7. There is a significant difference, of 13 GPa, in the pressures at the rear copper plate. The higher pressure of 53 GPa produces a lower plate velocity. This occurs because the reflected shock, R1, in the foam interacts with a reflected rarefaction, B, in the copper plate. It would be advantageous to remove this interaction in any future experiment and thereby remove one of the uncertainties in the calculation. This could be achieved simply by using a thinner sample of foam.

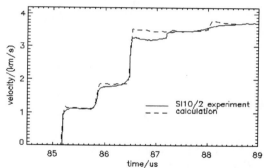

FIGURE 6. SI10/2: Comparison with a calculation using a Cubic Gruneisen EoS

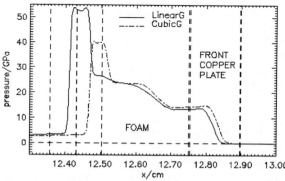

FIGURE 7. Comparison of pressures calculated using different Equations of State.

In the Gruneisen EoS referred to above, Γ is dependent on density but in the Osborne EoS gamma is also dependent on internal energy. Therefore by using the Osborne EoS it may be possible to vary gamma at higher energies and obtain better agreement with the reflected shock. However, several Osborne EoSs were derived which matched the initial velocity jump and shock transit times but the problem of matching the reflected shock persisted.

Carter and Marsh (5) noted that polyurethane undergoes a phase change at 21.7 GPa. The reflected shock in the foam is above this phase change pressure and therefore a multiphase EoS should give better agreement with the experimental data. Work has begun to derive a multi-phase EoS for the polyurethane foam. Fig. 8 shows the results from an initial attempt to derive such an EoS.

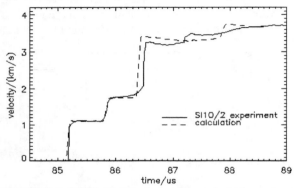

FIGURE 8. SI10/2: Comparison with a calculation using a multi-phase EoS.

Data at the phase change has been taken from Ref. 5. Gruneisen gamma is proportional to volume in both phases. It increases from 0.45 at solid density to 1.0 at the phase change, then decreases to 0.8 and then increases linearly with a gradient of 1.74. A Hugoniot has been calculated and it agrees reasonably well with the data published by Carter and Marsh but further work is needed to improve the match with the SI10 results and validate the phase boundary.

It is possible that the treatment of porosity in the above calculations would be improved by using a P-α model for the foam. However Kipp et al (6) determined the crush strength of polyurethane foam to be 2.1 MPa which is several orders of magnitude below the pressures in the SI10 experiments. This finding suggests that using this model may not affect the results significantly. Work is underway to calculate polyurethane foam using the Sandia CTH wave code in which P-α and P-λ (7) models have been implemented. The effect of using these models on the SI10 geometry will be assessed.

CONCLUSIONS

A 1D computational model has been designed that closely matches results from the SI10 PWL geometry. The previously published AWE EoS for polyurethane foam has been improved to give better agreement with experimental data. However more work is needed, especially at pressures above 20 GPa. The SI10 configuration will be used to provide additional data to further improve the foam EoS.

REFERENCES

1. Maw, J.R., Whitworth, N.J. and Holland, R.B.,"Multiple Shock Compression of Polyurethane and Syntactic Foams", in *Shock Compression of Condensed matter –1995*, pp133-136.
2. Dobratz, B.M. and Crawford, P.C., LLNL Explosives Handbook, UCRL-52997, January 1985.
3. Maw, J.R. and Whitworth, N.J., "Shock Compression and the Equation of State of Fully Dense and Porous Polyurethane", *Shock Compression of Condensed matter –1997*, pp111-114.
4. Wise, J.L. and Cox, D. E., SNL internal memorandum, 1996.
5. Carter, W.J., and Marsh, S.P., "Hugoniot Equation of State of Polymers", LA-13006-MS, UC-910, July 1995.
6. Kipp, M.E., Chhabilidas, L.C., Reinhart, W.D and Wong, M.K., "Polyurethane Foam Impact Experiments and Simulations", *Shock Compression of Condensed matter –1999*.
7. Grady, D.E., and Winfree, N.A., "A Computational Model for Polyurethane Foam", In *Fundamental Issues and Applications of Shock-Wave and High-Strain-Rate Phenomena*, 2001, Elsevier Science Ltd.

CP706, *Shock Compression of Condensed Matter - 2003*
edited by M. D. Furnish, Y. M. Gupta, and J. W. Forbes
© 2004 American Institute of Physics 0-7354-0181-0/04/$22.00

LONGITUDINAL AND LATERAL STRESS MEASUREMENTS IN SHOCK LOADED POLYETHER ETHER KETONE

J.C.F. Millett, G.T. Gray III*, N.K. Bourne

Royal Military College of Science, Cranfield University, Shrivenham, Swindon, SN6 8LA. United Kingdom.
**MST-8, Los Alamos National Laboratory, Los Alamos, NM 87545. U.S.A.*

Abstract. The shock response of poyether ether ketone (PEEK) has been investigated using manganin stress gauges mounted in longitudinal and lateral orientation to the impact axis. Measurements of the longitudinal stress with gauges at different positions within the shock assembly have determined the Hugoniot in terms of shock stress, shock velocity and particle velocity. It has been shown that the shock velocity has a simple linear response to particle velocity, in common with many but not all polymers. Measurements of lateral stress show a decrease behind the shock front, implying an increase in shear strength, possibly due to the viscoplastic nature of PEEK. Shear strength was also observed to increase with shock stress. A break in slope was observed at *ca.* 1.0 GPa, indicating a divergence between elastic and inelastic behaviour.

INTRODUCTION

An understanding of the high-strain-rate and shock-induced mechanical response of polymers is becoming increasing important, both as structural materials in their own right or as components in composite systems. This latter is of particular importance as polymers such as polychloro-trifluroethylene (Kel-F) and hydroxy-terminated polybutadiene (HTPB) are used as binders in plastic-bonded explosives. This paper examines the shock-induced response of polyether-ether ketone (PEEK), as part of a wider investigation of polymers.

PEEK is a relatively new polymer, with particular application in high temperature applications. Walley *et al.* [1] have tabulated its properties, giving a melting point of 334°C, a glass transition temperature of 143°C and a quasi-static yield strength of 100 MPa. Some high-strain-rate work has been performed [1, 2], showing that up to

strain-rates of 10^3 s^{-1}, the response of the yield stress is linear, but above that, the yield stress appears to drop. Similar results were also observed for polycarbonate and polysulphone.

EXPERIMENTAL

All shots were performed on the 5 m, 50 mm-bore single-stage gas gun at the Royal Military College of Science. 5 mm flyer plates of dural (aluminium alloy 6082-T6) and copper were fired in the velocity range 199 to 971 m s^{-1}. Velocities were measured via the sequential shorting of pairs of pins to an accuracy of 0.1%. Hugoniot measurements were made by mounting manganin gauges (MicroMeasurements type LM-SS-124CH-048) between 12.5 mm plates of PEEK, machined such that they were flat and parallel to better than 5 μm. A second gauge was supported on the front of the target assembly with a 1 mm coverplate of the same material as the flyer, such that the shock velocity could be determined from known gauge positions. Gauge calibrations were according to Rosenberg *et al* [3]. Lateral stresses were measured with manganin stress gauges (MicroMeasurements type J2M-SS-580SF-025) mounted 4 mm from the impact surface of a sectioned 12.5 mm plate of PEEK. Stress data was determined from the work of Rosenberg and Partom [4], using a modified analysis that requires no knowledge of the longitudinal stress [5]. Specimen configurations and gauge placements are presented in figure 1.

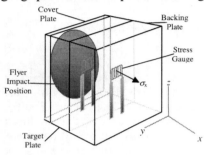

a. Hugoniot stress measurements.

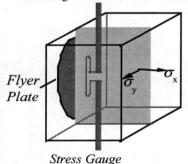

b. Lateral stress measurements.

Figure 1. Specimen configuration and gauge placement.

MATERIALS DATA

The properties of the PEEK investigated were – longitudinal sound speed (c_L)=2.47±0.03 mm μs^{-1}, shear sound speed (c_S)=1.06 mm μs^{-1}, density (ρ_0)=1.30±0.01 g cm^{-3} and Poisson's ratio (ν)=0.387. These were measured using a Panametrics 500PR pulse receiver with quartz transducers at a frequency of 5 MHz.

RESULTS AND DISCUSSION

Representative longitudinal gauge traces from PEEK are presented in figure 2.

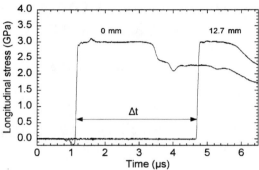

Figure 2. Longitudinal stress gauge traces from PEEK. 5 mm copper flyer plates at 674 m s^{-1}.

Both gauges record a maximum stress of *ca.* 3.0 GPa, indicating no attenuation as the shock wave moves through the target. The temporal spacing between the gauges (Δt), combined with the known gauge spacing (Δw of 12.7 mm) have been used to determine the shock velocity through U_s=$\Delta w/\Delta t$. The particle velocity (u_p) was determined from impedance matching techniques. The Hugoniot of PEEK is presented below in figure 3, both in terms of σ_x (shock stress) – u_p (particle velocity) and U_s-u_p.

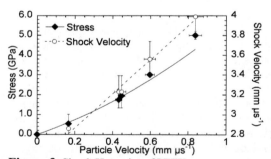

Figure 3. Shock Hugoniot of PEEK.

A simple linear regression has been fitted to the shock velocity data (the dotted line), assuming the standard relation U_s=c_0+Su_p, yielding –

$$U_s = 2.52 + 1.71u_p. \tag{1}$$

664

One feature from this data that immediately stands out is the difference between c_0 from figure 3 and equation 1, and the measured c_L, with c_0 being slightly greater. In metallic system, c0 generally equates with the bulk sound speed, and thus is lower than the measured longitudinal sound speed. However, the reverse is true in PEEK, a feature that both we have observed in other polymers such as epoxy resins [6] and polychloroprene [7] and others in epoxy [8] and polymethylmethacrylate [9], thus suggesting that such behaviour is characterisitic of polymeric materials. Also observe from figure 2 that the gauge labeled 12.7 mm, there is no break in slope that may indicate a Hugoniot Elastic Limit. Given that the elastic impedance of PEEK is similar to that of the epoxy adhesive and gauge backing, one would assume that the signal could rise quickly enough to resolve such changes in material response. However, as has already been demonstrated, the shock velocity in this material is always faster than the longitudinal sound speed, thus the elastic precursor will always be overtaken by the main shock, and thus no HEL will be seen in gauge traces.

This in turn, has been used to determine the hydrodynamic response (solid) through,

$$\sigma_x = \rho_0 U_s u_p \qquad (2)$$

It can be seen that the agreement between equation 2 and the measured stress data show increasing divergence, with the calculated stresses being the lower. A possible explanation presents itself when one considers that the shock stress is composed of hydrodynamic (P) and shear strength components (τ), thus,

$$\sigma_x = P + \frac{4}{3}\tau. \qquad (3)$$

Therefore the differences between the measured and calculated stresses would suggest that the materials shear strength increases with increasing impact stress. This issue is explored further in the following section.

In figure 4, traces from gauges mounted such that they are sensitive to the lateral component of stress are presented. The reason for these measurements is that from knowledge of the

longitudinal and lateral components of stress during shock loading, the material shear strength (τ) can be determined, thus,

$$2\tau = \sigma_x - \sigma_y. \qquad (4)$$

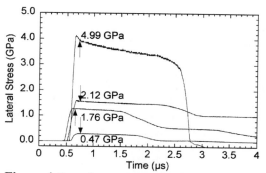

Figure 4. Lateral gauges traces in PEEK.

The most noticeable feature from these traces is the pronounced decrease in lateral stress behind the shock front for the 4.99 GPa shot and a gradual decrease for the remaining 3 peak pressures. If it is assumed that the longitudinal stress remains constant behind the shock (which from figure 2 is the case), this suggests that the shear strength increases behind the shock. Such behaviour has been observed in polymethylmethacrylate (PMMA) [10], and an epoxy resin [6], where it was suggested that this was a manifestation of the viscoplastic response of polymers. The data in figure 4 in the current study suggest that PEEK is responding in a similar manner.

The variation of shear strength with impact stress has been determined, using equation 4, and lateral stresses from figure 4 indicated by the arrows. The results are presented in figure 5.

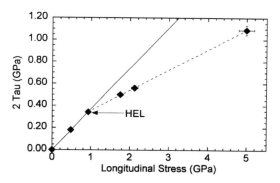

Figure 5. Shear strengths of PEEK.

665

Two straight lines have been fitted to this data. The solid line is according to the simple elastic response of materials,

$$2\tau = \frac{1-2\nu}{1-\nu}\sigma_x.$$ (4)

It can be seen that it passes through the lowest stress point. The second (dashed) is a simple linear regression fitted to the three highest stress points, and extended such that it intercepts with the elastic response. In doing so, we have assumed that this intercept corresponds to the Hugoniot Elastic Limit (HEL), yielding a value of ca. 1.0 GPa. As no such values for PEEK have appeared in the open literature, it is difficult to cross reference this result. However, we note that this method has been used for PMMA [10], and shown to provide good agreement with accepted values for the HEL of PMMA (0.75 GPa) from the literature [9, 11]. Therefore the data in this study suggest that an HEL of ca. 1.0 GPa is representative of PEEK.

CONCLUSIONS

The shock response of polyether ether ketone has been measured using manganin stress gauges mounted in longitudinal and lateral orientations. The relation between the shock velocity and the particle velocity has been shown to be linear, in common with many other materials. Lateral stress measurements have proven to be more revealing. The experimental data shows that the lateral stress decreases behind the shock front for PEEK, indicating an increase in shear strength. Such behaviour has been observed in other polymers, where it was suggested that this was due to the viscoplastic response of those materials. Finally, a break in slope of the shear strength – shock stress curve at 1 GPa has been suggested as the HEL of PEEK.

ACKNOWLEDGMENTS

The authors would like to thank Gary Cooper, Matt Eatwell, Gary Stevens and Paul Dicker of RMCS for technical assistance.

REFERENCES

1. Walley, S.M., Field, J.E., Greengrass, M. Wear, 1987. **114** 59.
2. Walley, S.M., Field, J.E.. DYMAT, 1994. **1** 211.
3. Rosenberg, Z., Yaziv, D., Partom, Y. J. Appl. Phys., 1980. **51** 3702.
4. Rosenberg, Z. Partom, Y. J. Appl. Phys., 1985. **58** 3072.
5. Millett, J.C.F., Bourne, N.K., Rosenberg, Z. J. Phys. D. Applied Physics, 1996. **29** 2466.
6. Millett, J.C.F., Bourne, N.K., Barnes, N.R.. J. Appl. Phys., 2002. **92** 6590.
7. Millett, J.C.F. Bourne, N.K. J. Appl. Phys., 2001. **89** 2576.
8. Munson, D.E., May, R.P. J. Appl. Phys., 1972. **43** 962.
9. Barker, L.M., Hollenbach, R.E. J. Appl. Phys., 1970. **41** 4208.
10. Millett, J.C.F., Bourne, N.K.. J. Appl. Phys., 2000. **88** 7037.
11. Schuler, K.W., Nunziato, J.W. J. Appl. Phys., 1976. **47** 2995.

CP706, *Shock Compression of Condensed Matter - 2003*
edited by M. D. Furnish, Y. M. Gupta, and J. W. Forbes
© 2004 American Institute of Physics 0-7354-0181-0/04/$22.00

HUGONIOT OF C$_{60}$ FULLERITE

V.V. Milyavskiy[1], A.V. Utkin[1,2], E.B. Zaretsky[3], A.Z. Zhuk[1], V.V. Yakushev[2], V.E. Fortov[1,2]

[1]*Institute for High Energy Densities RAS, Moscow, Russia*
[2]*Institute for Problem of Chemical Physics RAS, Chernogolovka, Russia*
[3]*Ben-Gurion University of the Negev, Beer-Sheva, Israel*

Abstract. Hugoniot of C$_{60}$ fullerite and the dependence of sound velocity versus pressure were measured. Experiments were carried out at pressures up to ~ 45 GPa. Hugoniot of the fullerite has several features. The transformation of C$_{60}$ to a dense carbon phase is observed at the transition onset pressure ~ 15 GPa. If shock pressures higher than ~ 33 GPa, Hugoniot of C$_{60}$ fullerite is determined by the thermodynamic properties of diamond-like carbon phase.

INTRODUCTION

Fullerenes were discovered in 1985 by Kroto et al. [1]. In 1990 Kratchmer et al. [2] proposed a technique for production of fullerenes in macroscopic quantities by electric arc evaporating of graphite electrodes. This discovery initiated numerous studies of their physical [3] and chemical [4] properties.

According to ref. [5,6], C$_{60}$ is a less stable allotrope of carbon. At ambient conditions, C$_{60}$ is a face centered cubic (fcc) molecular crystal. At static pressures of about 0.3-0.4 GPa reversible transformation to simple cubic (sc) modification occurs (see ref. 3). This transformation is accompanying by the small (~ 1%) decrease of the specific volume.

It is known that series of structural transformations take place at static compression of fullerene C$_{60}$ in conditions of increased temperature. These transformations consist in drawing together C$_{60}$ molecules and forming intermolecular covalent bonds (polymerization) [7,8]. Formation of 1D-, 2D-, 3D- polymerized and disordered states were experimentally observed. Disputable is the question about hardness of high-

pressure phases of C$_{60}$ (hpp). For example, in ref. [8] it is asserted that hardness some of hpp considerably exceeds that of diamond. On the other hand, the authors of ref. [7] consider that the mechanical characteristics of hpp are significantly lower than those of diamond.

Phase transitions of the C$_{60}$ fullerite under the shock loading have been investigated by means of the so-called recovery technique in details (see ref. [9,10] for example). This method utilizes shock loading of the starting material and the subsequent examination of the structure of the recovered material. Because of the reverberations of the waves between walls of the specimen container compression pulse usually has a step-like front (quasi-isentropic compression occurs).

Recovery experiments show that fullerite C$_{60}$ is stable up to pressure ~20 GPa under quasi-isentropic compression. Fullerite decomposes at lower pressure if temperature is higher (for example at compression in a high-porous metal matrix). Specimens recovered after dynamic compression of fullerite up to pressure P > 20 GPa, contained graphite, diamond and amorphous carbon. In some cases short range order of different types was detected in amorphous carbon. Traces of

C_{60} polymerized phases were detected in some experiments at pressures near 20 GPa.

EXPERIMENTAL PROCEDURE

In our experiments we used polycrystalline C_{60} specimens (with purity not less than 99,5% and thickness 1.5-4.5 mm). Specimens were prepared by high (1 GPa) hydrostatic pressure treatment using procedure, similar to that described in ref. [11]. The specimens had a density 1.64 g/cc. It is of about 97.5% of the crystalline density of C_{60} fullerite under normal conditions. Specimens were loaded by impacts of the metal plates accelerated up to 5.3 km/s.

At low pressure region (impact velocity ≤ 0.53 km/s) gas gun was used. The velocity of the interface between rear surface of the specimen and PMMA window was measured by laser velocity interferometer – VISAR [12]. To measure projectile velocity and shock velocity contact gauges were mounted on the front surface of the specimen.

At high-pressure experiments we used calibrated explosive projectile systems. In this case the specimen was installed on the metal screen. The velocity of the interface between the specimen and water window was measured by VISAR. In addition, ring piezoelectric gage was mounted around the specimen. At high pressures every experiment included two stages. In the first stage we did not install the specimen on the screen. This way we registered the pressure pulse on the rear surface of the screen. In the second stage we used the experimental assembly with specimen (Fig. 1).

To process the results we superposed signals of the piezoelectric gauges, corresponding to the first stage and to the second stage. After that we measured time intervals between arrival of the shock wave and release wave to the front and to the rear surfaces of the specimen. Thus, we calculated the velocity of shock wave front U_S and estimated the longitudinal sound velocity in shock-compressed specimen. To calculate pressure and specific volume we used standard relations conserving mass and momentum across the shock front [13].

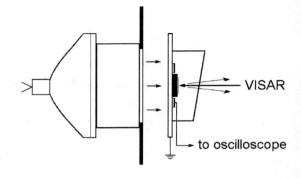

FIGURE 1. Scheme of an experimental assembly with use of explosive projectile system. The VISAR laser beam is Doppler shifted by the movement of the C_{60}/water interface.

RESULTS AND DISCUSSION

Results of measurements of shock compressibility of C_{60} fullerite are presented in Fig. 2-4. Hugoniot of porous diamond are shown, too. These Hugoniot were computed with use of equation of state (EOS) [14]. In Fig. 2-4 we also presented low-pressure Hugoniot of C_{60} (fcc), calculated with use of static compressibility data (see ref. [3]).

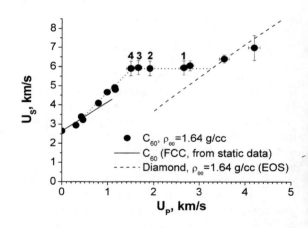

FIGURE 2. Hugoniot of C_{60} fullerite (U_S -U_P plane). U_P - mass velocity behind the shock wave front. Dotted lines - linear approximation.

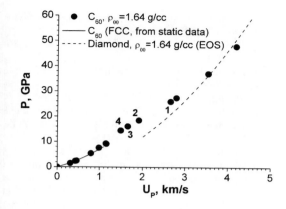

FIGURE 3. Hugoniot of C_{60} fullerite (P -U_P plane). P - pressure behind the shock wave front.

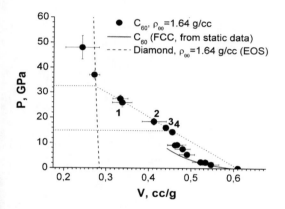

FIGURE 4. Hugoniot of C_{60} fullerite (P - V plane). V - specific volume behind the shock wave front. Horizontal dotted lines enframe the region of existence of two-wave structure of shock front.

The first offset of the Hugoniot from the calculated compression curve is observed at pressure $\sim$ 2.3 GPa and may correspond to structural transformation fcc $\rightarrow$ sc.

The break of the Hugoniot at pressure $P_H \sim 15$ GPa may correspond to the onset of phase transition to more dense phase of carbon. One cannot assert definitely, that an end-product of this phase transition is cubic diamond. However, good agreement of the sound velocity in shocked C_{60} with calculated sound velocity of compressed diamond at pressure >35 GPa (Fig. 5) testifies for the benefit of this assumption. It is also possible to propose an opportunity of two-stage phase

transition, for example: C_{60} $\rightarrow$ graphite (or amorphous $sp^2 + sp^3$ carbon) $\rightarrow$ diamond.

As follows from Fig. 4, the shock wave at $P_H < P < \sim 33$ GPa should lose stability. In this case two-wave structure of the shock front should be observed [13]. Similar features were detected in two experiments (points 1, 2 in Fig. 2-4). To find parameters of points 1, 2 we used measured values U_S in C_{60} and known velocities of impactors. Parameters of points 3, 4 (Fig. 2-4) correspond to the first shocks in the same experiments. These parameters were found from wave profiles monitored by VISAR with use of the procedure, similar to that described in ref. [15].

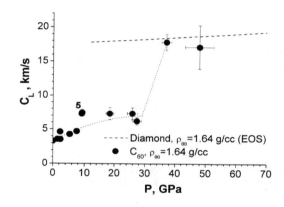

FIGURE 5. The experimental dependence of longitudinal (Euler) sound velocity C_L in shock-compressed fullerite specimens versus pressure P.

At pressure > 33 GPa Hugoniot of C_{60} fullerite is determined by the thermodynamic properties of diamond-like carbon phase (possible - cubic diamond).

An anomalous behavior of the sound velocity was detected at pressure $\sim$ 9 GPa (point 5 in Fig. 5). This fact may be explained either by existence of one more phase transition in this pressure region or by an exothermic reaction behind the shock front. The latter proposal is not so improbable if to take into account that enthalpy of formation of C_{60} fullerite is about 3270 kJ/kg [5,6]. This energy should be released during transformation of C_{60} to graphite. It is comparable, for example, with heat of explosion of ammonite (4312 kJ/kg).

CONCLUSIONS

Shock compressibility of C_{60} fullerite and sound velocities in shock-compressed fullerite were measured up to ~ 45 GPa.

For C_{60} fullerite a transformation to a dense carbon phase is observed with a transition onset pressure ~ 15 GPa, the stability limit of the fullerite structure under single-step (adiabatic) shock compression. The two-wave structure of the transition is overdriven to a single wave above ~ 33 GPa. If shock pressures higher than ~ 33 GPa, Hugoniot of C_{60} fullerite is determined by the thermodynamic properties of diamond-like carbon phase.

An anomalous behavior of the sound velocity was detected at pressure ~ 9 GPa. Additional experiments are necessary to clarify processes occurring at pressure region near 9 GPa.

ACKNOWLEDGEMENTS

We thank to L.G. Khvostantsev (IHPP RAS) for specimens fabrication and K.V. Khischenko (IHED RAS) for performing EOS calculations.

This work was supported by Russian Academy of Sciences. Additional funding was provided by RFBR (project no. 02-02-16582 and INTAS (YSF 01/1-40).

REFERENCES

1. Kroto, H. W., et al., *Nature* **318**, 162-163 (1985).
2. Kratchmer, W., et al., *Nature* **347**, 354-358 (1990).
3. Sandquist, B. "Fullerenes under high pressures", in *Advances in Physics* **48**, no. 1, 1-134 (1999).
4. Bezmel'nitsyn, V. N., et al., *Physics-Uspekhi* **41**, no. 11, 1091-1114 *(1998)*.
5. Kolesov, V. P., et al., *J. Chem. Thermodynamics* **28**, 1121-1125 (1996).
6. Lebedev, B. V., et al., *Thermochimica Acta* **299**, 127-131 (1997).
7. Brazhkin, V. V., and Lyapin, A. G. *Physics-Uspekhi* **39**, no. 8, 837-840 (1996).
8. Blank, V. D., et al., *Carbon* **36**, 319-343 (1998).
9. Yoo, C. S., and Nellis, W. J. *Science* **254**, 1489-1491 (1991).
10. Milyavskiy, V.V., et al., *Defect and Diffusion Forum* **208-209**, 161-174 (2002).
11. Milyavskiy, V.V., et al., *High Temp.* **39**, no. 5, 786-788, (2001).
12. Barker, L. M., and Hollenbach, R. E., *J. Appl. Phys.* **43**, 4669-4675 (1972).
13. Zeldovich, Ya. B., and Raizer, Yu. P., *Physics of Shock Waves and High-Temperature Hydrodynamic Phenomena*, Dover Publications Inc., New York, 2002, pp. 46-47.
14. Khishchenko, K. V., et al., "Shock compression, adiabatic expansion and multi-phase equation of state of carbon", in *Shock Compression of Condensed Matter-2001*, edited by M. D. Furnish et al., AIP Conference Proceedings 620, Melvile, New-York, 2002, pp. 759-762.
15. Erskine, D.J., and Nellis, W. J., *J. Appl. Phys.* **71**, 4882-4886 (1992).

CP706, *Shock Compression of Condensed Matter - 2003*
edited by M. D. Furnish, Y. M. Gupta, and J. W. Forbes
© 2004 American Institute of Physics 0-7354-0181-0/04/$22.00

THE TAYLOR IMPACT RESPONSE OF PTFE (TEFLON)

Philip J. Rae*, George T. Gray*, Dana M. Dattelbaum† and Neil K. Bourne**

**MST-8, MS-G755, LANL, Los Alamos, NM 87545*
†*MST-7, MS-E549, LANL, Los Alamos, NM 87545*
***Cranfield University, Defence Academy of the United Kingdom, Shrivenham, Swindon, SN6 8LA, U.K.*

Abstract. Whilst Polytetrafluoroethylene (PTFE) is an unusually ductile polymer, it undergoes an abrupt ductile-brittle transition at modest impact velocities. No previous explanation for this behaviour seems to have been presented. In this paper we examine the role of a pressure-induced phase transition in PTFE in the failure of Taylor cylinder samples. Whilst a phase transition occurs at approximately 0.65 GPa at 21°C, the transition pressure is inversely related to temperature. Varying the temperature of the fired Taylor cylinders shows that the phase transition is likely to be involved because the critical velocity increased for decreasing temperature, despite the material fracture toughness decreasing.

INTRODUCTION

Although P.T.F.E (polytetrafluoroethylene) is a commonly used material in industry and engineering, its mechanical response has received relatively little study in recent years. Early literature on PTFE describes a ductile, inert and stable polymer[1, 2], that had many potential uses in bearings, gaskets and electrical fittings. PTFE is unique among polymers in retaining some measure of ductility ($\approx$3-5 %) at liquid helium temperatures. It was therefore surprising that such a ductile polymer should undergo an abrupt ductile–brittle transition when impact loaded at quite modest rates. The earliest report of this transition occurred in 1958 in a report prepared by the University of Texas for the Sandia Corporation[3]. PTFE samples of various geometries were impacted by a 830 g projectile at 58 m s^{-1}. In most impact experiments the samples disintegrated into small pieces, whilst at quasi-static loading rates, the same geometries deformed gracefully in a ductile manner. Previous authors have commented on the sometimes 'brittle' nature of PTFE, but to our knowledge, no explanation has been postulated.

A small team at Los Alamos National Laboratory has been tasked with investigating the properties of PTFE with a view to developing a realistic constitutive model of the polymers behaviour over a wide range of strain-rates, stress-states and temperatures. As part of this study, Taylor cylinder[4] experiments were conducted, together with high-speed photography to observe the ballistic behaviour of PTFE. It was quickly realised that the ductile–brittle transition reported in the literature was exhibited in the Taylor test and an explanation for this response was sought.

MATERIAL

Central to the goal of developing a mechanical material model for PTFE was the use of pedigreed polymer material. Too many mechanical response papers, particularly in the field of polymers, relate that a sample of 'α' was tested with no further mention of exact chemistry, how it was processed, what its morphology was or how it was aged, appearing in the report. For this study a single $600 \times 600 \times 65$ mm^3 billet of DuPont 7A Teflon was sintered from a known powder batch, to a known pressing and heating schedule. Samples of the billet were cut to confirm the isotropic nature of the billet and various other physical properties were quantified, see Table 1

TABLE 1. Properties of the Teflon 7A tested. See Lehnert[5] for information on the different methods of estimating PTFE crystallinity.

Property	Value
Density (Immersion)	2.1583 kg m^{-3}
Density (Pycnometry)	2.1577 kg m^{-3}
Crystallinity (X-Ray Diffraction)	69%
Crystallinity (Density)	43%
Crystallinty (MDSC)	32%
Crystallinty (Infra-red)	73%

for details. The Taylor cylinders machined for these experiments were taken from the through-thickness (65 mm) direction of the billet.

PTFE is a complex material. It exhibits at least four phase changes depending on a combination of temperature and pressure[6]. Additionally, PTFE always contains a mix of amorphous and crystalline regions, it is not possible to manufacture fully amorphous or fully crystalline PTFE. At atmospheric pressure, below 19°C, PTFE has a triclinic crystalline structure (II). Above this temperature it undergoes a 1st order phase transition into a hexagonal structure exhibiting a 1.8% volume increase (IV). A second order transition occurs at 30°C into pseudo-hexagonal (I). From 30°C until melting (321°C for once melted material, 341°C for virgin moulding powder) a general relaxation of the crystalline structure occurs until, given enough time, a fully amorphous state is reached.

A pseudo-equilibrium pressure-induced phase transition has been reported in PTFE at ≈0.65 GPa at room temperature (III) [7, 8]. This transition is strongly temperature dependent, as shown in Figure 1, however considerable hysterisis was noted leading to large error bars. Recent work at Los Alamos using a diamond cell anvil and Raman Spectroscopy suggests the transition occurs at 0.65 GPa and exhibits around ±0.5 GPa of hysterisis.

EXPERIMENTAL

A helium gas driven gun was used for these Taylor experiments[4]. 50.8 mm (2 inch) long right-cylinders were fired at a 91 kg hardened steel anvil. The impact face of the anvil is polished to a mirror finish and the surface lubricated with a synthetic oil

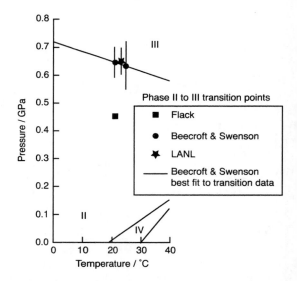

FIGURE 1. The strongly temperature related nature of the pressure-induced phase transition in PTFE.

containing colloidal PTFE. The sample is fired into an evacuated catcher chamber.

A three inch section of the barrel at the breech end is capable of being heated or cooled between +200 and -100°C using electrical heaters or liquid nitrogen cooling coils. In this way the impact behaviour of samples may be investigated with respect to temperature. For these PTFE tests, samples were heated or cooled to the desired temperature for at least one hour in an environmental chamber before being quickly loaded in to the temperature controlled section of the Taylor barrel.

Owing to the soft nature of PTFE and the low velocities used (120-150 m s^{-1}) the anvil remained well within the elastic regime and no visible damage to the impact face was caused. An Imacon 200 high-speed camera was used to photograph the shots. This camera is capable of taking up to 16 frames at a maximum rate corresponding to 200 million frames per second. The exposure time and inter-frame time (IFT) of each exposure are fully programmable. In these experiments 14 frames were used with a 500 ns exposure and a 15 μs IFT. The projectile velocity was measured using two laser beams spaced 32.28 mm apart.

To ensure accurate syncronisation of the photography with respect to the impact time, an up-down counter was employed. This device digitally counts up whilst the projectile is between the laser beams and starts to count down again once the projectile reaches the second beam. The camera is triggered when the count reaches zero. By setting the ratio of the up to down count frequency to the ratio of the distance between beams and the anvil, the camera is always triggered when the projectile is a set distance from impact, irrespective of the projectile velocity. The camera flash needs to be triggered at least 100 μs before impact, but because the flash lasts for approximately 1.2 milliseconds, a well illuminated scene is easily achieved with a simple delay generator.

RESULTS

Taylor samples were found to exhibit an abrupt ductile-brittle transition. Figure 2 shows the marked change in response for only a 1 m s^{-1} change in velocity at 21°C. Figure 3 shows the fracture threshold map plotted for 7A Teflon at 21°C. It was decided to see if the pressure-induced phase transition in PTFE might play a part in this behaviour.

A primitive dynamic finite element model (elastic, perfectly plastic) of a Taylor impact was run using the Lagrangian EPIC code[9]. An axisymmetric PTFE rod was simulated impacting an infinite smooth steel block at 135 m s^{-1}. A maximum compressive hydrostatic stress of $\approx$0.5 GPa was generated approximately 3 μs after impact. At later times, tensile stresses develop in the end of the rod. From this it was concluded that the magnitude of the stress was of the correct order to affect the material. Further shots were carried out to test this hypothesis making use of the strongly temperature sensitive nature of the phase transition.

The mechanical response of a polymer is affected by temperature. Generally, lowering the temperature increases the yield strength, but lowers the fracture toughness. This is the case with PTFE. Figure 4 shows the effect of temperature on the compressive properties of 7A Teflon whilst Figure 5 shows the fracture toughness (Gc) verses temperature. To limit the effect of mechanical property changes, additional Taylor shots at 1 and 40°C were undertaken. Table 2 shows the measured ductile-brittle transition velocity with respect to temperature.

TABLE 2. Measured ductile-brittle transition velocities vs. temperature.

Temperature / °C	Transition Velocity / m s^{-1}
1	139±2
21	134±1
40	131±1

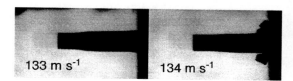

FIGURE 2. Two 7A Teflon Taylor images 165 μs after impact. Left: 133 m s^{-1}. Right: 134 m s^{-1}.

DISCUSSION

Definitive evidence of a phase transition during impact is difficult to obtain because flash X-Ray crystallography would be required. Post sample analysis is likely to be inconclusive because the phase transition is known to be reversible upon unloading. The ductile-brittle transition is certainly abrupt enough to be related to a phase transition. Given

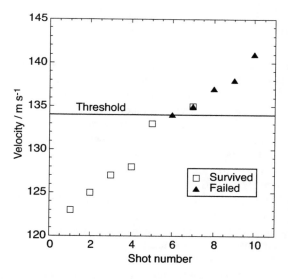

FIGURE 3. Fracture velocity threshold map for 7A Teflon at 21°C.

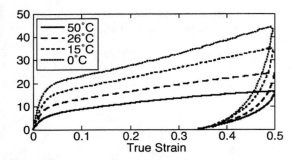

FIGURE 4. Stress/Strain response of 7A Teflon vs. temperature.

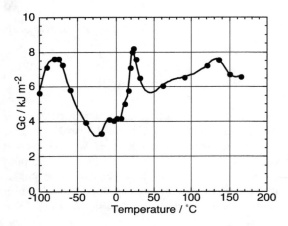

FIGURE 5. Fracture toughness (Gc) for PTFE vs. temperature. Re-drawn from Kisbenyi[10].

that the critical velocity of the Teflon 7A actually increased with lower temperature is further evidence because, as reported, the fracture toughness of Teflon would be decreased at temperatures higher and lower than the 19°C phase transition. Whilst the strength of Teflon is increased at lower temperatures, the strain-to-failure is reduced[11].

Further evidence that the variability in strength of the PTFE with temperature is not playing a part can be derived from high-speed photography. Samples that are going to fail in a brittle manner do so within the first 15 μs, and at radial strains smaller than those supported at later times in specimens that do not become brittle. It will be remembered from the computer model that the maximum hydrostatic stress was developed approximately 3 μs after impact. Similar room temperature Taylor experiments under take at

Cranfield University on un-pedigreed extruded PTFE material show a similar failure transition velocity to that found in Teflon 7A. It may therefore be supposed that the material processing does not, to a first order, effect the impact fracture behaviour. This would be consistent with a phase transition phenomena being involved.

In conclusion, it is felt that these Taylor shots present strong evidence in favour of a pressure-induced phase transition being responsible for the ductile-brittle transition found in PTFE.

ACKNOWLEDGMENTS

The authors gratefully acknowledge financial support for this research from the joint DOD/DOE MOU on the dynamic behaviour of polymers.

REFERENCES

1. Renfrew, M. M., and Lewis, E. E., *Industrial and Engineering Chemistry*, **38**, 870–877 (1946).
2. Sperati, C. A., and Starkweather, H. W., *Advances in Polymer Science*, **2**, 465–495 (1961).
3. Ripperger, E. A., Stress-strain characteristics of materials at high strain rates, part ii, Tech. Rep. AT(29-2)-621, Structural Mechanics Research Laboratory, UT (1958).
4. Maudlin, P. J., Bingert, J. F., and Gray III, G. T., *International Journal of Plasticity*, **19**, 483–515 (2003).
5. Lehnert, R. J., Hendra, P. J., Everall, N., and Clayden, N. J., *Polymer*, **38**, 1521–1535 (1997).
6. Weir, C. E., *Journal of Research of the National Bureau of Standards*, **50**, 95–97 (1953).
7. Beecroft, R. I., and Swenson, C. A., *Journal of Applied Physics*, **30**, 1793–1798 (1959).
8. Flack, H. D., *Journal of Polymer Science: Part A-2*, **10**, 1799–1809 (1972).
9. Johnson, G. R., Stryk, R. A., Holmquist, T. J., and Beissel, S. R., User instructions for the 1997 version of the epic code., Tech. Rep. WL-TR-1997-7037, Wright Laboratory, Armament Directorate, Eglin Airforce Base (1997).
10. Kisbenyi, M., Birch, M. W., Hodgkinson, J. M., and Williams, J. G., "The impact fracture toughness of PTFE", in *Proc 175th ACS (Organic Coatings and Plastics Chemistry)*, Anaheim, 1978, vol. 38, pp. 394–399.
11. Koo, G. P., *Cold drawing behaviour of Polytetrafluoroethylene*, Ph.D. thesis (1969).

CP706, *Shock Compression of Condensed Matter - 2003*
edited by M. D. Furnish, Y. M. Gupta, and J. W. Forbes
© 2004 American Institute of Physics 0-7354-0181-0/04/$22.00

MAGNETIC PARTICLE VELOCITY MEASUREMENTS OF SHOCKED TEFLON[†]

D. L. Robbins, S. A. Sheffield, and R. R. Alcon

Los Alamos National Laboratory, Los Alamos, NM 87545

Abstract. A series of shock compression experiments have been undertaken on Teflon using single- and two-stage gas-guns. Peak pressures in these experiments range from a few kbars to over 10 kbars, as well as one shot completed at 117 kbar. Multiple particle velocity wave profiles, at a number of Langrangian positions, are obtained for each experiment using *in-situ* magnetic gauges. Shock velocity is calculated from arrival times at both the particle velocity gauges and at embedded shock trackers. These direct measurements of particle and shock velocity are compared to previous shock compression results on Teflon. Particular attention is focused in the region below 10 kbar where evidence of a shock induced phase transition has been reported [1], based upon a cusp in the Hugoniot. The volume change for this transition is only ~ 2.2 % making its observation difficult. A two-wave structure on the shock front would be strong evidence of the shock-induced transition, but has not been observed in these initial low-pressure experiments. However, the Hugoniot data does show a subtle cusp between two of these shots at pressures of 6.4 and 7.9 kbar. The presence of the cusp is consistent with existing data, but appears at slightly higher pressure. Additionally, the *in-situ* particle velocity gauges show an evolving wave front, which is likely associated with Teflon's visco-elastic properties. The wave front is initially steep, but rounds significantly after the wave has propagated several millimeters into the target.

INTRODUCTION

Teflon's low-pressure (below 8 kbar) phase diagram exhibits multiple crystal structures. However, only one study [1] has attempted to measure shock-induced phase transitions in this region. In this study, Champion, reports a shock-induced phase transition occurring at 5.0 kbar, based upon a slight cusp in the Hugoniot data. Additionally, Teflon is known to undergo a pressure induced phase transition at a hydrostatic pressure of 5.4 kbar.[2,3] The calculated volume change associated with this transition is 2.2 %.[1] A focus of the current work is to confirm the presence of this shock-induced transition, by (preferably) direct observation of a two-wave structure or by generation of Hugoniot data with less scatter for a more distinct cusp. The small

volume change associated with this transition makes its observation

difficult. The present set of experiments employ embedded magnetic gauges, which allow direct measurement of particle velocity wave profiles at a number of Langrangian positions within the Teflon targets.

A second objective of these shock compression experiments is to help in the development of constitutive and hydrodynamic models of Teflon. To fulfill these objectives a series of experiments are being completed aiming to generate accurate equation of state data, measure any evidence of reaction or phase transitions, and obtain particle velocity wave profiles. Four measurements have been completed on a single-stage gas gun at pressures between 4 and 11 kbar. A fifth shot has been completed on two-stage gas-gun reaching a

---

[†] Work performed under the auspices of the U.S. Dept. of Energy.

pressure of 117 kbar. In addition to generating Hugoniot data for comparison to previous work the embedded gauge technique used for these shots yields a series of particle velocity-time profiles as a function of depth in the Teflon target. Particularly apparent from these profiles is the rounding in the shock front due to Teflon's visco-elastic properties.

EXPERIMENTAL

Materials: We have focused on the polymer polytetrafluoroethylene with the trade name Teflon. Two billets (24" x 24" x 2") were purchased from Balfor Industries Inc. for this work. The billets were made from Dupont resins designated "7A" and "7C". The billets have been well characterized with respect to density. This was accomplished by taking core samples across the billet and through its thickness; thereby determining any effects of the pressing and sintering process upon density. Targets for the gas-gun experiments (figure 1) were machined from ~ 2 inch cubes cut from the 7C billet. These wedges were labeled to allow mapping to their respective location within the original billet. Prior to assembly the density of each piece was measured.

Technique: Experiments were done on a 72-mm-bore single-stage gas gun capable of launching projectiles to 1.5 km/s or on a 50-mm-bore two-stage gas gun capable of launching projectiles to velocities of 3.1 km/s. Z-cut quartz impactors were used for experiments on the single stage gun and Kel-F 81 impactors were used on the two-stage gun experiments. The magnetic gauging system used in these experiments was developed by Vorthman and Wackerle [4,5] in the early 1980's and further refined by Sheffield and others[6]. An electromagnet was used in both guns to provide the magnetic field. The magnetic gauges consist of a 5 μm layer of aluminum sandwiched between two thin layers of 25 μm thick Teflon; the complete gauge thickness is about 60 μm. These gauges are glued between two Teflon target wedges, which had been machined to a 30-degree angle (figure 1). The gauge consists of 9 particle velocity gauges and 3 shock trackers (used to measure the shock velocity). When the gauge package is mounted at a 30-degree angle, individual gauge elements are at

nine different Lagrangian positions inside the sample and spaced at incremental depths between 0.7 and 1 mm. Additionally, a single (or double) gauge (stirrup gauge) is glued to the target's impact face, yielding the target/impactor interface particle velocity and also serving as the trigger for the suite of oscilloscopes required for the in-material gauges. The assembled target is shown in figure 2. The stirrup or impact gauge is shown on the target's face and the gauge package is seen exiting the target to the left.

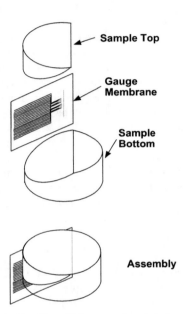

FIGURE 1. Details of the sample and the magnetic gauge package installation.

RESULTS

Four shots have been completed at pressures between 4 and 11 kbars and a fifth shot was completed at 117 kbar (Table 1). Particle velocities presented in the table were measured from the gauge located at the Teflon target's front impact face and represent the particle velocity at the impactor/target interface. Shock velocities were obtained by measuring the shock's arrival time at each of the particle velocity gauges as well as at each step on the shock tracker. Figures 3 and 4 show particle velocity wave profiles for two of these shots. The first profile in each of these plots

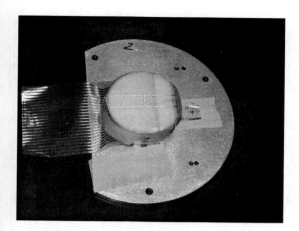

FIGURE 2. Target assembly showing the double stirrup gauge membrane glued to the front of the target. As can be seen on the right side of the sample, the sample edge is rounded, the leads of the stirrup gauge are glued to the rounded edge, and the extra volume is filled with glue to minimize lead spreading. The external leads from the embedded gauge package containing 9 particle velocity gauges and 3 shock trackers is shown coming out of the target on the left side.

corresponds to the gauge located at the Teflon's impact face.

Subsequent particle velocity wave profiles are spaced at nominal depths of 1 mm starting at 0.4

(figure 3) and 1 mm (figure 4) depths.

The embedded magnetic gauge technique used in this study gives much more insight into the evolution of the shock wave as it propagates through the target than was possible with the quartz gauge employed by Champion in the early '70s. In each of the low-pressure shots, considerable wave front rounding is observed from the *in-situ* gauges. This wave front rounding is an evolving process, which is observed with the embedded gauges yielding "snap shots" at every millimeter in the Teflon target. Little rounding is observed from the impact face gauge, but the wave front becomes more round through the first few gauges. This rounding or smearing out of the wave front increases until the wave has traveled ~ 2.5 mm into the target. Subsequent (deeper) gauges show the wave to be steady after this depth. This wave front rounding in the low-pressure shots is in contrast to the high-pressure shot (figure 4), where each profile is similar. The rounding is probably associated with Teflon's visco-elasticity, which is over-driven at higher pressures.

We hoped to exploit this embedded gauge technique to observe a two-wave structure. But, this definitive evidence of the shock-induced phase transition has not been observed.

The U_s-u_p Hugoniot points extracted from these

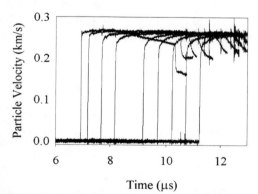

FIGURE 3. Dynamic particle velocity wave profiles for Teflon shock experiment completed at 10.8 kbar. The first waveform represents the particle velocity at the interface between the impactor and the Teflon target. Subsequent profiles represent particle velocities at nominal depths of 1 mm inside the Teflon target. Gauge 4 is missing due to a problem with the gauge connection prior to the shot.

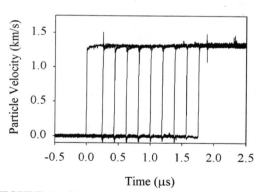

FIGURE 4. Dynamic particle velocity wave profiles for Teflon shock experiment completed at 117 kbar. The first waveform represents the particle velocity at the interface between the impactor and the Teflon target. The first internal gauge is positioned at a depth of ~ 1mm in the Teflon (deeper than the target shown in Figure 3 Subsequent gauges are located at nominal depths of 1 mm inside the Teflon target.

TABLE 1 Summary of shock compression experiments on Teflon.

Shot #	Impactor	Impactor Velocity (mm/μs)	Teflon Density (g/cm³)	Shock Velocity (mm/μs)	Particle Velocity (mm/μs)	Pressure (Kbar)
2S-108	Kel-F 81	2.65	2.163	4.17	1.3	117
1S-1249	Z-Quartz	0.146	2.166	1.67	0.123	4.45
1S-1251	Z-Quartz	0.206	2.1649	1.75	0.170	6.45
1S-1271	Z-Quartz	0.254	2.1652	1.77	0.205	7.86
1S-1272	Z-Quartz	0.328	2.1653	1.92	0.259	10.8

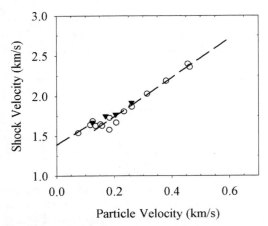

FIGURE 5. Low-pressure Hugoniot data for Teflon. Champion's data is shown with unfilled circles. Also shown are Champion's two linear fits above and below the reported phase transition. Data for the four low-pressure shots from the present work are shown with solid triangles.

four low pressure shots are plotted in figure 5 along with Champion's data. These new data are consistent with Champion's, but show a subtle cusp between shots 1251 and 1272 corresponding to pressures of 6.4 and 7.9 kbar, respectively. This pressure is slightly higher than that reported by Champion for the phase change (5 kbar), but is reasonably close given the scatter in his data (figure 5). The four Hugoniot points lie close to Champion's two linear fits corresponding to pressures above ($U = 1.393 + 2.217u$) and below ($U = 1.258 + 2.434u$) the phase transition.

CONCLUSIONS

Four low-pressure shock compression experiments have been completed between 4 and 11

kbars. The data is consistent with Champion's data and reported shock-induced phase transition. This initial data indicate a slightly higher pressure (~ 7 kbar) for the shock-induced phase transition than that reported by Champion [1]. Unfortunately, a two-wave structure was not observed in these shots. It is important to note that the phase transition under discussion has an associated volume change of only 2.2% making this observation difficult. The embedded magnetic gauges used in these experiments show an evolving wave front consistent with the visco-elastic behavior of Teflon. This wave front rounding becomes steady after the wave has propagated ~2.5 mm into the target and is not present in the high pressure shot completed at 117 kbar. Wave front rounding and the small volume change make observation of the shock-induced transition more difficult. Additional shots in the vicinity of the transition will add confidence to the presence (and pressure) of this transition.

REFERENCES

1. Champion, A.R., J Appl Phys. **42**, 5546, (1971).
2. Kennedy, G. C. and LaMori, P. N., J. Geophys. Res., **67**, 851 (1962)
3. Weir, C. E., J. Res. Natl. Bur. Std., **50**, 95 (1953).
4. Vorthman, J. E., Andrews, G., and Wackerle, J., "Reaction Rates from Electromagnetic Gauge Data," in *Proceedings of the Eighth Symposium (International) on Detonation*, Office of Naval Research, Report NSWC MP-86-194, p. 951 (1986).
5. Vorthman, J.E., "Facilities for the Study of Shock Induced Decomposition in High Explosive," in *Shock Waves in Conden-sed Matter -- 1981*, Eds. W. J. Nellis, L. Seaman, and R.A. Graham, AIP Conference Proceedings No. 78 (1982).
6. Sheffield, S. A., Gustavsen, R. L., and Alcon, R. R., In-Situ Magnetic Gauging Technique use at LANL-Method and Shock Information Obtained," in Shock Compression of Condensed Matter-1999, Eds. M. D. Furnish, L. C. Chhabildas, and R. S. Hisxon, AIP Conference Proceedings No 505 p 1043 (2000).

CP706, *Shock Compression of Condensed Matter - 2003*
edited by M. D. Furnish, Y. M. Gupta, and J. W. Forbes
2004 American Institute of Physics 0-7354-0181-0/04/$22.00

MEASUREMENT OF EQUATION OF STATE OF SILICONE ELASTOMER

R E Winter[1], G Whiteman[1], G S Haining[1], D A Salisbury[1], and K Tsembelis[2]

[1]AWE, Aldermaston, Reading, Berks, RG7 4PR, UK
[2]PCS, Cavendish Laboratory, Madingley Rd, Cambridge, CB3 0HE, UK

Abstract. Silicone Elastomer, ("Sylgard 184 ®"), samples were mounted between copper plates. Manganin stress gauges were placed within the front copper plate, halfway through the Sylgard and at the interface between the Sylgard and the rear copper plate. A series of experiments was performed in which the front plate was impacted by copper plates projected at a range of velocities. It was assumed that a Grüneisen Gamma form with a constant Γ could fit the Equation of State of the sample. A trial set of EoS parameters, including Gamma, was entered into a spreadsheet, then the state variables for the different stress jumps were calculated with the aid of a "Goalseek" function. This enabled the stresses and times for each jump to be calculated. Comparing these predictions with the experimentally determined parameters enabled optimum values of the EoS parameters to be identified.

INTRODUCTION

In 1980 Kondo, Yasumoto, Sugiura and Sawaoka (1) used the reverberation technique to determine the Hugoniot of fused quartz. When they calculated all the shock jumps using Hugoniots centred on the ground state they found that the pressure levels calculated for the later shock jumps were significantly higher than those determined experimentally. They obtained a better fit to their experiments by using a Grüneisen Gamma Equation of State (EOS) to calculate Hugoniots centred on the previous shock state. They pointed out that the method had the potential to derive the Grüneisen Gamma, (Γ), of the sample. This paper describes how the technique described by Kondo et al was used to determine Γ in the Grüneisen EOS of Silicone elastomer or "Sylgard 184 ®". We assume that the material can be fitted by a linear particle velocity shock velocity relation. Further, it is assumed that the material has negligible strength and therefore that stress and pressure in the shocked material are equal.

EXPERIMENTS

The experiments were conducted on the 50 mm gas gun at PCS, Cavendish Laboratory, Cambridge. A schematic of a typical experiment is shown in Fig. 1. Dimensions and impact velocities for all of the experiments are listed in Table 1. Following Kondo et al's notation, the elements of this system have been named from left to right: flyer plate, driver plate, delay plate, sample plate 1, sample plate 2 and reflector plate. Manganin gauges were mounted as shown.

A position (x) versus time (t) plot for SHOT 4 is shown in Fig. 2. The pressure (p) versus particle velocity (u_p) plot corresponding to Fig. 2 is shown in Fig. 3.

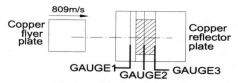

FIGURE 1. Shot 4 Schematic.

TABLE 1. Configurations of Shots.

Shot #	Plate thicknesses (mm)					m/s	Pos.wrt front of Syl (mm)		
	Flyer	Driver	Delay	Samp1	Samp2	Impact Velocity	Gauge1	Gauge2	Gauge3
1	10.00	3.00	0.00	3	3	506	0.00	3	6.00
4	10.00	3.00	2.00	3	3	809	-2.00	3	6.15
5	10.00	2.00	0.80	3	3	658	-0.80	3	6.15
6	10.00	2.00	0.85	3	3	655	-0.85	3	6.15
7	10.00	2.00	0.87	3	3	782	-0.87	3	6.15
8	10.00	2.00	0.86	3	3	305	0.86	3	6.15
10	7.95	2.75	0.00	2	2	671	0.00	2	6.75
10b	7.92	2.92	0.00	2	2	909	0.00	2	6.92

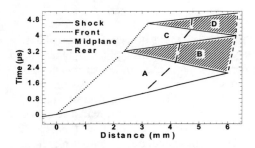

FIGURE 2. Position Vs. Time Plot for Shot 4.

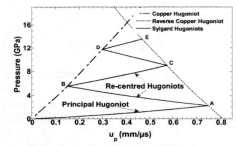

FIGURE 3. Pressure Vs. Particle Velocity Plot for Shot 4. Hugoniots are centred on their initial state.

The configuration was designed to maximise the period over which 1D conditions persisted. Figure 4 shows hydrocode calculations of Pressure vs. Time at the GAUGE 2 position in SHOT 4. A 1-D calculation with a very thick, (50 mm), flyer shows an ideal, but experimentally unattainable situation, in which many pressure steps asymptote to the copper-on-copper pressure. A 2-D calculation of this same geometry shows that pressure release from the perimeter of the impacted area arrives just before the 4th velocity jump. When, in a 1-D calculation the projectile length is reduced to 10 mm, a pressure release from the back of the flyer

arrives at the gauge at about the same time as the pressure release from the sides of the experiment. We conclude from these calculations that the first three jumps seen by GAUGE 2 are valid 1-D data but that the 4th Jump should be excluded from this analysis as it will probably be subject to 2-D effects.

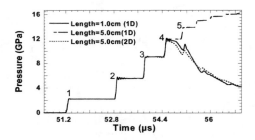

FIGURE 4. 1D and 2D Hydrocode Calculations at GAUGE 2/SHOT 4 showing that the pressure of the 4th jump has probably been affected by relief waves.

A gauge record for SHOT 4 is shown in Fig. 5. The pressures and times of each of the valid velocity jumps as measured in all of the experiments are listed in Table 2.

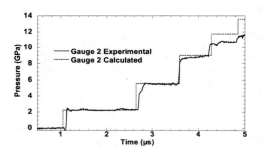

FIGURE 5. SHOT 4/GAUGE 2: Pressure vs. Time for $\Gamma=1.5$. Comparison of data and spreadsheet calculation.

THE PRINCIPAL HUGONIOT

We assume that, over the pressure range of interest, the Hugoniot of Sylgard can be fitted by a relation of the form

$$U_s = a + bu_p \quad\text{...}(1)$$

where U_s and u_p are the shock and particle velocities and a and b are constants which depend on the material.

680

TABLE 2. Pressures and Times of Valid Jumps. * The first jumps in each experiment were used to estimate the principal Hugoniot, the other jumps (below the line) were used to estimate Γ.

VALID JUMPS	MEASURED t(us)	MEASURED P(GPa)	CALC using GAMMA=1.5 t(us)	CALC using GAMMA=1.5 P(GPa)	Deltat/t	DeltaP/P	d
*s1-g2-j1	1.31	1.24	1.24	1.19	0.11	0.05	0.12
*s4-g2-j1	1.09	2.20	1.05	2.23	0.08	-0.01	0.08
*s5-g2-j1	1.16	1.92	1.13	1.68	0.04	0.14	0.15
*s6-g2-j1	1.17	1.82	1.14	1.67	0.06	0.09	0.11
*s7-g2-j1	1.10	2.41	1.06	2.13	0.07	0.14	0.15
*s8-g2-j1	1.46	0.60	1.43	0.63	0.04	-0.05	0.06
*s10-g2-j1	0.77	1.73	0.74	1.82	0.07	-0.05	0.09
*s10b-g2j1	0.66	2.85	0.65	2.80	0.02	0.02	0.03
s1-g2-j2	3.32	2.79	3.24	2.82	0.05	-0.01	0.05
s1-g2-j3	4.50	4.54	4.55	4.54	-0.02	0.00	0.02
s1-g2-j4	5.41	5.62	5.56		-0.05		0.05
s1-g3-j2	5.01	4.74	5.10	6.05	-0.04	-0.22	0.22
s4-g2-j2	2.69	5.46	2.64	5.59	0.04	-0.02	0.04
s4-g2-j3	3.56	8.79	3.59	9.00	-0.02	-0.02	0.03
s4-g2-j4	4.23	10.65	4.29		-0.03		0.03
s4-g3-j2	3.96	10.60	3.97	11.73	-0.01	-0.10	0.10
s5-g2-j2	2.96	4.24	2.91	4.12	0.04	0.03	0.05
s5-g2-j3	3.98	6.88	4.00	6.64	-0.01	0.04	0.04
s5-g2-j4	4.74	8.39	4.83		-0.03		0.03
s5-g3-j2	4.47	7.81	4.46	8.76	0.01	-0.11	0.11
s6-g2-j2	2.96	4.24	2.91	4.09	0.03	0.04	0.05
s6-g2-j3	3.97	6.86	4.01	6.60	-0.02	0.04	0.04
s6-g2-j4	4.73	8.58	4.84		-0.05		0.05
s6-g3-j2	4.43	8.16	4.47	8.70	-0.02	-0.06	0.07
s7-g2-j2	2.76	5.70	2.69	5.31	0.05	0.07	0.09
s7-g2-j3	3.66	8.65	3.65	8.56	0.00	0.01	0.01
s7-g2-j4	4.35	10.65	4.37		-0.01		0.01
s7-g3-j2	4.07	10.41	4.05	11.18	0.01	-0.07	0.07
s8-g2-j2	4.14	1.34	3.84	1.41	0.16	-0.05	0.17
s8-g2-j3	5.76	2.14	5.60	2.23	0.06	-0.04	0.07
s8-g2-j4	7.11	2.80	7.02		0.03		0.03
s8-g3-j2	6.59	2.20	6.37	3.00	0.07	-0.27	0.28
s10-g2-j2	1.97	4.90	1.88	4.88	0.10	0.00	0.10
s10-g2-j3	2.58	8.60	2.54	8.75	0.02	-0.02	0.03
s10-g2-j4	3.05	11.52	3.02		0.02		0.02
s10b-g2-j2	1.67	8.37	1.62	7.86	0.07	0.06	0.09
s10b-g2-j3	2.22	14.07	2.14	14.31	0.08	-0.02	0.08
s10b-g2-j4	2.61	18.96	2.50		0.09		0.09
				Average d			0.073

Since the Hugoniot of copper is known, the first jump in each of the 6 experiments listed in Table 2 may be used to estimate the principal Hugoniot of Sylgard. Figure 6 shows our data together with data from Ref. 2 plotted in the U_s-u_p plane. Fitting a straight line to all of the data gives $a=1.63$ $mm/\mu s$ and $b=1.66$.

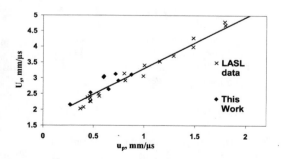

FIGURE 6. . U_s-u_P Plot using our data and LASL data (2).

The Hugoniots in the P-v and E-v planes are given by

$$p_h(v) = \frac{a^2(v_0 - v)}{(v_0 - b(v_0 - v))^2} \quad \dots\dots\dots\dots\dots\dots\dots\dots(2)$$

and

$$E_h(v) = \frac{1}{2} p_h(v)(v_0 - v) \dots\dots\dots\dots\dots\dots\dots(3)$$

where the initial state of the material, before the shock reaches it, is given by parameters: ρ_0, (initial density); v_0, (initial specific volume); p_0, (initial pressure); E_0, (initial internal energy) and u_0, (initial particle velocity), and the state after the shock has passed is given by U_s, (shock velocity), ρ, p, v, and u_p.

RE-CENTRED HUGONIOTS

We assume that the off-Hugoniot states can be fitted by a Grüneisen Gamma (EoS):

$$p = p_h(v) + \frac{\Gamma}{v}(E - E_h(v)) \quad\dots\dots\dots\dots\dots\dots\dots\dots(4)$$

where $p_h(v)$ and $E_h(v)$ are as given in Eqns 2 and 3 above and Γ is a constant which depends on the material.

It can be shown by combining the Rankine-Hugoniot equations with Eqn. 4 that the Hugoniot centred on any initial state v_1, p_1, E_1, is given by:

$$p = \frac{p_h(v) + \frac{\Gamma}{v}\left[\frac{p_1}{2}(v_1 - v) + E_1 - E_h(v)\right]}{1 - \frac{\Gamma}{2v}(v_1 - v)} \quad\dots\dots\dots\dots(5)$$

SPREADSHEET METHOD

The state variables at A, B, C etc. in Figs 2 and 3 were calculated using MS EXCEL®. Values for the density and the U_s-u_p constants of Sylgard are entered into the spreadsheet together with a trial values for Γ. A best guess for v^A, the volume at state A, is also entered. The principal Hugoniot is given by Eqn. 2 and the particle velocity at A is given by:

$$u_p{}^A = u_p{}^0 + \sqrt{(p^A - p_0)(v_0{}^{syl} - v^A)} \quad\dots\dots\dots\dots(6)$$

To calculate the state variables at A we must find the intersection between the principle Hugoniot and the reflection Hugoniot of the copper reflector plate, given by

$$p = (a_{Cu} + b_{Cu}(V_{imp} - u_p{}^A))(V_{imp} - u_p{}^A)/v_0{}^{Cu} \quad \dots\dots\dots(7)$$

where V_{imp} is the impact velocity of the flyer. The intersection is found by using the GOALSEEK function in MS EXCEL®.

Conditions at point B are obtained in a similar way by finding the intersection between the Sylgard Hugoniot centered on A (using Eqn. 5) and the principal Hugoniot of the Copper. A similar process gives Point C and subsequent points as required. The shock states at A, B, C etc., calculated assuming $\Gamma = 1.5$ are listed for SHOT 4 in Table 3.

TABLE 3. SHOT 4 GAUGE 2: The Shock States

States	V cc/g	P GPa	up mm/us	Us mm/us	E GPa.cc/g
0	0.96	0	0	0	0
A	0.71	2.23	0.75	2.87	0.28
B	0.61	5.59	0.15	-3.26	0.69
C	0.55	9.00	0.57	5.03	1.08
D	0.53	11.73	0.30	-4.93	1.36
E	0.51	13.61	0.47	6.20	1.55

Knowing the pressures and shock and particle velocities at the successive states allows the x vs. t plot shown in Fig. 2 to be constructed. Again, the MS EXCEL® GOALSEEK function is used to derive the arrival times of the shocks at the gauge positions by determining the intersections of the shock velocity and particle velocity paths.

The spreadsheet was used to determine the value of Γ which gave the best fit to the jumps in the second part of the list in Table 2. The table shows a comparison between measured times and pressures and those obtained using a chosen value of Γ. Table 2 also shows the differences between the calculated and observed pressures (Δp) and calculated and observed times (Δt). A measure of the difference between calculation and experiment for each valid shock jump is provided by:

$$d = \sqrt{\left(\frac{\Delta p}{p_{calc}}\right)^2 + \left(\frac{\Delta t}{t_{calc}}\right)^2}$$

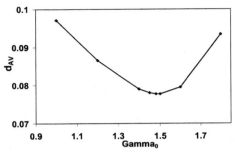

FIGURE 7. Plot of Average d, (d_{AV}), vs. Γ showing that the best fit to the experiments reported in this paper is obtained with $\Gamma = 1.5$.

Figure 7 shows a plot of d, averaged over all the valid jumps in all shots, (d_{AV}), plotted against Γ. It is seen that putting $\Gamma = 1.5$ minimizes d_{AV} and therefore provides the best fit to the body of experiments. It was assumed in the above analysis that Γ is independent of v. We investigated whether a better fit to the data could be obtained by putting

$$\Gamma(v) = \Gamma_0 + \Gamma_1 \frac{v}{v_{0s}}$$ and attempting to improve the fit by

trying a range of values of Γ_0 and Γ_1. However the conclusion of this exercise was that the best fit was obtained with $\Gamma_1 = 0$.

CONCLUSIONS

The reverberation technique has been used to find the Γ in the Grüneisen Gamma EoS for Sylgard 184 ® which gave the best fit to experimental data. The spreadsheet enables a large number of different values of Γ to be investigated quickly and generates illustrative diagrams such as x vs. t and P vs. u_p plots. It was found that the best fit to experiment was obtained by using the parameters a=1.63mm/μs, b=1.66, ρ_0=1.04 g/cm^3 and Γ=1.5.

REFERENCES

1. Kondo, K., Yasumoto, Y., Sugiura, H., and Sawaoka, *J. Appl. Phys.* **52**(2), February 1981
2. LASL Shock Hugoniot Data, Ed S P Marsh, *Univ of California Press*, 1980, p482

CHAPTER XI

MECHANICAL PROPERTIES OF COMPOSITES

CP706, *Shock Compression of Condensed Matter - 2003*
edited by M. D. Furnish, Y. M. Gupta, and J. W. Forbes
© 2004 American Institute of Physics 0-7354-0181-0/04/$22.00

SHOCK COMPRESSION AND RELEASE PROPERTIES OF ALUMINA-FILLED EPOXY

M. U. Anderson, R. E. Setchell, and D. E. Cox

Sandia National Laboratories, Albuquerque, NM, 87185

Abstract. Alumina-filled epoxies are used to encapsulate ferroelectric ceramics in shock-driven, pulsed power devices. Device performance is strongly influenced by the shock compression and release properties of the encapsulant, which must be adequately understood in order to develop a capability for numerically simulating the operation of these power sources. In previous studies, Hugoniot states and release velocities were measured in reverse-impact experiments using laser interferometry (VISAR) at stresses up to 5 GPa. In addition, wave profiles were obtained in transmitted-wave experiments at fixed impact conditions as a function of initial temperature. These experiments showed an extended wave structure having a rise time that increased with decreasing temperature. In recent studies, Hugoniot states and release velocities at stresses up to 10 GPa have been obtained in reverse-impact experiments. Transmitted-wave experiments have examined the effects of wave amplitude on the wave structure and also the evolution of this structure with increasing propagation distance. Unsteady wave propagation with strongly viscous behavior is observed over the range of shock conditions examined.

INTRODUCTION

Shock-induced depoling of ferroelectric ceramics has been used in pulsed power devices for many years. These ceramics are encapsulated using alumina-filled epoxies, and the stress history experienced by a ferroelectric element during device operation is strongly influenced by the shock compression and release properties of the encapsulant. Recent interest in numerically simulating the operation of pulsed power devices has motivated a number of new experimental and theoretical efforts to develop better models for the dynamic behavior of the materials involved. The shock response of alumina-filled epoxies had been examined in only a few previous studies. Munson et al. (1) examined the shock and release behavior of an epoxy using Epon 828 resin and Z hardener (2) with different volume fractions of alumina

powder added. Gas gun experiments on these materials used both laser interferometer (VISAR) and quartz gauge diagnostics. Transmitted wave profiles showed extended rise times and the distinct rounding near peak values characteristic of dispersive materials. Surprisingly high release wave velocities were observed as shock pressures increased. The experimental results were modeled using a rate-dependent formulation for a Maxwell solid. The data of Munson et al. (1) and the results of a pressure-shear loading experiment on filled Epon 828/Z by Chhabildas and Swegle (3) were modeled by Drumheller (4) using a general theory developed for immiscible mixtures. The Epon 828/Z material with the highest alumina content studied by Munson et al. (1) and modeled by Drumheller (4) was one of the materials examined in the first recent study of the shock response of encapsulants (5). Hugoniot states and release velocities were obtained at stresses up to 5 GPa in several filled and unfilled epoxies. Transmitted wave profiles were obtained at a fixed impact stress in unfilled and filled materials to compare wave structures. A subsequent study (6) examined differences in compression and release wave profiles in filled and unfilled epoxies resulting from initial temperature variations.

In the present study, one of the filled epoxies is examined in more detail. This material consists of Epon 828 resin, Z hardener, and an Alcoa tabular alumina mixed using 23.8, 4.8, and 71.4 weight percents, respectively. The alumina particles range in size from approximately 5 to 50 microns, and represent 43% by volume of the final material. The nominal density is 2.37 g/cm^3. For simplicity, this material will be denoted by ALOX in the following text and figures. Hugoniot states and release velocities were obtained in reverse-impact experiments, and transmitted wave experiments were used to examine the wave structure as wave amplitude and propagation distance were varied.

REVERSE-IMPACT EXPERIMENTS

Planar-impact experiments were conducted on a 63.5-mm diameter gas gun. In the reverse-impact configuration, ALOX samples backed by low-density carbon foam were mounted as projectile facings. The targets consisted of a 1.6-mm thick fused-silica buffer followed by a 12.7-mm thick fused-silica window. A diffusely reflecting film of aluminum was deposited at the interface, and a dual-delay VISAR system was used to record the particle velocity history at this location. The elastic properties of fused silica (7) were used in a method-of-characteristics calculation to find the corresponding particle velocity history at the impact interface. The final value of particle velocity at this interface, the corresponding axial stress given by fused silica properties, and the measured impact velocity provide a direct measurement of an ALOX Hugoniot state. A release wave is generated when the shock generated in the ALOX sample reaches the back interface with carbon foam, and the arrival time of this wave at the impact interface provides a release wave velocity corresponding to the shocked state. The highest states were achieved using sapphire for the

buffer/window assembly. Figure 1 shows the ALOX Hugoniot states obtained in these experiments, together with data from previous studies (1,8,9). The collective measurements can be accurately fit by a simple quadratic polynomial, corresponding to a linear relation between shock velocity and particle velocity given by:

$$U_s = 2.87 + 2.02 \, u_p \ (\text{km/s units}).$$

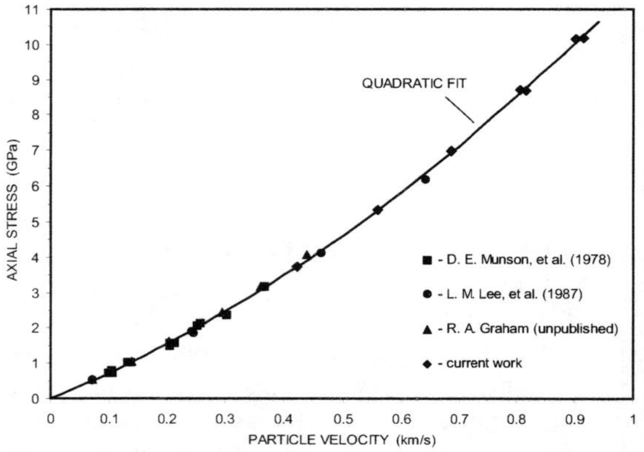

FIGURE 1. Summary of Hugoniot data.

The intercept value of 2.87 km/s is very close to the bulk sound speed of 2.88 km/s reported by Munson et al. (1) for the same material, and much less than their value of 3.40 km/s for the longitudinal sound speed. The large value of the particle velocity coefficient reflects the strong positive curvature in the fit to the Hugoniot data. Figure 2 shows ALOX release wave velocities obtained in the current study, together with values obtained previously. For comparison, a shock velocity curve obtained from the Hugoniot fit is also shown. Release speeds continue to increase to very high values, exceeding 8 km/s at a shock stress of 10 GPa.

TRANSMITTED WAVE EXPERIMENTS

In all transmitted-wave experiments, an ALOX sample mounted on carbon foam was impacted into a stationary ALOX target backed with a fused silica buffer and window. VISAR was used to obtain the transmitted wave profile at the ALOX/buffer interface using the same procedures followed in reverse-impact experiments. In a previous study (5), a profile recorded after a 1.6 GPa shock had propagated through a 6-mm-thick structure with a rise time of several hundred nano-

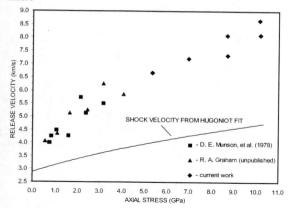

FIGURE 2. Summary of release wave velocities.

seconds, followed by dispersive rounding near the peak state. In contrast, a corresponding profile in unfilled epoxy showed a typical viscoelastic structure consisting of a sharp shock jump followed by similar rounding near the peak state. In the present study, the extended structure of transmitted waves in ALOX was investigated in several ways. Figure 3 shows transmitted wave profiles after propagation through 6-mm-thick ALOX samples for four different impact conditions, resulting in impact stresses from 0.61 to 5.51 GPa. Wave risetime decreases rapidly with increasing stress, approaching a shock jump at the highest level. This amplitude dependence is similar to trends seen in plastic deformation waves in a number of materials. In their investigation of viscosity in steady wave structures, Swegle and Grady (10) examined this

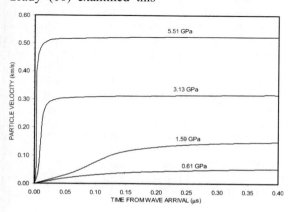

FIGURE 3. Transmitted wave profiles for different impact stresses. All ALOX samples were 6.0 mm thick.

data and found that peak strain rates varied with the fourth power of stress for most of these materials. The profiles shown in Fig. 3 were analyzed in the same manner as followed by Swegle and Grady, and the results are shown in Fig. 4. The three highest impact stresses had peak strain rates that roughly followed a fourth-power dependence, but the lowest stress case clearly deviated from this trend.

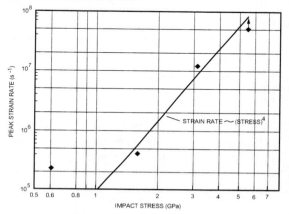

FIGURE 4. Peak strain rate versus impact stress from the profiles in Fig. 3.

Figure 5 shows transmitted wave profiles recorded after different propagation distances for a fixed impact stress of 1.6 GPa. The profiles have been shifted in time to display how the extended wave structure evolves as the wave propagates. The wave motion is unsteady over this range of distances, but appears to be approaching a steady condition. Swegle and Grady (10) identified a criterion for the propagation distance required for achieving a steady wave state as follows:

$$\delta = 3C_0 / 8S\dot{\eta} \qquad (1)$$

where C_0 and S are the coefficients in the linear relation between shock velocity and particle velocity, and the final variable is the peak strain rate. This expression reflects a balance between viscous wave spreading and wave steepening implied by positive curvature in the Hugoniot curve. Using a peak strain rate obtained from the final profile in Fig. 5 and the coefficients found from the Hugoniot curve, Eq. (1) gives $\delta \sim 12$ mm,

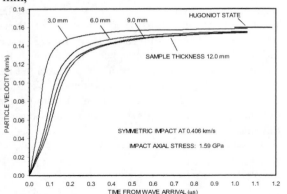

FIGURE 5. Transmitted wave profiles for increasing sample thickness.

which is the ALOX thickness in this case. Additional wave evolution features are shown in Fig. 6, where the same wave profiles are plotted against the time since impact divided by sample thickness. The "foot" of the wave is slowing with propagation distance, but part of the profile appears to be propagating at a nearly constant velocity close to the value predicted by the Hugoniot curve and steady-wave jump conditions.

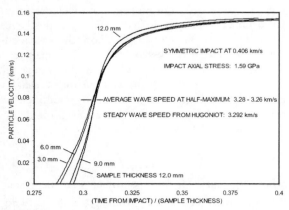

FIGURE 6. Additional timing features of the wave profiles shown in Fig. 5.

SUMMARY

The shock response of a particular filled epoxy has been examined in some detail. Hugoniot data at stresses up to 10 GPa are fit by a simple quadratic curve, and release velocities continue to rise to very high values. Transmitted wave experiments over a range of impact stresses and propagation distances show unsteady, strongly viscous behavior that appears to evolve towards steady wave conditions.

ACKNOWLEDGMENTS

The authors would like to thank Steve Montgomery of Sandia for the ALOX samples. Sandia is a multiprogram laboratory operated by Sandia Corporation, a Lockheed Martin Company, for the United States Department of Energy's National Nuclear Security Administration under Contract DE-AC04-94AL85000.

REFERENCES

1. Munson, D. E., Boade, R. R. and Schuler, K. W., *J. Appl Phys.* **49**, 4797-4807 (1978).
2. Registered products of Shell Chemical Company.
3. Chhabildas, L. C. and Swegle, J. W., *J. Appl. Phys.* **53**, 954-956 (1982).
4. Drumheller, D. S., *J. Appl. Phys.* **53**, 957-969 (1982).
5. Anderson, M. U., Setchell, R. E., and Cox, D. E., "Shock and Release Behavior of Filled and Unfilled Epoxies", in *Shock Compression of Condensed Matter – 1999*, edited by M. D. Furnish et al., AIP Conference Proceedings 505, New York, 2000, pp. 551-554.
6. Anderson, M. U., Setchell, R. E., and Cox, D. E., "Effects of Initial Temperature on the Shock and Release Behavior of Filled and Unfilled Epoxies", in *Shock Compression of Condensed Matter – 2001*, edited by M. D. Furnish et al., AIP Conference Proceedings 620, New York, 2002, pp. 669-672.
7. Barker, L. M. and Hollenbach, R. E., *J. Appl. Phys.* **41**, 4208-4226 (1970).
8. Lee, L. M., Jenrette, B. D., and Greb, A., Air Force Weapons Laboratory Report AFWL-TR-87-133 (1987).
9. Graham, R. A., private communication.
10. Swegle, J. W. and Grady, D. E., *J. Appl. Phys.* **58**, 692-701 (1985).

CP706, *Shock Compression of Condensed Matter - 2003*
edited by M. D. Furnish, Y. M. Gupta, and J. W. Forbes
© 2004 American Institute of Physics 0-7354-0181-0/04/$22.00

DYNAMIC LOADING OF A DESIGNER COMPOSITE

N.K. Bourne* and K.S. Vecchio

*Royal Military College of Science, Cranfield University, Shrivenham, Swindon, SN6 8LA, UK.
UC San Diego, MC-0411, Dept. of Mechanical and Aerospace Engineering,
9500 Gilman Drive, La Jolla, CA 92093-0411.*

Abstract. Advances in computer simulation and composite design capability have opened up the possibility of fitting a structure, and the material from which it is constructed, to the loading it may experience. One such class of materials is Metal-Intermetallic Laminate (MIL) composites, a designable material processed from titanium and aluminum sheets by reactive foil metallurgy. The material forms a unique microstructure that allows great flexibility in resistance to loading. Shock loading of a target has allowed assessment of shock profile and spall strength. A series of experiments has been designed to derive and validate materials' models. This process has been demonstrated with material properties derived allowing construction of models validated using impact tests. Inspection of the damaged composite suggests new avenues for better performance.

INTRODUCTION

MIL composites are produced through a simple reaction foil metallurgy approach that makes them inexpensive designer composites. The fabrication of metallic-intermetallic laminate composites has several key advantages that make it ideally suited for the production of novel armour materials. The metallic-intermetallic laminate composites may be fabricated from both Ti-Al metal foil combinations and Ti-(Al-metal matrix composite) foil combinations. Since the MIL composites are fabricated from metal foils, it is possible to directly replace the monolithic metal foils with metal-matrix composite foils. Further, inert non-metals, ceramics or sensors may be included for specific applications and performances.

The means to match the design of a material to the task for which it is intended opens a new series of opportunities for novel structure development. What is needed to rapidly converge on an efficient design is a synergy of experimental testing and metallugically flexible means to produce new material [1]. Such an approach is illustrated here for the design of a supported armour panel. Some of the

parameters for intermetallics build on a range of work done on various Ti-Al systems [2-4].

The experimental procedure aims to generate a simple elasto-plastic model for the constitutive response using a small number of experiments. The design of composite equations of state and constitutive models that correctly accounts for the properties of constituents is under development and this offers data to fuel that effort. For this simple, 1D laminate, where properties vary gradually between phases, it is found that response can be modelled using discrete layer of Ti and Al_3Ti.

The construction of a validated model for the composite has allowed an impact experiment to be designed that exploits its properties to best effect. The steps down this route are shown and then a validation experiment is given which illustrates how the derived model may be used to counter a specified threat.

EXPERIMENTAL PROCEDURE

Plate impact experiments were performed on a 50 mm bore, 5 m long single stage gas gun [5]. Flyer plates of dural (aluminium alloy 6082-T6) and copper of thickness 3 mm were fired at velocities between 150 and 820 m s⁻¹. Impact velocities were measured by the shorting of sequentially mounted pairs of pins, and the specimen was aligned to the flyer plate to better than 1 milliradian by an adjustable specimen mount. Longitudinal stresses were measured by embedding a manganin gauge (MicroMeasurements type LM-SS-125CH-048) between an 8.7 mm plates of the composite and a PMMA backing plate. The power supply was triggered by shorting pins ahead of the target. Additionally, a gauge was supported on the front surface of the target with a 1 mm plate of the same material as the flyer plate. In this way, that gauge would experience an embedded stress whilst the time difference, Δt, and the separation, Δw, of both gauges (thorough knowledge of the specimen dimensions is known) could be used to determine the shock velocity ($U_s = \Delta w / \Delta t$). Specimen configurations and gauge placements are shown in Fig. 1.

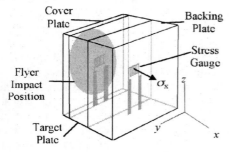

FIGURE 1. Specimen configuration and gauge placement.

The arrangement of target and impact plates leads to a complex series of interactions of shock and release waves in the composite. The initial wave loci are indicated in Fig. 2. The MIL target is represented as five 1.6 mm thick layers of Al₃Ti with 0.2 mm layers of Ti in between. The spall strength of the intermetallic is low so that failure occurs as the release waves interact between several layers at this stress level. Delamination is indeed observed at these positions in recovered specimens.

MATERIAL PROPERTIES

The MIL studied in this investigation had a density (ρ_0) of 3.48 g cm⁻³, a longitudinal sound speed (c_L) of 7.77 mm μs⁻¹, a shear wave speed (c_S) of 4.50 mm μs⁻¹ and a Poisson's ratio, ν (calculated

using the previous elastic wave speeds and the initial density) was 0.25. The acoustic properties were measured using 5 MHz quartz transducers, connected to a Panametrics 500 PR pulse receiver. These material properties were determined for a MIL composite containing approx. 80 vol.% Al₃Ti and 20 vol.% Ti (Ti-3Al-2.5V), with layer thickness of approx. 800 microns for the intermetallic and 200 microns for the remnant Ti layers.

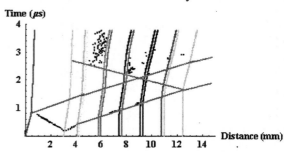

FIGURE 2. Representative x-t diagram for impact of 3 mm copper flyer at 606 m s⁻¹. Gauges 1 mm (*i.e.* between coverplate and specimen) and 9.7 mm from impact. Note shock waves and release interactions leading to multiple spall planes.

RESULTS AND DISCUSSION

In Figure 3, typical embedded gauge traces are presented from the MIL. In this example, a 3 mm copper flyer has been impacted onto the 8.7 mm target at a velocity of 606 m s⁻¹. This corresponds to the x-t diagram of Fig. 2. Note that in the first trace the signal rises to its maximum stress value with no evidence of a break in slope that might indicate an elastic precursor. The rise is not immediate, since the gauge is embedded between plates of insulation which are not matched in impedance to the MIL.

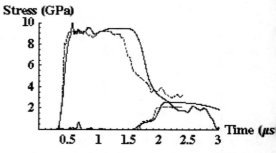

FIGURE 3. Representative embedded gauge traces at 1 mm from impact (*i.e.* between coverplate and specimen) and 9.7 mm from impact. 3 mm copper flyer at 606 m s⁻¹. Solid line experiment, dotted, model fillted from measured c_0 and S.

The second trace rises rapidly and shows dispersion between the elastic and plastic waves. The rise of the wave is rapid since the gauge mounting insulation is matched to the PMMA backing as can be seen from Fig. 3. The height of the first wave is the Hugoniot elastic limit (HEL) of the MIL and corresponds to *ca.* 1.5 GPa. The separation of the two traces may be used to determine the shock velocity at the stress level sampled which allows the U_s-u_p relation to be plotted for the composite (Fig. 4).

The wave is modelled using a constitutive equation derived from the measured value of the yield stress at this strain rate and the Hugoniot determined from a series of experiments (Fig. 4). The derived model is the dotted curve in Fig. 3 and the agreement may be seen to be close.

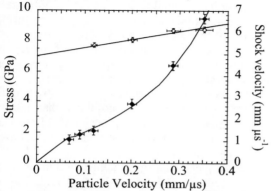

FIGURE 4. Shock velocity and stress v particle velocity for MIL composite. Best fit values for c_0 and S are 5.0 +3.4 u_p.

The Hugoniot for the MIL is plotted in stress–particle velocity space in Fig. 4 and shows no irregularities in slope. Further investigation is necessary to determine more detailed behaviour. The data is consistent with an HEL of 1.5 GPa determined from the back surface gauge.

In each of the traces recorded, some indication of the spall strength of the composite is recorded since the PMMA backing allows release to interact with that from the rear of the flyer. The measured value of this strength is small and so a value of 0.01 GPa was fixed as the minimum pressure that the Al3Ti interlayers could sustain in tension.

The parameters fixed by the plate impact experiments were inserted into a constitutive model and equations of state, in the Eulerian code Eden. This code was used to predict the performance of the composite in an impact experiment. The geometry adopted is an adaptation of the Taylor test with the MIL plate as the anvil placed onto a semi-infinite backing. Fig. 5 shows contours of pressure in an axi-symmetric simulation of such an experiment with a copper cylinder impacting at 300 m s^{-1}.

The simulation shows that the MIL surface is indented but that it resists delamination. There is very significant plastic flow for a pure Ti or Al block. The validation of this was a parallel series of experiments in which a 10 mm diameter rod was fired onto the MIL target. Fig. 6 shows such an experiment where a 10 mm diameter copper cylinder impacts at 217 m s^{-1}. The pictures were taken using a 68 frame DRS Hadland Ultra camera. A series of frames is shown showing the rod impacting from above and then rebounding from the surface of the plate. The frames are 20 μs apart and have an exposure time of 200 ns.

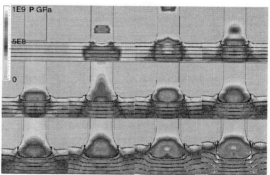

FIGURE 5. Impact of a copper cylinder onto the MIL block supported on a rigid block. Copper impacts at 300 m s^{-1}. Each frame is 4 μs apart running left to right, top to bottom.

In the first frames impact occurs and the darker material that is exhausted from beneath the cylinder is the lubricant placed onto the MIL to ensure negligible friction. Plastic deformation of the copper occurs in an identical manner to that observed when the same experiment is repeated on a semi-infinite, perfectly rigid anvil. There is, as predicted, no discernable deformation of the MIL anvil or delamination of the interlayers in the sequence.

The impact velocity was chosen such that the impact stress was *ca.* twice the elastic limit of the composite MIL. This gives the impact stress in this example to be *ca.* 3 GPa. Under these conditions, there is considerable plastic deformation of the copy cylinder as recorded by the camera. The last frames show the cylinder lifting off the anvil having suffered considerable plastic deformation.

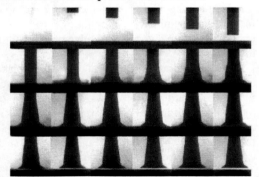

FIGURE 6. Taylor impact of a copper cylinder onto a MIL tile at 300 m s^{-1}. Rod diameter 10 mm. Frames are 20 µs apart and have exposure time 200 ns. Frames run left to right, top to bottom

Both target and rod were soft recovered for further examination. The two are shown in Fig. 7 after the experiment. There is small depression as predicted in the top face of the MIL target but otherwise there is little surface deformation despite impact pressures of initially around twice the yield stress. The impact site may be seen to be a lighter colour, indicative of the deposition of some copper onto the surface of the MIL as it deformed.

FIGURE 7. Recovered MIL plate and copper rod after the impact of Fig. 6. The copper rod has expanded on impact as seen in the sequence.

Work is going on at present to determine the state of the impacting rod and, more importantly, to assess the damage induced in the target which will be reported later. These dynamic recovery tests will be used to determine the constitutive behaviours of the various composite materials developed in the program.

CONCLUSIONS

An initial study has been reported for integrated design of a composite armour material. A method for the fast assessment, and mathematical description, of a metal composite has been described. The scheme has been proven for a layered metallic-intermetallic system (MIL). The physical parameters of interest were derived in a series of experiments that derived the Hugoniot elastic limit of the composite, its equation of state, and spall strength as a function of pressure.

These constants were used to construct a simple model for insertion into an Eulerian hydrocode. Predictions of an impact of a copper cylinder onto the laminate proved the validity of the approach. The designed laminate resisted plastic deformation under the impact loading as predicted. Further tests ongoing at intermediate rates, will provide further means to refine the constitutive description.

The method is proposed as a means of rapidly assessing the response of a laminate system in real time to influence design of further structures. It was not the intention of this study to construct more elaborate mathematical descriptions that require more extensive investigation. but to identify a fast, inexpensive route to integrate design in a cross-disciplinary approach.

Such a scheme has great application in a variety of environments and further work is underway to demonstrate such materials in fully designed structures.

REFERENCES

1. Vecchio, K. S., Andrade, U., Meyers, M. A., and Meyer, L. W., in *Shock Compression of Condensed Matter - 1991*, edited by S. C. Schmidt, R. D. Dick, J. W. Forbes, and D. G. Tasker (Elsevier, Amsterdam, 1992), p. 527-530.
2. Millett, J. C. F., Jones, I. P., Bourne, N. K., and Gray III, G. T., in *Shock Compression of Condensed Matter 2001*, edited by M. D. Furnish, N. N. Thadhani, and Y. Horie (American Institute of Physics, Melville, New York, 2002), p. 634-637.
3. Millett, J. C. F., Bourne, N. K., and Jones, I. P., *J. Appl. Phys.* **90**, 1188-1191 (2001).
4. Millett, J. C. F., Bourne, N. K., and Jones, I. P., *J. Appl. Phys.* **89**, 2566-2570 (2001).
5. Bourne, N. K., *Meas. Sci. Technol.* **14**, 273–278 (2003).

CP706, *Shock Compression of Condensed Matter - 2003*
edited by M. D. Furnish, Y. M. Gupta, and J. W. Forbes

THE SHOCK HUGONIOT OF TWO ALUMINA-EPOXY COMPOSITES

D. Deas, J.C.F. Millett*, N.K. Bourne*, K. Kos

AWE, Aldermaston, Reading, Berkshire, RG7 4PR, United Kingdom
*Royal Military College of Science, Cranfield University, Shrivenham, Swindon,
SN6 8LA. United Kingdom.

Abstract. The shock Hugoniots of alumina-epoxy composite materials at two different alumina loadings have been investigated, using manganin stress gauges. These have been determined in terms of both the shock stress and the shock velocity with regard to particle velocity. Results show that the shock velocity has a linear relationship with particle velocity, in common with many other materials. These in turn have been used to determine the shock stress, and show a high degree of agreement with the measured shock stresses.

INTRODUCTION

The use of epoxy matrix based composites is widespread as such materials combine low densities with usefully high strengths and stiffnesses, depending upon the nature of the additional phase. The shock response of such materials has been investigated, including glass fibre [1] and carbon fibre [2] based materials. However, given the nature of those composites (fibres in an epoxy matrix) the response is complicated by the directional nature of these materials. Simpler composite systems such as alumina particles in an epoxy matrix have also been studied [3]. In this situation, it can be assumed that the material is isotropic. Results showed that the shock pulses did not show evidence of an elastic precursor, due to the dispersive nature of the microstructure. It was also noted that these materials also displayed abnormally high release wave speeds, where release wave speed increased with increasing shock amplitude and particulate loading.

Particulate composites have been studied under shock loading conditions as they form the basis of many plastic-bonded explosives and propellants, see for example [4, 5]. Therefore there is an interest in investigating the response of such materials. Additionally, as no chemical interactions between the alumina-epoxy phases would be expected, in many ways it can be seen that such a composite system would be an ideal model system to study the high-strain-rate response of particulate composites, without the attendant safety risks and potential complications of chemical interactions between components.

In this article, we investigate the shock response of two different loadings of alumina particles in an epoxy resin, and compare to the known response of the epoxy matrix [6].

EXPERIMENTAL

All shots were performed on the 5 m, 50 mm bore single stage gas gun at the Royal Military College of Science. 60 mm diameter billets of composite were manufactured, with loadings of alumina to epoxy of 3 to 1 (fully loaded) and 1.5 to 1 (half loaded) by weight. These were sectioned into 10 mm thick discs and lapped such that they were flat and parallel to 5 optical fringes from a monochromatic light source. Manganin stress gauges (Micro-Measurements type LM-SS-125CH-

048) were embedded between the composite plates (the 10 mm position). In addition, a second gauge was supported upon the front of the specimen assembly with a 1 mm thick coverplate of either Dural (aluminium alloy 6082-T6) or copper, matched to the material of the flyer plate (the 0 mm position). This technique has been used successfully to determine the Hugoniot of other materials including epoxy [6] and polychloroprene [7]. Stresses were induced by the impact of 10 mm plates of either Dural or copper, in the velocity range 190 to 670 m s^{-1}.

Specimen configurations and gauge placements are shown below in figure 1.

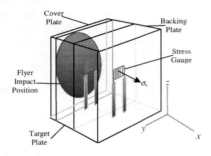

Figure 1. Specimen configuration and gauge placement.

MATERIALS DATA

The properties of the experimental materials in this study were –

Fully loaded. c_L=3.16±0.03 mm μs^{-1}, c_S=1.68±0.03 mm μs^{-1} and ρ_0=2.28±0.01 g/cm^3.
Half loaded. c_L=2.78±0.03 mm μs^{-1}, c_S=1.40±0.03 mm μs^{-1} and ρ_0=1.87±0.01 g/cm^3.

For comparison, the properties of the epoxy used and a polycrystalline alumina, AD995 were -
Epoxy. c_L=2.38±0.03 mm μs^{-1}, c_S=1.20±0.03 mm μs^{-1} and ρ_0=1.14±0.01 g cm^{-3}.
AD995. c_L=10.66±0.03 mm μs^{-1}, c_S=6.28±0.03 mm μs^{-1} and ρ_0=3.89±0.01 g cm^{-3}.

We have chosen the properties of a polycrystalline alumina due to the random distribution of the alumina particles in the composite.

All sound speed measurements were made using a Panametrics 500PR pulse receiver with quartz transducers operating at 5 MHz.

RESULTS AND DISCUSSION.

Typical gauge traces from both materials are presented in figure 2.

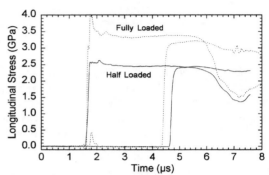

Figure 2. Stress gauge traces from fully and half loaded composites. 10 mm copper flyer plates at 490 m s^{-1}. Gauge traces are taken 0 mm and 10 mm from the impact surface.

As expected, from the densities and sound speeds discussed in the materials data section, the measured stresses for otherwise identical impact conditions, were greater in the fully loaded sample than in the half loaded. The temporal spacing between the traces at the 0 mm position and the 10 mm position have been used, along with the particle velocity (u_p) determined from impedance matching techniques, to obtain the shock velocity (U_s). The results are shown in figure 3, along with those for epoxy [6] for comparison.
Simple straight lines have been fitted to the data sets assuming the standard linear relationship between shock and particle velocities (U_s=c_0+Su_p), yielding the relations –

U_s=2.93 + 1.63u_p Fully loaded
U_s=2.63 + 1.66u_p Half loaded
U_s=2.58 + 1.47u_p Epoxy   1.

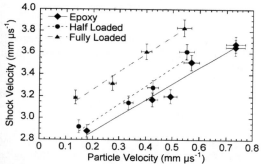

Figure 3. Shock Hugoniot of alumina-epoxy composites in U_s-u_p space.

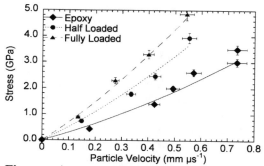

Figure 4. Shock Hugoniots of alumina-epoxy composites in σ_x-u_p space. The curve fits are according to equation 2, using the data from equation 1.

The values of c_0 increase with increasing alumina loading, in line with the increase in sound speed. Notice however, that the values of S in all three materials are similar, and thus it would seem possible that the role of the epoxy in defining the shock response of these materials at these alumina loading levels is the more significant. In Davison and Graham [8], it has been suggested that S is dependent upon the first derivative of the isentropic bulk modulus with pressure. One can therefore foresee a situation that as alumina loading increases, the global compressibility will decrease, as shown by the increasing values of c_0. However, as long as there remain significant amounts of epoxy in the material, the rate of change of compressibility with pressure could be relatively unchanged as such changes would be taken up by the epoxy itself. From figure 3 and equation 1, it would seem that this is possible.

The relations determined from equation 1 have been to determine the Hugoniot stress (σ_x) of each material according to –

$$\sigma_x = \rho_0 U_s u_p \qquad 2.$$

In doing so, it should be noted that the elastic response of the material has not been accounted for, and therefore is a measure of the hydrodynamic response. These calculations in turn are compared to the measured stresses, and the results presented in figure 4.

Looking at the two composite materials, the agreement between the calculated hydrodynamic response (equation 2) and the directly measured stresses is high. This suggests that the linear rela-

tionship between the shock and particle velocities assumed previously holds true. However, we note with interest that when a similar comparison is made with the epoxy Hugoniot data, an increasing divergence between measured stress and calculated hydrodynamic response is seen, with the measured stresses being the higher. The Hugoniot stress can be expressed as a function of the hydrostatic pressure, P and the shear strength, τ, thus,

$$\sigma_x = P + \frac{4}{3}\tau. \qquad 3.$$

We acknowledge that the stress calculated from equation 2 is the hydrodynamic response and thus will vary slightly from the hydrostatic pressure, which is generally calculated from extrapolation of ambient pressure bulk moduli data. However, it can be seen that failure to take into account the shear strength response of a material to shock loading may lead to a degree of disagreement between the calculated hydrodynamic response and the measured stress. Results from the two composites show that such a divergence does not occur, and thus suggests that the shear strength of these materials is near constant with increasing shock stress. We have confidence in this hypothesis given that in another polymer, polychloroprene [9] close agreement between measured shock stresses and the calculated hydrodynamic response was shown to correlate with a constant shear strength. Therefore, it would appear that the presence of alumina particles in an epoxy matrix has the effect of negating the overall hardening response of the epoxy in the overall material behaviour.

CONCLUSIONS

The shock Hugoniot of two alumina-epoxy particulate composites has been determined. Shock velocity and shock stress increase with increasing alumina loading. Both materials have been shown to have linear relationships between the shock and particle velocities. These in turn have been used to calculate the shock stresses with a high degree of agreement. This is different from the response of the epoxy matrix where of degree of divergence between the measured stresses and hydrodynamic pressures have been noted. This indicates that alumina particles have the overall effect of making the global response of these composites more 'elastic–perfectly plastic' than the base polymer.

Although the results have not been presented in this paper, there is ongoing work to model the equation of state of such composites using the known shock induced response of the constituent materials. Results have already shown that a simple rule of mixtures approach is inadequate, massively over predicting the Hugoniot in comparison of with the experimentally determined data. This is because mesoscale simulations have shown that in the time scales of the shock propagation, the material does not achieve either pressure or temperature equilibration between the individual phases of the composite. Work is in progress that will resolve these differences.

ACKNOWLEDGMENTS

We would like to thank Gary Cooper, Matt Eatwell and Paul Dicker of RMCS for valuable technical assistance.

REFERENCES

1. A.Z. Zhuk, G.I. Kanel, A.A. Lash,. J. Phys. IV France Colloq. C8 (DYMAT 94), 1994. **4**403.
2. W.R. Thissell, A.K. Zurek, F. Addessio, in *Shock Compression of Condensed Matter 1995*, S.C. Schmidt and W.C. Tao, Editors. 1996, American Institute of Physics: Woodbury, New York. p. 551.
3. D.E. Munson, R.R. Boade, K.W. Schuler. J. Appl. Phys., 1978. **49** 4797.
4. H.J. John Jr., F.E. Hudson, R. Robbs, in *Shock Compression of Condensed Matter 1997*, S.C. Schmidt, D.P. Dandekar, and J.W. Forbes, Editors. 1998, American Institute of Physics: Woodbury, New York. p. 603.
5. P.J. Miller, A.J. Lindfors, in *Shock Compression of Condensed Matter 1997*, S.C. Schmidt, D.P. Dandekar, and J.W. Forbes, Editors. 1998, American Institute of Physics: Woodbury, New York. 373.
6. J.C.F. Millett, N.K. Bourne, N.R. Barnes. J. Appl. Phys., 2002. **92** 6590.
7. J.C.F. Millett, N.K. Bourne, J. Appl. Phys., 2001. **89** 2576.
8. L. Davison, R.A. Graham, Physics Reports, 1979. **55** 255.
9. N.K. Bourne, J.C.F. Millett, Proc. R. Soc. Lond. A, 2003. **459** 567.

CP706, *Shock Compression of Condensed Matter - 2003*
edited by M. D. Furnish, Y. M. Gupta, and J. W. Forbes
© 2004 American Institute of Physics 0-7354-0181-0/04/$22.00

LATERAL STRESS MEASUREMENTS IN A SHOCK LOADED SILICON CARBIDE: SHEAR STRENGTH AND DELAYED FAILURE

M. Eatwell, J.C.F. Millett, N.K. Bourne

Royal Military College of Science, Cranfield University, Shrivenham, Swindon, SN6 8LA. United Kingdom.

Abstract. The shock response of a silicon carbide has been investigated using the methods of plate impact. Results show the presence of failure at the impact face, in common with other brittle materials, including glasses, alumina and other silicon carbides. Analysis of the measured shear strengths, both ahead of and behind the failure front, shows close agreement with the measured shear strengths in other silicon carbides. There is also close agreement with those values determined using a simple elastic analysis.

INTRODUCTION

The response of polycrystalline ceramics to high velocity impacts has been of interest for the past few decades. This has been due to their high compressive strengths that, in combination with low densities, make them attractive candidates for armour materials.

One particular feature common to many brittle materials under shock loading is that of the failure wave. This was first observed by Razorenov *et al* [1], who observed small reload signals superimposed on rear surface velocity measurements from a silicate glass. This was interpreted as a release from the rear of the target interacting with a moving front, behind which the shock impedance was reduced. It was suggested that this was due to material failure, and the feature dubbed the 'failure wave'. Subsequent work by other researchers showed that spall strength and shear strength significantly reduced behind the failure wave [2], referred hereto after as the failure front, and that the material became opaque, as revealed by high-speed photography [3]. Up to this point, the failure wave remained an interesting, but

relatively unimportant feature in the shock response of glass, but an investigation by Bourne *et al* [4] detected its presence in silicon carbide. Subsequent work also found failure waves in other ceramics, including alumina [5] and titanium diboride [6]. Unlike glasses, where it was observed that the failure front velocity was constant as it moved into the target [4, 7], in ceramics, it was observed that it only penetrated the first few millimeters of the target [5]. It was suggested that this be due to the presence of grain boundaries impeding the transmission of cracks across them [8].

In this paper, we present data on a variant of silicon carbide, SiC-N.

EXPERIMENTAL

All shots were performed on the 5 m, 50 mm bore single stage gas gun at the Royal Military College of Science. 10 mm flyer plates of copper and tungsten heavy alloy were fired in the velocity range 315 to 670 m s^{-1}. Velocities were measured via the sequential shorting of pairs of pins to an accuracy of 0.1%. Lateral stresses were measured with manganin stress gauges (MicroMeasurements type J2M-SS-580SF-025) mounted 2 mm from the impact surface of a sectioned 25 mm plate of SiC-N. Stress data was determined from the work of Rosenberg and Partom [9], using a modified analysis that requires no knowledge of the longitudinal stress [10]. Specimen configurations and gauge placements are presented in figure 1.

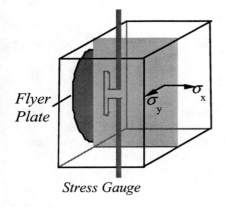

Figure 1. Specimen configuration and gauge placement.

As the Hugoniot Elastic Limit (HEL) was fixed at *ca.* 13 GPa, it was assumed that under the quoted impact conditions the material was behaving elastically, and thus impact stresses (σ_x) were calculated through,

$$\sigma_x = \rho_0 c_L u_p, \qquad 1.$$

where u_p is the particle velocity.

MATERIAL'S DATA

The properties of the silicon carbide investigated were longitudinal wave speed (c_L)=12.24±0.03 mm μs⁻¹, shear wave speed (c_S)=7.76 mm μs⁻¹, density (ρ_0)=3.20±0.01 g cm⁻³ and Poisson's ratio (ν)=0.164.

RESULTS AND DISCUSSION

Lateral gauge traces from SiC-N are presented in figure 2.

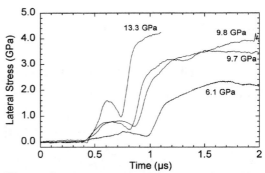

Figure 2. Lateral stress gauge traces from SiC-N. All gauges are 2 mm from the impact face.

It can be seen that in all cases, lateral stress rises to an initial value, then drops somewhat before increasing to a higher, near constant value. As the shear strength (τ) is half the difference between the orthogonal components of stress, thus,

$$2\tau = \sigma_x - \sigma_y. \qquad 2.$$

This indicates that the shear strength is initially high before the arrival of a feature that drops the shear strength to a lower value. This is consistent with the presence of a failure front, and thus it is reasonable to assume that this feature is present in SiC-N as well. The drop in lateral stress immediately prior to the arrival of the failure front has been observed in other materials such as silicon carbide pad B [4] and titanium diboride [6], although not to this extent. This remains an active area of investigation. Finally, observe that as impact stress increases, the failure wave arrives closer to the main shock front, that is the failure wave velocity increases with impact stress. We have not attempted to determine the failure wave velocity as evidence from other ceramics indicates that it slows as it progresses into the target until it is stopped entirely approximately 4 to 5 mm from the impact face [5].

From the literature, we observe there is perhaps some uncertainty [11, 12] as to the existence of failure fronts in ceramics. Therefore it would be desirable if there was independent evidence for them. In the case of silicate glasses, this has not a problem as failure fronts have been resolved using high-speed photography [3] as well as reload signals superimposed upon rear surface velocity traces [1]. However, we note with interest that reload signals, in a manner similar to glass *have* been observed in AD85 alumina by Rosenberg and Yeshurun [13]. Therefore it would seem likely that failure fronts do occur. Further, we would point out that failure fronts have not been observed in another brittle material, the igneous rocks gabbro [14] and dolerite [15]. One would argue that if the failure front is simply an artifact of the specimen configuration, then they should be present in all brittle materials. The fact that they are present in some, but not all, would lend credence to the failure front being a material's characteristic rather than one of the experimental technique.

Shear strengths have been determined using equation 2, and the lateral stresses ahead of and behind the failure wave. This in turn gives two values of shear strength, and these have been plotted against impact stress in figure 3.

The straight line fit is based upon a simple elastic response, thus,

$$2\tau = \frac{1-2\nu}{1-\nu}\sigma_x . \qquad 3.$$

Both unfailed and failed strengths increase monotonically, with the unfailed strengths lying slightly above the elastic fit.

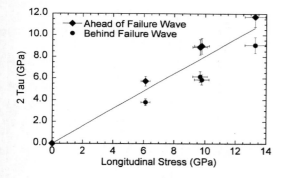

Figure 3. Shear strengths in SiC-N.

This may be an indication that the shear modulus of this material is pressure dependent, which we have not accounted for. We have also compared these results to similar measurements made in another silicon carbide [4], SiC-B. Also included are the results of Feng *et al.* [11], for that material. The results are presented in figure 4.

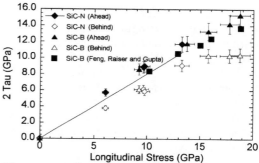

Figure 4. Shear strengths in SiC-B and SiC-N.

It can be seen that the results from SiC-N agree closely with our previous results from SiC-B [4], with both lying slightly above the calculated response, most likely for the reasons discussed previously. The results of Feng *et al* [11] appear more consistent with the calculated elastic response, although as they lie within our own errors, it is difficult to draw definite conclusions. Interestingly, the *failed* strengths of SiC-N and SiC-B appear identical, despite differences in processing route and microstructure. A similar response was seen in various grades of alumina (purities ranging from 85 to 97.5%) [8], where it was suggested that the failed strength was mainly controlled by the alumina grains themselves, with secondary features such as glassy phase and porosity making a much smaller contribution. Thus it seems that silicon carbide also behaves in this manner.

CONCLUSIONS

Plate impact experiments have been performed upon specimens of a silicon carbide, SiC-N. Measurements have been made using manganin stress gauges mounted in such an orientation so as to render them sensitive to the lateral components of stress, 2 mm from the impact face. The resultant traces show an initial rise, followed by a secondary increase to a final higher lateral stress, sometime after the first. This indicates that the shear strength has an initial high value before dropping significantly behind a second moving front behind the main shock wave. This is indicative of the failure wave, seen in other brittle materials such as glasses and some armour ceramics. Comparison with similar measurements made in another silicon carbide (SiC-

B) show close agreement between the data sets. This suggests that the strength of these materials is controlled by the silicon carbide grains themselves, with other features such as additional phases or porosity having a less significant effect. Comparison of the measured strengths ahead of the failure wave are similar to those calculated using a simple elastic analysis. Further work is in progress to determine how the failure wave penetrates into the bulk of the target.

ACKNOWLEDGMENTS

The authors would like to thank Gary Cooper and Gary Stevens of RMCS for technical assistance. We would also like to thank Dr. D.P. Dandekar of ARL, Aberdeen for his support and encouragement. The research reported has been made possible through the support and sponsorship of the U.S. Government through its European Research Office of the U.S. Army.

REFERENCES

1. Razorenov, S.V., Kanel, G.I., Fortov, V.E., Abasemov, M.M., High Press. Res., 1991. **6** 225.
2. Brar, N.S., S.J. Bless, Z. Rosenberg, Appl. Phys. Letts., 1991. **59** 3396.
3. Bourne, N.K., Rosenberg Z., Field, J.E. J. Appl. Phys., 1995. **78** 3736.
4. Bourne, N.K., Millett, J.C.F., Pickup, I. J. Appl. Phys., 1997. **81** 6019.
5. Bourne, N., Millett, J., Murray, N., Rosenberg, Z. J. Mech. Phys. Solids, 1998. **46** 1887.
6. Bourne, N.K. Gray III, G.T., Proc. R. Soc. Lond. A, 2002. **458** 1273.
7. Dandekar, D.P. Beaulieu, P.A., in *Metallurgical and materials applications of shock-wave and high-strain-rate phenomena*, L.E. Murr, K.P. Staudhammer, and M.A. Meyers, Editors. 1995, Elsevier Science BV. p. 211.
8. Millett, J.C.F. Bourne, N.K. J. Mater. Sci., 2001. **36** 3409.
9. Rosenberg, Z., Partom, Y. J. Appl. Phys., 1985. **58** 3072.
10. Millett, J.C.F., Bourne, N.K. Rosenberg, Z. J. Phys. D. Applied Physics, 1996. **29** 2466.
11. Feng, R., Raiser, G.F., Gupta, Y.M. . J. Appl. Phys., 1998. **83** 79.
12. Murray, N.H. Millett, J.C.F., Proud, W.G. Rosenberg, Z., in *Shock Compression of Condensed Matter - 1999*, M.D. Furnish, L.C. Chhabildas, and R.S. Hixson, Editors. 2000, AIP Press: Woodbury, NY. p. 581.
13. Rosenberg, Z. Yeshurun, Y. J. Appl. Phys., 1986. **60** 1844.
14. Millett, J.C.F., Tsembelis, K. Bourne, N.K. J. Appl. Phys., 2000. **87** 3678.
15. Tsembelis, K., Proud, W.G., Field, J.E. in *Shock Compression of Condensed Matter 2001*, M.D. Furnish, N.N. Thadhani, and Y. Horie, Editors. 2002, American Institute of Physics: Melville, New York. p. 1385.

CP706, *Shock Compression of Condensed Matter - 2003*
edited by M. D. Furnish, Y. M. Gupta, and J. W. Forbes
© 2004 American Institute of Physics 0-7354-0181-0/04/$22.00

EQUATION OF STATE PROPERTIES OF MODERN COMPOSITE MATERIALS: MODELING SHOCK, RELEASE AND SPALLATION

W. Riedel, H. Nahme, K. Thoma

Fraunhofer Ernst-Mach-Institut, Eckerstr. 4, D-79104 Freiburg, Germany

Abstract. As Finite-Element codes with explicit time integration schemes (Hydrocodes) are intensively used to simulate fiber reinforced composite structures under impact loading, reliable composite material models with thoroughly derived parameters are essential for successful simulations. The paper describes a closed approach to derive and validate the Hugoniot curve, release properties (Grüneisen Γ) and spall strength using a combined experimental and numerical methodology. The method is applicable to a wide range of materials, exemplary results are shown for a high strength Carbon-fiber / Epoxy laminate (CFRP) used in aeronautic structures and a low stiffness, highly deformable Aramid / Epoxy protection material.

INTRODUCTION

As hydrodynamic simulations become more sophisticated, it is necessary to develop characterization techniques for all classes of material behavior. Shock absorbing porous materials, polymers, composites, weaves or rubber type materials are only some examples of materials which have to be simulated to replicate realistic scenarios. It is therefore necessary to develop test methods which are widely applicable.

THIN PLATE CONFIGURATION

The inverse plate impact configuration is very well suited to derive shock and release data for a range of material types [1]. In this configuration a disc-shaped sample together with a backing plate impacts a target plate of known shock and release properties. In the present study steel target plates and aluminum backing plates have been used. With a VISAR (Velocity Interferometer System for Any Reflector, [2]) the velocity of the rear target surface was recorded. The wave propagation properties for a configuration with thin target and projectile backing plates and the resulting stress / particle-velocity states are illustrated by the Lagrange – and σ-u_p-diagrams given in Figure 1. The different σ-u_p-states are denoted by numbers for the target plate, letters for the composite material and roman numbers for the backing plate. Upon impact, composite and target plate are shocked to the Hugoniot state <1,b>. Reflection of the pressure wave at the target rear surface results in the released state <2> and provides a velocity increase to u_2 recorded with the VISAR. Velocity-time histories for two tests with different impact velocities are shown in Figure 2. From this first velocity step the shock state is derived using the equation of state properties of the steel target plate (e.g. bulk sound speed c_B and slope S). The free surface approximation $u_{p,1} \sim 0.5 u_2$ is well applicable for the compact steel target. Using the continuity conditions for stress and particle velocity across the material boundary steel-composite, the Hugoniot state <b> and the shock velocity U_s in the CFRP is derived from the velocity step u_2 according to

$$\sigma_{1,b} = \rho_{steel} c_{B,steel} \left(\frac{1}{2} u_2 \right) + \rho_{steel} S_{steel} \left(\frac{1}{2} u_2 \right)^2 \qquad (1)$$

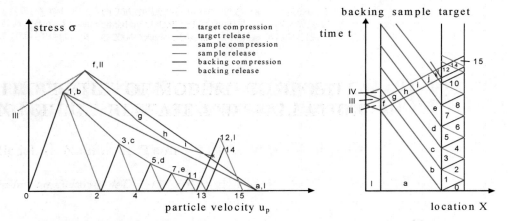

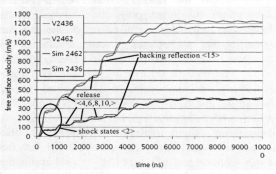

FIGURE 1. Lagrange and σ-u_p-diagram of a reverse plate impact to derive lower Hugoniot and release properties.

and

$$U_s = \frac{U_{s,steel}\, \rho_{steel}\, u_2}{\rho_{composite}\left(2v_0 - u_2\right)} \quad (2)$$

respectively.

The first release wave propagating back into the target is partially reflected at the steel-composite boundary as a pressure wave and partially transmitted as release wave. Subsequent reflections cause the stepwise velocity increase of the steel target plate observed with the VISAR. The transmitted release waves result in a stepwise unloading of the composite material.

FIGURE 2. Experimental velocity-time histories and replication of shock, release and reflected shock states in numerical simulations (see text below).

The points of the Hugoniot curve for the composite material determined from the first velocity step are

shown in terms of a shock velocity / particle velocity relationship (full symbols) in Figure 3. Except for two very low shock states we notice in reasonable approximation linear increasing U_s-u_p-characteristics for both the structural and the protective composites.

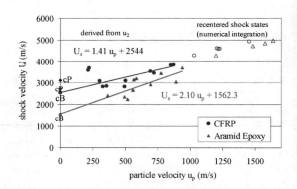

FIGURE 3. Hugoniot curve (U_s-u_p relation) of CFRP.

The original compressive wave in the composite meanwhile propagates upstream. Encountering the backing plate it is reflected as a pressure wave back through the sample and the target causing the distinct velocity change u_{15} of the target surface.

So far, the release isentrope has not been directly derived from the velocity steps u_4, u_6, u_8 etc. Different approximations for the Grueneisen coefficient have been tested and validated against the release states measured in the tests. The best reproduction of the subsequent release states <3,c>,

<5,d>, <7,e> is obtained using the simple approach given by Dugdale and MacDonald [3],(3).

$$\Gamma \approx 2S - 1 \qquad (3)$$

One-dimensional hydrocode simulations of the plane shock conditions using the described Hugoniot and release data show excellent agreement with respect to timing and level of the velocity plateaus associated with Hugoniot states <1,b>, release states <3,c>, <5,d>, <7,e> and states <f,II> caused by the reflected shock and resulting in the velocity u_{15} (see Figure 3, $v_0 = 339$ and 993 m/s respectively).

SHOCK REVERBERATION

So far the maximum shock pressures are limited to 5.3 GPa. With respect to higher impact velocities e.g. in space applications a con-figuration with a thin composite plate and thick target and backing plates was used to reach higher stress states (Fig. 4). The recorded velocity-time histories are displayed in Figure 5 together with the results of the numerical simulations.

Release waves from the free surfaces of target and backing take a long time to reach the sample plate and the initial pressure wave is reflected at the composite plate surfaces compressing the material in several steps to the states denoted by <II,c>, <3,d> etc. Those stress levels, caused by wave reflection at the interface composite-target, can be determined from the velocity steps u_4 and u_6 of the free target surface.

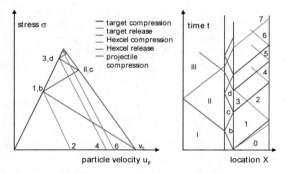

FIGURE 4. Shock reverberation configuration using thick witness and backing plates to derive higher Hugoniot states (elastic waves are omitted).

The shock state <1,b> caused by the initial pressure wave is derived as for the thin plate configuration.

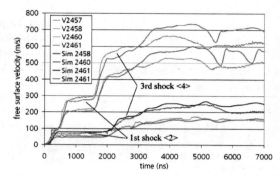

FIGURE 5. Velocity-time histories and replication of multiple shock states from numerical simulations of reverberation tests.

The reflected higher shock states were simulated extrapolating the Hugoniot curve of Figure 3. The third Hugoniot state <3,d> can be validated against the timing and velocity level of the measurement velocity signal u_4 (see Figure 5). We note good agreement of the simulations using this approximation. On this basis, the second shock state <II,c>, not accessible to VISAR measurements, can be evaluated from numerical simulations. For impact velocities up to 390 m/s, good agreement is observed also for the late times response of the impact process. Maximum pressures of 11.3 GPa were reached. The derived re-centered shock states are added to Figure 3.

In the case of the higher impact velocities (up to 1018 m/s), later states cannot be compared because of the target steel experiencing a crystal phase transition at about 13.5 GPa.

SPALLATION / DELAMINATION

The spallation configuration is the third plate impact technique to complete the shock and release properties with respect to dynamic through thickness failure. In this configuration the composite sample was impacted by a thin metal flyer. The ratios of plate thicknesses and shock velocities are adapted to obtain superposition of release waves from the target free surface and from the projectile rear surface to cause high tension stress and finally spallation. The delamination strength increases from the quasi static value of 45 MPa to around 95±25 MPa (Aramid / Epoxy) and 250±12 MPa (CFRP) at strain rates of 150000 s^{-1}.

Figure 6 shows a delaminated sample softly recovered for further examination. As delamination is a primary failure mechanism of composites, these results are very important for the calculation of impact damage.

3 cm

FIGURE 6. Dynamic delamination of a CFRP-sample

APPLICATIONS

The above CFRP material data has been employed in studies on the survivability of modern high strength aircraft components [4]. Figure 7 shows a wing structure with 9 co-bonded and bolted T spars after impact of 120 steel fragments (1.5g) (upper left) and loading by the blast of 0.3 kg high explosive at a distance of 50 cm.

High material costs for composites and the large range of loading scenarios (location, relative velocity, distance, fuel fill level) make numerical simulations a valuable tool for time and cost efficient parameter analysis of such configurations.

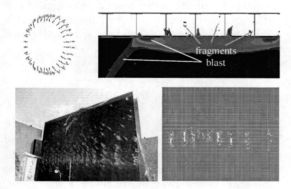

fragments
blast

FIGURE 7. Top: Fragment generation and impact onto wing structure (seen from above). Bottom: Side view of test configuration and simulated damage

The results of the coupled simulations are:

- Coupled calculations of the effects of 100 fragment impacts and a blast wave are feasible with parallelized hydrocodes [5].
- Location and distribution of fragments are well reproduced.
- Extent of delamination and structural degradation are simulated qualitatively correctly but quantitatively are too small in their total effects [4].

CONCLUSIONS

A suite of different plate impact experiments has been applied to derive low and extreme Hugoniot stresses, release states and dynamic delamination strength of two modern composite materials. By this methodology the area of the equation state surface relevant for shock and impact applications can be determined. The reverse plate impact technique allows the above analysis of any new material without any assumptions on its mechanical behaviour.

Numerical simulations of aeronautical applications show that the material data can well reproduce the momentum conservation during impact and shock loading. Next step will be the combination of nonlinear shock compressibility with state of the art composite failure models to accurately replicate strength decrease.

REFERENCES

1. Grady D.E., Furnish M.D.: *Shock and Release Wave Properties of MJ-2 Grout.* Sandia National Laboratories Rept., SAND88-1642, 1988
2. Barker, L. M., and Hollenback, R. E., "Laser Interferometer for Measuring High Velocities of any Reflective Surface", J. of Applied Physics 43, 4669, 1972.
3. Dugdale J.S., MacDonald D.: *The Thermal Expansion of Solids.* Physics Revues, Vol. 89, 1953, S. 832-834
4. W. Riedel, K. Thoma, A. Kurtz, P. Collins, L. Greaves, *Vulnerability of Composite Aircraft Components to Fragmenting Warheads*, 20th Int. Symp. Ballistics, 2002, pp. 702
5. N.N., AUTODYN, *Theory Manual*, Century Dynamics Ltd. Horsham, UK, 2003

CHAPTER XII

MECHANICAL PROPERTIES OF CERAMICS, GLASSES, IONIC SOLIDS, AND LIQUIDS

CP706, *Shock Compression of Condensed Matter - 2003*
edited by M. D. Furnish, Y. M. Gupta, and J. W. Forbes
© 2004 American Institute of Physics 0-7354-0181-0/04/$22.00

REEXAMINATION OF THE REQUIREMENTS TO DETECT THE FAILURE WAVE VELOCITY IN SiC USING PENETRATION EXPERIMENTS

C. E. Anderson, Jr.[1], D. L. Orphal[2], D. W. Templeton[3]

[1]*Southwest Research Institute, P. O. Drawer, San Antonio, TX 78240*
[2]*International Research Associates, 4450 Black Ave, Pleasanton, CA 94566*
[3]*U. S. Army TACOM-TARDEC, AMSTA-TR, Warren, MI 48397*

Abstract: Data for projectile penetration of silicon carbide (SiC) from two types of experiments were combined and analyzed in previous work [1-2]. Analysis of the data suggested the presence of the so-called "failure wave" phenomenon, interpreted as the apparent increase in the strength of SiC when the penetration velocity exceeds some critical value. These data are used as the basis for the design of a new set of experiments. The objectives of these new experiments are to remove ambiguities and uncertainties that exist in the analysis and interpretation of the original data sets and thereby more definitively detect and characterize the phenomena attributed to a "failure wave." The design requirements for experiments to achieve the high impact velocities necessary for investigating the physical phenomenon are described.

INTRODUCTION

Data from two data sets, long-rod tungsten projectiles and copper shaped-charge jets into SiC targets, were combined to examine the penetration resistance of SiC as a function of penetration velocity [1-2]. Analysis of the data suggested that there is an increase in the strength of the SiC above some threshold penetration velocity. Penetration velocity versus impact velocity is shown for the two data sets in Fig. 1. Also shown in the figure are the hydrodynamic limits for tungsten into SiC and copper into SiC:

$$u_{hydro} = v / \left[1 + \left(\rho_t / \rho_p \right)^{1/2} \right] \qquad (1)$$

where u is the penetration velocity, v is the impact velocity, ρ is the density, and the subscripts t and p indicate the target and projectile, respectively.

Kozhushko, *et al* [2], adjusted the penetration velocities to account, at least to first order, the differences in penetrator and target densities for the two data sets. The densities are given in Table 1. The adjusted penetration velocity is given by:

$$u_{adj} = u \left[1 + \left(\rho_t / \rho_p \right)^{1/2} \right] \qquad (2)$$

Using this adjustment, the hydrodynamic limit, from Eqn. (1) is

$$\left(u_{adj} \right)_{hydro} = v \qquad (3)$$

The results are shown in Fig. 2. It is noted that the tungsten into SiC data are approaching the hydrodynamic limit with increasing impact velocity, but that the copper into SiC data have the opposite trend.

Table 1. Physical Properties

	ρ_p (g/cm^3)	ρ_t (g/cm^3)
Ballistic Experiments	19.2	3.22
Shaped-Charge Expts.	8.9	3.0

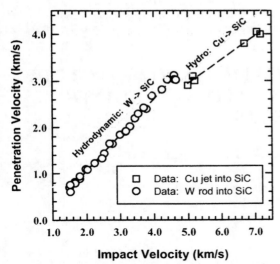

Figure 1. Penetration velocity vs. impact velocity.

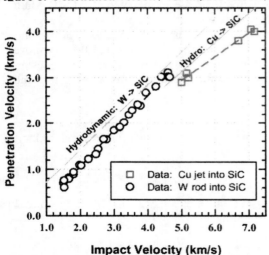

Figure 2. Adjusted penetration velocity vs. impact velocity.

An alternative representation of the data is shown in Fig. 3, where the difference between the hydrodynamic penetration velocity and the measured penetration velocity, normalized by the hydro-dynamic penetration velocity, $[(u_{hydro}-u)/u_{hydro}]$, is plotted versus the impact velocity. Two curves are drawn through the data. The dotted curve is an exponential least-squares regression that extrapolates the tungsten data to higher velocities. The dashed line is a quadratic least-squares regression fit to all the data. In this representation of the data, it is observed that the normalized difference in the penetration velocities begins to increase above an impact velocity of ~5 km/s.

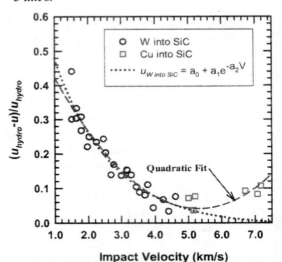

Figure 3. Normalized difference in penetration velocities vs. impact velocity.

ISSUES WITH PREVIOUS EXPERIMENTS

There are a number of issues with the previous experiments, which are discussed by the authors in Ref. [2]. These include, but are not necessarily limited to:

- The experiments used two different projectiles: a tungsten rod for the reverse ballistic tests and a portion of a copper shaped-charge jet for the highest velocity experiments;
- Different pedigrees of silicon carbide (different processing and densities);
- The reverse ballistic tungsten rod tests used a semi-infinite ceramic target while the shaped charge jet experiments used a finite-thickness ceramic target;
- Penetration velocity is inferred in the shaped charge experiments from arrival times at the front and back surface of the target;
- There is essentially no overlap in the experimental data sets.

DESIGN OF NEW EXPERIMENTS

Higher Impact Velocity. Our analysis resulted in a decision to resolve the issues associated with the previous research by conducting a new series of reverse ballistic experiments using a long-rod penetrator. Based on the original analysis [1-2] it was concluded that impact velocities up to at least 6.5 km/s are necessary to obtain penetration velocities sufficiently high that the failure kinetics of the ceramic might be revealed with confidence.

New Target Design. The requirement for very high impact velocities means that the ceramic targets had to have less mass than in the previous research. This required a redesign of the target to remove the titanium sleeve, reduce both the diameter and length of the ceramic, remove the cover plate, and reduce the thickness of the back plate. A series of numerical simulations was performed to determine the minimum target diameter required to avoid effects of the radial boundary on penetration. For SiC the final target design has a ceramic diameter of 15-20 mm (which is 20-27 projectile diameters) and a length of 40 mm.

New SiC Material. The previous experiments used SiC-B. It was decided to use SiC-N for the new experiments since it is a ceramic of choice for armor applications. In order to be sure that the new experiments reproduced the previous results, a number of experiments will be performed with both SiC-N and SiC-B.

New Penetrator Material and Diameter. The previous research used a tungsten long rod of diameter, $D = 0.762$ mm. The new experiments will use a gold rod with $D = 0.75$ mm. Gold was selected for the penetrator material in order to have a very high density but very low strength penetrator. The low strength for the penetrator essentially removes effects of projectile strength from analysis of the results.

Measurement Accuracy and Precision. Error analysis showed that resolving the expected transition required higher accuracy measurements than were possible in the previous research. The accuracy and precision necessary are indicated in Fig. 4. The improvements and procedures implemented to achieve the necessary experimental accuracy are summarized in [3].

Testing. Initial tests have been successfully performed to validate the function and precision of the flash X-ray timing and analysis procedures. Testing is currently being conducted to demonstrate that the changes in the experiment design described above, as compared to the earlier work [1-2], have no unexpected effects on the penetration velocity.

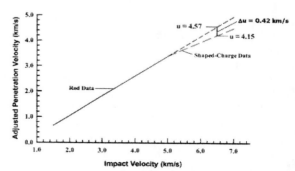

Figure 4. Schematic of measurement precision required for experiments.

HYPOTHESIS

Partom postulates the existence of a failure wave for planar impact of glass [4], based on results of Kanel [5] and others (see references in [4]). Above some threshold stress, damage is assumed to propagate at a "failure wave" velocity that is pressure dependent. Behind the failure wave, there is a two to threefold reduction in the shear modulus. With these assumptions, Partom can reasonably reproduce the experimental wave profiles of one-dimensional planar impact experiments.

For penetration, the hypothesis is that failure propagates ahead of the projectile, so that the projectile penetrates failed material. If, however, the projectile penetrates the material faster than the propagating failure wave, then the projectile must penetrate undamaged, and therefore stronger, material. Therefore, the increase in penetration resistance above some threshold penetration velocity has been attributed to the existence of a failure wave.

Although speculative at this point, we propose that the increase in penetration resistance is due to the kinetics of failure, instead of the propagation of a failure wave (which is described by the wave equation). That is, it takes a finite amount of time for material to undergo failure. If the projectile is penetrating with a velocity sufficiently high that full failure of the material directly in front of the projectile does not have time to occur, the material will respond as less than fully failed, and thus stronger, material.

A schematic of our current thinking regarding the failure mechanism is shown in Figs. 5 and 6. A damage variable, D^*, represents failure, with "0" being no failure and "1" being complete failure. In this context, these rod penetration experiments can be interpreted as showing that below u ~3.2 km/s, Fig. 6, the material in front of the projectile has sufficient time to completely fail and the projectile is moving into totally failed (comminuted) material ($D^* = 1$). However, once failure is initiated, by whatever mechanism, there is an "incubation" time for "full" damage to be realized (dislocation motion [6], microcracking, crack coalescence, comminution, etc.). At sufficiently high velocities—the data in [1-2] would suggest u greater than about 3.8 km/s—the projectile is penetrating material that has not had time to transition from intact to fully failed material; and consequently, the penetration resistance increases. An objective of the present work, in addition to the experimental efforts, is to investigate this hypothesis.

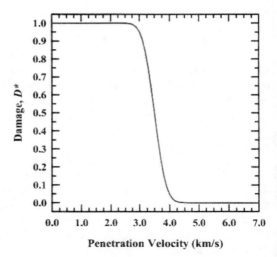

Figure 6. Schematic of damage at the projectile-target interface as a function of penetration velocity.

velocity in SiC using hypervelocity penetration experiments," *Shock Compression of Condensed Matter – 1999*, Edited by M. D. Furnish, *et al*, pp. 577-580, AIP, 2000.
2. A.A. Kozhushko, D.L. Orphal, A.B. Sinani, and R.R. Franzen, "Possible detection of failure wave velocity using hypervelocity penetration experiments," *Int. J. Impact Engng.*, **23**, pp. 467-475, 1999.
3. Th. Behner, V. Hohler, C. E. Anderson, Jr., and D.L. Orphal, "Accuracy and position require-ments for penetration experiments to detect the effect of failure kinetics in ceramics," *21st Int. Symp. on Ballistics*, Adelaide, Australia, submitted, April 2004.
4. Y. Partom, "Modeling failure waves in glass," *Int. J. Impact Engng.*, **21**(9), 791-799, 1998.
5. G. I. Kanel, S. V. Rasorenov, and V. E. Fortov, "The failure waves and spallations in homogeneous brittle materials," *Shock Com-pression of Condensed Matter – 1991*, Edited by S. C. Schmidt, *et al*, pp. 451-454, AIP, 1992.
6. J. Lankford, C. E. Anderson, Jr., A. J. Nagy, J. D. Walker, A. E. Nicholls, and R. A. Page, "Inelastic response of confined aluminum oxide under dynamic loading conditions," *J. Mat. Science*, **33**(6), 1619-1625, 1998.

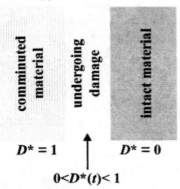

Figure 5. Schematic of damage kinetics process.

REFERENCES

1. D. L. Orphal, A. A. Kozhushko, and A. B. Sinani, "Possible detection of failure wave

CP706, *Shock Compression of Condensed Matter - 2003*
edited by M. D. Furnish, Y. M. Gupta, and J. W. Forbes
© 2004 American Institute of Physics 0-7354-0181-0/04/$22.00

THE EFFECT OF SHOCK RISE TIME ON STRENGTH OF ALUMINA IN 1-D STRESS AND 1-D STRAIN

Stephan J. Bless[1], Neil K. Bourne[2]

[1]Institute for Advanced Technology, University of Texas, Austin, TX 78759
[2]Cranfield University, Bedfordshire, U.K.

Abstract. Graded-density flyer plates were used in plate-impact and bar-impact tests with alumina targets. The stress wave ramp induced by these flyers results in data that are more useful for resolving strength parameters.

In recent years, we have seen increasing use of ceramics deployed as armor. Engineering practice has been enabled by better failure models for ceramics and the incorporation of these models into wave-propagation codes. Nevertheless, certain fundamental aspects of the impact behavior of ceramics remain obscure.

The cavity expansion approach to modeling ceramic armor, as applied by Satapathy and Bless, provides an essential framework for discussion of the properties of ceramics that are important for thick tile armor [1]. According to this model, the material around an advancing cavity can be divided into distinct regions; the far field experiences elastic deformation (compression in the radial direction, tension in the hoop direction, and zero mean stress). Where this material fails in hoop tension, there is a transition to a region of needle cracks in which the stress is thought to be effectively unidirectional (because there is no hoop strength). When the stress builds to a point where shear failure occurs, the ceramic comminutes to a rubble that surrounds the cavity. Consideration of the stress history of ceramic elements plus parameter studies shows the following [1]:

1. The most important strength property of intact ceramic is the strength under uniaxial stress loading.
2. The strength and dilation of comminuted material is also extremely important.

3. Penetration can be avoided altogether if a critical confining stress is applied, and this critical strength is mainly a function of the tensile strength of the intact ceramic.

Measurement of these critical ceramic properties is possible by means of plate-impact and bar-impact tests. The aim of the present work is to improve these test techniques through use of graded-density flyer plates. All data are for 98 percent dense alumina.

The bar-impact tests, as discussed in [2, 3] have the potential ability to measure 1-D stress strength (Y) and the properties of comminuted materials (at least for low confining stress). Thus, they are very relevant to modeling ballistic behavior. Moreover, conventional ceramic models that describe plate-impact tests and penetration experiments provide only poor agreement with bar-impact data [4]. On the other hand, the tests have had two problems historically. First, values of Y have been dependent on test configuration. This is illustrated for AD995 in Fig. 1, which shows that values between 2.8 and 4.2 GPa have been measured. Also shown in Fig. 1 are the strength values that would be predicted from the HEL using a Von-Mises and a Griffith [5] failure criteria. Were it not for the scatter, one might conclude that the Von-Mises criterion was valid, and this would be very useful, for it would allow values of Y to be inferred from standard plate-impact experiments.

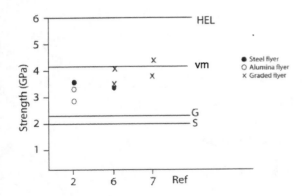

FIGURE 1. Strength of alumina measured in various bar-impact experiments. Plot data are from Refs. 2, 6, and 7.

The second problem with bar-impact tests is that stress recordings are almost always terminated before the arrival of the signal from recompression of failed material. This is illustrated in Fig. 2, which compares records from AlON and Al_2O_3 [3]. The difference in initial strength of the materials is resolvable, but possible variations in post-failure strength are not.

Use of graded-density flyer plates may be the best way to solve both of these problems. Encouraging initial results were obtained by [7].

Plate-impact tests are needed for measurements of compressive strength (values of Y derived from the HEL), tensile strength (e.g., spall stress), and post-failure behavior (e.g., failure wave analysis). It has been discussed by several authors, for example [8], that use of a ramped shock aids discrimination of strain-rate effects on compressive and tensile strength. We find here that they also greatly assist definition of spall strength.

The graded-density flyer plates used in this work are faced with 0.4 mm of PMMA, followed by 0.4 mm of Al and 0.4 mm of Cu, mounted on a substrate. These plates are fabricated in a special fixture developed at Cranfield University. Figure 3 illustrates hydrocode calculations of stress profiles in AD85 ceramic struck by 5-mm thick tungsten flyer plates with and without the graded-density packets. Impact velocities were 300 m/s. These experiments show that the strain rate is reduced from its usual value of 10^6/s to 10^4/s with the graded flyer. This lower rate is commensurate with

values in bar impacts and ballistic penetration, so strength measurements with graded flyers can be applied without ambiguity associated with possible rate dependence. However, the most striking new modification to conventional tests is that the location of the spall becomes dependent on the spall strength. This is illustrated in Fig. 4, which shows the predicted difference between 1-GPa and 0.1-GPa spall stress. Actual data that have been obtained to date on AD998 alumina are not in agreement with Fig. 4, which shows that the Pmin-type model used in the code cannot predict the spall induced by the graded flyer plate.

The use of graded flyer plates in the bar-impact test is illustrated in Fig. 5 (WC flyer plate) and Fig. 6 (graded flyer plate). The target bar was AD998 alumina. Peak stress was measured by means of fly-off disks that were affixed to the free end of the bars and whose motion was followed by framing cameras. In Fig. 5, the peak stress (computed from $Y=1/2 \ C_b \mu_{fs}$) is 3.2 GPa, and in Fig. 6, it is 3.5 GPa. This confirms the expectation that the more gradual loading associated with graded impactors allows the bar to propagate an elastic wave with an amplitude that has reached (or at least come closer to reaching) its value of Y.

The failure sequence, as followed by the camera, is also different. Using the graded impactor, there is no longer a zone of uncomminuted material adjacent to the free surface, as occurs in bar impacts on ceramics (this work) and brittle plastics [11].

In conclusion, use of graded impactors improves measurements of key ceramic properties in both the plate-impact and bar-impact configurations. In the plate-impact configuration, graded impactors can be used to better resolve strain-rate effects and spall strength. In the bar-impact geometry, the graded flyer increases the observed elastic wave amplitude, hopefully approaching the intrinsic value of Y for the ceramic. Used with manganin gauges in a witness bar, we believe that it will also allow observation of the stress transmitted through the comminuted material that faces the impact, thereby providing essential strength data for rubblized ceramic.

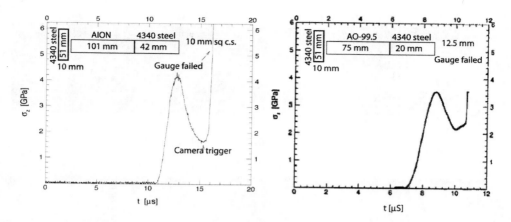

FIGURE 2. Comparison of manganin gauge records in AlON and Alumina [10]. Initial stress level is well resolved, but gauge breaks before recompression signal arrives.

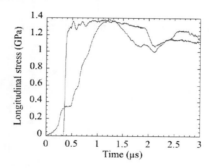

FIGURE 3. Comparison of computed stress records with (slow-rising pulse) and without (fast-rising pulse) graded-density flyer.

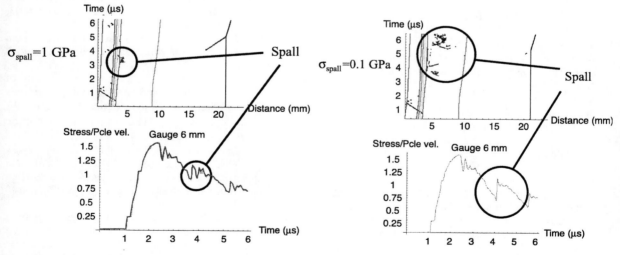

FIGURE 4. Effect of spall strength on gauge records. Gauge is 6 mm from impact surface and backed with PMMA. Aluminum flyer is 3-mm thick.

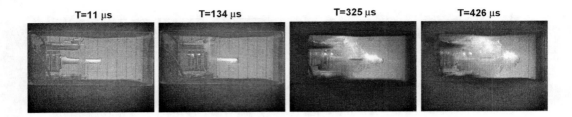

FIGURE 5. AD998 bar struck by WC flyer plate at 274 m/s.

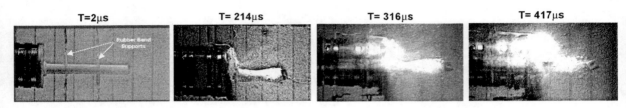

FIGURE 6. AD998 bar struck by WC flyer plate with graded-density packet at 266 m/s.

ACKNOWLEDGMENTS

The research reported in this document was performed in connection with Contract number DAAD17-01-D-0001 with the U.S. Army Research Laboratory.

Bar-impact tests at the IAT were conducted by Rod Russell and Tim Beno.

REFERENCES

1. Satapathy, S., Bless, S., "Cavity Expansion Resistance of Brittle Materials Obeying a Two-Curve Pressure Shear Behavior," *J. Appl. Phys.*, **88**, 4004–4012 (2000).
2. Simha, C. H. M., Bless, S. J., Bedford, A., "What is the Peak Stress in the Ceramic Bar Impact Experiment?" *Shock Compression of Condensed Matter—1999*, edited by M. D. Furnish, L. C. Chhabildas, R. S. Hixson, AIP Conference Proceedings 505, Melville, New York, 2000, pp. 615–618.
3. Brar, N. S., Bless, S. J., Rosenberg, Z., "Brittle Failure of Ceramic Rods under Dynamic Compression," *J. de Physique* **C3**, 607–612 (1988).
4. Bless, S., Russell, R., Beno, T., "Bar Impact Test to Characterize the Ballistic Properties of Ceramics," Army Symposium on Solid Mechanics, Charleston, SC, May 5–7, 2003.
5. Rosenberg, Z., "On the Relation between the Hugoniot Elastic Limit and the Yield Strength of Brittle Materials," *J. Appl. Phys.* **74**, 752–753 (1993).
6. Cazamias, J., Reinhart, B., Konrad, C., Chhabildas, L. C., Bless, S., "Bar Impact Tests on Alumina (AD995)," *Shock Compression of Condensed Matter—2001*, edited by M. D. Furnish et al., AIP Conference Proceedings 620, Melville, New York, 2002, pp. 787–790.
7. Chhabildas, L. C., Furnish, M. D., Grady, D. E., "Impact of Alumina Rods: A Computational and Experimental Study" *J. de Physique IV*, Colloque C3, **7**, 137–143 (1997).
8. Asay, J. R., "The Use of Shock-Structure Methods in Evaluating High-Pressure Material Properties," *Int. J. Impact Eng.* **20**, 27–61, 1997.
9. Kanel, G. I., Bless, S. J., "Compressive Fracture of Brittle Solids Under Shock-Wave Loading," *Ceramic Armor Materials by Design,* edited by James W. McCauley, et al., The American Ceramic Society, Ohio, 2002, pp. 197–216.
10. Cazamias, J., Bless, S., Fiske, P., "Shock Properties of AlON," EXPLOMET 2000 Int'l Conf. On Fundamental Issues and Appl. Shock-Wave & High Strain-Rate Phenomena, Albuquerque, NM June 19–23, 2000.
11. Russell, R., Bless, S., Beno, T., "Impact Induced Failure Phenomenology in Homalite Bars," *Shock Compression of Condensed Matter—2001*, edited by M. D. Furnish et al., AIP Conference Proceedings 620, Melville, New York, 2002, pp. 811–814

CP706, *Shock Compression of Condensed Matter - 2003*
edited by M. D. Furnish, Y. M. Gupta, and J. W. Forbes
© 2004 American Institute of Physics 0-7354-0181-0/04/$22.00

ESTIMATING THE BREAK-UP DIAMETER OF AN IMPULSIVELY DRIVEN INITIALLY SMOOTH FLUID CYLINDER

John P. Borg

Department of Mechanical Engineering, Marquette University
1515 W. Wisconsin Avenue, Milwaukee Wisconsin 53233

Abstract. This analysis pursues the underlying physics governing breakup of a fluid, which is initially contained in a smooth thin walled steel vessel, after being impacted by a high velocity aluminum sphere. It has been observed experimentally that the impact generates a radially expanding cohesive thin shell of liquid which stays intact to at least a diameter 8 times that of the original cylinder diameter. The cohesive nature of the shell is aided by the fact that the shell is expanding in a vacuum. If it is assumed that the shell will break once the height scale of the fastest growing instability exceeds the shell thickness then a linear Richtmyer-Meshkov stability analysis under predicts the break-up diameter. The difficulty with such an approach is determining the proper initial disturbance length scale for an initially smooth geometry. In this analysis it is assumed that the fracture process of the thin walled steel vessel imposes an initial disturbance on the expanding liquid. By combining this with the geometry of the cylinder an initial disturbance length scale can be determined. A breakup model is constructed by combining an initial disturbance derived from the characteristic fragment size of the vessel with a nonlinear Richtmyer-Meshkov growth rate model for the instability growth rate. This model extends the breakup diameter of the expanding liquid shell beyond 8 diameters.

INTRODUCTION

It is not clear at this time how a fluid initially contained in a smooth walled cylinder will evolve after being impacted, what mechanisms will govern the fluid evolution, or what drop distribution will result. It has been suggested that a variety of dynamic mechanisms will govern the fluid behavior including: Rayleigh-Taylor (RT), Richtmyer-Meshkov (RM) [1], Kelvin-Helmholtz (KH) instabilities, varicose or capillary waves [2], instabilities established through the strain history [3] or even dynamic spall. RT and RM instabilities which were discarded because of their rapid growth rate, not observed in experiments, will be reexamined in this work.

In typical RT or RM instability experiments there is a thin membrane which separates the two fluids. The effects of this membrane, which is typically made of solid material, are usually minimized or ignored [4,5]. The membrane is shaped in such a way that disturbances are initially imposed on the interface between the gas and the fluid. The membrane thickness is made small as compared to the total sheet thickness and the wave length of the imposed disturbance. By knowing the scale and wave number of the initially imposed disturbance the growth rate on the interface can be tracked. These experiments are not directly applicable to the smooth cylinders of interest here. A major difficulty in applying a classic stability analysis to the problem under investigation here is determining the disturbance spectra associated with an initially smooth interface. One must ascertain by what mechanisms disturbances are imposed on an initially smooth cylinder. It is very difficult to theoretically or experimentally ascertain the most unstable wave number [6].

In order to establish a working model the geometry is reduced to a simple configuration composed of a thin multi layered gas-solid-fluid sheet. Figure 1 is a pictorial representation of the reaction of the solid case and fluid when shock loaded; the actual dynamics are much more complicated. After impact, the sheet is loaded with a shockwave which will rupture the solid case and, in so doing, launch the fluid. The solid case will produce the fragments with a length scale, ε_1. This scale coupled with the geometry of the problem will indirectly impose a fluid scale length, ε_2, through which the fluid will evolve. Through the divergent nature of this geometry the fluid length scale is time dependent. By careful selection of the impedance of the materials either the fluid or the solid fragments will have the higher launch velocity, figure 1a or 1b.

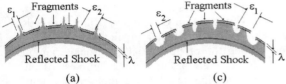

(a)   (c)

Figure 1. Post rupture of target and dispersal of liquid

In the absence of imposed disturbances it is believed that the membrane dynamics, in this case a thin solid sheet, will play an important role in the evolution of the liquid and correlate to the final drop size distribution from this event. Thus the fracture characteristics of the membrane play an important role in the evolution of the fluid. Although RT and RM instabilities are the leading candidates for the mechanism which will govern the break up of gas-solid-fluid sheet it will not be assumed that they are the only mechanism at work. Other instability mechanisms should be considered as the fluid evolves.

THEORETICAL FOUNDATION

Several theoretical models have been developed which describe the process of prompt dynamic fracture of solids [7,8]. In this context prompt relates to the high strain rates experienced by a material as a result of being explosively loaded. These theories balance a driving kinetic energy against a resisting material fragmentation energy to estimate the size of a fragment, ε_1, resulting from a dynamic event. Presented below are two such models [7,8]; the first of which the resisting fragmentation energy is fracture toughness, K_{co}, and the second is flow stress, where Y_0 is yield strength. The fragment size is then typically coupled with a statistical distribution to produce a theoretical distribution of fragments.

Fracture dominated
by fracture toughness:
$$\varepsilon_1 = \left(\frac{\sqrt{24}K_c}{\rho c \dot{\delta}}\right)^{2/3}, \quad (1)$$

Fracture dominated
by flow stress:
$$\varepsilon_1 = \left(\frac{1.2Y}{\rho \dot{\delta}^2}\right)^{1/2}, \quad (2)$$

where:
$$K_c = K_{c0}\left(1-\frac{T}{T_m}\right)^{n'} \quad \text{and} \quad Y = Y_0\left(1-\frac{T}{T_m}\right)^{n}\left(\frac{\dot{\delta}}{\dot{\delta}_0}\right)^{m}.$$
In the above equations T is the temperature, T_m is the melt temperature, $\dot{\delta}_0$ is a reference value of strain and n, m, and n' are constants [7,8].

Taylor developed a generalized analytical two-dimensional planar theory for finite thickness membrane which demonstrated that any initial disturbance is unconditionally unstable [9]. Richtmyer [1] then modified Taylor's original formulation so that the acceleration of the fluid interface was finite in time i.e. an impulse and compressibility effects were included. He concluded that the shock loaded interface initially grows linearly and then asymptotes to:

$$v_{asymptote} = \frac{dv(t)}{dt} \cong u_0^0 \, k \, a_0(+0)\frac{\rho_2^0 - \rho_1^0}{\rho_2^0 + \rho_1^0}, \quad (3)$$

where the superscript 0 denotes a post shock value. Richtmyer determined that using the post-shock densities produced much better results than did using the pre-shock values. Experiments have verified that Richtmyer's growth rate predictions are good in the region of short time for a gas-gas geometry with a single interface and a single disturbance [4,5]. However, Richtmyer's results are not accurate for long time evolution where nonlinear effects begin to dominate the flow.

An improvement to Richtmyer's formulation which included nonlinear effects for the growth rate of a disturbance was advanced by Zhang and

Sohn [10,11]. A nonlinear solution was sought by applying a perturbation technique to the governing equations which had been expanded using a Taylor's series approximation. The first order nonlinear description of the time varying growth rate is described as follows:

$$v = \frac{v_{lin}(t)}{1 + v_{lin}a_0k^2t + \max\left\{0, a_0^2k^2 - A^2 + 0.5\right\}v_{lin}^2k^2t^2}, \quad (4)$$

where v_{lin} is the compressible linear growth rate derived by Richtmyer [1], a_0 is the initial amplitude of the disturbance and t is time. If, following the work of Zhang and Sohn, $v_{lin}^\infty = \lim_{t\to\infty} v_{lin}(t)$ then the steady state compressible growth rate can be estimated as a function of time. Zhang found that including nonlinear effects greatly decreases the growth rate of instabilities as compared to the linear growth rate. Zhang showed that as time goes to infinity the growth rate approaches zero, not the constant growth put forth by Taylor or the nonzero asymptotic value put forth by Richtmyer. These results have been verified by Holmes [12] and Prasad [13] in which experimental data, numerical simulation and Zhang's nonlinear theory all compared favorably. All of these experiments were conducted using a air-SF_6 interface with a single disturbance in a planar 2D configuration.

RESULTS AND DISCUSSION

Associating the fracture dynamics, geometry and shock conditions of a thin gas-metal-fluid sheet in order to characterize the time varying growth rate of an initial perturbations driven by a RM instability is a novel idea. As a check, this idea will be applied to a previously published data set [14,15]. The data was collected at the University of Alabama Huntsville in the Aero Physics test range. The geometry consists of impacting a liquid filled steel right circular cylinder with an aluminum sphere at approximately 2.5 km/s.

The application of equations 1 and 2 bracket the experimentally recovered fragments. Equation 1 predicts a fragment size of $\varepsilon_1 = 0.587$ cm while equation 2 predicts a fragment size of $\varepsilon_1 = 4.6$ cm.

Figure 2 presents images from the UAH test series [14,15]. From these images a typical

fragment size in the radial direction is on the order of 1.2 cm. The fragments are not square and are oriented such that their long dimension lies along the shot direction. The observed particle velocity in the UAH test was approximately $u_p = 400$ m/s, the Atwood number was approximately A=0.99. The wave number associated with the steel fragments in figure 2 is $1/\varepsilon_2 = 2\pi/2.5 \times 10^{-5}$ m^{-1} and the initial amplitude of the perturbation will be on the order of the steel thickness $a(+0) = 1.5 \times 10^{-4}$ m. Thus the linear impulse growth rate was calculated using equation 3 to be $v \approx 15900$ m/s. Using this growth rate, the initial perturbation should have grown to a scale height of over 20 meters by the time the fluid had traveled 8 diameters. No such disturbances were observed in the experiments.

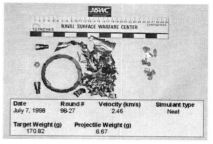

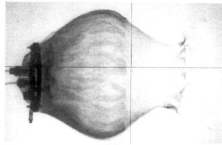

Figure 2. Theoretical Development: Rayleigh-Taylor and Richtmyer-Meshkov Instability

Obviously the linear growth rate over predicts the growth rate of the instability scale height. Applying the same data from the UAH test series to the nonlinear theory yields a scale height of the instability to be 1.2×10^{-3} cm at 8 diameters, a much more reasonable approximation to the phenomena observed at UAH [14,15]. A plot of the linear theory growth rate with time, shown in blue, is presented on figure 3. The nonlinear incompressible approximate theory, equation 4, is presented in figure 3 in black. A significant variation in the results can be seen as compared to the linear theory. In addition it is obvious from the initial conditions that the initial growth rate should be zero thus the time evolution of the growth rate was estimated and presented in figure 3 in red. This estimate of the full non-linear growth rate starts at zero, has an initial increase in disturbance growth rate, and the growth rate rapidly asymptotes to zero in less than a micro-second. Using this growth rate curve the liquid cylinder should grow to over 800 diameters before the scale of the instability exceeds the sheet thickness.

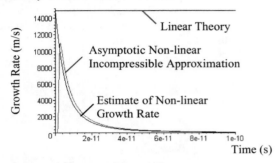

Figure 3. Growth rate of instabilities

CONCLUSIONS

Combining the fracture model, to establish an imposed scale length, and the nonlinear Richtmyer-Meshkov stability analysis better explains why liquid breakup was not observed in the UAH test. These nonlinear results also indicate that RM instabilities, once believe not to effect the governing dynamics because they were not observed within 8 diameters, could in fact play an import role in the late time development of the fluid evolution.

ACKNOWLEDGEMENTS

Funding was provided by the Wisconsin Space Grant Consortium. The author would also like to thank the Naval Surface Warfare Center, Dahlgren Division.

REFERENCES

1. Richtmyer, R. D., "Taylor Instability in Shock Acceleration of Compressible Fluids." Communications on Pure and Applied Mathematics, vol. XIII pg. 297-319, 1960.
2. Cawley, Robert "Software Design Report for DropGen 1.0", NSWCDD/TR-00/62, June 2000
3. Borg, J.P., Grady, D. and Cogar, J.R. "Instability and fragmentation of expanding liquid system", Int. J. of Impact Eng., **vol. 26**, pg 65-76 2001
4. Jones M.A. and Jacobs J.W. "A membraneless experiment for the study of Richtmyer-Meshkov instability of a shock-accelerated gas interface", Phys. Fluids 9 3078-3085, 1997.
5. Mikaelian, K. O., "Rayleigh-Taylor and Richtmyer-Meshkov Instabilities in Multilayer Fluids with Surface Tension" Phys. Rev. A 42 12, 1990.
6. Borg, J.P. "Application of Rayleigh-Taylor Instability Applied to an Expanding Shell", NSWCDD/TR-00/11, February 2000
7. Davison, L., Grady, D., and Shahinpoor, M., High-Pressure Shock Compression of Solids II: Dynamic Fracture and Fragmentation Springer-Verlag ISBN: 0-387-94402-8, 1996.
8. Grady, D. E. and Kipp, M. E. "Geometric statistics and dynamic fragmentation", J. Appl Phys. 58 (3) pg. 1210-12222, 1985
9. Taylor, G. I., "The Instability of Liquid Surfaces when Accelerated in a Direction Perpendicular to their Planes I", Proc. Royal. Soc. London. A **vol. 201** 192-196, 1950.
10. Zhang, Q and Sohn, S "An analytical nonlinear theory of Richtmyer-Meshkov instability". Physics Letters A 212 149-155, 1996
11. Zhang, Q and Graham, M. J. "A numerical study of Richtmyer-Meshkov instability driven by cylindrical shocks" Physics of Fluids 10 4 pg. 974-992, 1998.
12. Holmes, R. L., et. al. "Richtmyer-Meshkov instability growth: experiment, simulation and theory", J. Fluid Mech. **389** pg. 55-79, 1999.
13. Prasad, J.K., Rasheed, A., Kumar, S., and Sturtevant, B. "The late-time development of the Richtmyer-Meshkov instability", Physics of fluids **vol. 12**, number 8, 2000.
14. Ference, S.L., Borg, J.P. and Cogar, J.R. "Source Term Investigation of TBM Bulk Chemical Target Intercepts: Phase 1 Test Report", NSWCDD/TR 99-82, Dahlgren Virginia, 1999
15. Ference, S.L., Borg, J.P. and Cogar, J.R. "Source Term Investigation of TBM Bulk Chemical Target Intercepts: Phase I1 Test Report", NSWCDD/TR 99-83, Dahlgren Virginia, 1999.

CP706, *Shock Compression of Condensed Matter - 2003*
edited by M. D. Furnish, Y. M. Gupta, and J. W. Forbes
© 2004 American Institute of Physics 0-7354-0181-0/04/$22.00

NEW INSIGHTS INTO SHOCK PROPAGATION IN GLASS

N.K. Bourne and G.A. Cooper

Royal Military College of Science, Cranfield University, Shrivenham, Swindon, SN6 8LA, UK.

Abstract. There has been much interest over the last twenty years in failure of glasses triggered by the passage of shock fronts. The precise mechanisms for this failure (and their physical consequences) are as yet, incompletely understood. This has resulted in ability to model brittle materials in a first order manner that has considerable shortcomings. To advance requires full identification of all the operating mechanisms. This work presents results obtained by placing particle velocity gauges, similar to those developed in explosives investigations, into a fully filled glass. The new information gained allows greater understand of operating mechanisms. Features of the failure are discussed in relation to other measurements of the phenomenon previously presented.

INTRODUCTION

There has been much work investigating the response of various brittle materials to one-dimensional impact. In particular, various workers have been interested in planar shock-loading (plate-impact) to investigate the high strain-rate response. There are similarities and differences in the behaviours seen in amorphous glasses and polycrystalline ceramics. Glasses vary in microstructure from open-structured materials, such as fused silica and borosilicate, through the partially filled soda-lime, to the highly filled lead glasses. The filled glasses exhibit fast-rising elastic waves and have similar wave profiles to polycrystalline ceramics [1]. One such material is type D, extra dense flint (DEDF) that has a network structure disrupted by the presence of a large lead oxide phase that fill the matrix. In fact the microstructure is such that the amorphous layers sit in plates. It has been noted that many materials show decay of the elastic precursor with distance in plate impact. The kinetics of the deformation processes determines this distance, and the form of the stress history as the precursor steadies in different materials. DEDF shows pronounced precursor decay with target thickness [1].

Rasorenov *et al.* [2] were the first to observe delayed failure behind the elastic wave in glass, across a front which has been called a fracture or a failure wave. They noted that the wave appeared to travel at velocities in the 1.5–2.5 km s^{-1} range in the K19 (similar to soda-lime) glass. They inferred the presence of the wave by observing a small recompression signal caused by reflection of the elastic release off the failure front in their record of the free-surface velocity. This allowed calculation of front velocity and indicated that material behind the front was of lower impedance than that ahead of it. Later work [3,4] confirmed the existence of these waves by using more direct techniques of spall and lateral stress measurements ahead of and behind the failure front. The strength of glasses, deduced from lateral stress measurement, lies on a common curve for three glasses of varying density and microstructure tested [5]. The failed strength of these glasses crosses the constructed elastic line at 1.9 GPa [5]. It is also possible to show that these phenomena occur in other polycrystalline brittle materials [6].

It is still necessary to fully explain the observed effects and to track the state of all of the variables in the uniaxial strain conditions existing initially on impact. This paper aims to further understand the operating mechanisms in brittle failure.

EXPERIMENTAL PROCEDURE

The gauge package consisted of seven thin strips 25 μm thick, orientated orthogonally to the magnetic field. They are spaced at 2 mm intervals. Each lead ends at a tab, to which external cabling is attached, so that gauge signals can be taken to the scopes. As can be seen from the bottom of Fig. 1, there is an eighth gauge element, consisting of a castellated loop running the length of the gauge assembly. The shock tracker (reported by Alcon and Mulford [7]) is orientated such that the wave front propagates along its length. The result is a castellated voltage–time trace, the periodicity of which gives a direct measure of the time intervals between known Eulerian measurement stations.

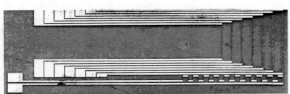

FIGURE 1. Particle velocity gauge constructed for the measurement of flow variables within a magnetic field.

All shots were performed on a 50 mm bore, 6 m long, single-stage gas gun. Particle velocity (u_p) gauge assemblies were placed in targets of DEDF and impacted with alumina flyers at a range of velocities. A magnetic field was generated using Helmholtz coils, configured such that they produced a uniform field over a volume of 20 mm^3, to a magnetic field strength of *ca.* 0.1 T. For similar reasons to others (see for example [7]), the gauge plane in the sample has been cut at an angle (usually 30°) to the impact face. The reasoning is that each gauge element is not directly behind another (which possibly impedes material flow) whilst minimizing the required machining on the sample itself. Therefore, a series of otherwise identical shots were performed with gauge planes orientated at 30°, 60° and 90° to the impact face.

In a parallel series already published [8], experiments were conducted with embedded strain gauges placed on cuts at 90° to the impact face. The constantan sensors were mounted 2 mm from the impact face and 2 mm from the rear of 15 mm thick targets. The technique had previously been used to measure strains in soda-lime glass [9].

The DEDF targets had a density of 5.18 g cm^{-3}, c_L = 3.49 mm μs^{-1}, c_S = 2.02 mm μs^{-1} and a Poisson's ratio of 0.25.

RESULTS AND DISCUSSION

To explain the mechanisms operating at the impact requires a means to track the flow field throughout the target. It was decided to mount particle velocity gauges with targets shocked to 4.2, 5.4 and 5.8 GPa. That is near, just above, and further beyond the Hugoniot elastic limit (HEL).

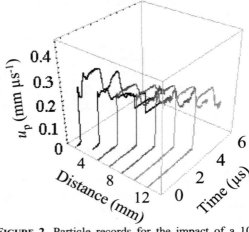

FIGURE 2. Particle records for the impact of a 10 mm alumina tile inducing a longitudinal stress of 4.2 GPa.

In Figs. 2, 3 and 4, particle velocity histories are shown, gathered from sensors embedded at 90° to the plane of the shock, and mounted in cut targets reassembled as discussed. The shots were then repeated with sensors in a plane at 30° to the impact face. One is shown in Fig 5 that is a repeat of that of Fig. 3. Here, whilst each gauge element still sits on an interface, it does not sit behind another sensor 2 mm in front of it.

There is not the space to fully analyse the wealth of information contained in these datasets. For example, wave speed data was also obtained from the shock tracker elements that gave the expected elastic and shock velocities. Thus highlights will be commented upon here, and a full journal paper will follow with complete analyses.

Fig. 2 shows a series of seven histories from each of the elements. The first sensor element rises rapidly to a value consistent with the levels seen coming after it. After 0.5 μs there is second rise to a level *ca.* 50 mm μs^{-1} above the first. Release comes into the gauge again in two steps separated by 0.5 μs. The pulse length is 2 μs consistent with the elastic wave speed of the 10 mm alumina flyer used to impact this target. The flyer behaves elastically in all of these tests.

The pulse at 3 mm shows a similar form, except that the second front arrives after 1.1 μs at this location. Taking into account that the second sensor has moved after the front has arrived at the first element, gives a velocity of 1.41 mm μs^{-1} for the front propagation speed. This compares with a Rayleigh wave speed for DEDF of 1.85 mm μs^{-1}.

The remaining sensors travel in the flow at 0.22 mm μs^{-1}

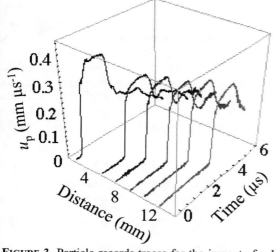

FIGURE 3. Particle records traces for the impact of a 10 mm alumina tile onto DEDF inducing 5.4 GPa.

Fig. 3 shows the impact of the same alumina flyer at a velocity of 490 m s^{-1}. The second gauge element was damaged, and there is thus a space at the 3 mm position. Similar features are seen to those noted earlier. There is a rise to a velocity of 0.3 mm μs^{-1} and then a second pulse takes u_p higher to 0.36 mm μs^{-1}. Note that the last six records have a round top and that the break from the elastic rise decays with distance into the target. This break is the HEL, and the rounded form of the pulse is typical of stress gauge histories recorded in this material. The level of the HEL is at 0.3 mm μs^{-1} on the impact face reducing to 0.23 mm μs^{-1} in the bulk. It reaches this value after a travel of 9 mm.

The gauge record for the highest impact velocity is shown in Fig. 4. Its form is qualitatively

similar to that of Fig. 3. The first trace rises to 0.33 mm μs^{-1} and the second pulse arrives after *ca.* 400 ns. At the second gauge, the pulse first breaks in slope at the HEL, and then rises again after 0.9 μs. The calculated travel speed of the front gives a velocity of *ca.* 1.8 mm μs^{-1}. There is precursor attenuation of the HEL as observed previously.

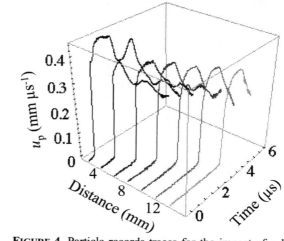

FIGURE 4. Particle records traces for the impact of a 10 mm alumina tile onto DEDF inducing 5.8 GPa.

DISCUSSION

Precursor Decay

The two shots above the HEL exhibit precursor decay which has been measured with stress gauges for this glass [1]. The decay distance is *ca.* 9 mm and the final level achieved is 0.20±0.05 mm μs^{-1}. This agrees precisely with 3.8 GPa measured previously (though the elastic impedance) [1].

Stress Levels Achieved

There are two levels reached at the first two sensors in each experiment (up to 3 mm into tile). The first corresponds to particle velocity assuming elastic behaviour and the peak of the five later sensors are at this level. The second, higher plateau corresponds with that calculated assuming the Hugoniot measured using 12 mm tiles. The glass is taken to this state by the passage of a front travelling at velocities consistent with those recorded for failure fronts, asymptoting to the Rayleigh wave speed at higher stresses. Thus this state is induced by cracks propagating from the *impact face*. Longitudinal release prevents fracture of material affecting the pulse at distances greater than 5 mm.

Origin of the interfacial failure

A series of shots repeated with a 30° instead of the 90° cut, show the *same* values of the final pulse level. The two histories on the impact face however show particle velocity *reduced* below this level rather than *increased* in the 90° case. This is consistent with the cut itself initiating the failure. An example sequence is shown in Fig. 5.

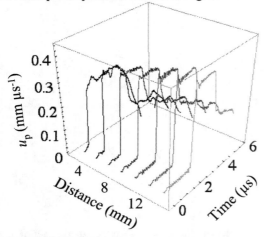

FIGURE 5 Particle records traces for the impact of a 10 mm alumina tile onto DEDF inducing 5.8 GPa.

Other work has measured the lateral and longitudinal stresses, as well as longitudinal strains [1,5,8]. The difference in the stresses as a function of time is twice the shear strength. Using this value connects the measured strength at the HEL (1.9 GPa) to the longitudinal stress via the Griffith's yield criterion. Substitution into the relation

$$\sigma_x^{HEL} = (1-\nu)/(1-2\nu)^2 \, 2\tau^{HEL}, \tag{1}$$

gives a value for *onset* of failure to be 5.6 GPa. This value represents the stress required to fail the material *immediately* at the impact face. This corresponds to a particle velocity of 0.31 mm μs^{-1}. This then defines a level at which failure can occur directly at the shock front immediately. However, over a longer time other flaws may fail the material driven primarily from the interface where their concentration is highest. The speed at which fracture originating at these larger flaws can travel is limited by the Rayleigh wave speed as observed previously [10].

The elastic-inelastic threshold, and the fracture phenomena noted above are *not* related. One is a material response to the passage of the shock and has a threshold stress of 6 GPa. The second is driven from the interface and has a threshold of 4 GPa. The latter effect results in the creation of lower impedance material that reduces the stress at the front behind which it travels.

CONCLUSIONS

A lead filled glass has been chosen for study since it is fully dense, has been studied extensively previously, and has been shown to show the various failure phenomena associated with brittle materials. This work builds on previous studies by introducing particle velocity sensors into the glass to both gain further flow information, but also to investigate the effect of the cut in the target produced to introduce the gauge. The cut is shown to have the effect of a controlled flaw and sends the material to its failed state. Above the HEL, the material also starts to fail at the shock front and the threshold for this behaviour corresponds to the Griffith's criteria for brittle fracture. This work has demonstrated that there are two distinct failure mechanisms, one originating from the bulk and the other from interfaces.

REFERENCES

1. Bourne, N. K., Millett, J. C. F., and Rosenberg, Z., *Proc. R. Soc. Lond. A* **452**, 1945-1951 (1996).
2. Rasorenov, S. V., Kanel, G. I., Fortov, V. E., and Abasehov, M. M., *High Press. Res.* **6**, 225-232 (1991).
3. Brar, N. S., Rosenberg, Z., and Bless, S. J., *J. Phys. IV France Colloq. C3 (DYMAT 91)* **1**, 639-644 (1991).
4. Brar, N. S. and Bless, S. J., *High Press. Res.* **10**, 773-784 (1992).
5. Bourne, N. K., Millett, J. C. F., and Field, J. E., *Proc. R. Soc. Lond. A* **455**, 1275-1282 (1999).
6. Bourne, N. K. and Millett, J. C. F., *J. Phys. IV France* **10**, 281-286 (2000).
7. Alcon, R. R., Sheffield, S. A., Martinez, A. R., and Gustavsen, R. L., in *Shock Compression of Condensed Matter 1997*, edited by S. C. Schmidt, D. P. Dandekar, and J. W. Forbes (American Institute of Physics, Woodbury, New York, 1998), p. 845-848.
8. Millett, J. C. F., Bourne, N. K., and Rosenberg, Z., *J. Appl. Phys.* **87**, 8457-8460 (2000).
9. Rosenberg, Z., Bourne, N. K., and Millett, J. C. F., *J. Appl. Phys.* **79**, 3971-3974 (1996).
10. Bourne, N. K., Millett, J. C. F., Rosenberg, Z., and Murray, N. H., *J. Mech. Phys. Solids* **46**, 1887-1908 (1998).

CP706, *Shock Compression of Condensed Matter - 2003*
edited by M. D. Furnish, Y. M. Gupta, and J. W. Forbes
© 2004 American Institute of Physics 0-7354-0181-0/04/$22.00

PROPAGATION OF WAVES THROUGH A FLOAT GLASS LAMINATE

N.K. Bourne, Z. Rosenberg*, J.C.F. Millett

Royal Military College of Science, Cranfield University, Shrivenham, Swindon, SN6 8LA, UK.
* RAFAEL, PO Box 2250, Haifa, Israel.

Abstract. Recent work has shown that failure of plates bonded together during impact has shaped the pulse transmitted through materials [Kanel, G. I., *et al.*, *J. Appl. Phys.* **92**, 5045-5052 (2002)]. Some of the features of their recorded pulse were noted by using particle interferometers to track the motion of interfaces. In this work high-speed photography has been used to monitor interfaces and multiple, embedded stress gauges were used to record the evolving stress level as it tracked through the laminate. Interpretation of the data was compared with the particle velocity work and related to the failure phenomena that occur at the interfaces.

INTRODUCTION

The response of glasses to shock has been much studied over the past decades [1-9]. Glasses may vary from the more open-structure materials such as fused silica and borosilicate, through the partially filled soda-lime (SL) to the highly filled lead glasses. The most commonly studied material is the partially filled soda-lime glass and it is this that is used here. One remarkable property, obtained using lateral stress gauges to measure the strengths of glasses ahead and behind the failure front (but behind the shock), is that these strengths lie on a common curve for glasses of varying density and microstructure [10].

Rasorenov *et al.* [8] observed the phenomenon of a delayed failure behind the elastic wave in glass, across a front which has been called a fracture, or more lately a failure wave. They noted that a front appeared to travel at velocities in the 1.5–2.5 km s^{-1} range in the K19 (similar to soda-lime) glass. They inferred its presence by observing a small recompression signal caused by reflection of the elastic release off the failure front in their record of the free-surface velocity, allowing calculation of velocity and indicating that material behind the front was of lower impedance than that ahead of it.

Later work confirmed these observations, and has shown that finite spall strength ahead, diminishes to zero behind the front, and that a substantial decrease in shear strength results [4,11].

Shocked silicate glasses have been recovered, and refractive index changes were noted [6]. There was found to be no permanent refractive index change up to 4 GPa, but beyond this point it increased up to a second threshold at 8 GPa.

In a previous paper, the effect of internal interfaces in soda-lime laminates was studied by measuring components of the stress field with simultaneous high-speed photography [12]. Previous work had identified the stress thresholds for onset of the failure at the lower extreme, and failure occurring within the shock front at the higher [13]. If the internal surface of the blocks is highly polished, there is a delay before more failure occurs from the interface. The front then propagates from the boundary at a constant velocity. Roughening the surface reduces the delay before the failure propagates. The sensors indicate that the second plate of the laminate behaves as if it is stronger than the first. Thus there are two effects of an internal interface; delayed failure and increased strength in the second tile.

Recently Kanel *et al.* [14] have published work in which they study the effects of shock propagation through an array of thin sheets of glass. They embedded stress gauges and collected particle velocity histories from several locations within an array at differing stress levels. They suggested that the failure is a process of crack propagation, and define conditions on several of the flow parameters for this (observed in 15).

The following work extends the approach of gauge measurements to further define the response of the laminate at three stress levels. The first is below the threshold for the loss of strength (3 GPa), the second is at *ca.* 5 GPa and the last is *ca.* 7 GPa.

EXPERIMENTAL PROCEDURE

Laminates were formed from 1.1 mm thick SL plates such that both faces of each target were lapped and polished to a defined surface finish. Manganin gauges (MicroMeasurements type LM-SS-125CH-048) were embedded between the faces of the tiles to measure the longitudinal component of the stress. After these were prepared, they were bonded together with a low viscosity epoxy to form a single assemblage containing gauges now placed at various positions in the direction of shock propagation. The impact surface of the blocks were lapped with a 25 μm alumina paste (removing at least seven times the average surface flaw size) and then polished to an optical finish using jeweller's rouge. The impact and internal surfaces were ground back and polished so as to ensure a surface flaw distribution typical of the bulk. The tiles were then reassembled so that each contained gauges; at the first, third, fifth and seventh tile interfaces. The targets were backed with thick glass blocks to ensure that no release entered the target during the times of interest. The samples were impacted by 10 mm thick, Al6082-T6 or copper flyer plates inducing longitudinal stresses of *ca.* 3, 5 and 7 GPa respectively. These stresses were chosen to be below and around previously determined thresholds for each glass.

In another set of experiments, high-speed photography was used to record the damage occurring across an internal interface during impact. The frames had exposure times of 50 ns, whilst the interframe time was 200 ns. The sequences were back-lit using a beam collimated before it passed through the target, so that light refracted from the shock region was not collected in the camera lens.

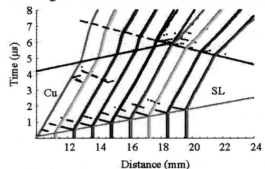

FIGURE 1. Distance-time plot for a typical impact. The plates are each 1.1 mm thick and are bonded with thin (0.1 mm) layers of epoxy.

The impact introduces a shock wave, travelling forward into the glass target and a second travelling back into the flyer plate (see Fig. 1). This shock also induces a failure front, which travels into the glass target from the interface at a velocity slower, or at that of the shock wave according to the impact stress level. This failure may be reinitiated at each interface it meets within the target.

The glass tested had a density of 2.49 g cm^{-3}, a longitudinal wave speed of 5.84 mm μs^{-1}, and a shear wave speed of 3.46 mm μs^{-1}. Microstructural examination of the bulk indicated randomly distributed flaws of three types. Sub-micron bubbles of average separation *ca.* 10 μm, larger bubbles of radius of order 50 μm and conchoidal cracks of size *ca.* 50 μm. Both these larger flaws were separated by 5-10 mm.

Plate impact experiments were performed on 50 mm bore, 5 and 6 m long, single stage gas guns. Impact velocities were measured to an accuracy of 0.5% by the shorting of sequentially mounted pairs of pins, and the specimen was aligned to the flyer plate to better than 1 milliradian by an adjustable specimen mount. The signals were recorded using a fast (1 GS s^{-1}) digital storage oscilloscope and transferred onto a micro-computer for data reduction.

RESULTS AND DISCUSSION

Fig. 2 shows a streak record of the interaction of a shock with an interface, constructed from strips of a framing sequence. The shock is transmitted and a reflected component from the glue layer at the interface may be seen. The damage fronts run from the interface back into the first plate and forward into the second. The front starts to propagate almost immediately the shock wave interacts with the region since this surface is roughened.

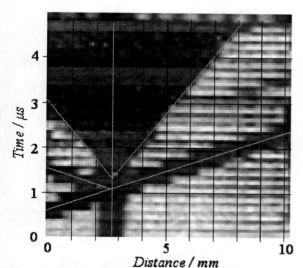

FIGURE 2. Constructed streak sequence showing the impact of a 10 mm thick copper flyer plate travelling at 503 m s⁻¹ onto a soda lime glass laminate.

Interpolating the velocity points gives a time of origin *ca.* 200 ns after the shock crosses the interface, but this is within error and the damage may initiate immediately it arrives. Cracks travel forward into the rear glass block and backwards into the impact one. The velocities of these two waves are different from one another with recorded speeds of 1.6±0.2 km s⁻¹ in the forward direction and 1.4±0.2 km s⁻¹ in the reverse

When a laminate containing many blocks is constructed, stress gauges may be embedded at each layer. Fig. 3 shows the results of impacts on aluminium and copper flyers of thickness 10 mm onto soda lime laminates. Gauges are placed at odd interfaces (i.e. the first, third, *etc.* that is 1.1 and 3.3 mm plus joint thickness *etc.* from the impact face). The gauge plus epoxy layer thickness is estimated to be 0.1 mm or less. Three stress levels were chosen, estimated to be at 3, 5, and 7 GPa calculated using the measured Hugoniot (where gauges were placed greater than 8 mm into a monolith).

It is instructive to separate out the sets of traces from each of the interfaces. Fig. 4 presents these. The time axis is expanded to more easily view the records. All the traces at the first level rise to the expected stress assuming elastic behaviour in the glass. The top of the trace is flat with dips due to release interactions from the 0.1 mm glue layers.

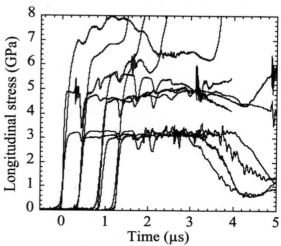

FIGURE 3. Longitudinal stress histories at each of four interfaces in a soda lime glass laminate (the first, third, fifth and seventh). The lower stress level is in the stress range before failure fronts are observed. The upper two are in this region. Note the rounded front that appears at the first location at the highest stress level.

At the second level, the gauge records show the rounded shape characteristic of that seen in polycrystalline materials. The expected elastic stress level is 5.45 GPa at the impact velocity recorded, which corresponds to that recorded by the first gauge. These stress level drops to *ca.* 5 GPa at sensors placed further into the target. The peak of the rounded envelope that surround the first records, arrives more rapidly at the highest stress level. The gauge does not survive at this stress for the entire record and is broken (level goes upward rapidly) not shorted. The release from the rear of the projectile comes into the flyer after 3.1 μs as expected for the lower stress. However, on the upper level, at the first gauge, a partial release occurs after 2 μs before this second release comes in as expected. It is believed that this release is not from the flyer, but could be the result of waves interacting from failed interfaces inside the target.

725

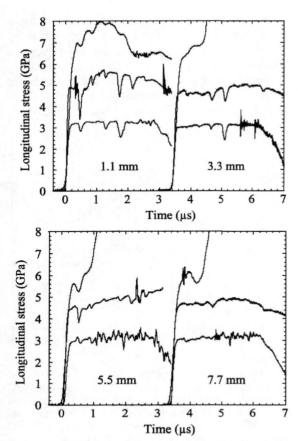

FIGURE 4. Expanded view of the gauge locations at 1.1, 3.3, 5.5 and 7.7 mm from the impact face.

CONCLUSIONS

Embedded stress gauges and high-speed photography have viewed the development of stress pulses transmitted through glass laminates. The following features have been noted.

(i)   The initial elastic level on the impact face attenuates with each glass layer to the Hugoniot value measured with thick monoliths.

(ii)   Levels below 4 GPa show no evidence of attenuation or pulse-rounding. Additionally, there is no observed stress reductions as one passes from sheet to sheet.

(iii)   Release wave speeds at the highest stress level appear to be as expected. However, partial release occurs ahead of that from the rear of the flyer which

is believed to result from interactions resulting from the failure of material at the interfaces.

(iv)   The envelope of the pulse is reminiscent of the wave shape from polycrystalline materials that this system mimics.

Further work will track other variables in the flow with a range of sensors.

REFERENCES

1.   Barker, L. M. and Hollenbach, R. E., *J. Appl. Phys.* **41**, 4208-4226 (1970).
2.   Bless, S. J., Brar, N. S., and Rosenberg, Z., in *Shock Waves in Condensed Matter 1987*, edited by S. C. Schmidt and N. C. Holmes (North Holland, Amsterdam, 1988), p. 309-312.
3.   Bourne, N. K., Millett, J. C. F., Rosenberg, Z., and Murray, N. H., *J. Mech. Phys. Solids* **46**, 1887-1908 (1998).
4.   Brar, N. S. and Bless, S. J., *High Press. Res.* **10**, 773-784 (1992).
5.   Dandekar, D. P., in *Shock Compression of Condensed Matter - 1997*, edited by S. C. Schmidt, D. P. Dandekar, and J. W. Forbes (American Institute of Physics, Woodbury, New York, 1998), p. 525-528.
6.   Gibbons, R. V. and Ahrens, T. J., *J. Geophys. Res.* **76**, 5489-5498 (1971).
7.   Kanel, G. I., Rasorenov, S. V., and Fortov, V. E., in *Shock Compression of Condensed Matter - 1991*, edited by S. C. Schmidt, R. D. Dick, J. W. Forbes, and D. G. Tasker (Elsevier, Amsterdam, 1992), p. 451-454.
8.   Rasorenov, S. V., Kanel, G. I., Fortov, V. E., and Abasehov, M. M., *High Press. Res.* **6**, 225-232 (1991).
9.   Rosenberg, Z., Yaziv, D., and Bless, S. J., *J. Appl. Phys.* **58**, 3249-3251 (1985).
10.   Bourne, N. K., Millett, J. C. F., and Field, J. E., *Proc. R. Soc. Lond. A* **455**, 1275-1282 (1999).
11.   Brar, N. S., Rosenberg, Z., and Bless, S. J., *J. Phys. IV France Colloq. C3 (DYMAT 91)* **1**, 639-644 (1991).
12.   Bourne, N. K. and Millett, J. C. F., *Proc. R. Soc. Lond. A.* **456**, 2673-2688 (2000).
13.   Bourne, N. K., Millett, J. C. F., Rosenberg, Z., and Murray, N. H., *J. Mech. Phys. Solids* **46**, 1887-1908 (1998).
14.   Kanel, G. I., Bogatch, A. A., Razorenov, S. V., and Chen, Z., *J. Appl. Phys.* **92**, 5045-5052 (2002).
15.   Bourne, N. K., Rosenberg, Z., and Field, J. E., *J. Appl. Phys.* **78**, 3736-3739 (1995).

CP706, *Shock Compression of Condensed Matter - 2003*
edited by M. D. Furnish, Y. M. Gupta, and J. W. Forbes
© 2004 American Institute of Physics 0-7354-0181-0/04/$22.00

CERAMIC BAR IMPACT EXPERIMENTS FOR IMPROVED MATERIAL MODEL

N. S. Brar[1], W. G. Proud[2], and A. M. Rajendran[3]

[1]Mechanical and Aerospace Engineering and Research Institute, University of Dayton, OH 45469-0182
[2]Cavendish Laboratory, University of Cambridge, Cambridge, UK
[3]US Army Research Office, Research Triangle Park, NC 27709

Abstract. Ceramic bar-on-bar (uniaxial stress) experiments are performed to extend uniaxial strain deformation states imposed in flyer plate impact experiments. A number of investigators engaged in modeling the bar-on-bar experiments have varying degrees of success in capturing the observed fracture modes in bars and correctly simulating the measured in-situ axial stress or free surface velocity histories. The difficulties encountered are related to uncertainties in understanding the dominant failure mechanisms as a function of different stress states imposed in bar impacts. Free surface velocity of the far end of the target AD998 bar were measured using a VISAR in a series of bar-on-bar impact experiments at nominal impact speeds of 100 m/s, 220 m/s, and 300 m/s. Velocity history data at an impact of 100 m/s show the material response as elastic. At higher impact velocities of 200 m/s and 300 m/s the velocity history data suggest an inelastic material response. A high-speed (Imacon) camera was employed to examine the fracture and failure of impactor and target bars. High speed photographs provide comprehensive data on geometry of damage and failure patterns as a function of time to check the validity of a particular constitutive material model for AD998 alumina used in numerical simulations of fracture and failure of the bars on impact.

INTRODUCTION

Bar impact has been widely used to study failure in brittle materials. Bar impact experiments are conducted by impacting a specimen bar, about 8-10 diameters long, either with a flyer plate or a bar impactor (bar-on-bar impact) of the same material or of a material of known Hugoniot. Strain rates produced in the bar target ($\sim 10^{3-4}$/s) bridge the gap between the strain rates achieved in split Hopkinson bar (10^2-10^3/s) and flyer plate impact tests (10^5-10^7/s). Bar impacts on ductile materials provide data on flow or failure stress at a strain rate of $\sim 10^4$/s [1-2]. Various investigators have conducted bar impact tests on brittle materials (e.g., rocks, concrete, glasses, ceramics) [3-8]. In the studies on ceramics, tests have been performed on both unconfined (bare) and laterally confined (sleeved with a ductile alloy) bars. Brar et al. performed impact tests on unconfined 12.7-mm diameter Coors 94% and 99.8% alumina bars to determine the amplitude of the propagating stress wave (in-situ stress) as a function of the impact stress generated by a steel flyer plate launched at 100-500 m/s [4]. They reported that a constant amplitude stress ($\sim$ static strength of the material) wave was produced when the impact stress was below or equal to the compressive yield stress of alumina. When the impact stress exceeded the yield stress, loading stress in the bar decayed with distance and time due to lateral release involving dilatancy.

Cosculluela et al. impacted 10-mm diameter T299 alumina (density=3.86 g/cm^3) bars with tungsten alloy impactor plates at velocities ranging form 80 to 700 m/s [5]. Lagrangian analysis of the velocity histories showed the highest stress amplitude wave that propagates in the alumina bar is 3.2 GPa, regardless of the impactor velocity. Wise and Grady performed impact tests on unconfined and confined (in close-fitting tantalum sleeve) Coors 99.5% alumina bars with aluminum flyer plates launched at velocities from 1035 to 2182 m/s [6]. They measured a maximum in-situ axial stress in unconfined bars of 3.15 GPa, irrespective of the impactor velocity, as observed by Cosculluela et al. [5] on similar alumina bars. Maximum in-situ axial stress in confined alumina bars is about twice (6.1 GPa) the value for the unconfined bars. This increase in measured axial stress was interpreted in terms of confinement, consistent with an expected upper dynamic limit equal to the HEL of alumina (6.2 GPa). Simha also reported a similar increase in measured in-situ axial stress (4.2 GPa) in 12.7-mm diameter confined (with a steel sleeve) Coors 99.5% alumina bars (6 diameters long) compared to that (3.7 GPa) in unconfined bars [7].

Chhabildas et al extended Wise and Grady's study to determine differences, if any, in measured in-situ axial stress in unconfined and sleeved (in shrink fit steel cylinder) 99.5% alumina bars impacted by a graded density flyer plate versus a single density (steel) impactor plate [8]. The measured in-situ axial stress in unconfined bars was slightly higher (3.5 GPa), when impacted with a graded density flyer plate compared to that (3.4 GPa) with a single density (steel) impactor. In sleeved bar targets, measured in-situ axial stress was in the range of 4.6 to 5.1 GPa, significantly higher than in the case of unconfined bars, in agreement with earlier studies.

Cosculluela et al. simulated free surface velocity histories at 8, 10, and 12 diameters away from the impact face of alumina (T299) bars impacted with steel plate impactors [5]. Simulated peak axial stress, 3.9 GPa, was greater than the measured value of 3.2 GPa. Espinosa and Brar simulated the alumina bar impacts using the multi-plane microcracking model [9]. Simha simulated velocity histories in 99.5% alumina bar impact experiments on unconfined and confined (sleeved) bars by Chabbildas et al. [8] These simulations matched the measured peak velocity only, missing all the detailed features, including the timings in initial rise of the profile. Grove and Rajendran developed R-G model for 99.5% alumina from flyer plate data and used model parameters to simulate the free-surface velocity profiles in bar impact experiments given in Reference 8 [9]. The simulated velocity histories in bar impact test configurations agreed with the measurements only for the steel sleeved bars. For unconfined bar targets simulated velocity histories were significantly lower than the measured data.

The objective of the present alumina bar-on-bar experiments is to generate velocity history data at impact velocities from 100-300 m/s using VISAR. These data are intended to develop improved material model and verify it through simulation of ballistic impact on alumina.

EXPERIMENTAL PROCEDURE

Schematic of the ceramic bar on bar experiment is shown in Figure 1. The experiments were performed at the 50-mm gas gun facility of the University of Cambridge on 12.7-mm diameter Coors AD-998 alumina bars at nominal impact speeds of 100 m/s, 200m/s, and 300m/s. Shot matrix is summarized in Table 1. Free surface velocity of the far end of the target bar was measured using VISAR. Fracture and failure modes of impactor and target bars are photographed using a high-speed (Imacon) camera (10^5 frames/s). Data on measured free surface velocity and high speed photographs are presented in this paper.

RESULTS AND DISCUSSION

Measured velocity history data for the six shots are shown in Figures 2-4 in three pairs: (i) shots 021106a and 0211106b at impact velocities of 112.2 m/s and 112.5 m/s, (ii) shots 021025b and 021028c at impact velocities of 220.8 m/s and 222.8 m/s, and (iii) shots 021022a and 021025a at impact velocities of 295.7 m/s and 298.6 m/s. Free surface velocities in the two shots at 112.2 m/s and 112.5 m/s agree within about 7%. Similar agreement between the free surface velocities for two shots at ~220 m/s and two at ~ 300 m/s is observed.

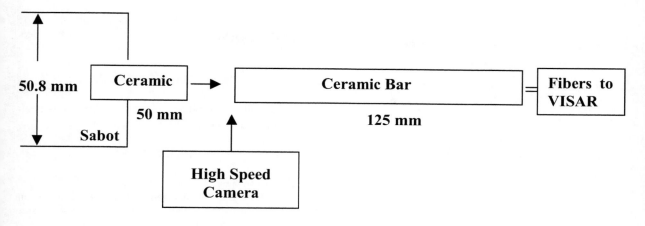

FIGURE 1.   Schematic of the ceramic bar-on-bar impact experiment

TABLE 1. Summary of AD998 Alumina Bar-on-Bar Impact Shots

Shot No	Shot ID	Proj. Mass (g)	Vel. (m s^{-1})	Inter-frame time (µs)
1	021022a	275	295.7	5
2	021025a	275	298.6	5
3	021025b	1006	220.8	5
4	021028c	1006	222.3	20
5	021106a	1006	112.2	20
6	021106b	1006	112.5	20

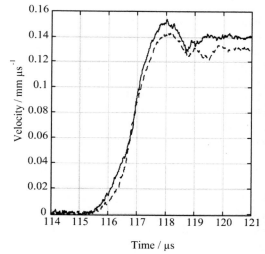

FIGURE 3. Measured free surface velocity histories at impact velocities of 220 m/s.

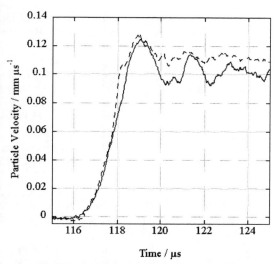

FIGURE 2. Measured free surface velocity histories at impact velocities of 112 m/s.

CONCLUSIONS

Velocity history data from ceramic bar-on-bar impact experiments obtained using VISAR presented in this paper are the first of its kind. In most of the earlier ceramic bar-on-bar impact experiments axial stress was measured with embedded manganin stress gauges. These stress-time data depended on the survival of the stress

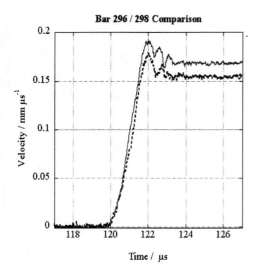

FIGURE 4. Measured velocity histories for impact velocities of 296 m/s and 298 m/s.

gauges and in many instances gauges failed prematurely resulting in incomplete data from the impact event. In cases where VISAR was employed to record velocity history of the far end of the target bar, impact stress was introduced using a flyer plate rather than a bar to suppress axial splitting during the impact event.

Measured velocity history data (particle velocity vs time) from the two shots at the lowest impact velocity of 112 m/s (Figure 2) agree within 7%. Average peak particle velocities of 105 m/s and 112 m/s from the two shots are approximately equal to the impact velocity, exhibiting an elastic response of AD-998 alumina. Measured particle velocities in the other four shots, two at ~220 m/s and the other two at ~295 m/s show a variation of ~10% between each pair of shots. These data further show that measured peak particle velocities are ~60% of the impact velocity, suggesting an inelastic response of AD-998. These observations are in agreement with earlier studies on the failure of impacted alumina bars through axial splitting.

We will be extending the present study to perform the bar impact experiments on metal sleeved (confined) bar targets with varying levels of confinement stress. The confinement stress would delay and inhibit the premature failure of the bars through axial splitting. Velocity history data from these experiments will provide ceramic bar impact data under one more impact geometry to the researchers engaged in generating material models for armor ceramics.

ACKNOWLEDGEMENTS

Funding was provided by the U.S. Army Research Office, under Grant DAAD19-01-1-0791.

REFERENCES

1. Rosenberg, Z., Mayseless, M., and Partom, Y. ," The use of manganin stress transducers in impulsively loaded long rod experiments," Trans. ASME, 51, pp. 202-4, 1984.
2. Rosenberg Z. and Bless, S. J., "Determination of dynamic yield strengths with embedded manganin gauges in plate-impact and long-rod experiments," Exp. Mech., 26, pp. 279-82, 1986.
3. Brar, N. S., Bless, S. J., and Rosenberg, Z., "Brittle failure of ceramic rods under dynamic compression," DYMAT 88, J. de Physique, 49, Suppl. 9, pp. C-3-607-12, 1988.
4. Cosculluela, A., Cagnoux, J., and Collombet, F., "Two types of experiments for studying uniaxial dynamic compression of alumina," Shock Compression of Condensed Matter-1991, Eds. S. C. Schmidt et al., Elsevier Science Publishers, pp. 951-54, 1992.
5. Wise, J. L. and Grady, D. E., "Dynamic multiaxial impact response of ceramic rods," High Pressure Science and Technology-1993, Eds. S. C. Schmidt et al., AIP Conference Proceedings 309, New York, pp. 777-80, 1992.
6. Chhabildas, L.C., Furnish, M.D., Reinhart, W.D., and Grady, D. E.," Impact of AD995 Alumina Rods," Shock Compression of Condensed Matter-1997, Eds. S. C. Schmidt et al., AIP Conference Proceedings 429, New York, pp. 505-08, 1998.
7. Espinosa, H.D. and Brar, N.S.,"Dynamic Failure Mechanisms of Ceramic Bars: Experiments and Numerical Simulations," J. Mech. Phys. Solids, 43, pp. 1615-38, 1995.
8. Simha, C. M., Bless, S. J., and Bedford, A.," Computational modeling of the penetration response of a high-purity ceramic,"Int. J. of Impact Eng., 27, pp. 65-86, 2002.
9. Grove, D.J. and Rajendran, A.M,"Modeling of microcrack density based damage evolution in ceramic rods, Shock Compression of Condensed Matter-1999, Eds. M. D. Furnish et al., AIP Conference Proceedings 505, pp. 619-22, 2000.

CP706, *Shock Compression of Condensed Matter - 2003*
edited by M. D. Furnish, Y. M. Gupta, and J. W. Forbes
2004 American Institute of Physics 0-7354-0181-0/04/$22.00

A RE-EXAMINATION OF TWO-STEP LATERAL STRESS HISTORY IN SILICON CARBIDE

Dattatraya P. Dandekar

U. S. Army Research Laboratory, Aberdeen Proving Ground, MD 21005

Abstract. The observed two-step lateral stress history in silicon carbide, SiC-B under plane shock wave propagation [N. K. Bourne, J. Millett, and I. Pickup, J. Appl. Phys. **81**, 6019 (1997)] is attributed to a delayed failure in SiC-B due to propagation of a slow moving front traveling behind the main shock wave. According to this attribution, the first lower magnitude, step corresponds to the lateral stress in intact shock compressed silicon carbide as a result of the fast moving plane shock wave. The second step of higher magnitude, observed after a few hundred nanoseconds, corresponds to the lateral stress in failed silicon carbide due to propagation of the slower moving front. The current analysis, takes into account additional relevant existing results dealing with shock response of SiC-B, and shows that the suggested explanation for the observed phenomenon remains in doubt.

INTRODUCTION

The growing body of investigations pertaining to impact-induced response of ceramics permits revisiting the explanation of the two-step structure observed in lateral stress history of silicon carbide under plane shock wave propagation reported by Bourne et al. [1] and Murray and Field [2]. Bourne et al. attribute the observed two-step lateral stress history in silicon carbide to a delayed failure in shocked silicon carbide due to propagation of a slow moving front traveling behind the main shock wave. Thus, they suggest that the first step of lower magnitude corresponds to the lateral stress developed in silicon carbide due to fast moving plane shock wave in intact material. The second step of higher magnitude, observed after a few hundred nanoseconds, corresponds to the lateral stress developed in failed silicon carbide due to propagation of the slower moving front behind the fast moving plane shock wave.

Silicon carbide, in particular, SiC-B manufactured by Cercom Inc., has been the subject of a large number of investigations.

Grady and Crawford [3], Feng et al. [4], Sekine and Kobayashi [5] investigated shock hugoniot of SiC-B. Dandekar [6] reported a review of hydrostatic and hydrodynamic compression of SiC-B. Bourne et al. [1], Feng et al. [7], and Murray and Field [2] determined shear strength of SiC-B through the lateral stress measurements. Bourne et al., and Murray and Field provide values of measured longitudinal and lateral stresses only. They do not report the concomitant volumetric compression for the measured longitudinal stress. The values of Hugoniot Elastic Limit (HEL) reported by the above investigators range from 11.7 to 18 GPa. Yuan et al. [8] measured longitudinal and shear wave velocities in shocked SiC-B to a peak stress value of 18 GPa to verify the reported values of shear strength from the lateral stress measurements reported by Feng et al. [7]. Dandekar [9] reported spall strength of five different varieties of silicon carbide including SiC-B. Feng et al. [7] did not observe a two-step structure in the lateral stress history in their experiments. Taking together all relevant existing results dealing with shock response of

Table I. Values of Measured Longitudinal Stress and Two Lateral Stresses [σ(Y1) and σ(Y2)] Calculated Shear Stresses [τ(1) and τ(2)] and Mean Stresses [<σ(1)> and <σ(2)>] and Strain in SiC-B. All Stresses are in GPa.

σ	σ(Y1)	σ(Y2)	τ(1)	τ(2)	<σ(1)>	<σ(2)>	μ(σ)
			Bourne et al. [1]				
16.7	3.3	3.3	6.7	6.7	7.77	7.77	0.0372
21.2	4.1	5.3	8.55	7.95	9.80	10.60	0.0506
23.4	6.7	7.3	8.35	8.05	12.27	12.67	0.0576
			Murray and Field [2]				
5.40	0.40	1.40	2.50	2.00	2.07	2.73	0.0114*
12.30	1.10	3.40	5.60	4.45	4.83	6.37	0.0254
16.50	3.30	3.30	6.60	6.60	7.70	7.70	0.0366
17.90	4.10	5.50	6.90	6.20	8.70	9.63	0.0406
20.90	6.10	7.50	7.40	6.70	11.03	11.97	0.0496
22.00	6.70	7.30	7.65	7.35	11.80	12.20	0.0531
34.00	17.20	18.24	8.40	7.88	22.80	23.49	0.0957

* This value is calculated using elastic deformation of SiC-B. The calculated value from Eq.1 is 0.0093.

SiC-B, this communication explores whether the observed two-step lateral stress history in SiC-B can be satisfactorily explained by the hypothesis suggested by Bourne et al. [1].

ANALYSIS

The present analysis requires longitudinal stress (σ) versus strain data to re-examine the lateral stress data. The strain (μ) in this work is defined as (ρ/ρ_0-1), where ρ is the mass density, and subscript 0 designates the ambient value. Bourne et al. [1] and Murray and Field [2] provide only the longitudinal stress data. However, Murray and Field show that the longitudinal stress versus particle velocity are consistent with the corresponding data reported by Grady and Crawford [3], and Feng et al. [4] Therefore, in the present analysis, we assume that strains for the measured longitudinal stresses reported in Ref.1 and 2 can be consistently calculated from the least square fit to SiC-B data given in Ref. 3 and 4. The maximum longitudinal stress generated in SiC-B was 40 GPa in Ref. 3 and 4. The maximum longitudinal stress attained in Ref. 1 and 2 are 23 and 34 GPa, respectively.

The least squares fit to longitudinal stress and strain data reported in Ref. 3 and 4 yield the following relation with a correlation coefficient

of 0.999. Stress is in the unit of GPa.

$$\mu = -0.0012 + 0.0018\sigma + 3.183 \times 10^{-5}\sigma^2. \quad (1)$$

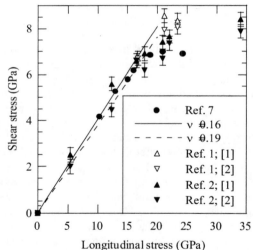

Figure 1. Shear stress versus longitudinal stress in SiC-B.

Table I gives the values of measured longitudinal and two lateral stresses [σ(Y1) and σ(Y2)], calculated pairs of shear [τ(1) and τ(2)] and mean stresses [<σ(1)> and <σ(2)>] corresponding to the two lateral stress values, and strain in SiC-B for the experiments reported in Ref. 1 and 2. Shear (τ), and mean $\langle\sigma\rangle$ stresses are given by $0.5\ [\sigma-\sigma\ (Y)]$, and $(\sigma +2\ \sigma\ (Y))/3$ respectively.

Fig. 1 shows shear stress versus longitudinal stress in SiC-B obtained by Feng et al. [7], Bourne et al. [1], and Murray and Field [2] In this figure, [1] and [2] refer to shear stress values obtained from the first (lower) and the second (higher) values of lateral stresses, corresponding to two-step structure in the lateral stress profiles observed in Ref. 1 and 2. Further, the effect of change in Poisson's ratio due to shock compression reported by Yuan et al. [8] from their compression and shear wave velocity measurements are indicated in this figure. They reported that Poisson's ratio (ν) of SiC-B increases with elastic compression from its ambient value of 0.16 to 0.192 at the HEL. The two lines corresponding to these two values of Poisson's ratios imply a decrease in the shear stress sustained by SiC-B under elastic compression. The termination of these lines at 20 GPa is arbitrary. This figure shows that the shear stress reported by Feng et al. [7] lie close to the

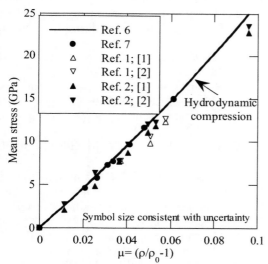

Figure 2. Mean stress versus μ for SiC-B from hugoniot and lateral stress measurements.

values indicated by the elastic compression of SiC-B to around 16 GPa. At higher values of longitudinal stress, the shear stress values

deviate from elastic loci and increase more slowly. Feng et al. [7] report the HEL of SiC-B to be 11.5 ± 0.4 GPa. Sekine and Kobayashi [5] report the value of the HEL of SiC-B to be 18 GPa. Bourne et al. report the HEL of SiC-B to be 15.7 ± 0.3 GPa. The values of first and second shear stress values reported by Murray and Field [2] lie close to the elastic loci to 16 GPa and at higher stresses these values increase slowly and deviate from the loci. The results of Bourne et al. [1] indicate that their two values of shear stresses continue to lie near and above the elastic loci to 21 GPa and deviate from it at longitudinal stress of 23 GPa.

Since the state of an intact or failed/damaged material under confinement is also characterized by its compression under confining stress, it is instructive to look at the compression of SiC-B in an undamaged state, represented by the first value of the lateral stress and the failed state represented by the presence of the second step, higher value of lateral stress reported in Ref. 1 and 2. Figure 2 shows compression of SiC-B represented in mean stress-strain plane. In this plot, hydrodynamic compression i.e., intact compression, of SiC-B reported by Dandekar [6] is based on the shock compression data of Sekine and Kobayashi [5]. Hydrodynamic compression of SiC-B is indistinguishable from the hydrostatic compression of SiC-B obtained from ultra sonic wave velocity measurements to 2 GPa. Dandekar [6] also showed that the mean stress-strain curve obtained by Yuan et al. [8] is indistinguishable from the hydrodynamic compression of SiC-B to around 60 GPa. The respective relations for hydrodynamic (P) given by Dandekar [6] and mean stress ($\langle\sigma\rangle$) compressions for SiC-B given by Feng et al. and Yuan et al. are,

$$P = 221\ \mu+398\ \mu^2 + 283\ \mu^3 \qquad (2)$$

$$\langle\sigma\rangle = 220.0\ \mu+361.3\ \mu^2 \qquad (3)$$

$$\langle\sigma\rangle = 1293\ \mu -1076.3\ \ln (1+\mu) \qquad (4)$$

The mean stress compression of SiC-B given by Feng et al. [7] differs slightly from those of Dandekar [6] and Yuan et al. [8]. For example, at μ=0.10 the values of pressure and mean stresses obtained from Eqs. (2) – (4) are 26.4, 25.6, and 26.7 GPa, respectively.

Fig. 2 shows that the mean stress-strain coordinates reported by Feng et al. [7] tend to lie on the hydrodynamic compression curve of SiC-B. However, the pair of mean stresses, computed from the first (lower) and second (higher) values in the two-step lateral stress profiles reported in Ref. 1 consistently lie below the intact/hydrodynamic compression of SiC-B (Fig.2). In other words, both of the mean stress-strain coordinates indicate SiC-B to be more compressible than it is under hydrodynamic compression. Further, the mean stress-strain coordinates attributed to intact SiC-B lie below the corresponding coordinates for the failed/damaged SiC-B indicating intact SiC-B to be more compressible than the damaged SiC-B. The mean stress-strain coordinates corresponding to the measured pair of lateral stresses reported in Ref. 2 show similar trend except that the mean stress-strain coordinates corresponding to the damaged state of SiC-B lie close to or on the hydrodynamic compression curve to 12 GPa. At 21 GPa, both mean stress-strain coordinates lie below the hydrodynamic compression curve of SiC-B. Thus, the characterization of the first level of lateral stress representing intact/undamaged SiC-B, and the subsequent second level of lateral stress observed after a few hundred nanoseconds, representing the failed/damaged state induced in SiC-B due to propagation of a failure front following the main shock wave in SiC-B as proposed by Bourne et al. [1] does not appear to hold.

CLOSURE

Recently, Murray et al. [10] have shown that the two step lateral stress profiles is not observed when either a ceramic buffer plate is mounted in front of the target containing lateral stress gages, or a ceramic-like flyer is used for impact. Thus, the presence of a two-step structure in lateral stress history may well be due to the experimental configuration. Further, Feng et al.

[11] suggested that the stress field can be disturbed by both stress concentration at the material joint and injection of soft flyer plate material into the epoxy layer used to bond a lateral stress gage in the target, which could lead to the observed two-step structure in a lateral stress profile. A third possibility is that in the absence of a ceramic buffer plate, it takes some time before both the shock induced longitudinal and lateral stress attain their respective final values. The observed trend in Fig. 2 tends to suggest this. But irrespective of the above, it appears that the explanation for the presence of a two-step structure in a lateral stress profile in SiC-B suggested by Bourne et al.[1] remains in doubt.

REFERENCES

1. N. J. Bourne, J. Millett, and I. Pickup, J. Appl. Phys. **81**, 6019 (1997).
2. N. H. Murray and J. E. Field, Cavendish Laboratory Report PCS/SP1053, University of Cambridge, Cambridge, UK (1999).
3. D. E Grady and D. A. Crawford, Sandia Technical Memorandum TMDG0593, Sandia national laboratories, Albuquerque, NM, 1993.
4. R. Feng, G. F. Raiser, and Y. M. Gupta, J. Appl. Phys. **79**, 1378 (1996).
5. T. sekine, and T. kobayashi, Phys. Rev. B **55**, 8034 (1997).
6. D. P Dandekar, Army Research Laboratory Technical Report, ARL-TR-2695 (2001).
7. R. Feng, G. F. Raiser, and Y. M. Gupta, J. Appl. Phys. **83**, 79 (1998).
8. G. Yuan, R. Feng, and Y. M. Gupta, J. Appl. Phys. **89**, 5372 (2001).
9. D. P. Dandekar and P. Bartkowski, in *Shock Wave and High-Strain-Rate Phenomena*, ed.K. P. Staudhammer, L. E. Murr, and M. A. Meyers (Elsevier, New York, 1991), p. 71.
10. N. H. Murray, J. C. F. Millett, W. G. Proud, and Z. Rosenberg, in *Shock compression of Condensed Matter-1999*, Ed. M. D. Furnish, L. C. Chhabildas, and R. S. Hixson, (American Institute of Physics, New York, 2000), p. 581.
11. R. Feng, Y. M. Gupta, and M. K. W. Wong, J. Appl. Phys. **82**, 2845 (1997).

CP706, *Shock Compression of Condensed Matter - 2003*
edited by M. D. Furnish, Y. M. Gupta, and J. W. Forbes
© 2004 American Institute of Physics 0-7354-0181-0/04/$22.00

IMPACT RESPONSE OF SINGLE CRYSTAL POTASSIUM CHLORIDE AT ELEVATED TEMPERATURES

R. Golkov, D. Kleiman and E. Zaretsky

*Department of Mechanical Engineering, Ben-Gurion University of the Negev,
P.O.B. 653, Beer-Sheva 84105, Israel*

Abstract. Two types of planar impact experiments were performed with [100]-oriented single crystals of potassium chloride (KCl) having initial temperatures ranged from 293 to 523 K. In the experiments of the first type the maximum impact strength did not exceed the pressure P_{tr} of B1-B2 transformation in KCl. In these experiments the aluminum buffers were placed between the impactor and the sample and the velocity of aluminum-KCl interface was measured by VISAR. The impact strength in the experiments of the second type (with copper buffers) was higher than P_{tr} and the VISAR was used for monitoring the free surface velocity of the KCl samples. Temperature dependencies of the longitudinal and the bulk sound velocities obtained in the experiments of the first type were used for the treatment of the results of the second-type experiments. The results show clearly the deceleration of the B1-B2 transformation kinetics and the increase of the KCl yield strength with temperature. Similar temperature-induced strengthening was observed recently in aluminum single crystals [G.I. Kanelet et al, *J. Appl. Phys.* 90, (2001) 136]. The value of the transformation entropy determined from the temperature dependence of P_{tr} is found to be close to zero.

INTRODUCTION

Kinetics of shock-induced B1 (NaCl) → B2 (CsCl) phase transition in potassium chloride (KCl) was studied in planar impact experiments using quartz gauge[1] and VISAR[2] diagnostic tools. Both the techniques reveal a two-stage character of the transformation. At the first, very early, stage the transformation process is extremely fast (characteristic time less than 1ns) and hardly-detectable. The kinetics of the second stage of the transformation was found slow enough to allow observing the decay of both the P1 (low-pressure phase) and P2 (high-pressure phase) waves with the propagation distance. The characteristic time (hundred ns) of this decay was found similar to the characteristic time of decay of the elastic precursor wave Pel[2]. The similarity leads us to the assumption that both the stress relaxation behind the elastic precursor and the transformation kinetics

are governed by the same dislocation-based mechanism. The main factor controlling the dislocation motion in pure crystals is the phonon drag.[3] Since the phonon drag grows with temperature the heating should enhance the strength of pure crystals. Such behavior was observed by Kanel et al.[4] in planar impact experiments with single crystals of pure aluminum.

The goal of the present work is to study the influence of the initial heating of the KCl single crystals on kinetics of the stress relaxation behind the elastic precursor wave and on the kinetics of B1→B2 phase transformation.

MATERIALS AND EXPERIMENTAL

The studied KCl samples were [100]-oriented single crystals of Graseby-Specac Ltd., UK 15-mm diameter. The longitudinal C_l and the transversal C_t

sound velocities measured prior to the impact experiments were found equal to C_l =4.53±0.01 km/s and C_t =1.83±0.021 km/s, respectively. The density of the samples was not measured and the initial density value was accepted equal to ρ_0 =1.989 g/cm³.

The resistive nichrome coil embedded into a 25-mm copper reel produced the initial heating of both the sample and the buffer. The impact velocity was measured with use of the sequentially shortened electrical pins. The parts of the experimental assembly were bonded by high temperature alumina-based cement.

Two types of planar impact experiments were performed with the crystals using the 58-mm gas gun. In the experiments of the first type (two shots) the maximum impact strength did not exceed the pressure P_{tr} of B1-B2 transformation in KCl. In these experiments the 3-mm aluminum buffers were placed between the 5-mm aluminum impactor and the sample and the velocity of aluminum-KCl interface was measured by VISAR. These experiments were performed with the KCl samples of 2-mm thickness having initial temperature 523±3 K. In the experiments of the second type (with 2-mm copper buffers) the impact strength was higher than P_{tr}: the 2- and 6-mm copper impactors were accelerated up to the velocity 570±10 m/s, and the VISAR was used for monitoring the free surface velocity of the gold sputtered (Scancoat 106) KCl samples. In these experiments the KCl samples of about 1-, 2-, and 3-mm thickness were tested at ambient (298 K) temperature and at elevated temperatures of 373 and 523 K. The temperature of the samples in all experiments with heating was controlled by two chromel-alumel thermocouples pressed into 1-mm pits at 1-mm depth from the sample surface.

EXPERIMENTAL RESULTS

The results of the impact experiments of the first type are shown in Fig. 1. The experiments utilize the departure of the refractive index of KCl from the Gladston-Dale law under compression and the contribution of this departure in the velocity value measured by VISAR.[5] Both the velocity profiles show the instant of the collision between the

aluminum buffer and the KCl sample (point I). Up to this instant the buffer surface is free and no intervention of KCl refractive index in the VISAR measurements takes place. Just after the instant I the apparent (VISAR-detected) interface velocity is higher than the real one. The arrival of the elastic wave at the sample interface, point Pel, results in the propagation of the elastic release wave through the sample and in some decrease of the apparent interface velocity. The arrival of the plastic wave (point P1) at the free surface results in further decrease of the apparent velocity. After point R (the arrival of the elastic release wave at the buffer-sample interface) the apparent velocity again starts to grow. It was shown by Johnes and Gupta[6] that the

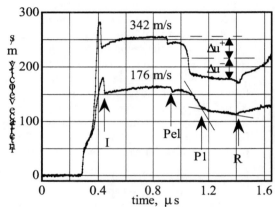

Figure 1. Velocity of the buffer-sample interface obtained in the first type experiments with impact velocities 176 and 342 m/s. The temperature of the KCl samples was 523±3 K.

apparent interface velocity excess Δu^+ caused by the presences of the compression wave in the window is very close to the apparent velocity decrease Δu^- caused by the presence of both the compression and the release waves in the sample. This allows us to find the value of the particle velocity u corresponding to the velocities $C_{el}(u)$ $C_{Pel}(u)$ and $C_{P1}(u)$ of propagation of the stress perturbation of finite amplitude.

The results of three (298, 373 and 523 K) series of the experiments of the second type are shown in Fig. 2a. Each of the recorded profiles is characterized by the three-wave structure containing elastic wave Pel,, plastic wave P1, compressing the low-pressure phase B1 up to the transformation pressure, and the wave P2, produced by the interaction between the reflection of the P1 wave from the free sample surface and oncoming phase transition wave.[7] Both Pel and P1 waves are decaying with the propagation distance (since the wave P2 may be distorted due to the arrival of the unloading wave from the sample lateral surface the P2 wave amplitudes do not considered here). As well, the velocity profiles show apparent increase of both Pel and P1 amplitudes with temperature, in particular, the heating of the KCl samples from 298 to 523 K results in some three-fold increase of the Pel amplitude, Fig. 2b (we consider the velocity values just behind the spike; the spikes amplitudes are hardly-reproducible).

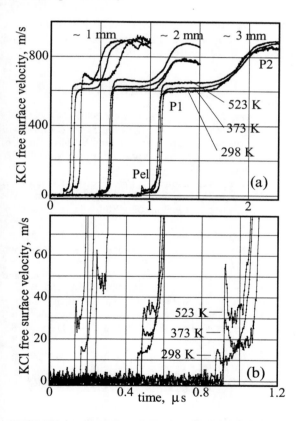

Figure 2. Free surface velocity profiles obtained after 570-m/s impacts with KCl samples of different thickness having different initial temperature (a) and elastic parts of the profiles (b).

DATA TREATMENT

The values of $C_{Pel}(573)$ and $C_{P1}(573)$ obtained in the first type experiments were used, together with the temperature dependencies of longitudinal and bulk speeds of sound,[8] for calculating the amplitudes σ_{HEL} and σ_{P1} of Pel and P1 waves, the specific volume V_{P1} behind the P1 wave, and the velocity C_{P2} of the propagation of the transformation wave P2 after 373K- and 523K- experiments.[2] The corresponding estimates for different impact temperatures are given in Tab. 1. Three leftmost columns contain the real transformation temperature calculated with use of the known Grüneisen's constant[8], and the maximum stress and temperature achieved in the phase P1 at the extension of the P1 Hugoniot after impact by heated copper buffer. The $\sigma_{HEL\infty}$ and $\sigma_{P1\infty}$ data attributed to the samples of infinite thickness (bold figures) are the parameters of the exponential best-fit of the amplitudes $\sigma_{HEL}-\sigma_{HEL\infty}$ and $\sigma_{P1}-\sigma_{P1\infty}$ decaying with propagation distance. The same fit yields the distances δ_{HEL} and δ_{P1} of the amplitudes logarithmic decay. Simplified linear kinetics accepted for the waves decay[9] allows one to estimate corresponding relaxation times $\tau_{HEL} = \delta_{HEL}/2C_{el}$ and $\tau_{P1} = \delta_{P1}/2C_{P1}$.

Assuming the invariability with temperature of the transformation-related change of the specific volume[10] $\Delta V=-4.25$ cm³mole⁻¹ and accepting $d\sigma_{P1\infty}/dT_{tr}$ as the temperature derivative dp_{tr}/dT_{tr} on the phase boundary yields the transformation entropy change $\Delta S=0.05\pm0.03$ Joule mole⁻¹ K⁻¹. This value is in satisfactory agreement with those suggested by Bassett,[10] $\Delta S=0.11\pm0.09$ Joule mole⁻¹ K⁻¹. The estimates of the characteristic times τ_{HEL} and τ_{P1} of the elastic and plastic waves decay are shown in Fig. 3 together with values of $\sigma_{HEL\infty}(T_{exp})$ and $\sigma_{P1\infty}(T_{exp})$. The strength of the KCl single crystal shows three-fold increase with temperature increasing from 300 to 523 K. As it was pointed by Kanel et al.[4] such strength behavior is expectable for pure single crystals under impact loading where a high-speed dislocation motion is controlled by scattering of the lattice phonons on the dislocation stress field. This "phonon drag" associates with the shear stress[11] $\tau = 0.1\omega_T v/C_t$ required for maintaining a constant dislocation velocity v in the

737

TABLE 1. Summary of the KCl shock data

T_{exp}, K-(shot)	C_{el}, km/s	C_{P1}, km/s	C_{P2}, km/s	σ_{HEL}, MPa	σ_{P1}, GPa	V_{P1}, cm^3/g	$V_0{}^a$, cm^3/g	T_{tr}, K	σ_{max}, GPa	T_{max}, K
298-1				65(3)	2.15(.02)	0.458(.001)				
298-2				39(2)	2.11(.02)	0.459(.001)				
298-3				32(2)	2.10(.02)	0.459(.001)				
298-∞	4.63(2)	3.48(5)	1.72(5)	**30(4)**	**2.09(.03)**	0.459(.002)	0.503	339(10)	3.92(.05)	364(10)
373-1				118(6)	2.20(.02)	0.461(.001)				
373-2				70(4)	2.14(.02)	0.462(.001)				
373-3				58(4)	2.12(.02)	0.462(.001)				
373-∞	4.51(2)	3.45(5)	1.75(5)	**53(6)**	**2.10(.03)**	0.462(.002)	0.507	426(15)	3.88(.05)	456(15)
523-1				170(9)	2.25(.02)	0.465(.001)				
523-2				130(5)	2.19(.02)	0.466(.001)				
523-3				110(5)	2.14(.02)	0.467(.001)				
523-∞	4.28(2)	3.39(5)	1.79(5)	**100(7)**	**2.10(.03)**	0.467(.002)	0.515	603(20)	3.80(.05)	643(20)

[a]The V_0 values uncertainty is less than 0.001. For other columns the uncertainty estimate is given in the parentheses.

material with phonon energy density ω_T. Accounting in that Debye temperature of KCl is about 230K one should expect that below this temperature the observed linear temperature dependence of KCl σ_{HEL} should be replaced by stronger one.

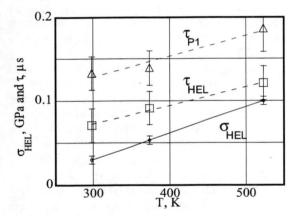

FIGURE 3. Temperature dependencies of the KCl σ_{HEL} strength and characteristic times of stress relaxation behind the Pel and P1 waves.

CONCLUSIONS

It is apparent from the same Fig. 3 that the temperature dependencies of relaxation times τ_{HEL} and τ_{P1} correlate with the crystal strength. The correlation makes it plausible to assume that the dislocation motion controls both the kinetics of the shear stress relaxation behind the elastic precursor front and the kinetics of the B1→B2 transformation. The latter case corresponds to the propagation of the dislocated boundary between old, B1, and new, B2, phases.

REFERENCES

1. Hayes, D. B., *J.Appl.Phys.*, **45**, 1208, 1974
   Ding, J.–L., and Hayes, D. B., in *Shock Compression of Condensed Matter-1999*, edited by M.D. Furnish, et al., AIP Conference Proceedings 505, Melwille, N-Y, 2000, p. 633
2. Zaretsky, E., in *Shock Compression of Condensed Matter-2001*, edited by M.D. Furnish, et al., AIP Conference Proceedings 620, Melwille, N-Y, 2002, p. 217
3. Kocks, U. S., Argon, A. S., and Ashby, M. F., Thermodynamics and kinetics of slip, Pergamon Press, Oxford, 1975, p. 45
4. Kanel, G. I., Razorenov, S. V., Baumung, K., and Singer, J., *J. Appl. Phys.* **90**, (2001) 136
5. Wackerle, J., Stacy, H. L., and Dallman, J. C., in High Speed Photography, Videography and Photonics V, Proc. SPIE Vol. 832,San-Diego, 1987, p.72
6. Jones, S. C., and Gupta, Y. M., *J.Appl.Phys.*, **88**, 5671, (2000).
7. Barker, L. M., and Hollenbach, R. E., *J.Appl.Phys.*, **45**, 4872, (1974).
8. Enck, F. D., *Phys. Rev*, **119**, 1873 (1960)
9. Duval, G.E. and Graham, R.A., *Rev. of Mod. Phys.*, **49**, 523 (1977)
10. Bassett, W. A., Takahashi, T., Mao, H.-K., and Weaver, J., S., *J.Appl.Phys.*, **39**, 319, (1968).
11. Leibfried, G., Z. fur Phys., **127**, 344 (1959)

CP706, *Shock Compression of Condensed Matter - 2003*
edited by M. D. Furnish, Y. M. Gupta, and J. W. Forbes
© 2004 American Institute of Physics 0-7354-0181-0/04/$22.00

A STUDY OF THE FAILURE WAVE PHENOMENON IN BRITTLE MATERIALS

G.I. Kanel [1], A.A. Bogach [2], S.V. Razorenov [2], A.S. Savinykh [2], Zhen Chen [3] and A. Rajendran [4]

[1] *Institute for High Energy Densities,IVTAN, Izhorskaya 13/19, Moscow, 125412 Russia*
[2] *Institute of Problems of Chemical Physics, Chernogolovka, Moscow region, 142432 Russia*
[3] *Department of Civil and Environmental Engineering, University of Missouri-Columbia, Columbia, Missouri, 65211-2200 U.S.A.*
[4] *U.S. Army Research Laboratory, ARO, RTP, NC 27709-2211*

Abstract. Shock-wave experiments with four glasses of different hardness, two ceramics, quartz and silicon single crystals have been carried out. Experiments with piles of thin sample plates confirm the appearance of the failure wave in elastically compressed fused quartz, K8 crown glass, and heavy flint glass, although the relationships between the Hugoniot elastic limits and the failure thresholds of these glasses are different. The failure waves were not recorded in quartz and silicon single crystals and polycrystalline alumina and boron carbide ceramics. The results show that the propagation speed of the failure wave in glass slightly depends on the stress above the failure threshold, and does not depend on the propagation distance. The process becomes unstable and stops at stresses near the failure threshold.

INTRODUCTION

The failure wave is a network of cracks that are nucleated on the surface and propagate with subsonic speed into the stressed body.[1] Our recent investigations[2] confirmed that the network of growing cracks in shock-compressed glass may indeed be considered as a wave which obeys the Rankine-Hugoniot conservation laws. It was shown that, when the failure wave is formed, shock compression of glass leads to a two-wave structure. Since the failure waves nucleate at each surface, the magnitude of the leading elastic wave in a pile of glass plates decreases as a result of its decomposition into two waves. The decrease of elastic wave amplitude repeats at each interface until the failure threshold is reached. As a result, for a sufficiently large number of plates in the pile, an elastic precursor wave having amplitude equal to the failure threshold is formed. Thus, the response of a layered assembly of thin brittle plates as compared to that of one thick plate is a simple way to diagnose nucleation of the failure process and determine the failure threshold.

The existing experimental data are not sufficient to say in which materials the failure wave may appear under shock compression and what are the threshold conditions. In this study we tested glasses of various hardness (soda lime glass, K8 crown glass, TF1 heavy flint glass, and fused quartz), brittle single crystals (quartz and silicon), and ceramics (alumina and boron carbide) in order to distinguish contributions of matter hardness, homogeneity, and surface state to nucleation of compressive fracture. The principal experiments were plane wave experiments with thick single sample plates and piles of plates of the same total thickness.

EXPERIMENTAL PROCEDURE

Since it was found earlier[2] that the shock-wave behavior of lapped glass plates is much more reproducible than that of as-received plates with mirror-like surfaces, the surfaces of sample plates were lapped. Plane shock waves were created by impacting the sample with a flyer plate which was launched by explosive facilities. The free-surface velocity profiles or the velocity histories of interface between the sample and water or PMMA window were recorded with the VISAR.

RESULTS AND DISCUSSION

A series of shock-wave experiments with soda lime glass plates of different thicknesses have been performed with a goal to measure the failure wave velocities at various stresses and propagation distances. The measured free surface velocity histories are presented in Fig. 1 where the time is normalized by the sample plate thickness. The wave profiles contain small recompression pulses which are due to the wave reflection from a failed layer inside the sample. The failure wave speed c_f was determined by means of measurement of the time interval t_r between arrivals of the initial compression wave and the recompression pulse fronts at the plate surface using the relationship

$$c_f = \frac{x}{t_x} = c_l \frac{2 - c_l t_r / \delta}{2 + c_l t_r / \delta} \qquad (1)$$

where δ is the glass plate thickness.

It follows from Eq. (1) that for constant speed of the failure wave the ratio t_r / δ should not depend on the plate thickness. The data in Fig. 1 show that the failure waves propagate with a constant speed,

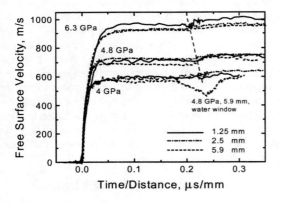

FIGURE 1. Free surface velocity histories of the soda lime glass plates of different thickness. One shot at 4.8 GPa and shorter load duration has been carried out with a water window.

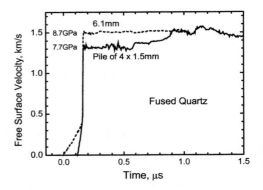

FIGURE 2. Results of experiments with thick plate and a pile of thin plates of fused quartz. Impact by aluminum flyer plate 4 mm in thickness with 1.5 km/s velocity.

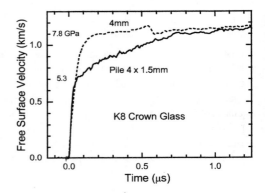

FIGURE 3. Free surface velocity histories of a thick plate and a pile of thin plates of K8 crown glass. Impact by aluminum flyer plate 7 mm in thickness with 1.15 km/s velocity.

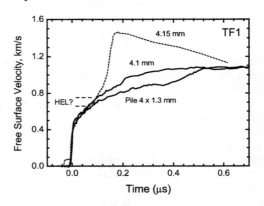

FIGURE 4. Free surface velocity histories of thick plates impacted to two different peak stresses and a pile of thin plates of TF1 heavy flint glass.

which depends on the stress. Using the average value of 5.3 km/s for the sound speed we could find the failure wave speed decreases from 1.58± 0.06 km/s at 6.3 GPa of compressive stress ahead of the failure front to 1.35±0.06 km/s at 4 GPa. The stress dependence of the failure wave speed explains its apparent deceleration that was found in the first observation[3] where the glass samples were loaded by decaying stress pulses. The process becomes unstable and obviously stops at stresses near the failure threshold.

Figures 2 to 4 present the results of experiments with thick plates and piles of thin plates of different glasses which demonstrate appearance of the failure waves in these materials. In the pile, the stress magnitude at elastic wave front should approach the failure threshold. Thus estimated failure thresholds are 7.7 GPa for fused quartz (Hugoniot elastic limit is 8.7 GPa) and 5.3 GPa for K8 crown glass (HEL ≈ 8 GPa). The failure threshold for soda lime glass[2] is 4 GPa (HEL ≈ 8 GPa). In other words, there are very different relationships between the failure threshold and the Hugoniot elastic limit for soda lime glass, fused quartz, and K8 crown glass. In the case of softest TF1 glass, both the failure threshold and the HEL are not so certain. It looks like the failure threshold is very close to the HEL value in this case.

Figure 5 presents results of 2 experiments with thick plates and 2 shots with piles of thin plates of alumina. The plates in the piles had surface roughness of order of a few micrometers that created thin gaps between them. As a result, the total time of propagation of the elastic precursor front through the pile is a sum of the time of wave propagation through the plates and the time of closing these gaps. The gaps become closed ahead the "plastic" compressive wave so the velocity of the later in a pile is the same as in a single plate. Thus, the delay in arrival of the elastic precursor front decreases the time interval between the elastic precursor front and the "plastic" compressive wave in the free surface velocity history. Unlike similar experiments with glasses, no evidence of the failure wave has been recorded. The experimental data demonstrate the increase, instead of the decrease, of the Hugoniot elastic limit in the pile as compared with the single plate. Probably this is an evidence of the precursor decay in alumina.

Figure 6 presents results of two shots with boron carbide plates of 5 mm and 6 mm in thickness, and one shot with a pile of 4 plates of 6.2 mm total thickness. Like in the experiments by Kipp and Grady,[4] the waveforms are oscillated. Whereas there are many reasons to expect certainly brittle behavior of B_4C at shock compression, the wave profiles don't show any evidence of appearance of the failure wave in this material. The amplitude of elastic precursor wave even increased in the pile as compared to the single plates, instead of expected decrease.

Figure 7 presents the results of two experiments with x-cut quartz plates 6 mm thick and two shots with piles of four quartz plates 1.5 mm thick each.

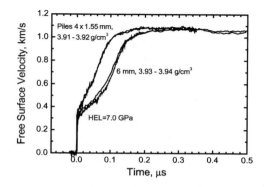

FIGURE 5. Free surface velocity histories of alumina samples composed of one plate 6 mm in thickness and 4 plates 1.55 mm thick each. Impact by aluminum flyer plate 2 mm in thickness at 1.9 km/s velocity.

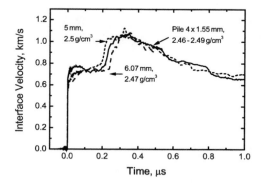

FIGURE 6. The interface velocity histories of boron carbide samples 5 mm and 6 mm in thickness and a pile of four B_4C plates of 1.55 mm thick each. Experiments with impact by aluminum flyer plate 2 mm in thickness at 1.9 km/s velocity using PMMA windows.

The waveforms are strongly oscillated, but one can certainly see that, again, the Hugoniot elastic limit is reproducibly higher for piles than that for single plates: the HEL is 5.3 GPa for 6-mm plates and 5.8 GPa for the piles of the same total thickness. No evidences of the failure wave phenomenon in quartz single crystals are seen in the wave profiles.

Figure 8 presents the results of four shots with silicon crystals. The velocity profiles for single plates distinctly show spike-like elastic precursor waves. The stress at elastic spike is 7.7 to 9.5 GPa; behind the elastic precursor front the stress drops down to 3.5–4.3 GPa. When the thick samples have been replaced by assemblies of two thinner silicon

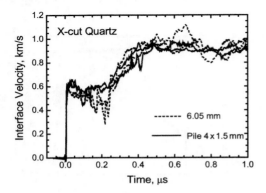

FIGURE 7. The interface velocity histories of thick x-cut quartz plates and piles of thin plates of the same total thickness. Measurements at interface between the sample and a water window. Impact by aluminum flyer plates with the velocity of 1.15 km/s.

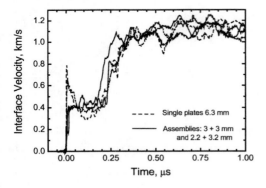

FIGURE 8. The interface velocity histories of thick silicon plates and assemblies of two thin silicon plates. Measurements at interface between the sample and a water window. Impact by aluminum flyer plates of 4 mm in thickness with the velocity of 1.5 km/s.

plates the shape of elastic precursor wave changed: instead of the spike the wave profiles exhibit almost rectangular precursor with 4.8 GPa stress at its front. By analogy with the glass piles, we could suppose that the observed decrease of HEL value in the assemble of plates is an evidence of the failure wave phenomenon in silicon crystals. However, it looks more possible that the effect is again just the result of delay in porous interlayer between the lapped plates.

CONCLUSION

Experiments with piles of thin plates confirm the appearance of the failure wave in elastically compressed glasses, although the relationships between the Hugoniot elastic limits and the failure thresholds are different for different kinds of glasses. For softest heavy flint glass, the failure threshold is closest to the HEL. The failure waves were not recorded in quartz and silicon, as well as in alumina and boron carbide ceramics. It is reasonable to conclude that failure waves are more likely to occur in materials which contain a higher concentration of brittle-failure-inducing flaws near the surface than in the interior. Another necessary condition might be a certain relationship between the stress components. Probably the failure waves may occur in single crystals and ceramics at divergent shock loading.

ACKNOWLEDGEMENT

The work was supported by the US Army Research Office under contract number N62558-02-M-6020, and by Russian Foundation for Basic Research under grant number 01-01-00436.

REFERENCES

1. G.I. Kanel, S.J. Bless. *Ceramics Transaction*, **134**, 197-216 (2002).
2. G.I. Kanel, A.A. Bogatch, S.V. Razorenov, Zhen Chen. *J. Appl. Phys.*, **92**(9), 5045-5052 (2002).
3. S.V. Rasorenov, G.I. Kanel, V.E. Fortov, M.M. Abasehov. *High Pressure Research*, **6**, pp. 225-232 (1991).
4. Kipp, M.E. and D.E.Grady. In: *Shock Compression of Condensed Matter 1989*. Eds: S.C.Schmidt, J.N. Johnson and L.W. Davison. Elsevier, pp. 377-380 (1990).

CP706, *Shock Compression of Condensed Matter - 2003*
edited by M. D. Furnish, Y. M. Gupta, and J. W. Forbes
© 2004 American Institute of Physics 0-7354-0181-0/04/$22.00

EQUATION OF STATE FOR A HIGH-DENSITY GLASS

A. E. Mattsson

Computational Materials & Molecular Biology MS 0196, Sandia National Laboratories, Albuquerque, NM 87185-0196

Abstract. Properties of relevance for the equation of state for a high-density glass are discussed. We review the effects of failure waves, comminuted phase, and compaction on the validity of the Mie-Grüneisen EOS. The specific heat and the Grüneisen parameter at standard conditions for a $\rho_0 = 5.085$ g/cm^3 glass ("Glass A") is then estimated to be 522 mJ/g/K and $0.1 - 0.3$, respectively. The latter value is substantially smaller than the value of 2.1751 given in the SESAME tables for a high-density glass with $\rho_0 = 5.46$ g/cm^3 [1]. The present unusual value of the Grüneisen parameter is confirmed from the volume dependence determined from fitting the Mie-Grüneisen EOS to shock data in Ref. [2].

INTRODUCTION

Brittle materials can exhibit failure waves. In glasses this phenomena is well studied; failure waves are found to exist in high- and low-density glasses. For example, the Hugoniot Elastic Limit (HEL) for DEDF glass ($\rho_0 = 5.18$ g/cm^3) is 4.3 GPa [3]. A slower failure wave is present for peak stresses from below the HEL to about 2 times the HEL. Thus, for certain peak states there is a complicated 3-wave structure with the elastic, shock and failure waves (there is evidence, though, that the failure wave is actually a diffusive process [4]). The Hugoniot relations have been shown to hold, to a good approximation, between the intact material and the comminuted material across the failure wave [3]. I also note that the failure wave not is due to tensile stress.

There are several studies of densification of silica glass [5] but not leaded glass. Silica has an open structure while high-density glasses with high lead content are considered "filled" materials and should not exhibit the large densification seen in silica. Note, however, in Fig. 1 that even though silica and low-density Pyrex glass exhibit a high degree of compaction, there also is noticeable compaction in the high-density glasses. I will discuss this issue further below in the context of composition. Note that the

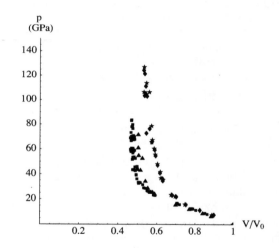

FIGURE 1. Hugoniots for 4 different glasses: Fused silica $\rho_0 = 2.204$ g/cm^3 (box), Pyrex $\rho_0 = 2.230$ g/cm^3 (triangle), $\rho_0 = 4.817$ g/cm^3 (diamond), and $\rho_0 = 5.085$ g/cm^3 (star). Data are from Ref. [2].

compaction in excess of comminution in the failure wave probably is due to a structural phase transition (e. g. Ref. [5]) in the grains of the comminuted material. Owing to its higher density the Mie-Grüneisen EOS should be valid for the comminuted glass.

Because of the complicated behavior of glasses

I did not model the cold curve, $p_c(V)$, directly. Instead my analysis is based on available shock data, the Hugoniot relations, the Mie-Grüneisen EOS and two models relating the Grüneisen parameter to the derivatives of the cold curve. I will compare my result with data derived from modelling the cold curve [1].

ROUGH ESTIMATE OF THE GRÜNEISEN PARAMETER

With knowledge of the linear thermal expansion, α, the bulk speed of sound, c_0, and the specific heat, c_V, the thermodynamic Grüneisen parameter, Γ, can be deduced from the relation $\Gamma = 3\,\alpha\,c_0^2\,/c_V$.

Estimate of specific heat

The specific heat can be estimated, by the Dulong-Petit law, to be $3k_B N$, where k_B is the Boltzmann's constant and N is the number of atoms per unit mass. This value is only dependent on the composition of the glass (larger lead content gives smaller number of atoms per unit mass). Since we do not know the exact composition of Glass A it needs to be estimated. Known compositions: (wt %)

- Fused silica, density 2.20 g/cm^3: SiO$_2$ (100).
- Glass ZF1, density 3.86 g/cm^3: SiO$_2$ (41.32), K$_2$O (7.00), As$_2$O$_3$ (0.50), PbO (51.18) [6].
- DEDF, density 5.18 g/cm^3: SiO$_2$ (27.4), K$_2$O (1.5), As$_2$O$_3$ (0.1), PbO (71.0) [3].

From the known compositions and densities above we can calculate the number of single atoms in a unit volume for each of these glasses. This number turns out to be almost the same regardless of composition. That is, the density difference between these three different glasses is mainly due to substitution of heavy PbO for light SiO$_2$ (trade 2 SiO$_2$ for 3 PbO). ZF1 and DEDF have exactly the same number of single atoms (within error-bars of the composition) while fused silica has 3% more atoms than these materials in the same volume. I assume the same total number of atoms in a volume in Glass A as in DEDF and ZF1. For simplicity I keep the K$_2$O and As$_2$O$_3$ content the same as in DEDF. The amount of PbO is, by far, the most important parameter for the density anyway. With these assumptions

the composition is calculated to be:

- Glass A, density $\rho_0 = 5.085$ g/cm^3: SiO$_2$ (28.4), K$_2$O (1.5), As$_2$O$_3$ (0.1), PbO (70.0).

The number of atoms in a unit volume for the high-density glasses above is 0.1065 N_A cm^{-3} (N_A is Avogadro's constant), which gives a specific heat of 522 mJ/g/K for Glass A.

Note that in the classical limit the specific heat only depends on the number of atoms in a unit volume of the material and not on type of atoms. Since this number is nearly constant in glasses, the specific heat scales linearly with the inverse of the density (or linearly with the specific volume). The specific heat of the high-density glass in Ref. [1], with density 5.46 g/cm^3, can thus be estimated to 486 mJ/g/K. The value extracted from the data in Ref. [1] itself, by thermodynamic relations, is approximately 530 mJ/g/K, which shows that using the linearity for estimating the specific heat in glasses is sufficiently accurate for these purposes.

Estimating the Grüneisen parameter

Data for 31 different glasses are available in Ref. [7]. Where possible I have made consistency checks between these data and data from other sources and found no inconsistencies. Using the above estimate for the specific heat, the DEDF bulk speed of sound, 2.60 km/s, and the range of linear thermal expansions from the data sheet [7], the Grüneisen parameter is estimated to $\Gamma = 0.02 - 0.38$. The large uncertainty in this value stems from the thermal expansion, which varies substantially between glasses. The 3 leaded glasses available in the data sheet have thermal expansions of $83 - 95 \times 10^{-7}$ K^{-1}, implying a Grüneisen parameter in the upper end of the interval ($\Gamma = 0.32 - 0.37$). Fused silica, as an extreme, has only 5.5×10^{-7} K^{-1} ($\Gamma = 0.02$), while Pyrex has the more ordinary value of 32.5×10^{-7} K^{-1} ($\Gamma = 0.13$). We clearly see, however, that the Grüneisen parameter in glass is significantly lower than in more ordinary materials, where it ranges from 1.0 to 3.0.

This range of Grüneisen parameter values is about a factor of ten smaller than the value presented in Ref. [1] for a high-density glass with 5.46 g/cm^3. The speed of sound derived from the data in Ref. [1] is 2.8 km/s, not very different from the speed of sound we have used. The full difference is instead due to a linear thermal expansion value of 481×10^{-7} K^{-1}

derived from the data in Ref. [1] (giving a $\Gamma = 2.13$, the value given in Ref. [1] being 2.1751). I note that none of the 31 glasses in Ref. [7] has a thermal expansion coefficient even a quarter as large as this value.

FURTHER DISCUSSION ON COMPACTION

Contrary to some opinions, I believe a structural phase transition contributes to the compaction of the high-density glasses. All glasses with known composition mentioned above have nearly the same number of atoms in a unit volume, which indicates that there are no large structural differences between high and low-density glasses. However, Pb atoms in a condensed phase occupies larger volume than Si/O atoms (in the variety of silica phases the nearest neighbor (nn) distance is approximately 1.75 Å while the nn distance in a Pb crystal is 3.50 Å). This implies that the Hugoniot curve stiffens, governed by the inter-atomic forces in the compact phase, at a larger relative volume for the leaded glasses than for the low-density glasses. This is indeed seen in Fig. 1.

The statement that there is no compaction in high density glasses [3] is based on a comparison of measurements of the longitudinal strain in soda-lime glass, Ref. [8] Fig. 6, and DEDF, Ref. [9] Fig. 5. I do not find this conclusion well supported.

MIE-GRÜNEISEN EOS

We neglect any contribution from the electrons. Using the Mie-Grüneisen EOS, standard thermodynamics relations, a model (Slater-Landau (SL) or Dugdale-MacDonald (DM) [10]) for connecting the Grüneisen parameter to the derivatives of the cold curve, and the Hugoniot relations, $p_H(V)$ can be written as a function of only the cold curve $p_c(V)$ and the volume, V_{0c}, where $p_c(V_{0c}) = 0$. The procedure is outlined in Ref. [10], Chapter 13. By assuming that

$$p_c(V) = \sum_{i=1}^{6} a_i \left(\frac{V_{0c}}{V} \right)^{\frac{i}{3}+1}, \qquad (1)$$

and fitting the parameters a_i to the $p_H(V)$ data given in Ref. [2], the cold curve, and thereby the Mie-Grüneisen EOS, can be determined. In Fig. 2 the cold curve and the resulting Hugoniot are shown together with the data to which they were fitted. Note that we

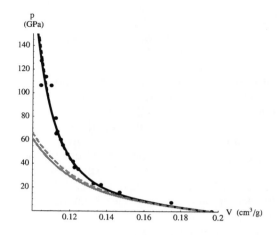

FIGURE 2. The resulting Hugoniot (black) and cold curve (gray) for Glass A from fitting to data in Ref. [2] (dots). The dashed cold curve results from the DM model and the solid from the SL model of the Grüneisen parameter. The dashed black line shows the Hugoniot derived from Eqn. 2.

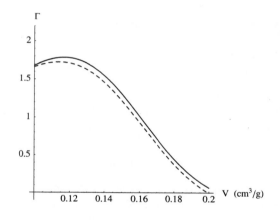

FIGURE 3. The Grüneisen parameter versus volume for Glass A. The solid curve corresponds to the SL model and the dashed to the DM model.

ignore the elastic region at low pressure. Fig. 3 shows the volume dependence of the Grüneisen parameter.

In the fitted curves, V_{0c} is chosen so that the bulk speed of sound is 2.6 km/s ($V_{0c} = 0.196300$ cm^3/g in the SL model, and 0.196600 cm^3/g in the DM model. The standard volume is $V_0 = 0.196657$ cm^3/g). The lower of the cold curves in Fig. 2 and the upper of the Grüneisen parameter curves in Fig. 3 are results using the SL model (solid lines), the other two corresponding curves (dashed) result from the DM model. In this case the SL model seems to work best since it gives a more sensible value for the internal energy E_0, which is the temperature dependent specific heat integrated from 0 to standard temperature. In the SL we obtain a value of E_0 that corresponds to a mean specific heat of 364 mJ/g/K in the $0 - 300$ K range while the DM model gives a value of only 156 mJ/g/K.

Note that this derivation of the Grüneisen parameter is independent of the rough estimate of the value at standard conditions made above. This indicates that the present very low value is internally consistent.

The cold curve corresponding to the SL model (full gray line in Fig. 2), the model for the Grüneisen parameter I recomend in this case, is obtained from Eqn. 1 with fitting parameters (a_1, a_2, a_3, a_4, a_5, a_6)= (-2989.05, 8360.78, -7125.9, 313.609, 2149.33, -708.758). From this cold curve, the Mie-Grüneisen EOS and thereby the Hugoniot, $p_H(V)$ can be determined and a relation between the shock wave velocity, u_s, and particle velocity, u_p, in the shock wave can be calculated. I found that at least a third degree polynomial was needed to model this curve,

$$u_s = 2.64 + 0.51u_p + 0.42u_p{}^2 - 0.05u_p{}^3 , \quad (2)$$

as is seen in Fig. 4. The Hugoniot derived from Eqn. 2 is shown as a black dashed line in Fig. 2.

CONCLUSION

For a high-density glass with $\rho_0 = 5.085$ g/cm^3, the Grüneisen parameter has an unusual behavior and its value at standard conditions is in the range 0.1-0.3.

ACKNOWLEDGMENTS

Sandia is a multiprogram laboratory operated by Sandia Corporation, a Lockheed Martin Company, for the United States Department of Energy under Contract DE-AC04-94AL85000.

REFERENCES

1. *SESAME database*, http://t1web.lanl.gov/home/t1/newweb_dir/t1sesame.html

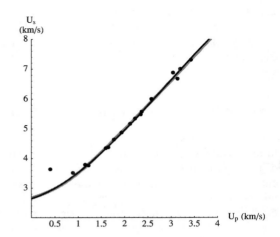

FIGURE 4. u_s vs. u_p for Glass A. The dots show data from Ref. [2]. The thin black line shows Eqn. 2, a fit to the thick gray line derived from the lower cold curve in Fig. 2.

2. Marsh, S.P., *LASL shock Hugoniot data*, University of California Press, Berkeley, 1980.
3. Millett, J.C.F., Bourne, N.K., and Rosenberg, Z., "Direct measurements of strain in a shock-loaded, lead filled glass", *J. Appl. Phys.* **87**, 8457-8460 (2000).
4. Feng, R., "Formation and propagation of failure in shocked glasses", *J. Appl. Phys.* **87**, 1693-1700 (2000).
5. Guido Della Valle, R., and Venuti, E., "High-pressure densification of silica glass: A molecular-dynamics simulation", *Phys. Rev. B* **54**, 3809-3816 (1996); Klug, D.D., Rousseau, R., Uehara, K., Bernasconi, M., Le Page, Y., and Tse, J.S., "Ab initio molecular dynamics study of the pressure-induced phase transformation in cristobalite", *Phys. Rev. B* **63**, 104106 (2001).
6. He, H., Jing, F., and Jin, X., "Evaluating the damage in shock compressed glass coupling with VISAR measurement", *Int. J. Imp. Eng.* **25**, 599-605 (2001).
7. *Materials Properties - Glass and Quartz/Ceramic*, data sheet extracted from http://www.mindrum.com/tech.html.
8. Bourne, N.K., and Rosenberg, Z., "The dynamic response of soda-lime glass", in *Shock Compression of Condensed Matter-1995*, edited by S.C. Schmidt and W.C. Tao, AIP Conference Proceedings 370, Woodbury, New York, 1996, pp. 567-572.
9. Bourne, N.K., Millett, J.C.F., and Rosenberg, Z., "Failure in a shocked high-density glass", *J. Appl. Phys.* **80**, 4328-4331 (1996).
10. Eliezer, S., Ghatak, A., and Hora, H., *Fundamentals of Equation of State*, World Scientific, New Jersey, 2002; and references therein.

CP706, *Shock Compression of Condensed Matter - 2003*
edited by M. D. Furnish, Y. M. Gupta, and J. W. Forbes
© 2004 American Institute of Physics 0-7354-0181-0/04/$22.00

DIAGNOSTICS OF DUCTILITY, FAILURE AND COMPACTION OF CERAMICS UNDER SHOCK COMPRESSION

V.E. Paris[1], E.B. Zaretsky[1], G.I. Kanel[2], and A.S. Savinykh[2]

[1]*Department of Mechanical Engineering, Ben-Gurion University of the Negev, P.O.B. 653, Beer-Sheva 84105, Israel*
[2]*Institute for High Energy Densities of Russian Academy of Sciences, IVTAN, Izhorskaya 13/19, Moscow, 125412 Russia*

Abstract. The presented experimental technique uses a radial sample pre-stressing in order to reveal whether the behavior of ceramics is brittle or ductile under uniaxial shock-wave compression. The controlled radial pre-stressing of ceramic samples was produced by a shrink-fit steel sleeve. The HEL increment caused by the pre-stressing was expected to be of about 2.5 times higher for materials whose brittle failure in compression is governed by Griffith's criterion than for those that obeyed the Von Mises criterion of ductile yielding. The experiments indicate certainly ductile behavior of hot-pressed alumina samples whereas the samples of hot-pressed high-density (2.5 g/cm^3) boron carbide exhibit compressive failure. The lateral pre-stressing of boron carbide samples of lower density (2.37 g/cm^3) results in a slight HEL decrease that is treated in terms of a compaction process.

INTRODUCTION

It is known that cracking of brittle solids caused by an axial compression is suppressed by applying a hydrostatic pressure [1]. The increase of the confining stress increases the stress deviator required for triggering the micro-cracking or failure and may result in a transition from the brittle failure to ductile yielding. It was shown by Heard and Cline [2] that the compressive strength of ceramics increases rapidly with pressure below the transition while above the transition the compressive strength is virtually constant. The different pressure sensitivity of the elastic limits corresponding to the different, brittle or ductile, deformation mode can be used for revealing the kind of the material response [3]: knowledge is especially important when ceramics or rocks are tested. For analyzing the HEL of brittle materials, Rosenberg [4] suggested to use, instead of the Tresca (or von Mises) yielding criterion for the difference between longitudinal σ_1 and radial σ_2 stress

$$\sigma_1 - \sigma_2 = Y_{duct}, \tag{1}$$

the Griffith's criterion of brittle failure

$$(\sigma_1 - \sigma_2)^2 = Y_{brit}(\sigma_1 + \sigma_2). \tag{1a}$$

The Griffith criterion is based on the assumption that the compressive failure starts when the tensile stress acting on the largest crack of the most vulnerable orientation exceeds some critical value causing spontaneous extension of the crack. In the case of the uniaxial strain, the stress σ_{HEL} corresponding to the onset of the ductile yielding is

$$\sigma_1 = \sigma_{HEL} = Y_{duct}(1-\nu)/(1-2\nu). \tag{2}$$

Here ν is the Poisson's ratio. For the material whose elastic response is terminated as soon as a threshold value Y_{brit} of the equivalent failure stress

TABLE 1. Mechanical properties of the studied Al_2O_3 and B_4C samples.

Material	C_l, km/s	C_s, km/s	ρ_0, g/cm^3	Poisson's ratio v	E, GPa
Al_2O_3 average[a]	10.03 (±0.01)	5.93 (±0.02)	3.745 (±0.005)	0.231 (±0.002)	324 (±2)
B_4C-A	13.91	8.83	2.493	0.163	452
B_4C-B	14.12	8.78	2.499	0.185	456
B_4C-C	14.15	8.77	2.494	0.188	456
B_4C-D	13.79	8.65	2.501	0.176	440
B_4C-E	14.09	8.74	2.497	0.187	453
B_4C-F	13.48	8.75	2.474	0.136	430
B_4C-G	13.67	8.75	2.495	0.153	440
B_4Caverage[a]	13.89 (±0.25)	8.75 (±0.05)	2.493 (±0.009)	0.17 (±0.02)	447 (±10)
B_4C-H	13.36	8.26	2.373	0.191	386
B_4C-I	12.84	8.19	2.368	0.157	368
B_4C-J	13.14	8.33	2.420	0.165	391

[a] Average values corresponding to four alumina samples and to seven dense B_4C samples.

is achieved the value of σ_{HEL} is

$$\sigma_1 = \sigma_{HEL} = Y_{brit}\,(1-v)/(1-2v)^2 . \qquad (2a)$$

Applying an additional compressive stress π in radial direction will change the expressions (1) and (1a). In the case of $\pi/Y \ll 1$ this give [3] for the derivatives of σ_{HEL} on π

$$\frac{d\sigma_{HEL}^{duct}}{d\pi} = \frac{(1-v)}{(1-2v)} \qquad (3)$$

$$\frac{d\sigma_{HEL}^{brit}}{d\pi} = \frac{(1-v)}{(1-2v)}(3-2v) \qquad (3a)$$

As apparent from (3) and (3a) the sensitivity of σ_{HEL} to the pre-stressing is in $(3-2v)$ times higher when the material response is brittle.

MATERIALS AND EXPERIMENTAL

The experiments were performed with hot pressed alumina (96% purity) and boron carbide (97% purity) samples (Microceramica Ltd., Haifa, Israel). The samples were precisely (with 2-μm tolerance) cut disks of 25-mm diameter and 5-mm thickness. The average measured properties of four alumina samples are given in Tab. 1. The properties of the boron carbide samples were found highly scattered; the densities, e.g., were found within the range from 2.215±0.005 to 2.500±0.005 g/cm^3 (the ultimate density of B_4C is 2.52 g/cm^3). The dependencies of both longitudinal C_l and transversal C_s sound velocities on the density were found very close to those of other authors. Average grain size in the studied B_4C samples was found to be ranged from 15 to 20 μm, close to the size of the B_4C grains in the samples studied by Gust and Royce [5]. Seven dense ($\rho_0 \approx 2.5$ g/cm^3) and three relatively porous (B_4C -H, -I, -J of Tab. 1) samples were chosen for the experiments.

Pre-stressing of the ceramic samples was provided by a shrink-fit steel ring as it was suggested by Chen and Ravichandran [6]. The rings were cut from rods of normalized 4340 steel. The inside diameter of the rings were machined to a diameter smaller than the diameter of the samples by $\delta = 0.1 \pm 0.005$ mm. Prior to the insertion of the ceramic discs, the rings were heated up to 600°C. For the 0.1-mm misfit the confining stress acting on the alumina and boron carbide samples were equal to 0.30 GPa and 0.32 GPa, respectively [3].

The planar impact loading of ceramic samples was produced by 1-mm flyer plates launched by a 57-mm gas gun. The copper (alumina samples) and tungsten (B_4C samples) flyer plates were accelerated up to the velocities of 500±10 and 580±11 m/s, respectively. The stronger shots with boron carbide samples (B$_4$C-D to B$_4$C-I samples of Tab. I) were performed with use of explosive facility accelerating 2-mm aluminum flyer plates up to the velocity of 1900±50 m/s. In the experiments with alumina ceramics the sample free surface velocity histories were recorded with a VISAR [7]. In the experiments with the boron carbide samples the velocity histories were recorded at the interface between the sample and the PMMA window.

RESULTS AND DISCUSSION

Figure 1 summarizes the results of experiments with pre-stressed and free alumina samples. Each shown waveform was obtained by averaging of the free surface velocity histories of two shots. The transition from elastic to inelastic response corresponds to $\sigma_{HEL} = \rho_0 C_l w_{HEL}/2 = 5.35$ GPa. The ramped velocity growth behind the elastic wave front is associated with strain hardening and stress relaxation effects. Assuming that the inelastic response of the alumina is ductile the confining stress $\pi = 0.3$ GPa should result, Eq. (3), in the increase of the HEL equal to $\Delta\sigma_{HEL} = 0.43$ GPa. The latter corresponds to the increase of the sample surface velocity by $\Delta w_{HEL} = 23$ m/s. The measured difference for the first 50 ns is 15 to 10 m/s.

The results of the experiments with boron carbide samples are shown in Fig. 2. Shock compression of the boron carbide differs from that of alumina. Even in the weak shots in which HEL of dense samples was not exceeded (right profiles in Fig. 2a) the specific response of each sample is evident. We associate these features with porosity of the samples. At stronger impact, right profiles in Fig. 2, this irregular process results in strong deformation heterogeneity proved by remarkable oscillations of the interface velocity. Increase of the porosity, Fig. 2b, decreases the HEL, smoothes the transition from elastic to inelastic deformation in the weak impact, and suppresses the velocity oscillations in strong shots.

In the shots where the HEL of the boron carbide samples was certainly exceeded the elastic compression wave consists of a shock discontinuity of less than 1-ns rise time followed by a gradual velocity increase towards the foot of the second,

inelastic, compression wave whose rise time varied from 40 to 70 ns.

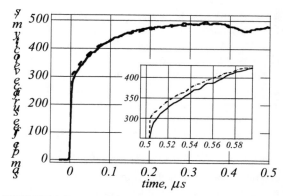

FIGURE 1. Free surface velocity data obtained with free of stress (solid lines) and pre-stressed (dashed lines) samples of alumina. Insert shows the difference between two average curves.

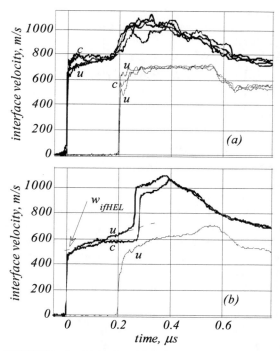

FIGURE 2. Velocity histories of sample-PMMA interfaces obtained with free (u) and pre-stressed (c) samples of boron carbide having density close to 2.5 (a) and 2.4 g/cm^3 (b). The velocity profiles obtained after weaker impacts (hair lines) are shifted to the right. The arrow points to the velocity related to the material HEL.

In the pre-stressed samples, independently of their initial density, a velocity plateau and, hence, the constant compressive stress, replaces the gradual increase in the part of waveform between the elastic precursor front and inelastic, "plastic", compression waves. In the following we define the HEL as the intersection of the upward extrapolation of the elastic shock discontinuity with the leftward extrapolation of the intermediate part of the profile as shown by the dashed line in Fig 2b. In accordance with this definition, the values of the interface velocities equal to $w_{ifHEL} = 700\pm10$ and $w_{ifHEL} = 775\pm10$ m/s have been related to the elastic-inelastic transitions in the free-of-stress and the pre-stressed samples, respectively. Using Barker and Hollenbach [8] data on shock Hugoniot of PMMA yields for the HEL of the free and pre-stressed B$_4$C samples, we compute the values $\sigma_{HEL} = 13.5\pm0.15$ GPa and $\sigma_{HEL} = 15.1\pm0.2$ GPa, respectively. So, the pre-stressing of the boron carbide samples equal to 0.32 GPa resulted in the average increase of the HEL of about $\Delta\sigma_{HEL} \approx 1.6\pm0.35$ GPa. Equation (3a) yields for the HEL increment the value of $\Delta\sigma_{HEL} \approx 1.06\pm0.03$ GPa. The discrepancy, 1.6 GPa vs. 1.06 GPa, seems due to the use of the Griffith's criterion, which, as is known [1], underestimates the values of Y_{brit}. Like in the case of alumina, the difference between the velocity profiles of free and pre-stressed samples decreases to zero after a lapse of about 150 ns.

In the lower-density samples of boron carbide containing a noticeable (about 5%) amount of pores or micro-cracks, Fig. 2b, the radial pre-stressing decreases the HEL. Compression of a porous ceramic is accompanied by two competitive processes: by the cracking which increases the porosity and by the compaction which results in the porosity decrease. Thus, the HEL of porous brittle material is determined by the lesser of the thresholds for either compressive failure or compaction. In the free-of-pores material (the density close to ultimate one) the compressive fracture may only increase the porosity. Note, that under hydrostatic compression no compressive failure can occur whereas the compaction of porous material is possible. The radial pre-stressing contributes to the increase of the hydrostatic stress associated with the shock compression decreasing an apparent compaction threshold of porous material.

The experiments with pre-stressed ceramic samples unambiguously show the ductile response of alumina and the brittle response of boron carbide to planar impact loading. The knowledge of the mode of the response of brittle material is important for calibrating its constitutive model. As well, in the case of the ductile shock response of brittle ceramics it is necessary to know the parameters of their brittle-to-ductile transition [2]. In the case of the alumina the lateral stress at the HEL in our experiments was varied between 1.6 and 1.9 GPa for free and pre-stressed samples, respectively. That means that in alumina the transition from brittle to ductile response occurs between the confining pressure 1.25 GPa, at which the alumina remains brittle [2], and 1.6 GPa. In the case of the boron carbide samples the highest confining pressure, corresponding to the highest detected w_{ifHEL} value and equal to 2.97 GPa, is apparently insufficient for the brittle-to-ductile transition.

REFERENCES

1. Kanel, G. I. and Bless, S. J. *Ceramics Transaction*, **134**, 197 (2002).
2. Heard, H.C. and Cline C.F., *J. Mat. Sci.*, **15** 1889 (1980).
3. Zaretsky, E.B. and Kanel, G.I., *Appl. Phys. Letters*, **81**, 1192 (2002).
4. Rosenberg, Z., *J. Appl. Phys.*, **76** 1543 (1994).
5. Gust W.H. and Royce, E.B., *J. Appl. Phys.*, **42**, 276 (1971).
6. Chen, W. and Ravichandran, G., *Int. J. of Fracture*, **101**, 141 (2000).
7. Barker, L.M. and Hollenbach, R.E., *J. Appl. Phys.* **43**, 4669 (1972)
8. Barker, L.M. and Hollenbach, R.E., *J. Appl. Phys.* **41**, 4208 (1970)

CP706, *Shock Compression of Condensed Matter - 2003*
edited by M. D. Furnish, Y. M. Gupta, and J. W. Forbes
2004 American Institute of Physics 0-7354-0181-0/04/$22.00

THE SHOCK BEHAVIOUR OF A SiO$_2$-Li$_2$O TRANSPARENT GLASS-CERAMIC ARMOUR MATERIAL

I.M. Pickup, J. C. F Millett* and N.K. Bourne*

Defence Acadamy of the UK, Cranfield University, Shrivenham, Swindon, SN6 8LA, UK.

Defence Science and Technology Laboratory, Porton Down, Salisbury, Wiltshire, SP4 OJQ, UK

Abstract. The dynamic behaviour of a transparent glass-ceramic material, Transarm, developed by Alstom UK for the UK MoD has been studied. Plate impact experiments have been used to measure the materials Hugoniot characteristics and failure behaviour. Longitudinal stresses have been measured using embedded and back surface mounted Manganin gauges. Above a threshold stress of *ca.* 4 GPa, the longitudinal stress histories exhibit a significant secondary rise, prior to attaining their Hugoniot stress. Lateral stresses were also measured by embedding Manganin gauges in longitudinal cuts. Significant secondary rises in stress were observed when the applied longitudinal stress exceeded the 4 GPa threshold, indicating the presence of a failure front. The dynamic shear strength of the glass has been measured using the longitudinal and lateral data. Even though significant strength drops have been measured before and behind the failure front, the material has a high post-failure strength compared to non- crystalline glasses.

INTRODUCTION

Transarm is a glass-ceramic [1], transparent armour material developed by ALSTOM UK Ltd. for the UK MOD [2]. The objective was to improve ballistic protection of transparent armour systems against small arm threats, offering superior performance to that of existing glass-polymer laminated systems. Controlled heat-treatment transformed a SiO$_2$-Li$_2$O based amorphous glass into a, primarily, polycrystalline glass-ceramic. The crystallites have an average grain size of *ca.* 50 nm, (*i.e.* about one tenth of the wavelength of visible light) resulting in good optical transparency. The objective of the studies presented here, was to provide basic data on shock properties of this development material to allow models to be developed and to compare the deviatoric strength and failure behaviour with conventional transparent armour materials.

EXPERIMENTAL

Plate impact experiments were conducted using a 50 mm diameter gas gun. Impact velocity was measured to an accuracy of 0.5% using a sequential pin-shorting method and tilt was fixed to be less than 1 mrad by means of an adjustable specimen mount. Impactor plates were made from lapped copper discs mounted onto a polycarbonate sabot with a relieved front surface in order that the rear of the flyer plate remained unconfined. Targets were flat to less than 2 μm across the surface. Longitudinal stresses were measured using Manganin stress gauges (MicroMeasurements type LM-SS-125CH-048). Lateral stresses were measured using Manganin stress gauges (MicroMeassurments type J2M-SS-580SF-025 - resistance 25 Ω). The gauges were placed at varying distances from the impact face as shown in Fig. 1. They had an active width of 240 μm.

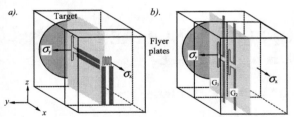

FIGURE 1. Experimental arrangement for lateral and longitudinal stress experiments. a) Longitudinal and lateral stress gauge mounting positions with rear PMMA plate. b). Sectioning for multiple lateral gauge measurements.

The lateral stress, σ_y, was used along with the longitudinal stress, σ_x, to calculate the shear strength of the material, τ, using the well-known relation

$$\tau = \frac{1}{2}\left(\sigma_x - \sigma_y\right). \tag{1}$$

This quantity has been shown to be a good indicator of the ballistic performance of the material [3]. This method of determining the shear strength has the advantage of being direct since no computation of the hydrostat is required.

Three types of plate impact (PI) experiment were performed;

Embedded gauge (EG): to measure longitudinal stress, σ_x, (in the direction of the shock). A Manganin stress gauge was embedded between a 50 x 50 x 6 mm thick front plate and a 50 x 50 x 8 mm back plate. This method allows accurate Hugoniot stress measurement without prior knowledge of the materials' impedance but reduces temporal resolution of the sensor;

Back surface (BS): The stress gauge was placed on the back surface of the glass target and sandwiched with a PMMA backing block. This allows a more detailed observation of the temporal development of the longitudinal stress history than the EG method (since the gauge is impedance matched) but the in-material value of the stress must be calculated

Lateral gauge (LG): The glass target was sectioned longitudinally and a lateral stress gauge was placed 2 mm from the impact surface. It measures the stress

perpendicular to the shock propagation direction, σ_y.

There were a total of 8 plate impact shots indicated in Table 1. Physical properties for Transarm are presented in Table 2.

Table 1. Summary of plate impact shots

Shot	Type	Tile thickness		Flyer plate		Velocity m/s
		Front	Back			
1	EG	6	8	Cu	3mm	374
2	EG	6	8	Cu	6mm	493
3	EG	6	8	Cu	3mm	722
4	BS	6	N/A	Cu	3mm	725
5	BS	8	N/A	Cu	6mm	727
6	LG	10	N/A	Cu	6mm	371
7	LG	10	N/A	Cu	6mm	504
8	LG	10	N/A	Cu	6mm	727

TABLE 2. Properties of Transarm

ρ_0 g cm^{-3}	c_L mm μs^{-1}	c_S mm μs^{-1}	v	HV Kg/mm^2
2.53	6.690	4.060	0.21	803

RESULTS

Longitudinal stress histories are presented for shots 1-5 in Fig. 2. The first three are from embedded gauge experiments, shots 5 and 6 are from back surface mounted gauges. Lateral stress histories from shots 6 to 8 are presented in Fig. 3. The configuration of each shot, including impact velocity and flyer plate type is indicated in Table 1.

In all the longitudinal experiments, above a threshold stress where the induced longitudinal stress exceeded 4 GPa, a pronounced secondary rise in the stress histories was observed. Both the embedded and the back surface mounted gauges exhibited this behaviour.

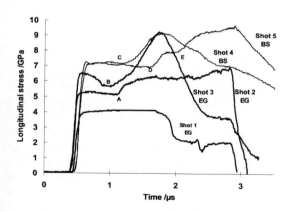

FIGURE 2. Longitudinal stress histories for embedded gauge (1-3) and back surface gauged shots (4-5)

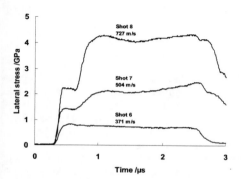

FIGURE 3. Lateral stress histories for shots 6-8

In the 8 mm thick tile (Shot 5) there was a 3rd rise. Such features are not usually so prominently observed in amorphous glasses like soda-lime or Pyrex [4]. However, Nahme and Hiltl [5] showed similar behaviour for one shot on a polycrystalline glass based on the Li_2O-Al_2O_3-SiO_2 system.

For the lowest velocity shot (1, 374m/s) there is no secondary rise and the history is typical of a material stressed below an elastic limit. The maximum stress measured, 4.10 GPa. is close to the value calculated from elastic impedance, 4.31 GPa. For the higher velocity, (embedded, shots 3 and 4) the initial break in slope occurred at significantly different stress levels. For shot 6 there was an apparent relaxation following the initial 'yield'. The stress at the initial break in slope was consistently higher for back surface mounted gauges (shots 5 and 6) than for the embedded gauge specimen (shot

3) when impacted at the same velocity (*ca.* 725 m/s). However, this would not appear to be a gauge calibration effect as the maximum stress values for all three experiments are the same.

Phenomenological propositions were considered to explain the structure of the longitudinal traces. These included a phase transition, a plastic wave or a failure front. The initial plateau stress for all 4 specimens impacted at stress levels above 4 GPa were below the calculated Hugoniot stress levels appropriate for the impact velocity. This does not preclude a phase change causing the secondary rise, but favours an inelastic transition.

Shots 4 and 5 were impacted at the same velocity and using the same diagnostic method. The tile thickness in Shot 4 was 6 mm compared to 8mm for shot 5. This allows a Lagrangian X-t analysis to test various wave interactions, Fig 4.

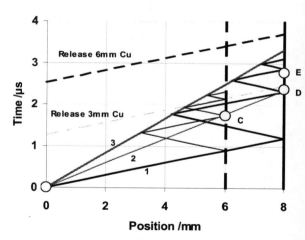

FIGURE 4. Lagrange X-t diagram for shots 4, (6mm thick tile) and 5 (8mm thick tile).

The three lines radiating from the origin are: 1) an elastic wave; 2) a proposed plastic wave; 3) a proposed failure front. The data points C, D and E mark the times of measured secondary and tertiary rises that are also indicated in Fig. 2 for shots 4 and 5. A plastic wave velocity of 3400 m/s was chosen to coincide with the rise time observed in the 6 mm thick specimen (c) and then extrapolated to the 8mm specimen surface line where it was also coincident with the secondary rise (D). Thus a plastic wave of this velocity could explain the observed rise in stress in both specimens.

A failure front with a velocity of 2400 m/s was selected so that the release wave from the back surface of the 8 mm tile would reflect off the failure front as a recompression wave and cause the rise indicated at point D. A significant impedance drop on the failed side of the front of *ca.* 60% would be required to produce the reload signal observed. A second reflection is indicated in Fig. 4 that is coincident with a reload at time E. The same approach was used for the 6 mm tile. The reload signal, observed at time C, is also coincident with the arrival of the reflected release wave in the X-t diagram. No second reload would be expected in the 6 mm tile due to unloading of the pulse from the back of the 3 mm flyer plate. The failure front velocity measured from the lateral stress data in Fig. 3 indicates a velocity of *ca.* 3000 m/s which is somewhat greater than that required to accommodate the observed reloads described above.

The dynamic shear strength of Transarm has been measured ahead of and behind the failure front from the longitudinal and lateral gauge experiments using Equation 1 and the methods described previously [6], Fig.5.

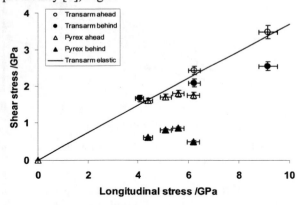

FIGURE 5. Dynamic shear strength of Transarm and borosilicate.

The measured shear strength of the material ahead of the failure front follows the calculated elastic strength up to the highest impact stresses. Above a threshold of 4 GPa there is a significant reduction in shear strength. Shear data for a borosilicate glass, Pyrex, is presented for comparison. The polycrystalline Transarm has significantly higher shear strength than Pyrex.

CONCLUSIONS

The shock properties of the glass-ceramic Transarm have been measured using plate impact techniques. The longitudinal stress histories exhibit very significant rises in slope. Phenomenological propositions were considered to explain the structure of the longitudinal traces based on inelastic fronts. Further experiments are required to confirm the propositions. The dynamic shear strength of Transarm has been determined. Its shear strength is significantly higher than standard transparent armours like soda-lime glass and borosilicate indicating its potential as a transparent armour against small arm threats.

Support from UK MoD, EGC Domain, is gratefully acknowledged.

REFERENCES

1. Pinckney l., 'Phase separated glasses and glass ceramics', Vol. **4** Engineered Materials Handbook, Ed. Scheider et al, ASM International, pp. 433-438, 1991
2. Hyde, A. R. and Darrant J. G., 'TRANSARM-improved transparent armour', in Personal Armour Systems Symposium, Colchester UK, 1998.
3. Bourne, N. K. and Millett, J. C. F., 'On impact upon brittle solids', DYMAT 2000, Internat. Conf. Mechanical and Physical Behaviour of Materials under Dynamic Loading, Journal de Physique IV, pp. 281-286, 2000.
4. Bourne, N. K., Rosenburg Z. and Millett, J. C. F, 'The plate impact response of three glasses', in Structures under Shock and Impact **IV,** Ed. Jones N. et al, pp 553-562, Computational Mechanics Publications, Southampton UK, 1996.
5. Hiltl M and Nahme H, 'Dynamic behaviour of a shock loaded glass-ceramic based on the $Li_2O-Al_2O_3-SiO_2$ system' J Phys IV France 7, Colloque C3, Supplement au Journal de physique III, 1997.
6. Pickup, I. M. and Barker, A. K., 'Deviatoric strength of silicon carbide subject to shock,' in Shock Compression of Condensed Matter-1999, edited by M. D. Furnish et al., AIP Press, *pp.573-576,1999.*

CP706, *Shock Compression of Condensed Matter - 2003*
edited by M. D. Furnish, Y. M. Gupta, and J. W. Forbes
© 2004 American Institute of Physics 0-7354-0181-0/04/$22.00

THE EFFECT OF STRUCTURE ON FAILURE FRONT VELOCITIES IN GLASS RODS

D.D. Radford[*], G.R. Willmott and J.E. Field

PCS Group, Cavendish Laboratory, Madingley Road, Cambridge, CB3 0HE, UK
[*]*Currently at University of Cambridge, Department of Engineering, Cambridge, CB2 1PZ, UK*

Abstract. Symmetric Taylor and reverse ballistics tests were used in conjunction with high-speed photography to examine the characteristics of failure in fused silica rods. Experiments were performed at impact velocities up to 800 m s^{-1} yielding impact pressures to approximately 7 GPa. Failure front velocities were strongly dependent on impact pressure consistent with results for borosilicate and soda-lime glasses [1]. One-dimensional strain waves were observed in some experiments before a one-dimensional stress regime was reached. For high impact pressures (> 2 GPa) the failure fronts followed almost immediately behind the incident shocks at a velocity near $\sqrt{2}\, c_S$ for fused silica, borosilicate and soda-lime glasses.

INTRODUCTION

Since Kanel [2] first introduced the idea of the so-called "failure wave" in shock loaded glass, there has been a significant number of studies into the failure of glasses and other brittle materials [3-12]. These investigations have generally involved examination of the failure process in a state of 1-D strain, as obtained in plate impact experiments.

Recently, the failure of glasses has been investigated for materials in a state of 1-D stress using the well-known Taylor impact test method [1,13-15]. In these studies, failure was observed in both the classic [16] and symmetric [17] Taylor test configurations using high-speed photography, stress and strain gauges, and VISAR. Results for borosilicate and soda-lime glasses have shown that failure front velocities increase for increasing impact pressure, approaching the elastic wave speeds above *ca.* 2 GPa. The time dependent nature of the failure process in brittle materials is consistent in 1-D strain [12] and 1-D stress [1], indicating that the physical mechanism of failure is related to the material structure.

The aim of the current study is to investigate failure in the silica based glass denoted "fused quartz" or "fused silica". This glass is the most open structured of the silica glasses, meaning that the spaces between interlocking SiO_2 tetrahedra are devoid of additives. As a result, fused quartz has a lower density than both borosilicate and soda-lime glasses. High-speed photography from Taylor impact experiments are used to determine the characteristics of failure in fused quartz and the results are compared to previous data for borosilicate and soda-lime glasses.

EXPERIMENTAL METHOD

In the current investigation, classic and symmetric Taylor impact experiments were performed on fused quartz rods in evacuated chambers using two single-stage light gas guns at the Cavendish Laboratory. Impact velocities, measured to an accuracy of 1 %, ranged from 0.18 to 1 km s^{-1}. The pressures induced at the impact plane were calculated by matching impedances, using the well-known shock responses of the materials.

TABLE 1. Material properties of the glasses studied. Wave velocities are for material in a state of 1-D stress.

Glass Type	ρ (± 0.01 g cm^{-3})	E (GPa)	μ (GPa)	ν	c_L (± 0.01 mm μs^{-1})	c_S (± 0.01 mm μs^{-1})
Fused Quartz	2.20	72.5	30.9	0.17	5.74	3.61
Borosilicate	2.23	73.1	30.4	0.20	5.72	3.49
Soda-Lime	2.49	73.3	29.8	0.23	5.43	3.22

In the classic configuration, the rod was impacted against the flat, polished surface of a hardened steel anvil with a measured hardness of 58.4 ± 0.7 (Rockwell C). In the symmetric configuration, the target and projectile were identical rods. In both configurations, the target was carefully aligned to ensure an axially symmetric impact.

The specimens were cut from industry-standard, high purity fused silica rods. Table 1 summarizes the material properties of the rods used in this investigation and the properties of borosilicate and soda-lime glasses used previously.

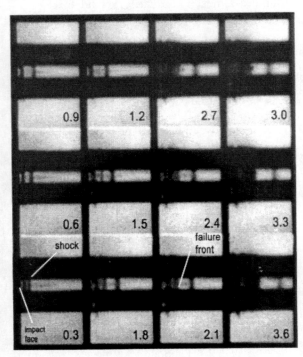

FIGURE 1. High-speed photographic sequence of a symmetric Taylor impact test using 25 mm diameter fused quartz rods (impact velocity is 534 m s^{-1}). Time (in microseconds) after impact is indicated in each frame.

Each cylindrical rod used in these experiments had a diameter of 10 or 25 mm and a length-to-diameter ratio of 10. The flat ends of each rod were polished. The sabots and supports used for alignment did not significantly affect observations of the failure front.

High-speed photographic sequences of the impacts were taken using a programmable image-converter camera (either a Hadland Ultra-8 or an Imacon Ultranac 501). Each impact was photographed over a sequence of between 8 and 15 frames. The exposure time for each frame was between 100 and 250 ns and inter-frame times ranged from 1 to 3 μs. The images were spatially calibrated using fiducial markers.

RESULTS AND DISCUSSION

Figure 1 shows a high-speed photographic sequence for a symmetric impact of 25 mm diameter rods. The flyer rod was travelling from the left to the right. The wide horizontal lines along the edges of the specimen are due to the refraction of light through the curved surface of the rods. In the first frame (0.3 μs after impact), two vertical dark lines are observed. The line to the left is the impact face and the line to the right is the shock wave. The shock can be seen propagating into the rod ahead of other features in subsequent frames. Another front, first observed in the frames labelled 1.5 μs and 1.8 μs, has become the failure front by 2.1 μs after impact. As in plate impact experiments, a 1-D state of strain exists until release waves from the outer edges converge along the axis of the specimen, so that there is a transition to a 1-D stress state. Several effects of this transition are observed. Firstly, the incident shock slows from a velocity approximately equal to the 1-D strain longitudinal wave speed in fused quartz (5.96 mm μs^{-1}) during the initial 2.1 μs after impact, to 5.3 ± 0.3 mm μs^{-1} after 3.9 μs. Secondly, the incident shock wave is not visible in later frames because the pressure gradient across the shock front decreases in the 1-D stress state. Finally, frames from 1.5 μs onwards clearly show that the advancement of the radial expansion corresponds with the leading edge of the failure front. Similar observations were made in other experiments in this study and by Willmott and Radford [1] for borosilicate and soda-lime glasses.

Figure 2 is a sequence showing a symmetric impact performed using 10 mm diameter rods. The projectile was travelling from the right to the left. 3 μs after impact, the failure front (dark portion of the rod) has moved into the target rod. Subsequent frames show further propagation of the failure front and the lateral expansion of the failed material. Following the technique described in [1], the failure front velocity in the rod was measured to be 4.51 ± 0.03 mm μs^{-1}.

Figure 3 shows the calculated failure front velocity as a function of impact pressure for all of the experiments performed on fused quartz rods in this investigation. Included in the figure are the longitudinal wave speed (c_L) and $\sqrt{2}$ times the shear wave speed ($\sqrt{2}\, c_S$). The failure front velocity

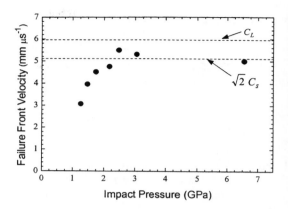

FIGURE 3. Measured failure front velocities as a function of impact pressure for fused quartz rods.

increases rapidly as the impact pressure increases from 1 GPa, and then asymptotically approaches a maximum value near $\sqrt{2}\, c_S$.

Classical fracture theories predict that the limiting velocity of a single mode I or II crack in an infinite medium is the surface (Rayleigh) wave speed (c_R), approximately 90% of the shear wave velocity. Rosakis et al. [18] have suggested that single shear cracks may propagate at speeds close to $\sqrt{2}$ of the shear wave speed. In practice, mode I cracks typically bifurcate at a velocity of approximately $0.5\, c_R$ [19].

At impact pressures > 2 GPa, the average failure front velocity for the fused quartz is 5.1 ± 0.3 mm μs^{-1}, which is within experimental error of $\sqrt{2}\, c_S$. Table 2 compares the data for fused quartz with combined data for borosilicate and soda-lime glasses [1,15] at impact pressures between 3.5 and 4 GPa. Trends in the failure front velocity are similar in all three materials. In practice, single cracks do not propagate faster than the shear wave speed [19], so the observed fracture pattern is caused by the nucleation of multiple cracks within uncomminuted glass behind the incident shock. It appears that under severe impact conditions, ensembles of cracks can approach group velocities near $\sqrt{2}\, c_S$ independent of the material.

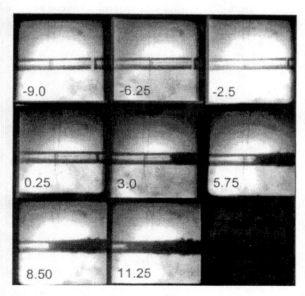

FIGURE 2. High-speed photographic sequence of a symmetric Taylor impact test using 10 mm diameter fused quartz rods (impact velocity is 306 m s^{-1}). Time (in microseconds) after impact is indicated in each frame.

TABLE 2. Experimentally determined failure front velocities for three silica-based glasses.

Glass Type	Failure Front Velocity (mm μs^{-1})	$\sqrt{2}\,c_S$ (mm μs^{-1})
Fused Quartz	5.1 ± 0.3	5.11
Borosilicate	4.8 ± 0.5	4.94
Soda-Lime	4.55 ± 0.5	4.55

CONCLUSIONS

Recent results from Taylor impact tests on borosilicate and soda-lime glasses and those presented in this paper for fused quartz show that failure front propagation in rods is irregular and the structure of the failure is material dependent. The results for fused quartz further demonstrate the dependence of the failure front velocity on impact pressure. It is believed that dynamic failure of glasses is not due to a continuous boundary sweeping through the material (failure wave), but is a time dependent process that evolves from inelastic deformation and subsequent nucleation of microcracks due to shear failure.

ACKNOWLEDGEMENTS

We acknowledge financial support of Qinetiq and are grateful to T. Andrews, P.D. Church, B. Goldthorpe, and I.G. Cullis for continued encouragement. We also thank J.M. Burley and W.G. Proud for assistance with experimentation. D.L.A. Cross and R.P. Flaxman provided valuable technical support.

REFERENCES

1. Willmott, G.R., and Radford, D.D., *Proc. R. Soc. Lond.*, submitted for review (June 2003).
2. Kanel, G.I., Rasorenov, S.V., and Fortov, V.E., "The Failure Waves and Spallations in Homogeneous Brittle Materials", in *Shock Compression of Condensed Matter - 1991*, edited by S.C. Schmidt et al., Elsevier, Amsterdam, 1992, pp. 451-454.
3. Bourne, N.K., Rosenberg, Z., Mebar, Y., Obara, T., and Field, J.E., *J. Phys. IV France* **4**, C8-635-C8-640 (1994).
4. Bourne, N.K., Millett, J.C.F., and Rosenberg, Z., *J. Appl. Phys.* **81**, 6670-6674 (1997).
5. Radford, D.D., and Tsembelis, K., *(to be published)*, (2003).
6. Brar, N.S., and Bless, S.J., *High Pressure Research* **10**, 773-784 (1992).
7. Clifton, R.J., *Appl. Mech. Rev.* **46**, 540-546 (1993).
8. Grady, D.E., "Dynamic Properties of Ceramic Materials," Sandia National Laboratories SAND94-3266 (1995).
9. Espinosa, H.D., Xu, Y., and Brar, N.S., *J. Am. Ceram. Soc.* **80**, 2061-2073 (1997).
10. Kondaurov, V.I., *J. Appl. Maths Mechs* **62**, 657-663 (1998).
11. Kanel, G.I., Bogatch, A.A., Razorenov, S.V., and Chen, Z., *J. Appl. Phys* **92**, 5045-5052 (2002).
12. Radford, D.D., Proud, W.G., and Field, J.E., "The Deviatoric Response of Three Dense Glasses Under Shock Loading Conditions" in *Shock Compression of Condensed Matter - 2001*, edited by M.D. Furnish et al., AIP Conference Proceedings 620, New York, 2001, pp. 807-810.
13. Bless, S.J., Brar, N.S. and Rosenberg, Z., "Failure of Ceramic and Glass Rods Under Dynamic Compression" in *Shock Compression of Condensed Matter - 1989*, edited by S.C. Schmidt et al., Elsevier, Amsterdam, 1990, pp. 939-942.
14. Bless, S.J., and Brar, N.S., *High-Pressure Science and Technology*, 1813-1816 (1994).
15. Murray, N.H., Bourne, N.K., Field, J.E., and Rosenberg, Z., "Symmetrical Taylor Impact of Glass Bars", in *Shock Compression of Condensed Matter - 1997*, edited by S. C. Schmidt et al., AIP Conference Proceedings 429, pp. 533-536.
16. Taylor, G.I., *J. Inst. Civil Engrs.* **26**, 486-519 (1946).
17. Erlich, D.C., Shockey, D.A., and Seaman, L., "Symmetric Rod Impact Technique for Dynamic Yield Determination" in *Shock Waves in Condensed Matter - 1981*, edited by W.J. Nellis et al., American Physical Society, New York, 1982, pp. 402-406.
18. Rosakis, A.J., Samudrala, O., and Coker, D., *Science* **284**, 1337-1340 (1999).
19. Field, J.E., *Contemp. Phys.* **12**, 1-31 (1971).

CP706, *Shock Compression of Condensed Matter - 2003*
edited by M. D. Furnish, Y. M. Gupta, and J. W. Forbes
© 2004 American Institute of Physics 0-7354-0181-0/04/$22.00

DYNAMIC STRENGTH OF AD995 ALUMINA
AT MBAR STRESS LEVELS

W. D. Reinhart, L. C. Chhabildas

Sandia National Laboratories, Department 1647, Albuquerque, New Mexico

Abstract: An investigation of the strength of AD995 alumina in the shocked state was assessed over the stress range of 26-120 GPa. Velocity interferometry was used to measure loading, unloading, and reloading profiles from the initial shocked state. These results show that alumina retains considerable strength at stress states exceeding 120 GPa. An important observation, as with some metals, is that there is a substantial increase in strength during reloading and a well-defined elastic behavior is observed. The unloading and reloading technique described also yields data to estimate a dynamic mean stress.

INTRODUCTION

There has been a resurgence in interest in using ceramics as an effective armor material as ceramics out perform metals for protective applications and are much lighter. Three important mechanical properties control the performance of ceramics: compressive, shear, and tensile/spall strengths. This paper will discuss plane shock wave experimental results that will estimate strengths of alumina in the shocked state.

Accumulated experimental results pertaining to plane shock wave response of polycrystalline ceramics indicate that whereas their compressive strengths are high and their tensile/spall strengths are always extremely low (typically one tenth of the HEL), their shear strengths are unpredictable. The reason for the observed variability in the shear strengths of ceramics is not known. The compressive strength of these materials do not degrade significantly even in the presence of moderate pore volume fraction or micro-cracks/-fissures. But their tensile strengths are very sensitive to this pore volume fraction, and the micro-cracks/-fissures. As a consequence, modeling of ceramics assumes that the roles of the pore volume fraction and the micro cracks are primarily to degrade its compressive strength. The complexity of the response of a ceramic to shock wave loading is conjectured to be controlled by inelastic deformation of a polycrystalline ceramic even below the HEL. The inelastic deformation is dominated by micro cracking, pore collapse, and intragranular plastic deformation, and anisotropy of the grains. Thus, the problem is to understand competing roles of localized plasticity and propagation of cracks and dominance of one over the other in the whole range of applicable compressive and tensile stresses. In order to understand the response of a ceramic to shock wave loading, it is essential to separate the effects of compression, shear and tension. The primary advantage of the results derived from plane shock wave experiments is that the measured strength properties of a material are the best values. In other words, the three strengths mentioned above is degraded in any other non-planar impact situation. Thus the value of the properties derived from plane shock wave experiments under single or multiple shocks can never be exceeded in any other impact loading conditions. Further, the strength parameters derived from shock wave

---

* This work was supported by the U.S. Department of Energy under contract DE-AC04-94AL8500000. Sandia is a multiprogram laboratory operated by Sandia Corporation a Lockheed Martin Company, for the United States Department of Energy.

experiments permit us to understand better the response of ceramics. These results are thus source of valuable information to develop material models, which can be used in simulation of ballistic events.

Shock experiments are confined uniaxial strain experiments and are generally referred to as the Hugoniot state of the material. The Hugoniot elastic limit identifies initial yield in ceramics and provides an estimate of the initial dynamic material strength. Assumptions of yield strength in the shocked state cannot be derived from the hugoniot elastic limit, unless the post shock strength does not vary with stress. Cumulative inelastic deformation with changing confining stress, however, can alter the strength of the material. To estimate the dynamic yield strength of the material at high stress, the Hugoniot state is generally compared to a hydrostat - which is either measured directly or determined by extrapolating the pressure-volume behavior of the material determined at lower hydrostatic pressures. Based on Von-Mises yield criteria the difference between the Hugoniot stress and the hydrostatic pressure curve is defined as two-thirds the dynamic yield strength. If this difference is (1) independent of the shock-loading stress then the material exhibits elastic-perfectly plastic behavior, (2) changing with increasing stress then the material exhibits a pressure-dependent yield strength. An increase in yield strength may be attributed to many factors such as rate-dependence and/or a pressure dependent yield behavior, while a decrease would be related to a softening behavior resulting from heterogeneous deformation process and or from damage resulting from shock compression. For an accurate application of this technique, the hydrostat has to be extremely well-defined, and therefore, has the potential for limited use at lower stresses. Alternately, a direct measurement of both longitudinal and transverse stress can be made. This technique has been used recently to determine the full stress state under shock compression in silicon carbide in excess of 20 GPa[1]. This technique is exploratory and promising and may have limited applicability in the higher stress regimes quoted in this paper. A technique is therefore needed determine the shock hydrostat (mean stress) for use at higher stresses.

The objective of the present study is to dynamically determine the mean stress for ceramics. This technique has been previously applied to investigate metals[2, 3], and in particular has been used extensively to characterize the strength properties of 6061-T6 aluminum and tungsten in the shocked state. The method employs re-shock and release experiments to be conducted from the same Hugoniot stress state to experimentally evaluate the departure of the initial loading stress state from an elastic-plastic behavior. The asymmetry in the reloading and release path is then used to determine the shock hydrostat. The mean value of the upper strength limit and the lower strength limit will define the dynamic hydrostat. These two bounds also define the strength of the material (undamaged material) at that mean stress while the difference between the hugoniot state and the lower strength limit will define the strength of the material (damaged material) in the shocked state.

MATERIAL DESCRIPTION

The aluminum oxide (Al_2O_3) used in this study is referred to as Coors AD995. Its composition consists of 99.5% alumina and the remainder of the material is aluminosilicate glass. The density of the material (Al_2O_3) was 3.89 g/cm^3 and the average longitudinal and shear wave speed was 10.56 km/s and 6.24 km/s respectively. This yields an estimate of 7.71 km/s, 0.234 and 231.7 GPa for the bulk wave velocity, Poisson's ratio and the bulk modulus, respectively.

EXPERIMENTAL TECHNIQUE

Compressive shock, reshock and release waves are produced in alumina oxide using a two-stage light gas gun. The experimental configuration used for this study on the two-stage light gas gun is shown in Figure 1. This gas gun utilizes a 28 mm bore diameter with projectile velocities approaching 8 km/s. In Figure 1, the projectile is faced with the ceramic Coors AD995 and is backed with either a plastic of low shock impedance for release experiments, or a high-shock impedance material (tantalum or copper), for reloading experiments. The target configuration in Figure 1, will have an alumina ceramic disk similar to that mounted on the projectile and a single crystal lithium-fluoride is bonded with epoxy to the back of the ceramic sample. The particle velocity histories resulting

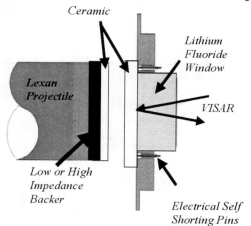

Figure 1. Experimental configuration for unloading and reloading experiments

from impact were measured at the target/lithium-fluoride window interface using a velocity interferometer system for any reflector (VISAR).

ANALYSIS

The impact conditions for the experiments in the current study are summarized in Table 1. The measured wave profiles shown in Figure 2 are used to determine the Hugoniot properties of the ceramic. In Figure 2, the leading edge of the elastic shock is used as a fiducial for the analysis in this study. For alumina, the leading edge of the elastic wave is assumed to traverse at the elastic longitudinal wave speed of 10.74 km/s, a value that has been determined on earlier studies in the alumina [Grady, 1995].

Table I: Summary of Impact Conditions

Exp	Vel. (km/s)	Targ. (mm)	Imp. (mm)	u_{pe} (km/s)	σ_e (GPa)	ε_e Strain
RL3	2.158	7.988	3.070	0.169	7.06	.0157
RS2	2.208	6.337	4.211	0.162	6.78	.0151
RS4	2.948	3.170	1.031	0.182	7.61	.0170
RL4	3.003	3.160	1.029	0.182	7.61	.0170
RS5	3.728	2.938	1.026	0.189	7.89	.0176
RL5	3.793	2.952	1.025	0.189	7.89	.0176
RS6	4.842	2.946	1.021	0.189	7.89	.0176
RL6	4.856	2.949	1.016	0.188	7.87	.0175
RS7	1.561	2.995	0.993	0.167	6.977	.0156
RL7	1.636	2.999	0.984	0.168	7.035	.0157

The Hugoniot elastic limit stress, (σ_{hel}), is determined using the relation:

$$\sigma_{hel} = (\rho_0 C_l u_e) \quad (1)$$

where ρ_0 is the initial density of the ceramic, C_l the elastic longitudinal wave speed, and u_e is the in-material particle velocity measurement prior to transition to a plastic wave. The in-material particle velocity is determined through the impedance matching relation:

$$u_e = u_w (Z_{w} + Z_m) / 2 Z_m \quad (2)$$

where u_w, Z_w, and Z_m are the measured velocity in the window material, and the shock impedances of the lithium-fluoride window and the ceramic material, respectively.

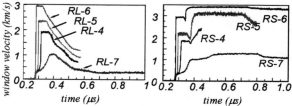

Figure 2. Window.interface particle velocity profiles for Al_2O_3 a) unloading; b) reloading experiments

As compression within the shock increases during the shock loading process, shear stresses will exceed the critical strength of the material (HEL) and plastic deformation occurs in the observed second wave. Because finite rise times are measured for the plastic wave, the plastic-wave velocity, U_{sp}, is taken at the center of the wave and the corresponding wave speed is given in Table 2. All the experiments were performed using symmetric-impact technique, therefore the particle (material) velocity, u_{ph}, behind the shock front is exactly one-half the impact velocity. The Hugoniot stress, σ_{ph}, and strain, ε_{ph}, behind the plastic-wave front are estimated using the following relations:

$$\sigma_{ph} = \sigma_e + \rho_0 U_{sp} (u_{ph} - u_e) \quad (3)$$

$$\varepsilon_{ph} = \varepsilon_e + (u_{ph} - u_e) / U_{sp} \quad (4)$$

The summary of the Hugoniot data is summarized in Tables 1 and 2 and plotted in Figures 3, 4, and 5.

OFF-HUGONIOT ANALYSIS:

All the experiments conducted in this investigation, evaluate the off-Hugoniot states of Coors AD995

Table 2: Hugoniot Summary

Exp. No.	U_{sp} (km/s)	u_{ph} (km/s)	σ_{ph} (GPa)	ε_{ph} Strain
RL3	8.94	1.079	38.70	0.1176
RS2	8.93	1.104	40.38	0.1178
RS4	9.49	1.474	55.30	0.1531
RL4	9.46	1.502	56.16	0.1564
RS5	9.96	1.864	72.79	0.1858
RL5	10.01	1.897	74.35	0.1883
RS6	10.52	2.421	99.26	0.2297
RL6	10.49	2.428	99.30	0.2310
RS7	8.227	0.7805	26.61	0.0901
RL7	8.218	0.8180	27.80	0.0947

alumina as it is allowed to release and reload from compression. An incremental form of the conservation equations given by the relations:

$$\sigma = \Sigma \rho_o\, c \Delta u \qquad (5)$$
$$\varepsilon = \Sigma\, \Delta u/c \qquad (6)$$

are used to estimate the final released or reshocked stress, σ, and strain, ε, respectively. The Lagrangian-wave velocity, c, corresponds to the current material particle velocity corresponding to the change, $\Sigma\, \Delta u$. The Lagrangian-wave velocity is estimated from the time difference between the arrival of the leading edge of the release/reshock wave at the window/sample interface and the time at which the effective shock arrives at the back surface of the impactor (finite rise times upon reshock are measured because an elastic-plastic wave is observed). The results of the companion reshock and release experiments conducted at an initial shock stresses from 38 to 99 GPa will be highlighted in this paper.

STRENGTH IN THE SHOCKED STATE

The results of companion reshock and release experiments conducted are shown in the stress vs. strain plane in Figure 6. In these experiments the leading edge of the reshock or release wave traverses at an elastic wave speed. The release path exhibits an elastic release from the initial

shocked state. The reloading path also shows precursor elastic recompression at all Hugoniot stresses, even at the Hugoniot state of 100 GPa to a final state of 120 GPa. The reshock final states always lie above the Hugoniot states. Similar

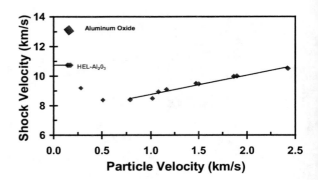

Figure 3. Shock velocity vs. particle velocity

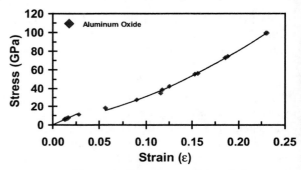

Figure 4. Stress vs. particle velocity

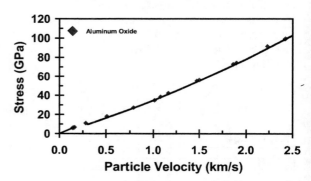

Figure 5. Stress vs. particle velocity

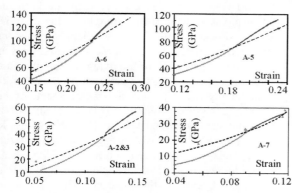

Figure 6. Results of reshock and unloading companion experiments in the stress vs. strain plane

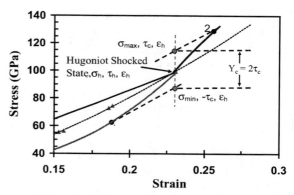

Figure 7. 100 GPa experiment is used to depict technique to estimate the shear stress state, τ_h, and the maximum strength, τ_c, the material can sustain.

results are obtained for the release experiments at all stresses. Note that the two experiments at the lower stresses up to 57 GPa, the recompression wave exhibits a distinct elastic precursor followed by ramped plastic wave, whereas the experiments at the higher stresses do not exhibit the distinct elastic-plastic behavior. In all companion reshock and release experiments reported in this study, the leading edge of the release-wave or the reshock wave, traverses at an elastic wave velocity thus providing evidence that the leading edge of the reloading wave is elastic and the material is reloading or unloading elastically.

This phenomena has been observed previously in 6061-T6 aluminum[2], tungsten[3], beryllium[5], copper[6] and was observed in previous studies of Coors AD995 alumina even at its Hugoniot elastic limit. The experimental technique in this study provides the ability to determine mean stress[2] at very high dynamic stresses.

In Figure 7 the method used for determining strength is graphically depicted. The relation between the hugoniot stress (σ_h), mean stress (**P**), and shear stress (τ_h) is given by:

$$\tau_h = \tfrac{3}{4} \left(\sigma_h - P \right) \tag{7}$$

equation 7 will yield the following relations,

$$\sigma_h = P + 4/3\, \tau_h \tag{8}$$
$$\sigma_{max} = P + 4/3\, \tau_c \tag{9}$$
$$\sigma_{min} = P - 4/3\, \tau_c \tag{10}$$

where τ_h is the shear stress of the material at the shocked or Hugoniot state of the material, τ_c is the maximum shear stress state that the material can sustain, and P is the stress of the material. σ_{max} and σ_{min} will be the stress states determined from the reshock and release experiments at the common Hugoniot strain ε_h.

As indicated in Figure 7, states 1 and 2 are defined as maximum or minimum shear stress states because the Lagrangian wave velocity transitions from elastic wave to the bulk wave at those states. Combining equations 8-10, the following expressions are obtained:

$$\tau_c + \tau_h = \tfrac{3}{4} \left(\sigma_h - \sigma_{min} \right) \tag{11}$$

$$\tau_c - \tau_h = \tfrac{3}{4} \left(\sigma_{max} - \sigma_h \right) \tag{12}$$

whereby the critical strength (Y_c) is defined by the expression,

$$Y_c = 2\tau_c = \tfrac{3}{4} \left(\sigma_{max} - \sigma_{min} \right) \tag{13}$$

Table 3 summarizes the results for the strength measurements obtained from this study.

Table 3: Strength summary--stress values (GPa)

Exp.	τ_{max}	τ_{min}	P	Y_c	$\tilde{\tau_c}\tau_h$	$\tau_c\tau_h$
A-6	111	83	97	21	9	12
A-5	89	61	75	21	10.5	10
A-4	63	46	54.5	12.8	6	7
A-2&3	46	32	39	10.5	5	6
A-7	29	20	24	7	2	5.5

Note: Rows are representation of combination of unloading (RL) and reloading (RS) experiments

763

Figure 8 describes the variation of the average shear stress τ_h, in the shocked state and the critical strength, Y_c of the alumina as determined from the unloading and reloading data.

CONCLUSIONS

Unloading, reloading experiments on Coors AD995 alumina shocked to stresses from 38 GPa

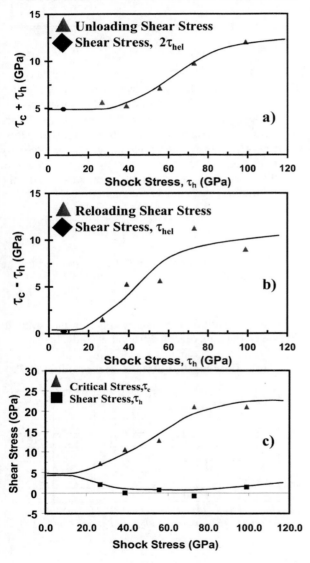

Figure 8. Measurement of strength of material upon release (a) and unloading (b). Variation of average shear strength in the shocked state (c).

to 100 GPa have been used to estimate both the shear stress and the shear strength of the material in the shocked state. The two main conclusions that can be derived from these measurements is that the yield strength of the material does increase with increasing stress – suggesting a pressure dependent yield, while the shears stress state of the material suggest a collapse towards the hydrostat over the stress regime of 38 to 100 GPa. Multiple factors such as the yield strength of the material being pressure dependent combined with shear stress losses in the shocked state are responsible for the behavior of the material. This technique also allows the determination of a dynamic hydrostat. For alumina the dynamic hydrostat and the shock Hugoniot overlay – mainly because the values of the remnant shear stress in the shocked state are very nearly zero.

REFERENCES

1. Feng, R., Gupta, Y. M., Yuan, G., *Dynamic Srength and Inelastic Deformation of Ceramics under Shock Wave Loading,* in Shock-waves in Condensed Matter – 1997, S. C. Schmidt, D. P. Dandekar, J. W. Forbes (eds.), AIP Proceedings 429, 483-488, (1998).
2. Asay, J. R. and Chhabildas, L. C. *Determination of Shear Strength of Shock-Compressed 6061- t6 Aluminum,* Shock Waves and High-Strain-Rate Phenomena in Metals, ed. Myers, M.M. and Murr, L. E. Plenum Pub. Corp, New York, NY (1981).
3. Chhabildas, L. C., Asay, J. R., Barker, L. M. *Shear strength of Tungsten Under Shock and Quasi-Isentropic Loading to 250 GPa*, Sandia National Laboratories Report, SAND88-0306, April (1988).
4. Grady, D. E. *Dynamic Properties of Ceramic Materials*, Sandia National Laboratories Report, SAND88-3266, February (1995).
5. Chhabildas, L. C., Wise, J. L., Asay, and J. R., *Reshock and Release Behavior of Beryllium,* Shock-waves in Condensed Matter, W. J. Nellis, L. Seaman, and R. A. Graham (eds.), AIP Conference Proceedings 78, 422-426 (1981).
6. Chhabildas, L. C., and Asay, and J. R., *Time-Resolved Wave Profile Measurements in Copper to Megabar Pressures,* in High Pressure in Research and Industry, C. M. Backman, T. Johanisson, and L. Tegner (eds.), V1, 183-189, (1982).

CP706, *Shock Compression of Condensed Matter - 2003*
edited by M. D. Furnish, Y. M. Gupta, and J. W. Forbes
© 2004 American Institute of Physics 0-7354-0181-0/04/$22.00

TENSION OF LIQUIDS BY SHOCK WAVES

Alexander V. Utkin, Vasiliy A. Sosikov, Andrey A. Bogach, Vladimir E. Fortov

Institute of Problems of Chemical Physics RAS, 142432, Chernogolovka, Russia

Abstract. The influences of strain rate and initial temperature on the negative pressure in distillate water, hexane, glycerol and methyl alcohol under shock waves have been investigated. The wave profiles were registered by laser interferometer VISAR. Shock waves were produced by aluminum plates accelerated by high explosive up to 600 m/s. At initial temperature 19 ^{0}C spall strength of water, hexane, and methyl alcohol is equal to 45, 15, and 47 MPa respectively and not depend on the strain rate in interval from 10^{-4} to 10^{-5} 1/s. A strong dependence of negative pressure on strain rate was observed only for glycerol. The reason is that the initial temperature of glycerol was equal to the freezing point, and in the vicinity of it the relaxation properties are usually very much more pronounced. To confirm this assumption the experiments with water an initial temperature 0.7 ^{0}C were made and strong influence of strain rate on spall strength was observed close to freezing temperature too. Moreover expansion isentropes intersected the melting curve at negative pressure and double metastable state was realized in water. Theory of homogeneous bubble nucleation was used to explain the experimental results.

INTRODUCTION

According to theoretical models, high negative pressure can be realized in liquids (~1 GPa [1]) if the cavitation process is controlled by the homogeneous nucleation of voids. However, significantly lower tensile strengths [2] are recorded under static experiments. Such discrepancies between experimental and theoretical values are generally attributed to heterogeneous void nucleation on impurities suspended in liquid. The tensile failure in liquids approximating by the theory of homogeneous nucleation can be achieved under dynamic tension. For this purpose we used the method of spall strength measurement [3], which has been applied to the investigation of cavitation in liquids [4-6]. The dependence of spall strength from the strain rate and initial temperature is useful information, which allows the mechanism of tensile failure to be found in a liquid. In this study we investigated the spall strength of water, hexane, methyl alcohol and glycerol over a wide range of shock waves duration. Experiments with water were made at different initial temperature.

SCHEME OF EXPERIMENTS

Figure 1 shows the experimental configuration. Shock waves were produced by aluminum plate (1) with a thickness from 0.2 up to 2 mm accelerated by high explosive up to 600 m/s. The 2 mm thick PMMA screen (2) was placed between the investigated liquid sample (3) and impactor. In some shots the shock waves were generated by the effect of high explosive products on the 20 mm thick copper screen. The wave profiles were registered by laser interferometer VISAR with 2 ns time resolution and an accuracy of the velocity determination was equal to ±2 m/s. The laser beam reflected from a 7 μm aluminum foil (4) placed between the liquid and the air. One-dimensional shock waves were provided by the geometric sizes of experimental setup.

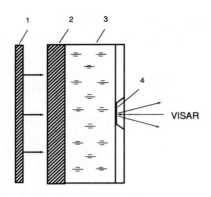

FIGURE 1. The scheme of experiments.

EXPERIMENTAL RESULTS

Distilled water at the initial temperature 19 °C was used in experiments. Figure 2 shows the history of free surface velocity as a function of time, when the shock wave amplitude was changed from 40 up to 430 MPa and the strain rate in the unloading part of incident shock pulse $\dot{\varepsilon} = (dW/dt)/(2c_0)$ was varied from $2.7 \cdot 10^4$ up to $1.4 \cdot 10^5$ s^{-1} (c_0 – is the initial sound velocity). When the shock front reaches the free surface, the velocity of the latter undergoes a jump from zero up to W_0, which is equal to double particle velocity. Reflection of the shock wave by the free surface produces a rarefaction wave, which interacts with the incident unloading wave. Tensile stresses arise in the interaction region and the damage occurs inside the liquid. Fracture of material allows the tensile stress to decrease to zero. As a result, a compression wave appears which propagates to the free surface and forms a so-called spall pulse in the free surface velocity profile. These details are observed on the velocity profiles presented on the figure 2. The tensile stress value P_s just before spalling is determined by the linear approximation [3]: $P_s=0.5\rho_0 c_0(W_0-W_m)$, where ρ_0 is the initial density, W_m is the free surface velocity just before the spall pulse. Figure 3 shows the spall strength P_s as function of $\dot{\varepsilon}$. The P_s is essentially independent of the strain rate in all measured interval, and is equal to 42 MPa as seen in Fig.3.

The initial density and sound velocity of the hexane used in experiments were equal to 0.66 g/cm^3 and 1.083 km/s, respectively, at the

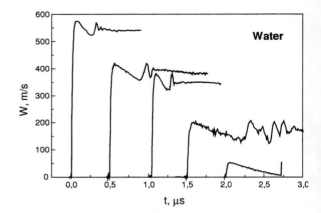

FIGURE 2. The history of free surface velocity in the shots with water.

temperature 19 °C. For the glycerol $\rho_0=1.26$ g/cm^3, $c_0=1.895$ km/s, and for the methyl alcohol $\rho_0=0.786$ g/cm^3, $c_0=1.730$ km/s. The results of experiments for these liquids are shown in Fig.3-5. The most interesting result is a strong dependence of the glycerol spall strength on the strain rate. It was not observed for water, methyl alcohol, and hexane.

In a qualitative sense the velocity profiles of hexane and methyl alcohol are similar to that of glycerol but in the glycerol the spall pulse and the velocity oscillations produced by the reverberation of spall pulse between the free surface and spall plane are more clearly defined. It is probable that the cavitation zone of glycerol has well-defined

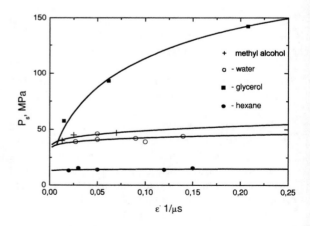

FIGURE 3. Tensile stress as a function of strain rate. Points are experimental results, solid lines are calculation.

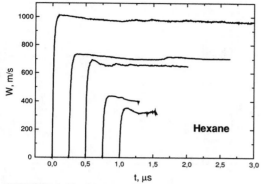

FIGURE 4. The history of free surface velocity in the shots with hexane.

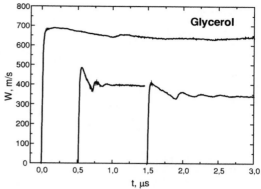

Figure 5. The history of free surface velocity in the shots with glycerol.

boundaries. Unlike hexane, methyl alcohol, and glycerol, free surface velocity profiles of the water (Fig.2) have very steep front of spall pulse (with a characteristic time ~30 ns) and its amplitude is essentially coincide with the velocity maximum. Since spall pulse steepness is determined by the damage rate [7], it means that the growth of cavitation in hexane, methyl alcohol, and glycerol is significantly below than in water. High amplitude of spall pulse in water probably is due to the growth of the cavitation zone in time and moving its boundary to the free surface. Most likely for this reason, and owing to absence of sharp boundary between the water and damage region, the oscillations of free surface velocity do not observed after spall pulse in Fig.2.

DISCUSSION

To explain of the experimental results let us consider the application of a theory of homogeneous bubble nucleation to the tensile failure in liquids. According to the classical nucleation concept [1], the number of critical void formed per unit volume per unit time J is given by

$$J = N_0 \frac{\sigma}{\eta} \sqrt{\frac{\sigma}{kT}} \exp\left(-\frac{16\pi\sigma^3}{3P_s^2 kT}\right), \qquad (1)$$

where N_0 is the number density of molecules, σ is the surface energy of the liquid, η is the viscosity, T is the temperature (^{0}K), k is Bolzmann constant. From Eq. (1) follows the dependence of spall strength on a strain rate [6, 7]:

$$P_s \approx A/\sqrt{\ln(B/\dot{\varepsilon})}. \qquad (2)$$

Here A and B are the constants which are the functions of T, σ, and η. As seen in Fig.3, the equation (2) describes well the experimental data for hexane (A=62 MPa, B=10^{13} s^{-1}), methyl alcohol (A=200 MPa, B=10^{13} s^{-1}), and water (A=110 MPa, B=10^{16} s^{-1}). As to glycerol, it is necessary to consider the influence of strain rate on the viscosity. Strong temperature dependence of the glycerol viscosity nearly the freezing point (291 K) [5] leads one to expect that temperature-frequency principle trues for the η. That is the change of viscosity under the increasing of strain rate identical to that at the temperature decreasing. To estimate the strain rate influence on the viscosity, a simple relaxation model was used: $\dot{\eta} \sim -\eta/\tau$, where τ is the characteristic time of glycerol. Since constant $B \sim 1/\eta \sim exp(1/(\dot{\varepsilon}\,\tau))$, the relationship (2) can be rewritten

$$P_s = A/\sqrt{\ln\left(B \exp(\beta/\dot{\varepsilon})/\dot{\varepsilon}\right)}, \qquad (3)$$

where $\beta \sim 1/\tau$. Figure 3 shows that the dependence (3) at A=970 MPa, B=10^{15} 1/s, β=5 µs^{-1} is in reasonably good agreement with the glycerol experimental data.

One would expect a strong dependence of spall strength from the strain rate in other liquids nearly freezing point. To test the validity of this assumption the experiments with water at initial temperature 0.7 ^{0}C were made. The experimental arrangement was similar that in Fig.1, but liquid sample (3) was surrounded by a vessel with a

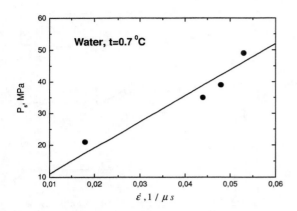

FIGURE 6. Tensile stress as a function of strain rate for water at low temperature.

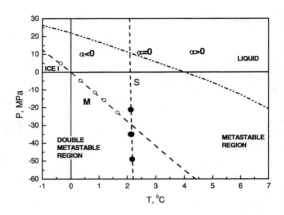

FIGURE 7. The phase diagram for water. M – melting curve [8], α- thermal expansivity, S – isentrope, solid points – experiments.

mixture of ice and water. The velocity profiles in this case are similar those presented in Fig.2. Figure 6 shows spall strength of water at low temperature and evidently that it is very sensitive to the stain rate, the linear dependence of P_s on $\dot\varepsilon$ is observed.

It is interesting to analyse the evolution of water state in a plane *P-T* during the shock wave compression and following expansion up to negative pressure. The phase diagram for water nearly 0 ^{0}C is shown in Fig.7. The melting temperature of ice I decreases with increasing pressure because the density of ice lower than that of liquid water [8]. This behavior is the reverse of normal melting behavior, and as result the isentrope of liqued water can intersect the melting curve under expansion. Experimental results are black points, pressure corresponds to the spall strengh of water and temperature was estimated from equation of state. After shock compression the pressure in water is equal to 300 MPa and increase in temperature is equal to 10 degree. In the rearfaction wave the pressure and temperature decrease along the isentrope S (Fig.7) which crosses melting curve at negative pressure. This implies that double metastable region is realized where the water can either freeze or boil.

CONCLUSION

A theory of homogeneous bubble nucleation explains the dependence of measured tensile strength from the strain rate in water, methyl alcohol, hexane, and glycerol, but in glycerol we have to take into account the influence of $\dot\varepsilon$ on the viscosity. The reason is that the initial temperature of glycerol is equal to the freezing point, and in the vicinity of it the relaxation properties are usually very much more pronounced. This assumption was confirmed by experiments with water at initial temperature 0.7 ^{0}C where strong influence of strain rate on spall strength was observed too. Moreover in water the expension isentropes intersected the melting curve at negative pressure and double metastable state was realized.

The work has been funded by Russian Academy of Science.

REFERENCES

1. Zeldovich, Ya.B., *Zhurnal Eksp. Tekh. Fiz.* 1942. **12**, No. 11/12, 525-537.
2. Skripov V.P. *Metastable Liquid*, Moscow, Nauka, 1972, pp.60-62.
3. Kanel, G.I., Razorenov, S.V., Utkin, A.V., Fortov, V.E., *Shock-Wave Phenomenon in Condensed Matter*, Moscow, Yanus-K, 1996, pp.150-168.
4. Erlich, D.C., Wooten, D.C., Crewdson, R.C., *J.Appl.Phys.* **42**, pp. 5495-5501 (1971).
5. Carlson, G.A., Levine, H.S., *J.Appl.Phys.*, **46** pp. 1594-1599 (1975).
6. Bogach, A.A., Utkin, A.V. *Prikl. Mekh. Tekh. Fiz.* **41**, No.4, pp. 198-225 (2000).
7. Utkin, A.V. *Prikl. Mekh. Tekh. Fiz.* **38**, No. 7, pp. 151-160 (1997).
8. Henderson, S.J., and Speedy R.J., *J.Phys.Chem.* **91**, 3069-3072 (1987).

A

Abe, T., 465
Ackland, G. J., 1436
Addessio, L. B., 469
Ahrens, T. J., 172, 1419, 1478, 1484
Aiken, A. C., 1377
Airijant, E., 1496
Akhavan, J., 99, 435
Akinci, A., 1500
Alcon, R. R., 651, 675, 973, 1033
Allen, R. M., 823
Aminov, Y. A., 913
Ananin, A. V., 851
Anderson, J. E., 1377
Anderson, Jr., C. E., 707, 1347
Anderson, M. U., 685
Anderson, W. A., 499
Anderson, W. E., 1191
Anderson, W. W., 37, 483, 573, 1289
Antoun, T. H., 1423, 1462
Arad, B., 1139
Aramovich Weaver, C., 1427
Armstrong, R. W., 779
Arnold, W., 1319
Arrigoni, M., 1369, 1373, 1393
Arrington Jr., C. A., 1377
Asay, B. W., 827, 855, 981, 1009, 1037, 1500
Asay, J. R., 3, 81, 617
Asimow, P. D., 95
Aslam, T. D., 831, 847
Atchison, W. L., 499, 1191
Auroux, E., 1393
Averin, A. N., 917
Avrillaud, G., 1209

B

Baer, B. J., 1249
Baer, M. R., 517
Bailey, J. E., 81
Baker, E. L., 375, 891
Baker, S. A., 839

Bandyopadhyay, A., 1079, 1094
Bardenhagen, S. G., 187
Barlow, A. J., 521, 625
Barradas, S., 1369, 1373
Barsoum, M. W., 77
Baskes, M. I., 281
Batalov, S. V., 859, 917
Batani, D., 1397
Bauer, F., 1121
Baumung, K., 645
Bdzil, J. B., 831
Beissel, S. R., 193
Belak, J. F., 1195
Belenovsky, Y. A., 917
Belmas, R., 951
Ben-Dor, G., 201, 1261
Benson, D. J., 343, 537, 1098
Benuzzi-Mounaix, A., 1397
Berthe, L., 1369, 1373
Bessette, G. C., 1323
Bezruchko, G. S., 29
Birnbaum, N., 271
Blanco, M., 251
Bless, S. J., 711
Bliss, D. E., 1175
Blottiau, P., 289
Blumenthal, W. R., 525
Boettger, J. C., 65
Bogatch, A. A., 739, 765
Bolis, C., 1369, 1373
Bonitz, M., 53
Bonner, M. P., 1470
Bonora, N., 439, 1355
Borg, J. P., 715
Borodina, T. I., 1082
Bour, N. K., 823
Bourne, N. K., 99, 371, 435, 461, 557, 561, 641, 663, 671, 689, 693, 697, 711, 719, 723, 751, 939, 1474
Boustie, M., 1369, 1373, 1393
Braga, D., 1013
Brar, N. S., 613, 727, 1151
Brennen, C. E., 69
Bringa, E. M., 225, 1195
Britan, A., 201, 1261
Brown, D. W., 525, 541
Browning, R. V., 921
Bry, A., 951

Bucholtz, S. M., 1163, 1405
Buelow, S. J., 1377
Buess, M. L., 929
Bulannikov, A. S., 73
Burkett, M. W., 529
Burtsev, V. V., 73
Busse, J. R., 827
Buy, F., 533
Byers, M. E., 483

C

Cai, J., 775
Cai, L., 33
Callahan, J. H., 843
Campos, J., 871, 887, 1013, 1021, 1500
Cao, B. Y., 537
Capellos, C., 783, 891
Carion, N., 1253
Carley, D., 521
Carney, J. R., 1143, 1281
Cart, E. J., 925
Carton, E. P., 1086, 1110
Caturla, M. J., 225, 1195
Caulder, S. M., 929
Cerreta, E. K., 499, 541
Chakravarty, A., 935
Chau, R., 127
Chavez, P. J., 1041
Chen, K. H., 1102
Chen, Q., 33
Chen, S. R., 633
Chen, Z., 739
Cherne, F. J., 229, 281
Chesnut, G. N., 37
Chhabildas, L. C., 759, 1323
Chidester, S. K., 1045, 1057
Chikov, B. M., 1496
Chisolm, E., 41, 45
Chitanvis, S. M., 319
Church, P. D., 1125, 1147
Clancy, S. P., 529
Clark, D. A., 473, 477
Clarke, S. A., 1401
Clements, B. E., 389, 443
Cline, C. F., 597
Cochrane, J. C., 499

A1

SUBJECT INDEX